HANDBOOK OF
LINEAR PARTIAL DIFFERENTIAL EQUATIONS for ENGINEERS and SCIENTISTS

Andrei D. Polyanin

CHAPMAN & HALL/CRC

A CRC Press Company
Boca Raton   London   New York   Washington, D.C.

Library of Congress Cataloging-in-Publication Data

Polianin, A. D. (Andrei Dmitrievich)
    Handbook of linear partial differential equations for engineers and scientists / by Andrei D. Polyanin
        p. cm.
    Includes bibliographical references and index.
    ISBN 1-58488-299-9
    1. Differential equations, Linear--Numerical solution--Handbooks, manuals, etc. I. Title.

QA377 .P568 2001
515'.354—dc21   2001052427
                                                                  CIP

This book contains information obtained from authentic and highly regarded sources. Reprinted material is quoted with permission, and sources are indicated. A wide variety of references are listed. Reasonable efforts have been made to publish reliable data and information, but the author and the publisher cannot assume responsibility for the validity of all materials or for the consequences of their use.

Apart from any fair dealing for the purpose of research or private study, or criticism or review, as permitted under the UK Copyright Designs and Patents Act, 1988, this publication may not be reproduced, stored or transmitted, in any form or by any means, electronic or mechanical, including photocopying, microfilming, and recording, or by any information storage or retrieval system, without the prior permission in writing of the publishers, or in the case of reprographic reproduction only in accordance with the terms of the licenses issued by the Copyright Licensing Agency in the UK, or in accordance with the terms of the license issued by the appropriate Reproduction Rights Organization outside the UK.

All rights reserved. Authorization to photocopy items for internal or personal use, or the personal or internal use of specific clients, may be granted by CRC Press LLC, provided that $1.50 per page photocopied is paid directly to Copyright Clearance Center, 222 Rosewood Drive, Danvers, MA 01923 USA. The fee code for users of the Transactional Reporting Service is ISBN 1-58488-299-9/02/$0.00+$1.50. The fee is subject to change without notice. For organizations that have been granted a photocopy license by the CCC, a separate system of payment has been arranged.

The consent of CRC Press LLC does not extend to copying for general distribution, for promotion, for creating new works, or for resale. Specific permission must be obtained in writing from CRC Press LLC for such copying.

Direct all inquiries to CRC Press LLC, 2000 N.W. Corporate Blvd., Boca Raton, Florida 33431.

Trademark Notice: Product or corporate names may be trademarks or registered trademarks, and are used only for identification and explanation, without intent to infringe.

Visit the CRC Press Web site at www.crcpress.com

© 2002 by Chapman & Hall/CRC

No claim to original U.S. Government works
International Standard Book Number 1-58488-299-9
Library of Congress Card Number 2001052427
Printed in the United States of America 1 2 3 4 5 6 7 8 9 0

FOREWORD

Linear partial differential equations arise in various fields of science and numerous applications, e.g., heat and mass transfer theory, wave theory, hydrodynamics, aerodynamics, elasticity, acoustics, electrostatics, electrodynamics, electrical engineering, diffraction theory, quantum mechanics, control theory, chemical engineering sciences, and biomechanics.

This book presents brief statements and exact solutions of more than 2000 linear equations and problems of mathematical physics. Nonstationary and stationary equations with constant and variable coefficients of parabolic, hyperbolic, and elliptic types are considered. A number of new solutions to linear equations and boundary value problems are described. Special attention is paid to equations and problems of general form that depend on arbitrary functions. Formulas for the effective construction of solutions to nonhomogeneous boundary value problems of various types are given. We consider second-order and higher-order equations as well as the corresponding boundary value problems. All in all, the handbook presents more equations and problems of mathematical physics than any other book currently available.

For the reader's convenience, the introduction outlines some definitions and basic equations, problems, and methods of mathematical physics. It also gives useful formulas that enable one to express solutions to stationary and nonstationary boundary value problems of general form in terms of the Green's function.

Two supplements are given at the end of the book. Supplement A lists properties of the most common special functions (the gamma function, Bessel functions, degenerate hypergeometric functions, Mathieu functions, etc.). Supplement B describes the methods of generalized and functional separation of variables for nonlinear partial differential equations. We give specific examples and an overview application of these methods to construct exact solutions for various classes of second-, third-, fourth-, and higher-order equations (in total, about 150 nonlinear equations with solutions are described). Special attention is paid to equations of heat and mass transfer theory, wave theory, and hydrodynamics as well as to mathematical physics equations of general form that involve arbitrary functions.

The equations in all chapters are in ascending order of complexity. Many sections can be read independently, which facilitates working with the material. An extended table of contents will help the reader find the desired equations and boundary value problems. We refer to specific equations using notation like "1.8.5.2," which means "Equation 2 in Subsection 1.8.5."

To extend the range of potential readers with diverse mathematical backgrounds, the author strove to avoid the use of special terminology wherever possible. For this reason, some results are presented schematically, in a simplified manner (without details), which is however quite sufficient in most applications.

Separate sections of the book can serve as a basis for practical courses and lectures on equations of mathematical physics.

The author thanks Alexei Zhurov for useful remarks on the manuscript.

The author hopes that the handbook will be useful for a wide range of scientists, university teachers, engineers, and students in various areas of mathematics, physics, mechanics, control, and engineering sciences.

Andrei D. Polyanin

BASIC NOTATION

Latin Characters

$\mathscr{E}$ fundamental solution
$\text{Im}[A]$ imaginary part of a complex quantity A
G Green's function
$\mathbb{R}^n$ n-dimensional Euclidean space, $\mathbb{R}^n = \{-\infty < x_k < \infty;\ k = 1,\ldots,n\}$
$\text{Re}[A]$ real part of a complex quantity A
r, φ, z cylindrical coordinates, $r = \sqrt{x^2 + y^2}$ and $x = r\cos\varphi,\ y = r\sin\varphi$
r, θ, φ spherical coordinates, $r = \sqrt{x^2 + y^2 + z^2}$ and $x = r\sin\theta\cos\varphi,\ y = \sin\theta\sin\varphi,\ z = r\cos\theta$
t time ($t \geq 0$)
w unknown function (dependent variable)
x, y, z space (Cartesian) coordinates
$x_1, \ldots, x_n$ Cartesian coordinates in n-dimensional space
$\mathbf{x}$ n-dimensional vector, $\mathbf{x} = \{x_1, \ldots, x_n\}$
$|\mathbf{x}|$ magnitude (length) of n-dimensional vector, $|\mathbf{x}| = \sqrt{x_1^2 + x_2^2 + \cdots + x_n^2}$
$\mathbf{y}$ n-dimensional vector, $\mathbf{y} = \{y_1, \ldots, y_n\}$

Greek Characters

Δ Laplace operator
Δ_2 two-dimensional Laplace operator, $\Delta_2 = \frac{\partial^2}{\partial x^2} + \frac{\partial^2}{\partial y^2}$
Δ_3 three-dimensional Laplace operator, $\Delta_3 = \frac{\partial^2}{\partial x^2} + \frac{\partial^2}{\partial y^2} + \frac{\partial^2}{\partial z^2}$
Δ_n n-dimensional Laplace operator, $\Delta_n = \sum_{k=1}^{n} \frac{\partial^2}{\partial x_k^2}$
$\delta(x)$ Dirac delta function; $\int_{-a}^{a} f(y)\delta(x-y)\,dy = f(x)$, where $f(x)$ is any continuous function, $a > 0$
δ_{nm} Kronecker delta, $\delta_{nm} = \begin{cases} 1 & \text{if } n = m, \\ 0 & \text{if } n \neq m \end{cases}$
$\vartheta(x)$ Heaviside unit step function, $\vartheta(x) = \begin{cases} 1 & \text{if } x \geq 0, \\ 0 & \text{if } x < 0 \end{cases}$

Brief Notation for Derivatives

$$\partial_t w = \frac{\partial w}{\partial t}, \quad \partial_x w = \frac{\partial w}{\partial x}, \quad \partial_{tt} w = \frac{\partial^2 w}{\partial t^2}, \quad \partial_{xx} w = \frac{\partial^2 w}{\partial x^2} \qquad \text{(partial derivatives)}$$

$$f'_x = \frac{df}{dx}, \quad f''_{xx} = \frac{d^2 f}{dx^2}, \quad f'''_{xxx} = \frac{d^3 f}{dx^3}, \quad f^{(n)}_x = \frac{d^n f}{dx^n} \qquad \text{(derivatives for } f = f(x)\text{)}$$

Special Functions (See Also Supplement A)

$\text{Ai}(x) = \dfrac{1}{\pi} \displaystyle\int_0^\infty \cos\!\left(\tfrac{1}{3}t^3 + xt\right) dt$ Airy function; $\text{Ai}(x) = \tfrac{1}{\pi}\sqrt{\tfrac{1}{3}x}\,K_{1/3}\!\left(\tfrac{2}{3}x^{3/2}\right)$

$\text{Ce}_{2n+p}(x, q) = \displaystyle\sum_{k=0}^{\infty} A_{2k+p}^{2n+p} \cosh[(2k+p)x]$ even modified Mathieu functions, where $p = 0, 1$; $\text{Ce}_{2n+p}(x, q) = \text{ce}_{2n+p}(ix, q)$

$\mathrm{ce}_{2n}(x,q) = \sum_{k=0}^{\infty} A_{2k}^{2n} \cos 2kx$	even π-periodic Mathieu functions; these satisfy the equation $y'' + (a - 2q\cos 2x)y = 0$, where $a = a_{2n}(q)$ are eigenvalues
$\mathrm{ce}_{2n+1}(x,q) = \sum_{k=0}^{\infty} A_{2k+1}^{2n+1} \cos[(2k+1)x]$	even 2π-periodic Mathieu functions; these satisfy the equation $y'' + (a - 2q\cos 2x)y = 0$, where $a = a_{2n+1}(q)$ are eigenvalues
$D_\nu = D_\nu(x)$	parabolic cylinder function (see Paragraph 7.3.4-1); it satisfies the equation $y'' + \left(\nu + \frac{1}{2} - \frac{1}{4}x^2\right)y = 0$
$\mathrm{erf}\, x = \frac{2}{\sqrt{\pi}} \int_0^x \exp(-\xi^2)\, d\xi$	error function
$\mathrm{erfc}\, x = \frac{2}{\sqrt{\pi}} \int_x^\infty \exp(-\xi^2)\, d\xi$	complementary error function
$H_n(x) = (-1)^n e^{x^2} \frac{d^n}{dx^n}\left(e^{-x^2}\right)$	Hermite polynomial
$H_\nu^{(1)}(x) = J_\nu(x) + iY_\nu(x)$	Hankel function of first kind, $i^2 = -1$
$H_\nu^{(2)}(x) = J_\nu(x) - iY_\nu(x)$	Hankel function of second kind, $i^2 = -1$
$F(a,b,c;x) = 1 + \sum_{n=1}^{\infty} \frac{(a)_n (b)_n}{(c)_n} \frac{x^n}{n!}$	hypergeometric function, $(a)_n = a(a+1)\ldots(a+n-1)$
$I_\nu(x) = \sum_{n=0}^{\infty} \frac{(x/2)^{\nu+2n}}{n!\, \Gamma(\nu+n+1)}$	modified Bessel function of first kind
$J_\nu(x) = \sum_{n=0}^{\infty} \frac{(-1)^n (x/2)^{\nu+2n}}{n!\, \Gamma(\nu+n+1)}$	Bessel function of first kind
$K_\nu(x) = \frac{\pi}{2} \frac{I_{-\nu}(x) - I_\nu(x)}{\sin(\pi\nu)}$	modified Bessel function of second kind
$L_n^s(x) = \frac{1}{n!} x^{-s} e^x \frac{d^n}{dx^n}\left(x^{n+s} e^{-x}\right)$	generalized Laguerre polynomial
$P_n(x) = \frac{1}{n!\, 2^n} \frac{d^n}{dx^n}(x^2 - 1)^n$	Legendre polynomial
$P_n^m(x) = (1-x^2)^{m/2} \frac{d^m}{dx^m} P_n(x)$	associated Legendre functions
$\mathrm{Se}_{2n+p}(x,q) = \sum_{k=0}^{\infty} B_{2k+p}^{2n+p} \sinh[(2k+p)x]$	odd modified Mathieu functions, where $p = 0, 1$; $\mathrm{Se}_{2n+p}(x,q) = -i\, \mathrm{se}_{2n+p}(ix, q)$
$\mathrm{se}_{2n}(x,q) = \sum_{k=0}^{\infty} B_{2k}^{2n} \sin 2kx$	odd π-periodic Mathieu functions; these satisfy the equation $y'' + (a - 2q\cos 2x)y = 0$, where $a = b_{2n}(q)$ are eigenvalues
$\mathrm{se}_{2n+1}(x,q) = \sum_{k=0}^{\infty} B_{2k+1}^{2n+1} \sin[(2k+1)x]$	odd 2π-periodic Mathieu functions; these satisfy the equation $y'' + (a - 2q\cos 2x)y = 0$, where $a = b_{2n+1}(q)$ are eigenvalues
$Y_\nu(x) = \frac{J_\nu(x)\cos(\pi\nu) - J_{-\nu}(x)}{\sin(\pi\nu)}$	Bessel function of second kind
$\gamma(\alpha, x) = \int_0^x e^{-\xi} \xi^{\alpha-1}\, d\xi$	incomplete gamma function
$\Gamma(\alpha) = \int_0^\infty e^{-\xi} \xi^{\alpha-1}\, d\xi$	gamma function
$\Phi(a,b;x) = 1 + \sum_{n=1}^{\infty} \frac{(a)_n}{(b)_n} \frac{x^n}{n!}$	degenerate hypergeometric function, $(a)_n = a(a+1)\ldots(a+n-1)$

AUTHOR

Andrei D. Polyanin, D.Sc., Ph.D., is a noted scientist of broad interests, who works in various areas of mathematics, mechanics, and chemical engineering sciences.

A. D. Polyanin graduated from the Department of Mechanics and Mathematics of the Moscow State University in 1974. He received his Ph.D. degree in 1981 and D.Sc. degree in 1986 at the Institute for Problems in Mechanics of the Russian (former USSR) Academy of Sciences. Since 1975, A. D. Polyanin has been a member of the staff of the Institute for Problems in Mechanics of the Russian Academy of Sciences.

Professor Polyanin has made important contributions to developing new exact and approximate analytical methods of the theory of differential equations, mathematical physics, integral equations, engineering mathematics, nonlinear mechanics, theory of heat and mass transfer, and chemical hydrodynamics. He obtained exact solutions for several thousand ordinary differential, partial differential, mathematical physics, and integral equations.

Professor Polyanin is an author of 27 books in English, Russian, German, and Bulgarian, as well as over 120 research papers and three patents. He has written a number of fundamental handbooks, including A. D. Polyanin and V. F. Zaitsev, *Handbook of Exact Solutions for Ordinary Differential Equations*, CRC Press, 1995; A. D. Polyanin and A. V. Manzhirov, *Handbook of Integral Equations*, CRC Press, 1998; and A. D. Polyanin, V. F. Zaitsev, and A. Moussiaux, *Handbook of First Order Partial Differential Equations*, Gordon and Breach, 2001.

In 1991, A. D. Polyanin was awarded a Chaplygin Prize of the USSR Academy of Sciences for his research in mechanics.

Address: Institute for Problems in Mechanics, RAS, 101 Vernadsky Avenue, Building 1, 117526 Moscow, Russia
E-mail: polyanin@ipmnet.ru

CONTENTS

Foreword iii

Basic Notation and Remarks v

Author vii

Introduction. Some Definitions, Formulas, Methods, and Solutions 1

0.1. Classification of Second-Order Partial Differential Equations 1
    0.1.1. Equations with Two Independent Variables 1
    0.1.2. Equations with Many Independent Variables 3

0.2. Basic Problems of Mathematical Physics 4
    0.2.1. Initial and Boundary Conditions. Cauchy Problem. Boundary Value Problems . 4
    0.2.2. First, Second, Third, and Mixed Boundary Value Problems 6

0.3. Properties and Particular Solutions of Linear Equations 7
    0.3.1. Homogeneous Linear Equations 7
    0.3.2. Nonhomogeneous Linear Equations 10

0.4. Separation of Variables Method 11
    0.4.1. General Description of the Separation of Variables Method 11
    0.4.2. Solution of Boundary Value Problems for Parabolic and Hyperbolic Equations . 15

0.5. Integral Transforms Method 17
    0.5.1. Main Integral Transforms 17
    0.5.2. Laplace Transform and Its Application in Mathematical Physics 18
    0.5.3. Fourier Transform and Its Application in Mathematical Physics 21

0.6. Representation of the Solution of the Cauchy Problem via the Fundamental Solution . . 23
    0.6.1. Cauchy Problem for Parabolic Equations 23
    0.6.2. Cauchy Problem for Hyperbolic Equations 24

0.7. Nonhomogeneous Boundary Value Problems with One Space Variable. Representation of Solutions via the Green's Function 25
    0.7.1. Problems for Parabolic Equations 25
    0.7.2. Problems for Hyperbolic Equations 27

0.8. Nonhomogeneous Boundary Value Problems with Many Space Variables. Representation of Solutions via the Green's Function 28
    0.8.1. Problems for Parabolic Equations 28
    0.8.2. Problems for Hyperbolic Equations 29
    0.8.3. Problems for Elliptic Equations 30
    0.8.4. Comparison of the Solution Structures for Boundary Value Problems for Equations of Various Types 31

0.9. Construction of the Green's Functions. General Formulas and Relations 32
    0.9.1. Green's Functions of Boundary Value Problems for Equations of Various Types in Bounded Domains 32
    0.9.2. Green's Functions Admitting Incomplete Separation of Variables 33
    0.9.3. Construction of Green's Functions via Fundamental Solutions 35

0.10.	Duhamel's Principles in Nonstationary Problems	38
	0.10.1. Problems for Homogeneous Linear Equations	38
	0.10.2. Problems for Nonhomogeneous Linear Equations	39
0.11.	Transformations Simplifying Initial and Boundary Conditions	40
	0.11.1. Transformations That Lead to Homogeneous Boundary Conditions	40
	0.11.2. Transformations That Lead to Homogeneous Initial and Boundary Conditions	41

1. Parabolic Equations with One Space Variable 43

1.1. Constant Coefficient Equations 43
    1.1.1. Heat Equation $\frac{\partial w}{\partial t} = a\frac{\partial^2 w}{\partial x^2}$ 43
    1.1.2. Equation of the Form $\frac{\partial w}{\partial t} = a\frac{\partial^2 w}{\partial x^2} + \Phi(x,t)$ 51
    1.1.3. Equation of the Form $\frac{\partial w}{\partial t} = a\frac{\partial^2 w}{\partial x^2} + bw + \Phi(x,t)$ 54
    1.1.4. Equation of the Form $\frac{\partial w}{\partial t} = a\frac{\partial^2 w}{\partial x^2} + b\frac{\partial w}{\partial x} + \Phi(x,t)$ 58
    1.1.5. Equation of the Form $\frac{\partial w}{\partial t} = a\frac{\partial^2 w}{\partial x^2} + b\frac{\partial w}{\partial x} + cw + \Phi(x,t)$ 62

1.2. Heat Equation with Axial or Central Symmetry and Related Equations 65
    1.2.1. Equation of the Form $\frac{\partial w}{\partial t} = a\left(\frac{\partial^2 w}{\partial r^2} + \frac{1}{r}\frac{\partial w}{\partial r}\right)$ 65
    1.2.2. Equation of the Form $\frac{\partial w}{\partial t} = a\left(\frac{\partial^2 w}{\partial r^2} + \frac{1}{r}\frac{\partial w}{\partial r}\right) + \Phi(r,t)$ 70
    1.2.3. Equation of the Form $\frac{\partial w}{\partial t} = a\left(\frac{\partial^2 w}{\partial r^2} + \frac{2}{r}\frac{\partial w}{\partial r}\right)$ 73
    1.2.4. Equation of the Form $\frac{\partial w}{\partial t} = a\left(\frac{\partial^2 w}{\partial r^2} + \frac{2}{r}\frac{\partial w}{\partial r}\right) + \Phi(r,t)$ 79
    1.2.5. Equation of the Form $\frac{\partial w}{\partial t} = \frac{\partial^2 w}{\partial x^2} + \frac{1-2\beta}{x}\frac{\partial w}{\partial x}$ 81
    1.2.6. Equation of the Form $\frac{\partial w}{\partial t} = \frac{\partial^2 w}{\partial x^2} + \frac{1-2\beta}{x}\frac{\partial w}{\partial x} + \Phi(x,t)$ 84

1.3. Equations Containing Power Functions and Arbitrary Parameters 85
    1.3.1. Equations of the Form $\frac{\partial w}{\partial t} = a\frac{\partial^2 w}{\partial x^2} + f(x,t)w$ 85
    1.3.2. Equations of the Form $\frac{\partial w}{\partial t} = a\frac{\partial^2 w}{\partial x^2} + f(x,t)\frac{\partial w}{\partial x}$ 90
    1.3.3. Equations of the Form $\frac{\partial w}{\partial t} = a\frac{\partial^2 w}{\partial x^2} + f(x,t)\frac{\partial w}{\partial x} + g(x,t)w + h(x,t)$ 93
    1.3.4. Equations of the Form $\frac{\partial w}{\partial t} = (ax+b)\frac{\partial^2 w}{\partial x^2} + f(x,t)\frac{\partial w}{\partial x} + g(x,t)w$ 95
    1.3.5. Equations of the Form $\frac{\partial w}{\partial t} = (ax^2+bx+c)\frac{\partial^2 w}{\partial x^2} + f(x,t)\frac{\partial w}{\partial x} + g(x,t)w$ 98
    1.3.6. Equations of the Form $\frac{\partial w}{\partial t} = f(x)\frac{\partial^2 w}{\partial x^2} + g(x,t)\frac{\partial w}{\partial x} + h(x,t)w$ 100
    1.3.7. Equations of the Form $\frac{\partial w}{\partial t} = f(x,t)\frac{\partial^2 w}{\partial x^2} + g(x,t)\frac{\partial w}{\partial x} + h(x,t)w$ 104
    1.3.8. Liquid-Film Mass Transfer Equation $(1-y^2)\frac{\partial w}{\partial x} = a\frac{\partial^2 w}{\partial y^2}$ 105
    1.3.9. Equations of the Form $f(x,y)\frac{\partial w}{\partial x} + g(x,y)\frac{\partial w}{\partial y} = \frac{\partial^2 w}{\partial y^2} + h(x,y)$ 107

1.4. Equations Containing Exponential Functions and Arbitrary Parameters 108
    1.4.1. Equations of the Form $\frac{\partial w}{\partial t} = a\frac{\partial^2 w}{\partial x^2} + f(x,t)w$ 108
    1.4.2. Equations of the Form $\frac{\partial w}{\partial t} = a\frac{\partial^2 w}{\partial x^2} + f(x,t)\frac{\partial w}{\partial x}$ 110
    1.4.3. Equations of the Form $\frac{\partial w}{\partial t} = a\frac{\partial^2 w}{\partial x^2} + f(x,t)\frac{\partial w}{\partial x} + g(x,t)w$ 113
    1.4.4. Equations of the Form $\frac{\partial w}{\partial t} = ax^n\frac{\partial^2 w}{\partial x^2} + f(x,t)\frac{\partial w}{\partial x} + g(x,t)w$ 114
    1.4.5. Equations of the Form $\frac{\partial w}{\partial t} = ae^{\beta x}\frac{\partial^2 w}{\partial x^2} + f(x,t)\frac{\partial w}{\partial x} + g(x,t)w$ 114
    1.4.6. Other Equations 117

1.5. Equations Containing Hyperbolic Functions and Arbitrary Parameters 117
    1.5.1. Equations Containing a Hyperbolic Cosine 117
    1.5.2. Equations Containing a Hyperbolic Sine 118
    1.5.3. Equations Containing a Hyperbolic Tangent 119
    1.5.4. Equations Containing a Hyperbolic Cotangent 120

1.6. Equations Containing Logarithmic Functions and Arbitrary Parameters 121
    1.6.1. Equations of the Form $\frac{\partial w}{\partial t} = a\frac{\partial^2 w}{\partial x^2} + f(x,t)\frac{\partial w}{\partial x} + g(x,t)w$ 121
    1.6.2. Equations of the Form $\frac{\partial w}{\partial t} = ax^k\frac{\partial^2 w}{\partial x^2} + f(x,t)\frac{\partial w}{\partial x} + g(x,t)w$ 122
1.7. Equations Containing Trigonometric Functions and Arbitrary Parameters 123
    1.7.1. Equations Containing a Cosine 123
    1.7.2. Equations Containing a Sine 124
    1.7.3. Equations Containing a Tangent 125
    1.7.4. Equations Containing a Cotangent 126
1.8. Equations Containing Arbitrary Functions 127
    1.8.1. Equations of the Form $\frac{\partial w}{\partial t} = a\frac{\partial^2 w}{\partial x^2} + f(x,t)w$ 127
    1.8.2. Equations of the Form $\frac{\partial w}{\partial t} = a\frac{\partial^2 w}{\partial x^2} + f(x,t)\frac{\partial w}{\partial x}$ 130
    1.8.3. Equations of the Form $\frac{\partial w}{\partial t} = a\frac{\partial^2 w}{\partial x^2} + f(x,t)\frac{\partial w}{\partial x} + g(x,t)w$ 133
    1.8.4. Equations of the Form $\frac{\partial w}{\partial t} = ax^n\frac{\partial^2 w}{\partial x^2} + f(x,t)\frac{\partial w}{\partial x} + g(x,t)w$ 136
    1.8.5. Equations of the Form $\frac{\partial w}{\partial t} = ae^{\beta x}\frac{\partial^2 w}{\partial x^2} + f(x,t)\frac{\partial w}{\partial x} + g(x,t)w$ 137
    1.8.6. Equations of the Form $\frac{\partial w}{\partial t} = f(x)\frac{\partial^2 w}{\partial x^2} + g(x,t)\frac{\partial w}{\partial x} + h(x,t)w$ 138
    1.8.7. Equations of the Form $\frac{\partial w}{\partial t} = f(t)\frac{\partial^2 w}{\partial x^2} + g(x,t)\frac{\partial w}{\partial x} + h(x,t)w$ 146
    1.8.8. Equations of the Form $\frac{\partial w}{\partial t} = f(x,t)\frac{\partial^2 w}{\partial x^2} + g(x,t)\frac{\partial w}{\partial x} + h(x,t)w$ 148
    1.8.9. Equations of the Form $s(x)\frac{\partial w}{\partial t} = \frac{\partial}{\partial x}\left[p(x)\frac{\partial w}{\partial x}\right] - q(x)w + \Phi(x,t)$ 151
1.9. Equations of Special Form 155
    1.9.1. Equations of the Diffusion (Thermal) Boundary Layer 155
    1.9.2. One-Dimensional Schrödinger Equation $i\hbar\frac{\partial w}{\partial t} = -\frac{\hbar^2}{2m}\frac{\partial^2 w}{\partial x^2} + U(x)w$ 157

2. Parabolic Equations with Two Space Variables 161
2.1. Heat Equation $\frac{\partial w}{\partial t} = a\Delta_2 w$ 161
    2.1.1. Boundary Value Problems in Cartesian Coordinates 161
    2.1.2. Problems in Polar Coordinates 173
    2.1.3. Axisymmetric Problems 179
2.2. Heat Equation with a Source $\frac{\partial w}{\partial t} = a\Delta_2 w + \Phi(x,y,t)$ 187
    2.2.1. Problems in Cartesian Coordinates 187
    2.2.2. Problems in Polar Coordinates 194
    2.2.3. Axisymmetric Problems 195
2.3. Other Equations 198
    2.3.1. Equations Containing Arbitrary Parameters 198
    2.3.2. Equations Containing Arbitrary Functions 200

3. Parabolic Equations with Three or More Space Variables 205
3.1. Heat Equation $\frac{\partial w}{\partial t} = a\Delta_3 w$ 205
    3.1.1. Problems in Cartesian Coordinates 205
    3.1.2. Problems in Cylindrical Coordinates 225
    3.1.3. Problems in Spherical Coordinates 250
3.2. Heat Equation with Source $\frac{\partial w}{\partial t} = a\Delta_3 w + \Phi(x,y,z,t)$ 254
    3.2.1. Problems in Cartesian Coordinates 254
    3.2.2. Problems in Cylindrical Coordinates 257
    3.2.3. Problems in Spherical Coordinates 260
3.3. Other Equations with Three Space Variables 261
    3.3.1. Equations Containing Arbitrary Parameters 261
    3.3.2. Equations Containing Arbitrary Functions 263
    3.3.3. Equations of the Form $\rho(x,y,z)\frac{\partial w}{\partial t} = \text{div}[a(x,y,z)\nabla w] - q(x,y,z)w + \Phi(x,y,z,t)$ 266

3.4. Equations with n Space Variables 268
3.4.1. Equations of the Form $\frac{\partial w}{\partial t} = a\Delta_n w + \Phi(x_1,\ldots,x_n,t)$ 268
3.4.2. Other Equations Containing Arbitrary Parameters 270
3.4.3. Equations Containing Arbitrary Functions 271

4. Hyperbolic Equations with One Space Variable 279
4.1. Constant Coefficient Equations 279
4.1.1. Wave Equation $\frac{\partial^2 w}{\partial t^2} = a^2 \frac{\partial^2 w}{\partial x^2}$ 279
4.1.2. Equations of the Form $\frac{\partial^2 w}{\partial t^2} = a^2 \frac{\partial^2 w}{\partial x^2} + \Phi(x,t)$ 284
4.1.3. Equation of the Form $\frac{\partial^2 w}{\partial t^2} = a^2 \frac{\partial^2 w}{\partial x^2} - bw + \Phi(x,t)$ 287
4.1.4. Equation of the Form $\frac{\partial^2 w}{\partial t^2} = a^2 \frac{\partial^2 w}{\partial x^2} - b\frac{\partial w}{\partial x} + \Phi(x,t)$ 291
4.1.5. Equation of the Form $\frac{\partial^2 w}{\partial t^2} = a^2 \frac{\partial^2 w}{\partial x^2} + b\frac{\partial w}{\partial x} + cw + \Phi(x,t)$ 293
4.2. Wave Equation with Axial or Central Symmetry 295
4.2.1. Equations of the Form $\frac{\partial^2 w}{\partial t^2} = a^2\big(\frac{\partial^2 w}{\partial r^2} + \frac{1}{r}\frac{\partial w}{\partial r}\big)$ 295
4.2.2. Equation of the Form $\frac{\partial^2 w}{\partial t^2} = a^2\big(\frac{\partial^2 w}{\partial r^2} + \frac{1}{r}\frac{\partial w}{\partial r}\big) + \Phi(r,t)$ 298
4.2.3. Equation of the Form $\frac{\partial^2 w}{\partial t^2} = a^2\big(\frac{\partial^2 w}{\partial r^2} + \frac{2}{r}\frac{\partial w}{\partial r}\big)$ 298
4.2.4. Equation of the Form $\frac{\partial^2 w}{\partial t^2} = a^2\big(\frac{\partial^2 w}{\partial r^2} + \frac{2}{r}\frac{\partial w}{\partial r}\big) + \Phi(r,t)$ 301
4.2.5. Equation of the Form $\frac{\partial^2 w}{\partial t^2} = a^2\big(\frac{\partial^2 w}{\partial r^2} + \frac{1}{r}\frac{\partial w}{\partial r}\big) - bw + \Phi(r,t)$ 302
4.2.6. Equation of the Form $\frac{\partial^2 w}{\partial t^2} = a^2\big(\frac{\partial^2 w}{\partial r^2} + \frac{2}{r}\frac{\partial w}{\partial r}\big) - bw + \Phi(r,t)$ 305
4.3. Equations Containing Power Functions and Arbitrary Parameters 308
4.3.1. Equations of the Form $\frac{\partial^2 w}{\partial t^2} = (ax+b)\frac{\partial^2 w}{\partial x^2} + c\frac{\partial w}{\partial x} + kw + \Phi(x,t)$ 308
4.3.2. Equations of the Form $\frac{\partial^2 w}{\partial t^2} = (ax^2+b)\frac{\partial^2 w}{\partial x^2} + cx\frac{\partial w}{\partial x} + kw + \Phi(x,t)$ 312
4.3.3. Other Equations 314
4.4. Equations Containing the First Time Derivative 320
4.4.1. Equations of the Form $\frac{\partial^2 w}{\partial t^2} + k\frac{\partial w}{\partial t} = a^2\frac{\partial^2 w}{\partial x^2} + b\frac{\partial w}{\partial x} + cw + \Phi(x,t)$ 320
4.4.2. Equations of the Form $\frac{\partial^2 w}{\partial t^2} + k\frac{\partial w}{\partial t} = f(x)\frac{\partial^2 w}{\partial x^2} + g(x)\frac{\partial w}{\partial x} + h(x)w + \Phi(x,t)$ 326
4.4.3. Other Equations 331
4.5. Equations Containing Arbitrary Functions 333
4.5.1. Equations of the Form $s(x)\frac{\partial^2 w}{\partial t^2} = \frac{\partial}{\partial x}\big[p(x)\frac{\partial w}{\partial x}\big] - q(x)w + \Phi(x,t)$ 333
4.5.2. Equations of the Form $\frac{\partial^2 w}{\partial t^2} + a(t)\frac{\partial w}{\partial t} = b(t)\big\{\frac{\partial}{\partial x}\big[p(x)\frac{\partial w}{\partial x}\big] - q(x)w\big\} + \Phi(x,t)$ 335
4.5.3. Other Equations 337

5. Hyperbolic Equations with Two Space Variables 341
5.1. Wave Equation $\frac{\partial^2 w}{\partial t^2} = a^2\Delta_2 w$ 341
5.1.1. Problems in Cartesian Coordinates 341
5.1.2. Problems in Polar Coordinates 346
5.1.3. Axisymmetric Problems 351
5.2. Nonhomogeneous Wave Equation $\frac{\partial^2 w}{\partial t^2} = a^2\Delta_2 w + \Phi(x,y,t)$ 355
5.2.1. Problems in Cartesian Coordinates 355
5.2.2. Problems in Polar Coordinates 357
5.2.3. Axisymmetric Problems 360
5.3. Equations of the Form $\frac{\partial^2 w}{\partial t^2} = a^2\Delta_2 w - bw + \Phi(x,y,t)$ 362
5.3.1. Problems in Cartesian Coordinates 362
5.3.2. Problems in Polar Coordinates 366
5.3.3. Axisymmetric Problems 371

5.4. Telegraph Equation $\frac{\partial^2 w}{\partial t^2} + k\frac{\partial w}{\partial t} = a^2\Delta_2 w - bw + \Phi(x,y,t)$ 376
    5.4.1. Problems in Cartesian Coordinates 376
    5.4.2. Problems in Polar Coordinates 381
    5.4.3. Axisymmetric Problems 386
5.5. Other Equations with Two Space Variables 391

6. Hyperbolic Equations with Three or More Space Variables 393

6.1. Wave Equation $\frac{\partial^2 w}{\partial t^2} = a^2\Delta_3 w$ 393
    6.1.1. Problems in Cartesian Coordinates 393
    6.1.2. Problems in Cylindrical Coordinates 399
    6.1.3. Problems in Spherical Coordinates 408
6.2. Nonhomogeneous Wave Equation $\frac{\partial^2 w}{\partial t^2} = a^2\Delta_3 w + \Phi(x,y,z,t)$ 412
    6.2.1. Problems in Cartesian Coordinates 412
    6.2.2. Problems in Cylindrical Coordinates 412
    6.2.3. Problems in Spherical Coordinates 413
6.3. Equations of the Form $\frac{\partial^2 w}{\partial t^2} = a^2\Delta_3 w - bw + \Phi(x,y,z,t)$ 414
    6.3.1. Problems in Cartesian Coordinates 414
    6.3.2. Problems in Cylindrical Coordinates 420
    6.3.3. Problems in Spherical Coordinates 429
6.4. Telegraph Equation $\frac{\partial^2 w}{\partial t^2} + k\frac{\partial w}{\partial t} = a^2\Delta_3 w - bw + \Phi(x,y,z,t)$ 434
    6.4.1. Problems in Cartesian Coordinates 434
    6.4.2. Problems in Cylindrical Coordinates 438
    6.4.3. Problems in Spherical Coordinates 448
6.5. Other Equations with Three Space Variables 453
    6.5.1. Equations Containing Arbitrary Parameters 453
    6.5.2. Equation of the Form $\rho(x,y,z)\frac{\partial^2 w}{\partial t^2} = \text{div}\big[a(x,y,z)\nabla w\big] - q(x,y,z)w + \Phi(x,y,z,t)$ 453
6.6. Equations with n Space Variables 455
    6.6.1. Wave Equation $\frac{\partial^2 w}{\partial t^2} = a^2\Delta_n w$ 456
    6.6.2. Nonhomogeneous Wave Equation $\frac{\partial^2 w}{\partial t^2} = a^2\Delta_n w + \Phi(x_1,\ldots,x_n,t)$ 457
    6.6.3. Equations of the Form $\frac{\partial^2 w}{\partial t^2} = a^2\Delta_n w - bw + \Phi(x_1,\ldots,x_n,t)$ 460
    6.6.4. Equations Containing the First Time Derivative 463

7. Elliptic Equations with Two Space Variables 467

7.1. Laplace Equation $\Delta_2 w = 0$ 467
    7.1.1. Problems in Cartesian Coordinate System 467
    7.1.2. Problems in Polar Coordinate System 472
    7.1.3. Other Coordinate Systems. Conformal Mappings Method 476
7.2. Poisson Equation $\Delta_2 w = -\Phi(x)$ 478
    7.2.1. Preliminary Remarks. Solution Structure 478
    7.2.2. Problems in Cartesian Coordinate System 480
    7.2.3. Problems in Polar Coordinate System 485
    7.2.4. Arbitrary Shape Domain. Conformal Mappings Method 489
7.3. Helmholtz Equation $\Delta_2 w + \lambda w = -\Phi(x)$ 490
    7.3.1. General Remarks, Results, and Formulas 490
    7.3.2. Problems in Cartesian Coordinate System 494
    7.3.3. Problems in Polar Coordinate System 503
    7.3.4. Other Orthogonal Coordinate Systems. Elliptic Domain 508

7.4.	Other Equations	510
	7.4.1. Stationary Schrödinger Equation $\Delta_2 w = f(x, y)w$	510
	7.4.2. Convective Heat and Mass Transfer Equations	512
	7.4.3. Equations of Heat and Mass Transfer in Anisotropic Media	518
	7.4.4. Other Equations Arising in Applications	526
	7.4.5. Equations of the Form $a(x)\frac{\partial^2 w}{\partial x^2} + \frac{\partial^2 w}{\partial y^2} + b(x)\frac{\partial w}{\partial x} + c(x)w = -\Phi(x,y)$	529

8. Elliptic Equations with Three or More Space Variables 533

8.1.	Laplace Equation $\Delta_3 w = 0$	533
	8.1.1. Problems in Cartesian Coordinates	533
	8.1.2. Problems in Cylindrical Coordinates	535
	8.1.3. Problems in Spherical Coordinates	537
	8.1.4. Other Orthogonal Curvilinear Systems of Coordinates	539
8.2.	Poisson Equation $\Delta_3 w + \Phi(x) = 0$	539
	8.2.1. Preliminary Remarks. Solution Structure	539
	8.2.2. Problems in Cartesian Coordinates	544
	8.2.3. Problems in Cylindrical Coordinates	554
	8.2.4. Problems in Spherical Coordinates	558
8.3.	Helmholtz Equation $\Delta_3 w + \lambda w = -\Phi(x)$	561
	8.3.1. General Remarks, Results, and Formulas	561
	8.3.2. Problems in Cartesian Coordinates	567
	8.3.3. Problems in Cylindrical Coordinates	580
	8.3.4. Problems in Spherical Coordinates	588
	8.3.5. Other Orthogonal Curvilinear Coordinates	591
8.4.	Other Equations with Three Space Variables	593
	8.4.1. Equations Containing Arbitrary Functions	593
	8.4.2. Equations of the Form div $[a(x, y, z)\nabla w] - q(x, y, z)w = -\Phi(x, y, z)$	595
8.5.	Equations with n Space Variables	597
	8.5.1. Laplace Equation $\Delta_n w = 0$	597
	8.5.2. Other Equations	598

9. Higher-Order Partial Differential Equations 601

9.1.	Third-Order Partial Differential Equations	601
9.2.	Fourth-Order One-Dimensional Nonstationary Equations	602
	9.2.1. Equations of the Form $\frac{\partial w}{\partial t} + a^2 \frac{\partial^4 w}{\partial x^4} = \Phi(x, t)$	602
	9.2.2. Equations of the Form $\frac{\partial^2 w}{\partial t^2} + a^2 \frac{\partial^4 w}{\partial x^4} = 0$	605
	9.2.3. Equations of the Form $\frac{\partial^2 w}{\partial t^2} + a^2 \frac{\partial^4 w}{\partial x^4} = \Phi(x, t)$	606
	9.2.4. Equations of the Form $\frac{\partial^2 w}{\partial t^2} + a^2 \frac{\partial^4 w}{\partial x^4} + kw = \Phi(x, t)$	608
	9.2.5. Other Equations	611
9.3.	Two-Dimensional Nonstationary Fourth-Order Equations	613
	9.3.1. Equations of the Form $\frac{\partial w}{\partial t} + a^2 \left(\frac{\partial^4 w}{\partial x^4} + \frac{\partial^4 w}{\partial y^4}\right) = \Phi(x, y, t)$	613
	9.3.2. Two-Dimensional Equations of the Form $\frac{\partial^2 w}{\partial t^2} + a^2 \Delta\Delta w = 0$	615
	9.3.3. Three- and n-Dimensional Equations of the Form $\frac{\partial^2 w}{\partial t^2} + a^2 \Delta\Delta w = 0$	617
	9.3.4. Equations of the Form $\frac{\partial^2 w}{\partial t^2} + a^2 \Delta\Delta w + kw = \Phi(x, y, t)$	619
	9.3.5. Equations of the Form $\frac{\partial^2 w}{\partial t^2} + a^2 \left(\frac{\partial^4 w}{\partial x^4} + \frac{\partial^4 w}{\partial y^4}\right) + kw = \Phi(x, y, t)$	620
9.4.	Fourth-Order Stationary Equations	621
	9.4.1. Biharmonic Equation $\Delta\Delta w = 0$	621
	9.4.2. Equations of the Form $\Delta\Delta w = \Phi(x, y)$	625

	9.4.3. Equations of the Form $\Delta\Delta w - \lambda w = \Phi(x, y)$	626
	9.4.4. Equations of the Form $\frac{\partial^4 w}{\partial x^4} + \frac{\partial^4 w}{\partial y^4} = \Phi(x, y)$	628
	9.4.5. Equations of the Form $\frac{\partial^4 w}{\partial x^4} + \frac{\partial^4 w}{\partial y^4} + kw = \Phi(x, y)$	629
	9.4.6. Stokes Equation (Axisymmetric Flows of Viscous Fluids)	630
9.5.	Higher-Order Linear Equations with Constant Coefficients	633
	9.5.1. Fundamental Solutions. Cauchy Problem	633
	9.5.2. Elliptic Equations	635
	9.5.3. Hyperbolic Equations	637
	9.5.4. Regular Equations. Number of Initial Conditions in the Cauchy Problem	637
	9.5.5. Some Special-Type Equations	640
9.6.	Higher-Order Linear Equations with Variable Coefficients	644
	9.6.1. Equations Containing the First Time Derivative	644
	9.6.2. Equations Containing the Second Time Derivative	648
	9.6.3. Nonstationary Problems with Many Space Variables	649
	9.6.4. Some Special-Type Equations	651

Supplement A. Special Functions and Their Properties **655**

A.1.	Some Symbols and Coefficients	655
	A.1.1. Factorials	655
	A.1.2. Binomial Coefficients	655
	A.1.3. Pochhammer Symbol	656
	A.1.4. Bernoulli Numbers	656
A.2.	Error Functions and Exponential Integral	656
	A.2.1. Error Function and Complementary Error Function	656
	A.2.2. Exponential Integral	656
	A.2.3. Logarithmic Integral	657
A.3.	Sine Integral and Cosine Integral. Fresnel Integrals	657
	A.3.1. Sine Integral	657
	A.3.2. Cosine Integral	658
	A.3.3. Fresnel Integrals	658
A.4.	Gamma and Beta Functions	659
	A.4.1. Gamma Function	659
	A.4.2. Beta Function	660
A.5.	Incomplete Gamma and Beta Functions	660
	A.5.1. Incomplete Gamma Function	660
	A.5.2. Incomplete Beta Function	661
A.6.	Bessel Functions	661
	A.6.1. Definitions and Basic Formulas	661
	A.6.2. Integral Representations and Asymptotic Expansions	663
	A.6.3. Zeros and Orthogonality Properties of Bessel Functions	664
	A.6.4. Hankel Functions (Bessel Functions of the Third Kind)	665
A.7.	Modified Bessel Functions	666
	A.7.1. Definitions. Basic Formulas	666
	A.7.2. Integral Representations and Asymptotic Expansions	667
A.8.	Airy Functions	668
	A.8.1. Definition and Basic Formulas	668
	A.8.2. Power Series and Asymptotic Expansions	668

A.9. Degenerate Hypergeometric Functions 669
    A.9.1. Definitions and Basic Formulas 669
    A.9.2. Integral Representations and Asymptotic Expansions 671
A.10. Hypergeometric Functions 672
    A.10.1. Definition and Some Formulas 672
    A.10.2. Basic Properties and Integral Representations 672
A.11. Whittaker Functions 672
A.12. Legendre Polynomials and Legendre Functions 674
    A.12.1. Definitions. Basic Formulas 674
    A.12.2. Zeros of Legendre Polynomials and the Generating Function 674
    A.12.3. Associated Legendre Functions 674
A.13. Parabolic Cylinder Functions 675
    A.13.1. Definitions. Basic Formulas 675
    A.13.2. Integral Representations and Asymptotic Expansions 675
A.14. Mathieu Functions 675
    A.14.1. Definitions and Basic Formulas 675
A.15. Modified Mathieu Functions 677
A.16. Orthogonal Polynomials 677
    A.16.1. Laguerre Polynomials and Generalized Laguerre Polynomials 678
    A.16.2. Chebyshev Polynomials and Functions 679
    A.16.3. Hermite Polynomial 679
    A.16.4. Jacobi Polynomials 680

Supplement B. Methods of Generalized and Functional Separation of Variables in Nonlinear Equations of Mathematical Physics ... 681

B.1. Introduction 681
    B.1.1. Preliminary Remarks 681
    B.1.2. Simple Cases of Variable Separation in Nonlinear Equations 682
    B.1.3. Examples of Nontrivial Variable Separation in Nonlinear Equations 683
B.2. Methods of Generalized Separation of Variables 685
    B.2.1. Structure of Generalized Separable Solutions 685
    B.2.2. Solution of Functional Differential Equations by Differentiation 686
    B.2.3. Solution of Functional Differential Equations by Splitting 688
    B.2.4. Simplified Scheme for Constructing Exact Solutions of Equations with Quadratic Nonlinearities 691
B.3. Methods of Functional Separation of Variables 693
    B.3.1. Structure of Functional Separable Solutions 693
    B.3.2. Special Functional Separable Solutions 693
    B.3.3. Differentiation Method 696
    B.3.4. Splitting Method. Reduction to a Functional Equation with Two Variables ... 700
    B.3.5. Some Functional Equations and Their Solutions. Exact Solutions of Heat and Wave Equations 701
B.4. First-Order Nonlinear Equations 706
    B.4.1. Preliminary Remarks 706
    B.4.2. Individual Equations 706
B.5. Second-Order Nonlinear Equations 709
    B.5.1. Parabolic Equations 709
    B.5.2. Hyperbolic Equations 721
    B.5.3. Elliptic Equations 727
    B.5.4. Equations Containing Mixed Derivatives 733

	B.5.5. General Form Equations	736
B.6.	Third-Order Nonlinear Equations	739
	B.6.1. Stationary Hydrodynamic Boundary Layer Equations	739
	B.6.2. Nonstationary Hydrodynamic Boundary Layer Equations	741
B.7.	Fourth-Order Nonlinear Equations	749
	B.7.1. Stationary Hydrodynamic Equations (Navier–Stokes Equations)	749
	B.7.2. Nonstationary Hydrodynamic Equations	752
B.8.	Higher-Order Nonlinear Equations	757
	B.8.1. Equations of the Form $\frac{\partial w}{\partial t} = F\left(x, t, w, \frac{\partial w}{\partial x}, \ldots, \frac{\partial^n w}{\partial x^n}\right)$	757
	B.8.2. Equations of the Form $\frac{\partial^2 w}{\partial t^2} = F\left(x, t, w, \frac{\partial w}{\partial x}, \ldots, \frac{\partial^n w}{\partial x^n}\right)$	761
	B.8.3. Other Equations	765
References	....	769
Index	....	777

Introduction

Some Definitions, Formulas, Methods, and Solutions

0.1. Classification of Second-Order Partial Differential Equations

0.1.1. Equations with Two Independent Variables

0.1.1-1. Examples of equations encountered in applications.

Three basic types of partial differential equations are distinguished—*parabolic*, *hyperbolic*, and *elliptic*. The solutions of the equations pertaining to each of the types have their own characteristic qualitative differences.

The simplest example of a *parabolic* equation is the heat equation

$$\frac{\partial w}{\partial t} - \frac{\partial^2 w}{\partial x^2} = 0, \tag{1}$$

where the variables t and x play the role of time and the spatial coordinate, respectively. Note that equation (1) contains only one highest derivative term.

The simplest example of a *hyperbolic* equation is the wave equation

$$\frac{\partial^2 w}{\partial t^2} - \frac{\partial^2 w}{\partial x^2} = 0, \tag{2}$$

where the variables t and x play the role of time and the spatial coordinate, respectively. Note that the highest derivative terms in equation (2) differ in sign.

The simplest example of an *elliptic* equation is the Laplace equation

$$\frac{\partial^2 w}{\partial x^2} + \frac{\partial^2 w}{\partial y^2} = 0, \tag{3}$$

where x and y play the role of the spatial coordinates. Note that the highest derivative terms in equation (3) have like signs.

Any linear partial differential equation of the second-order with two independent variables can be reduced, by appropriate manipulations, to a simpler equation which has one of the three highest derivative combinations specified above in examples (1), (2), and (3).

0.1.1-2. Types of equations. Characteristic equations.

Consider a second-order partial differential equation with two independent variables which has the general form

$$a(x,y)\frac{\partial^2 w}{\partial x^2} + 2b(x,y)\frac{\partial^2 w}{\partial x \partial y} + c(x,y)\frac{\partial^2 w}{\partial y^2} = F\left(x, y, w, \frac{\partial w}{\partial x}, \frac{\partial w}{\partial y}\right), \tag{4}$$

where a, b, c are some functions of x and y that have continuous derivatives up to the second-order inclusive.*

Given a point (x, y), equation (4) is said to be

$$\begin{aligned} &\text{parabolic} && \text{if } b^2 - ac = 0, \\ &\text{hyperbolic} && \text{if } b^2 - ac > 0, \\ &\text{elliptic} && \text{if } b^2 - ac < 0 \end{aligned}$$

at this point.

In order to reduce equation (4) to a canonical form, one should first write out the characteristic equation

$$a\,dy^2 - 2b\,dx\,dy + c\,dx^2 = 0,$$

which splits into two equations

$$a\,dy - \left(b + \sqrt{b^2 - ac}\right) dx = 0, \tag{5}$$

and

$$a\,dy - \left(b - \sqrt{b^2 - ac}\right) dx = 0, \tag{6}$$

and find their general integrals.

0.1.1-3. Canonical form of parabolic equations (case $b^2 - ac = 0$).

In this case, equations (5) and (6) coincide and have a common general integral,

$$\varphi(x, y) = C.$$

By passing from x, y to new independent variables ξ, η in accordance with the relations

$$\xi = \varphi(x, y), \qquad \eta = \eta(x, y),$$

where $\eta = \eta(x, y)$ is any twice differentiable function that satisfies the condition of nondegeneracy of the Jacobian $\frac{D(\xi, \eta)}{D(x, y)}$ in the given domain, we reduce equation (4) to the canonical form

$$\frac{\partial^2 w}{\partial \eta^2} = F_1\left(\xi, \eta, w, \frac{\partial w}{\partial \xi}, \frac{\partial w}{\partial \eta}\right). \tag{7}$$

As η, one can take $\eta = x$ or $\eta = y$.

It is apparent that, just as the heat equation (1), the transformed equation (7) has only one highest-derivative term.

Remark. In the degenerate case where the function F_1 does not depend on the derivative $\partial_\xi w$, equation (7) is an ordinary differential equation for the variable η, in which ξ serves as a parameter.

0.1.1-4. Canonical form of hyperbolic equations (case $b^2 - ac > 0$).

The general integrals

$$\varphi(x, y) = C_1, \qquad \psi(x, y) = C_2$$

of equations (5) and (6) are real and different. These integrals determine two different families of real characteristics.

---

* The right-hand side of equation (4) may be nonlinear. The classification and the procedure of reducing such equations to a canonical form are only determined by the left-hand side of the equation.

By passing from x, y to new independent variables ξ, η in accordance with the relations

$$\xi = \varphi(x, y), \qquad \eta = \psi(x, y),$$

we reduce equation (4) to

$$\frac{\partial^2 w}{\partial \xi \partial \eta} = F_2\left(\xi, \eta, w, \frac{\partial w}{\partial \xi}, \frac{\partial w}{\partial \eta}\right).$$

This is the so-called first canonical form of a hyperbolic equation.

The transformation

$$\xi = t + z, \qquad \eta = t - z$$

brings the above equation to another canonical form,

$$\frac{\partial^2 w}{\partial t^2} - \frac{\partial^2 w}{\partial z^2} = F_3\left(t, z, w, \frac{\partial w}{\partial t}, \frac{\partial w}{\partial z}\right),$$

where $F_3 = 4F_2$. This is the so-called second canonical form of a hyperbolic equation. Apart from notation, the left-hand side of the last equation coincides with that of the wave equation (2).

0.1.1-5. Canonical form of elliptic equations (case $b^2 - ac < 0$).

In this case the general integrals of equations (5) and (6) are complex conjugate; these determine two families of complex characteristics.

Let the general integral of equation (5) have the form

$$\varphi(x, y) + i\psi(x, y) = C, \qquad i^2 = -1,$$

where $\varphi(x, y)$ and $\psi(x, y)$ are real-valued functions.

By passing from x, y to new independent variables ξ, η in accordance with the relations

$$\xi = \varphi(x, y), \qquad \eta = \psi(x, y),$$

we reduce equation (4) to the canonical form

$$\frac{\partial^2 w}{\partial \xi^2} + \frac{\partial^2 w}{\partial \eta^2} = F_4\left(\xi, \eta, w, \frac{\partial w}{\partial \xi}, \frac{\partial w}{\partial \eta}\right).$$

Apart from notation, the left-hand side of the last equation coincides with that of the Laplace equation (3).

0.1.2. Equations with Many Independent Variables

Consider a second-order partial differential equation with n independent variables $x_1, \ldots, x_n$ that has the form

$$\sum_{i,j=1}^{n} a_{ij}(\mathbf{x}) \frac{\partial^2 w}{\partial x_i \partial x_j} = F\left(\mathbf{x}, w, \frac{\partial w}{\partial x_1}, \ldots, \frac{\partial w}{\partial x_n}\right), \qquad (8)$$

where a_{ij} are some functions that have continuous derivatives with respect to all variables to the second-order inclusive, and $\mathbf{x} = \{x_1, \ldots, x_n\}$. [The right-hand side of equation (8) may be nonlinear. The left-hand side only is required for the classification of this equation.]

At a point $\mathbf{x} = \mathbf{x}_0$, the following quadratic form is assigned to equation (8):

$$Q = \sum_{i,j=1}^{n} a_{ij}(\mathbf{x}_0) \xi_i \xi_j. \qquad (9)$$

TABLE 1
Classification of equations with many independent variables

Type of equation (8) at a point $\mathbf{x} = \mathbf{x}_0$	Coefficients of the canonical form (11)
Parabolic (in the broad sense)	At least one coefficient of the c_i is zero
Hyperbolic (in the broad sense)	All c_i are nonzero and some c_i differ in sign
Elliptic	All c_i are nonzero and have like signs

By an appropriate linear nondegenerate transformation

$$\xi_i = \sum_{k=1}^{n} \beta_{ik} \eta_k \qquad (i = 1, \ldots, n) \tag{10}$$

the quadratic form (9) can be reduced to the canonical form

$$Q = \sum_{i=1}^{n} c_i \eta_i^2, \tag{11}$$

where the coefficients c_i assume the values 1, −1, and 0. The number of negative and zero coefficients in (11) does not depend on the way in which the quadratic form is reduced to the canonical form.

Table 1 presents the basic criteria according to which the equations with many independent variables are classified.

Suppose all coefficients of the highest derivatives in (8) are constant, $a_{ij} = \text{const}$. By introducing the new independent variables $y_1, \ldots, y_n$ in accordance with the formulas $y_i = \sum_{k=1}^{n} \beta_{ik} x_k$, where the β_{ik} are the coefficients of the linear transformation (10), we reduce equation (8) to the canonical form

$$\sum_{i=1}^{n} c_i \frac{\partial^2 w}{\partial y_i^2} = F_1\left(\mathbf{y}, w, \frac{\partial w}{\partial y_1}, \ldots, \frac{\partial w}{\partial y_n}\right). \tag{12}$$

Here, the coefficients c_i are the same as in the quadratic form (11), and $\mathbf{y} = \{y_1, \ldots, y_n\}$.

Remark 1. Among the parabolic equations, it is conventional to distinguish the parabolic equations in the narrow sense, i.e., the equations for which only one of the coefficients, c_k, is zero, while the other c_i is the same, and in this case the right-hand side of equation (12) must contain the first-order partial derivative with respect to y_k.

Remark 2. In turn, the hyperbolic equations are divided into normal hyperbolic equations—for which all c_i but one have like signs—and ultrahyperbolic equations—for which there are two or more positive c_i and two or more negative c_i.

Specific equations of parabolic, elliptic, and hyperbolic types will be discussed further in Subsection 0.2.

⊙ *References for Section* 0.1: V. M. Babich, M. B. Kapilevich, S. G. Mikhlin, et al. (1964), S. J. Farlow (1982), D. Colton (1988), E. Zauderer (1989), A. N. Tikhonov and A. A. Samarskii (1990), I. G. Petrovsky (1991), W. A. Strauss (1992), R. B. Guenther and J. W. Lee (1996), D. Zwillinger (1998).

0.2. Basic Problems of Mathematical Physics

0.2.1. Initial and Boundary Conditions. Cauchy Problem. Boundary Value Problems

Every equation of mathematical physics governs infinitely many qualitatively similar phenomena or processes. This follows from the fact that differential equations have infinitely many particular solutions. The specific solution that describes the physical phenomenon under study is separated from the set of particular solutions of the given differential equation by means of the initial and boundary conditions.

Throughout this section, we consider linear equations in the n-dimensional Euclidean space $\mathbb{R}^n$ or in an open domain $V \in \mathbb{R}^n$ (exclusive of the boundary) with a sufficiently smooth boundary $S = \partial V$.

0.2.1-1. Parabolic equations. Initial and boundary conditions.

In general, a linear second-order partial differential equation of the parabolic type with n independent variables can be written as

$$\frac{\partial w}{\partial t} - L_{\mathbf{x},t}[w] = \Phi(\mathbf{x}, t), \tag{1}$$

where

$$L_{\mathbf{x},t}[w] \equiv \sum_{i,j=1}^{n} a_{ij}(\mathbf{x}, t) \frac{\partial^2 w}{\partial x_i \partial x_j} + \sum_{i=1}^{n} b_i(\mathbf{x}, t) \frac{\partial w}{\partial x_i} + c(\mathbf{x}, t) w, \tag{2}$$

$$\mathbf{x} = \{x_1, \dots, x_n\}, \quad \sum_{i,j=1}^{n} a_{ij}(\mathbf{x}, t) \xi_i \xi_j \geq \sigma \sum_{i=1}^{n} \xi_i^2, \quad \sigma > 0.$$

Parabolic equations govern unsteady thermal, diffusion, and other phenomena dependent on time t.

Equation (1) is called homogeneous if $\Phi(\mathbf{x}, t) \equiv 0$.

Cauchy problem ($t \geq 0$, $\mathbf{x} \in \mathbb{R}^n$). Find a function w that satisfies equation (1) for $t > 0$ and the initial condition

$$w = f(\mathbf{x}) \quad \text{at} \quad t = 0. \tag{3}$$

Boundary value problem* ($t \geq 0$, $\mathbf{x} \in V$). Find a function w that satisfies equation (1) for $t > 0$, the initial condition (3), and the boundary condition

$$\Gamma_{\mathbf{x},t}[w] = g(\mathbf{x}, t) \quad \text{at} \quad \mathbf{x} \in S \quad (t > 0). \tag{4}$$

In general, $\Gamma_{\mathbf{x},t}$ is a first-order linear differential operator in the space variables $\mathbf{x}$ with coefficient dependent on $\mathbf{x}$ and t. The basic types of boundary conditions are described below in Subsection 0.2.2.

The initial condition (3) is called homogeneous if $f(\mathbf{x}) \equiv 0$. The boundary condition (4) is called homogeneous if $g(\mathbf{x}, t) \equiv 0$.

0.2.1-2. Hyperbolic equations. Initial and boundary conditions.

Consider a second-order linear partial differential equation of the hyperbolic type with n independent variables of the general form

$$\frac{\partial^2 w}{\partial t^2} + \varphi(\mathbf{x}, t) \frac{\partial w}{\partial t} - L_{\mathbf{x},t}[w] = \Phi(\mathbf{x}, t), \tag{5}$$

where the linear differential operator $L_{\mathbf{x},t}$ is defined by (2). Hyperbolic equations govern unsteady wave processes, which depend on time t.

Equation (5) is said to be homogeneous if $\Phi(\mathbf{x}, t) \equiv 0$.

Cauchy problem ($t \geq 0$, $\mathbf{x} \in \mathbb{R}^n$). Find a function w that satisfies equation (5) for $t > 0$ and the initial conditions

$$\begin{aligned} w &= f_0(\mathbf{x}) \quad \text{at} \quad t = 0, \\ \partial_t w &= f_1(\mathbf{x}) \quad \text{at} \quad t = 0. \end{aligned} \tag{6}$$

---

* *Boundary value problems* for parabolic and hyperbolic equations are sometimes called *mixed* or *initial-boundary value problems*.

Boundary value problem ($t \geq 0$, $\mathbf{x} \in V$). Find a function w that satisfies equation (5) for $t > 0$, the initial conditions (6), and boundary condition (4).

The initial conditions (6) are called homogeneous if $f_0(\mathbf{x}) \equiv 0$ and $f_1(\mathbf{x}) \equiv 0$.

Goursat problem. On the characteristics of a hyperbolic equation with two independent variables, the values of the unknown function w are prescribed.

0.2.1-3. Elliptic equations. Boundary conditions.

In general, a second-order linear partial differential equation of elliptic type with n independent variables can be written as

$$-L_{\mathbf{x}}[w] = \Phi(\mathbf{x}), \tag{7}$$

where

$$L_{\mathbf{x}}[w] \equiv \sum_{i,j=1}^{n} a_{ij}(\mathbf{x}) \frac{\partial^2 w}{\partial x_i \partial x_j} + \sum_{i=1}^{n} b_i(\mathbf{x}) \frac{\partial w}{\partial x_i} + c(\mathbf{x}) w, \tag{8}$$

$$\sum_{i,j=1}^{n} a_{ij}(\mathbf{x}) \xi_i \xi_j \geq \sigma \sum_{i=1}^{n} \xi_i^2, \quad \sigma > 0.$$

Elliptic equations govern steady-state thermal, diffusion, and other phenomena independent of time t.

Equation (7) is said to be homogeneous if $\Phi(\mathbf{x}) \equiv 0$.

Boundary value problem. Find a function w that satisfies equation (7) and the boundary condition

$$\Gamma_{\mathbf{x}}[w] = g(\mathbf{x}) \quad \text{at} \quad \mathbf{x} \in S. \tag{9}$$

In general, $\Gamma_{\mathbf{x}}$ is a first-order linear differential operator in the space variables $\mathbf{x}$. The basic types of boundary conditions are described below in Subsection 0.2.2.

The boundary condition (9) is called homogeneous if $g(\mathbf{x}) \equiv 0$. The boundary value problem (7)–(9) is said to be homogeneous if $\Phi \equiv 0$ and $g \equiv 0$.

0.2.2. First, Second, Third, and Mixed Boundary Value Problems

For any (parabolic, hyperbolic, and elliptic) second-order partial differential equations, it is conventional to distinguish four basic types of boundary value problems, depending on the form of the boundary conditions (4) [see also the analogous equation (9)]. For simplicity, here we confine ourselves to the case where the coefficients a_{ij} of equations (1) and (5) have the special form

$$a_{ij}(\mathbf{x}, t) = a(\mathbf{x}, t) \delta_{ij}, \quad \delta_{ij} = \begin{cases} 1 & \text{if } i = j, \\ 0 & \text{if } i \neq j. \end{cases}$$

This situation is rather frequent in applications; such coefficients are used to describe various phenomena (processes) in isotropic media.

First boundary value problem. The function $w(\mathbf{x}, t)$ takes prescribed values at the boundary S of the domain,

$$w(\mathbf{x}, t) = g_1(\mathbf{x}, t) \quad \text{for} \quad \mathbf{x} \in S. \tag{10}$$

Second boundary value problem. The derivative along the (outward) normal is prescribed at the boundary S of the domain,

$$\frac{\partial w}{\partial N} = g_2(\mathbf{x}, t) \quad \text{for} \quad \mathbf{x} \in S. \tag{11}$$

In heat transfer problems, where w is temperature, the left-hand side of the boundary condition (11) is proportional to the heat flux per unit area of the surface S.

Third boundary value problem. A linear relationship between the unknown function and its normal derivative is prescribed at the boundary S of the domain,

$$\frac{\partial w}{\partial N} + k(\mathbf{x}, t)w = g_3(\mathbf{x}, t) \quad \text{for} \quad \mathbf{x} \in S. \tag{12}$$

Usually, it is assumed that $k(\mathbf{x}, t) = \text{const}$. In mass transfer problems, where w is concentration, the boundary condition (12) with $g_3 \equiv 0$ describes a surface chemical reaction of the first order.

Mixed boundary value problems. Conditions of various types, listed above, are set at different portions of the boundary S.

If $g_1 \equiv 0$, $g_2 \equiv 0$, or $g_3 \equiv 0$, the respective boundary conditions (10), (11), (12) are said to be homogeneous.

⊙ *References for Section* 0.2: V. M. Babich, M. B. Kapilevich, S. G. Mikhlin, et al. (1964), M. A. Pinsky (1984), R. Leis (1986), R. Haberman (1987), A. A. Dezin (1987), A. G. Mackie (1989), A. N. Tikhonov and A. A. Samarskii (1990), I. Stakgold (2000).

0.3. Properties and Particular Solutions of Linear Equations

0.3.1. Homogeneous Linear Equations

0.3.1-1. Preliminary remarks.

For brevity, in this paragraph a homogeneous linear partial differential equation will be written as

$$\mathfrak{L}[w] = 0. \tag{1}$$

For second-order linear parabolic and hyperbolic equations, the linear differential operator $\mathfrak{L}[w]$ is defined by the left-hand side of equations (1) and (5) from Subsection 0.2.1, respectively. It is assumed that equation (1) is an arbitrary homogeneous linear partial differential equation of any order in the variables $t, x_1, \ldots, x_n$ with sufficiently smooth coefficients.

A linear operator $\mathfrak{L}$ possesses the properties

$$\mathfrak{L}[w_1 + w_2] = \mathfrak{L}[w_1] + \mathfrak{L}[w_2],$$
$$\mathfrak{L}[Aw] = A\mathfrak{L}[w], \quad A = \text{const}.$$

An arbitrary homogeneous linear equation (1) has a trivial solution, $w \equiv 0$.

A function w is called a classical solution of equation (1) if w, when substituted into (1), turns the equation into an identity and if all partial derivatives of w that occur in (1) are continuous; the notion of a classical solution is directly linked to the range of the independent variables. In what follows, we usually write "solution" instead of "classical solution" for brevity.

0.3.1-2. Usage of particular solutions for the construction of other particular solutions.

Below are some properties of particular solutions of homogeneous linear equations.

$1°$. Let $w_1 = w_1(\mathbf{x}, t), w_2 = w_2(\mathbf{x}, t), \ldots, w_k = w_k(\mathbf{x}, t)$ be any particular solutions of the homogeneous equation (1). Then the linear combination

$$w = A_1 w_1 + A_2 w_2 + \cdots + A_k w_k \tag{2}$$

with arbitrary constants $A_1, A_2, \ldots, A_k$ is also a solution of equation (1); in physics, this property is known as the *principle of linear superposition*.

Suppose $\{w_k\}$ is an infinite sequence of solutions of equation (1). Then the series $\sum_{k=1}^{\infty} w_k$, irrespective of its convergence, is called a formal solution of (1). If the solutions w_k are classical, the series is uniformly convergent, and the sum of the series has all the necessary particular derivatives, then the sum of the series is a classical solution of equation (1).

2°. Let the coefficients of the differential operator $\mathcal{L}$ be independent of time t. If equation (1) has a particular solution $\widetilde{w} = \widetilde{w}(\mathbf{x}, t)$, then the partial derivatives of $\widetilde{w}$ with respect to time,*

$$\frac{\partial \widetilde{w}}{\partial t}, \quad \frac{\partial^2 \widetilde{w}}{\partial t^2}, \quad \ldots, \quad \frac{\partial^k \widetilde{w}}{\partial t^k}, \quad \ldots,$$

are also solutions of equation (1)

3°. Let the coefficients of the differential operator $\mathcal{L}$ be independent of the space variables $x_1, \ldots, x_n$. If equation (1) has a particular solution $\widetilde{w} = \widetilde{w}(\mathbf{x}, t)$, then the partial derivatives of $\widetilde{w}$ with respect to the space coordinates,

$$\frac{\partial \widetilde{w}}{\partial x_1}, \quad \frac{\partial \widetilde{w}}{\partial x_2}, \quad \frac{\partial \widetilde{w}}{\partial x_3}, \quad \ldots, \quad \frac{\partial^2 \widetilde{w}}{\partial x_1^2}, \quad \frac{\partial^2 \widetilde{w}}{\partial x_1 \partial x_2}, \quad \ldots, \quad \frac{\partial^{k+m} \widetilde{w}}{\partial x_2^k \partial x_3^m}, \quad \ldots,$$

are also solutions of equation (1)

If the coefficients of $\mathcal{L}$ are independent of only one space coordinate, say x_1, and equation (1) has a particular solution $\widetilde{w} = \widetilde{w}(\mathbf{x}, t)$, then the partial derivatives

$$\frac{\partial \widetilde{w}}{\partial x_1}, \quad \frac{\partial^2 \widetilde{w}}{\partial x_1^2}, \quad \ldots, \quad \frac{\partial^k \widetilde{w}}{\partial x_1^k}, \quad \ldots$$

are also solutions of equation (1).

4°. Let the coefficients of $\mathcal{L}$ be constant and let equation (1) have a particular solution $\widetilde{w} = \widetilde{w}(\mathbf{x}, t)$. Then any particular derivatives of $\widetilde{w}$ with respect to time and the space coordinates (inclusive mixed derivatives),

$$\frac{\partial \widetilde{w}}{\partial t}, \quad \frac{\partial \widetilde{w}}{\partial x_1}, \quad \ldots, \quad \frac{\partial^2 \widetilde{w}}{\partial x_2^2}, \quad \frac{\partial^2 \widetilde{w}}{\partial t \partial x_1}, \quad \ldots, \quad \frac{\partial^k \widetilde{w}}{\partial x_3^k}, \quad \ldots,$$

are solutions of equation (1).

5°. Suppose equation (1) has a particular solution dependent on a parameter μ, $\widetilde{w} = \widetilde{w}(\mathbf{x}, t; \mu)$, and the coefficients of $\mathcal{L}$ are independent of μ (but can depend on time and the space coordinates). Then, by differentiating $\widetilde{w}$ with respect to μ, one obtains other solutions of equation (1),

$$\frac{\partial \widetilde{w}}{\partial \mu}, \quad \frac{\partial^2 \widetilde{w}}{\partial \mu^2}, \quad \ldots, \quad \frac{\partial^k \widetilde{w}}{\partial \mu^k}, \quad \ldots$$

Let some constants $\mu_1, \ldots, \mu_k$ belong to the range of the parameter μ. Then the sum

$$w = A_1 \widetilde{w}(\mathbf{x}, t; \mu_1) + \cdots + A_k \widetilde{w}(\mathbf{x}, t; \mu_k), \tag{3}$$

where $A_1, \ldots, A_k$ are arbitrary constants, is also a solution of the homogeneous linear equation (1). The number of terms in sum (3) can be both finite and infinite.

6°. Another effective way of constructing solutions involves the following. The particular solution $\widetilde{w}(\mathbf{x}, t; \mu)$, which depends on the parameter μ (as before, it is assumed that the coefficients of $\mathcal{L}$ are independent of μ), is first multiplied by an arbitrary function $\varphi(\mu)$. Then the resulting expression is integrated with respect to μ over some interval $[\alpha, \beta]$. Thus, one obtains a new function,

$$\int_\alpha^\beta \widetilde{w}(\mathbf{x}, t; \mu) \varphi(\mu) \, d\mu,$$

which is also a solution of the original homogeneous linear equation.

The properties listed in Items 1°–6° enable one to use known particular solutions to construct other particular solutions of homogeneous linear equations of mathematical physics.

---

* Here and in what follows, it is assumed that the particular solution $\widetilde{w}$ is differentiable sufficiently many times with respect to t and $x_1, \ldots, x_n$ (or the parameters).

TABLE 2
Homogeneous linear partial differential equations that admit separable solutions

No	Form of equation (1)	Form of particular solutions
1	Equation coefficients are constant	$w(\mathbf{x}, t) = A \exp(\lambda t + \beta_1 x_1 + \cdots + \beta_n x_n)$, $\lambda, \beta_1, \ldots, \beta_n$ are related by an algebraic equation
2	Equation coefficients are independent of time t	$w(\mathbf{x}, t) = e^{\lambda t}\psi(\mathbf{x})$, λ is an arbitrary constant, $\mathbf{x} = \{x_1, \ldots, x_n\}$
3	Equation coefficients are independent of the coordinates $x_1, \ldots, x_n$	$w(\mathbf{x}, t) = \exp(\beta_1 x_1 + \cdots + \beta_n x_n)\psi(t)$, $\beta_1, \ldots, \beta_n$ are arbitrary constants
4	Equation coefficients are independent of the coordinates $x_1, \ldots, x_k$	$w(\mathbf{x}, t) = \exp(\beta_1 x_1 + \cdots + \beta_k x_k)\psi(t, x_{k+1}, \ldots, x_n)$, $\beta_1, \ldots, \beta_k$ are arbitrary constants
5	$\mathfrak{L}[w] = L_t[w] + L_\mathbf{x}[w]$, operator L_t depends on only t, operator $L_\mathbf{x}$ depends on only $\mathbf{x}$	$w(\mathbf{x}, t) = \varphi(t)\psi(\mathbf{x})$, $\varphi(t)$ satisfies the equation $L_t[\varphi] + \lambda\varphi = 0$, $\psi(\mathbf{x})$ satisfies the equation $L_\mathbf{x}[\psi] - \lambda\psi = 0$
6	$\mathfrak{L}[w] = L_t[w] + L_1[w] + \cdots + L_n[w]$, operator L_t depends on only t, operator L_k depends on only x_k	$w(\mathbf{x}, t) = \varphi(t)\psi_1(x_1)\ldots\psi_n(x_n)$, $\varphi(t)$ satisfies the equation $L_t[\varphi] + \lambda\varphi = 0$, $\psi_k(x_k)$ satisfies the equation $L_k[\psi_k] + \beta_k\psi_k = 0$, $\lambda + \beta_1 + \cdots + \beta_n = 0$
7	$\mathfrak{L}[w] = f_0(x_1)L_t[w] + \sum_{k=1}^{n} f_k(x_1)L_k[w]$, operator L_t depends on only t, operator L_k depends on only x_k	$w(\mathbf{x}, t) = \varphi(t)\psi_1(x_1)\ldots\psi_n(x_n)$, $L_t[\varphi] + \lambda\varphi = 0$, $L_k[\psi_k] + \beta_k\psi_k = 0, \quad k = 2, \ldots, n$, $f_1(x_1)L_1[\psi_1] - \left[\lambda f_0(x_1) + \sum_{k=2}^{n}\beta_k f_k(x_1)\right]\psi_1 = 0$
8	$\mathfrak{L}[w] = \dfrac{\partial w}{\partial t} + L_{1,t}[w] + \cdots + L_{n,t}[w]$, where $L_{k,t}[w] = \sum_{s=0}^{m_k} f_{ks}(x_k, t)\dfrac{\partial^s w}{\partial x_k^s}$	$w(\mathbf{x}, t) = \psi_1(x_1, t)\psi_2(x_2, t)\ldots\psi_n(x_n, t)$, $\dfrac{\partial \psi_k}{\partial t} + L_{k,t}[\psi_k] = \lambda_k(t)\psi_k, \quad k = 1, \ldots, n$, $\lambda_1(t) + \lambda_2(t) + \cdots + \lambda_n(t) = 0$

0.3.1-3. Separable solutions.

Many homogeneous linear partial differential equations have solutions that can be represented as the product of functions depending on different arguments. Such solutions are referred to as separable solutions.

Table 2 presents the most commonly encountered types of homogeneous linear differential equations with many independent variables that admit exact separable solutions. Linear combinations of particular solutions that correspond to different values of the separation parameters, $\lambda, \beta_1, \ldots, \beta_n$, are also solutions of the equations in question. For brevity, the word "operator" is used to denote "linear differential operator."

For a constant coefficient equation (see the first row in Table 2), the separation parameters must satisfy the algebraic equation
$$D(\lambda, \beta_1, \ldots, \beta_n) = 0, \tag{4}$$
which results from substituting the solution into the equation (1). In physical applications, equation (4) is usually referred to as a variance equation. Any n of the $n+1$ separation parameters in (4) can be treated as arbitrary.

Note that constant coefficient equations also admit more sophisticated solutions; see the second and third rows, the last column.

The eighth row of Table 2 presents the case of incomplete separation of variables where the solution is separated with respect to the space variables $x_1, \ldots, x_n$ but is not separated with respect to time t.

Remark. For stationary equations, which do not depend on t, one should set $\lambda = 0$, $L_t[w] \equiv 0$, and $\varphi(t) \equiv 1$ in rows 1, 6, and 7 of Table 2.

0.3.1-4. Solutions in the form of infinite series in t.

1°. The equation
$$\frac{\partial w}{\partial t} = M[w],$$
where M is an arbitrary linear differential operator of the second (or any) order that only depends on the space variables, has the formal series solution
$$w(\mathbf{x}, t) = f(\mathbf{x}) + \sum_{k=1}^{\infty} \frac{t^k}{k!} M^k[f(\mathbf{x})], \qquad M^k[f] = M\left[M^{k-1}[f]\right],$$
where $f(\mathbf{x})$ is an arbitrary infinitely differentiable function. This solution satisfies the initial condition $w(\mathbf{x}, 0) = f(\mathbf{x})$.

2°. The equation
$$\frac{\partial^2 w}{\partial t^2} = M[w],$$
where M is a linear differential operator, just as in Item 1°, has a formal solution represented by the sum of two series as
$$w(\mathbf{x}, t) = \sum_{k=0}^{\infty} \frac{t^{2k}}{(2k)!} M^k[f(\mathbf{x})] + \sum_{k=0}^{\infty} \frac{t^{2k+1}}{(2k+1)!} M^k[g(\mathbf{x})],$$
where $f(\mathbf{x})$ and $g(\mathbf{x})$ are arbitrary infinitely differentiable functions. This solution satisfies the initial conditions $w(\mathbf{x}, 0) = f(\mathbf{x})$ and $\partial_t w(\mathbf{x}, 0) = g(\mathbf{x})$.

0.3.2. Nonhomogeneous Linear Equations

0.3.2-1. Simplest properties of nonhomogeneous linear equations.

For brevity, we write a nonhomogeneous linear partial differential equation in the form
$$\mathfrak{L}[w] = \Phi(\mathbf{x}, t), \tag{5}$$
where the linear differential operator $\mathfrak{L}$ is defined above, see the beginning of Paragraph 0.3.1-1.

Below are the simplest properties of particular solutions of the nonhomogeneous equation (5).

1°. If $\widetilde{w}_\Phi(\mathbf{x}, t)$ is a particular solution of the nonhomogeneous equation (5) and $\widetilde{w}_0(\mathbf{x}, t)$ is a particular solution of the corresponding homogeneous equation (1), then the sum
$$A\widetilde{w}_0(\mathbf{x}, t) + \widetilde{w}_\Phi(\mathbf{x}, t),$$
where A is an arbitrary constant, is also a solution of the nonhomogeneous equation (5). The following, more general statement holds: The general solution of the nonhomogeneous equation (5) is the sum of the general solution of the corresponding homogeneous equation (1) and any particular solution of the nonhomogeneous equation (5).

2°. Suppose w_1 and w_2 are solutions of nonhomogeneous linear equations with the same left-hand side and different right-hand sides, i.e.,
$$\mathfrak{L}[w_1] = \Phi_1(\mathbf{x}, t), \qquad \mathfrak{L}[w_2] = \Phi_2(\mathbf{x}, t).$$
Then the function $w = w_1 + w_2$ is a solution of the equation
$$\mathfrak{L}[w] = \Phi_1(\mathbf{x}, t) + \Phi_2(\mathbf{x}, t).$$

0.3.2-2. Fundamental and particular solutions of stationary equations.

Consider the second-order linear stationary (time-independent) nonhomogeneous equation

$$L_{\mathbf{x}}[w] = \Phi(\mathbf{x}). \tag{6}$$

Here, $L_{\mathbf{x}}$ is a linear differential operator of the second (or any) order of general form whose coefficients are dependent on $\mathbf{x}$, where $\mathbf{x} \in \mathbb{R}^n$.

A distribution $\mathscr{E} = \mathscr{E}(\mathbf{x}, \mathbf{y})$ that satisfies the equation with a special right-hand side

$$L_{\mathbf{x}}[\mathscr{E}] = \delta(\mathbf{x} - \mathbf{y}) \tag{7}$$

is called a fundamental solution corresponding to the operator $L_{\mathbf{x}}$. In (7), $\delta(\mathbf{x})$ is an n-dimensional Dirac delta function and the vector quantity $\mathbf{y} = \{y_1, \ldots, y_n\}$ appears in equation (7) as an n-dimensional free parameter. It is assumed that $\mathbf{y} \in \mathbb{R}^n$.

The n-dimensional Dirac delta function possesses the following basic properties:

1. $\delta(\mathbf{x}) = \delta(x_1)\delta(x_2)\ldots\delta(x_n),$
2. $\int_{\mathbb{R}^n} \Phi(\mathbf{y})\delta(\mathbf{x} - \mathbf{y})\,d\mathbf{y} = \Phi(\mathbf{x}),$

where $\delta(x_k)$ is the one-dimensional Dirac delta function, $\Phi(\mathbf{x})$ is an arbitrary continuous function, and $d\mathbf{y} = dy_1 \ldots dy_n$.

For constant coefficient equations, a fundamental solution always exists; it can be found by means of the n-dimensional Fourier transform (see Paragraph 0.5.3-2).

The fundamental solution $\mathscr{E} = \mathscr{E}(\mathbf{x}, \mathbf{y})$ can be used to construct a particular solution of the linear stationary nonhomogeneous equation (6) for arbitrary continuous $\Phi(\mathbf{x})$; this particular solution is expressed as follows:

$$w(\mathbf{x}) = \int_{\mathbb{R}^n} \Phi(\mathbf{y})\mathscr{E}(\mathbf{x}, \mathbf{y})\,d\mathbf{y}. \tag{8}$$

Remark 1. The fundamental solution $\mathscr{E}$ is not unique; it is defined up to an additive term $w_0 = w_0(\mathbf{x})$ which is an arbitrary solution of the homogeneous equation $L_{\mathbf{x}}[w_0] = 0$.

Remark 2. For constant coefficient differential equations, the fundamental solution possesses the property $\mathscr{E}(\mathbf{x}, \mathbf{y}) = \mathscr{E}(\mathbf{x} - \mathbf{y})$.

Remark 3. The right-hand sides of equations (6) and (7) are often prefixed with the minus sign. In this case, formula (8) remains valid.

Remark 4. Particular solutions of linear nonstationary nonhomogeneous equations can be expressed in terms of the fundamental solution of the Cauchy problem; see Section 0.6.

⊙ *References for Section* 0.3: G. A. Korn and T. M. Korn (1968), W. Miller, Jr. (1977), R. P. Kanwal (1983), L. Hörmander (1983, 1990), V. S. Vladimirov (1988), A. N. Tikhonov and A. A. Samarskii (1990), D. Zwillinger (1998), A. D. Polyanin, A. V. Vyazmin, A. I. Zhurov, and D. A. Kazenin (1998).

0.4. Separation of Variables Method

0.4.1. General Description of the Separation of Variables Method

0.4.1-1. Scheme of solving linear boundary value problems by separation of variables.

Many linear problems of mathematical physics can be solved by separation of variables. Figure 1 depicts the scheme of application of this method to solve boundary value problems for second-order homogeneous linear equations of the parabolic and hyperbolic type* with homogeneous boundary

---

* The separation of variables method is also used to solve linear boundary value problems for elliptic equations.

conditions and nonhomogeneous initial conditions. For simplicity, problems with two independent variables x and t, are considered, with $x_1 \leq x \leq x_2$ and $t \geq 0$.

The scheme presented in Fig. 1 applies to boundary value problems for second-order linear homogeneous partial differential equations of the form

$$\alpha(t)\frac{\partial^2 w}{\partial t^2} + \beta(t)\frac{\partial w}{\partial t} = a(x)\frac{\partial^2 w}{\partial x^2} + b(x)\frac{\partial w}{\partial x} + \bigl[c(x) + \gamma(t)\bigr]w \tag{1}$$

with homogeneous linear boundary conditions,

$$\begin{aligned} s_1\partial_x w + k_1 w &= 0 \quad \text{at} \quad x = x_1, \\ s_2\partial_x w + k_2 w &= 0 \quad \text{at} \quad x = x_2, \end{aligned} \tag{2}$$

and arbitrary initial conditions,

$$w = f_0(x) \quad \text{at} \quad t = 0, \tag{3}$$
$$\partial_t w = f_1(x) \quad \text{at} \quad t = 0. \tag{4}$$

For parabolic equations, which correspond to $\alpha(t) \equiv 0$ in (1), only the initial condition (3) is set.

Below we consider the basic stages of the method of separation of variables in more detail. We assume that the coefficients of equation (1) and boundary conditions (2) meet the following requirements:

$$\alpha(t), \ \beta(t), \ \gamma(t), \ a(x), \ b(x), \ c(x) \ \text{are continuous functions},$$
$$\alpha(t) \geq 0, \quad 0 < a(x) < \infty, \quad |s_1| + |k_1| > 0, \quad |s_2| + |k_2| > 0.$$

0.4.1-2. Search for particular solutions. Derivation of equations and boundary conditions.

The approach is based on searching for particular solutions of equation (1) in the product form

$$w(x, t) = \varphi(x)\psi(t). \tag{5}$$

After separation of the variables and elementary manipulations, one arrives at the following linear ordinary differential equations for the functions $\varphi = \varphi(x)$ and $\psi = \psi(t)$:

$$a(x)\varphi''_{xx} + b(x)\varphi'_x + [\lambda + c(x)]\varphi = 0, \tag{6}$$
$$\alpha(t)\psi''_{tt} + \beta(t)\psi'_t + [\lambda - \gamma(t)]\psi = 0. \tag{7}$$

These equations contain a free parameter λ called the separation constant. With the notation adopted in Fig. 1, equations (6) and (7) can be rewritten as follows: $\varphi F_1(x, \varphi, \varphi'_x, \varphi''_{xx}) + \lambda\varphi = 0$ and $\psi F_2(t, \psi, \psi'_t, \psi''_{tt}) + \lambda\psi = 0$.

Substituting (5) into (2) yields the boundary conditions for $\varphi = \varphi(x)$:

$$\begin{aligned} s_1\varphi'_x + k_1\varphi &= 0 \quad \text{at} \quad x = x_1, \\ s_2\varphi'_x + k_2\varphi &= 0 \quad \text{at} \quad x = x_2. \end{aligned} \tag{8}$$

The homogeneous linear ordinary differential equation (6) in conjunction with the homogeneous linear boundary conditions (8) make up an eigenvalue problem.

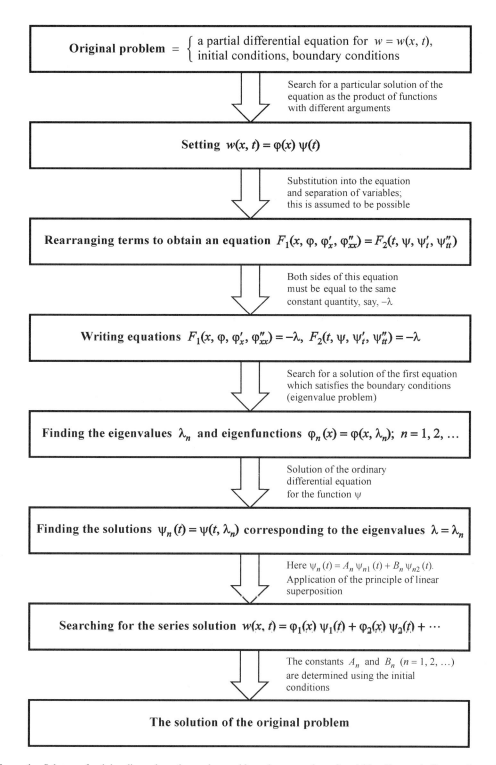

Figure 1. Scheme of solving linear boundary value problems by separation of variables (for parabolic equations, the function F_2 does not depend on ψ''_{tt}, and $B_n = 0$).

0.4.1-3. Solution of eigenvalue problems. Orthogonality of eigenfunctions.

Suppose $\widetilde{\varphi}_1 = \widetilde{\varphi}_1(x, \lambda)$ and $\widetilde{\varphi}_2 = \widetilde{\varphi}_2(x, \lambda)$ are linearly independent particular solutions of equation (6). Then the general solution of this equation can be represented as the linear combination

$$\varphi = C_1\widetilde{\varphi}_1(x, \lambda) + C_2\widetilde{\varphi}_2(x, \lambda), \tag{9}$$

where C_1 and C_2 are arbitrary constants.

Substituting solution (9) into the boundary conditions (8) yields the following homogeneous linear algebraic system of equations for C_1 and C_2:

$$\begin{aligned}\varepsilon_{11}(\lambda)C_1 + \varepsilon_{12}(\lambda)C_2 &= 0,\\ \varepsilon_{21}(\lambda)C_1 + \varepsilon_{22}(\lambda)C_2 &= 0,\end{aligned} \tag{10}$$

where $\varepsilon_{ij}(\lambda) = \left[s_i(\widetilde{\varphi}_j)'_x + k_i\widetilde{\varphi}_j\right]_{x=x_i}$. For system (10) to have nontrivial solutions, its determinant must be zero; we have

$$\varepsilon_{11}(\lambda)\varepsilon_{22}(\lambda) - \varepsilon_{12}(\lambda)\varepsilon_{21}(\lambda) = 0. \tag{11}$$

Solving the transcendental equation (11) for λ, one obtains the eigenvalues $\lambda = \lambda_n$, where $n = 1, 2, \ldots$ For these values of λ, there are nontrivial solutions of equation (6),

$$\varphi_n(x) = \varepsilon_{12}(\lambda_n)\widetilde{\varphi}_1(x, \lambda_n) - \varepsilon_{11}(\lambda_n)\widetilde{\varphi}_2(x, \lambda_n), \tag{12}$$

which are called eigenfunctions (these functions are defined up to a constant multiplier).

To facilitate the further analysis, we represent equation (6) in the form

$$[p(x)\varphi'_x]'_x + [\lambda\rho(x) - q(x)]\varphi = 0, \tag{13}$$

where

$$p(x) = \exp\left[\int \frac{b(x)}{a(x)}\,dx\right], \quad q(x) = -\frac{c(x)}{a(x)}\exp\left[\int \frac{b(x)}{a(x)}\,dx\right], \quad \rho(x) = \frac{1}{a(x)}\exp\left[\int \frac{b(x)}{a(x)}\,dx\right]. \tag{14}$$

It follows from the adopted assumptions (see the end of Paragraph 0.4.1-1) that $p(x)$, $p'_x(x)$, $q(x)$, and $\rho(x)$ are continuous functions, with $p(x) > 0$ and $\rho(x) > 0$.

The eigenvalue problem (13), (8) is known to possess the following properties:

1. All eigenvalues $\lambda_1, \lambda_2, \ldots$ are real, and $\lambda_n \to \infty$ as $n \to \infty$; consequently, the number of negative eigenvalues is finite.

2. The system of eigenfunctions $\varphi_1(x), \varphi_2(x), \ldots$ is orthogonal on the interval $x_1 \leq x \leq x_2$ with weight $\rho(x)$, i.e.,

$$\int_{x_1}^{x_2} \rho(x)\varphi_n(x)\varphi_m(x)\,dx = 0 \quad \text{for} \quad n \neq m. \tag{15}$$

3. If

$$q(x) \geq 0, \quad s_1k_1 \leq 0, \quad s_2k_2 \geq 0, \tag{16}$$

there are no negative eigenvalues. If $q \equiv 0$ and $k_1 = k_2 = 0$, the least eigenvalue is $\lambda_1 = 0$ and the corresponding eigenfunction is $\varphi_1 = \text{const}$. Otherwise, all eigenvalues are positive, provided that conditions (16) are satisfied; the first inequality in (16) is satisfied if $c(x) \leq 0$.

Subsection 1.8.9 presents some estimates for the eigenvalues λ_n and eigenfunctions $\varphi_n(x)$.

0.4.2. Solution of Boundary Value Problems for Parabolic and Hyperbolic Equations

0.4.2-1. Solution of boundary value problems for parabolic equations.

For parabolic equations, one should set $\alpha(t) \equiv 0$ in (1) and (7). In addition, we assume that $\beta(t) > 0$ and $\gamma(t) < \min \lambda_n$.

First we search for the solutions of equation (7) corresponding to the eigenvalues $\lambda = \lambda_n$ and satisfying the normalizing conditions $\psi_n(0) = 1$ to obtain

$$\psi_n(t) = \exp\left[\int_0^t \frac{\gamma(\xi) - \lambda_n}{\beta(\xi)} d\xi\right]. \tag{17}$$

Then the solution of the original nonstationary boundary value problem (1)–(3) for the parabolic equation is sought in the form

$$w(x, t) = \sum_{n=1}^{\infty} A_n \varphi_n(x) \psi_n(t), \tag{18}$$

where the A_n are arbitrary constants and the functions $w_n(x, t) = \varphi_n(x) \psi_n(t)$ are particular solutions (5) satisfying the boundary conditions (2). By the principle of linear superposition, series (18) is also a solution of the original partial differential equation which satisfies the boundary conditions.

To determine the coefficients A_n, we substitute series (18) into the initial condition (3), thus obtaining

$$\sum_{n=1}^{\infty} A_n \varphi_n(x) = f_0(x).$$

Multiplying this equation by $\rho(x)\varphi_n(x)$, integrating the resulting relation with respect to x over the interval $x_1 \le x \le x_2$, and taking into account the properties (15), we find

$$A_n = \frac{1}{\|\varphi_n\|^2} \int_{x_1}^{x_2} \rho(x)\varphi_n(x)f_0(x)\, dx, \quad \|\varphi_n\|^2 = \int_{x_1}^{x_2} \rho(x)\varphi_n^2(x)\, dx. \tag{19}$$

The weight function $\rho(x)$ is defined in (14).

Relations (18), (12), (17), and (19) give a formal solution of the nonstationary boundary value problem (1)–(3) if $\alpha(t) \equiv 0$.

Example 1. Let $\beta(t) = 1$ and $\gamma(t) = 0$. Substituting these values into (17) yields

$$\psi_n(t) = \exp(-\lambda_n t). \tag{20}$$

If the function $f_0(x)$ is twice continuously differentiable and the compatibility conditions (see Paragraph 0.4.2-3) are satisfied, then series (18) is convergent and admits termwise differentiation, once with respect to t and twice with respect to x. In this case, relations (18), (12), (19), and (20) give the classical smooth solution of problem (1)–(3). [If $f_0(x)$ is not as smooth as indicated or if the compatibility conditions are not met, then series (18) may converge to a discontinuous function, thus giving only a generalized solution.]

0.4.2-2. Solution of boundary value problems for hyperbolic equations.

For hyperbolic equations, the solution of the boundary value problem (1)–(4) is sought in the form

$$w(x, t) = \sum_{n=1}^{\infty} \varphi_n(x) \left[A_n \psi_{n1}(t) + B_n \psi_{n2}(t)\right]. \tag{21}$$

Here, A_n and B_n are arbitrary constants. The functions $\psi_{n1}(t)$ and $\psi_{n2}(t)$ are particular solutions of the linear equation (7) for ψ (with $\lambda = \lambda_n$) which satisfy the conditions

$$\psi_{n1}(0) = 1, \quad \psi'_{n1}(0) = 0; \quad \psi_{n2}(0) = 0, \quad \psi'_{n2}(0) = 1. \tag{22}$$

Substituting solution (21) into the initial conditions (3)–(4) yields

$$\sum_{n=1}^{\infty} A_n \varphi_n(x) = f_0(x), \quad \sum_{n=1}^{\infty} B_n \varphi_n(x) = f_1(x).$$

Multiplying these equations by $\rho(x)\varphi_n(x)$, integrating the resulting relations with respect to x on the interval $x_1 \leq x \leq x_2$, and taking into account the properties (15), we obtain the coefficients of series (21) in the form

$$A_n = \frac{1}{\|\varphi_n\|^2} \int_{x_1}^{x_2} \rho(x)\varphi_n(x) f_0(x)\, dx, \quad B_n = \frac{1}{\|\varphi_n\|^2} \int_{x_1}^{x_2} \rho(x)\varphi_n(x) f_1(x)\, dx. \tag{23}$$

The quantity $\|\varphi_n\|$ is defined in (19).

Relations (21), (12), and (23) give a formal solution of the nonstationary boundary value problem (1)–(4) for $\alpha(t) > 0$.

Example 2. Let $\alpha(t) = 1$, $\beta(t) = \gamma(t) = 0$, and $\lambda_n > 0$. The solutions of (7) satisfying conditions (22) are expressed as

$$\psi_{n1}(t) = \cos(\sqrt{\lambda_n}\, t), \quad \psi_{n2}(t) = \frac{1}{\sqrt{\lambda_n}} \sin(\sqrt{\lambda_n}\, t). \tag{24}$$

If $f_0(x)$ and $f_1(x)$ have three and two continuous derivatives, respectively, and the compatibility conditions are met (see Paragraph 0.4.2-3), then series (21) is convergent and admits double termwise differentiation. In this case, formulas (21), (12), (23), and (24) give the classical smooth solution of problem (1)–(4).

0.4.2-3. Conditions of compatibility of initial and boundary conditions.

Parabolic equations, $\alpha(t) \equiv 0$. Suppose the function w has a continuous derivative with respect to t and two continuous derivatives with respect to x and is a solution of problem (1)–(3). Then the boundary conditions (2) and the initial condition (3) must be consistent; namely, the following compatibility conditions must hold:

$$[s_1 f_0' + k_1 f_0]_{x=x_1} = 0, \quad [s_2 f_0' + k_2 f_0]_{x=x_2} = 0. \tag{25}$$

If $s_1 = 0$ or $s_2 = 0$, then the additional compatibility conditions

$$\begin{aligned}{}[a(x) f_0'' + b(x) f_0']_{x=x_1} = 0 \quad \text{if} \quad s_1 = 0, \\ [a(x) f_0'' + b(x) f_0']_{x=x_2} = 0 \quad \text{if} \quad s_2 = 0 \end{aligned} \tag{26}$$

must also hold; the primes denote the derivatives with respect to x.

Hyperbolic equations. Suppose w is a twice continuously differentiable solution of problem (1)–(4). Then conditions (25) and (26) must hold. In addition, the following conditions of compatibility of the boundary conditions (2) and initial condition (4) must be satisfied:

$$[s_1 f_1' + k_1 f_1]_{x=x_1} = 0, \quad [s_2 f_1' + k_2 f_1]_{x=x_2} = 0.$$

0.4.2-4. Linear nonhomogeneous equations with nonhomogeneous boundary conditions.

Parabolic equations, $\alpha(t) \equiv 0$. The solution of the boundary value problem for the parabolic linear homogeneous equation (1) subject to the homogeneous linear boundary conditions (2) and nonhomogeneous initial condition (3) is given by relations (18), (12), (17), and (19). This solution can be rewritten in the form

$$w(x,t) = \int_{x_1}^{x_2} G(x,y,t,0) f_0(y)\, dy.$$

Here, $G(x, y, t, \tau)$ is the Green's function, which is expressed as

$$G(x, y, t, \tau) = \rho(y) \sum_{n=1}^{\infty} \frac{\varphi_n(x)\varphi_n(y)}{\|\varphi_n\|^2} \psi_n(t, \tau), \tag{27}$$

where $\psi_n = \psi_n(t, \tau)$ is the solution of equation (7) with $\alpha(t) \equiv 0$ and $\lambda = \lambda_n$ which satisfies the initial condition

$$\psi_n = 1 \quad \text{at} \quad t = \tau.$$

The function $\psi_n(t, \tau)$ can be calculated by formula (17) with the lower limit of integration equal to τ (rather than zero).

The simplest way to obtain the solutions of more general boundary value problems for the corresponding nonhomogeneous linear equations with nonhomogeneous boundary and initial conditions is to take advantage of formula (6) from Subsection 0.7.1 and use the Green's function (27).

Hyperbolic equations. The solution of the boundary value problem for the hyperbolic linear homogeneous equation (1) subject to the homogeneous linear boundary conditions (2) and semihomogeneous initial conditions (3)–(4) with $f_0(x) \equiv 0$ is given by relations (21), (12), and (23) with $A_n = 0$ ($n = 1, 2, \dots$). This solution can be rewritten in the form

$$w(x, t) = \int_{x_1}^{x_2} G(x, y, t, 0) f_1(y)\, dy.$$

Here, $G(x, y, t, \tau)$ is the Green's function defined by relation (27), where $\psi_n = \psi_n(t, \tau)$ is the solution of equation (7) for ψ with $\lambda = \lambda_n$ which satisfies the initial conditions

$$\psi_n = 0 \quad \text{at} \quad t = \tau, \qquad \psi_n' = 1 \quad \text{at} \quad t = \tau.$$

In the special case $\alpha(t) = 1$, $\beta(t) = \gamma(t) = 0$, we have $\psi_n(t, \tau) = \lambda_n^{-1/2} \sin\!\left[\lambda_n^{1/2}(t - \tau)\right]$.

The simplest way to obtain the solutions of more general boundary value problems for the corresponding nonhomogeneous linear equations with nonhomogeneous boundary and initial conditions is to take advantage of formula (14) from Subsection 0.7.2 and use the Green's function (27).

⊙ *References for Section* 0.4: V. M. Babich, M. B. Kapilevich, S. G. Mikhlin, et al. (1964), E. Butkov (1968), E. C. Zachmanoglou and D. W. Thoe (1986), T. U.-Myint and L. Debnath (1987), A. N. Tikhonov and A. A. Samarskii (1990), R. B. Guenther and J. W. Lee (1996), D. Zwillinger (1998), I. Stakgold (2000), A. D. Polyanin (2001a).

0.5. Integral Transforms Method

0.5.1. Main Integral Transforms

Various integral transforms are widely used to solve linear problems of mathematical physics.

An integral transform is defined as

$$\widetilde{f}(\lambda) = \int_a^b \varphi(x, \lambda) f(x)\, dx.$$

The function $\widetilde{f}(\lambda)$ is called the transform of the function $f(x)$ and $\varphi(x, \lambda)$ is called the kernel of the integral transform. The function $f(x)$ is called the inverse transform of $\widetilde{f}(\lambda)$. The limits of integration a and b are real numbers (usually, $a = 0$, $b = \infty$ or $a = -\infty$, $b = \infty$).

Corresponding inversion formulas, which have the form

$$f(x) = \int_{\mathcal{L}} \psi(x, \lambda) \widetilde{f}(\lambda)\, d\lambda$$

make it possible to recover $f(x)$ if $\widetilde{f}(\lambda)$ is given. The integration path $\mathcal{L}$ can lie either on the real axis or in the complex plane.

The most commonly used integral transforms are listed in Table 3 (for the constraints imposed on the functions and parameters occurring in the integrand, see the references given at the end of Section 0.5).

TABLE 3
Main integral transforms

Integral Transform	Definition	Inversion Formula
Laplace transform	$\tilde{f}(p) = \int_0^\infty e^{-pt} f(t)\, dt$	$f(t) = \dfrac{1}{2\pi i} \int_{c-i\infty}^{c+i\infty} e^{pt} \tilde{f}(p)\, dp$
Fourier transform	$\tilde{f}(u) = \dfrac{1}{\sqrt{2\pi}} \int_{-\infty}^\infty e^{-iux} f(x)\, dx$	$f(x) = \dfrac{1}{\sqrt{2\pi}} \int_{-\infty}^\infty e^{iux} \tilde{f}(u)\, du$
Fourier sine transform	$\tilde{f}_s(u) = \sqrt{\dfrac{2}{\pi}} \int_0^\infty \sin(xu) f(x)\, dx$	$f(x) = \sqrt{\dfrac{2}{\pi}} \int_0^\infty \sin(xu) \tilde{f}_s(u)\, du$
Fourier cosine transform	$\tilde{f}_c(u) = \sqrt{\dfrac{2}{\pi}} \int_0^\infty \cos(xu) f(x)\, dx$	$f(x) = \sqrt{\dfrac{2}{\pi}} \int_0^\infty \cos(xu) \tilde{f}_c(u)\, du$
Mellin transform	$\hat{f}(s) = \int_0^\infty x^{s-1} f(x)\, dx$	$f(x) = \dfrac{1}{2\pi i} \int_{c-i\infty}^{c+i\infty} x^{-s} \hat{f}(s)\, ds$
Hankel transform	$\hat{f}_\nu(u) = \int_0^\infty x J_\nu(xu) f(x)\, dx$	$f(x) = \int_0^\infty u J_\nu(xu) \hat{f}_\nu(u)\, du$
Meijer transform	$\hat{f}(s) = \sqrt{\dfrac{2}{\pi}} \int_0^\infty \sqrt{sx}\, K_\nu(sx) f(x)\, dx$	$f(x) = \dfrac{1}{i\sqrt{2\pi}} \int_{c-i\infty}^{c+i\infty} \sqrt{sx}\, I_\nu(sx) \hat{f}(s)\, ds$

Notation: $i^2 = -1$, $J_\mu(x)$ and $Y_\mu(x)$ are the Bessel functions of the first and the second kind, respectively; $I_\mu(x)$ and $K_\mu(x)$ are the modified Bessel functions of the first and the second kind.

The Laplace transform and the Fourier transform are in most common use. These integral transforms are briefly described below.

0.5.2. Laplace Transform and Its Application in Mathematical Physics

0.5.2-1. The Laplace transform. The inverse Laplace transform.

The Laplace transform of an arbitrary (complex-valued) function $f(t)$ of a real variable t ($t \geq 0$) is defined by

$$\tilde{f}(p) = \int_0^\infty e^{-pt} f(t)\, dt, \tag{1}$$

where $p = s + i\sigma$ is a complex variable, $i^2 = -1$.

The Laplace transform exists for any continuous or piecewise-continuous function satisfying the condition $|f(t)| < M e^{\sigma_0 t}$ with some $M > 0$ and $\sigma_0 \geq 0$. In what follows, σ_0 often means the greatest lower bound of the possible values of σ_0 in this condition. For any $f(t)$, the transform $\tilde{f}(p)$ is defined in the half-plane $\operatorname{Re} p > \sigma_0$ and is analytic there.

Given the transform $\tilde{f}(p)$, the function $f(t)$ can be found by means of the inverse Laplace transform

$$f(t) = \frac{1}{2\pi i} \int_{c-i\infty}^{c+i\infty} \tilde{f}(p) e^{pt}\, dp, \tag{2}$$

where the integration path is parallel to the imaginary axis and lies to the right of all singularities of $\tilde{f}(p)$, which corresponds to $c > \sigma_0$.

The integral in (2) is understood in the sense of the Cauchy principal value:

$$\int_{c-i\infty}^{c+i\infty} \tilde{f}(p) e^{pt}\, dp = \lim_{\omega \to \infty} \int_{c-i\omega}^{c+i\omega} \tilde{f}(p) e^{pt}\, dp.$$

In the domain $t < 0$, formula (2) gives $f(t) \equiv 0$.

TABLE 4
Main properties of the Laplace transform

No	Function	Laplace transform	Operation
1	$af_1(t) + bf_2(t)$	$a\widetilde{f}_1(p) + b\widetilde{f}_2(p)$	Linearity
2	$f(t/a)$, $a > 0$	$a\widetilde{f}(ap)$	Scaling
3	$t^n f(t)$; $n = 1, 2, \ldots$	$(-1)^n \widetilde{f}_p^{(n)}(p)$	Differentiation of the transform
4	$e^{at} f(t)$	$\widetilde{f}(p - a)$	Shift in the complex plane
5	$f'_t(t)$	$p\widetilde{f}(p) - f(+0)$	Differentiation
6	$f_t^{(n)}(t)$	$p^n \widetilde{f}(p) - \sum_{k=1}^{n} p^{n-k} f_t^{(k-1)}(+0)$	Differentiation
7	$t^m f_t^{(n)}(t)$, $m \geq n$	$(-1)^m \left[p^n \widetilde{f}(p)\right]_p^{(m)}$	Differentiation
8	$\int_0^t f(\tau)\, d\tau$	$\dfrac{1}{p}\widetilde{f}(p)$	Integration
9	$\int_0^t f_1(\tau) f_2(t - \tau)\, d\tau$	$\widetilde{f}_1(p)\widetilde{f}_2(p)$	Convolution

Formula (2) holds for continuous functions. If $f(t)$ has a (finite) jump discontinuity at a point $t = t_0 > 0$, then the left-hand side of (2) is equal to $\frac{1}{2}[f(t_0 - 0) + f(t_0 + 0)]$ at this point (for $t_0 = 0$, the first term in the square brackets must be omitted).

We will briefly denote the Laplace transform (1) and the inverse Laplace transform (2) as

$$\widetilde{f}(p) = \mathcal{L}\{f(t)\}, \quad f(t) = \mathcal{L}^{-1}\{\widetilde{f}(p)\}.$$

0.5.2-2. Main properties of the Laplace transform.

The main properties of the correspondence between functions and their Laplace transforms are gathered in Table 4. The Laplace transforms of some functions are listed in Table 5.

There are a number of books that contain detailed tables of direct and inverse Laplace transforms elsewhere; see the references at the end of Section 0.5. Such tables are convenient to use in solving linear differential equations.

Note the important case in which the transform is a rational function of the form

$$\widetilde{f}(p) = \frac{R(p)}{Q(p)},$$

where $Q(p)$ and $R(p)$ are polynomials in the variable p and the degree of $Q(p)$ exceeds that of $R(p)$. Assume that the zeros of the denominator are simple, i.e., $Q(p) \equiv \text{const}\,(p - \lambda_1)(p - \lambda_2) \ldots (p - \lambda_n)$. Then the inverse transform can be determined by the formula

$$f(t) = \sum_{k=1}^{n} \frac{R(\lambda_k)}{Q'(\lambda_k)} \exp(\lambda_k t),$$

where the primes denote the derivatives.

TABLE 5

The Laplace transforms of some functions

No	Function, $f(t)$	Laplace transform, $\tilde{f}(p)$	Remarks
1	1	$\dfrac{1}{p}$	
2	t^n	$\dfrac{n!}{p^{n+1}}$	$n = 1, 2, \ldots$
3	t^a	$\Gamma(a+1)p^{-a-1}$	$a > -1$
4	e^{-at}	$(p+a)^{-1}$	
5	$t^a e^{-bt}$	$\Gamma(a+1)(p+b)^{-a-1}$	$a > -1$
6	$\sinh(at)$	$\dfrac{a}{p^2 - a^2}$	
7	$\cosh(at)$	$\dfrac{p}{p^2 - a^2}$	
8	$\ln t$	$-\dfrac{1}{p}(\ln p + \mathcal{C})$	$\mathcal{C} = 0.5772\ldots$ is the Euler constant
9	$\sin(at)$	$\dfrac{a}{p^2 + a^2}$	
10	$\cos(at)$	$\dfrac{p}{p^2 + a^2}$	
11	$\operatorname{erfc}\left(\dfrac{a}{2\sqrt{t}}\right)$	$\dfrac{1}{p}\exp(-a\sqrt{p})$	$a \geq 0$
12	$J_0(at)$	$\dfrac{1}{\sqrt{p^2 + a^2}}$	$J_0(x)$ is the Bessel function

0.5.2-3. Solving linear problems of mathematical physics by the Laplace transform.

Figure 2 shows schematically how one can utilize the Laplace transforms to solve boundary value problems for linear parabolic or hyperbolic equations with two independent variables in the case where the equation coefficients are independent of t.

It is significant that with the Laplace transform, the original problems for a partial differential equation is reduced to a simpler problem for an ordinary differential equation with parameter p; the derivatives with respect to t are replaced by appropriate algebraic expressions taking into account the initial conditions (see property 5 or 6 in Table 4).

Example 1. Consider the following problem for the heat equation:

$$\partial_t w = \partial_{xx} w, \quad (x > 0,\ t > 0),$$
$$w = 0 \quad \text{at} \quad t = 0 \quad \text{(initial condition)},$$
$$w = w_0 \quad \text{at} \quad x = 0 \quad \text{(boundary condition)},$$
$$w \to 0 \quad \text{at} \quad x \to \infty \quad \text{(boundary condition)}.$$

We apply the Laplace transform with respect to t. Setting $\tilde{w} = \mathcal{L}\{w\}$ and taking into account the relations

$$\mathcal{L}\{\partial_t w\} = p\tilde{w} - w|_{t=0} = p\tilde{w} \quad \text{(used are property 5 of Table 4 and the initial condition)},$$
$$\mathcal{L}\{w_0\} = w_0 \mathcal{L}\{1\} = w_0/p \quad \text{(used are property 1 of Table 4 and the relation } \mathcal{L}\{1\} = 1/p\text{)},$$

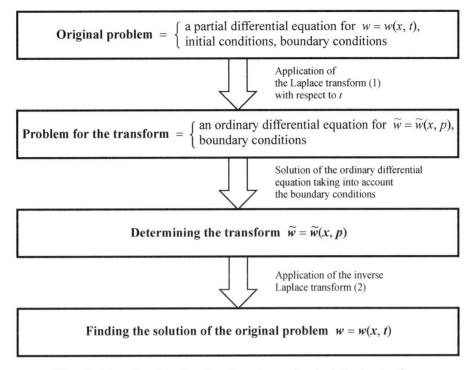

Figure 2. Scheme for solving linear boundary value problems by the Laplace transform.

we arrive at the following problem for a second-order linear ordinary differential equation with parameter p:

$$\widetilde{w}''_{xx} - p\widetilde{w} = 0,$$
$$\widetilde{w} = w_0/p \quad \text{at} \quad x = 0 \quad \text{(boundary condition)},$$
$$\widetilde{w} \to 0 \quad \text{at} \quad x \to \infty \quad \text{(boundary condition)}.$$

Integrating the equation yields the general solution $\widetilde{w} = A_1(p)e^{-x\sqrt{p}} + A_2(p)e^{x\sqrt{p}}$. Using the boundary conditions, we determine the constants, $A_1(p) = w_0/p$ and $A_2(p) = 0$. Thus, we have

$$\widetilde{w} = \frac{w_0}{p} e^{-x\sqrt{p}}.$$

Let us apply the inverse Laplace transform to both sides of this relation. We refer to Table 5, row 11 with $a = x$ to find the inverse transform of the right-hand side. Finally, we obtain the solution of the original problem in the form

$$w = w_0 \operatorname{erfc}\left(\frac{x}{2\sqrt{t}}\right).$$

0.5.3. Fourier Transform and Its Application in Mathematical Physics

0.5.3-1. The Fourier transform and its properties.

The Fourier transform is defined as follows:

$$\widetilde{f}(u) = \frac{1}{\sqrt{2\pi}} \int_{-\infty}^{\infty} f(x)e^{-iux}\, dx, \qquad i^2 = -1. \tag{3}$$

This relation is meaningful for any function $f(x)$ absolutely integrable on the interval $(\infty, +\infty)$. We will briefly write $\widetilde{f}(u) = \mathcal{F}\{f(x)\}$ to denote the Fourier transform (3).

Given $\widetilde{f}(u)$, the function $f(x)$ can be found by means of the inverse Fourier transform

$$f(x) = \frac{1}{\sqrt{2\pi}} \int_{-\infty}^{\infty} \widetilde{f}(u)e^{iux}\, du, \tag{4}$$

TABLE 6
Main properties of the Fourier transform

No	Function	Fourier transform	Operation
1	$af_1(x) + bf_2(x)$	$a\widetilde{f}_1(u) + b\widetilde{f}_2(u)$	Linearity
2	$f(x/a)$, $a > 0$	$a\widetilde{f}(au)$	Scaling
3	$x^n f(x)$; $n = 1, 2, \ldots$	$i^n \widetilde{f}_u^{(n)}(u)$	Differentiation of the transform
4	$f''_{xx}(x)$	$-u^2 \widetilde{f}(u)$	Differentiation
5	$f_x^{(n)}(x)$	$(iu)^n \widetilde{f}(u)$	Differentiation
6	$\int_{-\infty}^{\infty} f_1(\xi) f_2(x-\xi)\, d\xi$	$\widetilde{f}_1(u) \widetilde{f}_2(u)$	Convolution

where the integral is understood in the sense of the Cauchy principal value. We will briefly write $f(x) = \mathcal{F}^{-1}\{\widetilde{f}(u)\}$ to denote the inverse Fourier transform (4).

The inversion formula (4) holds for continuous functions. If $f(x)$ has a (finite) jump discontinuity at a point $x = x_0$, then the left-hand side of (4) is equal to $\frac{1}{2}\bigl[f(x_0 - 0) + f(x_0 + 0)\bigr]$ at this point.

The main properties of the correspondence between functions and their Fourier transforms are gathered in Table 6.

0.5.3-2. Solving linear problems of mathematical physics by the Fourier transform.

The Fourier transform is usually employed to solve boundary value problems for linear partial differential equations whose coefficients are independent of the space variable x, $-\infty < x < \infty$.

The scheme for solving linear boundary value problems with the help of the Fourier transform is similar to that used in solving problems with help of the Laplace transform. With the Fourier transform, the derivatives with respect to x in the equation are replaced by appropriate algebraic expressions; see property 4 or 5 in Table 6. In the case of two independent variables, the problem for a partial differential equation is reduced to a simpler problem for an ordinary differential equation with parameter u. On solving the latter problem, one determines the transform. After that, by applying the inverse Fourier transform, one obtains the solution of the original boundary value problem.

Example 2. Consider the following Cauchy problem for the heat equation:
$$\partial_t w = \partial_{xx} w \qquad (-\infty < x < \infty),$$
$$w = f(x) \quad \text{at} \quad t = 0 \quad \text{(initial condition)}.$$

We apply the Fourier transform with respect to the space variable x. Setting $\widetilde{w} = \mathcal{F}\{w\}$ and taking into account the relation $\mathcal{F}\{\partial_{xx} w\} = -u^2 \widetilde{w}$ (see property 4 of Table 6), we arrive at the following problem for a first-order linear ordinary differential equation with parameter u:
$$\widetilde{w}'_t + u^2 \widetilde{w} = 0,$$
$$w = \widetilde{f}(u) \quad \text{at} \quad t = 0,$$

where $\widetilde{f}(u)$ is defined by (3). On solving this problem for the transform $\widetilde{w}$, we find
$$\widetilde{w} = \widetilde{f}(u) e^{-u^2 t}.$$

Let us apply the inversion formula to both sides of this equation. After some calculations, we obtain the solution of the original problem in the form
$$w = \frac{1}{\sqrt{2\pi}} \int_{-\infty}^{\infty} \widetilde{f}(u) e^{-u^2 t} e^{iux}\, du = \frac{1}{2\pi} \int_{-\infty}^{\infty} \left[\int_{-\infty}^{\infty} f(\xi) e^{-iu\xi}\, d\xi\right] e^{-u^2 t + iux}\, du$$
$$= \frac{1}{2\pi} \int_{-\infty}^{\infty} f(\xi)\, d\xi \int_{-\infty}^{\infty} e^{-u^2 t + iu(x-\xi)}\, du = \frac{1}{\sqrt{2\pi t}} \int_{-\infty}^{\infty} f(\xi) \exp\left[-\frac{(x-\xi)^2}{4t}\right] d\xi.$$

At the last stage we used the relation $\int_{-\infty}^{\infty} \exp(-a^2 u^2 + bu)\, du = \frac{\sqrt{\pi}}{|a|} \exp\left(\frac{b^2}{4a^2}\right)$.

The Fourier transform admits n-dimensional generalization:

$$\widetilde{f}(\mathbf{u}) = \frac{1}{(2\pi)^{n/2}} \int_{\mathbb{R}^n} f(\mathbf{x}) e^{-i(\mathbf{u} \cdot \mathbf{x})}\, d\mathbf{x}, \qquad (\mathbf{u} \cdot \mathbf{x}) = u_1 x_1 + \cdots + u_n x_n, \tag{5}$$

where $f(\mathbf{x}) = f(x_1, \ldots, x_n)$, $\widetilde{f}(\mathbf{u}) = \widetilde{f}(u_1, \ldots, u_n)$, and $d\mathbf{x} = dx_1 \ldots dx_n$.

The corresponding inversion formula is

$$f(\mathbf{x}) = \frac{1}{(2\pi)^{n/2}} \int_{\mathbb{R}^n} \widetilde{f}(\mathbf{u}) e^{i(\mathbf{u} \cdot \mathbf{x})}\, d\mathbf{u}, \qquad d\mathbf{u} = du_1 \ldots du_n.$$

The Fourier transform (5) is frequently used in the theory of linear partial differential equations with constant coefficients ($\mathbf{x} \in \mathbb{R}^n$).

⊙ *References for Section* 0.5: H. Bateman and A. Erdélyi (1954), V. A. Ditkin and A. P. Prudnikov (1965), J. W. Miles (1971), B. Davis (1978), Yu. A. Brychkov and A. P. Prudnikov (1989), L. Hörmander (1990), J. R. Hanna and J. H. Rowland (1990), W. H. Beyer (1991), D. Zwillinger (1998), A. D. Polyanin and A. V. Manzhirov (1998).

0.6. Representation of the Solution of the Cauchy Problem via the Fundamental Solution

0.6.1. Cauchy Problem for Parabolic Equations

0.6.1-1. General formula for the solution of the Cauchy problem.

Let $\mathbf{x} = \{x_1, \ldots, x_n\}$ and $\mathbf{y} = \{y_1, \ldots, y_n\}$, where $\mathbf{x} \in \mathbb{R}^n$ and $\mathbf{y} \in \mathbb{R}^n$.

Consider a nonhomogeneous linear equation of the parabolic type with an arbitrary right-hand side,

$$\frac{\partial w}{\partial t} - L_{\mathbf{x},t}[w] = \Phi(\mathbf{x}, t), \tag{1}$$

where the second-order linear differential operator $L_{\mathbf{x},t}$ is defined by relation (2) from Subsection 0.2.1.

The solution of the Cauchy problem for equation (1) with an arbitrary initial condition,

$$w = f(\mathbf{x}) \quad \text{at} \quad t = 0,$$

can be represented as the sum of two integrals,

$$w(\mathbf{x}, t) = \int_0^t \int_{\mathbb{R}^n} \Phi(\mathbf{y}, \tau) \mathscr{E}(\mathbf{x}, \mathbf{y}, t, \tau)\, d\mathbf{y}\, d\tau + \int_{\mathbb{R}^n} f(\mathbf{y}) \mathscr{E}(\mathbf{x}, \mathbf{y}, t, 0)\, d\mathbf{y}, \qquad d\mathbf{y} = dy_1 \ldots dy_n.$$

Here, $\mathscr{E} = \mathscr{E}(\mathbf{x}, \mathbf{y}, t, \tau)$ is the fundamental solution of the Cauchy problem that satisfies, for $t > \tau \geq 0$, the homogeneous linear equation

$$\frac{\partial \mathscr{E}}{\partial t} - L_{\mathbf{x},t}[\mathscr{E}] = 0 \tag{2}$$

with the nonhomogeneous initial condition of special form

$$\mathscr{E} = \delta(\mathbf{x} - \mathbf{y}) \quad \text{at} \quad t = \tau. \tag{3}$$

The quantities τ and $\mathbf{y}$ appear in problem (2)–(3) as free parameters, and $\delta(\mathbf{x}) = \delta(x_1) \ldots \delta(x_n)$ is the n-dimensional Dirac delta function.

Remark 1. If the coefficients of the differential operator $L_{\mathbf{x},t}$ in (2) are independent of time t, then the fundamental solution of the Cauchy problem depends on only three arguments, $\mathscr{E}(\mathbf{x}, \mathbf{y}, t, \tau) = \mathscr{E}(\mathbf{x}, \mathbf{y}, t - \tau)$.

Remark 2. If the differential operator $L_{\mathbf{x},t}$ has constant coefficients, then the fundamental solution of the Cauchy problem depends on only two arguments, $\mathscr{E}(\mathbf{x}, \mathbf{y}, t, \tau) = \mathscr{E}(\mathbf{x} - \mathbf{y}, t - \tau)$.

0.6.1-2. The fundamental solution allowing incomplete separation of variables.

Consider the special case where the differential operator $L_{\mathbf{x},t}$ in equation (1) can be represented as the sum

$$L_{\mathbf{x},t}[w] = L_{1,t}[w] + \cdots + L_{n,t}[w], \tag{4}$$

where each term depends on a single space coordinate and time,

$$L_{k,t}[w] \equiv a_k(x_k, t)\frac{\partial^2 w}{\partial x_k^2} + b_k(x_k, t)\frac{\partial w}{\partial x_k} + c_k(x_k, t)w, \qquad k = 1, \ldots, n.$$

Equations of this form are often encountered in applications. The fundamental solution of the Cauchy problem for the n-dimensional equation (1) with operator (4) can be represented in the product form

$$\mathscr{E}(\mathbf{x}, \mathbf{y}, t, \tau) = \prod_{k=1}^{n} \mathscr{E}_k(x_k, y_k, t, \tau), \tag{5}$$

where $\mathscr{E}_k = \mathscr{E}_k(x_k, y_k, t, \tau)$ are the fundamental solutions satisfying the one-dimensional equations

$$\frac{\partial \mathscr{E}_k}{\partial t} - L_{k,t}[\mathscr{E}_k] = 0 \qquad (k = 1, \ldots, n)$$

with the initial conditions

$$\mathscr{E}_k = \delta(x_k - y_k) \quad \text{at} \quad t = \tau.$$

In this case, the fundamental solution of the Cauchy problem (5) admits incomplete separation of variables; the fundamental solution is separated in the space variables $x_1, \ldots, x_n$ but not in time t.

0.6.2. Cauchy Problem for Hyperbolic Equations

Consider a nonhomogeneous linear equation of the hyperbolic type with an arbitrary right-hand side,

$$\frac{\partial^2 w}{\partial t^2} + \varphi(\mathbf{x}, t)\frac{\partial w}{\partial t} - L_{\mathbf{x},t}[w] = \Phi(\mathbf{x}, t), \tag{6}$$

where the second-order linear differential operator $L_{\mathbf{x},t}$ is defined by relation (2) from Subsection 0.2.1, with $\mathbf{x} \in \mathbb{R}^n$.

The solution of the Cauchy problem for equation (6) with general initial conditions,

$$w = f_0(\mathbf{x}) \quad \text{at} \quad t = 0,$$
$$\partial_t w = f_1(\mathbf{x}) \quad \text{at} \quad t = 0,$$

can be represented as the sum

$$w(\mathbf{x}, t) = \int_0^t \int_{\mathbb{R}^n} \Phi(\mathbf{y}, \tau)\mathscr{E}(\mathbf{x}, \mathbf{y}, t, \tau)\, d\mathbf{y}\, d\tau - \int_{\mathbb{R}^n} f_0(\mathbf{y})\left[\frac{\partial}{\partial \tau}\mathscr{E}(\mathbf{x}, \mathbf{y}, t, \tau)\right]_{\tau=0} d\mathbf{y}$$
$$+ \int_{\mathbb{R}^n} \left[f_1(\mathbf{y}) + f_0(\mathbf{y})\varphi(\mathbf{y}, 0)\right]\mathscr{E}(\mathbf{x}, \mathbf{y}, t, 0)\, d\mathbf{y}, \qquad d\mathbf{y} = dy_1 \ldots dy_n.$$

Here, $\mathscr{E} = \mathscr{E}(\mathbf{x}, \mathbf{y}, t, \tau)$ is the fundamental solution of the Cauchy problem that satisfies, for $t > \tau \geq 0$, the homogeneous linear equation

$$\frac{\partial^2 \mathscr{E}}{\partial t^2} + \varphi(\mathbf{x}, t)\frac{\partial \mathscr{E}}{\partial t} - L_{\mathbf{x},t}[\mathscr{E}] = 0 \tag{7}$$

with the semihomogeneous initial conditions of special form

$$\mathscr{E} = 0 \qquad \text{at} \quad t = \tau,$$
$$\partial_t \mathscr{E} = \delta(\mathbf{x} - \mathbf{y}) \qquad \text{at} \quad t = \tau. \tag{8}$$

The quantities τ and $\mathbf{y}$ appear in problem (7)–(8) as free parameters ($\mathbf{y} \in \mathbb{R}^n$).

Remark 1. If the coefficients of the differential operator $L_{\mathbf{x},t}$ in (7) are independent of time t, then the fundamental solution of the Cauchy problem depends on only three arguments, $\mathscr{E}(\mathbf{x}, \mathbf{y}, t, \tau) = \mathscr{E}(\mathbf{x}, \mathbf{y}, t - \tau)$. Here, $\frac{\partial}{\partial \tau} \mathscr{E}(\mathbf{x}, \mathbf{y}, t, \tau)\big|_{\tau=0} = -\frac{\partial}{\partial t} \mathscr{E}(\mathbf{x}, \mathbf{y}, t)$.

Remark 2. If the differential operator $L_{\mathbf{x},t}$ has constant coefficients, then the fundamental solution of the Cauchy problem depends on only two arguments, $\mathscr{E}(\mathbf{x}, \mathbf{y}, t, \tau) = \mathscr{E}(\mathbf{x} - \mathbf{y}, t - \tau)$.

⊙ *References for Section 0.6:* V. M. Babich, M. B. Kapilevich, S. G. Mikhlin, et al. (1964), G. E. Shilov (1965), A. D. Polyanin (2000a, 2000b, 2000c, 2001a).

0.7. Nonhomogeneous Boundary Value Problems with One Space Variable. Representation of Solutions via the Green's Function

0.7.1. Problems for Parabolic Equations

0.7.1-1. Statement of the problem ($t \geq 0$, $x_1 \leq x \leq x_2$).

In general, a nonhomogeneous linear differential equation of the parabolic type with variable coefficients in one dimension can be written as

$$\frac{\partial w}{\partial t} - L_{x,t}[w] = \Phi(x,t), \tag{1}$$

where

$$L_{x,t}[w] \equiv a(x,t)\frac{\partial^2 w}{\partial x^2} + b(x,t)\frac{\partial w}{\partial x} + c(x,t)w, \qquad a(x,t) > 0. \tag{2}$$

Consider the nonstationary boundary value problem for equation (1) with an initial condition of general form,

$$w = f(x) \quad \text{at} \quad t = 0, \tag{3}$$

and arbitrary nonhomogeneous linear boundary conditions,

$$s_1 \frac{\partial w}{\partial x} + k_1(t)w = g_1(t) \quad \text{at} \quad x = x_1, \tag{4}$$

$$s_2 \frac{\partial w}{\partial x} + k_2(t)w = g_2(t) \quad \text{at} \quad x = x_2. \tag{5}$$

By appropriately choosing the coefficients s_1, s_2 and the functions $k_1 = k_1(t)$, $k_2 = k_2(t)$ in (4) and (5), we obtain the first, second, third, and mixed boundary value problems for equation (1).

0.7.1-2. Representation of the problem solution in terms of the Green's function.

The solution of the nonhomogeneous linear boundary value problem (1)–(5) can be represented as

$$w(x,t) = \int_0^t \int_{x_1}^{x_2} \Phi(y,\tau) G(x,y,t,\tau)\, dy\, d\tau + \int_{x_1}^{x_2} f(y) G(x,y,t,0)\, dy$$

$$+ \int_0^t g_1(\tau) a(x_1,\tau) \Lambda_1(x,t,\tau)\, d\tau + \int_0^t g_2(\tau) a(x_2,\tau) \Lambda_2(x,t,\tau)\, d\tau. \tag{6}$$

Here, $G(x,y,t,\tau)$ is the Green's function that satisfies, for $t > \tau \geq 0$, the homogeneous equation

$$\frac{\partial G}{\partial t} - L_{x,t}[G] = 0 \tag{7}$$

TABLE 7
Expressions of the functions $\Lambda_1(x,t,\tau)$ and $\Lambda_2(x,t,\tau)$
involved in the integrands of the last two terms in solution (6)

Type of problem	Form of boundary conditions	Functions $\Lambda_m(x,t,\tau)$
First boundary value problem $(s_1=s_2=0,\ k_1=k_2=1)$	$w=g_1(t)$ at $x=x_1$ $w=g_2(t)$ at $x=x_2$	$\Lambda_1(x,t,\tau)=\partial_y G(x,y,t,\tau)\big|_{y=x_1}$ $\Lambda_2(x,t,\tau)=-\partial_y G(x,y,t,\tau)\big|_{y=x_2}$
Second boundary value problem $(s_1=s_2=1,\ k_1=k_2=0)$	$\partial_x w=g_1(t)$ at $x=x_1$ $\partial_x w=g_2(t)$ at $x=x_2$	$\Lambda_1(x,t,\tau)=-G(x,x_1,t,\tau)$ $\Lambda_2(x,t,\tau)=G(x,x_2,t,\tau)$
Third boundary value problem $(s_1=s_2=1,\ k_1<0,\ k_2>0)$	$\partial_x w+k_1 w=g_1(t)$ at $x=x_1$ $\partial_x w+k_2 w=g_2(t)$ at $x=x_2$	$\Lambda_1(x,t,\tau)=-G(x,x_1,t,\tau)$ $\Lambda_2(x,t,\tau)=G(x,x_2,t,\tau)$
Mixed boundary value problem $(s_1=k_2=0,\ s_2=k_1=1)$	$w=g_1(t)$ at $x=x_1$ $\partial_x w=g_2(t)$ at $x=x_2$	$\Lambda_1(x,t,\tau)=\partial_y G(x,y,t,\tau)\big|_{y=x_1}$ $\Lambda_2(x,t,\tau)=G(x,x_2,t,\tau)$
Mixed boundary value problem $(s_1=k_2=1,\ s_2=k_1=0)$	$\partial_x w=g_1(t)$ at $x=x_1$ $w=g_2(t)$ at $x=x_2$	$\Lambda_1(x,t,\tau)=-G(x,x_1,t,\tau)$ $\Lambda_2(x,t,\tau)=-\partial_y G(x,y,t,\tau)\big|_{y=x_2}$

with the nonhomogeneous initial condition of special form

$$G = \delta(x-y) \quad \text{at} \quad t=\tau \tag{8}$$

and the homogeneous boundary conditions

$$s_1 \frac{\partial G}{\partial x} + k_1(t)G = 0 \quad \text{at} \quad x=x_1, \tag{9}$$

$$s_2 \frac{\partial G}{\partial x} + k_2(t)G = 0 \quad \text{at} \quad x=x_2. \tag{10}$$

The quantities y and τ appear in problem (7)–(10) as free parameters ($x_1 \leq y \leq x_2$), and $\delta(x)$ is the Dirac delta function.

The initial condition (8) implies the limit relation

$$f(x) = \lim_{t\to\tau} \int_{x_1}^{x_2} f(y)G(x,y,t,\tau)\,dy$$

for any continuous function $f=f(x)$.

The functions $\Lambda_1(x,t,\tau)$ and $\Lambda_2(x,t,\tau)$ involved in the integrands of the last two terms in solution (6) can be expressed in terms of the Green's function $G(x,y,t,\tau)$. The corresponding formulas for $\Lambda_m(x,t,\tau)$ are given in Table 7 for the basic types of boundary value problems.

It is significant that the Green's function G and the functions Λ_1, Λ_2 are independent of the functions Φ, f, g_1, and g_2 that characterize various nonhomogeneities of the boundary value problem.

If the coefficients of equation (1)–(2) and the coefficients k_1, k_2 in the boundary conditions (4) and (5) are independent of time t, i.e., the conditions

$$a = a(x), \quad b = b(x), \quad c = c(x), \quad k_1 = \text{const}, \quad k_2 = \text{const} \tag{11}$$

hold, then the Green's function depends on only three arguments,

$$G(x,y,t,\tau) = G(x,y,t-\tau).$$

In this case, the functions Λ_m depend on only two arguments, $\Lambda_m = \Lambda_m(x, t - \tau)$, $m = 1, 2$.

Formula (6) also remains valid for the problem with boundary conditions of the third kind if $k_1 = k_1(t)$ and $k_2 = k_2(t)$. Here, the relation between Λ_m ($m = 1, 2$) and the Green's function G is the same as that in the case of constants k_1 and k_2; the Green's function itself is now different.

The condition that the solution must vanish at infinity, $w \to 0$ as $x \to \infty$, is often set for the first, second, and third boundary value problems that are considered on the interval $x_1 \le x < \infty$. In this case, the solution is calculated by formula (6) with $\Lambda_2 = 0$ and Λ_1 specified in Table 7.

0.7.2. Problems for Hyperbolic Equations

0.7.1-2. Statement of the problem ($t \ge 0$, $x_1 \le x \le x_2$).

In general, a one-dimensional nonhomogeneous linear differential equation of hyperbolic type with variable coefficients is written as

$$\frac{\partial^2 w}{\partial t^2} + \varphi(x, t) \frac{\partial w}{\partial t} - L_{x,t}[w] = \Phi(x, t), \tag{12}$$

where the operator $L_{x,t}[w]$ is defined by (2).

Consider the nonstationary boundary value problem for equation (12) with the initial conditions

$$\begin{aligned} w &= f_0(x) \quad \text{at} \quad t = 0, \\ \partial_t w &= f_1(x) \quad \text{at} \quad t = 0 \end{aligned} \tag{13}$$

and arbitrary nonhomogeneous linear boundary conditions (4)–(5).

0.7.2-2. Representation of the problem solution in terms of the Green's function.

The solution of problem (12), (13), (4), (5) can be represented as the sum

$$\begin{aligned} w(x, t) = &\int_0^t \int_{x_1}^{x_2} \Phi(y, \tau) G(x, y, t, \tau) \, dy \, d\tau \\ &- \int_{x_1}^{x_2} f_0(y) \left[\frac{\partial}{\partial \tau} G(x, y, t, \tau) \right]_{\tau=0} dy + \int_{x_1}^{x_2} \left[f_1(y) + f_0(y) \varphi(y, 0) \right] G(x, y, t, 0) \, dy \\ &+ \int_0^t g_1(\tau) a(x_1, \tau) \Lambda_1(x, t, \tau) \, d\tau + \int_0^t g_2(\tau) a(x_2, \tau) \Lambda_2(x, t, \tau) \, d\tau. \end{aligned} \tag{14}$$

Here, the Green's function $G(x, y, t, \tau)$ is determined by solving the homogeneous equation

$$\frac{\partial^2 G}{\partial t^2} + \varphi(x, t) \frac{\partial G}{\partial t} - L_{x,t}[G] = 0 \tag{15}$$

with the semihomogeneous initial conditions

$$G = 0 \qquad \text{at} \quad t = \tau, \tag{16}$$

$$\partial_t G = \delta(x - y) \qquad \text{at} \quad t = \tau, \tag{17}$$

and the homogeneous boundary conditions (9) and (10). The quantities y and τ appear in problem (15)–(17), (9), (10) as free parameters ($x_1 \le y \le x_2$), and $\delta(x)$ is the Dirac delta function.

The functions $\Lambda_1(x, t, \tau)$ and $\Lambda_2(x, t, \tau)$ involved in the integrands of the last two terms in solution (14) can be expressed via the Green's function $G(x, y, t, \tau)$. The corresponding formulas for $\Lambda_m(x, t, \tau)$ are given in Table 7 for the basic types of boundary value problems.

It is significant that the Green's function G and Λ_1, Λ_2 are independent of the functions Φ, f_0, f_1, g_1, and g_2 that characterize various nonhomogeneities of the boundary value problem.

If the coefficients of equation (12) and the coefficients k_1, k_2 in the boundary conditions (4) and (5) are independent of time t, then the Green's function depends on only three arguments, $G(x, y, t, \tau) = G(x, y, t - \tau)$. In this case, one can set $\frac{\partial}{\partial \tau} G(x, y, t, \tau)\big|_{\tau=0} = -\frac{\partial}{\partial t} G(x, y, t)$ in solution (14).

⊙ *References for Section* 0.7: V. M. Babich, M. B. Kapilevich, S. G. Mikhlin, et al. (1964), E. Butkov (1968), A. G. Butkovskiy (1982), E. Zauderer (1989), A. D. Polyanin (2000a, 2000b, 2000c, 2001a).

0.8. Nonhomogeneous Boundary Value Problems with Many Space Variables. Representation of Solutions via the Green's Function

0.8.1. Problems for Parabolic Equations

0.8.1-1. Statement of the problem.

In general, a nonhomogeneous linear differential equation of the parabolic type in n space variables has the form

$$\frac{\partial w}{\partial t} - L_{\mathbf{x},t}[w] = \Phi(\mathbf{x}, t), \tag{1}$$

where

$$L_{\mathbf{x},t}[w] \equiv \sum_{i,j=1}^{n} a_{ij}(\mathbf{x}, t) \frac{\partial^2 w}{\partial x_i \partial x_j} + \sum_{i=1}^{n} b_i(\mathbf{x}, t) \frac{\partial w}{\partial x_i} + c(\mathbf{x}, t) w,$$

$$\mathbf{x} = \{x_1, \ldots, x_n\}, \quad \sum_{i,j=1}^{n} a_{ij}(\mathbf{x}, t) \xi_i \xi_j \geq \sigma \sum_{i=1}^{n} \xi_i^2, \quad \sigma > 0. \tag{2}$$

Let V be some simply connected domain in $\mathbb{R}^n$ with a sufficiently smooth boundary $S = \partial V$. We consider the nonstationary boundary value problem for equation (1) in the domain V with an arbitrary initial condition,

$$w = f(\mathbf{x}) \quad \text{at} \quad t = 0, \tag{3}$$

and nonhomogeneous linear boundary conditions,

$$\Gamma_{\mathbf{x},t}[w] = g(\mathbf{x}, t) \quad \text{for} \quad \mathbf{x} \in S. \tag{4}$$

In the general case, $\Gamma_{\mathbf{x},t}$ is a first-order linear differential operator in the space coordinates with coefficients dependent on $\mathbf{x}$ and t.

0.8.1-2. Representation of the problem solution in terms of the Green's function.

The solution of the nonhomogeneous linear boundary value problem (1)–(4) can be represented as the sum

$$w(\mathbf{x}, t) = \int_0^t \int_V \Phi(\mathbf{y}, \tau) G(\mathbf{x}, \mathbf{y}, t, \tau) \, dV_y \, d\tau + \int_V f(\mathbf{y}) G(\mathbf{x}, \mathbf{y}, t, 0) \, dV_y$$
$$+ \int_0^t \int_S g(\mathbf{y}, \tau) H(\mathbf{x}, \mathbf{y}, t, \tau) \, dS_y \, d\tau, \tag{5}$$

where $G(\mathbf{x}, \mathbf{y}, t, \tau)$ is the Green's function; for $t > \tau \geq 0$, it satisfies the homogeneous equation

$$\frac{\partial G}{\partial t} - L_{\mathbf{x},t}[G] = 0 \tag{6}$$

with the nonhomogeneous initial condition of special form

$$G = \delta(\mathbf{x} - \mathbf{y}) \quad \text{at} \quad t = \tau \tag{7}$$

and the homogeneous boundary condition

$$\Gamma_{\mathbf{x},t}[G] = 0 \quad \text{for} \quad \mathbf{x} \in S. \tag{8}$$

The vector $\mathbf{y} = \{y_1, \ldots, y_n\}$ appears in problem (6)–(8) as an n-dimensional free parameter ($\mathbf{y} \in V$), and $\delta(\mathbf{x} - \mathbf{y}) = \delta(x_1 - y_1) \ldots \delta(x_n - y_n)$ is the n-dimensional Dirac delta function. The Green's

TABLE 8

The form of the function $H(\mathbf{x}, \mathbf{y}, t, \tau)$ for the basic types of nonstationary boundary value problems

Type of problem	Form of boundary condition (4)	Function $H(\mathbf{x}, \mathbf{y}, t, \tau)$
1st boundary value problem	$w = g(\mathbf{x}, t)$ for $\mathbf{x} \in S$	$H(\mathbf{x}, \mathbf{y}, t, \tau) = -\frac{\partial G}{\partial M_y}(\mathbf{x}, \mathbf{y}, t, \tau)$
2nd boundary value problem	$\frac{\partial w}{\partial M_x} = g(\mathbf{x}, t)$ for $\mathbf{x} \in S$	$H(\mathbf{x}, \mathbf{y}, t, \tau) = G(\mathbf{x}, \mathbf{y}, t, \tau)$
3rd boundary value problem	$\frac{\partial w}{\partial M_x} + kw = g(\mathbf{x}, t)$ for $\mathbf{x} \in S$	$H(\mathbf{x}, \mathbf{y}, t, \tau) = G(\mathbf{x}, \mathbf{y}, t, \tau)$

function G is independent of the functions Φ, f, and g that characterize various nonhomogeneities of the boundary value problem. In (5), the integration is everywhere performed with respect to $\mathbf{y}$, with $dV_y = dy_1 \ldots dy_n$.

The function $H(\mathbf{x}, \mathbf{y}, t, \tau)$ involved in the integrand of the last term in solution (5) can be expressed via the Green's function $G(\mathbf{x}, \mathbf{y}, t, \tau)$. The corresponding formulas for $H(\mathbf{x}, \mathbf{y}, t, \tau)$ are given in Table 8 for the three basic types of boundary value problems; in the third boundary value problem, the coefficient k can depend on $\mathbf{x}$ and t. The boundary conditions of the second and third kind, as well as the solution of the first boundary value problem, involve operators of differentiation along the conormal of operator (2); these operators act as follows:

$$\frac{\partial G}{\partial M_x} \equiv \sum_{i,j=1}^{n} a_{ij}(\mathbf{x}, t) N_j \frac{\partial G}{\partial x_i}, \qquad \frac{\partial G}{\partial M_y} \equiv \sum_{i,j=1}^{n} a_{ij}(\mathbf{y}, \tau) N_j \frac{\partial G}{\partial y_i}, \tag{9}$$

where $\mathbf{N} = \{N_1, \ldots, N_n\}$ is the unit outward normal to the surface S. In the special case where $a_{ii}(\mathbf{x}, t) = 1$ and $a_{ij}(\mathbf{x}, t) = 0$ for $i \neq j$, operator (9) coincides with the ordinary operator of differentiation along the outward normal to S.

If the coefficient of equation (6) and the boundary condition (8) are independent of t, then the Green's function depends on only three arguments, $G(\mathbf{x}, \mathbf{y}, t, \tau) = G(\mathbf{x}, \mathbf{y}, t - \tau)$.

Remark. Let S_i ($i = 1, \ldots, p$) be different portions of the surface S such that $S = \sum_{i=1}^{p} S_i$ and let boundary conditions of various types be set on the S_i,

$$\Gamma_{\mathbf{x},t}^{(i)}[w] = g_i(\mathbf{x}, t) \quad \text{for} \quad \mathbf{x} \in S_i, \quad i = 1, \ldots, p. \tag{10}$$

Then formula (5) remains valid but the last term in (5) must be replaced by the sum

$$\sum_{i=1}^{p} \int_{0}^{t} \int_{S_i} g_i(\mathbf{y}, \tau) H_i(\mathbf{x}, \mathbf{y}, t, \tau) \, dS_y \, d\tau. \tag{11}$$

0.8.2. Problems for Hyperbolic Equations

0.8.2-1. Statement of the problem.

The general nonhomogeneous linear differential hyperbolic equation in n space variables can be written as

$$\frac{\partial^2 w}{\partial t^2} + \varphi(\mathbf{x}, t) \frac{\partial w}{\partial t} - L_{\mathbf{x},t}[w] = \Phi(\mathbf{x}, t), \tag{12}$$

where the operator $L_{\mathbf{x},t}[w]$ is explicitly defined in (2).

We consider the nonstationary boundary value problem for equation (12) in the domain V with arbitrary initial conditions,

$$w = f_0(\mathbf{x}) \quad \text{at} \quad t = 0, \tag{13}$$

$$\partial_t w = f_1(\mathbf{x}) \quad \text{at} \quad t = 0, \tag{14}$$

and the nonhomogeneous linear boundary condition (4).

0.8.2-2. Representation of the problem solution in terms of the Green's function.

The solution of the nonhomogeneous linear boundary value problem (12)–(14), (4) can be represented as the sum

$$w(\mathbf{x},t) = \int_0^t \int_V \Phi(\mathbf{y},\tau) G(\mathbf{x},\mathbf{y},t,\tau)\, dV_y\, d\tau - \int_V f_0(\mathbf{y}) \left[\frac{\partial}{\partial \tau} G(\mathbf{x},\mathbf{y},t,\tau)\right]_{\tau=0} dV_y$$

$$+ \int_V \big[f_1(\mathbf{y}) + f_0(\mathbf{y})\varphi(\mathbf{y},0)\big] G(\mathbf{x},\mathbf{y},t,0)\, dV_y + \int_0^t \int_S g(\mathbf{y},\tau) H(\mathbf{x},\mathbf{y},t,\tau)\, dS_y\, d\tau. \qquad (15)$$

Here, $G(\mathbf{x},\mathbf{y},t,\tau)$ is the Green's function; for $t > \tau \geq 0$ it satisfies the homogeneous equation

$$\frac{\partial^2 G}{\partial t^2} + \varphi(\mathbf{x},t)\frac{\partial G}{\partial t} - L_{\mathbf{x},t}[G] = 0 \qquad (16)$$

with the semihomogeneous initial conditions

$$G = 0 \quad \text{at} \quad t = \tau,$$
$$\partial_t G = \delta(\mathbf{x}-\mathbf{y}) \quad \text{at} \quad t = \tau,$$

and the homogeneous boundary condition (8).

If the coefficients of equation (16) and the boundary condition (8) are independent of time t, then the Green's function depends on only three arguments, $G(\mathbf{x},\mathbf{y},t,\tau) = G(\mathbf{x},\mathbf{y},t-\tau)$. In this case, one can set $\frac{\partial}{\partial \tau} G(\mathbf{x},\mathbf{y},t,\tau)\big|_{\tau=0} = -\frac{\partial}{\partial t} G(\mathbf{x},\mathbf{y},t)$ in solution (15).

The function $H(\mathbf{x},\mathbf{y},t,\tau)$ involved in the integrand of the last term in solution (15) can be expressed via the Green's function $G(\mathbf{x},\mathbf{y},t,\tau)$. The corresponding formulas for H are given in Table 8 for the three basic types of boundary value problems; in the third boundary value problem, the coefficient k can depend on $\mathbf{x}$ and t.

Remark. Let S_i ($i = 1,\ldots,p$) be different portions of the surface S such that $S = \sum_{i=1}^{p} S_i$ and let boundary conditions of various types (10) be set on the S_i. Then formula (15) remains valid but the last term in (15) must be replaced by the sum (11).

0.8.3. Problems for Elliptic Equations

0.8.3-1. Statement of the problem.

In general, a nonhomogeneous linear elliptic equation can be written as

$$-L_{\mathbf{x}}[w] = \Phi(\mathbf{x}), \qquad (17)$$

where

$$L_{\mathbf{x}}[w] \equiv \sum_{i,j=1}^{n} a_{ij}(\mathbf{x})\frac{\partial^2 w}{\partial x_i \partial x_j} + \sum_{i=1}^{n} b_i(\mathbf{x})\frac{\partial w}{\partial x_i} + c(\mathbf{x})w. \qquad (18)$$

Two-dimensional problems correspond to $n = 2$ and three-dimensional problems, to $n = 3$.

We consider equation (17)–(18) in a domain V and assume that the equation is subject to the general linear boundary condition

$$\Gamma_{\mathbf{x}}[w] = g(\mathbf{x}) \quad \text{for} \quad \mathbf{x} \in S. \qquad (19)$$

The solution of the stationary problem (17)–(19) can be obtained by passing in (5) to the limit as $t \to \infty$. To this end, one should start with equation (1) whose coefficients are independent of t and take the homogeneous initial condition (3), with $f(\mathbf{x}) = 0$, and the stationary boundary condition (4).

TABLE 9
The form of the function $H(\mathbf{x}, \mathbf{y})$ involved in the integrand of the last term
in solution (20) for the basic types of stationary boundary value problems

Type of problem	Form of boundary condition (19)	Function $H(\mathbf{x}, \mathbf{y})$
1st boundary value problem	$w = g(\mathbf{x})$ for $\mathbf{x} \in S$	$H(\mathbf{x}, \mathbf{y}) = -\frac{\partial G}{\partial M_y}(\mathbf{x}, \mathbf{y})$
2nd boundary value problem	$\frac{\partial w}{\partial M_x} = g(\mathbf{x})$ for $\mathbf{x} \in S$	$H(\mathbf{x}, \mathbf{y}) = G(\mathbf{x}, \mathbf{y})$
3rd boundary value problem	$\frac{\partial w}{\partial M_x} + kw = g(\mathbf{x})$ for $\mathbf{x} \in S$	$H(\mathbf{x}, \mathbf{y}) = G(\mathbf{x}, \mathbf{y})$

0.8.3-2. Representation of the problem solution in terms of the Green's function.

The solution of the linear boundary value problem (17)–(19) can be represented as the sum

$$w(\mathbf{x}) = \int_V \Phi(\mathbf{y}) G(\mathbf{x}, \mathbf{y}) \, dV_y + \int_S g(\mathbf{y}) H(\mathbf{x}, \mathbf{y}) \, dS_y. \qquad (20)$$

Here, the Green's function $G(\mathbf{x}, \mathbf{y})$ satisfies the nonhomogeneous equation of special form

$$-L_{\mathbf{x}}[G] = \delta(\mathbf{x} - \mathbf{y}) \qquad (21)$$

with the homogeneous boundary condition

$$\Gamma_{\mathbf{x}}[G] = 0 \quad \text{for} \quad \mathbf{x} \in S. \qquad (22)$$

The vector $\mathbf{y} = \{y_1, \dots, y_n\}$ appears in problem (21), (22) as an n-dimensional free parameter ($\mathbf{y} \in V$). Note that G is independent of the functions Φ and g characterizing various nonhomogeneities of the original boundary value problem.

The function $H(\mathbf{x}, \mathbf{y})$ involved in the integrand of the second term in solution (20) can be expressed via the Green's function $G(\mathbf{x}, \mathbf{y})$. The corresponding formulas for H are given in Table 9 for the three basic types of boundary value problems. The boundary conditions of the second and third kind, as well as the solution of the first boundary value problem, involve operators of differentiation along the conormal of operator (18); these operators are defined by (9); in this case, the coefficients a_{ij} depend on only $\mathbf{x}$.

Remark. For the second boundary value problem with $c(\mathbf{x}) \equiv 0$, the thus defined Green's function must not necessarily exist; see Remark 2 in Paragraph 8.2.1-2.

0.8.4. Comparison of the Solution Structures for Boundary Value Problems for Equations of Various Types

Table 10 lists brief formulations of boundary value problems for second-order equations of elliptic, parabolic, and hyperbolic types. The coefficients of the differential operators $L_{\mathbf{x}}$ and $\Gamma_{\mathbf{x}}$ in the space variables $x_1, \dots, x_n$ are assumed to be independent of time t; these operators are the same for the problems under consideration.

Below are the respective general formulas defining the solutions of these problems with zero initial conditions ($f = f_0 = f_1 = 0$):

$$w_0(\mathbf{x}) = \int_V \Phi(\mathbf{y}) G_0(\mathbf{x}, \mathbf{y}) \, dV_y + \int_S g(\mathbf{y}) \mathcal{H}[G_0(\mathbf{x}, \mathbf{y})] \, dS_y,$$

$$w_1(\mathbf{x}, t) = \int_0^t \int_V \Phi(\mathbf{y}, \tau) G_1(\mathbf{x}, \mathbf{y}, t - \tau) \, dV_y \, d\tau + \int_0^t \int_S g(\mathbf{y}, \tau) \mathcal{H}[G_1(\mathbf{x}, \mathbf{y}, t - \tau)] \, dS_y \, d\tau,$$

$$w_2(\mathbf{x}, t) = \int_0^t \int_V \Phi(\mathbf{y}, \tau) G_2(\mathbf{x}, \mathbf{y}, t - \tau) \, dV_y \, d\tau + \int_0^t \int_S g(\mathbf{y}, \tau) \mathcal{H}[G_2(\mathbf{x}, \mathbf{y}, t - \tau)] \, dS_y \, d\tau,$$

TABLE 10
Formulations of boundary value problems for equations of various types

Type of equation	Form of equation	Initial conditions	Boundary conditions
Elliptic	$-L_\mathbf{x}[w] = \Phi(\mathbf{x})$	not set	$\Gamma_\mathbf{x}[w] = g(\mathbf{x})$ for $\mathbf{x} \in S$
Parabolic	$\partial_t w - L_\mathbf{x}[w] = \Phi(\mathbf{x}, t)$	$w = f(\mathbf{x})$ at $t = 0$	$\Gamma_\mathbf{x}[w] = g(\mathbf{x}, t)$ for $\mathbf{x} \in S$
Hyperbolic	$\partial_{tt} w - L_\mathbf{x}[w] = \Phi(\mathbf{x}, t)$	$w = f_0(\mathbf{x})$ at $t = 0$, $\partial_t w = f_1(\mathbf{x})$ at $t = 0$	$\Gamma_\mathbf{x}[w] = g(\mathbf{x}, t)$ for $\mathbf{x} \in S$

where the G_n are the Green's functions, the subscripts 0, 1, and 2 refer to the elliptic, parabolic, and hyperbolic problem, respectively. All solutions involve the same operator $\mathcal{H}[G]$; it is explicitly defined in Subsections 0.8.1–0.8.3 (see also Section 0.7) for different boundary conditions.

It is apparent that the solutions of the parabolic and hyperbolic problems with zero initial conditions have the same structure. The structure of the solution to the problem for a parabolic equation differs from that for an elliptic equation by the additional integration with respect to t.

⊙ *References for Section* 0.8: P. M. Morse and H. Feshbach (1953), V. M. Babich, M. B. Kapilevich, S. G. Mikhlin, et al. (1964), E. Butkov (1968), A. G. Butkovskiy (1979, 1982), E. Zauderer (1989), A. N. Tikhonov and A. A. Samarskii (1990), A. D. Polyanin (2000a, 2000c, 2001a).

0.9. Construction of the Green's Functions. General Formulas and Relations

0.9.1. Green's Functions of Boundary Value Problems for Equations of Various Types in Bounded Domains

0.9.1-1. Expressions of the Green's function in terms of infinite series.

Table 11 lists the Green's functions of boundary value problems for second-order equations of various types in a bounded domain V. It is assumed that $L_\mathbf{x}$ is a second-order linear self-adjoint differential operator (e.g., see Zwillinger, 1998) in the space variables $x_1, \ldots, x_n$, and $\Gamma_\mathbf{x}$ is a zeroth- or first-order linear boundary operator that can define a boundary condition of the first, second, or third kind; the coefficients of the operators $L_\mathbf{x}$ and $\Gamma_\mathbf{x}$ can depend on the space variables but are independent of time t. The coefficients λ_k and the functions $u_k(\mathbf{x})$ are determined by solving the homogeneous eigenvalue problem

$$L_\mathbf{x}[u] + \lambda u = 0, \tag{1}$$

$$\Gamma_\mathbf{x}[u] = 0 \quad \text{for} \quad \mathbf{x} \in S. \tag{2}$$

It is apparent from Table 11 that, given the Green's function in the problem for a parabolic (or hyperbolic) equation, one can easily construct the Green's functions of the corresponding problems for elliptic and hyperbolic (or parabolic) equations. In particular, the Green's function of the problem for an elliptic equation can be expressed via the Green's function of the problem for a parabolic equation as follows:

$$G_0(\mathbf{x}, \mathbf{y}) = \int_0^\infty G_1(\mathbf{x}, \mathbf{y}, t)\, dt. \tag{3}$$

Here, the fact that all λ_k are positive is taken into account; for the second boundary value problem, it is assumed that $\lambda = 0$ is not an eigenvalue of problem (1)–(2).

TABLE 11
The Green's functions of boundary value problems for equations of various types in bounded domains. In all problems, the operators $L_\mathbf{x}$ and $\Gamma_\mathbf{x}$ are the same; $\mathbf{x} = \{x_1, \ldots, x_n\}$

Equation	Initial and boundary conditions	Green's function
Elliptic equation $-L_\mathbf{x}[w] = \Phi(\mathbf{x})$	$\Gamma_\mathbf{x}[w] = g(\mathbf{x})$ for $\mathbf{x} \in S$ (no initial condition required)	$G(\mathbf{x}, \mathbf{y}) = \sum_{k=1}^{\infty} \dfrac{u_k(\mathbf{x}) u_k(\mathbf{y})}{\|u_k\|^2 \lambda_k}, \quad \lambda_k \neq 0$
Parabolic equation $\partial_t w - L_\mathbf{x}[w] = \Phi(\mathbf{x}, t)$	$w = f(\mathbf{x})$ at $t = 0$ $\Gamma_\mathbf{x}[w] = g(\mathbf{x}, t)$ for $\mathbf{x} \in S$	$G(\mathbf{x}, \mathbf{y}, t) = \sum_{k=1}^{\infty} \dfrac{u_k(\mathbf{x}) u_k(\mathbf{y})}{\|u_k\|^2} \exp(-\lambda_k t)$
Hyperbolic equation $\partial_{tt} w - L_\mathbf{x}[w] = \Phi(\mathbf{x}, t)$	$w = f_0(\mathbf{x})$ at $t = 0$ $w = f_1(\mathbf{x})$ at $t = 0$ $\Gamma_\mathbf{x}[w] = g(\mathbf{x}, t)$ for $\mathbf{x} \in S$	$G(\mathbf{x}, \mathbf{y}, t) = \sum_{k=1}^{\infty} \dfrac{u_k(\mathbf{x}) u_k(\mathbf{y})}{\|u_k\|^2 \sqrt{\lambda_k}} \sin(t \sqrt{\lambda_k})$

0.9.1-2. Some remarks and generalizations.

Remark 1. Formula (3) can also be used if the domain V is infinite. In this case, one should make sure that the integral on the right-hand side is convergent.

Remark 2. Suppose the equations given in the first column of Table 11 contain $-L_\mathbf{x}[w] - \beta w$ instead of $-L_\mathbf{x}[w]$, with β being a free parameter. Then the λ_k in the expressions of the Green's function in the third column of Table 11 must be replaced by $\lambda_k - \beta$; just as previously, the λ_k and $u_k(\mathbf{x})$ were determined by solving the eigenvalue problem (1)–(2).

Remark 3. The formulas for the Green's functions presented in Table 11 will also hold for boundary value problems described by equations of the fourth or higher order in the space variables; provided that the eigenvalue problem for equation (1) subject to appropriate boundary conditions is self-adjoint.

0.9.2. Green's Functions Admitting Incomplete Separation of Variables

0.9.2-1. Boundary value problems for rectangular domains.

1°. Consider the parabolic equation

$$\frac{\partial w}{\partial t} = L_{1,t}[w] + \cdots + L_{n,t}[w] + \Phi(\mathbf{x}, t), \qquad (4)$$

where each term $L_{m,t}[w]$ depends on only one space variable, x_m, and time t:

$$L_{m,t}[w] \equiv a_m(x_m, t) \frac{\partial^2 w}{\partial x_m^2} + b_m(x_m, t) \frac{\partial w}{\partial x_m} + c_m(x_m, t) w, \qquad m = 1, \ldots, n.$$

For equation (4) we set the initial condition of general form

$$w = f(\mathbf{x}) \quad \text{at} \quad t = 0. \qquad (5)$$

Consider the domain $V = \{\alpha_m \leq x_m \leq \beta_m, \ m = 1, \ldots, n\}$ which is an n-dimensional parallelepiped. We set the following boundary conditions at the faces of the parallelepiped:

$$s_m^{(1)} \frac{\partial w}{\partial x_m} + k_m^{(1)}(t) w = g_m^{(1)}(\mathbf{x}, t) \quad \text{at} \quad x_m = \alpha_m,$$
$$s_m^{(2)} \frac{\partial w}{\partial x_m} + k_m^{(2)}(t) w = g_m^{(2)}(\mathbf{x}, t) \quad \text{at} \quad x_m = \beta_m. \qquad (6)$$

By appropriately choosing the coefficients $s_m^{(1)}$, $s_m^{(2)}$ and functions $k_m^{(1)} = k_m^{(1)}(t)$, $k_m^{(2)} = k_m^{(2)}(t)$, we can obtain the boundary conditions of the first, second, or third kind. For infinite domains, the boundary conditions corresponding to $\alpha_m = -\infty$ or $\beta_m = \infty$ are omitted.

2°. The Green's function of the nonstationary n-dimensional boundary value problem (4)–(6) can be represented in the product form

$$G(\mathbf{x}, \mathbf{y}, t, \tau) = \prod_{m=1}^{n} G_m(x_m, y_m, t, \tau), \qquad (7)$$

where the Green's functions $G_m = G_m(x_m, y_m, t, \tau)$ satisfy the one-dimensional equations

$$\frac{\partial G_m}{\partial t} - L_{m,t}[G_m] = 0 \qquad (m = 1, \ldots, n)$$

with the initial conditions

$$G_m = \delta(x_m - y_m) \quad \text{at} \quad t = \tau$$

and the homogeneous boundary conditions

$$s_m^{(1)} \frac{\partial G_m}{\partial x_m} + k_m^{(1)}(t) G_m = 0 \quad \text{at} \quad x_m = \alpha_m,$$

$$s_m^{(2)} \frac{\partial G_m}{\partial x_m} + k_m^{(2)}(t) G_m = 0 \quad \text{at} \quad x_m = \beta_m.$$

Here, y_m and τ are free parameters ($\alpha_m \leq y_m \leq \beta_m$ and $t \geq \tau \geq 0$), and $\delta(x)$ is the Dirac delta function.

It can be seen that the Green's function (7) admits incomplete separation of variables; it separates in the space variables $x_1, \ldots, x_n$ but not in time t.

0.9.2-2. Boundary value problems for a cylindrical domain with arbitrary cross-section.

1°. Consider the parabolic equation

$$\frac{\partial w}{\partial t} = L_{\mathbf{x},t}[w] + M_{z,t}[w] + \Phi(\mathbf{x}, z, t), \qquad (8)$$

where $L_{\mathbf{x},t}$ is an arbitrary second-order linear differential operator in $x_1, \ldots, x_n$ with coefficients dependent on $\mathbf{x}$ and t, and $M_{z,t}$ is an arbitrary second-order linear differential operator in z with coefficients dependent on z and t.

For equation (8) we set the general initial condition (5), where $f(\mathbf{x})$ must be replaced by $f(\mathbf{x}, z)$.

We assume that the space variables belong to a cylindrical domain $V = \{\mathbf{x} \in D, z_1 \leq z \leq z_2\}$ with arbitrary cross-section D. We set the boundary conditions*

$$\begin{aligned}
\Gamma_1[w] &= g_1(\mathbf{x}, t) & \text{at} \quad z &= z_1 & (\mathbf{x} \in D), \\
\Gamma_2[w] &= g_2(\mathbf{x}, t) & \text{at} \quad z &= z_2 & (\mathbf{x} \in D), \\
\Gamma_3[w] &= g_3(\mathbf{x}, z, t) & \text{for} \quad \mathbf{x} &\in \partial D & (z_1 \leq z \leq z_2),
\end{aligned} \qquad (9)$$

where the linear boundary operators Γ_k ($k = 1, 2, 3$) can define boundary conditions of the first, second, or third kind; in the last case, the coefficients of the differential operators Γ_k can be dependent on t.

---

* If $z_1 = -\infty$ or $z_2 = \infty$, the corresponding boundary condition is to be omitted.

2°. The Green's function of problem (8)–(9), (5) can be represented in the product form

$$G(\mathbf{x}, \mathbf{y}, z, \zeta, t, \tau) = G_L(\mathbf{x}, \mathbf{y}, t, \tau) G_M(z, \zeta, t, \tau), \tag{10}$$

where $G_L = G_L(\mathbf{x}, \mathbf{y}, t, \tau)$ and $G_M = G_M(z, \zeta, t, \tau)$ are auxiliary Green's functions; these can be determined from the following two simpler problems with fewer independent variables:

Problem on the cross-section D:

$$\begin{cases} \dfrac{\partial G_L}{\partial t} = L_{\mathbf{x},t}[G_L] & \text{for } \mathbf{x} \in D, \\ G_L = \delta(\mathbf{x} - \mathbf{y}) & \text{at } t = \tau, \\ \Gamma_3[G_L] = 0 & \text{for } \mathbf{x} \in \partial D, \end{cases}$$

Problem on the interval $z_1 \leq z \leq z_2$:

$$\begin{cases} \dfrac{\partial G_M}{\partial t} = M_{z,t}[G_M] & \text{for } z_1 < z < z_2, \\ G_M = \delta(z - \zeta) & \text{at } t = \tau, \\ \Gamma_k[G_M] = 0 & \text{at } z = z_k \ (k = 1, 2). \end{cases}$$

Here, $\mathbf{y}$, ζ, and τ are free parameters ($\mathbf{y} \in D$, $z_1 \leq \zeta \leq z_2$, $t \geq \tau \geq 0$).

It can be seen that the Green's function (10) admits incomplete separation of variables; it separates in the space variables $\mathbf{x}$ and z but not in time t.

0.9.3. Construction of Green's Functions via Fundamental Solutions

0.9.3-1. Elliptic equations. Fundamental solution.

Consider the elliptic equation

$$L_{\mathbf{x}}[w] + \frac{\partial^2 w}{\partial z^2} = \Phi(\mathbf{x}, z), \tag{11}$$

where $\mathbf{x} = \{x_1, \ldots, x_n\} \in \mathbb{R}^n$, $z \in \mathbb{R}^1$, and $L_{\mathbf{x}}[w]$ is a linear differential operator that depends on $x_1, \ldots, x_n$ but is independent of z. For subsequent analysis it is significant that the homogeneous equation (with $\Phi \equiv 0$) does not change under the replacement of z by $-z$ and z by $z + \text{const}$.

Let $\mathscr{E} = \mathscr{E}(\mathbf{x}, \mathbf{y}, z - \zeta)$ be a fundamental solution of equation (11), which means that

$$L_{\mathbf{x}}[\mathscr{E}] + \frac{\partial^2 \mathscr{E}}{\partial z^2} = \delta(\mathbf{x} - \mathbf{y}) \delta(z - \zeta).$$

Here, $\mathbf{y} = \{y_1, \ldots, y_n\} \in \mathbb{R}^n$ and $\zeta \in \mathbb{R}^1$ are free parameters.

The fundamental solution of equation (11) is an even function in the last argument, i.e.,

$$\mathscr{E}(\mathbf{x}, \mathbf{y}, z) = \mathscr{E}(\mathbf{x}, \mathbf{y}, -z).$$

Below, Paragraphs 0.9.3-2 and 0.9.3-3 present relations that permit one to express the Green's functions of some boundary value problems for equation (11) via its fundamental solution.

0.9.3-2. Domain: $\mathbf{x} \in \mathbb{R}^n$, $0 \leq z < \infty$. Boundary value problems for elliptic equations.

1°. *First boundary value problem.* The boundary condition:

$$w = f(\mathbf{x}) \quad \text{at} \quad z = 0.$$

Green's function:

$$G(\mathbf{x}, \mathbf{y}, z, \zeta) = \mathscr{E}(\mathbf{x}, \mathbf{y}, z - \zeta) - \mathscr{E}(\mathbf{x}, \mathbf{y}, z + \zeta).$$

Domain of the free parameters: $\mathbf{y} \in \mathbb{R}^n$ and $0 \leq \zeta < \infty$.

2°. *Second boundary value problem.* The boundary condition:
$$\partial_z w = f(\mathbf{x}) \quad \text{at} \quad z = 0.$$
Green's function:
$$G(\mathbf{x}, \mathbf{y}, z, \zeta) = \mathscr{E}(\mathbf{x}, \mathbf{y}, z - \zeta) + \mathscr{E}(\mathbf{x}, \mathbf{y}, z + \zeta).$$

3°. *Third boundary value problem.* The boundary condition:
$$\partial_z w - kw = f(\mathbf{x}) \quad \text{at} \quad z = 0.$$
Green's function:
$$\begin{aligned}G(\mathbf{x}, \mathbf{y}, z, \zeta) &= \mathscr{E}(\mathbf{x}, \mathbf{y}, z - \zeta) + \mathscr{E}(\mathbf{x}, \mathbf{y}, z + \zeta) - 2k \int_0^\infty e^{-ks} \mathscr{E}(\mathbf{x}, \mathbf{y}, z + \zeta + s)\, ds \\ &= \mathscr{E}(\mathbf{x}, \mathbf{y}, z - \zeta) + \mathscr{E}(\mathbf{x}, \mathbf{y}, z + \zeta) - 2k \int_{z+\zeta}^\infty e^{-k(\sigma - z - \zeta)} \mathscr{E}(\mathbf{x}, \mathbf{y}, \sigma)\, d\sigma.\end{aligned}$$

0.9.3-3. Domain: $\mathbf{x} \in \mathbb{R}^n$, $0 \leq z \leq l$. Boundary value problems for elliptic equations.

1°. *First boundary value problem.* Boundary conditions:
$$w = f_1(\mathbf{x}) \quad \text{at} \quad z = 0, \qquad w = f_2(\mathbf{x}) \quad \text{at} \quad z = l.$$
Green's function:
$$G(\mathbf{x}, \mathbf{y}, z, \zeta) = \sum_{n=-\infty}^{\infty} \bigl[\mathscr{E}(\mathbf{x}, \mathbf{y}, z - \zeta + 2nl) - \mathscr{E}(\mathbf{x}, \mathbf{y}, z + \zeta + 2nl) \bigr]. \tag{12}$$
Domain of the free parameters: $\mathbf{y} \in \mathbb{R}^n$ and $0 \leq \zeta \leq l$.

2°. *Second boundary value problem.* Boundary conditions:
$$\partial_z w = f_1(\mathbf{x}) \quad \text{at} \quad z = 0, \qquad \partial_z w = f_2(\mathbf{x}) \quad \text{at} \quad z = l.$$
Green's function:
$$G(\mathbf{x}, \mathbf{y}, z, \zeta) = \sum_{n=-\infty}^{\infty} \bigl[\mathscr{E}(\mathbf{x}, \mathbf{y}, z - \zeta + 2nl) + \mathscr{E}(\mathbf{x}, \mathbf{y}, z + \zeta + 2nl) \bigr]. \tag{13}$$

3°. *Mixed boundary value problem.* The unknown function and its derivative are prescribed at the left and right end, respectively:
$$w = f_1(\mathbf{x}) \quad \text{at} \quad z = 0, \qquad \partial_z w = f_2(\mathbf{x}) \quad \text{at} \quad z = l.$$
Green's function:
$$G(\mathbf{x}, \mathbf{y}, z, \zeta) = \sum_{n=-\infty}^{\infty} (-1)^n \bigl[\mathscr{E}(\mathbf{x}, \mathbf{y}, z - \zeta + 2nl) - \mathscr{E}(\mathbf{x}, \mathbf{y}, z + \zeta + 2nl) \bigr]. \tag{14}$$

4°. *Mixed boundary value problem.* The derivative and the unknown function itself are prescribed at the left and right end, respectively:
$$\partial_z w = f_1(\mathbf{x}) \quad \text{at} \quad z = 0, \qquad w = f_2(\mathbf{x}) \quad \text{at} \quad z = l.$$
Green's function:
$$G(\mathbf{x}, \mathbf{y}, z, \zeta) = \sum_{n=-\infty}^{\infty} (-1)^n \bigl[\mathscr{E}(\mathbf{x}, \mathbf{y}, z - \zeta + 2nl) + \mathscr{E}(\mathbf{x}, \mathbf{y}, z + \zeta + 2nl) \bigr]. \tag{15}$$

Remark. One should make sure that series (12)–(15) are convergent; in particular, for the three-dimensional Laplace equation, series (12), (14), and (15) are convergent and series (13) is divergent.

TABLE 12

Representation of the Green's functions of some nonstationary boundary value problems in terms of the fundamental solution of the Cauchy problem

Boundary value problems	Boundary conditions	Green's functions
First problem $\mathbf{x} \in \mathbb{R}^n$, $z \in \mathbb{R}^1$	$G=0$ at $z=0$	$G(\mathbf{x},\mathbf{y},z,\zeta,t,\tau) = \mathscr{E}(\mathbf{x},\mathbf{y},z-\zeta,t,\tau) - \mathscr{E}(\mathbf{x},\mathbf{y},z+\zeta,t,\tau)$
Second problem $\mathbf{x} \in \mathbb{R}^n$, $z \in \mathbb{R}^1$	$\partial_z G = 0$ at $z=0$	$G(\mathbf{x},\mathbf{y},z,\zeta,t,\tau) = \mathscr{E}(\mathbf{x},\mathbf{y},z-\zeta,t,\tau) + \mathscr{E}(\mathbf{x},\mathbf{y},z+\zeta,t,\tau)$
Third problem $\mathbf{x} \in \mathbb{R}^n$, $z \in \mathbb{R}^1$	$\partial_z G - kG = 0$ at $z=0$	$G(\mathbf{x},\mathbf{y},z,\zeta,t,\tau) = \mathscr{E}(\mathbf{x},\mathbf{y},z-\zeta,t,\tau) + \mathscr{E}(\mathbf{x},\mathbf{y},z+\zeta,t,\tau)$ $-2k \int_0^\infty e^{-ks} \mathscr{E}(\mathbf{x},\mathbf{y},z+\zeta+s,t,\tau)\, ds$
First problem $\mathbf{x} \in \mathbb{R}^n$, $0 \le z \le l$	$G=0$ at $z=0$, $G=0$ at $z=l$	$G(\mathbf{x},\mathbf{y},z,\zeta,t,\tau) = \sum_{n=-\infty}^{\infty} \big[\mathscr{E}(\mathbf{x},\mathbf{y},z-\zeta+2nl,t,\tau)$ $- \mathscr{E}(\mathbf{x},\mathbf{y},z+\zeta+2nl,t,\tau) \big]$
Second problem $\mathbf{x} \in \mathbb{R}^n$, $0 \le z \le l$	$\partial_z G = 0$ at $z=0$, $\partial_z G = 0$ at $z=l$	$G(\mathbf{x},\mathbf{y},z,\zeta,t,\tau) = \sum_{n=-\infty}^{\infty} \big[\mathscr{E}(\mathbf{x},\mathbf{y},z-\zeta+2nl,t,\tau)$ $+ \mathscr{E}(\mathbf{x},\mathbf{y},z+\zeta+2nl,t,\tau) \big]$
Mixed problem $\mathbf{x} \in \mathbb{R}^n$, $0 \le z \le l$	$G=0$ at $z=0$, $\partial_z G = 0$ at $z=l$	$G(\mathbf{x},\mathbf{y},z,\zeta,t,\tau) = \sum_{n=-\infty}^{\infty} (-1)^n \big[\mathscr{E}(\mathbf{x},\mathbf{y},z-\zeta+2nl,t,\tau)$ $- \mathscr{E}(\mathbf{x},\mathbf{y},z+\zeta+2nl,t,\tau) \big]$
Mixed problem $\mathbf{x} \in \mathbb{R}^n$, $0 \le z \le l$	$\partial_z G = 0$ at $z=0$, $G=0$ at $z=l$	$G(\mathbf{x},\mathbf{y},z,\zeta,t,\tau) = \sum_{n=-\infty}^{\infty} (-1)^n \big[\mathscr{E}(\mathbf{x},\mathbf{y},z-\zeta+2nl,t,\tau)$ $+ \mathscr{E}(\mathbf{x},\mathbf{y},z+\zeta+2nl,t,\tau) \big]$

0.9.3-4. Boundary value problems for parabolic equations.

Let $\mathbf{x} \in \mathbb{R}^n$, $z \in \mathbb{R}^1$, and $t \ge 0$. Consider the parabolic equation

$$\frac{\partial w}{\partial t} = L_{\mathbf{x},t}[w] + \frac{\partial^2 w}{\partial z^2} + \Phi(\mathbf{x}, z, t), \tag{16}$$

where $L_{\mathbf{x},t}[w]$ is a linear differential operator that depends on $x_1, \ldots, x_n$ and t but is independent of z.

Let $\mathscr{E} = \mathscr{E}(\mathbf{x}, \mathbf{y}, z - \zeta, t, \tau)$ be a fundamental solution of the Cauchy problem for equation (16), i.e.,

$$\frac{\partial \mathscr{E}}{\partial t} = L_{\mathbf{x},t}[\mathscr{E}] + \frac{\partial^2 \mathscr{E}}{\partial z^2} \quad \text{for} \quad t > \tau,$$
$$\mathscr{E} = \delta(\mathbf{x} - \mathbf{y})\delta(z - \zeta) \quad \text{at} \quad t = \tau.$$

Here, $\mathbf{y} \in \mathbb{R}^n$, $\zeta \in \mathbb{R}^1$, and $\tau \ge 0$ are free parameters.

The fundamental solution of the Cauchy problem possesses the property

$$\mathscr{E}(\mathbf{x}, \mathbf{y}, z, t, \tau) = \mathscr{E}(\mathbf{x}, \mathbf{y}, -z, t, \tau).$$

Table 12 presents formulas that permit one to express the Green's functions of some nonstationary boundary value problems for equation (16) via the fundamental solution of the Cauchy problem.

⊙ *References for Section* 0.9: V. M. Babich, M. B. Kapilevich, S. G. Mikhlin, et al. (1964), B. M. Budak, A. A. Samarskii, and A. N. Tikhonov (1980), A. D. Polyanin (2000b, 2001a).

0.10. Duhamel's Principles in Nonstationary Problems

0.10.1. Problems for Homogeneous Linear Equations

0.10.1-1. Parabolic equations with two independent variables.

Consider the problem for the homogeneous linear equation of parabolic type

$$\frac{\partial w}{\partial t} = a(x)\frac{\partial^2 w}{\partial x^2} + b(x)\frac{\partial w}{\partial x} + c(x)w \tag{1}$$

with the homogeneous initial condition

$$w = 0 \quad \text{at} \quad t = 0 \tag{2}$$

and the boundary conditions

$$s_1\partial_x w + k_1 w = g(t) \quad \text{at} \quad x = x_1, \tag{3}$$
$$s_2\partial_x w + k_2 w = 0 \quad \text{at} \quad x = x_2. \tag{4}$$

By appropriately choosing the values of the coefficients s_1, s_2, k_1, and k_2 in (3) and (4), one can obtain the first, second, third, and mixed boundary value problems for equation (1).

The solution of problem (1)–(4) with the nonstationary boundary condition (3) at $x = x_1$ can be expressed by the formula (Duhamel's first principle)

$$w(x,t) = \frac{\partial}{\partial t}\int_0^t u(x, t-\tau)g(\tau)\,d\tau = \int_0^t \frac{\partial u}{\partial t}(x, t-\tau)g(\tau)\,d\tau \tag{5}$$

in terms of the solution $u(x,t)$ of the auxiliary problem for equation (1) with the initial and boundary conditions (2) and (4), for u instead of w, and the following simpler stationary boundary condition at $x = x_1$:

$$s_1\partial_x u + k_1 u = 1 \quad \text{at} \quad x = x_1. \tag{6}$$

Remark. A similar formula also holds for the homogeneous boundary condition at $x = x_1$ and a nonhomogeneous nonstationary boundary condition at $x = x_2$.

0.10.1-2. Hyperbolic equations with two independent variables.

Consider the problem for the homogeneous linear hyperbolic equation

$$\frac{\partial^2 w}{\partial t^2} + \varphi(x)\frac{\partial w}{\partial t} = a(x)\frac{\partial^2 w}{\partial x^2} + b(x)\frac{\partial w}{\partial x} + c(x)w \tag{7}$$

with the homogeneous initial conditions

$$\begin{aligned} w = 0 \quad &\text{at} \quad t = 0, \\ \partial_t w = 0 \quad &\text{at} \quad t = 0, \end{aligned} \tag{8}$$

and the boundary conditions (3) and (4).

The solution of problem (7), (8), (3), (4) with the nonstationary boundary condition (3) at $x = x_1$ can be expressed by formula (5) in terms of the solution $u(x,t)$ of the auxiliary problem for equation (7) with the initial conditions (8) and boundary condition (4), for u instead of w, and the simpler stationary boundary condition (6) at $x = x_1$.

In this case, the remark made in Paragraph 0.10.1-1 remains valid.

0.10.1-3. Second-order equations with several independent variables.

Duhamel's first principle can also be used to solve homogeneous linear equations of the parabolic or hyperbolic type with many space variables,

$$\frac{\partial^k w}{\partial t^k} = \sum_{i,j=1}^{n} a_{ij}(\mathbf{x}) \frac{\partial^2 w}{\partial x_i \partial x_j} + \sum_{i=1}^{n} b_i(\mathbf{x}) \frac{\partial w}{\partial x_i} + c(\mathbf{x}) w, \tag{9}$$

where $k = 1, 2$ and $\mathbf{x} = \{x_1, \ldots, x_n\}$.

Let V be some bounded domain in $\mathbb{R}^n$ with a sufficiently smooth surface $S = \partial V$. The solution of the boundary value problem for equation (9) in V with the homogeneous initial conditions (2) if $k = 1$ or (8) if $k = 2$, and the nonhomogeneous linear boundary condition

$$\Gamma_\mathbf{x}[w] = g(t) \quad \text{for} \quad \mathbf{x} \in S, \tag{10}$$

is given by

$$w(\mathbf{x}, t) = \frac{\partial}{\partial t} \int_0^t u(\mathbf{x}, t - \tau) g(\tau) \, d\tau = \int_0^t \frac{\partial u}{\partial t}(\mathbf{x}, t - \tau) g(\tau) \, d\tau.$$

Here, $u(\mathbf{x}, t)$ is the solution of the auxiliary problem for equation (9) with the same initial conditions, (2) or (8), for u instead of w, and the simpler stationary boundary condition

$$\Gamma_\mathbf{x}[u] = 1 \quad \text{for} \quad \mathbf{x} \in S.$$

Note that (10) can represent a boundary condition of the first, second, or third kind; the coefficients of the operator $\Gamma_\mathbf{x}$ are assumed to be independent of t.

0.10.2. Problems for Nonhomogeneous Linear Equations

0.10.2-1. Parabolic equations.

The solution of the nonhomogeneous linear equation

$$\frac{\partial w}{\partial t} = \sum_{i,j=1}^{n} a_{ij}(\mathbf{x}) \frac{\partial^2 w}{\partial x_i \partial x_j} + \sum_{i=1}^{n} b_i(\mathbf{x}) \frac{\partial w}{\partial x_i} + c(\mathbf{x}) w + \Phi(\mathbf{x}, t)$$

with the homogeneous initial condition (2) and the homogeneous boundary condition

$$\Gamma_\mathbf{x}[w] = 0 \quad \text{for} \quad \mathbf{x} \in S \tag{11}$$

can be represented in the form (Duhamel's second principle)

$$w(\mathbf{x}, t) = \int_0^t U(\mathbf{x}, t - \tau, \tau) \, d\tau. \tag{12}$$

Here, $U(\mathbf{x}, t, \tau)$ is the solution of the auxiliary problem for the homogeneous equation

$$\frac{\partial U}{\partial t} = \sum_{i,j=1}^{n} a_{ij}(\mathbf{x}) \frac{\partial^2 U}{\partial x_i \partial x_j} + \sum_{i=1}^{n} b_i(\mathbf{x}) \frac{\partial U}{\partial x_i} + c(\mathbf{x}) U$$

with the boundary condition (11), in which w must be substituted by U, and the nonhomogeneous initial condition

$$U = \Phi(\mathbf{x}, \tau) \quad \text{at} \quad t = 0,$$

where τ is a parameter.

Note that (11) can represent a boundary condition of the first, second, or third kind; the coefficients of the operator $\Gamma_\mathbf{x}$ are assumed to be independent of t.

0.10.2-2. Hyperbolic equations.

The solution of the nonhomogeneous linear equation

$$\frac{\partial^2 w}{\partial t^2} + \varphi(\mathbf{x})\frac{\partial w}{\partial t} = \sum_{i,j=1}^{n} a_{ij}(\mathbf{x})\frac{\partial^2 w}{\partial x_i \partial x_j} + \sum_{i=1}^{n} b_i(\mathbf{x})\frac{\partial w}{\partial x_i} + c(\mathbf{x})w + \Phi(\mathbf{x},t)$$

with the homogeneous initial conditions (8) and homogeneous boundary condition (11) can be expressed by formula (12) in terms of the solution $U = U(\mathbf{x}, t, \tau)$ of the auxiliary problem for the homogeneous equation

$$\frac{\partial^2 U}{\partial t^2} + \varphi(\mathbf{x})\frac{\partial U}{\partial t} = \sum_{i,j=1}^{n} a_{ij}(\mathbf{x})\frac{\partial^2 U}{\partial x_i \partial x_j} + \sum_{i=1}^{n} b_i(\mathbf{x})\frac{\partial U}{\partial x_i} + c(\mathbf{x})U$$

with the homogeneous initial and boundary conditions, (2) and (11), where w must be replaced by U, and the nonhomogeneous initial condition

$$\partial_t U = \Phi(\mathbf{x}, \tau) \quad \text{at} \quad t = 0,$$

where τ is a parameter.

Note that (11) can represent a boundary condition of the first, second, or third kind.

⊙ *References for Section* 0.10: E. Butkov (1968), S. J. Farlow (1982), E. Zauderer (1989), R. Courant and D. Hilbert (1989), D. Zwillinger (1998).

0.11. Transformations Simplifying Initial and Boundary Conditions

0.11.1. Transformations That Lead to Homogeneous Boundary Conditions

A linear problem with arbitrary nonhomogeneous boundary conditions,

$$\Gamma_{\mathbf{x},t}^{(k)}[w] = g_k(\mathbf{x}, t) \quad \text{for} \quad \mathbf{x} \in S_k, \tag{1}$$

can be reduced to a linear problem with homogeneous boundary conditions. To this end, one should perform the change of variable

$$w(\mathbf{x}, t) = \psi(\mathbf{x}, t) + u(\mathbf{x}, t), \tag{2}$$

where u is a new unknown function and ψ is any function that satisfies the nonhomogeneous boundary conditions (1),

$$\Gamma_{\mathbf{x},t}^{(k)}[\psi] = g_k(\mathbf{x}, t) \quad \text{for} \quad \mathbf{x} \in S_k. \tag{3}$$

Table 13 gives examples of such transformations for linear boundary value problems with one space variable for parabolic and hyperbolic equations. In the third boundary value problem, it is assumed that $k_1 < 0$ and $k_2 > 0$.

Note that the selection of the function ψ is of a purely algebraic nature and is not connected with the equation in question; there are infinitely many suitable functions ψ that satisfy condition (3). Transformations of the form (2) can often be used at the first stage of solving boundary value problems.

TABLE 13
Simple transformations of the form $w(x,t) = \psi(x,t) + u(x,t)$ that lead to homogeneous boundary conditions in problems with one space variables ($0 \leq x \leq l$)

No	Problems	Boundary conditions	Function $\psi(x,t)$
1	First boundary value problem	$w = g_1(t)$ at $x = 0$ $w = g_2(t)$ at $x = l$	$\psi(x,t) = g_1(t) + \dfrac{x}{l}\left[g_2(t) - g_1(t)\right]$
2	Second boundary value problem	$\partial_x w = g_1(t)$ at $x = 0$ $\partial_x w = g_2(t)$ at $x = l$	$\psi(x,t) = x g_1(t) + \dfrac{x^2}{2l}\left[g_2(t) - g_1(t)\right]$
3	Third boundary value problem	$\partial_x w + k_1 w = g_1(t)$ at $x = 0$ $\partial_x w + k_2 w = g_2(t)$ at $x = l$	$\psi(x,t) = \dfrac{(k_2 x - 1 - k_2 l)g_1(t) + (1 - k_1 x)g_2(t)}{k_2 - k_1 - k_1 k_2 l}$
4	Mixed boundary value problem	$w = g_1(t)$ at $x = 0$ $\partial_x w = g_2(t)$ at $x = l$	$\psi(x,t) = g_1(t) + x g_2(t)$
5	Mixed boundary value problem	$\partial_x w = g_1(t)$ at $x = 0$ $w = g_2(t)$ at $x = l$	$\psi(x,t) = (x - l)g_1(t) + g_2(t)$

0.11.2. Transformations That Lead to Homogeneous Initial and Boundary Conditions

A linear problem with nonhomogeneous initial and boundary conditions can be reduced to a linear problem with homogeneous initial and boundary conditions. To this end, one should introduce a new dependent variable u by formula (2), where the function ψ must satisfy nonhomogeneous initial and boundary conditions.

Below we specify some simple functions ψ that can be used in transformation (2) to obtain boundary value problems with homogeneous initial and boundary conditions. To be specific, we consider a parabolic equation with one space variable and the general initial condition

$$w = f(x) \quad \text{at} \quad t = 0. \tag{4}$$

1. First boundary value problem: the initial condition is (4) and the boundary conditions are given in row 1 of Table 13. Suppose that the initial and boundary conditions are compatible, i.e., $f(0) = g_1(0)$ and $f(l) = g_2(0)$. Then, in transformation (2), one can take

$$\psi(x,t) = f(x) + g_1(t) - g_1(0) + \frac{x}{l}\left[g_2(t) - g_1(t) + g_1(0) - g_2(0)\right].$$

2. Second boundary value problem: the initial condition is (4) and the boundary conditions are given in row 2 of Table 13. Suppose that the initial and boundary conditions are compatible, i.e., $f'(0) = g_1(0)$ and $f'(l) = g_2(0)$. Then, in transformation (2), one can set

$$\psi(x,t) = f(x) + x\left[g_1(t) - g_1(0)\right] + \frac{x^2}{2l}\left[g_2(t) - g_1(t) + g_1(0) - g_2(0)\right].$$

3. Third boundary value problem: the initial condition is (4) and the boundary conditions are given in row 3 of Table 13. If the initial and boundary conditions are compatible, then, in transformation (2), one can take

$$\psi(x,t) = f(x) + \frac{(k_2 x - 1 - k_2 l)[g_1(t) - g_1(0)] + (1 - k_1 x)[g_2(t) - g_2(0)]}{k_2 - k_1 - k_1 k_2 l} \quad (k_1 < 0, \ k_2 > 0).$$

4. *Mixed boundary value problem*: the initial condition is (4) and the boundary conditions are given in row 4 of Table 13. Suppose that the initial and boundary conditions are compatible, i.e., $f(0) = g_1(0)$ and $f'(l) = g_2(0)$. Then, in transformation (2), one can set

$$\psi(x,t) = f(x) + g_1(t) - g_1(0) + x\bigl[g_2(t) - g_2(0)\bigr].$$

5. *Mixed boundary value problem*: the initial condition is (4) and the boundary conditions are given in row 5 of Table 13. Suppose that the initial and boundary conditions are compatible, i.e., $f'(0) = g_1(0)$ and $f(l) = g_2(0)$. Then, in transformation (2), one can take

$$\psi(x,t) = f(x) + (x - l)\bigl[g_1(t) - g_1(0)\bigr] + g_2(t) - g_2(0).$$

⊙ *References for Section* 0.11: V. M. Babich, M. B. Kapilevich, S. G. Mikhlin, et al. (1964), A. D. Polyanin, A. V. Vyazmin, A. I. Zhurov, and D. A. Kazenin (1998).

Chapter 1

Parabolic Equations with One Space Variable

1.1. Constant Coefficient Equations

1.1.1. Heat Equation $\frac{\partial w}{\partial t} = a \frac{\partial^2 w}{\partial x^2}$

This equation is often encountered in the theory of heat and mass transfer. It describes one-dimensional unsteady thermal processes in quiescent media or solids with constant thermal diffusivity. A similar equation is used in studying corresponding one-dimensional unsteady mass-exchange processes with constant diffusivity.

1.1.1-1. Particular solutions (A, B, and μ are arbitrary constants).

$$w(x) = Ax + B,$$
$$w(x,t) = A(x^2 + 2at) + B,$$
$$w(x,t) = A(x^3 + 6atx) + B,$$
$$w(x,t) = A(x^4 + 12atx^2 + 12a^2t^2) + B,$$
$$w(x,t) = A(x^5 + 20atx^3 + 60a^2t^2x) + B,$$
$$w(x,t) = A(x^6 + 30atx^4 + 180a^2t^2x^2 + 120a^3t^3) + B,$$
$$w(x,t) = A(x^7 + 42atx^5 + 420a^2t^2x^3 + 840a^3t^3x) + B,$$
$$w(x,t) = x^{2n} + \sum_{k=1}^{n} \frac{(2n)(2n-1)\ldots(2n-2k+1)}{k!}(at)^k x^{2n-2k},$$
$$w(x,t) = x^{2n+1} + \sum_{k=1}^{n} \frac{(2n+1)(2n)\ldots(2n-2k+2)}{k!}(at)^k x^{2n-2k+1},$$
$$w(x,t) = A\exp(a\mu^2 t \pm \mu x) + B,$$
$$w(x,t) = A\frac{1}{\sqrt{t}}\exp\left(-\frac{x^2}{4at}\right) + B,$$
$$w(x,t) = A\frac{x}{t^{3/2}}\exp\left(-\frac{x^2}{4at}\right) + B,$$
$$w(x,t) = A\exp(-a\mu^2 t)\cos(\mu x) + B,$$
$$w(x,t) = A\exp(-a\mu^2 t)\sin(\mu x) + B,$$
$$w(x,t) = A\exp(-\mu x)\cos(\mu x - 2a\mu^2 t) + B,$$
$$w(x,t) = A\exp(-\mu x)\sin(\mu x - 2a\mu^2 t) + B,$$
$$w(x,t) = A\operatorname{erf}\left(\frac{x}{2\sqrt{at}}\right) + B,$$

$$w(x,t) = A\,\mathrm{erfc}\left(\frac{x}{2\sqrt{at}}\right) + B,$$

$$w(x,t) = A\left[\sqrt{\frac{t}{\pi}}\exp\left(-\frac{x^2}{4at}\right) - \frac{x}{2\sqrt{a}}\,\mathrm{erfc}\left(\frac{x}{2\sqrt{at}}\right)\right] + B,$$

where n is a positive integer, $\mathrm{erf}\,z \equiv \dfrac{2}{\sqrt{\pi}}\displaystyle\int_0^z \exp(-\xi^2)\,d\xi$ the error function (probability integral), and $\mathrm{erfc}\,z = 1 - \mathrm{erf}\,z$ the complementary error function (complementary probability integral).

Fundamental solution:

$$\mathscr{E}(x,t) = \frac{1}{2\sqrt{\pi a t}}\exp\left(-\frac{x^2}{4at}\right).$$

⊙ *References*: H. S. Carslaw and J. C. Jaeger (1984), A. D. Polyanin, A. V. Vyazmin, A. I. Zhurov, and D. A. Kazenin (1998).

1.1.1-2. Formulas allowing the construction of particular solutions.

Suppose $w = w(x,t)$ is a solution of the heat equation. Then the functions

$$w_1 = Aw(\pm\lambda x + C_1,\ \lambda^2 t + C_2),$$

$$w_2 = A\exp(\lambda x + a\lambda^2 t)w(x + 2a\lambda t + C_1,\ t + C_2),$$

$$w_3 = \frac{A}{\sqrt{|\delta + \beta t|}}\exp\left[-\frac{\beta x^2}{4a(\delta + \beta t)}\right]w\left(\pm\frac{x}{\delta + \beta t},\ \frac{\gamma + \lambda t}{\delta + \beta t}\right), \qquad \lambda\delta - \beta\gamma = 1,$$

where A, C_1, C_2, β, δ, and λ are arbitrary constants, are also solutions of this equation. The last formula with $\beta = 1$, $\gamma = -1$, $\delta = \lambda = 0$ was obtained with the Appell transformation.

⊙ *References*: W. Miller, Jr. (1977), P. J. Olver (1986).

1.1.1-3. Infinite series solutions.

A solution involving an arbitrary function of the space variable:

$$w(x,t) = f(x) + \sum_{n=1}^{\infty}\frac{(at)^n}{n!}f_x^{(2n)}(x), \qquad f_x^{(m)}(x) = \frac{d^m}{dx^m}f(x),$$

where $f(x)$ is any infinitely differentiable function. This solution satisfies the initial condition $w(x,0) = f(x)$. The sum is finite if $f(x)$ is a polynomial.

Solutions involving arbitrary functions of time:

$$w(x,t) = g(t) + \sum_{n=1}^{\infty}\frac{1}{a^n(2n)!}x^{2n}g_t^{(n)}(t),$$

$$w(x,t) = xh(t) + x\sum_{n=1}^{\infty}\frac{1}{a^n(2n+1)!}x^{2n}h_t^{(n)}(t),$$

where $g(t)$ and $h(t)$ are infinitely differentiable functions. The sums are finite if $g(t)$ and $h(t)$ are polynomials. The first solution satisfies the boundary condition of the first kind $w(0,t) = g(t)$ and the second solution to the boundary condition of the second kind $\partial_x w(0,t) = h(t)$.

⊙ *Reference*: H. S. Carslaw and J. C. Jaeger (1984).

TABLE 14
Transformations of the form $\xi = f(x,t)$, $w = g(\xi,t)u(\xi,t)$ for which the equation $\partial_t w - \partial_{xx} w = 0$ admits multiplicatively separable particular solutions with $u(\xi,t) = \varphi(t)\psi(\xi)$

No	Function $\xi = f(x,t)$	Factor $g = g(\xi,t)$	Function $\varphi = \varphi(t)$, λ is any	Equation for $\psi = \psi(\xi)$
1	$\xi = \dfrac{x}{\sqrt{t}}$	$g = 1$	$\varphi = t^\lambda$	$\psi''_{\xi\xi} + \frac{1}{2}\xi\psi'_\xi - \lambda\psi = 0$
2	$\xi = x - \frac{1}{2}t^2$	$g = \exp\!\left(-\frac{1}{2}\xi t\right)$	$\varphi = \exp\!\left(-\frac{1}{12}t^3 + \lambda t\right)$	$\psi''_{\xi\xi} + \left(\frac{1}{2}\xi - \lambda\right)\psi = 0$
3	$\xi = \dfrac{x}{\sqrt{1+t^2}}$	$g = \exp\!\left(-\frac{1}{4}\xi^2 t\right)$	$\varphi = \dfrac{\exp(\lambda \arctan t)}{(1+t^2)^{1/4}}$	$\psi''_{\xi\xi} + \left(\frac{1}{4}\xi^2 - \lambda\right)\psi = 0$

1.1.1-4. Transformations allowing separation of variables.

Table 14 presents transformations that reduce the heat equation to separable equations (the identity transformation with $\xi = x$, $g = 1$ is omitted).

Remark. In general, the solution of the equation for ψ in the first row of Table 14 is expressed in terms of degenerate hypergeometric functions. In the special case $\lambda = \frac{1}{2}n$ ($n = 0, 1, 2, \ldots$), the equation admits solutions of the form $\psi(\xi) = (i/2)^n H_n(i\xi/2)$, where $H_n(z)$ is the nth Hermite polynomial, $i^2 = -1$. The solution of the equation for ψ in the second row of Table 14 is expressed in terms of Bessel functions, and that in the third row, in terms of parabolic cylinder functions.

⊙ *References*: E. Kalnins and W. Miller, Jr. (1974), W. Miller, Jr. (1977).

1.1.1-5. Domain: $-\infty < x < \infty$. Cauchy problem.

An initial condition is prescribed:
$$w = f(x) \quad \text{at} \quad t = 0.$$

Solution:
$$w(x,t) = \frac{1}{2\sqrt{\pi a t}} \int_{-\infty}^{\infty} \exp\!\left[-\frac{(x-\xi)^2}{4at}\right] f(\xi)\,d\xi.$$

Example 1. The initial temperatures in the domains $|x| < x_0$ and $|x| > x_0$ are constant and equal to w_1 and w_2, respectively, i.e.,
$$f(x) = \begin{cases} w_1 & \text{for } |x| < x_0, \\ w_2 & \text{for } |x| > x_0. \end{cases}$$

Solution:
$$w = \frac{1}{2}(w_1 - w_2)\left[\mathrm{erf}\!\left(\frac{x_0 - x}{2\sqrt{at}}\right) + \mathrm{erf}\!\left(\frac{x_0 + x}{2\sqrt{at}}\right)\right] + w_2.$$

⊙ *Reference*: H. S. Carslaw and J. C. Jaeger (1984).

1.1.1-6. Domain: $0 \leq x < \infty$. First boundary value problem.

The following conditions are prescribed:
$$w = f(x) \quad \text{at} \quad t = 0 \quad \text{(initial condition)},$$
$$w = g(t) \quad \text{at} \quad x = 0 \quad \text{(boundary condition)}.$$

Solution:

$$w(x,t) = \frac{1}{2\sqrt{\pi a t}} \int_0^\infty \left\{ \exp\left[-\frac{(x-\xi)^2}{4at}\right] - \exp\left[-\frac{(x+\xi)^2}{4at}\right] \right\} f(\xi)\, d\xi$$
$$+ \frac{x}{2\sqrt{\pi a}} \int_0^t \exp\left[-\frac{x^2}{4a(t-\tau)}\right] \frac{g(\tau)\, d\tau}{(t-\tau)^{3/2}}.$$

Example 2. The initial temperature is linearly dependent on the space coordinate, $f(x) = w_0 + bx$. The temperature at the boundary is zero, $g(t) = 0$.
Solution:
$$w = w_0 \operatorname{erf}\left(\frac{x}{2\sqrt{at}}\right) + bx.$$

The case of uniform initial temperature with $f(x) = w_0$ corresponds to the value $b = 0$.

Example 3. The initial temperature is zero, $f(x) = 0$. The temperature at the boundary increases linearly with time, $g(t) = At$.
Solution:
$$w = At\left[\left(1 + \frac{x^2}{2at}\right) \operatorname{erfc}\left(\frac{x}{2\sqrt{at}}\right) - \frac{x}{\sqrt{\pi a t}} \exp\left(-\frac{x^2}{4at}\right)\right].$$

⊙ *References*: A. G. Butkovskiy (1979), H. S. Carslaw and J. C. Jaeger (1984).

1.1.1-7. Domain: $0 \leq x < \infty$. Second boundary value problem.

The following conditions are prescribed:
$$w = f(x) \quad \text{at} \quad t = 0 \quad \text{(initial condition)},$$
$$\partial_x w = g(t) \quad \text{at} \quad x = 0 \quad \text{(boundary condition)}.$$

Solution:
$$w(x,t) = \frac{1}{2\sqrt{\pi a t}} \int_0^\infty \left\{ \exp\left[-\frac{(x-\xi)^2}{4at}\right] + \exp\left[-\frac{(x+\xi)^2}{4at}\right] \right\} f(\xi)\, d\xi$$
$$- \sqrt{\frac{a}{\pi}} \int_0^t \exp\left[-\frac{x^2}{4a(t-\tau)}\right] \frac{g(\tau)}{\sqrt{t-\tau}}\, d\tau.$$

Example 4. The initial temperature is zero, $f(x) = 0$. A constant thermal flux is maintained at the boundary all the time, $g(t) = -Q$.
Solution:
$$w = 2Q\sqrt{\frac{at}{\pi}} \exp\left(-\frac{x^2}{4at}\right) - Qx \operatorname{erfc}\left(\frac{x}{2\sqrt{at}}\right).$$

⊙ *References*: A. G. Butkovskiy (1979), H. S. Carslaw and J. C. Jaeger (1984).

1.1.1-8. Domain: $0 \leq x < \infty$. Third boundary value problem.

The following conditions are prescribed:
$$w = f(x) \quad \text{at} \quad t = 0 \quad \text{(initial condition)},$$
$$\partial_x w - kw = g(t) \quad \text{at} \quad x = 0 \quad \text{(boundary condition)}.$$

Solution:
$$w(x,t) = \int_0^\infty f(\xi) G(x,\xi,t)\, d\xi - a \int_0^t g(\tau) G(x,0,t-\tau)\, d\tau,$$

where
$$G(x,\xi,t) = \frac{1}{2\sqrt{\pi a t}} \left\{ \exp\left[-\frac{(x-\xi)^2}{4at}\right] + \exp\left[-\frac{(x+\xi)^2}{4at}\right] - 2k \int_0^\infty \exp\left[-\frac{(x+\xi+\eta)^2}{4at} - k\eta\right] d\eta \right\}.$$

The improper integral may be calculated by the formula

$$\int_0^\infty \exp\left[-\frac{(x+\xi+\eta)^2}{4at} - k\eta\right] d\eta = \sqrt{\pi a t}\, \exp\left[ak^2 t + k(x+\xi)\right] \operatorname{erfc}\left(\frac{x+\xi}{2\sqrt{at}} + k\sqrt{at}\right).$$

Example 5. The initial temperature is uniform, $f(x) = w_0$. The temperature of the contacting medium is zero, $g(t) = 0$.
Solution:

$$w = w_0\left[\operatorname{erf}\left(\frac{x}{2\sqrt{at}}\right) + \exp(kx + ak^2 t)\operatorname{erfc}\left(\frac{x}{2\sqrt{at}} + k\sqrt{at}\right)\right].$$

⊙ *References*: B. M. Budak, A. A. Samarskii, and A. N. Tikhonov (1980), H. S. Carslaw and J. C. Jaeger (1984).

1.1.1-9. Domain: $0 \leq x \leq l$. First boundary value problem.

The following conditions are prescribed:

$$w = f(x) \quad \text{at} \quad t = 0 \quad \text{(initial condition)},$$
$$w = g_1(t) \quad \text{at} \quad x = 0 \quad \text{(boundary condition)},$$
$$w = g_2(t) \quad \text{at} \quad x = l \quad \text{(boundary condition)}.$$

Solution:

$$w(x,t) = \frac{2}{l} \sum_{n=1}^\infty \sin\left(\frac{n\pi x}{l}\right) \exp\left(-\frac{an^2\pi^2 t}{l^2}\right) M_n(t),$$

where

$$M_n(t) = \int_0^l f(\xi)\sin\left(\frac{n\pi\xi}{l}\right) d\xi + \frac{an\pi}{l}\int_0^t \exp\left(\frac{an^2\pi^2\tau}{l^2}\right)\left[g_1(\tau) - (-1)^n g_2(\tau)\right] d\tau.$$

Remark. Using the relations [see Prudnikov, Brychkov, and Marichev (1986)]

$$\sum_{n=1}^\infty \frac{\sin n\xi}{n} = \frac{\pi - \xi}{2} \quad (0 < \xi < 2\pi); \qquad \sum_{n=1}^\infty (-1)^{n-1}\frac{\sin n\xi}{n} = \frac{\xi}{2} \quad (-\pi < \xi < \pi),$$

one can transform the solution to

$$w(x,t) = g_1(t) + \frac{x}{l}[g_2(t) - g_1(t)] + \frac{2}{l}\sum_{n=1}^\infty \sin(\lambda_n x)\exp(-a\lambda_n^2 t)R_n(t), \quad \lambda_n = \frac{n\pi}{l},$$

where

$$R_n(t) = \int_0^l f(\xi)\sin(\lambda_n \xi)\,d\xi - \frac{1}{\lambda_n}\exp(a\lambda_n^2 t)[g_1(t) - (-1)^n g_2(t)] + a\lambda_n \int_0^t \exp(a\lambda_n^2 \tau)[g_1(\tau) - (-1)^n g_2(\tau)]\,d\tau$$

$$= \int_0^l f(\xi)\sin(\lambda_n \xi)\,d\xi - \frac{1}{\lambda_n}[g_1(0) - (-1)^n g_2(0)] - \frac{1}{\lambda_n}\int_0^t \exp(a\lambda_n^2 \tau)[g_1'(\tau) - (-1)^n g_2'(\tau)]\,d\tau.$$

Note that another representation of the solution is given in Paragraph 1.1.2 5.

Example 6. The initial temperature is uniform, $f(x) = w_0$. Both ends are maintained at zero temperature, $g_1(t) = g_2(t) = 0$.
Solution:

$$w = \frac{4w_0}{\pi}\sum_{n=0}^\infty \frac{1}{(2n+1)}\sin\left[\frac{(2n+1)\pi x}{l}\right]\exp\left[-\frac{a(2n+1)^2\pi^2 t}{l^2}\right].$$

Example 7. The initial temperature is zero, $f(x) = 0$. The ends are maintained at uniform temperatures, $g_1(t) = w_1$ and $g_2(t) = w_2$.
Solution:

$$w = w_1 + (w_2 - w_1)\frac{x}{l} + \frac{2}{\pi}\sum_{n=0}^\infty \frac{(-1)^n w_2 - w_1}{n}\sin\left(\frac{n\pi x}{l}\right)\exp\left(-\frac{an^2\pi^2 t}{l^2}\right).$$

⊙ *References*: H. S. Carslaw and J. C. Jaeger (1984), A. D. Polyanin, A. V. Vyazmin, A. I. Zhurov, and D. A. Kazenin (1998).

1.1.1-10. Domain: $0 \leq x \leq l$. Second boundary value problem.

The following conditions are prescribed:

$$w = f(x) \quad \text{at} \quad t = 0 \quad \text{(initial condition)},$$
$$\partial_x w = g_1(t) \quad \text{at} \quad x = 0 \quad \text{(boundary condition)},$$
$$\partial_x w = g_2(t) \quad \text{at} \quad x = l \quad \text{(boundary condition)}.$$

Solution:

$$w(x,t) = \int_0^l f(\xi) G(x,\xi,t)\, d\xi - a \int_0^t g_1(\tau) G(x,0,t-\tau)\, d\tau + a \int_0^t g_2(\tau) G(x,l,t-\tau)\, d\tau,$$

where

$$G(x,\xi,t) = \frac{1}{l} + \frac{2}{l} \sum_{n=1}^{\infty} \cos\left(\frac{n\pi x}{l}\right) \cos\left(\frac{n\pi \xi}{l}\right) \exp\left(-\frac{an^2\pi^2 t}{l^2}\right).$$

⊙ *References*: B. M. Budak, A. A. Samarskii, and A. N. Tikhonov (1980), H. S. Carslaw and J. C. Jaeger (1984).

1.1.1-11. Domain: $0 \leq x \leq l$. Third boundary value problem ($k_1 > 0$ and $k_2 > 0$).

The following conditions are prescribed:

$$w = f(x) \quad \text{at} \quad t = 0 \quad \text{(initial condition)},$$
$$\partial_x w - k_1 w = g_1(t) \quad \text{at} \quad x = 0 \quad \text{(boundary condition)},$$
$$\partial_x w + k_2 w = g_2(t) \quad \text{at} \quad x = l \quad \text{(boundary condition)},$$

Solution:

$$w(x,t) = \int_0^l f(\xi) G(x,\xi,t)\, d\xi - a \int_0^t g_1(\tau) G(x,0,t-\tau)\, d\tau + a \int_0^t g_2(\tau) G(x,l,t-\tau)\, d\tau,$$

where

$$G(x,\xi,t) = \sum_{n=1}^{\infty} \frac{1}{\|y_n\|^2} y_n(x) y_n(\xi) \exp(-a\mu_n^2 t),$$

$$y_n(x) = \cos(\mu_n x) + \frac{k_1}{\mu_n}\sin(\mu_n x), \quad \|y_n\|^2 = \frac{k_2}{2\mu_n^2}\frac{\mu_n^2 + k_1^2}{\mu_n^2 + k_2^2} + \frac{k_1}{2\mu_n^2} + \frac{l}{2}\left(1 + \frac{k_1^2}{\mu_n^2}\right).$$

Here, the μ_n are positive roots of the transcendental equation $\dfrac{\tan(\mu l)}{\mu} = \dfrac{k_1 + k_2}{\mu^2 - k_1 k_2}.$

⊙ *References*: A. G. Butkovskiy (1979), H. S. Carslaw and J. C. Jaeger (1984).

1.1.1-12. Domain: $0 \leq x \leq l$. Mixed boundary value problems.

1°. The following conditions are prescribed:

$$w = f(x) \quad \text{at} \quad t = 0 \quad \text{(initial condition)},$$
$$w = g_1(t) \quad \text{at} \quad x = 0 \quad \text{(boundary condition)},$$
$$\partial_x w = g_2(t) \quad \text{at} \quad x = l \quad \text{(boundary condition)}.$$

Solution:

$$w(x,t) = \int_0^l f(\xi) G(x,\xi,t)\, d\xi + a \int_0^t g_1(\tau) \Lambda(x,t-\tau)\, d\tau + a \int_0^t g_2(\tau) G(x,l,t-\tau)\, d\tau,$$

where

$$G(x,\xi,t) = \frac{2}{l}\sum_{n=0}^{\infty}\sin\left[\frac{\pi(2n+1)x}{2l}\right]\sin\left[\frac{\pi(2n+1)\xi}{2l}\right]\exp\left[-\frac{a\pi^2(2n+1)^2 t}{4l^2}\right],$$

$$\Lambda(x,t) = \frac{\partial}{\partial\xi}G(x,\xi,t)\bigg|_{\xi=0}.$$

2°. The following conditions are prescribed:

$$w = f(x) \quad \text{at} \quad t = 0 \quad \text{(initial condition)},$$
$$\partial_x w = g_1(t) \quad \text{at} \quad x = 0 \quad \text{(boundary condition)},$$
$$w = g_2(t) \quad \text{at} \quad x = l \quad \text{(boundary condition)}.$$

Solution:

$$w(x,t) = \int_0^l f(\xi)G(x,\xi,t)\,d\xi - a\int_0^t g_1(\tau)G(x,0,t-\tau)\,d\tau - a\int_0^t g_2(\tau)H(x,t-\tau)\,d\tau,$$

where

$$G(x,\xi,t) = \frac{2}{l}\sum_{n=0}^{\infty}\cos\left[\frac{\pi(2n+1)x}{2l}\right]\cos\left[\frac{\pi(2n+1)\xi}{2l}\right]\exp\left[-\frac{a\pi^2(2n+1)^2 t}{4l^2}\right],$$

$$H(x,t) = \frac{\partial}{\partial\xi}G(x,\xi,t)\bigg|_{\xi=l}.$$

Note that Paragraph 1.1.2-8 also gives other forms of representation of solutions to mixed boundary value problems.

Example 8. The initial temperature is zero, $f(x) = 0$. The left end is heat insulated, and the right end is maintained at a constant temperature, $g_1(t) = 0$ and $g_2(t) = A$.

Solution:

$$w = A + \frac{4A}{\pi}\sum_{n=0}^{\infty}\frac{(-1)^{n+1}}{2n+1}\cos\left[\frac{\pi(2n+1)x}{2l}\right]\exp\left[-\frac{a\pi^2(2n+1)^2 t}{4l^2}\right].$$

⊙ *References*: B. M. Budak, A. A. Samarskii, and A. N. Tikhonov (1980), A. V. Bitsadze and D. F. Kalinichenko (1985).

1.1.1-13. Problems without initial conditions.

In applications, problems are encountered in which the process is studied at a time instant fairly remote from the initial instant and, in this case, the initial conditions do not practically affect the distribution of the desired quantity at the observation instant. In such problems, no initial condition is stated, and the boundary conditions are assumed to be prescribed for all preceding time instants, $\infty < t$. However, in addition, the boundedness condition in the entire domain is imposed on the solution.

As an example, consider the first boundary value problem for the half-space $0 \le x < \infty$ with the boundary condition

$$w = g(t) \quad \text{at} \quad x = 0.$$

Solution:

$$w(x,t) = \frac{x}{2\sqrt{\pi a}}\int_{-\infty}^{t}\frac{g(\tau)}{(t-\tau)^{3/2}}\exp\left[-\frac{x^2}{4a(t-\tau)}\right]d\tau.$$

Example 9. The temperature at the boundary is a harmonic function of time, i.e.,

$$g(t) = w_0\cos(\omega t + \beta).$$

Solution:

$$w = w_0\exp\left(-\sqrt{\frac{\omega}{2a}}\,x\right)\cos\left(-\sqrt{\frac{\omega}{2a}}\,x + \omega t + \beta\right).$$

⊙ *References*: V. M. Babich, M. B. Kapilevich, S. G. Mikhlin, et al. (1964), A. N. Tikhonov and A. A. Samarskii (1990).

1.1.1-14. Conjugate heat and mass transfer problems.

In such problems, one deals with two (or more) domains, V_1 and V_2, with interface S. The domains are filled by different media. Each of the media is characterized by its own thermal conductivity, λ_1 and λ_2, and thermal diffusivity, a_1 and a_2. The processes in each of the media are described by an appropriate (different) equations of heat and mass transfer. The thermal equilibrium conditions express the equality of the temperatures and of the thermal fluxes at the interface. Below we consider a typical example of a conjugate problem (a more detailed analysis of such problems is beyond the scope of this handbook).

Consider two semi-infinite solids (two semi-infinite quiescent media) the temperature distributions, in which $w_1 = w_1(x,t)$ and $w_2 = w_2(x,t)$ are governed by the equations

$$\frac{\partial w_1}{\partial t} = a_1 \frac{\partial^2 w_1}{\partial x^2} \quad \text{(in the range } 0 < x < \infty\text{)},$$

$$\frac{\partial w_2}{\partial t} = a_2 \frac{\partial^2 w_2}{\partial x^2} \quad \text{(in the range } -\infty < x < 0\text{)}.$$

Each of the solids has its own temperature profile at the initial instant $t=0$, and at the interface $x=0$ conjugate boundary solutions are imposed, specifically,

$$w_1 = f_1(x) \quad \text{at} \quad t = 0 \quad \text{(initial condition)},$$
$$w_2 = f_2(x) \quad \text{at} \quad t = 0 \quad \text{(initial condition)},$$
$$w_1 = w_2 \quad \text{at} \quad x = 0 \quad \text{(boundary condition)},$$
$$\lambda_1 \partial_x w_1 = \lambda_2 \partial_x w_2 \quad \text{at} \quad x = 0 \quad \text{(boundary condition)}.$$

Solution:

$$w_1(x,t) = \frac{1}{2\sqrt{\pi a_1 t}} \int_0^\infty f_1(\xi) \left\{ \exp\left[-\frac{(x-\xi)^2}{4a_1 t}\right] + \exp\left[-\frac{(x+\xi)^2}{4a_1 t}\right] \right\} d\xi$$
$$- \sqrt{\frac{a_1}{\pi \lambda_1^2}} \int_0^t \exp\left[-\frac{x^2}{4a_1(t-\tau)}\right] \frac{g(\tau)\,d\tau}{\sqrt{t-\tau}},$$

$$w_2(x,t) = \frac{1}{2\sqrt{\pi a_2 t}} \int_0^\infty f_2(-\xi) \left\{ \exp\left[-\frac{(x-\xi)^2}{4a_2 t}\right] + \exp\left[-\frac{(x+\xi)^2}{4a_2 t}\right] \right\} d\xi$$
$$+ \sqrt{\frac{a_2}{\pi \lambda_2^2}} \int_0^t \exp\left[-\frac{x^2}{4a_2(t-\tau)}\right] \frac{g(\tau)\,d\tau}{\sqrt{t-\tau}}.$$

The function $g(t)$ is given by

$$g(t) = \frac{\lambda_1 \lambda_2}{\pi(\lambda_1 \sqrt{a_2} + \lambda_2 \sqrt{a_1})} \frac{d}{dt} \int_0^t \frac{F(\tau)\,d\tau}{\sqrt{\tau(t-\tau)}},$$

where

$$F(t) = \frac{1}{\sqrt{a_1}} \int_0^\infty f_1(\xi) \exp\left(-\frac{\xi^2}{4a_1 t}\right) d\xi - \frac{1}{\sqrt{a_2}} \int_0^\infty f_2(-\xi) \exp\left(-\frac{\xi^2}{4a_2 t}\right) d\xi.$$

Example 10. The initial temperatures are uniform, $f_1(x) = A$ and $f_2(x) = B$.
Solution:
$$\frac{w_1(x,t) - B}{A - B} = \frac{K}{1+K}\left[1 + \frac{1}{K}\operatorname{erf}\left(\frac{x}{2\sqrt{a_1 t}}\right)\right],$$
$$\frac{w_2(x,t) - B}{A - B} = \frac{K}{1+K}\operatorname{erfc}\left(\frac{|x|}{2\sqrt{a_1 t}}\right),$$

where the quantity $K = \frac{\lambda_1}{\lambda_2}\sqrt{\frac{a_2}{a_1}}$ characterizes the thermal activity of the first medium with respect to the second medium.

⊙ *References*: A. V. Lykov (1967), H. S. Carslaw and J. C. Jaeger (1984).

1.1.2. Equation of the Form $\frac{\partial w}{\partial t} = a\frac{\partial^2 w}{\partial x^2} + \Phi(x,t)$

This sort of equation describes one-dimensional unsteady thermal processes in quiescent media or solids with constant thermal diffusivity in the presence of a volume thermal source dependent on the space coordinate and time.

1.1.2-1. Domain: $-\infty < x < \infty$. Cauchy problem.

An initial condition is prescribed:
$$w = f(x) \quad \text{at} \quad t = 0.$$

Solution:
$$w(x,t) = \int_{-\infty}^{\infty} f(\xi)G(x,\xi,t)\,d\xi + \int_0^t \int_{-\infty}^{\infty} \Phi(\xi,\tau)G(x,\xi,t-\tau)\,d\xi\,d\tau,$$

where
$$G(x,\xi,t) = \frac{1}{2\sqrt{\pi at}} \exp\left[-\frac{(x-\xi)^2}{4at}\right].$$

⊙ *References*: A. G. Butkovskiy (1979), H. S. Carslaw and J. C. Jaeger (1984).

1.1.2-2. Domain: $0 \le x < \infty$. First boundary value problem.

The following conditions are prescribed:
$$w = f(x) \quad \text{at} \quad t = 0 \quad \text{(initial condition)},$$
$$w = g(t) \quad \text{at} \quad x = 0 \quad \text{(boundary condition)}.$$

Solution:
$$w(x,t) = \int_0^{\infty} f(\xi)G(x,\xi,t)\,d\xi + \int_0^t g(\tau)H(x,t-\tau)\,d\tau + \int_0^t \int_0^{\infty} \Phi(\xi,\tau)G(x,\xi,t-\tau)\,d\xi\,d\tau,$$

where
$$G(x,\xi,t) = \frac{1}{2\sqrt{\pi at}}\left\{\exp\left[-\frac{(x-\xi)^2}{4at}\right] - \exp\left[-\frac{(x+\xi)^2}{4at}\right]\right\}, \quad H(x,t) = \frac{x}{2\sqrt{\pi a}\,t^{3/2}} \exp\left(-\frac{x^2}{4at}\right).$$

⊙ *References*: A. G. Butkovskiy (1979), H. S. Carslaw and J. C. Jaeger (1984).

1.1.2-3. Domain: $0 \le x < \infty$. Second boundary value problem.

The following conditions are prescribed:
$$w = f(x) \quad \text{at} \quad t = 0 \quad \text{(initial condition)},$$
$$\partial_x w = g(t) \quad \text{at} \quad x = 0 \quad \text{(boundary condition)}.$$

Solution:
$$w(x,t) = \int_0^{\infty} G(x,\xi,t)f(\xi)\,d\xi - a\int_0^t g(\tau)G(x,0,t-\tau)\,d\tau + \int_0^t \int_0^{\infty} \Phi(\xi,\tau)G(x,\xi,t-\tau)\,d\xi\,d\tau,$$

where
$$G(x,\xi,t) = \frac{1}{2\sqrt{\pi at}}\left\{\exp\left[-\frac{(x-\xi)^2}{4at}\right] + \exp\left[-\frac{(x+\xi)^2}{4at}\right]\right\}, \quad G(x,0,t) = \frac{1}{\sqrt{\pi at}} \exp\left(-\frac{x^2}{4at}\right).$$

⊙ *References*: A. G. Butkovskiy (1979), H. S. Carslaw and J. C. Jaeger (1984).

1.1.2-4. Domain: $0 \leq x < \infty$. Third boundary value problem.

The following conditions are prescribed:

$$w = f(x) \quad \text{at} \quad t = 0 \quad \text{(initial condition)},$$
$$\partial_x w - kw = g(t) \quad \text{at} \quad x = 0 \quad \text{(boundary condition)}.$$

Solution:

$$w(x,t) = \int_0^\infty f(\xi) G(x,\xi,t)\,d\xi - a \int_0^t g(\tau) G(x,0,t-\tau)\,d\tau$$
$$+ \int_0^t \int_0^\infty \Phi(\xi,\tau) G(x,\xi,t-\tau)\,d\xi\,d\tau,$$

where

$$G(x,\xi,t) = \frac{1}{2\sqrt{\pi a t}} \left\{ \exp\left[-\frac{(x-\xi)^2}{4at}\right] + \exp\left[-\frac{(x+\xi)^2}{4at}\right] - 2k \int_0^\infty \exp\left[-\frac{(x+\xi+\eta)^2}{4at} - k\eta\right] d\eta \right\}.$$

⊙ *References*: A. G. Butkovskiy (1979), B. M. Budak, A. A. Samarskii, and A. N. Tikhonov (1980).

1.1.2-5. Domain: $0 \leq x \leq l$. First boundary value problem.

The following conditions are prescribed:

$$w = f(x) \quad \text{at} \quad t = 0 \quad \text{(initial condition)},$$
$$w = g_1(t) \quad \text{at} \quad x = 0 \quad \text{(boundary condition)},$$
$$w = g_2(t) \quad \text{at} \quad x = l \quad \text{(boundary condition)}.$$

Solution:

$$w(x,t) = \int_0^l f(\xi) G(x,\xi,t)\,d\xi + \int_0^t \int_0^l \Phi(\xi,\tau) G(x,\xi,t-\tau)\,d\xi\,d\tau$$
$$+ a \int_0^t g_1(\tau) H_1(x,t-\tau)\,d\tau - a \int_0^t g_2(\tau) H_2(x,t-\tau)\,d\tau.$$

Two forms of representation of the Green's function:

$$G(x,\xi,t) = \frac{2}{l} \sum_{n=1}^\infty \sin\left(\frac{n\pi x}{l}\right) \sin\left(\frac{n\pi \xi}{l}\right) \exp\left(-\frac{an^2\pi^2 t}{l^2}\right)$$
$$= \frac{1}{2\sqrt{\pi a t}} \sum_{n=-\infty}^\infty \left\{ \exp\left[-\frac{(x-\xi+2nl)^2}{4at}\right] - \exp\left[-\frac{(x+\xi+2nl)^2}{4at}\right] \right\}.$$

The first series converges rapidly at large t and the second series at small t. The functions H_1 and H_2 are expressed in terms of the Green's function as

$$H_1(x,t) = \frac{\partial}{\partial \xi} G(x,\xi,t)\Big|_{\xi=0}, \quad H_2(x,t) = \frac{\partial}{\partial \xi} G(x,\xi,t)\Big|_{\xi=l}.$$

⊙ *References*: A. G. Butkovskiy (1979), B. M. Budak, A. A. Samarskii, and A. N. Tikhonov (1980), H. S. Carslaw and J. C. Jaeger (1984).

1.1.2-6. Domain: $0 \leq x \leq l$. Second boundary value problem.

The following conditions are prescribed:

$$w = f(x) \quad \text{at} \quad t = 0 \quad \text{(initial condition)},$$
$$\partial_x w = g_1(t) \quad \text{at} \quad x = 0 \quad \text{(boundary condition)},$$
$$\partial_x w = g_2(t) \quad \text{at} \quad x = l \quad \text{(boundary condition)}.$$

Solution:

$$w(x,t) = \int_0^l f(\xi) G(x,\xi,t)\, d\xi + \int_0^t \int_0^l \Phi(\xi,\tau) G(x,\xi,t-\tau)\, d\xi\, d\tau$$
$$- a \int_0^t g_1(\tau) G(x,0,t-\tau)\, d\tau + a \int_0^t g_2(\tau) G(x,l,t-\tau)\, d\tau.$$

Two forms of representation of the Green's function:

$$G(x,\xi,t) = \frac{1}{l} + \frac{2}{l} \sum_{n=1}^{\infty} \cos\left(\frac{n\pi x}{l}\right) \cos\left(\frac{n\pi \xi}{l}\right) \exp\left(-\frac{an^2\pi^2 t}{l^2}\right)$$
$$= \frac{1}{2\sqrt{\pi a t}} \sum_{n=-\infty}^{\infty} \left\{ \exp\left[-\frac{(x-\xi+2nl)^2}{4at}\right] + \exp\left[-\frac{(x+\xi+2nl)^2}{4at}\right] \right\}.$$

The first series converges rapidly at large t and the second series at small t.

⊙ *References*: A. G. Butkovskiy (1979), B. M. Budak, A. A. Samarskii, and A. N. Tikhonov (1980).

1.1.2-7. Domain: $0 \leq x \leq l$. Third boundary value problem ($k_1 > 0$, $k_2 > 0$).

The following conditions are prescribed:

$$w = f(x) \quad \text{at} \quad t = 0 \quad \text{(initial condition)},$$
$$\partial_x w - k_1 w = g_1(t) \quad \text{at} \quad x = 0 \quad \text{(boundary condition)},$$
$$\partial_x w + k_2 w = g_2(t) \quad \text{at} \quad x = l \quad \text{(boundary condition)}.$$

The solution is given by the formula presented in Paragraph 1.1.1-11 with the additional term

$$\int_0^t \int_0^l \Phi(\xi,\tau) G(x,\xi,t-\tau)\, d\xi\, d\tau,$$

which takes into account the nonhomogeneity of the equation.

⊙ *References*: A. G. Butkovskiy (1979), H. S. Carslaw and J. C. Jaeger (1984).

1.1.2-8. Domain: $0 \leq x \leq l$. Mixed boundary value problems.

1°. The following conditions are prescribed:

$$w = f(x) \quad \text{at} \quad t = 0 \quad \text{(initial condition)},$$
$$w = g_1(t) \quad \text{at} \quad x = 0 \quad \text{(boundary condition)},$$
$$\partial_x w = g_2(t) \quad \text{at} \quad x = l \quad \text{(boundary condition)}.$$

Solution:

$$w(x,t) = \int_0^l f(\xi)G(x,\xi,t)\,d\xi + \int_0^t \int_0^l \Phi(\xi,\tau)G(x,\xi,t-\tau)\,d\xi\,d\tau$$
$$+ a\int_0^t g_1(\tau)\left[\frac{\partial}{\partial \xi}G(x,\xi,t-\tau)\right]_{\xi=0}d\tau + a\int_0^t g_2(\tau)G(x,l,t-\tau)\,d\tau.$$

Two forms of representation of the Green's function:

$$G(x,\xi,t) = \frac{2}{l}\sum_{n=0}^\infty \sin\left[\frac{\pi(2n+1)x}{2l}\right]\sin\left[\frac{\pi(2n+1)\xi}{2l}\right]\exp\left[-\frac{a\pi^2(2n+1)^2 t}{4l^2}\right]$$
$$= \frac{1}{2\sqrt{\pi a t}}\sum_{n=-\infty}^\infty (-1)^n\left\{\exp\left[-\frac{(x-\xi+2nl)^2}{4at}\right] - \exp\left[-\frac{(x+\xi+2nl)^2}{4at}\right]\right\}.$$

The first series converges rapidly at large t and the second series at small t.

2°. The following conditions are prescribed:

$$w = f(x) \quad \text{at} \quad t=0 \quad \text{(initial condition)},$$
$$\partial_x w = g_1(t) \quad \text{at} \quad x=0 \quad \text{(boundary condition)},$$
$$w = g_2(t) \quad \text{at} \quad x=l \quad \text{(boundary condition)}.$$

Solution:

$$w(x,t) = \int_0^l f(\xi)G(x,\xi,t)\,d\xi + \int_0^t \int_0^l \Phi(\xi,\tau)G(x,\xi,t-\tau)\,d\xi\,d\tau$$
$$- a\int_0^t g_1(\tau)G(x,0,t-\tau)\,d\tau - a\int_0^t g_2(\tau)\left[\frac{\partial}{\partial \xi}G(x,\xi,t-\tau)\right]_{\xi=l}d\tau.$$

Two forms of representation of the Green's function:

$$G(x,\xi,t) = \frac{2}{l}\sum_{n=0}^\infty \cos\left[\frac{\pi(2n+1)x}{2l}\right]\cos\left[\frac{\pi(2n+1)\xi}{2l}\right]\exp\left[-\frac{a\pi^2(2n+1)^2 t}{4l^2}\right]$$
$$= \frac{1}{2\sqrt{\pi a t}}\sum_{n=-\infty}^\infty (-1)^n\left\{\exp\left[-\frac{(x-\xi+2nl)^2}{4at}\right] + \exp\left[-\frac{(x+\xi+2nl)^2}{4at}\right]\right\}.$$

The first series converges rapidly at large t and the second series at small t.

⊙ *References*: B. M. Budak, A. A. Samarskii, and A. N. Tikhonov (1980), A. V. Bitsadze and D. F. Kalinichenko (1985).

1.1.3. Equation of the Form $\frac{\partial w}{\partial t} = a\frac{\partial^2 w}{\partial x^2} + bw + \Phi(x,t)$

Homogeneous equations of this form describe one-dimensional unsteady mass transfer in a quiescent medium with a first-order volume chemical reaction; the cases $b < 0$ and $b > 0$ correspond to absorption and release of substance, respectively. A similar equation is used to analyze appropriate one-dimensional thermal processes in which volume heat release ($b > 0$) proportional to temperature occurs in the medium. Furthermore, this equation governs heat transfer in a one-dimensional rod whose lateral surface exchanges heat with the ambient medium having constant temperature; $b > 0$ if the temperature of the medium is greater than that of the rod, and $b < 0$ otherwise.

1.1.3-1. Homogeneous equation ($\Phi \equiv 0$).

1°. Particular solutions:

$$w(x) = Ae^{\lambda x} + Be^{-\lambda x}, \quad \lambda = \sqrt{-b/a},$$

$$w(x,t) = (Ax + B)e^{bt},$$

$$w(x,t) = \left[A(x^2 + 2at) + B\right]e^{bt},$$

$$w(x,t) = \left[A(x^3 + 6atx) + B\right]e^{bt},$$

$$w(x,t) = \left[A(x^4 + 12atx^2 + 12a^2t^2) + B\right]e^{bt},$$

$$w(x,t) = \left[A(x^5 + 20atx^3 + 60a^2t^2x) + B\right]e^{bt},$$

$$w(x,t) = \left[A(x^6 + 30atx^4 + 180a^2t^2x^2 + 120a^3t^3) + B\right]e^{bt},$$

$$w(x,t) = A\exp\left[(a\mu^2 + b)t \pm \mu x\right] + Be^{bt},$$

$$w(x,t) = A\frac{1}{\sqrt{t}}\exp\left(-\frac{x^2}{4at} + bt\right) + Be^{bt},$$

$$w(x,t) = A\frac{x}{t^{3/2}}\exp\left(-\frac{x^2}{4at} + bt\right) + Be^{bt},$$

$$w(x,t) = A\exp\left[(b - a\mu^2)t\right]\cos(\mu x) + Be^{bt},$$

$$w(x,t) = A\exp\left[(b - a\mu^2)t\right]\sin(\mu x) + Be^{bt},$$

$$w(x,t) = A\exp(-\mu x + bt)\cos(\mu x - 2a\mu^2 t) + Be^{bt},$$

$$w(x,t) = A\exp(-\mu x + bt)\sin(\mu x - 2a\mu^2 t) + Be^{bt},$$

$$w(x,t) = A\exp(-\mu x)\cos(\beta x - 2a\beta\mu t), \quad \beta = \sqrt{\mu^2 + b/a},$$

$$w(x,t) = A\exp(-\mu x)\sin(\beta x - 2a\beta\mu t), \quad \beta = \sqrt{\mu^2 + b/a},$$

$$w(x,t) = Ae^{bt}\operatorname{erf}\left(\frac{x}{2\sqrt{at}}\right) + Be^{bt},$$

$$w(x,t) = Ae^{bt}\operatorname{erfc}\left(\frac{x}{2\sqrt{at}}\right) + Be^{bt},$$

where A, B, and μ are arbitrary constants.

2°. Fundamental solution:

$$\mathscr{E}(x,t) = \frac{1}{2\sqrt{\pi at}}\exp\left(-\frac{x^2}{4at} + bt\right).$$

1.1.3-2. Reduction to the heat equation. Remarks on the Green's functions.

The substitution $w(x,t) = e^{bt}u(x,t)$ leads to the nonhomogeneous heat equation

$$\frac{\partial u}{\partial t} = a\frac{\partial^2 u}{\partial x^2} + e^{-bt}\Phi(x,t),$$

which is discussed in Subsection 1.1.2 in detail. The initial condition for the new variable u remains the same, and the nonhomogeneous part in the boundary conditions is multiplied by e^{-bt}. Taking this into account, one can easily solve the original equation subject to the initial and boundary conditions considered in Subsection 1.1.2.

In all the boundary value problems that are dealt with in the current subsection, the Green's function can be represented in the form

$$G_b(x, \xi, t) = e^{bt}G_0(x, \xi, t),$$

where $G_0(x, \xi, t)$ is the Green's function for the heat equation that corresponds to $b = 0$.

1.1.3-3. Domain: $-\infty < x < \infty$. Cauchy problem.

An initial condition is prescribed:

$$w = f(x) \quad \text{at} \quad t = 0.$$

Solution:

$$w(x,t) = \int_{-\infty}^{\infty} f(\xi) G(x,\xi,t)\, d\xi + \int_0^t \int_{-\infty}^{\infty} \Phi(\xi,\tau) G(x,\xi,t-\tau)\, d\xi\, d\tau,$$

where

$$G(x,\xi,t) = \frac{1}{2\sqrt{\pi a t}} \exp\left[-\frac{(x-\xi)^2}{4at} + bt\right].$$

⊙ *Reference*: A. G. Butkovskiy (1979).

1.1.3-4. Domain: $0 \leq x < \infty$. First boundary value problem.

The following conditions are prescribed:

$$w = f(x) \quad \text{at} \quad t = 0 \quad \text{(initial condition)},$$
$$w = g(t) \quad \text{at} \quad x = 0 \quad \text{(boundary condition)}.$$

Solution:

$$w(x,t) = \int_0^{\infty} f(\xi) G(x,\xi,t)\, d\xi + \int_0^t g(\tau) H(x,t-\tau)\, d\tau + \int_0^t \int_0^{\infty} \Phi(\xi,\tau) G(x,\xi,t-\tau)\, d\xi\, d\tau,$$

where

$$G(x,\xi,t) = \frac{e^{bt}}{2\sqrt{\pi a t}} \left\{ \exp\left[-\frac{(x-\xi)^2}{4at}\right] - \exp\left[-\frac{(x+\xi)^2}{4at}\right] \right\}, \quad H(x,t) = \frac{x e^{bt}}{2\sqrt{\pi a}\, t^{3/2}} \exp\left(-\frac{x^2}{4at}\right).$$

⊙ *Reference*: A. D. Polyanin, A. V. Vyazmin, A. I. Zhurov, and D. A. Kazenin (1998).

1.1.3-5. Domain: $0 \leq x < \infty$. Second boundary value problem.

The following conditions are prescribed:

$$w = f(x) \quad \text{at} \quad t = 0 \quad \text{(initial condition)},$$
$$\partial_x w = g(t) \quad \text{at} \quad x = 0 \quad \text{(boundary condition)}.$$

Solution:

$$w(x,t) = \int_0^{\infty} G(x,\xi,t) f(\xi)\, d\xi - a \int_0^t g(\tau) G(x,0,t-\tau)\, d\tau + \int_0^t \int_0^{\infty} \Phi(\xi,\tau) G(x,\xi,t-\tau)\, d\xi\, d\tau,$$

where

$$G(x,\xi,t) = \frac{e^{bt}}{2\sqrt{\pi a t}} \left\{ \exp\left[-\frac{(x-\xi)^2}{4at}\right] + \exp\left[-\frac{(x+\xi)^2}{4at}\right] \right\}.$$

1.1.3-6. Domain: $0 \leq x < \infty$. Third boundary value problem.

The following conditions are prescribed:

$$w = f(x) \quad \text{at} \quad t = 0 \quad \text{(initial condition)},$$
$$\partial_x w - kw = g(t) \quad \text{at} \quad x = 0 \quad \text{(boundary condition)}.$$

The solution $w(x,t)$ is determined by the formula in Paragraph 1.1.3-5 where

$$G(x,\xi,t) = \frac{e^{bt}}{2\sqrt{\pi a t}} \left\{ \exp\left[-\frac{(x-\xi)^2}{4at}\right] + \exp\left[-\frac{(x+\xi)^2}{4at}\right] - 2k \int_0^{\infty} \exp\left[-\frac{(x+\xi+\eta)^2}{4at} - k\eta\right] d\eta \right\}.$$

1.1.3-7. Domain: $0 \leq x \leq l$. First boundary value problem.

The following conditions are prescribed:

$$w = f(x) \quad \text{at} \quad t = 0 \quad \text{(initial condition)},$$
$$w = g_1(t) \quad \text{at} \quad x = 0 \quad \text{(boundary condition)},$$
$$w = g_2(t) \quad \text{at} \quad x = l \quad \text{(boundary condition)}.$$

Solution:

$$w(x,t) = \int_0^l f(\xi) G(x,\xi,t)\, d\xi + \int_0^t \int_0^l \Phi(\xi,\tau) G(x,\xi,t-\tau)\, d\xi\, d\tau$$
$$+ a \int_0^t g_1(\tau) H_1(x, t-\tau)\, d\tau - a \int_0^t g_2(\tau) H_2(x, t-\tau)\, d\tau,$$

where

$$G(x,\xi,t) = \frac{2}{l} e^{bt} \sum_{n=1}^{\infty} \sin\left(\frac{n\pi x}{l}\right) \sin\left(\frac{n\pi \xi}{l}\right) \exp\left(-\frac{a n^2 \pi^2 t}{l^2}\right),$$
$$H_1(x,t) = \frac{\partial}{\partial \xi} G(x,\xi,t)\Big|_{\xi=0}, \quad H_2(x,t) = \frac{\partial}{\partial \xi} G(x,\xi,t)\Big|_{\xi=l}.$$

⦿ *Reference*: A. G. Butkovskiy (1979).

1.1.3-8. Domain: $0 \leq x \leq l$. Second boundary value problem.

The following conditions are prescribed:

$$w = f(x) \quad \text{at} \quad t = 0 \quad \text{(initial condition)},$$
$$\partial_x w = g_1(t) \quad \text{at} \quad x = 0 \quad \text{(boundary condition)},$$
$$\partial_x w = g_2(t) \quad \text{at} \quad x = l \quad \text{(boundary condition)}.$$

Solution:

$$w(x,t) = \int_0^l f(\xi) G(x,\xi,t)\, d\xi + \int_0^t \int_0^l \Phi(\xi,\tau) G(x,\xi,t-\tau)\, d\xi\, d\tau$$
$$- a \int_0^t g_1(\tau) G(x,0,t-\tau)\, d\tau + a \int_0^t g_2(\tau) G(x,l,t-\tau)\, d\tau,$$

where

$$G(x,\xi,t) = e^{bt}\left[\frac{1}{l} + \frac{2}{l}\sum_{n=1}^{\infty} \cos\left(\frac{n\pi x}{l}\right) \cos\left(\frac{n\pi \xi}{l}\right) \exp\left(-\frac{a n^2 \pi^2 t}{l^2}\right)\right].$$

⦿ *Reference*: A. G. Butkovskiy (1979).

1.1.3-9. Domain: $0 \leq x \leq l$. Third boundary value problem.

The following conditions are prescribed:

$$w = f(x) \quad \text{at} \quad t = 0 \quad \text{(initial condition)},$$
$$\partial_x w - k_1 w = g_1(t) \quad \text{at} \quad x = 0 \quad \text{(boundary condition)},$$
$$\partial_x w + k_2 w = g_2(t) \quad \text{at} \quad x = l \quad \text{(boundary condition)}.$$

The solution $w(x, t)$ is determined by the formula in Paragraph 1.1.3-8 where

$$G(x, \xi, t) = e^{bt} \sum_{n=1}^{\infty} \frac{1}{\|y_n\|^2} y_n(x) y_n(\xi) \exp(-a\mu_n^2 t),$$

$$y_n(x) = \cos(\mu_n x) + \frac{k_1}{\mu_n} \sin(\mu_n x), \quad \|y_n\|^2 = \frac{k_2}{2\mu_n^2} \frac{\mu_n^2 + k_1^2}{\mu_n^2 + k_2^2} + \frac{k_1}{2\mu_n^2} + \frac{l}{2}\left(1 + \frac{k_1^2}{\mu_n^2}\right).$$

Here, the μ_n are positive roots of the transcendental equation $\dfrac{\tan(\mu l)}{\mu} = \dfrac{k_1 + k_2}{\mu^2 - k_1 k_2}$.

⊙ *Reference*: A. G. Butkovskiy (1979).

1.1.3-10. Domain: $0 \le x \le l$. Mixed boundary value problem.

The following conditions are prescribed:

$$\begin{aligned} w &= f(x) & \text{at} \quad & t = 0 & \text{(initial condition)}, \\ w &= g_1(t) & \text{at} \quad & x = 0 & \text{(boundary condition)}, \\ \partial_x w &= g_2(t) & \text{at} \quad & x = l & \text{(boundary condition)}. \end{aligned}$$

Solution:

$$w(x, t) = \int_0^l f(\xi) G(x, \xi, t)\, d\xi + \int_0^t \int_0^l \Phi(\xi, \tau) G(x, \xi, t - \tau)\, d\xi\, d\tau$$

$$+ a \int_0^t g_1(\tau) \Lambda(x, t - \tau)\, d\tau + a \int_0^t g_2(\tau) G(x, l, t - \tau)\, d\tau,$$

where

$$G(x, \xi, t) = \frac{2}{l} e^{bt} \sum_{n=0}^{\infty} \sin\left[\frac{\pi(2n+1)x}{2l}\right] \sin\left[\frac{\pi(2n+1)\xi}{2l}\right] \exp\left[-\frac{a\pi^2(2n+1)^2 t}{4l^2}\right],$$

$$\Lambda(x, t) = \frac{\partial}{\partial \xi} G(x, \xi, t)\Big|_{\xi=0}.$$

1.1.4. Equation of the Form $\dfrac{\partial w}{\partial t} = a\dfrac{\partial^2 w}{\partial x^2} + b\dfrac{\partial w}{\partial x} + \Phi(x, t)$

This equation is encountered in one-dimensional nonstationary problems of convective mass transfer in a continuous medium that moves with a constant velocity; the case $\Phi \equiv 0$ means that there is no absorption or release of substance.

1.1.4-1. Homogeneous equation ($\Phi \equiv 0$).

1°. Particular solutions:

$$\begin{aligned} w(x) &= A e^{-\lambda x} + B, \quad \lambda = b/a, \\ w(x, t) &= Ax + Abt + B, \\ w(x, t) &= A(x + bt)^2 + 2Aat + B, \\ w(x, t) &= A(x + bt)^3 + 6Aatx + B, \\ w(x, t) &= A \exp\left[(a\mu^2 + b\mu)t + \mu x\right] + B, \\ w(x, t) &= A \frac{1}{\sqrt{t}} \exp\left[-\frac{(x + bt)^2}{4at}\right] + B, \end{aligned}$$

$$w(x,t) = A\exp(-a\mu^2 t)\cos(\mu x + b\mu t) + B,$$
$$w(x,t) = A\exp(-a\mu^2 t)\sin(\mu x + b\mu t) + B,$$
$$w(x,t) = A\exp(-\mu x)\cos\bigl[\beta x + \beta(b - 2a\mu)t\bigr] + B, \qquad \beta = \sqrt{\mu^2 - (b/a)\mu},$$
$$w(x,t) = A\exp(-\mu x)\sin\bigl[\beta x + \beta(b - 2a\mu)t\bigr] + B, \qquad \beta = \sqrt{\mu^2 - (b/a)\mu},$$
$$w(x,t) = A\,\mathrm{erf}\!\left(\frac{x + bt}{2\sqrt{at}}\right) + B,$$
$$w(x,t) = A\,\mathrm{erfc}\!\left(\frac{x + bt}{2\sqrt{at}}\right) + B,$$

where A, B, and μ are arbitrary constants.

2°. Fundamental solution:
$$\mathscr{E}(x,t) = \frac{1}{2\sqrt{\pi a t}}\exp\!\left[-\frac{(x+bt)^2}{4at}\right].$$

1.1.4-2. Reduction to the heat equation. Remarks on the Green's function.

1°. The substitution
$$w(x,t) = \exp(\beta t + \mu x)u(x,t), \qquad \beta = -\frac{b^2}{4a}, \qquad \mu = -\frac{b}{2a}$$

leads to the nonhomogeneous heat equation
$$\frac{\partial u}{\partial t} = a\frac{\partial^2 u}{\partial x^2} + \exp(-\beta t - \mu x)\Phi(x,t),$$

which is considered in Subsection 1.1.2 in detail.

2°. On passing from t, x to the new variables t, $z = x + bt$, we obtain the nonhomogeneous heat equation
$$\frac{\partial w}{\partial t} = a\frac{\partial^2 w}{\partial z^2} + \Phi(z - bt, t),$$

which is treated in Subsection 1.1.2.

3°. For all first boundary value problems, the Green's function can be represented as
$$G_b(x,\zeta,t) = \exp\!\left[\frac{b}{2a}(\zeta - x) - \frac{b^2}{4a}t\right]G_0(x,\zeta,t),$$

where $G_0(x,\xi,t)$ is the Green's function for the heat equation that corresponds to $b = 0$.

1.1.4-3. Domain: $-\infty < x < \infty$. Cauchy problem.

An initial condition is prescribed:
$$w = f(x) \quad \text{at} \quad t = 0.$$

Solution:
$$w(x,t) = \int_{-\infty}^{\infty} f(\xi)G(x,\xi,t)\,d\xi + \int_0^t\!\int_{-\infty}^{\infty} \Phi(\xi,\tau)G(x,\xi,t-\tau)\,d\xi\,d\tau,$$

where
$$G(x,\xi,t) = \frac{1}{2\sqrt{\pi a t}}\exp\!\left[\frac{b(\xi - x)}{2a} - \frac{b^2 t}{4a} - \frac{(x-\xi)^2}{4at}\right].$$

1.1.4-4. Domain: $0 \le x < \infty$. First boundary value problem.

The following conditions are prescribed:

$$w = f(x) \quad \text{at} \quad t = 0 \quad \text{(initial condition)},$$
$$w = g(t) \quad \text{at} \quad x = 0 \quad \text{(boundary condition)}.$$

Solution:

$$w(x,t) = \int_0^\infty f(\xi) G(x,\xi,t)\,d\xi + a \int_0^t g(\tau) \Lambda(x, t-\tau)\,d\tau$$
$$+ \int_0^t \int_0^\infty \Phi(\xi,\tau) G(x,\xi,t-\tau)\,d\xi\,d\tau,$$

where

$$G(x,\xi,t) = \frac{1}{2\sqrt{\pi a t}} \exp\left[\frac{b(\xi - x)}{2a} - \frac{b^2 t}{4a}\right] \left\{ \exp\left[-\frac{(x-\xi)^2}{4at}\right] - \exp\left[-\frac{(x+\xi)^2}{4at}\right] \right\},$$

$$\Lambda(x,t) = \frac{\partial}{\partial \xi} G(x,\xi,t) \bigg|_{\xi=0}.$$

1.1.4-5. Domain: $0 \le x < \infty$. Second boundary value problem.

The following conditions are prescribed:

$$w = f(x) \quad \text{at} \quad t = 0 \quad \text{(initial condition)},$$
$$\partial_x w = g(t) \quad \text{at} \quad x = 0 \quad \text{(boundary condition)}.$$

Solution:

$$w(x,t) = \int_0^\infty G(x,\xi,t) f(\xi)\,d\xi - a \int_0^t g(\tau) G(x,0,t-\tau)\,d\tau$$
$$+ \int_0^t \int_0^\infty \Phi(\xi,\tau) G(x,\xi,t-\tau)\,d\xi\,d\tau,$$

where

$$G(x,\xi,t) = \frac{1}{2\sqrt{\pi a t}} \exp\left[\frac{b(\xi - x)}{2a} - \frac{b^2 t}{4a}\right] \left\{ \exp\left[-\frac{(x-\xi)^2}{4at}\right] + \exp\left[-\frac{(x+\xi)^2}{4at}\right] \right\}.$$

1.1.4-6. Domain: $0 \le x < \infty$. Third boundary value problem.

The following conditions are prescribed:

$$w = f(x) \quad \text{at} \quad t = 0 \quad \text{(initial condition)},$$
$$\partial_x w - kw = g(t) \quad \text{at} \quad x = 0 \quad \text{(boundary condition)}.$$

The solution $w(x,t)$ is determined by the formula in Paragraph 1.1.4-5 where

$$G(x,\xi,t) = \frac{1}{2\sqrt{\pi a t}} \exp\left[\frac{b(\xi - x)}{2a} - \frac{b^2 t}{4a}\right] \left\{ \exp\left[-\frac{(x-\xi)^2}{4at}\right] + \exp\left[-\frac{(x+\xi)^2}{4at}\right] \right.$$
$$\left. - 2s \int_0^\infty \exp\left[-\frac{(x+\xi+\eta)^2}{4at} - s\eta\right] d\eta \right\}, \qquad s = k + \frac{b}{2a}.$$

1.1.4-7. Domain: $0 \leq x \leq l$. First boundary value problem.

The following conditions are prescribed:

$$w = f(x) \quad \text{at} \quad t = 0 \quad \text{(initial condition)},$$
$$w = g_1(t) \quad \text{at} \quad x = 0 \quad \text{(boundary condition)},$$
$$w = g_2(t) \quad \text{at} \quad x = l \quad \text{(boundary condition)}.$$

Solution:

$$w(x,t) = \int_0^l f(\xi) G(x,\xi,t)\,d\xi + \int_0^t \int_0^l \Phi(\xi,\tau) G(x,\xi,t-\tau)\,d\xi\,d\tau$$
$$+ a \int_0^t g_1(\tau) H_1(x,t-\tau)\,d\tau - a \int_0^t g_2(\tau) H_2(x,t-\tau)\,d\tau,$$

where

$$G(x,\xi,t) = \frac{2}{l} \exp\left[\frac{b}{2a}(\xi-x) - \frac{b^2}{4a}t\right] \sum_{n=1}^\infty \sin\left(\frac{\pi n x}{l}\right) \sin\left(\frac{\pi n \xi}{l}\right) \exp\left(-\frac{a\pi^2 n^2}{l^2}t\right),$$

$$H_1(x,t) = \frac{\partial}{\partial \xi} G(x,\xi,t)\bigg|_{\xi=0}, \quad H_2(x,t) = \frac{\partial}{\partial \xi} G(x,\xi,t)\bigg|_{\xi=l}.$$

⊙ *Reference*: A. G. Butkovskiy (1979).

1.1.4-8. Domain: $0 \leq x \leq l$. Second boundary value problem.

The following conditions are prescribed:

$$w = f(x) \quad \text{at} \quad t = 0 \quad \text{(initial condition)},$$
$$\partial_x w = g_1(t) \quad \text{at} \quad x = 0 \quad \text{(boundary condition)},$$
$$\partial_x w = g_2(t) \quad \text{at} \quad x = l \quad \text{(boundary condition)}.$$

Solution:

$$w(x,t) = \int_0^l f(\xi) G(x,\xi,t)\,d\xi + \int_0^t \int_0^l \Phi(\xi,\tau) G(x,\xi,t-\tau)\,d\xi\,d\tau$$
$$- a \int_0^t g_1(\tau) G(x,0,t-\tau)\,d\tau + a \int_0^t g_2(\tau) G(x,l,t-\tau)\,d\tau,$$

where

$$G(x,\xi,t) = \frac{b}{a(e^{bl/a}-1)} \exp\left(\frac{b\xi}{a}\right) + \frac{2}{l} \exp\left[\frac{b(\xi-x)}{2a} - \frac{b^2 t}{4a}\right] \sum_{n=1}^\infty \frac{y_n(x) y_n(\xi)}{1+\mu_n^2} \exp\left(-\frac{a\pi^2 n^2}{l^2}t\right),$$

$$y_n(x) = \cos\left(\frac{\pi n x}{l}\right) + \mu_n \sin\left(\frac{\pi n x}{l}\right), \quad \mu_n = \frac{bl}{2a\pi n}.$$

⊙ *Reference*: A. G. Butkovskiy (1979).

1.1.4-9. Domain: $0 \leq x \leq l$. Third boundary value problem.

The following conditions are prescribed:

$$w = f(x) \quad \text{at} \quad t = 0 \quad \text{(initial condition)},$$
$$\partial_x w - k_1 w = g_1(t) \quad \text{at} \quad x = 0 \quad \text{(boundary condition)},$$
$$\partial_x w + k_2 w = g_2(t) \quad \text{at} \quad x = l \quad \text{(boundary condition)}.$$

The solution $w(x, t)$ is determined by the formula in Paragraph 1.1.4-8 where

$$G(x, \xi, t) = \exp\left[\frac{b(\xi - x)}{2a} - \frac{b^2 t}{4a}\right] \sum_{n=1}^{\infty} \frac{1}{B_n} y_n(x) y_n(\xi) \exp(-a\mu_n^2 t),$$

$$y_n(x) = \cos(\mu_n x) + \frac{2ak_1 + b}{2a\mu_n} \sin(\mu_n x),$$

$$B_n = \frac{2ak_2 - b}{4a\mu_n^2} \frac{4a^2 \mu_n^2 + (2ak_1 + b)^2}{4a^2 \mu_n^2 + (2ak_2 - b)^2} + \frac{2ak_1 + b}{4a\mu_n^2} + \frac{l}{2} + \frac{l(2ak_1 + b)^2}{8a^2 \mu_n^2},$$

and the μ_n are positive roots of the transcendental equation

$$\frac{\tan(\mu l)}{\mu} = \frac{4a^2(k_1 + k_2)}{4a^2 \mu^2 - (2ak_1 + b)(2ak_2 - b)}.$$

1.1.5. Equation of the Form $\dfrac{\partial w}{\partial t} = a \dfrac{\partial^2 w}{\partial x^2} + b \dfrac{\partial w}{\partial x} + cw + \Phi(x, t)$

For $\Phi \equiv 0$, this equation describes one-dimensional unsteady convective mass transfer with a first-order volume chemical reaction in a continuous medium that moves with a constant velocity. A similar equation is used for the analysis of the corresponding one-dimensional thermal processes in a moving medium with volume heat release proportional to temperature.

1.1.5-1. Homogeneous equation ($\Phi \equiv 0$).

1°. Particular solutions:

$$w(x, t) = e^{ct}(Ax + Abt + B),$$
$$w(x, t) = e^{ct}\left[A(x + bt)^2 + 2Aat + B\right],$$
$$w(x, t) = e^{ct}\left[A(x + bt)^3 + 6Aatx + B\right],$$
$$w(x, t) = Ae^{-\lambda x + ct} + Be^{ct}, \quad \lambda = b/a,$$
$$w(x, t) = A\exp\left[(a\mu^2 + b\mu + c)t + \mu x\right] + Be^{ct},$$
$$w(x, t) = A\frac{1}{\sqrt{t}} \exp\left[-\frac{(x + bt)^2}{4at} + ct\right] + Be^{ct},$$
$$w(x, t) = A\exp(ct - a\mu^2 t)\cos(\mu x + b\mu t) + Be^{ct},$$
$$w(x, t) = A\exp(ct - a\mu^2 t)\sin(\mu x + b\mu t) + Be^{ct},$$
$$w(x, t) = A\exp(-\mu x)\cos\left[\beta x + \beta(b - 2a\mu)t\right], \quad \beta = \sqrt{\mu^2 - (b/a)\mu + c/a},$$
$$w(x, t) = A\exp(-\mu x)\sin\left[\beta x + \beta(b - 2a\mu)t\right], \quad \beta = \sqrt{\mu^2 - (b/a)\mu + c/a},$$
$$w(x, t) = Ae^{ct}\operatorname{erf}\left(\frac{x + bt}{2\sqrt{at}}\right) + Be^{ct},$$
$$w(x, t) = Ae^{ct}\operatorname{erfc}\left(\frac{x + bt}{2\sqrt{at}}\right) + Be^{ct},$$

where A, B, and μ are arbitrary constants.

2°. Fundamental solution:
$$\mathscr{E}(x,t) = \frac{1}{2\sqrt{\pi at}} \exp\left[-\frac{(x+bt)^2}{4at} + ct\right].$$

1.1.5-2. Reduction to the heat equation. Remarks on the Green's functions.

1°. The substitution
$$w(x,t) = \exp(\beta t + \mu x)u(x,t), \qquad \beta = c - \frac{b^2}{4a}, \quad \mu = -\frac{b}{2a}$$
leads to the nonhomogeneous heat equation
$$\frac{\partial u}{\partial t} = a\frac{\partial^2 u}{\partial x^2} + \exp(-\beta t - \mu x)\Phi(x,t),$$
which is considered in Subsection 1.1.2 in detail.

2°. The transformation
$$w(x,t) = e^{ct}v(z,t), \quad z = x + bt,$$
leads to the nonhomogeneous heat equation
$$\frac{\partial v}{\partial t} = a\frac{\partial^2 v}{\partial z^2} + e^{-ct}\Phi(z - bt, t),$$
which it treated in Subsection 1.1.2.

3°. For all first boundary value problems, the Green's function can be represented as
$$G_{b,c}(x,\xi,t) = \exp\left[\frac{b}{2a}(\xi - x) + \left(c - \frac{b^2}{4a}\right)t\right]G_{0,0}(x,\xi,t),$$
where $G_{0,0}(x,\xi,t)$ is the Green's function for the heat equation that corresponds to $b = c = 0$.

1.1.5-3. Domain: $-\infty < x < \infty$. Cauchy problem.

An initial condition is prescribed:
$$w = f(x) \quad \text{at} \quad t = 0.$$

Solution:
$$w(x,t) = \int_{-\infty}^{\infty} f(\xi)G(x,\xi,t)\,d\xi + \int_0^t \int_{-\infty}^{\infty} \Phi(\xi,\tau)G(x,\xi,t-\tau)\,d\xi\,d\tau,$$
where
$$G(x,\xi,t) = \frac{1}{2\sqrt{\pi at}} \exp\left[\frac{b}{2a}(\xi - x) + \left(c - \frac{b^2}{4a}\right)t - \frac{(x-\xi)^2}{4at}\right].$$

1.1.5-4. Domain: $0 \leq x < \infty$. First boundary value problem.

The following conditions are prescribed:
$$w = f(x) \quad \text{at} \quad t = 0 \quad \text{(initial condition)},$$
$$w = g(t) \quad \text{at} \quad x = 0 \quad \text{(boundary condition)}.$$

Solution:
$$w(x,t) = \int_0^{\infty} f(\xi)G(x,\xi,t)\,d\xi + a\int_0^t g(\tau)\Lambda(x,t-\tau)\,d\tau + \int_0^t \int_0^{\infty} \Phi(\xi,\tau)G(x,\xi,t-\tau)\,d\xi\,d\tau,$$
where
$$G(x,\xi,t) = \frac{1}{2\sqrt{\pi at}} \exp\left[\frac{b(\xi - x)}{2a} + \left(c - \frac{b^2}{4a}\right)t\right]\left\{\exp\left[-\frac{(x-\xi)^2}{4at}\right] - \exp\left[-\frac{(x+\xi)^2}{4at}\right]\right\},$$
$$\Lambda(x,t) = \frac{\partial}{\partial \xi}G(x,\xi,t)\bigg|_{\xi=0}.$$

1.1.5-5. Domain: $0 \leq x < \infty$. Second boundary value problem.

The following conditions are prescribed:

$$w = f(x) \quad \text{at} \quad t = 0 \quad \text{(initial condition)},$$
$$\partial_x w = g(t) \quad \text{at} \quad x = 0 \quad \text{(boundary condition)}.$$

Solution:

$$w(x,t) = \int_0^\infty G(x,\xi,t) f(\xi)\, d\xi - a \int_0^t g(\tau) G(x,0,t-\tau)\, d\tau + \int_0^t \int_0^\infty \Phi(\xi,\tau) G(x,\xi,t-\tau)\, d\xi\, d\tau,$$

where

$$G(x,\xi,t) = \frac{1}{2\sqrt{\pi a t}} \exp\left[\frac{b(\xi-x)}{2a} + \left(c - \frac{b^2}{4a}\right)t\right] \left\{\exp\left[-\frac{(x-\xi)^2}{4at}\right] + \exp\left[-\frac{(x+\xi)^2}{4at}\right]\right\}.$$

1.1.5-6. Domain: $0 \leq x < \infty$. Third boundary value problem.

The following conditions are prescribed:

$$w = f(x) \quad \text{at} \quad t = 0 \quad \text{(initial condition)},$$
$$\partial_x w - k w = g(t) \quad \text{at} \quad x = 0 \quad \text{(boundary condition)}.$$

The solution $w(x,t)$ is determined by the formula in Paragraph 1.1.5-5 where

$$G(x,\xi,t) = \frac{1}{2\sqrt{\pi a t}} \exp\left[\frac{b(\xi-x)}{2a} + \left(c - \frac{b^2}{4a}\right)t\right] \left\{\exp\left[-\frac{(x-\xi)^2}{4at}\right] + \exp\left[-\frac{(x+\xi)^2}{4at}\right]\right.$$
$$\left. - 2s \int_0^\infty \exp\left[-\frac{(x+\xi+\eta)^2}{4at} - s\eta\right] d\eta \right\}, \qquad s = k + \frac{b}{2a}.$$

1.1.5-7. Domain: $0 \leq x \leq l$. First boundary value problem.

The following conditions are prescribed:

$$w = f(x) \quad \text{at} \quad t = 0 \quad \text{(initial condition)},$$
$$w = g_1(t) \quad \text{at} \quad x = 0 \quad \text{(boundary condition)},$$
$$w = g_2(t) \quad \text{at} \quad x = l \quad \text{(boundary condition)}.$$

Solution:

$$w(x,t) = \int_0^l f(\xi) G(x,\xi,t)\, d\xi + \int_0^t \int_0^l \Phi(\xi,\tau) G(x,\xi,t-\tau)\, d\xi\, d\tau$$
$$+ a \int_0^t g_1(\tau) H_1(x,t-\tau)\, d\tau - a \int_0^t g_2(\tau) H_2(x,t-\tau)\, d\tau,$$

where

$$G(x,\xi,t) = \frac{2}{l} \exp\left[\frac{b(\xi-x)}{2a} + \left(c - \frac{b^2}{4a}\right)t\right] \sum_{n=1}^\infty \sin\left(\frac{\pi n x}{l}\right) \sin\left(\frac{\pi n \xi}{l}\right) \exp\left(-\frac{a \pi^2 n^2}{l^2} t\right),$$

$$H_1(x,t) = \frac{\partial}{\partial \xi} G(x,\xi,t)\Big|_{\xi=0}, \qquad H_2(x,t) = \frac{\partial}{\partial \xi} G(x,\xi,t)\Big|_{\xi=l}.$$

⊙ *Reference*: A. G. Butkovskiy (1979).

1.1.5-8. Domain: $0 \leq x \leq l$. Second boundary value problem.

The following conditions are prescribed:

$$w = f(x) \quad \text{at} \quad t = 0 \quad \text{(initial condition)},$$
$$\partial_x w = g_1(t) \quad \text{at} \quad x = 0 \quad \text{(boundary condition)},$$
$$\partial_x w = g_2(t) \quad \text{at} \quad x = l \quad \text{(boundary condition)}.$$

Solution:

$$w(x,t) = \int_0^l f(\xi) G(x,\xi,t)\,d\xi + \int_0^t \int_0^l \Phi(\xi,\tau) G(x,\xi,t-\tau)\,d\xi\,d\tau$$
$$- a\int_0^t g_1(\tau) G(x,0,t-\tau)\,d\tau + a\int_0^t g_2(\tau) G(x,l,t-\tau)\,d\tau,$$

where

$$G(x,\xi,t) = A \exp\left(\frac{b\xi}{a} + ct\right) + \frac{2}{l}\exp\left[\frac{b(\xi-x)}{2a} + \left(c - \frac{b^2}{4a}\right)t\right] \sum_{n=1}^{\infty} \frac{y_n(x)y_n(\xi)}{1+\mu_n^2} \exp\left(-\frac{a\pi^2 n^2}{l^2}t\right),$$

$$A = \frac{b}{a(e^{bl/a} - 1)}, \quad y_n(x) = \cos\left(\frac{\pi n x}{l}\right) + \mu_n \sin\left(\frac{\pi n x}{l}\right), \quad \mu_n = \frac{bl}{2a\pi n}.$$

⊙ *Reference*: A. G. Butkovskiy (1979).

1.1.5-9. Domain: $0 \leq x \leq l$. Third boundary value problem.

The following conditions are prescribed:

$$w = f(x) \quad \text{at} \quad t = 0 \quad \text{(initial condition)},$$
$$\partial_x w - k_1 w = g_1(t) \quad \text{at} \quad x = 0 \quad \text{(boundary condition)},$$
$$\partial_x w + k_2 w = g_2(t) \quad \text{at} \quad x = l \quad \text{(boundary condition)}.$$

The solution $w(x,t)$ is determined by the formula in Paragraph 1.1.5-8 where

$$G(x,\xi,t) = \exp\left[\frac{b(\xi-x)}{2a} + \left(c - \frac{b^2}{4a}\right)t\right] \sum_{n=1}^{\infty} \frac{1}{B_n} y_n(x) y_n(\xi) \exp(-a\mu_n^2 t),$$

$$y_n(x) = \cos(\mu_n x) + \frac{2ak_1 + b}{2a\mu_n}\sin(\mu_n x),$$

$$B_n = \frac{2ak_2 - b}{4a\mu_n^2} \frac{4a^2\mu_n^2 + (2ak_1+b)^2}{4a^2\mu_n^2 + (2ak_2-b)^2} + \frac{2ak_1+b}{4a\mu_n^2} + \frac{l}{2} + \frac{l(2ak_1+b)^2}{8a^2\mu_n^2},$$

and the μ_n are positive roots of the transcendental equation

$$\frac{\tan(\mu l)}{\mu} = \frac{4a^2(k_1 + k_2)}{4a^2\mu^2 - (2ak_1+b)(2ak_2-b)}.$$

1.2. Heat Equation with Axial or Central Symmetry and Related Equations

1.2.1. Equation of the Form $\dfrac{\partial w}{\partial t} = a\left(\dfrac{\partial^2 w}{\partial r^2} + \dfrac{1}{r}\dfrac{\partial w}{\partial r}\right)$

This is a sourceless heat equation that describes one-dimensional unsteady thermal processes having axial symmetry. It is often represented in the equivalent form

$$\frac{\partial w}{\partial t} = \frac{a}{r}\frac{\partial}{\partial r}\left(r\frac{\partial w}{\partial r}\right).$$

A similar equation is used for the analysis of the corresponding one-dimensional unsteady diffusion processes.

1.2.1-1. Particular solutions (A, B, and μ are arbitrary constants).

$$w(r) = A + B \ln r,$$

$$w(r,t) = A + B(r^2 + 4at),$$

$$w(r,t) = A + B(r^4 + 16atr^2 + 32a^2t^2),$$

$$w(r,t) = A + B\left(r^{2n} + \sum_{k=1}^{n} \frac{4^k [n(n-1)\ldots(n-k+1)]^2}{k!}(at)^k r^{2n-2k}\right),$$

$$w(r,t) = A + B\left(4at \ln r + r^2 \ln r - r^2\right),$$

$$w(r,t) = A + \frac{B}{t} \exp\left(-\frac{r^2}{4at}\right),$$

$$w(r,t) = A + B \int_1^{\zeta} e^{-z} \frac{dz}{z}, \quad \zeta = \frac{r^2}{4at},$$

$$w(r,t) = A + B \exp(-a\mu^2 t) J_0(\mu r),$$

$$w(r,t) = A + B \exp(-a\mu^2 t) Y_0(\mu r),$$

$$w(r,t) = A + \frac{B}{t} \exp\left(-\frac{r^2 + \mu^2}{4t}\right) I_0\left(\frac{\mu r}{2t}\right),$$

$$w(r,t) = A + \frac{B}{t} \exp\left(-\frac{r^2 + \mu^2}{4t}\right) K_0\left(\frac{\mu r}{2t}\right),$$

where n is an arbitrary positive integer, $J_0(z)$ and $Y_0(z)$ are the Bessel functions, and $I_0(z)$ and $K_0(z)$ are the modified Bessel functions.

Suppose $w = w(r,t)$ is a solution of the original equation. Then the functions

$$w_1 = Aw(\pm \lambda r, \lambda^2 t + C),$$

$$w_2 = \frac{A}{\delta + \beta t} \exp\left[-\frac{\beta r^2}{4a(\delta + \beta t)}\right] w\left(\pm \frac{r}{\delta + \beta t}, \frac{\gamma + \lambda t}{\delta + \beta t}\right), \quad \lambda \delta - \beta \gamma = 1,$$

where A, C, β, δ, and λ are arbitrary constants, are also solutions of this equation. The second formula usually may be encountered with $\beta = 1$, $\gamma = -1$, and $\delta = \lambda = 0$.

⊙ *Reference*: H. S. Carslaw and J. C. Jaeger (1984), A. D. Polyanin, A. V. Vyazmin, A. I. Zhurov, and D. A. Kazenin (1998).

1.2.1-2. Particular solutions in the form of an infinite series.

A solution containing an arbitrary function of the space variable:

$$w(r,t) = f(r) + \sum_{n=1}^{\infty} \frac{(at)^n}{n!} \boldsymbol{L}^n[f(r)], \quad \boldsymbol{L} \equiv \frac{d^2}{dr^2} + \frac{1}{r}\frac{d}{dr},$$

where $f(r)$ is any infinitely differentiable function. This solution satisfies the initial condition $w(r,0) = f(r)$. The sum is finite if $f(r)$ is a polynomial that contains only even powers.

A solution containing an arbitrary function of time:

$$w(r,t) = g(t) + \sum_{n=1}^{\infty} \frac{1}{(4a)^n (n!)^2} r^{2n} g_t^{(n)}(t),$$

where $g(t)$ is any infinitely differentiable function. This solution is bounded at $r = 0$ and possesses the properties

$$w(0,t) = g(t), \quad \partial_r w(0,t) = 0.$$

1.2.1-3. Domain: $0 \leq r \leq R$. First boundary value problem.

The following conditions are prescribed:

$$w = f(r) \quad \text{at} \quad t = 0 \quad \text{(initial condition)},$$
$$w = g(t) \quad \text{at} \quad r = R \quad \text{(boundary condition)},$$
$$|w| \neq \infty \quad \text{at} \quad r = 0 \quad \text{(boundedness condition)}.$$

Solution:
$$w(r,t) = \int_0^R f(\xi) G(r,\xi,t)\, d\xi - a \int_0^t g(\tau) \Lambda(r, t-\tau)\, d\tau.$$

Here,
$$G(r,\xi,t) = \sum_{n=1}^{\infty} \frac{2\xi}{R^2 J_1^2(\mu_n)} J_0\!\left(\mu_n \frac{r}{R}\right) J_0\!\left(\mu_n \frac{\xi}{R}\right) \exp\!\left(-\frac{a\mu_n^2 t}{R^2}\right), \quad \Lambda(r,t) = \frac{\partial}{\partial \xi} G(r,\xi,t)\bigg|_{\xi=R},$$

where the μ_n are positive zeros of the Bessel function, $J_0(\mu) = 0$. Below are the numerical values of the first ten roots:

$\mu_1 = 2.4048, \quad \mu_2 = 5.5201, \quad \mu_3 = 8.6537, \quad \mu_4 = 11.7915, \quad \mu_5 = 14.9309,$
$\mu_6 = 18.0711, \quad \mu_7 = 21.2116, \quad \mu_8 = 24.3525, \quad \mu_9 = 27.4935, \quad \mu_{10} = 30.6346.$

The zeroes of the Bessel function $J_0(\mu)$ may be approximated by the formula
$$\mu_n = 2.4 + 3.13(n-1) \qquad (n = 1, 2, 3, \ldots),$$
which is accurate within 0.3%. As $n \to \infty$, we have $\mu_{n+1} - \mu_n \to \pi$.

Example 1. The initial temperature of the cylinder is uniform, $f(r) = w_0$, and its lateral surface is maintained all the time at a constant temperature, $g(t) = w_R$.
Solution:
$$\frac{w(r,t) - w_R}{w_0 - w_R} = \sum_{n=1}^{\infty} \frac{2}{\mu_n J_1(\mu_n)} \exp\!\left(-\mu_n^2 \frac{at}{R^2}\right) J_0\!\left(\mu_n \frac{r}{R}\right).$$

⊙ *Reference*: H. S. Carslaw and J. C. Jaeger (1984).

1.2.1-4. Domain: $0 \leq r \leq R$. Second boundary value problem.

The following conditions are prescribed:

$$w = f(r) \quad \text{at} \quad t = 0 \quad \text{(initial condition)},$$
$$\partial_r w = g(t) \quad \text{at} \quad r = R \quad \text{(boundary condition)},$$
$$|w| \neq \infty \quad \text{at} \quad r = 0 \quad \text{(boundedness condition)}.$$

Solution:
$$w(r,t) = \int_0^R f(\xi) G(r,\xi,t)\, d\xi + a \int_0^t g(\tau) G(r, R, t-\tau)\, d\tau.$$

Here,
$$G(r,\xi,t) = \frac{2}{R^2}\xi + \frac{2}{R^2} \sum_{n=1}^{\infty} \frac{\xi}{J_0^2(\mu_n)} J_0\!\left(\frac{\mu_n r}{R}\right) J_0\!\left(\frac{\mu_n \xi}{R}\right) \exp\!\left(-\frac{a\mu_n^2 t}{R^2}\right),$$

where the μ_n are positive zeros of the first-order Bessel function, $J_1(\mu) = 0$. Below are the numerical values of the first ten roots:

$\mu_1 = 3.8317, \quad \mu_2 = 7.0156, \quad \mu_3 = 10.1735, \quad \mu_4 = 13.3237, \quad \mu_5 = 16.4706,$
$\mu_6 = 19.6159, \quad \mu_7 = 22.7601, \quad \mu_8 = 25.9037, \quad \mu_9 = 29.0468, \quad \mu_{10} = 32.1897.$

As $n \to \infty$, we have $\mu_{n+1} - \mu_n \to \pi$.

Example 2. The initial temperature of the cylinder is uniform, $f(r) = w_0$. The lateral surface is maintained at constant thermal flux, $g(t) = g_R$.
Solution:
$$w(r,t) = w_0 + g_R R\left[2\frac{at}{R^2} - \frac{1}{4} + \frac{r^2}{2R^2} - \sum_{n=1}^{\infty} \frac{2}{\mu_n^2 J_0(\mu_n)} \exp\!\left(-\mu_n^2 \frac{at}{R^2}\right) J_0\!\left(\mu_n \frac{r}{R}\right)\right].$$

⊙ *Reference*: H. S. Carslaw and J. C. Jaeger (1984).

1.2.1-5. Domain: $0 \leq r \leq R$. Third boundary value problem.

The following conditions are prescribed:

$$w = f(r) \quad \text{at} \quad t = 0 \quad \text{(initial condition)},$$
$$\partial_r w + kw = g(t) \quad \text{at} \quad r = R \quad \text{(boundary condition)},$$
$$|w| \neq \infty \quad \text{at} \quad r = 0 \quad \text{(boundedness condition)}.$$

Solution:

$$w(r,t) = \int_0^R f(\xi) G(r,\xi,t)\, d\xi + a \int_0^t g(\tau) G(r,R,t-\tau)\, d\tau.$$

Here,

$$G(r,\xi,t) = \frac{2}{R^2} \sum_{n=1}^\infty \frac{\mu_n^2 \xi}{(k^2 R^2 + \mu_n^2) J_0^2(\mu_n)} J_0\!\left(\frac{\mu_n r}{R}\right) J_0\!\left(\frac{\mu_n \xi}{R}\right) \exp\!\left(-\frac{a\mu_n^2 t}{R^2}\right),$$

where the μ_n are positive roots of the transcendental equation

$$\mu J_1(\mu) - kR J_0(\mu) = 0.$$

The numerical values of the first six roots μ_n can be found in Carslaw and Jaeger (1984).

Example 3. The initial temperature of the cylinder is uniform, $f(r) = w_0$. The temperature of the environment is also uniform and is equal to w_R, which corresponds to $g(t) = kw_R$.

Solution:

$$\frac{w(r,t) - w_0}{w_R - w_0} = 1 - \sum_{n=1}^\infty A_n \exp\!\left(-\frac{a\mu_n^2 t}{R^2}\right) J_0\!\left(\frac{\mu_n r}{R}\right), \quad A_n = \frac{2kR}{(k^2 R^2 + \mu_n^2) J_0(\mu_n)}.$$

⦿ *References*: A. V. Lykov (1967), B. M. Budak, A. A. Samarskii, and A. N. Tikhonov (1980), H. S. Carslaw and J. C. Jaeger (1984).

1.2.1-6. Domain: $R_1 \leq r \leq R_2$. First boundary value problem.

The following conditions are prescribed:

$$w = f(r) \quad \text{at} \quad t = 0 \quad \text{(initial condition)},$$
$$w = g_1(t) \quad \text{at} \quad r = R_1 \quad \text{(boundary condition)},$$
$$w = g_2(t) \quad \text{at} \quad r = R_2 \quad \text{(boundary condition)}.$$

Solution:

$$w(r,t) = \int_{R_1}^{R_2} f(\xi) G(r,\xi,t)\, d\xi + a \int_0^t g_1(\tau) \Lambda_1(r,t-\tau)\, d\tau - a \int_0^t g_2(\tau) \Lambda_2(r,t-\tau)\, d\tau.$$

Here,

$$G(r,\xi,t) = \frac{\pi^2}{2R_1^2} \sum_{n=1}^\infty \frac{\mu_n^2 J_0^2(s\mu_n) \xi}{J_0^2(\mu_n) - J_0^2(s\mu_n)} \Psi_n(r) \Psi_n(\xi) \exp\!\left(-\frac{a\mu_n^2 t}{R_1^2}\right),$$

$$\Psi_n(r) = Y_0(\mu_n) J_0\!\left(\frac{\mu_n r}{R_1}\right) - J_0(\mu_n) Y_0\!\left(\frac{\mu_n r}{R_1}\right), \quad s = \frac{R_2}{R_1},$$

$$\Lambda_1(r,t) = \frac{\partial}{\partial \xi} G(r,\xi,t)\bigg|_{\xi=R_1}, \quad \Lambda_2(r,t) = \frac{\partial}{\partial \xi} G(r,\xi,t)\bigg|_{\xi=R_2},$$

where $J_0(z)$ and $Y_0(z)$ are the Bessel functions; the μ_n are positive roots of the transcendental equation

$$J_0(\mu) Y_0(s\mu) - J_0(s\mu) Y_0(\mu) = 0.$$

The numerical values of the first five roots $\mu_n = \mu_n(s)$ range in the interval $1.4 \leq s \leq 4.0$ and can be found in Carslaw and Jaeger (1984). See also Abramowitz and Stegun (1964).

Example 4. The initial temperature of the hollow cylinder is zero, and its interior and exterior surfaces are held all the time at constant temperatures, $g_1(t) = w_1$ and $g_2(t) = w_2$.

Solution:

$$w(r,t) = \frac{1}{\ln s}\left(w_1 \ln \frac{R_2}{r} + w_2 \ln \frac{r}{R_1}\right) - \pi \sum_{n=1}^{\infty} \frac{J_0(\mu_n)[w_2 J_0(\mu_n) - w_1 J_0(s\mu_n)]}{J_0^2(\mu_n) - J_0^2(s\mu_n)} \exp\left(-\frac{a\mu_n^2 t}{R_1^2}\right) \Psi_n(r).$$

⊙ *Reference*: H. S. Carslaw and J. C. Jaeger (1984).

1.2.1-7. Domain: $R_1 \leq r \leq R_2$. Second boundary value problem.

The following conditions are prescribed:

$$\begin{aligned} w &= f(r) & \text{at} \quad t &= 0 & \text{(initial condition)}, \\ \partial_r w &= g_1(t) & \text{at} \quad r &= R_1 & \text{(boundary condition)}, \\ \partial_r w &= g_2(t) & \text{at} \quad r &= R_2 & \text{(boundary condition)}. \end{aligned}$$

Solution:

$$w(r,t) = \int_{R_1}^{R_2} f(\xi) G(r,\xi,t)\,d\xi - a \int_0^t g_1(\tau) G(r,R_1,t-\tau)\,d\tau + a \int_0^t g_2(\tau) G(r,R_2,t-\tau)\,d\tau.$$

Here,

$$G(r,\xi,t) = \frac{2\xi}{R_2^2 - R_1^2} + \frac{\pi^2}{2R_1^2}\sum_{n=1}^{\infty} \frac{\mu_n^2 J_1^2(s\mu_n)\xi}{J_1^2(\mu_n) - J_1^2(s\mu_n)} \Psi_n(r)\Psi_n(\xi)\exp\left(-\frac{a\mu_n^2 t}{R_1^2}\right),$$

$$\Psi_n(r) = Y_1(\mu_n) J_0\left(\frac{\mu_n r}{R_1}\right) - J_1(\mu_n) Y_0\left(\frac{\mu_n r}{R_1}\right), \quad s = \frac{R_2}{R_1},$$

where $J_k(z)$ and $Y_k(z)$ are the Bessel functions ($k = 0, 1$), and the μ_n are positive roots of the transcendental equation

$$J_1(\mu)Y_1(s\mu) - J_1(s\mu)Y_1(\mu) = 0.$$

The numerical values of the first five roots $\mu_n = \mu_n(s)$ can be found in Abramowitz and Stegun (1964).

⊙ *References*: A. V. Lykov (1967), H. S. Carslaw and J. C. Jaeger (1984).

1.2.1-8. Domain: $R_1 \leq r \leq R_2$. Third boundary value problem.

The following conditions are prescribed:

$$\begin{aligned} w &= f(r) & \text{at} \quad t &= 0 & \text{(initial condition)}, \\ \partial_r w - k_1 w &= g_1(t) & \text{at} \quad r &= R_1 & \text{(boundary condition)}, \\ \partial_r w + k_2 w &= g_2(t) & \text{at} \quad r &= R_2 & \text{(boundary condition)}. \end{aligned}$$

Solution:

$$w(r,t) = \int_{R_1}^{R_2} f(\xi) G(r,\xi,t)\,d\xi - a \int_0^t g_1(\tau) G(r,R_1,t-\tau)\,d\tau + a \int_0^t g_2(\tau) G(r,R_2,t-\tau)\,d\tau.$$

Here,

$$G(r,\xi,t) = \frac{\pi^2}{2}\sum_{n=1}^{\infty} \frac{\lambda_n^2}{B_n}[k_2 J_0(\lambda_n R_2) - \lambda_n J_1(\lambda_n R_2)]^2 \xi H_n(r) H_n(\xi) \exp(-a\lambda_n^2 t),$$

where

$$B_n = (\lambda_n^2 + k_2^2)[k_1 J_0(\lambda_n R_1) + \lambda_n J_1(\lambda_n R_1)]^2 - (\lambda_n^2 + k_1^2)[k_2 J_0(\lambda_n R_2) - \lambda_n J_1(\lambda_n R_2)]^2,$$
$$H_n(r) = [k_1 Y_0(\lambda_n R_1) + \lambda_n Y_1(\lambda_n R_1)] J_0(\lambda_n r) - [k_1 J_0(\lambda_n R_1) + \lambda_n J_1(\lambda_n R_1)] Y_0(\lambda_n r),$$

and the λ_n are positive roots of the transcendental equation

$$[k_1 J_0(\lambda R_1) + \lambda J_1(\lambda R_1)][k_2 Y_0(\lambda R_2) - \lambda Y_1(\lambda R_2)]$$
$$- [k_2 J_0(\lambda R_2) - \lambda J_1(\lambda R_2)][k_1 Y_0(\lambda R_1) + \lambda Y_1(\lambda R_1)] = 0.$$

⊙ *References*: A. V. Lykov (1967), H. S. Carslaw and J. C. Jaeger (1984).

1.2.1-9. Domain: $0 \leq r < \infty$. Cauchy type problem.

This problem is encountered in the theory of diffusion wake behind a drop or a solid particle.
    Given the initial condition

$$w = f(r) \quad \text{at} \quad t = 0,$$

the equation has the following bounded solution:

$$w(r,t) = \frac{1}{2a} \int_0^\infty \frac{\xi}{t} \exp\left(-\frac{r^2 + \xi^2}{4at}\right) I_0\left(\frac{r\xi}{2at}\right) f(\xi)\, d\xi,$$

where $I_0(\xi)$ is the modified Bessel function.

⊙ *References*: W. G. L. Sutton (1943), B. M. Budak, A. A. Samarskii, and A. N. Tikhonov (1980), Yu. P. Gupalo, A. D. Polyanin, and Yu. S. Ryazantsev (1985).

1.2.1-10. Domain: $R \leq r < \infty$. Third boundary value problem.

The following conditions are prescribed:

$$w = f(r) \quad \text{at} \quad t = 0 \quad \text{(initial condition)},$$
$$\partial_r w - kw = g(t) \quad \text{at} \quad r = R \quad \text{(boundary condition)},$$
$$|w| \neq \infty \quad \text{at} \quad r \to \infty \quad \text{(boundedness condition)}.$$

Solution:

$$w(r,t) = \int_R^\infty f(\xi) G(r,\xi,t)\, d\xi - a \int_0^t g(\tau) G(r,R,t-\tau)\, d\tau,$$

where

$$G(r,\xi,t) = \xi \int_0^\infty \exp(-au^2 t) F(r,u) F(\xi,u) u\, du,$$
$$F(r,u) = \frac{J_0(ur)[uY_1(uR) + kY_0(uR)] - Y_0(ur)[uJ_1(uR) + kJ_0(uR)]}{\sqrt{[uJ_1(uR) + kJ_0(uR)]^2 + [uY_1(uR) + kY_0(uR)]^2}}.$$

⊙ *Reference*: H. S. Carslaw and J. C. Jaeger (1984).

1.2.2. Equation of the Form $\dfrac{\partial w}{\partial t} = a\left(\dfrac{\partial^2 w}{\partial r^2} + \dfrac{1}{r}\dfrac{\partial w}{\partial r}\right) + \Phi(r,t)$

This equation is encountered in plane problems of heat conduction with heat release (the function Φ is proportional to the amount of heat released per unit time in the volume under consideration). The equation describes one-dimensional unsteady thermal processes having axial symmetry.

1.2.2-1. Domain: $0 \leq r \leq R$. First boundary value problem.

The following conditions are prescribed:

$$w = f(r) \quad \text{at} \quad t = 0 \quad \text{(initial condition)},$$
$$w = g(t) \quad \text{at} \quad r = R \quad \text{(boundary condition)},$$
$$|w| \neq \infty \quad \text{at} \quad r = 0 \quad \text{(boundedness condition)}.$$

Solution:
$$w(r,t) = \int_0^R f(\xi) G(r,\xi,t)\, d\xi - a \int_0^t g(\tau) \Lambda(r, t-\tau)\, d\tau + \int_0^t \int_0^R \Phi(\xi,\tau) G(r,\xi,t-\tau)\, d\xi\, d\tau,$$

where

$$G(r,\xi,t) = \sum_{n=1}^{\infty} \frac{2\xi}{R^2 J_1^2(\mu_n)} J_0\!\left(\mu_n \frac{r}{R}\right) J_0\!\left(\mu_n \frac{\xi}{R}\right) \exp\!\left(-\frac{a\mu_n^2 t}{R^2}\right), \quad \Lambda(r,t) = \left.\frac{\partial}{\partial \xi} G(r,\xi,t)\right|_{\xi=R}.$$

Here, the μ_n are positive zeros of the Bessel function, $J_0(\mu_n) = 0$. The numerical values of the first ten roots μ_n are given in Paragraph 1.2.1-3.

1.2.2-2. Domain: $0 \leq r \leq R$. Second boundary value problem.

The following conditions are prescribed:

$$w = f(r) \quad \text{at} \quad t = 0 \quad \text{(initial condition)},$$
$$\partial_r w = g(t) \quad \text{at} \quad r = R \quad \text{(boundary condition)},$$
$$|w| \neq \infty \quad \text{at} \quad r = 0 \quad \text{(boundedness condition)}.$$

Solution:
$$w(r,t) = \int_0^R f(\xi) G(r,\xi,t)\, d\xi + a \int_0^t g(\tau) G(r,R,t-\tau)\, d\tau + \int_0^t \int_0^R \Phi(\xi,\tau) G(r,\xi,t-\tau)\, d\xi\, d\tau.$$

Here,

$$G(r,\xi,t) = \frac{2}{R^2}\xi + \frac{2}{R^2} \sum_{n=1}^{\infty} \frac{\xi}{J_0^2(\mu_n)} J_0\!\left(\frac{\mu_n r}{R}\right) J_0\!\left(\frac{\mu_n \xi}{R}\right) \exp\!\left(-\frac{a\mu_n^2 t}{R^2}\right),$$

where the μ_n are positive zeros of the first-order Bessel function, $J_1(\mu) = 0$. The numerical values of the first ten roots μ_n can be found in Paragraph 1.2.1-4.

⊙ *References*: A. V. Lykov (1967), H. S. Carslaw and J. C. Jaeger (1984).

1.2.2-3. Domain: $0 \leq r \leq R$. Third boundary value problem.

The following conditions are prescribed:

$$w = f(r) \quad \text{at} \quad t = 0 \quad \text{(initial condition)},$$
$$\partial_r w + kw = g(t) \quad \text{at} \quad r = R \quad \text{(boundary condition)},$$
$$|w| \neq \infty \quad \text{at} \quad r = 0 \quad \text{(boundedness condition)}.$$

Solution:
$$w(r,t) = \int_0^R f(\xi) G(r,\xi,t)\, d\xi + a \int_0^t g(\tau) G(r,R,t-\tau)\, d\tau + \int_0^t \int_0^R \Phi(\xi,\tau) G(r,\xi,t-\tau)\, d\xi\, d\tau.$$

Here,

$$G(r,\xi,t) = \frac{2}{R^2} \sum_{n=1}^{\infty} \frac{\mu_n^2 \xi}{(k^2 R^2 + \mu_n^2) J_0^2(\mu_n)} J_0\!\left(\frac{\mu_n r}{R}\right) J_0\!\left(\frac{\mu_n \xi}{R}\right) \exp\!\left(-\frac{a\mu_n^2 t}{R^2}\right),$$

where the μ_n are positive roots of the transcendental equation

$$\mu J_1(\mu) - kR J_0(\mu) = 0.$$

The numerical values of the first six roots μ_n can be found in Carslaw and Jaeger (1984).

⊙ *References*: A. V. Lykov (1967), H. S. Carslaw and J. C. Jaeger (1984).

1.2.2-4. Domain: $R_1 \leq r \leq R_2$. First boundary value problem.

The following conditions are prescribed:

$$w = f(r) \quad \text{at} \quad t = 0 \quad \text{(initial condition)},$$
$$w = g_1(t) \quad \text{at} \quad r = R_1 \quad \text{(boundary condition)},$$
$$w = g_2(t) \quad \text{at} \quad r = R_2 \quad \text{(boundary condition)}.$$

Solution:

$$w(r,t) = \int_{R_1}^{R_2} f(\xi) G(r,\xi,t)\, d\xi + \int_0^t \int_{R_1}^{R_2} \Phi(\xi,\tau) G(r,\xi,t-\tau)\, d\xi\, d\tau$$
$$+ a \int_0^t g_1(\tau) \Lambda_1(r,t-\tau)\, d\tau - a \int_0^t g_2(\tau) \Lambda_2(r,t-\tau)\, d\tau.$$

Here,

$$G(r,\xi,t) = \sum_{n=1}^{\infty} A_n \xi \Psi_n(r) \Psi_n(\xi) \exp\left(-\frac{a\mu_n^2 t}{R_1^2}\right), \quad \Lambda_1(r,t) = \frac{\partial G}{\partial \xi}\bigg|_{\xi=R_1}, \quad \Lambda_2(r,t) = \frac{\partial G}{\partial \xi}\bigg|_{\xi=R_2},$$

$$A_n = \frac{\pi^2 \mu_n^2 J_0^2(s\mu_n)}{2R_1^2 [J_0^2(\mu_n) - J_0^2(s\mu_n)]}, \quad \Psi_n(r) = Y_0(\mu_n) J_0\left(\frac{\mu_n r}{R_1}\right) - J_0(\mu_n) Y_0\left(\frac{\mu_n r}{R_1}\right), \quad s = \frac{R_2}{R_1},$$

where $J_0(z)$ and $Y_0(z)$ are the Bessel functions, the μ_n are positive roots of the transcendental equation

$$J_0(\mu) Y_0(s\mu) - J_0(s\mu) Y_0(\mu) = 0.$$

The numerical values of the first five roots $\mu_n = \mu_n(s)$ can be found in Carslaw and Jaeger (1984).

1.2.2-5. Domain: $R_1 \leq r \leq R_2$. Second boundary value problem.

The following conditions are prescribed:

$$w = f(r) \quad \text{at} \quad t = 0 \quad \text{(initial condition)},$$
$$\partial_r w = g_1(t) \quad \text{at} \quad r = R_1 \quad \text{(boundary condition)},$$
$$\partial_r w = g_2(t) \quad \text{at} \quad r = R_2 \quad \text{(boundary condition)}.$$

Solution:

$$w(r,t) = \int_{R_1}^{R_2} f(\xi) G(r,\xi,t)\, d\xi + \int_0^t \int_{R_1}^{R_2} \Phi(\xi,\tau) G(r,\xi,t-\tau)\, d\xi\, d\tau$$
$$- a \int_0^t g_1(\tau) G(r,R_1,t-\tau)\, d\tau + a \int_0^t g_2(\tau) G(r,R_2,t-\tau)\, d\tau.$$

Here,

$$G(r,\xi,t) = \frac{2\xi}{R_2^2 - R_1^2} + \frac{\pi^2}{2R_1^2} \sum_{n=1}^{\infty} \frac{\mu_n^2 J_1^2(s\mu_n) \xi}{J_1^2(\mu_n) - J_1^2(s\mu_n)} \Psi_n(r) \Psi_n(\xi) \exp\left(-\frac{a\mu_n^2 t}{R_1^2}\right),$$

$$\Psi_n(r) = Y_1(\mu_n) J_0\left(\frac{\mu_n r}{R_1}\right) - J_1(\mu_n) Y_0\left(\frac{\mu_n r}{R_1}\right), \quad s = \frac{R_2}{R_1},$$

where $J_k(z)$ and $Y_k(z)$ are the Bessel functions of order $k = 0, 1$ and the μ_n are positive roots of the transcendental equation

$$J_1(\mu) Y_1(s\mu) - J_1(s\mu) Y_1(\mu) = 0.$$

The numerical values of the first five roots $\mu_n = \mu_n(s)$ can be found in Abramowitz and Stegun (1964).

1.2.2-6. Domain: $R_1 \leq r \leq R_2$. Third boundary value problem.

The following conditions are prescribed:

$$w = f(r) \quad \text{at} \quad t = 0 \quad \text{(initial condition)},$$
$$\partial_r w - k_1 w = g_1(t) \quad \text{at} \quad r = R_1 \quad \text{(boundary condition)},$$
$$\partial_r w + k_2 w = g_2(t) \quad \text{at} \quad r = R_2 \quad \text{(boundary condition)}.$$

The solution is given by the formula from Paragraph 1.2.1-8 with the additional term

$$\int_0^t \int_{R_1}^{R_2} \Phi(\xi, \tau) G(r, \xi, t - \tau)\, d\xi\, d\tau,$$

which takes into account the nonhomogeneity of the equation.

1.2.2-7. Domain: $0 \leq r < \infty$. Cauchy type problem.

The bounded solution of this equation subject to the initial condition

$$w = f(r) \quad \text{at} \quad t = 0$$

is given by the relations

$$w(r, t) = \int_0^\infty G(r, \xi, t) f(\xi)\, d\xi + \int_0^t \int_0^\infty G(r, \xi, t - \tau) \Phi(\xi, \tau)\, d\xi\, d\tau,$$

$$G(r, \xi, t) = \frac{\xi}{2at} \exp\left(-\frac{r^2 + \xi^2}{4at}\right) I_0\left(\frac{r\xi}{2at}\right),$$

where $I_0(z)$ is the modified Bessel function.

⊙ *References*: W. G. L. Sutton (1943), B. M. Budak, A. A. Samarskii, and A. N. Tikhonov (1980).

1.2.2-8. Domain: $R \leq r < \infty$. Third boundary value problem.

The following conditions are prescribed:

$$w = f(r) \quad \text{at} \quad t = 0 \quad \text{(initial condition)},$$
$$\partial_r w - kw = g(t) \quad \text{at} \quad r = R \quad \text{(boundary condition)},$$
$$|w| \neq \infty \quad \text{at} \quad r \to \infty \quad \text{(boundedness condition)}.$$

The solution is given by the formula from Paragraph 1.2.1-10 with the additional term

$$\int_0^t \int_R^\infty \Phi(\xi, \tau) G(r, \xi, t - \tau)\, d\xi\, d\tau,$$

which takes into account the nonhomogeneity of the equation.

1.2.3. Equation of the Form $\dfrac{\partial w}{\partial t} = a\left(\dfrac{\partial^2 w}{\partial r^2} + \dfrac{2}{r}\dfrac{\partial w}{\partial r}\right)$

This is a sourceless heat equation that describes unsteady heat processes with central symmetry. It is often represented in the equivalent form

$$\frac{\partial w}{\partial t} = \frac{a}{r^2} \frac{\partial}{\partial r}\left(r^2 \frac{\partial w}{\partial r}\right).$$

A similar equation is used for the analysis of the corresponding one-dimensional unsteady diffusion processes.

1.2.3-1. Particular solutions (A, B, and μ are arbitrary constants).

$$w(r) = A + B\frac{1}{r},$$

$$w(r,t) = A + B(r^2 + 6at),$$

$$w(r,t) = A + B(r^4 + 20atr^2 + 60a^2t^2),$$

$$w(r,t) = A + B\left[r^{2n} + \sum_{k=1}^{n} \frac{(2n+1)(2n)\ldots(2n-2k+2)}{k!}(at)^k r^{2n-2k}\right],$$

$$w(r,t) = A + 2aB\frac{t}{r} + Br,$$

$$w(r,t) = Ar^{-1}\exp(a\mu^2 t \pm \mu r) + B,$$

$$w(r,t) = A + \frac{B}{t^{3/2}}\exp\left(-\frac{r^2}{4at}\right),$$

$$w(r,t) = A + \frac{B}{r\sqrt{t}}\exp\left(-\frac{r^2}{4at}\right),$$

$$w(r,t) = Ar^{-1}\exp(-a\mu^2 t)\cos(\mu r) + B,$$

$$w(r,t) = Ar^{-1}\exp(-a\mu^2 t)\sin(\mu r) + B,$$

$$w(r,t) = Ar^{-1}\exp(-\mu r)\cos(\mu r - 2a\mu^2 t) + B,$$

$$w(r,t) = Ar^{-1}\exp(-\mu r)\sin(\mu r - 2a\mu^2 t) + B,$$

$$w(r,t) = \frac{A}{r}\operatorname{erf}\left(\frac{r}{2\sqrt{at}}\right) + B,$$

$$w(r,t) = \frac{A}{r}\operatorname{erfc}\left(\frac{r}{2\sqrt{at}}\right) + B,$$

where n is an arbitrary positive integer.

1.2.3-2. Reduction to a constant coefficient equation. Some formulas.

$1°$. The substitution $u(r,t) = rw(r,t)$ brings the original equation with variable coefficients to the constant coefficient equation

$$\frac{\partial u}{\partial t} = a\frac{\partial^2 u}{\partial r^2},$$

which is discussed in Subsection 1.1.1 in detail.

$2°$. Suppose $w = w(r,t)$ is a solution of the original equation. Then the functions

$$w_1 = Aw(\pm\lambda r, \lambda^2 t + C),$$

$$w_2 = \frac{A}{|\delta + \beta t|^{3/2}}\exp\left[-\frac{\beta r^2}{4a(\delta + \beta t)}\right]w\left(\pm\frac{r}{\delta + \beta t}, \frac{\gamma + \lambda t}{\delta + \beta t}\right), \qquad \lambda\delta - \beta\gamma = 1,$$

where A, C, β, δ, and λ are arbitrary constants, are also solutions of this equation. The second formula may usually be encountered with $\beta = 1$, $\gamma = -1$, and $\delta = \lambda = 0$.

1.2.3-3. Infinite series particular solutions.

A solution containing an arbitrary function of the space variable:

$$w(r,t) = f(r) + \sum_{n=1}^{\infty} \frac{(at)^n}{n!}\boldsymbol{L}^n[f(r)], \qquad \boldsymbol{L} \equiv \frac{d^2}{dr^2} + \frac{2}{r}\frac{d}{dr},$$

where $f(r)$ is any infinitely differentiable function. This solution satisfies the initial condition $w(r, 0) = f(r)$. The sum is finite if $f(r)$ is a polynomial that contains only even powers.

A solution containing an arbitrary function of time:

$$w(r,t) = g(t) + \sum_{n=1}^{\infty} \frac{1}{a^n(2n+1)!} r^{2n} g_t^{(n)}(t),$$

where $g(t)$ is any infinitely differentiable function. This solution is bounded at $r = 0$ and possesses the properties

$$w(0,t) = g(t), \qquad \partial_r w(0,t) = 0.$$

1.2.3-4. Domain: $0 \le r \le R$. First boundary value problem.

The following conditions are prescribed:

$$\begin{aligned} w &= f(r) & \text{at} \quad t &= 0 & \text{(initial condition)}, \\ w &= g(t) & \text{at} \quad r &= R & \text{(boundary condition)}, \\ |w| &\neq \infty & \text{at} \quad r &= 0 & \text{(boundedness condition)}. \end{aligned}$$

Solution:

$$w(r,t) = \frac{2}{R} \sum_{n=1}^{\infty} \frac{1}{r} \sin\left(\frac{\pi n r}{R}\right) \exp\left(-\frac{a\pi^2 n^2 t}{R^2}\right) M_n(t),$$

where

$$M_n(t) = \int_0^R \xi f(\xi) \sin\left(\frac{\pi n \xi}{R}\right) d\xi - (-1)^n a\pi n \int_0^t g(\tau) \exp\left(\frac{a\pi^2 n^2 \tau}{R^2}\right) d\tau.$$

Remark. Using the relation [see Prudnikov, Brychkov, and Marichev (1986)]

$$\sum_{n=1}^{\infty} (-1)^{n-1} \frac{\sin nz}{n} = \frac{z}{2} \qquad (-\pi < z < \pi),$$

we rewrite the solution in the form

$$w(r,t) = g(t) + \frac{2}{Rr} \sum_{n=1}^{\infty} \sin(\lambda_n r) \exp(-a\lambda_n^2 t) H_n(t), \qquad \lambda_n = \frac{\pi n}{R},$$

where

$$\begin{aligned} H_n(t) &= \int_0^R \xi f(\xi) \sin(\lambda_n \xi) \, d\xi + (-1)^n \frac{R}{\lambda_n} g(t) \exp(a\lambda_n^2 t) - (-1)^n a\pi n \int_0^t g(\tau) \exp(a\lambda_n^2 \tau) \, d\tau \\ &= \int_0^R \xi f(\xi) \sin(\lambda_n \xi) \, d\xi + (-1)^n \frac{R}{\lambda_n} g(0) + (-1)^n \frac{R}{\lambda_n} \int_0^t g_\tau'(\tau) \exp(a\lambda_n^2 \tau) \, d\tau. \end{aligned}$$

Example 1. The initial temperature is uniform, $f(r) = w_0$, and the surface of the sphere is maintained at constant temperature, $g(t) = w_R$.
Solution:

$$\frac{w(r,t) - w_R}{w_0 - w_R} = \frac{2R}{\pi r} \sum_{n=1}^{\infty} \frac{(-1)^{n+1}}{n} \sin\left(\frac{\pi n r}{R}\right) \exp\left(-\frac{a\pi^2 n^2 t}{R^2}\right).$$

The average temperature $\overline{w}$ depends on time t as follows:

$$\frac{\overline{w} - w_R}{w_0 - w_R} = \frac{6}{\pi^2} \sum_{n=1}^{\infty} \frac{1}{n^2} \exp\left(-\frac{a\pi^2 n^2 t}{R^2}\right), \qquad \overline{w} = \frac{1}{V} \int_v w \, dv,$$

where V is the volume of the sphere of radius R.

⊙ *Reference*: H. S. Carslaw and J. C. Jaeger (1984).

1.2.3-5. Domain: $0 \leq r \leq R$. Second boundary value problem.

The following conditions are prescribed:

$$w = f(r) \quad \text{at} \quad t = 0 \quad \text{(initial condition)},$$
$$\partial_r w = g(t) \quad \text{at} \quad r = R \quad \text{(boundary condition)},$$
$$|w| \neq \infty \quad \text{at} \quad r = 0 \quad \text{(boundedness condition)}.$$

Solution:

$$w(r,t) = \int_0^R f(\xi) G(r,\xi,t)\, d\xi + a \int_0^t g(\tau) G(r,R,t-\tau)\, d\tau,$$

where

$$G(r,\xi,t) = \frac{3\xi^2}{R^3} + \frac{2\xi}{Rr} \sum_{n=1}^{\infty} \frac{\mu_n^2 + 1}{\mu_n^2} \sin\left(\frac{\mu_n r}{R}\right) \sin\left(\frac{\mu_n \xi}{R}\right) \exp\left(-\frac{a\mu_n^2 t}{R^2}\right).$$

Here, the μ_n are positive roots of the transcendental equation $\tan\mu - \mu = 0$. The first five roots are

$$\mu_1 = 4.4934, \quad \mu_2 = 7.7253, \quad \mu_3 = 10.9041, \quad \mu_4 = 14.0662, \quad \mu_5 = 17.2208.$$

Example 2. The initial temperature of the sphere is uniform, $f(r) = w_0$. The thermal flux at the sphere surface is a maintained constant, $g(t) = g_R$.
Solution:

$$w(r,t) = w_0 + g_R R \left[\frac{3at}{R^2} + \frac{5r^2 - 3R^2}{10R^2} - \sum_{n=1}^{\infty} \frac{2R}{\mu_n^3 \cos(\mu_n)} \frac{1}{r} \sin\left(\frac{\mu_n r}{R}\right) \exp\left(-\frac{a\mu_n^2 t}{R^2}\right) \right].$$

⦿ *References*: A. V. Lykov (1967), A. V. Bitsadze and D. F. Kalinichenko (1985).

1.2.3-6. Domain: $0 \leq r \leq R$. Third boundary value problem.

The following conditions are prescribed:

$$w = f(r) \quad \text{at} \quad t = 0 \quad \text{(initial condition)},$$
$$\partial_r w + kw = g(t) \quad \text{at} \quad r = R \quad \text{(boundary condition)},$$
$$|w| \neq \infty \quad \text{at} \quad r = 0 \quad \text{(boundedness condition)}.$$

Solution:

$$w(r,t) = \int_0^R f(\xi) G(r,\xi,t)\, d\xi + a \int_0^t g(\tau) G(r,R,t-\tau)\, d\tau.$$

Here,

$$G(r,\xi,t) = \frac{2\xi}{Rr} \sum_{n=1}^{\infty} \frac{\mu_n^2 + (kR-1)^2}{\mu_n^2 + kR(kR-1)} \sin\left(\frac{\mu_n r}{R}\right) \sin\left(\frac{\mu_n \xi}{R}\right) \exp\left(-\frac{a\mu_n^2 t}{R^2}\right),$$

where the μ_n are positive roots of the transcendental equation

$$\mu \cot\mu + kR - 1 = 0.$$

The numerical values of the first six roots μ_n can be found in Carslaw and Jaeger (1984).

Example 3. The initial temperature of the sphere is uniform, $f(r) = w_0$. The temperature of the ambient medium is zero, $g(t) = 0$.
Solution:

$$w(r,t) = \frac{2kR^2 w_0}{r} \sum_{n=1}^{\infty} \frac{\sin\mu_n [\mu_n^2 + (kR-1)^2]}{\mu_n^2 [\mu_n^2 + kR(kR-1)]} \sin\left(\frac{\mu_n r}{R}\right) \exp\left(-\frac{a\mu_n^2 t}{R^2}\right),$$

⦿ *Reference*: H. S. Carslaw and J. C. Jaeger (1984).

1.2. Heat Equation with Axial or Central Symmetry and Related Equations

1.2.3-7. Domain: $R_1 \leq r \leq R_2$. First boundary value problem.

The following conditions are prescribed:

$$w = f(r) \quad \text{at} \quad t = 0 \quad \text{(initial condition)},$$
$$w = g_1(t) \quad \text{at} \quad r = R_1 \quad \text{(boundary condition)},$$
$$w = g_2(t) \quad \text{at} \quad r = R_2 \quad \text{(boundary condition)}.$$

Solution:

$$w(r,t) = \int_{R_1}^{R_2} f(\xi) G(r,\xi,t) \, d\xi + a \int_0^t g_1(\tau) \Lambda_1(r, t-\tau) \, d\tau - a \int_0^t g_2(\tau) \Lambda_2(r, t-\tau) \, d\tau,$$

where

$$G(r,\xi,t) = \frac{2\xi}{(R_2 - R_1)r} \sum_{n=1}^{\infty} \sin\left[\frac{\pi n(r - R_1)}{R_2 - R_1}\right] \sin\left[\frac{\pi n(\xi - R_1)}{R_2 - R_1}\right] \exp\left[-\frac{\pi^2 n^2 at}{(R_2 - R_1)^2}\right],$$

$$\Lambda_1(r,t) = \frac{\partial}{\partial \xi} G(r,\xi,t) \bigg|_{\xi = R_1}, \quad \Lambda_2(r,t) = \frac{\partial}{\partial \xi} G(r,\xi,t) \bigg|_{\xi = R_2}.$$

Example 4. The initial temperature is zero. The temperatures of the interior and exterior surfaces of the spherical layer are maintained constants, $g_1(t) = w_1$ and $g_2(t) = w_2$.

Solution:

$$w(r,t) = \frac{R_1 w_1}{r} + \frac{(r - R_1)(R_2 w_2 - R_1 w_1)}{r(R_2 - R_1)} + \frac{2}{r} \sum_{n=1}^{\infty} \frac{(-1)^n R_2 w_2 - R_1 w_1}{\pi n} \sin\left[\frac{\pi n(r - R_1)}{R_2 - R_1}\right] \exp\left[-\frac{\pi^2 n^2 at}{(R_2 - R_1)^2}\right].$$

⊙ *Reference*: H. S. Carslaw and J. C. Jaeger (1984).

1.2.3-8. Domain: $R_1 \leq r \leq R_2$. Second boundary value problem.

The following conditions are prescribed:

$$w = f(r) \quad \text{at} \quad t = 0 \quad \text{(initial condition)},$$
$$\partial_r w = g_1(t) \quad \text{at} \quad r = R_1 \quad \text{(boundary condition)},$$
$$\partial_r w = g_2(t) \quad \text{at} \quad r = R_2 \quad \text{(boundary condition)}.$$

Solution:

$$w(r,t) = \int_{R_1}^{R_2} f(\zeta) G(r,\zeta,t) \, d\zeta - a \int_0^t g_1(\tau) G(r, R_1, t-\tau) \, d\tau + a \int_0^t g_2(\tau) G(r, R_2, t-\tau) \, d\tau.$$

Here,

$$G(r,\xi,t) = \frac{3\xi^2}{R_2^3 - R_1^3} + \frac{2\xi}{(R_2 - R_1)r} \sum_{n=1}^{\infty} \frac{(1 + R_2^2 \lambda_n^2) \Psi_n(r) \Psi_n(\xi) \exp(-a\lambda_n^2 t)}{\lambda_n^2 \left[R_1^2 + R_2^2 + R_1 R_2 (1 + R_1 R_2 \lambda_n^2)\right]},$$

$$\Psi_n(r) = \sin[\lambda_n(r - R_1)] + R_1 \lambda_n \cos[\lambda_n(r - R_1)],$$

where the λ_n are positive roots of the transcendental equation

$$(\lambda^2 R_1 R_2 + 1) \tan[\lambda(R_2 - R_1)] - \lambda(R_2 - R_1) = 0.$$

⊙ *Reference*: H. S. Carslaw and J. C. Jaeger (1984).

1.2.3-9. Domain: $R_1 \leq r \leq R_2$. Third boundary value problem.

The following conditions are prescribed:

$$w = f(r) \quad \text{at} \quad t = 0 \quad \text{(initial condition)},$$
$$\partial_r w - k_1 w = g_1(t) \quad \text{at} \quad r = R_1 \quad \text{(boundary condition)},$$
$$\partial_r w + k_2 w = g_2(t) \quad \text{at} \quad r = R_2 \quad \text{(boundary condition)}.$$

Solution:

$$w(r,t) = \int_{R_1}^{R_2} f(\xi) G(r,\xi,t)\, d\xi - a \int_0^t g_1(\tau) G(r,R_1,t-\tau)\, d\tau + a \int_0^t g_2(\tau) G(r,R_2,t-\tau)\, d\tau.$$

Here,

$$G(r,\xi,t) = \frac{2\xi}{r} \sum_{n=1}^{\infty} \frac{(b_2^2 + R_2^2 \lambda_n^2) \Psi_n(r) \Psi_n(\xi) \exp(-a\lambda_n^2 t)}{(R_2 - R_1)(b_1^2 + R_1^2 \lambda_n^2)(b_2^2 + R_2^2 \lambda_n^2) + (b_1 R_2 + b_2 R_1)(b_1 b_2 + R_1 R_2 \lambda_n^2)},$$

$$\Psi_n(r) = b_1 \sin[\lambda_n(r - R_1)] + R_1 \lambda_n \cos[\lambda_n(r - R_1)], \quad b_1 = k_1 R_1 + 1, \quad b_2 = k_2 R_2 - 1,$$

where the λ_n are positive roots of the transcendental equation

$$(b_1 b_2 - R_1 R_2 \lambda^2) \sin[\lambda(R_2 - R_1)] + \lambda(R_1 b_2 + R_2 b_1) \cos[\lambda(R_2 - R_1)] = 0.$$

⊙ *Reference*: H. S. Carslaw and J. C. Jaeger (1984).

1.2.3-10. Domain: $0 \leq r < \infty$. Cauchy type problem.

The bounded solution of this equation subject to the initial condition

$$w = f(r) \quad \text{at} \quad t = 0$$

has the form

$$w(r,t) = \frac{1}{2r\sqrt{\pi a t}} \int_0^\infty \xi \left\{ \exp\left[-\frac{(r-\xi)^2}{4at}\right] - \exp\left[-\frac{(r+\xi)^2}{4at}\right] \right\} f(\xi)\, d\xi.$$

⊙ *Reference*: A. G. Butkovskiy (1979).

1.2.3-11. Domain: $R \leq r < \infty$. First boundary value problem.

The following conditions are prescribed:

$$w = f(r) \quad \text{at} \quad t = 0 \quad \text{(initial condition)},$$
$$w = g(t) \quad \text{at} \quad r = R \quad \text{(boundary condition)}.$$

Solution:

$$w(r,t) = \frac{1}{2r\sqrt{\pi a t}} \int_R^\infty \xi f(\xi) \left\{ \exp\left[-\frac{(r-\xi)^2}{4at}\right] - \exp\left[-\frac{(r+\xi - 2R)^2}{4at}\right] \right\} d\xi$$
$$+ \frac{2R}{r\sqrt{\pi}} \int_z^\infty g\left(t - \frac{(r-R)^2}{4a\tau^2}\right) \exp(-\tau^2)\, d\tau, \quad z = \frac{r-R}{2\sqrt{at}}.$$

Example 5. The temperature of the ambient medium is uniform at the initial instant $t = 0$ and the boundary of the domain is held at constant temperature, that is, $f(r) = w_0$ and $g(t) = w_R$.
Solution:

$$\frac{w - w_0}{w_R - w_0} = \frac{R}{r} \operatorname{erfc}\left(\frac{r - R}{2\sqrt{at}}\right),$$

where $\operatorname{erfc} z = \dfrac{2}{\sqrt{\pi}} \int_z^\infty \exp(-\xi^2)\, d\xi$ is the error function.

⊙ *Reference*: H. S. Carslaw and J. C. Jaeger (1984).

1.2.4. Equation of the Form $\frac{\partial w}{\partial t} = a\left(\frac{\partial^2 w}{\partial r^2} + \frac{2}{r}\frac{\partial w}{\partial r}\right) + \Phi(r,t)$

This equation is encountered in heat conduction problems with heat release; the function Φ is proportional to the amount of heat released per unit time in the volume under consideration. The equation describes one-dimensional unsteady thermal processes having central symmetry.

The substitution $u(r,t) = rw(r,t)$ brings the original nonhomogeneous equation with variable coefficients to the nonhomogeneous constant coefficient equation

$$\frac{\partial u}{\partial t} = a\frac{\partial^2 u}{\partial r^2} + r\Phi(r,t),$$

which is considered in Subsection 1.1.2.

1.2.4-1. Domain: $0 \leq r \leq R$. First boundary value problem.

The following conditions are prescribed:

$$w = f(r) \quad \text{at} \quad t = 0 \quad \text{(initial condition)},$$
$$w = g(t) \quad \text{at} \quad r = R \quad \text{(boundary condition)},$$
$$w \neq \infty \quad \text{at} \quad r = 0 \quad \text{(boundedness condition)}.$$

Solution:

$$w(r,t) = \int_0^R f(\xi) G(r,\xi,t)\, d\xi - a \int_0^t g(\tau) \Lambda(r, t-\tau)\, d\tau + \int_0^t \int_0^R \Phi(\xi,\tau) G(r,\xi,t-\tau)\, d\xi\, d\tau,$$

where

$$G(r,\xi,t) = \frac{2\xi}{Rr} \sum_{n=1}^{\infty} \sin\left(\frac{n\pi r}{R}\right) \sin\left(\frac{n\pi \xi}{R}\right) \exp\left(-\frac{an^2\pi^2 t}{R^2}\right), \quad \Lambda(r,t) = \frac{\partial}{\partial \xi} G(r,\xi,t)\bigg|_{\xi=R}.$$

1.2.4-2. Domain: $0 \leq r \leq R$. Second boundary value problem.

The following conditions are prescribed:

$$w = f(r) \quad \text{at} \quad t = 0 \quad \text{(initial condition)},$$
$$\partial_r w = g(t) \quad \text{at} \quad r = R \quad \text{(boundary condition)},$$
$$w \neq \infty \quad \text{at} \quad r = 0 \quad \text{(boundedness condition)}.$$

Solution:

$$w(r,t) = \int_0^R f(\xi) G(r,\xi,t)\, d\xi + a \int_0^t g(\tau) G(r,R,t-\tau)\, d\tau + \int_0^t \int_0^R \Phi(\xi,\tau) G(r,\xi,t-\tau)\, d\xi\, d\tau,$$

where

$$G(r,\xi,t) = \frac{3\xi^2}{R^3} + \frac{2\xi}{Rr} \sum_{n=1}^{\infty} \frac{\mu_n^2 + 1}{\mu_n^2} \sin\left(\frac{\mu_n r}{R}\right) \sin\left(\frac{\mu_n \xi}{R}\right) \exp\left(-\frac{a\mu_n^2 t}{R^2}\right).$$

Here, the μ_n are positive roots of the transcendental equation $\tan \mu - \mu = 0$. The values of the first five roots μ_n can be found in Paragraph 1.2.3-5.

⊙ *References*: A. V. Lykov (1967), A. V. Bitsadze and D. F. Kalinichenko (1985).

1.2.4-3. Domain: $0 \leq r \leq R$. Third boundary value problem.

The following conditions are prescribed:

$$w = f(r) \quad \text{at} \quad t = 0 \quad \text{(initial condition)},$$
$$\partial_r w + kw = g(t) \quad \text{at} \quad r = R \quad \text{(boundary condition)},$$
$$w \neq \infty \quad \text{at} \quad r = 0 \quad \text{(boundedness condition)}.$$

Solution:

$$w(r,t) = \int_0^R f(\xi) G(r,\xi,t)\, d\xi + a \int_0^t g(\tau) G(r,R,t-\tau)\, d\tau + \int_0^t \int_0^R \Phi(\xi,\tau) G(r,\xi,t-\tau)\, d\xi\, d\tau.$$

Here,

$$G(r,\xi,t) = \frac{2\xi}{Rr} \sum_{n=1}^{\infty} \frac{\mu_n^2 + (kR-1)^2}{\mu_n^2 + kR(kR-1)} \sin\left(\frac{\mu_n r}{R}\right) \sin\left(\frac{\mu_n \xi}{R}\right) \exp\left(-\frac{a\mu_n^2 t}{R^2}\right),$$

where the μ_n are positive roots of the transcendental equation $\mu \cot \mu + kR - 1 = 0$.

⊙ *Reference*: H. S. Carslaw and J. C. Jaeger (1984).

1.2.4-4. Domain: $R_1 \leq r \leq R_2$. First boundary value problem.

The following conditions are prescribed:

$$w = f(r) \quad \text{at} \quad t = 0 \quad \text{(initial condition)},$$
$$w = g_1(t) \quad \text{at} \quad r = R_1 \quad \text{(boundary condition)},$$
$$w = g_2(t) \quad \text{at} \quad r = R_2 \quad \text{(boundary condition)}.$$

Solution:

$$w(r,t) = \int_{R_1}^{R_2} f(\xi) G(r,\xi,t)\, d\xi + \int_0^t \int_{R_1}^{R_2} \Phi(\xi,\tau) G(r,\xi,t-\tau)\, d\xi\, d\tau$$
$$+ a \int_0^t g_1(\tau) \Lambda_1(r,t-\tau)\, d\tau - a \int_0^t g_2(\tau) \Lambda_2(r,t-\tau)\, d\tau,$$

where

$$G(r,\xi,t) = \frac{2\xi}{(R_2-R_1)r} \sum_{n=1}^{\infty} \sin\left[\frac{\pi n(r-R_1)}{R_2-R_1}\right] \sin\left[\frac{\pi n(\xi-R_1)}{R_2-R_1}\right] \exp\left[-\frac{\pi^2 n^2 a t}{(R_2-R_1)^2}\right],$$

$$\Lambda_1(r,t) = \frac{\partial}{\partial \xi} G(r,\xi,t)\bigg|_{\xi=R_1}, \quad \Lambda_2(r,t) = \frac{\partial}{\partial \xi} G(r,\xi,t)\bigg|_{\xi=R_2}.$$

⊙ *Reference*: H. S. Carslaw and J. C. Jaeger (1984).

1.2.4-5. Domain: $R_1 \leq r \leq R_2$. Second boundary value problem.

The following conditions are prescribed:

$$w = f(r) \quad \text{at} \quad t = 0 \quad \text{(initial condition)},$$
$$\partial_r w = g_1(t) \quad \text{at} \quad r = R_1 \quad \text{(boundary condition)},$$
$$\partial_r w = g_2(t) \quad \text{at} \quad r = R_2 \quad \text{(boundary condition)}.$$

Solution:

$$w(r,t) = \int_{R_1}^{R_2} f(\xi) G(r,\xi,t)\, d\xi + \int_0^t \int_{R_1}^{R_2} \Phi(\xi,\tau) G(r,\xi,t-\tau)\, d\xi\, d\tau$$
$$- a \int_0^t g_1(\tau) G(r,R_1,t-\tau)\, d\tau + a \int_0^t g_2(\tau) G(r,R_2,t-\tau)\, d\tau,$$

where the function $G(r,\xi,t)$ is the same as in Paragraph 1.2.3-8.

⊙ *Reference*: H. S. Carslaw and J. C. Jaeger (1984).

1.2.4-6. Domain: $R_1 \leq r \leq R_2$. Third boundary value problem.

The following conditions are prescribed:

$$w = f(r) \quad \text{at} \quad t = 0 \quad \text{(initial condition)},$$
$$\partial_r w - k_1 w = g_1(t) \quad \text{at} \quad r = R_1 \quad \text{(boundary condition)},$$
$$\partial_r w + k_2 w = g_2(t) \quad \text{at} \quad r = R_2 \quad \text{(boundary condition)}.$$

The solution is given by the formula of Paragraph 1.2.3-9 with the additional term

$$\int_0^t \int_{R_1}^{R_2} \Phi(\xi, \tau) G(r, \xi, t - \tau) \, d\xi \, d\tau,$$

which takes into account the nonhomogeneity of the equation.

1.2.4-7. Domain: $0 \leq r < \infty$. Cauchy type problem.

The bounded solution of this equation subject to the initial condition

$$w = f(r) \quad \text{at} \quad t = 0$$

is given by

$$w(r,t) = \int_0^\infty f(\xi) G(r, \xi, t) \, d\xi + \int_0^t \int_0^\infty \Phi(\xi, \tau) G(r, \xi, t - \tau) \, d\xi \, d\tau,$$

$$G(r, \xi, t) = \frac{\xi}{2r\sqrt{\pi a t}} \left\{ \exp\left[-\frac{(r-\xi)^2}{4at}\right] - \exp\left[-\frac{(r+\xi)^2}{4at}\right] \right\}.$$

⦿ *Reference*: A. G. Butkovskiy (1979).

1.2.4-8. Domain: $R \leq r < \infty$. First boundary value problem.

The following conditions are prescribed:

$$w = f(r) \quad \text{at} \quad t = 0 \quad \text{(initial condition)},$$
$$w = g(t) \quad \text{at} \quad r = R \quad \text{(boundary condition)}.$$

Solution:

$$w(r,t) = \int_R^\infty f(\xi) G(r, \xi, t) \, d\xi + \int_0^t \int_R^\infty \Phi(\xi, \tau) G(r, \xi, t - \tau) \, d\xi \, d\tau + a \int_0^t g(\tau) \Lambda(r, t - \tau) \, d\tau,$$

where

$$G(r, \xi, t) = \frac{\xi}{2r\sqrt{\pi a t}} \left\{ \exp\left[-\frac{(r-\xi)^2}{4at}\right] - \exp\left[-\frac{(r+\xi-2R)^2}{4at}\right] \right\}, \quad \Lambda(r,t) = \frac{\partial}{\partial \xi} G(r, \xi, t) \bigg|_{\xi=R}.$$

⦿ *Reference*: H. S. Carslaw and J. C. Jaeger (1984).

1.2.5. Equation of the Form $\dfrac{\partial w}{\partial t} = \dfrac{\partial^2 w}{\partial x^2} + \dfrac{1 - 2\beta}{x} \dfrac{\partial w}{\partial x}$

This dimensionless equation is encountered in problems of the diffusion boundary layer. For $\beta = 0$, $\beta = \frac{1}{2}$, or $\beta = -\frac{1}{2}$, see the equations of Subsections 1.2.1, 1.1.1, or 1.2.3, respectively.

1.2.5-1. Particular solutions (A, B, and μ are arbitrary constants).

$$w(x) = A + Bx^{2\beta},$$

$$w(x,t) = A + 4(1-\beta)Bt + Bx^2,$$

$$w(x,t) = A + 16(2-\beta)(1-\beta)Bt^2 + 8(2-\beta)Btx^2 + Bx^4,$$

$$w(x,t) = x^{2n} + \sum_{p=1}^{n} \frac{4^p}{p!} S_{n,p} S_{n-\beta,p} t^p x^{2(n-p)}, \quad S_{q,p} = q(q-1)\ldots(q-p+1),$$

$$w(x,t) = A + 4(1+\beta)Btx^{2\beta} + Bx^{2\beta+2},$$

$$w(x,t) = A + Bt^{\beta-1} \exp\left(-\frac{x^2}{4t}\right),$$

$$w(x,t) = A + B\frac{x^{2\beta}}{t^{\beta+1}} \exp\left(-\frac{x^2}{4t}\right),$$

$$w(x,t) = A + B\gamma\left(\beta, \frac{x^2}{4t}\right),$$

$$w(x,t) = A + B\exp(-\mu^2 t)x^\beta J_\beta(\mu x),$$

$$w(x,t) = A + B\exp(-\mu^2 t)x^\beta Y_\beta(\mu x),$$

$$w(x,t) = A + B\frac{x^\beta}{t} \exp\left(-\frac{x^2+\mu^2}{4t}\right) I_\beta\left(\frac{\mu x}{2t}\right),$$

$$w(x,t) = A + B\frac{x^\beta}{t} \exp\left(-\frac{x^2+\mu^2}{4t}\right) I_{-\beta}\left(\frac{\mu x}{2t}\right),$$

$$w(x,t) = A + B\frac{x^\beta}{t} \exp\left(-\frac{x^2+\mu^2}{4t}\right) K_\beta\left(\frac{\mu x}{2t}\right),$$

where n is an arbitrary positive integer, $\gamma(\beta, z)$ is the incomplete gamma function, $J_\beta(z)$ and $Y_\beta(z)$ are the Bessel functions, and $I_\beta(z)$ and $K_\beta(z)$ are the modified Bessel functions.

⊙ *References*: W. G. L. Sutton (1943), A. D. Polyanin (2001a).

1.2.5-2. Infinite series solutions.

A solution containing an arbitrary function of the space variable:

$$w(x,t) = f(x) + \sum_{n=1}^{\infty} \frac{1}{n!} t^n \boldsymbol{L}^n[f(x)], \quad \boldsymbol{L} \equiv \frac{d^2}{dx^2} + \frac{1-2\beta}{x}\frac{d}{dx},$$

where $f(x)$ is any infinitely differentiable function. This solution satisfies the initial condition $w(x,0) = f(x)$. The sum is finite if $f(x)$ is a polynomial that contains only even powers.

A solution containing an arbitrary function of time:

$$w(x,t) = g(t) + \sum_{n=1}^{\infty} \frac{1}{4^n n! (1-\beta)(2-\beta)\ldots(n-\beta)} x^{2n} g_t^{(n)}(t),$$

where $g(t)$ any infinitely differentiable function. This solution is bounded at $x = 0$ and possesses the properties

$$w(0,t) = g(t), \qquad \partial_x w(0,t) = 0.$$

1.2.5-3. Formulas and transformations for constructing particular solutions.

Suppose $w = w(x, t)$ is a solution of the original equation. Then the functions

$$w_1 = Aw(\pm \lambda x, \lambda^2 t + C),$$

$$w_2 = A|a + bt|^{\beta-1} \exp\left[-\frac{bx^2}{4(a+bt)}\right] w\left(\pm \frac{x}{a+bt}, \frac{c+kt}{a+bt}\right), \qquad ak - bc = 1,$$

where A, C, a, b, and c are arbitrary constants, are also solutions of this equation. The second formula usually may be encountered with $a = k = 0$, $b = 1$, and $c = -1$.

The substitution $w = x^{2\beta} u(x,t)$ brings the equation with parameter β to an equation of the same type with parameter $-\beta$:

$$\frac{\partial u}{\partial t} = \frac{\partial^2 u}{\partial x^2} + \frac{1+2\beta}{x} \frac{\partial u}{\partial x}.$$

1.2.5-4. Domain: $0 \leq x < \infty$. First boundary value problem.

The following conditions are prescribed:

$$w = f(x) \quad \text{at} \quad t = 0 \quad \text{(initial condition)},$$
$$w = g(t) \quad \text{at} \quad x = 0 \quad \text{(boundary condition)}.$$

Solution for $0 < \beta < 1$:

$$w(x,t) = \frac{x^\beta}{2t} \int_0^\infty f(\xi) \xi^{1-\beta} \exp\left(-\frac{x^2+\xi^2}{4t}\right) I_\beta\left(\frac{\xi x}{2t}\right) d\xi$$
$$+ \frac{x^{2\beta}}{2^{2\beta+1} \Gamma(\beta+1)} \int_0^t g(\tau) \exp\left[-\frac{x^2}{4(t-\tau)}\right] \frac{d\tau}{(t-\tau)^{1+\beta}}.$$

Example. For $f(x) = w_0$ and $g(t) = w_1$, where $f(x) = \text{const}$ and $g(t) = \text{const}$, we have

$$w = \frac{(w_0 - w_1)}{\Gamma(\beta)} \gamma\left(\beta, \frac{x^2}{4t}\right) + w_1, \qquad \gamma(\beta, z) = \int_0^z \xi^{\beta-1} e^{-\xi} d\xi.$$

Here, $\gamma(\beta, z)$ is the incomplete gamma function and $\Gamma(\beta) = \gamma(\beta, \infty)$ is the gamma function.

⊙ *Reference*: W. G. L. Sutton (1943).

1.2.5-5. Domain: $0 \leq x < \infty$. Second boundary value problem.

The following conditions are prescribed:

$$w = f(x) \quad \text{at} \quad t = 0 \quad \text{(initial condition)},$$
$$\left(x^{1-2\beta} \partial_x w\right) = g(t) \quad \text{at} \quad x = 0 \quad \text{(boundary condition)}.$$

Solution for $0 < \beta < 1$:

$$w(x,t) = \frac{x^\beta}{2t} \int_0^\infty f(\xi) \xi^{1-\beta} \exp\left(-\frac{x^2+\xi^2}{4t}\right) I_{-\beta}\left(\frac{\xi x}{2t}\right) d\xi$$
$$- \frac{2^{2\beta-1}}{\Gamma(1-\beta)} \int_0^t g(\tau) \exp\left[-\frac{x^2}{4(t-\tau)}\right] \frac{d\tau}{(t-\tau)^{1-\beta}}.$$

1.2.5-6. Domain: $0 \le x < \infty$. Third boundary value problem.

The following conditions are prescribed:
$$w = 0 \quad \text{at} \quad t = 0 \quad \text{(initial condition)},$$
$$\left[x^{1-2\beta}\partial_x w + k(w_0 - w)\right] = 0 \quad \text{at} \quad x = 0 \quad \text{(boundary condition)}.$$

Solution for $0 < \beta < 1$:
$$w(x,t) = w_0 \frac{2^{2\beta-1}k}{\Gamma(1-\beta)} \int_0^t \varphi(\tau) \exp\left[-\frac{x^2}{4(t-\tau)}\right] \frac{d\tau}{(t-\tau)^{1-\beta}},$$

where the function $\varphi(t)$ is given as the series
$$\varphi(t) = \sum_{n=0}^{\infty} \frac{(-\mu t^\beta)^n}{\Gamma(n\beta+1)}, \qquad \mu = \frac{2^{2\beta-1}k\Gamma(\beta)}{\Gamma(1-\beta)},$$

which is convergent for any t.

⊙ *Reference*: W. G. L. Sutton (1943).

1.2.6. Equation of the Form $\dfrac{\partial w}{\partial t} = \dfrac{\partial^2 w}{\partial x^2} + \dfrac{1-2\beta}{x}\dfrac{\partial w}{\partial x} + \Phi(x,t)$

This equation is encountered in problems of a diffusion boundary layer with sources/sinks of substance. For $\beta = 0$, $\beta = \frac{1}{2}$, or $\beta = -\frac{1}{2}$, see the equations of Subsections 1.2.2, 1.1.2, or 1.2.4, respectively.

1.2.6-1. Domain: $0 \le x < \infty$. First boundary value problem.

The following conditions are prescribed:
$$w = f(x) \quad \text{at} \quad t = 0 \quad \text{(initial condition)},$$
$$w = g(t) \quad \text{at} \quad x = 0 \quad \text{(boundary condition)}.$$

Solution for $0 < \beta < 1$:
$$w(x,t) = \frac{x^\beta}{2t} \int_0^\infty f(\xi)\xi^{1-\beta} \exp\left(-\frac{x^2+\xi^2}{4t}\right) I_\beta\left(\frac{\xi x}{2t}\right) d\xi$$
$$+ \frac{x^{2\beta}}{2^{2\beta+1}\Gamma(\beta+1)} \int_0^t g(\tau) \exp\left[-\frac{x^2}{4(t-\tau)}\right] \frac{d\tau}{(t-\tau)^{1+\beta}}$$
$$+ \frac{1}{2} \int_0^t \int_0^\infty \Phi(\xi,\tau) \frac{x^\beta \xi^{1-\beta}}{t-\tau} \exp\left[-\frac{x^2+\xi^2}{4(t-\tau)}\right] I_\beta\left(\frac{\xi x}{2(t-\tau)}\right) d\xi\, d\tau.$$

1.2.6-2. Domain: $0 \le x < \infty$. Second boundary value problem.

The following conditions are prescribed:
$$w = f(x) \quad \text{at} \quad t = 0 \quad \text{(initial condition)},$$
$$\left(x^{1-2\beta}\partial_x w\right) = g(t) \quad \text{at} \quad x = 0 \quad \text{(boundary condition)}.$$

Solution for $0 < \beta < 1$:
$$w(x,t) = \frac{x^\beta}{2t} \int_0^\infty f(\xi)\xi^{1-\beta} \exp\left(-\frac{x^2+\xi^2}{4t}\right) I_{-\beta}\left(\frac{\xi x}{2t}\right) d\xi$$
$$- \frac{2^{2\beta-1}}{\Gamma(1-\beta)} \int_0^t g(\tau) \exp\left[-\frac{x^2}{4(t-\tau)}\right] \frac{d\tau}{(t-\tau)^{1-\beta}}$$
$$+ \frac{1}{2} \int_0^t \int_0^\infty \Phi(\xi,\tau) \frac{x^\beta \xi^{1-\beta}}{t-\tau} \exp\left[-\frac{x^2+\xi^2}{4(t-\tau)}\right] I_{-\beta}\left(\frac{\xi x}{2(t-\tau)}\right) d\xi\, d\tau.$$

⊙ *Reference for Subsection* 1.2.6: W. G. L. Sutton (1943).

1.3. Equations Containing Power Functions and Arbitrary Parameters

1.3.1. Equations of the Form $\frac{\partial w}{\partial t} = a\frac{\partial^2 w}{\partial x^2} + f(x,t)w$

1.3.1-1. The function f depends on the space coordinate x alone.

Such equations are encountered in problems of heat and mass transfer with heat release (or volume chemical reaction). The one-dimensional Schrödinger equation can be reduced to this form by the change of variable $t \to -iht$ [the function $-f(x)$ describes the potential against the space coordinate; see Subsection 1.9.2].

1. $\dfrac{\partial w}{\partial t} = a\dfrac{\partial^2 w}{\partial x^2} + (bx + c)w.$

This equation is a special case of equation 1.8.9 with $s(x)=1$, $p(x)=a$, $q(x)=-bx-c$, and $\Phi(x,t)=0$. Also, it is a special case of equation 1.8.1.6 with $f(t)=b$ and $g(t)=c$.

1°. Particular solutions (A and μ are arbitrary constants):

$$w(x,t) = A\exp\left(btx + \tfrac{1}{3}ab^2t^3 + ct\right),$$

$$w(x,t) = A(x + abt^2)\exp\left(btx + \tfrac{1}{3}ab^2t^3 + ct\right),$$

$$w(x,t) = A\exp\left[x(bt+\mu) + \tfrac{1}{3}ab^2t^3 + ab\mu t^2 + (a\mu^2+c)t\right],$$

$$w(x,t) = A\exp(-\mu t)\sqrt{\xi}\,J_{1/3}\left(\frac{2}{3b\sqrt{a}}\xi^{3/2}\right), \quad \xi = bx+c+\mu,$$

$$w(x,t) = A\exp(-\mu t)\sqrt{\xi}\,Y_{1/3}\left(\frac{2}{3b\sqrt{a}}\xi^{3/2}\right), \quad \xi = bx+c+\mu,$$

where $J_{1/3}(z)$ and $Y_{1/3}(z)$ are the Bessel functions of the first and second kind of order $1/3$.

2°. The transformation

$$w(x,t) = u(z,t)\exp\left(btx + \tfrac{1}{3}ab^2t^3 + ct\right), \quad z = x + abt^2$$

leads to a constant coefficient equation, $\partial_t u = a\partial_{zz}u$, which is considered in Subsection 1.1.1.

3°. Domain: $-\infty < x < \infty$. Cauchy problem.
An initial condition is prescribed:

$$w = f(x) \quad \text{at} \quad t = 0.$$

Solution:

$$w(x,t) = \frac{1}{2\sqrt{\pi a t}}\exp\left(btx + \tfrac{1}{3}ab^2t^3 + ct\right)\int_{-\infty}^{\infty}\exp\left[-\frac{(x+abt^2-\xi)^2}{4at}\right]f(\xi)\,d\xi.$$

⊙ *Reference*: A. D. Polyanin, A. V. Vyazmin, A. I. Zhurov, and D. A. Kazenin (1998), see also W. Miller, Jr. (1977).

2. $\dfrac{\partial w}{\partial t} = a\dfrac{\partial^2 w}{\partial x^2} - (bx^2 + c)w, \quad b > 0.$

This equation is a special case of equation 1.8.9 with $s(x)=1$, $p(x)=a$, $q(x)=bx^2+c$, and $\Phi(x,t)=0$. In addition, it is a special case of equation 1.8.1.7 with $f(t)=-c$.

1°. Particular solutions (A and μ are arbitrary constants):

$$w(x,t) = A\exp\left[(\sqrt{ab}-c)t + \frac{\sqrt{b}}{2\sqrt{a}}x^2\right],$$

$$w(x,t) = A\exp\left[-(\sqrt{ab}+c)t - \frac{\sqrt{b}}{2\sqrt{a}}x^2\right],$$

$$w(x,t) = A\exp\left(-\mu t - \frac{\sqrt{b}}{2\sqrt{a}}x^2\right)\Phi\left(\frac{c-\mu}{4\sqrt{ab}} + \frac{1}{4}, \frac{1}{2}; \sqrt{\frac{b}{a}}x^2\right),$$

$$w(x,t) = A\exp\left(-\mu t - \frac{\sqrt{b}}{2\sqrt{a}}x^2\right)x\Phi\left(\frac{c-\mu}{4\sqrt{ab}} + \frac{3}{4}, \frac{3}{2}; \sqrt{\frac{b}{a}}x^2\right),$$

where $\Phi(\alpha,\beta;z) = 1 + \sum_{m=1}^{\infty}\frac{\alpha(\alpha+1)\ldots(\alpha+m-1)}{\beta(\beta+1)\ldots(\beta+m-1)}\frac{z^m}{m!}$ is the degenerate hypergeometric function.

2°. In quantum mechanics the following particular solution is encountered:

$$w(x,t) = e^{-[c+\sqrt{ab}\,(2n+1)]t}\psi_n(\xi), \quad \psi_n(\xi) = \frac{1}{\pi^{1/4}\sqrt{2^n n!\,x_0}}e^{-\frac{1}{2}\xi^2}H_n(\xi), \quad \xi = \frac{x}{x_0}, \quad x_0 = \left(\frac{a}{b}\right)^{1/4},$$

where $H_n(\xi) = (-1)^n e^{\xi^2}\frac{d^n}{d\xi^n}\left(e^{-\xi^2}\right)$ are the Hermite polynomials, $n = 0, 1, 2, \ldots$ These solutions satisfy the normalization condition

$$\int_{-\infty}^{\infty}|\psi_n(\xi)|^2 d\xi = 1.$$

3°. The transformation (A is any number)

$$w(x,t) = u(z,\tau)\exp\left[\frac{\sqrt{b}}{2\sqrt{a}}x^2 + (\sqrt{ab}-c)t\right], \quad z = x\exp(2\sqrt{ab}\,t), \quad \tau = \frac{\sqrt{a}}{4\sqrt{b}}\exp(4\sqrt{ab}\,t) + A,$$

leads to the constant coefficient equation $\partial_\tau u = \partial_{zz}u$, which is considered in Subsection 1.1.1.

⊙ *Reference*: A. D. Polyanin, A. V. Vyazmin, A. I. Zhurov, and D. A. Kazenin (1998), see also W. Miller, Jr. (1977).

3. $\dfrac{\partial w}{\partial t} = a\dfrac{\partial^2 w}{\partial x^2} + (bx^2 - c)w, \quad b > 0.$

This is a special case of equation 1.8.9 with $s(x) = 1$, $p(x) = a$, $q(x) = c - bx^2$, and $\Phi(x,t) = 0$. The transformation

$$w(x,t) = \frac{1}{\sqrt{|\cos(2\sqrt{ab}\,t)|}}\exp\left[\frac{\sqrt{b}}{2\sqrt{a}}x^2\tan(2\sqrt{ab}\,t) - ct\right]u(z,\tau),$$

$$z = \frac{x}{\cos(2\sqrt{ab}\,t)}, \quad \tau = \frac{\sqrt{a}}{2\sqrt{b}}\tan(2\sqrt{ab}\,t),$$

leads to the constant coefficient equation $\partial_\tau u = \partial_{zz}u$, which is considered in Subsection 1.1.1.

4. $\dfrac{\partial w}{\partial t} = a\dfrac{\partial^2 w}{\partial x^2} + (bx^2 + cx + k)w.$

The substitution $z = x + c/(2b)$ leads to an equation of the form 1.3.1.2 (for $b < 0$) or 1.3.1.3 (for $b > 0$).

1.3. Equations Containing Power Functions and Arbitrary Parameters

5. $\dfrac{\partial w}{\partial t} = a\dfrac{\partial^2 w}{\partial x^2} + \left(b + cx^{-2}\right)w.$

1°. Particular solutions

$$w(x,t) = e^{(b-a\mu^2)t}\sqrt{x}\left[AJ_\nu(\mu x) + BY_\nu(\mu x)\right], \qquad \nu^2 = \dfrac{1}{4} - \dfrac{c}{a},$$

where $J_\nu(z)$ and $Y_\nu(z)$ are the Bessel functions; A, B, and μ are arbitrary constants.

2°. Domain: $0 \le x < \infty$. First boundary value problem.
The following conditions are prescribed:

$$\begin{aligned} w &= f(x) &&\text{at}\quad t = 0 &&\text{(initial condition)},\\ w &= 0 &&\text{at}\quad x = 0 &&\text{(boundary condition)}. \end{aligned}$$

Solution:

$$w(x,t) = \dfrac{e^{bt}}{2at}\int_0^\infty \sqrt{x\xi}\,\exp\!\left(-\dfrac{x^2+\xi^2}{4at}\right) I_\nu\!\left(\dfrac{\xi x}{2at}\right) f(\xi)\,d\xi, \qquad \nu^2 = \dfrac{1}{4} - \dfrac{c}{a},$$

where $-\tfrac{3}{4}a < c < \tfrac{1}{4}a$.

3°. The transformation

$$w(x,t) = e^{bt} x^k u(x,\tau), \qquad \tau = at,$$

where k is a root of the quadratic equation $ak^2 - ak + c = 0$, leads to an equation of the form 1.2.5:

$$\dfrac{\partial u}{\partial \tau} = \dfrac{\partial^2 u}{\partial x^2} + \dfrac{2k}{x}\dfrac{\partial u}{\partial x}.$$

See also Miller, Jr. (1977).

6. $\dfrac{\partial w}{\partial t} = a\dfrac{\partial^2 w}{\partial x^2} + \left(-bx^2 + c + kx^{-2}\right)w, \qquad b > 0.$

This is a special case of equation 1.8.1.2 with $f(x) = -bx^2 - c + kx^{-2}$. The transformation ($A$ is any number)

$$w(x,t) = u(z,\tau)\exp\!\left[\dfrac{\sqrt{b}}{2\sqrt{a}}x^2 + (\sqrt{ab}+c)t\right], \quad z = x\exp(2\sqrt{ab}\,t), \quad \tau = \dfrac{1}{4\sqrt{ab}}\exp(4\sqrt{ab}\,t) + A,$$

leads to an equation of the form 1.3.1.5:

$$\dfrac{\partial u}{\partial \tau} = \dfrac{\partial^2 u}{\partial z^2} + \dfrac{k}{a}z^{-2}u.$$

See also Miller, Jr. (1977).

7. $\dfrac{\partial w}{\partial t} = a\dfrac{\partial^2 w}{\partial x^2} + (bx^2 - c + kx^{-2})w, \qquad b > 0.$

This is a special case of equation 1.8.1.2 with $f(x) = bx^2 - c + kx^{-2}$. The transformation

$$w(x,t) = \dfrac{1}{\sqrt{|\cos(2\sqrt{ab}\,t)|}}\exp\!\left[\dfrac{\sqrt{b}}{2\sqrt{a}}x^2\tan(2\sqrt{ab}\,t) - ct\right] u(z,\tau),$$

$$z = \dfrac{x}{\cos(2\sqrt{ab}\,t)}, \qquad \tau = \dfrac{\sqrt{a}}{2\sqrt{b}}\tan(2\sqrt{ab}\,t),$$

leads to an equation of the form 1.3.1.5:

$$\dfrac{\partial u}{\partial \tau} = \dfrac{\partial^2 u}{\partial z^2} + \dfrac{k}{a}z^{-2}u.$$

1.3.1-2. The function f depends on time t alone.

8. $\dfrac{\partial w}{\partial t} = a\dfrac{\partial^2 w}{\partial x^2} + (bt+c)w.$

This is a special case of equation 1.8.1.1 with $f(t) = bt + c$.

1°. Particular solutions (A, B, and μ are arbitrary constants):
$$w(x,t) = (Ax+B)\exp\left(\tfrac{1}{2}bt^2 + ct\right),$$
$$w(x,t) = A(x^2+2at)\exp\left(\tfrac{1}{2}bt^2 + ct\right),$$
$$w(x,t) = A\exp\left[\mu x + \tfrac{1}{2}bt^2 + (c+a\mu^2)t\right],$$
$$w(x,t) = A\exp\left[\tfrac{1}{2}bt^2 + (c-a\mu^2)t\right]\cos(\mu x),$$
$$w(x,t) = A\exp\left[\tfrac{1}{2}bt^2 + (c-a\mu^2)t\right]\sin(\mu x).$$

2°. The substitution $w(x,t) = u(x,t)\exp\left(\tfrac{1}{2}bt^2 + ct\right)$ leads to a constant coefficient equation, $\partial_t u = a\partial_{xx}u$, which is considered in Subsection 1.1.1.

9. $\dfrac{\partial w}{\partial t} = a\dfrac{\partial^2 w}{\partial x^2} + bt^k w.$

This is a special case of equation 1.8.1.1 with $f(t) = bt^k$.

1°. Particular solutions (A, B, and μ are arbitrary constants):
$$w(x,t) = (Ax+B)\exp\left(\dfrac{b}{k+1}t^{k+1}\right),$$
$$w(x,t) = A(x^2+2at)\exp\left(\dfrac{b}{k+1}t^{k+1}\right),$$
$$w(x,t) = A\exp\left(\mu x + a\mu^2 t + \dfrac{b}{k+1}t^{k+1}\right),$$
$$w(x,t) = A\exp\left(\dfrac{b}{k+1}t^{k+1} - a\mu^2 t\right)\cos(\mu x),$$
$$w(x,t) = A\exp\left(\dfrac{b}{k+1}t^{k+1} - a\mu^2 t\right)\sin(\mu x).$$

2°. The substitution $w(x,t) = u(x,t)\exp\left(\tfrac{b}{k+1}t^{k+1}\right)$ leads to a constant coefficient equation, $\partial_t u = a\partial_{xx}u$, which is considered in Subsection 1.1.1.

1.3.1-3. The function f depends on both x and t.

10. $\dfrac{\partial w}{\partial t} = a\dfrac{\partial^2 w}{\partial x^2} + (bx + ct + d)w.$

This is a special case of equation 1.8.1.6 with $f(t) = b$ and $g(t) = ct + d$.

1°. Particular solutions (A and μ are arbitrary constants):
$$w(x,t) = A\exp\left(btx + \tfrac{1}{3}ab^2 t^3 + \tfrac{1}{2}ct^2 + dt\right),$$
$$w(x,t) = A(x+abt^2)\exp\left(btx + \tfrac{1}{3}ab^2 t^3 + \tfrac{1}{2}ct^2 + dt\right),$$
$$w(x,t) = A\exp(\tfrac{1}{2}ct^2 - \mu t)\sqrt{\xi}\,J_{1/3}\!\left(\dfrac{2}{3b\sqrt{a}}\xi^{3/2}\right), \quad \xi = bx + d + \mu,$$
$$w(x,t) = A\exp(\tfrac{1}{2}ct^2 - \mu t)\sqrt{\xi}\,Y_{1/3}\!\left(\dfrac{2}{3b\sqrt{a}}\xi^{3/2}\right), \quad \xi = bx + d + \mu,$$

where $J_{1/3}(\xi)$ and $Y_{1/3}(\xi)$ are the Bessel functions of the first and second kind.

2°. The transformation
$$w(x,t) = u(z,t)\exp\left(btx + \tfrac{1}{3}ab^2 t^3 + \tfrac{1}{2}ct^2 + dt\right), \quad z = x + abt^2$$
leads to a constant coefficient equation, $\partial_t u = a\partial_{zz}u$, which is considered in Subsection 1.1.1.

11. $\dfrac{\partial w}{\partial t} = a\dfrac{\partial^2 w}{\partial x^2} + x(bt + c)w.$

This is a special case of equation 1.8.1.3 with $f(t) = bt + c$.

1°. Particular solutions (A and μ are arbitrary constants):

$$w(x,t) = A\exp\left[x\left(\tfrac{1}{2}bt^2 + ct\right) + a\left(\tfrac{1}{20}b^2 t^5 + \tfrac{1}{4}bct^4 + \tfrac{1}{3}c^2 t^3\right)\right],$$

$$w(x,t) = A\left[x + a\left(\tfrac{1}{3}bt^3 + ct^2\right)\right]\exp\left[x\left(\tfrac{1}{2}bt^2 + ct\right) + a\phi(t)\right],$$

$$w(x,t) = A\exp\left[x\left(\tfrac{1}{2}bt^2 + ct + \mu\right) + a\mu\left(\tfrac{1}{3}bt^3 + ct^2 + \mu t\right) + a\phi(t)\right],$$

where $\phi(t) = \tfrac{1}{20}b^2 t^5 + \tfrac{1}{4}bct^4 + \tfrac{1}{3}c^2 t^3$.

2°. The transformation

$$w(x,t) = u(z,t)\exp\left[x\left(\tfrac{1}{2}bt^2 + ct\right) + a\left(\tfrac{1}{20}b^2 t^5 + \tfrac{1}{4}bct^4 + \tfrac{1}{3}c^2 t^3\right)\right], \quad z = x + a\left(\tfrac{1}{3}bt^3 + ct^2\right)$$

leads to a constant coefficient equation, $\partial_t u = a\partial_{zz} u$, which is considered in Subsection 1.1.1.

12. $\dfrac{\partial w}{\partial t} = a\dfrac{\partial^2 w}{\partial x^2} + (bxt + cx + dt + e)w.$

This is a special case of equation 1.8.1.6 with $f(t) = bt + c$ and $g(t) = dt + e$.

1°. Particular solution:

$$w(x,t) = \exp\left[x\left(\tfrac{1}{2}bt^2 + ct\right) + a\left(\tfrac{1}{20}b^2 t^5 + \tfrac{1}{4}bct^4 + \tfrac{1}{3}c^2 t^3\right) + \tfrac{1}{2}dt^2 + et\right].$$

2°. The transformation

$$w(x,t) = u(z,t)\exp\left[x\left(\tfrac{1}{2}bt^2 + ct\right) + a\left(\tfrac{1}{20}b^2 t^5 + \tfrac{1}{4}bct^4 + \tfrac{1}{3}c^2 t^3\right) + \tfrac{1}{2}dt^2 + et\right], \quad z = x + a(bt^2 + 2ct)$$

leads to a constant coefficient equation, $\partial_t u = a\partial_{zz} u$, which is considered in Subsection 1.1.1.

13. $\dfrac{\partial w}{\partial t} = a\dfrac{\partial^2 w}{\partial x^2} + (-bx^2 + ct + d)w.$

This is a special case of equation 1.8.1.7 with $f(t) = ct + d$.

1°. Particular solutions (A is an arbitrary constant):

$$w(x,t) = A\exp\left[\tfrac{1}{2}\sqrt{\tfrac{b}{a}}\, x^2 + \tfrac{1}{2}ct^2 + \left(\sqrt{ab} + d\right)t\right],$$

$$w(x,t) = Ax\exp\left[\tfrac{1}{2}\sqrt{\tfrac{b}{a}}\, x^2 + \tfrac{1}{2}ct^2 + \left(3\sqrt{ab} + d\right)t\right].$$

2°. The transformation (A is any number)

$$w(x,t) = u(z,\tau)\exp\left[\tfrac{1}{2}\sqrt{\tfrac{b}{a}}\, x^2 + \tfrac{1}{2}ct^2 + \left(\sqrt{ab} + d\right)t\right],$$

$$z = x\exp\left(2\sqrt{ab}\, t\right), \quad \tau = \tfrac{1}{4}\sqrt{\tfrac{b}{a}}\exp\left(4\sqrt{ab}\, t\right) + A$$

leads to the constant coefficient equation $\partial_\tau u = \partial_{zz} u$, which is considered in Subsection 1.1.1.

14. $\dfrac{\partial w}{\partial t} = a\dfrac{\partial^2 w}{\partial x^2} + x(-bx + ct + d)w.$

This is a special case of equation 1.8.1.8 with $f(t) = ct + d$.

15. $\dfrac{\partial w}{\partial t} = a\dfrac{\partial^2 w}{\partial x^2} + bxt^k w.$

This is a special case of equation 1.8.1.3 with $f(t) = bt^k$.

1°. Particular solutions (A and μ are arbitrary constants):

$$w(x,t) = A\exp\left[\dfrac{b}{k+1}xt^{k+1} + \dfrac{ab^2}{(k+1)^2(2k+3)}t^{2k+3}\right],$$

$$w(x,t) = A\left[x + \dfrac{2ab}{(k+1)(k+2)}t^{k+2}\right]\exp\left[\dfrac{b}{k+1}xt^{k+1} + \dfrac{ab^2}{(k+1)^2(2k+3)}t^{2k+3}\right],$$

$$w(x,t) = A\exp\left[\dfrac{b}{k+1}xt^{k+1} + \mu x + \dfrac{ab^2}{(k+1)^2(2k+3)}t^{2k+3} + \dfrac{2ab\mu}{(k+1)(k+2)}t^{k+2} + a\mu^2 t\right].$$

2°. The transformation

$$w(x,t) = u(z,t)\exp\left[\dfrac{b}{k+1}xt^{k+1} + \dfrac{ab^2}{(k+1)^2(2k+3)}t^{2k+3}\right], \qquad z = x + \dfrac{2ab}{(k+1)(k+2)}t^{k+2}$$

leads to a constant coefficient equation, $\partial_t u = a\partial_{zz} u$, which is considered in Subsection 1.1.1.

16. $\dfrac{\partial w}{\partial t} = a\dfrac{\partial^2 w}{\partial x^2} + (bx^2 t^n + cxt^m + dt^k)w.$

This is a special case of equation 1.8.7.5 with $n(t) = a$, $f(t) = g(t) = 0$, $h(t) = bt^n$, $s(t) = ct^m$, and $p(t) = dt^k$.

1.3.2. Equations of the Form $\dfrac{\partial w}{\partial t} = a\dfrac{\partial^2 w}{\partial x^2} + f(x,t)\dfrac{\partial w}{\partial x}$

1. $\dfrac{\partial w}{\partial t} = a\dfrac{\partial^2 w}{\partial x^2} + (bt + c)\dfrac{\partial w}{\partial x}.$

This is a special case of equation 1.8.2.1 with $f(t) = bt + c$.

1°. Particular solutions (A, B, and μ are arbitrary constants):

$$w(x,t) = 2Ax + A(bt^2 + 2ct) + B,$$

$$w(x,t) = A\left(x + \tfrac{1}{2}bt^2 + ct\right)^2 + 2aAt + B,$$

$$w(x,t) = A\exp\left[\mu x + \tfrac{1}{2}\mu bt^2 + (a\mu^2 + \mu c)t\right] + B.$$

2°. The substitution $z = x + \tfrac{1}{2}bt^2 + ct$ leads to a constant coefficient equation, $\partial_t w = a\partial_{zz} w$, which is considered in Subsection 1.1.1.

2. $\dfrac{\partial w}{\partial t} = a\dfrac{\partial^2 w}{\partial x^2} + bx\dfrac{\partial w}{\partial x}.$

This equation is a special case of equation 1.8.2.2 with $f(x) = bx$ and a special case of equation 1.8.2.3 with $f(t) = b$.

1°. Particular solutions (A, B, and μ are arbitrary constants):

$$w(x) = A\int \exp\left(-\dfrac{b}{2a}x^2\right)dx + B,$$

$$w(x,t) = Axe^{bt} + B,$$

$$w(x,t) = Abx^2 e^{2bt} + Aae^{2bt} + B,$$

$$w(x,t) = A\exp\left(2b\mu xe^{bt} + 2ab\mu^2 e^{2bt}\right) + B.$$

2°. On passing from t, x to the new variables (A and B are any numbers)

$$\tau = \frac{A^2}{2b} e^{2bt} + B, \quad z = Axe^{bt},$$

for the function $w(\tau, z)$ we obtain a constant coefficient equation, $\partial_\tau w = a \partial_{zz} w$, which is considered in Subsection 1.1.1.

3°. Domain: $-\infty < x < \infty$. Cauchy problem.
An initial condition is prescribed:

$$w = f(x) \quad \text{at} \quad t = 0.$$

Solution:

$$w(x,t) = \left[\frac{2\pi a}{b} \left(e^{2bt} - 1 \right) \right]^{-1/2} \int_{-\infty}^{\infty} \exp\left[-\frac{b(xe^{bt} - \xi)^2}{2a(e^{2bt} - 1)} \right] f(\xi) \, d\xi.$$

⊙ *References*: W. Miller, Jr. (1977), A. D. Polyanin, A. V. Vyazmin, A. I. Zhurov, and D. A. Kazenin (1998).

3. $\dfrac{\partial w}{\partial t} = a \dfrac{\partial^2 w}{\partial x^2} + (bt^2 + c) \dfrac{\partial w}{\partial x}.$

This is a special case of equation 1.8.2.1 with $f(t) = bt^2 + c$.

1°. Particular solutions (A, B, and μ are arbitrary constants):

$$w(x,t) = A\left(x + \tfrac{1}{3}bt^3 + ct\right) + B,$$
$$w(x,t) = A\left(x + \tfrac{1}{3}bt^3 + ct\right)^2 + 2aAt + B,$$
$$w(x,t) = A\exp\left[\mu x + \tfrac{1}{3}\mu bt^3 + \mu(a\mu + c)t\right] + B.$$

2°. On passing from t, x to the new variables t, $z = x + \tfrac{1}{3}bt^3 + ct$, we obtain a constant coefficient equation, $\partial_t w = a\partial_{zz} w$, which is considered in Subsection 1.1.1.

4. $\dfrac{\partial w}{\partial t} = a \dfrac{\partial^2 w}{\partial x^2} + x(bt + c) \dfrac{\partial w}{\partial x}.$

This is a special case of equation 1.8.2.3 with $f(t) = bt + c$.

1°. Particular solutions (A, B, and μ are arbitrary constants):

$$w(x,t) = Ax \exp\left(\tfrac{1}{2}bt^2 + ct\right) + B,$$
$$w(x,t) = Ax^2 \exp(bt^2 + 2ct) + 2Aa \int \exp(bt^2 + 2ct) \, dt + B,$$
$$w(x,t) = A \exp\left[\mu x \exp\left(\tfrac{1}{2}bt^2 + ct\right) + a\mu^2 \int \exp(bt^2 + 2ct) \, dt\right] + B.$$

2°. On passing from t, x to the new variables (A is any number)

$$\tau = \int \exp(bt^2 + 2ct) \, dt + A, \quad z = x \exp\left(\tfrac{1}{2}bt^2 + ct\right),$$

for the function $w(\tau, z)$ we obtain a constant coefficient equation, $\partial_\tau w = a\partial_{zz} w$, which is considered in Subsection 1.1.1.

5. $\dfrac{\partial w}{\partial t} = \dfrac{\partial^2 w}{\partial x^2} + (ax + bt + c) \dfrac{\partial w}{\partial x}.$

This is a special case of equation 1.8.2.7 with $f(t) = a$ and $g(t) = bt + c$. See also equation 1.8.2.4.

6. $\dfrac{\partial w}{\partial t} = a\dfrac{\partial^2 w}{\partial x^2} + b\dfrac{x}{t}\dfrac{\partial w}{\partial x}.$

Ilkovič's equation. It describes heat transfer to the surface of a growing drop that flows out of a thin capillary into a fluid solution (the mass rate of flow of the fluid moving in the capillary is assumed constant). This equation is a special case of equation 1.8.2.3 with $f(t) = b/t$.

1°. Particular solutions (A, B, and μ are arbitrary constants):

$$w(x,t) = Axt^b + B,$$
$$w(x,t) = A(2b+1)x^2 t^{2b} + 2Aat^{2b+1} + B,$$
$$w(x,t) = A\exp\left(\mu x t^b + \dfrac{a\mu^2}{2b+1}t^{2b+1}\right) + B.$$

2°. On passing from t, x to the new variables

$$\tau = \dfrac{1}{2b+1}t^{2b+1}, \quad z = xt^b,$$

for the function $w(\tau, z)$ we obtain a constant coefficient equation, $\partial_\tau w = a\partial_{zz} w$, which is considered in Subsection 1.1.1.

3°. The solution of the original equation in the important special case where the drop surface has a time-invariant temperature w_s and the heat exchange occurs with an infinite medium having an initial temperature w_0, namely,

$$w = w_0 \quad \text{at} \quad t = 0 \quad \text{(initial condition)},$$
$$w = w_s \quad \text{at} \quad x = 0 \quad \text{(boundary condition)},$$
$$w \to w_0 \quad \text{at} \quad x \to \infty \quad \text{(boundary condition)},$$

is expressed in terms of the error function as follows:

$$\dfrac{w - w_s}{w_0 - w_s} = \operatorname{erf}\left(\dfrac{\sqrt{2b+1}}{2\sqrt{a}}\dfrac{x}{\sqrt{t}}\right), \quad \operatorname{erf}\xi = \dfrac{2}{\sqrt{\pi}}\int_0^\xi \exp(-\zeta^2)\,d\zeta.$$

⊙ *Reference*: Yu. P. Gupalo, A. D. Polyanin, and Yu. S. Ryazantsev (1985).

7. $\dfrac{\partial w}{\partial t} = a\dfrac{\partial^2 w}{\partial x^2} + (bt^k x + ct^m)\dfrac{\partial w}{\partial x}.$

This is a special case of equation 1.8.2.7 with $f(t) = bt^k$ and $g(t) = ct^m$.

8. $\dfrac{\partial w}{\partial t} = a\dfrac{\partial^2 w}{\partial x^2} + \left(cx + \dfrac{b}{x}\right)\dfrac{\partial w}{\partial x}.$

On passing from t, x to the new variables z, τ by the formulas

$$z = xe^{ct}, \quad \tau = \dfrac{a}{2c}e^{2ct} + \text{const},$$

we obtain a simpler equation of the form 1.2.5:

$$\dfrac{\partial w}{\partial \tau} = \dfrac{\partial^2 w}{\partial z^2} + \dfrac{\mu}{z}\dfrac{\partial w}{\partial z}, \quad \mu = \dfrac{b}{a}.$$

For $\mu = 1$ and $\mu = 2$, see also equations from Subsections 1.2.1 and 1.2.3.

9. $\dfrac{\partial w}{\partial t} = a\dfrac{\partial^2 w}{\partial x^2} + \left(ct^n x + \dfrac{b}{x}\right)\dfrac{\partial w}{\partial x}.$

This is a special case of equation 1.8.2.6 with $f(t) = ct^n$.

1.3.3. Equations of the Form $\frac{\partial w}{\partial t} = a\frac{\partial^2 w}{\partial x^2} + f(x,t)\frac{\partial w}{\partial x} + g(x,t)w + h(x,t)$

1. $\dfrac{\partial w}{\partial t} = a\dfrac{\partial^2 w}{\partial x^2} + b\dfrac{\partial w}{\partial x} + (cx+d)w.$

This is a special case of equation 1.8.7.4 with $n(t) = a$, $f(t) = 0$, $g(t) = b$, $h(t) = c$, and $s(t) = d$.

1°. Particular solutions (A and μ are arbitrary constants):

$$w(x,t) = A\exp\left[ctx - \frac{b}{2a}x + \frac{1}{3}ac^2t^3 + \left(d - \frac{b^2}{4a}\right)t\right],$$

$$w(x,t) = A(x+act^2)\exp\left[ctx - \frac{b}{2a}x + \frac{1}{3}ac^2t^3 + \left(d - \frac{b^2}{4a}\right)t\right],$$

$$w(x,t) = A\exp\left[x\left(ct + \mu - \frac{b}{2a}\right) + \frac{1}{3}ac^2t^3 + ac\mu t^2 + \left(a\mu^2 + d - \frac{b^2}{4a}\right)t\right],$$

$$w(x,t) = A\exp\left(-\mu t - \frac{b}{2a}x\right)\sqrt{\xi}\, J_{1/3}\!\left(\frac{2}{3c\sqrt{a}}\xi^{3/2}\right), \quad \xi = cx + \mu + d - \frac{b^2}{4a},$$

$$w(x,t) = A\exp\left(-\mu t - \frac{b}{2a}x\right)\sqrt{\xi}\, Y_{1/3}\!\left(\frac{2}{3c\sqrt{a}}\xi^{3/2}\right), \quad \xi = cx + \mu + d - \frac{b^2}{4a},$$

where $J_{1/3}(\xi)$ and $Y_{1/3}(\xi)$ are the Bessel functions of the first and second kind of order 1/3, respectively.

2°. The transformation

$$w(x,t) = u(z,t)\exp\left[ctx - \frac{b}{2a}x + \frac{1}{3}ac^2t^3 + \left(d - \frac{b^2}{4a}\right)t\right], \quad z = x + act^2$$

leads to a constant coefficient equation, $\partial_t u = a\partial_{zz}u$, which is considered in Subsection 1.1.1.

2. $\dfrac{\partial w}{\partial t} = a\dfrac{\partial^2 w}{\partial x^2} + bx\dfrac{\partial w}{\partial x} + (cx+d)w.$

This is a special case of equation 1.8.7.4 with $n(t) = a$, $f(t) = b$, $g(t) = 0$, $h(t) = c$, and $s(t) = d$.

1°. Particular solutions (A and μ are arbitrary constants):

$$w(x,t) = A\exp\left[-\frac{c}{b}x + \left(d + \frac{ac^2}{b^2}\right)t\right],$$

$$w(x,t) = A\left(x - \frac{2ac}{b^2}\right)\exp\left[-\frac{c}{b}x + \left(b + d + \frac{ac^2}{b^2}\right)t\right],$$

$$w(x,t) = A\exp\left[\frac{a\mu^2}{2b}e^{2bt} + \mu e^{bt}\left(x - \frac{2ac}{b^2}\right) - \frac{c}{b}x + \left(d + \frac{ac^2}{b^2}\right)t\right].$$

See 1.3.4.7 for more complicated solutions.

2°. The transformation

$$w(x,t) = u(z,\tau)\exp\left[-\frac{c}{b}x + \left(d + \frac{ac^2}{b^2}\right)t\right], \quad \tau = \frac{a}{2b}e^{2bt}, \quad z = e^{bt}\left(x - \frac{2ac}{b^2}\right),$$

leads to a constant coefficient equation, $\partial_\tau u = \partial_{zz}u$, which is considered in Subsection 1.1.1.

3. $\dfrac{\partial w}{\partial t} = a\dfrac{\partial^2 w}{\partial x^2} + (bx+c)\dfrac{\partial w}{\partial x} + (dx+e)w.$

For $b=0$, see equation 1.3.3.1. For $b \neq 0$, the substitution $z = x + c/b$ leads to an equation of the form 1.3.3.2:
$$\dfrac{\partial w}{\partial t} = a\dfrac{\partial^2 w}{\partial z^2} + bz\dfrac{\partial w}{\partial z} + (dz+k)w, \qquad k = e - \dfrac{cd}{b}.$$

4. $\dfrac{\partial w}{\partial t} = \dfrac{\partial^2 w}{\partial x^2} + 2(ax+b)\dfrac{\partial w}{\partial x} + (a^2 x^2 + 2abx + c)w.$

This equation is a special case of equation 1.8.6.5 and a special case of equation 1.8.7.5. The substitution $w(x,t) = u(x,t)\exp\!\left(-\tfrac{1}{2}ax^2 - bx\right)$ leads to a constant coefficient equation of the form 1.1.3 with $\Phi \equiv 0$, namely, $\partial_t u = \partial_{xx} u + (c - a - b^2)u$.

5. $\dfrac{\partial w}{\partial t} = \dfrac{\partial^2 w}{\partial x^2} + (ax+b)\dfrac{\partial w}{\partial x} + (cx^2 + dx + e)w.$

This equation is a special case of equation 1.8.6.5 and a special case of equation 1.8.7.5.

1°. The substitution
$$w(x,t) = u(x,t)\exp\!\left(\tfrac{1}{2}Ax^2\right),$$
where A is a root of the quadratic equation $A^2 + aA + c = 0$, yields an equation of the form 1.8.7.4,
$$\dfrac{\partial u}{\partial t} = \dfrac{\partial^2 u}{\partial x^2} + \bigl[(2A+a)x + b\bigr]\dfrac{\partial u}{\partial x} + \bigl[(Ab+d)x + A + e\bigr]u,$$
which is reduced to a constant coefficient equation.

2°. The substitution
$$w(x,t) = u(x,t)\exp\!\left(\tfrac{1}{2}Ax^2 + Bx + Ct\right)$$
leads to an equation of the analogous form:
$$\dfrac{\partial u}{\partial t} = \dfrac{\partial^2 u}{\partial x^2} + \bigl[(2A+a)x + 2B + b\bigr]\dfrac{\partial u}{\partial x} \\ + \bigl[(A^2 + Aa + c)x^2 + (2AB + Ab + Ba + d)x + B^2 + Bb + A - C + e\bigr]u.$$

By appropriately choosing the coefficients A, B, and C, one can simplify the original equation in various ways.

6. $\dfrac{\partial w}{\partial t} = \dfrac{\partial^2 w}{\partial x^2} + (ax + bt + c)\dfrac{\partial w}{\partial x} + (sx^2 + ptx + qt^2 + kx + lt + m)w.$

This is a special case of equation 1.8.7.5.

7. $\dfrac{\partial w}{\partial t} = a\dfrac{\partial^2 w}{\partial x^2} + (bt^k x + ct^m)\dfrac{\partial w}{\partial x} + st^n w.$

This is a special case of equation 1.8.3.6 with $f(t) = bt^k$, $g(t) = ct^m$, and $h(t) = st^n$.

8. $\dfrac{\partial w}{\partial t} = a\dfrac{\partial^2 w}{\partial x^2} + \dfrac{b}{x}\dfrac{\partial w}{\partial x} + \left(c + \dfrac{k}{x^2}\right)w.$

1°. The transformation
$$w(x,t) = e^{ct} x^\lambda u(x,\tau), \qquad \tau = at,$$
where λ is a root of the quadratic equation $a\lambda^2 + (b-a)\lambda + k = 0$, leads to an equation of the form 1.2.5:
$$\dfrac{\partial u}{\partial \tau} = \dfrac{\partial^2 u}{\partial x^2} + \left(2\lambda + \dfrac{b}{a}\right)\dfrac{1}{x}\dfrac{\partial u}{\partial x}.$$

2°. If $w(x, t)$ is a solution of the original equation, then the functions

$$w_1 = Ae^{c(1-a^2)\tau} w(\pm ax, a^2\tau), \quad \tau = t + B,$$

$$w_2 = A\tau^{\lambda-1} \exp\left(-\frac{x^2}{4a\tau} + c\tau + \frac{c}{a^2\tau}\right) w\left(\pm\frac{x}{a\tau}, -\frac{1}{a^2\tau}\right), \quad \lambda = \frac{1}{2} - \frac{b}{2a},$$

where A and B are arbitrary constants, are also solutions of this equation.

9. $\quad \dfrac{\partial w}{\partial t} = a\dfrac{\partial^2 w}{\partial x^2} + \dfrac{b}{x}\dfrac{\partial w}{\partial x} + \left(c + \dfrac{k}{x^2}\right)w + \Phi(x, t).$

The transformation

$$w(x, t) = e^{ct} x^\lambda u(x, \tau), \quad \tau = at,$$

where λ is a root of the quadratic equation $a\lambda^2 + (b-a)\lambda + k = 0$, leads to an equation of the form 1.2.6:

$$\frac{\partial u}{\partial \tau} = \frac{\partial^2 u}{\partial x^2} + \left(2\lambda + \frac{b}{a}\right)\frac{1}{x}\frac{\partial u}{\partial x} + \Psi(x, \tau), \quad \Psi(x, \tau) = \frac{1}{a}e^{-ct}x^{-\lambda}\Phi(x, t).$$

1.3.4. Equations of the Form $\dfrac{\partial w}{\partial t} = (ax + b)\dfrac{\partial^2 w}{\partial x^2} + f(x, t)\dfrac{\partial w}{\partial x} + g(x, t)w$

1. $\quad \dfrac{\partial w}{\partial t} = ax\dfrac{\partial^2 w}{\partial x^2}.$

This is a special case of equation 1.8.6.1 with $f(x) = ax$. See also equation 1.3.6.6 with $n = 0$.

1°. Particular solutions (A, B, C, and μ are arbitrary constants):

$$w(x) = Ax + B,$$

$$w(x, t) = 2Aatx + Ax^2 + B,$$

$$w(x, t) = Aa^2t^2x + Aatx^2 + \tfrac{1}{6}Ax^3 + B,$$

$$w(x, t) = 2Aa^3t^3x + 3Aa^2t^2x^2 + Aatx^3 + \tfrac{1}{12}Ax^4 + B,$$

$$w(x, t) = x^n + \sum_{k=1}^{n-1} \frac{[n(n-1)\ldots(n-k)]^2}{n(n-k)k!}(at)^k x^{n-k},$$

$$w(x, t) = A\exp\left(-\frac{x}{at + B}\right) + C,$$

$$w(x, t) = \frac{Ax}{(at + B)^2}\exp\left(-\frac{x}{at + B}\right) + C,$$

$$w(x, t) = Aat + A(x\ln x - x) + B,$$

$$w(x, t) = Aa^2t^2 + 2Aat(x\ln x - x) + A(x^2\ln x - \tfrac{5}{2}x^2) + B,$$

$$w(x, t) = e^{\mu t}\sqrt{x}\left[AJ_1\left(\frac{2}{\sqrt{a}}\sqrt{-\mu x}\right) + BY_1\left(\frac{2}{\sqrt{a}}\sqrt{-\mu x}\right)\right] \quad \text{for } \mu < 0,$$

$$w(x, t) = e^{\mu t}\sqrt{x}\left[AI_1\left(\frac{2}{\sqrt{a}}\sqrt{\mu x}\right) + BK_1\left(\frac{2}{\sqrt{a}}\sqrt{\mu x}\right)\right] \quad \text{for } \mu > 0,$$

where $J_1(z)$ and $Y_1(z)$ are the Bessel functions, $I_1(z)$ and $K_1(z)$ are the modified Bessel functions.

2°. A solution containing an arbitrary function of the space variable:

$$w(x, t) = f(x) + \sum_{n=1}^{\infty} \frac{(at)^n}{n!}L^n[f(x)], \quad L \equiv x\frac{d^2}{dx^2},$$

where $f(x)$ is any infinitely differentiable function. This solution satisfies the initial condition $w(x, 0) = f(x)$. The sum is finite if $f(x)$ is a polynomial.

3°. A solution containing an arbitrary function of time:
$$w(x,t) = A + xg(t) + \sum_{n=2}^{\infty} \frac{1}{n[(n-1)!]^2 a^{n-1}} x^n g_t^{(n-1)}(t),$$

where $g(t)$ is any infinitely differentiable function and A is an arbitrary number. This solution possesses the properties
$$w(0,t) = A, \qquad \partial_x w(0,t) = g(t).$$

4°. Suppose $w = w(x,t)$ is a solution of the original equation. Then the functions
$$w_1 = Aw(\lambda x, \lambda t + C),$$
$$w_2 = A \exp\left[-\frac{\beta x}{a(\delta + \beta t)}\right] w\left(\frac{x}{(\delta + \beta t)^2}, \frac{\gamma + \lambda t}{\delta + \beta t}\right), \qquad \lambda\delta - \beta\gamma = 1,$$

where A, C, β, δ, and λ are arbitrary constants, are also solutions of the equation.

2. $\dfrac{\partial w}{\partial t} = ax \dfrac{\partial^2 w}{\partial x^2} + (-bx + c)w.$

The transformation
$$w(x,t) = u(z,\tau) \exp\left(\sqrt{\frac{b}{a}}\, x + ct\right), \quad z = x \exp(2\sqrt{ab}\, t), \quad \tau = \frac{1}{2}\sqrt{\frac{a}{b}} \exp(2\sqrt{ab}\, t)$$

leads to a simpler equation of the form 1.3.4.1:
$$\frac{\partial u}{\partial \tau} = z \frac{\partial^2 u}{\partial z^2}.$$

3. $\dfrac{\partial w}{\partial t} = ax \dfrac{\partial^2 w}{\partial x^2} + bxt^n w.$

This is a special case of equation 1.8.8.1 with $f(t) = a$, $g(t) = 0$, $h(t) = bt^n$, and $s(t) = 0$.

4. $\dfrac{\partial w}{\partial t} = ax \dfrac{\partial^2 w}{\partial x^2} + (bt^n x + ct^m)w.$

This is a special case of equation 1.8.8.1 with $f(t) = a$, $g(t) = 0$, $h(t) = bt^n$, and $s(t) = ct^m$.

5. $\dfrac{\partial w}{\partial t} = a\left[(x+b)\dfrac{\partial^2 w}{\partial x^2} + \dfrac{\partial w}{\partial x}\right].$

This equation describes heat transfer in a quiescent medium (solid body) in the case of thermal diffusivity as a linear function of the space coordinate.

1°. The original equation can be rewritten in a form more suitable for applications,
$$\frac{\partial w}{\partial t} = a \frac{\partial}{\partial x}\left[(x+b)\frac{\partial w}{\partial x}\right].$$

2°. The substitution $x = \frac{1}{4}z^2 - b$ leads to the equation
$$\frac{\partial w}{\partial t} = a\left(\frac{\partial^2 w}{\partial z^2} + \frac{1}{z}\frac{\partial w}{\partial z}\right),$$

which is considered in Subsection 1.2.1.

6. $\dfrac{\partial w}{\partial t} = ax\dfrac{\partial^2 w}{\partial x^2} + bx\dfrac{\partial w}{\partial x} + (cx+d)w.$

This is a special case of equation 1.8.8.1 with $f(t) = a$, $g(t) = b$, $h(t) = c$, and $s(t) = d$.

7. $\dfrac{\partial w}{\partial t} = (a_2 x + b_2)\dfrac{\partial^2 w}{\partial x^2} + (a_1 x + b_1)\dfrac{\partial w}{\partial x} + (a_0 x + b_0)w.$

This is a special case of equation 1.8.6.5 with $f(x) = a_2 x + b_2$, $g(x) = a_1 x + b_1$, $h(x) = a_0 x + b_0$, and $\Phi \equiv 0$.

Particular solutions of the original equation are presented in Table 15, where the function

$$\mathcal{J}(a,b;x) = C_1 \Phi(a,b;x) + C_2 \Psi(a,b;x), \qquad C_1, C_2 \text{ are any numbers,}$$

is an arbitrary solution of the degenerate hypergeometric equation

$$x y''_{xx} + (b-x)y'_x - ay = 0,$$

and the function

$$Z_\nu(x) = C_1 J_\nu(x) + C_2 Y_\nu(x), \qquad C_1, C_2 \text{ are any numbers,}$$

is an arbitrary solution of the Bessel equation

$$x^2 y''_{xx} + x y'_x + (x^2 - \nu^2)y = 0.$$

TABLE 15
Particular solutions of equation 1.3.4.7 for different values of the determining parameters (μ is an arbitrary number)

Particular solution: $w(x,t) = \exp(hx - \mu t) F(\xi)$, where $\xi = (x+\gamma)/p$					
Constraints	h	p	γ	F	Parameters
$a_2 \neq 0$, $D \neq 0$	$\dfrac{D - a_1}{2 a_2}$	$-\dfrac{a_2}{A(h)}$	$\dfrac{b_2}{a_2}$	$\mathcal{J}(a,b;\xi)$	$a = B(h)/A(h)$, $b = (a_2 b_1 - a_1 b_2)a_2^{-2}$
$a_2 = 0$, $a_1 \neq 0$	$-\dfrac{a_0}{a_1}$	1	$\dfrac{2 b_2 h + b_1}{a_1}$	$\mathcal{J}\!\left(a, \tfrac{1}{2}; k\xi^2\right)$	$a = B(h)/(2 a_1)$, $k = -a_1/(2 b_2)$
$a_2 \neq 0$, $a_1^2 = 4 a_0 a_2$	$-\dfrac{a_1}{2 a_2}$	a_2	$\dfrac{b_2}{a_2}$	$\xi^\alpha Z_{2\alpha}(\beta\sqrt{\xi})$	$\alpha = \dfrac{1}{2} - \dfrac{2 b_2 h + b_1}{2 a_2}$, $\beta = 2\sqrt{B(h)}$
$a_2 = a_1 = 0$, $a_0 \neq 0$	$-\dfrac{b_1}{2 b_2}$	1	$\dfrac{4(b_0 + \mu)b_2 - b_1^2}{4 a_0 b_2}$	$\xi^{1/2} Z_{1/3}(k\xi^{3/2})$	$k = \dfrac{2}{3}\left(\dfrac{a_0}{b_2}\right)^{1/2}$
Notation: $D^2 = a_1^2 - 4 a_0 a_2$, $A(h) = 2 a_2 h + a_1$, $B(h) = b_2 h^2 + b_1 h + b_0 + \mu$					

For the degenerate hypergeometric functions $\Phi(a,b;x)$ and $\Psi(a,b;x)$, see Supplement A.9 and books by Bateman and Erdélyi (1953, Vol. 1) and Abramowitz and Stegun (1964).

For the Bessel functions $J_\nu(x)$ and $Y_\nu(x)$, see Supplement A.6 and Bateman and Erdélyi (1953, Vol. 2), Abramowitz and Stegun (1964), Nikiforov and Uvarov (1988), and Temme (1996).

Remark. For $b_2 = 0$ the original equation is a special case of equation 1.8.7.4 and can be reduced to the constant coefficient heat equation that is considered in Subsection 1.1.1. In this case, a number of solutions are not displayed in Table 15.

1.3.5. Equations of the Form $\frac{\partial w}{\partial t}=(ax^2+bx+c)\frac{\partial^2 w}{\partial x^2}+f(x,t)\frac{\partial w}{\partial x}+g(x,t)w$

1. $\quad \dfrac{\partial w}{\partial t} = ax^2 \dfrac{\partial^2 w}{\partial x^2} + bx \dfrac{\partial w}{\partial x} + cw.$

1°. Particular solutions (A, B, and μ are arbitrary constants):

$$w(x,t) = \bigl(A\ln|x| + B\bigr)|x|^n \exp\bigl[(c - an^2)t\bigr],$$
$$w(x,t) = A\bigl(2at + \ln^2|x|\bigr)|x|^n \exp\bigl[(c - an^2)t\bigr],$$
$$w(x,t) = A|x|^\mu \exp\bigl[(c + a\mu^2 - 2an\mu)t\bigr],$$

where $n = \tfrac{1}{2}(a-b)/a$.

2°. The transformation

$$w(x,t) = |x|^n \exp\bigl[(c - an^2)t\bigr] u(z,t), \quad z = \ln|x|, \quad n = \frac{a-b}{2a},$$

leads to a constant coefficient equation, $\partial_t u = a\partial_{zz} u$, which is considered in Subsection 1.1.1.

2. $\quad \dfrac{\partial w}{\partial t} = ax^2 \dfrac{\partial^2 w}{\partial x^2} + (bt^n + c)w.$

This is a special case of equation 1.8.4.2 with $f(t) = bt^n + c$.

The transformation

$$w(x,t) = u(z,t) \exp\!\left(\frac{b}{n+1}t^{n+1} + ct\right), \quad z = \ln|x|$$

leads to a constant coefficient equation of the form 1.1.4:

$$\frac{\partial u}{\partial \tau} = a\frac{\partial^2 u}{\partial z^2} - a\frac{\partial u}{\partial z}.$$

3. $\quad \dfrac{\partial w}{\partial t} = ax^2 \dfrac{\partial^2 w}{\partial x^2} + bx \dfrac{\partial w}{\partial x} + (cx^k + s)w.$

This is a special case of equation 1.8.6.5 with $f(x) = ax^2$, $g(x) = bx$, $h(x) = cx^k + s$, and $\Phi \equiv 0$. For $c = 0$, see equation 1.3.5.1.

Particular solutions for $c \neq 0$:

$$w(x,t) = Ae^{-\mu t} x^{\frac{a-b}{2a}} J_\nu\!\left(\frac{2}{k}\sqrt{\frac{c}{a}}\, x^{\frac{k}{2}}\right), \quad \nu = \frac{1}{ak}\sqrt{(a-b)^2 - 4a(s+\mu)},$$
$$w(x,t) = Ae^{-\mu t} x^{\frac{a-b}{2a}} Y_\nu\!\left(\frac{2}{k}\sqrt{\frac{c}{a}}\, x^{\frac{k}{2}}\right), \quad \nu = \frac{1}{ak}\sqrt{(a-b)^2 - 4a(s+\mu)},$$

where A and μ are arbitrary constants, $J_\nu(z)$ and $Y_\nu(z)$ are the Bessel functions.

4. $\quad \dfrac{\partial w}{\partial t} = a_2 x^2 \dfrac{\partial^2 w}{\partial x^2} + (a_1 x^2 + b_1 x)\dfrac{\partial w}{\partial x} + (a_0 x^2 + b_0 x + c_0)w.$

This is a special case of equation 1.8.6.5 with $f(x) = a_2 x^2$, $g(x) = a_1 x^2 + b_1 x$, $h(x) = a_0 x^2 + b_0 x + c_0$, and $\Phi \equiv 0$.

1°. Particular solutions for $a_1^2 \neq 4a_0a_2$:

$$w(x,t) = A\exp(-\nu t + \mu x)x^k \Phi\left(\alpha, 2k + \frac{b_1}{a_2}; -\gamma x\right),$$
$$w(x,t) = A\exp(-\nu t + \mu x)x^k \Psi\left(\alpha, 2k + \frac{b_1}{a_2}; -\gamma x\right),$$
(1)

where A and ν are arbitrary constants,

$$\mu = \frac{\sqrt{a_1^2 - 4a_0a_2} - a_1}{2a_2}, \quad \alpha = \frac{(b_1 + 2a_2k)\mu + b_0 + a_1k}{2a_2\mu + a_1}, \quad \gamma = 2\mu + \frac{a_1}{a_2},$$

$k = k(\nu)$ is a root of the quadratic equation $a_2k^2 + (b_1 - a_2)k + c_0 + \nu = 0$, and $\Phi(\alpha, \beta; z)$ and $\Psi(\alpha, \beta; z)$ are the degenerate hypergeometric functions. [For the degenerate hypergeometric functions, see Supplement A.9 and the books by Abramowitz and Stegun (1964) and Bateman and Erdélyi (1953, Vol. 1)].

2°. Particular solutions for $a_1^2 = 4a_0a_2$:

$$w(x,t) = A\exp\left(-\nu t - \frac{a_1}{2a_2}x\right)x^k \xi^m J_{2m}(2\sqrt{p\xi}), \quad \xi = \frac{x}{a_2},$$
$$w(x,t) = A\exp\left(-\nu t - \frac{a_1}{2a_2}x\right)x^k \xi^m Y_{2m}(2\sqrt{p\xi}), \quad \xi = \frac{x}{a_2},$$
(2)

where A and ν are arbitrary constants,

$$m = \frac{1}{2} - k - \frac{b_1}{2a_2}, \quad p = -\frac{a_1}{2a_2}(b_1 + 2a_2k) + b_0 + a_1k = 0,$$

$k = k(\nu)$ is a root of the quadratic equation $a_2k^2 + (b_1 - a_2)k + c_0 + \nu = 0$, and $J_m(z)$ and $Y_m(z)$ are the Bessel functions. [For the Bessel functions, see Supplement A.6 and the books by Abramowitz and Stegun (1964) and Bateman and Erdélyi (1953, Vol. 2)].

Remark. In solutions (1) and (2), the parameter k can be regarded as arbitrary, and then $\nu = -a_2k^2 - (b_1 - a_2)k - c_0$.

5. $\dfrac{\partial w}{\partial t} = a_2x^2\dfrac{\partial^2 w}{\partial x^2} + (a_1x^{k+1} + b_1x)\dfrac{\partial w}{\partial x} + (a_0x^{2k} + b_0x^k + c_0)w.$

This is a special case of equation 1.8.6.5 with $f(x) = a_2x^2$, $g(x) = a_1x^{k+1} + b_1x$, $h(x) = a_0x^{2k} + b_0x^k + c_0$, and $\Phi \equiv 0$.

The substitution $\xi = x^k$ leads to an equation of the form 1.3.5.4:

$$\frac{\partial w}{\partial t} = a_2k^2\xi^2 \frac{\partial^2 w}{\partial \xi^2} + k(a_1\xi^2 + \beta\xi)\frac{\partial w}{\partial \xi} + (a_0\xi^2 + b_0\xi + c_0)w,$$

where $\beta = b_1 + a_2(k-1)$.

6. $\dfrac{\partial w}{\partial t} = (ax^2 + b)\dfrac{\partial^2 w}{\partial x^2} + ax\dfrac{\partial w}{\partial x} + cw.$

The substitution $z = \displaystyle\int \frac{dx}{\sqrt{ax^2 + b}}$ leads to the constant coefficient equation

$$\frac{\partial w}{\partial t} = \frac{\partial^2 w}{\partial z^2} + cw,$$

which is considered in Subsection 1.1.3.

1.3.6. Equations of the Form $\dfrac{\partial w}{\partial t} = f(x)\dfrac{\partial^2 w}{\partial x^2} + g(x,t)\dfrac{\partial w}{\partial x} + h(x,t)w$

1. $\dfrac{\partial w}{\partial t} = ax^3\dfrac{\partial^2 w}{\partial x^2} + bxt^k\dfrac{\partial w}{\partial x} + ct^m w.$

This is a special case of equation 1.8.8.7 with $n = 3$, $f(t) = a$, $g(t) = bt^k$, and $h(t) = ct^m$.

2. $\dfrac{\partial w}{\partial t} = ax^4\dfrac{\partial^2 w}{\partial x^2} + bw.$

This is a special case of equation 1.3.7.6 with $n = m = 0$.

$1°$. Particular solutions (A, B, and μ are arbitrary constants):

$$w(x,t) = e^{bt}(Ax + B),$$

$$w(x,t) = e^{bt}\left(2Aatx + \dfrac{A}{x} + B\right),$$

$$w(x,t) = Ax\exp\left[(b + a\mu^2)t + \dfrac{\mu}{x}\right].$$

$2°$. The transformation $w(x,t) = xe^{bt}u(\xi,t)$, $\xi = 1/x$ leads to a constant coefficient equation, $\partial_t u = a\partial_{\xi\xi}u$, which is considered in Subsection 1.1.1.

3. $\dfrac{\partial w}{\partial t} = (x^2 + a^2)^2\dfrac{\partial^2 w}{\partial x^2} + \Phi(x,t).$

Domain: $0 \le x \le l$. First boundary value problem.
The following conditions are prescribed:

$$\begin{aligned} w &= f(x) & \text{at} \quad t &= 0 & \text{(initial condition)}, \\ w &= g(t) & \text{at} \quad x &= 0 & \text{(boundary condition)}, \\ w &= h(t) & \text{at} \quad x &= l & \text{(boundary condition)}. \end{aligned}$$

Solution:

$$w(x,t) = \int_0^t\!\!\int_0^l G(x,\xi,t-\tau)\Phi(\xi,\tau)\,d\xi\,d\tau + \int_0^l G(x,\xi,t)f(\xi)\,d\xi$$

$$+ a^4\int_0^t g(\tau)\Lambda_1(x,t-\tau)\,d\tau - (a^2 + l^2)^2\int_0^t h(\tau)\Lambda_2(x,t-\tau)\,d\tau.$$

Here, the Green's function G is given by

$$G(x,\xi,t) = \sum_{n=1}^{\infty}\dfrac{y_n(x)y_n(\xi)\exp(-\mu_n^2 t)}{\|y_n\|^2(\xi^2 + a^2)^2},\qquad \mu_n^2 = \left[\dfrac{\pi n a}{\arctan(l/a)}\right]^2 - a^2,$$

$$y_n(x) = \sqrt{x^2 + a^2}\sin\left[\pi n\dfrac{\arctan(x/a)}{\arctan(l/a)}\right],\qquad \|y_n\|^2 = \dfrac{\arctan(l/a)}{2a},$$

and the functions Λ_1 and Λ_2 are expressed via the Green's function as follows:

$$\Lambda_1(x,t) = \left.\dfrac{\partial}{\partial\xi}G(x,\xi,t)\right|_{\xi=0},\qquad \Lambda_2(x,t) = \left.\dfrac{\partial}{\partial\xi}G(x,\xi,t)\right|_{\xi=l}.$$

⊙ *Reference*: A. G. Butkovskiy (1979).

4. $\dfrac{\partial w}{\partial t} = (x - a_1)^2(x - a_2)^2 \dfrac{\partial^2 w}{\partial x^2} - bw, \qquad a_1 \neq a_2.$

The transformation

$$w(x, t) = (x - a_2)e^{-bt}u(\xi, \tau), \quad \xi = \ln\left|\dfrac{x - a_1}{x - a_2}\right|, \quad \tau = (a_1 - a_2)^2 t$$

leads to a constant coefficient equation,

$$\dfrac{\partial u}{\partial \tau} = \dfrac{\partial^2 u}{\partial \xi^2} - \dfrac{\partial u}{\partial \xi},$$

which is considered in Subsection 1.1.4.

5. $\dfrac{\partial w}{\partial t} = (a_2 x^2 + a_1 x + a_0)^2 \dfrac{\partial^2 w}{\partial x^2} + bw.$

The transformation

$$w(x, t) = \exp\left[(a_2 a_0 - \tfrac{1}{4}a_1^2 + b)t\right] \sqrt{|a_2 x^2 + a_1 x + a_0|}\, u(\xi, t), \quad \xi = \int \dfrac{dx}{a_2 x^2 + a_1 x + a_0}$$

leads to a constant coefficient equation, $\partial_t u = \partial_{\xi\xi} u$, which is considered in Subsection 1.1.1.

6. $\dfrac{\partial w}{\partial t} = a x^{1-n} \dfrac{\partial^2 w}{\partial x^2}.$

This equation is encountered in diffusion boundary layer problems (see equation 1.9.1.3) and is a special case of 1.8.6.1 with $f(x) = ax^{1-n}$. In addition, it is a special case of equation 1.3.5.1 with $n = -1$ and is an equation of the form 1.3.6.2 for $n = -3$ (in both cases the equation is reduced to a constant coefficient equation). For $n = 0$, see equation 1.3.4.1.

$1°$. Particular solutions (A, B, and μ are arbitrary constants):

$$w(x) = Ax + B,$$

$$w(x, t) = Aan(n+1)t + Ax^{n+1} + B,$$

$$w(x, t) = Aa(n+1)(n+2)tx + Ax^{n+2} + B,$$

$$w(x, t) = A\left[an(n+1)t^2 + 2tx^{n+1} + \dfrac{x^{2n+2}}{a(n+1)(2n+1)}\right] + B,$$

$$w(x, t) = A\left[a(n+1)(n+2)t^2 x + 2tx^{n+2} + \dfrac{x^{2n+3}}{a(n+1)(2n+3)}\right] + B,$$

$$w(x, t) = A + Bt^{-\frac{n}{n+1}} \exp\left[-\dfrac{x^{n+1}}{a(n+1)^2 t}\right],$$

$$w(x, t) = A + Bxt^{-\frac{n+2}{n+1}} \exp\left[-\dfrac{x^{n+1}}{a(n+1)^2 t}\right],$$

$$w(x, t) = e^{\mu t}\sqrt{x}\left[AJ_{\frac{1}{2q}}\left(\dfrac{\sqrt{-\mu}}{\sqrt{a}\,q} x^q\right) + BY_{\frac{1}{2q}}\left(\dfrac{\sqrt{-\mu}}{\sqrt{a}\,q} x^q\right)\right] \qquad \text{for } \mu < 0,$$

$$w(x, t) = e^{\mu t}\sqrt{x}\left[AI_{\frac{1}{2q}}\left(\dfrac{\sqrt{\mu}}{\sqrt{a}\,q} x^q\right) + BK_{\frac{1}{2q}}\left(\dfrac{\sqrt{\mu}}{\sqrt{a}\,q} x^q\right)\right] \qquad \text{for } \mu > 0,$$

where $q = \tfrac{1}{2}(n+1)$, $J_\nu(z)$ and $Y_\nu(z)$ are the Bessel functions, and $I_\nu(z)$ and $K_\nu(z)$ are the modified Bessel functions.

Suppose $2/(n+1) = 2m+1$, where m is an integer. We have the following particular solutions (A, B, and μ are arbitrary constants):

$$w(x,t) = e^{\mu t} x (x^{1-2q} D)^{m+1} \left[A \exp\left(\frac{\sqrt{\mu}}{\sqrt{a}\,q} x^q\right) + B \exp\left(-\frac{\sqrt{\mu}}{\sqrt{a}\,q} x^q\right) \right] \quad \text{for } m \geq 0,$$

$$w(x,t) = e^{\mu t} x (x^{1-2q} D)^{-m} \left[A \exp\left(\frac{\sqrt{\mu}}{\sqrt{a}\,q} x^q\right) + B \exp\left(-\frac{\sqrt{\mu}}{\sqrt{a}\,q} x^q\right) \right] \quad \text{for } m < 0,$$

where $D = \dfrac{d}{dx}$, $q = \dfrac{n+1}{2} = \dfrac{1}{2m+1}$.

2°. Suppose $w = w(x,t)$ is a solution of the original equation. Then the functions

$$w_1 = Aw(\lambda x, \lambda^{n+1} t + C),$$

$$w_2 = \frac{A}{|\delta + \beta t|^{nk}} \exp\left[-\frac{\beta k^2 x^{n+1}}{a(\delta + \beta t)}\right] w\left(\frac{x}{(\delta + \beta t)^{2k}}, \frac{\gamma + \lambda t}{\delta + \beta t}\right), \quad k = \frac{1}{n+1}, \quad \lambda\delta - \beta\gamma = 1,$$

where A, C, β, δ, and λ are arbitrary constants, are also solutions of the equation.

3°. Domain: $0 \leq x < \infty$. First boundary value problem.
The following conditions are prescribed:

$$w = w_0 \quad \text{at} \quad t = 0 \quad \text{(initial condition)},$$
$$w = w_1 \quad \text{at} \quad x = 0 \quad \text{(boundary condition)},$$
$$w \to w_0 \quad \text{at} \quad x \to \infty \quad \text{(boundary condition)},$$

where $w_0 = \text{const}$ and $w_1 = \text{const}$.
Solution:

$$\frac{w - w_1}{w_0 - w_1} = \frac{1}{\Gamma(k)} \gamma\left(k, \frac{k^2 x^{n+1}}{at}\right), \quad k = \frac{1}{n+1},$$

where $\Gamma(k) = \gamma(k, \infty)$ is the gamma function and $\gamma(k, \zeta) = \displaystyle\int_0^\zeta \zeta^{k-1} e^{-\zeta} d\zeta$ is the incomplete gamma function.

4°. The transformation

$$\tau = \tfrac{1}{4} a(n+1)^2 t, \quad \xi = x^{\frac{n+1}{2}}$$

leads to the equation

$$\frac{\partial w}{\partial \tau} = \frac{\partial^2 w}{\partial \xi^2} + \frac{1-2k}{\xi} \frac{\partial w}{\partial \xi}, \quad k = \frac{1}{n+1},$$

which is considered in Subsection 1.2.5.

5°. Two discrete transformations are worth mentioning. They preserve the form of the original equation, but the parameter n is changed.
5.1. The point transformation

$$z = \frac{1}{x}, \quad u = \frac{w}{x} \quad (\text{transformation } \mathcal{F})$$

leads to a similar equation,

$$\frac{\partial u}{\partial t} = a z^{n+3} \frac{\partial^2 u}{\partial z^2}. \tag{1}$$

The transformation $\mathcal{F}$ changes the equation parameter in accordance with the rule $n \stackrel{\mathcal{F}}{\Longrightarrow} -n-2$. The second application of the transformation $\mathcal{F}$ leads to the original equation.

5.2. Using the Bäcklund transformation (see 1.8.6.1, Item 5.2)

$$\xi = x^n, \quad w = \frac{\partial v}{\partial \xi} \quad \text{(transformation } \mathcal{H}\text{)}$$

and integrating the resulting equation with respect to ξ, we obtain

$$\frac{\partial v}{\partial t} = an^2 \xi^{\frac{n-1}{n}} \frac{\partial^2 v}{\partial \xi^2}. \tag{2}$$

The transformation $\mathcal{H}$ changes the equation parameter in accordance with the rule $n \stackrel{\mathcal{H}}{\Longrightarrow} \frac{1}{n}$. The second application of the transformation $\mathcal{H}$ leads to the original equation.

The composition of transformations $\mathcal{G} = \mathcal{H} \circ \mathcal{F}$ changes the equation parameter in accordance with the rule $n \stackrel{\mathcal{G}}{\Longrightarrow} -\frac{1}{n+2}$.

The original equation reduces to a constant coefficient equation for $n = -3$ (see 1.3.6.2). Substituting $n = -3$ into (2) yields the equation

$$\frac{\partial v}{\partial t} = A\xi^{4/3} \frac{\partial^2 v}{\partial \xi^2},$$

which also can be reduced to a constant coefficient equation.

Likewise, using the transformations $\mathcal{F}$, $\mathcal{G}$, and $\mathcal{H}$, one may find some other equations of the given type that are reduced to a constant coefficient heat equation.

7. $\dfrac{\partial w}{\partial t} = ax^n \dfrac{\partial^2 w}{\partial x^2} + bx \dfrac{\partial w}{\partial x}.$

This is a special case of equation 1.8.4.5 with $f(t) = b$. On passing from t, x to the new variables

$$z = xe^{bt}, \quad \tau = \frac{a}{b(2-n)} e^{b(2-n)t} + \text{const},$$

we obtain an equation of the form 1.3.6.6:

$$\frac{\partial w}{\partial \tau} = z^n \frac{\partial^2 w}{\partial z^2}.$$

8. $\dfrac{\partial w}{\partial t} = a\left(x^n \dfrac{\partial^2 w}{\partial x^2} + nx^{n-1} \dfrac{\partial w}{\partial x}\right).$

This equation describes heat transfer in a quiescent medium (solid body) in the case where thermal diffusivity is a power law function of the coordinate. The equation can be rewritten in the form

$$\frac{\partial w}{\partial t} = a\frac{\partial}{\partial x}\left(x^n \frac{\partial w}{\partial x}\right),$$

which is more customary for applications.

1°. For $n = 2$, see equation 1.3.5.1. For $n \neq 2$, by passing from t, x to the new variables $\tau = \frac{1}{4}a(2-n)^2 t$, $z = x^{\frac{2-n}{2}}$, we obtain an equation of the form 1.2.5:

$$\frac{\partial w}{\partial \tau} = \frac{\partial^2 w}{\partial z^2} + \frac{n}{2-n} \frac{1}{z} \frac{\partial w}{\partial z}.$$

2°. The transformation

$$w(x,t) = x^{1-n} u(\xi, t), \quad \xi = x^{3-2n}$$

leads to a similar equation

$$\frac{\partial u}{\partial t} = b\frac{\partial}{\partial \xi}\left(\xi^{\frac{4-3n}{3-2n}}\frac{\partial u}{\partial \xi}\right), \qquad b = a(3-2n)^2.$$

9. $\quad \dfrac{\partial w}{\partial t} = a\left(x^{2m}\dfrac{\partial^2 w}{\partial x^2} + mx^{2m-1}\dfrac{\partial w}{\partial x}\right).$

This is a special case of equation 1.8.4.7 with $f(t) = g(t) = 0$.

The substitution

$$\xi = \begin{cases} \dfrac{1}{1-m}x^{1-m} & \text{if } m \neq 1, \\ \ln|x| & \text{if } m = 1, \end{cases}$$

leads to a constant coefficient equation, $\partial_t w = a\partial_{\xi\xi} w$, which is considered in Subsection 1.1.1.

10. $\quad \dfrac{\partial w}{\partial t} = ax^n\dfrac{\partial^2 w}{\partial x^2} + x(bt^m + c)\dfrac{\partial w}{\partial x}.$

This is a special case of equation 1.8.4.5 with $f(t) = bt^m + c$.

1.3.7. Equations of the Form $\dfrac{\partial w}{\partial t} = f(x,t)\dfrac{\partial^2 w}{\partial x^2} + g(x,t)\dfrac{\partial w}{\partial x} + h(x,t)w$

1. $\quad \dfrac{\partial w}{\partial t} = axt\dfrac{\partial^2 w}{\partial x^2} + (bx + ct^n)w.$

This is a special case of equation 1.8.8.1 with $f(t) = at$, $g(t) = 0$, $h(t) = b$, and $s(t) = ct^n$.

2. $\quad \dfrac{\partial w}{\partial t} = axt^k\dfrac{\partial^2 w}{\partial x^2} + bxt^m\dfrac{\partial w}{\partial x} + ct^n w.$

This is a special case of equation 1.8.8.1 with $f(t) = at^k$, $g(t) = bt^m$, $h(t) = 0$, and $s(t) = ct^n$.

3. $\quad \dfrac{\partial w}{\partial t} = x\left(at^k\dfrac{\partial^2 w}{\partial x^2} + bt^m\dfrac{\partial w}{\partial x} + ct^n w\right).$

This is a special case of equation 1.8.8.1 with $f(t) = at^k$, $g(t) = bt^m$, $h(t) = ct^n$, and $s(t) = 0$.

4. $\quad \dfrac{\partial w}{\partial t} = ax^2 t^k\dfrac{\partial^2 w}{\partial x^2} + bt^m x\dfrac{\partial w}{\partial x} + ct^n w.$

This is a special case of equation 1.8.8.2 with $f(t) = at^k$, $g(t) = bt^m$, and $h(t) = ct^n$.

5. $\quad \dfrac{\partial w}{\partial t} = ax^3 t^m\dfrac{\partial^2 w}{\partial x^2} + bxt^k\dfrac{\partial w}{\partial x} + ct^l w.$

This is a special case of equation 1.8.8.7 with $n = 3$, $f(t) = at^m$, $g(t) = bt^k$, and $h(t) = ct^l$.

6. $\quad \dfrac{\partial w}{\partial t} = ax^4 t^n\dfrac{\partial^2 w}{\partial x^2} + bt^m w.$

This is a special case of equation 1.8.8.4 with $f(t) = at^n$ and $g(t) = bt^m$.

1°. Particular solutions (A, B, and λ are arbitrary constants):

$$w(x,t) = (Ax + B)\exp\left(\frac{b}{m+1}t^{m+1}\right),$$

$$w(x,t) = A\left(\frac{2a}{n+1}t^{n+1}x + \frac{1}{x}\right)\exp\left(\frac{b}{m+1}t^{m+1}\right),$$

$$w(x,t) = Ax\exp\left(\frac{b}{m+1}t^{m+1} + \frac{a\lambda^2}{n+1}t^{n+1} + \frac{\lambda}{x}\right).$$

2°. The transformation

$$w(x,t) = x\exp\left(\frac{b}{m+1}t^{m+1}\right)u(\xi,\tau), \quad \xi = \frac{1}{x}, \quad \tau = \frac{a}{n+1}t^{n+1}$$

leads to a constant coefficient equation, $\partial_\tau u = \partial_{\xi\xi} u$, which is considered in Subsection 1.1.1.

7. $\dfrac{\partial w}{\partial t} = at^n \dfrac{\partial^2 w}{\partial x^2} + (bt^m x + ct^i)\dfrac{\partial w}{\partial x} + (dt^l x + et^p)w.$

This is a special case of equation 1.8.7.4 (hence, it can be reduced to a constant coefficient equation, which is considered in Subsection 1.1.1).

8. $\dfrac{\partial w}{\partial t} = at^n \dfrac{\partial^2 w}{\partial x^2} + (bt^m x + ct^i)\dfrac{\partial w}{\partial x} + (dt^l x^2 + et^p x + st^q)w.$

This is a special case of equation 1.8.7.5.

9. $\dfrac{\partial w}{\partial t} = ax^n t^m \dfrac{\partial^2 w}{\partial x^2} + bxt^k \dfrac{\partial w}{\partial x} + ct^p w.$

This is a special case of equation 1.8.8.7 with $f(t) = at^m$, $g(t) = bt^k$, and $h(t) = ct^p$.

1.3.8. Liquid-Film Mass Transfer Equation $(1-y^2)\dfrac{\partial w}{\partial x} = a\dfrac{\partial^2 w}{\partial y^2}$

This equation describes steady-state heat and mass transfer in a fluid film with a parabolic velocity profile. The variables have the following physical meanings: w is a dimensionless temperature (concentration); x and y are dimensionless coordinates measured, respectively, along and across the film ($y = 0$ corresponds to the free surface of the film and $y = 1$ to the solid surface the film flows down), and Pe $= 1/a$ is the Peclet number. Mixed boundary conditions are usually encountered in practical applications.

1.3.8-1. Particular solutions (A, B, k, and λ are arbitrary constants).

$$w(x,y) = kx - \frac{k}{12a}y^4 + \frac{k}{2a}y^2 + Ay + B,$$
$$w(x,y) = A\exp(-a\lambda^2 x)\exp(-\tfrac{1}{2}\lambda y^2)\,\Phi\!\left(\tfrac{1}{4} - \tfrac{1}{4}\lambda,\ \tfrac{1}{2};\ \lambda y^2\right),$$
$$w(x,y) = A\exp(-a\lambda^2 x)y\exp(-\tfrac{1}{2}\lambda y^2)\,\Phi\!\left(\tfrac{3}{4} - \tfrac{1}{4}\lambda,\ \tfrac{3}{2};\ \lambda y^2\right),$$

where $\Phi(\alpha,\beta;z) = 1 + \sum_{m=1}^{\infty} \dfrac{\alpha(\alpha+1)\ldots(\alpha+m-1)}{\beta(\beta+1)\ldots(\beta+m-1)}\dfrac{z^m}{m!}$ is the degenerate hypergeometric function.

1.3.8-2. Mass exchange between fluid film and gas.

The mass exchange between a fluid film and the gas above the free surface, provided that the admixture concentration at the film surface is constant and there is no mass transfer through the solid surface, meets the boundary conditions

$$\begin{aligned} w &= 0 \quad \text{at} \quad x = 0 \quad (0 < y < 1),\\ w &= 1 \quad \text{at} \quad y = 0 \quad (x > 0),\\ \partial_y w &= 0 \quad \text{at} \quad y = 1 \quad (x > 0). \end{aligned}$$

The solution of the original equation subject to these boundary conditions is given by

$$w(x,y) = 1 - \sum_{m=1}^{\infty} A_m \exp(-a\lambda_m^2 x) F_m(y), \tag{1}$$

$$F_m(y) = y \exp\left(-\tfrac{1}{2}\lambda_m y^2\right) \Phi\left(\tfrac{3}{4} - \tfrac{1}{4}\lambda_m, \tfrac{3}{2}; \lambda_m y^2\right),$$

where the function F_m and the coefficients A_m and λ_m are independent of the parameter a. The eigenvalues λ_m are solutions of the transcendental equation

$$\lambda_m \Phi\left(\tfrac{3}{4} - \tfrac{1}{4}\lambda_m, \tfrac{3}{2}; \lambda_m\right) - \Phi\left(\tfrac{3}{4} - \tfrac{1}{4}\lambda_m, \tfrac{1}{2}; \lambda_m\right) = 0.$$

The series coefficients A_m are calculated from

$$A_m = \frac{\int_0^1 (1-y^2) F_m(y)\, dy}{\int_0^1 (1-y^2)[F_m(y)]^2\, dy}, \qquad \text{where} \quad m = 1, 2, \ldots$$

Table 16 shows the first ten eigenvalues λ_m and coefficients A_m.

TABLE 16
Eigenvalues λ_m and coefficients A_m in solution (1)

m	λ_m	A_m	m	λ_m	A_m
1	2.2631	1.3382	6	22.3181	−0.1873
2	6.2977	−0.5455	7	26.3197	0.1631
3	10.3077	0.3589	8	30.3209	−0.1449
4	14.3128	−0.2721	9	34.3219	0.1306
5	18.3159	0.2211	10	38.3227	−0.1191

The solution asymptotics as $ax \to 0$ is given by

$$w = \operatorname{erfc}\left(\frac{y}{2\sqrt{ax}}\right),$$

where $\operatorname{erfc} z = \int_z^\infty \exp(-\xi^2)\, d\xi$ is the complementary error function.

1.3.8-3. Dissolution of a plate by a laminar fluid film.

The dissolution of a plate by a laminar fluid film, provided that the concentration at the solid surface is constant and there is no mass flux from the film into the gas, satisfies the boundary conditions

$$\begin{aligned} w &= 0 \quad \text{at} \quad x = 0 \quad (0 < y < 1), \\ \partial_y w &= 0 \quad \text{at} \quad y = 0 \quad (x > 0), \\ w &= 1 \quad \text{at} \quad y = 1 \quad (x > 0). \end{aligned}$$

The solution of the original equation subject to these boundary conditions is given by

$$w(x,y) = 1 - \sum_{m=0}^{\infty} A_m \exp(-a\lambda_m^2 x) G_m(y), \tag{2}$$

$$G_m(y) = \exp\left(-\tfrac{1}{2}\lambda_m y^2\right) \Phi\left(\tfrac{1}{4} - \tfrac{1}{4}\lambda_m, \tfrac{1}{2}; \lambda_m y^2\right),$$

where the functions G_m and the constants A_m and λ_m are independent of the parameter a.

The eigenvalues λ_m are solutions of the transcendental equation
$$\Phi\left(\tfrac{1}{4} - \tfrac{1}{4}\lambda_m, \tfrac{1}{2}; \lambda_m\right) = 0.$$

The following approximate relation is convenient to calculate λ_m:
$$\lambda_m = 4m + 1.68 \qquad (m = 0, 1, 2, \dots). \tag{3}$$

The maximum error of this formula is less than 0.2%.

The coefficients A_m are approximated by
$$A_0 = 1.2, \qquad A_m = (-1)^m 2.27\, \lambda_m^{-7/6} \qquad \text{for} \quad m = 1, 2, 3, \dots,$$
where the eigenvalues λ_m are defined by (3). The maximum error of the expressions for A_m is less than 0.1%.

The solution asymptotics as $ax \to 0$ is given by
$$w = \frac{1}{\Gamma\left(\tfrac{1}{3}\right)}\Gamma\left(\tfrac{1}{3}, \tfrac{2}{9}\zeta\right), \qquad \zeta = \frac{(1-y)^3}{ax},$$
where $\Gamma(\alpha, z) = \displaystyle\int_z^\infty e^{-\xi}\xi^{\alpha-1}\,d\xi$ is the incomplete gamma function, $\Gamma(\alpha) = \Gamma(\alpha, 0)$ is the gamma function, and $\Gamma\left(\tfrac{1}{3}\right) \approx 2.679$.

⊙ *References for Subsection* 1.3.8: Z. Rotem and J. E. Neilson (1966), E. J. Davis (1973), A. D. Polyanin, A. M. Kutepov, A. V. Vyazmin, and D. A. Kazenin (2001).

1.3.9. Equations of the Form $f(x,y)\dfrac{\partial w}{\partial x} + g(x,y)\dfrac{\partial w}{\partial y} = \dfrac{\partial^2 w}{\partial y^2} + h(x,y)$

1. $ax^n \dfrac{\partial w}{\partial x} + bx^k y \dfrac{\partial w}{\partial y} = \dfrac{\partial^2 w}{\partial y^2}.$

This is a special case of equation 1.9.1.1 with $f(x) = ax^n$ and $g(x) = bx^k$.

The transformation
$$t = \frac{1}{a}\int \exp\left[-\frac{2b}{a(k-n+1)}x^{k-n+1}\right]\frac{dx}{x^n}, \qquad z = y\exp\left[-\frac{b}{a(k-n+1)}x^{k-n+1}\right]$$
leads to a constant coefficient equation, $\partial_t w = \partial_{zz} w$, which is considered in Subsection 1.1.1.

2. $ax^n \dfrac{\partial w}{\partial x} + bx^k y \dfrac{\partial w}{\partial y} = \dfrac{\partial^2 w}{\partial y^2} - cx^m w.$

This is a special case of equation 1.9.1.2 with $f(x) = ax^n$, $g(x) = bx^k$, and $h(x) = cx^m$.

3. $ax^m y^{n-1} \dfrac{\partial w}{\partial x} + bx^k y^n \dfrac{\partial w}{\partial y} = \dfrac{\partial^2 w}{\partial y^2}.$

This is a special case of equation 1.9.1.3 with $f(x) = ax^m$ and $g(x) = bx^k$.

The transformation
$$t = \frac{1}{a}\int \exp\left[-\frac{b(n+1)}{a(k-m+1)}x^{k-m+1}\right]\frac{dx}{x^m}, \qquad z = y\exp\left[-\frac{b}{a(k-m+1)}x^{k-m+1}\right]$$
leads to a simpler equation of the form 1.3.6.6:
$$\frac{\partial w}{\partial t} = z^{1-n}\frac{\partial^2 w}{\partial z^2}.$$

4. $a\left(\dfrac{y}{\sqrt{x}}\right)^n \dfrac{\partial w}{\partial x} + \dfrac{b}{\sqrt{x}}\left(\dfrac{y}{\sqrt{x}}\right)^k \dfrac{\partial w}{\partial y} = \dfrac{\partial^2 w}{\partial y^2}.$

This is a special case of equation 1.9.1.4 with $f(z) = az^n$ and $g(z) = bz^k$.

▶ See Subsection 1.9.1 for other equations of this form.

1.4. Equations Containing Exponential Functions and Arbitrary Parameters

1.4.1. Equations of the Form $\frac{\partial w}{\partial t} = a\frac{\partial^2 w}{\partial x^2} + f(x,t)w$

1. $\quad \dfrac{\partial w}{\partial t} = a\dfrac{\partial^2 w}{\partial x^2} + (be^{\beta t} + c)w.$

This is a special case of equation 1.8.1.1 with $f(t) = be^{\beta t} + c$.

1°. Particular solutions (A, B, and λ are arbitrary constants):

$$w(x,t) = (Ax + B)\exp\left(\frac{b}{\beta}e^{\beta t} + ct\right),$$

$$w(x,t) = A(x^2 + 2at)\exp\left(\frac{b}{\beta}e^{\beta t} + ct\right),$$

$$w(x,t) = A\exp\left(\lambda x + a\lambda^2 t + \frac{b}{\beta}e^{\beta t} + ct\right).$$

2°. The substitution $w(x,t) = u(x,t)\exp\left(\dfrac{b}{\beta}e^{\beta t} + ct\right)$ leads to a constant coefficient equation, $\partial_t u = a\partial_{xx} u$, which is considered in Subsection 1.1.1.

2. $\quad \dfrac{\partial w}{\partial t} = a\dfrac{\partial^2 w}{\partial x^2} + (be^{\beta t} + cx)w.$

This is a special case of equation 1.8.1.6 with $f(t) = c$ and $g(t) = be^{\beta t}$.

1°. Particular solutions (A and λ are arbitrary constants):

$$w(x,t) = A\exp\left(cxt + \frac{b}{\beta}e^{\beta t} + \frac{1}{3}ac^2 t^3\right),$$

$$w(x,t) = A(x + act^2)\exp\left(cxt + \frac{b}{\beta}e^{\beta t} + \frac{1}{3}ac^2 t^3\right),$$

$$w(x,t) = A\exp\left[x(ct + \lambda) + \frac{b}{\beta}e^{\beta t} + \frac{1}{3}ac^2 t^3 + ac\lambda t^2 + a\lambda^2 t\right].$$

2°. The transformation

$$w(x,t) = u(z,t)\exp\left(cxt + \frac{b}{\beta}e^{\beta t} + \frac{1}{3}ac^2 t^3\right), \quad z = x + act^2$$

leads to a constant coefficient equation, $\partial_t u = a\partial_{zz} u$, which is considered in Subsection 1.1.1.

3. $\quad \dfrac{\partial w}{\partial t} = a\dfrac{\partial^2 w}{\partial x^2} + (be^{\beta x} - c)w.$

This is a special case of equation 1.8.1.2 with $f(x) = be^{\beta x} - c$.

Particular solutions (A and λ are arbitrary constants):

$$w(x,t) = A\exp(-\lambda t)J_\nu\left(\frac{2\sqrt{b}}{\beta\sqrt{a}}e^{\beta x/2}\right), \quad \nu = \frac{2}{\beta}\sqrt{\frac{c-\lambda}{a}},$$

$$w(x,t) = A\exp(-\lambda t)Y_\nu\left(\frac{2\sqrt{b}}{\beta\sqrt{a}}e^{\beta x/2}\right), \quad \nu = \frac{2}{\beta}\sqrt{\frac{c-\lambda}{a}},$$

where $J_\nu(\xi)$ and $Y_\nu(\xi)$ are the Bessel functions of the first and second kind, respectively.

4. $\dfrac{\partial w}{\partial t} = a\dfrac{\partial^2 w}{\partial x^2} + (bxe^{\beta t} + c)w.$

This is a special case of equation 1.8.1.6 with $f(t) = be^{\beta t}$ and $g(t) = c$.

1°. Particular solutions (A and λ are arbitrary constants):

$$w(x,t) = A\exp\left(\frac{b}{\beta}xe^{\beta t} + \frac{ab^2}{2\beta^3}e^{2\beta t} + ct\right),$$

$$w(x,t) = A\left(x + \frac{2ab}{\beta^2}e^{\beta t}\right)\exp\left(\frac{b}{\beta}xe^{\beta t} + \frac{ab^2}{2\beta^3}e^{2\beta t} + ct\right),$$

$$w(x,t) = A\exp\left[x\left(\frac{b}{\beta}e^{\beta t} + \lambda\right) + \frac{ab^2}{2\beta^3}e^{2\beta t} + \frac{2ab\lambda}{\beta^2}e^{\beta t} + (a\lambda^2 + c)t\right].$$

2°. The transformation

$$w(x,t) = u(z,t)\exp\left(\frac{b}{\beta}xe^{\beta t} + \frac{ab^2}{2\beta^3}e^{2\beta t} + ct\right), \quad z = x + \frac{2ab}{\beta^2}e^{\beta t}$$

leads to a constant coefficient equation, $\partial_t u = a\partial_{zz} u$, which is considered in Subsection 1.1.1.

5. $\dfrac{\partial w}{\partial t} = a\dfrac{\partial^2 w}{\partial x^2} + x(be^{\beta t} + c)w.$

This is a special case of equation 1.8.1.3 with $f(t) = be^{\beta t} + c$.

1°. Particular solutions (A and λ are arbitrary constants):

$$w(x,t) = A\exp\left[x\left(\frac{b}{\beta}e^{\beta t} + ct\right) + a\phi(t)\right],$$

$$w(x,t) = A\left[x + a\left(\frac{2b}{\beta^2}e^{\beta t} + ct^2\right)\right]\exp\left[x\left(\frac{b}{\beta}e^{\beta t} + ct\right) + a\phi(t)\right],$$

$$w(x,t) = A\exp\left[x\left(\frac{b}{\beta}e^{\beta t} + ct + \lambda\right) + a\lambda\left(\frac{2b}{\beta^2}e^{\beta t} + ct^2 + \lambda t\right) + a\phi(t)\right],$$

where $\phi(t) = \tfrac{1}{2}b^2\beta^{-3}e^{2\beta t} + 2bc\beta^{-3}(\beta t - 1)e^{\beta t} + \tfrac{1}{3}c^2 t^3$.

2°. The transformation

$$w(x,t) = u(z,t)\exp\left[x\left(\frac{b}{\beta}e^{\beta t} + ct\right) + a\phi(t)\right], \quad z = x + a\left(\frac{2b}{\beta^2}e^{\beta t} + ct^2\right)$$

leads to a constant coefficient equation, $\partial_t u = a\partial_{zz} u$, which is considered in Subsection 1.1.1.

6. $\dfrac{\partial w}{\partial t} = a\dfrac{\partial^2 w}{\partial x^2} + x^2(be^{\beta t} + c)w.$

This is a special case of equation 1.8.7.5 with $n(t) = a$, $f(t) = g(t) = s(t) = p(t) = 0$, and $h(t) = be^{\beta t} + c$.

7. $\dfrac{\partial w}{\partial t} = a\dfrac{\partial^2 w}{\partial x^2} + (be^{\beta t} + ce^{\lambda t})w.$

This is a special case of equation 1.8.1.1 with $f(t) = be^{\beta t} + ce^{\lambda t}$.

1°. Particular solutions (A, B, and ν are arbitrary constants):

$$w(x,t) = (Ax + B)\exp\left(\frac{b}{\beta}e^{\beta t} + \frac{c}{\lambda}e^{\lambda t}\right),$$

$$w(x,t) = A(x^2 + 2at)\exp\left(\frac{b}{\beta}e^{\beta t} + \frac{c}{\lambda}e^{\lambda t}\right),$$

$$w(x,t) = A\exp\left(\nu x + a\nu^2 t + \frac{b}{\beta}e^{\beta t} + \frac{c}{\lambda}e^{\lambda t}\right).$$

2°. The substitution $w(x,t) = u(x,t)\exp\left(\frac{b}{\beta}e^{\beta t} + \frac{c}{\lambda}e^{\lambda t}\right)$ leads to a constant coefficient equation, $\partial_t u = a\partial_{xx}u$, which is considered in Subsection 1.1.1.

8. $\dfrac{\partial w}{\partial t} = a\dfrac{\partial^2 w}{\partial x^2} + (be^{\beta x} + ce^{\lambda t} + d)w.$

The substitution $w(x,t) = u(x,t)\exp\left(\frac{c}{\lambda}e^{\lambda t}\right)$ leads to an equation of the form 1.4.1.3:

$$\frac{\partial u}{\partial t} = a\frac{\partial^2 u}{\partial x^2} + (be^{\beta x} + d)u.$$

9. $\dfrac{\partial w}{\partial t} = a\dfrac{\partial^2 w}{\partial x^2} + (be^{\beta x + \lambda t} + c)w.$

For $\beta = 0$, see equation 1.4.1.1; for $\lambda = 0$, see equation 1.4.1.3.

For $\beta \neq 0$, the transformation

$$w(x,t) = u(z,t)e^{\mu x}, \quad z = x + \frac{\lambda}{\beta}t, \quad \text{where} \quad \mu = \frac{\lambda}{2a\beta},$$

leads to an equation of the form 1.4.1.3:

$$\frac{\partial u}{\partial t} = a\frac{\partial^2 u}{\partial z^2} + (be^{\beta z} + c + a\mu^2)u.$$

1.4.2. Equations of the Form $\dfrac{\partial w}{\partial t} = a\dfrac{\partial^2 w}{\partial x^2} + f(x,t)\dfrac{\partial w}{\partial x}$

1. $\dfrac{\partial w}{\partial t} = a\dfrac{\partial^2 w}{\partial x^2} + (be^{\beta t} + c)\dfrac{\partial w}{\partial x}.$

This is a special case of equation 1.8.2.1 with $f(t) = be^{\beta t} + c$.

1°. Particular solutions (A, B, and λ are arbitrary constants):

$$w(x,t) = Ax + A\left(\frac{b}{\beta}e^{\beta t} + ct\right) + B,$$

$$w(x,t) = A\left(x + \frac{b}{\beta}e^{\beta t} + ct\right)^2 + 2aAt + B,$$

$$w(x,t) = A\exp\left[\lambda x + \lambda\frac{b}{\beta}e^{\beta t} + (a\lambda^2 + c\lambda)t\right] + B.$$

2°. On passing from t, x to the new variables t, $z = x + \dfrac{b}{\beta}e^{\beta t} + ct$, we obtain a constant coefficient equation, $\partial_t w = a\partial_{zz}w$, which is considered in Subsection 1.1.1.

2. $\dfrac{\partial w}{\partial t} = a\dfrac{\partial^2 w}{\partial x^2} + (be^{\beta x} + c)\dfrac{\partial w}{\partial x}.$

This is a special case of equation 1.8.2.2 with $f(x) = be^{\beta x} + c$.

1°. Particular solutions (A and λ are arbitrary constants):

$$w(x,t) = A\exp(-\lambda t + k\beta x)\,\Phi\!\left(k,\; 2k+1+\dfrac{c}{a\beta};\; -\dfrac{b}{a\beta}e^{\beta x}\right),$$
$$w(x,t) = A\exp(-\lambda t + k\beta x)\,\Psi\!\left(k,\; 2k+1+\dfrac{c}{a\beta};\; -\dfrac{b}{a\beta}e^{\beta x}\right),$$
(1)

where $k = k(\lambda)$ is a root of the quadratic equation $a\beta^2 k^2 + c\beta k + \lambda = 0$; $\Phi(\alpha,\nu;z)$ and $\Psi(\alpha,\nu;z)$ are degenerate hypergeometric functions. [Regarding the degenerate hypergeometric functions, see Supplement A.9 and the books by Abramowitz and Stegun (1964) and Bateman and Erdélyi (1953, Vol. 1)].

Remark. In solutions (1), the parameter k can be considered arbitrary, and then $\lambda = -a\beta^2 k^2 - c\beta k$.

2°. Other particular solutions (A and B are arbitrary constants):

$$w(x) = A + B\int F(x)\,dx, \quad F(x) = \exp\!\left(-\dfrac{b}{a\beta}e^{\beta x} - \dfrac{c}{a}x\right),$$
$$w(x,t) = Aat + A\int F(x)\!\left(\int \dfrac{dx}{F(x)}\right) dx,$$
$$w(x,t) = AatG(x) + A\int F(x)\!\left(\int \dfrac{G(x)\,dx}{F(x)}\right) dx, \quad G(x) = \int F(x)\,dx.$$

3°. The substitution $z = e^{\beta x}$ leads to an equation of the form 1.3.5.4:

$$\dfrac{\partial w}{\partial t} = a\beta^2 z^2\dfrac{\partial^2 w}{\partial z^2} + \beta z(bz + c + a\beta)\dfrac{\partial w}{\partial z}.$$

3. $\dfrac{\partial w}{\partial t} = a\dfrac{\partial^2 w}{\partial x^2} + x(be^{\beta t} + c)\dfrac{\partial w}{\partial x}.$

This is a special case of equation 1.8.2.3 with $f(t) = be^{\beta t} + c$.

1°. Particular solutions (A, B, and λ are arbitrary constants):

$$w(x,t) = AxF(t) + B, \quad F(t) = \exp\!\left(\dfrac{b}{\beta}e^{\beta t} + ct\right),$$
$$w(x,t) = Ax^2 F^2(t) + 2Aa\int F^2(t)\,dt + B,$$
$$w(x,t) = A\exp\!\left[\lambda x F(t) + a\lambda^2\int F^2(t)\,dt\right] + B.$$

2°. On passing from t, x to the new variables (A is any number)

$$\tau = \int F^2(t)\,dt + A, \quad z = xF(t), \quad \text{where}\quad F(t) = \exp\!\left(\dfrac{b}{\beta}e^{\beta t} + ct\right),$$

for the function $w(\tau, z)$ we obtain a constant coefficient equation, $\partial_\tau w = a\partial_{zz} w$, which is considered in Subsection 1.1.1.

4. $\dfrac{\partial w}{\partial t} = a\dfrac{\partial^2 w}{\partial x^2} + (be^{\beta t} + ce^{\lambda t})\dfrac{\partial w}{\partial x}.$

This is a special case of equation 1.8.2.1 with $f(t) = be^{\beta t} + ce^{\lambda t}$.

1°. Particular solutions (A, B, and μ are arbitrary constants):

$$w(x,t) = Ax + A\left(\dfrac{b}{\beta}e^{\beta t} + \dfrac{c}{\lambda}e^{\lambda t}\right) + B,$$

$$w(x,t) = A\left(x + \dfrac{b}{\beta}e^{\beta t} + \dfrac{c}{\lambda}e^{\lambda t}\right)^2 + 2aAt + B,$$

$$w(x,t) = A\exp\left(\mu x + \mu\dfrac{b}{\beta}e^{\beta t} + \mu\dfrac{c}{\lambda}e^{\lambda t} + a\mu^2 t\right) + B.$$

2°. On passing from t, x to the new variables t, $z = x + \dfrac{b}{\beta}e^{\beta t} + \dfrac{c}{\lambda}e^{\lambda t}$, we obtain a constant coefficient equation, $\partial_t w = a\partial_{zz} w$, which is considered in Subsection 1.1.1.

5. $\dfrac{\partial w}{\partial t} = a\dfrac{\partial^2 w}{\partial x^2} + (be^{\beta t + \lambda x} + c)\dfrac{\partial w}{\partial x}.$

For $\beta = 0$, see equation 1.4.2.2; for $\lambda = 0$, see equation 1.4.2.1.

For $\lambda \neq 0$, the substitution $z = x + (\beta/\lambda)t$ leads to an equation of the form 1.4.2.2:

$$\dfrac{\partial w}{\partial t} = a\dfrac{\partial^2 w}{\partial z^2} + \left(be^{\lambda z} + c - \dfrac{\beta}{\lambda}\right)\dfrac{\partial w}{\partial z}.$$

6. $\dfrac{\partial w}{\partial t} = a\dfrac{\partial^2 w}{\partial x^2} + (be^{\beta t} + cx)\dfrac{\partial w}{\partial x}.$

This is a special case of equation 1.8.1.6 with $f(t) = c$ and $g(t) = be^{\beta t}$.

7. $\dfrac{\partial w}{\partial t} = a\dfrac{\partial^2 w}{\partial x^2} + (bxe^{\beta t} + c)\dfrac{\partial w}{\partial x}.$

This is a special case of equation 1.8.1.6 with $f(t) = be^{\beta t}$ and $g(t) = c$.

8. $\dfrac{\partial w}{\partial t} = a\dfrac{\partial^2 w}{\partial x^2} + (bxe^{\beta t} + ce^{\lambda t})\dfrac{\partial w}{\partial x}.$

This is a special case of equation 1.8.1.6 with $f(t) = be^{\beta t}$ and $g(t) = ce^{\lambda t}$.

9. $\dfrac{\partial w}{\partial t} = a\dfrac{\partial^2 w}{\partial x^2} + x(be^{\beta t} + ce^{\lambda t})\dfrac{\partial w}{\partial x}.$

This is a special case of equation 1.8.2.3 with $f(t) = be^{\beta t} + ce^{\lambda t}$.

1°. Particular solutions (A, B, and μ are arbitrary constants):

$$w(x,t) = AxF(t) + B, \qquad F(t) = \exp\left(\dfrac{b}{\beta}e^{\beta t} + \dfrac{c}{\lambda}e^{\lambda t}\right),$$

$$w(x,t) = Ax^2 F^2(t) + 2Aa\int F^2(t)\,dt + B,$$

$$w(x,t) = A\exp\left[\mu x F(t) + a\mu^2 \int F^2(t)\,dt\right] + B.$$

2°. On passing from t, x to the new variables (A is any number)

$$\tau = \int F^2(t)\,dt + A, \quad z = xF(t), \qquad \text{where} \quad F(t) = \exp\left(\dfrac{b}{\beta}e^{\beta t} + \dfrac{c}{\lambda}e^{\lambda t}\right),$$

for the function $w(\tau, z)$ we obtain a constant coefficient equation, $\partial_\tau w = a\partial_{zz} w$, which is considered in Subsection 1.1.1.

10. $\dfrac{\partial w}{\partial t} = a\dfrac{\partial^2 w}{\partial x^2} + \left(cxe^{\beta t} + \dfrac{b}{x}\right)\dfrac{\partial w}{\partial x}.$

This is a special case of equation 1.8.2.6 with $f(t) = ce^{\beta t}$.

1.4.3. Equations of the Form $\dfrac{\partial w}{\partial t} = a\dfrac{\partial^2 w}{\partial x^2} + f(x,t)\dfrac{\partial w}{\partial x} + g(x,t)w$

1. $\dfrac{\partial w}{\partial t} = a\dfrac{\partial^2 w}{\partial x^2} + b\dfrac{\partial w}{\partial x} + (ce^{\beta t} + s)w.$

The substitution
$$w(x,t) = u(x,t)\exp\left[-\dfrac{b}{2a}x + \dfrac{c}{\beta}e^{\beta t} + \left(s - \dfrac{b^2}{4a}\right)t\right]$$
leads to a constant coefficient equation, $\partial_t u = a\partial_{xx} u$, which is considered in Subsection 1.1.1.

2. $\dfrac{\partial w}{\partial t} = a\dfrac{\partial^2 w}{\partial x^2} + b\dfrac{\partial w}{\partial x} + (ce^{\beta x} + s)w.$

The substitution $w(x,t) = u(x,t)\exp\left(-\dfrac{b}{2a}x\right)$ leads to an equation of the form 1.4.1.3:
$$\dfrac{\partial u}{\partial t} = a\dfrac{\partial^2 u}{\partial x^2} + \left(ce^{\beta x} + s - \dfrac{b^2}{4a}\right)u.$$

3. $\dfrac{\partial w}{\partial t} = a\dfrac{\partial^2 w}{\partial x^2} + b\dfrac{\partial w}{\partial x} + (ce^{\beta t} + se^{\mu t})w.$

The substitution
$$w(x,t) = u(x,t)\exp\left(-\dfrac{b}{2a}x + \dfrac{c}{\beta}e^{\beta t} + \dfrac{s}{\mu}e^{\mu t} - \dfrac{b^2}{4a}t\right)$$
leads to a constant coefficient equation, $\partial_t u = a\partial_{xx} u$, which is considered in Subsection 1.1.1.

4. $\dfrac{\partial w}{\partial t} = a\dfrac{\partial^2 w}{\partial x^2} + b\dfrac{\partial w}{\partial x} + (ce^{\beta x} + se^{\mu t})w.$

The substitution $w(x,t) = u(x,t)\exp\left(\dfrac{s}{\mu}e^{\mu t}\right)$ leads to an equation of the form 1.4.3.2:
$$\dfrac{\partial u}{\partial t} = a\dfrac{\partial^2 u}{\partial x^2} + b\dfrac{\partial u}{\partial x} + ce^{\beta x}u.$$

5. $\dfrac{\partial w}{\partial t} = a\dfrac{\partial^2 w}{\partial x^2} + \beta x\dfrac{\partial w}{\partial x} + ce^{2\beta t}w.$

On passing from t, x to the new variables (A and B are any numbers)
$$\tau = \dfrac{A^2}{2\beta}e^{2\beta t} + B, \quad z = Axe^{\beta t},$$
for the function $w(\tau, z)$ we obtain a constant coefficient equation of the form 1.1.3:
$$\dfrac{\partial w}{\partial \tau} = a\dfrac{\partial^2 w}{\partial z^2} + cA^{-2}w.$$

6. $\dfrac{\partial w}{\partial t} = a_2\dfrac{\partial^2 w}{\partial x^2} + \left(a_1 e^{\beta x} + b_1\right)\dfrac{\partial w}{\partial x} + \left(a_0 e^{2\beta x} + b_0 e^{\beta x} + c_0\right)w.$

The substitution $z = e^{\beta x}$ leads to an equation of the form 1.3.5.4:
$$\dfrac{\partial w}{\partial t} = a_2\beta^2 z^2\dfrac{\partial^2 w}{\partial z^2} + \beta z\left(a_1 z + b_1 + a_2\beta\right)\dfrac{\partial w}{\partial z} + \left(a_0 z^2 + b_0 z + c_0\right)w.$$

1.4.4. Equations of the Form $\dfrac{\partial w}{\partial t} = ax^n \dfrac{\partial^2 w}{\partial x^2} + f(x,t)\dfrac{\partial w}{\partial x} + g(x,t)w$

1. $\dfrac{\partial w}{\partial t} = ax \dfrac{\partial^2 w}{\partial x^2} + (be^{\beta t} + cx)w.$

   This is a special case of equation 1.8.8.1 with $f(t) = a$, $g(t) = 0$, $h(t) = c$, and $s(t) = be^{\beta t}$.

2. $\dfrac{\partial w}{\partial t} = ax \dfrac{\partial^2 w}{\partial x^2} + (bxe^{\beta t} + c)w.$

   This is a special case of equation 1.8.8.1 with $f(t) = a$, $g(t) = 0$, $h(t) = be^{\beta t}$, and $s(t) = c$.

3. $\dfrac{\partial w}{\partial t} = ax \dfrac{\partial^2 w}{\partial x^2} + bx \dfrac{\partial w}{\partial x} + (ce^{\beta t} + d)w.$

   This is a special case of equation 1.8.8.1 with $f(t) = a$, $g(t) = b$, $h(t) = 0$, and $s(t) = ce^{\beta t} + d$.

4. $\dfrac{\partial w}{\partial t} = ax^2 \dfrac{\partial^2 w}{\partial x^2} + (be^{\beta t} + c)w.$

   This is a special case of equation 1.8.8.2 with $f(t) = a$, $g(t) = 0$, and $h(t) = be^{\beta t} + c$.

5. $\dfrac{\partial w}{\partial t} = ax^4 \dfrac{\partial^2 w}{\partial x^2} + (be^{\beta t} + c)w.$

   This is a special case of equation 1.8.8.4 with $f(t) = a$ and $g(t) = be^{\beta t} + c$.

6. $\dfrac{\partial w}{\partial t} = ax^n \dfrac{\partial^2 w}{\partial x^2} + x(be^{\beta t} + c) \dfrac{\partial w}{\partial x}.$

   This is a special case of equation 1.8.4.5 with $f(t) = be^{\beta t} + c$.

7. $\dfrac{\partial w}{\partial t} = ax^n \dfrac{\partial^2 w}{\partial x^2} + bxe^{\beta t} \dfrac{\partial w}{\partial x} + ce^{\mu t} w.$

   This is a special case of equation 1.8.8.7 with $f(t) = a$, $g(t) = be^{\beta t}$, and $h(t) = ce^{\mu t}$.

1.4.5. Equations of the Form $\dfrac{\partial w}{\partial t} = ae^{\beta x} \dfrac{\partial^2 w}{\partial x^2} + f(x,t)\dfrac{\partial w}{\partial x} + g(x,t)w$

1. $\dfrac{\partial w}{\partial t} = ae^{\beta x} \dfrac{\partial^2 w}{\partial x^2}.$

   This is a special case of equation 1.8.6.1 with $f(x) = ae^{\beta x}$.

   Particular solutions (A, B, and μ are arbitrary constants):

   $$w(x) = Ax + B,$$
   $$w(x,t) = A(a\beta^2 t + e^{-\beta x}) + B,$$
   $$w(x,t) = A(a\beta^3 tx + \beta x e^{-\beta x} + 2e^{-\beta x}) + B,$$
   $$w(x,t) = A(2a^2\beta^4 t^2 + 4a\beta^2 t e^{-\beta x} + e^{-2\beta x}) + B,$$
   $$w(x,t) = A\exp(-\mu t)\, J_0\!\left(\dfrac{2}{\beta}\sqrt{\dfrac{\mu}{a}}\exp(-\tfrac{1}{2}\beta x)\right),$$
   $$w(x,t) = A\exp(-\mu t)\, Y_0\!\left(\dfrac{2}{\beta}\sqrt{\dfrac{\mu}{a}}\exp(-\tfrac{1}{2}\beta x)\right),$$

   where $J_0(\xi)$ and $Y_0(\xi)$ are the Bessel functions.

2. $\dfrac{\partial w}{\partial t} = ae^{\beta x}\dfrac{\partial^2 w}{\partial x^2} + bw.$

1°. Particular solutions (A, B, and μ are arbitrary constants):

$$w(x,t) = e^{bt}(Ax+B),$$
$$w(x,t) = Ae^{bt}\left(a\beta^2 t + e^{-\beta x}\right) + Be^{bt},$$
$$w(x,t) = Ae^{bt}\left(a\beta^3 tx + \beta x e^{-\beta x} + 2e^{-\beta x}\right) + Be^{bt},$$
$$w(x,t) = Ae^{bt}\left(2a^2\beta^4 t^2 + 4a\beta^2 t e^{-\beta x} + e^{-2\beta x}\right),$$
$$w(x,t) = A\exp(-\mu t)\, J_0\left(\dfrac{2}{\beta}\sqrt{\dfrac{\mu+b}{a}}\,\exp(-\tfrac{1}{2}\beta x)\right),$$
$$w(x,t) = A\exp(-\mu t)\, Y_0\left(\dfrac{2}{\beta}\sqrt{\dfrac{\mu+b}{a}}\,\exp(-\tfrac{1}{2}\beta x)\right),$$

where $J_0(\xi)$ and $Y_0(\xi)$ are the Bessel functions.

2°. The substitution $w(x,t) = e^{bt}u(x,t)$ leads to an equation of the form 1.4.5.1: $\partial_t u = ae^{\beta x}\partial_{xx}u$.

3. $\dfrac{\partial w}{\partial t} = ae^{\beta x}\dfrac{\partial^2 w}{\partial x^2} + (be^{\mu t} + c)w.$

1°. Particular solutions (A, B, μ are arbitrary constants):

$$w(x) = (Ax+B)\exp\left(\dfrac{b}{\mu}e^{\mu t} + ct\right),$$
$$w(x,t) = A\left(a\beta^2 t + e^{-\beta x}\right)\exp\left(\dfrac{b}{\mu}e^{\mu t} + ct\right),$$
$$w(x,t) = A\left(a\beta^3 tx + \beta x e^{-\beta x} + 2e^{-\beta x}\right)\exp\left(\dfrac{b}{\mu}e^{\mu t} + ct\right),$$
$$w(x,t) = A\left(2a^2\beta^4 t^2 + 4a\beta^2 t e^{-\beta x} + e^{-2\beta x}\right)\exp\left(\dfrac{b}{\mu}e^{\mu t} + ct\right),$$
$$w(x,t) = A\exp\left[\dfrac{b}{\mu}e^{\mu t} + (c-\mu)t\right]J_0\left(\dfrac{2}{\beta}\sqrt{\dfrac{\mu}{a}}\,\exp(-\tfrac{1}{2}\beta x)\right),$$
$$w(x,t) = A\exp\left[\dfrac{b}{\mu}e^{\mu t} + (c-\mu)t\right]Y_0\left(\dfrac{2}{\beta}\sqrt{\dfrac{\mu}{a}}\,\exp(\tfrac{1}{2}\beta x)\right).$$

2°. The substitution $w(x,t) = \exp\!\left(\dfrac{b}{\mu}e^{\mu t} + ct\right)u(x,t)$ leads to an equation of the form 1.4.5.1: $\partial_t u = ae^{\beta x}\partial_{xx}u$.

4. $\dfrac{\partial w}{\partial t} = a\left(e^{\beta x}\dfrac{\partial^2 w}{\partial x^2} + \beta e^{\beta x}\dfrac{\partial w}{\partial x}\right).$

This equation describes heat transfer in a quiescent medium (solid body) in the case where thermal diffusivity is an exponential function of the coordinate. The equation can be rewritten in the divergence form

$$\dfrac{\partial w}{\partial t} = a\dfrac{\partial}{\partial x}\left(e^{\beta x}\dfrac{\partial w}{\partial x}\right)$$

that is more customary for applications.

1°. Particular solutions (A, B, C, and μ are arbitrary constants):

$$w(x,t) = A\exp\left(-\frac{e^{-\beta x}}{a\beta^2 t + C}\right) + B,$$

$$w(x,t) = Aa\beta^2 t - A(\beta x + 1)e^{-\beta x} + B,$$

$$w(x,t) = 2Aa\beta^2 t e^{-\beta x} + Ae^{-2\beta x} + B,$$

$$w(x,t) = Aa^2\beta^4 t^2 - 2Aa\beta^2 t(\beta x + 1)e^{-\beta x} - A(\beta x + \tfrac{5}{2})e^{-2\beta x} + B,$$

$$w(x,t) = Aa^2\beta^4 t^2 e^{-\beta x} + Aa\beta^2 t e^{-2\beta x} + \tfrac{1}{6}Ae^{-3\beta x} + B,$$

$$w(x,t) = 2Aa^3\beta^6 t^3 e^{-\beta x} + 3Aa^2\beta^4 t^2 e^{-2\beta x} + Aa\beta^2 t e^{-3\beta x} + \tfrac{1}{12}Ae^{-4\beta x} + B,$$

$$w(x,t) = e^{-n\beta x} + \sum_{k=1}^{n-1} \frac{[n(n-1)\dots(n-k)]^2}{n(n-k)k!}(a\beta^2 t)^k e^{(k-n)\beta x},$$

$$w(x,t) = e^{\mu t - \frac{1}{2}\beta x}\left[AJ_1\left(\frac{2\sqrt{-\mu}}{\beta\sqrt{a}}e^{-\frac{1}{2}\beta x}\right) + BY_1\left(\frac{2\sqrt{-\mu}}{\beta\sqrt{a}}e^{-\frac{1}{2}\beta x}\right)\right] \quad \text{for } \mu < 0,$$

$$w(x,t) = e^{\mu t - \frac{1}{2}\beta x}\left[AI_1\left(\frac{2\sqrt{\mu}}{\beta\sqrt{a}}e^{-\frac{1}{2}\beta x}\right) + BK_1\left(\frac{2\sqrt{\mu}}{\beta\sqrt{a}}e^{-\frac{1}{2}\beta x}\right)\right] \quad \text{for } \mu > 0,$$

where $J_1(z)$ and $Y_1(z)$ are the Bessel functions, $I_1(z)$ and $K_1(z)$ are the modified Bessel functions.

2°. A solution containing an arbitrary function of the space variable:

$$w(x,t) = f(x) + \sum_{n=1}^{\infty} \frac{(at)^n}{n!} \boldsymbol{L}^n[f(x)], \qquad \boldsymbol{L} \equiv \frac{d}{dx}\left(e^{\beta x}\frac{d}{dx}\right),$$

where $f(x)$ is any infinitely differentiable function. This solution satisfies the initial condition $w(x,0) = f(x)$.

3°. A solution containing an arbitrary function of time:

$$w(x,t) = A + e^{-\beta x}g(t) + \sum_{n=2}^{\infty} \frac{1}{n[(n-1)!]^2(a\beta^2)^{n-1}} e^{-\beta n x} g_t^{(n-1)}(t),$$

where $g(t)$ is any infinitely differentiable function. If $g(t)$ is a polynomial, then the series has finitely many terms.

4°. The transformation (C_1, C_2, and C_3 are any numbers)

$$w(x,t) = u(\xi,\tau)\exp\left[-\frac{e^{-\beta x}}{a\beta^2(t+C_1)}\right], \qquad \xi = x - \frac{1}{\beta}\ln\frac{C_2}{(t+C_1)^2}, \qquad \tau = C_3 - \frac{C_2}{t+C_1},$$

leads to the same equation, up to the notation,

$$\frac{\partial u}{\partial \tau} = a\frac{\partial}{\partial \xi}\left(e^{\beta\xi}\frac{\partial u}{\partial \xi}\right).$$

5°. The substitution $z = e^{-\beta x}$ leads to an equation of the form 1.3.4.1:

$$\frac{\partial w}{\partial t} = a\beta^2 z \frac{\partial^2 w}{\partial z^2}.$$

6°. A series solution of the original equation (under constant values of w at the boundary and at the initial instant) can be found in Lykov (1967).

5. $\dfrac{\partial w}{\partial t} = ae^{2\beta x}\dfrac{\partial^2 w}{\partial x^2} + \sqrt{a}\,e^{\beta x}\left(\sqrt{a}\,\beta e^{\beta x} + b\right)\dfrac{\partial w}{\partial x}.$

This is a special case of equation 1.8.5.2 with $f(t)=b$ and $g(t)=0$. The substitution $\xi = \dfrac{1}{\sqrt{a}\,\beta}(1-e^{-\beta x})$ leads to a constant coefficient equation of the form 1.1.4 with $\Phi \equiv 0$:

$$\dfrac{\partial w}{\partial t} = \dfrac{\partial^2 w}{\partial \xi^2} + b\dfrac{\partial w}{\partial \xi}.$$

6. $\dfrac{\partial w}{\partial t} = ae^{2\beta x}\dfrac{\partial^2 w}{\partial x^2} + \sqrt{a}\,e^{\beta x}\left(\sqrt{a}\,\beta e^{\beta x} + be^{\mu t}\right)\dfrac{\partial w}{\partial x} + ce^{\nu t} w.$

This is a special case of equation 1.8.5.2 with $f(t) = be^{\mu t}$ and $g(t) = ce^{\nu t}$.

1.4.6. Other Equations

1. $\dfrac{\partial w}{\partial t} = ae^{\beta t}\dfrac{\partial^2 w}{\partial x^2} + be^{\mu t}\dfrac{\partial w}{\partial x} + ce^{\nu t} w.$

This is a special case of equation 1.8.7.3 with $f(t) = ae^{\beta t}$, $g(t) = be^{\mu t}$, and $h(t) = ce^{\nu t}$.

2. $\dfrac{\partial w}{\partial t} = ae^{\beta t}\dfrac{\partial^2 w}{\partial x^2} + bxe^{\mu t}\dfrac{\partial w}{\partial x} + cxe^{\nu t} w.$

This is a special case of equation 1.8.7.4 with $n(t) = ae^{\beta t}$, $f(t) = be^{\mu t}$, $g(t) = 0$, $h(t) = ce^{\nu t}$, and $s(t) = 0$.

3. $\dfrac{\partial w}{\partial t} = ae^{\beta x + \mu t}\dfrac{\partial^2 w}{\partial x^2} + be^{\nu t} w.$

The transformation

$$w(x,t) = \exp\left(\dfrac{b}{\nu}e^{\nu t}\right) u(x,\tau), \quad \tau = \dfrac{a}{\mu} e^{\mu t}$$

leads to an equation of the form 1.4.5.1: $\partial_\tau u = e^{\beta x}\partial_{xx} u$.

1.5. Equations Containing Hyperbolic Functions and Arbitrary Parameters

1.5.1. Equations Containing a Hyperbolic Cosine

1. $\dfrac{\partial w}{\partial t} = a\dfrac{\partial^2 w}{\partial x^2} + (b\cosh^k \omega t + c)w.$

This is a special case of equation 1.8.1.1 with $f(t) = b\cosh^k \omega t + c$.

2. $\dfrac{\partial w}{\partial t} = a\dfrac{\partial^2 w}{\partial x^2} + (b\cosh^k \omega t + cx)w.$

This is a special case of equation 1.8.1.6 with $f(t) = c$ and $g(t) = b\cosh^k \omega t$.

3. $\dfrac{\partial w}{\partial t} = a\dfrac{\partial^2 w}{\partial x^2} + x(b\cosh^k \omega t + c)w.$

This is a special case of equation 1.8.1.3 with $f(t) = b\cosh^k \omega t + c$.

4. $\dfrac{\partial w}{\partial t} = a\dfrac{\partial^2 w}{\partial x^2} + (c\cosh^k \omega t - bx^2)w.$

This is a special case of equation 1.8.1.7 with $f(t) = c\cosh^k \omega t$.

5. $\dfrac{\partial w}{\partial t} = a\dfrac{\partial^2 w}{\partial x^2} + (b\cosh^k \omega t + c)\dfrac{\partial w}{\partial x}.$

This is a special case of equation 1.8.2.1 with $f(t) = b\cosh^k \omega t + c$.

6. $\dfrac{\partial w}{\partial t} = a\dfrac{\partial^2 w}{\partial x^2} + x(b\cosh^k \omega t + c)\dfrac{\partial w}{\partial x}.$

This is a special case of equation 1.8.2.3 with $f(t) = b\cosh^k \omega t + c$.

7. $\dfrac{\partial w}{\partial t} = a\dfrac{\partial^2 w}{\partial x^2} + b\dfrac{\partial w}{\partial x} + (c\cosh^k \omega t + s)w.$

This is a special case of equation 1.8.3.1 with $f(t) = b$ and $g(t) = c\cosh^k \omega t + s$.

8. $\dfrac{\partial w}{\partial t} = a\dfrac{\partial^2 w}{\partial x^2} + \left(cx\cosh^k \omega t + \dfrac{b}{x}\right)\dfrac{\partial w}{\partial x}.$

This is a special case of equation 1.8.2.6 with $f(t) = c\cosh^k \omega t$.

9. $\dfrac{\partial w}{\partial t} = ax\dfrac{\partial^2 w}{\partial x^2} + (bx\cosh^k \omega t + c)w.$

This is a special case of equation 1.8.4.1 with $f(t) = b\cosh^k \omega t$ and $g(t) = c$.

10. $\dfrac{\partial w}{\partial t} = ax^2\dfrac{\partial^2 w}{\partial x^2} + (b\cosh^k \omega t + c)w.$

This is a special case of equation 1.8.4.2 with $f(t) = b\cosh^k \omega t + c$.

11. $\dfrac{\partial w}{\partial t} = ax^4\dfrac{\partial^2 w}{\partial x^2} + (b\cosh^k \omega t + c)w.$

This is a special case of equation 1.8.8.4 with $f(t) = a$ and $g(t) = b\cosh^k \omega t + c$.

12. $\dfrac{\partial w}{\partial t} = ax^n\dfrac{\partial^2 w}{\partial x^2} + x(b\cosh^k \omega t + c)\dfrac{\partial w}{\partial x}.$

This is a special case of equation 1.8.4.5 with $f(t) = b\cosh^k \omega t + c$.

1.5.2. Equations Containing a Hyperbolic Sine

1. $\dfrac{\partial w}{\partial t} = a\dfrac{\partial^2 w}{\partial x^2} + (b\sinh^k \omega t + c)w.$

This is a special case of equation 1.8.1.1 with $f(t) = b\sinh^k \omega t + c$.

2. $\dfrac{\partial w}{\partial t} = a\dfrac{\partial^2 w}{\partial x^2} + (b\sinh^k \omega t + cx)w.$

This is a special case of equation 1.8.1.6 with $f(t) = c$ and $g(t) = b\sinh^k \omega t$.

3. $\dfrac{\partial w}{\partial t} = a\dfrac{\partial^2 w}{\partial x^2} + x(b\sinh^k \omega t + c)w.$

This is a special case of equation 1.8.1.3 with $f(t) = b\sinh^k \omega t + c$.

4. $\dfrac{\partial w}{\partial t} = a\dfrac{\partial^2 w}{\partial x^2} + (c\sinh^k \omega t - bx^2)w.$

This is a special case of equation 1.8.1.7 with $f(t) = c\sinh^k \omega t$.

5. $\dfrac{\partial w}{\partial t} = a\dfrac{\partial^2 w}{\partial x^2} + (b\sinh^k \omega t + c)\dfrac{\partial w}{\partial x}.$

This is a special case of equation 1.8.2.1 with $f(t) = b\sinh^k \omega t + c$.

6. $\dfrac{\partial w}{\partial t} = a\dfrac{\partial^2 w}{\partial x^2} + x(b\sinh^k \omega t + c)\dfrac{\partial w}{\partial x}.$

This is a special case of equation 1.8.2.3 with $f(t) = b\sinh^k \omega t + c$.

7. $\dfrac{\partial w}{\partial t} = a\dfrac{\partial^2 w}{\partial x^2} + b\dfrac{\partial w}{\partial x} + (c\sinh^k \omega t + s)w.$

This is a special case of equation 1.8.3.1 with $f(t) = b$ and $g(t) = c\sinh^k \omega t + s$.

8. $\dfrac{\partial w}{\partial t} = a\dfrac{\partial^2 w}{\partial x^2} + \left(cx\sinh^k \omega t + \dfrac{b}{x}\right)\dfrac{\partial w}{\partial x}.$

This is a special case of equation 1.8.2.6 with $f(t) = c\sinh^k \omega t$.

9. $\dfrac{\partial w}{\partial t} = ax\dfrac{\partial^2 w}{\partial x^2} + (bx\sinh^k \omega t + c)w.$

This is a special case of equation 1.8.4.1 with $f(t) = b\sinh^k \omega t$ and $g(t) = c$.

10. $\dfrac{\partial w}{\partial t} = ax^2\dfrac{\partial^2 w}{\partial x^2} + (b\sinh^k \omega t + c)w.$

This is a special case of equation 1.8.4.2 with $f(t) = b\sinh^k \omega t + c$.

11. $\dfrac{\partial w}{\partial t} = ax^4\dfrac{\partial^2 w}{\partial x^2} + (b\sinh^k \omega t + c)w.$

This is a special case of equation 1.8.8.4 with $f(t) = a$ and $g(t) = b\sinh^k \omega t + c$.

12. $\dfrac{\partial w}{\partial t} = ax^n\dfrac{\partial^2 w}{\partial x^2} + x(b\sinh^k \omega t + c)\dfrac{\partial w}{\partial x}.$

This is a special case of equation 1.8.4.5 with $f(t) = b\sinh^k \omega t + c$.

1.5.3. Equations Containing a Hyperbolic Tangent

1. $\dfrac{\partial w}{\partial t} = a\dfrac{\partial^2 w}{\partial x^2} + (b\tanh^k \omega t + c)w.$

This is a special case of equation 1.8.1.1 with $f(t) = b\tanh^k \omega t + c$.

2. $\dfrac{\partial w}{\partial t} = a\dfrac{\partial^2 w}{\partial x^2} + (b\tanh^k \omega t + cx)w.$

This is a special case of equation 1.8.1.6 with $f(t) = c$ and $g(t) = b\tanh^k \omega t$.

3. $\dfrac{\partial w}{\partial t} = a\dfrac{\partial^2 w}{\partial x^2} + x(b\tanh^k \omega t + c)w.$

This is a special case of equation 1.8.1.3 with $f(t) = b\tanh^k \omega t + c$.

4. $\dfrac{\partial w}{\partial t} = a\dfrac{\partial^2 w}{\partial x^2} + (c\tanh^k \omega t - bx^2)w.$

This is a special case of equation 1.8.1.7 with $f(t) = c\tanh^k \omega t$.

5. $\dfrac{\partial w}{\partial t} = a\dfrac{\partial^2 w}{\partial x^2} + (b\tanh^k \omega t + c)\dfrac{\partial w}{\partial x}.$

This is a special case of equation 1.8.2.1 with $f(t) = b\tanh^k \omega t + c$.

6. $\dfrac{\partial w}{\partial t} = a\dfrac{\partial^2 w}{\partial x^2} + x(b\tanh^k \omega t + c)\dfrac{\partial w}{\partial x}.$

This is a special case of equation 1.8.2.3 with $f(t) = b\tanh^k \omega t + c$.

7. $\dfrac{\partial w}{\partial t} = a\dfrac{\partial^2 w}{\partial x^2} + b\dfrac{\partial w}{\partial x} + (c\tanh^k \omega t + s)w.$

This is a special case of equation 1.8.3.1 with $f(t) = b$ and $g(t) = c\tanh^k \omega t + s$.

8. $\dfrac{\partial w}{\partial t} = a\dfrac{\partial^2 w}{\partial x^2} + \left(cx\tanh^k \omega t + \dfrac{b}{x}\right)\dfrac{\partial w}{\partial x}.$

This is a special case of equation 1.8.2.6 with $f(t) = c\tanh^k \omega t$.

9. $\dfrac{\partial w}{\partial t} = ax\dfrac{\partial^2 w}{\partial x^2} + (bx\tanh^k \omega t + c)w.$

This is a special case of equation 1.8.4.1 with $f(t) = b\tanh^k \omega t$ and $g(t) = c$.

10. $\dfrac{\partial w}{\partial t} = ax^2\dfrac{\partial^2 w}{\partial x^2} + (b\tanh^k \omega t + c)w.$

This is a special case of equation 1.8.4.2 with $f(t) = b\tanh^k \omega t + c$.

11. $\dfrac{\partial w}{\partial t} = ax^4\dfrac{\partial^2 w}{\partial x^2} + (b\tanh^k \omega t + c)w.$

This is a special case of equation 1.8.8.4 with $f(t) = a$ and $g(t) = b\tanh^k \omega t + c$.

12. $\dfrac{\partial w}{\partial t} = ax^n\dfrac{\partial^2 w}{\partial x^2} + x(b\tanh^k \omega t + c)\dfrac{\partial w}{\partial x}.$

This is a special case of equation 1.8.4.5 with $f(t) = b\tanh^k \omega t + c$.

1.5.4. Equations Containing a Hyperbolic Cotangent

1. $\dfrac{\partial w}{\partial t} = a\dfrac{\partial^2 w}{\partial x^2} + (b\coth^k \omega t + c)w.$

This is a special case of equation 1.8.1.1 with $f(t) = b\coth^k \omega t + c$.

2. $\dfrac{\partial w}{\partial t} = a\dfrac{\partial^2 w}{\partial x^2} + (b\coth^k \omega t + cx)w.$

This is a special case of equation 1.8.1.6 with $f(t) = c$ and $g(t) = b\coth^k \omega t$.

3. $\dfrac{\partial w}{\partial t} = a\dfrac{\partial^2 w}{\partial x^2} + x(b\coth^k \omega t + c)w.$

This is a special case of equation 1.8.1.3 with $f(t) = b\coth^k \omega t + c$.

4. $\dfrac{\partial w}{\partial t} = a\dfrac{\partial^2 w}{\partial x^2} + (c\coth^k \omega t - bx^2)w.$

This is a special case of equation 1.8.1.7 with $f(t) = c\coth^k \omega t$.

5. $\dfrac{\partial w}{\partial t} = a\dfrac{\partial^2 w}{\partial x^2} + (b\coth^k \omega t + c)\dfrac{\partial w}{\partial x}.$

This is a special case of equation 1.8.2.1 with $f(t) = b\coth^k \omega t + c$.

6. $\dfrac{\partial w}{\partial t} = a\dfrac{\partial^2 w}{\partial x^2} + x(b\coth^k \omega t + c)\dfrac{\partial w}{\partial x}.$

This is a special case of equation 1.8.2.3 with $f(t) = b\coth^k \omega t + c$.

7. $\dfrac{\partial w}{\partial t} = a\dfrac{\partial^2 w}{\partial x^2} + b\dfrac{\partial w}{\partial x} + (c\coth^k \omega t + s)w.$

This is a special case of equation 1.8.3.1 with $f(t) = b$ and $g(t) = c\coth^k \omega t + s$.

8. $\dfrac{\partial w}{\partial t} = a\dfrac{\partial^2 w}{\partial x^2} + \left(cx\coth^k \omega t + \dfrac{b}{x}\right)\dfrac{\partial w}{\partial x}.$

This is a special case of equation 1.8.2.6 with $f(t) = c\coth^k \omega t$.

9. $\dfrac{\partial w}{\partial t} = ax\dfrac{\partial^2 w}{\partial x^2} + (bx\coth^k \omega t + c)w.$

This is a special case of equation 1.8.4.1 with $f(t) = b\coth^k \omega t$ and $g(t) = c$.

10. $\dfrac{\partial w}{\partial t} = ax^2\dfrac{\partial^2 w}{\partial x^2} + (b\coth^k \omega t + c)w.$

This is a special case of equation 1.8.4.2 with $f(t) = b\coth^k \omega t + c$.

11. $\dfrac{\partial w}{\partial t} = ax^4\dfrac{\partial^2 w}{\partial x^2} + (b\coth^k \omega t + c)w.$

This is a special case of equation 1.8.8.4 with $f(t) = a$ and $g(t) = b\coth^k \omega t + c$.

12. $\dfrac{\partial w}{\partial t} = ax^n\dfrac{\partial^2 w}{\partial x^2} + x(b\coth^k \omega t + c)\dfrac{\partial w}{\partial x}.$

This is a special case of equation 1.8.4.5 with $f(t) = b\coth^k \omega t + c$.

1.6. Equations Containing Logarithmic Functions and Arbitrary Parameters

1.6.1. Equations of the Form $\dfrac{\partial w}{\partial t} = a\dfrac{\partial^2 w}{\partial x^2} + f(x,t)\dfrac{\partial w}{\partial x} + g(x,t)w$

1. $\dfrac{\partial w}{\partial t} = a\dfrac{\partial^2 w}{\partial x^2} + (b\ln t + c)w.$

This is a special case of equation 1.8.1.1 with $f(t) = b\ln t + c$.

The substitution $w(x,t) = u(x,t)\exp(bt\ln t - bt + ct)$ leads to the constant coefficient equation $\partial_t u = a\partial^2_{xx} u$, which is considered in Subsection 1.1.1.

2. $\dfrac{\partial w}{\partial t} = a\dfrac{\partial^2 w}{\partial x^2} + (bx + c\ln t)w.$

This is a special case of equation 1.8.1.6 with $f(t) = b$ and $g(t) = c\ln t$.

3. $\dfrac{\partial w}{\partial t} = a\dfrac{\partial^2 w}{\partial x^2} + x(b\ln^k t + c)w.$

This is a special case of equation 1.8.1.3 with $f(t) = b\ln^k t + c$.

4. $\dfrac{\partial w}{\partial t} = a\dfrac{\partial^2 w}{\partial x^2} + (-bx^2 + c\ln^k t)w.$

This is a special case of equation 1.8.1.7 with $f(t) = c\ln^k t$.

5. $\dfrac{\partial w}{\partial t} = a\dfrac{\partial^2 w}{\partial x^2} + (b\ln t + c)\dfrac{\partial w}{\partial x}.$

The change of variable $z = x + bt\ln t - bt + ct$ leads to the constant coefficient equation $\partial_t w = a\partial^2_{zz} w$ that is considered in Subsection 1.1.1.

6. $\dfrac{\partial w}{\partial t} = a\dfrac{\partial^2 w}{\partial x^2} + x(b\ln t + c)\dfrac{\partial w}{\partial x}.$

This is a special case of equation 1.8.2.3 with $f(t) = b\ln t + c$.

On passing from t, x to the new variables (A and B are any numbers)

$$\tau = \int F^2(t)\,dt + A, \quad z = xF(t), \quad \text{where} \quad F(t) = B\exp(bt\ln t - bt + ct),$$

we arrive at the constant coefficient equation $\partial_\tau w = a\partial_{zz} w$ for $w(\tau, z)$; this equation is considered in Subsection 1.1.1.

1.6.2. Equations of the Form $\dfrac{\partial w}{\partial t} = ax^k\dfrac{\partial^2 w}{\partial x^2} + f(x,t)\dfrac{\partial w}{\partial x} + g(x,t)w$

1. $\dfrac{\partial w}{\partial t} = ax\dfrac{\partial^2 w}{\partial x^2} + (bx + c\ln t)w.$

This is a special case of equation 1.8.8.1 with $f(t) = a$, $g(t) = 0$, $h(t) = b$, and $s(t) = c\ln t$.

2. $\dfrac{\partial w}{\partial t} = ax\dfrac{\partial^2 w}{\partial x^2} + (bx\ln t + c)w.$

This is a special case of equation 1.8.8.1 with $f(t) = a$, $g(t) = 0$, $h(t) = b\ln t$, and $s(t) = c$.

3. $\dfrac{\partial w}{\partial t} = ax\dfrac{\partial^2 w}{\partial x^2} + bx\dfrac{\partial w}{\partial x} + (c\ln t + d)w.$

This is a special case of equation 1.8.8.1 with $f(t) = a$, $g(t) = b$, $h(t) = 0$, and $s(t) = c\ln t + d$.

4. $\dfrac{\partial w}{\partial t} = ax^2\dfrac{\partial^2 w}{\partial x^2} + (b\ln t + c)w.$

This is a special case of equation 1.8.8.2 with $f(t) = a$, $g(t) = 0$, and $h(t) = b\ln t + c$.

5. $\dfrac{\partial w}{\partial t} = ax^2\dfrac{\partial^2 w}{\partial x^2} + b\ln x\, w.$

This is a special case of equation 1.8.8.3 with $n(t) = a$, $f(t) = g(t) = h(t) = p(t) = 0$, and $s(t) = b$.

6. $\dfrac{\partial w}{\partial t} = ax^2 \dfrac{\partial^2 w}{\partial x^2} + bt^k \ln x\, w.$

This is a special case of equation 1.8.8.3 with $n(t) = a$, $f(t) = g(t) = h(t) = p(t) = 0$, and $s(t) = bt^k$.

7. $\dfrac{\partial w}{\partial t} = ax^2 \dfrac{\partial^2 w}{\partial x^2} + b \ln^2 x\, w.$

This is a special case of equation 1.8.8.3 with $n(t) = a$, $f(t) = g(t) = s(t) = p(t) = 0$, and $h(t) = b$.
See also equation 1.8.6.5.

8. $\dfrac{\partial w}{\partial t} = ax^2 \dfrac{\partial^2 w}{\partial x^2} + bt^k \ln^2 x\, w.$

This is a special case of equation 1.8.8.3 with $n(t) = a$, $f(t) = g(t) = s(t) = p(t) = 0$, and $h(t) = bt^k$.

9. $\dfrac{\partial w}{\partial t} = ax^2 \dfrac{\partial^2 w}{\partial x^2} + (b \ln^2 x + c \ln x \ln t + d \ln^2 t)w.$

This is a special case of equation 1.8.8.3 with $n(t) = a$, $f(t) = g(t) = 0$, $h(t) = b$, $s(t) = c \ln t$, and $p(t) = d \ln^2 t$.

10. $\dfrac{\partial w}{\partial t} = ax^4 \dfrac{\partial^2 w}{\partial x^2} + (b \ln t + c)w.$

This is a special case of equation 1.8.8.4 with $f(t) = a$ and $g(t) = b \ln t + c$.

11. $\dfrac{\partial w}{\partial t} = ax^n \dfrac{\partial^2 w}{\partial x^2} + (b \ln t + c)w.$

This is a special case of equation 1.8.8.7 with $f(t) = a$, $g(t) = 0$, and $h(t) = b \ln t + c$.

12. $\dfrac{\partial w}{\partial t} = ax^n \dfrac{\partial^2 w}{\partial x^2} + x(b \ln t + c)\dfrac{\partial w}{\partial x}.$

This is a special case of equation 1.8.4.5 with $f(t) = b \ln t + c$.

1.7. Equations Containing Trigonometric Functions and Arbitrary Parameters

1.7.1. Equations Containing a Cosine

1. $\dfrac{\partial w}{\partial t} = a \dfrac{\partial^2 w}{\partial x^2} + (b \cos^k \omega t + c)w.$

This is a special case of equation 1.8.1.1 with $f(t) = b \cos^k \omega t + c$.

2. $\dfrac{\partial w}{\partial t} = a \dfrac{\partial^2 w}{\partial x^2} + (b \cos^k \omega t + cx)w.$

This is a special case of equation 1.8.1.6 with $f(t) = c$ and $g(t) = b \cos^k \omega t$.

3. $\dfrac{\partial w}{\partial t} = a \dfrac{\partial^2 w}{\partial x^2} + x(b \cos^k \omega t + c)w.$

This is a special case of equation 1.8.1.3 with $f(t) = b \cos^k \omega t + c$.

4. $\dfrac{\partial w}{\partial t} = a \dfrac{\partial^2 w}{\partial x^2} + (c \cos^k \omega t - bx^2)w.$

This is a special case of equation 1.8.1.7 with $f(t) = c \cos^k \omega t$.

5. $\dfrac{\partial w}{\partial t} = a\dfrac{\partial^2 w}{\partial x^2} + (b\cos^k \omega t + c)\dfrac{\partial w}{\partial x}$.

This is a special case of equation 1.8.2.1 with $f(t) = b\cos^k \omega t + c$.

6. $\dfrac{\partial w}{\partial t} = a\dfrac{\partial^2 w}{\partial x^2} + x(b\cos^k \omega t + c)\dfrac{\partial w}{\partial x}$.

This is a special case of equation 1.8.2.3 with $f(t) = b\cos^k \omega t + c$.

7. $\dfrac{\partial w}{\partial t} = a\dfrac{\partial^2 w}{\partial x^2} + b\dfrac{\partial w}{\partial x} + (c\cos^k \omega t + s)w$.

This is a special case of equation 1.8.3.1 with $f(t) = b$ and $g(t) = c\cos^k \omega t + s$.

8. $\dfrac{\partial w}{\partial t} = a\dfrac{\partial^2 w}{\partial x^2} + \left(cx\cos^k \omega t + \dfrac{b}{x}\right)\dfrac{\partial w}{\partial x}$.

This is a special case of equation 1.8.2.6 with $f(t) = c\cos^k \omega t$.

9. $\dfrac{\partial w}{\partial t} = ax\dfrac{\partial^2 w}{\partial x^2} + (bx\cos^k \omega t + c)w$.

This is a special case of equation 1.8.4.1 with $f(t) = b\cos^k \omega t$ and $g(t) = c$.

10. $\dfrac{\partial w}{\partial t} = ax^2\dfrac{\partial^2 w}{\partial x^2} + (b\cos^k \omega t + c)w$.

This is a special case of equation 1.8.4.2 with $f(t) = b\cos^k \omega t + c$.

11. $\dfrac{\partial w}{\partial t} = ax^4\dfrac{\partial^2 w}{\partial x^2} + (b\cos^k \omega t + c)w$.

This is a special case of equation 1.8.8.4 with $f(t) = a$ and $g(t) = b\cos^k \omega t + c$.

12. $\dfrac{\partial w}{\partial t} = ax^n\dfrac{\partial^2 w}{\partial x^2} + x(b\cos^k \omega t + c)\dfrac{\partial w}{\partial x}$.

This is a special case of equation 1.8.4.5 with $f(t) = b\cos^k \omega t + c$.

1.7.2. Equations Containing a Sine

1. $\dfrac{\partial w}{\partial t} = a\dfrac{\partial^2 w}{\partial x^2} + (b\sin^k \omega t + c)w$.

This is a special case of equation 1.8.1.1 with $f(t) = b\sin^k \omega t + c$.

2. $\dfrac{\partial w}{\partial t} = a\dfrac{\partial^2 w}{\partial x^2} + (b\sin^k \omega t + cx)w$.

This is a special case of equation 1.8.1.6 with $f(t) = c$ and $g(t) = b\sin^k \omega t$.

3. $\dfrac{\partial w}{\partial t} = a\dfrac{\partial^2 w}{\partial x^2} + x(b\sin^k \omega t + c)w$.

This is a special case of equation 1.8.1.3 with $f(t) = b\sin^k \omega t + c$.

4. $\dfrac{\partial w}{\partial t} = a\dfrac{\partial^2 w}{\partial x^2} + (c\sin^k \omega t - bx^2)w$.

This is a special case of equation 1.8.1.7 with $f(t) = c\sin^k \omega t$.

5. $\dfrac{\partial w}{\partial t} = a\dfrac{\partial^2 w}{\partial x^2} + (b\sin^k \omega t + c)\dfrac{\partial w}{\partial x}$.

This is a special case of equation 1.8.2.1 with $f(t) = b\sin^k \omega t + c$.

6. $\dfrac{\partial w}{\partial t} = a\dfrac{\partial^2 w}{\partial x^2} + x(b\sin^k \omega t + c)\dfrac{\partial w}{\partial x}$.

This is a special case of equation 1.8.2.3 with $f(t) = b\sin^k \omega t + c$.

7. $\dfrac{\partial w}{\partial t} = a\dfrac{\partial^2 w}{\partial x^2} + b\dfrac{\partial w}{\partial x} + (c\sin^k \omega t + s)w$.

This is a special case of equation 1.8.3.1 with $f(t) = b$ and $g(t) = c\sin^k \omega t + s$.

8. $\dfrac{\partial w}{\partial t} = a\dfrac{\partial^2 w}{\partial x^2} + \left(cx\sin^k \omega t + \dfrac{b}{x}\right)\dfrac{\partial w}{\partial x}$.

This is a special case of equation 1.8.2.6 with $f(t) = c\sin^k \omega t$.

9. $\dfrac{\partial w}{\partial t} = ax\dfrac{\partial^2 w}{\partial x^2} + (bx\sin^k \omega t + c)w$.

This is a special case of equation 1.8.4.1 with $f(t) = b\sin^k \omega t$ and $g(t) = c$.

10. $\dfrac{\partial w}{\partial t} = ax^2\dfrac{\partial^2 w}{\partial x^2} + (b\sin^k \omega t + c)w$.

This is a special case of equation 1.8.4.2 with $f(t) = b\sin^k \omega t + c$.

11. $\dfrac{\partial w}{\partial t} = ax^4\dfrac{\partial^2 w}{\partial x^2} + (b\sin^k \omega t + c)w$.

This is a special case of equation 1.8.8.4 with $f(t) = a$ and $g(t) = b\sin^k \omega t + c$.

12. $\dfrac{\partial w}{\partial t} = ax^n\dfrac{\partial^2 w}{\partial x^2} + x(b\sin^k \omega t + c)\dfrac{\partial w}{\partial x}$.

This is a special case of equation 1.8.4.5 with $f(t) = b\sin^k \omega t + c$.

1.7.3. Equations Containing a Tangent

1. $\dfrac{\partial w}{\partial t} = a\dfrac{\partial^2 w}{\partial x^2} + (b\tan^k \omega t + c)w$.

This is a special case of equation 1.8.1.1 with $f(t) = b\tan^k \omega t + c$.

2. $\dfrac{\partial w}{\partial t} = a\dfrac{\partial^2 w}{\partial x^2} + (b\tan^k \omega t + cx)w$.

This is a special case of equation 1.8.1.6 with $f(t) = c$ and $g(t) = b\tan^k \omega t$.

3. $\dfrac{\partial w}{\partial t} = a\dfrac{\partial^2 w}{\partial x^2} + x(b\tan^k \omega t + c)w$.

This is a special case of equation 1.8.1.3 with $f(t) = b\tan^k \omega t + c$.

4. $\dfrac{\partial w}{\partial t} = a\dfrac{\partial^2 w}{\partial x^2} + (c\tan^k \omega t - bx^2)w$.

This is a special case of equation 1.8.1.7 with $f(t) = c\tan^k \omega t$.

5. $\dfrac{\partial w}{\partial t} = a\dfrac{\partial^2 w}{\partial x^2} + (b\tan^k \omega t + c)\dfrac{\partial w}{\partial x}$.

This is a special case of equation 1.8.2.1 with $f(t) = b\tan^k \omega t + c$.

6. $\dfrac{\partial w}{\partial t} = a\dfrac{\partial^2 w}{\partial x^2} + x(b\tan^k \omega t + c)\dfrac{\partial w}{\partial x}$.

This is a special case of equation 1.8.2.3 with $f(t) = b\tan^k \omega t + c$.

7. $\dfrac{\partial w}{\partial t} = a\dfrac{\partial^2 w}{\partial x^2} + b\dfrac{\partial w}{\partial x} + (c\tan^k \omega t + s)w$.

This is a special case of equation 1.8.3.1 with $f(t) = b$ and $g(t) = c\tan^k \omega t + s$.

8. $\dfrac{\partial w}{\partial t} = a\dfrac{\partial^2 w}{\partial x^2} + \left(cx\tan^k \omega t + \dfrac{b}{x}\right)\dfrac{\partial w}{\partial x}$.

This is a special case of equation 1.8.2.6 with $f(t) = c\tan^k \omega t$.

9. $\dfrac{\partial w}{\partial t} = ax\dfrac{\partial^2 w}{\partial x^2} + (bx\tan^k \omega t + c)w$.

This is a special case of equation 1.8.4.1 with $f(t) = b\tan^k \omega t$ and $g(t) = c$.

10. $\dfrac{\partial w}{\partial t} = ax^2\dfrac{\partial^2 w}{\partial x^2} + (b\tan^k \omega t + c)w$.

This is a special case of equation 1.8.4.2 with $f(t) = b\tan^k \omega t + c$.

11. $\dfrac{\partial w}{\partial t} = ax^4\dfrac{\partial^2 w}{\partial x^2} + (b\tan^k \omega t + c)w$.

This is a special case of equation 1.8.8.4 with $f(t) = a$ and $g(t) = b\tan^k \omega t + c$.

12. $\dfrac{\partial w}{\partial t} = ax^n\dfrac{\partial^2 w}{\partial x^2} + x(b\tan^k \omega t + c)\dfrac{\partial w}{\partial x}$.

This is a special case of equation 1.8.4.5 with $f(t) = b\tan^k \omega t + c$.

1.7.4. Equations Containing a Cotangent

1. $\dfrac{\partial w}{\partial t} = a\dfrac{\partial^2 w}{\partial x^2} + (b\cot^k \omega t + c)w$.

This is a special case of equation 1.8.1.1 with $f(t) = b\cot^k \omega t + c$.

2. $\dfrac{\partial w}{\partial t} = a\dfrac{\partial^2 w}{\partial x^2} + (b\cot^k \omega t + cx)w$.

This is a special case of equation 1.8.1.6 with $f(t) = c$ and $g(t) = b\cot^k \omega t$.

3. $\dfrac{\partial w}{\partial t} = a\dfrac{\partial^2 w}{\partial x^2} + x(b\cot^k \omega t + c)w$.

This is a special case of equation 1.8.1.3 with $f(t) = b\cot^k \omega t + c$.

4. $\dfrac{\partial w}{\partial t} = a\dfrac{\partial^2 w}{\partial x^2} + (c\cot^k \omega t - bx^2)w$.

This is a special case of equation 1.8.1.7 with $f(t) = c\cot^k \omega t$.

5. $\dfrac{\partial w}{\partial t} = a\dfrac{\partial^2 w}{\partial x^2} + (b\cot^k \omega t + c)\dfrac{\partial w}{\partial x}.$

This is a special case of equation 1.8.2.1 with $f(t) = b\cot^k \omega t + c$.

6. $\dfrac{\partial w}{\partial t} = a\dfrac{\partial^2 w}{\partial x^2} + x(b\cot^k \omega t + c)\dfrac{\partial w}{\partial x}.$

This is a special case of equation 1.8.2.3 with $f(t) = b\cot^k \omega t + c$.

7. $\dfrac{\partial w}{\partial t} = a\dfrac{\partial^2 w}{\partial x^2} + b\dfrac{\partial w}{\partial x} + (c\cot^k \omega t + s)w.$

This is a special case of equation 1.8.3.1 with $f(t) = b$ and $g(t) = c\cot^k \omega t + s$.

8. $\dfrac{\partial w}{\partial t} = a\dfrac{\partial^2 w}{\partial x^2} + \left(cx\cot^k \omega t + \dfrac{b}{x}\right)\dfrac{\partial w}{\partial x}.$

This is a special case of equation 1.8.2.6 with $f(t) = c\cot^k \omega t$.

9. $\dfrac{\partial w}{\partial t} = ax\dfrac{\partial^2 w}{\partial x^2} + (bx\cot^k \omega t + c)w.$

This is a special case of equation 1.8.4.1 with $f(t) = b\cot^k \omega t$ and $g(t) = c$.

10. $\dfrac{\partial w}{\partial t} = ax^2\dfrac{\partial^2 w}{\partial x^2} + (b\cot^k \omega t + c)w.$

This is a special case of equation 1.8.4.2 with $f(t) = b\cot^k \omega t + c$.

11. $\dfrac{\partial w}{\partial t} = ax^4\dfrac{\partial^2 w}{\partial x^2} + (b\cot^k \omega t + c)w.$

This is a special case of equation 1.8.8.4 with $f(t) = a$ and $g(t) = b\cot^k \omega t + c$.

12. $\dfrac{\partial w}{\partial t} = ax^n\dfrac{\partial^2 w}{\partial x^2} + x(b\cot^k \omega t + c)\dfrac{\partial w}{\partial x}.$

This is a special case of equation 1.8.4.5 with $f(t) = b\cot^k \omega t + c$.

1.8. Equations Containing Arbitrary Functions

1.8.1. Equations of the Form $\dfrac{\partial w}{\partial t} = a\dfrac{\partial^2 w}{\partial x^2} + f(x,t)w$

1. $\dfrac{\partial w}{\partial t} = a\dfrac{\partial^2 w}{\partial x^2} + f(t)w.$

1°. Particular solutions (A, B, and λ are arbitrary constants):

$$w(x,t) = (Ax + B)\exp\left[\int f(t)\,dt\right],$$

$$w(x,t) = A(x^2 + 2at)\exp\left[\int f(t)\,dt\right],$$

$$w(x,t) = A\exp\left[\lambda x + a\lambda^2 t + \int f(t)\,dt\right],$$

$$w(x,t) = A\cos(\lambda x)\exp\left[-a\lambda^2 t + \int f(t)\,dt\right],$$

$$w(x,t) = A\sin(\lambda x)\exp\left[-a\lambda^2 t + \int f(t)\,dt\right].$$

2°. The substitution $w(x,t) = u(x,t)\exp\left[\int f(t)\,dt\right]$ leads to a constant coefficient equation, $\partial_t u = a\partial_{xx}u$, which is considered in Subsection 1.1.1.

2. $\dfrac{\partial w}{\partial t} = a\dfrac{\partial^2 w}{\partial x^2} + f(x)w.$

This is a special case of equation 1.8.9 with $s(x) \equiv 1$, $p(x) = a = \text{const}$, $q(x) = -f(x)$, and $\Phi \equiv 0$.

3. $\dfrac{\partial w}{\partial t} = a\dfrac{\partial^2 w}{\partial x^2} + xf(t)w.$

1°. Particular solutions (A and λ are arbitrary constants):

$$w(x,t) = A\exp\left[xF(t) + a\int F^2(t)\,dt\right], \quad F(t) = \int f(t)\,dt,$$

$$w(x,t) = A\left[x + 2a\int F(t)\,dt\right]\exp\left[xF(t) + a\int F^2(t)\,dt\right],$$

$$w(x,t) = A\exp\left[xF(t) + \lambda x + a\lambda^2 t + a\int F^2(t)\,dt + 2a\lambda\int F(t)\,dt\right].$$

2°. The transformation

$$w(x,t) = u(z,t)\exp\left[xF(t) + a\int F^2(t)\,dt\right], \quad z = x + 2a\int F(t)\,dt,$$

where $F(t) = \int f(t)\,dt$, leads to a constant coefficient equation, $\partial_t u = a\partial_{zz}u$, which is considered in Subsection 1.1.1.

4. $\dfrac{\partial w}{\partial t} = a\dfrac{\partial^2 w}{\partial x^2} + x^2 f(t)w.$

This is a special case of equation 1.8.7.5.

5. $\dfrac{\partial w}{\partial t} = a\dfrac{\partial^2 w}{\partial x^2} + [f(x) + g(t)]w.$

1°. There are particular solutions in the product form (λ is an arbitrary constant)

$$w(x,t) = \exp\left[\lambda t + \int g(t)\,dt\right]\varphi(x),$$

where the function $\varphi = \varphi(x)$ is determined by the ordinary differential equation

$$a\varphi''_{xx} + [f(x) - \lambda]\varphi = 0.$$

2°. The substitution $w(x,t) = u(x,t)\exp\left[\int g(t)\,dt\right]$ leads to an equation of the form 1.8.1.2:

$$\dfrac{\partial u}{\partial t} = a\dfrac{\partial^2 u}{\partial x^2} + f(x)u.$$

6. $\dfrac{\partial w}{\partial t} = a\dfrac{\partial^2 w}{\partial x^2} + [xf(t) + g(t)]w.$

1°. Particular solutions (A and λ are arbitrary constants):

$$w(x,t) = A\exp\left[xF(t) + a\int F^2(t)\,dt + \int g(t)\,dt\right], \quad F(t) = \int f(t)\,dt,$$

$$w(x,t) = A\left[x + 2a\int F(t)\,dt\right]\exp\left[xF(t) + a\int F^2(t)\,dt + \int g(t)\,dt\right],$$

$$w(x,t) = \exp\left[xF(t) + \lambda x + a\lambda^2 t + 2a\lambda\int F(t)\,dt + a\int F^2(t)\,dt + \int g(t)\,dt\right].$$

2°. The transformation

$$w(x,t) = u(z,t)\exp\left[xF(t) + a\int F^2(t)\,dt + \int g(t)\,dt\right], \quad z = x + 2a\int F(t)\,dt,$$

where $F(t) = \int f(t)\,dt$, leads to a constant coefficient equation, $\partial_t u = a\partial_{zz}u$, which is considered in Subsection 1.1.1.

7. $\dfrac{\partial w}{\partial t} = a\dfrac{\partial^2 w}{\partial x^2} + \left[-bx^2 + f(t)\right]w.$

1°. Particular solutions (A is an arbitrary constant):

$$w(x,t) = A\exp\left[\frac{1}{2}\sqrt{\frac{b}{a}}x^2 + \sqrt{ab}\,t + \int f(t)\,dt\right],$$

$$w(x,t) = Ax\exp\left[\frac{1}{2}\sqrt{\frac{b}{a}}x^2 + 3\sqrt{ab}\,t + \int f(t)\,dt\right].$$

2°. The transformation (C is any number)

$$w(x,t) = u(z,\tau)\exp\left[\frac{1}{2}\sqrt{\frac{b}{a}}x^2 + \sqrt{ab}\,t + \int f(t)\,dt\right],$$

$$z = x\exp(2\sqrt{ab}\,t), \quad \tau = \frac{1}{4}\sqrt{\frac{a}{b}}\exp(4\sqrt{ab}\,t) + C$$

leads to a constant coefficient equation, $\partial_\tau u = \partial_{zz} u$, which is considered in Subsection 1.1.1.

8. $\dfrac{\partial w}{\partial t} = a\dfrac{\partial^2 w}{\partial x^2} + x\left[-bx + f(t)\right]w.$

1°. Particular solution (A and B are arbitrary constants):

$$w(x,t) = \exp\left[\frac{1}{2}\sqrt{\frac{b}{a}}x^2 + xF(t) + \sqrt{ab}\,t + a\int_A^t F^2(\tau)\,d\tau\right],$$

$$F(t) = \exp(2\sqrt{ab}\,t)\int_B^t f(\tau)\exp(-2\sqrt{ab}\,\tau)\,d\tau.$$

2°. The transformation

$$w(x,t) = \exp\left(\frac{1}{2}\sqrt{\frac{b}{a}}x^2\right)u(z,\tau), \quad z = x\exp(2\sqrt{ab}\,t), \quad \tau = \frac{1}{4\sqrt{ab}}\exp(4\sqrt{ab}\,t)$$

leads to an equation of the form 1.8.1.6:

$$\frac{\partial u}{\partial \tau} = a\frac{\partial^2 u}{\partial z^2} + \left[z\Phi(\tau) + \frac{1}{4\tau}\right]u,$$

where $\Psi(\tau) = \dfrac{1}{(n\tau)^{3/2}}f\left(\dfrac{\ln\tau + \ln n}{n}\right)$, $n - 4\sqrt{ab}$.

9. $\dfrac{\partial w}{\partial t} = a\dfrac{\partial^2 w}{\partial x^2} + \left[x^2 f(t) + xg(t) + h(t)\right]w.$

This is a special case of equation 1.8.7.5.

10. $\dfrac{\partial w}{\partial t} = a\dfrac{\partial^2 w}{\partial x^2} + f(x - bt)w.$

On passing from t, x to the new variables $t, \xi = x - bt$, we obtain an equation of the form 1.8.6.5:

$$\frac{\partial w}{\partial t} = a\frac{\partial^2 w}{\partial \xi^2} + b\frac{\partial w}{\partial \xi} + f(\xi)w.$$

1.8.2. Equations of the Form $\frac{\partial w}{\partial t} = a\frac{\partial^2 w}{\partial x^2} + f(x,t)\frac{\partial w}{\partial x}$

1. $\dfrac{\partial w}{\partial t} = a\dfrac{\partial^2 w}{\partial x^2} + f(t)\dfrac{\partial w}{\partial x}.$

This equation describes heat transfer in a moving medium where the velocity of motion is an arbitrary function of time.

1°. Particular solutions (A, B, and μ are arbitrary constants):

$$w(x,t) = Ax + A\int f(t)\,dt + B,$$

$$w(x,t) = A\left[x + \int f(t)\,dt\right]^2 + 2aAt + B,$$

$$w(x,t) = A\exp\left[\lambda x + a\lambda^2 t + \lambda \int f(t)\,dt\right] + B.$$

2°. On passing from t, x to the new variables t, $z = x + \int f(t)\,dt$, we obtain a constant coefficient equation, $\partial_t w = a\partial_{zz} w$, which is considered in Subsection 1.1.1.

2. $\dfrac{\partial w}{\partial t} = a\dfrac{\partial^2 w}{\partial x^2} + f(x)\dfrac{\partial w}{\partial x}.$

This is a special case of equation 1.8.6.4. This equation describes heat transfer in a moving medium where the velocity of motion is an arbitrary function of the coordinate.

1°. The equation has particular solutions of the form

$$w(x,t) = e^{-\lambda t} u(x),$$

where the function $u = u(x)$ is determined by solving the following ordinary differential equation with parameter λ:

$$au''_{xx} + f(x)u'_x + \lambda u = 0.$$

Other particular solutions (A and B are arbitrary constants):

$$w(x) = A + B\int F(x)\,dx, \quad F(x) = \exp\left[-\frac{1}{a}\int f(x)\,dx\right],$$

$$w(x,t) = Aat + A\int\left(\int \frac{dx}{F(x)}\right)F(x)\,dx + B,$$

$$w(x,t) = Aat\Phi(x) + A\int\left(\int \frac{\Phi(x)\,dx}{F(x)}\right)F(x)\,dx, \quad \Phi(x) = \int F(x)\,dx.$$

More sophisticated solutions are specified below.

2°. The original equation admits particular solutions of the form

$$w_n(x,t) = \sum_{i=0}^{n} t^i \varphi_{n,i}(x) \tag{1}$$

for any $f(x)$. Substituting the expression of (1) into the original equation and matching the coefficients of like powers of t, we arrive at the following system of ordinary differential equations for $\varphi_{n,i} = \varphi_{n,i}(x)$:

$$a\varphi''_{n,n} + f(x)\varphi'_{n,n} = 0,$$
$$a\varphi''_{n,i} + f(x)\varphi'_{n,i} = (i+1)\varphi_{n,i+1}; \quad i = 0, 1, \ldots, n-1,$$

where the prime denotes the derivative with respect to x. Integrating these equations successively in order of decreasing number i, we obtain the solution (A and B are any numbers):

$$\varphi_{n,n}(x) = A + B \int F(x)\,dx, \quad F(x) = \exp\left[-\frac{1}{a}\int f(x)\,dx\right], \tag{2}$$

$$\varphi_{n,i}(x) = n(n-1)\ldots(i+1)\boldsymbol{L}_f^{n-i}[\varphi_{n,n}(x)]; \quad i = 0, 1, \ldots, n-1.$$

Here, the integral operator $\boldsymbol{L}_f$ is introduced as follows:

$$\boldsymbol{L}_f[y(x)] \equiv \frac{1}{a}\int F(x)\left(\int \frac{y(x)\,dx}{F(x)}\right)dx. \tag{3}$$

The powers of the operator are defined by $\boldsymbol{L}_f^i[y(x)] = \boldsymbol{L}_f\big[\boldsymbol{L}_f^{i-1}[y(x)]\big]$.

Formulas (1)–(3) give an exact analytical solution of the original equation for arbitrary $f(x)$.

A linear combination of particular solutions

$$w(x,t) = \sum_{n=0}^{N} C_n w_n(x,t) \quad (C_n \text{ are arbitrary constants})$$

is also a particular solution of the original equation.

3. $\dfrac{\partial w}{\partial t} = a\dfrac{\partial^2 w}{\partial x^2} + xf(t)\dfrac{\partial w}{\partial x}.$

Generalized Ilkovič equation. The equation describes mass transfer to the surface of a growing drop that flows out of a thin capillary into a fluid solution (the mass rate of flow of the fluid moving through the capillary is an arbitrary function of time).

1°. Particular solutions (A, B, and λ are arbitrary constants):

$$w(x,t) = AxF(t) + B, \quad F(t) = \exp\left[\int f(t)\,dt\right],$$

$$w(x,t) = Ax^2 F^2(t) + 2Aa\int F^2(t)\,dt + B,$$

$$w(x,t) = A\exp\left[\lambda x F(t) + a\lambda^2 \int F^2(t)\,dt\right] + B.$$

2°. On passing from t, x to the new variables (A is any number)

$$\tau = \int F^2(t)\,dt + A, \quad z = xF(t), \quad \text{where} \quad F(t) = \exp\left[\int f(t)\,dt\right],$$

for the function $w(\tau, z)$ we obtain a constant coefficient equation, $\partial_\tau w = a\partial_{zz}w$, which is considered in Subsection 1.1.1.

3°. Consider the special case where the heat exchange occurs with a semiinfinite medium; the medium has a uniform temperature w_0 at the initial instant $t = 0$ and the boundary $x = 0$ is maintained at a constant temperature w_1 all the time. In this case, the original equation subject to the initial and boundary conditions

$$w = w_0 \quad \text{at} \quad t = 0 \quad \text{(initial condition)},$$
$$w = w_1 \quad \text{at} \quad x = 0 \quad \text{(boundary condition)},$$
$$w \to w_0 \quad \text{at} \quad x \to \infty \quad \text{(boundary condition)}$$

has the solution

$$\frac{w - w_1}{w_0 - w_1} = \operatorname{erf}\left(\frac{z}{2\sqrt{a\tau}}\right), \quad \operatorname{erf}\xi = \frac{2}{\sqrt{\pi}}\int_0^\xi \exp(-\zeta^2)\,d\zeta,$$

$$\tau = \int_0^t F^2(\zeta)\,d\zeta, \quad z = xF(t), \quad F(t) = \exp\left[\int f(t)\,dt\right],$$

where $\operatorname{erf}\xi$ is the error function.

4. $\dfrac{\partial w}{\partial t} = a\dfrac{\partial^2 w}{\partial x^2} + f(x - bt)\dfrac{\partial w}{\partial x}.$

1°. Particular solutions (A and B are arbitrary constants):
$$w(x,t) = A + B\int F(z)\,dz, \quad F(z) = \exp\left[-\dfrac{1}{a}\int f(z)\,dz - \dfrac{b}{a}z\right],$$
$$w(x,t) = Aat + A\int\left(\int \dfrac{dz}{F(z)}\right)F(z)\,dz,$$
$$w(x,t) = Aat\Phi(z) + A\int\left(\int\dfrac{\Phi(z)\,dz}{F(z)}\right)F(z)\,dx, \quad \Phi(z) = \int F(z)\,dz,$$

where $z = x - bt$.

2°. On passing from t, x to the new variables t, $z = x - bt$, we obtain a separable equation of the form 1.8.2.2:
$$\dfrac{\partial w}{\partial t} = a\dfrac{\partial^2 w}{\partial z^2} + [f(z) + b]\dfrac{\partial w}{\partial z}.$$

5. $\dfrac{\partial w}{\partial t} = a\dfrac{\partial^2 w}{\partial x^2} + \dfrac{1}{\sqrt{t}}f\!\left(\dfrac{x}{\sqrt{t}}\right)\dfrac{\partial w}{\partial x}.$

1°. On passing from t, x to the new variables $\tau = \ln t$, $\xi = x/\sqrt{t}$, we obtain a separable equation of the form 1.8.2.2:
$$\dfrac{\partial w}{\partial \tau} = a\dfrac{\partial^2 w}{\partial \xi^2} + [f(\xi) + \tfrac{1}{2}\xi]\dfrac{\partial w}{\partial \xi}.$$

2°. Consider the special case where the heat exchange occurs with a semiinfinite medium; the medium has a uniform temperature w_0 at the initial instant $t = 0$ and the boundary $x = 0$ is maintained at a constant temperature w_1 all the time. In this case, the original equation subject to the initial and boundary conditions

$$w = w_0 \quad \text{at} \quad t = 0 \quad \text{(initial condition)},$$
$$w = w_1 \quad \text{at} \quad x = 0 \quad \text{(boundary condition)},$$
$$w \to w_0 \quad \text{at} \quad x \to \infty \quad \text{(boundary condition)}$$

has the solution
$$\dfrac{w - w_0}{w_1 - w_0} = \dfrac{\int_\xi^\infty \exp[-\Phi(\xi)]\,d\xi}{\int_0^\infty \exp[-\Phi(\xi)]\,d\xi}, \quad \Phi(\xi) = \dfrac{1}{4a}\xi^2 + \dfrac{1}{a}\int_0^\xi f(\xi)\,d\xi,$$

where $\xi = x/\sqrt{t}$.

6. $\dfrac{\partial w}{\partial t} = a\dfrac{\partial^2 w}{\partial x^2} + \left[xf(t) + \dfrac{b}{x}\right]\dfrac{\partial w}{\partial x}.$

1°. Particular solutions (A and B are arbitrary constants):
$$w(x,t) = Ax^n \exp\left[n\int f(t)\,dt\right] + B, \quad n = 1 - \dfrac{b}{a},$$
$$w(x,t) = Ax^2 F(t) + 2A(a+b)\int F(t)\,dt + B, \quad F(t) = \exp\left[2\int f(t)\,dt\right].$$

2°. On passing from t, x to the new variables (A is any number)
$$\tau = \int F^2(t)\,dt + A, \quad z = xF(t), \quad \text{where} \quad F(t) = \exp\left[\int f(t)\,dt\right],$$

for the function $w(\tau, z)$ we obtain a simpler equation
$$\dfrac{\partial w}{\partial \tau} = a\dfrac{\partial^2 w}{\partial z^2} + \dfrac{b}{z}\dfrac{\partial w}{\partial z},$$

which is considered in Subsections 1.2.1, 1.2.3, and 1.2.5.

7. $\dfrac{\partial w}{\partial t} = a\dfrac{\partial^2 w}{\partial x^2} + [xf(t) + g(t)]\dfrac{\partial w}{\partial x}.$

The transformation (A, B, and C are any numbers)

$$\tau = \int F^2(t)\, dt + A, \qquad z = xF(t) + \int g(t)F(t)\, dt + C, \qquad F(t) = B\exp\left[\int f(t)\, dt\right],$$

leads to a constant coefficient equation, $\partial_\tau w = a\partial_{zz} w$, which is considered in Subsection 1.1.1.

8. $\dfrac{\partial w}{\partial t} = a\dfrac{\partial^2 w}{\partial x^2} + \dfrac{f(t)}{x}\dfrac{\partial w}{\partial x}.$

Particular solutions:

$$w = \dfrac{A}{\sqrt{t}}\exp\left[-\dfrac{x^2}{4at} - \dfrac{1}{2a}\int\dfrac{f(t)}{t}\, dt\right] + B,$$

$$w = x^2 + 2at + 2\int f(t)\, dt + A,$$

$$w = x^4 + p(t)x^2 + q(t), \qquad p(t) = 12at + 4\int f(t)\, dt + A, \qquad q(t) = 2\int [a + f(t)]p(t)\, dt + B,$$

where A and B are arbitrary constants. The second and third solutions are special cases of a solution having the form

$$w = x^{2n} + A_{2n-2}(t)x^{2n-2} + \cdots + A_2(t)x^2 + A_0(t),$$

which contains n arbitrary constants.

9. $\dfrac{\partial w}{\partial t} = a\dfrac{\partial^2 w}{\partial x^2} + \left[xf(t) + \dfrac{g(t)}{x}\right]\dfrac{\partial w}{\partial x}.$

On passing from t, x to the new variables (A is any number)

$$\tau = \int F^2(t)\, dt + A, \qquad z = xF(t), \qquad \text{where} \quad F(t) = \exp\left[\int f(t)\, dt\right],$$

for the function $w(\tau, z)$ we obtain a simpler equation of the form 1.8.2.8:

$$\dfrac{\partial w}{\partial \tau} = a\dfrac{\partial^2 w}{\partial z^2} + \dfrac{\varphi(\tau)}{z}\dfrac{\partial w}{\partial z}.$$

The function $\varphi = \varphi(\tau)$ is defined parametrically as

$$\varphi = \dfrac{g(t)}{F(t)}, \qquad \tau = \int F^2(t)\, dt + A.$$

1.8.3. Equations of the Form $\dfrac{\partial w}{\partial t} = a\dfrac{\partial^2 w}{\partial x^2} + f(x,t)\dfrac{\partial w}{\partial x} + g(x,t)w$

1. $\dfrac{\partial w}{\partial t} = a\dfrac{\partial^2 w}{\partial x^2} + f(t)\dfrac{\partial w}{\partial x} + g(t)w.$

This is a special case of equation 1.8.7.3.

1°. Particular solutions (A, B, and λ are arbitrary constants):

$$w(x,t) = \left[Ax + AF(t) + B\right] \exp\left[\int g(t)\,dt\right], \quad F(t) = \int f(t)\,dt,$$

$$w(x,t) = A\left\{\left[x + F(t)\right]^2 + 2at\right\} \exp\left[\int g(t)\,dt\right],$$

$$w(x,t) = A\exp\left[a\lambda^2 t + \int g(t)\,dt \pm \lambda F(t) \pm \lambda x\right],$$

$$w(x,t) = A\exp\left[-a\lambda^2 t + \int g(t)\,dt\right] \cos\left[\lambda x + \lambda F(t)\right],$$

$$w(x,t) = A\exp\left[-a\lambda^2 t + \int g(t)\,dt\right] \sin\left[\lambda x + \lambda F(t)\right].$$

2°. The transformation

$$w(x,t) = u(z,t)\exp\left[\int g(t)\,dt\right], \quad z = x + \int f(t)\,dt$$

leads to a constant coefficient equation, $\partial_t u = a\partial_{zz}u$, which is considered in Subsection 1.1.1.

2. $\dfrac{\partial w}{\partial t} = a\dfrac{\partial^2 w}{\partial x^2} + f(x)\dfrac{\partial w}{\partial x} + g(t)w.$

1°. Particular solutions (A and B are arbitrary constants):

$$w(x,t) = \left[A + B\int F(x)\,dx\right]G(t),$$

$$w(x,t) = A\left[at + \int F(x)\left(\int \frac{dx}{F(x)}\right)dx\right]G(t),$$

$$w(x,t) = A\left[at\Psi(x) + \int F(x)\left(\int \frac{\Psi(x)\,dx}{F(x)}\right)dx\right]G(t).$$

The following notation is used here:

$$G(t) = \exp\left[\int g(t)\,dt\right], \quad F(x) = \exp\left[-\frac{1}{a}\int f(x)\,dx\right], \quad \Psi(x) = \int F(x)\,dx.$$

2°. The substitution $w(x,t) = u(x,t)\exp\left[\int g(t)\,dt\right]$ leads to an equation of the form 1.8.2.2:

$$\frac{\partial u}{\partial t} = a\frac{\partial^2 u}{\partial x^2} + f(x)\frac{\partial u}{\partial x}.$$

3. $\dfrac{\partial w}{\partial t} = a\dfrac{\partial^2 w}{\partial x^2} + f(x)\dfrac{\partial w}{\partial x} + g(x)w.$

This is a special case of equation 1.8.6.5.

4. $\dfrac{\partial w}{\partial t} = a\dfrac{\partial^2 w}{\partial x^2} + xf(t)\dfrac{\partial w}{\partial x} + g(t)w.$

The transformation (A, B, and C are any numbers)

$$\tau = \int F^2(t)\,dt + A, \quad z = xF(t) + C, \quad w(t,x) = u(\tau,z)\exp\left[\int g(t)\,dt\right],$$

where $F(t) = B\exp\left[\int f(t)\,dt\right]$, leads to a constant coefficient equation, $\partial_\tau u = a\partial_{zz}u$, which is considered in Subsection 1.1.1.

5. $\dfrac{\partial w}{\partial t} = a\dfrac{\partial^2 w}{\partial x^2} + \left[xf(t) + \dfrac{b}{x}\right]\dfrac{\partial w}{\partial x} + g(t)w.$

The substitution $w(x,t) = u(x,t)\exp\left[\int g(t)\,dt\right]$ leads to an equation of the form 1.8.2.6:

$$\dfrac{\partial u}{\partial t} = a\dfrac{\partial^2 u}{\partial x^2} + \left[xf(t) + \dfrac{b}{x}\right]\dfrac{\partial u}{\partial x}.$$

For the special case $b = 0$, see equation 1.8.2.3.

6. $\dfrac{\partial w}{\partial t} = a\dfrac{\partial^2 w}{\partial x^2} + [xf(t) + g(t)]\dfrac{\partial w}{\partial x} + h(t)w.$

The transformation (A, B, and C are any numbers)

$$\tau = \int F^2(t)\,dt + A, \quad z = xF(t) + \int g(t)F(t)\,dt + C, \quad w(t,x) = u(\tau, z)\exp\left[\int h(t)\,dt\right],$$

where $F(t) = B\exp\left[\int f(t)\,dt\right]$, leads to a constant coefficient equation, $\partial_\tau u = a\partial_{zz}u$, which is considered in Subsection 1.1.1.

7. $\dfrac{\partial w}{\partial t} = a\dfrac{\partial^2 w}{\partial x^2} + [xf(t) + g(t)]\dfrac{\partial w}{\partial x} + [xh(t) + s(t)]w.$

This is a special case of equation 1.8.7.4 with $n(t) = a$.

8. $\dfrac{\partial w}{\partial t} = a\dfrac{\partial^2 w}{\partial x^2} + [xf(t) + g(t)]\dfrac{\partial w}{\partial x} + [x^2 h(t) + xs(t) + p(t)]w.$

This is a special case of equation 1.8.7.5 with $n(t) = a$.

9. $\dfrac{\partial w}{\partial t} = a\dfrac{\partial^2 w}{\partial x^2} + \dfrac{f(t)}{x}\dfrac{\partial w}{\partial x} + g(t)w.$

1°. Particular solutions (A, B, and C are arbitrary constants):

$$w = \dfrac{A}{\sqrt{t}}\exp\left[-\dfrac{x^2}{4at} - \dfrac{1}{2a}\int \dfrac{f(t)}{t}\,dt + \int g(t)\,dt\right],$$

$$w = A\exp\left[\int g(t)\,dt\right]\left[x^2 + 2at + 2\int f(t)\,dt + B\right],$$

$$w = A\exp\left[\int g(t)\,dt\right]\left[x^4 + p(t)x^2 + q(t)\right],$$

where

$$p(t) = 12at + 4\int f(t)\,dt + B, \quad q(t) = 2\int [a + f(t)]p(t)\,dt + C.$$

2°. The substitution $w = \exp\left[\int g(t)\,dt\right]u(x,t)$ leads to an equation of the form 1.8.2.8 for $u = u(x,t)$.

10. $\dfrac{\partial w}{\partial t} = a\dfrac{\partial^2 w}{\partial x^2} + \left[xf(t) + \dfrac{g(t)}{x}\right]\dfrac{\partial w}{\partial x} + h(t)w.$

The substitution $w = \exp\left[\int h(t)\,dt\right]u(x,t)$ leads to an equation of the form 1.8.2.9 for $u = u(x,t)$.

1.8.4. Equations of the Form $\dfrac{\partial w}{\partial t} = ax^n \dfrac{\partial^2 w}{\partial x^2} + f(x,t)\dfrac{\partial w}{\partial x} + g(x,t)w$

1. $\dfrac{\partial w}{\partial t} = ax\dfrac{\partial^2 w}{\partial x^2} + [xf(t) + g(t)]w.$

This is a special case of equation 1.8.8.1.

2. $\dfrac{\partial w}{\partial t} = ax^2\dfrac{\partial^2 w}{\partial x^2} + f(t)w.$

This is a special case of equation 1.8.8.2. The transformation

$$w(x,t) = u(z,t)\exp\left[\int f(t)\,dt\right], \quad z = \ln|x|$$

leads to a constant coefficient equation of the form 1.1.4:

$$\frac{\partial u}{\partial \tau} = a\frac{\partial^2 u}{\partial z^2} - a\frac{\partial u}{\partial z}.$$

3. $\dfrac{\partial w}{\partial t} = ax^2\dfrac{\partial^2 w}{\partial x^2} + \ln x\, f(t)w.$

This is a special case of equation 1.8.8.3.

4. $\dfrac{\partial w}{\partial t} = ax^2\dfrac{\partial^2 w}{\partial x^2} + \ln^2 x\, f(t)w.$

This is a special case of equation 1.8.8.3.

5. $\dfrac{\partial w}{\partial t} = ax^n\dfrac{\partial^2 w}{\partial x^2} + xf(t)\dfrac{\partial w}{\partial x}.$

1°. Particular solutions (A and B are arbitrary constants):

$$w(x,t) = Ax\exp\left[\int f(t)\,dt\right] + B,$$

$$w(x,t) = Ax^{2-n}F(t) + Aa(n-1)(n-2)\int F(t)\,dt,$$

where $F(t) = \exp\left[(2-n)\int f(t)\,dt\right].$

2°. On passing from t, x to the new variables

$$z = xF(t), \quad \tau = a\int F^{2-n}(t)\,dt, \quad F(t) = \exp\left[\int f(t)\,dt\right],$$

we obtain an equation of the form 1.3.6.6:

$$\frac{\partial w}{\partial \tau} = z^n \frac{\partial^2 w}{\partial z^2}.$$

6. $\dfrac{\partial w}{\partial t} = ax^n\dfrac{\partial^2 w}{\partial x^2} + xf(t)\dfrac{\partial w}{\partial x} + bw.$

The substitution $w(x,t) = e^{bt}u(x,t)$ leads to an equation of the form 1.8.4.5:

$$\frac{\partial u}{\partial t} = ax^n\frac{\partial^2 u}{\partial x^2} + xf(t)\frac{\partial u}{\partial x}.$$

7. $\dfrac{\partial w}{\partial t} = ax^{2n}\dfrac{\partial^2 w}{\partial x^2} + \sqrt{a}\, x^n\left[\sqrt{a}\, nx^{n-1} + f(t)\right]\dfrac{\partial w}{\partial x} + g(t)w.$

The substitution
$$\xi = \dfrac{1}{\sqrt{a}}\begin{cases} \dfrac{x^{1-n}}{1-n} & \text{if } n \neq 1, \\ \ln|x| & \text{if } n = 1 \end{cases}$$
leads to a special case of equation 1.8.7.3, namely,
$$\dfrac{\partial w}{\partial t} = \dfrac{\partial^2 w}{\partial \xi^2} + f(t)\dfrac{\partial w}{\partial \xi} + g(t)w.$$

1.8.5. Equations of the Form $\dfrac{\partial w}{\partial t} = ae^{\beta x}\dfrac{\partial^2 w}{\partial x^2} + f(x,t)\dfrac{\partial w}{\partial x} + g(x,t)w$

1. $\dfrac{\partial w}{\partial t} = ae^{\beta x}\dfrac{\partial^2 w}{\partial x^2} + f(t)w.$

The substitution $w(x,t) = u(x,t)\exp\left[\int f(t)\,dt\right]$ leads to an equation of the form 1.4.5.1:
$$\dfrac{\partial u}{\partial t} = ae^{\beta x}\dfrac{\partial^2 u}{\partial x^2}.$$

2. $\dfrac{\partial w}{\partial t} = ae^{2\beta x}\dfrac{\partial^2 w}{\partial x^2} + \sqrt{a}\, e^{\beta x}\left[\sqrt{a}\,\beta e^{\beta x} + f(t)\right]\dfrac{\partial w}{\partial x} + g(t)w.$

The substitution $\xi = \dfrac{1}{\beta\sqrt{a}}\left(1 - e^{-\beta x}\right)$ leads to a special case of equation 1.8.7.3:
$$\dfrac{\partial w}{\partial t} = \dfrac{\partial^2 w}{\partial \xi^2} + f(t)\dfrac{\partial w}{\partial \xi} + g(t)w.$$

3. $\dfrac{\partial w}{\partial t} = ae^{\beta x}\dfrac{\partial^2 w}{\partial x^2} + f(x)\dfrac{\partial w}{\partial x}.$

This is a special case of equation 1.8.6.4.

1°. The equation has particular solutions of the form
$$w(x,t) = e^{-\lambda t}u(x), \tag{1}$$
where the function $u(x)$ is determined by solving the following linear ordinary differential equation with parameter λ:
$$ae^{\beta x}u''_{xx} + f(x)u'_x + \lambda u = 0. \tag{2}$$

2°. Other particular solutions (A and B are arbitrary constants):
$$w(x) = A + B\int F(x)\,dx, \quad F(x) = \exp\left[-\dfrac{1}{a}\int e^{-\beta x}f(x)\,dx\right],$$
$$w(x,t) = Aat + A\int F(x)\left(\int \dfrac{dx}{e^{\beta x}F(x)}\right)dx,$$
$$w(x,t) = Aat\Phi(x) + A\int F(x)\left(\int \dfrac{\Phi(x)\,dx}{e^{\beta x}F(x)}\right)dx, \quad \Phi(x) = \int F(x)\,dx.$$

4. $\dfrac{\partial w}{\partial t} = ae^{\beta x}\dfrac{\partial^2 w}{\partial x^2} + f(x)\dfrac{\partial w}{\partial x} + g(t)w.$

This is a special case of equation 1.8.6.6.

The substitution $w(x,t) = u(x,t)\exp\left[\int g(t)\,dt\right]$ leads to an equation of the form 1.8.5.3:
$$\dfrac{\partial u}{\partial t} = ae^{\beta x}\dfrac{\partial^2 u}{\partial x^2} + f(x)\dfrac{\partial u}{\partial x}.$$

1.8.6. Equations of the Form $\dfrac{\partial w}{\partial t} = f(x)\dfrac{\partial^2 w}{\partial x^2} + g(x,t)\dfrac{\partial w}{\partial x} + h(x,t)w$

1. $\dfrac{\partial w}{\partial t} = f(x)\dfrac{\partial^2 w}{\partial x^2}.$

This is an equation of the form 1.8.9 with $s(x) = 1/f(x)$, $p(x) \equiv 1$, $q(x) \equiv 0$, and $\Phi(x,t) \equiv 0$.

1°. The equation has particular solutions of the form

$$w(x,t) = e^{-\lambda t} u(x), \qquad (1)$$

where the function $u(x)$ is determined by solving the following linear ordinary differential equation with parameter λ:

$$f(x) u''_{xx} + \lambda u = 0. \qquad (2)$$

The procedure for constructing solutions to specific boundary value problems for the original equations with the help of particular solutions of the form (1) is described in detail in Section 0.4 (Example 1).

The main problem here is to investigate the auxiliary equation (2), which is far from always admitting a closed-form solution; therefore, recourse to numerical solution methods is often necessary. Many specific solvable equations of the form (2) can be found in the handbooks by Murphy (1960), Kamke (1977), and Polyanin and Zaitsev (1995).

2°. Particular solutions (A, B, and x_0 are arbitrary constants):

$$w(x) = Ax + B,$$

$$w(x,t) = At + AF(x), \quad F(x) = \int_{x_0}^{x} \dfrac{x-\xi}{f(\xi)} d\xi,$$

$$w(x,t) = Atx + AG(x), \quad G(x) = \int_{x_0}^{x} \dfrac{x-\xi}{f(\xi)} \xi \, d\xi,$$

$$w(x,t) = At^2 + 2AtF(x) + 2A \int_{x_0}^{x} \dfrac{x-\xi}{f(\xi)} F(\xi) \, d\xi,$$

$$w(x,t) = At^2 x + 2AtG(x) + 2A \int_{x_0}^{x} \dfrac{x-\xi}{f(\xi)} G(\xi) \, d\xi.$$

More sophisticated solutions are specified below in Item 3°.

3°. For any function $f(x)$, the original equation admits exact analytical solutions of the form

$$w_n(x,t) = t^n + \sum_{i=0}^{n-1} t^i \varphi_{n,i}(x). \qquad (3)$$

Substituting expression (3) into the original equation and matching the coefficients of like powers of t, we arrive at the following system of ordinary differential equations for $\varphi_{n,i} = \varphi_{n,i}(x)$:

$$f(x) \varphi''_{n,i} = (i+1) \varphi_{n,i+1},$$
$$i = 0, 1, \ldots, n-1; \quad \varphi_{n,n} \equiv 1,$$

where the prime stands for the differentiation with respect to x. Integrating these equation successively in order of decreasing number i, we obtain

$$\varphi_{n,i}(x) = n(n-1)\ldots(i+1) \boldsymbol{L}_f^{n-i}[1]. \qquad (4)$$

Here, the integral operator $\boldsymbol{L}_f$ is introduced as follows:

$$\boldsymbol{L}_f[y(x)] \equiv \int \left(\int \dfrac{y(x)}{f(x)} dx \right) dx = \int_{x_0}^{x} \dfrac{x-\xi}{f(\xi)} y(\xi)\, d\xi + Ax + B, \qquad (5)$$

where x_0, A, and B are arbitrary constants. The powers of the operator are defined by the usual relation $\boldsymbol{L}_f^i[y(x)] = \boldsymbol{L}_f\big[\boldsymbol{L}_f^{i-1}[y(x)]\big]$; generally speaking, the constants A and B are not the same in repeated actions of $\boldsymbol{L}_f$ in this formula.

Formulas (3) and (4) determine an exact analytical solution of the original equation for arbitrary $f(x)$.

A linear combination of particular solutions (3),

$$w(x,t) = \sum_{n=0}^{N} C_n w_n(x,t) \qquad (C_n \text{ are arbitrary constants})$$

is also a particular solution of the original equation.

The original equation also admits other exact analytical solutions, specifically,

$$w_n(x,t) = t^n x + \sum_{i=0}^{n-1} t^i \phi_{n,i}(x), \qquad \phi_{n,i}(x) = n(n-1)\ldots(i+1)\boldsymbol{L}_f^{n-i}[x],$$

where n is a positive integer and the operator $\boldsymbol{L}_f$ is given by relation (5). A linear combination of these solutions with a linear combination of solutions (3) is also a solution of the original equation.

For the structure of other particular solutions, see equation 1.8.6.5 (the remark in Item 3°).

4°. The equation admits the following infinite-series solution that contains an arbitrary function of the coordinate:

$$w(x,t) = \Theta(x) + \sum_{n=1}^{\infty} \frac{1}{n!} t^n \boldsymbol{L}^n[\Theta(x)], \qquad \boldsymbol{L} \equiv f(x)\frac{d^2}{dx^2},$$

where $\Theta(x)$ is any infinitely differentiable function. This solution satisfies the initial condition $w(x,0) = \Theta(x)$.

5°. Below are two discrete transformations that preserve the form of the original equation; the function f is subject to changes.

5.1. The transformation

$$z = \frac{1}{x}, \quad u = \frac{w}{x} \qquad \text{(point transformation)}$$

leads to a similar equation

$$\frac{\partial u}{\partial t} = z^4 f\left(\frac{1}{z}\right) \frac{\partial^2 u}{\partial z^2}.$$

5.2. First, we perform the change of variable

$$\xi = \int \frac{dx}{f(x)}$$

to obtain the equation

$$\frac{\partial w}{\partial t} = \frac{\partial}{\partial \xi}\left[F(\xi)\frac{\partial w}{\partial \xi}\right],$$

where the function $F = F(\xi)$ is defined parametrically as

$$F = \frac{1}{f(x)}, \quad \xi = \int \frac{dx}{f(x)}.$$

Introducing the new dependent variable $v = v(\xi, t)$ by the formula

$$w = \frac{\partial v}{\partial \xi} \qquad \text{(Bäcklund transformation)}$$

and integrating the resulting equation with respect to ξ, we arrive at the desired equation

$$\frac{\partial v}{\partial t} = F(\xi)\frac{\partial^2 v}{\partial \xi^2}.$$

(Here, the function v is defined up to an arbitrary additive term that depends on t.)

For power-law and exponential functions the above transformation acts as follows:

$$f(x) = bx^n \quad \Longrightarrow \quad F(\xi) = A\xi^{\frac{n}{n-1}},$$
$$f(x) = be^{-\beta x} \quad \Longrightarrow \quad F(\xi) = \beta\xi,$$

where $A = \dfrac{1}{b}[b(1-n)]^{\frac{n}{n-1}}$.

2. $\dfrac{\partial w}{\partial t} = f(x)\dfrac{\partial^2 w}{\partial x^2} + \Phi(x,t).$

This is an equation of the form 1.8.9 with $s(x) = 1/f(x)$, $p(x) \equiv 1$, and $q(x) \equiv 0$. For $\Phi(x,t) \equiv 0$, see equation 1.8.6.1.

1°. For

$$\Phi(x,t) = g_n(x)t^n \qquad (n = 0, 1, 2, \dots)$$

and arbitrary functions $f(x)$ and $g_n(x)$, the original equation has a particular solution of the form

$$\bar{w}_n(x,t) = \sum_{i=0}^{n} t^i \psi_{n,i}(x). \tag{1}$$

The functions $\psi_{n,i} = \psi_{n,i}(x)$ are calculated by the formulas

$$\psi_{n,i}(x) = \begin{cases} -\boldsymbol{L}_f[g_n(x)] & \text{if } i = n, \\ -n(n-1)\dots(i+1)\boldsymbol{L}_f^{n-i+1}[g_n(x)] & \text{if } i = 0, 1, \dots, n-1 \end{cases} \tag{2}$$

with the aid of the integral operator $\boldsymbol{L}_f$ that is defined by relation (5) in equation 1.8.6.1.

2°. If the nonhomogeneous part of the equation can be represented in the form

$$\Phi(x,t) = \sum_{n=1}^{N} g_n(x)t^n,$$

then there is a particular solution that is the sum of particular solutions of the form (1):

$$\bar{w}(x,t) = \sum_{n=1}^{N} \bar{w}_n(x,t).$$

For example, if

$$\Phi(x,t) = g(x)t + h(x),$$

where $g(x)$ and $h(x)$ are arbitrary functions, the original equation has a solution of the form

$$\bar{w}(x,t) = -t\psi(x) - \int_{x_0}^{x} \frac{\psi(\xi) + h(\xi)}{f(\xi)}(x-\xi)\,d\xi,$$

$$\psi(x) = \int_{x_0}^{x} \frac{g(\xi)}{f(\xi)}(x-\xi)\,d\xi, \quad x_0 \text{ is any}.$$

For the structure of particular solutions for other $\Phi(x,t)$, see equation 1.8.6.5, Item 3°.

By summing different solutions of the homogeneous equation (see equation 1.8.6.1) and any particular solution of the nonhomogeneous equation, one can obtain a wide class of particular solutions of the nonhomogeneous equation.

3. $\dfrac{\partial w}{\partial t} = \dfrac{\partial}{\partial x}\left[f(x)\dfrac{\partial w}{\partial x}\right].$

This is a special case of equation 1.8.6.4 with $g(x) = f'_x(x)$. The equation describes heat transfer in a quiescent medium (solid body) in the case where the thermal diffusivity $f(x)$ is a coordinate dependent function.

1°. Particular solutions (A and B are arbitrary constants):

$$w(x) = A + B\int \dfrac{dx}{f(x)},$$

$$w(x,t) = At + A\int \dfrac{x\,dx}{f(x)} + B,$$

$$w(x,t) = At\varphi(x) + A\int\left(\int \varphi(x)\,dx\right)\dfrac{dx}{f(x)} + B, \quad \varphi(x) = \int \dfrac{dx}{f(x)},$$

$$w(x,t) = At^2 + 2At\psi(x) + 2A\int\left(\int \psi(x)\,dx\right)\dfrac{dx}{f(x)} + B, \quad \psi(x) = \int \dfrac{x\,dx}{f(x)},$$

$$w(x,t) = At^2\varphi(x) + 2AtI(x) + 2A\int\left(\int I(x)\,dx\right)\dfrac{dx}{f(x)} + B, \quad I(x) = \int\left(\int \varphi(x)\,dx\right)\dfrac{dx}{f(x)}.$$

2°. A solution in the form of an infinite series:

$$w(x,t) = \Theta(x) + \sum_{n=1}^{\infty} \dfrac{1}{n!} t^n \boldsymbol{L}^n[\Theta(x)], \quad \boldsymbol{L} \equiv \dfrac{d}{dx}\left[f(x)\dfrac{d}{dx}\right].$$

It contains an arbitrary function of the space variable, $\Theta = \Theta(x)$. This solution satisfies the initial condition $w(x,0) = \Theta(x)$.

3°. The transformation

$$w(x,t) = \varphi(x)u(\xi,t), \quad \xi = -\int \varphi^2(x)\,dx, \quad \varphi(x) = \int \dfrac{dx}{f(x)},$$

leads to the analogous equation

$$\dfrac{\partial u}{\partial t} = \dfrac{\partial}{\partial \xi}\left[F(\xi)\dfrac{\partial u}{\partial \xi}\right],$$

where the function $F(\xi)$ is defined parametrically as

$$F(\xi) = f(x)\psi^4(x), \quad \xi = -\int \varphi^2(x)\,dx, \quad \varphi(x) = \int \dfrac{dx}{f(x)}.$$

4°. The substitution $z = \int \dfrac{dx}{f(x)}$ leads to an equation of the form 1.8.6.1:

$$\dfrac{\partial w}{\partial t} = g(z)\dfrac{\partial^2 w}{\partial z^2},$$

where the function $q(z)$ is defined parametrically as

$$g(z) = \dfrac{1}{f(x)}, \quad z = \int \dfrac{dx}{f(x)}.$$

4. $\dfrac{\partial w}{\partial t} = f(x)\dfrac{\partial^2 w}{\partial x^2} + g(x)\dfrac{\partial w}{\partial x}.$

1°. This equation can be rewritten in the form

$$s(x)\dfrac{\partial w}{\partial t} = \dfrac{\partial}{\partial x}\left[p(x)\dfrac{\partial w}{\partial x}\right], \quad (1)$$

where

$$s(x) = \dfrac{1}{f(x)}\exp\left[\int \dfrac{g(x)}{f(x)}\,dx\right], \quad p(x) = \exp\left[\int \dfrac{g(x)}{f(x)}\,dx\right].$$

For solutions of equation (1), see Subsection 1.8.9 with $q(x) \equiv 0$.

2°. There are particular solutions of the form

$$w(x,t) = e^{-\lambda t} u(x), \tag{2}$$

where the function $u(x)$ is identified by solving the following linear ordinary differential equation with parameter λ:

$$f(x) u''_{xx} + g(x) u'_x + \lambda u = 0. \tag{3}$$

A procedure for constructing solutions to specific boundary value problems for the original equation with the aid of particular solutions (2) is described in detail in Subsection 0.4.1. A good deal of specific solvable equations of the form (3) can be found in Murphy (1960), Kamke (1977), and Polyanin and Zaitsev (1995).

3°. Other particular solutions (A and B are arbitrary constants):

$$w(x) = A + B \int F(x)\, dx, \quad F(x) = \exp\left[-\int \frac{g(x)}{f(x)}\, dx\right],$$

$$w(x,t) = At + A \int F(x) \left(\int \frac{dx}{f(x) F(x)}\right) dx,$$

$$w(x,t) = At\Phi(x) + A \int F(x) \left(\int \frac{\Phi(x)\, dx}{f(x) F(x)}\right) dx, \quad \Phi(x) = \int F(x)\, dx.$$

More sophisticated solutions are presented below in Item 4°.

4°. For any $f(x)$ and $g(x)$, the original equation admits particular solutions of the form

$$w_n(x,t) = \sum_{i=0}^{n} t^i \varphi_{n,i}(x). \tag{4}$$

Substituting expression (4) into the original equation and matching the coefficients of like powers of t, we arrive at the following system of ordinary differential equations for $\varphi_{n,i} = \varphi_{n,i}(x)$:

$$f(x)\varphi''_{n,n} + g(x)\varphi'_{n,n} = 0,$$
$$f(x)\varphi''_{n,i} + g(x)\varphi'_{n,i} = (i+1)\varphi_{n,i+1}, \quad i = 0, 1, \ldots, n-1,$$

where the prime stands for the differentiation with respect to x. Integrating these equation successively in order of decreasing number i, we obtain (A and B are any numbers)

$$\varphi_{n,n}(x) = A + B \int F(x)\, dx, \quad F(x) = \exp\left[-\int \frac{g(x)}{f(x)}\, dx\right],$$
$$\varphi_{n,i}(x) = n(n-1)\ldots(i+1) \boldsymbol{L}_f^{n-i}[\varphi_{n,n}(x)]; \quad i = 0, 1, \ldots, n-1. \tag{5}$$

Here, the integral operator $\boldsymbol{L}_f$ is introduced as follows:

$$\boldsymbol{L}_f[y(x)] \equiv \int F(x) \left(\int \frac{y(x)\, dx}{f(x) F(x)}\right) dx. \tag{6}$$

The powers of the operator are defined as $\boldsymbol{L}_f^i[y(x)] = \boldsymbol{L}_f\!\left[\boldsymbol{L}_f^{i-1}[y(x)]\right]$.

Formulas (4)–(6) determine an exact analytical solution of the original equation for arbitrary $f(x)$.

A linear combination of particular solutions (4),

$$w(x,t) = \sum_{n=0}^{N} C_n w_n(x,t) \quad (C_n \text{ are arbitrary numbers}),$$

is also a particular solution of the homogeneous equation.

For the structure of other particular solutions, see equation 1.8.6.5 (the remark in Item 3°).

5°. The substitution $\xi = \int \varphi(x)\,dx$, $\varphi(x) = \exp\left[-\int \dfrac{g(x)}{f(x)}dx\right]$ leads to an equation of the form 1.8.6.1:

$$\frac{\partial w}{\partial t} = F(\xi)\frac{\partial^2 w}{\partial \xi^2},$$

where the function $F = F(\xi)$ is determined by eliminating x from the relations

$$F = f(x)\varphi^2(x), \quad \xi = \int \varphi(x)\,dx.$$

6°. An infinite series solution containing an arbitrary function of the coordinate:

$$w(x,t) = \Theta(x) + \sum_{n=1}^{\infty} \frac{1}{n!} t^n \boldsymbol{L}^n[\Theta(x)], \quad \boldsymbol{L} \equiv f(x)\frac{d^2}{dx^2} + g(x)\frac{d}{dx},$$

where $\Theta(x)$ is any infinitely differentiable function. This solution satisfies the initial condition $w(x,0) = \Theta(x)$.

5. $\dfrac{\partial w}{\partial t} = f(x)\dfrac{\partial^2 w}{\partial x^2} + g(x)\dfrac{\partial w}{\partial x} + h(x)w + \Phi(x,t).$

1°. This equation can be rewritten in the form

$$s(x)\frac{\partial w}{\partial t} = \frac{\partial}{\partial x}\left[p(x)\frac{\partial w}{\partial x}\right] - q(x)w + s(x)\Phi(x,t), \tag{1}$$

where

$$s(x) = \frac{1}{f(x)}\exp\left[\int \frac{g(x)}{f(x)}dx\right], \quad p(x) = \exp\left[\int \frac{g(x)}{f(x)}dx\right], \quad q(x) = -\frac{h(x)}{f(x)}\exp\left[\int \frac{g(x)}{f(x)}dx\right].$$

For solutions of equation (1), see Subsection 1.8.9.

2°. Consider the homogeneous equation, i.e., the case $\Phi(x,t) \equiv 0$.

2.1. There are particular solutions of the form

$$w(x,t) = e^{-\lambda t}u(x),$$

where the function $u(x)$ is determined by solving the following linear ordinary differential equation with parameter λ:

$$f(x)u''_{xx} + g(x)u'_x + \big[h(x) + \lambda\big]u = 0.$$

2.2. Suppose we know a nontrivial particular solution $w_0 = w_0(x)$ of the ordinary differential equation

$$f(x)w''_0 + g(x)w'_0 + h(x)w_0 = 0 \tag{2}$$

that corresponds to the stationary case $(\partial_t w \equiv 0)$. Then the functions

$$w(x) = Aw_0 + Bw_0 \int \frac{F}{w_0^2}dx, \quad F = \exp\left(-\int \frac{g}{f}dx\right),$$

$$w(x,t) = Atw_0 + Aw_0 \int \frac{F}{w_0^2}\left(\int \frac{w_0^2}{fF}dx\right)dx,$$

$$w(x,t) = Atw_0\Psi + Aw_0 \int \frac{F}{w_0^2}\left(\int \frac{w_0^2\Psi}{fF}dx\right)dx, \quad \Psi = \int \frac{F}{w_0^2}dx,$$

where A and B are arbitrary constants, are also particular solutions of the original equation.

By performing the change of variable $w(x,t) = w_0(x)u(x,t)$, we arrive at the simpler equation

$$\frac{\partial u}{\partial t} = f(x)\frac{\partial^2 u}{\partial x^2} + \left[2f(x)\frac{w_0'(x)}{w_0(x)} + g(x)\right]\frac{\partial u}{\partial x}.$$

It determines a wide class of more complicated analytical solutions of the original equation.

It follows, with reference to the results of Item 4° from 1.8.6.4, that any nontrivial particular solution of the auxiliary linear ordinary differential equation (2) generates infinitely many particular solutions of the original partial differential equation.

For the structure of other particular solutions, see the remark at the end of Item 3°.

2.3. Let a particular nonstationary solution $w_0 = w_0(x,t)$ ($\partial_t w_0 \neq 0$) of the homogeneous equation be known. Then the functions

$$w_n(x,t) = \frac{\partial^n w_0}{\partial t^n}(x,t),$$

obtained by differentiating the solution w_0 with respect to t, are also particular solutions of the equation in question.

In addition, a new particular solution can be sought in the form

$$\bar{w}(x,t) = \int_{t_0}^{t} w_0(x,\tau)\,d\tau + \phi(x), \tag{3}$$

where the unknown function $\phi(x)$ is determined on substituting expression (3) into the original equation. On constructing solution (3), one can use the above approach to construct another solution, and so on.

2.4. Case $h(x) = h = \text{const}$. Particular solutions (A and B are arbitrary constants):

$$w(x,t) = e^{ht}\left[A + B\int F(x)\,dx\right], \quad F(x) = \exp\left[-\int \frac{g(x)}{f(x)}\,dx\right],$$

$$w(x,t) = Ae^{ht}\left[t + \int F(x)\left(\int \frac{dx}{f(x)F(x)}\right)dx\right],$$

$$w(x,t) = Ae^{ht}\left[t\Psi(x) + \int F(x)\left(\int \frac{\Psi(x)\,dx}{f(x)F(x)}\right)dx\right], \quad \Psi(x) = \int F(x)\,dx.$$

The substitution $w(x,t) = e^{ht}v(x,t)$ leads to an equation of the form 1.8.6.4:

$$\frac{\partial v}{\partial t} = f(x)\frac{\partial^2 v}{\partial x^2} + g(x)\frac{\partial v}{\partial x}.$$

3°. The structure of particular solutions $\bar{w}(x,t)$ of the nonhomogeneous equation 1.8.6.5 for some functions $\Phi(x,t)$ is presented in Table 17.

Remark. The homogeneous equation (with $\Phi \equiv 0$) admits all the particular solutions specified in Table 17. In this case, n should be assumed an integer and β and λ arbitrary numbers.

6. $\quad \dfrac{\partial w}{\partial t} = f(x)\dfrac{\partial^2 w}{\partial x^2} + g(x)\dfrac{\partial w}{\partial x} + h(t)w.$

1°. Particular solutions (A and B are arbitrary constants):

$$w(x,t) = \left[A + B\int F(x)\,dx\right]H(t),$$

$$w(x,t) = A\left[t + \int F(x)\left(\int \frac{dx}{f(x)F(x)}\right)dx\right]H(t),$$

$$w(x,t) = A\left[t\Psi(x) + \int F(x)\left(\int \frac{\Psi(x)\,dx}{f(x)F(x)}\right)dx\right]H(t),$$

TABLE 17
Structure of particular solutions of linear nonhomogeneous equations of special form

No	Functions $\Phi(x,t)$	Form of particular solutions $\bar{w}(x,t)$	Remarks
1	$\varphi(x)t^n$	$\sum_{m=0}^{n} \psi_m(x)t^m$	n is an integer; the equations for $\psi_m(x)$ are solved consecutively, starting with $m=n$
2	$\varphi(x)e^{\beta t}$	$\psi(x)e^{\beta t}$	$\psi(x)$ is governed by a single equation
3	$\varphi(x)t^n e^{\beta t}$	$e^{\beta t}\sum_{m=0}^{n} \psi_m(x)t^m$	n is an integer; the equations for $\psi_m(x)$ are solved consecutively, starting with $m=n$
4	$\varphi(x)\sinh(\beta t)$	$\psi(x)e^{\beta t} + \chi(x)e^{-\beta t}$	the equations for $\psi(x)$ and $\chi(x)$ are independent
5	$\varphi(x)\cosh(\beta t)$	$\psi(x)e^{\beta t} + \chi(x)e^{-\beta t}$	the equations for $\psi(x)$ and $\chi(x)$ are independent
6	$\varphi(x)\sin(\beta t)$	$\psi(x)\sin(\beta t) + \chi(x)\cos(\beta t)$	$\psi(x)$ and $\chi(x)$ are determined by a system of equations
7	$\varphi(x)\cos(\beta t)$	$\psi(x)\sin(\beta t) + \chi(x)\cos(\beta t)$	$\psi(x)$ and $\chi(x)$ are determined by a system of equations
8	$\varphi(x)e^{\lambda t}\sin(\beta t)$	$\psi(x)e^{\lambda t}\sin(\beta t) + \chi(x)e^{\lambda t}\cos(\beta t)$	$\psi(x)$ and $\chi(x)$ are determined by a system of equations
9	$\varphi(x)e^{\lambda t}\cos(\beta t)$	$\psi(x)e^{\lambda t}\sin(\beta t) + \chi(x)e^{\lambda t}\cos(\beta t)$	$\psi(x)$ and $\chi(x)$ are determined by a system of equations

where

$$H(t) = \exp\left[\int h(t)\,dt\right], \quad F(x) = \exp\left[-\int \frac{g(x)}{f(x)}dx\right], \quad \Psi(x) = \int F(x)\,dx.$$

2°. The substitution $w(x,t) = u(x,t)\exp\left[\int h(t)\,dt\right]$ leads to an equation of the form 1.8.6.4:

$$\frac{\partial u}{\partial t} = f(x)\frac{\partial^2 u}{\partial x^2} + g(x)\frac{\partial u}{\partial x}.$$

7. $\dfrac{\partial w}{\partial t} = f(x)\dfrac{\partial^2 w}{\partial x^2} + g(x)\dfrac{\partial w}{\partial x} + [h_1(x) + h_2(t)]w.$

The substitution $w(x,t) = u(x,t)\exp\left[\int h_2(t)\,dt\right]$ leads to an equation of the form 1.8.6.5:

$$\frac{\partial u}{\partial t} = f(x)\frac{\partial^2 u}{\partial x^2} + g(x)\frac{\partial u}{\partial x} + h_1(x)u.$$

8. $\dfrac{\partial w}{\partial t} = f^2(x)\dfrac{\partial^2 w}{\partial x^2} + f(x)[f'_x(x) + g(t)]\dfrac{\partial w}{\partial x} + h(t)w.$

The change of variable $\xi = \displaystyle\int \dfrac{dx}{f(x)}$ leads to a special case of equation 1.8.7.3:

$$\dfrac{\partial w}{\partial t} = \dfrac{\partial^2 w}{\partial \xi^2} + g(t)\dfrac{\partial w}{\partial \xi} + h(t)w.$$

9. $\dfrac{\partial w}{\partial t} = f^2\dfrac{\partial^2 w}{\partial x^2} + f(f'_x + 2g + \varphi)\dfrac{\partial w}{\partial x} + (fg'_x + g^2 + g\varphi + \psi)w,$

where $f = f(x)$, $g = g(x)$, $\varphi = \varphi(t)$, $\psi = \psi(t)$.

The transformation

$$w(x,t) = u(\xi, t)\exp\!\left(-\int \dfrac{g}{f}dx\right), \quad \xi = \int \dfrac{dx}{f(x)}$$

leads to a special case of equation 1.8.7.3:

$$\dfrac{\partial u}{\partial t} = \dfrac{\partial^2 u}{\partial \xi^2} + \varphi(t)\dfrac{\partial u}{\partial \xi} + \psi(t)u.$$

1.8.7. Equations of the Form $\dfrac{\partial w}{\partial t} = f(t)\dfrac{\partial^2 w}{\partial x^2} + g(x,t)\dfrac{\partial w}{\partial x} + h(x,t)w$

1. $\dfrac{\partial w}{\partial t} = f(t)\dfrac{\partial^2 w}{\partial x^2} + g(t)\dfrac{\partial w}{\partial x}.$

1°. Particular solutions (A, B, and λ are arbitrary constants):

$$w(x,t) = Ax + A\int g(t)\,dt + B,$$

$$w(x,t) = A\!\left[x + \int g(t)\,dt\right]^2 + 2A\int f(t)\,dt + B,$$

$$w(x,t) = A\exp\!\left[\lambda x + \lambda^2 \int f(t)\,dt + \lambda \int g(t)\,dt\right].$$

2°. On passing from t, x to the new variables (A and B are any numbers)

$$\tau = \int f(t)\,dt + A, \quad z = x + \int g(t)\,dt + B,$$

for the function $w(\tau, z)$ we obtain a constant coefficient equation, $\partial_\tau w = \partial_{zz} w$, which is considered in Subsection 1.1.1.

2. $\dfrac{\partial w}{\partial t} = f(t)\dfrac{\partial^2 w}{\partial x^2} + xg(t)\dfrac{\partial w}{\partial x}.$

1°. Particular solutions (A, B, and λ are arbitrary constants):

$$w(x,t) = AxG(t) + B, \quad G(t) = \exp\!\left[\int g(t)\,dt\right],$$

$$w(x,t) = Ax^2 G^2(t) + 2A\int f(t)G^2(t)\,dt + B,$$

$$w(x,t) = A\exp\!\left[\lambda x G(t) + \lambda^2 \int f(t)G^2(t)\,dt\right].$$

2°. On passing from t, x to the new variables (A is any number)

$$\tau = \int f(t)G^2(t)\,dt + A, \quad z = xG(t), \quad \text{where } G(t) = \exp\!\left[\int g(t)\,dt\right],$$

for the function $w(\tau, z)$ we obtain a constant coefficient equation, $\partial_\tau w = \partial_{zz} w$, which is considered in Subsection 1.1.1.

3. $\dfrac{\partial w}{\partial t} = f(t)\dfrac{\partial^2 w}{\partial x^2} + g(t)\dfrac{\partial w}{\partial x} + h(t)w.$

This is a special case of equation 1.8.7.4.

The transformation

$$w(x,t) = u(z,\tau)\exp\left[\int h(t)\,dt\right], \quad z = x + \int g(t)\,dt, \quad \tau = \int f(t)\,dt$$

leads to a constant coefficient equation, $\partial_\tau u = \partial_{zz} u$, which is considered in Subsection 1.1.1.

4. $\dfrac{\partial w}{\partial t} = n(t)\dfrac{\partial^2 w}{\partial x^2} + \big[xf(t)+g(t)\big]\dfrac{\partial w}{\partial x} + \big[xh(t)+s(t)\big]w.$

Let us perform the transformation

$$w(x,t) = \exp\big[x\alpha(t) + \beta(t)\big]u(z,\tau), \quad \tau = \varphi(t), \quad z = x\psi(t) + \chi(t), \tag{1}$$

where the unknown functions $\alpha(t)$, $\beta(t)$, $\varphi(t)$, $\psi(t)$, and $\chi(t)$ are chosen so that the resulting equation is as simple as possible. For the new dependent variable $u(z,\tau)$ we have

$$\varphi'_t \frac{\partial u}{\partial \tau} = n\psi^2 \frac{\partial^2 u}{\partial z^2} + \big[x(f\psi - \psi'_t) + 2n\psi\alpha + g\psi - \chi'_t\big]\frac{\partial u}{\partial z}$$
$$+ \big[x(f\alpha + h - \alpha'_t) + n\alpha^2 + g\alpha + s - \beta'_t\big]u.$$

Let the unknown functions satisfy the system of ordinary differential equations

$$\varphi'_t = n\psi^2, \tag{2}$$
$$\psi'_t = f\psi, \tag{3}$$
$$\chi'_t = 2n\alpha\psi + g\psi, \tag{4}$$
$$\alpha'_t = f\alpha + h, \tag{5}$$
$$\beta'_t = n\alpha^2 + g\alpha + s. \tag{6}$$

Then the original equation can be reduced with the transformation (1)–(6) to the constant coefficient equation

$$\frac{\partial u}{\partial \tau} = \frac{\partial^2 u}{\partial z^2},$$

which is discussed in Subsection 1.1.1 in detail.

System (2)–(6) can be solved successively. To this end, we start with equation (3), for example, in order (3) → (2) → (5) → (6) → (4). As a result, we obtain

$$\psi = C_1 \exp\left(\int f\,dt\right), \quad C_1 \neq 0,$$

$$\varphi = \int n\psi^2\,dt + C_2,$$

$$\alpha = \psi \int \frac{h}{\psi}\,dt + C_3\psi,$$

$$\beta = \int \left(n\alpha^2 + g\alpha + s\right)dt + C_4,$$

$$\chi = \int (2n\alpha + g)\psi\,dt + C_5,$$

where C_1, C_2, C_3, C_4, and C_5 are arbitrary constants.

Remark. Likewise, one can simplify the nonhomogeneous equation with an additional term $\Phi(x,t)$ on the right-hand side.

5. $\dfrac{\partial w}{\partial t} = n(t)\dfrac{\partial^2 w}{\partial x^2} + \big[xf(t) + g(t)\big]\dfrac{\partial w}{\partial x} + \big[x^2 h(t) + xs(t) + p(t)\big]w.$

The substitution $w(x,t) = \exp\!\big[\varphi(t)x^2\big]u(x,t)$ leads to an equation

$$\begin{aligned}\dfrac{\partial u}{\partial t} &= n\dfrac{\partial^2 u}{\partial x^2} + \big[x(4n\varphi + f) + g\big]\dfrac{\partial u}{\partial x} \\ &\quad + [x^2(h + 2f\varphi + 4n\varphi^2 - \varphi_t') + x(s + 2g\varphi) + p + 2n\varphi]u.\end{aligned} \qquad (1)$$

We choose the function $\varphi = \varphi(t)$ so that it is a (particular) solution of the Riccati ordinary differential equation

$$\varphi_t' = 4n\varphi^2 + 2f\varphi + h. \qquad (2)$$

Then the transformed equation (1) becomes an equation of the form 1.8.7.4:

$$\dfrac{\partial u}{\partial t} = n\dfrac{\partial^2 u}{\partial x^2} + \big[x(4n\varphi + f) + g\big]\dfrac{\partial u}{\partial x} + [x(s + 2g\varphi) + p + 2n\varphi]u.$$

A number of specific solvable Riccati equations (2) can be found in Murphy (1960), Kamke (1977), and Polyanin and Zaitsev (1995).

In the special case where

$$n = af, \qquad h = bf, \qquad \text{with} \quad a, b = \text{const}, \ f = f(t),$$

the roots of the quadratic equation $4a\varphi^2 + 2\varphi + b = 0$ ($\varphi = \text{const}$) are particular solutions of equation (2).

1.8.8. Equations of the Form $\dfrac{\partial w}{\partial t} = f(x,t)\dfrac{\partial^2 w}{\partial x^2} + g(x,t)\dfrac{\partial w}{\partial x} + h(x,t)w$

1. $\dfrac{\partial w}{\partial t} = xf(t)\dfrac{\partial^2 w}{\partial x^2} + xg(t)\dfrac{\partial w}{\partial x} + \big[xh(t) + s(t)\big]w.$

Let us perform the transformation

$$\tau = \varphi(t), \quad z = x\psi(t), \quad w(x,t) = u(z,\tau)\exp\!\left[x\alpha(t) + \int s(t)\,dt\right], \qquad (1)$$

where the unknown functions $\varphi(t)$, $\psi(t)$, and $\alpha(t)$ are chosen so that the resulting equation is as simple as possible. For the new dependent variable $u(z,\tau)$ we have

$$\varphi_t' \dfrac{\partial u}{\partial \tau} = zf\psi\dfrac{\partial^2 u}{\partial z^2} + \dfrac{z}{\psi}\big(2f\psi\alpha + g\psi - \psi_t'\big)\dfrac{\partial u}{\partial z} + \dfrac{z}{\psi}\big(f\alpha^2 + g\alpha + h - \alpha_t'\big)u.$$

Let the unknown functions satisfy the system of ordinary differential equations

$$\varphi_t' = f\psi, \qquad (2)$$
$$\psi_t' = 2f\alpha\psi + g\psi, \qquad (3)$$
$$\alpha_t' = f\alpha^2 + g\alpha + h. \qquad (4)$$

Then the original equation can be reduced with the transformation (1)–(4) to a constant coefficient equation of the form 1.3.4.1:

$$\dfrac{\partial u}{\partial \tau} = z\dfrac{\partial^2 u}{\partial z^2}.$$

Let us solve system (2)–(4) successively, starting with equation (4) in the order (4) → (3) → (2).

The Riccati equation (4) can be solved separately. A lot of specific solvable Riccati equations can be found in Murphy (1960), Kamke (1977), and Polyanin and Zaitsev (1995).

Suppose a solution $\alpha = \alpha(t)$ of equation (4) is known. Then the solutions of equations (2) and (3) can be found in the form

$$\psi(t) = C_1 \exp\left[\int (2f\alpha + g)\, dt\right], \quad \varphi(t) = \int f\psi\, dt + C_2,$$

where C_1 and C_2 are arbitrary constants.

Remark. The transformation (1)–(4) can also be used to simplify the nonhomogeneous equation with an additional term $\Phi(x,t)$ on the right-hand side.

2. $\dfrac{\partial w}{\partial t} = x^2 f(t) \dfrac{\partial^2 w}{\partial x^2} + x g(t) \dfrac{\partial w}{\partial x} + h(t) w.$

The substitution $x = \pm e^\xi$ leads to an equation of the form 1.8.7.3:

$$\frac{\partial w}{\partial t} = f(t) \frac{\partial^2 w}{\partial \xi^2} + \left[g(t) - f(t)\right] \frac{\partial w}{\partial \xi} + h(t) w.$$

3. $\dfrac{\partial w}{\partial t} = x^2 n(t) \dfrac{\partial^2 w}{\partial x^2} + x\left[f(t)\ln x + g(t)\right] \dfrac{\partial w}{\partial x} + \left[h(t)\ln^2 x + s(t)\ln x + p(t)\right] w.$

The substitution $z = \ln x$ leads to an equation of the form 1.8.7.5:

$$\frac{\partial w}{\partial t} = n(t) \frac{\partial^2 w}{\partial z^2} + \left[z f(t) + g(t) - n(t)\right] \frac{\partial w}{\partial z} + \left[z^2 h(t) + z s(t) + p(t)\right] w.$$

4. $\dfrac{\partial w}{\partial t} = x^4 f(t) \dfrac{\partial^2 w}{\partial x^2} + g(t) w.$

1°. Particular solutions (A, B, and λ are arbitrary constants):

$$w(x,t) = (Ax + B)\exp\left[\int g(t)\, dt\right],$$

$$w(x,t) = \left[2Ax \int f(t)\, dt + Bx + \frac{A}{x}\right] \exp\left[\int g(t)\, dt\right],$$

$$w(x,t) = Ax \exp\left[\lambda^2 \int f(t)\, dt + \int g(t)\, dt + \frac{\lambda}{x}\right].$$

2°. The transformation

$$w(x,t) = x \exp\left[\int g(t)\, dt\right] u(\xi, \tau), \quad \xi = \frac{1}{x}, \quad \tau = \int f(t)\, dt$$

leads to a constant coefficient equation, $\partial_\tau u = \partial_{\xi\xi} u$, which is considered in Subsection 1.1.1.

5. $\dfrac{\partial w}{\partial t} = (x - a_1)^2 (x - a_2)^2 f(t) \dfrac{\partial^2 w}{\partial x^2} + g(t) w, \quad a_1 \neq a_2.$

The transformation

$$w(x,t) = (x - a_2) \exp\left[\int g(t)\, dt\right] u(\xi, \tau), \quad \xi = \ln\left|\frac{x - a_1}{x - a_2}\right|, \quad \tau = (a_1 - a_2)^2 \int f(t)\, dt$$

leads to a constant coefficient equation

$$\frac{\partial u}{\partial \tau} = \frac{\partial^2 u}{\partial \xi^2} - \frac{\partial u}{\partial \xi},$$

which is considered in Subsection 1.1.4.

6. $\dfrac{\partial w}{\partial t} = (ax^2 + bx + c)^2 f(t) \dfrac{\partial^2 w}{\partial x^2} + g(t)w.$

The transformation
$$w(x,t) = |ax^2 + bx + c|^{1/2} \exp\left[\int g(t)\,dt\right] u(\xi,\tau), \quad \xi = \int \dfrac{dx}{ax^2 + bx + c}, \quad \tau = \int f(t)\,dt$$
leads to a constant coefficient equation
$$\dfrac{\partial u}{\partial \tau} = \dfrac{\partial^2 u}{\partial \xi^2} + \left(ac - \tfrac{1}{4}b^2\right)u,$$
which is considered in Subsection 1.1.3.

7. $\dfrac{\partial w}{\partial t} = x^n f(t) \dfrac{\partial^2 w}{\partial x^2} + xg(t)\dfrac{\partial w}{\partial x} + h(t)w.$

The transformation
$$w(x,t) = u(z,\tau) \exp\left[\int h(t)\,dt\right], \quad z = xG(t), \quad \tau = \int f(t)G^{2-n}(t)\,dt,$$
where $G(t) = \exp\left[\int g(t)\,dt\right]$, leads to an equation of the form 1.3.6.6:
$$\dfrac{\partial u}{\partial \tau} = z^n \dfrac{\partial^2 u}{\partial z^2}.$$

8. $\dfrac{\partial w}{\partial t} = e^{\beta x} f(t) \dfrac{\partial^2 w}{\partial x^2} + g(t)w.$

The transformation
$$w(x,t) = \exp\left[\int g(t)\,dt\right] u(x,\tau), \quad \tau = \int f(t)\,dt$$
leads to an equation of the form 1.4.5.1:
$$\dfrac{\partial u}{\partial \tau} = e^{\beta x} \dfrac{\partial^2 u}{\partial x^2}.$$

9. $\dfrac{\partial w}{\partial t} = f(x)g(t) \dfrac{\partial^2 w}{\partial x^2} + h(t)w.$

1°. Particular solutions (A, B, and x_0 are arbitrary constants):
$$w(x,t) = (Ax + B)H(t),$$
$$w(x,t) = A\big[M(t) + F(x)\big]H(t),$$
$$w(x,t) = A\big[M(t)x + \Psi(x)\big]H(t),$$
$$w(x,t) = A\left[M^2(t) + 2M(t)F(x) + 2\int_{x_0}^{x} \dfrac{x-\xi}{f(\xi)} F'(\xi)\,d\xi\right] H(t),$$
$$w(x,t) = A\left[M^2(t)x + 2M(t)\Psi(x) + 2\int_{x_0}^{x} \dfrac{x-\xi}{f(\xi)} \Psi(\xi)\,d\xi\right] H(t).$$

Here we use the shorthand notation
$$H(t) = \exp\left[\int h(t)\,dt\right], \quad M(t) = \int g(t)\,dt, \quad F(x) = \int_{x_0}^{x} \dfrac{x-\xi}{f(\xi)}\,d\xi, \quad \Psi(x) = \int_{x_0}^{x} \dfrac{x-\xi}{f(\xi)} \xi\,d\xi.$$

2°. The transformation
$$w(x,t) = \exp\left[\int h(t)\,dt\right] u(x,\tau), \quad \tau = \int g(t)\,dt$$
leads to a simpler equation
$$\dfrac{\partial u}{\partial \tau} = f(x) \dfrac{\partial^2 u}{\partial x^2}.$$

A wide class of exact analytical solutions to this equation is specified in 1.8.6.1.

1.8.9. Equations of the Form $s(x)\dfrac{\partial w}{\partial t} = \dfrac{\partial}{\partial x}\left[p(x)\dfrac{\partial w}{\partial x}\right] - q(x)w + \Phi(x,t)$

Equations of this form are often encountered in heat and mass transfer theory and chemical engineering sciences. Throughout this subsection, we assume that the functions s, p, p'_x, and q are continuous and $s > 0$, $p > 0$, and $x_1 \le x \le x_2$.

1.8.9-1. General formulas for solving linear nonhomogeneous boundary value problems.

The solution of the equation in question under the initial condition
$$w = f(x) \quad \text{at} \quad t = 0 \tag{1}$$
and the arbitrary linear nonhomogeneous boundary conditions
$$\begin{aligned}a_1\partial_x w + b_1 w &= g_1(t) \quad \text{at} \quad x = x_1, \\ a_2\partial_x w + b_2 w &= g_2(t) \quad \text{at} \quad x = x_2,\end{aligned} \tag{2}$$
can be represented as the sum
$$\begin{aligned}w(x,t) &= \int_0^t \int_{x_1}^{x_2} \Phi(\xi,\tau)\mathcal{G}(x,\xi,t-\tau)\,d\xi\,d\tau + \int_{x_1}^{x_2} s(\xi)f(\xi)\mathcal{G}(x,\xi,t)\,d\xi \\ &\quad + p(x_1)\int_0^t g_1(\tau)\Lambda_1(x,t-\tau)\,d\tau + p(x_2)\int_0^t g_2(\tau)\Lambda_2(x,t-\tau)\,d\tau.\end{aligned} \tag{3}$$
Here, the modified Green's function is given by
$$\mathcal{G}(x,\xi,t) = \sum_{n=1}^{\infty} \frac{y_n(x)y_n(\xi)}{\|y_n\|^2}\exp(-\lambda_n t), \qquad \|y_n\|^2 = \int_{x_1}^{x_2} s(x)y_n^2(x)\,dx, \tag{4}$$
where the λ_n and $y_n(x)$ are the eigenvalues and corresponding eigenfunctions of the following Sturm–Liouville problem for a second-order linear ordinary differential equation:
$$\begin{aligned}[p(x)y'_x]'_x + [\lambda s(x) - q(x)]y &= 0, \\ a_1 y'_x + b_1 y &= 0 \quad \text{at} \quad x = x_1, \\ a_2 y'_x + b_2 y &= 0 \quad \text{at} \quad x = x_2.\end{aligned} \tag{5}$$
The functions $\Lambda_1(x,t)$ and $\Lambda_2(x,t)$ that occur in the integrands of the last two terms in solution (3) are expressed via the Green's function (4). Appropriate formulas will be given below in Paragraphs 1.8.9-3–1.8.9-7 when considering specific boundary value problems.

1.8.9-2. General properties of the Sturm–Liouville problem (5).

1°. There are infinitely many eigenvalues. All eigenvalues are real and different and can be ordered so that $\lambda_1 < \lambda_2 < \lambda_3 < \cdots$, with $\lambda_n \to \infty$ as $n \to \infty$ (therefore, there can exist only finitely many negative eigenvalues). Each eigenvalue is of multiplicity 1.

2°. The eigenfunctions are determined up to a constant multiplier. Each eigenfunction $y_n(x)$ has exactly $n-1$ zeros in the open interval (x_1, x_2).

3°. Eigenfunctions $y_n(x)$ and $y_m(x)$, $n \ne m$, are orthogonal with weight $s(x)$ on the interval $x_1 \le x \le x_2$:
$$\int_{x_1}^{x_2} s(x)y_n(x)y_m(x)\,dx = 0 \quad \text{for} \quad n \ne m.$$

4°. An arbitrary function $F(x)$ that has a continuous derivative and satisfies the boundary conditions of the Sturm–Liouville problem can be expanded into an absolutely and uniformly convergent series in eigenfunctions:
$$F(x) = \sum_{n=1}^{\infty} F_n y_n(x), \qquad F_n = \frac{1}{\|y_n\|^2}\int_{x_1}^{x_2} s(x)F(x)y_n(x)\,dx,$$
where the norm $\|y_n\|^2$ is defined in (4).

5°. If the conditions
$$q(x) \geq 0, \quad a_1 b_1 \leq 0, \quad a_2 b_2 \geq 0 \tag{6}$$
are satisfied, there are no negative eigenvalues. If $q \equiv 0$ and $b_1 = b_2 = 0$, then $\lambda_1 = 0$ is the least eigenvalue, to which there corresponds the eigenfunction $\varphi_1 = \text{const}$. Otherwise, all eigenvalues are positive, provided that conditions (6) are satisfied.

6°. The following asymptotic relation holds for large eigenvalues as $n \to \infty$:
$$\lambda_n = \frac{\pi^2 n^2}{\Delta^2} + O(1), \quad \Delta = \int_{x_1}^{x_2} \sqrt{\frac{s(x)}{p(x)}}\, dx. \tag{7}$$

Special, boundary value condition-dependent properties of the Sturm–Liouville problem are presented in Paragraphs 1.8.9-3 through 1.8.9-7.

Remark. Equation (5) can be reduced to one with $p(x) \equiv 1$ and $s(x) \equiv 1$ by the change of variables
$$\zeta = \int \sqrt{\frac{s(x)}{p(x)}}\, dx, \quad u(\zeta) = \left[p(x)s(x)\right]^{1/4} y(x).$$
The boundary conditions transform into boundary conditions of the same type.

1.8.9-3. First boundary value problem: the case of $a_1 = a_2 = 0$ and $b_1 = b_2 = 1$.

The solution of the first boundary value problem with the initial condition (1) and the boundary conditions
$$w = g_1(t) \quad \text{at} \quad x = x_1,$$
$$w = g_2(t) \quad \text{at} \quad x = x_2$$
is given by formulas (3)–(4) with
$$\Lambda_1(x,t) = \frac{\partial}{\partial \xi} \mathcal{G}(x,\xi,t) \bigg|_{\xi = x_1}, \quad \Lambda_2(x,t) = -\frac{\partial}{\partial \xi} \mathcal{G}(x,\xi,t) \bigg|_{\xi = x_2}.$$

Some special properties of the Sturm–Liouville problem are worth mentioning.

1°. For $n \to \infty$, the asymptotic relation (7) can be used to estimate eigenvalues λ_n. The corresponding eigenfunctions $y_n(x)$ satisfy the asymptotic relation
$$\frac{y_n(x)}{\|y_n\|} = \left[\frac{4}{\Delta^2 p(x) s(x)}\right]^{1/4} \sin\left[\frac{\pi n}{\Delta} \int_{x_1}^{x} \sqrt{\frac{s(x)}{p(x)}}\, dx\right] + O\left(\frac{1}{n}\right), \quad \Delta = \int_{x_1}^{x_2} \sqrt{\frac{s(x)}{p(x)}}\, dx.$$

2°. For $q \geq 0$, the following upper estimate (Rayleigh principle) holds for the least eigenvalue:
$$\lambda_1 \leq \frac{\int_{x_1}^{x_2} \left[p(x)(z'_x)^2 + q(x) z^2\right] dx}{\int_{x_1}^{x_2} s(x) z^2\, dx}, \tag{8}$$
where $z = z(x)$ is any twice differentiable function that satisfies the conditions $z(x_1) = z(x_2) = 0$. The equality in (8) is attained for $z = y_1(x)$, where $y_1(x)$ is the eigenfunction of the Sturm–Liouville problem corresponding to the eigenvalue λ_1. To obtain particular estimates, one may set $z = (x - x_1)(x_2 - x)$ or $z = \sin[\pi(x - x_1)/(x_2 - x_1)]$ in (8).

3°. Suppose
$$0 < p_{\min} \le p(x) \le p_{\max}, \quad 0 < q_{\min} \le q(x) \le q_{\max}, \quad 0 < s_{\min} \le s(x) \le s_{\max}.$$
Then the following double-ended estimate holds for the eigenvalues:
$$\frac{p_{\min}}{s_{\max}} \frac{\pi^2 n^2}{(x_2-x_1)^2} + \frac{q_{\min}}{s_{\max}} \le \lambda_n \le \frac{p_{\max}}{s_{\min}} \frac{\pi^2 n^2}{(x_2-x_1)^2} + \frac{q_{\max}}{s_{\min}}.$$

4°. In engineering calculations, the approximate formula
$$\lambda_n = \frac{\pi^2 n^2}{\Delta^2} + \frac{1}{x_2-x_1}\int_{x_1}^{x_2} \frac{q(x)}{s(x)}\,dx, \quad \text{where} \quad \Delta = \int_{x_1}^{x_2} \sqrt{\frac{s(x)}{p(x)}}\,dx, \tag{9}$$
may be used to determine eigenvalues. This formula is exact if $p(x)s(x)=\text{const}$ and $q(x)/s(x)=\text{const}$ (in particular, for constant $p=p_0$, $q=q_0$, and $s=s_0$) and provides correct asymptotics (7) for any $p(x)$, $q(x)$, and $s(x)$. Furthermore, for $p(x)=\text{const}$ and $s(x)=\text{const}$, relation (9) gives two correct first terms as $n\to\infty$; the same holds true if $p(x)s(x)=\text{const}$.

5°. Suppose $p(x)=s(x)=1$ and the function $q=q(x)$ has a continuous derivative. Then the following asymptotic relations hold for eigenvalues λ_n and eigenfunctions $y_n(x)$ as $n\to\infty$:
$$\sqrt{\lambda_n} = \frac{\pi n}{x_2-x_1} + \frac{1}{\pi n}Q(x_1,x_2) + O\left(\frac{1}{n^2}\right),$$
$$y_n(x) = \sin\frac{\pi n(x-x_1)}{x_2-x_1} - \frac{1}{\pi n}\big[(x_1-x)Q(x,x_2)+(x_2-x)Q(x_1,x)\big]\cos\frac{\pi n(x-x_1)}{x_2-x_1} + O\left(\frac{1}{n^2}\right),$$
where
$$Q(u,v) = \frac{1}{2}\int_u^v q(x)\,dx. \tag{10}$$

1.8.9-4. Second boundary value problem: the case of $a_1 = a_2 = 1$ and $b_1 = b_2 = 0$.

The solution of the second boundary value problem with the initial condition (1) and the boundary conditions
$$\partial_x w = g_1(t) \quad \text{at} \quad x = x_1,$$
$$\partial_x w = g_2(t) \quad \text{at} \quad x = x_2$$
is given by formulas (3)–(4) with
$$\Lambda_1(x,t) = -\mathcal{G}(x,x_1,t), \quad \Lambda_2(x,t) = \mathcal{G}(x,x_2,t).$$

Some special properties of the Sturm–Liouville problem are worth mentioning.

1°. For $q > 0$, the upper estimate (8) holds for the least eigenvalue, with $z = z(x)$ being any twice differentiable function that satisfies the conditions $z'_x(x_1) = z'_x(x_2) = 0$. The equality in (8) is attained for $z = y_1(x)$, where $y_1(x)$ is the eigenfunction of the Sturm–Liouville problem corresponding to the eigenvalue λ_1.

2°. Suppose $p(x)=s(x)=1$ and the function $q=q(x)$ has a continuous derivative. Then the following asymptotic relations hold for eigenvalues λ_n and eigenfunctions $y_n(x)$ as $n\to\infty$:
$$\sqrt{\lambda_n} = \frac{\pi(n-1)}{x_2-x_1} + \frac{1}{\pi(n-1)}Q(x_1,x_2) + O\left(\frac{1}{n^2}\right),$$
$$y_n(x) = \cos\frac{\pi(n-1)(x-x_1)}{x_2-x_1} + \frac{1}{\pi(n-1)}\big[(x_1-x)Q(x,x_2) + (x_2-x)Q(x_1,x)\big]\sin\frac{\pi(n-1)(x-x_1)}{x_2-x_1} + O\left(\frac{1}{n^2}\right),$$
where the function $Q(u,v)$ is defined by (10).

1.8.9-5. Third boundary value problem: the case of $a_1 = a_2 = 1$, $b_1 \neq 0$, and $b_2 \neq 0$.

The solution of the third boundary value problem with the initial condition (1) and boundary conditions (2), with $a_1 = a_2 = 1$, is given by relations (3)–(4) in which

$$\Lambda_1(x,t) = -\mathcal{G}(x, x_1, t), \quad \Lambda_2(x,t) = \mathcal{G}(x, x_2, t).$$

Suppose $p(x) = s(x) = 1$ and the function $q = q(x)$ has a continuous derivative. Then the following asymptotic relations hold for eigenvalues λ_n and eigenfunctions $y_n(x)$ as $n \to \infty$:

$$\sqrt{\lambda_n} = \frac{\pi(n-1)}{x_2 - x_1} + \frac{1}{\pi(n-1)}\Big[Q(x_1, x_2) - b_1 + b_2\Big] + O\!\left(\frac{1}{n^2}\right),$$

$$y_n(x) = \cos\frac{\pi(n-1)(x-x_1)}{x_2 - x_1} + \frac{1}{\pi(n-1)}\Big\{(x_1 - x)\big[Q(x, x_2) + b_2\big]$$

$$+ (x_2 - x)\big[Q(x_1, x) - b_1\big]\Big\} \sin\frac{\pi(n-1)(x-x_1)}{x_2 - x_1} + O\!\left(\frac{1}{n^2}\right),$$

where $Q(u,v)$ is defined by (10).

1.8.9-6. Mixed boundary value problem: the case of $a_1 = b_2 = 0$ and $a_2 = b_1 = 1$.

The solution of the mixed boundary value problem with the initial condition (1) and the boundary conditions

$$w = g_1(t) \quad \text{at} \quad x = x_1,$$
$$\partial_x w = g_2(t) \quad \text{at} \quad x = x_2$$

is given by relations (3)–(4) with

$$\Lambda_1(x,t) = \frac{\partial}{\partial \xi}\mathcal{G}(x, \xi, t)\bigg|_{\xi = x_1}, \quad \Lambda_2(x,t) = \mathcal{G}(x, x_2, t).$$

Below are some special properties of the Sturm–Liouville problem.

$1°$. For $q \geq 0$, the upper estimate (8) holds for the least eigenvalue, with $z = z(x)$ being any twice differentiable function that satisfies the conditions $z(x_1) = 0$ and $z'_x(x_2) = 0$. The equality in (8) is attained for $z = y_1(x)$, where $y_1(x)$ is the eigenfunction corresponding to the eigenvalue λ_1.

$2°$. Suppose $p(x) = s(x) = 1$ and the function $q = q(x)$ has a continuous derivative. Then the following asymptotic relations hold for eigenvalues λ_n and eigenfunctions $y_n(x)$ as $n \to \infty$:

$$\sqrt{\lambda_n} = \frac{\pi(2n-1)}{2(x_2-x_1)} + \frac{1}{\pi(2n-1)}Q(x_1, x_2) + O\!\left(\frac{1}{n^2}\right),$$

$$y_n(x) = \sin\frac{\pi(2n-1)(x-x_1)}{2(x_2-x_1)} - \frac{2}{\pi(2n-1)}\Big[(x_1 - x)Q(x, x_2)$$

$$+ (x_2 - x)Q(x_1, x)\Big] \cos\frac{\pi(2n-1)(x-x_1)}{2(x_2-x_1)} + O\!\left(\frac{1}{n^2}\right),$$

where $Q(u,v)$ is defined by (10).

1.8.9-7. Mixed boundary value problem: the case of $a_1 = b_2 = 1$ and $a_2 = b_1 = 0$.

The solution of the mixed boundary value problem with the initial condition (1) and the boundary conditions

$$\partial_x w = g_1(t) \quad \text{at} \quad x = x_1,$$
$$w = g_2(t) \quad \text{at} \quad x = x_2$$

is given by formulas (3)–(4) in which

$$\Lambda_1(x,t) = -\mathcal{G}(x, x_1, t), \quad \Lambda_2(x,t) = -\frac{\partial}{\partial \xi}\mathcal{G}(x, \xi, t)\bigg|_{\xi = x_2}.$$

Below are some special properties of the Sturm–Liouville problem.

1°. For $q \geq 0$, the upper estimate (8) holds for the least eigenvalue, with $z = z(x)$ being any twice differentiable function that satisfies the conditions $z'_x(x_1) = 0$ and $z(x_2) = 0$. The equality in (8) is attained for $z = y_1(x)$, where $y_1(x)$ is the eigenfunction corresponding to the eigenvalue λ_1.

2°. Suppose $p(x) = s(x) = 1$ and the function $q = q(x)$ has a continuous derivative. Then the following asymptotic relations hold for eigenvalues λ_n and eigenfunctions $y_n(x)$ as $n \to \infty$:

$$\sqrt{\lambda_n} = \frac{\pi(2n-1)}{2(x_2-x_1)} + \frac{2}{\pi(2n-1)} Q(x_1, x_2) + O\left(\frac{1}{n^2}\right),$$

$$y_n(x) = \cos\frac{\pi(2n-1)(x-x_1)}{2(x_2-x_1)} + \frac{2}{\pi(2n-1)}\big[(x_1-x)Q(x, x_2)$$

$$+ (x_2-x)Q(x_1, x)\big] \sin\frac{\pi(2n-1)(x-x_1)}{2(x_2-x_1)} + O\left(\frac{1}{n^2}\right),$$

where the function $Q(u, v)$ is defined by relation (10).

⊙ *References for Subsection 1.8.9:* V. M. Babich, M. B. Kapilevich, S. G. Mikhlin, et al. (1964), E. Kamke (1977), V. A. Marchenko (1986), V. S. Vladimirov (1988), B. M. Levitan, I. S. Sargsyan (1988), L. D. Akulenko and S. V. Nesterov (1997), A. D. Polyanin (2001a).

1.9. Equations of Special Form

1.9.1. Equations of the Diffusion (Thermal) Boundary Layer

1. $f(x)\dfrac{\partial w}{\partial x} + g(x)y\dfrac{\partial w}{\partial y} = \dfrac{\partial^2 w}{\partial y^2}.$

This equation is encountered in diffusion boundary layer problems (mass exchange of drops and bubbles with flow).

The transformation (A and B are any numbers)

$$t = \int \frac{h^2(x)}{f(x)}\,dx + A, \quad z = yh(x), \quad \text{where} \quad h(x) = B\exp\left[-\int\frac{g(x)}{f(x)}\,dx\right],$$

leads to a constant coefficient equation, $\partial_t w = \partial_{zz} w$, which is considered in Subsection 1.1.1.

⊙ *References:* V. G. Levich (1962), A. D. Polyanin and V. V. Dilman (1994), A. D. Polyanin, A. M. Kutepov, A. V. Vyazmin, and D. A. Kazenin (2001).

2. $f(x)\dfrac{\partial w}{\partial x} + g(x)y\dfrac{\partial w}{\partial y} = \dfrac{\partial^2 w}{\partial y^2} - h(x)w.$

This equation is encountered in diffusion boundary layer problems with a first-order volume chemical reaction (usually $h \equiv \text{const}$).

The transformation (A and B are any numbers)

$$w(x, y) = u(t, z)\exp\left[-\int\frac{h(x)}{f(x)}dx\right], \quad t = \int\frac{\varphi^2(x)}{f(x)}dx + A, \quad z = y\varphi(x),$$

where $\varphi(x) = B\exp\left[-\int\dfrac{g(x)}{f(x)}dx\right]$, leads to a constant coefficient equation, $\partial_t u = \partial_{zz} u$, which is considered in Subsection 1.1.1.

⊙ *Reference:* Yu. P. Gupalo, A. D. Polyanin, and Yu. S. Ryazantsev (1985).

3. $f(x)y^{n-1}\dfrac{\partial w}{\partial x} + g(x)y^n \dfrac{\partial w}{\partial y} = \dfrac{\partial^2 w}{\partial y^2}.$

This equation is encountered in diffusion boundary layer problems (mass exchange of solid particles, drops, and bubbles with flow).

The transformation (A and B are any numbers)

$$t = \int \dfrac{h^{n+1}(x)}{f(x)}\,dx + A, \quad z = yh(x), \quad \text{where} \quad h(x) = B\exp\left[-\int \dfrac{g(x)}{f(x)}\,dx\right],$$

leads to a simpler equation of the form 1.3.6.6:

$$\dfrac{\partial w}{\partial t} = z^{1-n}\dfrac{\partial^2 w}{\partial z^2}.$$

⊙ *References*: V. G. Levich (1962), A. D. Polyanin and V. V. Dilman (1994), A. D. Polyanin, A. V. Vyazmin, A. I. Zhurov, and D. A. Kazenin (1998), A. D. Polyanin, A. M. Kutepov, A. V. Vyazmin, and D. A. Kazenin (2001).

4. $f\!\left(\dfrac{y}{\sqrt{x}}\right)\dfrac{\partial w}{\partial x} + \dfrac{1}{\sqrt{x}}g\!\left(\dfrac{y}{\sqrt{x}}\right)\dfrac{\partial w}{\partial y} = \dfrac{\partial^2 w}{\partial y^2}.$

This is a generalization of the problem of thermal boundary layer on a flat plate.

1°. By passing from x, y to the new variables $t = \ln x$, $\xi = y/\sqrt{x}$, we arrive at the separable equation

$$f(\xi)\dfrac{\partial w}{\partial t} + \left[g(\xi) - \tfrac{1}{2}\xi f(\xi)\right]\dfrac{\partial w}{\partial \xi} = \dfrac{\partial^2 w}{\partial \xi^2}.$$

There are particular solutions of the form

$$w(t,\xi) = Ae^{\beta t}\phi(\xi),$$

where the function $\phi(\xi)$ satisfies the ordinary differential equation

$$\phi''_{\xi\xi} = \left[g(\xi) - \tfrac{1}{2}\xi f(\xi)\right]\phi'_\xi + \beta f(\xi)\phi.$$

2°. The solution of the original equation with the boundary conditions

$$x = 0, \ w = w_0; \quad y = 0, \ w = w_1; \quad y \to \infty, \ w \to w_0$$

(w_0 and w_1 are some constants) is given by

$$\dfrac{w - w_0}{w_1 - w_0} = \dfrac{\int_\xi^\infty \exp[-\Psi(\xi)]\,d\xi}{\int_0^\infty \exp[-\Psi(\xi)]\,d\xi}, \qquad \Psi(\xi) = \int_0^\xi \left[\tfrac{1}{2}\xi f(\xi) - g(\xi)\right]d\xi,$$

where $\xi = y/\sqrt{x}$. It is assumed that the inequality $\xi f(\xi) > 2g(\xi)$ holds for $\xi > 0$.

3°. The equation of thermal boundary layer on a flat plate corresponds to

$$f(\xi) = \Pr F'_\xi(\xi), \qquad g(\xi) = \tfrac{1}{2}\Pr\!\left[\xi F'_\xi(\xi) - F(\xi)\right],$$

where $F(\xi)$ is Blasius' solution in the problem of translational flow past a flat plate and Pr is the Prandtl number (x is the coordinate measured along the plate and y is the transverse coordinate to the plate surface). In this case the formulas in Item 2° transform into Polhausen's solution. See Schlichting (1981) for details.

⊙ *Reference*: A. D. Polyanin, A. V. Vyazmin, A. I. Zhurov, and D. A. Kazenin (1998).

5. $f(x)\dfrac{\partial w}{\partial x} + \left[g(x)y - \dfrac{b}{y}\right]\dfrac{\partial w}{\partial y} = \dfrac{\partial^2 w}{\partial y^2}$.

For $b = 1$, equations of this sort govern the concentration distribution in the internal region of the diffusion wake behind a moving particle or drop.

The transformation (A and B are any numbers)

$$t = \int \dfrac{h^2(x)}{f(x)}\,dx + A, \quad z = yh(x), \quad \text{where} \quad h(x) = B\exp\left[-\int \dfrac{g(x)}{f(x)}\,dx\right],$$

leads to an equation of the form 1.2.5:

$$\dfrac{\partial w}{\partial t} = \dfrac{\partial^2 w}{\partial z^2} + \dfrac{b}{z}\dfrac{\partial w}{\partial z}.$$

⊙ *References*: Yu. P. Gupalo, A. D. Polyanin, and Yu. S. Ryazantsev (1985), A. D. Polyanin, A. V. Vyazmin, A. I. Zhurov, and D. A. Kazenin (1998).

6. $f(x)\dfrac{\partial w}{\partial x} + \left[g(x)y - \dfrac{b}{y}\right]\dfrac{\partial w}{\partial y} = \dfrac{\partial^2 w}{\partial y^2} + h(x)w$.

The substitution $w(x,y) = u(x,y)\exp\left[\int \dfrac{h(x)}{f(x)}\,dx\right]$ leads to an equation of the form 1.9.1.5:

$$f(x)\dfrac{\partial u}{\partial x} + \left[g(x)y - \dfrac{b}{y}\right]\dfrac{\partial u}{\partial y} = \dfrac{\partial^2 u}{\partial y^2}.$$

⊙ *Reference*: A. D. Polyanin, A. V. Vyazmin, A. I. Zhurov, and D. A. Kazenin (1998).

7. $f(x)y^{n-1}\dfrac{\partial w}{\partial x} + \left[g(x)y^n - \dfrac{b}{y}\right]\dfrac{\partial w}{\partial y} = \dfrac{\partial^2 w}{\partial y^2}$.

The transformation (A and B are any numbers)

$$t = \int \dfrac{h^{n+1}(x)}{f(x)}\,dx + A, \quad \xi = \dfrac{2}{n+1}\left[yh(x)\right]^{\frac{n+1}{2}},$$

where $h(x) = B\exp\left[-\int \dfrac{g(x)}{f(x)}\,dx\right]$, leads to the equation

$$\dfrac{\partial w}{\partial t} = \dfrac{\partial^2 w}{\partial \xi^2} + \dfrac{1-2\beta}{\xi}\dfrac{\partial w}{\partial \xi}, \qquad \beta = \dfrac{1-b}{n+1},$$

which is considered in Subsection 1.2.5 (see also equations in Subsections 1.2.1 and 1.2.3).

1.9.2. One-Dimensional Schrödinger Equation

$$i\hbar\dfrac{\partial w}{\partial t} = -\dfrac{\hbar^2}{2m}\dfrac{\partial^2 w}{\partial x^2} + U(x)w$$

1.9.2-1. Eigenvalue problem. Cauchy problem.

Schrödinger's equation is the basic equation of quantum mechanics; w is the wave function, $i^2 = -1$, $\hbar$ is Planck's constant, m is the mass of the particle, and $U(x)$ is the potential energy of the particle in the force field.

1°. In discrete spectrum problems, the particular solutions are sought in the form

$$w(x,t) = \exp\left(-\frac{iE_n}{\hbar}t\right)\psi_n(x),$$

where the eigenfunctions ψ_n and the respective energies E_n have to be determined by solving the eigenvalue problem

$$\frac{d^2\psi_n}{dx^2} + \frac{2m}{\hbar^2}[E_n - U(x)]\psi_n = 0,$$

$$\psi_n \to 0 \text{ at } x \to \pm\infty, \qquad \int_{-\infty}^{\infty} |\psi_n|^2 \, dx = 1.$$

(1)

The last relation is the normalizing condition for ψ_n.

2°. In the cases where the eigenfunctions $\psi_n(x)$ form an orthonormal basis in $L_2(\mathbb{R})$, the solution of the Cauchy problem for Schrödinger's equation with the initial condition

$$w = f(x) \quad \text{at} \quad t = 0 \qquad (2)$$

is given by

$$w(x,t) = \int_{-\infty}^{\infty} G(x,\xi,t) f(\xi)\, d\xi, \qquad G(x,\xi,t) = \sum_{n=0}^{\infty} \psi_n(x)\psi_n(\xi) \exp\left(-\frac{iE_n}{\hbar}t\right).$$

Various potentials $U(x)$ are considered below and particular solutions of the boundary value problem (1) or the Cauchy problem for Schrödinger's equation are presented. In some cases, nonnormalized eigenfunctions $\Psi_n(x)$ are given instead of normalized eigenfunctions $\psi_n(x)$; the former differ from the latter by a constant multiplier.

1.9.2-2. Free particle: $U(x) = 0$.

The solution of the Cauchy problem with the initial condition (2) is given by

$$w(x,t) = \frac{1}{2\sqrt{i\pi\tau}} \int_{-\infty}^{\infty} \exp\left[-\frac{(x-\xi)^2}{4i\tau}\right] f(\xi)\, d\xi, \qquad \tau = \frac{\hbar t}{2m}, \qquad \sqrt{ia} = \begin{cases} e^{\pi i/4}\sqrt{|a|} & \text{if } a > 0, \\ e^{-\pi i/4}\sqrt{|a|} & \text{if } a < 0. \end{cases}$$

⊙ *Reference*: W. Miller, Jr. (1977).

1.9.2-3. Linear potential (motion in a uniform external field): $U(x) = ax$.

Solution of the Cauchy problem with the initial condition (2):

$$w(x,t) = \frac{1}{2\sqrt{i\pi\tau}} \exp\left(-ib\tau x - \tfrac{1}{3}ib^2\tau^3\right) \int_{-\infty}^{\infty} \exp\left[-\frac{(x+b\tau^2-\xi)^2}{4i\tau}\right] f(\xi)\, d\xi, \qquad \tau = \frac{\hbar t}{2m}, \qquad b = \frac{2am}{\hbar^2}.$$

See also Miller, Jr. (1977).

1.9.2-4. Linear harmonic oscillator: $U(x) = \tfrac{1}{2}m\omega^2 x^2$.

Eigenvalues:

$$E_n = \hbar\omega\left(n + \tfrac{1}{2}\right), \qquad n = 0, 1, \ldots$$

Normalized eigenfunctions:

$$\psi_n(x) = \frac{1}{\pi^{1/4}\sqrt{2^n n!\, x_0}} \exp\left(-\tfrac{1}{2}\xi^2\right) H_n(\xi), \qquad \xi = \frac{x}{x_0}, \qquad x_0 = \sqrt{\frac{\hbar}{m\omega}},$$

where $H_n(\xi)$ are the Hermite polynomials.

The functions $\psi_n(x)$ form an orthonormal basis in $L_2(\mathbb{R})$.

⊙ *References*: S. G. Krein (1964), W. Miller, Jr. (1977), A. N. Tikhonov and A. A. Samarskii (1990).

1.9.2-5. Isotropic free particle: $U(x) = a/x^2$.

Here, the variable $x \geq 0$ plays the role of the radial coordinate, and $a > 0$. The equation with $U(x) = a/x^2$ results from Schrödinger's equation for a free particle with n space coordinates if one passes to spherical (cylindrical) coordinates and separates the angular variables.

The solution of Schrödinger's equation satisfying the initial condition (2) has the form

$$w(x,t) = \frac{\exp\left[-\tfrac{1}{2}i\pi(\mu+1)\operatorname{sign} t\right]}{2|\tau|} \int_0^\infty \sqrt{xy}\, \exp\left(i\frac{x^2+y^2}{4\tau}\right) J_\mu\left(\frac{xy}{2|\tau|}\right) f(y)\, dy,$$

$$\tau = \frac{\hbar t}{2m}, \quad \mu = \sqrt{\frac{2am}{\hbar^2} + \frac{1}{4}} \geq 1,$$

where $J_\mu(\xi)$ is the Bessel function.

⊙ *Reference*: W. Miller, Jr. (1977).

1.9.2-6. Isotropic harmonic oscillator: $U(x) = \tfrac{1}{2}m\omega^2 x^2 + ax^{-2}$.

Here, the variable $x \geq 0$ plays the role of the radial coordinate, and $a > 0$. The equation with this $U(x)$ results from Schrödinger's equation for a harmonic oscillator with n space coordinates if one passes to spherical (cylindrical) coordinates and separates the angular variables.

Eigenvalues:

$$E_n = -\hbar\omega(2n+\mu+1), \quad \mu = \sqrt{\frac{2am}{\hbar^2}+\frac{1}{4}} \geq 1, \quad n = 0, 1, \dots$$

Normalized eigenfunctions:

$$\psi_n(x) = \sqrt{\frac{2n!}{\Gamma(n+1+\mu)x_0}}\, \xi^{\frac{2\mu+1}{2}} \exp\left(-\tfrac{1}{2}\xi^2\right) L_n^\mu(\xi^2), \quad \xi = \frac{x}{x_0}, \quad x_0 = \sqrt{\frac{\hbar}{m\omega}},$$

where $L_n^\mu(z)$ is the nth generalized Laguerre polynomial with parameter μ. The norm $|\psi_n(x)|^2$ refers to the semiaxis $x \geq 0$.

The functions $\psi_n(x)$ form an orthonormal basis in $L_2(\mathbb{R}_+)$.

⊙ *Reference*: W. Miller, Jr. (1977).

1.9.2-7. Morse potential: $U(x) = U_0(e^{-2x/a} - 2e^{-x/a})$.

Eigenvalues:

$$E_n = -U_0\left[1 - \frac{1}{\beta}(n+\tfrac{1}{2})\right]^2, \quad \beta = \frac{a\sqrt{2mU_0}}{\hbar}, \quad 0 \leq n < \beta - 2.$$

Eigenfunctions:

$$\psi_n(x) = \xi^s e^{-\xi/2}\Phi(-n, 2s+1, \xi), \quad \xi = 2\beta e^{-x/a}, \quad s = \frac{a\sqrt{-2mE_n}}{\hbar},$$

where $\Phi(a, b, \xi)$ is the degenerate hypergeometric function.

In this case the number of eigenvalues (energy levels) E_n and eigenfunctions ψ_n is finite: $n = 0, 1, \dots, n_{\max}$.

⊙ *References*: S. G. Krein (1964), L. D. Landau and E. M. Lifshitz (1974).

1.9.2-8. Potential with a hyperbolic function: $U(x) = -U_0 \cosh^{-2}(x/a)$.

Eigenvalues:

$$E_n = -\frac{\hbar^2}{2ma^2}(s-n)^2, \quad s = \frac{1}{2}\left(-1 + \sqrt{1 + \frac{8mU_0 a^2}{\hbar^2}}\right), \quad 0 \le n < s.$$

Eigenfunctions:

$$\psi_n(x) = \begin{cases} \left(\cosh\dfrac{x}{a}\right)^{-2s} F\left(\dfrac{\beta-s}{2}, -\dfrac{\beta+s}{2}, \dfrac{1}{2}, -\sinh^2\dfrac{x}{a}\right) & \text{for even } n, \\ \sinh\dfrac{x}{a}\left(\cosh\dfrac{x}{a}\right)^{-2s} F\left(\dfrac{1+\beta-s}{2}, \dfrac{1-\beta-s}{2}, \dfrac{3}{2}, -\sinh^2\dfrac{x}{a}\right) & \text{for odd } n, \end{cases}$$

where $F(a,b,c,\xi)$ is the hypergeometric function and $\beta = \dfrac{a}{\hbar}\sqrt{-2mE_n}$.

The number of eigenvalues (energy levels) E_n and eigenfunctions ψ_n is finite in this case: $n = 0, 1, \ldots, n_{\max}$.

⊙ *Reference*: S. G. Krein (1964).

1.9.2-9. Potential with a trigonometric function: $U(x) = U_0 \cot^2(\pi x/a)$.

Eigenvalues:

$$E_n = \frac{\pi^2 \hbar^2}{2ma^2}(n^2 + 2ns - s), \quad s = \frac{1}{2}\left(-1 + \sqrt{1 + \frac{8mU_0 a^2}{\pi^2 \hbar^2}}\right), \quad n = 0, 1, \ldots$$

Eigenfunctions:

$$\psi_n(x) = \begin{cases} \cos\dfrac{\pi x}{a}\left(\sin\dfrac{\pi x}{a}\right)^{-2s} F\left(\dfrac{1-n-s}{2}, \dfrac{n+1}{2}, \dfrac{3}{2}, \cos^2\dfrac{\pi x}{a}\right) & \text{for even } n, \\ \left(\sin\dfrac{\pi x}{a}\right)^{-2s} F\left(-\dfrac{n+s}{2}, \dfrac{n}{2}, \dfrac{1}{2}, \cos^2\dfrac{\pi x}{a}\right) & \text{for odd } n, \end{cases}$$

where $F(a,b,c,\xi)$ is the hypergeometric function.

In particular, if $a = \pi\hbar/\sqrt{2m}$, $U_0 = 2$, and $n = k-1$, we have

$$E_k = k^2 - 2, \quad \psi_k(x) = k\cos\frac{k\pi x}{a} - \sin\frac{k\pi x}{a}\cot\frac{\pi x}{a}, \quad k = 1, 2, \ldots$$

⊙ *References*: S. G. Krein (1964), L. D. Landau and E. M. Lifshitz (1974).

Chapter 2

Parabolic Equations with Two Space Variables

2.1. Heat Equation $\frac{\partial w}{\partial t} = a\Delta_2 w$

2.1.1. Boundary Value Problems in Cartesian Coordinates

In rectangular Cartesian coordinates, the two-dimensional sourceless heat equation has the form

$$\frac{\partial w}{\partial t} = a\left(\frac{\partial^2 w}{\partial x^2} + \frac{\partial^2 w}{\partial y^2}\right).$$

It governs two-dimensional unsteady heat transfer processes in quiescent media or solid bodies with constant thermal diffusivity a. A similar equation is used to study analogous two-dimensional unsteady mass transfer phenomena with constant diffusivity; in this case the equation is called a diffusion equation.

2.1.1-1. Particular solutions:

$$w(x,y) = Axy + C_1 x + C_2 y + C_3,$$
$$w(x,y,t) = Ax^2 + By^2 + 2a(A+B)t,$$
$$w(x,y,t) = A(x^2 + 2at)(y^2 + 2at) + B,$$
$$w(x,y,t) = A\exp\left[k_1 x + k_2 y + (k_1^2 + k_2^2)at\right] + B,$$
$$w(x,y,t) = A\cos(k_1 x + C_1)\cos(k_2 y + C_2)\exp\left[-(k_1^2 + k_2^2)at\right],$$
$$w(x,y,t) = A\cos(k_1 x + C_1)\sinh(k_2 y + C_2)\exp\left[-(k_1^2 - k_2^2)at\right],$$
$$w(x,y,t) = A\cos(k_1 x + C_1)\cosh(k_2 y + C_2)\exp\left[-(k_1^2 - k_2^2)at\right],$$
$$w(x,y,t) = A\exp(-\mu x - \lambda y)\cos(\mu x - 2a\mu^2 t + C_1)\cos(\lambda y - 2a\lambda^2 t + C_2),$$
$$w(x,y,t) = \frac{A}{t-t_0}\exp\left[-\frac{(x-x_0)^2 + (y-y_0)^2}{4a(t-t_0)}\right],$$
$$w(x,y,t) = A\operatorname{erf}\left(\frac{x-x_0}{2\sqrt{at}}\right)\operatorname{erf}\left(\frac{y-y_0}{2\sqrt{at}}\right) + B,$$

where A, B, C_1, C_2, C_3, k_1, k_2, x_0, y_0, and t_0 are arbitrary constants.

Fundamental solution:

$$\mathscr{E}(x,y,t) = \frac{1}{4\pi at}\exp\left(-\frac{x^2+y^2}{4at}\right).$$

2.1.1-2. Formulas to construct particular solutions. Remarks on the Green's functions.

$1°$. Apart from usual separable solutions $w(x,y,t) = f_1(x)f_2(y)f_3(t)$, the equation in question has more sophisticated solutions in the product form

$$w(x,y,t) = u(x,t)v(y,t),$$

where $u = u(x,t)$ and $v = v(y,t)$ are solutions of the one-dimensional heat equations

$$\frac{\partial u}{\partial t} = a\frac{\partial^2 u}{\partial x^2}, \quad \frac{\partial v}{\partial t} = a\frac{\partial^2 v}{\partial y^2},$$

considered in Subsection 1.1.1.

2°. Suppose $w = w(x,y,t)$ is a solution of the heat equation. Then the functions

$$w_1 = Aw(\pm\lambda x + C_1, \pm\lambda y + C_2, \lambda^2 t + C_3),$$
$$w_2 = Aw(x\cos\beta - y\sin\beta + C_1, x\sin\beta + y\cos\beta + C_2, t + C_3),$$
$$w_3 = A\exp[\lambda_1 x + \lambda_2 y + a(\lambda_1^2 + \lambda_2^2)t]w(x + 2a\lambda_1 t + C_1, y + 2a\lambda_2 t + C_2, t + C_3),$$
$$w_4 = \frac{A}{\delta + \beta t}\exp\left[-\frac{\beta(x^2 + y^2)}{4a(\delta + \beta t)}\right]w\left(\frac{x}{\delta + \beta t}, \frac{y}{\delta + \beta t}, \frac{\gamma + \lambda t}{\delta + \beta t}\right), \quad \lambda\delta - \beta\gamma = 1,$$

where A, C_1, C_2, C_3, β, δ, λ, λ_1 and λ_2 are arbitrary constants, are also solutions of this equation. The signs at λ's in the formula for w_1 are taken arbitrarily, independently of each other.

⊙ *Reference*: W. Miller, Jr. (1977).

3°. For all two-dimensional boundary value problems discussed in Subsection 2.1.1, the Green's function can be represented in the product form

$$G(x,y,\xi,\eta,t) = G_1(x,\xi,t)G_2(y,\eta,t),$$

where $G_1(x,\xi,t)$ and $G_2(y,\eta,t)$ are the Green's functions of the corresponding one-dimensional boundary value problems (these functions are specified in Subsections 1.1.1 and 1.1.2).

Example 1. The Green's function of the first boundary value problem for a semiinfinite strip ($0 \le x \le l$, $0 \le y < \infty$), considered in Subsection 2.1.1-12, is the product of two one-dimensional Green's functions. The first Green's function is that of the first boundary value problem on a closed interval ($0 \le x \le l$) presented in Subsection 1.1.2-5. The second Green's function is that of the first boundary value problem on a semiinfinite interval ($0 \le y < \infty$) presented in Subsection 1.1.2-2, where x and ξ must be renamed y and η, respectively.

2.1.1-3. Transformations that allow separation of variables.

Table 18 lists possible transformations that allow reduction of the two-dimensional heat equation to a separable equation. All transformations of the independent variables have the form $(x,y,t) \longmapsto (\xi,\eta,t)$. The transformations that can be obtained by interchange of independent variables, $x \rightleftarrows y$, are omitted.

The anharmonic oscillator functions are solutions of the second-order ordinary differential equation $F''_{zz} + (az^4 + bz^2 + c)F = 0$. The Ince polynomials are the 2π-periodic solutions of the Whittaker–Hill equation $F''_{zz} + k\sin 2z\, F'_z + (a - bk\cos 2z)F = 0$; see Arscott (1964, 1967).

⊙ *Reference*: W. Miller, Jr. (1977).

2.1.1-4. Domain: $-\infty < x < \infty$, $-\infty < y < \infty$. Cauchy problem.

An initial condition is prescribed:

$$w = f(x,y) \quad \text{at} \quad t = 0.$$

Solution:

$$w(x,y,t) = \frac{1}{4\pi a t}\int_{-\infty}^{\infty}\int_{-\infty}^{\infty} f(\xi,\eta)\exp\left[-\frac{(x-\xi)^2 + (y-\eta)^2}{4at}\right]d\xi\, d\eta.$$

Example 2. The initial temperature is piecewise-constant and equal to w_1 in the domain $|x| < x_0$, $|y| < y_0$ and w_2 in the domain $|x| > x_0$, $|y| > y_0$, specifically,

$$f(x,y) = \begin{cases} w_1 & \text{for } |x| < x_0, |y| < y_0, \\ w_2 & \text{for } |x| > x_0, |y| > y_0. \end{cases}$$

TABLE 18

Transformations $(x, y, t) \longmapsto (\xi, \eta, t)$ that allow solutions with $\mathcal{R}$-separated variables, $w = \exp[\mathcal{R}(\xi, \eta, t)] f(\xi) g(\eta) h(t)$, for the two-dimensional heat equation $\partial_t w = \partial_{xx} w + \partial_{yy} w$. Everywhere, the function $h(t)$ is exponential

No	Transformations	Factor $\exp \mathcal{R}$	Function $f(\xi)$	Function $g(\eta)$
1	$x = \xi,$ $y = \eta$	$\mathcal{R} = 0$	Exponential function	Exponential function
2	$x = \xi,$ $y = \eta\sqrt{|t|}$	$\mathcal{R} = 0$	Exponential function	Hermite function
3	$x = \xi\sqrt{|t|},$ $y = \eta\sqrt{|t|}$	$\mathcal{R} = 0$	Hermite function	Hermite function
4	$x = \frac{1}{2}(\xi^2 - \eta^2),$ $y = \xi\eta$	$\mathcal{R} = 0$	Parabolic cylinder function	Parabolic cylinder function
5	$x = \xi \cos\eta,$ $y = \xi \sin\eta$	$\mathcal{R} = 0$	Bessel function	Exponential function
6	$x = \cosh\xi \cos\eta,$ $y = \sinh\xi \sin\eta$	$\mathcal{R} = 0$	Modified Mathieu function	Mathieu function
7	$x = \sqrt{|t|}\,\xi \cos\eta,$ $y = \sqrt{|t|}\,\xi \sin\eta$	$\mathcal{R} = 0$	Laguerre function	Exponential function
8	$x = \sqrt{|t|}\cosh\xi \cos\eta,$ $y = \sqrt{|t|}\sinh\xi \sin\eta$	$\mathcal{R} = 0$	Ince polynomial	Ince polynomial
9	$x = \xi,$ $y = \eta + at^2$	$\mathcal{R} = -a\eta t$	Exponential function	Airy function
10	$x = \xi,$ $y = \eta t + b/t$	$\mathcal{R} = -\frac{1}{4}\eta^2 t + \frac{1}{2}b\eta/t$	Exponential function	Airy function
11	$x = \xi,$ $y = \eta\sqrt{1 + t^2}$	$\mathcal{R} = -\frac{1}{4}\eta^2 t$	Exponential function	Parabolic cylinder function
12	$x = \xi,$ $y = \eta\sqrt{|1 - t^2|}$	$\mathcal{R} = -\frac{1}{4}\varepsilon\eta^2 t,$ $\varepsilon = \mathrm{sign}(1 - t^2)$	Exponential function	Hermite function
13	$x = \xi t,$ $y = \eta t$	$\mathcal{R} = -\frac{1}{4}(\xi^2 + \eta^2)t$	Exponential function	Exponential function
14	$x = \xi + at^2,$ $y = \eta + bt^2$	$\mathcal{R} = -(a\xi + b\eta)t$	Airy function	Airy function
15	$x = \xi t + a/t,$ $y = \eta t + b/t$	$\mathcal{R} = -\frac{1}{4}(\xi^2 + \eta^2)t$ $+\frac{1}{2}(a\xi + b\eta)/t$	Airy function	Airy function
16	$x = \frac{1}{2}(\xi^2 - \eta^2)t,$ $y = \xi\eta t$	$\mathcal{R} = -\frac{1}{16}(\xi^2 + \eta^2)^2 t$	Parabolic cylinder function	Parabolic cylinder function
17	$x = \xi\sqrt{1 + t^2},$ $y = \eta\sqrt{1 + t^2}$	$\mathcal{R} = -\frac{1}{4}(\xi^2 + \eta^2)t$	Parabolic cylinder function	Parabolic cylinder function
18	$x = \xi\sqrt{|1 - t^2|},$ $y = \eta\sqrt{|1 - t^2|}$	$\mathcal{R} = -\frac{1}{4}\varepsilon(\xi^2 + \eta^2)t,$ $\varepsilon = \mathrm{sign}(1 - t^2)$	Hermite function	Hermite function

TABLE 18
(continued)

No	Transformations	Factor exp $\mathcal{R}$	Function $f(\xi)$	Function $g(\eta)$
19	$x = \frac{1}{2}(\xi^2 - \eta^2) + at^2$, $y = \xi\eta$	$\mathcal{R} = -\frac{1}{2}a(\xi^2 - \eta^2)t$	Anharmonic oscillator function	Anharmonic oscillator function
20	$x = \frac{1}{2}(\xi^2 - \eta^2)t + a/t$, $y = \xi\eta t$	$\mathcal{R} = -\frac{1}{16}(\xi^2 + \eta^2)^2 t$ $+ \frac{1}{4}a(\xi^2 - \eta^2)/t$	Anharmonic oscillator function	Anharmonic oscillator function
21	$x = \xi t \cos\eta$, $y = \xi t \sin\eta$	$\mathcal{R} = -\frac{1}{4}\xi^2 t$	Bessel function	Exponential function
22	$x = t \cosh\xi \cos\eta$, $y = t \sinh\xi \sin\eta$	$\mathcal{R} = -\frac{1}{4}(\sinh^2\xi + \cos^2\eta)t$	Modified Mathieu function	Mathieu function
23	$x = \sqrt{1+t^2}\,\xi\cos\eta$, $y = \sqrt{1+t^2}\,\xi\sin\eta$	$\mathcal{R} = -\frac{1}{4}\xi^2 t$	Whittaker function	Exponential function
24	$x = \sqrt{\|1-t^2\|}\,\xi\cos\eta$, $y = \sqrt{\|1-t^2\|}\,\xi\sin\eta$	$\mathcal{R} = -\frac{1}{4}\varepsilon\xi^2 t$, $\varepsilon = \text{sign}(1-t^2)$	Laguerre function	Exponential function
25	$x = \sqrt{1+t^2}\,\cosh\xi\cos\eta$, $y = \sqrt{1+t^2}\,\sinh\xi\sin\eta$	$\mathcal{R} = -\frac{1}{4}(\sinh^2\xi + \cos^2\eta)t$	Ince polynomial	Ince polynomial
26	$x = \sqrt{\|1-t^2\|}\,\cosh\xi\cos\eta$, $y = \sqrt{\|1-t^2\|}\,\sinh\xi\sin\eta$	$\mathcal{R} = -\frac{1}{4}\varepsilon(\sinh^2\xi + \cos^2\eta)t$, $\varepsilon = \text{sign}(1-t^2)$	Ince polynomial	Ince polynomial

Solution:
$$w = \frac{1}{4}(w_1 - w_2)\left[\text{erf}\left(\frac{x_0 - x}{2\sqrt{at}}\right) + \text{erf}\left(\frac{x_0 + x}{2\sqrt{at}}\right)\right]\left[\text{erf}\left(\frac{y_0 - y}{2\sqrt{at}}\right) + \text{erf}\left(\frac{y_0 + y}{2\sqrt{at}}\right)\right] + w_2.$$

If the initial temperature distribution $f(x, y)$ is an infinitely differentiable function in both arguments, then the solution can be represented in the series form

$$w(x, y, t) = f(x, y) + \sum_{n=1}^{\infty} \frac{(at)^n}{n!} \boldsymbol{L}^n[f(x,y)], \quad \boldsymbol{L} \equiv \frac{\partial^2}{\partial x^2} + \frac{\partial^2}{\partial y^2}.$$

Such a representation is useful for small t.

⊙ *Reference*: H. S. Carslaw and J. C. Jaeger (1984).

2.1.1-5. Domain: $0 \leq x < \infty$, $-\infty < y < \infty$. First boundary value problem.

A half-plane is considered. The following conditions are prescribed:

$$w = f(x, y) \quad \text{at} \quad t = 0 \quad \text{(initial condition)},$$
$$w = g(y, t) \quad \text{at} \quad x = 0 \quad \text{(boundary condition)}.$$

Solution:
$$w(x, y, t) = \int_0^\infty \int_{-\infty}^\infty f(\xi, \eta) G(x, y, \xi, \eta, t)\, d\eta\, d\xi$$
$$+ a \int_0^t \int_{-\infty}^\infty g(\eta, \tau)\left[\frac{\partial}{\partial \xi} G(x, y, \xi, \eta, t - \tau)\right]_{\xi=0} d\eta\, d\tau,$$

where
$$G(x,y,\xi,\eta,t) = \frac{1}{4\pi at}\left\{\exp\left[-\frac{(x-\xi)^2+(y-\eta)^2}{4at}\right] - \exp\left[-\frac{(x+\xi)^2+(y-\eta)^2}{4at}\right]\right\}.$$

⊙ *Reference*: H. S. Carslaw and J. C. Jaeger (1984).

2.1.1-6. Domain: $0 \leq x < \infty$, $-\infty < y < \infty$. Second boundary value problem.

A half-plane is considered. The following conditions are prescribed:
$$w = f(x,y) \quad \text{at} \quad t=0 \quad \text{(initial condition)},$$
$$\partial_x w = g(y,t) \quad \text{at} \quad x=0 \quad \text{(boundary condition)}.$$

Solution:
$$w(x,y,t) = \int_0^\infty \int_{-\infty}^\infty f(\xi,\eta)G(x,y,\xi,\eta,t)\,d\eta\,d\xi - a\int_0^t \int_{-\infty}^\infty g(\eta,\tau)G(x,y,0,\eta,t-\tau)\,d\eta\,d\tau,$$

where
$$G(x,y,\xi,\eta,t) = \frac{1}{4\pi at}\left\{\exp\left[-\frac{(x-\xi)^2+(y-\eta)^2}{4at}\right] + \exp\left[-\frac{(x+\xi)^2+(y-\eta)^2}{4at}\right]\right\}.$$

2.1.1-7. Domain: $0 \leq x < \infty$, $-\infty < y < \infty$. Third boundary value problem.

A half-plane is considered. The following conditions are prescribed:
$$w = f(x,y) \quad \text{at} \quad t=0 \quad \text{(initial condition)},$$
$$\partial_x w - kw = g(y,t) \quad \text{at} \quad x=0 \quad \text{(boundary condition)}.$$

The solution $w(x,y,t)$ is determined by the formula in Paragraph 2.1.1-6 where
$$G(x,y,\xi,\eta,t) = \frac{1}{4\pi at}\exp\left[-\frac{(y-\eta)^2}{4at}\right]\left\{\exp\left[-\frac{(x-\xi)^2}{4at}\right] + \exp\left[-\frac{(x+\xi)^2}{4at}\right]\right.$$
$$\left. - 2k\int_0^\infty \exp\left[-\frac{(x+\xi+s)^2}{4at} - ks\right]ds\right\}.$$

2.1.1-8. Domain: $0 \leq x < \infty$, $0 \leq y < \infty$. First boundary value problem.

A quadrant of the plane is considered. The following conditions are prescribed:
$$w = f(x,y) \quad \text{at} \quad t=0 \quad \text{(initial condition)},$$
$$w = g_1(y,t) \quad \text{at} \quad x=0 \quad \text{(boundary condition)},$$
$$w = g_2(x,t) \quad \text{at} \quad y=0 \quad \text{(boundary condition)}.$$

Solution:
$$w(x,y,t) = \int_0^\infty \int_0^\infty f(\xi,\eta)G(x,y,\xi,\eta,t)\,d\xi\,d\eta$$
$$+ a\int_0^t \int_0^\infty g_1(\eta,\tau)\left[\frac{\partial}{\partial \xi}G(x,y,\xi,\eta,t-\tau)\right]_{\xi=0}d\eta\,d\tau$$
$$+ a\int_0^t \int_0^\infty g_2(\xi,\tau)\left[\frac{\partial}{\partial \eta}G(x,y,\xi,\eta,t-\tau)\right]_{\eta=0}d\xi\,d\tau,$$

where
$$G(x,y,\xi,\eta,t) = \frac{1}{4\pi at}\left\{\exp\left[-\frac{(x-\xi)^2}{4at}\right]-\exp\left[-\frac{(x+\xi)^2}{4at}\right]\right\}\left\{\exp\left[-\frac{(y-\eta)^2}{4at}\right]-\exp\left[-\frac{(y+\eta)^2}{4at}\right]\right\}.$$

Example 3. The initial temperature is uniform, $f(x,y) = w_0$. The boundary is maintained at zero temperature, $g_1(y,t) = g_2(x,t) = 0$.
Solution:
$$w = w_0\,\mathrm{erf}\left(\frac{x}{2\sqrt{at}}\right)\mathrm{erf}\left(\frac{y}{2\sqrt{at}}\right).$$

⊙ *References*: A. G. Butkovskiy (1979), H. S. Carslaw and J. C. Jaeger (1984).

2.1.1-9. Domain: $0 \leq x < \infty$, $0 \leq y < \infty$. Second boundary value problem.

A quadrant of the plane is considered. The following conditions are prescribed:
$$w = f(x,y) \quad \text{at} \quad t = 0 \quad \text{(initial condition)},$$
$$\partial_x w = g_1(y,t) \quad \text{at} \quad x = 0 \quad \text{(boundary condition)},$$
$$\partial_y w = g_2(x,t) \quad \text{at} \quad y = 0 \quad \text{(boundary condition)}.$$

Solution:
$$w(x,y,t) = \int_0^\infty\!\!\int_0^\infty f(\xi,\eta)G(x,y,\xi,\eta,t)\,d\xi\,d\eta$$
$$- a\int_0^t\!\!\int_0^\infty g_1(\eta,\tau)G(x,y,0,\eta,t-\tau)\,d\eta\,d\tau - a\int_0^t\!\!\int_0^\infty g_2(\xi,\tau)G(x,y,\xi,0,t-\tau)\,d\xi\,d\tau,$$
where
$$G(x,y,\xi,\eta,t) = \frac{1}{4\pi at}\left\{\exp\left[-\frac{(x-\xi)^2}{4at}\right]+\exp\left[-\frac{(x+\xi)^2}{4at}\right]\right\}\left\{\exp\left[-\frac{(y-\eta)^2}{4at}\right]+\exp\left[-\frac{(y+\eta)^2}{4at}\right]\right\}.$$

2.1.1-10. Domain: $0 \leq x < \infty$, $0 \leq y < \infty$. Third boundary value problem.

A quadrant of the plane is considered. The following conditions are prescribed:
$$w = f(x,y) \quad \text{at} \quad t = 0 \quad \text{(initial condition)},$$
$$\partial_x w - k_1 w = g_1(y,t) \quad \text{at} \quad x = 0 \quad \text{(boundary condition)},$$
$$\partial_y w - k_2 w = g_2(x,t) \quad \text{at} \quad y = 0 \quad \text{(boundary condition)}.$$

The solution $w(x,y,t)$ is determined by the formula in Paragraph 2.1.1-9 where
$$G(x,y,\xi,\eta,t) = \frac{1}{4\pi at}\left\{\exp\left[-\frac{(x-\xi)^2}{4at}\right]+\exp\left[-\frac{(x+\xi)^2}{4at}\right]\right.$$
$$\left. - 2k_1\sqrt{\pi at}\,\exp\!\left[ak_1^2 t + k_1(x+\xi)\right]\mathrm{erfc}\!\left(\frac{x+\xi}{2\sqrt{at}}+k_1\sqrt{at}\right)\right\}$$
$$\times\left\{\exp\left[-\frac{(y-\eta)^2}{4at}\right]+\exp\left[-\frac{(y+\eta)^2}{4at}\right]\right.$$
$$\left. - 2k_2\sqrt{\pi at}\,\exp\!\left[ak_2^2 t + k_2(y+\eta)\right]\mathrm{erfc}\!\left(\frac{y+\eta}{2\sqrt{at}}+k_2\sqrt{at}\right)\right\}.$$

Example 4. The initial temperature is constant, $f(x,y) = w_0$. The temperature of the environment is zero, $g_1(y,t) = g_2(x,t) = 0$.
Solution:
$$w = w_0\left[\mathrm{erf}\left(\frac{x}{2\sqrt{at}}\right)+\exp(k_1 x + ak_1^2 t)\,\mathrm{erfc}\left(\frac{x}{2\sqrt{at}}+k_1\sqrt{at}\right)\right]$$
$$\times\left[\mathrm{erf}\left(\frac{y}{2\sqrt{at}}\right)+\exp(k_2 y + ak_2^2 t)\,\mathrm{erfc}\left(\frac{y}{2\sqrt{at}}+k_2\sqrt{at}\right)\right].$$

⊙ *Reference*: H. S. Carslaw and J. C. Jaeger (1984).

2.1.1-11. Domain: $0 \leq x < \infty$, $0 \leq y < \infty$. Mixed boundary value problems.

1°. A quadrant of the plane is considered. The following conditions are prescribed:

$$w = f(x, y) \quad \text{at} \quad t = 0 \quad \text{(initial condition),}$$
$$w = g_1(y, t) \quad \text{at} \quad x = 0 \quad \text{(boundary condition),}$$
$$\partial_y w = g_2(x, t) \quad \text{at} \quad y = 0 \quad \text{(boundary condition).}$$

Solution:

$$w(x, y, t) = \int_0^\infty \int_0^\infty f(\xi, \eta) G(x, y, \xi, \eta, t) \, d\xi \, d\eta$$
$$+ a \int_0^t \int_0^\infty g_1(\eta, \tau) \left[\frac{\partial}{\partial \xi} G(x, y, \xi, \eta, t - \tau)\right]_{\xi=0} d\eta \, d\tau$$
$$- a \int_0^t \int_0^\infty g_2(\xi, \tau) G(x, y, \xi, 0, t - \tau) \, d\xi \, d\tau,$$

where

$$G(x, y, \xi, \eta, t) = \frac{1}{4\pi a t} \left\{\exp\left[-\frac{(x-\xi)^2}{4at}\right] - \exp\left[-\frac{(x+\xi)^2}{4at}\right]\right\} \left\{\exp\left[-\frac{(y-\eta)^2}{4at}\right] + \exp\left[-\frac{(y+\eta)^2}{4at}\right]\right\}.$$

2°. A quadrant of the plane is considered. The following conditions are prescribed:

$$w = f(x, y) \quad \text{at} \quad t = 0 \quad \text{(initial condition),}$$
$$\partial_x w - kw = g_1(y, t) \quad \text{at} \quad x = 0 \quad \text{(boundary condition),}$$
$$w = g_2(x, t) \quad \text{at} \quad y = 0 \quad \text{(boundary condition).}$$

Solution:

$$w(x, y, t) = \int_0^\infty \int_0^\infty f(\xi, \eta) G(x, y, \xi, \eta, t) \, d\xi \, d\eta$$
$$- a \int_0^t \int_0^\infty g_1(\eta, \tau) G(x, y, 0, \eta, t - \tau) \, d\eta \, d\tau$$
$$+ a \int_0^t \int_0^\infty g_2(\xi, \tau) \left[\frac{\partial}{\partial \eta} G(x, y, \xi, \eta, t - \tau)\right]_{\eta=0} d\xi \, d\tau,$$

where

$$G(x, y, \xi, \eta, t) = \frac{1}{4\pi a t} \left\{\exp\left[-\frac{(y-\eta)^2}{4at}\right] - \exp\left[-\frac{(y+\eta)^2}{4at}\right]\right\} \left\{\exp\left[-\frac{(x-\xi)^2}{4at}\right] + \exp\left[-\frac{(x+\xi)^2}{4at}\right]\right.$$
$$\left. - 2k\sqrt{\pi a t} \, \exp[ak^2 t + k(x+\xi)] \, \text{erfc}\left(\frac{x+\xi}{2\sqrt{at}} + k\sqrt{at}\right)\right\}.$$

Example 5. The initial temperature is uniform, $f(x, y) = w_0$. Heat exchange with the environment of zero temperature occurs at one side and the other side is maintained at zero temperature: $g_1(y, t) = g_2(x, t) = 0$.

Solution:

$$w = w_0 \left[\text{erf}\left(\frac{x}{2\sqrt{at}}\right) + \exp(kx + ak^2 t) \, \text{erfc}\left(\frac{x}{2\sqrt{at}} + k\sqrt{at}\right)\right] \text{erf}\left(\frac{y}{2\sqrt{at}}\right).$$

2.1.1-12. Domain: $0 \leq x \leq l$, $0 \leq y < \infty$. First boundary value problem.

A semiinfinite strip is considered. The following conditions are prescribed:

$$w = f(x, y) \quad \text{at} \quad t = 0 \quad \text{(initial condition)},$$
$$w = g_1(y, t) \quad \text{at} \quad x = 0 \quad \text{(boundary condition)},$$
$$w = g_2(y, t) \quad \text{at} \quad x = l \quad \text{(boundary condition)},$$
$$w = g_3(x, t) \quad \text{at} \quad y = 0 \quad \text{(boundary condition)}.$$

Solution:

$$w(x, y, t) = \int_0^\infty \int_0^l f(\xi, \eta) G(x, y, \xi, \eta, t) \, d\xi \, d\eta$$
$$+ a \int_0^t \int_0^\infty g_1(\eta, \tau) \left[\frac{\partial}{\partial \xi} G(x, y, \xi, \eta, t - \tau) \right]_{\xi=0} d\eta \, d\tau$$
$$- a \int_0^t \int_0^\infty g_2(\eta, \tau) \left[\frac{\partial}{\partial \xi} G(x, y, \xi, \eta, t - \tau) \right]_{\xi=l} d\eta \, d\tau$$
$$+ a \int_0^t \int_0^l g_3(\xi, \tau) \left[\frac{\partial}{\partial \eta} G(x, y, \xi, \eta, t - \tau) \right]_{\eta=0} d\xi \, d\tau,$$

where

$$G(x, y, \xi, \eta, t) = G_1(x, \xi, t) G_2(y, \eta, t),$$
$$G_1(x, \xi, t) = \frac{2}{l} \sum_{n=1}^{\infty} \sin\left(\frac{n\pi x}{l}\right) \sin\left(\frac{n\pi \xi}{l}\right) \exp\left(-\frac{a n^2 \pi^2 t}{l^2}\right),$$
$$G_2(y, \eta, t) = \frac{1}{2\sqrt{\pi a t}} \left\{ \exp\left[-\frac{(y-\eta)^2}{4at}\right] - \exp\left[-\frac{(y+\eta)^2}{4at}\right] \right\}.$$

Example 6. The initial temperature is uniform, $f(x, y) = w_0$. The boundary is maintained at zero temperature, $g_1(y, t) = g_2(y, t) = g_3(x, t) = 0$.

Solution:

$$w = \frac{4w_0}{\pi} \operatorname{erf}\left(\frac{y}{2\sqrt{at}}\right) \sum_{n=0}^{\infty} \frac{1}{2n+1} \sin\left[\frac{(2n+1)\pi x}{l}\right] \exp\left[-\frac{\pi^2 (2n+1)^2 at}{l^2}\right].$$

⊙ *Reference*: H. S. Carslaw and J. C. Jaeger (1984).

2.1.1-13. Domain: $0 \leq x \leq l$, $0 \leq y < \infty$. Second boundary value problem.

A semiinfinite strip is considered. The following conditions are prescribed:

$$w = f(x, y) \quad \text{at} \quad t = 0 \quad \text{(initial condition)},$$
$$\partial_x w = g_1(y, t) \quad \text{at} \quad x = 0 \quad \text{(boundary condition)},$$
$$\partial_x w = g_2(y, t) \quad \text{at} \quad x = l \quad \text{(boundary condition)},$$
$$\partial_y w = g_3(x, t) \quad \text{at} \quad y = 0 \quad \text{(boundary condition)}.$$

Solution:

$$w(x, y, t) = \int_0^\infty \int_0^l f(\xi, \eta) G(x, y, \xi, \eta, t) \, d\xi \, d\eta$$
$$- a \int_0^t \int_0^\infty g_1(\eta, \tau) G(x, y, 0, \eta, t - \tau) \, d\eta \, d\tau$$
$$+ a \int_0^t \int_0^\infty g_2(\eta, \tau) G(x, y, l, \eta, t - \tau) \, d\eta \, d\tau$$
$$- a \int_0^t \int_0^l g_3(\xi, \tau) G(x, y, \xi, 0, t - \tau) \, d\xi \, d\tau,$$

where
$$G(x,y,\xi,\eta,t) = G_1(x,\xi,t)G_2(y,\eta,t),$$
$$G_1(x,\xi,t) = \frac{1}{l} + \frac{2}{l}\sum_{n=1}^{\infty}\cos\left(\frac{n\pi x}{l}\right)\cos\left(\frac{n\pi\xi}{l}\right)\exp\left(-\frac{an^2\pi^2 t}{l^2}\right),$$
$$G_2(y,\eta,t) = \frac{1}{2\sqrt{\pi at}}\left\{\exp\left[-\frac{(y-\eta)^2}{4at}\right] + \exp\left[-\frac{(y+\eta)^2}{4at}\right]\right\}.$$

2.1.1-14. Domain: $0 \le x \le l$, $0 \le y < \infty$. Third boundary value problem.

A semiinfinite strip is considered. The following conditions are prescribed:

$$\begin{aligned}w &= f(x,y) &\text{at}\quad& t=0 &\text{(initial condition)},\\ \partial_x w - k_1 w &= g_1(y,t) &\text{at}\quad& x=0 &\text{(boundary condition)},\\ \partial_x w + k_2 w &= g_2(y,t) &\text{at}\quad& x=l &\text{(boundary condition)},\\ \partial_y w - k_3 w &= g_3(x,t) &\text{at}\quad& y=0 &\text{(boundary condition)}.\end{aligned}$$

The solution $w(x,y,t)$ is determined by the formula in Paragraph 2.1.1-13 where the Green's function $G(x,y,\xi,\eta,t)$ is the product of the Green's function of Subsection 1.1.1-11 and that of Subsection 1.1.1-8; one should replace x, ξ, and k by y, η and k_3, respectively, in the last Green's function.

2.1.1-15. Domain: $0 \le x \le l$, $0 \le y < \infty$. Mixed boundary value problems.

1°. A semiinfinite strip is considered. The following conditions are prescribed:

$$\begin{aligned}w &= f(x,y) &\text{at}\quad& t=0 &\text{(initial condition)},\\ w &= g_1(y,t) &\text{at}\quad& x=0 &\text{(boundary condition)},\\ w &= g_2(y,t) &\text{at}\quad& x=l &\text{(boundary condition)},\\ \partial_y w &= g_3(x,t) &\text{at}\quad& y=0 &\text{(boundary condition)}.\end{aligned}$$

Solution:
$$\begin{aligned}w(x,y,t) =& \int_0^\infty\int_0^l f(\xi,\eta)G(x,y,\xi,\eta,t)\,d\xi\,d\eta\\ &+ a\int_0^t\int_0^\infty g_1(\eta,\tau)\left[\frac{\partial}{\partial\xi}G(x,y,\xi,\eta,t-\tau)\right]_{\xi=0}d\eta\,d\tau\\ &- a\int_0^t\int_0^\infty g_2(\eta,\tau)\left[\frac{\partial}{\partial\xi}G(x,y,\xi,\eta,t-\tau)\right]_{\xi=l}d\eta\,d\tau\\ &- a\int_0^t\int_0^l g_3(\xi,\tau)G(x,y,\xi,0,t-\tau)\,d\xi\,d\tau,\end{aligned}$$

where
$$G(x,y,\xi,\eta,t) = G_1(x,\xi,t)G_2(y,\eta,t),$$
$$G_1(x,\xi,t) = \frac{2}{l}\sum_{n=1}^{\infty}\sin\left(\frac{n\pi x}{l}\right)\sin\left(\frac{n\pi\xi}{l}\right)\exp\left(-\frac{an^2\pi^2 t}{l^2}\right),$$
$$G_2(y,\eta,t) = \frac{1}{2\sqrt{\pi at}}\left\{\exp\left[-\frac{(y-\eta)^2}{4at}\right] + \exp\left[-\frac{(y+\eta)^2}{4at}\right]\right\}.$$

2°. A semiinfinite strip is considered. The following conditions are prescribed:

$$w = f(x,y) \quad \text{at} \quad t = 0 \quad \text{(initial condition)},$$
$$\partial_x w = g_1(y,t) \quad \text{at} \quad x = 0 \quad \text{(boundary condition)},$$
$$\partial_x w = g_2(y,t) \quad \text{at} \quad x = l \quad \text{(boundary condition)},$$
$$w = g_3(x,t) \quad \text{at} \quad y = 0 \quad \text{(boundary condition)}.$$

Solution:

$$w(x,y,t) = \int_0^\infty \int_0^l f(\xi,\eta) G(x,y,\xi,\eta,t)\, d\xi\, d\eta$$
$$- a \int_0^t \int_0^\infty g_1(\eta,\tau) G(x,y,0,\eta,t-\tau)\, d\eta\, d\tau$$
$$+ a \int_0^t \int_0^\infty g_2(\eta,\tau) G(x,y,l,\eta,t-\tau)\, d\eta\, d\tau$$
$$+ a \int_0^t \int_0^l g_3(\xi,\tau) \left[\frac{\partial}{\partial \eta} G(x,y,\xi,\eta,t-\tau)\right]_{\eta=0} d\xi\, d\tau,$$

where

$$G(x,y,\xi,\eta,t) = G_1(x,\xi,t) G_2(y,\eta,t),$$
$$G_1(x,\xi,t) = \frac{1}{l} + \frac{2}{l} \sum_{n=1}^\infty \cos\left(\frac{n\pi x}{l}\right) \cos\left(\frac{n\pi \xi}{l}\right) \exp\left(-\frac{an^2\pi^2 t}{l^2}\right),$$
$$G_2(y,\eta,t) = \frac{1}{2\sqrt{\pi a t}} \left\{\exp\left[-\frac{(y-\eta)^2}{4at}\right] - \exp\left[-\frac{(y+\eta)^2}{4at}\right]\right\}.$$

2.1.1-16. Domain: $0 \le x \le l_1$, $0 \le y \le l_2$. First boundary value problem.

A rectangle is considered. The following conditions are prescribed:

$$w = f(x,y) \quad \text{at} \quad t = 0 \quad \text{(initial condition)},$$
$$w = g_1(y,t) \quad \text{at} \quad x = 0 \quad \text{(boundary condition)},$$
$$w = g_2(y,t) \quad \text{at} \quad x = l_1 \quad \text{(boundary condition)},$$
$$w = g_3(x,t) \quad \text{at} \quad y = 0 \quad \text{(boundary condition)},$$
$$w = g_4(x,t) \quad \text{at} \quad y = l_2 \quad \text{(boundary condition)}.$$

Solution:

$$w(x,y,t) = \int_0^{l_1} \int_0^{l_2} f(\xi,\eta) G(x,y,\xi,\eta,t)\, d\eta\, d\xi$$
$$+ a \int_0^t \int_0^{l_2} g_1(\eta,\tau) \left[\frac{\partial}{\partial \xi} G(x,y,\xi,\eta,t-\tau)\right]_{\xi=0} d\eta\, d\tau$$
$$- a \int_0^t \int_0^{l_2} g_2(\eta,\tau) \left[\frac{\partial}{\partial \xi} G(x,y,\xi,\eta,t-\tau)\right]_{\xi=l_1} d\eta\, d\tau$$
$$+ a \int_0^t \int_0^{l_1} g_3(\xi,\tau) \left[\frac{\partial}{\partial \eta} G(x,y,\xi,\eta,t-\tau)\right]_{\eta=0} d\xi\, d\tau$$
$$- a \int_0^t \int_0^{l_1} g_4(\xi,\tau) \left[\frac{\partial}{\partial \eta} G(x,y,\xi,\eta,t-\tau)\right]_{\eta=l_2} d\xi\, d\tau,$$

where
$$G(x,y,\xi,\eta,t) = \frac{4}{l_1 l_2} \sum_{n=1}^{\infty} \sum_{m=1}^{\infty} \sin\frac{n\pi x}{l_1} \sin\frac{n\pi \xi}{l_1} \sin\frac{m\pi y}{l_2} \sin\frac{m\pi \eta}{l_2} \exp\left[-\pi^2\left(\frac{n^2}{l_1^2} + \frac{m^2}{l_2^2}\right)at\right].$$

Example 7. The initial temperature is uniform, $f(x,y) = w_0$. The boundary is maintained at zero temperature, $g_1(y,t) = g_2(y,t) = g_3(x,t) = g_4(x,t) = 0$.

Solution:
$$w = \frac{16 w_0}{\pi^2} \left\{ \sum_{n=0}^{\infty} \frac{1}{2n+1} \sin\left[\frac{(2n+1)\pi x}{l_1}\right] \exp\left[-\frac{\pi^2 (2n+1)^2 at}{l_1^2}\right] \right\}$$
$$\times \left\{ \sum_{m=0}^{\infty} \frac{1}{2m+1} \sin\left[\frac{(2m+1)\pi y}{l_2}\right] \exp\left[-\frac{\pi^2 (2m+1)^2 at}{l_2^2}\right] \right\}.$$

⊙ *Reference*: H. S. Carslaw and J. C. Jaeger (1984).

2.1.1-17. Domain: $0 \le x \le l_1$, $0 \le y \le l_2$. Second boundary value problem.

A rectangle is considered. The following conditions are prescribed:

$$w = f(x,y) \quad \text{at} \quad t = 0 \quad \text{(initial condition)},$$
$$\partial_x w = g_1(y,t) \quad \text{at} \quad x = 0 \quad \text{(boundary condition)},$$
$$\partial_x w = g_2(y,t) \quad \text{at} \quad x = l_1 \quad \text{(boundary condition)},$$
$$\partial_y w = g_3(x,t) \quad \text{at} \quad y = 0 \quad \text{(boundary condition)},$$
$$\partial_y w = g_4(x,t) \quad \text{at} \quad y = l_2 \quad \text{(boundary condition)}.$$

Solution:
$$w(x,y,t) = \int_0^{l_1} \int_0^{l_2} f(\xi,\eta) G(x,y,\xi,\eta,t) \, d\eta \, d\xi$$
$$- a \int_0^t \int_0^{l_2} g_1(\eta,\tau) G(x,y,0,\eta,t-\tau) \, d\eta \, d\tau$$
$$+ a \int_0^t \int_0^{l_2} g_2(\eta,\tau) G(x,y,l_1,\eta,t-\tau) \, d\eta \, d\tau$$
$$- a \int_0^t \int_0^{l_1} g_3(\xi,\tau) G(x,y,\xi,0,t-\tau) \, d\xi \, d\tau$$
$$+ a \int_0^t \int_0^{l_1} g_4(\xi,\tau) G(x,y,\xi,l_2,t-\tau) \, d\xi \, d\tau,$$

where
$$G(x,y,\xi,\eta,t) = \frac{1}{l_1 l_2} \left[1 + 2 \sum_{n=1}^{\infty} \exp\left(-\frac{\pi^2 n^2 at}{l_1^2}\right) \cos\frac{n\pi x}{l_1} \cos\frac{n\pi \xi}{l_1} \right]$$
$$\times \left[1 + 2 \sum_{m=1}^{\infty} \exp\left(-\frac{\pi^2 m^2 at}{l_2^2}\right) \cos\frac{m\pi y}{l_2} \cos\frac{m\pi \eta}{l_2} \right].$$

⊙ *Reference*: H. S. Carslaw and J. C. Jaeger (1984).

2.1.1-18. Domain: $0 \le x \le l_1$, $0 \le y \le l_2$. Third boundary value problem.

A rectangle is considered. The following conditions are prescribed:

$$w = f(x,y) \quad \text{at} \quad t = 0 \quad \text{(initial condition)},$$
$$\partial_x w - k_1 w = g_1(y,t) \quad \text{at} \quad x = 0 \quad \text{(boundary condition)},$$
$$\partial_x w + k_2 w = g_2(y,t) \quad \text{at} \quad x = l_1 \quad \text{(boundary condition)},$$
$$\partial_y w - k_3 w = g_3(x,t) \quad \text{at} \quad y = 0 \quad \text{(boundary condition)},$$
$$\partial_y w + k_4 w = g_4(x,t) \quad \text{at} \quad y = l_2 \quad \text{(boundary condition)}.$$

The solution $w(x, y, t)$ is determined by the formula in Paragraph 2.1.1-17 where

$$G(x,y,\xi,\eta,t) = \left\{\sum_{n=1}^{\infty} \frac{\varphi_n(x)\varphi_n(\xi)}{\|\varphi_n\|^2} \exp(-a\mu_n^2 t)\right\}\left\{\sum_{m=1}^{\infty} \frac{\psi_m(y)\psi_m(\eta)}{\|\psi_m\|^2} \exp(-a\lambda_m^2 t)\right\},$$

$$\varphi_n(x) = \cos(\mu_n x) + \frac{k_1}{\mu_n}\sin(\mu_n x), \quad \|\varphi_n\|^2 = \frac{k_2}{2\mu_n^2}\frac{\mu_n^2 + k_1^2}{\mu_n^2 + k_2^2} + \frac{k_1}{2\mu_n^2} + \frac{l_1}{2}\left(1 + \frac{k_1^2}{\mu_n^2}\right),$$

$$\psi_m(y) = \cos(\lambda_m y) + \frac{k_3}{\lambda_m}\sin(\lambda_m y), \quad \|\psi_m\|^2 = \frac{k_4}{2\lambda_m^2}\frac{\lambda_m^2 + k_3^2}{\lambda_m^2 + k_4^2} + \frac{k_3}{2\lambda_m^2} + \frac{l_2}{2}\left(1 + \frac{k_3^2}{\lambda_m^2}\right).$$

Here, the μ_n and λ_m are positive roots of the transcendental equations

$$\frac{\tan(\mu l_1)}{\mu} = \frac{k_1 + k_2}{\mu^2 - k_1 k_2}, \quad \frac{\tan(\lambda l_2)}{\lambda} = \frac{k_3 + k_4}{\lambda^2 - k_3 k_4}.$$

2.1.1-19. Domain: $0 \le x \le l_1$, $0 \le y \le l_2$. Mixed boundary value problems.

1°. A rectangle is considered. The following conditions are prescribed:

$$\begin{aligned}
w &= f(x, y) &&\text{at} && t = 0 && \text{(initial condition)}, \\
w &= g_1(y, t) &&\text{at} && x = 0 && \text{(boundary condition)}, \\
w &= g_2(y, t) &&\text{at} && x = l_1 && \text{(boundary condition)}, \\
\partial_y w &= g_3(x, t) &&\text{at} && y = 0 && \text{(boundary condition)}, \\
\partial_y w &= g_4(x, t) &&\text{at} && y = l_2 && \text{(boundary condition)}.
\end{aligned}$$

Solution:

$$\begin{aligned}
w(x,y,t) = &\int_0^{l_1}\int_0^{l_2} f(\xi,\eta) G(x,y,\xi,\eta,t)\, d\eta\, d\xi \\
&+ a\int_0^t\int_0^{l_2} g_1(\eta,\tau)\left[\frac{\partial}{\partial \xi}G(x,y,\xi,\eta,t-\tau)\right]_{\xi=0} d\eta\, d\tau \\
&- a\int_0^t\int_0^{l_2} g_2(\eta,\tau)\left[\frac{\partial}{\partial \xi}G(x,y,\xi,\eta,t-\tau)\right]_{\xi=l_1} d\eta\, d\tau \\
&- a\int_0^t\int_0^{l_1} g_3(\xi,\tau) G(x,y,\xi,0,t-\tau)\, d\xi\, d\tau \\
&+ a\int_0^t\int_0^{l_1} g_4(\xi,\tau) G(x,y,\xi,l_2,t-\tau)\, d\xi\, d\tau,
\end{aligned}$$

where

$$G(x,y,\xi,\eta,t) = \frac{4}{l_1 l_2}\left[\sum_{n=1}^{\infty} \sin\frac{n\pi x}{l_1}\sin\frac{n\pi \xi}{l_1}\exp\left(-\frac{\pi^2 n^2 a t}{l_1^2}\right)\right]$$

$$\times \left[\frac{1}{2} + \sum_{m=1}^{\infty} \cos\frac{m\pi y}{l_2}\cos\frac{m\pi \eta}{l_2}\exp\left(-\frac{\pi^2 m^2 a t}{l_2^2}\right)\right].$$

2°. A rectangle is considered. The following conditions are prescribed:

$$\begin{aligned}
w &= f(x, y) &&\text{at} && t = 0 && \text{(initial condition)}, \\
w &= g_1(y, t) &&\text{at} && x = 0 && \text{(boundary condition)}, \\
\partial_x w &= g_2(y, t) &&\text{at} && x = l_1 && \text{(boundary condition)}, \\
w &= g_3(x, t) &&\text{at} && y = 0 && \text{(boundary condition)}, \\
\partial_y w &= g_4(x, t) &&\text{at} && y = l_2 && \text{(boundary condition)}.
\end{aligned}$$

Solution:

$$w(x,y,t) = \int_0^{l_1}\int_0^{l_2} f(\xi,\eta) G(x,y,\xi,\eta,t)\,d\eta\,d\xi$$
$$+ a\int_0^t\int_0^{l_2} g_1(\eta,\tau)\left[\frac{\partial}{\partial \xi} G(x,y,\xi,\eta,t-\tau)\right]_{\xi=0} d\eta\,d\tau$$
$$+ a\int_0^t\int_0^{l_2} g_2(\eta,\tau) G(x,y,l_1,\eta,t-\tau)\,d\eta\,d\tau$$
$$+ a\int_0^t\int_0^{l_1} g_3(\xi,\tau)\left[\frac{\partial}{\partial \eta} G(x,y,\xi,\eta,t-\tau)\right]_{\eta=0} d\xi\,d\tau$$
$$+ a\int_0^t\int_0^{l_1} g_4(\xi,\tau) G(x,y,\xi,l_2,t-\tau)\,d\xi\,d\tau,$$

where

$$G(x,y,\xi,\eta,t) = \frac{4}{l_1 l_2}\left\{\sum_{n=0}^{\infty} \sin\left[\frac{\pi(2n+1)x}{2l_1}\right]\sin\left[\frac{\pi(2n+1)\xi}{2l_1}\right]\exp\left[-\frac{a\pi^2(2n+1)^2 t}{4l_1^2}\right]\right\}$$
$$\times \left\{\sum_{m=0}^{\infty} \sin\left[\frac{\pi(2m+1)y}{2l_2}\right]\sin\left[\frac{\pi(2m+1)\eta}{2l_2}\right]\exp\left[-\frac{a\pi^2(2m+1)^2 t}{4l_2^2}\right]\right\}.$$

2.1.2. Problems in Polar Coordinates

The sourceless heat equation with two space variables in the polar coordinate system r, φ has the form

$$\frac{\partial w}{\partial t} = a\left(\frac{\partial^2 w}{\partial r^2} + \frac{1}{r}\frac{\partial w}{\partial r} + \frac{1}{r^2}\frac{\partial^2 w}{\partial \varphi^2}\right), \qquad r = \sqrt{x^2 + y^2}.$$

One-dimensional problems with axial symmetry that have solutions of the form $w = w(r,t)$ are considered in Subsection 1.2.1.

2.1.2-1. Domain: $0 \le r < \infty$, $0 \le \varphi \le 2\pi$. Cauchy problem.

An initial condition is prescribed:

$$w = f(r,\varphi) \quad \text{at} \quad t = 0.$$

Solution:

$$w(r,\varphi,t) = \frac{1}{4\pi at}\int_0^{2\pi}\int_0^{\infty} \xi \exp\left[-\frac{r^2 + \xi^2 - 2r\xi\cos(\varphi-\eta)}{4at}\right] f(\xi,\eta)\,d\xi\,d\eta.$$

2.1.2-2. Domain: $0 \le r \le R$, $0 \le \varphi \le 2\pi$. First boundary value problem.

A circle is considered. The following conditions are prescribed:

$$w = f(r,\varphi) \quad \text{at} \quad t = 0 \quad \text{(initial condition)},$$
$$w = g(\varphi,t) \quad \text{at} \quad r = R \quad \text{(boundary condition)}.$$

Solution:

$$w(r,\varphi,t) = \int_0^{2\pi}\int_0^R f(\xi,\eta) G(r,\varphi,\xi,\eta,t)\xi\,d\xi\,d\eta$$
$$- aR\int_0^t\int_0^{2\pi} g(\eta,\tau)\left[\frac{\partial}{\partial \xi} G(r,\varphi,\xi,\eta,t-\tau)\right]_{\xi=R} d\eta\,d\tau.$$

Here,

$$G(r,\varphi,\xi,\eta,t) = \frac{1}{\pi R^2} \sum_{n=0}^{\infty} \sum_{m=1}^{\infty} \frac{A_n}{[J_n'(\mu_{nm}R)]^2} J_n(\mu_{nm}r) J_n(\mu_{nm}\xi) \cos[n(\varphi-\eta)] \exp(-\mu_{nm}^2 at),$$

$$A_0 = 1, \quad A_n = 2 \quad (n = 1, 2, \ldots),$$

where $J_n(\xi)$ are the Bessel functions (the prime denotes the derivative with respect to the argument), and μ_{nm} are positive roots of the transcendental equation $J_n(\mu R) = 0$.

⊙ *Reference*: H. S. Carslaw and J. C. Jaeger (1984).

2.1.2-3. Domain: $0 \le r \le R$, $0 \le \varphi \le 2\pi$. Second boundary value problem.

A circle is considered. The following conditions are prescribed:

$$w = f(r,\varphi) \quad \text{at} \quad t = 0 \quad \text{(initial condition)},$$
$$\partial_r w = g(\varphi, t) \quad \text{at} \quad r = R \quad \text{(boundary condition)}.$$

Solution:

$$w(r,\varphi,t) = \int_0^{2\pi} \int_0^R f(\xi,\eta) G(r,\varphi,\xi,\eta,t)\xi \, d\xi \, d\eta + aR \int_0^t \int_0^{2\pi} g(\eta,\tau) G(r,\varphi,R,\eta,t-\tau) \, d\eta \, d\tau.$$

Here,

$$G(r,\varphi,\xi,\eta,t) = \frac{1}{\pi R^2} + \frac{1}{\pi} \sum_{n=0}^{\infty} \sum_{m=1}^{\infty} \frac{A_n \mu_{nm}^2 J_n(\mu_{nm}r) J_n(\mu_{nm}\xi)}{(\mu_{nm}^2 R^2 - n^2)[J_n(\mu_{nm}R)]^2} \cos[n(\varphi-\eta)] \exp(-\mu_{nm}^2 at),$$

$$A_0 = 1, \quad A_n = 2 \quad (n = 1, 2, \ldots),$$

where $J_n(\xi)$ are the Bessel functions, and μ_{nm} are positive roots of the transcendental equation $J_n'(\mu R) = 0$.

2.1.2-4. Domain: $0 \le r \le R$, $0 \le \varphi \le 2\pi$. Third boundary value problem.

A circle is considered. The following conditions are prescribed:

$$w = f(r,\varphi) \quad \text{at} \quad t = 0 \quad \text{(initial condition)},$$
$$\partial_r w + kw = g(\varphi,t) \quad \text{at} \quad r = R \quad \text{(boundary condition)}.$$

The solution $w(r,\varphi,t)$ is determined by the formula in Paragraph 2.1.2-3 where

$$G(r,\varphi,\xi,\eta,t) = \frac{1}{\pi} \sum_{n=0}^{\infty} \sum_{m=1}^{\infty} \frac{A_n \mu_{nm}^2 J_n(\mu_{nm}r) J_n(\mu_{nm}\xi)}{(\mu_{nm}^2 R^2 + k^2 R^2 - n^2)[J_n(\mu_{nm}R)]^2} \cos[n(\varphi-\eta)] \exp(-\mu_{nm}^2 at),$$

$$A_0 = 1, \quad A_n = 2 \quad (n = 1, 2, \ldots).$$

Here, $J_n(\xi)$ are the Bessel functions, and μ_{nm} are positive roots of the transcendental equation

$$\mu J_n'(\mu R) + k J_n(\mu R) = 0.$$

⊙ *Reference*: H. S. Carslaw and J. C. Jaeger (1984).

2.1.2-5. Domain: $R_1 \leq r \leq R_2$, $0 \leq \varphi \leq 2\pi$. First boundary value problem.

An annular domain is considered. The following conditions are prescribed:

$$w = f(r, \varphi) \quad \text{at} \quad t = 0 \quad \text{(initial condition)},$$
$$w = g_1(\varphi, t) \quad \text{at} \quad r = R_1 \quad \text{(boundary condition)},$$
$$w = g_2(\varphi, t) \quad \text{at} \quad r = R_2 \quad \text{(boundary condition)}.$$

Solution:

$$w(r, \varphi, t) = \int_0^{2\pi} \int_{R_1}^{R_2} f(\xi, \eta) G(r, \varphi, \xi, \eta, t) \xi \, d\xi \, d\eta$$

$$+ a R_1 \int_0^t \int_0^{2\pi} g_1(\eta, \tau) \left[\frac{\partial}{\partial \xi} G(r, \varphi, \xi, \eta, t - \tau) \right]_{\xi = R_1} d\eta \, d\tau$$

$$- a R_2 \int_0^t \int_0^{2\pi} g_2(\eta, \tau) \left[\frac{\partial}{\partial \xi} G(r, \varphi, \xi, \eta, t - \tau) \right]_{\xi = R_2} d\eta \, d\tau.$$

Here,

$$G(r, \varphi, \xi, \eta, t) = \frac{\pi}{2} \sum_{n=0}^{\infty} \sum_{m=1}^{\infty} A_n B_{nm} Z_n(\mu_{nm} r) Z_n(\mu_{nm} \xi) \cos[n(\varphi - \eta)] \exp(-\mu_{nm}^2 a t),$$

$$A_n = \begin{cases} 1/2 & \text{if } n = 0, \\ 1 & \text{if } n \neq 0, \end{cases} \quad B_{nm} = \frac{\mu_{nm}^2 J_n^2(\mu_{nm} R_2)}{J_n^2(\mu_{nm} R_1) - J_n^2(\mu_{nm} R_2)},$$

$$Z_n(\mu_{nm} r) = J_n(\mu_{nm} R_1) Y_n(\mu_{nm} r) - Y_n(\mu_{nm} R_1) J_n(\mu_{nm} r),$$

where $J_n(r)$ and $Y_n(r)$ are the Bessel functions, and μ_{nm} are positive roots of the transcendental equation

$$J_n(\mu R_1) Y_n(\mu R_2) - Y_n(\mu R_1) J_n(\mu R_2) = 0.$$

⊙ *Reference*: B. M. Budak, A. A. Samarskii, and A. N. Tikhonov (1980).

2.1.2-6. Domain: $R_1 \leq r \leq R_2$, $0 \leq \varphi \leq 2\pi$. Second boundary value problem.

An annular domain is considered. The following conditions are prescribed:

$$w = f(r, \varphi) \quad \text{at} \quad t = 0 \quad \text{(initial condition)},$$
$$\partial_r w = g_1(\varphi, t) \quad \text{at} \quad r = R_1 \quad \text{(boundary condition)},$$
$$\partial_r w = g_2(\varphi, t) \quad \text{at} \quad r = R_2 \quad \text{(boundary condition)}.$$

Solution:

$$w(r, \varphi, t) = \int_0^{2\pi} \int_{R_1}^{R_2} f(\xi, \eta) G(r, \varphi, \xi, \eta, t) \xi \, d\xi \, d\eta$$

$$- a R_1 \int_0^t \int_0^{2\pi} g_1(\eta, \tau) G(r, \varphi, R_1, \eta, t - \tau) \, d\eta \, d\tau$$

$$+ a R_2 \int_0^t \int_0^{2\pi} g_2(\eta, \tau) G(r, \varphi, R_2, \eta, t - \tau) \, d\eta \, d\tau.$$

Here,

$$G(r, \varphi, \xi, \eta, t) = \frac{1}{\pi(R_2^2 - R_1^2)} + \frac{1}{\pi} \sum_{n=0}^{\infty} \sum_{m=1}^{\infty} \frac{A_n \mu_{nm}^2 Z_n(\mu_{nm} r) Z_n(\mu_{nm} \xi) \cos[n(\varphi - \eta)] \exp(-\mu_{nm}^2 a t)}{(\mu_{nm}^2 R_2^2 - n^2) Z_n^2(\mu_{nm} R_2) - (\mu_{nm}^2 R_1^2 - n^2) Z_n^2(\mu_{nm} R_1)},$$

$$Z_n(\mu_{nm} r) = J_n'(\mu_{nm} R_1) Y_n(\mu_{nm} r) - Y_n'(\mu_{nm} R_1) J_n(\mu_{nm} r),$$

where $A_0 = 1$ and $A_n = 2$ for $n = 1, 2, \ldots$; $J_n(r)$ and $Y_n(r)$ are the Bessel functions, and μ_{nm} are positive roots of the transcendental equation

$$J_n'(\mu R_1) Y_n'(\mu R_2) - Y_n'(\mu R_1) J_n'(\mu R_2) = 0.$$

2.1.2-7. Domain: $R_1 \leq r \leq R_2$, $0 \leq \varphi \leq 2\pi$. Third boundary value problem.

An annular domain is considered. The following conditions are prescribed:

$$w = f(r, \varphi) \quad \text{at} \quad t = 0 \quad \text{(initial condition)},$$
$$\partial_r w - k_1 w = g_1(\varphi, t) \quad \text{at} \quad r = R_1 \quad \text{(boundary condition)},$$
$$\partial_r w + k_2 w = g_2(\varphi, t) \quad \text{at} \quad r = R_2 \quad \text{(boundary condition)}.$$

Solution:

$$w(r, \varphi, t) = \int_0^{2\pi} \int_{R_1}^{R_2} f(\xi, \eta) G(r, \varphi, \xi, \eta, t) \xi \, d\xi \, d\eta$$
$$- a R_1 \int_0^t \int_0^{2\pi} g_1(\eta, \tau) G(r, \varphi, R_1, \eta, t - \tau) \, d\eta \, d\tau$$
$$+ a R_2 \int_0^t \int_0^{2\pi} g_2(\eta, \tau) G(r, \varphi, R_2, \eta, t - \tau) \, d\eta \, d\tau.$$

Here,

$$G(r, \varphi, \xi, \eta, t) = \frac{1}{\pi} \sum_{n=0}^{\infty} \sum_{m=1}^{\infty} \frac{A_n \mu_{nm}^2 Z_n(\mu_{nm} r) Z_n(\mu_{nm} \xi) \cos[n(\varphi - \eta)] \exp(-\mu_{nm}^2 a t)}{(k_2^2 R_2^2 + \mu_{nm}^2 R_2^2 - n^2) Z_n^2(\mu_{nm} R_2) - (k_1^2 R_1^2 + \mu_{nm}^2 R_1^2 - n^2) Z_n^2(\mu_{nm} R_1)},$$

$$Z_n(\mu_{nm} r) = [\mu_{nm} J'_n(\mu_{nm} R_1) - k_1 J_n(\mu_{nm} R_1)] Y_n(\mu_{nm} r)$$
$$- [\mu_{nm} Y'_n(\mu_{nm} R_1) - k_1 Y_n(\mu_{nm} R_1)] J_n(\mu_{nm} r),$$

where $A_0 = 1$ and $A_n = 2$ for $n = 1, 2, \ldots$; $J_n(r)$ and $Y_n(r)$ are the Bessel functions, and μ_{nm} are positive roots of the transcendental equation

$$[\mu J'_n(\mu R_1) - k_1 J_n(\mu R_1)] [\mu Y'_n(\mu R_2) + k_2 Y_n(\mu R_2)]$$
$$= [\mu Y'_n(\mu R_1) - k_1 Y_n(\mu R_1)] [\mu J'_n(\mu R_2) + k_2 J_n(\mu R_2)].$$

⊙ *Reference*: B. M. Budak, A. A. Samarskii, and A. N. Tikhonov (1980).

2.1.2-8. Domain: $0 \leq r < \infty$, $0 \leq \varphi \leq \varphi_0$. First boundary value problem.

A wedge domain is considered. The following conditions are prescribed:

$$w = f(r, \varphi) \quad \text{at} \quad t = 0 \quad \text{(initial condition)},$$
$$w = g_1(r, t) \quad \text{at} \quad \varphi = 0 \quad \text{(boundary condition)},$$
$$w = g_2(r, t) \quad \text{at} \quad \varphi = \varphi_0 \quad \text{(boundary condition)}.$$

Solution:

$$w(r, \varphi, t) = \int_0^{\varphi_0} \int_0^{\infty} f(\xi, \eta) G(r, \varphi, \xi, \eta, t) \xi \, d\xi \, d\eta$$
$$+ a \int_0^t \int_0^{\infty} g_1(\xi, \tau) \frac{1}{\xi} \left[\frac{\partial}{\partial \eta} G(r, \varphi, \xi, \eta, t - \tau) \right]_{\eta=0} d\xi \, d\tau$$
$$- a \int_0^t \int_0^{\infty} g_2(\xi, \tau) \frac{1}{\xi} \left[\frac{\partial}{\partial \eta} G(r, \varphi, \xi, \eta, t - \tau) \right]_{\eta=\varphi_0} d\xi \, d\tau.$$

Here,

$$G(r, \varphi, \xi, \eta, t) = \frac{1}{a \varphi_0 t} \exp\left(-\frac{r^2 + \xi^2}{4at}\right) \sum_{n=1}^{\infty} I_{n\pi/\varphi_0}\left(\frac{r\xi}{2at}\right) \sin\left(\frac{n\pi\varphi}{\varphi_0}\right) \sin\left(\frac{n\pi\eta}{\varphi_0}\right),$$

where $I_\nu(r)$ are the modified Bessel functions.

⊙ *Reference*: B. M. Budak, A. A. Samarskii, and A. N. Tikhonov (1980).

2.1.2-9. Domain: $0 \le r < \infty$, $0 \le \varphi \le \varphi_0$. Second boundary value problem.

A wedge domain is considered. The following conditions are prescribed:

$$w = f(r, \varphi) \quad \text{at} \quad t = 0 \quad \text{(initial condition)},$$
$$r^{-1}\partial_\varphi w = g_1(r, t) \quad \text{at} \quad \varphi = 0 \quad \text{(boundary condition)},$$
$$r^{-1}\partial_\varphi w = g_2(r, t) \quad \text{at} \quad \varphi = \varphi_0 \quad \text{(boundary condition)}.$$

Solution:
$$w(r, \varphi, t) = \int_0^{\varphi_0} \int_0^\infty f(\xi, \eta) G(r, \varphi, \xi, \eta, t) \xi \, d\xi \, d\eta$$
$$- a \int_0^t \int_0^\infty g_1(\xi, \tau) G(r, \varphi, \xi, 0, t - \tau) \, d\xi \, d\tau$$
$$+ a \int_0^t \int_0^\infty g_2(\xi, \tau) G(r, \varphi, \xi, \varphi_0, t - \tau) \, d\xi \, d\tau.$$

Here,
$$G(r, \varphi, \xi, \eta, t) = \frac{1}{a\varphi_0 t} \exp\!\left(-\frac{r^2 + \xi^2}{4at}\right)\!\left[\frac{1}{2} I_0\!\left(\frac{r\xi}{2at}\right) + \sum_{n=1}^\infty I_{n\pi/\varphi_0}\!\left(\frac{r\xi}{2at}\right) \cos\!\left(\frac{n\pi\varphi}{\varphi_0}\right) \cos\!\left(\frac{n\pi\eta}{\varphi_0}\right)\right],$$

where $I_\nu(r)$ are the modified Bessel functions.

⊙ *Reference*: B. M. Budak, A. A. Samarskii, and A. N. Tikhonov (1980).

2.1.2-10. Domain: $0 \le r \le R$, $0 \le \varphi \le \varphi_0$. First boundary value problem.

A circular sector is considered. The following conditions are prescribed:

$$w = f(r, \varphi) \quad \text{at} \quad t = 0 \quad \text{(initial condition)},$$
$$w = g_1(\varphi, t) \quad \text{at} \quad r = R \quad \text{(boundary condition)},$$
$$w = g_2(r, t) \quad \text{at} \quad \varphi = 0 \quad \text{(boundary condition)},$$
$$w = g_3(r, t) \quad \text{at} \quad \varphi = \varphi_0 \quad \text{(boundary condition)}.$$

Solution:
$$w(r, \varphi, t) = \int_0^{\varphi_0}\!\int_0^R f(\xi, \eta) G(r, \varphi, \xi, \eta, t) \xi \, d\xi \, d\eta$$
$$- aR \int_0^t\!\int_0^{\varphi_0} g_1(\eta, \tau) \left[\frac{\partial}{\partial \xi} G(r, \varphi, \xi, \eta, t - \tau)\right]_{\xi=R} d\eta \, d\tau$$
$$+ a \int_0^t\!\int_0^R g_2(\xi, \tau) \frac{1}{\xi}\!\left[\frac{\partial}{\partial \eta} G(r, \varphi, \xi, \eta, t - \tau)\right]_{\eta=0} d\xi \, d\tau$$
$$- a \int_0^t\!\int_0^R g_3(\xi, \tau) \frac{1}{\xi}\!\left[\frac{\partial}{\partial \eta} G(r, \varphi, \xi, \eta, t - \tau)\right]_{\eta=\varphi_0} d\xi \, d\tau.$$

Here,
$$G(r, \varphi, \xi, \eta, t) = \frac{4}{R^2 \varphi_0} \sum_{n=1}^\infty \sum_{m=1}^\infty \frac{J_{n\pi/\varphi_0}(\mu_{nm} r) J_{n\pi/\varphi_0}(\mu_{nm}\xi)}{[J'_{n\pi/\varphi_0}(\mu_{nm} R)]^2} \sin\!\left(\frac{n\pi\varphi}{\varphi_0}\right) \sin\!\left(\frac{n\pi\eta}{\varphi_0}\right) \exp(-\mu_{nm}^2 at),$$

where the $J_{n\pi/\varphi_0}(r)$ are the Bessel functions, and the μ_{nm} are positive roots of the transcendental equation $J_{n\pi/\varphi_0}(\mu R) = 0$.

Example. The initial temperature is uniform, $f(r, \varphi) = w_0$. The boundary is maintained at zero temperature, $g_1(\varphi, t) = g_2(r, t) = g_3(r, t) = 0$.

Solution:

$$w = \frac{8w_0}{\pi R^2} \sum_{n=0}^{\infty} \frac{1}{2n+1} \sin(s_n \varphi) \sum_{m=1}^{\infty} \exp(-\mu_{nm}^2 at) \frac{J_{s_n}(\mu_{nm} r)}{[J'_{s_n}(\mu_{nm} R)]^2} \int_0^R J_{s_n}(\mu_{nm}\xi)\xi \, d\xi, \quad s_n = \frac{(2n+1)\pi}{\varphi_0},$$

where the μ_{nm} are positive roots of the transcendental equation $J_{s_n}(\mu R) = 0$.

⊙ *References*: B. M. Budak, A. A. Samarskii, and A. N. Tikhonov (1980), H. S. Carslaw and J. C. Jaeger (1984).

2.1.2-11. Domain: $0 \leq r \leq R$, $0 \leq \varphi \leq \varphi_0$. **Second boundary value problem.**

A circular sector is considered. The following conditions are prescribed:

$$w = f(r, \varphi) \quad \text{at} \quad t = 0 \quad \text{(initial condition)},$$
$$\partial_r w = g_1(\varphi, t) \quad \text{at} \quad r = R \quad \text{(boundary condition)},$$
$$r^{-1}\partial_\varphi w = g_2(r, t) \quad \text{at} \quad \varphi = 0 \quad \text{(boundary condition)},$$
$$r^{-1}\partial_\varphi w = g_3(r, t) \quad \text{at} \quad \varphi = \varphi_0 \quad \text{(boundary condition)}.$$

Solution:

$$w(r, \varphi, t) = \int_0^{\varphi_0} \int_0^R f(\xi, \eta) G(r, \varphi, \xi, \eta, t) \xi \, d\xi \, d\eta$$
$$+ aR \int_0^t \int_0^{\varphi_0} g_1(\eta, \tau) G(r, \varphi, R, \eta, t - \tau) \, d\eta \, d\tau$$
$$- a \int_0^t \int_0^R g_2(\xi, \tau) G(r, \varphi, \xi, 0, t - \tau) \, d\xi \, d\tau$$
$$+ a \int_0^t \int_0^R g_3(\xi, \tau) G(r, \varphi, \xi, \varphi_0, t - \tau) \, d\xi \, d\tau.$$

Here,

$$G(r, \varphi, \xi, \eta, t) = \frac{2}{R^2 \varphi_0} + 4\varphi_0 \sum_{n=0}^{\infty} \sum_{m=1}^{\infty} \frac{\mu_{nm}^2 J_{n\pi/\varphi_0}(\mu_{nm} r) J_{n\pi/\varphi_0}(\mu_{nm}\xi)}{(R^2 \varphi_0^2 \mu_{nm}^2 - n^2 \pi^2)[J_{n\pi/\varphi_0}(\mu_{nm} R)]^2}$$
$$\times \cos\left(\frac{n\pi\varphi}{\varphi_0}\right) \cos\left(\frac{n\pi\eta}{\varphi_0}\right) \exp(-\mu_{nm}^2 at),$$

where the $J_{n\pi/\varphi_0}(r)$ are the Bessel functions, and the μ_{nm} are positive roots of the transcendental equation $J'_{n\pi/\varphi_0}(\mu R) = 0$.

⊙ *Reference*: B. M. Budak, A. A. Samarskii, and A. N. Tikhonov (1980).

2.1.2-12. Domain: $0 \leq r \leq R$, $0 \leq \varphi \leq \varphi_0$. **Mixed boundary value problem.**

A circular sector is considered. The following conditions are prescribed:

$$w = f(r, \varphi) \quad \text{at} \quad t = 0 \quad \text{(initial condition)},$$
$$\partial_r w - kw = g(\varphi, t) \quad \text{at} \quad r = R \quad \text{(boundary condition)},$$
$$\partial_\varphi w = 0 \quad \text{at} \quad \varphi = 0 \quad \text{(boundary condition)},$$
$$\partial_\varphi w = 0 \quad \text{at} \quad \varphi = \varphi_0 \quad \text{(boundary condition)}.$$

Solution:

$$w(r, \varphi, t) = \int_0^{\varphi_0} \int_0^R f(\xi, \eta) G(r, \varphi, \xi, \eta, t) \xi \, d\xi \, d\eta$$
$$+ aR \int_0^t \int_0^{\varphi_0} g(\eta, \tau) G(r, \varphi, R, \eta, t - \tau) \, d\eta \, d\tau.$$

Here,

$$G(r,\varphi,\xi,\eta,t) = \sum_{n=0}^{\infty}\sum_{m=1}^{\infty} A_{nm} J_{s_n}(\mu_{nm}r) J_{s_n}(\mu_{nm}\xi) \cos(s_n\varphi)\cos(s_n\eta)\exp(-\mu_{nm}^2 at),$$

$$s_n = \frac{n\pi}{\varphi_0}, \qquad A_{nm} = \frac{4\mu_{nm}^2}{\varphi_0(\mu_{nm}^2 R^2 + k^2 R^2 - s_n^2)\left[J_{s_n}(\mu_{nm}R)\right]^2},$$

where $J_{s_n}(r)$ are the Bessel functions, and μ_{nm} are positive roots of the transcendental equation

$$\mu J'_{s_n}(\mu R) + k J_{s_n}(\mu R) = 0.$$

⊙ *Reference*: B. M. Budak, A. A. Samarskii, and A. N. Tikhonov (1980).

2.1.2-13. Domain: $R_1 \le r \le R_2$, $0 \le \varphi \le \varphi_0$. Different boundary value problems.

Some problems for this domain were studied in Budak, Samarskii, and Tikhonov (1980).

2.1.3. Axisymmetric Problems

In the case of angular symmetry, the two-dimensional sourceless heat equation in the cylindrical coordinate system has the form

$$\frac{\partial w}{\partial t} = a\left(\frac{\partial^2 w}{\partial r^2} + \frac{1}{r}\frac{\partial w}{\partial r} + \frac{\partial^2 w}{\partial z^2}\right), \qquad r = \sqrt{x^2 + y^2}.$$

This equation governs two-dimensional unsteady thermal processes in quiescent media or solid bodies (bounded by coordinate surfaces of the cylindrical system) in the case where the initial and boundary conditions are independent of the angular coordinate. A similar equation is used to study analogous two-dimensional unsteady mass transfer phenomena.

2.1.3-1. Particular solutions. Remarks on the Green's functions.

$1°$. Apart from usual separable solutions $w(r,z,t) = f_1(r)f_2(z)f_3(t)$, the equation in question has more sophisticated solutions in the product form

$$w(r,z,t) = u(r,t)v(z,t),$$

where $u = u(r,t)$ and $v = v(z,t)$ are solutions of the simpler one-dimensional equations

$$\frac{\partial u}{\partial t} = a\left(\frac{\partial^2 u}{\partial r^2} + \frac{1}{r}\frac{\partial u}{\partial r}\right) \qquad \text{(see Subsection 1.2.1-1 for particular solutions of this equation),}$$

$$\frac{\partial v}{\partial t} = a\frac{\partial^2 v}{\partial z^2} \qquad \text{(see Subsection 1.1.1-1 for particular solutions of this equation).}$$

$2°$. For all two-dimensional boundary value problems considered in Subsection 2.1.3, the Green's function can be represented in the product form

$$G(r,z,\xi,\eta,t) = G_1(r,\xi,t)G_2(z,\eta,t),$$

where $G_1(r,\xi,t)$ and $G_2(z,\eta,t)$ are the Green's functions of appropriate one-dimensional boundary value problems.

2.1.3-2. Domain: $0 \leq r \leq R$, $0 \leq z < \infty$. First boundary value problem.

A semiinfinite circular cylinder is considered. The following conditions are prescribed:

$$w = f(r, z) \quad \text{at} \quad t = 0 \quad \text{(initial condition)},$$
$$w = g_1(z, t) \quad \text{at} \quad r = R \quad \text{(boundary condition)},$$
$$w = g_2(r, t) \quad \text{at} \quad z = 0 \quad \text{(boundary condition)}.$$

Solution:

$$w(r, z, t) = 2\pi \int_0^\infty \int_0^R \xi f(\xi, \eta) G(r, z, \xi, \eta, t) \, d\xi \, d\eta$$
$$- 2\pi a R \int_0^t \int_0^\infty g_1(\eta, \tau) \left[\frac{\partial}{\partial \xi} G(r, z, \xi, \eta, t - \tau) \right]_{\xi=R} d\eta \, d\tau$$
$$+ 2\pi a \int_0^t \int_0^R \xi g_2(\xi, \tau) \left[\frac{\partial}{\partial \eta} G(r, z, \xi, \eta, t - \tau) \right]_{\eta=0} d\xi \, d\tau.$$

Here,

$$G(r, z, \xi, \eta, t) = G_1(r, \xi, t) G_2(z, \eta, t),$$

$$G_1(r, \xi, t) = \frac{1}{\pi R^2} \sum_{n=1}^\infty \frac{1}{J_1^2(\mu_n)} J_0\left(\frac{\mu_n r}{R}\right) J_0\left(\frac{\mu_n \xi}{R}\right) \exp\left(-\frac{a\mu_n^2 t}{R^2}\right),$$

$$G_2(z, \eta, t) = \frac{1}{2\sqrt{\pi a t}} \left\{ \exp\left[-\frac{(z-\eta)^2}{4at}\right] - \exp\left[-\frac{(z+\eta)^2}{4at}\right] \right\},$$

where the μ_n are positive zeros of the Bessel function, $J_0(\mu_n) = 0$.

Example 1. The initial temperature is the same at every point of the cylinder, $f(r, z) = w_0$. The lateral surface and the end face are maintained at zero temperature, $g_1(r, t) = g_2(z, t) = 0$.
Solution:

$$w(r, z, t) = \frac{2w_0}{R} \operatorname{erf}\left(\frac{z}{2\sqrt{at}}\right) \sum_{n=1}^\infty \frac{J_0(\lambda_n r)}{\lambda_n J_1(\lambda_n R)} \exp(-\lambda_n^2 a t), \qquad \lambda_n = \frac{\mu_n}{R}.$$

Example 2. The initial temperature of the cylinder is everywhere zero, $f(r, z) = 0$. The lateral surface $r = R$ is maintained at a constant temperature w_0, and the end face $z = 0$ at zero temperature.
Solution:

$$w(r, z, t) = w_0 - \frac{w_0}{R} \sum_{n=1}^\infty \frac{J_0(\lambda_n r)}{\lambda_n J_1(\lambda_n R)} \left[2\exp(\lambda_n^2 a t) \operatorname{erf}\left(\frac{z}{2\sqrt{at}}\right) \right.$$
$$\left. + \exp(\lambda_n z) \operatorname{erfc}\left(\frac{z}{2\sqrt{at}} + \lambda_n \sqrt{at}\right) + \exp(-\lambda_n z) \operatorname{erfc}\left(\frac{z}{2\sqrt{at}} - \lambda_n \sqrt{at}\right) \right],$$

where the λ_n are positive zeros of the Bessel function, $J_0(\lambda R) = 0$.

⊙ *References*: B. M. Budak, A. A. Samarskii, and A. N. Tikhonov (1980), H. S. Carslaw and J. C. Jaeger (1984).

2.1.3-3. Domain: $0 \leq r \leq R$, $0 \leq z < \infty$. Second boundary value problem.

A semiinfinite circular cylinder is considered. The following conditions are prescribed:

$$w = f(r, z) \quad \text{at} \quad t = 0 \quad \text{(initial condition)},$$
$$\partial_r w = g_1(z, t) \quad \text{at} \quad r = R \quad \text{(boundary condition)},$$
$$\partial_z w = g_2(r, t) \quad \text{at} \quad z = 0 \quad \text{(boundary condition)}.$$

Solution:

$$w(r,z,t) = 2\pi \int_0^\infty \int_0^R \xi f(\xi,\eta) G(r,z,\xi,\eta,t)\, d\xi\, d\eta$$
$$+ 2\pi aR \int_0^t \int_0^\infty g_1(\eta,\tau) G(r,z,R,\eta,t-\tau)\, d\eta\, d\tau$$
$$- 2\pi a \int_0^t \int_0^R \xi g_2(\xi,\tau) G(r,z,\xi,0,t-\tau)\, d\xi\, d\tau.$$

Here,

$$G(r,z,\xi,\eta,t) = G_1(r,\xi,t) G_2(z,\eta,t),$$

$$G_1(r,\xi,t) = \frac{1}{\pi R^2} + \frac{1}{\pi R^2} \sum_{n=1}^\infty \frac{1}{J_0^2(\mu_n)} J_0\!\left(\frac{\mu_n r}{R}\right) J_0\!\left(\frac{\mu_n \xi}{R}\right) \exp\!\left(-\frac{a\mu_n^2 t}{R^2}\right),$$

$$G_2(z,\eta,t) = \frac{1}{2\sqrt{\pi at}} \left\{ \exp\!\left[-\frac{(z-\eta)^2}{4at}\right] + \exp\!\left[-\frac{(z+\eta)^2}{4at}\right] \right\},$$

where the μ_n are positive zeros of the first-order Bessel function, $J_1(\mu_n) = 0$.

2.1.3-4. Domain: $0 \le r \le R$, $0 \le z < \infty$. Third boundary value problem.

A semiinfinite circular cylinder is considered. The following conditions are prescribed:

$$w = f(r,z) \quad \text{at} \quad t = 0 \quad \text{(initial condition)},$$
$$\partial_r w + k_1 w = g_1(z,t) \quad \text{at} \quad r = R \quad \text{(boundary condition)},$$
$$\partial_z w - k_2 w = g_2(r,t) \quad \text{at} \quad z = 0 \quad \text{(boundary condition)}.$$

The solution $w(r,z,t)$ is determined by the formula in Paragraph 2.1.3-3 where

$$G(r,z,\xi,\eta,t) = G_1(r,\xi,t) G_2(z,\eta,t),$$

$$G_1(r,\xi,t) = \frac{1}{\pi R^2} \sum_{n=1}^\infty \frac{\mu_n^2}{(k_1^2 R^2 + \mu_n^2) J_0^2(\mu_n)} J_0\!\left(\frac{\mu_n r}{R}\right) J_0\!\left(\frac{\mu_n \xi}{R}\right) \exp\!\left(-\frac{a\mu_n^2 t}{R^2}\right),$$

$$G_2(z,\eta,t) = \frac{1}{2\sqrt{\pi at}} \left\{ \exp\!\left[\frac{(z-\eta)^2}{4at}\right] + \exp\!\left[\frac{(z+\eta)^2}{4at}\right] - 2k_2 \int_0^\infty \exp\!\left[-\frac{(z+\eta+s)^2}{4at} - k_2 s\right] ds \right\}.$$

Here, $J_0(\mu)$ is the zeroth Bessel function and the μ_n are positive roots of the transcendental equation

$$\mu J_1(\mu) - k_1 R J_0(\mu) = 0.$$

Example 3. The initial temperature is the same at every point of the cylinder, $f(r,z) = w_0$. At the lateral surface and the end face, heat exchange of the cylinder with the zero temperature environment occurs, $g_1(z,t) = g_2(r,t) = 0$.
Solution:

$$w(r,z,t) = \frac{2w_0 k_1}{R} \left[\operatorname{erf}\!\left(\frac{z}{2\sqrt{at}}\right) + \exp(k_2 z + k_2^2 at)\operatorname{erfc}\!\left(\frac{z}{2\sqrt{at}} + k_2\sqrt{at}\right) \right] \sum_{n=1}^\infty \frac{J_0(\nu_n r)\exp(-\nu_n^2 at)}{(k_1^2 + \nu_n^2) J_0(\nu_n R)},$$

where the ν_n are positive roots of the transcendental equation $\nu J_1(\nu R) - k_1 J_0(\nu R) = 0$.

⊙ *Reference*: H. S. Carslaw and J. C. Jaeger (1984).

2.1.3-5. Domain: $0 \le r \le R$, $0 \le z < \infty$. Mixed boundary value problems.

1°. A semiinfinite circular cylinder is considered. The following conditions are prescribed:

$$w = f(r,z) \quad \text{at} \quad t = 0 \quad \text{(initial condition)},$$
$$w = g_1(z,t) \quad \text{at} \quad r = R \quad \text{(boundary condition)},$$
$$\partial_z w = g_2(r,t) \quad \text{at} \quad z = 0 \quad \text{(boundary condition)}.$$

Solution:

$$w(r,z,t) = 2\pi \int_0^\infty \int_0^R \xi f(\xi,\eta) G(r,z,\xi,\eta,t)\,d\xi\,d\eta$$
$$- 2\pi a R \int_0^t \int_0^\infty g_1(\eta,\tau) \left[\frac{\partial}{\partial \xi} G(r,z,\xi,\eta,t-\tau)\right]_{\xi=R} d\eta\,d\tau$$
$$- 2\pi a \int_0^t \int_0^R \xi g_2(\xi,\tau) G(r,z,\xi,0,t-\tau)\,d\xi\,d\tau.$$

Here,

$$G(r,z,\xi,\eta,t) = G_1(r,\xi,t) G_2(z,\eta,t),$$
$$G_1(r,\xi,t) = \frac{1}{\pi R^2} \sum_{n=1}^\infty \frac{1}{J_1^2(\mu_n)} J_0\!\left(\frac{\mu_n r}{R}\right) J_0\!\left(\frac{\mu_n \xi}{R}\right) \exp\!\left(-\frac{a\mu_n^2 t}{R^2}\right),$$
$$G_2(z,\eta,t) = \frac{1}{2\sqrt{\pi a t}} \left\{\exp\!\left[-\frac{(z-\eta)^2}{4at}\right] + \exp\!\left[-\frac{(z+\eta)^2}{4at}\right]\right\},$$

where the μ_n are positive zeros of the Bessel function, $J_0(\mu_n) = 0$.

2°. A semiinfinite circular cylinder is considered. The following conditions are prescribed:

$$w = f(r,z) \quad \text{at} \quad t = 0 \quad \text{(initial condition)},$$
$$\partial_r w = g_1(z,t) \quad \text{at} \quad r = R \quad \text{(boundary condition)},$$
$$w = g_2(r,t) \quad \text{at} \quad z = 0 \quad \text{(boundary condition)}.$$

Solution:

$$w(r,z,t) = 2\pi \int_0^\infty \int_0^R \xi f(\xi,\eta) G(r,z,\xi,\eta,t)\,d\xi\,d\eta$$
$$+ 2\pi a R \int_0^t \int_0^\infty g_1(\eta,\tau) G(r,z,R,\eta,t-\tau)\,d\eta\,d\tau$$
$$+ 2\pi a \int_0^t \int_0^R \xi g_2(\xi,\tau) \left[\frac{\partial}{\partial \eta} G(r,z,\xi,\eta,t-\tau)\right]_{\eta=0} d\xi\,d\tau.$$

Here,

$$G(r,z,\xi,\eta,t) = G_1(r,\xi,t) G_2(z,\eta,t),$$
$$G_1(r,\xi,t) = \frac{1}{\pi R^2} + \frac{1}{\pi R^2} \sum_{n=1}^\infty \frac{1}{J_0^2(\mu_n)} J_0\!\left(\frac{\mu_n r}{R}\right) J_0\!\left(\frac{\mu_n \xi}{R}\right) \exp\!\left(-\frac{a\mu_n^2 t}{R^2}\right),$$
$$G_2(z,\eta,t) = \frac{1}{2\sqrt{\pi a t}} \left\{\exp\!\left[-\frac{(z-\eta)^2}{4at}\right] - \exp\!\left[-\frac{(z+\eta)^2}{4at}\right]\right\},$$

where the μ_n are positive zeros of the first-order Bessel function, $J_1(\mu_n) = 0$.

3°. A semiinfinite circular cylinder is considered. The following conditions are prescribed:

$$w = f(r,z) \quad \text{at} \quad t = 0 \quad \text{(initial condition)},$$
$$\partial_r w + kw = g_1(z,t) \quad \text{at} \quad r = R \quad \text{(boundary condition)},$$
$$w = g_2(r,t) \quad \text{at} \quad z = 0 \quad \text{(boundary condition)}.$$

The solution $w(r, z, t)$ is determined by the formula in Paragraph 2.1.3-5, Item 2° where

$$G(r, z, \xi, \eta, t) = G_1(r, \xi, t) G_2(z, \eta, t),$$

$$G_1(r, \xi, t) = \frac{1}{\pi R^2} \sum_{n=1}^{\infty} \frac{\mu_n^2}{(k^2 R^2 + \mu_n^2) J_0^2(\mu_n)} J_0\left(\frac{\mu_n r}{R}\right) J_0\left(\frac{\mu_n \xi}{R}\right) \exp\left(-\frac{a\mu_n^2 t}{R^2}\right),$$

$$G_2(z, \eta, t) = \frac{1}{2\sqrt{\pi a t}} \left\{ \exp\left[-\frac{(z-\eta)^2}{4at}\right] - \exp\left[-\frac{(z+\eta)^2}{4at}\right] \right\},$$

where the μ_n are positive roots of the transcendental equation

$$\mu J_1(\mu) - kR J_0(\mu) = 0.$$

Example 4. The initial temperature is the same at every point of the cylinder, $f(r,z) = w_0$. Heat exchange of the cylinder with the zero temperature environment occurs at the lateral surface, $g_1(z,t) = 0$. The end face is maintained at zero temperature, $g_2(r,t) = 0$.
Solution:

$$w(r, z, t) = \frac{2w_0 k}{R} \operatorname{erf}\left(\frac{z}{2\sqrt{at}}\right) \sum_{n=1}^{\infty} \frac{J_0(\lambda_n r)}{(k^2 + \lambda_n^2) J_0(\lambda_n R)} \exp(-\lambda_n^2 a t), \qquad \lambda_n = \frac{\mu_n}{R}.$$

Example 5. The initial temperature of the cylinder is everywhere zero, $f(r,z) = 0$. Heat exchange of the cylinder with the zero temperature environment occurs at the lateral surface, $g_1(z,t) = 0$. The end face is maintained at a constant temperature, $g_2(r,t) = w_0$.
Solution:

$$w(r,z,t) = \frac{kw_0}{R} \sum_{n=1}^{\infty} \frac{J_0(\lambda_n r)}{(\lambda_n^2 + k^2) J_0(\lambda_n R)} \bigg[2 \exp(-\lambda_n z) $$
$$+ \exp(\lambda_n z) \operatorname{erfc}\left(\lambda_n \sqrt{at} + \frac{z}{2\sqrt{at}}\right) - \exp(-\lambda_n z) \operatorname{erfc}\left(\lambda_n \sqrt{at} - \frac{z}{2\sqrt{at}}\right) \bigg], \qquad \lambda_n = \frac{\mu_n}{R}.$$

⊙ *Reference*: H. S. Carslaw and J. C. Jaeger (1984).

2.1.3-6. Domain: $0 \leq r \leq R$, $0 \leq z \leq l$. First boundary value problem.

A circular cylinder of finite length is considered. The following conditions are prescribed:

$$w = f(r,z) \quad \text{at} \quad t = 0 \quad \text{(initial condition)},$$
$$w = g_1(z,t) \quad \text{at} \quad r = R \quad \text{(boundary condition)},$$
$$w = g_2(r,t) \quad \text{at} \quad z = 0 \quad \text{(boundary condition)},$$
$$w = g_3(r,t) \quad \text{at} \quad z = l \quad \text{(boundary condition)}.$$

Solution:

$$w(r, z, t) = 2\pi \int_0^l \int_0^R \xi f(\xi, \eta) G(r, z, \xi, \eta, t) \, d\xi \, d\eta$$
$$- 2\pi a R \int_0^t \int_0^l g_1(\eta, \tau) \left[\frac{\partial}{\partial \xi} G(r, z, \xi, \eta, t - \tau) \right]_{\xi = R} d\eta \, d\tau$$
$$+ 2\pi a \int_0^t \int_0^R \xi g_2(\xi, \tau) \left[\frac{\partial}{\partial \eta} G(r, z, \xi, \eta, t - \tau) \right]_{\eta = 0} d\xi \, d\tau$$
$$- 2\pi a \int_0^t \int_0^R \xi g_3(\xi, \tau) \left[\frac{\partial}{\partial \eta} G(r, z, \xi, \eta, t - \tau) \right]_{\eta = l} d\xi \, d\tau.$$

Here,

$$G(r, z, \xi, \eta, t) = G_1(r, \xi, t) G_2(z, \eta, t),$$

$$G_1(r, \xi, t) = \frac{1}{\pi R^2} \sum_{n=1}^{\infty} \frac{1}{J_1^2(\mu_n)} J_0\left(\frac{\mu_n r}{R}\right) J_0\left(\frac{\mu_n \xi}{R}\right) \exp\left(-\frac{a\mu_n^2 t}{R^2}\right),$$

$$G_2(z, \eta, t) = \frac{2}{l} \sum_{n=1}^{\infty} \sin\left(\frac{n\pi z}{l}\right) \sin\left(\frac{n\pi \eta}{l}\right) \exp\left(-\frac{an^2\pi^2 t}{l^2}\right),$$

where the μ_n are positive zeros of the Bessel function, $J_0(\mu_n) = 0$.

Example 6. The initial temperature is the same at every point of the cylinder, $f(r,z) = w_0$. The lateral surface and the end faces are maintained at zero temperature, $g_1(z,t) = g_2(r,t) = g_3(r,t) = 0$.

Solution:

$$w = \frac{8w_0}{\pi} \left\{ \sum_{n=0}^{\infty} \frac{1}{2n+1} \sin\left[\frac{(2n+1)\pi z}{l}\right] \exp\left[-\frac{a(2n+1)^2\pi^2 t}{l^2}\right] \right\} \left\{ \sum_{n=1}^{\infty} \frac{1}{\mu_n J_1(\mu_n)} J_0\left(\frac{\mu_n r}{R}\right) \exp\left(-\frac{\mu_n^2 a t}{R^2}\right) \right\},$$

where μ_n are positive zeros of the Bessel function, $J_0(\mu_n) = 0$.

⊙ *Reference*: H. S. Carslaw and J. C. Jaeger (1984).

2.1.3-7. Domain: $0 \leq r \leq R$, $0 \leq z \leq l$. Second boundary value problem.

A circular cylinder of finite length is considered. The following conditions are prescribed:

$$\begin{aligned}
w &= f(r, z) & \text{at} \quad t &= 0 & \text{(initial condition)},\\
\partial_r w &= g_1(z, t) & \text{at} \quad r &= R & \text{(boundary condition)},\\
\partial_z w &= g_2(r, t) & \text{at} \quad z &= 0 & \text{(boundary condition)},\\
\partial_z w &= g_3(r, t) & \text{at} \quad z &= l & \text{(boundary condition)}.
\end{aligned}$$

Solution:

$$w(r, z, t) = 2\pi \int_0^l \int_0^R \xi f(\xi, \eta) G(r, z, \xi, \eta, t) \, d\xi \, d\eta$$
$$+ 2\pi a R \int_0^t \int_0^l g_1(\eta, \tau) G(r, z, R, \eta, t - \tau) \, d\eta \, d\tau$$
$$- 2\pi a \int_0^t \int_0^R \xi g_2(\xi, \tau) G(r, z, \xi, 0, t - \tau) \, d\xi \, d\tau$$
$$+ 2\pi a \int_0^t \int_0^R \xi g_3(\xi, \tau) G(r, z, \xi, l, t - \tau) \, d\xi \, d\tau.$$

Here,

$$G(r, z, \xi, \eta, t) = G_1(r, \xi, t) G_2(z, \eta, t),$$

$$G_1(r, \xi, t) = \frac{1}{\pi R^2} + \frac{1}{\pi R^2} \sum_{n=1}^{\infty} \frac{1}{J_0^2(\mu_n)} J_0\left(\frac{\mu_n r}{R}\right) J_0\left(\frac{\mu_n \xi}{R}\right) \exp\left(-\frac{a\mu_n^2 t}{R^2}\right),$$

$$G_2(z, \eta, t) = \frac{1}{l} + \frac{2}{l} \sum_{n=1}^{\infty} \cos\left(\frac{n\pi z}{l}\right) \cos\left(\frac{n\pi \eta}{l}\right) \exp\left(-\frac{an^2\pi^2 t}{l^2}\right),$$

where the μ_n are positive zeros of the first-order Bessel function, $J_1(\mu_n) = 0$.

2.1.3-8. Domain: $0 \le r \le R$, $0 \le z \le l$. Third boundary value problem.

A circular cylinder of finite length is considered. The following conditions are prescribed:

$$w = f(r, z) \quad \text{at} \quad t = 0 \quad \text{(initial condition)},$$
$$\partial_r w + k_1 w = g_1(z, t) \quad \text{at} \quad r = R \quad \text{(boundary condition)},$$
$$\partial_z w - k_2 w = g_2(r, t) \quad \text{at} \quad z = 0 \quad \text{(boundary condition)},$$
$$\partial_z w + k_3 w = g_3(r, t) \quad \text{at} \quad z = l \quad \text{(boundary condition)}.$$

The solution $w(r, z, t)$ is determined by the formula in Paragraph 2.1.3-7 where

$$G(r, z, \xi, \eta, t) = G_1(r, \xi, t) G_2(z, \eta, t),$$

$$G_1(r, \xi, t) = \frac{1}{\pi R^2} \sum_{n=1}^{\infty} \frac{\mu_n^2}{(k_1^2 R^2 + \mu_n^2) J_0^2(\mu_n)} J_0\left(\frac{\mu_n r}{R}\right) J_0\left(\frac{\mu_n \xi}{R}\right) \exp\left(-\frac{a\mu_n^2 t}{R^2}\right),$$

$$G_2(z, \eta, t) = \sum_{m=1}^{\infty} \frac{\varphi_m(z)\varphi_m(\eta)}{\|\varphi_m\|^2} \exp(-a\lambda_m^2 t), \quad \varphi_m(z) = \cos(\lambda_m z) + \frac{k_2}{\lambda_m} \sin(\lambda_m z),$$

$$\|\varphi_m\|^2 = \frac{k_3}{2\lambda_m^2} \frac{\lambda_m^2 + k_2^2}{\lambda_m^2 + k_3^2} + \frac{k_2}{2\lambda_m^2} + \frac{l}{2}\left(1 + \frac{k_2^2}{\lambda_m^2}\right),$$

and the μ_n and λ_m are positive roots of the transcendental equations

$$\mu J_1(\mu) - k_1 R J_0(\mu) = 0, \qquad \frac{\tan(\lambda l)}{\lambda} = \frac{k_2 + k_3}{\lambda^2 - k_2 k_3}.$$

2.1.3-9. Domain: $0 \le r \le R$, $0 \le z \le l$. Mixed boundary value problems.

1°. A circular cylinder of finite length is considered. The following conditions are prescribed:

$$w = f(r, z) \quad \text{at} \quad t = 0 \quad \text{(initial condition)},$$
$$w = g_1(z, t) \quad \text{at} \quad r = R \quad \text{(boundary condition)},$$
$$\partial_z w = g_2(r, t) \quad \text{at} \quad z = 0 \quad \text{(boundary condition)},$$
$$\partial_z w = g_3(r, t) \quad \text{at} \quad z = l \quad \text{(boundary condition)}.$$

Solution:

$$w(r, z, t) = 2\pi \int_0^l \int_0^R \xi f(\xi, \eta) G(r, z, \xi, \eta, t) \, d\xi \, d\eta$$
$$- 2\pi a R \int_0^t \int_0^l g_1(\eta, \tau) \left[\frac{\partial}{\partial \zeta} G(r, z, \xi, \eta, t - \tau)\right]_{\xi = R} d\eta \, d\tau$$
$$- 2\pi a \int_0^t \int_0^R \xi g_2(\xi, \tau) G(r, z, \xi, 0, t - \tau) \, d\xi \, d\tau$$
$$+ 2\pi a \int_0^t \int_0^R \xi g_3(\xi, \tau) G(r, z, \xi, l, t - \tau) \, d\xi \, d\tau.$$

Here,

$$G(r, z, \xi, \eta, t) = G_1(r, \xi, t) G_2(z, \eta, t),$$

$$G_1(r, \xi, t) = \frac{1}{\pi R^2} \sum_{n=1}^{\infty} \frac{1}{J_1^2(\mu_n)} J_0\left(\frac{\mu_n r}{R}\right) J_0\left(\frac{\mu_n \xi}{R}\right) \exp\left(-\frac{a\mu_n^2 t}{R^2}\right),$$

$$G_2(z, \eta, t) = \frac{1}{l} + \frac{2}{l} \sum_{n=1}^{\infty} \cos\left(\frac{n\pi z}{l}\right) \cos\left(\frac{n\pi \eta}{l}\right) \exp\left(-\frac{an^2\pi^2 t}{l^2}\right),$$

where the μ_n are positive zeros of the Bessel function, $J_0(\mu_n) = 0$.

2°. A circular cylinder of finite length is considered. The following conditions are prescribed:

$$w = f(r, z) \quad \text{at} \quad t = 0 \quad \text{(initial condition)},$$
$$\partial_r w = g_1(z, t) \quad \text{at} \quad r = R \quad \text{(boundary condition)},$$
$$w = g_2(r, t) \quad \text{at} \quad z = 0 \quad \text{(boundary condition)},$$
$$w = g_3(r, t) \quad \text{at} \quad z = l \quad \text{(boundary condition)}.$$

Solution:

$$w(r, z, t) = 2\pi \int_0^l \int_0^R \xi f(\xi, \eta) G(r, z, \xi, \eta, t) \, d\xi \, d\eta$$
$$+ 2\pi a R \int_0^t \int_0^l g_1(\eta, \tau) G(r, z, R, \eta, t - \tau) \, d\eta \, d\tau$$
$$+ 2\pi a \int_0^t \int_0^R \xi g_2(\xi, \tau) \left[\frac{\partial}{\partial \eta} G(r, z, \xi, \eta, t - \tau) \right]_{\eta=0} d\xi \, d\tau$$
$$- 2\pi a \int_0^t \int_0^R \xi g_3(\xi, \tau) \left[\frac{\partial}{\partial \eta} G(r, z, \xi, \eta, t - \tau) \right]_{\eta=l} d\xi \, d\tau.$$

Here,

$$G(r, z, \xi, \eta, t) = G_1(r, \xi, t) G_2(z, \eta, t),$$
$$G_1(r, \xi, t) = \frac{1}{\pi R^2} + \frac{1}{\pi R^2} \sum_{n=1}^{\infty} \frac{1}{J_0^2(\mu_n)} J_0\left(\frac{\mu_n r}{R}\right) J_0\left(\frac{\mu_n \xi}{R}\right) \exp\left(-\frac{a \mu_n^2 t}{R^2}\right),$$
$$G_2(z, \eta, t) = \frac{2}{l} \sum_{n=1}^{\infty} \sin\left(\frac{n \pi z}{l}\right) \sin\left(\frac{n \pi \eta}{l}\right) \exp\left(-\frac{a n^2 \pi^2 t}{l^2}\right),$$

where the μ_n are positive zeros of the first-order Bessel function, $J_1(\mu_n) = 0$.

2.1.3-10. Domain: $R_1 \leq r \leq R_2$, $0 \leq z \leq l$. First boundary value problem.

A hollow circular cylinder of finite length is considered. The following conditions are prescribed:

$$w = f(r, z) \quad \text{at} \quad t = 0 \quad \text{(initial condition)},$$
$$w = g_1(z, t) \quad \text{at} \quad r = R_1 \quad \text{(boundary condition)},$$
$$w = g_2(z, t) \quad \text{at} \quad r = R_2 \quad \text{(boundary condition)},$$
$$w = g_3(r, t) \quad \text{at} \quad z = 0 \quad \text{(boundary condition)},$$
$$w = g_4(r, t) \quad \text{at} \quad z = l \quad \text{(boundary condition)}.$$

Solution:

$$w(r, z, t) = 2\pi \int_0^l \int_{R_1}^{R_2} \xi f(\xi, \eta) G(r, z, \xi, \eta, t) \, d\xi \, d\eta$$
$$+ 2\pi a R_1 \int_0^t \int_0^l g_1(\eta, \tau) \left[\frac{\partial}{\partial \xi} G(r, z, \xi, \eta, t - \tau) \right]_{\xi=R_1} d\eta \, d\tau$$
$$- 2\pi a R_2 \int_0^t \int_0^l g_2(\eta, \tau) \left[\frac{\partial}{\partial \xi} G(r, z, \xi, \eta, t - \tau) \right]_{\xi=R_2} d\eta \, d\tau$$
$$+ 2\pi a \int_0^t \int_{R_1}^{R_2} \xi g_3(\xi, \tau) \left[\frac{\partial}{\partial \eta} G(r, z, \xi, \eta, t - \tau) \right]_{\eta=0} d\xi \, d\tau$$
$$- 2\pi a \int_0^t \int_{R_1}^{R_2} \xi g_4(\xi, \tau) \left[\frac{\partial}{\partial \eta} G(r, z, \xi, \eta, t - \tau) \right]_{\eta=l} d\xi \, d\tau.$$

Here,
$$G(r, z, \xi, \eta, t) = G_1(z, \eta, t)G_2(r, \xi, t),$$
$$G_1(z, \eta, t) = \frac{2}{l} \sum_{n=1}^{\infty} \sin\left(\frac{n\pi z}{l}\right) \sin\left(\frac{n\pi \eta}{l}\right) \exp\left(-\frac{an^2\pi^2 t}{l^2}\right),$$
$$G_2(r, \xi, t) = \frac{\pi}{4R_1^2} \sum_{n=1}^{\infty} \frac{\mu_n^2 J_0^2(s\mu_n)}{J_0^2(\mu_n) - J_0^2(s\mu_n)} \Psi_n(r)\Psi_n(\xi) \exp\left(-\frac{a\mu_n^2 t}{R_1^2}\right),$$
$$\Psi_n(r) = Y_0(\mu_n) J_0\left(\frac{\mu_n r}{R_1}\right) - J_0(\mu_n) Y_0\left(\frac{\mu_n r}{R_1}\right), \qquad s = \frac{R_2}{R_1},$$

where $J_0(\mu)$ and $Y_0(\mu)$ are the Bessel functions, the μ_n are positive roots of the transcendental equation
$$J_0(\mu)Y_0(s\mu) - J_0(s\mu)Y_0(\mu) = 0.$$

2.1.3-11. Domain: $R_1 \leq r \leq R_2$, $0 \leq z \leq l$. Second boundary value problem.

A hollow circular cylinder of finite length is considered. The following conditions are prescribed:

$$w = f(r, z) \quad \text{at} \quad t = 0 \quad \text{(initial condition)},$$
$$\partial_r w = g_1(z, t) \quad \text{at} \quad r = R_1 \quad \text{(boundary condition)},$$
$$\partial_r w = g_2(z, t) \quad \text{at} \quad r = R_2 \quad \text{(boundary condition)},$$
$$\partial_z w = g_3(r, t) \quad \text{at} \quad z = 0 \quad \text{(boundary condition)},$$
$$\partial_z w = g_4(r, t) \quad \text{at} \quad z = l \quad \text{(boundary condition)}.$$

Solution:
$$w(r, z, t) = 2\pi \int_0^l \int_{R_1}^{R_2} \xi f(\xi, \eta) G(r, z, \xi, \eta, t) \, d\xi \, d\eta$$
$$- 2\pi a R_1 \int_0^t \int_0^l g_1(\eta, \tau) G(r, z, R_1, \eta, t - \tau) \, d\eta \, d\tau$$
$$+ 2\pi a R_2 \int_0^t \int_0^l g_2(\eta, \tau) G(r, z, R_2, \eta, t - \tau) \, d\eta \, d\tau$$
$$- 2\pi a \int_0^t \int_{R_1}^{R_2} \xi g_3(\xi, \tau) G(r, z, \xi, 0, t - \tau) \, d\xi \, d\tau$$
$$+ 2\pi a \int_0^t \int_{R_1}^{R_2} \xi g_4(\xi, \tau) G(r, z, \xi, l, t - \tau) \, d\xi \, d\tau.$$

Here,
$$G(r, z, \xi, \eta, t) = G_1(z, \eta, t)G_2(r, \xi, t),$$
$$G_1(z, \eta, t) = \frac{1}{l} + \frac{2}{l} \sum_{n=1}^{\infty} \cos\left(\frac{n\pi z}{l}\right) \cos\left(\frac{n\pi \eta}{l}\right) \exp\left(-\frac{an^2\pi^2 t}{l^2}\right),$$
$$G_2(r, \xi, t) = \frac{1}{\pi(R_2^2 - R_1^2)} + \frac{\pi}{4R_1^2} \sum_{n=1}^{\infty} \frac{\mu_n^2 J_1^2(s\mu_n)}{J_1^2(\mu_n) - J_1^2(s\mu_n)} \Psi_n(r)\Psi_n(\xi) \exp\left(-\frac{a\mu_n^2 t}{R_1^2}\right),$$
$$\Psi_n(r) = Y_1(\mu_n) J_0\left(\frac{\mu_n r}{R_1}\right) - J_1(\mu_n) Y_0\left(\frac{\mu_n r}{R_1}\right), \qquad s = \frac{R_2}{R_1},$$

where $J_k(\mu)$ and $Y_k(\mu)$ are the Bessel functions of order $k = 0, 1$ and the μ_n are positive roots of the transcendental equation
$$J_1(\mu)Y_1(s\mu) - J_1(s\mu)Y_1(\mu) = 0.$$

2.1.3-12. Domain: $R_1 \leq r \leq R_2$, $0 \leq z \leq l$. Third boundary value problem.

A hollow circular cylinder of finite length is considered. The following conditions are prescribed:

$$w = f(r, z) \quad \text{at} \quad t = 0 \quad \text{(initial condition)},$$
$$\partial_r w - k_1 w = g_1(z, t) \quad \text{at} \quad r = R_1 \quad \text{(boundary condition)},$$
$$\partial_r w + k_2 w = g_2(z, t) \quad \text{at} \quad r = R_2 \quad \text{(boundary condition)},$$
$$\partial_z w - k_3 w = g_3(r, t) \quad \text{at} \quad z = 0 \quad \text{(boundary condition)},$$
$$\partial_z w + k_4 w = g_4(r, t) \quad \text{at} \quad z = l \quad \text{(boundary condition)}.$$

For the solution of this problem, see Subsection 2.2.3-4 with $\Phi \equiv 0$.

2.2. Heat Equation with a Source $\dfrac{\partial w}{\partial t} = a \Delta_2 w + \Phi(x, y, t)$

2.2.1. Problems in Cartesian Coordinates

In the rectangular Cartesian coordinate system, the heat equation has the form

$$\frac{\partial w}{\partial t} = a \left(\frac{\partial^2 w}{\partial x^2} + \frac{\partial^2 w}{\partial y^2} \right) + \Phi(x, y, t).$$

It governs two-dimensional unsteady thermal processes in quiescent media or solids with constant thermal diffusivity in the cases where there are volume thermal sources or sinks.

2.2.1-1. Domain: $-\infty < x < \infty$, $-\infty < y < \infty$. Cauchy problem.

An initial condition is prescribed:

$$w = f(x, y) \quad \text{at} \quad t = 0.$$

Solution:

$$w(x, y, t) = \int_{-\infty}^{\infty} \int_{-\infty}^{\infty} f(\xi, \eta) G(x, y, \xi, \eta, t) \, d\xi \, d\eta$$
$$+ \int_0^t \int_{-\infty}^{\infty} \int_{-\infty}^{\infty} \Phi(\xi, \eta, \tau) G(x, y, \xi, \eta, t - \tau) \, d\xi \, d\eta \, d\tau,$$

where

$$G(x, y, \xi, \eta, t) = \frac{1}{4\pi a t} \exp\left[-\frac{(x-\xi)^2 + (y-\eta)^2}{4at} \right].$$

⊙ *Reference*: A. G. Butkovskiy (1979).

2.2.1-2. Domain: $0 \leq x < \infty$, $-\infty < y < \infty$. First boundary value problem.

A half-plane is considered. The following conditions are prescribed:

$$w = f(x, y) \quad \text{at} \quad t = 0 \quad \text{(initial condition)},$$
$$w = g(y, t) \quad \text{at} \quad x = 0 \quad \text{(boundary condition)}.$$

Solution:

$$w(x, y, t) = \int_0^{\infty} \int_{-\infty}^{\infty} f(\xi, \eta) G(x, y, \xi, \eta, t) \, d\eta \, d\xi$$
$$+ a \int_0^t \int_{-\infty}^{\infty} g(\eta, \tau) \left[\frac{\partial}{\partial \xi} G(x, y, \xi, \eta, t - \tau) \right]_{\xi=0} d\eta \, d\tau$$
$$+ \int_0^t \int_0^{\infty} \int_{-\infty}^{\infty} \Phi(\xi, \eta, \tau) G(x, y, \xi, \eta, t - \tau) \, d\eta \, d\xi \, d\tau,$$

where
$$G(x,y,\xi,\eta,t) = \frac{1}{4\pi at}\left\{\exp\left[-\frac{(x-\xi)^2+(y-\eta)^2}{4at}\right] - \exp\left[-\frac{(x+\xi)^2+(y-\eta)^2}{4at}\right]\right\}.$$

⊙ *References*: A. G. Butkovskiy (1979), H. S. Carslaw and J. C. Jaeger (1984).

2.2.1-3. Domain: $0 \leq x < \infty$, $-\infty < y < \infty$. Second boundary value problem.

A half-plane is considered. The following conditions are prescribed:

$$w = f(x,y) \quad \text{at} \quad t = 0 \quad \text{(initial condition)},$$
$$\partial_x w = g(y,t) \quad \text{at} \quad x = 0 \quad \text{(boundary condition)}.$$

Solution:
$$w(x,y,t) = \int_0^\infty \int_{-\infty}^\infty f(\xi,\eta)G(x,y,\xi,\eta,t)\,d\eta\,d\xi - a\int_0^t\int_{-\infty}^\infty g(\eta,\tau)G(x,y,0,\eta,t-\tau)\,d\eta\,d\tau$$
$$+ \int_0^t\int_0^\infty\int_{-\infty}^\infty \Phi(\xi,\eta,\tau)G(x,y,\xi,\eta,t-\tau)\,d\eta\,d\xi\,d\tau,$$

where
$$G(x,y,\xi,\eta,t) = \frac{1}{4\pi at}\left\{\exp\left[-\frac{(x-\xi)^2+(y-\eta)^2}{4at}\right] + \exp\left[-\frac{(x+\xi)^2+(y-\eta)^2}{4at}\right]\right\}.$$

2.2.1-4. Domain: $0 \leq x < \infty$, $-\infty < y < \infty$. Third boundary value problem.

A half-plane is considered. The following conditions are prescribed:

$$w = f(x,y) \quad \text{at} \quad t = 0 \quad \text{(initial condition)},$$
$$\partial_x w - kw = g(y,t) \quad \text{at} \quad x = 0 \quad \text{(boundary condition)}.$$

The solution $w(x,y,t)$ is determined by the formula in Paragraph 2.2.1-3 where

$$G(x,y,\xi,\eta,t) = \frac{1}{4\pi at}\exp\left[-\frac{(y-\eta)^2}{4at}\right]\left\{\exp\left[-\frac{(x-\xi)^2}{4at}\right] + \exp\left[-\frac{(x+\xi)^2}{4at}\right]\right.$$
$$\left. - 2k\int_0^\infty \exp\left[-\frac{(x+\xi+s)^2}{4at} - ks\right]ds\right\}.$$

2.2.1-5. Domain: $0 \leq x < \infty$, $0 \leq y < \infty$. First boundary value problem.

A quadrant of the plane is considered. The following conditions are prescribed:

$$w = f(x,y) \quad \text{at} \quad t = 0 \quad \text{(initial condition)},$$
$$w = g_1(y,t) \quad \text{at} \quad x = 0 \quad \text{(boundary condition)},$$
$$w = g_2(x,t) \quad \text{at} \quad y = 0 \quad \text{(boundary condition)}.$$

Solution:
$$w(x,y,t) = \int_0^\infty \int_0^\infty f(\xi,\eta)G(x,y,\xi,\eta,t)\,d\xi\,d\eta$$
$$+ a\int_0^t\int_0^\infty g_1(\eta,\tau)\left[\frac{\partial}{\partial\xi}G(x,y,\xi,\eta,t-\tau)\right]_{\xi=0}d\eta\,d\tau$$
$$+ a\int_0^t\int_0^\infty g_2(\xi,\tau)\left[\frac{\partial}{\partial\eta}G(x,y,\xi,\eta,t-\tau)\right]_{\eta=0}d\xi\,d\tau$$
$$+ \int_0^t\int_0^\infty\int_0^\infty \Phi(\xi,\eta,\tau)G(x,y,\xi,\eta,t-\tau)\,d\xi\,d\eta\,d\tau,$$

where

$$G(x,y,\xi,\eta,t) = \frac{1}{4\pi at}\left\{\exp\left[-\frac{(x-\xi)^2}{4at}\right] - \exp\left[-\frac{(x+\xi)^2}{4at}\right]\right\}\left\{\exp\left[-\frac{(y-\eta)^2}{4at}\right] - \exp\left[-\frac{(y+\eta)^2}{4at}\right]\right\}.$$

⊙ *Reference*: H. S. Carslaw and J. C. Jaeger (1984).

2.2.1-6. Domain: $0 \le x < \infty$, $0 \le y < \infty$. Second boundary value problem.

A quadrant of the plane is considered. The following conditions are prescribed:

$$w = f(x,y) \quad \text{at} \quad t = 0 \quad \text{(initial condition)},$$
$$\partial_x w = g_1(y,t) \quad \text{at} \quad x = 0 \quad \text{(boundary condition)},$$
$$\partial_y w = g_2(x,t) \quad \text{at} \quad y = 0 \quad \text{(boundary condition)}.$$

Solution:

$$w(x,y,t) = \int_0^\infty \int_0^\infty f(\xi,\eta) G(x,y,\xi,\eta,t)\, d\xi\, d\eta$$
$$- a \int_0^t \int_0^\infty g_1(\eta,\tau) G(x,y,0,\eta,t-\tau)\, d\eta\, d\tau$$
$$- a \int_0^t \int_0^\infty g_2(\xi,\tau) G(x,y,\xi,0,t-\tau)\, d\xi\, d\tau$$
$$+ \int_0^t \int_0^\infty \int_0^\infty \Phi(\xi,\eta,\tau) G(x,y,\xi,\eta,t-\tau)\, d\xi\, d\eta\, d\tau,$$

where

$$G(x,y,\xi,\eta,t) = \frac{1}{4\pi at}\left\{\exp\left[-\frac{(x-\xi)^2}{4at}\right] + \exp\left[-\frac{(x+\xi)^2}{4at}\right]\right\}\left\{\exp\left[-\frac{(y-\eta)^2}{4at}\right] + \exp\left[-\frac{(y+\eta)^2}{4at}\right]\right\}.$$

2.2.1-7. Domain: $0 \le x < \infty$, $0 \le y < \infty$. Third boundary value problem.

A quadrant of the plane is considered. The following conditions are prescribed:

$$w = f(x,y) \quad \text{at} \quad t = 0 \quad \text{(initial condition)},$$
$$\partial_x w - k_1 w = g_1(y,t) \quad \text{at} \quad x = 0 \quad \text{(boundary condition)},$$
$$\partial_y w - k_2 w = g_2(x,t) \quad \text{at} \quad y = 0 \quad \text{(boundary condition)}.$$

The solution $w(x,y,t)$ is determined by the formula in Paragraph 2.2.1-6 where

$$G(x,y,\xi,\eta,t) = \frac{1}{4\pi at}\left\{\exp\left[-\frac{(x-\xi)^2}{4at}\right] + \exp\left[-\frac{(x+\xi)^2}{4at}\right]\right.$$
$$\left. - 2k_1\sqrt{\pi at}\, \exp\left[ak_1^2 t + k_1(x+\xi)\right] \operatorname{erfc}\left(\frac{x+\xi}{2\sqrt{at}} + k_1\sqrt{at}\right)\right\}$$
$$\times \left\{\exp\left[-\frac{(y-\eta)^2}{4at}\right] + \exp\left[-\frac{(y+\eta)^2}{4at}\right]\right.$$
$$\left. - 2k_2\sqrt{\pi at}\, \exp\left[ak_2^2 t + k_2(y+\eta)\right] \operatorname{erfc}\left(\frac{y+\eta}{2\sqrt{at}} + k_2\sqrt{at}\right)\right\}.$$

2.2.1-8. Domain: $0 \leq x < \infty$, $0 \leq y < \infty$. Mixed boundary value problem.

A quadrant of the plane is considered. The following conditions are prescribed:

$$w = f(x, y) \quad \text{at} \quad t = 0 \quad \text{(initial condition)},$$
$$w = g_1(y, t) \quad \text{at} \quad x = 0 \quad \text{(boundary condition)},$$
$$\partial_y w = g_2(x, t) \quad \text{at} \quad y = 0 \quad \text{(boundary condition)}.$$

Solution:

$$w(x, y, t) = \int_0^\infty \int_0^\infty f(\xi, \eta) G(x, y, \xi, \eta, t) \, d\xi \, d\eta$$
$$+ a \int_0^t \int_0^\infty g_1(\eta, \tau) \left[\frac{\partial}{\partial \xi} G(x, y, \xi, \eta, t-\tau) \right]_{\xi=0} d\eta \, d\tau$$
$$- a \int_0^t \int_0^\infty g_2(\xi, \tau) G(x, y, \xi, 0, t-\tau) \, d\xi \, d\tau$$
$$+ \int_0^t \int_0^\infty \int_0^\infty \Phi(\xi, \eta, \tau) G(x, y, \xi, \eta, t-\tau) \, d\xi \, d\eta \, d\tau,$$

where

$$G(x, y, \xi, \eta, t) = \frac{1}{4\pi a t} \left\{ \exp\left[-\frac{(x-\xi)^2}{4at}\right] - \exp\left[-\frac{(x+\xi)^2}{4at}\right] \right\} \left\{ \exp\left[-\frac{(y-\eta)^2}{4at}\right] + \exp\left[-\frac{(y+\eta)^2}{4at}\right] \right\}.$$

2.2.1-9. Domain: $0 \leq x \leq l$, $0 \leq y < \infty$. First boundary value problem.

A semiinfinite strip is considered. The following conditions are prescribed:

$$w = f(x, y) \quad \text{at} \quad t = 0 \quad \text{(initial condition)},$$
$$w = g_1(y, t) \quad \text{at} \quad x = 0 \quad \text{(boundary condition)},$$
$$w = g_2(y, t) \quad \text{at} \quad x = l \quad \text{(boundary condition)},$$
$$w = g_3(x, t) \quad \text{at} \quad y = 0 \quad \text{(boundary condition)}.$$

The solution is given by the formula of Subsection 2.1.1-12 with the additional term

$$\int_0^t \int_0^l \int_0^\infty \Phi(\xi, \eta, \tau) G(x, y, \xi, \eta, t-\tau) \, d\eta \, d\xi \, d\tau,$$

which takes into account the equation's nonhomogeneity.

2.2.1-10. Domain: $0 \leq x \leq l$, $0 \leq y < \infty$. Second boundary value problem.

A semiinfinite strip is considered. The following conditions are prescribed:

$$w = f(x, y) \quad \text{at} \quad t = 0 \quad \text{(initial condition)},$$
$$\partial_x w = g_1(y, t) \quad \text{at} \quad x = 0 \quad \text{(boundary condition)},$$
$$\partial_x w = g_2(y, t) \quad \text{at} \quad x = l \quad \text{(boundary condition)},$$
$$\partial_y w = g_3(x, t) \quad \text{at} \quad y = 0 \quad \text{(boundary condition)}.$$

Solution:

$$w(x, y, t) = \int_0^\infty \int_0^l f(\xi, \eta) G(x, y, \xi, \eta, t) \, d\xi \, d\eta + \int_0^t \int_0^\infty \int_0^l \Phi(\xi, \eta, \tau) G(x, y, \xi, \eta, t-\tau) \, d\xi \, d\eta \, d\tau$$
$$- a \int_0^t \int_0^\infty g_1(\eta, \tau) G(x, y, 0, \eta, t-\tau) \, d\eta \, d\tau + a \int_0^t \int_0^\infty g_2(\eta, \tau) G(x, y, l, \eta, t-\tau) \, d\eta \, d\tau$$
$$- a \int_0^t \int_0^l g_3(\xi, \tau) G(x, y, \xi, 0, t-\tau) \, d\xi \, d\tau,$$

where
$$G(x,y,\xi,\eta,t) = G_1(x,\xi,t)G_2(y,\eta,t),$$
$$G_1(x,\xi,t) = \frac{1}{l} + \frac{2}{l}\sum_{n=1}^{\infty} \cos\left(\frac{n\pi x}{l}\right)\cos\left(\frac{n\pi\xi}{l}\right)\exp\left(-\frac{an^2\pi^2 t}{l^2}\right),$$
$$G_2(y,\eta,t) = \frac{1}{2\sqrt{\pi a t}}\left\{\exp\left[-\frac{(y-\eta)^2}{4at}\right] + \exp\left[-\frac{(y+\eta)^2}{4at}\right]\right\}.$$

2.2.1-11. Domain: $0 \le x \le l$, $0 \le y < \infty$. Third boundary value problem.

A semiinfinite strip is considered. The following conditions are prescribed:

$$\begin{aligned}
w &= f(x,y) & \text{at} \quad t &= 0 & &\text{(initial condition)}, \\
\partial_x w - k_1 w &= g_1(y,t) & \text{at} \quad x &= 0 & &\text{(boundary condition)}, \\
\partial_x w + k_2 w &= g_2(y,t) & \text{at} \quad x &= l & &\text{(boundary condition)}, \\
\partial_y w - k_3 w &= g_3(x,t) & \text{at} \quad y &= 0 & &\text{(boundary condition)}.
\end{aligned}$$

The solution $w(x,y,t)$ is determined by the formula in Paragraph 2.2.1-10 where the Green's function $G(x,y,\xi,\eta,t)$ is the product of the Green's function of Subsection 1.1.1-11 and that of Subsection 1.1.1-8; x, ξ, and k in the last Green's function must be replaced by y, η, and k_3, respectively.

2.2.1-12. Domain: $0 \le x \le l$, $0 \le y < \infty$. Mixed boundary value problem.

A semiinfinite strip is considered. The following conditions are prescribed:

$$\begin{aligned}
w &= f(x,y) & \text{at} \quad t &= 0 & &\text{(initial condition)}, \\
w &= g_1(y,t) & \text{at} \quad x &= 0 & &\text{(boundary condition)}, \\
w &= g_2(y,t) & \text{at} \quad x &= l & &\text{(boundary condition)}, \\
\partial_y w &= g_3(x,t) & \text{at} \quad y &= 0 & &\text{(boundary condition)}.
\end{aligned}$$

The solution is given by the formula of Subsection 2.1.1-15 (Item 1°) with the additional term

$$\int_0^t \int_0^l \int_0^\infty \Phi(\xi,\eta,\tau)G(x,y,\xi,\eta,t-\tau)\,d\eta\,d\xi\,d\tau,$$

which takes into account the equation's nonhomogeneity.

2.2.1-13. Domain: $0 \le x \le l_1$, $0 \le y \le l_2$. First boundary value problem.

A rectangle is considered. The following conditions are prescribed:

$$\begin{aligned}
w &= f(x,y) & \text{at} \quad t &= 0 & &\text{(initial condition)}, \\
w &= g_1(y,t) & \text{at} \quad x &= 0 & &\text{(boundary condition)}, \\
w &= g_2(y,t) & \text{at} \quad x &= l_1 & &\text{(boundary condition)}, \\
w &= g_3(x,t) & \text{at} \quad y &= 0 & &\text{(boundary condition)}, \\
w &= g_4(x,t) & \text{at} \quad y &= l_2 & &\text{(boundary condition)}.
\end{aligned}$$

The solution is given by the formula of Subsection 2.1.1-16 with the additional term

$$\int_0^t \int_0^{l_1} \int_0^{l_2} \Phi(\xi,\eta,\tau)G(x,y,\xi,\eta,t-\tau)\,d\eta\,d\xi\,d\tau,$$

which takes into account the equation's nonhomogeneity.

⊙ *Reference*: H. S. Carslaw and J. C. Jaeger (1984).

2.2.1-14. Domain: $0 \leq x \leq l_1$, $0 \leq y \leq l_2$. Second boundary value problem.

A rectangle is considered. The following conditions are prescribed:

$$w = f(x,y) \quad \text{at} \quad t = 0 \quad \text{(initial condition)},$$
$$\partial_x w = g_1(y,t) \quad \text{at} \quad x = 0 \quad \text{(boundary condition)},$$
$$\partial_x w = g_2(y,t) \quad \text{at} \quad x = l_1 \quad \text{(boundary condition)},$$
$$\partial_y w = g_3(x,t) \quad \text{at} \quad y = 0 \quad \text{(boundary condition)},$$
$$\partial_y w = g_4(x,t) \quad \text{at} \quad y = l_2 \quad \text{(boundary condition)}.$$

Solution:

$$w(x,y,t) = \int_0^{l_1}\int_0^{l_2} f(\xi,\eta) G(x,y,\xi,\eta,t)\,d\eta\,d\xi + \int_0^t\int_0^{l_1}\int_0^{l_2} \Phi(\xi,\eta,\tau) G(x,y,\xi,\eta,t-\tau)\,d\eta\,d\xi\,d\tau$$
$$- a\int_0^t\int_0^{l_2} g_1(\eta,\tau) G(x,y,0,\eta,t-\tau)\,d\eta\,d\tau + a\int_0^t\int_0^{l_2} g_2(\eta,\tau) G(x,y,l_1,\eta,t-\tau)\,d\eta\,d\tau$$
$$- a\int_0^t\int_0^{l_1} g_3(\xi,\tau) G(x,y,\xi,0,t-\tau)\,d\xi\,d\tau + a\int_0^t\int_0^{l_1} g_4(\xi,\tau) G(x,y,\xi,l_2,t-\tau)\,d\xi\,d\tau,$$

where

$$G(x,y,\xi,\eta,t) = \frac{1}{l_1 l_2}\left[1 + 2\sum_{n=1}^{\infty} \exp\left(-\frac{\pi^2 n^2 at}{l_1^2}\right) \cos\frac{n\pi x}{l_1} \cos\frac{n\pi\xi}{l_1}\right]$$
$$\times \left[1 + 2\sum_{m=1}^{\infty} \exp\left(-\frac{\pi^2 m^2 at}{l_2^2}\right) \cos\frac{m\pi y}{l_2} \cos\frac{m\pi\eta}{l_2}\right].$$

⊙ *Reference*: H. S. Carslaw and J. C. Jaeger (1984).

2.2.1-15. Domain: $0 \leq x \leq l_1$, $0 \leq y \leq l_2$. Third boundary value problem.

A rectangle is considered. The following conditions are prescribed:

$$w = f(x,y) \quad \text{at} \quad t = 0 \quad \text{(initial condition)},$$
$$\partial_x w - k_1 w = g_1(y,t) \quad \text{at} \quad x = 0 \quad \text{(boundary condition)},$$
$$\partial_x w + k_2 w = g_2(y,t) \quad \text{at} \quad x = l_1 \quad \text{(boundary condition)},$$
$$\partial_y w - k_3 w = g_3(x,t) \quad \text{at} \quad y = 0 \quad \text{(boundary condition)},$$
$$\partial_y w + k_4 w = g_4(x,t) \quad \text{at} \quad y = l_2 \quad \text{(boundary condition)}.$$

The solution $w(x,y,t)$ is determined by the formula in Paragraph 2.2.1-14 where

$$G(x,y,\xi,\eta,t) = \left\{\sum_{n=1}^{\infty} \frac{\varphi_n(x)\varphi_n(\xi)}{\|\varphi_n\|^2} \exp(-a\mu_n^2 t)\right\}\left\{\sum_{m=1}^{\infty} \frac{\psi_m(y)\psi_m(\eta)}{\|\psi_m\|^2} \exp(-a\lambda_m^2 t)\right\},$$

$$\varphi_n(x) = \cos(\mu_n x) + \frac{k_1}{\mu_n}\sin(\mu_n x), \quad \|\varphi_n\|^2 = \frac{k_2}{2\mu_n^2}\frac{\mu_n^2 + k_1^2}{\mu_n^2 + k_2^2} + \frac{k_1}{2\mu_n^2} + \frac{l_1}{2}\left(1 + \frac{k_1^2}{\mu_n^2}\right),$$

$$\psi_m(y) = \cos(\lambda_m y) + \frac{k_3}{\lambda_m}\sin(\lambda_m y), \quad \|\psi_m\|^2 = \frac{k_4}{2\lambda_m^2}\frac{\lambda_m^2 + k_3^2}{\lambda_m^2 + k_4^2} + \frac{k_3}{2\lambda_m^2} + \frac{l_2}{2}\left(1 + \frac{k_3^2}{\lambda_m^2}\right).$$

Here, the μ_n and λ_m are positive roots of the transcendental equations

$$\frac{\tan(\mu l_1)}{\mu} = \frac{k_1 + k_2}{\mu^2 - k_1 k_2}, \quad \frac{\tan(\lambda l_2)}{\lambda} = \frac{k_3 + k_4}{\lambda^2 - k_3 k_4}.$$

2.2.1-16. Domain: $0 \le x \le l_1$, $0 \le y \le l_2$. Mixed boundary value problem.

A rectangle is considered. The following conditions are prescribed:

$$w = f(x, y) \quad \text{at} \quad t = 0 \quad \text{(initial condition)},$$
$$w = g_1(y, t) \quad \text{at} \quad x = 0 \quad \text{(boundary condition)},$$
$$w = g_2(y, t) \quad \text{at} \quad x = l_1 \quad \text{(boundary condition)},$$
$$\partial_y w = g_3(x, t) \quad \text{at} \quad y = 0 \quad \text{(boundary condition)},$$
$$\partial_y w = g_4(x, t) \quad \text{at} \quad y = l_2 \quad \text{(boundary condition)}.$$

The solution is given by the formula of Subsection 2.1.1-19 (Item 1°) with the additional term

$$\int_0^t \int_0^{l_1} \int_0^{l_2} \Phi(\xi, \eta, \tau) G(x, y, \xi, \eta, t - \tau) \, d\eta \, d\xi \, d\tau,$$

which takes into account the equation's nonhomogeneity.

2.2.2. Problems in Polar Coordinates

The heat equation with a volume source in the polar coordinate system r, φ is written as

$$\frac{\partial w}{\partial t} = a \left(\frac{\partial^2 w}{\partial r^2} + \frac{1}{r} \frac{\partial w}{\partial r} + \frac{1}{r^2} \frac{\partial^2 w}{\partial \varphi^2} \right) + \Phi(r, \varphi, t).$$

Solutions of the form $w = w(r, t)$ that are independent of the angular coordinate φ and govern plane thermal processes with central symmetry, are presented in Subsection 1.2.2.

2.2.2-1. Domain: $0 \le r < \infty$, $0 \le \varphi \le 2\pi$. Cauchy problem.

An initial condition is prescribed:

$$w = f(r, \varphi) \quad \text{at} \quad t = 0.$$

Solution:

$$w(r, \varphi, t) = \int_0^{2\pi} \int_0^\infty f(\xi, \eta) G(r, \varphi, \xi, \eta, t) \, \xi \, d\xi \, d\eta$$
$$+ \int_0^t \int_0^{2\pi} \int_0^\infty \Phi(\xi, \eta, \tau) G(r, \varphi, \xi, \eta, t - \tau) \, \xi \, d\xi \, d\eta \, d\tau,$$

where

$$G(r, \varphi, \xi, \eta, t) = \frac{1}{4\pi a t} \exp\left[-\frac{r^2 + \xi^2 - 2r\xi \cos(\varphi - \eta)}{4at} \right].$$

2.2.2-2. Domain: $0 \le r \le R$, $0 \le \varphi \le 2\pi$. Different boundary value problems.

1°. The solution of the first boundary value problem for a circle of radius R is given by the formula from Subsection 2.1.2-2 with the additional term

$$\int_0^t \int_0^{2\pi} \int_0^R \Phi(\xi, \eta, \tau) G(r, \varphi, \xi, \eta, t - \tau) \xi \, d\xi \, d\eta \, d\tau, \tag{1}$$

which allows for the equation's nonhomogeneity.

2°. The solution of the second boundary value problem for a circle is given by the formula in Paragraph 2.1.2-3 with the additional term (1).

3°. The solution of the third boundary value problem for a circle is given by the formula in Paragraph 2.1.2-4 with the additional term (1).

2.2.2-3. Domain: $R_1 \leq r \leq R_2$, $0 \leq \varphi \leq 2\pi$. Different boundary value problems.

1°. The solution of the first boundary value problem for an annular domain is given by the formula in Paragraph 2.1.2-5 with the additional term

$$\int_0^t \int_0^{2\pi} \int_{R_1}^{R_2} \Phi(\xi, \eta, \tau) G(r, \varphi, \xi, \eta, t - \tau) \xi \, d\xi \, d\eta \, d\tau, \tag{2}$$

which allows for the equation's nonhomogeneity.

2°. The solution of the third boundary value problem for an annular domain is given by the formula in Paragraph 2.1.2-7 with the additional term (2).

2.2.2-4. Domain: $0 \leq r < \infty$, $0 \leq \varphi \leq \varphi_0$. Different boundary value problems.

1°. The solution of the first boundary value problem for a wedge domain is given by the formula in Paragraph 2.1.2-8 with the additional term

$$\int_0^t \int_0^{\varphi_0} \int_0^{\infty} \Phi(\xi, \eta, \tau) G(r, \varphi, \xi, \eta, t - \tau) \xi \, d\xi \, d\eta \, d\tau, \tag{3}$$

which allows for the equation's nonhomogeneity.

2°. The solution of the second boundary value problem for a wedge domain is given by the formula in Paragraph 2.1.2-9 with the additional term (3).

2.2.2-5. Domain: $0 \leq r \leq R$, $0 \leq \varphi \leq \varphi_0$. Different boundary value problems.

1°. The solution of the first boundary value problem for a sector of a circle is given by the formula of Paragraph 2.1.2-10 with the additional term

$$\int_0^t \int_0^{\varphi_0} \int_0^R \Phi(\xi, \eta, \tau) G(r, \varphi, \xi, \eta, t - \tau) \xi \, d\xi \, d\eta \, d\tau, \tag{4}$$

which allows for the equation's nonhomogeneity.

2°. The solution of the mixed boundary value problem for a sector of a circle is given by the formula of Paragraph 2.1.2-11 with the additional term (4).

2.2.3. Axisymmetric Problems

In the case of axial symmetry, the heat equation in the cylindrical coordinate system is written as

$$\frac{\partial w}{\partial t} = a \left(\frac{\partial^2 w}{\partial r^2} + \frac{1}{r} \frac{\partial w}{\partial r} + \frac{\partial^2 w}{\partial z^2} \right) + \Phi(r, z, t),$$

provided there are heat sources or sinks.

One-dimensional axisymmetric problems that have solutions of the form $w = w(r, t)$ can be found in Subsection 1.2.2.

2.2.3-1. Domain: $0 \leq r \leq R$, $0 \leq z < \infty$. Different boundary value problems.

1°. The solution to the first boundary value problem for a semiinfinite circular cylinder of radius R is given by the formula of Subsection 2.1.3-2 with the term

$$2\pi \int_0^t \int_0^{\infty} \int_0^R \xi \Phi(\xi, \eta, \tau) G(r, z, \xi, \eta, t - \tau) \, d\xi \, d\eta \, d\tau \tag{1}$$

added; this term takes into account the nonhomogeneity of the equation.

2°. The solution to the second boundary value problem for a semiinfinite circular cylinder is given by the formula of Subsection 2.1.3-3 with the additional term (1).

3°. The solution to the third boundary value problem for a semiinfinite circular cylinder is given by the formula of Subsection 2.1.3-4 with the additional term (1).

4°. The solutions to various mixed boundary value problems for a semiinfinite circular cylinder are defined by formulas of Subsection 2.1.3-5 with additional terms of the form (1).

2.2.3-2. Domain: $0 \leq r \leq R$, $0 \leq z \leq l$. Different boundary value problems.

1°. The solution to the first boundary value problem for a circular cylinder of radius R and length l is given by the formula of Subsection 2.1.3-6 with the term

$$2\pi \int_0^t \int_0^l \int_0^R \xi \Phi(\xi, \eta, \tau) G(r, z, \xi, \eta, t - \tau) \, d\xi \, d\eta \, d\tau, \tag{2}$$

added; this term takes into account the nonhomogeneity of the equation.

2°. The solution to the second boundary value problem for a finite circular cylinder is given by the formula of Subsection 2.1.3-7 with the additional term (2).

3°. The solution to the third boundary value problem for a finite circular cylinder is given by the formula of Subsection 2.1.3-8 with the additional term (2).

4°. The solutions to various mixed boundary value problems for a finite circular cylinder are defined by formulas of Subsection 2.1.3-9 with additional terms of the form (2).

2.2.3-3. Domain: $R_1 \leq r \leq R_2$, $0 \leq z \leq l$. First and second boundary value problems.

1°. The solution to the first boundary value problem for a hollow circular cylinder of interior radius R_1, exterior radius R_2, and length l is given by the formula of Subsection 2.1.3-10 with the term

$$2\pi \int_0^t \int_0^l \int_{R_1}^{R_2} \xi \Phi(\xi, \eta, \tau) G(r, z, \xi, \eta, t - \tau) \, d\xi \, d\eta \, d\tau \tag{3}$$

added; this term takes into account the equation's nonhomogeneity.

2°. The solution to the second boundary value problem for a finite hollow circular cylinder is given by the formula of Subsection 2.1.3-11 with the additional term (3).

2.2.3-4. Domain: $R_1 \leq r \leq R_2$, $0 \leq z \leq l$. Third boundary value problem.

A hollow circular cylinder of finite length is considered. The following conditions are prescribed:

$$\begin{aligned}
w &= f(r, z) & \text{at} \quad t &= 0 & \text{(initial condition)}, \\
\partial_r w - k_1 w &= g_1(z, t) & \text{at} \quad r &= R_1 & \text{(boundary condition)}, \\
\partial_r w + k_2 w &= g_2(z, t) & \text{at} \quad r &= R_2 & \text{(boundary condition)}, \\
\partial_z w - k_3 w &= g_3(r, t) & \text{at} \quad z &= 0 & \text{(boundary condition)}, \\
\partial_z w + k_4 w &= g_4(r, t) & \text{at} \quad z &= l & \text{(boundary condition)}.
\end{aligned}$$

Solution:

$$w(r,z,t) = 2\pi \int_0^l \int_{R_1}^{R_2} \xi f(\xi,\eta) G(r,z,\xi,\eta,t)\, d\xi\, d\eta$$

$$- 2\pi a R_1 \int_0^t \int_0^l g_1(\eta,\tau) G(r,z,R_1,\eta,t-\tau)\, d\eta\, d\tau$$

$$+ 2\pi a R_2 \int_0^t \int_0^l g_2(\eta,\tau) G(r,z,R_2,\eta,t-\tau)\, d\eta\, d\tau$$

$$- 2\pi a \int_0^t \int_{R_1}^{R_2} \xi g_3(\xi,\tau) G(r,z,\xi,0,t-\tau)\, d\xi\, d\tau$$

$$+ 2\pi a \int_0^t \int_{R_1}^{R_2} \xi g_4(\xi,\tau) G(r,z,\xi,l,t-\tau)\, d\xi\, d\tau$$

$$+ 2\pi \int_0^t \int_0^l \int_{R_1}^{R_2} \xi \Phi(\xi,\eta,\tau) G(r,\xi,z,\eta,t-\tau)\, d\xi\, d\eta\, d\tau.$$

Here, the Green's function is given by

$$G(r,z,\xi,\eta,t) = G_1(r,\xi,t) G_2(z,\eta,t),$$

$$G_1(r,\xi,t) = \frac{\pi}{4} \sum_{n=1}^{\infty} \frac{\lambda_n^2}{B_n} \left[k_2 J_0(\lambda_n R_2) - \lambda_n J_1(\lambda_n R_2)\right]^2 H_n(r) H_n(\xi) \exp(-\lambda_n^2 a t),$$

$$G_2(z,\eta,t) = \sum_{m=1}^{\infty} \frac{\varphi_m(z) \varphi_m(\eta)}{\|\varphi_m\|^2} \exp(-\mu_m^2 a t),$$

where

$$B_n = (\lambda_n^2 + k_2^2)\left[k_1 J_0(\lambda_n R_1) + \lambda_n J_1(\lambda_n R_1)\right]^2 - (\lambda_n^2 + k_1^2)\left[k_2 J_0(\lambda_n R_2) - \lambda_n J_1(\lambda_n R_2)\right]^2,$$

$$H_n(r) = \left[k_1 Y_0(\lambda_n R_1) + \lambda_n Y_1(\lambda_n R_1)\right] J_0(\lambda_n r) - \left[k_1 J_0(\lambda_n R_1) + \lambda_n J_1(\lambda_n R_1)\right] Y_0(\lambda_n r),$$

$$\varphi_m(z) = \mu_m \cos(\mu_m z) + k_3 \sin(\mu_m z), \quad \|\varphi_m\|^2 = \frac{k_4}{2} \frac{\mu_m^2 + k_3^2}{\mu_m^2 + k_4^2} + \frac{k_3}{2} + \frac{l}{2}(\mu_m^2 + k_3^2),$$

$J_0(\lambda)$, $J_1(\lambda)$, $Y_0(\lambda)$, and $Y_1(\lambda)$ are the Bessel functions, the λ_n are positive roots of the transcendental equation

$$\left[k_1 J_0(\lambda R_1) + \lambda J_1(\lambda R_1)\right]\left[k_2 Y_0(\lambda R_2) - \lambda Y_1(\lambda R_2)\right]$$
$$- \left[k_2 J_0(\lambda R_2) - \lambda J_1(\lambda R_2)\right]\left[k_1 Y_0(\lambda R_1) + \lambda Y_1(\lambda R_1)\right] = 0,$$

and the μ_m are positive roots of the transcendental equation

$$\frac{\tan \mu l}{\mu} = \frac{k_3 + k_4}{\mu^2 - k_3 k_4}.$$

⊙ *Reference*: A. G. Butkovskiy (1979).

2.2.3-5. Domain: $R_1 \leq r \leq R_2$, $0 \leq z \leq l$. Mixed boundary value problem.

A hollow circular cylinder of finite length is considered. The following conditions are prescribed:

$$w = f(r,z) \quad \text{at} \quad t = 0 \quad \text{(initial condition)},$$
$$\partial_r w - k_1 w = g_1(z,t) \quad \text{at} \quad r = R_1 \quad \text{(boundary condition)},$$
$$\partial_r w + k_2 w = g_2(z,t) \quad \text{at} \quad r = R_2 \quad \text{(boundary condition)},$$
$$w = g_3(r,t) \quad \text{at} \quad z = 0 \quad \text{(boundary condition)},$$
$$w = g_4(r,t) \quad \text{at} \quad z = l \quad \text{(boundary condition)}.$$

Solution:

$$w(r,z,t) = 2\pi \int_0^l \int_{R_1}^{R_2} \xi f(\xi,\eta) G(r,z,\xi,\eta,t)\, d\xi\, d\eta$$

$$- 2\pi a R_1 \int_0^t \int_0^l g_1(\eta,\tau) G(r,z,R_1,\eta,t-\tau)\, d\eta\, d\tau$$

$$+ 2\pi a R_2 \int_0^t \int_0^l g_2(\eta,\tau) G(r,z,R_2,\eta,t-\tau)\, d\eta\, d\tau$$

$$+ 2\pi a \int_0^t \int_{R_1}^{R_2} \xi g_3(\xi,\tau) \left[\frac{\partial}{\partial \eta} G(r,z,\xi,\eta,t-\tau)\right]_{\eta=0} d\xi\, d\tau$$

$$- 2\pi a \int_0^t \int_{R_1}^{R_2} \xi g_4(\xi,\tau) \left[\frac{\partial}{\partial \eta} G(r,z,\xi,\eta,t-\tau)\right]_{\eta=l} d\xi\, d\tau$$

$$+ 2\pi \int_0^t \int_0^l \int_{R_1}^{R_2} \xi \Phi(\xi,\eta,\tau) G(r,\xi,z,\eta,t-\tau)\, d\xi\, d\eta\, d\tau.$$

Here,

$$G(r,\xi,z,\eta,t) = G_1(r,\xi,t) \frac{2}{l} \sum_{m=1}^{\infty} \sin\left(\frac{\pi m z}{l}\right) \sin\left(\frac{\pi m \eta}{l}\right) \exp\left(-\frac{\pi^2 m^2 a t}{l^2}\right),$$

where the expression of $G_1(r,\xi,t)$ is specified in Subsection 2.2.3-4.

▶ *Subsection 3.2.2 presents solutions of other boundary value problems; a more general, three-dimensional equation is discussed there.*

2.3. Other Equations

2.3.1. Equations Containing Arbitrary Parameters

1. $\dfrac{\partial w}{\partial t} = a\left(\dfrac{\partial^2 w}{\partial x^2} + \dfrac{\partial^2 w}{\partial y^2}\right) + (bx + cy + k)w.$

The transformation

$$w(x,y,t) = u(\xi,\eta,t) \exp\left[(bx+cy+k)t + \tfrac{1}{3}a(b^2+c^2)t^3\right], \quad \xi = x + abt^2, \quad \eta = y + act^2$$

leads to the two-dimensional heat equation $\partial_t u = a(\partial_{\xi\xi} u + \partial_{\eta\eta} u)$.

See also Niederer (1973) and Boyer (1974).

2. $\dfrac{\partial w}{\partial t} = a\left(\dfrac{\partial^2 w}{\partial x^2} + \dfrac{\partial^2 w}{\partial y^2}\right) - (bx^2 + by^2 + k)w, \qquad b > 0.$

The transformation (C is an arbitrary constant)

$$w(x,y,t) = u(\xi,\eta,\tau) \exp\left[\frac{1}{2}\sqrt{\frac{b}{a}}(x^2+y^2) + (2\sqrt{ab}-k)t\right],$$

$$\xi = x\exp(2\sqrt{ab}\,t), \quad \eta = y\exp(2\sqrt{ab}\,t), \quad \tau = \frac{1}{4\sqrt{ab}}\exp(4\sqrt{ab}\,t) + C$$

leads to the two-dimensional heat equation $\partial_\tau u = a(\partial_{\xi\xi} u + \partial_{\eta\eta} u)$.

See also Niederer (1973) and Boyer (1974).

3. $\dfrac{\partial w}{\partial t} = a\left(\dfrac{\partial^2 w}{\partial x^2} + \dfrac{\partial^2 w}{\partial y^2}\right) + (bx^2 + by^2 - k)w, \qquad b > 0.$

The transformation
$$w(x,y,t) = \dfrac{1}{\cos(2\sqrt{ab}\,t)} \exp\left[\dfrac{\sqrt{b}}{2\sqrt{a}} \tan(2\sqrt{ab}\,t)(x^2 + y^2) - kt\right] u(\xi,\eta,\tau),$$
$$\xi = \dfrac{x}{\cos(2\sqrt{ab}\,t)}, \quad \eta = \dfrac{y}{\cos(2\sqrt{ab}\,t)}, \quad \tau = \dfrac{\sqrt{a}}{2\sqrt{b}}\tan(2\sqrt{ab}\,t)$$

leads to the two-dimensional heat equation $\partial_\tau u = \partial_{\xi\xi} u + \partial_{\eta\eta} u$.

See also Niederer (1973) and Boyer (1974).

4. $\dfrac{\partial w}{\partial t} = \dfrac{\partial^2 w}{\partial x^2} + \dfrac{\partial^2 w}{\partial y^2} + (ax^{-2} + by^{-2})w.$

This is a special case of equation 2.3.2.7. Boyer (1976) showed that this equation admits the separation of variables into 25 systems of coordinates for $ab = 0$ and 15 systems of coordinates for $ab \neq 0$.

5. $\dfrac{\partial w}{\partial t} = a\left(\dfrac{\partial^2 w}{\partial x^2} + \dfrac{\partial^2 w}{\partial y^2}\right) + (bt^n x + ct^m y + st^k)w.$

This is a special case of equation 2.3.2.2 with $f(t) = bt^n$, $g(t) = ct^m$, and $h(t) = st^k$.

6. $\dfrac{\partial w}{\partial t} = a\left(\dfrac{\partial^2 w}{\partial x^2} + \dfrac{\partial^2 w}{\partial y^2}\right) + \left[-b(x^2 + y^2) + c_1 t^{n_1} x + c_2 t^{n_2} y + st^k\right]w.$

This is a special case of equation 2.3.2.3 with $f(t) = c_1 t^{n_1}$, $g(t) = c_2 t^{n_2}$, and $h(t) = st^k$.

7. $\dfrac{\partial w}{\partial t} = a\left(\dfrac{\partial^2 w}{\partial x^2} + \dfrac{\partial^2 w}{\partial y^2}\right) + b_1 \dfrac{\partial w}{\partial x} + b_2 \dfrac{\partial w}{\partial y} + cw.$

This equation describes an unsteady temperature (concentration) field in a medium moving with a constant velocity, provided there is volume release (absorption) of heat proportional to temperature.

The substitution
$$w(x,y,t) = \exp(A_1 x + A_2 y + Bt)U(x,y,t),$$
$$A_1 = -\dfrac{b_1}{2a}, \quad A_2 = -\dfrac{b_2}{2a}, \quad B = c - \dfrac{b_1^2 + b_2^2}{4a},$$

leads to the two-dimensional heat equation $\partial_t U = a\Delta_2 U$ that is considered in Subsection 2.1.1.

8. $\dfrac{\partial w}{\partial t} = a\left(\dfrac{\partial^2 w}{\partial x^2} + \dfrac{\partial^2 w}{\partial y^2}\right) + b_1 \dfrac{\partial w}{\partial x} + b_2 \dfrac{\partial w}{\partial y} + (c_1 x + c_2 y + k)w.$

The transformation
$$w(x,y,t) = \exp\left[(c_1 x + c_2 y)t + \tfrac{1}{3}a(c_1^2 + c_2^2)t^3 + \tfrac{1}{2}(b_1 c_1 + b_2 c_2)t^2 + kt\right] U(\xi,\eta,t),$$
$$\xi = x + ac_1 t^2 + b_1 t, \quad \eta = y + ac_2 t^2 + b_2 t$$

leads to the two-dimensional heat equation $\partial_t U = a(\partial_{\xi\xi} U + \partial_{\eta\eta} U)$ that is considered in Subsection 2.1.1.

9. $\dfrac{\partial w}{\partial t} = a\left(\dfrac{\partial^2 w}{\partial x^2} + \dfrac{\partial^2 w}{\partial y^2}\right) + b_1 t^{n_1} \dfrac{\partial w}{\partial x} + b_2 t^{n_2} \dfrac{\partial w}{\partial y} + (c_1 t^{m_1} x + c_2 t^{m_2} y + st^k)w.$

This is a special case of equation 2.3.2.5. The equation can be reduced to the two-dimensional heat equation treated in Subsection 2.1.1.

10. $i\hbar \dfrac{\partial w}{\partial t} + \dfrac{\hbar^2}{2m}\left(\dfrac{\partial^2 w}{\partial x^2} + \dfrac{\partial^2 w}{\partial y^2}\right) = 0.$

Two-dimensional Schrödinger equation, $i^2 = -1$.

Fundamental solution:
$$\mathcal{E}(x,y,t) = -\dfrac{im}{2\pi\hbar^2 t}\exp\left[\dfrac{im}{2\hbar t}(x^2+y^2) - i\dfrac{\pi}{2}\right].$$

⊙ *Reference*: V. S. Vladimirov, V. P. Mikhailov, A. A. Vasharin, et al. (1974).

2.3.2. Equations Containing Arbitrary Functions

1. $\dfrac{\partial w}{\partial t} = a\left(\dfrac{\partial^2 w}{\partial x^2} + \dfrac{\partial^2 w}{\partial y^2}\right) + f(t)w.$

This equation describes two-dimensional thermal phenomena in quiescent media or solids with constant thermal diffusivities in the case of unsteady volume heat release proportional to temperature.

The substitution $w(x,y,t) = \exp\left[\int f(t)\,dt\right]U(x,y,t)$ leads to the two-dimensional heat equation $\partial_t U = a(\partial_{xx}U + \partial_{yy}U)$ treated in Subsection 2.1.1.

2. $\dfrac{\partial w}{\partial t} = a\left(\dfrac{\partial^2 w}{\partial x^2} + \dfrac{\partial^2 w}{\partial y^2}\right) + [xf(t) + yg(t) + h(t)]w.$

The transformation
$$w(x,y,t) = u(\xi,\eta,t)\exp\left[xF(t) + yG(t) + H(t) + a\int F^2(t)\,dt + a\int G^2(t)\,dt\right],$$
$$\xi = x + 2a\int F(t)\,dt, \quad \eta = y + 2a\int G(t)\,dt,$$
where
$$F(t) = \int f(t)\,dt, \quad G(t) = \int g(t)\,dt, \quad H(t) = \int h(t)\,dt,$$
leads to the two-dimensional heat equation $\partial_t u = a(\partial_{\xi\xi}u + \partial_{\eta\eta}u)$.

3. $\dfrac{\partial w}{\partial t} = a\left(\dfrac{\partial^2 w}{\partial x^2} + \dfrac{\partial^2 w}{\partial y^2}\right) + [-b(x^2+y^2) + f(t)x + g(t)y + h(t)]w.$

1°. Case $b > 0$. The transformation
$$w(x,y,t) = u(\xi,\eta,\tau)\exp\left[\dfrac{1}{2}\sqrt{\dfrac{b}{a}}(x^2+y^2)\right],$$
$$\xi = x\exp(2\sqrt{ab}\,t), \quad \eta = y\exp(2\sqrt{ab}\,t), \quad \tau = \dfrac{1}{4\sqrt{ab}}\exp(4\sqrt{ab}\,t)$$
leads to an equation of the form 2.3.2.2:
$$\dfrac{\partial u}{\partial t} = a\left(\dfrac{\partial^2 u}{\partial \xi^2} + \dfrac{\partial^2 u}{\partial \eta^2}\right) + [F(\tau)\xi + G(\tau)\eta + H(\tau)]u,$$
$F(\tau) = \dfrac{1}{(c\tau)^{3/2}}f\left(\dfrac{\ln(c\tau)}{c}\right), \quad G(\tau) = \dfrac{1}{(c\tau)^{3/2}}g\left(\dfrac{\ln(c\tau)}{c}\right), \quad H(\tau) = \dfrac{1}{c\tau}h\left(\dfrac{\ln(c\tau)}{c}\right) + \dfrac{1}{2\tau}, \quad c = 4\sqrt{ab}.$

2°. Case $b < 0$. The transformation
$$w(x,y,t) = v(\bar\xi,\bar\eta,\bar\tau)\exp\left[\dfrac{\sqrt{-b}}{2\sqrt{a}}\tan(2\sqrt{-ab}\,t)(x^2+y^2)\right],$$
$$\bar\xi = \dfrac{x}{\cos(2\sqrt{-ab}\,t)}, \quad \bar\eta = \dfrac{y}{\cos(2\sqrt{-ab}\,t)}, \quad \bar\tau = \dfrac{1}{2\sqrt{-ab}}\tan(2\sqrt{-ab}\,t)$$
also leads to an equation of the form 2.3.2.2 (the transformed equation is not written out here).

4. $\dfrac{\partial w}{\partial t} = a_1(t)\dfrac{\partial^2 w}{\partial x^2} + a_2(t)\dfrac{\partial^2 w}{\partial y^2} + \Phi(x,y,t).$

This is a special case of equation 2.3.2.8. Let $0 < a_1(t) < \infty$ and $0 < a_2(t) < \infty$.

For the first, second, third, and mixed boundary value problems treated in rectangular, finite, or infinite domains ($x_1 \le x \le x_2$, $y_1 \le y \le y_2$), the Green's function can be represented in the product form

$$G(x,y,\xi,\eta,t,\tau) = G_1(x,\xi,T_1)G_2(y,\eta,T_2),$$

$$T_1 = \int_\tau^t a_1(\eta)\,d\eta, \quad T_2 = \int_\tau^t a_2(\eta)\,d\eta.$$

Here, $G_1 = G_1(x,\xi,t)$ is the auxiliary Green's function that corresponds to the one-dimensional heat equation for $a_1(t) = 1$, $a_2(t) = 0$, and $\Phi(x,y,t) = 0$ with homogeneous boundary conditions at $x = x_1$ and $x = x_2$ (the G_1's for various boundary value problems can be found in Subsections 1.1.1 and 1.1.2). Similarly, $G_2 = G_2(y,\eta,t)$ is the auxiliary Green's function that corresponds to the one-dimensional heat equation for $a_1(t) = 0$, $a_2(t) = 1$, and $\Phi(x,y,t) = 0$ with homogeneous boundary conditions at $y = y_1$ and $y = y_2$. Note that the Green's functions G_1 and G_2 are introduced for $\tau = 0$.

See Subsection 0.8.1 for solution of various boundary value problems with the help of the Green's function.

Example 1. Domain: $-\infty < x < \infty$, $-\infty < y < \infty$. Cauchy problem.
An initial condition is prescribed:
$$w = f(x,y) \quad \text{at} \quad t = 0.$$

Solution:

$$w(x,y,t) = \int_0^t\!\!\int_{-\infty}^{\infty}\!\!\int_{-\infty}^{\infty} \Phi(\xi,\eta,\tau)G(x,y,\xi,\eta,t,\tau)\,d\xi\,d\eta\,d\tau + \int_{-\infty}^{\infty}\!\!\int_{-\infty}^{\infty} f(\xi,\eta)G(x,y,\xi,\eta,t,0)\,d\xi\,d\eta,$$

where

$$G(x,y,\xi,\eta,t,\tau) = \dfrac{1}{4\pi\sqrt{T_1 T_2}} \exp\!\left[-\dfrac{(x-\xi)^2}{4T_1} - \dfrac{(y-\eta)^2}{4T_2}\right], \quad T_1 = \int_\tau^t a_1(\eta)\,d\eta, \quad T_2 = \int_\tau^t a_2(\eta)\,d\eta.$$

Example 2. Domain: $0 \le x < \infty$, $0 \le y < \infty$. Second boundary value problem.
The following conditions are prescribed:
$$w = f(x,y) \quad \text{at} \quad t = 0 \quad \text{(initial condition)},$$
$$\partial_x w = g_1(y,t) \quad \text{at} \quad x = 0 \quad \text{(boundary condition)},$$
$$\partial_y w = g_2(x,t) \quad \text{at} \quad y = 0 \quad \text{(boundary condition)}.$$

Solution:

$$w(x,y,t) = \int_0^t\!\!\int_0^\infty\!\!\int_0^\infty \Phi(\xi,\eta,\tau)G(x,y,\xi,\eta,t,\tau)\,d\xi\,d\eta\,d\tau + \int_0^\infty\!\!\int_0^\infty f(\xi,\eta)G(x,y,\xi,\eta,t,0)\,d\xi\,d\eta$$
$$- \int_0^t\!\!\int_0^\infty a_1(\tau)g_1(\eta,\tau)G(x,y,0,\eta,t,\tau)\,d\eta\,d\tau - \int_0^t\!\!\int_0^\infty a_2(\tau)g_2(\xi,\tau)G(x,y,\xi,0,t,\tau)\,d\xi\,d\tau,$$

where

$$G(x,y,\xi,\eta,t,\tau) = G_1(x,\xi,T_1)G_2(y,\eta,T_2),$$

$$G_1(x,\xi,T_1) = \dfrac{1}{2\sqrt{\pi T_1}}\left\{\exp\!\left[-\dfrac{(x-\xi)^2}{4T_1}\right] + \exp\!\left[-\dfrac{(x+\xi)^2}{4T_1}\right]\right\}, \quad T_1 = \int_\tau^t a_1(\eta)\,d\eta,$$

$$G_2(y,\eta,T_2) = \dfrac{1}{2\sqrt{\pi T_2}}\left\{\exp\!\left[-\dfrac{(y-\eta)^2}{4T_2}\right] + \exp\!\left[-\dfrac{(y+\eta)^2}{4T_2}\right]\right\}, \quad T_2 = \int_\tau^t a_2(\eta)\,d\eta.$$

Example 3. Domain: $0 \le x \le l_1$, $0 \le y \le l_2$. First boundary value problem.

The following conditions are prescribed:

$$w = f(x,y) \quad \text{at} \quad t = 0 \quad \text{(initial condition)},$$
$$w = g_1(y,t) \quad \text{at} \quad x = 0 \quad \text{(boundary condition)},$$
$$w = g_2(y,t) \quad \text{at} \quad x = l_1 \quad \text{(boundary condition)},$$
$$w = h_1(x,t) \quad \text{at} \quad y = 0 \quad \text{(boundary condition)},$$
$$w = h_2(x,t) \quad \text{at} \quad y = l_2 \quad \text{(boundary condition)}.$$

Solution:

$$w(x,y,t) = \int_0^t \int_0^{l_1} \int_0^{l_2} \Phi(\xi,\eta,\tau) G(x,y,\xi,\eta,t,\tau)\, d\eta\, d\xi\, d\tau + \int_0^{l_1} \int_0^{l_2} f(\xi,\eta) G(x,y,\xi,\eta,t,0)\, d\eta\, d\xi$$

$$+ \int_0^t \int_0^{l_2} a_1(\tau) g_1(\eta,\tau) \left[\frac{\partial}{\partial \xi} G(x,y,\xi,\eta,t,\tau)\right]_{\xi=0} d\eta\, d\tau - \int_0^t \int_0^{l_2} a_1(\tau) g_2(\eta,\tau) \left[\frac{\partial}{\partial \xi} G(x,y,\xi,\eta,t,\tau)\right]_{\xi=l_1} d\eta\, d\tau$$

$$+ \int_0^t \int_0^{l_1} a_2(\tau) h_1(\xi,\tau) \left[\frac{\partial}{\partial \eta} G(x,y,\xi,\eta,t,\tau)\right]_{\eta=0} d\xi\, d\tau - \int_0^t \int_0^{l_1} a_2(\tau) h_2(\xi,\tau) \left[\frac{\partial}{\partial \eta} G(x,y,\xi,\eta,t,\tau)\right]_{\eta=l_2} d\xi\, d\tau,$$

where

$$G(x,y,\xi,\eta,t,\tau) = G_1(x,\xi,T_1) G_2(y,\eta,T_2),$$

$$G_1(x,\xi,T_1) = \frac{2}{l_1} \sum_{n=1}^{\infty} \sin\left(\frac{n\pi x}{l_1}\right) \sin\left(\frac{n\pi \xi}{l_1}\right) \exp\left(-\frac{n^2\pi^2 T_1}{l_1^2}\right), \qquad T_1 = \int_\tau^t a_1(\eta)\, d\eta,$$

$$G_2(y,\eta,T_2) = \frac{2}{l_2} \sum_{n=1}^{\infty} \sin\left(\frac{n\pi y}{l_2}\right) \sin\left(\frac{n\pi \eta}{l_2}\right) \exp\left(-\frac{n^2\pi^2 T_2}{l_2^2}\right), \qquad T_2 = \int_\tau^t a_2(\eta)\, d\eta.$$

5. $\dfrac{\partial w}{\partial t} = a_1(t)\dfrac{\partial^2 w}{\partial x^2} + a_2(t)\dfrac{\partial^2 w}{\partial y^2} + [b_1(t)x + c_1(t)]\dfrac{\partial w}{\partial x}$
$$+ [b_2(t)y + c_2(t)]\dfrac{\partial w}{\partial y} + [s_1(t)x + s_2(t)y + p(t)]w.$$

The transformation

$$w(x,y,t) = \exp\bigl[f_1(t)x + f_2(t)y + g(t)\bigr] u(\xi,\eta,t), \quad \xi = h_1(t)x + r_1(t), \quad \eta = h_2(t)y + r_2(t),$$

where

$$h_k(t) = A_k \exp\left[\int b_k(t)\, dt\right],$$

$$f_k(t) = h_k(t) \int \frac{s_k(t)}{h_k(t)}\, dt + B_k h_k(t),$$

$$r_k(t) = \int \bigl[2a_k(t) f_k(t) + c_k(t)\bigr] h_k(t)\, dt + C_k,$$

$$g(t) = \int \bigl[a_1(t) f_1^2(t) + a_2(t) f_2^2(t) + c_1(t) f_1(t) + c_2(t) f_2(t) + p(t)\bigr] dt + D,$$

($k = 1, 2$; A_k, B_k, C_k, and D are arbitrary constants), leads to an equation of the form 2.3.2.4:

$$\frac{\partial u}{\partial t} = a_1(t) h_1^2(t) \frac{\partial^2 u}{\partial \xi^2} + a_2(t) h_2^2(t) \frac{\partial^2 u}{\partial \eta^2}.$$

6. $\dfrac{\partial w}{\partial t} = a_1(t)\dfrac{\partial^2 w}{\partial x^2} + a_2(t)\dfrac{\partial^2 w}{\partial y^2} + [b_1(t)x + c_1(t)]\dfrac{\partial w}{\partial x} + [b_2(t)y + c_2(t)]\dfrac{\partial w}{\partial y}$
$$+ [s_1(t)x^2 + s_2(t)y^2 + p_1(t)x + p_2(t)y + q(t)]w.$$

The substitution

$$w(x,y,t) = \exp\bigl[f_1(t)x^2 + f_2(t)y^2\bigr] u(x,y,t),$$

where the functions $f_1 = f_1(t)$ and $f_2 = f_2(t)$ are solutions of the Riccati equations

$$f_1' = 4a_1(t) f_1^2 + 2b_1(t) f_1 + s_1(t),$$
$$f_2' = 4a_2(t) f_2^2 + 2b_2(t) f_2 + s_2(t),$$

leads to an equation of the form 2.3.2.5 for $u = u(x,y,t)$.

7. $\dfrac{\partial w}{\partial t} = a_1(x)\dfrac{\partial^2 w}{\partial x^2} + a_2(y)\dfrac{\partial^2 w}{\partial y^2} + b_1(x)\dfrac{\partial w}{\partial x} + b_2(y)\dfrac{\partial w}{\partial y} + [c_1(x) + c_2(y)]w + \Phi(x,y,t)$.

Domain: $x_1 \le x \le x_2$, $y_1 \le y \le y_2$. Different boundary value problems:

$$w = f(x,y) \quad \text{at} \quad t = 0 \quad \text{(initial condition)},$$
$$s_1 \partial_x w - k_1 w = g_1(y,t) \quad \text{at} \quad x = x_1 \quad \text{(boundary condition)},$$
$$s_2 \partial_x w + k_2 w = g_2(y,t) \quad \text{at} \quad x = x_2 \quad \text{(boundary condition)},$$
$$s_3 \partial_y w - k_3 w = g_3(x,t) \quad \text{at} \quad y = y_1 \quad \text{(boundary condition)},$$
$$s_4 \partial_y w + k_4 w = g_4(x,t) \quad \text{at} \quad y = y_2 \quad \text{(boundary condition)}.$$

By choosing appropriate parameters s_n, k_n ($n = 1, 2, 3, 4$), one obtains the first, second, third, or mixed boundary value problem. If the domain is infinite, say, $x_2 = \infty$, the corresponding boundary condition should be omitted; this is also valid for $x_1 = -\infty$, $y_1 = -\infty$, or $y_2 = \infty$.

The Green's function admits incomplete separation of variables; specifically, it can be represented in the product form

$$G(x,y,\xi,\eta,t) = G_1(x,\xi,t)G_2(y,\eta,t).$$

Here, $G_1 = G_1(x,\xi,t)$ and $G_2 = G_2(y,\eta,t)$ are auxiliary Green's functions that are determined by solving the following simpler one-dimensional problems with homogeneous boundary conditions:

$$\dfrac{\partial G_1}{\partial t} = a_1(x)\dfrac{\partial^2 G_1}{\partial x^2} + b_1(x)\dfrac{\partial G_1}{\partial x} + c_1(x)G_1, \qquad \dfrac{\partial G_2}{\partial t} = a_2(y)\dfrac{\partial^2 G_2}{\partial y^2} + b_2(y)\dfrac{\partial G_2}{\partial y} + c_2(y)G_2,$$

$$G_1 = \delta(x - \xi) \quad \text{at} \quad t = 0, \qquad\qquad G_2 = \delta(y - \eta) \quad \text{at} \quad t = 0,$$
$$s_1 \partial_x G_1 - k_1 G_1 = 0 \quad \text{at} \quad x = x_1, \qquad s_3 \partial_y G_2 - k_3 G_2 = 0 \quad \text{at} \quad y = y_1,$$
$$s_2 \partial_x G_1 + k_2 G_1 = 0 \quad \text{at} \quad x = x_2, \qquad s_4 \partial_y G_2 + k_4 G_2 = 0 \quad \text{at} \quad y = y_2,$$

where ξ and η are free parameters, and $\delta(x)$ is the Dirac delta function.

The equation for G_1 coincides with equation 1.8.6.5, which is reduced to the equation of Subsection 1.8.9 (where the expression of the Green's function can also be found). In the general case, the equation for G_2 differs from the equation for G_1 in only notation.

8. $\dfrac{\partial w}{\partial t} = a_1(x,t)\dfrac{\partial^2 w}{\partial x^2} + a_2(y,t)\dfrac{\partial^2 w}{\partial y^2} + b_1(x,t)\dfrac{\partial w}{\partial x}$
$\qquad\qquad\qquad\qquad + b_2(y,t)\dfrac{\partial w}{\partial y} + [c_1(x,t) + c_2(y,t)]w + \Phi(x,y,t)$.

Suppose this equation is subject to the same initial and boundary conditions as equation 2.3.2.7. Then the Green's function for this problem can be represented in the product form

$$G(x,y,\xi,\eta,t,\tau) = G_1(x,\xi,t,\tau)G_2(y,\eta,t,\tau).$$

Here, $G_1 = G_1(x,\xi,t,\tau)$ and $G_2 = G_2(y,\eta,t,\tau)$ are auxiliary Green's functions that are determined by solving the following simpler boundary value problems with homogeneous boundary conditions:

$$\dfrac{\partial G_1}{\partial t} = a_1(x,t)\dfrac{\partial^2 G_1}{\partial x^2} + b_1(x,t)\dfrac{\partial G_1}{\partial x} + c_1(x,t)G_1, \qquad \dfrac{\partial G_2}{\partial t} = a_2(y,t)\dfrac{\partial^2 G_2}{\partial y^2} + b_2(y,t)\dfrac{\partial G_2}{\partial y} + c_2(y,t)G_2,$$

$$G_1 = \delta(x - \xi) \quad \text{at} \quad t = \tau, \qquad\qquad G_2 = \delta(y - \eta) \quad \text{at} \quad t = \tau,$$
$$s_1 \partial_x G_1 - k_1 G_1 = 0 \quad \text{at} \quad x = x_1, \qquad s_3 \partial_y G_2 - k_3 G_2 = 0 \quad \text{at} \quad y = y_1,$$
$$s_2 \partial_x G_1 + k_2 G_1 = 0 \quad \text{at} \quad x = x_2, \qquad s_4 \partial_y G_2 + k_4 G_2 = 0 \quad \text{at} \quad y = y_2,$$

where ξ, η, and τ are free parameters, and $\delta(x)$ is the Dirac delta function, $t \ge \tau$.

See Subsection 0.8.1 for the solution of boundary value problems with the help of the Green's function.

Chapter 3

Parabolic Equations with Three or More Space Variables

3.1. Heat Equation $\frac{\partial w}{\partial t} = a\Delta_3 w$

3.1.1. Problems in Cartesian Coordinates

The three-dimensional sourceless heat equation in the rectangular Cartesian system of coordinates has the form

$$\frac{\partial w}{\partial t} = a\left(\frac{\partial^2 w}{\partial x^2} + \frac{\partial^2 w}{\partial y^2} + \frac{\partial^2 w}{\partial z^2}\right).$$

It governs three-dimensional thermal phenomena in quiescent media or solids with constant thermal diffusivity. A similar equation is used to study the corresponding three-dimensional unsteady mass-exchange processes with constant diffusivity.

3.1.1-1. Particular solutions:

$$w(x, y, z, t) = Ax^2 + By^2 + Cz^2 + 2a(A + B + C)t,$$
$$w(x, y, z, t) = A(x^2 + 2at)(y^2 + 2at)(z^2 + 2at) + B,$$
$$w(x, y, z, t) = A\exp\left[k_1 x + k_2 y + k_3 z + (k_1^2 + k_2^2 + k_3^2)at\right] + B,$$
$$w(x, y, z, t) = A\cos(k_1 x + C_1)\cos(k_2 y + C_2)\cos(k_3 z + C_3)\exp\left[-(k_1^2 + k_2^2 + k_3^2)at\right],$$
$$w(x, y, z, t) = A\cos(k_1 x + C_1)\cos(k_2 y + C_2)\sinh(k_3 z + C_3)\exp\left[-(k_1^2 + k_2^2 - k_3^2)at\right],$$
$$w(x, y, z, t) = A\cos(k_1 x + C_1)\cos(k_2 y + C_2)\cosh(k_3 z + C_3)\exp\left[-(k_1^2 + k_2^2 - k_3^2)at\right],$$
$$w(x, y, z, t) = A\exp(-k_1 x - k_2 y - k_3 z)\cos(k_1 x - 2ak_1^2 t)\cos(k_2 y - 2ak_2^2 t)\cos(k_3 z - 2ak_3^2 t),$$
$$w(x, y, z, t) = \frac{A}{(t - t_0)^{3/2}}\exp\left[-\frac{(x - x_0)^2 + (y - y_0)^2 + (z - z_0)^2}{4a(t - t_0)}\right],$$
$$w(x, y, z, t) = A\operatorname{erf}\left(\frac{x - x_0}{2\sqrt{at}}\right)\operatorname{erf}\left(\frac{y - y_0}{2\sqrt{at}}\right)\operatorname{erf}\left(\frac{z - z_0}{2\sqrt{at}}\right) + B,$$

where A, B, C, C_1, C_2, C_3, k_1, k_2, k_3, x_0, y_0, z_0, and t_0 are arbitrary constants.
Fundamental solution:

$$\mathcal{E}(x, y, z, t) = \frac{1}{8(\pi at)^{3/2}}\exp\left(-\frac{x^2 + y^2 + z^2}{4at}\right).$$

3.1.1-2. Formulas to construct particular solutions. Remarks on the Green's functions.

1°. Apart from usual solutions with separated variables,

$$w(x, y, z, t) = f_1(x)f_2(y)f_3(z)f_4(t),$$

the equation in question admits more sophisticated solutions in the product form

$$w(x,y,z,t) = u_1(x,t)u_2(y,t)u_3(z,t),$$

where the functions $u_1 = u_1(x,t)$, $u_2 = u_2(y,t)$, and $u_3 = u_3(y,t)$ are solutions of the one-dimensional heat equations

$$\frac{\partial u_1}{\partial t} = a\frac{\partial^2 u_1}{\partial x^2}, \quad \frac{\partial u_2}{\partial t} = a\frac{\partial^2 u_2}{\partial y^2}, \quad \frac{\partial u_3}{\partial t} = a\frac{\partial^2 u_3}{\partial z^2},$$

treated in Subsection 1.1.1.

2°. Suppose $w = w(x,y,z,t)$ is a solution of the three-dimensional heat equation. Then the functions

$$w_1 = Aw(\pm\lambda x + C_1, \pm\lambda y + C_2, \pm\lambda z + C_3, \lambda^2 t + C_4),$$

$$w_2 = A\exp\left[\lambda_1 x + \lambda_2 y + \lambda_3 z + (\lambda_1^2 + \lambda_2^2 + \lambda_3^2)at\right]w(x + 2a\lambda_1 t, y + 2a\lambda_2 t, z + 2a\lambda_3 t, t),$$

$$w_3 = \frac{A}{|\delta + \beta t|^{3/2}}\exp\left[-\frac{\beta(x^2 + y^2 + z^2)}{4a(\delta + \beta t)}\right]w\left(\frac{x}{\delta + \beta t}, \frac{y}{\delta + \beta t}, \frac{z}{\delta + \beta t}, \frac{\gamma + \lambda t}{\delta + \beta t}\right), \quad \lambda\delta - \beta\gamma = 1,$$

where A, C_1, C_2, C_3, C_4, λ, λ_1, λ_2, λ_3, β, and δ are arbitrary constants, are also solutions of this equation. The signs at λ in the formula for w_1 can be taken independently of one another.

3°. For the three-dimensional boundary value problems considered in Subsection 3.1.1, the Green's function can be represented in the product form

$$G(x,y,z,\xi,\eta,\zeta,t) = G_1(x,\xi,t)G_2(y,\eta,t)G_3(z,\zeta,t),$$

where $G_1(x,\xi,t)$, $G_2(y,\eta,t)$, $G_3(z,\zeta,t)$ are the Green's functions of the corresponding one-dimensional boundary value problems; these functions can be found in Subsections 1.1.1 and 1.1.2.

Example 1. The Green's function of the mixed boundary value problem for a semiinfinite layer ($-\infty < x < \infty$, $0 \le y < \infty$, $0 \le z < l$) presented in Paragraph 3.1.1-14 is the product of three one-dimensional Green's functions from Paragraph 1.1.2-1 (Cauchy problem for $-\infty < x < \infty$), Paragraph 1.1.2-2 (first boundary value problem for $0 \le y < \infty$), and Paragraph 1.1.2-6 (second boundary value problem for $0 \le z < l$), in which one needs to carry out obvious renaming of variables.

3.1.1-3. Domain: $-\infty < x < \infty$, $-\infty < y < \infty$, $-\infty < z < \infty$. Cauchy problem.

An initial condition is prescribed:

$$w = f(x,y,z) \quad \text{at} \quad t = 0.$$

Solution:

$$w(x,y,z,t) = \frac{1}{8(\pi at)^{3/2}}\int_{-\infty}^{\infty}\int_{-\infty}^{\infty}\int_{-\infty}^{\infty} f(\xi,\eta,\zeta)\exp\left[-\frac{(x-\xi)^2 + (y-\eta)^2 + (z-\zeta)^2}{4at}\right]d\xi\,d\eta\,d\zeta.$$

Example 2. The initial temperature is constant and is equal to w_1 in the domain $|x| < x_0, |y| < y_0, |z| < z_0$ and is equal to w_2 in the domain $|x| > x_0, |y| > y_0, |z| > z_0$; specifically,

$$f(x,y,z) = \begin{cases} w_1 & \text{for } |x| < x_0, |y| < y_0, |z| < z_0, \\ w_2 & \text{for } |x| > x_0, |y| > y_0, |z| > z_0. \end{cases}$$

Solution:

$$w = \frac{1}{8}(w_1 - w_2)\left[\operatorname{erf}\left(\frac{x_0 - x}{2\sqrt{at}}\right) + \operatorname{erf}\left(\frac{x_0 + x}{2\sqrt{at}}\right)\right]$$
$$\times \left[\operatorname{erf}\left(\frac{y_0 - y}{2\sqrt{at}}\right) + \operatorname{erf}\left(\frac{y_0 + y}{2\sqrt{at}}\right)\right]\left[\operatorname{erf}\left(\frac{z_0 - z}{2\sqrt{at}}\right) + \operatorname{erf}\left(\frac{z_0 + z}{2\sqrt{at}}\right)\right] + w_2.$$

⊙ *Reference*: H. S. Carslaw and J. C. Jaeger (1984).

3.1.1-4. Domain: $0 \le x < \infty$, $-\infty < y < \infty$, $-\infty < z < \infty$. First boundary value problem.

A half-space is considered. The following conditions are prescribed:

$$w = f(x, y, z) \quad \text{at} \quad t = 0 \quad \text{(initial condition)},$$
$$w = g(y, z, t) \quad \text{at} \quad x = 0 \quad \text{(boundary condition)}.$$

Solution:

$$w(x, y, z, t) = \int_{-\infty}^{\infty} \int_{-\infty}^{\infty} \int_{0}^{\infty} f(\xi, \eta, \zeta) G(x, y, z, \xi, \eta, \zeta, t) \, d\xi \, d\eta \, d\zeta$$
$$+ a \int_{0}^{t} \int_{-\infty}^{\infty} \int_{-\infty}^{\infty} g(\eta, \zeta, \tau) \left[\frac{\partial}{\partial \xi} G(x, y, z, \xi, \eta, \zeta, t - \tau) \right]_{\xi=0} d\eta \, d\zeta \, d\tau,$$

where

$$G(x, y, z, \xi, \eta, \zeta, t) = \frac{1}{8(\pi a t)^{3/2}} \left\{ \exp\left[-\frac{(x-\xi)^2}{4at}\right] - \exp\left[-\frac{(x+\xi)^2}{4at}\right] \right\} \exp\left[-\frac{(y-\eta)^2 + (z-\zeta)^2}{4at}\right].$$

⊙ *References*: A. G. Butkovskiy (1979), H. S. Carslaw and J. C. Jaeger (1984).

3.1.1-5. Domain: $0 \le x < \infty$, $-\infty < y < \infty$, $-\infty < z < \infty$. Second boundary value problem.

A half-space is considered. The following conditions are prescribed:

$$w = f(x, y, z) \quad \text{at} \quad t = 0 \quad \text{(initial condition)},$$
$$\partial_x w = g(y, z, t) \quad \text{at} \quad x = 0 \quad \text{(boundary condition)}.$$

Solution:

$$w(x, y, z, t) = \int_{-\infty}^{\infty} \int_{-\infty}^{\infty} \int_{0}^{\infty} f(\xi, \eta, \zeta) G(x, y, z, \xi, \eta, \zeta, t) \, d\xi \, d\eta \, d\zeta$$
$$- a \int_{0}^{t} \int_{-\infty}^{\infty} \int_{-\infty}^{\infty} g(\eta, \zeta, \tau) G(x, y, z, 0, \eta, \zeta, t - \tau) \, d\eta \, d\zeta \, d\tau,$$

where

$$G(x, y, z, \xi, \eta, \zeta, t) = \frac{1}{8(\pi a t)^{3/2}} \left\{ \exp\left[-\frac{(x-\xi)^2}{4at}\right] + \exp\left[-\frac{(x+\xi)^2}{4at}\right] \right\} \exp\left[-\frac{(y-\eta)^2 + (z-\zeta)^2}{4at}\right].$$

⊙ *Reference*: A. G. Butkovskiy (1979).

3.1.1-6. Domain: $0 \le x < \infty$, $-\infty < y < \infty$, $-\infty < z < \infty$. Third boundary value problem.

A half-space is considered. The following conditions are prescribed:

$$w = f(x, y, z) \quad \text{at} \quad t = 0 \quad \text{(initial condition)},$$
$$\partial_x w - kw = g(y, z, t) \quad \text{at} \quad x = 0 \quad \text{(boundary condition)}.$$

The solution $w(x, y, z, t)$ is determined by the formula in Paragraph 3.1.1-5 where

$$G(x, y, z, \xi, \eta, \zeta, t) = \frac{1}{8(\pi a t)^{3/2}} \exp\left[-\frac{(y-\eta)^2 + (z-\zeta)^2}{4at}\right] \left\{ \exp\left[-\frac{(x-\xi)^2}{4at}\right] + \exp\left[-\frac{(x+\xi)^2}{4at}\right] \right.$$
$$\left. - 2k\sqrt{\pi a t} \exp[k^2 at + k(x+\xi)] \operatorname{erfc}\left(\frac{x+\xi}{2\sqrt{at}} + k\sqrt{at}\right) \right\}.$$

⊙ *Reference*: H. S. Carslaw and J. C. Jaeger (1984).

3.1.1-7. Domain: $-\infty < x < \infty$, $-\infty < y < \infty$, $0 \leq z \leq l$. First boundary value problem.

An infinite layer is considered. The following conditions are prescribed:

$$w = f(x,y,z) \quad \text{at} \quad t=0 \quad \text{(initial condition)},$$
$$w = g_1(x,y,t) \quad \text{at} \quad z=0 \quad \text{(boundary condition)},$$
$$w = g_2(x,y,t) \quad \text{at} \quad z=l \quad \text{(boundary condition)}.$$

Solution:

$$w(x,y,z,t) = \int_0^l \int_{-\infty}^{\infty} \int_{-\infty}^{\infty} f(\xi,\eta,\zeta) G(x,y,z,\xi,\eta,\zeta,t) \, d\xi \, d\eta \, d\zeta$$
$$+ a \int_0^t \int_{-\infty}^{\infty} \int_{-\infty}^{\infty} g_1(\xi,\eta,\tau) \left[\frac{\partial}{\partial \zeta} G(x,y,z,\xi,\eta,\zeta,t-\tau)\right]_{\zeta=0} d\xi \, d\eta \, d\tau$$
$$- a \int_0^t \int_{-\infty}^{\infty} \int_{-\infty}^{\infty} g_2(\xi,\eta,\tau) \left[\frac{\partial}{\partial \zeta} G(x,y,z,\xi,\eta,\zeta,t-\tau)\right]_{\zeta=l} d\xi \, d\eta \, d\tau,$$

where

$$G(x,y,z,\xi,\eta,\zeta,t) = \frac{1}{2\pi alt} \exp\left[-\frac{(x-\xi)^2 + (y-\eta)^2}{4at}\right] \sum_{n=1}^{\infty} \sin\frac{n\pi z}{l} \sin\frac{n\pi \zeta}{l} \exp\left(-\frac{n^2\pi^2 at}{l^2}\right),$$

or

$$G(x,y,z,\xi,\eta,\zeta,t) = \frac{1}{8(\pi at)^{3/2}} \exp\left[-\frac{(x-\xi)^2 + (y-\eta)^2}{4at}\right]$$
$$\times \sum_{n=-\infty}^{\infty} \left\{\exp\left[-\frac{(2nl+z-\zeta)^2}{4at}\right] - \exp\left[-\frac{(2nl+z+\zeta)^2}{4at}\right]\right\}.$$

⊙ *Reference*: H. S. Carslaw and J. C. Jaeger (1984).

3.1.1-8. Domain: $-\infty < x < \infty$, $-\infty < y < \infty$, $0 \leq z \leq l$. Second boundary value problem.

An infinite layer is considered. The following conditions are prescribed:

$$w = f(x,y,z) \quad \text{at} \quad t=0 \quad \text{(initial condition)},$$
$$\partial_z w = g_1(x,y,t) \quad \text{at} \quad z=0 \quad \text{(boundary condition)},$$
$$\partial_z w = g_2(x,y,t) \quad \text{at} \quad z=l \quad \text{(boundary condition)}.$$

Solution:

$$w(x,y,z,t) = \int_0^l \int_{-\infty}^{\infty} \int_{-\infty}^{\infty} f(\xi,\eta,\zeta) G(x,y,z,\xi,\eta,\zeta,t) \, d\xi \, d\eta \, d\zeta$$
$$- a \int_0^t \int_{-\infty}^{\infty} \int_{-\infty}^{\infty} g_1(\xi,\eta,\tau) G(x,y,z,\xi,\eta,0,t-\tau) \, d\xi \, d\eta \, d\tau$$
$$+ a \int_0^t \int_{-\infty}^{\infty} \int_{-\infty}^{\infty} g_2(\xi,\eta,\tau) G(x,y,z,\xi,\eta,l,t-\tau) \, d\xi \, d\eta \, d\tau,$$

where

$$G(x,y,z,\xi,\eta,\zeta,t) = \frac{1}{4\pi alt} \exp\left[-\frac{(x-\xi)^2 + (y-\eta)^2}{4at}\right]$$
$$\times \left[1 + 2\sum_{n=1}^{\infty} \cos\frac{n\pi z}{l} \cos\frac{n\pi \zeta}{l} \exp\left(-\frac{n^2\pi^2 at}{l^2}\right)\right],$$

or

$$G(x,y,z,\xi,\eta,\zeta,t) = \frac{1}{(2\sqrt{\pi a t})^3} \exp\left[-\frac{(x-\xi)^2+(y-\eta)^2}{4at}\right]$$
$$\times \sum_{n=-\infty}^{\infty} \left\{\exp\left[-\frac{(z-\zeta+2nl)^2}{4at}\right] + \exp\left[-\frac{(z+\zeta+2nl)^2}{4at}\right]\right\}.$$

⊙ *Reference*: H. S. Carslaw and J. C. Jaeger (1984).

3.1.1-9. Domain: $-\infty < x < \infty$, $-\infty < y < \infty$, $0 \le z \le l$. Third boundary value problem.

An infinite layer is considered. The following conditions are prescribed:

$$w = f(x,y,z) \quad \text{at} \quad t=0 \quad \text{(initial condition)},$$
$$\partial_z w - k_1 w = g_1(x,y,t) \quad \text{at} \quad z=0 \quad \text{(boundary condition)},$$
$$\partial_z w + k_2 w = g_2(x,y,t) \quad \text{at} \quad z=l \quad \text{(boundary condition)}.$$

The solution $w(x,y,z,t)$ is determined by the formula in Paragraph 3.1.1-8 where

$$G(x,y,z,\xi,\eta,\zeta,t) = \frac{1}{4\pi at} \exp\left[-\frac{(x-\xi)^2+(y-\eta)^2}{4at}\right] \sum_{n=1}^{\infty} \frac{\varphi_n(z)\varphi_n(\zeta)}{\|\varphi_n\|^2} \exp(-a\mu_n^2 t),$$

$$\varphi_n(z) = \cos(\mu_n z) + \frac{k_1}{\mu_n}\sin(\mu_n z), \quad \|\varphi_n\|^2 = \frac{k_2}{2\mu_n^2}\frac{\mu_n^2+k_1^2}{\mu_n^2+k_2^2} + \frac{k_1}{2\mu_n^2} + \frac{l}{2}\left(1+\frac{k_1^2}{\mu_n^2}\right).$$

Here, the μ_n are positive roots of the transcendental equation $\dfrac{\tan(\mu l)}{\mu} = \dfrac{k_1+k_2}{\mu^2-k_1 k_2}$.

3.1.1-10. Domain: $-\infty < x < \infty$, $-\infty < y < \infty$, $0 \le z \le l$. Mixed boundary value problem.

An infinite layer is considered. The following conditions are prescribed:

$$w = f(x,y,z) \quad \text{at} \quad t=0 \quad \text{(initial condition)},$$
$$w = g_1(x,y,t) \quad \text{at} \quad z=0 \quad \text{(boundary condition)},$$
$$\partial_z w = g_2(x,y,t) \quad \text{at} \quad z=l \quad \text{(boundary condition)}.$$

Solution:

$$w(x,y,z,t) = \int_0^l \int_{-\infty}^{\infty} \int_{-\infty}^{\infty} f(\xi,\eta,\zeta) G(x,y,z,\xi,\eta,\zeta,t)\, d\xi\, d\eta\, d\zeta$$
$$+ a\int_0^t \int_{-\infty}^{\infty} \int_{-\infty}^{\infty} g_1(\xi,\eta,\tau)\left[\frac{\partial}{\partial\zeta}G(x,y,z,\xi,\eta,\zeta,t-\tau)\right]_{\zeta=0} d\xi\, d\eta\, d\tau$$
$$+ a\int_0^t \int_{-\infty}^{\infty} \int_{-\infty}^{\infty} g_2(\xi,\eta,\tau) G(x,y,z,\xi,\eta,l,t-\tau)\, d\xi\, d\eta\, d\tau,$$

where

$$G(x,y,z,\xi,\eta,\zeta,t) = \frac{1}{2\pi alt} \exp\left[-\frac{(x-\xi)^2+(y-\eta)^2}{4at}\right]$$
$$\times \sum_{n=0}^{\infty} \sin\left[\frac{(2n+1)\pi z}{2l}\right] \sin\left[\frac{(2n+1)\pi\zeta}{2l}\right] \exp\left[-\frac{(2n+1)^2\pi^2 at}{4l^2}\right],$$

or

$$G(x,y,z,\xi,\eta,\zeta,t) = \frac{1}{(2\sqrt{\pi a t})^3} \exp\left[-\frac{(x-\xi)^2+(y-\eta)^2}{4at}\right]$$
$$\times \sum_{n=-\infty}^{\infty} (-1)^n \left\{\exp\left[-\frac{(z-\zeta+2nl)^2}{4at}\right] - \exp\left[-\frac{(z+\zeta+2nl)^2}{4at}\right]\right\}.$$

⊙ *Reference*: A. G. Butkovskiy (1979).

3.1.1-11. Domain: $-\infty < x < \infty$, $0 \le y < \infty$, $0 \le z \le l$. First boundary value problem.

A semiinfinite layer is considered. The following conditions are prescribed:

$$w = f(x, y, z) \quad \text{at} \quad t = 0 \quad \text{(initial condition)},$$
$$w = g_1(x, z, t) \quad \text{at} \quad y = 0 \quad \text{(boundary condition)},$$
$$w = g_2(x, y, t) \quad \text{at} \quad z = 0 \quad \text{(boundary condition)},$$
$$w = g_3(x, y, t) \quad \text{at} \quad z = l \quad \text{(boundary condition)}.$$

Solution:

$$w(x, y, z, t) = \int_0^l \int_0^\infty \int_{-\infty}^\infty f(\xi, \eta, \zeta) G(x, y, z, \xi, \eta, \zeta, t)\, d\xi\, d\eta\, d\zeta$$
$$+ a \int_0^t \int_0^l \int_{-\infty}^\infty g_1(\xi, \zeta, \tau) \left[\frac{\partial}{\partial \eta} G(x, y, z, \xi, \eta, \zeta, t-\tau)\right]_{\eta=0} d\xi\, d\zeta\, d\tau$$
$$+ a \int_0^t \int_0^\infty \int_{-\infty}^\infty g_2(\xi, \eta, \tau) \left[\frac{\partial}{\partial \zeta} G(x, y, z, \xi, \eta, \zeta, t-\tau)\right]_{\zeta=0} d\xi\, d\eta\, d\tau$$
$$- a \int_0^t \int_0^\infty \int_{-\infty}^\infty g_3(\xi, \eta, \tau) \left[\frac{\partial}{\partial \zeta} G(x, y, z, \xi, \eta, \zeta, t-\tau)\right]_{\zeta=l} d\xi\, d\eta\, d\tau,$$

where

$$G(x, y, z, \xi, \eta, \zeta, t) = \frac{1}{2\pi a l t} \exp\left[-\frac{(x-\xi)^2}{4at}\right] \left\{\exp\left[-\frac{(y-\eta)^2}{4at}\right] - \exp\left[-\frac{(y+\eta)^2}{4at}\right]\right\}$$
$$\times \sum_{n=1}^\infty \sin\frac{n\pi z}{l} \sin\frac{n\pi \zeta}{l} \exp\left(-\frac{n^2\pi^2 at}{l^2}\right).$$

3.1.1-12. Domain: $-\infty < x < \infty$, $0 \le y < \infty$, $0 \le z \le l$. Second boundary value problem.

A semiinfinite layer is considered. The following conditions are prescribed:

$$w = f(x, y, z) \quad \text{at} \quad t = 0 \quad \text{(initial condition)},$$
$$\partial_y w = g_1(x, z, t) \quad \text{at} \quad y = 0 \quad \text{(boundary condition)},$$
$$\partial_z w = g_2(x, y, t) \quad \text{at} \quad z = 0 \quad \text{(boundary condition)},$$
$$\partial_z w = g_3(x, y, t) \quad \text{at} \quad z = l \quad \text{(boundary condition)}.$$

Solution:

$$w(x, y, z, t) = \int_0^l \int_0^\infty \int_{-\infty}^\infty f(\xi, \eta, \zeta) G(x, y, z, \xi, \eta, \zeta, t)\, d\xi\, d\eta\, d\zeta$$
$$- a \int_0^t \int_0^l \int_{-\infty}^\infty g_1(\xi, \zeta, \tau) G(x, y, z, \xi, 0, \zeta, t-\tau)\, d\xi\, d\zeta\, d\tau$$
$$- a \int_0^t \int_0^\infty \int_{-\infty}^\infty g_2(\xi, \eta, \tau) G(x, y, z, \xi, \eta, 0, t-\tau)\, d\xi\, d\eta\, d\tau$$
$$+ a \int_0^t \int_0^\infty \int_{-\infty}^\infty g_3(\xi, \eta, \tau) G(x, y, z, \xi, \eta, l, t-\tau)\, d\xi\, d\eta\, d\tau,$$

where

$$G(x, y, z, \xi, \eta, \zeta, t) = \frac{1}{4\pi a l t} \exp\left[-\frac{(x-\xi)^2}{4at}\right] \left\{\exp\left[-\frac{(y-\eta)^2}{4at}\right] + \exp\left[-\frac{(y+\eta)^2}{4at}\right]\right\}$$
$$\times \left[1 + 2\sum_{n=1}^\infty \cos\frac{n\pi z}{l} \cos\frac{n\pi \zeta}{l} \exp\left(-\frac{n^2\pi^2 at}{l^2}\right)\right].$$

3.1.1-13. Domain: $-\infty < x < \infty$, $0 \le y < \infty$, $0 \le z \le l$. **Third boundary value problem.**

A semiinfinite layer is considered. The following conditions are prescribed:

$$w = f(x,y,z) \quad \text{at} \quad t = 0 \quad \text{(initial condition)},$$
$$\partial_y w - k_1 w = g_1(x,z,t) \quad \text{at} \quad y = 0 \quad \text{(boundary condition)},$$
$$\partial_z w - k_2 w = g_2(x,y,t) \quad \text{at} \quad z = 0 \quad \text{(boundary condition)},$$
$$\partial_z w + k_3 w = g_3(x,y,t) \quad \text{at} \quad z = l \quad \text{(boundary condition)}.$$

The solution $w(x,y,z,t)$ is determined by the formula in Paragraph 3.1.1-12 where

$$G(x,y,z,\xi,\eta,\zeta,t) = \frac{1}{4\pi at} \exp\left[-\frac{(x-\xi)^2}{4at}\right] H(y,\eta,t) \sum_{n=1}^{\infty} \frac{\varphi_n(z)\varphi_n(\zeta)}{\|\varphi_n\|^2} \exp(-a\mu_n^2 t),$$

$$H(y,\eta,t) = \exp\left[-\frac{(y-\eta)^2}{4at}\right] + \exp\left[-\frac{(y+\eta)^2}{4at}\right] - 2k_1 \int_0^{\infty} \exp\left[-\frac{(y+\eta+s)^2}{4at} - k_1 s\right] ds.$$

Here,

$$\varphi_n(z) = \cos(\mu_n z) + \frac{k_2}{\mu_n}\sin(\mu_n z), \quad \|\varphi_n\|^2 = \frac{k_3}{2\mu_n^2}\frac{\mu_n^2 + k_2^2}{\mu_n^2 + k_3^2} + \frac{k_2}{2\mu_n^2} + \frac{l}{2}\left(1 + \frac{k_2^2}{\mu_n^2}\right);$$

the μ_n are positive roots of the transcendental equation $\dfrac{\tan(\mu l)}{\mu} = \dfrac{k_2 + k_3}{\mu^2 - k_2 k_3}$.

3.1.1-14. Domain: $-\infty < x < \infty$, $0 \le y < \infty$, $0 \le z \le l$. **Mixed boundary value problems.**

1°. A semiinfinite layer is considered. The following conditions are prescribed:

$$w = f(x,y,z) \quad \text{at} \quad t = 0 \quad \text{(initial condition)},$$
$$w = g_1(x,z,t) \quad \text{at} \quad y = 0 \quad \text{(boundary condition)},$$
$$\partial_z w = g_2(x,y,t) \quad \text{at} \quad z = 0 \quad \text{(boundary condition)},$$
$$\partial_z w = g_3(x,y,t) \quad \text{at} \quad z = l \quad \text{(boundary condition)}.$$

Solution:

$$w(x,y,z,t) = \int_0^l \int_0^{\infty} \int_{-\infty}^{\infty} f(\xi,\eta,\zeta) G(x,y,z,\xi,\eta,\zeta,t)\, d\xi\, d\eta\, d\zeta$$
$$+ a \int_0^t \int_0^l \int_{-\infty}^{\infty} g_1(\xi,\zeta,\tau) \left[\frac{\partial}{\partial \eta} G(x,y,z,\xi,\eta,\zeta,t-\tau)\right]_{\eta=0} d\xi\, d\zeta\, d\tau$$
$$- a \int_0^t \int_0^{\infty} \int_{-\infty}^{\infty} g_2(\xi,\eta,\tau) G(x,y,z,\xi,\eta,0,t-\tau)\, d\xi\, d\eta\, d\tau$$
$$+ a \int_0^t \int_0^{\infty} \int_{-\infty}^{\infty} g_3(\xi,\eta,\tau) G(x,y,z,\xi,\eta,l,t-\tau)\, d\xi\, d\eta\, d\tau,$$

where

$$G(x,y,z,\xi,\eta,\zeta,t) = \frac{1}{4\pi alt} \exp\left[-\frac{(x-\xi)^2}{4at}\right] \left\{\exp\left[-\frac{(y-\eta)^2}{4at}\right] - \exp\left[-\frac{(y+\eta)^2}{4at}\right]\right\}$$
$$\times \left[1 + 2\sum_{n=1}^{\infty} \cos\frac{n\pi z}{l} \cos\frac{n\pi \zeta}{l} \exp\left(-\frac{n^2\pi^2 at}{l^2}\right)\right].$$

2°. A semiinfinite layer is considered. The following conditions are prescribed:

$$w = f(x, y, z) \quad \text{at} \quad t = 0 \quad \text{(initial condition)},$$
$$\partial_y w = g_1(x, z, t) \quad \text{at} \quad y = 0 \quad \text{(boundary condition)},$$
$$w = g_2(x, y, t) \quad \text{at} \quad z = 0 \quad \text{(boundary condition)},$$
$$w = g_3(x, y, t) \quad \text{at} \quad z = l \quad \text{(boundary condition)}.$$

Solution:

$$w(x, y, z, t) = \int_0^l \int_0^\infty \int_{-\infty}^\infty f(\xi, \eta, \zeta) G(x, y, z, \xi, \eta, \zeta, t) \, d\xi \, d\eta \, d\zeta$$
$$- a \int_0^t \int_0^l \int_{-\infty}^\infty g_1(\xi, \zeta, \tau) G(x, y, z, \xi, 0, \zeta, t - \tau) \, d\xi \, d\zeta \, d\tau$$
$$+ a \int_0^t \int_0^\infty \int_{-\infty}^\infty g_2(\xi, \eta, \tau) \left[\frac{\partial}{\partial \zeta} G(x, y, z, \xi, \eta, \zeta, t - \tau) \right]_{\zeta=0} d\xi \, d\eta \, d\tau$$
$$- a \int_0^t \int_0^\infty \int_{-\infty}^\infty g_3(\xi, \eta, \tau) \left[\frac{\partial}{\partial \zeta} G(x, y, z, \xi, \eta, \zeta, t - \tau) \right]_{\zeta=l} d\xi \, d\eta \, d\tau,$$

where

$$G(x, y, z, \xi, \eta, \zeta, t) = \frac{1}{2\pi a l t} \exp\left[-\frac{(x - \xi)^2}{4at} \right] \left\{ \exp\left[-\frac{(y - \eta)^2}{4at} \right] + \exp\left[-\frac{(y + \eta)^2}{4at} \right] \right\}$$
$$\times \sum_{n=1}^\infty \sin \frac{n \pi z}{l} \sin \frac{n \pi \zeta}{l} \exp\left(-\frac{n^2 \pi^2 a t}{l^2} \right).$$

3.1.1-15. Domain: $0 \le x < \infty$, $0 \le y < \infty$, $0 \le z < \infty$. First boundary value problem.

An octant is considered. The following conditions are prescribed:

$$w = f(x, y, z) \quad \text{at} \quad t = 0 \quad \text{(initial condition)},$$
$$w = g_1(y, z, t) \quad \text{at} \quad x = 0 \quad \text{(boundary condition)},$$
$$w = g_2(x, z, t) \quad \text{at} \quad y = 0 \quad \text{(boundary condition)},$$
$$w = g_3(x, y, t) \quad \text{at} \quad z = 0 \quad \text{(boundary condition)}.$$

Solution:

$$w(x, y, z, t) = \int_0^\infty \int_0^\infty \int_0^\infty G(x, y, z, \xi, \eta, \zeta, t) f(\xi, \eta, \zeta) \, d\xi \, d\eta \, d\zeta$$
$$+ a \int_0^t \int_0^\infty \int_0^\infty g_1(\eta, \zeta, \tau) \left[\frac{\partial}{\partial \xi} G(x, y, z, \xi, \eta, \zeta, t - \tau) \right]_{\xi=0} d\eta \, d\zeta \, d\tau$$
$$+ a \int_0^t \int_0^\infty \int_0^\infty g_2(\xi, \zeta, \tau) \left[\frac{\partial}{\partial \eta} G(x, y, z, \xi, \eta, \zeta, t - \tau) \right]_{\eta=0} d\xi \, d\zeta \, d\tau$$
$$+ a \int_0^t \int_0^\infty \int_0^\infty g_3(\xi, \eta, \tau) \left[\frac{\partial}{\partial \zeta} G(x, y, z, \xi, \eta, \zeta, t - \tau) \right]_{\zeta=0} d\xi \, d\eta \, d\tau,$$

where

$$G(x, y, z, \xi, \eta, \zeta, t) = \frac{1}{(2\sqrt{\pi a t})^3} H(x, \xi, t) H(y, \eta, t) H(z, \zeta, t),$$
$$H(x, \xi, t) = \exp\left[-\frac{(x - \xi)^2}{4at} \right] - \exp\left[-\frac{(x + \xi)^2}{4at} \right].$$

Example 3. The initial temperature is uniform, $f(x,y,z) = w_0$. The faces are maintained at zero temperature, $g_1 = g_2 = g_3 = 0$.
Solution:
$$w = w_0 \operatorname{erf}\left(\frac{x}{2\sqrt{at}}\right) \operatorname{erf}\left(\frac{y}{2\sqrt{at}}\right) \operatorname{erf}\left(\frac{z}{2\sqrt{at}}\right).$$

⊙ *Reference*: H. S. Carslaw and J. C. Jaeger (1984).

3.1.1-16. Domain: $0 \le x < \infty$, $0 \le y < \infty$, $0 \le z < \infty$. Second boundary value problem.

An octant is considered. The following conditions are prescribed:

$$\begin{aligned}
w &= f(x,y,z) & \text{at} \quad t &= 0 & \text{(initial condition)}, \\
\partial_x w &= g_1(y,z,t) & \text{at} \quad x &= 0 & \text{(boundary condition)}, \\
\partial_y w &= g_2(x,z,t) & \text{at} \quad y &= 0 & \text{(boundary condition)}, \\
\partial_z w &= g_3(x,y,t) & \text{at} \quad z &= 0 & \text{(boundary condition)}.
\end{aligned}$$

Solution:
$$\begin{aligned}
w(x,y,z,t) &= \int_0^\infty \int_0^\infty \int_0^\infty G(x,y,z,\xi,\eta,\zeta,t) f(\xi,\eta,\zeta) \, d\xi \, d\eta \, d\zeta \\
&\quad - a \int_0^t \int_0^\infty \int_0^\infty g_1(\eta,\zeta,\tau) G(x,y,z,0,\eta,\zeta,t-\tau) \, d\eta \, d\zeta \, d\tau \\
&\quad - a \int_0^t \int_0^\infty \int_0^\infty g_2(\xi,\zeta,\tau) G(x,y,z,\xi,0,\zeta,t-\tau) \, d\xi \, d\zeta \, d\tau \\
&\quad - a \int_0^t \int_0^\infty \int_0^\infty g_3(\xi,\eta,\tau) G(x,y,z,\xi,\eta,0,t-\tau) \, d\xi \, d\eta \, d\tau,
\end{aligned}$$

where
$$G(x,y,z,\xi,\eta,\zeta,t) = \frac{1}{\left(2\sqrt{\pi a t}\right)^3} H(x,\xi,t) H(y,\eta,t) H(z,\zeta,t),$$

$$H(x,\xi,t) = \exp\left[-\frac{(x-\xi)^2}{4at}\right] + \exp\left[-\frac{(x+\xi)^2}{4at}\right].$$

3.1.1-17. Domain: $0 \le x < \infty$, $0 \le y < \infty$, $0 \le z < \infty$. Third boundary value problem.

An octant is considered. The following conditions are prescribed:

$$\begin{aligned}
w &= f(x,y,z) & \text{at} \quad t &= 0 & \text{(initial condition)}, \\
\partial_x w - k_1 w &= g_1(y,z,t) & \text{at} \quad x &= 0 & \text{(boundary condition)}, \\
\partial_y w - k_2 w &= g_2(x,z,t) & \text{at} \quad y &= 0 & \text{(boundary condition)}, \\
\partial_z w - k_3 w &= g_3(x,y,t) & \text{at} \quad z &= 0 & \text{(boundary condition)}.
\end{aligned}$$

The solution $w(x,y,z,t)$ is determined by the formula in Paragraph 3.1.1-16 where

$$G(x,y,z,\xi,\eta,\zeta,t) = \frac{1}{\left(2\sqrt{\pi a t}\right)^3} H(x,\xi,t;k_1) H(y,\eta,t;k_2) H(z,\zeta,t;k_3),$$

$$\begin{aligned}
H(x,\xi,t;k) &= \exp\left[-\frac{(x-\xi)^2}{4at}\right] + \exp\left[-\frac{(x+\xi)^2}{4at}\right] \\
&\quad - 2k\sqrt{\pi a t} \exp[ak^2 t + k(x+\xi)] \operatorname{erfc}\left(\frac{x+\xi}{2\sqrt{at}} + k\sqrt{at}\right).
\end{aligned}$$

Example 4. The initial temperature is uniform, $f(x,y,z) = w_0$. The temperature of the contacting media is zero, $g_1 = g_2 = g_3 = 0$.

Solution:
$$w = w_0 \left[\operatorname{erf}\left(\frac{x}{2\sqrt{at}}\right) + \exp(k_1 x + k_1^2 at)\operatorname{erfc}\left(\frac{x}{2\sqrt{at}} + k_1\sqrt{at}\right) \right]$$
$$\times \left[\operatorname{erf}\left(\frac{y}{2\sqrt{at}}\right) + \exp(k_2 y + k_2^2 at)\operatorname{erfc}\left(\frac{y}{2\sqrt{at}} + k_2\sqrt{at}\right) \right]$$
$$\times \left[\operatorname{erf}\left(\frac{z}{2\sqrt{at}}\right) + \exp(k_3 z + k_3^2 at)\operatorname{erfc}\left(\frac{z}{2\sqrt{at}} + k_3\sqrt{at}\right) \right].$$

⊙ *Reference*: H. S. Carslaw and J. C. Jaeger (1984).

3.1.1-18. Domain: $0 \le x < \infty$, $0 \le y < \infty$, $0 \le z < \infty$. Mixed boundary value problems.

1°. An octant is considered. The following conditions are prescribed:

$$w = f(x, y, z) \quad \text{at} \quad t = 0 \quad \text{(initial condition)},$$
$$w = g_1(y, z, t) \quad \text{at} \quad x = 0 \quad \text{(boundary condition)},$$
$$\partial_y w = g_2(x, z, t) \quad \text{at} \quad y = 0 \quad \text{(boundary condition)},$$
$$\partial_z w = g_3(x, y, t) \quad \text{at} \quad z = 0 \quad \text{(boundary condition)}.$$

Solution:
$$w(x, y, z, t) = \int_0^\infty \int_0^\infty \int_0^\infty G(x, y, z, \xi, \eta, \zeta, t) f(\xi, \eta, \zeta)\, d\xi\, d\eta\, d\zeta$$
$$+ a \int_0^t \int_0^\infty \int_0^\infty g_1(\eta, \zeta, \tau) \left[\frac{\partial}{\partial \xi} G(x, y, z, \xi, \eta, \zeta, t - \tau)\right]_{\xi=0} d\eta\, d\zeta\, d\tau$$
$$- a \int_0^t \int_0^\infty \int_0^\infty g_2(\xi, \zeta, \tau) G(x, y, z, \xi, 0, \zeta, t - \tau)\, d\xi\, d\zeta\, d\tau$$
$$- a \int_0^t \int_0^\infty \int_0^\infty g_3(\xi, \eta, \tau) G(x, y, z, \xi, \eta, 0, t - \tau)\, d\xi\, d\eta\, d\tau,$$

where
$$G(x, y, z, \xi, \eta, \zeta, t) = \frac{1}{(2\sqrt{\pi at})^3} \left\{ \exp\left[-\frac{(x-\xi)^2}{4at}\right] - \exp\left[-\frac{(x+\xi)^2}{4at}\right] \right\} H(y, \eta, t) H(z, \zeta, t),$$
$$H(y, \eta, t) = \exp\left[-\frac{(y-\eta)^2}{4at}\right] + \exp\left[-\frac{(y+\eta)^2}{4at}\right].$$

2°. An octant is considered. The following conditions are prescribed:

$$w = f(x, y, z) \quad \text{at} \quad t = 0 \quad \text{(initial condition)},$$
$$w = g_1(y, z, t) \quad \text{at} \quad x = 0 \quad \text{(boundary condition)},$$
$$w = g_2(x, z, t) \quad \text{at} \quad y = 0 \quad \text{(boundary condition)},$$
$$\partial_z w = g_3(x, y, t) \quad \text{at} \quad z = 0 \quad \text{(boundary condition)}.$$

Solution:
$$w(x, y, z, t) = \int_0^\infty \int_0^\infty \int_0^\infty G(x, y, z, \xi, \eta, \zeta, t) f(\xi, \eta, \zeta)\, d\xi\, d\eta\, d\zeta$$
$$+ a \int_0^t \int_0^\infty \int_0^\infty g_1(\eta, \zeta, \tau) \left[\frac{\partial}{\partial \xi} G(x, y, z, \xi, \eta, \zeta, t - \tau)\right]_{\xi=0} d\eta\, d\zeta\, d\tau$$
$$+ a \int_0^t \int_0^\infty \int_0^\infty g_2(\xi, \zeta, \tau) \left[\frac{\partial}{\partial \eta} G(x, y, z, \xi, \eta, \zeta, t - \tau)\right]_{\eta=0} d\xi\, d\zeta\, d\tau$$
$$- a \int_0^t \int_0^\infty \int_0^\infty g_3(\xi, \eta, \tau) G(x, y, z, \xi, \eta, 0, t - \tau)\, d\xi\, d\eta\, d\tau,$$

where

$$G(x, y, z, \xi, \eta, \zeta, t) = \frac{1}{(2\sqrt{\pi at})^3} H(x, \xi, t) H(y, \eta, t) \left\{ \exp\left[-\frac{(z-\zeta)^2}{4at}\right] + \exp\left[-\frac{(z+\zeta)^2}{4at}\right] \right\},$$

$$H(x, \xi, t) = \exp\left[-\frac{(x-\xi)^2}{4at}\right] - \exp\left[-\frac{(x+\xi)^2}{4at}\right].$$

3.1.1-19. Domain: $0 \leq x \leq l_1$, $0 \leq y \leq l_2$, $-\infty < z < \infty$. First boundary value problem.

An infinite cylindrical domain of a rectangular cross-section is considered. The following conditions are prescribed:

$$\begin{aligned}
w &= f(x, y, z) & \text{at} \quad t &= 0 & \text{(initial condition)}, \\
w &= g_1(y, z, t) & \text{at} \quad x &= 0 & \text{(boundary condition)}, \\
w &= g_2(y, z, t) & \text{at} \quad x &= l_1 & \text{(boundary condition)}, \\
w &= g_3(x, z, t) & \text{at} \quad y &= 0 & \text{(boundary condition)}, \\
w &= g_4(x, z, t) & \text{at} \quad y &= l_2 & \text{(boundary condition)}.
\end{aligned}$$

Solution:

$$\begin{aligned}
w(x, y, z, t) &= \int_{-\infty}^{\infty} \int_0^{l_2} \int_0^{l_1} f(\xi, \eta, \zeta) G(x, y, z, \xi, \eta, \zeta, t) \, d\xi \, d\eta \, d\zeta \\
&+ a \int_0^t \int_{-\infty}^{\infty} \int_0^{l_2} g_1(\eta, \zeta, \tau) \left[\frac{\partial}{\partial \xi} G(x, y, z, \xi, \eta, \zeta, t - \tau)\right]_{\xi=0} d\eta \, d\zeta \, d\tau \\
&- a \int_0^t \int_{-\infty}^{\infty} \int_0^{l_2} g_2(\eta, \zeta, \tau) \left[\frac{\partial}{\partial \xi} G(x, y, z, \xi, \eta, \zeta, t - \tau)\right]_{\xi=l_1} d\eta \, d\zeta \, d\tau \\
&+ a \int_0^t \int_{-\infty}^{\infty} \int_0^{l_1} g_3(\xi, \zeta, \tau) \left[\frac{\partial}{\partial \eta} G(x, y, z, \xi, \eta, \zeta, t - \tau)\right]_{\eta=0} d\xi \, d\zeta \, d\tau \\
&- a \int_0^t \int_{-\infty}^{\infty} \int_0^{l_1} g_4(\xi, \zeta, \tau) \left[\frac{\partial}{\partial \eta} G(x, y, z, \xi, \eta, \zeta, t - \tau)\right]_{\eta=l_2} d\xi \, d\zeta \, d\tau,
\end{aligned}$$

where

$$G(x, y, z, \xi, \eta, \zeta, t) = \frac{1}{2\sqrt{\pi at}} \exp\left[-\frac{(z-\zeta)^2}{4at}\right] H_1(x, \xi, t) H_2(y, \eta, t),$$

$$H_1(x, \xi, t) = \frac{2}{l_1} \sum_{n=1}^{\infty} \sin\left(\frac{\pi n x}{l_1}\right) \sin\left(\frac{\pi n \xi}{l_1}\right) \exp\left(-\frac{\pi^2 n^2 a t}{l_1^2}\right),$$

$$H_2(y, \eta, t) = \frac{2}{l_2} \sum_{n=1}^{\infty} \sin\left(\frac{\pi n y}{l_2}\right) \sin\left(\frac{\pi n \eta}{l_2}\right) \exp\left(-\frac{\pi^2 n^2 a t}{l_2^2}\right).$$

3.1.1-20. Domain: $0 \leq x \leq l_1$, $0 \leq y \leq l_2$, $-\infty < z < \infty$. Second boundary value problem.

An infinite cylindrical domain of a rectangular cross-section is considered. The following conditions are prescribed:

$$\begin{aligned}
w &= f(x, y, z) & \text{at} \quad t &= 0 & \text{(initial condition)}, \\
\partial_x w &= g_1(y, z, t) & \text{at} \quad x &= 0 & \text{(boundary condition)}, \\
\partial_x w &= g_2(y, z, t) & \text{at} \quad x &= l_1 & \text{(boundary condition)}, \\
\partial_y w &= g_3(x, z, t) & \text{at} \quad y &= 0 & \text{(boundary condition)}, \\
\partial_y w &= g_4(x, z, t) & \text{at} \quad y &= l_2 & \text{(boundary condition)}.
\end{aligned}$$

Solution:

$$w(x,y,z,t) = \int_{-\infty}^{\infty}\int_0^{l_2}\int_0^{l_1} f(\xi,\eta,\zeta)G(x,y,z,\xi,\eta,\zeta,t)\,d\xi\,d\eta\,d\zeta$$

$$- a\int_0^t\int_{-\infty}^{\infty}\int_0^{l_2} g_1(\eta,\zeta,\tau)G(x,y,z,0,\eta,\zeta,t-\tau)\,d\eta\,d\zeta\,d\tau$$

$$+ a\int_0^t\int_{-\infty}^{\infty}\int_0^{l_2} g_2(\eta,\zeta,\tau)G(x,y,z,l_1,\eta,\zeta,t-\tau)\,d\eta\,d\zeta\,d\tau$$

$$- a\int_0^t\int_{-\infty}^{\infty}\int_0^{l_1} g_3(\xi,\zeta,\tau)G(x,y,z,\xi,0,\zeta,t-\tau)\,d\xi\,d\zeta\,d\tau$$

$$+ a\int_0^t\int_{-\infty}^{\infty}\int_0^{l_1} g_4(\xi,\zeta,\tau)G(x,y,z,\xi,l_2,\zeta,t-\tau)\,d\xi\,d\zeta\,d\tau,$$

where

$$G(x,y,z,\xi,\eta,\zeta,t) = \frac{1}{2\sqrt{\pi a t}}\exp\left[-\frac{(z-\zeta)^2}{4at}\right]H_1(x,\xi,t)H_2(y,\eta,t),$$

$$H_1(x,\xi,t) = \frac{1}{l_1}\left[1 + 2\sum_{n=1}^{\infty}\cos\left(\frac{\pi n x}{l_1}\right)\cos\left(\frac{\pi n \xi}{l_1}\right)\exp\left(-\frac{\pi^2 n^2 a t}{l_1^2}\right)\right],$$

$$H_2(y,\eta,t) = \frac{1}{l_2}\left[1 + 2\sum_{n=1}^{\infty}\cos\left(\frac{\pi n y}{l_2}\right)\cos\left(\frac{\pi n \eta}{l_2}\right)\exp\left(-\frac{\pi^2 n^2 a t}{l_2^2}\right)\right].$$

3.1.1-21. Domain: $0 \le x \le l_1$, $0 \le y \le l_2$, $-\infty < z < \infty$. Third boundary value problem.

An infinite cylindrical domain of a rectangular cross-section is considered. The following conditions are prescribed:

$$w = f(x,y,z) \quad \text{at} \quad t = 0 \quad \text{(initial condition)},$$
$$\partial_x w - k_1 w = g_1(y,z,t) \quad \text{at} \quad x = 0 \quad \text{(boundary condition)},$$
$$\partial_x w + k_2 w = g_2(y,z,t) \quad \text{at} \quad x = l_1 \quad \text{(boundary condition)},$$
$$\partial_y w - k_3 w = g_3(x,z,t) \quad \text{at} \quad y = 0 \quad \text{(boundary condition)},$$
$$\partial_y w + k_4 w = g_4(x,z,t) \quad \text{at} \quad y = l_2 \quad \text{(boundary condition)}.$$

The solution $w(x,y,z,t)$ is determined by the formula in Paragraph 3.1.1-20 where

$$G(x,y,z,\xi,\eta,\zeta,t) = \frac{1}{2\sqrt{\pi a t}}\exp\left[-\frac{(z-\zeta)^2}{4at}\right]H_1(x,\xi,t)H_2(y,\eta,t),$$

$$H_1(x,\xi,t) = \sum_{n=1}^{\infty}\frac{\varphi_n(x)\varphi_n(\xi)}{\|\varphi_n\|^2}\exp(-a\mu_n^2 t), \quad H_2(y,\eta,t) = \sum_{m=1}^{\infty}\frac{\psi_m(y)\psi_m(\eta)}{\|\psi_m\|^2}\exp(-a\lambda_m^2 t).$$

Here,

$$\varphi_n(x) = \cos(\mu_n x) + \frac{k_1}{\mu_n}\sin(\mu_n x), \quad \|\varphi_n\|^2 = \frac{k_2}{2\mu_n^2}\frac{\mu_n^2 + k_1^2}{\mu_n^2 + k_2^2} + \frac{k_1}{2\mu_n^2} + \frac{l_1}{2}\left(1 + \frac{k_1^2}{\mu_n^2}\right),$$

$$\psi_m(y) = \cos(\lambda_m y) + \frac{k_3}{\lambda_m}\sin(\lambda_m y), \quad \|\psi_m\|^2 = \frac{k_4}{2\lambda_m^2}\frac{\lambda_m^2 + k_3^2}{\lambda_m^2 + k_4^2} + \frac{k_3}{2\lambda_m^2} + \frac{l_2}{2}\left(1 + \frac{k_3^2}{\lambda_m^2}\right);$$

the μ_n and λ_m are positive roots of the transcendental equations

$$\frac{\tan(\mu l_1)}{\mu} = \frac{k_1 + k_2}{\mu^2 - k_1 k_2}, \quad \frac{\tan(\lambda l_2)}{\lambda} = \frac{k_3 + k_4}{\lambda^2 - k_3 k_4}.$$

3.1.1-22. Domain: $0 \leq x \leq l_1$, $0 \leq y \leq l_2$, $-\infty < z < \infty$. Mixed boundary value problem.

An infinite cylindrical domain of a rectangular cross-section is considered. The following conditions are prescribed:

$$w = f(x, y, z) \quad \text{at} \quad t = 0 \quad \text{(initial condition)},$$
$$w = g_1(y, z, t) \quad \text{at} \quad x = 0 \quad \text{(boundary condition)},$$
$$w = g_2(y, z, t) \quad \text{at} \quad x = l_1 \quad \text{(boundary condition)},$$
$$\partial_y w = g_3(x, z, t) \quad \text{at} \quad y = 0 \quad \text{(boundary condition)},$$
$$\partial_y w = g_4(x, z, t) \quad \text{at} \quad y = l_2 \quad \text{(boundary condition)}.$$

Solution:

$$w(x, y, z, t) = \int_{-\infty}^{\infty} \int_0^{l_2} \int_0^{l_1} f(\xi, \eta, \zeta) G(x, y, z, \xi, \eta, \zeta, t)\, d\xi\, d\eta\, d\zeta$$
$$+ a \int_0^t \int_{-\infty}^{\infty} \int_0^{l_2} g_1(\eta, \zeta, \tau) \left[\frac{\partial}{\partial \xi} G(x, y, z, \xi, \eta, \zeta, t - \tau)\right]_{\xi=0} d\eta\, d\zeta\, d\tau$$
$$- a \int_0^t \int_{-\infty}^{\infty} \int_0^{l_2} g_2(\eta, \zeta, \tau) \left[\frac{\partial}{\partial \xi} G(x, y, z, \xi, \eta, \zeta, t - \tau)\right]_{\xi=l_1} d\eta\, d\zeta\, d\tau$$
$$- a \int_0^t \int_{-\infty}^{\infty} \int_0^{l_1} g_3(\xi, \zeta, \tau) G(x, y, z, \xi, 0, \zeta, t - \tau)\, d\xi\, d\zeta\, d\tau$$
$$+ a \int_0^t \int_{-\infty}^{\infty} \int_0^{l_1} g_4(\xi, \zeta, \tau) G(x, y, z, \xi, l_2, \zeta, t - \tau)\, d\xi\, d\zeta\, d\tau,$$

where

$$G(x, y, z, \xi, \eta, \zeta, t) = \frac{2}{l_1 l_2 \sqrt{\pi a t}} \exp\left[-\frac{(z-\zeta)^2}{4at}\right] \left[\sum_{n=1}^{\infty} \sin\left(\frac{\pi n x}{l_1}\right) \sin\left(\frac{\pi n \xi}{l_1}\right) \exp\left(-\frac{\pi^2 n^2 a t}{l_1^2}\right)\right]$$
$$\times \left[\frac{1}{2} + \sum_{m=1}^{\infty} \cos\left(\frac{\pi m x}{l_2}\right) \cos\left(\frac{\pi m \xi}{l_2}\right) \exp\left(-\frac{\pi^2 m^2 a t}{l_2^2}\right)\right].$$

3.1.1-23. Domain: $0 \leq x \leq l_1$, $0 \leq y \leq l_2$, $0 \leq z < \infty$. First boundary value problem.

A semiinfinite cylindrical domain of a rectangular cross-section is considered. The following conditions are prescribed:

$$w = f(x, y, z) \quad \text{at} \quad t = 0 \quad \text{(initial condition)},$$
$$w = g_1(y, z, t) \quad \text{at} \quad x = 0 \quad \text{(boundary condition)},$$
$$w = g_2(y, z, t) \quad \text{at} \quad x = l_1 \quad \text{(boundary condition)},$$
$$w = g_3(x, z, t) \quad \text{at} \quad y = 0 \quad \text{(boundary condition)},$$
$$w = g_4(x, z, t) \quad \text{at} \quad y = l_2 \quad \text{(boundary condition)},$$
$$w = g_5(x, y, t) \quad \text{at} \quad z = 0 \quad \text{(boundary condition)}.$$

Solution:
$$w(x,y,z,t) = \int_0^\infty \int_0^{l_2} \int_0^{l_1} f(\xi,\eta,\zeta) G(x,y,z,\xi,\eta,\zeta,t)\, d\xi\, d\eta\, d\zeta$$
$$+ a \int_0^t \int_0^\infty \int_0^{l_2} g_1(\eta,\zeta,\tau) \left[\frac{\partial}{\partial \xi} G(x,y,z,\xi,\eta,\zeta,t-\tau)\right]_{\xi=0} d\eta\, d\zeta\, d\tau$$
$$- a \int_0^t \int_0^\infty \int_0^{l_2} g_2(\eta,\zeta,\tau) \left[\frac{\partial}{\partial \xi} G(x,y,z,\xi,\eta,\zeta,t-\tau)\right]_{\xi=l_1} d\eta\, d\zeta\, d\tau$$
$$+ a \int_0^t \int_0^\infty \int_0^{l_1} g_3(\xi,\zeta,\tau) \left[\frac{\partial}{\partial \eta} G(x,y,z,\xi,\eta,\zeta,t-\tau)\right]_{\eta=0} d\xi\, d\zeta\, d\tau$$
$$- a \int_0^t \int_0^\infty \int_0^{l_1} g_4(\xi,\zeta,\tau) \left[\frac{\partial}{\partial \eta} G(x,y,z,\xi,\eta,\zeta,t-\tau)\right]_{\eta=l_2} d\xi\, d\zeta\, d\tau$$
$$+ a \int_0^t \int_0^{l_2} \int_0^{l_1} g_5(\xi,\eta,\tau) \left[\frac{\partial}{\partial \zeta} G(x,y,z,\xi,\eta,\zeta,t-\tau)\right]_{\zeta=0} d\xi\, d\eta\, d\tau,$$

where
$$G(x,y,z,\xi,\eta,\zeta,t) = G_1(x,\xi,t;l_1)\, G_1(y,\eta,t;l_2)\, G_2(z,\zeta,t),$$
$$G_1(x,\xi,t;l) = \frac{2}{l} \sum_{n=1}^\infty \sin\left(\frac{\pi n x}{l}\right) \sin\left(\frac{\pi n \xi}{l}\right) \exp\left(-\frac{\pi^2 n^2 a t}{l^2}\right),$$
$$G_2(z,\zeta,t) = \frac{1}{2\sqrt{\pi a t}} \left\{ \exp\left[-\frac{(z-\zeta)^2}{4at}\right] - \exp\left[-\frac{(z+\zeta)^2}{4at}\right] \right\}.$$

3.1.1-24. Domain: $0 \leq x \leq l_1$, $0 \leq y \leq l_2$, $0 \leq z < \infty$. Second boundary value problem.

A semiinfinite cylindrical domain of a rectangular cross-section is considered. The following conditions are prescribed:

$$\begin{aligned}
w &= f(x,y,z) & \text{at} \quad t &= 0 & &\text{(initial condition)}, \\
\partial_x w &= g_1(y,z,t) & \text{at} \quad x &= 0 & &\text{(boundary condition)}, \\
\partial_x w &= g_2(y,z,t) & \text{at} \quad x &= l_1 & &\text{(boundary condition)}, \\
\partial_y w &= g_3(x,z,t) & \text{at} \quad y &= 0 & &\text{(boundary condition)}, \\
\partial_y w &= g_4(x,z,t) & \text{at} \quad y &= l_2 & &\text{(boundary condition)}, \\
\partial_z w &= g_5(x,y,t) & \text{at} \quad z &= 0 & &\text{(boundary condition)}.
\end{aligned}$$

Solution:
$$w(x,y,z,t) = \int_0^\infty \int_0^{l_2} \int_0^{l_1} f(\xi,\eta,\zeta) G(x,y,z,\xi,\eta,\zeta,t)\, d\xi\, d\eta\, d\zeta$$
$$- a \int_0^t \int_0^\infty \int_0^{l_2} g_1(\eta,\zeta,\tau) G(x,y,z,0,\eta,\zeta,t-\tau)\, d\eta\, d\zeta\, d\tau$$
$$+ a \int_0^t \int_0^\infty \int_0^{l_2} g_2(\eta,\zeta,\tau) G(x,y,z,l_1,\eta,\zeta,t-\tau)\, d\eta\, d\zeta\, d\tau$$
$$- a \int_0^t \int_0^\infty \int_0^{l_1} g_3(\xi,\zeta,\tau) G(x,y,z,\xi,0,\zeta,t-\tau)\, d\xi\, d\zeta\, d\tau$$
$$+ a \int_0^t \int_0^\infty \int_0^{l_1} g_4(\xi,\zeta,\tau) G(x,y,z,\xi,l_2,\zeta,t-\tau)\, d\xi\, d\zeta\, d\tau$$
$$- a \int_0^t \int_0^{l_2} \int_0^{l_1} g_5(\xi,\eta,\tau) G(x,y,z,\xi,\eta,0,t-\tau)\, d\xi\, d\eta\, d\tau,$$

where

$$G(x,y,z,\xi,\eta,\zeta,t) = \frac{1}{2\sqrt{\pi at}} \left\{ \exp\left[-\frac{(z-\zeta)^2}{4at}\right] + \exp\left[-\frac{(z+\zeta)^2}{4at}\right] \right\} G_1(x,\xi,t) G_2(y,\eta,t),$$

$$G_1(x,\xi,t) = \frac{1}{l_1}\left[1 + 2\sum_{n=1}^{\infty} \cos\left(\frac{\pi n x}{l_1}\right) \cos\left(\frac{\pi n \xi}{l_1}\right) \exp\left(-\frac{\pi^2 n^2 at}{l_1^2}\right)\right],$$

$$G_2(y,\eta,t) = \frac{1}{l_2}\left[1 + 2\sum_{n=1}^{\infty} \cos\left(\frac{\pi n y}{l_2}\right) \cos\left(\frac{\pi n \eta}{l_2}\right) \exp\left(-\frac{\pi^2 n^2 at}{l_2^2}\right)\right].$$

3.1.1-25. Domain: $0 \le x \le l_1$, $0 \le y \le l_2$, $0 \le z < \infty$. Third boundary value problem.

A semiinfinite cylindrical domain of a rectangular cross-section is considered. The following conditions are prescribed:

$$\begin{aligned}
w = f(x,y,z) &\quad \text{at} \quad t = 0 \quad \text{(initial condition)},\\
\partial_x w - k_1 w = g_1(y,z,t) &\quad \text{at} \quad x = 0 \quad \text{(boundary condition)},\\
\partial_x w + k_2 w = g_2(y,z,t) &\quad \text{at} \quad x = l_1 \quad \text{(boundary condition)},\\
\partial_y w - k_3 w = g_3(x,z,t) &\quad \text{at} \quad y = 0 \quad \text{(boundary condition)},\\
\partial_y w + k_4 w = g_4(x,z,t) &\quad \text{at} \quad y = l_2 \quad \text{(boundary condition)},\\
\partial_z w - k_5 w = g_5(x,y,t) &\quad \text{at} \quad z = 0 \quad \text{(boundary condition)}.
\end{aligned}$$

The solution $w(x,y,z,t)$ is determined by the formula in Paragraph 3.1.1-24 where

$$G(x,y,z,\xi,\eta,\zeta,t) = H_1(x,\xi,t) H_2(y,\eta,t) H_3(z,\zeta,t),$$

$$H_3(z,\zeta,t) = \frac{1}{2\sqrt{\pi at}} \left\{ \exp\left[-\frac{(z-\zeta)^2}{4at}\right] + \exp\left[-\frac{(z+\zeta)^2}{4at}\right] \right\}$$
$$- k_5 \exp\left[k_5^2 at + k_5(z+\zeta)\right] \operatorname{erfc}\left(\frac{z+\zeta}{2\sqrt{at}} + k_5\sqrt{at}\right),$$

and the functions $H_1(x,\xi,t)$ and $H_2(y,\eta,t)$ can be found in Paragraph 3.1.1-21.

3.1.1-26. Domain: $0 \le x \le l_1$, $0 \le y \le l_2$, $0 \le z < \infty$. Mixed boundary value problems.

1°. A semiinfinite cylindrical domain of a rectangular cross-section is considered. The following conditions are prescribed:

$$\begin{aligned}
w = f(x,y,z) &\quad \text{at} \quad t = 0 \quad \text{(initial condition)},\\
w = g_1(y,z,t) &\quad \text{at} \quad x = 0 \quad \text{(boundary condition)},\\
w = g_2(y,z,t) &\quad \text{at} \quad x = l_1 \quad \text{(boundary condition)},\\
w = g_3(x,z,t) &\quad \text{at} \quad y = 0 \quad \text{(boundary condition)},\\
w = g_4(x,z,t) &\quad \text{at} \quad y = l_2 \quad \text{(boundary condition)},\\
\partial_z w = g_5(x,y,t) &\quad \text{at} \quad z = 0 \quad \text{(boundary condition)}.
\end{aligned}$$

Solution:

$$w(x,y,z,t) = \int_0^\infty \int_0^{l_2} \int_0^{l_1} f(\xi,\eta,\zeta) G(x,y,z,\xi,\eta,\zeta,t)\, d\xi\, d\eta\, d\zeta$$

$$+ a \int_0^t \int_0^\infty \int_0^{l_2} g_1(\eta,\zeta,\tau) \left[\frac{\partial}{\partial \xi} G(x,y,z,\xi,\eta,\zeta,t-\tau)\right]_{\xi=0} d\eta\, d\zeta\, d\tau$$

$$- a \int_0^t \int_0^\infty \int_0^{l_2} g_2(\eta,\zeta,\tau) \left[\frac{\partial}{\partial \xi} G(x,y,z,\xi,\eta,\zeta,t-\tau)\right]_{\xi=l_1} d\eta\, d\zeta\, d\tau$$

$$+ a \int_0^t \int_0^\infty \int_0^{l_1} g_3(\xi,\zeta,\tau) \left[\frac{\partial}{\partial \eta} G(x,y,z,\xi,\eta,\zeta,t-\tau)\right]_{\eta=0} d\xi\, d\zeta\, d\tau$$

$$- a \int_0^t \int_0^\infty \int_0^{l_1} g_4(\xi,\zeta,\tau) \left[\frac{\partial}{\partial \eta} G(x,y,z,\xi,\eta,\zeta,t-\tau)\right]_{\eta=l_2} d\xi\, d\zeta\, d\tau$$

$$- a \int_0^t \int_0^{l_2} \int_0^{l_1} g_5(\xi,\eta,\tau) G(x,y,z,\xi,\eta,0,t-\tau)\, d\xi\, d\eta\, d\tau,$$

where

$$G(x,y,z,\xi,\eta,\zeta,t) = \frac{1}{2\sqrt{\pi a t}} \left\{ \exp\left[-\frac{(z-\zeta)^2}{4at}\right] + \exp\left[-\frac{(z+\zeta)^2}{4at}\right] \right\} H(x,\xi,t;l_1) H(y,\eta,t;l_2),$$

$$H(x,\xi,t;l) = \frac{2}{l} \sum_{n=1}^\infty \sin\left(\frac{\pi n x}{l}\right) \sin\left(\frac{\pi n \xi}{l}\right) \exp\left(-\frac{\pi^2 n^2 a t}{l^2}\right).$$

2°. A semiinfinite cylindrical domain of a rectangular cross-section is considered. The following conditions are prescribed:

$$\begin{aligned}
w &= f(x,y,z) & \text{at}\quad & t = 0 & \text{(initial condition)}, \\
\partial_x w &= g_1(y,z,t) & \text{at}\quad & x = 0 & \text{(boundary condition)}, \\
\partial_x w &= g_2(y,z,t) & \text{at}\quad & x = l_1 & \text{(boundary condition)}, \\
\partial_y w &= g_3(x,z,t) & \text{at}\quad & y = 0 & \text{(boundary condition)}, \\
\partial_y w &= g_4(x,z,t) & \text{at}\quad & y = l_2 & \text{(boundary condition)}, \\
w &= g_5(x,y,t) & \text{at}\quad & z = 0 & \text{(boundary condition)}.
\end{aligned}$$

Solution:

$$w(x,y,z,t) = \int_0^\infty \int_0^{l_2} \int_0^{l_1} f(\xi,\eta,\zeta) G(x,y,z,\xi,\eta,\zeta,t)\, d\xi\, d\eta\, d\zeta$$

$$- a \int_0^t \int_0^\infty \int_0^{l_2} g_1(\eta,\zeta,\tau) G(x,y,z,0,\eta,\zeta,t-\tau)\, d\eta\, d\zeta\, d\tau$$

$$+ a \int_0^t \int_0^\infty \int_0^{l_2} g_2(\eta,\zeta,\tau) G(x,y,z,l_1,\eta,\zeta,t-\tau)\, d\eta\, d\zeta\, d\tau$$

$$- a \int_0^t \int_0^\infty \int_0^{l_1} g_3(\xi,\zeta,\tau) G(x,y,z,\xi,0,\zeta,t-\tau)\, d\xi\, d\zeta\, d\tau$$

$$+ a \int_0^t \int_0^\infty \int_0^{l_1} g_4(\xi,\zeta,\tau) G(x,y,z,\xi,l_2,\zeta,t-\tau)\, d\xi\, d\zeta\, d\tau$$

$$+ a \int_0^t \int_0^{l_2} \int_0^{l_1} g_5(\xi,\eta,\tau) \left[\frac{\partial}{\partial \zeta} G(x,y,z,\xi,\eta,\zeta,t-\tau)\right]_{\zeta=0} d\xi\, d\eta\, d\tau,$$

where
$$G(x,y,z,\xi,\eta,\zeta,t) = \frac{1}{2\sqrt{\pi at}}\left\{\exp\left[-\frac{(z-\zeta)^2}{4at}\right] - \exp\left[-\frac{(z+\zeta)^2}{4at}\right]\right\}H(x,\xi,t;l_1)H(y,\eta,t;l_2),$$
$$H(x,\xi,t;l) = \frac{1}{l}\left[1 + 2\sum_{n=1}^{\infty}\cos\left(\frac{\pi nx}{l}\right)\cos\left(\frac{\pi n\xi}{l}\right)\exp\left(-\frac{\pi^2 n^2 at}{l^2}\right)\right].$$

3.1.1-27. Domain: $0 \leq x \leq l_1$, $0 \leq y \leq l_2$, $0 \leq z \leq l_3$. First boundary value problem.

A rectangular parallelepiped is considered. The following conditions are prescribed:

$$\begin{aligned}
w &= f(x,y,z) & \text{at} \quad t &= 0 & \text{(initial condition)}, \\
w &= g_1(y,z,t) & \text{at} \quad x &= 0 & \text{(boundary condition)}, \\
w &= g_2(y,z,t) & \text{at} \quad x &= l_1 & \text{(boundary condition)}, \\
w &= g_3(x,z,t) & \text{at} \quad y &= 0 & \text{(boundary condition)}, \\
w &= g_4(x,z,t) & \text{at} \quad y &= l_2 & \text{(boundary condition)}, \\
w &= g_5(x,y,t) & \text{at} \quad z &= 0 & \text{(boundary condition)}, \\
w &= g_6(x,y,t) & \text{at} \quad z &= l_3 & \text{(boundary condition)}.
\end{aligned}$$

Solution:
$$\begin{aligned}
w(x,y,z,t) &= \int_0^{l_3}\int_0^{l_2}\int_0^{l_1} f(\xi,\eta,\zeta)G(x,y,z,\xi,\eta,\zeta,t)\,d\xi\,d\eta\,d\zeta \\
&+ a\int_0^t\int_0^{l_3}\int_0^{l_2} g_1(\eta,\zeta,\tau)\left[\frac{\partial}{\partial\xi}G(x,y,z,\xi,\eta,\zeta,t-\tau)\right]_{\xi=0} d\eta\,d\zeta\,d\tau \\
&- a\int_0^t\int_0^{l_3}\int_0^{l_2} g_2(\eta,\zeta,\tau)\left[\frac{\partial}{\partial\xi}G(x,y,z,\xi,\eta,\zeta,t-\tau)\right]_{\xi=l_1} d\eta\,d\zeta\,d\tau \\
&+ a\int_0^t\int_0^{l_3}\int_0^{l_1} g_3(\xi,\zeta,\tau)\left[\frac{\partial}{\partial\eta}G(x,y,z,\xi,\eta,\zeta,t-\tau)\right]_{\eta=0} d\xi\,d\zeta\,d\tau \\
&- a\int_0^t\int_0^{l_3}\int_0^{l_1} g_4(\xi,\zeta,\tau)\left[\frac{\partial}{\partial\eta}G(x,y,z,\xi,\eta,\zeta,t-\tau)\right]_{\eta=l_2} d\xi\,d\zeta\,d\tau \\
&+ a\int_0^t\int_0^{l_2}\int_0^{l_1} g_5(\xi,\eta,\tau)\left[\frac{\partial}{\partial\zeta}G(x,y,z,\xi,\eta,\zeta,t-\tau)\right]_{\zeta=0} d\xi\,d\eta\,d\tau \\
&- a\int_0^t\int_0^{l_2}\int_0^{l_1} g_6(\xi,\eta,\tau)\left[\frac{\partial}{\partial\zeta}G(x,y,z,\xi,\eta,\zeta,t-\tau)\right]_{\zeta=l_3} d\xi\,d\eta\,d\tau,
\end{aligned}$$

where
$$G(x,y,z,\xi,\eta,\zeta,t) = G_1(x,\xi,t)G_2(y,\eta,t)G_3(z,\zeta,t),$$
$$G_1(x,\xi,t) = \frac{2}{l_1}\sum_{n=1}^{\infty}\sin\left(\frac{\pi nx}{l_1}\right)\sin\left(\frac{\pi n\xi}{l_1}\right)\exp\left(-\frac{\pi^2 n^2 at}{l_1^2}\right),$$
$$G_2(y,\eta,t) = \frac{2}{l_2}\sum_{n=1}^{\infty}\sin\left(\frac{\pi ny}{l_2}\right)\sin\left(\frac{\pi n\eta}{l_2}\right)\exp\left(-\frac{\pi^2 n^2 at}{l_2^2}\right),$$
$$G_3(z,\zeta,t) = \frac{2}{l_3}\sum_{n=1}^{\infty}\sin\left(\frac{\pi nz}{l_3}\right)\sin\left(\frac{\pi n\zeta}{l_3}\right)\exp\left(-\frac{\pi^2 n^2 at}{l_3^2}\right).$$

⊙ *Reference*: H. S. Carslaw and J. C. Jaeger (1984).

3.1.1-28. Domain: $0 \leq x \leq l_1$, $0 \leq y \leq l_2$, $0 \leq z \leq l_3$. **Second boundary value problem.**

A rectangular parallelepiped is considered. The following conditions are prescribed:

$$w = f(x, y, z) \quad \text{at} \quad t = 0 \quad \text{(initial condition)},$$
$$\partial_x w = g_1(y, z, t) \quad \text{at} \quad x = 0 \quad \text{(boundary condition)},$$
$$\partial_x w = g_2(y, z, t) \quad \text{at} \quad x = l_1 \quad \text{(boundary condition)},$$
$$\partial_y w = g_3(x, z, t) \quad \text{at} \quad y = 0 \quad \text{(boundary condition)},$$
$$\partial_y w = g_4(x, z, t) \quad \text{at} \quad y = l_2 \quad \text{(boundary condition)},$$
$$\partial_z w = g_5(x, y, t) \quad \text{at} \quad z = 0 \quad \text{(boundary condition)},$$
$$\partial_z w = g_6(x, y, t) \quad \text{at} \quad z = l_3 \quad \text{(boundary condition)}.$$

Solution:

$$w(x, y, z, t) = \int_0^{l_3}\int_0^{l_2}\int_0^{l_1} f(\xi, \eta, \zeta) G(x, y, z, \xi, \eta, \zeta, t)\, d\xi\, d\eta\, d\zeta$$
$$- a \int_0^t \int_0^{l_3} \int_0^{l_2} g_1(\eta, \zeta, \tau) G(x, y, z, 0, \eta, \zeta, t - \tau)\, d\eta\, d\zeta\, d\tau$$
$$+ a \int_0^t \int_0^{l_3} \int_0^{l_2} g_2(\eta, \zeta, \tau) G(x, y, z, l_1, \eta, \zeta, t - \tau)\, d\eta\, d\zeta\, d\tau$$
$$- a \int_0^t \int_0^{l_3} \int_0^{l_1} g_3(\xi, \zeta, \tau) G(x, y, z, \xi, 0, \zeta, t - \tau)\, d\xi\, d\zeta\, d\tau$$
$$+ a \int_0^t \int_0^{l_3} \int_0^{l_1} g_4(\xi, \zeta, \tau) G(x, y, z, \xi, l_2, \zeta, t - \tau)\, d\xi\, d\zeta\, d\tau$$
$$- a \int_0^t \int_0^{l_2} \int_0^{l_1} g_5(\xi, \eta, \tau) G(x, y, z, \xi, \eta, 0, t - \tau)\, d\xi\, d\eta\, d\tau$$
$$+ a \int_0^t \int_0^{l_2} \int_0^{l_1} g_6(\xi, \eta, \tau) G(x, y, z, \xi, \eta, l_3, t - \tau)\, d\xi\, d\eta\, d\tau,$$

where

$$G(x, y, z, \xi, \eta, \zeta, t) = G_1(x, \xi, t) G_2(y, \eta, t) G_3(z, \zeta, t),$$
$$G_1(x, \xi, t) = \frac{1}{l_1}\left[1 + 2 \sum_{n=1}^{\infty} \cos\left(\frac{\pi n x}{l_1}\right) \cos\left(\frac{\pi n \xi}{l_1}\right) \exp\left(-\frac{\pi^2 n^2 a t}{l_1^2}\right)\right],$$
$$G_2(y, \eta, t) = \frac{1}{l_2}\left[1 + 2 \sum_{n=1}^{\infty} \cos\left(\frac{\pi n y}{l_2}\right) \cos\left(\frac{\pi n \eta}{l_2}\right) \exp\left(-\frac{\pi^2 n^2 a t}{l_2^2}\right)\right],$$
$$G_3(z, \zeta, t) = \frac{1}{l_3}\left[1 + 2 \sum_{n=1}^{\infty} \cos\left(\frac{\pi n z}{l_3}\right) \cos\left(\frac{\pi n \zeta}{l_3}\right) \exp\left(-\frac{\pi^2 n^2 a t}{l_3^2}\right)\right].$$

3.1.1-29. Domain: $0 \leq x \leq l_1$, $0 \leq y \leq l_2$, $0 \leq z \leq l_3$. **Third boundary value problem.**

A rectangular parallelepiped is considered. The following conditions are prescribed:

$$w = f(x, y, z) \quad \text{at} \quad t = 0 \quad \text{(initial condition)},$$
$$\partial_x w - k_1 w = g_1(y, z, t) \quad \text{at} \quad x = 0 \quad \text{(boundary condition)},$$
$$\partial_x w + k_2 w = g_2(y, z, t) \quad \text{at} \quad x = l_1 \quad \text{(boundary condition)},$$
$$\partial_y w - k_3 w = g_3(x, z, t) \quad \text{at} \quad y = 0 \quad \text{(boundary condition)},$$
$$\partial_y w + k_4 w = g_4(x, z, t) \quad \text{at} \quad y = l_2 \quad \text{(boundary condition)},$$
$$\partial_z w - k_5 w = g_5(x, y, t) \quad \text{at} \quad z = 0 \quad \text{(boundary condition)},$$
$$\partial_z w + k_6 w = g_6(x, y, t) \quad \text{at} \quad z = l_3 \quad \text{(boundary condition)}.$$

The solution $w(x, y, z, t)$ is determined by the formula in Paragraph 3.1.1-28 where

$$G(x, y, z, \xi, \eta, \zeta, t) = H_1(x, \xi, t) H_2(y, \eta, t) H_3(z, \zeta, t).$$

The functions $H_1(x, \xi, t)$ and $H_2(y, \eta, t)$ can be found in Paragraph 3.1.1-21, and the function $H_3(z, \zeta, t)$ is given by

$$H_3(z, \zeta, t) = \sum_{n=1}^{\infty} \frac{\rho_n(z) \rho_n(\zeta)}{\|\rho_n\|^2} \exp(-a \nu_n^2 t),$$

$$\rho_n(x) = \cos(\nu_n x) + \frac{k_5}{\nu_n} \sin(\nu_n x), \quad \|\rho_n\|^2 = \frac{k_6}{2\nu_n^2} \frac{\nu_n^2 + k_5^2}{\nu_n^2 + k_6^2} + \frac{k_5}{2\nu_n^2} + \frac{l_3}{2}\left(1 + \frac{k_5^2}{\nu_n^2}\right),$$

where the ν_n are positive roots of the transcendental equation $\dfrac{\tan(\nu l_3)}{\nu} = \dfrac{k_5 + k_6}{\nu^2 - k_5 k_6}$.

3.1.1-30. Domain: $0 \le x \le l_1$, $0 \le y \le l_2$, $0 \le z \le l_3$. Mixed boundary value problems.

1°. A rectangular parallelepiped is considered. The following conditions are prescribed:

$$\begin{aligned}
w &= f(x, y, z) & \text{at} \quad & t = 0 & \text{(initial condition)}, \\
w &= g_1(y, z, t) & \text{at} \quad & x = 0 & \text{(boundary condition)}, \\
w &= g_2(y, z, t) & \text{at} \quad & x = l_1 & \text{(boundary condition)}, \\
w &= g_3(x, z, t) & \text{at} \quad & y = 0 & \text{(boundary condition)}, \\
w &= g_4(x, z, t) & \text{at} \quad & y = l_2 & \text{(boundary condition)}, \\
\partial_z w &= g_5(x, y, t) & \text{at} \quad & z = 0 & \text{(boundary condition)}, \\
\partial_z w &= g_6(x, y, t) & \text{at} \quad & z = l_3 & \text{(boundary condition)}.
\end{aligned}$$

Solution:

$$\begin{aligned}
w(x, y, z, t) = & \int_0^{l_3}\!\!\int_0^{l_2}\!\!\int_0^{l_1} f(\xi, \eta, \zeta) G(x, y, z, \xi, \eta, \zeta, t)\, d\xi\, d\eta\, d\zeta \\
& + a \int_0^t\!\!\int_0^{l_3}\!\!\int_0^{l_2} g_1(\eta, \zeta, \tau) \left[\frac{\partial}{\partial \xi} G(x, y, z, \xi, \eta, \zeta, t - \tau) \right]_{\xi=0} d\eta\, d\zeta\, d\tau \\
& - a \int_0^t\!\!\int_0^{l_3}\!\!\int_0^{l_2} g_2(\eta, \zeta, \tau) \left[\frac{\partial}{\partial \xi} G(x, y, z, \xi, \eta, \zeta, t - \tau) \right]_{\xi=l_1} d\eta\, d\zeta\, d\tau \\
& + a \int_0^t\!\!\int_0^{l_3}\!\!\int_0^{l_1} g_3(\xi, \zeta, \tau) \left[\frac{\partial}{\partial \eta} G(x, y, z, \xi, \eta, \zeta, t - \tau) \right]_{\eta=0} d\xi\, d\zeta\, d\tau \\
& - a \int_0^t\!\!\int_0^{l_3}\!\!\int_0^{l_1} g_4(\xi, \zeta, \tau) \left[\frac{\partial}{\partial \eta} G(x, y, z, \xi, \eta, \zeta, t - \tau) \right]_{\eta=l_2} d\xi\, d\zeta\, d\tau \\
& - a \int_0^t\!\!\int_0^{l_2}\!\!\int_0^{l_1} g_5(\xi, \eta, \tau) G(x, y, z, \xi, \eta, 0, t - \tau)\, d\xi\, d\eta\, d\tau \\
& + a \int_0^t\!\!\int_0^{l_2}\!\!\int_0^{l_1} g_6(\xi, \eta, \tau) G(x, y, z, \xi, \eta, l_3, t - \tau)\, d\xi\, d\eta\, d\tau,
\end{aligned}$$

where
$$G(x,y,z,\xi,\eta,\zeta,t) = G_1(x,\xi,t)G_2(y,\eta,t)G_3(z,\zeta,t),$$
$$G_1(x,\xi,t) = \frac{2}{l_1}\sum_{n=1}^{\infty}\sin\left(\frac{\pi n x}{l_1}\right)\sin\left(\frac{\pi n \xi}{l_1}\right)\exp\left(-\frac{\pi^2 n^2 a t}{l_1^2}\right),$$
$$G_2(y,\eta,t) = \frac{2}{l_2}\sum_{k=1}^{\infty}\sin\left(\frac{\pi k y}{l_2}\right)\sin\left(\frac{\pi k \eta}{l_2}\right)\exp\left(-\frac{\pi^2 k^2 a t}{l_2^2}\right),$$
$$G_3(z,\zeta,t) = \frac{1}{l_3} + \frac{2}{l_3}\sum_{m=1}^{\infty}\cos\left(\frac{\pi m z}{l_3}\right)\cos\left(\frac{\pi m \zeta}{l_3}\right)\exp\left(-\frac{\pi^2 m^2 a t}{l_3^2}\right).$$

2°. A rectangular parallelepiped is considered. The following conditions are prescribed:

$$\begin{aligned}
w &= f(x,y,z) & \text{at} \quad t &= 0 & &\text{(initial condition)}, \\
w &= g_1(y,z,t) & \text{at} \quad x &= 0 & &\text{(boundary condition)}, \\
w &= g_2(y,z,t) & \text{at} \quad x &= l_1 & &\text{(boundary condition)}, \\
\partial_y w &= g_3(x,z,t) & \text{at} \quad y &= 0 & &\text{(boundary condition)}, \\
\partial_y w &= g_4(x,z,t) & \text{at} \quad y &= l_2 & &\text{(boundary condition)}, \\
\partial_z w &= g_5(x,y,t) & \text{at} \quad z &= 0 & &\text{(boundary condition)}, \\
\partial_z w &= g_6(x,y,t) & \text{at} \quad z &= l_3 & &\text{(boundary condition)}.
\end{aligned}$$

Solution:
$$\begin{aligned}
w(x,y,z,t) &= \int_0^{l_3}\int_0^{l_2}\int_0^{l_1} f(\xi,\eta,\zeta)G(x,y,z,\xi,\eta,\zeta,t)\,d\xi\,d\eta\,d\zeta \\
&+ a\int_0^t\int_0^{l_3}\int_0^{l_2} g_1(\eta,\zeta,\tau)\left[\frac{\partial}{\partial \xi}G(x,y,z,\xi,\eta,\zeta,t-\tau)\right]_{\xi=0} d\eta\,d\zeta\,d\tau \\
&- a\int_0^t\int_0^{l_3}\int_0^{l_2} g_2(\eta,\zeta,\tau)\left[\frac{\partial}{\partial \xi}G(x,y,z,\xi,\eta,\zeta,t-\tau)\right]_{\xi=l_1} d\eta\,d\zeta\,d\tau \\
&- a\int_0^t\int_0^{l_3}\int_0^{l_2} g_3(\xi,\zeta,\tau)G(x,y,z,\xi,0,\zeta,t-\tau)\,d\xi\,d\zeta\,d\tau \\
&+ a\int_0^t\int_0^{l_3}\int_0^{l_1} g_4(\xi,\zeta,\tau)G(x,y,z,\xi,l_2,\zeta,t-\tau)\,d\xi\,d\zeta\,d\tau \\
&- a\int_0^t\int_0^{l_2}\int_0^{l_1} g_5(\xi,\eta,\tau)G(x,y,z,\xi,\eta,0,t-\tau)\,d\xi\,d\eta\,d\tau \\
&+ a\int_0^t\int_0^{l_2}\int_0^{l_1} g_6(\xi,\eta,\tau)G(x,y,z,\xi,\eta,l_3,t-\tau)\,d\xi\,d\eta\,d\tau,
\end{aligned}$$

where
$$G(x,y,z,\xi,\eta,\zeta,t) = G_1(x,\xi,t)G_2(y,\eta,t)G_3(z,\zeta,t),$$
$$G_1(x,\xi,t) = \frac{2}{l_1}\sum_{n=1}^{\infty}\sin\left(\frac{\pi n x}{l_1}\right)\sin\left(\frac{\pi n \xi}{l_1}\right)\exp\left(-\frac{\pi^2 n^2 a t}{l_1^2}\right),$$
$$G_2(y,\eta,t) = \frac{1}{l_2} + \frac{2}{l_2}\sum_{k=1}^{\infty}\cos\left(\frac{\pi k y}{l_2}\right)\cos\left(\frac{\pi k \eta}{l_2}\right)\exp\left(-\frac{\pi^2 k^2 a t}{l_2^2}\right),$$
$$G_3(z,\zeta,t) = \frac{1}{l_3} + \frac{2}{l_3}\sum_{m=1}^{\infty}\cos\left(\frac{\pi m z}{l_3}\right)\cos\left(\frac{\pi m \zeta}{l_3}\right)\exp\left(-\frac{\pi^2 m^2 a t}{l_3^2}\right).$$

3.1.2. Problems in Cylindrical Coordinates

The three-dimensional sourceless heat equation in the cylindrical coordinate system has the form

$$\frac{\partial w}{\partial t} = a\left[\frac{1}{r}\frac{\partial}{\partial r}\left(r\frac{\partial w}{\partial r}\right) + \frac{1}{r^2}\frac{\partial^2 w}{\partial \varphi^2} + \frac{\partial^2 w}{\partial z^2}\right], \qquad r = \sqrt{x^2+y^2}.$$

It is used to describe nonsymmetric unsteady processes in moving media or solids with cylindrical or plane boundaries. A similar equation is used to study the corresponding three-dimensional unsteady mass-exchange processes with constant diffusivity.

One-dimensional problems with axial symmetry that have solutions of the form $w = w(r,t)$ are discussed in Subsection 1.2.1. Two-dimensional problems whose solutions have the form $w = w(r, \varphi, t)$ or $w = w(r, z, t)$ are considered in Subsections 2.1.2 and 2.1.3.

3.1.2-1. Remarks on the Green's functions.

For the three-dimensional problems dealt with in Subsection 3.1.2, the Green's function can be represented in the product form

$$G(r, \varphi, z, \xi, \eta, \zeta, t) = G_1(r, \varphi, \xi, \eta, t)G_2(z, \zeta, t),$$

where $G_1(r, \varphi, \xi, \eta, t)$ is the Green's function of the two-dimensional boundary value problem (such functions are presented in Subsection 2.1.2), and $G_2(z, \zeta, t)$ is the Green's function of the corresponding one-dimensional boundary value problem (such functions can be found in Subsections 1.1.1 and 1.1.2).

Example. The Green's function of the first boundary value problem for a semiinfinite circular cylinder ($0 \leq r \leq R$, $0 \leq \varphi \leq 2\pi$, $0 \leq z < \infty$) of Paragraph 3.1.2-5 is the product of the two-dimensional Green's function of the first boundary value problem of Paragraph 2.1.2-2 ($0 \leq r \leq R$, $0 \leq \varphi \leq 2\pi$) and the one-dimensional Green's function of the first boundary value problem of Paragraph 1.1.2-2 ($0 \leq z < \infty$), in which one should perform obvious renaming of variables.

General formulas that enable one to obtain solutions of basic boundary value problems with the help of the Green's function can be found in Subsection 0.8.1.

3.1.2-2. Domain: $0 \leq r \leq R$, $0 \leq \varphi \leq 2\pi$, $-\infty < z < \infty$. First boundary value problem.

An infinite circular cylinder is considered. The following conditions are prescribed:

$$w = f(r, \varphi, z) \quad \text{at} \quad t = 0 \quad \text{(initial condition)},$$
$$w = g(\varphi, z, t) \quad \text{at} \quad r = R \quad \text{(boundary condition)}.$$

Solution:

$$w(r, \varphi, z, t) = \int_{-\infty}^{\infty}\int_0^{2\pi}\int_0^R \xi f(\xi, \eta, \zeta) G(r, \varphi, z, \xi, \eta, \zeta, t)\, d\xi\, d\eta\, d\zeta$$
$$- aR\int_0^t \int_{-\infty}^{\infty}\int_0^{2\pi} g(\eta, \zeta, \tau)\left[\frac{\partial}{\partial \xi}G(r, \varphi, z, \xi, \eta, \zeta, t-\tau)\right]_{\xi=R} d\eta\, d\zeta\, d\tau.$$

Here,

$$G(r, \varphi, z, \xi, \eta, \zeta, t) = G_1(r, \varphi, \xi, \eta, t)G_2(z, \zeta, t),$$

$$G_1(r, \varphi, \xi, \eta, t) = \frac{1}{\pi R^2}\sum_{n=0}^{\infty}\sum_{m=1}^{\infty}\frac{A_n}{[J'_n(\mu_{nm}R)]^2}J_n(\mu_{nm}r)J_n(\mu_{nm}\xi)\cos[n(\varphi-\eta)]\exp(-\mu_{nm}^2 at),$$

$$G_2(z, \zeta, t) = \frac{1}{2\sqrt{\pi a t}}\exp\left[-\frac{(z-\zeta)^2}{4at}\right], \qquad A_n = \begin{cases} 1 & \text{for } n=0, \\ 2 & \text{for } n=1, 2, \ldots, \end{cases}$$

where the $J_n(\xi)$ are the Bessel functions (the prime denotes the derivative with respect to the argument) and the μ_{nm} are positive roots of the transcendental equation $J_n(\mu R) = 0$.

⦿ *Reference*: B. M. Budak, A. A. Samarskii, and A. N. Tikhonov (1980).

3.1.2-3. Domain: $0 \leq r \leq R$, $0 \leq \varphi \leq 2\pi$, $-\infty < z < \infty$. Second boundary value problem.

An infinite circular cylinder is considered. The following conditions are prescribed:

$$w = f(r, \varphi, z) \quad \text{at} \quad t = 0 \quad \text{(initial condition)},$$
$$\partial_r w = g(\varphi, z, t) \quad \text{at} \quad r = R \quad \text{(boundary condition)}.$$

Solution:

$$w(r, \varphi, z, t) = \int_{-\infty}^{\infty} \int_{0}^{2\pi} \int_{0}^{R} \xi f(\xi, \eta, \zeta) G(r, \varphi, z, \xi, \eta, \zeta, t) \, d\xi \, d\eta \, d\zeta$$
$$+ aR \int_{0}^{t} \int_{-\infty}^{\infty} \int_{0}^{2\pi} g(\eta, \zeta, \tau) G(r, \varphi, z, R, \eta, \zeta, t - \tau) \, d\eta \, d\zeta \, d\tau.$$

Here,

$$G(r, \varphi, z, \xi, \eta, \zeta, t) = G_1(r, \varphi, \xi, \eta, t) G_2(z, \zeta, t),$$

$$G_1(r, \varphi, \xi, \eta, t) = \frac{1}{\pi R^2} + \frac{1}{\pi} \sum_{n=0}^{\infty} \sum_{m=1}^{\infty} \frac{A_n \mu_{nm}^2 J_n(\mu_{nm} r) J_n(\mu_{nm} \xi)}{(\mu_{nm}^2 R^2 - n^2)[J_n(\mu_{nm} R)]^2} \cos[n(\varphi - \eta)] \exp(-\mu_{nm}^2 a t),$$

$$G_2(z, \zeta, t) = \frac{1}{2\sqrt{\pi a t}} \exp\left[-\frac{(z - \zeta)^2}{4at}\right], \qquad A_n = \begin{cases} 1 & \text{for } n = 0, \\ 2 & \text{for } n = 1, 2, \ldots, \end{cases}$$

where the $J_n(\xi)$ are the Bessel functions and the μ_{nm} are positive roots of the transcendental equation $J_n'(\mu R) = 0$.

⊙ *Reference*: B. M. Budak, A. A. Samarskii, and A. N. Tikhonov (1980).

3.1.2-4. Domain: $0 \leq r \leq R$, $0 \leq \varphi \leq 2\pi$, $-\infty < z < \infty$. Third boundary value problem.

An infinite circular cylinder is considered. The following conditions are prescribed:

$$w = f(r, \varphi, z) \quad \text{at} \quad t = 0 \quad \text{(initial condition)},$$
$$\partial_r w + kw = g(\varphi, z, t) \quad \text{at} \quad r = R \quad \text{(boundary condition)}.$$

The solution $w(r, \varphi, z, t)$ is determined by the formula in Paragraph 3.1.2-3 where

$$G(r, \varphi, z, \xi, \eta, \zeta, t) = G_1(r, \varphi, \xi, \eta, t) G_2(z, \zeta, t),$$

$$G_1(r, \varphi, \xi, \eta, t) = \frac{1}{\pi} \sum_{n=0}^{\infty} \sum_{m=1}^{\infty} \frac{A_n \mu_{nm}^2 J_n(\mu_{nm} r) J_n(\mu_{nm} \xi)}{(\mu_{nm}^2 R^2 + k^2 R^2 - n^2)[J_n(\mu_{nm} R)]^2} \cos[n(\varphi - \eta)] \exp(-\mu_{nm}^2 a t),$$

$$G_2(z, \zeta, t) = \frac{1}{2\sqrt{\pi a t}} \exp\left[-\frac{(z - \zeta)^2}{4at}\right], \qquad A_n = \begin{cases} 1 & \text{for } n = 0, \\ 2 & \text{for } n = 1, 2, \ldots \end{cases}$$

Here, the $J_n(\xi)$ are the Bessel functions and the μ_{nm} are positive roots of the transcendental equation

$$\mu J_n'(\mu R) + k J_n(\mu R) = 0.$$

⊙ *Reference*: B. M. Budak, A. A. Samarskii, and A. N. Tikhonov (1980).

3.1.2-5. Domain: $0 \le r \le R$, $0 \le \varphi \le 2\pi$, $0 \le z < \infty$. **First boundary value problem.**

A semiinfinite circular cylinder is considered. The following conditions are prescribed:

$$w = f(r, \varphi, z) \quad \text{at} \quad t = 0 \quad \text{(initial condition)},$$
$$w = g_1(\varphi, z, t) \quad \text{at} \quad r = R \quad \text{(boundary condition)},$$
$$w = g_2(r, \varphi, t) \quad \text{at} \quad z = 0 \quad \text{(boundary condition)}.$$

Solution:

$$w(r, \varphi, z, t) = \int_0^\infty \int_0^{2\pi} \int_0^R \xi f(\xi, \eta, \zeta) G(r, \varphi, z, \xi, \eta, \zeta, t) \, d\xi \, d\eta \, d\zeta$$
$$- aR \int_0^t \int_0^\infty \int_0^{2\pi} g_1(\eta, \zeta, \tau) \left[\frac{\partial}{\partial \xi} G(r, \varphi, z, \xi, \eta, \zeta, t - \tau) \right]_{\xi=R} d\eta \, d\zeta \, d\tau$$
$$+ a \int_0^t \int_0^{2\pi} \int_0^R \xi g_2(\xi, \eta, \tau) \left[\frac{\partial}{\partial \zeta} G(r, \varphi, z, \xi, \eta, \zeta, t - \tau) \right]_{\zeta=0} d\xi \, d\eta \, d\tau.$$

Here,

$$G(r, \varphi, z, \xi, \eta, \zeta, t) = G_1(r, \varphi, \xi, \eta, t) G_2(z, \zeta, t),$$

$$G_1(r, \varphi, \xi, \eta, t) = \frac{1}{\pi R^2} \sum_{n=0}^\infty \sum_{m=1}^\infty \frac{A_n}{[J_n'(\mu_{nm} R)]^2} J_n(\mu_{nm} r) J_n(\mu_{nm} \xi) \cos[n(\varphi - \eta)] \exp(-\mu_{nm}^2 at),$$

$$G_2(z, \zeta, t) = \frac{1}{2\sqrt{\pi a t}} \left\{ \exp\left[-\frac{(z - \zeta)^2}{4at}\right] - \exp\left[-\frac{(z + \zeta)^2}{4at}\right] \right\}, \quad A_n = \begin{cases} 1 & \text{for } n = 0, \\ 2 & \text{for } n = 1, 2, \ldots, \end{cases}$$

where the $J_n(\xi)$ are the Bessel functions (the prime denotes the derivative with respect to the argument) and the μ_{nm} are positive roots of the transcendental equation $J_n(\mu R) = 0$.

⊙ *Reference*: B. M. Budak, A. A. Samarskii, and A. N. Tikhonov (1980).

3.1.2-6. Domain: $0 \le r \le R$, $0 \le \varphi \le 2\pi$, $0 \le z < \infty$. **Second boundary value problem.**

A semiinfinite circular cylinder is considered. The following conditions are prescribed:

$$w = f(r, \varphi, z) \quad \text{at} \quad t = 0 \quad \text{(initial condition)},$$
$$\partial_r w = g_1(\varphi, z, t) \quad \text{at} \quad r = R \quad \text{(boundary condition)},$$
$$\partial_z w = g_2(r, \varphi, t) \quad \text{at} \quad z = 0 \quad \text{(boundary condition)}.$$

Solution:

$$w(r, \varphi, z, t) = \int_0^\infty \int_0^{2\pi} \int_0^R \xi f(\xi, \eta, \zeta) G(r, \varphi, z, \xi, \eta, \zeta, t) \, d\xi \, d\eta \, d\zeta$$
$$+ aR \int_0^t \int_0^\infty \int_0^{2\pi} g_1(\eta, \zeta, \tau) G(r, \varphi, z, R, \eta, \zeta, t - \tau) \, d\eta \, d\zeta \, d\tau$$
$$- a \int_0^t \int_0^{2\pi} \int_0^R \xi g_2(\xi, \eta, \tau) G(r, \varphi, z, \xi, \eta, 0, t - \tau) \, d\xi \, d\eta \, d\tau.$$

Here,

$$G(r, \varphi, z, \xi, \eta, \zeta, t) = G_1(r, \varphi, \xi, \eta, t) G_2(z, \zeta, t),$$

$$G_1(r, \varphi, \xi, \eta, t) = \frac{1}{\pi R^2} + \frac{1}{\pi} \sum_{n=0}^\infty \sum_{m=1}^\infty \frac{A_n \mu_{nm}^2 J_n(\mu_{nm} r) J_n(\mu_{nm} \xi)}{(\mu_{nm}^2 R^2 - n^2)[J_n(\mu_{nm} R)]^2} \cos[n(\varphi - \eta)] \exp(-\mu_{nm}^2 at),$$

$$G_2(z, \zeta, t) = \frac{1}{2\sqrt{\pi a t}} \left\{ \exp\left[-\frac{(z - \zeta)^2}{4at}\right] + \exp\left[-\frac{(z + \zeta)^2}{4at}\right] \right\}, \quad A_n = \begin{cases} 1 & \text{for } n = 0, \\ 2 & \text{for } n = 1, 2, \ldots, \end{cases}$$

where the $J_n(\xi)$ are the Bessel functions and the μ_{nm} are positive roots of the transcendental equation $J_n'(\mu R) = 0$.

⊙ *Reference*: B. M. Budak, A. A. Samarskii, and A. N. Tikhonov (1980).

3.1.2-7. Domain: $0 \le r \le R$, $0 \le \varphi \le 2\pi$, $0 \le z < \infty$. Third boundary value problem.

A semiinfinite circular cylinder is considered. The following conditions are prescribed:

$$w = f(r, \varphi, z) \quad \text{at} \quad t = 0 \quad \text{(initial condition)},$$
$$\partial_r w + k_1 w = g(\varphi, z, t) \quad \text{at} \quad r = R \quad \text{(boundary condition)},$$
$$\partial_z w - k_2 w = g_2(r, \varphi, t) \quad \text{at} \quad z = 0 \quad \text{(boundary condition)}.$$

The solution $w(r, \varphi, z, t)$ is determined by the formula in Paragraph 3.1.2-6 where

$$G(r, \varphi, z, \xi, \eta, \zeta, t) = G_1(r, \varphi, \xi, \eta, t) G_2(z, \zeta, t),$$

$$G_1(r, \varphi, \xi, \eta, t) = \frac{1}{\pi} \sum_{n=0}^{\infty} \sum_{m=1}^{\infty} \frac{A_n \mu_{nm}^2 J_n(\mu_{nm} r) J_n(\mu_{nm} \xi)}{(\mu_{nm}^2 R^2 + k_1^2 R^2 - n^2)[J_n(\mu_{nm} R)]^2} \cos[n(\varphi - \eta)] \exp(-\mu_{nm}^2 a t),$$

$$G_2(z, \zeta, t) = \frac{1}{2\sqrt{\pi a t}} \left\{ \exp\left[-\frac{(z-\zeta)^2}{4at}\right] + \exp\left[-\frac{(z+\zeta)^2}{4at}\right] - 2k_2 \int_0^\infty \exp\left[-\frac{(z+\zeta+s)^2}{4at} - k_2 s\right] ds \right\}.$$

Here, $A_0 = 1$ and $A_n = 2$ for $n = 1, 2, \ldots$; the $J_n(\xi)$ are the Bessel functions and the μ_{nm} are positive roots of the transcendental equation

$$\mu J_n'(\mu R) + k_1 J_n(\mu R) = 0.$$

3.1.2-8. Domain: $0 \le r \le R$, $0 \le \varphi \le 2\pi$, $0 \le z < \infty$. Mixed boundary value problems.

1°. A semiinfinite circular cylinder is considered. The following conditions are prescribed:

$$w = f(r, \varphi, z) \quad \text{at} \quad t = 0 \quad \text{(initial condition)},$$
$$w = g_1(\varphi, z, t) \quad \text{at} \quad r = R \quad \text{(boundary condition)},$$
$$\partial_z w = g_2(r, \varphi, t) \quad \text{at} \quad z = 0 \quad \text{(boundary condition)}.$$

Solution:

$$w(r, \varphi, z, t) = \int_0^\infty \int_0^{2\pi} \int_0^R \xi f(\xi, \eta, \zeta) G(r, \varphi, z, \xi, \eta, \zeta, t) \, d\xi \, d\eta \, d\zeta$$
$$- aR \int_0^t \int_0^\infty \int_0^{2\pi} g_1(\eta, \zeta, \tau) \left[\frac{\partial}{\partial \xi} G(r, \varphi, z, \xi, \eta, \zeta, t - \tau)\right]_{\xi=R} d\eta \, d\zeta \, d\tau$$
$$- a \int_0^t \int_0^{2\pi} \int_0^R \xi g_2(\xi, \eta, \tau) G(r, \varphi, z, \xi, \eta, 0, t-\tau) \, d\xi \, d\eta \, d\tau.$$

Here,

$$G(r, \varphi, z, \xi, \eta, \zeta, t) = G_1(r, \varphi, \xi, \eta, t) G_2(z, \zeta, t),$$

$$G_1(r, \varphi, \xi, \eta, t) = \frac{1}{\pi R^2} \sum_{n=0}^{\infty} \sum_{m=1}^{\infty} \frac{A_n}{[J_n'(\mu_{nm} R)]^2} J_n(\mu_{nm} r) J_n(\mu_{nm} \xi) \cos[n(\varphi - \eta)] \exp(-\mu_{nm}^2 a t),$$

$$G_2(z, \zeta, t) = \frac{1}{2\sqrt{\pi a t}} \left\{ \exp\left[-\frac{(z-\zeta)^2}{4at}\right] + \exp\left[-\frac{(z+\zeta)^2}{4at}\right] \right\}, \quad A_n = \begin{cases} 1 & \text{for } n = 0, \\ 2 & \text{for } n = 1, 2, \ldots, \end{cases}$$

where the $J_n(\xi)$ are the Bessel functions (the prime denotes the derivative with respect to the argument) and the μ_{nm} are positive roots of the transcendental equation $J_n(\mu R) = 0$.

2°. A semiinfinite circular cylinder is considered. The following conditions are prescribed:

$$w = f(r, \varphi, z) \quad \text{at} \quad t = 0 \quad \text{(initial condition)},$$
$$\partial_r w = g_1(\varphi, z, t) \quad \text{at} \quad r = R \quad \text{(boundary condition)},$$
$$w = g_2(r, \varphi, t) \quad \text{at} \quad z = 0 \quad \text{(boundary condition)}.$$

Solution:

$$w(r, \varphi, z, t) = \int_0^\infty \int_0^{2\pi} \int_0^R \xi f(\xi, \eta, \zeta) G(r, \varphi, z, \xi, \eta, \zeta, t) \, d\xi \, d\eta \, d\zeta$$
$$+ aR \int_0^t \int_0^\infty \int_0^{2\pi} g_1(\eta, \zeta, \tau) G(r, \varphi, z, R, \eta, \zeta, t-\tau) \, d\eta \, d\zeta \, d\tau$$
$$+ a \int_0^t \int_0^{2\pi} \int_0^R \xi g_2(\xi, \eta, \tau) \left[\frac{\partial}{\partial \zeta} G(r, \varphi, z, \xi, \eta, \zeta, t-\tau) \right]_{\zeta=0} d\xi \, d\eta \, d\tau.$$

Here,

$$G(r, \varphi, z, \xi, \eta, \zeta, t) = G_1(r, \varphi, \xi, \eta, t) G_2(z, \zeta, t),$$

$$G_1(r, \varphi, \xi, \eta, t) = \frac{1}{\pi R^2} + \frac{1}{\pi} \sum_{n=0}^\infty \sum_{m=1}^\infty \frac{A_n \mu_{nm}^2 J_n(\mu_{nm} r) J_n(\mu_{nm} \xi)}{(\mu_{nm}^2 R^2 - n^2)[J_n(\mu_{nm} R)]^2} \cos[n(\varphi-\eta)] \exp(-\mu_{nm}^2 at),$$

$$G_2(z, \zeta, t) = \frac{1}{2\sqrt{\pi a t}} \left\{ \exp\left[-\frac{(z-\zeta)^2}{4at}\right] - \exp\left[-\frac{(z+\zeta)^2}{4at}\right] \right\}, \quad A_n = \begin{cases} 1 & \text{for } n = 0, \\ 2 & \text{for } n = 1, 2, \ldots, \end{cases}$$

where the $J_n(\xi)$ are the Bessel functions and the μ_{nm} are positive roots of the transcendental equation $J_n'(\mu R) = 0$.

3.1.2-9. Domain: $0 \le r \le R$, $0 \le \varphi \le 2\pi$, $0 \le z \le l$. First boundary value problem.

A circular cylinder of finite length is considered. The following conditions are prescribed:

$$w = f(r, \varphi, z) \quad \text{at} \quad t = 0 \quad \text{(initial condition)},$$
$$w = g_1(\varphi, z, t) \quad \text{at} \quad r = R \quad \text{(boundary condition)},$$
$$w = g_2(r, \varphi, t) \quad \text{at} \quad z = 0 \quad \text{(boundary condition)},$$
$$w = g_3(r, \varphi, t) \quad \text{at} \quad z = l \quad \text{(boundary condition)}.$$

Solution:

$$w(r, \varphi, z, t) = \int_0^l \int_0^{2\pi} \int_0^R \xi f(\xi, \eta, \zeta) G(r, \varphi, z, \xi, \eta, \zeta, t) \, d\xi \, d\eta \, d\zeta$$
$$- aR \int_0^t \int_0^l \int_0^{2\pi} g_1(\eta, \zeta, \tau) \left[\frac{\partial}{\partial \xi} G(r, \varphi, z, \xi, \eta, \zeta, t-\tau) \right]_{\xi=R} d\eta \, d\zeta \, d\tau$$
$$+ a \int_0^t \int_0^{2\pi} \int_0^R \xi g_2(\xi, \eta, \tau) \left[\frac{\partial}{\partial \zeta} G(r, \varphi, z, \xi, \eta, \zeta, t-\tau) \right]_{\zeta=0} d\xi \, d\eta \, d\tau$$
$$- a \int_0^t \int_0^{2\pi} \int_0^R \xi g_3(\xi, \eta, \tau) \left[\frac{\partial}{\partial \zeta} G(r, \varphi, z, \xi, \eta, \zeta, t-\tau) \right]_{\zeta=l} d\xi \, d\eta \, d\tau.$$

Here,

$$G(r, \varphi, z, \xi, \eta, \zeta, t) = G_1(r, \varphi, \xi, \eta, t) G_2(z, \zeta, t),$$

$$G_1(r, \varphi, \xi, \eta, t) = \frac{1}{\pi R^2} \sum_{n=0}^\infty \sum_{m=1}^\infty \frac{A_n}{[J_n'(\mu_{nm} R)]^2} J_n(\mu_{nm} r) J_n(\mu_{nm} \xi) \cos[n(\varphi-\eta)] \exp(-\mu_{nm}^2 at),$$

$$G_2(z, \zeta, t) = \frac{2}{l} \sum_{n=1}^\infty \sin\left(\frac{n\pi z}{l}\right) \sin\left(\frac{n\pi \zeta}{l}\right) \exp\left(-\frac{an^2\pi^2 t}{l^2}\right), \quad A_n = \begin{cases} 1 & \text{for } n = 0, \\ 2 & \text{for } n = 1, 2, \ldots, \end{cases}$$

where the $J_n(\xi)$ are the Bessel functions (the prime denotes the derivative with respect to the argument) and the μ_{nm} are positive roots of the transcendental equation $J_n(\mu R) = 0$.

3.1.2-10. Domain: $0 \le r \le R$, $0 \le \varphi \le 2\pi$, $0 \le z \le l$. Second boundary value problem.

A circular cylinder of finite length is considered. The following conditions are prescribed:

$$w = f(r, \varphi, z) \quad \text{at} \quad t = 0 \quad \text{(initial condition)},$$
$$\partial_r w = g_1(\varphi, z, t) \quad \text{at} \quad r = R \quad \text{(boundary condition)},$$
$$\partial_z w = g_2(r, \varphi, t) \quad \text{at} \quad z = 0 \quad \text{(boundary condition)},$$
$$\partial_z w = g_3(r, \varphi, t) \quad \text{at} \quad z = l \quad \text{(boundary condition)}.$$

Solution:

$$w(r, \varphi, z, t) = \int_0^l \int_0^{2\pi} \int_0^R \xi f(\xi, \eta, \zeta) G(r, \varphi, z, \xi, \eta, \zeta, t) \, d\xi \, d\eta \, d\zeta$$
$$+ aR \int_0^t \int_0^l \int_0^{2\pi} g_1(\eta, \zeta, \tau) G(r, \varphi, z, R, \eta, \zeta, t - \tau) \, d\eta \, d\zeta \, d\tau$$
$$- a \int_0^t \int_0^{2\pi} \int_0^R \xi g_2(\xi, \eta, \tau) G(r, \varphi, z, \xi, \eta, 0, t - \tau) \, d\xi \, d\eta \, d\tau$$
$$+ a \int_0^t \int_0^{2\pi} \int_0^R \xi g_3(\xi, \eta, \tau) G(r, \varphi, z, \xi, \eta, l, t - \tau) \, d\xi \, d\eta \, d\tau.$$

Here,

$$G(r, \varphi, z, \xi, \eta, \zeta, t) = G_1(r, \varphi, \xi, \eta, t) G_2(z, \zeta, t),$$

$$G_1(r, \varphi, \xi, \eta, t) = \frac{1}{\pi R^2} + \frac{1}{\pi} \sum_{n=0}^{\infty} \sum_{m=1}^{\infty} \frac{A_n \mu_{nm}^2 J_n(\mu_{nm} r) J_n(\mu_{nm} \xi)}{(\mu_{nm}^2 R^2 - n^2)[J_n(\mu_{nm} R)]^2} \cos[n(\varphi - \eta)] \exp(-\mu_{nm}^2 a t),$$

$$G_2(z, \zeta, t) = \frac{1}{l} + \frac{2}{l} \sum_{n=1}^{\infty} \cos\left(\frac{n\pi x}{l}\right) \cos\left(\frac{n\pi \xi}{l}\right) \exp\left(-\frac{an^2\pi^2 t}{l^2}\right), \quad A_n = \begin{cases} 1 & \text{for } n = 0, \\ 2 & \text{for } n = 1, 2, \ldots, \end{cases}$$

where the $J_n(\xi)$ are the Bessel functions and the μ_{nm} are positive roots of the transcendental equation $J_n'(\mu R) = 0$.

3.1.2-11. Domain: $0 \le r \le R$, $0 \le \varphi \le 2\pi$, $0 \le z \le l$. Third boundary value problem.

A circular cylinder of finite length is considered. The following conditions are prescribed:

$$w = f(r, \varphi, z) \quad \text{at} \quad t = 0 \quad \text{(initial condition)},$$
$$\partial_r w + k_1 w = g(\varphi, z, t) \quad \text{at} \quad r = R \quad \text{(boundary condition)},$$
$$\partial_z w - k_2 w = g_2(r, \varphi, t) \quad \text{at} \quad z = 0 \quad \text{(boundary condition)},$$
$$\partial_z w + k_3 w = g_3(r, \varphi, t) \quad \text{at} \quad z = l \quad \text{(boundary condition)}.$$

The solution $w(r, \varphi, z, t)$ is determined by the formula in Paragraph 3.1.2-10 where

$$G(r, \varphi, z, \xi, \eta, \zeta, t) = G_1(r, \varphi, \xi, \eta, t) \sum_{s=1}^{\infty} \frac{h_s(z) h_s(\zeta)}{\|h_s\|^2} \exp(-a\lambda_s^2 t),$$

$$G_1(r, \varphi, \xi, \eta, t) = \frac{1}{\pi} \sum_{n=0}^{\infty} \sum_{m=1}^{\infty} \frac{A_n \mu_{nm}^2 J_n(\mu_{nm} r) J_n(\mu_{nm} \xi)}{(\mu_{nm}^2 R^2 + k_1^2 R^2 - n^2)[J_n(\mu_{nm} R)]^2} \cos[n(\varphi - \eta)] \exp(-\mu_{nm}^2 a t),$$

$$h_s(z) = \cos(\lambda_s z) + \frac{k_2}{\lambda_s} \sin(\lambda_s z), \quad \|h_s\|^2 = \frac{k_3}{2\lambda_s^2} \frac{\lambda_s^2 + k_2^2}{\lambda_s^2 + k_3^2} + \frac{k_2}{2\lambda_s^2} + \frac{l}{2}\left(1 + \frac{k_2^2}{\lambda_s^2}\right).$$

Here, $A_0 = 1$ and $A_n = 2$ for $n = 1, 2, \ldots$; the $J_n(\xi)$ are the Bessel functions; and the μ_{nm} and λ_s are positive roots of the transcendental equations

$$\mu J_n'(\mu R) + k_1 J_n(\mu R) = 0, \qquad \frac{\tan(\lambda l)}{\lambda} = \frac{k_2 + k_3}{\lambda^2 - k_2 k_3}.$$

3.1.2-12. Domain: $0 \le r \le R$, $0 \le \varphi \le 2\pi$, $0 \le z \le l$. Mixed boundary value problems.

1°. A circular cylinder of finite length is considered. The following conditions are prescribed:

$$\begin{aligned}
w &= f(r, \varphi, z) & \text{at} \quad t &= 0 & \text{(initial condition)}, \\
w &= g_1(\varphi, z, t) & \text{at} \quad r &= R & \text{(boundary condition)}, \\
\partial_z w &= g_2(r, \varphi, t) & \text{at} \quad z &= 0 & \text{(boundary condition)}, \\
\partial_z w &= g_3(r, \varphi, t) & \text{at} \quad z &= l & \text{(boundary condition)}.
\end{aligned}$$

Solution:

$$\begin{aligned}
w(r, \varphi, z, t) = &\int_0^l \int_0^{2\pi} \int_0^R \xi f(\xi, \eta, \zeta) G(r, \varphi, z, \xi, \eta, \zeta, t) \, d\xi \, d\eta \, d\zeta \\
&- aR \int_0^t \int_0^l \int_0^{2\pi} g_1(\eta, \zeta, \tau) \left[\frac{\partial}{\partial \xi} G(r, \varphi, z, \xi, \eta, \zeta, t - \tau) \right]_{\xi=R} d\eta \, d\zeta \, d\tau \\
&- a \int_0^t \int_0^{2\pi} \int_0^R \xi g_2(\xi, \eta, \tau) G(r, \varphi, z, \xi, \eta, 0, t-\tau) \, d\xi \, d\eta \, d\tau \\
&+ a \int_0^t \int_0^{2\pi} \int_0^R \xi g_3(\xi, \eta, \tau) G(r, \varphi, z, \xi, \eta, l, t-\tau) \, d\xi \, d\eta \, d\tau.
\end{aligned}$$

Here,

$$G(r, \varphi, z, \xi, \eta, \zeta, t) = G_1(r, \varphi, \xi, \eta, t) G_2(z, \zeta, t),$$

$$G_1(r, \varphi, \xi, \eta, t) = \frac{1}{\pi R^2} \sum_{n=0}^{\infty} \sum_{m=1}^{\infty} \frac{A_n}{[J_n'(\mu_{nm} R)]^2} J_n(\mu_{nm} r) J_n(\mu_{nm} \xi) \cos[n(\varphi - \eta)] \exp(-\mu_{nm}^2 a t),$$

$$G_2(z, \zeta, t) = \frac{1}{l} + \frac{2}{l} \sum_{n=1}^{\infty} \cos\left(\frac{n\pi z}{l}\right) \cos\left(\frac{n\pi \zeta}{l}\right) \exp\left(-\frac{an^2\pi^2 t}{l^2}\right), \qquad A_n = \begin{cases} 1 & \text{for } n=0, \\ 2 & \text{for } n=1, 2, \ldots, \end{cases}$$

where the $J_n(\xi)$ are the Bessel functions (the prime denotes the derivative with respect to the argument) and the μ_{nm} are positive roots of the transcendental equation $J_n(\mu R) = 0$.

2°. A circular cylinder of finite length is considered. The following conditions are prescribed:

$$\begin{aligned}
w &= f(r, \varphi, z) & \text{at} \quad t &= 0 & \text{(initial condition)}, \\
\partial_r w &= g_1(\varphi, z, t) & \text{at} \quad r &= R & \text{(boundary condition)}, \\
w &= g_2(r, \varphi, t) & \text{at} \quad z &= 0 & \text{(boundary condition)}, \\
w &= g_3(r, \varphi, t) & \text{at} \quad z &= l & \text{(boundary condition)}.
\end{aligned}$$

Solution:

$$\begin{aligned}
w(r, \varphi, z, t) = &\int_0^l \int_0^{2\pi} \int_0^R \xi f(\xi, \eta, \zeta) G(r, \varphi, z, \xi, \eta, \zeta, t) \, d\xi \, d\eta \, d\zeta \\
&+ aR \int_0^t \int_0^l \int_0^{2\pi} g_1(\eta, \zeta, \tau) G(r, \varphi, z, R, \eta, \zeta, t - \tau) \, d\eta \, d\zeta \, d\tau \\
&+ a \int_0^t \int_0^{2\pi} \int_0^R \xi g_2(\xi, \eta, \tau) \left[\frac{\partial}{\partial \zeta} G(r, \varphi, z, \xi, \eta, \zeta, t - \tau) \right]_{\zeta=0} d\xi \, d\eta \, d\tau \\
&- a \int_0^t \int_0^{2\pi} \int_0^R \xi g_3(\xi, \eta, \tau) \left[\frac{\partial}{\partial \zeta} G(r, \varphi, z, \xi, \eta, \zeta, t - \tau) \right]_{\zeta=l} d\xi \, d\eta \, d\tau.
\end{aligned}$$

Here,

$$G(r,\varphi,z,\xi,\eta,\zeta,t) = G_1(r,\varphi,\xi,\eta,t)G_2(z,\zeta,t),$$

$$G_1(r,\varphi,\xi,\eta,t) = \frac{1}{\pi R^2} + \frac{1}{\pi}\sum_{n=0}^{\infty}\sum_{m=1}^{\infty}\frac{A_n\mu_{nm}^2 J_n(\mu_{nm}r)J_n(\mu_{nm}\xi)}{(\mu_{nm}^2 R^2 - n^2)[J_n(\mu_{nm}R)]^2}\cos[n(\varphi-\eta)]\exp(-\mu_{nm}^2 at),$$

$$G_2(z,\zeta,t) = \frac{2}{l}\sum_{n=1}^{\infty}\sin\left(\frac{n\pi z}{l}\right)\sin\left(\frac{n\pi \zeta}{l}\right)\exp\left(-\frac{an^2\pi^2 t}{l^2}\right),\quad A_n = \begin{cases} 1 & \text{for } n=0, \\ 2 & \text{for } n=1,2,\dots, \end{cases}$$

where the $J_n(\xi)$ are the Bessel functions and the μ_{nm} are positive roots of the transcendental equation $J_n'(\mu R) = 0$.

3.1.2-13. Domain: $R_1 \le r \le R_2$, $0 \le \varphi \le 2\pi$, $-\infty < z < \infty$. First boundary value problem.

An infinite hollow circular cylinder is considered. The following conditions are prescribed:

$$w = f(r,\varphi,z) \quad \text{at} \quad t = 0 \quad \text{(initial condition)},$$
$$w = g_1(\varphi,z,t) \quad \text{at} \quad r = R_1 \quad \text{(boundary condition)},$$
$$w = g_2(\varphi,z,t) \quad \text{at} \quad r = R_2 \quad \text{(boundary condition)}.$$

Solution:

$$w(r,\varphi,z,t) = \int_{-\infty}^{\infty}\int_{0}^{2\pi}\int_{R_1}^{R_2} f(\xi,\eta,\zeta)G(r,\varphi,z,\xi,\eta,\zeta,t)\xi\, d\xi\, d\eta\, d\zeta$$
$$+ aR_1\int_{0}^{t}\int_{-\infty}^{\infty}\int_{0}^{2\pi} g_1(\eta,\zeta,\tau)\left[\frac{\partial}{\partial \xi}G(r,\varphi,z,\xi,\eta,\zeta,t-\tau)\right]_{\xi=R_1} d\eta\, d\zeta\, d\tau$$
$$- aR_2\int_{0}^{t}\int_{-\infty}^{\infty}\int_{0}^{2\pi} g_2(\eta,\zeta,\tau)\left[\frac{\partial}{\partial \xi}G(r,\varphi,z,\xi,\eta,\zeta,t-\tau)\right]_{\xi=R_2} d\eta\, d\zeta\, d\tau.$$

Here,

$$G(r,\varphi,z,\xi,\eta,\zeta,t) = \frac{1}{2\sqrt{\pi at}}\exp\left[-\frac{(z-\zeta)^2}{4at}\right] G_1(r,\varphi,\xi,\eta,t),$$

$$G_1(r,\varphi,\xi,\eta,t) = \frac{\pi}{2}\sum_{n=0}^{\infty}\sum_{m=1}^{\infty} A_n B_{nm} Z_n(\mu_{nm}r)Z_n(\mu_{nm}\xi)\cos[n(\varphi-\eta)]\exp(-\mu_{nm}^2 at),$$

$$A_n = \begin{cases} 1/2 & \text{for } n=0, \\ 1 & \text{for } n\ne 0, \end{cases}\quad B_{nm} = \frac{\mu_{nm}^2 J_n^2(\mu_{nm}R_2)}{J_n^2(\mu_{nm}R_1) - J_n^2(\mu_{nm}R_2)},$$

$$Z_n(\mu_{nm}r) = J_n(\mu_{nm}R_1)Y_n(\mu_{nm}r) - Y_n(\mu_{nm}R_1)J_n(\mu_{nm}r),$$

where the $J_n(r)$ and $Y_n(r)$ are the Bessel functions and the μ_{nm} are positive roots of the transcendental equation

$$J_n(\mu R_1)Y_n(\mu R_2) - Y_n(\mu R_1)J_n(\mu R_2) = 0.$$

3.1.2-14. Domain: $R_1 \le r \le R_2$, $0 \le \varphi \le 2\pi$, $-\infty < z < \infty$. Second boundary value problem.

An infinite hollow circular cylinder is considered. The following conditions are prescribed:

$$w = f(r,\varphi,z) \quad \text{at} \quad t = 0 \quad \text{(initial condition)},$$
$$\partial_r w = g_1(\varphi,z,t) \quad \text{at} \quad r = R_1 \quad \text{(boundary condition)},$$
$$\partial_r w = g_2(\varphi,z,t) \quad \text{at} \quad r = R_2 \quad \text{(boundary condition)}.$$

Solution:

$$w(r,\varphi,z,t) = \int_{-\infty}^{\infty}\int_0^{2\pi}\int_{R_1}^{R_2} f(\xi,\eta,\zeta)G(r,\varphi,z,\xi,\eta,\zeta,t)\xi\,d\xi\,d\eta\,d\zeta$$

$$-aR_1\int_0^t\int_{-\infty}^{\infty}\int_0^{2\pi} g_1(\eta,\zeta,\tau)G(r,\varphi,z,R_1,\eta,\zeta,t-\tau)\,d\eta\,d\zeta\,d\tau$$

$$+aR_2\int_0^t\int_{-\infty}^{\infty}\int_0^{2\pi} g_2(\eta,\zeta,\tau)G(r,\varphi,z,R_2,\eta,\zeta,t-\tau)\,d\eta\,d\zeta\,d\tau.$$

Here,

$$G(r,\varphi,z,\xi,\eta,\zeta,t) = \frac{1}{2\sqrt{\pi a t}}\exp\left[-\frac{(z-\zeta)^2}{4at}\right]G_1(r,\varphi,\xi,\eta,t),$$

$$G_1(r,\varphi,\xi,\eta,t) = \frac{1}{\pi(R_2^2-R_1^2)} + \frac{1}{\pi}\sum_{n=0}^{\infty}\sum_{m=1}^{\infty}\frac{A_n\mu_{nm}^2 Z_n(\mu_{nm}r)Z_n(\mu_{nm}\xi)\cos[n(\varphi-\eta)]\exp(-\mu_{nm}^2 at)}{(\mu_{nm}^2 R_2^2 - n^2)Z_n^2(\mu_{nm}R_2) - (\mu_{nm}^2 R_1^2 - n^2)Z_n^2(\mu_{nm}R_1)},$$

$$Z_n(\mu_{nm}r) = J_n'(\mu_{nm}R_1)Y_n(\mu_{nm}r) - Y_n'(\mu_{nm}R_1)J_n(\mu_{nm}r),$$

where $A_0 = 1$ and $A_n = 2$ for $n = 1, 2, \ldots$; the $J_n(r)$ and $Y_n(r)$ are the Bessel functions (the prime denotes the derivative with respect to the argument); and the μ_{nm} are positive roots of the transcendental equation

$$J_n'(\mu R_1)Y_n'(\mu R_2) - Y_n'(\mu R_1)J_n'(\mu R_2) = 0.$$

3.1.2-15. Domain: $R_1 \leq r \leq R_2$, $0 \leq \varphi \leq 2\pi$, $-\infty < z < \infty$. Third boundary value problem.

An infinite hollow circular cylinder is considered. The following conditions are prescribed:

$$w = f(r,\varphi,z) \quad \text{at} \quad t = 0 \quad \text{(initial condition)},$$
$$\partial_r w - k_1 w = g_1(\varphi,z,t) \quad \text{at} \quad r = R_1 \quad \text{(boundary condition)},$$
$$\partial_r w + k_2 w = g_2(\varphi,z,t) \quad \text{at} \quad r = R_2 \quad \text{(boundary condition)}.$$

The solution $w(r,\varphi,z,t)$ is given by relations in Paragraph 3.1.2-14 in which

$$G(r,\varphi,z,\xi,\eta,\zeta,t) = \frac{1}{2\sqrt{\pi a t}}\exp\left[-\frac{(z-\zeta)^2}{4at}\right]G_1(r,\varphi,\xi,\eta,t),$$

$$G_1(r,\varphi,\xi,\eta,t) = \frac{1}{\pi}\sum_{n=0}^{\infty}\sum_{m=1}^{\infty}\frac{A_n\mu_{nm}^2 Z_n(\mu_{nm}r)Z_n(\mu_{nm}\xi)\cos[n(\varphi-\eta)]\exp(-\mu_{nm}^2 at)}{(k_2^2 R_2^2 + \mu_{nm}^2 R_2^2 - n^2)Z_n^2(\mu_{nm}R_2) - (k_1^2 R_1^2 + \mu_{nm}^2 R_1^2 - n^2)Z_n^2(\mu_{nm}R_1)},$$

$$Z_n(\mu_{nm}r) = \left[\mu_{nm}J_n'(\mu_{nm}R_1) - k_1 J_n(\mu_{nm}R_1)\right]Y_n(\mu_{nm}r)$$
$$- \left[\mu_{nm}Y_n'(\mu_{nm}R_1) - k_1 Y_n(\mu_{nm}R_1)\right]J_n(\mu_{nm}r).$$

Here, $A_0 = 1$ and $A_n = 2$ for $n = 1, 2, \ldots$; $J_n(r)$ and $Y_n(r)$ are the Bessel functions; and the μ_{nm} are positive roots of the transcendental equation

$$\left[\mu J_n'(\mu R_1) - k_1 J_n(\mu R_1)\right]\left[\mu Y_n'(\mu R_2) + k_2 Y_n(\mu R_2)\right]$$
$$= \left[\mu Y_n'(\mu R_1) - k_1 Y_n(\mu R_1)\right]\left[\mu J_n'(\mu R_2) + k_2 J_n(\mu R_2)\right].$$

3.1.2-16. Domain: $R_1 \leq r \leq R_2$, $0 \leq \varphi \leq 2\pi$, $0 \leq z < \infty$. **First boundary value problem.**

A semiinfinite hollow circular cylinder is considered. The following conditions are prescribed:

$$w = f(r, \varphi, z) \quad \text{at} \quad t = 0 \quad \text{(initial condition)},$$
$$w = g_1(\varphi, z, t) \quad \text{at} \quad r = R_1 \quad \text{(boundary condition)},$$
$$w = g_2(\varphi, z, t) \quad \text{at} \quad r = R_2 \quad \text{(boundary condition)},$$
$$w = g_3(r, \varphi, t) \quad \text{at} \quad z = 0 \quad \text{(boundary condition)}.$$

Solution:

$$w(r, \varphi, z, t) = \int_0^\infty \int_0^{2\pi} \int_{R_1}^{R_2} f(\xi, \eta, \zeta) G(r, \varphi, z, \xi, \eta, \zeta, t) \xi \, d\xi \, d\eta \, d\zeta$$

$$+ a R_1 \int_0^t \int_0^\infty \int_0^{2\pi} g_1(\eta, \zeta, \tau) \left[\frac{\partial}{\partial \xi} G(r, \varphi, z, \xi, \eta, \zeta, t - \tau) \right]_{\xi = R_1} d\eta \, d\zeta \, d\tau$$

$$- a R_2 \int_0^t \int_0^\infty \int_0^{2\pi} g_2(\eta, \zeta, \tau) \left[\frac{\partial}{\partial \xi} G(r, \varphi, z, \xi, \eta, \zeta, t - \tau) \right]_{\xi = R_2} d\eta \, d\zeta \, d\tau$$

$$+ a \int_0^t \int_0^{2\pi} \int_{R_1}^{R_2} g_3(\xi, \eta, \tau) \left[\frac{\partial}{\partial \zeta} G(r, \varphi, z, \xi, \eta, \zeta, t - \tau) \right]_{\zeta = 0} \xi \, d\xi \, d\eta \, d\tau.$$

Here,

$$G(r, \varphi, z, \xi, \eta, \zeta, t) = \frac{1}{2\sqrt{\pi a t}} \left\{ \exp\left[-\frac{(z - \zeta)^2}{4at} \right] - \exp\left[-\frac{(z + \zeta)^2}{4at} \right] \right\} G_1(r, \varphi, \xi, \eta, t),$$

$$G_1(r, \varphi, \xi, \eta, t) = \frac{\pi}{2} \sum_{n=0}^{\infty} \sum_{m=1}^{\infty} A_n B_{nm} Z_n(\mu_{nm} r) Z_n(\mu_{nm} \xi) \cos[n(\varphi - \eta)] \exp(-\mu_{nm}^2 a t),$$

$$A_n = \begin{cases} 1/2 & \text{for } n = 0, \\ 1 & \text{for } n \neq 0, \end{cases} \quad B_{nm} = \frac{\mu_{nm}^2 J_n^2(\mu_{nm} R_2)}{J_n^2(\mu_{nm} R_1) - J_n^2(\mu_{nm} R_2)},$$

$$Z_n(\mu_{nm} r) = J_n(\mu_{nm} R_1) Y_n(\mu_{nm} r) - Y_n(\mu_{nm} R_1) J_n(\mu_{nm} r),$$

where the $J_n(r)$ and $Y_n(r)$ are the Bessel functions and the μ_{nm} are positive roots of the transcendental equation

$$J_n(\mu R_1) Y_n(\mu R_2) - Y_n(\mu R_1) J_n(\mu R_2) = 0.$$

3.1.2-17. Domain: $R_1 \leq r \leq R_2$, $0 \leq \varphi \leq 2\pi$, $0 \leq z < \infty$. **Second boundary value problem.**

A semiinfinite hollow circular cylinder is considered. The following conditions are prescribed:

$$w = f(r, \varphi, z) \quad \text{at} \quad t = 0 \quad \text{(initial condition)},$$
$$\partial_r w = g_1(\varphi, z, t) \quad \text{at} \quad r = R_1 \quad \text{(boundary condition)},$$
$$\partial_r w = g_2(\varphi, z, t) \quad \text{at} \quad r = R_2 \quad \text{(boundary condition)},$$
$$\partial_z w = g_3(r, \varphi, t) \quad \text{at} \quad z = 0 \quad \text{(boundary condition)}.$$

Solution:

$$w(r, \varphi, z, t) = \int_0^\infty \int_0^{2\pi} \int_{R_1}^{R_2} f(\xi, \eta, \zeta) G(r, \varphi, z, \xi, \eta, \zeta, t) \xi \, d\xi \, d\eta \, d\zeta$$

$$- a R_1 \int_0^t \int_0^\infty \int_0^{2\pi} g_1(\eta, \zeta, \tau) G(r, \varphi, z, R_1, \eta, \zeta, t - \tau) \, d\eta \, d\zeta \, d\tau$$

$$+ a R_2 \int_0^t \int_0^\infty \int_0^{2\pi} g_2(\eta, \zeta, \tau) G(r, \varphi, z, R_2, \eta, \zeta, t - \tau) \, d\eta \, d\zeta \, d\tau$$

$$- a \int_0^t \int_0^{2\pi} \int_{R_1}^{R_2} g_3(\xi, \eta, \tau) G(r, \varphi, z, \xi, \eta, 0, t - \tau) \xi \, d\xi \, d\eta \, d\tau.$$

Here,

$$G(r,\varphi,z,\xi,\eta,\zeta,t)=\frac{1}{2\sqrt{\pi at}}\left\{\exp\left[-\frac{(z-\zeta)^2}{4at}\right]+\exp\left[-\frac{(z+\zeta)^2}{4at}\right]\right\}G_1(r,\varphi,\xi,\eta,t),$$

$$G_1(r,\varphi,\xi,\eta,t)=\frac{1}{\pi(R_2^2-R_1^2)}+\frac{1}{\pi}\sum_{n=0}^{\infty}\sum_{m=1}^{\infty}\frac{A_n\mu_{nm}^2 Z_n(\mu_{nm}r)Z_n(\mu_{nm}\xi)\cos[n(\varphi-\eta)]\exp(-\mu_{nm}^2 at)}{(\mu_{nm}^2 R_2^2-n^2)Z_n^2(\mu_{nm}R_2)-(\mu_{nm}^2 R_1^2-n^2)Z_n^2(\mu_{nm}R_1)},$$

$$Z_n(\mu_{nm}r)=J_n'(\mu_{nm}R_1)Y_n(\mu_{nm}r)-Y_n'(\mu_{nm}R_1)J_n(\mu_{nm}r),$$

where $A_0 = 1$ and $A_n = 2$ for $n = 1, 2, \ldots$; the $J_n(r)$ and $Y_n(r)$ are the Bessel functions; and the μ_{nm} are positive roots of the transcendental equation

$$J_n'(\mu R_1)Y_n'(\mu R_2) - Y_n'(\mu R_1)J_n'(\mu R_2) = 0.$$

3.1.2-18. Domain: $R_1 \leq r \leq R_2$, $0 \leq \varphi \leq 2\pi$, $0 \leq z < \infty$. Third boundary value problem.

A semiinfinite hollow circular cylinder is considered. The following conditions are prescribed:

$$\begin{aligned}
w &= f(r,\varphi,z) &&\text{at} \quad t=0 &&\text{(initial condition)}, \\
\partial_r w - k_1 w &= g_1(\varphi,z,t) &&\text{at} \quad r=R_1 &&\text{(boundary condition)}, \\
\partial_r w + k_2 w &= g_2(\varphi,z,t) &&\text{at} \quad r=R_2 &&\text{(boundary condition)}, \\
\partial_z w - k_3 w &= g_3(r,\varphi,t) &&\text{at} \quad z=0 &&\text{(boundary condition)}.
\end{aligned}$$

The solution $w(r,\varphi,z,t)$ is determined by the formula in Paragraph 3.1.2-17 where

$$G(r,\varphi,z,\xi,\eta,\zeta,t)=G_1(z,\zeta,t)G_2(r,\varphi,\xi,\eta,t),$$

$$G_1(z,\zeta,t)=\frac{1}{2\sqrt{\pi at}}\left\{\exp\left[-\frac{(z-\zeta)^2}{4at}\right]+\exp\left[-\frac{(z+\zeta)^2}{4at}\right]-2k_3\int_0^{\infty}\exp\left[-\frac{(z+\zeta+s)^2}{4at}-k_3 s\right]ds\right\},$$

$$G_2(r,\varphi,\xi,\eta,t)=\frac{1}{\pi}\sum_{n=0}^{\infty}\sum_{m=1}^{\infty}\frac{A_n\mu_{nm}^2 Z_n(\mu_{nm}r)Z_n(\mu_{nm}\xi)\cos[n(\varphi-\eta)]\exp(-\mu_{nm}^2 at)}{(k_2^2 R_2^2+\mu_{nm}^2 R_2^2-n^2)Z_n^2(\mu_{nm}R_2)-(k_1^2 R_1^2+\mu_{nm}^2 R_1^2-n^2)Z_n^2(\mu_{nm}R_1)},$$

$$Z_n(\mu_{nm}r) = \left[\mu_{nm}J_n'(\mu_{nm}R_1)-k_1 J_n(\mu_{nm}R_1)\right]Y_n(\mu_{nm}r)$$
$$-\left[\mu_{nm}Y_n'(\mu_{nm}R_1)-k_1 Y_n(\mu_{nm}R_1)\right]J_n(\mu_{nm}r).$$

Here, $A_0 = 1$ and $A_n = 2$ for $n = 1, 2, \ldots$; the $J_n(r)$ and $Y_n(r)$ are the Bessel functions; and the μ_{nm} are positive roots of the transcendental equation

$$\left[\mu J_n'(\mu R_1) - k_1 J_n(\mu R_1)\right]\left[\mu Y_n'(\mu R_2) + k_2 Y_n(\mu R_2)\right]$$
$$= \left[\mu Y_n'(\mu R_1) - k_1 Y_n(\mu R_1)\right]\left[\mu J_n'(\mu R_2) + k_2 J_n(\mu R_2)\right].$$

3.1.2-19. Domain: $R_1 \leq r \leq R_2$, $0 \leq \varphi \leq 2\pi$, $0 \leq z < \infty$. Mixed boundary value problems.

1°. A semiinfinite hollow circular cylinder is considered. The following conditions are prescribed:

$$\begin{aligned}
w &= f(r,\varphi,z) &&\text{at} \quad t=0 &&\text{(initial condition)}, \\
w &= g_1(\varphi,z,t) &&\text{at} \quad r=R_1 &&\text{(boundary condition)}, \\
w &= g_2(\varphi,z,t) &&\text{at} \quad r=R_2 &&\text{(boundary condition)}, \\
\partial_z w &= g_3(r,\varphi,t) &&\text{at} \quad z=0 &&\text{(boundary condition)}.
\end{aligned}$$

Solution:
$$w(r,\varphi,z,t) = \int_0^\infty \int_0^{2\pi} \int_{R_1}^{R_2} f(\xi,\eta,\zeta) G(r,\varphi,z,\xi,\eta,\zeta,t) \xi \, d\xi \, d\eta \, d\zeta$$
$$+ aR_1 \int_0^t \int_0^\infty \int_0^{2\pi} g_1(\eta,\zeta,\tau) \left[\frac{\partial}{\partial \xi} G(r,\varphi,z,\xi,\eta,\zeta,t-\tau) \right]_{\xi=R_1} d\eta \, d\zeta \, d\tau$$
$$- aR_2 \int_0^t \int_0^\infty \int_0^{2\pi} g_2(\eta,\zeta,\tau) \left[\frac{\partial}{\partial \xi} G(r,\varphi,z,\xi,\eta,\zeta,t-\tau) \right]_{\xi=R_2} d\eta \, d\zeta \, d\tau$$
$$- a \int_0^t \int_0^{2\pi} \int_{R_1}^{R_2} g_3(\xi,\eta,\tau) G(r,\varphi,z,\xi,\eta,0,t-\tau) \xi \, d\xi \, d\eta \, d\tau.$$

Here,
$$G(r,\varphi,z,\xi,\eta,\zeta,t) = \frac{1}{2\sqrt{\pi at}} \left\{ \exp\left[-\frac{(z-\zeta)^2}{4at}\right] + \exp\left[-\frac{(z+\zeta)^2}{4at}\right] \right\} G_1(r,\varphi,\xi,\eta,t),$$
$$G_1(r,\varphi,\xi,\eta,t) = \frac{\pi}{2} \sum_{n=0}^\infty \sum_{m=1}^\infty A_n B_{nm} Z_n(\mu_{nm}r) Z_n(\mu_{nm}\xi) \cos[n(\varphi-\eta)] \exp(-\mu_{nm}^2 at),$$
$$A_n = \begin{cases} 1/2 & \text{for } n=0, \\ 1 & \text{for } n \neq 0, \end{cases} \quad B_{nm} = \frac{\mu_{nm}^2 J_n^2(\mu_{nm}R_2)}{J_n^2(\mu_{nm}R_1) - J_n^2(\mu_{nm}R_2)},$$
$$Z_n(\mu_{nm}r) = J_n(\mu_{nm}R_1) Y_n(\mu_{nm}r) - Y_n(\mu_{nm}R_1) J_n(\mu_{nm}r),$$

where the $J_n(r)$ and $Y_n(r)$ are the Bessel functions and the μ_{nm} are positive roots of the transcendental equation
$$J_n(\mu R_1) Y_n(\mu R_2) - Y_n(\mu R_1) J_n(\mu R_2) = 0.$$

2°. A semiinfinite hollow circular cylinder is considered. The following conditions are prescribed:

$$\begin{aligned} w &= f(r,\varphi,z) & \text{at} \quad t=0 & \quad \text{(initial condition)}, \\ \partial_r w &= g_1(\varphi,z,t) & \text{at} \quad r=R_1 & \quad \text{(boundary condition)}, \\ \partial_r w &= g_2(\varphi,z,t) & \text{at} \quad r=R_2 & \quad \text{(boundary condition)}, \\ w &= g_3(r,\varphi,t) & \text{at} \quad z=0 & \quad \text{(boundary condition)}. \end{aligned}$$

Solution:
$$w(r,\varphi,z,t) = \int_0^\infty \int_0^{2\pi} \int_{R_1}^{R_2} f(\xi,\eta,\zeta) G(r,\varphi,z,\xi,\eta,\zeta,t) \xi \, d\xi \, d\eta \, d\zeta$$
$$- aR_1 \int_0^t \int_0^\infty \int_0^{2\pi} g_1(\eta,\zeta,\tau) G(r,\varphi,z,R_1,\eta,\zeta,t-\tau) \, d\eta \, d\zeta \, d\tau$$
$$+ aR_2 \int_0^t \int_0^\infty \int_0^{2\pi} g_2(\eta,\zeta,\tau) G(r,\varphi,z,R_2,\eta,\zeta,t-\tau) \, d\eta \, d\zeta \, d\tau$$
$$+ a \int_0^t \int_0^{2\pi} \int_{R_1}^{R_2} g_3(\xi,\eta,\tau) \left[\frac{\partial}{\partial \zeta} G(r,\varphi,z,\xi,\eta,\zeta,t-\tau) \right]_{\zeta=0} \xi \, d\xi \, d\eta \, d\tau.$$

Here,
$$G(r,\varphi,z,\xi,\eta,\zeta,t) = \frac{1}{2\sqrt{\pi at}} \left\{ \exp\left[-\frac{(z-\zeta)^2}{4at}\right] - \exp\left[-\frac{(z+\zeta)^2}{4at}\right] \right\} G_1(r,\varphi,\xi,\eta,t),$$
$$G_1(r,\varphi,\xi,\eta,t) = \frac{1}{\pi(R_2^2 - R_1^2)} + \frac{1}{\pi} \sum_{n=0}^\infty \sum_{m=1}^\infty \frac{A_n \mu_{nm}^2 Z_n(\mu_{nm}r) Z_n(\mu_{nm}\xi) \cos[n(\varphi-\eta)] \exp(-\mu_{nm}^2 at)}{(\mu_{nm}^2 R_2^2 - n^2) Z_n^2(\mu_{nm}R_2) - (\mu_{nm}^2 R_1^2 - n^2) Z_n^2(\mu_{nm}R_1)},$$
$$Z_n(\mu_{nm}r) = J_n'(\mu_{nm}R_1) Y_n(\mu_{nm}r) - Y_n'(\mu_{nm}R_1) J_n(\mu_{nm}r),$$

where $A_0 = 1$ and $A_n = 2$ for $n = 1, 2, \ldots$; the $J_n(r)$ and $Y_n(r)$ are the Bessel functions (the prime denotes the derivative with respect to the argument); and the μ_{nm} are positive roots of the transcendental equation

$$J'_n(\mu R_1) Y'_n(\mu R_2) - Y'_n(\mu R_1) J'_n(\mu R_2) = 0.$$

3.1.2-20. Domain: $R_1 \le r \le R_2$, $0 \le \varphi \le 2\pi$, $0 \le z \le l$. First boundary value problem.

A circular cylinder of finite length is considered. The following conditions are prescribed:

$$\begin{aligned}
w &= f(r, \varphi, z) & \text{at} \quad t &= 0 & \text{(initial condition)}, \\
w &= g_1(\varphi, z, t) & \text{at} \quad r &= R_1 & \text{(boundary condition)}, \\
w &= g_2(\varphi, z, t) & \text{at} \quad r &= R_2 & \text{(boundary condition)}, \\
w &= g_3(r, \varphi, t) & \text{at} \quad z &= 0 & \text{(boundary condition)}, \\
w &= g_4(r, \varphi, t) & \text{at} \quad z &= l & \text{(boundary condition)}.
\end{aligned}$$

Solution:

$$w(r, \varphi, z, t) = \int_0^l \int_0^{2\pi} \int_{R_1}^{R_2} f(\xi, \eta, \zeta) G(r, \varphi, z, \xi, \eta, \zeta, t) \xi \, d\xi \, d\eta \, d\zeta$$

$$+ a R_1 \int_0^t \int_0^l \int_0^{2\pi} g_1(\eta, \zeta, \tau) \left[\frac{\partial}{\partial \xi} G(r, \varphi, z, \xi, \eta, \zeta, t - \tau) \right]_{\xi=R_1} d\eta \, d\zeta \, d\tau$$

$$- a R_2 \int_0^t \int_0^l \int_0^{2\pi} g_2(\eta, \zeta, \tau) \left[\frac{\partial}{\partial \xi} G(r, \varphi, z, \xi, \eta, \zeta, t - \tau) \right]_{\xi=R_2} d\eta \, d\zeta \, d\tau$$

$$+ a \int_0^t \int_0^{2\pi} \int_{R_1}^{R_2} g_3(\xi, \eta, \tau) \left[\frac{\partial}{\partial \zeta} G(r, \varphi, z, \xi, \eta, \zeta, t - \tau) \right]_{\zeta=0} \xi \, d\xi \, d\eta \, d\tau$$

$$- a \int_0^t \int_0^{2\pi} \int_{R_1}^{R_2} g_4(\xi, \eta, \tau) \left[\frac{\partial}{\partial \zeta} G(r, \varphi, z, \xi, \eta, \zeta, t - \tau) \right]_{\zeta=l} \xi \, d\xi \, d\eta \, d\tau.$$

Here,

$$G(r, \varphi, z, \xi, \eta, \zeta, t) = G_1(r, \varphi, \xi, \eta, t) \left[\frac{2}{l} \sum_{n=1}^{\infty} \sin\left(\frac{n\pi z}{l}\right) \sin\left(\frac{n\pi \zeta}{l}\right) \exp\left(-\frac{a n^2 \pi^2 t}{l^2}\right) \right],$$

$$G_1(r, \varphi, \xi, \eta, t) = \frac{\pi}{2} \sum_{n=0}^{\infty} \sum_{m=1}^{\infty} A_n B_{nm} Z_n(\mu_{nm} r) Z_n(\mu_{nm} \xi) \cos[n(\varphi - \eta)] \exp(-\mu_{nm}^2 a t),$$

$$A_n = \begin{cases} 1/2 & \text{for } n = 0, \\ 1 & \text{for } n \ne 0, \end{cases} \qquad B_{nm} = \frac{\mu_{nm}^2 J_n^2(\mu_{nm} R_2)}{J_n^2(\mu_{nm} R_1) - J_n^2(\mu_{nm} R_2)},$$

$$Z_n(\mu_{nm} r) = J_n(\mu_{nm} R_1) Y_n(\mu_{nm} r) - Y_n(\mu_{nm} R_1) J_n(\mu_{nm} r),$$

where the $J_n(r)$ and $Y_n(r)$ are the Bessel functions and the μ_{nm} are positive roots of the transcendental equation

$$J_n(\mu R_1) Y_n(\mu R_2) - Y_n(\mu R_1) J_n(\mu R_2) = 0.$$

3.1.2-21. Domain: $R_1 \le r \le R_2$, $0 \le \varphi \le 2\pi$, $0 \le z \le l$. Second boundary value problem.

A circular cylinder of finite length is considered. The following conditions are prescribed:

$$\begin{aligned}
w &= f(r, \varphi, z) & \text{at} \quad t &= 0 & \text{(initial condition)}, \\
\partial_r w &= g_1(\varphi, z, t) & \text{at} \quad r &= R_1 & \text{(boundary condition)}, \\
\partial_r w &= g_2(\varphi, z, t) & \text{at} \quad r &= R_2 & \text{(boundary condition)}, \\
\partial_z w &= g_3(r, \varphi, t) & \text{at} \quad z &= 0 & \text{(boundary condition)}, \\
\partial_z w &= g_4(r, \varphi, t) & \text{at} \quad z &= l & \text{(boundary condition)}.
\end{aligned}$$

Solution:

$$w(r,\varphi,z,t) = \int_0^l \int_0^{2\pi} \int_{R_1}^{R_2} f(\xi,\eta,\zeta) G(r,\varphi,z,\xi,\eta,\zeta,t) \xi \, d\xi \, d\eta \, d\zeta$$

$$- aR_1 \int_0^t \int_0^l \int_0^{2\pi} g_1(\eta,\zeta,\tau) G(r,\varphi,z,R_1,\eta,\zeta,t-\tau) \, d\eta \, d\zeta \, d\tau$$

$$+ aR_2 \int_0^t \int_0^l \int_0^{2\pi} g_2(\eta,\zeta,\tau) G(r,\varphi,z,R_2,\eta,\zeta,t-\tau) \, d\eta \, d\zeta \, d\tau$$

$$- a \int_0^t \int_0^{2\pi} \int_{R_1}^{R_2} g_3(\xi,\eta,\tau) G(r,\varphi,z,\xi,\eta,0,t-\tau) \xi \, d\xi \, d\eta \, d\tau$$

$$+ a \int_0^t \int_0^{2\pi} \int_{R_1}^{R_2} g_4(\xi,\eta,\tau) G(r,\varphi,z,\xi,\eta,l,t-\tau) \xi \, d\xi \, d\eta \, d\tau.$$

Here,

$$G(r,\varphi,z,\xi,\eta,\zeta,t) = G_1(r,\varphi,\xi,\eta,t) \left[\frac{1}{l} + \frac{2}{l} \sum_{n=1}^{\infty} \cos\left(\frac{n\pi z}{l}\right) \cos\left(\frac{n\pi \zeta}{l}\right) \exp\left(-\frac{an^2\pi^2 t}{l^2}\right) \right],$$

$$G_1(r,\varphi,\xi,\eta,t) = \frac{1}{\pi(R_2^2 - R_1^2)} + \frac{1}{\pi} \sum_{n=0}^{\infty} \sum_{m=1}^{\infty} \frac{A_n \mu_{nm}^2 Z_n(\mu_{nm} r) Z_n(\mu_{nm} \xi) \cos[n(\varphi-\eta)] \exp(-\mu_{nm}^2 at)}{(\mu_{nm}^2 R_2^2 - n^2) Z_n^2(\mu_{nm} R_2) - (\mu_{nm}^2 R_1^2 - n^2) Z_n^2(\mu_{nm} R_1)},$$

$$Z_n(\mu_{nm} r) = J_n'(\mu_{nm} R_1) Y_n(\mu_{nm} r) - Y_n'(\mu_{nm} R_1) J_n(\mu_{nm} r),$$

where $A_0 = 1$ and $A_n = 2$ for $n = 1, 2, \ldots$; the $J_n(r)$ and $Y_n(r)$ are the Bessel functions (the prime denotes the derivative with respect to the argument); and the μ_{nm} are positive roots of the transcendental equation

$$J_n'(\mu R_1) Y_n'(\mu R_2) - Y_n'(\mu R_1) J_n'(\mu R_2) = 0.$$

3.1.2-22. Domain: $R_1 \leq r \leq R_2$, $0 \leq \varphi \leq 2\pi$, $0 \leq z \leq l$. **Third boundary value problem.**

A circular cylinder of finite length is considered. The following conditions are prescribed:

$$w = f(r,\varphi,z) \quad \text{at} \quad t = 0 \quad \text{(initial condition)},$$
$$\partial_r w - k_1 w = g_1(\varphi,z,t) \quad \text{at} \quad r = R_1 \quad \text{(boundary condition)},$$
$$\partial_r w + k_2 w = g_2(\varphi,z,t) \quad \text{at} \quad r = R_2 \quad \text{(boundary condition)},$$
$$\partial_z w - k_3 w = g_3(r,\varphi,t) \quad \text{at} \quad z = 0 \quad \text{(boundary condition)},$$
$$\partial_z w + k_4 w = g_4(r,\varphi,t) \quad \text{at} \quad z = l \quad \text{(boundary condition)}.$$

The solution $w(r,\varphi,z,t)$ is determined by the formula in Paragraph 3.1.2-21 where

$$G(r,\varphi,z,\xi,\eta,\zeta,t) = G_1(r,\varphi,\xi,\eta,t) G_2(z,\zeta,t).$$

Here, the first factor has the form

$$G_1(r,\varphi,\xi,\eta,t) = \frac{1}{\pi} \sum_{n=0}^{\infty} \sum_{m=1}^{\infty} \frac{A_n \mu_{nm}^2 Z_n(\mu_{nm} r) Z_n(\mu_{nm} \xi) \cos[n(\varphi-\eta)] \exp(-\mu_{nm}^2 at)}{(k_2^2 R_2^2 + \mu_{nm}^2 R_2^2 - n^2) Z_n^2(\mu_{nm} R_2) - (k_1^2 R_1^2 + \mu_{nm}^2 R_1^2 - n^2) Z_n^2(\mu_{nm} R_1)},$$

$$Z_n(\mu_{nm} r) = \left[\mu_{nm} J_n'(\mu_{nm} R_1) - k_1 J_n(\mu_{nm} R_1) \right] Y_n(\mu_{nm} r)$$
$$\qquad - \left[\mu_{nm} Y_n'(\mu_{nm} R_1) - k_1 Y_n(\mu_{nm} R_1) \right] J_n(\mu_{nm} r),$$

where $A_0 = 1$ and $A_n = 2$ for $n = 1, 2, \ldots$; the $J_n(r)$ and $Y_n(r)$ are the Bessel functions; and the μ_{nm} are positive roots of the transcendental equation

$$[\mu J_n'(\mu R_1) - k_1 J_n(\mu R_1)][\mu Y_n'(\mu R_2) + k_2 Y_n(\mu R_2)]$$
$$= [\mu Y_n'(\mu R_1) - k_1 Y_n(\mu R_1)][\mu J_n'(\mu R_2) + k_2 J_n(\mu R_2)].$$

The second factor is given by

$$G_2(z, \zeta, t) = \sum_{s=1}^{\infty} \frac{h_s(z) h_s(\zeta)}{\|h_s\|^2} \exp(-a\lambda_s^2 t),$$

$$h_s(z) = \cos(\lambda_s z) + \frac{k_3}{\lambda_s} \sin(\lambda_s z), \quad \|h_s\|^2 = \frac{k_4}{2\lambda_s^2} \frac{\lambda_s^2 + k_3^2}{\lambda_s^2 + k_4^2} + \frac{k_3}{2\lambda_s^2} + \frac{l}{2}\left(1 + \frac{k_3^2}{\lambda_s^2}\right),$$

where the λ_s are positive roots of the transcendental equation $\dfrac{\tan(\lambda l)}{\lambda} = \dfrac{k_3 + k_4}{\lambda^2 - k_3 k_4}.$

3.1.2-23. Domain: $R_1 \leq r \leq R_2$, $0 \leq \varphi \leq 2\pi$, $0 \leq z \leq l$. Mixed boundary value problems.

1°. A circular cylinder of finite length is considered. The following conditions are prescribed:

$$\begin{aligned}
w &= f(r, \varphi, z) &&\text{at} \quad t = 0 &&\text{(initial condition)}, \\
w &= g_1(\varphi, z, t) &&\text{at} \quad r = R_1 &&\text{(boundary condition)}, \\
w &= g_2(\varphi, z, t) &&\text{at} \quad r = R_2 &&\text{(boundary condition)}, \\
\partial_z w &= g_3(r, \varphi, t) &&\text{at} \quad z = 0 &&\text{(boundary condition)}, \\
\partial_z w &= g_4(r, \varphi, t) &&\text{at} \quad z = l &&\text{(boundary condition)}.
\end{aligned}$$

Solution:

$$\begin{aligned}
w(r, \varphi, z, t) = &\int_0^l \int_0^{2\pi} \int_{R_1}^{R_2} f(\xi, \eta, \zeta) G(r, \varphi, z, \xi, \eta, \zeta, t) \xi \, d\xi \, d\eta \, d\zeta \\
&+ aR_1 \int_0^t \int_0^l \int_0^{2\pi} g_1(\eta, \zeta, \tau) \left[\frac{\partial}{\partial \xi} G(r, \varphi, z, \xi, \eta, \zeta, t - \tau)\right]_{\xi = R_1} d\eta \, d\zeta \, d\tau \\
&- aR_2 \int_0^t \int_0^l \int_0^{2\pi} g_2(\eta, \zeta, \tau) \left[\frac{\partial}{\partial \xi} G(r, \varphi, z, \xi, \eta, \zeta, t - \tau)\right]_{\xi = R_2} d\eta \, d\zeta \, d\tau \\
&- a \int_0^t \int_0^{2\pi} \int_{R_1}^{R_2} g_3(\xi, \eta, \tau) G(r, \varphi, z, \xi, \eta, 0, t - \tau) \xi \, d\xi \, d\eta \, d\tau \\
&+ a \int_0^t \int_0^{2\pi} \int_{R_1}^{R_2} g_4(\xi, \eta, \tau) G(r, \varphi, z, \xi, \eta, l, t - \tau) \xi \, d\xi \, d\eta \, d\tau.
\end{aligned}$$

Here,

$$G(r, \varphi, z, \xi, \eta, \zeta, t) = G_1(r, \varphi, \xi, \eta, t) \left[\frac{1}{l} + \frac{2}{l} \sum_{n=1}^{\infty} \cos\left(\frac{n\pi z}{l}\right) \cos\left(\frac{n\pi \zeta}{l}\right) \exp\left(-\frac{an^2\pi^2 t}{l^2}\right)\right],$$

$$G_1(r, \varphi, \xi, \eta, t) = \frac{\pi}{2} \sum_{n=0}^{\infty} \sum_{m=1}^{\infty} A_n B_{nm} Z_n(\mu_{nm} r) Z_n(\mu_{nm} \xi) \cos[n(\varphi - \eta)] \exp(-\mu_{nm}^2 a t),$$

$$A_n = \begin{cases} 1/2 & \text{for } n = 0, \\ 1 & \text{for } n \neq 0, \end{cases} \quad B_{nm} = \frac{\mu_{nm}^2 J_n^2(\mu_{nm} R_2)}{J_n^2(\mu_{nm} R_1) - J_n^2(\mu_{nm} R_2)},$$

$$Z_n(\mu_{nm} r) = J_n(\mu_{nm} R_1) Y_n(\mu_{nm} r) - Y_n(\mu_{nm} R_1) J_n(\mu_{nm} r),$$

where the $J_n(r)$ and $Y_n(r)$ are the Bessel functions and the μ_{nm} are positive roots of the transcendental equation
$$J_n(\mu R_1)Y_n(\mu R_2) - Y_n(\mu R_1)J_n(\mu R_2) = 0.$$

2°. A circular cylinder of finite length is considered. The following conditions are prescribed:

$$w = f(r, \varphi, z) \quad \text{at} \quad t = 0 \quad \text{(initial condition)},$$
$$\partial_r w = g_1(\varphi, z, t) \quad \text{at} \quad r = R_1 \quad \text{(boundary condition)},$$
$$\partial_r w = g_2(\varphi, z, t) \quad \text{at} \quad r = R_2 \quad \text{(boundary condition)},$$
$$w = g_3(r, \varphi, t) \quad \text{at} \quad z = 0 \quad \text{(boundary condition)},$$
$$w = g_4(r, \varphi, t) \quad \text{at} \quad z = l \quad \text{(boundary condition)}.$$

Solution:
$$w(r, \varphi, z, t) = \int_0^l \int_0^{2\pi} \int_{R_1}^{R_2} f(\xi, \eta, \zeta) G(r, \varphi, z, \xi, \eta, \zeta, t) \xi\, d\xi\, d\eta\, d\zeta$$
$$- aR_1 \int_0^t \int_0^l \int_0^{2\pi} g_1(\eta, \zeta, \tau) G(r, \varphi, z, R_1, \eta, \zeta, t-\tau)\, d\eta\, d\zeta\, d\tau$$
$$+ aR_2 \int_0^t \int_0^l \int_0^{2\pi} g_2(\eta, \zeta, \tau) G(r, \varphi, z, R_2, \eta, \zeta, t-\tau)\, d\eta\, d\zeta\, d\tau$$
$$+ a \int_0^t \int_0^{2\pi} \int_{R_1}^{R_2} g_3(\xi, \eta, \tau) \left[\frac{\partial}{\partial \zeta} G(r, \varphi, z, \xi, \eta, \zeta, t-\tau)\right]_{\zeta=0} \xi\, d\xi\, d\eta\, d\tau$$
$$- a \int_0^t \int_0^{2\pi} \int_{R_1}^{R_2} g_4(\xi, \eta, \tau) \left[\frac{\partial}{\partial \zeta} G(r, \varphi, z, \xi, \eta, \zeta, t-\tau)\right]_{\zeta=l} \xi\, d\xi\, d\eta\, d\tau.$$

Here,
$$G(r, \varphi, z, \xi, \eta, \zeta, t) = G_1(r, \varphi, \xi, \eta, t)\left[\frac{2}{l}\sum_{n=1}^{\infty} \sin\left(\frac{n\pi z}{l}\right) \sin\left(\frac{n\pi \zeta}{l}\right) \exp\left(-\frac{an^2\pi^2 t}{l^2}\right)\right],$$
$$G_1(r, \varphi, \xi, \eta, t) = \frac{1}{\pi(R_2^2 - R_1^2)} + \frac{1}{\pi}\sum_{n=0}^{\infty}\sum_{m=1}^{\infty} \frac{A_n \mu_{nm}^2 Z_n(\mu_{nm} r) Z_n(\mu_{nm} \xi) \cos[n(\varphi-\eta)] \exp(-\mu_{nm}^2 a t)}{(\mu_{nm}^2 R_2^2 - n^2) Z_n^2(\mu_{nm} R_2) - (\mu_{nm}^2 R_1^2 - n^2) Z_n^2(\mu_{nm} R_1)},$$
$$Z_n(\mu_{nm} r) = J_n'(\mu_{nm} R_1) Y_n(\mu_{nm} r) - Y_n'(\mu_{nm} R_1) J_n(\mu_{nm} r),$$

where $A_0 = 1$ and $A_n = 2$ for $n = 1, 2, \ldots$; the $J_n(r)$ and $Y_n(r)$ are the Bessel functions (the prime denotes the derivative with respect to the argument); and the μ_{nm} are positive roots of the transcendental equation
$$J_n'(\mu R_1) Y_n'(\mu R_2) - Y_n'(\mu R_1) J_n'(\mu R_2) = 0.$$

3.1.2-24. Domain: $0 \leq r < \infty$, $0 \leq \varphi \leq \varphi_0$, $-\infty < z < \infty$. **First boundary value problem.**

A dihedral angle is considered. The following conditions are prescribed:

$$w = f(r, \varphi, z) \quad \text{at} \quad t = 0 \quad \text{(initial condition)},$$
$$w = g_1(r, z, t) \quad \text{at} \quad \varphi = 0 \quad \text{(boundary condition)},$$
$$w = g_2(r, z, t) \quad \text{at} \quad \varphi = \varphi_0 \quad \text{(boundary condition)}.$$

Solution:
$$w(r, \varphi, z, t) = \int_{-\infty}^{\infty} \int_0^{\varphi_0} \int_0^{\infty} f(\xi, \eta, \zeta) G(r, \varphi, z, \xi, \eta, \zeta, t) \xi\, d\xi\, d\eta\, d\zeta$$
$$+ a \int_0^t \int_{-\infty}^{\infty} \int_0^{\infty} g_1(\xi, \zeta, \tau) \frac{1}{\xi}\left[\frac{\partial}{\partial \eta} G(r, \varphi, z, \xi, \eta, \zeta, t-\tau)\right]_{\eta=0} d\xi\, d\zeta\, d\tau$$
$$- a \int_0^t \int_{-\infty}^{\infty} \int_0^{\infty} g_2(\xi, \zeta, \tau) \frac{1}{\xi}\left[\frac{\partial}{\partial \eta} G(r, \varphi, z, \xi, \eta, \zeta, t-\tau)\right]_{\eta=\varphi_0} d\xi\, d\zeta\, d\tau.$$

Here,
$$G(r,\varphi,z,\xi,\eta,\zeta,t) = \frac{1}{2\sqrt{\pi at}} \exp\left[-\frac{(z-\zeta)^2}{4at}\right] G_1(r,\varphi,\xi,\eta,t),$$
$$G_1(r,\varphi,\xi,\eta,t) = \frac{1}{a\varphi_0 t} \exp\left(-\frac{r^2+\xi^2}{4at}\right) \sum_{n=1}^{\infty} I_{n\pi/\varphi_0}\left(\frac{r\xi}{2at}\right) \sin\left(\frac{n\pi\varphi}{\varphi_0}\right) \sin\left(\frac{n\pi\eta}{\varphi_0}\right),$$
where the $I_\nu(r)$ are the modified Bessel functions.

3.1.2-25. Domain: $0 \le r < \infty$, $0 \le \varphi \le \varphi_0$, $-\infty < z < \infty$. Second boundary value problem.

A dihedral angle is considered. The following conditions are prescribed:
$$w = f(r,\varphi,z) \quad \text{at} \quad t = 0 \quad \text{(initial condition)},$$
$$r^{-1}\partial_\varphi w = g_1(r,z,t) \quad \text{at} \quad \varphi = 0 \quad \text{(boundary condition)},$$
$$r^{-1}\partial_\varphi w = g_2(r,z,t) \quad \text{at} \quad \varphi = \varphi_0 \quad \text{(boundary condition)}.$$

Solution:
$$w(r,\varphi,z,t) = \int_{-\infty}^{\infty} \int_0^{\varphi_0} \int_0^{\infty} f(\xi,\eta,\zeta) G(r,\varphi,z,\xi,\eta,\zeta,t) \xi\, d\xi\, d\eta\, d\zeta$$
$$- a \int_0^t \int_{-\infty}^{\infty} \int_0^{\infty} g_1(\xi,\zeta,\tau) G(r,\varphi,z,\xi,0,\zeta,t-\tau)\, d\xi\, d\zeta\, d\tau$$
$$+ a \int_0^t \int_{-\infty}^{\infty} \int_0^{\infty} g_2(\xi,\zeta,\tau) G(r,\varphi,z,\xi,\varphi_0,\zeta,t-\tau)\, d\xi\, d\zeta\, d\tau.$$

Here,
$$G(r,\varphi,z,\xi,\eta,\zeta,t) = \frac{1}{2\sqrt{\pi at}} \exp\left[-\frac{(z-\zeta)^2}{4at}\right] G_1(r,\varphi,\xi,\eta,t),$$
$$G_1(r,\varphi,\xi,\eta,t) = \frac{1}{a\varphi_0 t} \exp\left(-\frac{r^2+\xi^2}{4at}\right) \left[\frac{1}{2} I_0\left(\frac{r\xi}{2at}\right) + \sum_{n=1}^{\infty} I_{n\pi/\varphi_0}\left(\frac{r\xi}{2at}\right) \cos\left(\frac{n\pi\varphi}{\varphi_0}\right) \cos\left(\frac{n\pi\eta}{\varphi_0}\right)\right],$$
where the $I_\nu(r)$ are the modified Bessel functions.

3.1.2-26. Domain: $0 \le r < \infty$, $0 \le \varphi \le \varphi_0$, $0 \le z < \infty$. First boundary value problem.

The upper half of a dihedral angle is considered. The following conditions are prescribed:
$$w = f(r,\varphi,z) \quad \text{at} \quad t = 0 \quad \text{(initial condition)},$$
$$w = g_1(r,z,t) \quad \text{at} \quad \varphi = 0 \quad \text{(boundary condition)},$$
$$w = g_2(r,z,t) \quad \text{at} \quad \varphi = \varphi_0 \quad \text{(boundary condition)},$$
$$w = g_3(r,\varphi,t) \quad \text{at} \quad z = 0 \quad \text{(boundary condition)}.$$

Solution:
$$w(r,\varphi,z,t) = \int_0^{\infty} \int_0^{\varphi_0} \int_0^{\infty} f(\xi,\eta,\zeta) G(r,\varphi,z,\xi,\eta,\zeta,t) \xi\, d\xi\, d\eta\, d\zeta$$
$$+ a \int_0^t \int_0^{\infty} \int_0^{\infty} g_1(\xi,\zeta,\tau) \frac{1}{\xi}\left[\frac{\partial}{\partial\eta} G(r,\varphi,z,\xi,\eta,\zeta,t-\tau)\right]_{\eta=0} d\xi\, d\zeta\, d\tau$$
$$- a \int_0^t \int_0^{\infty} \int_0^{\infty} g_2(\xi,\zeta,\tau) \frac{1}{\xi}\left[\frac{\partial}{\partial\eta} G(r,\varphi,z,\xi,\eta,\zeta,t-\tau)\right]_{\eta=\varphi_0} d\xi\, d\zeta\, d\tau$$
$$+ a \int_0^t \int_0^{\varphi_0} \int_0^{\infty} g_3(\xi,\eta,\tau) \left[\frac{\partial}{\partial\zeta} G(r,\varphi,z,\xi,\eta,\zeta,t-\tau)\right]_{\zeta=0} \xi\, d\xi\, d\eta\, d\tau.$$

Here,

$$G(r,\varphi,z,\xi,\eta,\zeta,t) = \frac{1}{2\sqrt{\pi a t}}\left\{\exp\left[-\frac{(z-\zeta)^2}{4at}\right] - \exp\left[-\frac{(z+\zeta)^2}{4at}\right]\right\}G_1(r,\varphi,\xi,\eta,t),$$

$$G_1(r,\varphi,\xi,\eta,t) = \frac{1}{a\varphi_0 t}\exp\left(-\frac{r^2+\xi^2}{4at}\right)\sum_{n=1}^{\infty} I_{n\pi/\varphi_0}\left(\frac{r\xi}{2at}\right)\sin\left(\frac{n\pi\varphi}{\varphi_0}\right)\sin\left(\frac{n\pi\eta}{\varphi_0}\right),$$

where the $I_\nu(r)$ are the modified Bessel functions.

3.1.2-27. Domain: $0 \le r < \infty$, $0 \le \varphi \le \varphi_0$, $0 \le z < \infty$. Second boundary value problem.

The upper half of a dihedral angle is considered. The following conditions are prescribed:

$$\begin{aligned}
w &= f(r,\varphi,z) & \text{at} \quad t &= 0 & \text{(initial condition)}, \\
r^{-1}\partial_\varphi w &= g_1(r,z,t) & \text{at} \quad \varphi &= 0 & \text{(boundary condition)}, \\
r^{-1}\partial_\varphi w &= g_2(r,z,t) & \text{at} \quad \varphi &= \varphi_0 & \text{(boundary condition)}, \\
\partial_z w &= g_3(r,\varphi,t) & \text{at} \quad z &= 0 & \text{(boundary condition)}.
\end{aligned}$$

Solution:

$$\begin{aligned}
w(r,\varphi,z,t) = &\int_0^\infty \int_0^{\varphi_0} \int_0^\infty f(\xi,\eta,\zeta) G(r,\varphi,z,\xi,\eta,\zeta,t)\xi\, d\xi\, d\eta\, d\zeta \\
&- a\int_0^t \int_0^\infty \int_0^\infty g_1(\xi,\zeta,\tau) G(r,\varphi,z,\xi,0,\zeta,t-\tau)\, d\xi\, d\zeta\, d\tau \\
&+ a\int_0^t \int_0^\infty \int_0^\infty g_2(\xi,\zeta,\tau) G(r,\varphi,z,\xi,\varphi_0,\zeta,t-\tau)\, d\xi\, d\zeta\, d\tau \\
&- a\int_0^t \int_0^{\varphi_0} \int_0^\infty g_3(\xi,\eta,\tau) G(r,\varphi,z,\xi,\eta,0,t-\tau)\xi\, d\xi\, d\eta\, d\tau.
\end{aligned}$$

Here,

$$G(r,\varphi,z,\xi,\eta,\zeta,t) = \frac{1}{2\sqrt{\pi a t}}\left\{\exp\left[-\frac{(z-\zeta)^2}{4at}\right] + \exp\left[-\frac{(z+\zeta)^2}{4at}\right]\right\}G_1(r,\varphi,\xi,\eta,t),$$

$$G_1(r,\varphi,\xi,\eta,t) = \frac{1}{a\varphi_0 t}\exp\left(-\frac{r^2+\xi^2}{4at}\right)\left[\frac{1}{2}I_0\left(\frac{r\xi}{2at}\right) + \sum_{n=1}^{\infty} I_{n\pi/\varphi_0}\left(\frac{r\xi}{2at}\right)\cos\left(\frac{n\pi\varphi}{\varphi_0}\right)\cos\left(\frac{n\pi\eta}{\varphi_0}\right)\right],$$

where the $I_\nu(r)$ are the modified Bessel functions.

3.1.2-28. Domain: $0 \le r < \infty$, $0 \le \varphi \le \varphi_0$, $0 \le z < \infty$. Mixed boundary value problems.

1°. The upper half of a dihedral angle is considered. The following conditions are prescribed:

$$\begin{aligned}
w &= f(r,\varphi,z) & \text{at} \quad t &= 0 & \text{(initial condition)}, \\
w &= g_1(r,z,t) & \text{at} \quad \varphi &= 0 & \text{(boundary condition)}, \\
w &= g_2(r,z,t) & \text{at} \quad \varphi &= \varphi_0 & \text{(boundary condition)}, \\
\partial_z w &= g_3(r,\varphi,t) & \text{at} \quad z &= 0 & \text{(boundary condition)}.
\end{aligned}$$

Solution:

$$w(r,\varphi,z,t) = \int_0^\infty \int_0^{\varphi_0} \int_0^\infty f(\xi,\eta,\zeta) G(r,\varphi,z,\xi,\eta,\zeta,t)\xi\,d\xi\,d\eta\,d\zeta$$
$$+ a\int_0^t \int_0^\infty \int_0^\infty g_1(\xi,\zeta,\tau) \frac{1}{\xi}\left[\frac{\partial}{\partial \eta} G(r,\varphi,z,\xi,\eta,\zeta,t-\tau)\right]_{\eta=0} d\xi\,d\zeta\,d\tau$$
$$- a\int_0^t \int_0^\infty \int_0^\infty g_2(\xi,\zeta,\tau) \frac{1}{\xi}\left[\frac{\partial}{\partial \eta} G(r,\varphi,z,\xi,\eta,\zeta,t-\tau)\right]_{\eta=\varphi_0} d\xi\,d\zeta\,d\tau$$
$$- a\int_0^t \int_0^{\varphi_0} \int_0^\infty g_3(\xi,\eta,\tau) G(r,\varphi,z,\xi,\eta,0,t-\tau)\xi\,d\xi\,d\eta\,d\tau.$$

Here,

$$G(r,\varphi,z,\xi,\eta,\zeta,t) = \frac{1}{2\sqrt{\pi a t}}\left\{\exp\left[-\frac{(z-\zeta)^2}{4at}\right] + \exp\left[-\frac{(z+\zeta)^2}{4at}\right]\right\} G_1(r,\varphi,\xi,\eta,t),$$

$$G_1(r,\varphi,\xi,\eta,t) = \frac{1}{a\varphi_0 t}\exp\left(-\frac{r^2+\xi^2}{4at}\right) \sum_{n=1}^\infty I_{n\pi/\varphi_0}\left(\frac{r\xi}{2at}\right) \sin\left(\frac{n\pi\varphi}{\varphi_0}\right) \sin\left(\frac{n\pi\eta}{\varphi_0}\right),$$

where the $I_\nu(r)$ are the modified Bessel functions.

2°. The upper half of a dihedral angle is considered. The following conditions are prescribed:

$$w = f(r,\varphi,z) \quad \text{at} \quad t = 0 \qquad \text{(initial condition)},$$
$$r^{-1}\partial_\varphi w = g_1(r,z,t) \quad \text{at} \quad \varphi = 0 \qquad \text{(boundary condition)},$$
$$r^{-1}\partial_\varphi w = g_2(r,z,t) \quad \text{at} \quad \varphi = \varphi_0 \qquad \text{(boundary condition)},$$
$$w = g_3(r,\varphi,t) \quad \text{at} \quad z = 0 \qquad \text{(boundary condition)}.$$

Solution:

$$w(r,\varphi,z,t) = \int_0^\infty \int_0^{\varphi_0} \int_0^\infty f(\xi,\eta,\zeta) G(r,\varphi,z,\xi,\eta,\zeta,t)\xi\,d\xi\,d\eta\,d\zeta$$
$$- a\int_0^t \int_0^\infty \int_0^\infty g_1(\xi,\zeta,\tau) G(r,\varphi,z,\xi,0,\zeta,t-\tau)\,d\xi\,d\zeta\,d\tau$$
$$+ a\int_0^t \int_0^\infty \int_0^\infty g_2(\xi,\zeta,\tau) G(r,\varphi,z,\xi,\varphi_0,\zeta,t-\tau)\,d\xi\,d\zeta\,d\tau$$
$$+ a\int_0^t \int_0^{\varphi_0} \int_0^\infty g_3(\xi,\eta,\tau) \left[\frac{\partial}{\partial \zeta} G(r,\varphi,z,\xi,\eta,\zeta,t-\tau)\right]_{\zeta=0} \xi\,d\xi\,d\eta\,d\tau.$$

Here,

$$G(r,\varphi,z,\xi,\eta,\zeta,t) = \frac{1}{2\sqrt{\pi a t}}\left\{\exp\left[-\frac{(z-\zeta)^2}{4at}\right] - \exp\left[-\frac{(z+\zeta)^2}{4at}\right]\right\} G_1(r,\varphi,\xi,\eta,t),$$

$$G_1(r,\varphi,\xi,\eta,t) = \frac{1}{a\varphi_0 t}\exp\left(-\frac{r^2+\xi^2}{4at}\right) \left[\frac{1}{2} I_0\left(\frac{r\xi}{2at}\right) + \sum_{n=1}^\infty I_{n\pi/\varphi_0}\left(\frac{r\xi}{2at}\right) \cos\left(\frac{n\pi\varphi}{\varphi_0}\right) \cos\left(\frac{n\pi\eta}{\varphi_0}\right)\right],$$

where the $I_\nu(r)$ are the modified Bessel functions.

3.1.2-29. Domain: $0 \leq r < \infty$, $0 \leq \varphi \leq \varphi_0$, $0 \leq z \leq l$. First boundary value problem.

A wedge domain of finite thickness is considered. The following conditions are prescribed:

$$w = f(r, \varphi, z) \quad \text{at} \quad t = 0 \quad \text{(initial condition)},$$
$$w = g_1(r, z, t) \quad \text{at} \quad \varphi = 0 \quad \text{(boundary condition)},$$
$$w = g_2(r, z, t) \quad \text{at} \quad \varphi = \varphi_0 \quad \text{(boundary condition)},$$
$$w = g_3(r, \varphi, t) \quad \text{at} \quad z = 0 \quad \text{(boundary condition)},$$
$$w = g_4(r, \varphi, t) \quad \text{at} \quad z = l \quad \text{(boundary condition)}.$$

Solution:

$$w(r, \varphi, z, t) = \int_0^l \int_0^{\varphi_0} \int_0^\infty f(\xi, \eta, \zeta) G(r, \varphi, z, \xi, \eta, \zeta, t) \xi \, d\xi \, d\eta \, d\zeta$$
$$+ a \int_0^t \int_0^l \int_0^\infty g_1(\xi, \zeta, \tau) \frac{1}{\xi} \left[\frac{\partial}{\partial \eta} G(r, \varphi, z, \xi, \eta, \zeta, t - \tau) \right]_{\eta=0} d\xi \, d\zeta \, d\tau$$
$$- a \int_0^t \int_0^l \int_0^\infty g_2(\xi, \zeta, \tau) \frac{1}{\xi} \left[\frac{\partial}{\partial \eta} G(r, \varphi, z, \xi, \eta, \zeta, t - \tau) \right]_{\eta=\varphi_0} d\xi \, d\zeta \, d\tau$$
$$+ a \int_0^t \int_0^{\varphi_0} \int_0^\infty g_3(\xi, \eta, \tau) \left[\frac{\partial}{\partial \zeta} G(r, \varphi, z, \xi, \eta, \zeta, t - \tau) \right]_{\zeta=0} \xi \, d\xi \, d\eta \, d\tau$$
$$- a \int_0^t \int_0^{\varphi_0} \int_0^\infty g_4(\xi, \eta, \tau) \left[\frac{\partial}{\partial \zeta} G(r, \varphi, z, \xi, \eta, \zeta, t - \tau) \right]_{\zeta=l} \xi \, d\xi \, d\eta \, d\tau.$$

Here,

$$G(r, \varphi, z, \xi, \eta, \zeta, t) = G_1(r, \varphi, \xi, \eta, t) \left[\frac{2}{l} \sum_{n=1}^\infty \sin\left(\frac{n\pi z}{l}\right) \sin\left(\frac{n\pi \zeta}{l}\right) \exp\left(-\frac{an^2\pi^2 t}{l^2}\right) \right],$$

$$G_1(r, \varphi, \xi, \eta, t) = \frac{1}{a\varphi_0 t} \exp\left(-\frac{r^2 + \xi^2}{4at}\right) \sum_{n=1}^\infty I_{n\pi/\varphi_0}\left(\frac{r\xi}{2at}\right) \sin\left(\frac{n\pi \varphi}{\varphi_0}\right) \sin\left(\frac{n\pi \eta}{\varphi_0}\right),$$

where the $I_\nu(r)$ are the modified Bessel functions.

3.1.2-30. Domain: $0 \leq r < \infty$, $0 \leq \varphi \leq \varphi_0$, $0 \leq z \leq l$. Second boundary value problem.

A wedge domain of finite thickness is considered. The following conditions are prescribed:

$$w = f(r, \varphi, z) \quad \text{at} \quad t = 0 \quad \text{(initial condition)},$$
$$r^{-1}\partial_\varphi w = g_1(r, z, t) \quad \text{at} \quad \varphi = 0 \quad \text{(boundary condition)},$$
$$r^{-1}\partial_\varphi w = g_2(r, z, t) \quad \text{at} \quad \varphi = \varphi_0 \quad \text{(boundary condition)},$$
$$\partial_z w = g_3(r, \varphi, t) \quad \text{at} \quad z = 0 \quad \text{(boundary condition)},$$
$$\partial_z w = g_4(r, \varphi, t) \quad \text{at} \quad z = l \quad \text{(boundary condition)}.$$

Solution:

$$w(r,\varphi,z,t) = \int_0^l \int_0^{\varphi_0} \int_0^\infty f(\xi,\eta,\zeta) G(r,\varphi,z,\xi,\eta,\zeta,t) \xi \, d\xi \, d\eta \, d\zeta$$
$$- a \int_0^t \int_0^l \int_0^\infty g_1(\xi,\zeta,\tau) G(r,\varphi,z,\xi,0,\zeta,t-\tau) \, d\xi \, d\zeta \, d\tau$$
$$+ a \int_0^t \int_0^l \int_0^\infty g_2(\xi,\zeta,\tau) G(r,\varphi,z,\xi,\varphi_0,\zeta,t-\tau) \, d\xi \, d\zeta \, d\tau$$
$$- a \int_0^t \int_0^{\varphi_0} \int_0^\infty g_3(\xi,\eta,\tau) G(r,\varphi,z,\xi,\eta,0,t-\tau) \xi \, d\xi \, d\eta \, d\tau$$
$$+ a \int_0^t \int_0^{\varphi_0} \int_0^\infty g_4(\xi,\eta,\tau) G(r,\varphi,z,\xi,\eta,l,t-\tau) \xi \, d\xi \, d\eta \, d\tau.$$

Here,

$$G(r,\varphi,z,\xi,\eta,\zeta,t) = G_1(r,\varphi,\xi,\eta,t) G_2(z,\zeta,t),$$

$$G_1(r,\varphi,\xi,\eta,t) = \frac{1}{a\varphi_0 t} \exp\left(-\frac{r^2+\xi^2}{4at}\right)\left[\frac{1}{2} I_0\left(\frac{r\xi}{2at}\right) + \sum_{n=1}^\infty I_{n\pi/\varphi_0}\left(\frac{r\xi}{2at}\right) \cos\left(\frac{n\pi\varphi}{\varphi_0}\right) \cos\left(\frac{n\pi\eta}{\varphi_0}\right)\right],$$

$$G_2(z,\zeta,t) = \frac{1}{l} + \frac{2}{l} \sum_{n=1}^\infty \cos\left(\frac{n\pi z}{l}\right) \cos\left(\frac{n\pi\zeta}{l}\right) \exp\left(-\frac{an^2\pi^2 t}{l^2}\right),$$

where the $I_\nu(r)$ are the modified Bessel functions.

3.1.2-31. Domain: $0 \le r < \infty$, $0 \le \varphi \le \varphi_0$, $0 \le z \le l$. Mixed boundary value problems.

1°. A wedge domain of finite thickness is considered. The following conditions are prescribed:

$$\begin{aligned}
w &= f(r,\varphi,z) & \text{at} \quad & t = 0 & \text{(initial condition)}, \\
w &= g_1(r,z,t) & \text{at} \quad & \varphi = 0 & \text{(boundary condition)}, \\
w &= g_2(r,z,t) & \text{at} \quad & \varphi = \varphi_0 & \text{(boundary condition)}, \\
\partial_z w &= g_3(r,\varphi,t) & \text{at} \quad & z = 0 & \text{(boundary condition)}, \\
\partial_z w &= g_4(r,\varphi,t) & \text{at} \quad & z = l & \text{(boundary condition)}.
\end{aligned}$$

Solution:

$$w(r,\varphi,z,t) = \int_0^l \int_0^{\varphi_0} \int_0^\infty f(\xi,\eta,\zeta) G(r,\varphi,z,\xi,\eta,\zeta,t) \xi \, d\xi \, d\eta \, d\zeta$$
$$+ a \int_0^t \int_0^l \int_0^\infty g_1(\xi,\zeta,\tau) \frac{1}{\xi}\left[\frac{\partial}{\partial \eta} G(r,\varphi,z,\xi,\eta,\zeta,t-\tau)\right]_{\eta=0} d\xi \, d\zeta \, d\tau$$
$$- a \int_0^t \int_0^l \int_0^\infty g_2(\xi,\zeta,\tau) \frac{1}{\xi}\left[\frac{\partial}{\partial \eta} G(r,\varphi,z,\xi,\eta,\zeta,t-\tau)\right]_{\eta=\varphi_0} d\xi \, d\zeta \, d\tau$$
$$- a \int_0^t \int_0^{\varphi_0} \int_0^\infty g_3(\xi,\eta,\tau) G(r,\varphi,z,\xi,\eta,0,t-\tau) \xi \, d\xi \, d\eta \, d\tau$$
$$+ a \int_0^t \int_0^{\varphi_0} \int_0^\infty g_4(\xi,\eta,\tau) G(r,\varphi,z,\xi,\eta,l,t-\tau) \xi \, d\xi \, d\eta \, d\tau.$$

Here,

$$G(r,\varphi,z,\xi,\eta,\zeta,t) = G_1(r,\varphi,\xi,\eta,t) G_2(z,\zeta,t),$$

$$G_1(r,\varphi,\xi,\eta,t) = \frac{1}{a\varphi_0 t} \exp\left(-\frac{r^2+\xi^2}{4at}\right) \sum_{n=1}^{\infty} I_{n\pi/\varphi_0}\left(\frac{r\xi}{2at}\right) \sin\left(\frac{n\pi\varphi}{\varphi_0}\right) \sin\left(\frac{n\pi\eta}{\varphi_0}\right),$$

$$G_2(z,\zeta,t) = \frac{1}{l} + \frac{2}{l}\sum_{n=1}^{\infty} \cos\left(\frac{n\pi z}{l}\right)\cos\left(\frac{n\pi\zeta}{l}\right)\exp\left(-\frac{an^2\pi^2 t}{l^2}\right),$$

where the $I_\nu(r)$ are the modified Bessel functions.

2°. A wedge domain of finite thickness is considered. The following conditions are prescribed:

$$\begin{aligned} w &= f(r,\varphi,z) &&\text{at } t=0 &&\text{(initial condition)},\\ r^{-1}\partial_\varphi w &= g_1(r,z,t) &&\text{at } \varphi=0 &&\text{(boundary condition)},\\ r^{-1}\partial_\varphi w &= g_2(r,z,t) &&\text{at } \varphi=\varphi_0 &&\text{(boundary condition)},\\ w &= g_3(r,\varphi,t) &&\text{at } z=0 &&\text{(boundary condition)},\\ w &= g_4(r,\varphi,t) &&\text{at } z=l &&\text{(boundary condition)}. \end{aligned}$$

Solution:

$$\begin{aligned} w(r,\varphi,z,t) &= \int_0^l \int_0^{\varphi_0}\int_0^\infty f(\xi,\eta,\zeta)G(r,\varphi,z,\xi,\eta,\zeta,t)\xi\,d\xi\,d\eta\,d\zeta\\ &\quad - a\int_0^t\int_0^l\int_0^\infty g_1(\xi,\zeta,\tau)G(r,\varphi,z,\xi,0,\zeta,t-\tau)\,d\xi\,d\zeta\,d\tau\\ &\quad + a\int_0^t\int_0^l\int_0^\infty g_2(\xi,\zeta,\tau)G(r,\varphi,z,\xi,\varphi_0,\zeta,t-\tau)\,d\xi\,d\zeta\,d\tau\\ &\quad + a\int_0^t\int_0^{\varphi_0}\int_0^\infty g_3(\xi,\eta,\tau)\left[\frac{\partial}{\partial\zeta}G(r,\varphi,z,\xi,\eta,\zeta,t-\tau)\right]_{\zeta=0}\xi\,d\xi\,d\eta\,d\tau\\ &\quad - a\int_0^t\int_0^{\varphi_0}\int_0^\infty g_4(\xi,\eta,\tau)\left[\frac{\partial}{\partial\zeta}G(r,\varphi,z,\xi,\eta,\zeta,t-\tau)\right]_{\zeta=l}\xi\,d\xi\,d\eta\,d\tau. \end{aligned}$$

Here,

$$G(r,\varphi,z,\xi,\eta,\zeta,t) = G_1(r,\varphi,\xi,\eta,t)\left[\frac{2}{l}\sum_{n=1}^\infty \sin\left(\frac{n\pi z}{l}\right)\sin\left(\frac{n\pi\zeta}{l}\right)\exp\left(-\frac{an^2\pi^2 t}{l^2}\right)\right],$$

$$G_1(r,\varphi,\xi,\eta,t) = \frac{1}{a\varphi_0 t}\exp\left(-\frac{r^2+\xi^2}{4at}\right)\left[\frac{1}{2}I_0\left(\frac{r\xi}{2at}\right) + \sum_{n=1}^\infty I_{n\pi/\varphi_0}\left(\frac{r\xi}{2at}\right)\cos\left(\frac{n\pi\varphi}{\varphi_0}\right)\cos\left(\frac{n\pi\eta}{\varphi_0}\right)\right],$$

where the $I_\nu(r)$ are the modified Bessel functions.

3.1.2-32. Domain: $0 \leq r \leq R$, $0 \leq \varphi \leq \varphi_0$, $-\infty < z < \infty$. First boundary value problem.

An infinite cylindrical sector is considered. The following conditions are prescribed:

$$\begin{aligned} w &= f(r,\varphi,z) &&\text{at } t=0 &&\text{(initial condition)},\\ w &= g_1(\varphi,z,t) &&\text{at } r=R &&\text{(boundary condition)},\\ w &= g_2(r,z,t) &&\text{at } \varphi=0 &&\text{(boundary condition)},\\ w &= g_3(r,z,t) &&\text{at } \varphi=\varphi_0 &&\text{(boundary condition)}. \end{aligned}$$

Solution:

$$w(r,\varphi,z,t) = \int_{-\infty}^{\infty}\int_0^{\varphi_0}\int_0^R f(\xi,\eta,\zeta)G(r,\varphi,z,\xi,\eta,\zeta,t)\xi\,d\xi\,d\eta\,d\zeta$$

$$- aR\int_0^t\int_{-\infty}^{\infty}\int_0^{\varphi_0} g_1(\eta,\zeta,\tau)\left[\frac{\partial}{\partial \xi}G(r,\varphi,z,\xi,\eta,\zeta,t-\tau)\right]_{\xi=R} d\eta\,d\zeta\,d\tau$$

$$+ a\int_0^t\int_{-\infty}^{\infty}\int_0^R g_2(\xi,\zeta,\tau)\frac{1}{\xi}\left[\frac{\partial}{\partial \eta}G(r,\varphi,z,\xi,\eta,\zeta,t-\tau)\right]_{\eta=0} d\xi\,d\zeta\,d\tau$$

$$- a\int_0^t\int_{-\infty}^{\infty}\int_0^R g_3(\xi,\zeta,\tau)\frac{1}{\xi}\left[\frac{\partial}{\partial \eta}G(r,\varphi,z,\xi,\eta,\zeta,t-\tau)\right]_{\eta=\varphi_0} d\xi\,d\zeta\,d\tau.$$

Here,

$$G(r,\varphi,z,\xi,\eta,\zeta,t) = \frac{1}{2\sqrt{\pi a t}}\exp\left[-\frac{(z-\zeta)^2}{4at}\right]G_1(r,\varphi,\xi,\eta,t),$$

$$G_1(r,\varphi,\xi,\eta,t) = \frac{4}{R^2\varphi_0}\sum_{n=1}^{\infty}\sum_{m=1}^{\infty}\frac{J_{n\pi/\varphi_0}(\mu_{nm}r)J_{n\pi/\varphi_0}(\mu_{nm}\xi)}{[J'_{n\pi/\varphi_0}(\mu_{nm}R)]^2}\sin\left(\frac{n\pi\varphi}{\varphi_0}\right)\sin\left(\frac{n\pi\eta}{\varphi_0}\right)\exp(-\mu_{nm}^2 a t),$$

where the $J_{n\pi/\varphi_0}(r)$ are the Bessel functions and the μ_{nm} are positive roots of the transcendental equation $J_{n\pi/\varphi_0}(\mu R) = 0$.

3.1.2-33. Domain: $0 \le r \le R$, $0 \le \varphi \le \varphi_0$, $0 \le z < \infty$. First boundary value problem.

A semiinfinite cylindrical sector is considered. The following conditions are prescribed:

$$w = f(r,\varphi,z) \quad \text{at} \quad t = 0 \quad \text{(initial condition)},$$
$$w = g_1(\varphi,z,t) \quad \text{at} \quad r = R \quad \text{(boundary condition)},$$
$$w = g_2(r,z,t) \quad \text{at} \quad \varphi = 0 \quad \text{(boundary condition)},$$
$$w = g_3(r,z,t) \quad \text{at} \quad \varphi = \varphi_0 \quad \text{(boundary condition)},$$
$$w = g_4(r,\varphi,t) \quad \text{at} \quad z = 0 \quad \text{(boundary condition)}.$$

Solution:

$$w(r,\varphi,z,t) = \int_0^{\infty}\int_0^{\varphi_0}\int_0^R f(\xi,\eta,\zeta)G(r,\varphi,z,\xi,\eta,\zeta,t)\xi\,d\xi\,d\eta\,d\zeta$$

$$- aR\int_0^t\int_0^{\infty}\int_0^{\varphi_0} g_1(\eta,\zeta,\tau)\left[\frac{\partial}{\partial \xi}G(r,\varphi,z,\xi,\eta,\zeta,t-\tau)\right]_{\xi=R} d\eta\,d\zeta\,d\tau$$

$$+ a\int_0^t\int_0^{\infty}\int_0^R g_2(\xi,\zeta,\tau)\frac{1}{\xi}\left[\frac{\partial}{\partial \eta}G(r,\varphi,z,\xi,\eta,\zeta,t-\tau)\right]_{\eta=0} d\xi\,d\zeta\,d\tau$$

$$- a\int_0^t\int_0^{\infty}\int_0^R g_3(\xi,\zeta,\tau)\frac{1}{\xi}\left[\frac{\partial}{\partial \eta}G(r,\varphi,z,\xi,\eta,\zeta,t-\tau)\right]_{\eta=\varphi_0} d\xi\,d\zeta\,d\tau$$

$$+ a\int_0^t\int_0^{\varphi_0}\int_0^R g_4(\xi,\eta,\tau)\left[\frac{\partial}{\partial \zeta}G(r,\varphi,z,\xi,\eta,\zeta,t-\tau)\right]_{\zeta=0} \xi\,d\xi\,d\eta\,d\tau.$$

Here,

$$G(r,\varphi,z,\xi,\eta,\zeta,t) = \frac{1}{2\sqrt{\pi a t}}\left\{\exp\left[-\frac{(z-\zeta)^2}{4at}\right] - \exp\left[-\frac{(z+\zeta)^2}{4at}\right]\right\}G_1(r,\varphi,\xi,\eta,t),$$

$$G_1(r,\varphi,\xi,\eta,t) = \frac{4}{R^2\varphi_0}\sum_{n=1}^{\infty}\sum_{m=1}^{\infty}\frac{J_{n\pi/\varphi_0}(\mu_{nm}r)J_{n\pi/\varphi_0}(\mu_{nm}\xi)}{[J'_{n\pi/\varphi_0}(\mu_{nm}R)]^2}\sin\left(\frac{n\pi\varphi}{\varphi_0}\right)\sin\left(\frac{n\pi\eta}{\varphi_0}\right)\exp(-\mu_{nm}^2 a t),$$

where the $J_{n\pi/\varphi_0}(r)$ are the Bessel functions and the μ_{nm} are positive roots of the transcendental equation $J_{n\pi/\varphi_0}(\mu R) = 0$.

3.1.2-34. Domain: $0 \le r \le R$, $0 \le \varphi \le \varphi_0$, $0 \le z < \infty$. Mixed boundary value problem.

A semiinfinite cylindrical sector is considered. The following conditions are prescribed:

$$\begin{aligned}
w &= f(r,\varphi,z) & \text{at} \quad & t = 0 & \text{(initial condition)}, \\
w &= g_1(\varphi,z,t) & \text{at} \quad & r = R & \text{(boundary condition)}, \\
w &= g_2(r,z,t) & \text{at} \quad & \varphi = 0 & \text{(boundary condition)}, \\
w &= g_3(r,z,t) & \text{at} \quad & \varphi = \varphi_0 & \text{(boundary condition)}, \\
\partial_z w &= g_4(r,\varphi,t) & \text{at} \quad & z = 0 & \text{(boundary condition)}.
\end{aligned}$$

Solution:

$$\begin{aligned}
w(r,\varphi,z,t) =\ & \int_0^\infty \int_0^{\varphi_0} \int_0^R f(\xi,\eta,\zeta) G(r,\varphi,z,\xi,\eta,\zeta,t) \xi\, d\xi\, d\eta\, d\zeta \\
& - aR \int_0^t \int_0^\infty \int_0^{\varphi_0} g_1(\eta,\zeta,\tau) \left[\frac{\partial}{\partial \xi} G(r,\varphi,z,\xi,\eta,\zeta,t-\tau)\right]_{\xi=R} d\eta\, d\zeta\, d\tau \\
& + a \int_0^t \int_0^\infty \int_0^R g_2(\xi,\zeta,\tau) \frac{1}{\xi} \left[\frac{\partial}{\partial \eta} G(r,\varphi,z,\xi,\eta,\zeta,t-\tau)\right]_{\eta=0} d\xi\, d\zeta\, d\tau \\
& - a \int_0^t \int_0^\infty \int_0^R g_3(\xi,\zeta,\tau) \frac{1}{\xi} \left[\frac{\partial}{\partial \eta} G(r,\varphi,z,\xi,\eta,\zeta,t-\tau)\right]_{\eta=\varphi_0} d\xi\, d\zeta\, d\tau \\
& - a \int_0^t \int_0^{\varphi_0} \int_0^R g_4(\xi,\eta,\tau) G(r,\varphi,z,\xi,\eta,0,t-\tau) \xi\, d\xi\, d\eta\, d\tau.
\end{aligned}$$

Here,

$$G(r,\varphi,z,\xi,\eta,\zeta,t) = \frac{1}{2\sqrt{\pi a t}} \left\{ \exp\left[-\frac{(z-\zeta)^2}{4at}\right] + \exp\left[-\frac{(z+\zeta)^2}{4at}\right] \right\} G_1(r,\varphi,\xi,\eta,t),$$

$$G_1(r,\varphi,\xi,\eta,t) = \frac{4}{R^2 \varphi_0} \sum_{n=1}^\infty \sum_{m=1}^\infty \frac{J_{n\pi/\varphi_0}(\mu_{nm} r) J_{n\pi/\varphi_0}(\mu_{nm} \xi)}{[J'_{n\pi/\varphi_0}(\mu_{nm} R)]^2} \sin\left(\frac{n\pi\varphi}{\varphi_0}\right) \sin\left(\frac{n\pi\eta}{\varphi_0}\right) \exp(-\mu_{nm}^2 a t),$$

where the $J_{n\pi/\varphi_0}(r)$ are the Bessel functions and the μ_{nm} are positive roots of the transcendental equation $J_{n\pi/\varphi_0}(\mu R) = 0$.

3.1.2-35. Domain: $0 \le r \le R$, $0 \le \varphi \le \varphi_0$, $0 \le z \le l$. First boundary value problem.

A cylindrical sector of finite thickness is considered. The following conditions are prescribed:

$$\begin{aligned}
w &= f(r,\varphi,z) & \text{at} \quad & t = 0 & \text{(initial condition)}, \\
w &= g_1(\varphi,z,t) & \text{at} \quad & r = R & \text{(boundary condition)}, \\
w &= g_2(r,z,t) & \text{at} \quad & \varphi = 0 & \text{(boundary condition)}, \\
w &= g_3(r,z,t) & \text{at} \quad & \varphi = \varphi_0 & \text{(boundary condition)}, \\
w &= g_4(r,\varphi,t) & \text{at} \quad & z = 0 & \text{(boundary condition)}, \\
w &= g_5(r,\varphi,t) & \text{at} \quad & z = l & \text{(boundary condition)}.
\end{aligned}$$

Solution:

$$w(r,\varphi,z,t) = \int_0^l \int_0^{\varphi_0} \int_0^R f(\xi,\eta,\zeta) G(r,\varphi,z,\xi,\eta,\zeta,t)\xi\,d\xi\,d\eta\,d\zeta$$

$$- aR \int_0^t \int_0^l \int_0^{\varphi_0} g_1(\eta,\zeta,\tau)\left[\frac{\partial}{\partial \xi}G(r,\varphi,z,\xi,\eta,\zeta,t-\tau)\right]_{\xi=R} d\eta\,d\zeta\,d\tau$$

$$+ a \int_0^t \int_0^l \int_0^R g_2(\xi,\zeta,\tau)\frac{1}{\xi}\left[\frac{\partial}{\partial \eta}G(r,\varphi,z,\xi,\eta,\zeta,t-\tau)\right]_{\eta=0} d\xi\,d\zeta\,d\tau$$

$$- a \int_0^t \int_0^l \int_0^R g_3(\xi,\zeta,\tau)\frac{1}{\xi}\left[\frac{\partial}{\partial \eta}G(r,\varphi,z,\xi,\eta,\zeta,t-\tau)\right]_{\eta=\varphi_0} d\xi\,d\zeta\,d\tau$$

$$+ a \int_0^t \int_0^{\varphi_0} \int_0^R g_4(\xi,\eta,\tau)\left[\frac{\partial}{\partial \zeta}G(r,\varphi,z,\xi,\eta,\zeta,t-\tau)\right]_{\zeta=0} \xi\,d\xi\,d\eta\,d\tau$$

$$- a \int_0^t \int_0^{\varphi_0} \int_0^R g_5(\xi,\eta,\tau)\left[\frac{\partial}{\partial \zeta}G(r,\varphi,z,\xi,\eta,\zeta,t-\tau)\right]_{\zeta=l} \xi\,d\xi\,d\eta\,d\tau.$$

Here,

$$G(r,\varphi,z,\xi,\eta,\zeta,t) = G_1(r,\varphi,\xi,\eta,t) G_2(z,\zeta,t),$$

$$G_1(r,\varphi,\xi,\eta,t) = \frac{4}{R^2 \varphi_0} \sum_{n=1}^{\infty} \sum_{m=1}^{\infty} \frac{J_{n\pi/\varphi_0}(\mu_{nm}r) J_{n\pi/\varphi_0}(\mu_{nm}\xi)}{[J'_{n\pi/\varphi_0}(\mu_{nm}R)]^2} \sin\left(\frac{n\pi\varphi}{\varphi_0}\right) \sin\left(\frac{n\pi\eta}{\varphi_0}\right) \exp(-\mu_{nm}^2 a t),$$

$$G_2(z,\zeta,t) = \frac{2}{l} \sum_{n=1}^{\infty} \sin\left(\frac{n\pi z}{l}\right) \sin\left(\frac{n\pi \zeta}{l}\right) \exp\left(-\frac{a n^2 \pi^2 t}{l^2}\right),$$

where the $J_{n\pi/\varphi_0}(r)$ are the Bessel functions and the μ_{nm} are positive roots of the transcendental equation $J_{n\pi/\varphi_0}(\mu R) = 0$.

3.1.2-36. Domain: $0 \le r \le R$, $0 \le \varphi \le \varphi_0$, $0 \le z \le l$. Mixed boundary value problem.

A cylindrical sector of finite thickness is considered. The following conditions are prescribed:

$$\begin{array}{llll}
w = f(r,\varphi,z) & \text{at} & t=0 & \text{(initial condition)}, \\
w = g_1(\varphi,z,t) & \text{at} & r=R & \text{(boundary condition)}, \\
w = g_2(r,z,t) & \text{at} & \varphi=0 & \text{(boundary condition)}, \\
w = g_3(r,z,t) & \text{at} & \varphi=\varphi_0 & \text{(boundary condition)}, \\
\partial_z w = g_4(r,\varphi,t) & \text{at} & z=0 & \text{(boundary condition)}, \\
\partial_z w = g_5(r,\varphi,t) & \text{at} & z=l & \text{(boundary condition)}.
\end{array}$$

Solution:

$$w(r,\varphi,z,t) = \int_0^l \int_0^{\varphi_0} \int_0^R f(\zeta,\eta,\zeta) G(r,\varphi,z,\xi,\eta,\zeta,t)\xi\,d\xi\,d\eta\,d\zeta$$

$$- aR \int_0^t \int_0^l \int_0^{\varphi_0} g_1(\eta,\zeta,\tau)\left[\frac{\partial}{\partial \xi}G(r,\varphi,z,\xi,\eta,\zeta,t-\tau)\right]_{\xi=R} d\eta\,d\zeta\,d\tau$$

$$+ a \int_0^t \int_0^l \int_0^R g_2(\xi,\zeta,\tau)\frac{1}{\xi}\left[\frac{\partial}{\partial \eta}G(r,\varphi,z,\xi,\eta,\zeta,t-\tau)\right]_{\eta=0} d\xi\,d\zeta\,d\tau$$

$$- a \int_0^t \int_0^l \int_0^R g_3(\xi,\zeta,\tau)\frac{1}{\xi}\left[\frac{\partial}{\partial \eta}G(r,\varphi,z,\xi,\eta,\zeta,t-\tau)\right]_{\eta=\varphi_0} d\xi\,d\zeta\,d\tau$$

$$- a \int_0^t \int_0^{\varphi_0} \int_0^R g_4(\xi,\eta,\tau) G(r,\varphi,z,\xi,\eta,0,t-\tau) \xi\,d\xi\,d\eta\,d\tau$$

$$+ a \int_0^t \int_0^{\varphi_0} \int_0^R g_5(\xi,\eta,\tau) G(r,\varphi,z,\xi,\eta,l,t-\tau) \xi\,d\xi\,d\eta\,d\tau.$$

Here,

$$G(r,\varphi,z,\xi,\eta,\zeta,t) = G_1(r,\varphi,\xi,\eta,t)\left[\frac{1}{l} + \frac{2}{l}\sum_{n=1}^{\infty}\cos\left(\frac{n\pi z}{l}\right)\cos\left(\frac{n\pi\zeta}{l}\right)\exp\left(-\frac{an^2\pi^2 t}{l^2}\right)\right],$$

$$G_1(r,\varphi,\xi,\eta,t) = \frac{4}{R^2\varphi_0}\sum_{n=1}^{\infty}\sum_{m=1}^{\infty}\frac{J_{n\pi/\varphi_0}(\mu_{nm}r)J_{n\pi/\varphi_0}(\mu_{nm}\xi)}{[J'_{n\pi/\varphi_0}(\mu_{nm}R)]^2}\sin\left(\frac{n\pi\varphi}{\varphi_0}\right)\sin\left(\frac{n\pi\eta}{\varphi_0}\right)\exp(-\mu_{nm}^2 at),$$

where the $J_{n\pi/\varphi_0}(r)$ are the Bessel functions and the μ_{nm} are positive roots of the transcendental equation $J_{n\pi/\varphi_0}(\mu R) = 0$.

3.1.3. Problems in Spherical Coordinates

The heat equation in the spherical coordinate system has the form

$$\frac{\partial w}{\partial t} = a\left[\frac{1}{r^2}\frac{\partial}{\partial r}\left(r^2\frac{\partial w}{\partial r}\right) + \frac{1}{r^2\sin\theta}\frac{\partial}{\partial\theta}\left(\sin\theta\frac{\partial w}{\partial\theta}\right) + \frac{1}{r^2\sin^2\theta}\frac{\partial^2 w}{\partial\varphi^2}\right], \qquad r = \sqrt{x^2+y^2+z^2}.$$

This representation is convenient to describe three-dimensional heat and mass exchange phenomena in domains bounded by coordinate surfaces of the spherical coordinate system.

One-dimensional problems with central symmetry that have solutions of the form $w = w(r,t)$ are discussed in Subsection 1.2.3.

3.1.3-1. Domain: $0 \leq r \leq R$, $0 \leq \theta \leq \pi$, $0 \leq \varphi \leq 2\pi$. First boundary value problem.

A spherical domain is considered. The following conditions are prescribed:

$$w = f(r,\theta,\varphi) \quad \text{at} \quad t=0 \quad \text{(initial condition)},$$
$$w = g(\theta,\varphi,t) \quad \text{at} \quad r=R \quad \text{(boundary condition)}.$$

Solution:

$$w(r,\theta,\varphi,t) = \int_0^{2\pi}\int_0^{\pi}\int_0^R f(\xi,\eta,\zeta)G(r,\theta,\varphi,\xi,\eta,\zeta,t)\xi^2\sin\eta\, d\xi\, d\eta\, d\zeta$$
$$- aR^2\int_0^t\int_0^{2\pi}\int_0^{\pi} g(\eta,\zeta,\tau)\left[\frac{\partial}{\partial\xi}G(r,\theta,\varphi,\xi,\eta,\zeta,t-\tau)\right]_{\xi=R}\sin\eta\, d\eta\, d\zeta\, d\tau,$$

where

$$G(r,\theta,\varphi,\xi,\eta,\zeta,t) = \frac{1}{2\pi R^2\sqrt{r\xi}}\sum_{n=0}^{\infty}\sum_{m=1}^{\infty}\sum_{k=0}^{n} A_k B_{nmk} J_{n+1/2}(\lambda_{nm}r)J_{n+1/2}(\lambda_{nm}\xi)$$
$$\times P_n^k(\cos\theta)P_n^k(\cos\eta)\cos[k(\varphi-\zeta)]\exp(-\lambda_{nm}^2 at),$$

$$A_k = \begin{cases} 1 & \text{for } k=0, \\ 2 & \text{for } k\neq 0, \end{cases} \qquad B_{nmk} = \frac{(2n+1)(n-k)!}{(n+k)!\left[J'_{n+1/2}(\lambda_{nm}R)\right]^2}.$$

Here, the $J_{n+1/2}(r)$ are the Bessel functions, the $P_n^k(\mu)$ are the associated Legendre functions expressed in terms of the Legendre polynomials $P_n(\mu)$ as follows:

$$P_n^k(\mu) = (1-\mu^2)^{k/2}\frac{d^k}{d\mu^k}P_n(\mu), \qquad P_n(\mu) = \frac{1}{n!\,2^n}\frac{d^n}{d\mu^n}(\mu^2-1)^n;$$

and the λ_{nm} are positive roots of the transcendental equation $J_{n+1/2}(\lambda R) = 0$.

⊙ *References*: B. M. Budak, A. A. Samarskii, and A. N. Tikhonov (1980), H. S. Carslaw and J. C. Jaeger (1984).

3.1.3-2. Domain: $0 \leq r \leq R$, $0 \leq \theta \leq \pi$, $0 \leq \varphi \leq 2\pi$. Second boundary value problem.

A spherical domain is considered. The following conditions are prescribed:

$$w = f(r, \theta, \varphi) \quad \text{at} \quad t = 0 \quad \text{(initial condition)},$$
$$\partial_r w = g(\theta, \varphi, t) \quad \text{at} \quad r = R \quad \text{(boundary condition)}.$$

Solution:

$$w(r,\theta,\varphi,t) = \int_0^{2\pi}\int_0^\pi\int_0^R f(\xi,\eta,\zeta)G(r,\theta,\varphi,\xi,\eta,\zeta,t)\xi^2 \sin\eta\, d\xi\, d\eta\, d\zeta$$
$$+ aR^2 \int_0^t\int_0^{2\pi}\int_0^\pi g(\eta,\zeta,\tau)G(r,\theta,\varphi,R,\eta,\zeta,t-\tau)\sin\eta\, d\eta\, d\zeta\, d\tau,$$

where

$$G(r,\theta,\varphi,\xi,\eta,\zeta,t) = \frac{3}{4\pi R^3} + \frac{1}{2\pi\sqrt{r\xi}}\sum_{n=0}^\infty\sum_{m=1}^\infty\sum_{k=0}^n A_k B_{nmk} J_{n+1/2}(\lambda_{nm}r)J_{n+1/2}(\lambda_{nm}\xi)$$
$$\times P_n^k(\cos\theta)P_n^k(\cos\eta)\cos[k(\varphi-\zeta)]\exp(-\lambda_{nm}^2 at),$$

$$A_k = \begin{cases} 1 & \text{for } k = 0, \\ 2 & \text{for } k \neq 0, \end{cases} \quad B_{nmk} = \frac{\lambda_{nm}^2(2n+1)(n-k)!}{(n+k)!\left[R^2\lambda_{nm}^2 - n(n+1)\right]\left[J_{n+1/2}(\lambda_{nm}R)\right]^2}.$$

Here, the $J_{n+1/2}(r)$ are the Bessel functions, the $P_n^k(\mu)$ are the associated Legendre functions (see Paragraph 3.1.3-1), and the λ_{nm} are positive roots of the transcendental equation

$$2\lambda R J'_{n+1/2}(\lambda R) - J_{n+1/2}(\lambda R) = 0.$$

3.1.3-3. Domain: $0 \leq r \leq R$, $0 \leq \theta \leq \pi$, $0 \leq \varphi \leq 2\pi$. Third boundary value problem.

A spherical domain is considered. The following conditions are prescribed:

$$w = f(r,\theta,\varphi) \quad \text{at} \quad t = 0 \quad \text{(initial condition)},$$
$$\partial_r w + kw = g(\theta,\varphi,t) \quad \text{at} \quad r = R \quad \text{(boundary condition)}.$$

The solution $w(r,\theta,\varphi,t)$ is determined by the formula in Paragraph 3.1.3-2 where

$$G(r,\theta,\varphi,\xi,\eta,\zeta,t) = \frac{1}{2\pi\sqrt{r\xi}}\sum_{n=0}^\infty\sum_{m=1}^\infty\sum_{s=0}^n A_s B_{nms} J_{n+1/2}(\lambda_{nm}r)J_{n+1/2}(\lambda_{nm}\zeta)$$
$$\times P_n^s(\cos\theta)P_n^s(\cos\eta)\cos[s(\varphi-\zeta)]\exp(-\lambda_{nm}^2 at),$$

$$A_s = \begin{cases} 1 & \text{for } s = 0, \\ 2 & \text{for } s \neq 0, \end{cases} \quad B_{nms} = \frac{\lambda_{nm}^2(2n+1)(n-s)!}{(n+s)!\left[R^2\lambda_{nm}^2 + (kR+n)(kR-n-1)\right]\left[J_{n+1/2}(\lambda_{nm}R)\right]^2}.$$

Here, the $J_{n+1/2}(r)$ are the Bessel functions, the $P_n^s(\mu)$ are the associated Legendre functions (see Paragraph 3.1.3-1), and the λ_{nm} are positive roots of the transcendental equation

$$\lambda R J'_{n+1/2}(\lambda R) + \left(kR - \tfrac{1}{2}\right)J_{n+1/2}(\lambda R) = 0.$$

⊙ *Reference*: B. M. Budak, A. A. Samarskii, and A. N. Tikhonov (1980).

3.1.3-4. Domain: $R_1 \leq r \leq R_2$, $0 \leq \theta \leq \pi$, $0 \leq \varphi \leq 2\pi$. First boundary value problem.

A spherical layer is considered. The following conditions are prescribed:

$$w = f(r,\theta,\varphi) \quad \text{at} \quad t=0 \quad \text{(initial condition)},$$
$$w = g_1(\theta,\varphi,t) \quad \text{at} \quad r=R_1 \quad \text{(boundary condition)},$$
$$w = g_2(\theta,\varphi,t) \quad \text{at} \quad r=R_2 \quad \text{(boundary condition)}.$$

Solution:

$$w(r,\theta,\varphi,t) = \int_0^{2\pi}\int_0^{\pi}\int_{R_1}^{R_2} f(\xi,\eta,\zeta)G(r,\theta,\varphi,\xi,\eta,\zeta,t)\xi^2 \sin\eta \, d\xi \, d\eta \, d\zeta$$
$$+ aR_1^2 \int_0^t\int_0^{2\pi}\int_0^{\pi} g_1(\eta,\zeta,\tau)\left[\frac{\partial}{\partial\xi}G(r,\theta,\varphi,\xi,\eta,\zeta,t-\tau)\right]_{\xi=R_1} \sin\eta \, d\eta \, d\zeta \, d\tau$$
$$- aR_2^2 \int_0^t\int_0^{2\pi}\int_0^{\pi} g_2(\eta,\zeta,\tau)\left[\frac{\partial}{\partial\xi}G(r,\theta,\varphi,\xi,\eta,\zeta,t-\tau)\right]_{\xi=R_2} \sin\eta \, d\eta \, d\zeta \, d\tau,$$

where

$$G(r,\theta,\varphi,\xi,\eta,\zeta,t) = \frac{\pi}{8\sqrt{r\xi}} \sum_{n=0}^{\infty}\sum_{m=1}^{\infty}\sum_{k=0}^{n} A_k B_{nmk} Z_{n+1/2}(\lambda_{nm}r) Z_{n+1/2}(\lambda_{nm}\xi)$$
$$\times P_n^k(\cos\theta) P_n^k(\cos\eta) \cos[k(\varphi-\zeta)] \exp(-\lambda_{nm}^2 at).$$

Here,

$$Z_{n+1/2}(\lambda_{nm}r) = J_{n+1/2}(\lambda_{nm}R_1)Y_{n+1/2}(\lambda_{nm}r) - Y_{n+1/2}(\lambda_{nm}R_1)J_{n+1/2}(\lambda_{nm}r),$$

$$A_k = \begin{cases} 1 & \text{for } k=0, \\ 2 & \text{for } k \neq 0, \end{cases} \quad B_{nmk} = \frac{\lambda_{nm}^2(2n+1)(n-k)! \, J_{n+1/2}^2(\lambda_{nm}R_2)}{(n+k)! \left[J_{n+1/2}^2(\lambda_{nm}R_1) - J_{n+1/2}^2(\lambda_{nm}R_2)\right]},$$

where the $J_{n+1/2}(r)$ are the Bessel functions, the $P_n^k(\mu)$ are the associated Legendre functions expressed in terms of the Legendre polynomials $P_n(\mu)$ as follows:

$$P_n^k(\mu) = (1-\mu^2)^{k/2}\frac{d^k}{d\mu^k}P_n(\mu), \qquad P_n(\mu) = \frac{1}{n!\,2^n}\frac{d^n}{d\mu^n}(\mu^2-1)^n;$$

and the λ_{nm} are positive roots of the transcendental equation

$$Z_{n+1/2}(\lambda R_2) = 0.$$

⊙ *Reference*: B. M. Budak, A. A. Samarskii, and A. N. Tikhonov (1980).

3.1.3-5. Domain: $R_1 \leq r \leq R_2$, $0 \leq \theta \leq \pi$, $0 \leq \varphi \leq 2\pi$. Second boundary value problem.

A spherical layer is considered. The following conditions are prescribed:

$$w = f(r,\theta,\varphi) \quad \text{at} \quad t=0 \quad \text{(initial condition)},$$
$$\partial_r w = g_1(\theta,\varphi,t) \quad \text{at} \quad r=R_1 \quad \text{(boundary condition)},$$
$$\partial_r w = g_2(\theta,\varphi,t) \quad \text{at} \quad r=R_2 \quad \text{(boundary condition)}.$$

Solution:

$$w(r,\theta,\varphi,t) = \int_0^{2\pi}\int_0^{\pi}\int_{R_1}^{R_2} f(\xi,\eta,\zeta)G(r,\theta,\varphi,\xi,\eta,\zeta,t)\xi^2 \sin\eta \, d\xi \, d\eta \, d\zeta$$
$$- aR_1^2 \int_0^t\int_0^{2\pi}\int_0^{\pi} g_1(\eta,\zeta,\tau)G(r,\theta,\varphi,R_1,\eta,\zeta,t-\tau) \sin\eta \, d\eta \, d\zeta \, d\tau$$
$$+ aR_2^2 \int_0^t\int_0^{2\pi}\int_0^{\pi} g_2(\eta,\zeta,\tau)G(r,\theta,\varphi,R_2,\eta,\zeta,t-\tau) \sin\eta \, d\eta \, d\zeta \, d\tau,$$

where
$$G(r,\theta,\varphi,\xi,\eta,\zeta,t) = \frac{3}{4\pi(R_2^3 - R_1^3)} + \frac{1}{4\pi\sqrt{r\xi}} \sum_{n=0}^{\infty}\sum_{m=1}^{\infty}\sum_{k=0}^{n} \frac{A_k}{B_{nmk}} Z_{n+1/2}(\lambda_{nm}r)Z_{n+1/2}(\lambda_{nm}\xi)$$
$$\times P_n^k(\cos\theta)P_n^k(\cos\eta)\cos[k(\varphi-\zeta)]\exp(-\lambda_{nm}^2 at).$$

Here,
$$A_k = \begin{cases} 1 & \text{for } k = 0, \\ 2 & \text{for } k \neq 0, \end{cases} \quad B_{nmk} = \frac{(n+k)!}{(2n+1)(n-k)!}\int_{R_1}^{R_2} rZ_{n+1/2}^2(\lambda_{nm}r)\,dr,$$

$$Z_{n+1/2}(\lambda r) = \left[\lambda J'_{n+1/2}(\lambda R_1) - \frac{1}{2R_1}J_{n+1/2}(\lambda R_1)\right]Y_{n+1/2}(\lambda r)$$
$$- \left[\lambda Y'_{n+1/2}(\lambda R_1) - \frac{1}{2R_1}Y_{n+1/2}(\lambda R_1)\right]J_{n+1/2}(\lambda r),$$

where the $J_{n+1/2}(r)$ and $Y_{n+1/2}(r)$ are the Bessel functions, the $P_n^k(\mu)$ are the associated Legendre functions (see Paragraph 3.1.3-4), and the λ_{nm} are positive roots of the transcendental equation

$$\lambda Z'_{n+1/2}(\lambda R_2) - \frac{1}{2R_2}Z_{n+1/2}(\lambda R_2) = 0.$$

The integrals that determine the coefficients B_{nmk} can be expressed in terms of the Bessel functions and their derivatives; see Budak, Samarskii, and Tikhonov (1980).

3.1.3-6. Domain: $R_1 \leq r \leq R_2$, $0 \leq \theta \leq \pi$, $0 \leq \varphi \leq 2\pi$. Third boundary value problem.

A spherical layer is considered. The following conditions are prescribed:

$$w = f(r,\theta,\varphi) \quad \text{at} \quad t = 0 \quad \text{(initial condition)},$$
$$\partial_r w - k_1 w = g_1(\theta,\varphi,t) \quad \text{at} \quad r = R_1 \quad \text{(boundary condition)},$$
$$\partial_r w + k_2 w = g_2(\theta,\varphi,t) \quad \text{at} \quad r = R_2 \quad \text{(boundary condition)}.$$

The solution $w(r,\theta,\varphi,t)$ is determined by the formula in Paragraph 3.1.3-5 where
$$G(r,\theta,\varphi,\xi,\eta,\zeta,t) = \frac{1}{4\pi\sqrt{r\xi}} \sum_{n=0}^{\infty}\sum_{m=1}^{\infty}\sum_{s=0}^{n} \frac{A_s}{B_{nms}} Z_{n+1/2}(\lambda_{nm}r)Z_{n+1/2}(\lambda_{nm}\xi)$$
$$\times P_n^s(\cos\theta)P_n^s(\cos\eta)\cos[s(\varphi-\zeta)]\exp(-\lambda_{nm}^2 at).$$

Here,
$$A_s = \begin{cases} 1 & \text{for } s = 0, \\ 2 & \text{for } s \neq 0, \end{cases} \quad B_{nms} = \frac{(n+s)!}{(2n+1)(n-s)!}\int_{R_1}^{R_2} rZ_{n+1/2}^2(\lambda_{nm}r)\,dr,$$

$$Z_{n+1/2}(\lambda r) = \left[\lambda J'_{n+1/2}(\lambda R_1) - \left(k_1 + \frac{1}{2R_1}\right)J_{n+1/2}(\lambda R_1)\right]Y_{n+1/2}(\lambda r)$$
$$- \left[\lambda Y'_{n+1/2}(\lambda R_1) - \left(k_1 + \frac{1}{2R_1}\right)Y_{n+1/2}(\lambda R_1)\right]J_{n+1/2}(\lambda r),$$

where the $J_{n+1/2}(r)$ and $Y_{n+1/2}(r)$ are the Bessel functions, the $P_n^s(\mu)$ are the associated Legendre functions (see Paragraph 3.1.3-4), and the λ_{nm} are positive roots of the transcendental equation

$$\lambda Z'_{n+1/2}(\lambda R_2) + \left(k_2 - \frac{1}{2R_2}\right)Z_{n+1/2}(\lambda R_2) = 0.$$

The integrals that determine the coefficients B_{nms} can be expressed in terms of the Bessel functions and their derivatives.

⊙ *Reference*: B. M. Budak, A. A. Samarskii, and A. N. Tikhonov (1980).

3.1.3-7. Domain: $0 \leq r < \infty$, $0 \leq \theta \leq \theta_0$, $0 \leq \varphi \leq 2\pi$. **First boundary value problem.**

A cone is considered. The following conditions are prescribed:

$$w = f(r, \theta, \varphi) \quad \text{at} \quad t = 0 \quad \text{(initial condition)},$$
$$w = g(r, \varphi, t) \quad \text{at} \quad \theta = \theta_0 \quad \text{(boundary condition)}.$$

Solution:

$$w(r, \theta, \varphi, t) = \int_0^{2\pi} \int_0^{\theta_0} \int_0^{\infty} f(\xi, \eta, \zeta) G(r, \theta, \varphi, \xi, \eta, \zeta, t) \xi^2 \sin \eta \, d\xi \, d\eta \, d\zeta$$
$$- a \int_0^t \int_0^{2\pi} \int_0^{\infty} g(\xi, \zeta, \tau) \left[\sin \eta \frac{\partial}{\partial \eta} G(r, \theta, \varphi, \xi, \eta, \zeta, t - \tau) \right]_{\eta=\theta_0} d\xi \, d\zeta \, d\tau,$$

where

$$G(r, \theta, \varphi, \xi, \eta, \zeta, t) = \frac{1}{4\pi a t \sqrt{r\xi}} \sum_{m=0}^{\infty} \sum_{\nu} \frac{A_m (2\nu + 1)}{B_{m\nu}} \exp\left(-\frac{r^2 + \xi^2}{4at}\right) I_{\nu+1/2}\left(\frac{r\xi}{2at}\right)$$
$$\times P_\nu^{-m}(\cos \theta) P_\nu^{-m}(\cos \eta) \cos[m(\varphi - \zeta)],$$

$$A_m = \begin{cases} 1 & \text{for } m = 0, \\ 2 & \text{for } m \neq 0, \end{cases} \quad B_{m\nu} = \left[(1 - \mu)^2 \frac{d}{d\mu} P_\nu^{-m}(\mu) \frac{d}{d\nu} P_\nu^{-m}(\mu) \right]_{\mu = \cos \theta_0}.$$

Here, $P_\nu^{-m}(\mu)$ is the modified Legendre function expressed as

$$P_\nu^{-m}(\mu) = \frac{1}{\Gamma(1 + m)} \left(\frac{1 - \mu}{1 + \mu}\right)^{m/2} F\left(-\nu, \nu + 1, 1 + m; \tfrac{1}{2} - \tfrac{1}{2}\mu\right),$$

where $F(a, b, c; \mu)$ is the Gaussian hypergeometric function and $\Gamma(z)$ is the gamma function. The summation with respect to ν is performed over all roots of the equation $P_\nu^{-m}(\cos \theta_0) = 0$ that are greater than $-1/2$.

⊙ *Reference*: H. S. Carslaw and J. C. Jaeger (1984).

3.2. Heat Equation with Source $\dfrac{\partial w}{\partial t} = a\Delta_3 w + \Phi(x, y, z, t)$

3.2.1. Problems in Cartesian Coordinates

In the Cartesian coordinate system, the three-dimensional heat equation with a volume source has the form

$$\frac{\partial w}{\partial t} = a\left(\frac{\partial^2 w}{\partial x^2} + \frac{\partial^2 w}{\partial y^2} + \frac{\partial^2 w}{\partial z^2}\right) + \Phi(x, y, z, t).$$

It describes three-dimensional unsteady thermal phenomena in quiescent media or solids with constant thermal diffusivity. A similar equation is used to study the corresponding three-dimensional mass transfer processes with constant diffusivity.

3.2.1-1. Domain: $-\infty < x < \infty$, $-\infty < y < \infty$, $-\infty < z < \infty$. **Cauchy problem.**

An initial condition is prescribed:

$$w = f(x, y, z) \quad \text{at} \quad t = 0.$$

Solution:

$$w(x,y,z,t) = \int_{-\infty}^{\infty}\int_{-\infty}^{\infty}\int_{-\infty}^{\infty} f(\xi,\eta,\zeta) G(x,y,z,\xi,\eta,\zeta,t)\, d\xi\, d\eta\, d\zeta$$
$$+ \int_{0}^{t}\int_{-\infty}^{\infty}\int_{-\infty}^{\infty}\int_{-\infty}^{\infty} \Phi(\xi,\eta,\zeta,\tau) G(x,y,z,\xi,\eta,\zeta,t-\tau)\, d\xi\, d\eta\, d\zeta\, d\tau,$$

where

$$G(x,y,z,\xi,\eta,\zeta,t) = \frac{1}{8(\pi a t)^{3/2}} \exp\left[-\frac{(x-\xi)^2 + (y-\eta)^2 + (z-\zeta)^2}{4at}\right].$$

⊙ *References*: A. G. Butkovskiy (1979).

3.2.1-2. Domain: $0 \le x < \infty$, $-\infty < y < \infty$, $-\infty < z < \infty$. Different boundary value problems.

1°. The solution of the first boundary value problem for a half-space is given by the formula in Paragraph 3.1.1-4 with the additional term

$$\int_{0}^{t}\int_{-\infty}^{\infty}\int_{-\infty}^{\infty}\int_{0}^{\infty} \Phi(\xi,\eta,\zeta,\tau) G(x,y,z,\xi,\eta,\zeta,t-\tau)\, d\xi\, d\eta\, d\zeta\, d\tau, \tag{1}$$

which allows for the equation's nonhomogeneity.

2°. The solution of the second boundary value problem for a half-space is given by the formula in Paragraph 3.1.1-5 with the additional term (1).

3°. The solution of the third boundary value problem for a half-space is given by the formula in Paragraph 3.1.1-6 with the additional term (1).

⊙ *References*: A. G. Butkovskiy (1979), H. S. Carslaw and J. C. Jaeger (1984).

3.2.1-3. Domain: $-\infty < x < \infty$, $-\infty < y < \infty$, $0 \le z \le l$. Different boundary value problems.

1°. The solution of the first boundary value problem for an infinite layer is given by the formula in Paragraph 3.1.1-7 with the additional term

$$\int_{0}^{t}\int_{0}^{l}\int_{-\infty}^{\infty}\int_{-\infty}^{\infty} \Phi(\xi,\eta,\zeta,\tau) G(x,y,z,\xi,\eta,\zeta,t-\tau)\, d\xi\, d\eta\, d\zeta\, d\tau, \tag{2}$$

which allows for the equation's nonhomogeneity.

2°. The solution of the second boundary value problem for an infinite layer is given by the formula in Paragraph 3.1.1-8 with the additional term (2).

3°. The solution of the third boundary value problem for an infinite layer is given by the formula in Paragraph 3.1.1-9 with the additional term (2).

4°. The solution of a mixed boundary value problem for an infinite layer is given by the formula in Paragraph 3.1.1-10 with the additional term (2).

3.2.1-4. Domain: $-\infty < x < \infty$, $0 \le y < \infty$, $0 \le z \le l$. Different boundary value problems.

1°. The solution of the first boundary value problem for a semiinfinite layer is given by the formula in Paragraph 3.1.1-11 with the additional term

$$\int_{0}^{t}\int_{0}^{l}\int_{0}^{\infty}\int_{-\infty}^{\infty} \Phi(\xi,\eta,\zeta,\tau) G(x,y,z,\xi,\eta,\zeta,t-\tau)\, d\xi\, d\eta\, d\zeta\, d\tau, \tag{3}$$

which allows for the equation's nonhomogeneity.

2°. The solution of the second boundary value problem for a semiinfinite layer is given by the formula in Paragraph 3.1.1-12 with the additional term (3).

3°. The solution of the third boundary value problem for a semiinfinite layer is given by the formula in Paragraph 3.1.1-13 with the additional term (3).

4°. The solutions of mixed boundary value problems for a semiinfinite layer are given by the formulas in Paragraph 3.1.1-14 with additional terms of the form (3).

⊙ *References*: A. G. Butkovskiy (1979), H. S. Carslaw and J. C. Jaeger (1984).

3.2.1-5. Domain: $0 \le x < \infty$, $0 \le y < \infty$, $0 \le z < \infty$. Different boundary value problems.

1°. The solution of the first boundary value problem for the first octant is given by the formula in Paragraph 3.1.1-15 with the additional term

$$\int_0^t \int_0^\infty \int_0^\infty \int_0^\infty \Phi(\xi, \eta, \zeta, \tau) G(x, y, z, \xi, \eta, \zeta, t - \tau) \, d\xi \, d\eta \, d\zeta \, d\tau, \tag{4}$$

which allows for the equation's nonhomogeneity.

2°. The solution of the second boundary value problem for the first octant is given by the formula in Paragraph 3.1.1-16 with the additional term (4).

3°. The solution of the third boundary value problem for the first octant is given by the formula in Paragraph 3.1.1-17 with the additional term (4).

4°. The solutions of mixed boundary value problems for the first octant are given by the formulas in Paragraph 3.1.1-18 with additional terms of the form (4).

3.2.1-6. Domain: $0 \le x \le l_1$, $0 \le y \le l_2$, $-\infty < z < \infty$. Different boundary value problems.

1°. The solution of the first boundary value problem in an infinite rectangular domain is given by the formula in Paragraph 3.1.1-19 with the additional term

$$\int_0^t \int_{-\infty}^\infty \int_0^{l_2} \int_0^{l_1} \Phi(\xi, \eta, \zeta, \tau) G(x, y, z, \xi, \eta, \zeta, t - \tau) \, d\xi \, d\eta \, d\zeta \, d\tau, \tag{5}$$

which allows for the equation's nonhomogeneity.

2°. The solution of the second boundary value problem in an infinite rectangular domain is given by the formula in Paragraph 3.1.1-20 with the additional term (5).

3°. The solution of the third boundary value problem in an infinite rectangular domain is given by the formula in Paragraph 3.1.1-21 with the additional term (5).

4°. The solution of a mixed boundary value problem in an infinite rectangular domain is given by the formula in Paragraph 3.1.1-22 with the additional term (5).

3.2.1-7. Domain: $0 \le x \le l_1$, $0 \le y \le l_2$, $0 \le z < \infty$. Different boundary value problems.

1°. The solution of the first boundary value problem in a semiinfinite rectangular domain is given by the formula of Paragraph 3.1.1-23 with the additional term

$$\int_0^t \int_0^\infty \int_0^{l_2} \int_0^{l_1} \Phi(\xi, \eta, \zeta, \tau) G(x, y, z, \xi, \eta, \zeta, t - \tau) \, d\xi \, d\eta \, d\zeta \, d\tau, \tag{6}$$

which allows for the equation's nonhomogeneity.

2°. The solution of the second boundary value problem in a semiinfinite rectangular domain is given by the formula in Paragraph 3.1.1-24 with the additional term (6).

3°. The solution of the third boundary value problem in a semiinfinite rectangular domain is given by the formula in Paragraph 3.1.1-25 with the additional term (6).

4°. The solutions of mixed boundary value problems in a semiinfinite rectangular domain are given by the formulas in Paragraph 3.1.1-26 with additional terms of the form (6).

3.2.1-8. Domain: $0 \le x \le l_1$, $0 \le y \le l_2$, $0 \le z \le l_3$. Different boundary value problems.

1°. The solution of the first boundary value problem for a rectangular parallelepiped is given by the formula in Paragraph 3.1.1-27 with the additional term

$$\int_0^t \int_0^{l_3} \int_0^{l_2} \int_0^{l_1} \Phi(\xi, \eta, \zeta, \tau) G(x, y, z, \xi, \eta, \zeta, t-\tau) \, d\xi \, d\eta \, d\zeta \, d\tau, \tag{7}$$

which allows for the equation's nonhomogeneity.

2°. The solution of the second boundary value problem for a rectangular parallelepiped is given by the formula in Paragraph 3.1.1-28 with the additional term (7).

3°. The solution of the third boundary value problem for a rectangular parallelepiped is given by the formula in Paragraph 3.1.1-29 with the additional term (7).

4°. The solutions of mixed boundary value problems for a rectangular parallelepiped are given by the formulas in Paragraph 3.1.1-30 with additional terms of the form (7).

⊙ *References*: A. G. Butkovskiy (1979), H. S. Carslaw and J. C. Jaeger (1984).

3.2.2. Problems in Cylindrical Coordinates

In the cylindrical coordinate system, the heat equation with a volume source is written as

$$\frac{\partial w}{\partial t} = a \left[\frac{1}{r} \frac{\partial}{\partial r} \left(r \frac{\partial w}{\partial r} \right) + \frac{1}{r^2} \frac{\partial^2 w}{\partial \varphi^2} + \frac{\partial^2 w}{\partial z^2} \right] + \Phi(r, \varphi, z, t).$$

This representation is used to describe nonsymmetric unsteady thermal (diffusion) processes in quiescent media or solids bounded by cylindrical surfaces and planes.

3.2.2-1. Domain: $0 \le r \le R$, $0 \le \varphi \le 2\pi$, $-\infty < z < \infty$. Different boundary value problems.

1°. The solution of the first boundary value problem for an infinite circular cylinder is given by the formula in Paragraph 3.1.2-2 with the additional term

$$\int_0^t \int_{-\infty}^{\infty} \int_0^{2\pi} \int_0^R \Phi(\xi, \eta, \zeta, \tau) G(r, \varphi, z, \xi, \eta, \zeta, t-\tau) \xi \, d\xi \, d\eta \, d\zeta \, d\tau, \tag{1}$$

which allows for the equation's nonhomogeneity.

2°. The solution of the second boundary value problem for an infinite circular cylinder is given by the formula in Paragraph 3.1.2-3 with the additional term (1).

3°. The solution of the third boundary value problem for an infinite circular cylinder is the sum of the solution presented in Paragraph 3.1.2-4 and expression (1).

3.2.2-2. Domain: $0 \le r \le R$, $0 \le \varphi \le 2\pi$, $0 \le z < \infty$. Different boundary value problems.

1°. The solution of the first boundary value problem for a semiinfinite circular cylinder is given by the formula in Paragraph 3.1.2-5 with the additional term

$$\int_0^t \int_0^{\infty} \int_0^{2\pi} \int_0^R \Phi(\xi, \eta, \zeta, \tau) G(r, \varphi, z, \xi, \eta, \zeta, t-\tau) \xi \, d\xi \, d\eta \, d\zeta \, d\tau, \tag{2}$$

which allows for the equation's nonhomogeneity.

2°. The solution of the second boundary value problem for a semiinfinite circular cylinder is given by the formula in Paragraph 3.1.2-6 with the additional term (2).

3°. The solution of the third boundary value problem for a semiinfinite circular cylinder is the sum of the solution presented in Paragraph 3.1.2-7 and expression (2).

4°. The solutions of mixed boundary value problems for a semiinfinite circular cylinder are given by the formulas in Paragraph 3.1.2-8 with additional terms of the form (2).

> **3.2.2-3. Domain: $0 \le r \le R$, $0 \le \varphi \le 2\pi$, $0 \le z \le l$. Different boundary value problems.**

1°. The solution of the first boundary value problem for a circular cylinder of finite length is given by the formula in Paragraph 3.1.2-9 with the additional term

$$\int_0^t \int_0^l \int_0^{2\pi} \int_0^R \Phi(\xi,\eta,\zeta,\tau) G(r,\varphi,z,\xi,\eta,\zeta,t-\tau)\xi\, d\xi\, d\eta\, d\zeta\, d\tau, \tag{3}$$

which allows for the equation's nonhomogeneity.

2°. The solution of the second boundary value problem for a circular cylinder of finite length is given by the formula in Paragraph 3.1.2-10 with the additional term (3).

3°. The solution of the third boundary value problem for a circular cylinder of finite length is the sum of the solution presented in Paragraph 3.1.2-11 and expression (3).

4°. The solutions of mixed boundary value problems for a circular cylinder of finite length are given by the formulas in Paragraph 3.1.2-12 with additional terms of the form (3).

> **3.2.2-4. Domain: $R_1 \le r \le R_2$, $0 \le \varphi \le 2\pi$, $-\infty < z < \infty$. Different boundary value problems.**

1°. The solution of the first boundary value problem for an infinite hollow cylinder is given by the formula in Paragraph 3.1.2-13 with the additional term

$$\int_0^t \int_{-\infty}^{\infty} \int_0^{2\pi} \int_{R_1}^{R_2} \Phi(\xi,\eta,\zeta,\tau) G(r,\varphi,z,\xi,\eta,\zeta,t-\tau)\xi\, d\xi\, d\eta\, d\zeta\, d\tau, \tag{4}$$

which allows for the equation's nonhomogeneity.

2°. The solution of the second boundary value problem for an infinite hollow cylinder is given by the formula in Paragraph 3.1.2-14 with the additional term (4).

3°. The solution of the third boundary value problem for an infinite hollow cylinder is the sum of the solution presented in Paragraph 3.1.2-15 and expression (4).

> **3.2.2-5. Domain: $R_1 \le r \le R_2$, $0 \le \varphi \le 2\pi$, $0 \le z < \infty$. Different boundary value problems.**

1°. The solution of the first boundary value problem for a semiinfinite hollow cylinder is given by the formula in Paragraph 3.1.2-16 with the additional term

$$\int_0^t \int_0^{\infty} \int_0^{2\pi} \int_{R_1}^{R_2} \Phi(\xi,\eta,\zeta,\tau) G(r,\varphi,z,\xi,\eta,\zeta,t-\tau)\xi\, d\xi\, d\eta\, d\zeta\, d\tau, \tag{5}$$

which allows for the equation's nonhomogeneity.

2°. The solution of the second boundary value problem for a semiinfinite hollow cylinder is given by the formula in Paragraph 3.1.2-17 with the additional term (5).

3°. The solution of the third boundary value problem for a semiinfinite hollow cylinder is the sum of the solution presented in Paragraph 3.1.2-18 and expression (5).

4°. The solutions of mixed boundary value problems for a semiinfinite hollow cylinder are given by the formulas in Paragraph 3.1.2-19 with additional terms of the form (5).

3.2. Heat equation with source $\frac{\partial w}{\partial t} = a\Delta_3 w + \Phi(x,y,z,t)$

3.2.2-6. Domain: $R_1 \leq r \leq R_2$, $0 \leq \varphi \leq 2\pi$, $0 \leq z \leq l$. Different boundary value problems.

1°. The solution of the first boundary value problem for a hollow cylinder of finite length is given by the formula in Paragraph 3.1.2-20 with the additional term

$$\int_0^t \int_0^l \int_0^{2\pi} \int_{R_1}^{R_2} \Phi(\xi,\eta,\zeta,\tau) G(r,\varphi,z,\xi,\eta,\zeta,t-\tau)\xi\, d\xi\, d\eta\, d\zeta\, d\tau, \tag{6}$$

which allows for the equation's nonhomogeneity.

2°. The solution of the second boundary value problem for a hollow cylinder of finite length is given by the formula in Paragraph 3.1.2-21 with the additional term (6).

3°. The solution of the third boundary value problem for a hollow cylinder of finite length is the sum of the solution specified in Paragraph 3.1.2-22 and expression (6).

4°. The solutions of mixed boundary value problems for a hollow cylinder of finite length are given by the formulas in Paragraph 3.1.2-23 with additional terms of the form (6).

3.2.2-7. Domain: $0 \leq r < \infty$, $0 \leq \varphi \leq \varphi_0$, $-\infty < z < \infty$. Different boundary value problems.

1°. The solution of the first boundary value problem for an infinite wedge domain is given by the formula in Paragraph 3.1.2-24 with the additional term

$$\int_0^t \int_{-\infty}^{\infty} \int_0^{\varphi_0} \int_0^{\infty} \Phi(\xi,\eta,\zeta,\tau) G(r,\varphi,z,\xi,\eta,\zeta,t-\tau)\xi\, d\xi\, d\eta\, d\zeta\, d\tau, \tag{7}$$

which allows for the equation's nonhomogeneity.

2°. The solution of the second boundary value problem for an infinite wedge domain is given by the formula in Paragraph 3.1.2-25 with the additional term (7).

3.2.2-8. Domain: $0 \leq r < \infty$, $0 \leq \varphi \leq \varphi_0$, $0 \leq z < \infty$. Different boundary value problems.

1°. The solution of the first boundary value problem for a semiinfinite wedge domain is given by the formula in Paragraph 3.1.2-26 with the additional term

$$\int_0^t \int_0^{\infty} \int_0^{\varphi_0} \int_0^{\infty} \Phi(\xi,\eta,\zeta,\tau) G(r,\varphi,z,\xi,\eta,\zeta,t-\tau)\xi\, d\xi\, d\eta\, d\zeta\, d\tau, \tag{8}$$

which allows for the equation's nonhomogeneity.

2°. The solution of the second boundary value problem for a semiinfinite wedge domain is given by the formula in Paragraph 3.1.2-27 with the additional term (8).

3°. The solutions of mixed boundary value problems for a semiinfinite wedge domain are given by the formulas in Paragraph 3.1.2-28 with additional terms of the form (8).

3.2.2-9. Domain: $0 \leq r < \infty$, $0 \leq \varphi \leq \varphi_0$, $0 \leq z \leq l$. Different boundary value problems.

1°. The solution of the first boundary value problem for a wedge domain of finite height is given by the formula in Paragraph 3.1.2-29 with the additional term

$$\int_0^t \int_0^l \int_0^{\varphi_0} \int_0^{\infty} \Phi(\xi,\eta,\zeta,\tau) G(r,\varphi,z,\xi,\eta,\zeta,t-\tau)\xi\, d\xi\, d\eta\, d\zeta\, d\tau, \tag{9}$$

which allows for the equation's nonhomogeneity.

2°. The solution of the second boundary value problem for a wedge domain of finite height is given by the formula in Paragraph 3.1.2-30 with the additional term (9).

3°. The solutions of mixed boundary value problems for a wedge domain of finite height are given by the formulas in Paragraph 3.1.2-31 with additional terms of the form (9).

3.2.2-10. Different boundary value problems for a cylindrical sector.

1°. The solution of the first boundary value problem for an unbounded cylindrical sector ($0 \le r \le R$, $0 \le \varphi \le \varphi_0$, $-\infty < z < \infty$) is given by the formula in Paragraph 3.1.2-32 with the additional term

$$\int_0^t \int_{-\infty}^\infty \int_0^{\varphi_0} \int_0^R \Phi(\xi, \eta, \zeta, \tau) G(r, \varphi, z, \xi, \eta, \zeta, t-\tau) \xi \, d\xi \, d\eta \, d\zeta \, d\tau,$$

which allows for the equation's nonhomogeneity.

2°. The solution of the first boundary value problem for a semibounded cylindrical sector ($0 \le r \le R$, $0 \le \varphi \le \varphi_0$, $0 \le z < \infty$) is given by the formula in Paragraph 3.1.2-33 with the additional term

$$\int_0^t \int_0^\infty \int_0^{\varphi_0} \int_0^R \Phi(\xi, \eta, \zeta, \tau) G(r, \varphi, z, \xi, \eta, \zeta, t-\tau) \xi \, d\xi \, d\eta \, d\zeta \, d\tau, \tag{10}$$

which allows for the equation's nonhomogeneity.

3°. The solution of the mixed boundary value problem for a semibounded cylindrical sector ($0 \le r \le R$, $0 \le \varphi \le \varphi_0$, $0 \le z < \infty$) is given by the formula in Paragraph 3.1.2-34 with the additional term (10).

4°. The solution of the first boundary value problem for a cylindrical sector of finite height ($0 \le r \le R$, $0 \le \varphi \le \varphi_0$, $0 \le z \le l$) is given by the formula in Paragraph 3.1.2-35 with the additional term

$$\int_0^t \int_0^l \int_0^{\varphi_0} \int_0^R \Phi(\xi, \eta, \zeta, \tau) G(r, \varphi, z, \xi, \eta, \zeta, t-\tau) \xi \, d\xi \, d\eta \, d\zeta \, d\tau, \tag{11}$$

which allows for the equation's nonhomogeneity.

3°. The solution of a mixed boundary value problem for a cylindrical sector of finite height is given by the formula in Paragraph 3.1.2-36 with the additional term (11).

3.2.3. Problems in Spherical Coordinates

In the spherical coordinate system, the heat equation with a volume source has the form

$$\frac{\partial w}{\partial t} = a \left[\frac{1}{r^2} \frac{\partial}{\partial r} \left(r^2 \frac{\partial w}{\partial r} \right) + \frac{1}{r^2 \sin\theta} \frac{\partial}{\partial \theta} \left(\sin\theta \frac{\partial w}{\partial \theta} \right) + \frac{1}{r^2 \sin^2\theta} \frac{\partial^2 w}{\partial \varphi^2} \right] + \Phi(r, \theta, \varphi, t).$$

One-dimensional problems with central symmetry that have solutions of the form $w = w(r,t)$ are discussed in Subsection 1.2.4.

3.2.3-1. Domain: $0 \le r \le R$, $0 \le \theta \le \pi$, $0 \le \varphi \le 2\pi$. Different boundary value problems.

1°. The solution of the first boundary value problem for a spherical domain is given by the formula in Paragraph 3.1.3-1 with the additional term

$$\int_0^t \int_0^{2\pi} \int_0^\pi \int_0^R \Phi(\xi, \eta, \zeta, \tau) G(r, \theta, \varphi, \xi, \eta, \zeta, t-\tau) \xi^2 \sin\eta \, d\xi \, d\eta \, d\zeta \, d\tau, \tag{1}$$

which allows for the equation's nonhomogeneity.

2°. The solution of the second boundary value problem for a spherical domain is given by the formula in Paragraph 3.1.3-2 with the additional term (1).

3°. The solution of the third boundary value problem for a spherical domain is the sum of the solution specified in Paragraph 3.1.3-3 and expression (1).

3.2.3-2. Domain: $R_1 \leq r \leq R_2$, $0 \leq \theta \leq \pi$, $0 \leq \varphi \leq 2\pi$. Different boundary value problems.

1°. The solution of the first boundary value problem for a spherical layer is given by the formula in Paragraph 3.1.3-4 with the additional term

$$\int_0^t \int_0^{2\pi} \int_0^\pi \int_{R_1}^{R_2} \Phi(\xi,\eta,\zeta,\tau) G(r,\theta,\varphi,\xi,\eta,\zeta,t-\tau)\xi^2 \sin\eta \, d\xi \, d\eta \, d\zeta \, d\tau, \tag{2}$$

which allows for the equation's nonhomogeneity.

2°. The solution of the second boundary value problem for a spherical layer is given by the formula in Paragraph 3.1.3-5 with the additional term (2).

3°. The solution of the third boundary value problem for a spherical layer is the sum of the solution specified in Paragraph 3.1.3-6, and expression (2).

3.2.3-3. Domain: $0 \leq r < \infty$, $0 \leq \theta \leq \theta_0$, $0 \leq \varphi \leq 2\pi$. First boundary value problem.

The solution of the first boundary value problem for an infinite cone is given by the formula in Paragraph 3.1.3-7 with the additional term

$$\int_0^t \int_0^{2\pi} \int_0^{\theta_0} \int_0^\infty \Phi(\xi,\eta,\zeta,\tau) G(r,\theta,\varphi,\xi,\eta,\zeta,t-\tau)\xi^2 \sin\eta \, d\xi \, d\eta \, d\zeta \, d\tau,$$

which allows for the equation's nonhomogeneity.

3.3. Other Equations with Three Space Variables

3.3.1. Equations Containing Arbitrary Parameters

1. $\dfrac{\partial w}{\partial t} = a\left(\dfrac{\partial^2 w}{\partial x^2} + \dfrac{\partial^2 w}{\partial y^2} + \dfrac{\partial^2 w}{\partial z^2}\right) + (b_1 x + b_2 y + b_3 z + c)w.$

The transformation

$$w(x,y,z,t) = \exp\left[(b_1 x + b_2 y + b_3 z)t + \tfrac{1}{3}a(b_1^2 + b_2^2 + b_3^2)t^3 + ct\right] u(\xi,\eta,\zeta,t),$$
$$\xi = x + ab_1 t^2, \quad \eta = y + ab_2 t^2, \quad \zeta = z + ab_3 t^2$$

leads to the three-dimensional heat equation $\partial_t u = a(\partial_{\xi\xi} u + \partial_{\eta\eta} u + \partial_{\zeta\zeta} u)$ that is dealt with in Subsection 3.1.1.

2. $\dfrac{\partial w}{\partial t} = a\left(\dfrac{\partial^2 w}{\partial x^2} + \dfrac{\partial^2 w}{\partial y^2} + \dfrac{\partial^2 w}{\partial z^2}\right) - \left[b(x^2 + y^2 + z^2) + c\right]w, \qquad b > 0.$

The transformation (A is any number)

$$w(x,y,z,t) = \exp\left[\dfrac{\sqrt{ab}}{2a}(x^2 + y^2 + z^2) + \left(3\sqrt{ab} - c\right)t\right] u(\xi,\eta,\zeta,\tau),$$
$$\xi = x\exp(2\sqrt{ab}\,t), \quad \eta = y\exp(2\sqrt{ab}\,t), \quad \zeta = z\exp(2\sqrt{ab}\,t), \quad \tau = \dfrac{1}{4\sqrt{ab}}\exp(4\sqrt{ab}\,t) + A$$

leads to the three-dimensional heat equation $\partial_\tau u = a(\partial_{\xi\xi} u + \partial_{\eta\eta} u + \partial_{\zeta\zeta} u)$ that is dealt with in Subsection 3.1.1.

3. $\dfrac{\partial w}{\partial t} = a\left(\dfrac{\partial^2 w}{\partial x^2} + \dfrac{\partial^2 w}{\partial y^2} + \dfrac{\partial^2 w}{\partial z^2}\right) + \left[-b(x^2 + y^2 + z^2) + c_1 x + c_2 y + c_3 z + s\right]w.$

This is a special case of equation 3.3.2.3 with $f_k(t) = c_k$, $g(t) = s$.

4. $\dfrac{\partial w}{\partial t} = a\left(\dfrac{\partial^2 w}{\partial x^2} + \dfrac{\partial^2 w}{\partial y^2} + \dfrac{\partial^2 w}{\partial z^2}\right) + b_1\dfrac{\partial w}{\partial x} + b_2\dfrac{\partial w}{\partial y} + b_3\dfrac{\partial w}{\partial z} + cw.$

This equation governs the nonstationary temperature (concentration) field in a medium moving with a constant velocity, provided there is volume release (absorption) of heat proportional to temperature (concentration).

The substitution

$$w(x,y,z,t) = \exp\bigl(A_1 x + A_2 y + A_3 z + Bt\bigr) U(x,y,z,t),$$

where

$$A_1 = -\frac{b_1}{2a}, \quad A_2 = -\frac{b_2}{2a}, \quad A_3 = -\frac{b_3}{2a}, \quad B = c - \frac{1}{4a}\bigl(b_1^2 + b_2^2 + b_3^2\bigr),$$

leads to the three-dimensional heat equation $\partial_t U = a\Delta_3 U$ that is considered in Subsection 3.1.1.

5. $\dfrac{\partial w}{\partial t} = a\left(\dfrac{\partial^2 w}{\partial x^2} + \dfrac{\partial^2 w}{\partial y^2} + \dfrac{\partial^2 w}{\partial z^2}\right) - by\dfrac{\partial w}{\partial x}.$

This equation is encountered in problems of convective heat and mass transfer in a simple shear flow.

Fundamental solution:

$$\mathscr{E}(x,y,z,\xi,\eta,\zeta,t) = \frac{1}{(4\pi at)^{3/2}\bigl(1+\tfrac{1}{12}b^2 t^2\bigr)^{1/2}} \exp\left\{-\frac{\bigl[x-\xi-\tfrac{1}{2}bt(y+\eta)\bigr]^2}{4at\bigl(1+\tfrac{1}{12}b^2 t^2\bigr)} - \frac{(y-\eta)^2+(z-\zeta)^2}{4at}\right\}.$$

⊙ *Reference*: E. A. Novikov (1958).

6. $\dfrac{\partial w}{\partial t} = a\left(\dfrac{\partial^2 w}{\partial x^2} + \dfrac{\partial^2 w}{\partial y^2} + \dfrac{\partial^2 w}{\partial z^2}\right) + b_1 x\dfrac{\partial w}{\partial x} + b_2 y\dfrac{\partial w}{\partial x} + b_3 z\dfrac{\partial w}{\partial z} + \Phi(x,y,z,t).$

This equation is encountered in problems of convective heat and mass transfer in a linear shear flow.

Domain: $-\infty < x < \infty$, $-\infty < y < \infty$, $-\infty < z < \infty$. Cauchy problem.

An initial condition is prescribed:

$$w = f(x,y,z) \quad \text{at} \quad t = 0.$$

Solution:

$$w(x,y,z,t) = \int_{-\infty}^{\infty}\int_{-\infty}^{\infty}\int_{-\infty}^{\infty} f(\xi,\eta,\zeta) G(x,y,z,\xi,\eta,\zeta,t)\,dx\,dy\,dz$$
$$+ \int_0^t\int_{-\infty}^{\infty}\int_{-\infty}^{\infty}\int_{-\infty}^{\infty} \Phi(\xi,\eta,\zeta,\tau) G(x,y,z,\xi,\eta,\zeta,t-\tau)\,dx\,dy\,dz\,d\tau,$$

where

$$G(x,y,z,\xi,\eta,\zeta,t) = H(x,\xi,t;b_1) H(y,\eta,t;b_2) H(z,\zeta,t;b_3),$$

$$H(x,\xi,t;b) = \left[\frac{2\pi a}{b}(e^{2bt}-1)\right]^{-1/2} \exp\left[-\frac{b\bigl(xe^{bt}-\xi\bigr)^2}{2a(e^{2bt}-1)}\right].$$

7. $\dfrac{\partial w}{\partial t} = a\left(\dfrac{\partial^2 w}{\partial x^2} + \dfrac{\partial^2 w}{\partial y^2} + \dfrac{\partial^2 w}{\partial z^2}\right) + (b_1 x + c_1)\dfrac{\partial w}{\partial x} + (b_2 y + c_2)\dfrac{\partial w}{\partial y}$
$\qquad\qquad\qquad\qquad\qquad\qquad\qquad\qquad + (b_3 z + c_3)\dfrac{\partial w}{\partial z} + (s_1 x + s_2 y + s_3 z + p)w.$

This is a special case of equation 3.3.2.5.

8. $i\hbar \dfrac{\partial w}{\partial t} + \dfrac{\hbar^2}{2m}\left(\dfrac{\partial^2 w}{\partial x^2} + \dfrac{\partial^2 w}{\partial y^2} + \dfrac{\partial^2 w}{\partial z^2}\right) = 0.$

Three-dimensional Schrödinger equation, $i^2 = -1$.

Fundamental solution:

$$\mathscr{E}(x,y,z,t) = -\dfrac{i}{\hbar}\left(\dfrac{m}{2\pi\hbar t}\right)^{3/2} \exp\left[i\dfrac{m}{2\hbar t}(x^2+y^2+z^2) - i\dfrac{3\pi}{4}\right].$$

⊙ *Reference*: V. S. Vladimirov, V. P. Mikhailov, A. A. Vasharin, et al. (1974).

9. $\dfrac{\partial w}{\partial t} = \dfrac{\partial}{\partial x}\left(ax^n \dfrac{\partial w}{\partial x}\right) + \dfrac{\partial}{\partial y}\left(by^m \dfrac{\partial w}{\partial y}\right) + \dfrac{\partial}{\partial z}\left(cz^k \dfrac{\partial w}{\partial z}\right).$

This equation describes unsteady heat and mass transfer processes in inhomogeneous (anisotropic) media. It admits separable solutions, as well as solutions with incomplete separation of variables (see Subsection 0.9.2-1). In addition, for $n \ne 2$, $m \ne 2$, $k \ne 2$ there are particular solutions of the form

$$w = w(\xi, t), \qquad \xi^2 = 4\left[\dfrac{x^{2-n}}{a(2-n)^2} + \dfrac{y^{2-m}}{b(2-m)^2} + \dfrac{z^{2-k}}{c(2-k)^2}\right],$$

where the function $w(\xi, t)$ is determined by the one-dimensional nonstationary equation

$$\dfrac{\partial w}{\partial t} = \dfrac{\partial^2 w}{\partial \xi^2} + \dfrac{A}{\xi}\dfrac{\partial w}{\partial \xi}, \qquad A = 2\left(\dfrac{1}{2-n} + \dfrac{1}{2-m} + \dfrac{1}{2-k}\right) - 1.$$

For solutions of this equation, see Subsections 1.2.1, 1.2.3, and 1.2.5.

3.3.2. Equations Containing Arbitrary Functions

1. $\dfrac{\partial w}{\partial t} = a\left(\dfrac{\partial^2 w}{\partial x^2} + \dfrac{\partial^2 w}{\partial y^2} + \dfrac{\partial^2 w}{\partial z^2}\right) + f(t)w.$

This equation describes three-dimensional unsteady thermal phenomena in quiescent media or solids with constant thermal diffusivity, provided there is unsteady volume heat release proportional to temperature.

The substitution $w(x,y,z,t) = \exp\left[\int f(t)\,dt\right] U(x,y,z,t)$ leads to the usual heat equation $\partial_t U = a\Delta_3 U$ that is dealt with in Subsection 3.1.1.

2. $\dfrac{\partial w}{\partial t} = a\left(\dfrac{\partial^2 w}{\partial x^2} + \dfrac{\partial^2 w}{\partial y^2} + \dfrac{\partial^2 w}{\partial z^2}\right) + \left[xf_1(t) + yf_2(t) + zf_3(t) + g(t)\right]w.$

The transformation

$$w(x,y,z,t) = \exp\left\{xF_1(t) + yF_2(t) + zF_3(t) + a\int\left[F_1^2(t) + F_2^2(t) + F_3^2(t)\right]dt + \int g(t)\,dt\right\} u(\xi, \eta, \zeta, t),$$

$$\xi = x + 2a\int F_1(t)\,dt, \quad \eta = y + 2a\int F_2(t)\,dt, \quad \zeta = z + 2a\int F_3(t)\,dt, \quad F_k(t) = \int f_k(t)\,dt,$$

leads to the three-dimensional heat equation $\partial_t u = a\left(\partial_{\xi\xi}u + \partial_{\eta\eta}u + \partial_{\zeta\zeta}u\right)$ that is dealt with in Subsection 3.1.1.

3. $\dfrac{\partial w}{\partial t} = a\left(\dfrac{\partial^2 w}{\partial x^2} + \dfrac{\partial^2 w}{\partial y^2} + \dfrac{\partial^2 w}{\partial z^2}\right) + \left[-b(x^2+y^2+z^2) + x f_1(t) + y f_2(t) + z f_3(t) + g(t)\right] w.$

1°. Case $b > 0$. The transformation

$$w(x,y,z,t) = u(\xi, \eta, \zeta, \tau) \exp\left[\dfrac{\sqrt{ab}}{2a}(x^2+y^2+z^2)\right],$$

$\xi = x \exp(2\sqrt{ab}\, t), \quad \eta = y \exp(2\sqrt{ab}\, t), \quad \zeta = z \exp(2\sqrt{ab}\, t), \quad \tau = \dfrac{1}{4\sqrt{ab}} \exp(4\sqrt{ab}\, t) + A,$

where A is an arbitrary constant, leads to an equation of the form 3.3.2.2:

$$\dfrac{\partial u}{\partial t} = a\left(\dfrac{\partial^2 u}{\partial \xi^2} + \dfrac{\partial^2 u}{\partial \eta^2} + \dfrac{\partial^2 u}{\partial \zeta^2}\right) + \left[\xi F_1(\tau) + \eta F_2(\tau) + \zeta F_3(\tau) + G(\tau)\right] u,$$

$F_k(\tau) = \dfrac{1}{(c\tau)^{3/2}} f_k\left(\dfrac{\ln(c\tau)}{c}\right), \quad G(\tau) = \dfrac{1}{c\tau} g\left(\dfrac{\ln(c\tau)}{c}\right) + \dfrac{3}{4\tau}, \quad c = 4\sqrt{ab}, \quad k = 1, 2, 3.$

2°. Case $b < 0$. The transformation

$$w(x,y,z,t) = v(\xi_1, \eta_1, \zeta_1, \tau_1) \exp\left[\dfrac{\sqrt{-ab}}{2a}(x^2+y^2+z^2)\tan(2\sqrt{-ab}\, t)\right],$$

$\xi_1 = \dfrac{x}{\cos(2\sqrt{-ab}\, t)}, \quad \eta_1 = \dfrac{y}{\cos(2\sqrt{-ab}\, t)}, \quad \zeta_1 = \dfrac{z}{\cos(2\sqrt{-ab}\, t)}, \quad \tau_1 = \dfrac{1}{2\sqrt{-ab}} \tan(2\sqrt{-ab}\, t)$

also leads to an equation of the form 3.3.2.2 for $v = v(\xi_1, \eta_1, \zeta_1, \tau_1)$ (the equation for v is not written out here).

4. $\dfrac{\partial w}{\partial t} = a_1(t) \dfrac{\partial^2 w}{\partial x^2} + a_2(t) \dfrac{\partial^2 w}{\partial y^2} + a_3(t) \dfrac{\partial^2 w}{\partial z^2} + \Phi(x, y, z, t).$

Here, $0 < a_k(t) < \infty$; $k = 1, 2, 3$.

Domain: $-\infty < x < \infty$, $-\infty < y < \infty$, $-\infty < z < \infty$. Cauchy problem.
An initial condition is prescribed:

$$w = f(x, y, z) \quad \text{at} \quad t = 0.$$

Solution:

$$w(x, y, z, t) = \int_0^t \int_{-\infty}^{\infty} \int_{-\infty}^{\infty} \int_{-\infty}^{\infty} \Phi(\xi, \eta, \zeta, \tau) G(x, y, z, \xi, \eta, \zeta, t, \tau) \, d\xi \, d\eta \, d\zeta \, d\tau$$
$$+ \int_{-\infty}^{\infty} \int_{-\infty}^{\infty} \int_{-\infty}^{\infty} f(\xi, \eta, \zeta) G(x, y, z, \xi, \eta, \zeta, t, 0) \, d\xi \, d\eta \, d\zeta,$$

where

$$G(x, y, z, \xi, \eta, \zeta, t, \tau) = \dfrac{1}{8\pi^{3/2} \sqrt{T_1 T_2 T_3}} \exp\left[-\dfrac{(x-\xi)^2}{4T_1} - \dfrac{(y-\eta)^2}{4T_2} - \dfrac{(z-\zeta)^2}{4T_3}\right],$$

$$T_1 = \int_\tau^t a_1(\sigma) \, d\sigma, \quad T_2 = \int_\tau^t a_2(\sigma) \, d\sigma, \quad T_3 = \int_\tau^t a_3(\sigma) \, d\sigma.$$

See also the more general equation 3.4.3.3, where other boundary value problems are considered.

5. $\dfrac{\partial w}{\partial t} = a_1(t)\dfrac{\partial^2 w}{\partial x^2} + a_2(t)\dfrac{\partial^2 w}{\partial y^2} + a_3(t)\dfrac{\partial^2 w}{\partial z^2} + \bigl[b_1(t)x + c_1(t)\bigr]\dfrac{\partial w}{\partial x}$
$+ \bigl[b_2(t)y + c_2(t)\bigr]\dfrac{\partial w}{\partial y} + \bigl[b_3(t)z + c_3(t)\bigr]\dfrac{\partial w}{\partial z} + \bigl[s_1(t)x + s_2(t)y + s_3(t)z + p(t)\bigr]w.$

The transformation
$$w(x, y, z, t) = \exp\bigl[f_1(t)x + f_2(t)y + f_3(t)z + g(t)\bigr]u(\xi, \eta, \zeta, t),$$
$$\xi = h_1(t)x + r_1(t), \quad \eta = h_2(t)y + r_2(t), \quad \zeta = h_3(t)z + r_3(t),$$

where
$$h_k(t) = A_k \exp\!\left[\int b_k(t)\,dt\right],$$
$$f_k(t) = h_k(t)\int \dfrac{s_k(t)}{h_k(t)}\,dt + B_k h_k(t),$$
$$r_k(t) = \int \bigl[2a_k(t)f_k(t) + c_k(t)\bigr] h_k(t)\,dt + C_k,$$
$$g(t) = \int \sum_{k=1}^{3}\bigl[a_k(t)f_k^2(t) + c_k(t)f_k(t)\bigr]\,dt + \int p(t)\,dt + D,$$

($k = 1, 2, 3$; A_k, B_k, C_k, D are arbitrary constants) leads to an equation of the form 3.3.2.4:
$$\dfrac{\partial u}{\partial t} = a_1(t)h_1^2(t)\dfrac{\partial^2 u}{\partial \xi^2} + a_2(t)h_2^2(t)\dfrac{\partial^2 u}{\partial \eta^2} + a_3(t)h_3^2(t)\dfrac{\partial^2 u}{\partial \zeta^2}.$$

6. $\dfrac{\partial w}{\partial t} = a_1(t)\dfrac{\partial^2 w}{\partial x^2} + a_2(t)\dfrac{\partial^2 w}{\partial y^2} + a_3(t)\dfrac{\partial^2 w}{\partial z^2} + \bigl[b_1(t)x + c_1(t)\bigr]\dfrac{\partial w}{\partial x} + \bigl[b_2(t)y + c_2(t)\bigr]\dfrac{\partial w}{\partial y}$
$+ \bigl[b_3(t)z + c_3(t)\bigr]\dfrac{\partial w}{\partial z} + \bigl[s_1(t)x^2 + s_2(t)y^2 + s_3(t)z^2 + p_1(t)x + p_2(t)y + p_3(t)z + q(t)\bigr]w.$

The substitution
$$w(x, y, z, t) = \exp\bigl[f_1(t)x^2 + f_2(t)y^2 + f_3(t)z^2\bigr]u(x, y, z, t),$$
where the functions $f_k = f_k(t)$ are solutions of the respective Riccati equations
$$f_k' = 4a_k(t)f_k^2 + 2b_k(t)f_k + s_k(t) \qquad (k = 1, 2, 3),$$
leads to an equation of the form 3.3.2.5 for $u = u(x, y, z, t)$.

7. $\dfrac{\partial w}{\partial t} + \displaystyle\sum_{n=1}^{2}\bigl[f_n(t) + g_n(t)x_3\bigr]\dfrac{\partial w}{\partial x_n} - a\dfrac{\partial w}{\partial x_3} = \sum_{n,m=1}^{2} K_{nm}(t)\dfrac{\partial^2 w}{\partial x_n \partial x_m} + K_{33}(t)\dfrac{\partial^2 w}{\partial x_3^2}.$

Equation of turbulent diffusion. It describes the diffusion of an admixture in a horizontal stream whose velocity components are linear functions of the height.

Fundamental solution:
$$\mathscr{E}(x_1, x_2, x_3, \xi_1, \xi_2, \xi_3) = \dfrac{1}{(4\pi)^{3/2}\sqrt{\det |T|}}\exp\!\left[-\dfrac{1}{4}\sum_{i,j=1}^{3} T_{ij}^{-1}(t)y_i y_j\right], \qquad T_{ij}(t) = \int_0^t S_{ij}(\tau)\,d\tau.$$

Here, the following notation is used ($n, m = 1, 2$):
$$y_n = x_n - \xi_n - F_n(t) - x_3 G_n(t) - a\int_0^t (t - \tau)g_n(\tau)\,d\tau, \qquad y_3 = x_3 - \xi_3 + at,$$
$$S_{nm}(t) = K_{nm}(t) + K_{33}(t)G_n(t)G_m(t), \qquad S_{n3}(t) = S_{3n}(t) = -K_{33}(t)G_n(t),$$
$$S_{33}(t) = K_{33}(t), \qquad F_n(t) = \int_0^t f_n(t)\,dt, \qquad G_n(t) = \int_0^t g_n(t)\,dt,$$

$\det |T|$ is the determinant of the matrix $\mathbf{T}$ with entries $T_{ij}(t)$, $T_{ij}^{-1}(t)$ are the entries of the inverse of $\mathbf{T}$. The inequalities $T_{11}(t) > 0$, $T_{11}(t)T_{22}(t) - T_{12}^2(t) > 0$, and $\det |T| > 0$ are assumed to hold.

⊙ *Reference*: E. A. Novikov (1958).

3.3.3. Equations of the Form
$$\rho(x,y,z)\frac{\partial w}{\partial t} = \text{div}[a(x,y,z)\nabla w] - q(x,y,z)w + \Phi(x,y,z,t)$$

Equations of this form are often encountered in the theory of heat and mass transfer. For brevity, the following notation is used:

$$\text{div}[a(\mathbf{r})\nabla w] = \frac{\partial}{\partial x}\left[a(\mathbf{r})\frac{\partial w}{\partial x}\right] + \frac{\partial}{\partial y}\left[a(\mathbf{r})\frac{\partial w}{\partial y}\right] + \frac{\partial}{\partial z}\left[a(\mathbf{r})\frac{\partial w}{\partial z}\right], \qquad \mathbf{r} = \{x,y,z\}.$$

The problems presented in this subsection are assumed to refer to a simply connected bounded domain V with smooth boundary S. It is also assumed that $\rho(\mathbf{r}) > 0$, $a(\mathbf{r}) > 0$, and $q(\mathbf{r}) \geq 0$.

3.3.3-1. First boundary value problem.

The following conditions are prescribed:

$$w = f(\mathbf{r}) \quad \text{at} \quad t = 0 \quad \text{(initial condition)},$$
$$w = g(\mathbf{r},t) \quad \text{for} \quad \mathbf{r} \in S \quad \text{(boundary condition)}.$$

Solution:

$$w(\mathbf{r},t) = \int_0^t \int_V \Phi(\boldsymbol{\xi},\tau)\mathcal{G}(\mathbf{r},\boldsymbol{\xi},t-\tau)\, dV_{\boldsymbol{\xi}}\, d\tau + \int_V f(\boldsymbol{\xi})\rho(\boldsymbol{\xi})\mathcal{G}(\mathbf{r},\boldsymbol{\xi},t)\, dV_{\boldsymbol{\xi}}$$
$$- \int_0^t \int_S g(\boldsymbol{\xi},\tau)a(\boldsymbol{\xi})\left[\frac{\partial}{\partial N_{\boldsymbol{\xi}}}\mathcal{G}(\mathbf{r},\boldsymbol{\xi},t-\tau)\right] dS_{\boldsymbol{\xi}}\, d\tau. \qquad (1)$$

Here, the modified Green's function is given by

$$\mathcal{G}(\mathbf{r},\boldsymbol{\xi},t) = \sum_{n=1}^{\infty} \frac{u_n(\mathbf{r})u_n(\boldsymbol{\xi})}{\|u_n\|^2} \exp(-\lambda_n t), \quad \|u_n\|^2 = \int_V \rho(\mathbf{r})u_n^2(\mathbf{r})\, dV, \quad \boldsymbol{\xi} = \{\xi_1,\xi_2,\xi_3\}, \qquad (2)$$

where the λ_n and $u_n(\mathbf{r})$ are the eigenvalues and corresponding eigenfunctions of the Sturm–Liouville problem for the following elliptic second-order equation with a homogeneous boundary condition of the first kind:

$$\text{div}[a(\mathbf{r})\nabla u] - q(\mathbf{r})u + \lambda\rho(\mathbf{r})u = 0, \qquad (3)$$
$$u = 0 \quad \text{for} \quad \mathbf{r} \in S. \qquad (4)$$

The integration in solution (1) is carried out with respect to ξ_1, ξ_2, ξ_3; $\frac{\partial}{\partial N_{\boldsymbol{\xi}}}$ denotes the derivative along the outward normal to the surface S with respect to ξ_1, ξ_2, and ξ_3.

General properties of the Sturm–Liouville problem (3)–(4):

$1°$. There are countably many eigenvalues. All eigenvalues are real and can be ordered so that $\lambda_1 \leq \lambda_2 \leq \lambda_3 \leq \cdots$, with $\lambda_n \to \infty$ as $n \to \infty$; consequently, there can exist only finitely many negative eigenvalues.

$2°$. For $\rho(\mathbf{r}) > 0$, $a(\mathbf{r}) > 0$, and $q(\mathbf{r}) \geq 0$, all eigenvalues are positive: $\lambda_n > 0$.

$3°$. The eigenfunctions are defined up to a constant multiplier. Any two eigenfunctions $u_n(\mathbf{r})$ and $u_m(\mathbf{r})$ corresponding to different eigenvalues λ_n and λ_m are orthogonal with weight $\rho(\mathbf{r})$ in the domain V:

$$\int_V \rho(\mathbf{r})u_n(\mathbf{r})u_m(\mathbf{r})\, dV = 0 \quad \text{for} \quad n \neq m.$$

4°. An arbitrary function $F(\mathbf{r})$ that is twice continuously differentiable and satisfies the boundary condition of the Sturm–Liouville problem ($F = 0$ for $\mathbf{r} \in S$) can be expanded into an absolutely and uniformly convergent series in the eigenvalues:

$$F(\mathbf{r}) = \sum_{n=1}^{\infty} F_n u_n(\mathbf{r}), \quad F_n = \frac{1}{\|u_n\|^2} \int_V F(\mathbf{r}) \rho(\mathbf{r}) u_n(\mathbf{r}) \, dV,$$

where the formula for $\|u_n\|^2$ is given in (2).

Remark. In a three-dimensional problem, to each eigenvalue λ_n there generally correspond finitely many linearly independent eigenfunctions $u_n^{(1)}, u_n^{(2)}, \ldots, u_n^{(m)}$. These function can always be replaced by their linear combinations

$$\bar{u}_n^{(k)} = A_{k,1} u_n^{(1)} + \cdots + A_{k,k-1} u_n^{(k-1)} + u_n^{(k)}, \quad k = 1, 2, \ldots, m,$$

such that $\bar{u}_n^{(1)}, \bar{u}_n^{(2)}, \ldots, \bar{u}_n^{(m)}$ are now pairwise orthogonal. Thus, without loss of generality, we assume that all eigenfunctions are orthogonal.

3.3.3-2. Second boundary value problem.

The following conditions are prescribed:

$$w = f(\mathbf{r}) \quad \text{at} \quad t = 0 \quad \text{(initial condition)},$$

$$\frac{\partial w}{\partial N} = g(\mathbf{r}, t) \quad \text{for} \quad \mathbf{r} \in S \quad \text{(boundary condition)}.$$

Solution:

$$w(\mathbf{r}, t) = \int_0^t \int_V \Phi(\boldsymbol{\xi}, \tau) \mathcal{G}(\mathbf{r}, \boldsymbol{\xi}, t - \tau) \, dV_\xi \, d\tau + \int_V f(\boldsymbol{\xi}) \rho(\boldsymbol{\xi}) \mathcal{G}(\mathbf{r}, \boldsymbol{\xi}, t) \, dV_\xi$$
$$+ \int_0^t \int_S g(\boldsymbol{\xi}, \tau) a(\boldsymbol{\xi}) \mathcal{G}(\mathbf{r}, \boldsymbol{\xi}, t - \tau) \, dS_\xi \, d\tau. \tag{5}$$

Here, the modified Green's function is given by (2), where the λ_n and $u_n(\mathbf{r})$ are the eigenvalues and corresponding eigenfunctions of the Sturm–Liouville problem for the elliptic second-order equation (3) with a homogeneous boundary condition of the second kind,

$$\frac{\partial u}{\partial N} = 0 \quad \text{for} \quad \mathbf{r} \in S. \tag{6}$$

For $q(\mathbf{r}) > 0$ the general properties of the eigenvalue problem (3), (6) are the same as for the first boundary value problem (with all $\lambda_n > 0$).

3.3.3-3. Third boundary value problem.

The following conditions are prescribed:

$$w = f(\mathbf{r}) \quad \text{at} \quad t = 0 \quad \text{(initial condition)},$$

$$\frac{\partial w}{\partial N} + k(\mathbf{r}) w = g(\mathbf{r}, t) \quad \text{for} \quad \mathbf{r} \in S \quad \text{(boundary condition)}.$$

The solution of the third boundary value problem is given by formulas (5) and (2), where the λ_n and $u_n(\mathbf{r})$ are the eigenvalues and corresponding eigenfunctions of the Sturm–Liouville problem for the second-order elliptic equation (3) with a homogeneous boundary condition of the third kind,

$$\frac{\partial u}{\partial N} + k(\mathbf{r}) u = 0 \quad \text{for} \quad \mathbf{r} \in S. \tag{7}$$

For $q(\mathbf{r}) \geq 0$ and $k(\mathbf{r}) > 0$ the general properties of the eigenvalue problem (3), (7) are the same as for the first boundary value problem (see Paragraph 3.3.3-1).

Let $k(\mathbf{r}) = k = \text{const}$. Denote the Green's functions of the second and third boundary value problems by $G_2(\mathbf{r}, \boldsymbol{\xi}, t)$ and $G_3(\mathbf{r}, \boldsymbol{\xi}, t, k)$, respectively. If $q(\mathbf{r}) > 0$, then the following limiting relation holds: $G_2(\mathbf{r}, \boldsymbol{\xi}, t) = \lim_{k \to 0} G_3(\mathbf{r}, \boldsymbol{\xi}, t, k)$.

⊙ *References for Subsection* 3.3.3: V. S. Vladimirov (1988), A. D. Polyanin (2000a, 2000c).

3.4. Equations with n Space Variables

3.4.1. Equations of the Form $\frac{\partial w}{\partial t} = a\Delta_n w + \Phi(x_1, \ldots, x_n, t)$

This is an n-dimensional nonhomogeneous heat equation. In the Cartesian system of coordinates, it is represented as

$$\frac{\partial w}{\partial t} = a \sum_{k=1}^{n} \frac{\partial^2 w}{\partial x_k^2} + \Phi(\mathbf{x}, t), \qquad \mathbf{x} = \{x_1, \ldots, x_n\}.$$

The solutions of various problems for this equation can be constructed on the basis of incomplete separation of variables (see Paragraphs 0.6.1-2 and 0.9.2-1) taking into account the results of Subsections 1.1.1 and 1.1.2. Some examples of solving such problems can be found below in Paragraphs 3.4.1-2 through 3.4.1-4.

3.4.1-1. Homogeneous equation ($\Phi \equiv 0$).

1°. Particular solutions:

$$w(\mathbf{x}, t) = A \exp\left(\sum_{m=1}^{n} k_m x_m + at \sum_{m=1}^{n} k_m^2\right),$$

$$w(\mathbf{x}, t) = A \exp\left(-at \sum_{m=1}^{n} k_m^2\right) \prod_{m=1}^{n} \cos(k_m x_m + C_m),$$

$$w(\mathbf{x}, t) = A \exp\left(-\sum_{m=1}^{n} k_m x_m\right) \prod_{m=1}^{n} \cos(k_m x_m - 2ak_m^2 t + C_m),$$

$$w(\mathbf{x}, t) = \frac{A}{(t-t_0)^{n/2}} \exp\left[-\frac{1}{4a(t-t_0)} \sum_{m=1}^{n} (x_m - C_m)^2\right],$$

$$w(\mathbf{x}, t) = A \prod_{m=1}^{n} \operatorname{erf}\left(\frac{x_m - C_m}{2\sqrt{at}}\right),$$

where $\mathbf{x} = \{x_1, \ldots, x_n\}$; A, k_m, C_m, and t_0 are arbitrary constants.

2°. Fundamental solution:

$$\mathscr{E}(\mathbf{x}, t) = \frac{1}{(2\sqrt{\pi at})^n} \exp\left(-\frac{|\mathbf{x}|^2}{4at}\right), \qquad |\mathbf{x}|^2 = \sum_{k=1}^{n} x_k^2.$$

3°. Suppose $w = w(x_1, \ldots, x_n, t)$ is a solution of the homogeneous equation. Then the functions

$$w_1 = Aw(\pm \lambda x_1 + C_1, \ldots, \pm \lambda x_n + C_n, \lambda^2 t + C_{n+1}),$$

$$w_2 = A \exp\left(\sum_{k=1}^{n} \lambda_k x_k + at \sum_{k=1}^{n} \lambda_k^2\right) w(x_1 + 2a\lambda_1 t + C_1, \ldots, x_n + 2a\lambda_n t + C_n, t + C_{n+1}),$$

$$w_3 = \frac{A}{|\delta + \beta t|^{n/2}} \exp\left[-\frac{\beta}{4a(\delta + \beta t)} \sum_{k=1}^{n} x_k^2\right] w\left(\frac{x_1}{\delta + \beta t}, \ldots, \frac{x_n}{\delta + \beta t}, \frac{\gamma + \lambda t}{\delta + \beta t}\right), \qquad \lambda\delta - \beta\gamma = 1,$$

where A, $C_1, \ldots, C_{n+1}$, λ, $\lambda_1, \ldots, \lambda_n$ β, and δ are arbitrary constants, are also solutions of the equation. The signs at λ in the formula for w_1 can be taken independently of one another.

3.4.1-2. Domain: $\mathbb{R}^n = \{-\infty < x_k < \infty;\ k = 1, \ldots, n\}$. Cauchy problem.

An initial condition is prescribed:
$$w = f(\mathbf{x}) \quad \text{at} \quad t = 0.$$

Solution:
$$w(\mathbf{x}, t) = \frac{1}{(2\sqrt{\pi a t})^n} \int_{\mathbb{R}^n} f(\mathbf{y}) \exp\left(-\frac{|\mathbf{x}-\mathbf{y}|^2}{4at}\right) d\mathbf{y} + \int_0^t \int_{\mathbb{R}^n} \frac{\Phi(\mathbf{y}, \tau)}{(2\sqrt{\pi a(t-\tau)})^n} \exp\left(-\frac{|\mathbf{x}-\mathbf{y}|^2}{4a(t-\tau)}\right) d\mathbf{y}\, d\tau,$$

where $\mathbf{y} = \{y_1, \ldots, y_n\}$, $|\mathbf{x}-\mathbf{y}| = \sqrt{(x_1-y_1)^2 + \cdots + (x_n-y_n)^2}$, $d\mathbf{y} = dy_1\, dy_2 \ldots dy_n$.

⊙ *Reference*: V. S. Vladimirov (1988).

3.4.1-3. Domain: $V = \{0 \leq x_k \leq l_k;\ k = 1, \ldots, n\}$. First boundary value problem.

The following conditions are prescribed:
$$w = f(\mathbf{x}) \quad \text{at} \quad t = 0 \quad \text{(initial condition)},$$
$$w = g_k(\mathbf{x}, t) \quad \text{at} \quad x_k = 0 \quad \text{(boundary conditions)},$$
$$w = h_k(\mathbf{x}, t) \quad \text{at} \quad x_k = l_k \quad \text{(boundary conditions)}.$$

Solution:
$$w(\mathbf{x}, t) = \int_0^t \int_V \Phi(\mathbf{y}, \tau) G(\mathbf{x}, \mathbf{y}, t-\tau)\, d\mathbf{y}\, d\tau + \int_V f(\mathbf{y}) G(\mathbf{x}, \mathbf{y}, t)\, d\mathbf{y}$$
$$+ a \sum_{k=1}^n \int_0^t \int_{S^{(k)}} \left[g_k(\mathbf{y}, \tau) \frac{\partial}{\partial y_k} G(\mathbf{x}, \mathbf{y}, t-\tau) \right]_{y_k=0} dS_y^{(k)}\, d\tau$$
$$- a \sum_{k=1}^n \int_0^t \int_{S^{(k)}} \left[h_k(\mathbf{y}, \tau) \frac{\partial}{\partial y_k} G(\mathbf{x}, \mathbf{y}, t-\tau) \right]_{y_k=l_k} dS_y^{(k)}\, d\tau,$$

where the following notation is used:
$$dS_y^{(k)} = dy_1 \ldots dy_{k-1}\, dy_{k+1} \ldots dy_n, \quad S^{(k)} = \{0 \leq y_m \leq l_m \text{ for } m = 1, \ldots, k-1, k+1, \ldots, n\}.$$

The Green's function can be represented in the product form
$$G(\mathbf{x}, \mathbf{y}, t) = \prod_{k=1}^n G_k(x_k, y_k, t), \tag{1}$$

where the $G_k(x_k, y_k, t)$ are the Green's functions of the respective one-dimensional boundary value problems (see Paragraph 1.1.2-5):
$$G_k(x_k, y_k, t) = \frac{2}{l_k} \sum_{m=1}^\infty \sin\left(\frac{m\pi x}{l_k}\right) \sin\left(\frac{m\pi \xi}{l_k}\right) \exp\left(-\frac{a m^2 \pi^2 t}{l_k^2}\right).$$

3.4.1-4. Domain: $V = \{0 \leq x_k \leq l_k;\ k = 1, \ldots, n\}$. Second boundary value problem.

The following conditions are prescribed:
$$w = f(\mathbf{x}) \quad \text{at} \quad t = 0 \quad \text{(initial condition)},$$
$$\partial_{x_k} w = g_k(\mathbf{x}, t) \quad \text{at} \quad x_k = 0 \quad \text{(boundary conditions)},$$
$$\partial_{x_k} w = h_k(\mathbf{x}, t) \quad \text{at} \quad x_k = l_k \quad \text{(boundary conditions)}.$$

Solution:

$$w(\mathbf{x},t) = \int_0^t \int_V \Phi(\mathbf{y},\tau) G(\mathbf{x},\mathbf{y},t-\tau)\,d\mathbf{y}\,d\tau + \int_V f(\mathbf{y}) G(\mathbf{x},\mathbf{y},t)\,d\mathbf{y}$$
$$- a \sum_{k=1}^n \int_0^t \int_{S^{(k)}} \left[g_k(\mathbf{y},\tau) G(\mathbf{x},\mathbf{y},t-\tau)\right]_{y_k=0} dS_y^{(k)}\,d\tau$$
$$+ a \sum_{k=1}^n \int_0^t \int_{S^{(k)}} \left[h_k(\mathbf{y},\tau) G(\mathbf{x},\mathbf{y},t-\tau)\right]_{y_k=l_k} dS_y^{(k)}\,d\tau. \tag{2}$$

The Green's function can be represented as the product (1) of the corresponding one-dimensional Green's functions of the form (see Paragraph 1.1.2-6)

$$G_k(x_k,y_k,t) = \frac{1}{l_k} + \frac{2}{l_k} \sum_{m=1}^\infty \cos\left(\frac{m\pi x}{l_k}\right)\cos\left(\frac{m\pi \xi}{l_k}\right)\exp\left(-\frac{am^2\pi^2 t}{l_k^2}\right).$$

3.4.2. Other Equations Containing Arbitrary Parameters

1. $\dfrac{\partial w}{\partial t} = a \sum_{k=1}^n \dfrac{\partial^2 w}{\partial x_k^2} + \left(c + \sum_{k=1}^n b_k x_k\right) w.$

This is a special case of equation 3.4.3.1. The transformation

$$w(x_1,\ldots,x_n,t) = \exp\left(t \sum_{k=1}^n b_k x_k + \tfrac{1}{3} at^3 \sum_{k=1}^n b_k^2 + ct\right) u(\xi_1,\ldots,\xi_n,t), \qquad \xi_k = x_k + ab_k t^2$$

leads to the n-dimensional heat equation $\partial_t u = a \sum_{k=1}^n \partial_{\xi_k \xi_k} u$ that is dealt with in Subsection 3.4.1.

2. $\dfrac{\partial w}{\partial t} = a \sum_{k=1}^n \dfrac{\partial^2 w}{\partial x_k^2} - \left(c + b \sum_{k=1}^n x_k^2\right) w, \qquad b > 0.$

The transformation (A is any number)

$$w(x_1,\ldots,x_n,t) = u(\xi_1,\ldots,\xi_n,\tau) \exp\left[\frac{1}{2}\sqrt{\frac{b}{a}} \sum_{k=1}^n x_k^2 + \left(n\sqrt{ab} - c\right) t\right],$$

$$\xi_1 = x_1 \exp\left(2\sqrt{ab}\,t\right), \quad \ldots, \quad \xi_n = x_n \exp\left(2\sqrt{ab}\,t\right), \quad \tau = \frac{1}{4\sqrt{ab}} \exp\left(4\sqrt{ab}\,t\right) + A$$

leads to the n-dimensional heat equation $\partial_\tau u = a \sum_{k=1}^n \partial_{\xi_k \xi_k} u$ that is dealt with in Subsection 3.4.1.

3. $\dfrac{\partial w}{\partial t} = a \sum_{k=1}^n \dfrac{\partial^2 w}{\partial x_k^2} + \left(-b \sum_{k=1}^n x_k^2 + \sum_{k=1}^n c_k x_k + s\right) w.$

This is a special case of equation 3.4.3.2 with $f_k(t) = c_k$ and $g(t) = s$.

4. $\dfrac{\partial w}{\partial t} = a \sum_{k=1}^n \dfrac{\partial^2 w}{\partial x_k^2} + \sum_{k=1}^n b_k \dfrac{\partial w}{\partial x_k} + cw.$

The substitution

$$w(x_1,\ldots,x_n,t) = \exp\left(At - \frac{1}{2a}\sum_{k=1}^n b_k x_k\right) U(x_1,\ldots,x_n,t), \quad \text{where} \quad A = c - \frac{1}{4a}\sum_{k=1}^n b_k^2,$$

leads to the n-dimensional heat equation $\partial_t U = a \Delta_n U$ that is dealt with in Subsection 3.4.1.

5. $\dfrac{\partial w}{\partial t} = a \sum_{k=1}^{n} \dfrac{\partial^2 w}{\partial x_k^2} + \sum_{k=1}^{n} (b_k x_k + c_k) \dfrac{\partial w}{\partial x_k} + \left(\sum_{k=1}^{n} s_k x_k + p \right) w.$

This is a special case of equation 3.4.3.4.

6. $i\hbar \dfrac{\partial w}{\partial t} + \dfrac{\hbar^2}{2m} \sum_{k=1}^{n} \dfrac{\partial^2 w}{\partial x_k^2} = 0.$

This is the n-dimensional Schrödinger equation, $i^2 = -1$.

Fundamental solution:

$$\mathscr{E}(\mathbf{x}, t) = -\dfrac{i}{\hbar} \left(\dfrac{m}{2\pi\hbar t} \right)^{n/2} \exp\left(i \dfrac{m}{2\hbar t} |\mathbf{x}|^2 - i \dfrac{\pi n}{4} \right), \qquad |\mathbf{x}|^2 = x_1^2 + \cdots + x_n^2.$$

⊙ *Reference*: V. S. Vladimirov, V. P. Mikhailov, A. A. Vasharin, et al. (1974).

3.4.3. Equations Containing Arbitrary Functions

1. $\dfrac{\partial w}{\partial t} = a \sum_{k=1}^{n} \dfrac{\partial^2 w}{\partial x_k^2} + \left[\sum_{k=1}^{n} x_k f_k(t) + g(t) \right] w.$

The transformation

$$w(x_1, \ldots, x_n, t) = \exp\left[\sum_{k=1}^{n} x_k F_k(t) + a \sum_{k=1}^{n} \int F_k^2(t) \, dt + G(t) \right] u(\xi_1, \ldots, \xi_n, t),$$

$$\xi_k = x_k + 2a \int F_k(t) \, dt, \qquad F_k(t) = \int f_k(t) \, dt, \qquad G(t) = \int g(t) \, dt,$$

leads to the n-dimensional heat equation $\partial_t u = a \sum_{k=1}^{n} \partial_{\xi_k \xi_k} u$ that is discussed in Subsection 3.4.1.

2. $\dfrac{\partial w}{\partial t} = a \sum_{k=1}^{n} \dfrac{\partial^2 w}{\partial x_k^2} + \left[-b \sum_{k=1}^{n} x_k^2 + \sum_{k=1}^{n} x_k f_k(t) + g(t) \right] w.$

$1°$. Case $b > 0$. The transformation

$$w(x_1, \ldots, x_n, t) = u(\xi_1, \ldots, \xi_n, \tau) \exp\left(\dfrac{1}{2} \sqrt{\dfrac{b}{a}} \sum_{k=1}^{n} x_k^2 \right),$$

$$\xi_1 = x_1 \exp(2\sqrt{ab}\, t), \quad \ldots, \quad \xi_n = x_n \exp(2\sqrt{ab}\, t), \quad \tau = \dfrac{1}{4\sqrt{ab}} \exp(4\sqrt{ab}\, t) + C,$$

where C is an arbitrary constant, leads to an equation of the form 3.4.3.1:

$$\dfrac{\partial u}{\partial t} = a \sum_{k=1}^{n} \dfrac{\partial^2 u}{\partial \xi_k^2} + \left[\sum_{k=1}^{n} \xi_k F_k(\tau) + G(\tau) \right] u,$$

$$F_k(\tau) = \dfrac{1}{(s\tau)^{3/2}} f_k\left(\dfrac{\ln(s\tau)}{s} \right), \qquad G(\tau) = \dfrac{1}{s\tau} g\left(\dfrac{\ln(s\tau)}{s} \right) + \dfrac{n}{4\tau}, \qquad s = 4\sqrt{ab}.$$

$2°$. Case $b < 0$. The transformation

$$w(x_1, \ldots, x_n, t) = v(z_1, \ldots, z_n, \tau) \exp\left[\dfrac{\sqrt{-b}}{2\sqrt{a}} \tan(2\sqrt{-ab}\, t) \sum_{k=1}^{n} x_k^2 \right],$$

$$z_1 = \dfrac{x_1}{\cos(2\sqrt{-ab}\, t)}, \quad \ldots, \quad z_n = \dfrac{x_n}{\cos(2\sqrt{-ab}\, t)}, \quad \tau = \dfrac{1}{2\sqrt{-ab}} \tan(2\sqrt{-ab}\, t)$$

also leads to an equation of the form 3.4.3.1 (this equation is not specified here).

3. $\dfrac{\partial w}{\partial t} = \sum_{k=1}^{n} a_k(t) \dfrac{\partial^2 w}{\partial x_k^2} + \Phi(x_1, \ldots, x_n, t).$

The solutions of various problems for this equation can be constructed on the basis of incomplete separation of variables (see Paragraphs 0.6.1-2 and 0.9.2-1) taking into account the results of Subsections 1.1.1 and 1.1.2. Some examples of solving such problems are given below. It is assumed that $0 < a_k(t) < \infty$, $k = 1, \ldots, n$.

1°. Domain: $\mathbb{R}^n = \{-\infty < x_k < \infty;\ k = 1, \ldots, n\}$. Cauchy problem.
An initial condition is prescribed:
$$w = f(\mathbf{x}) \quad \text{at} \quad t = 0.$$

Solution:
$$w(\mathbf{x}, t) = \int_0^t \int_{\mathbb{R}^n} \Phi(\mathbf{y}, \tau) G(\mathbf{x}, \mathbf{y}, t, \tau)\, d\mathbf{y}\, d\tau + \int_{\mathbb{R}^n} f(\mathbf{y}) G(\mathbf{x}, \mathbf{y}, t, 0)\, d\mathbf{y},$$

where
$$G(\mathbf{x}, \mathbf{y}, t, \tau) = \dfrac{1}{2^n \pi^{n/2} \sqrt{T_1 T_2 \ldots T_n}} \exp\left[-\sum_{k=1}^{n} \dfrac{(x_k - y_k)^2}{4 T_k}\right], \quad T_k = \int_\tau^t a_k(\eta)\, d\eta,$$
$$\mathbf{x} = \{x_1, \ldots, x_n\}, \quad \mathbf{y} = \{y_1, \ldots, y_n\}, \quad d\mathbf{y} = dy_1\, dy_2 \ldots dy_n.$$

2°. Domain: $V = \{0 \leq x_k \leq l_k;\ k = 1, \ldots, n\}$. First boundary value problem.
The following conditions are prescribed:
$$\begin{aligned} w &= f(\mathbf{x}) & \text{at} & \quad t = 0 & \text{(initial condition)}, \\ w &= g_k(\mathbf{x}, t) & \text{at} & \quad x_k = 0 & \text{(boundary conditions)}, \\ w &= h_k(\mathbf{x}, t) & \text{at} & \quad x_k = l_k & \text{(boundary conditions)}. \end{aligned}$$

Solution:
$$\begin{aligned} w(\mathbf{x}, t) &= \int_0^t \int_V \Phi(\mathbf{y}, \tau) G(\mathbf{x}, \mathbf{y}, t, \tau)\, d\mathbf{y}\, d\tau + \int_V f(\mathbf{y}) G(\mathbf{x}, \mathbf{y}, t)\, d\mathbf{y} \\ &+ \sum_{k=1}^{n} \int_0^t \int_{S^{(k)}} a_k(\tau) \left[g_k(\mathbf{y}, \tau) \dfrac{\partial}{\partial y_k} G(\mathbf{x}, \mathbf{y}, t, \tau) \right]_{y_k = 0} dS_y^{(k)}\, d\tau \\ &- \sum_{k=1}^{n} \int_0^t \int_{S^{(k)}} a_k(\tau) \left[h_k(\mathbf{y}, \tau) \dfrac{\partial}{\partial y_k} G(\mathbf{x}, \mathbf{y}, t, \tau) \right]_{y_k = l_k} dS_y^{(k)}\, d\tau, \end{aligned}$$

where the following notation is used:
$$dS_y^{(k)} = dy_1 \ldots dy_{k-1}\, dy_{k+1} \ldots dy_n, \quad S^{(k)} = \{0 \leq y_m \leq l_m\ \text{for}\ m = 1, \ldots, k-1, k+1, \ldots, n\}.$$

The Green's function can be represented in the product form
$$G(\mathbf{x}, \mathbf{y}, t, \tau) = \prod_{k=1}^{n} G_k(x_k, y_k, t, \tau), \tag{1}$$

where the $G_k(x_k, y_k, t, \tau)$ are the Green's functions of the respective boundary value problems,
$$G_k(x_k, y_k, t, \tau) = \dfrac{2}{l_k} \sum_{m=1}^{\infty} \sin\left(\dfrac{m\pi x_k}{l_k}\right) \sin\left(\dfrac{m\pi y_k}{l_k}\right) \exp\left(-\dfrac{m^2 \pi^2 T_k}{l_k^2}\right), \quad T_k = \int_\tau^t a_k(\sigma)\, d\sigma. \tag{2}$$

3°. Domain: $V = \{0 \leq x_k \leq l_k;\ k = 1, \ldots, n\}$. Second boundary value problem.

The following conditions are prescribed:

$$w = f(\mathbf{x}) \quad \text{at} \quad t=0 \quad \text{(initial condition)},$$
$$\partial_{x_k} w = g_k(\mathbf{x},t) \quad \text{at} \quad x_k=0 \quad \text{(boundary conditions)},$$
$$\partial_{x_k} w = h_k(\mathbf{x},t) \quad \text{at} \quad x_k=l_k \quad \text{(boundary conditions)}.$$

Solution:

$$w(\mathbf{x},t) = \int_0^t \int_V \Phi(\mathbf{y},\tau) G(\mathbf{x},\mathbf{y},t,\tau)\, d\mathbf{y}\, d\tau + \int_V f(\mathbf{y}) G(\mathbf{x},\mathbf{y},t)\, d\mathbf{y}$$
$$- \sum_{k=1}^n \int_0^t \int_{S^{(k)}} a_k(\tau) \big[g_k(\mathbf{y},\tau) G(\mathbf{x},\mathbf{y},t,\tau)\big]_{y_k=0}\, dS_y^{(k)}\, d\tau$$
$$+ \sum_{k=1}^n \int_0^t \int_{S^{(k)}} a_k(\tau) \big[h_k(\mathbf{y},\tau) G(\mathbf{x},\mathbf{y},t,\tau)\big]_{y_k=l_k}\, dS_y^{(k)}\, d\tau.$$

The Green's function can be represented as the product (1) of the corresponding one-dimensional Green's functions

$$G_k(x_k,y_k,t,\tau) = \frac{1}{l_k} + \frac{2}{l_k} \sum_{m=1}^\infty \cos\!\left(\frac{m\pi x_k}{l_k}\right) \cos\!\left(\frac{m\pi y_k}{l_k}\right) \exp\!\left(-\frac{m^2\pi^2 T_k}{l_k^2}\right), \quad T_k = \int_\tau^t a_k(\sigma)\,d\sigma.$$

⊙ *Reference*: A. D. Polyanin (2000a, 2000b).

4. $\dfrac{\partial w}{\partial t} = \displaystyle\sum_{k=1}^n a_k(t)\dfrac{\partial^2 w}{\partial x_k^2} + \sum_{k=1}^n [b_k(t)x_k + c_k(t)]\dfrac{\partial w}{\partial x_k} + \left[\sum_{k=1}^n s_k(t)x_k + p(t)\right] w.$

Let us perform the transformation

$$w(x_1,\ldots,x_n,t) = \exp\!\left[\sum_{k=1}^n f_k(t)x_k + g(t)\right] u(z_1,\ldots,z_n,t), \quad z_k = h_k(t)x_k + r_k(t),$$

where the functions $f_k(t)$, $g(t)$, $h_k(t)$, and $r_k(t)$ are given by (A_k, B_k, C_k, and D are arbitrary constants):

$$h_k(t) = A_k \exp\!\left[\int b_k(t)\,dt\right],$$
$$f_k(t) = h_k(t) \int \frac{s_k(t)}{h_k(t)}\,dt + B_k h_k(t),$$
$$r_k(t) = \int \big[2a_k(t) f_k(t) + c_k(t)\big] h_k(t)\,dt + C_k,$$
$$g(t) = \int \left[p(t) + \sum_{k=1}^n a_k(t) f_k^2(t) + \sum_{k=1}^n c_k(t) f_k(t)\right] dt + D.$$

As a result, we arrive at an equation of the form 3.4.3.3 for the new dependent variable $u = u(z_1,\ldots,z_n,t)$:

$$\frac{\partial u}{\partial t} = \sum_{k=1}^n a_k(t) h_k^2(t) \frac{\partial^2 u}{\partial z_k^2}.$$

5. $\dfrac{\partial w}{\partial t} = \displaystyle\sum_{k=1}^n a_k(t)\dfrac{\partial^2 w}{\partial x_k^2} + \sum_{k=1}^n [b_k(t)x_k + c_k(t)]\dfrac{\partial w}{\partial x_k} + \left[\sum_{k=1}^n s_k(t)x_k^2 + \sum_{k=1}^n p_k(t)x_k + q(t)\right] w.$

The substitution

$$w(x_1,\ldots,x_n,t) = \exp\!\left[\sum_{k=1}^n f_k(t)x_k^2\right] u(x_1,\ldots,x_n,t),$$

where the functions $f_k = f_k(t)$ are solutions of the Riccati equation

$$f_k' = 4a_k(t) f_k^2 + 2b_k(t) f_k + s_k(t) \quad (k=1,\ldots,n),$$

leads to an equation of the form 3.4.3.4 for $u = u(x_1,\ldots,x_n,t)$.

6. $\dfrac{\partial w}{\partial t} - \sum_{k=1}^{n}\left[a_k(x_k,t)\dfrac{\partial^2 w}{\partial x_k^2} + b_k(x_k,t)\dfrac{\partial w}{\partial x_k} + c_k(x_k,t)w\right] = \Phi(x_1,\ldots,x_n,t).$

Here, $0 < a_k(x_k,t) < \infty$ for all k. We introduce the notation $\mathbf{x} = \{x_1,\ldots,x_n\}$, $\mathbf{y} = \{y_1,\ldots,y_n\}$ and consider the domain $V = \{\alpha_k \le x_k \le \beta_k,\ k = 1,\ldots,n\}$, which is an n-dimensional parallelepiped.

1°. **First boundary value problem.** The following conditions are prescribed:

$$w = f(\mathbf{x}) \quad \text{at} \quad t = 0 \qquad \text{(initial condition)},$$
$$w = g_k(\mathbf{x},t) \quad \text{at} \quad x_k = \alpha_k \qquad \text{(boundary conditions)},$$
$$w = h_k(\mathbf{x},t) \quad \text{at} \quad x_k = \beta_k \qquad \text{(boundary conditions)}.$$

Solution:

$$w(\mathbf{x},t) = \int_0^t \int_V \Phi(\mathbf{y},\tau) G(\mathbf{x},\mathbf{y},t,\tau)\,d\mathbf{y}\,d\tau + \int_V f(\mathbf{y}) G(\mathbf{x},\mathbf{y},t,0)\,d\mathbf{y}$$
$$+ \sum_{k=1}^n \int_0^t \int_{S^{(k)}} a_k(\alpha_k,\tau)\left[g_k(\mathbf{y},\tau)\dfrac{\partial}{\partial y_k} G(\mathbf{x},\mathbf{y},t,\tau)\right]_{y_k=\alpha_k} dS_y^{(k)}\,d\tau$$
$$- \sum_{k=1}^n \int_0^t \int_{S^{(k)}} a_k(\beta_k,\tau)\left[h_k(\mathbf{y},\tau)\dfrac{\partial}{\partial y_k} G(\mathbf{x},\mathbf{y},t,\tau)\right]_{y_k=\beta_k} dS_y^{(k)}\,d\tau,$$

where

$$d\mathbf{y} = dy_1\,dy_2\ldots dy_n, \qquad dS_y^{(k)} = dy_1\ldots dy_{k-1}\,dy_{k+1}\ldots dy_n,$$
$$S^{(k)} = \{\alpha_m \le y_m \le \beta_m \text{ for } m = 1,\ldots,k-1,k+1,\ldots,n\}.$$

The Green's function can be represented in the product form

$$G(\mathbf{x},\mathbf{y},t,\tau) = \prod_{k=1}^n G_k(x_k,y_k,t,\tau). \qquad (1)$$

Here, the $G_k = G_k(x_k,y_k,t,\tau)$ are auxiliary Green's functions that, for $t > \tau \ge 0$, satisfy the one-dimensional linear homogeneous equations

$$\dfrac{\partial G_k}{\partial t} - a_k(x_k,t)\dfrac{\partial^2 G_k}{\partial x_k^2} - b_k(x_k,t)\dfrac{\partial G_k}{\partial x_k} - c_k(x_k,t)G_k = 0 \qquad (k=1,\ldots,n) \qquad (2)$$

with nonhomogeneous initial conditions of a special form,

$$G_k = \delta(x_k - y_k) \quad \text{at} \quad t = \tau, \qquad (3)$$

and homogeneous boundary conditions of the first kind,

$$G_k = 0 \quad \text{at} \quad x_k = \alpha_k,$$
$$G_k = 0 \quad \text{at} \quad x_k = \beta_k.$$

In determining the function G_k, the quantities y_k and τ play the role of parameters; $\delta(x)$ is the Dirac delta function.

2°. **The second and third boundary value problems.** The following conditions are prescribed:

$$w = f(\mathbf{x}) \quad \text{at} \quad t = 0 \qquad \text{(initial condition)},$$
$$\partial_{x_k} w - s_k w = g_k(\mathbf{x},t) \quad \text{at} \quad x_k = \alpha_k \qquad \text{(boundary conditions)},$$
$$\partial_{x_k} w + p_k w = h_k(\mathbf{x},t) \quad \text{at} \quad x_k = \beta_k \qquad \text{(boundary conditions)}.$$

The second boundary value problem corresponds to $s_k = p_k = 0$.

Solution:

$$w(\mathbf{x},t) = \int_0^t \int_V \Phi(\mathbf{y},\tau) G(\mathbf{x},\mathbf{y},t,\tau)\, d\mathbf{y}\, d\tau + \int_V f(\mathbf{y}) G(\mathbf{x},\mathbf{y},t,0)\, d\mathbf{y}$$
$$- \sum_{k=1}^n \int_0^t \int_{S^{(k)}} a_k(\alpha_k,\tau) \big[g_k(\mathbf{y},\tau) G(\mathbf{x},\mathbf{y},t,\tau)\big]_{y_k=\alpha_k} dS_y^{(k)}\, d\tau$$
$$+ \sum_{k=1}^n \int_0^t \int_{S^{(k)}} a_k(\beta_k,\tau) \big[h_k(\mathbf{y},\tau) G(\mathbf{x},\mathbf{y},t,\tau)\big]_{y_k=\beta_k} dS_y^{(k)}\, d\tau.$$

The Green's function can be represented as the product (1) of the corresponding one-dimensional Green's functions satisfying the linear equations (2) with the initial conditions (3) and the homogeneous boundary conditions

$$\partial_{x_k} G_k - s_k G_k = 0 \quad \text{at} \quad x_k = \alpha_k,$$
$$\partial_{x_k} G_k + p_k G_k = 0 \quad \text{at} \quad x_k = \beta_k.$$

⊙ *Reference*: A. D. Polyanin (2000a, 2000b).

7. $$\frac{\partial w}{\partial t} = \sum_{i,j=1}^n \frac{\partial}{\partial x_i}\left[a_{ij}(x_1,\ldots,x_n)\frac{\partial w}{\partial x_j}\right] - q(x_1,\ldots,x_n)w + \Phi(x_1,\ldots,x_n,t).$$

The problems considered below are assume to refer to a bounded domain V with smooth surface S. We introduce the brief notation $\mathbf{x} = \{x_1,\ldots,x_n\}$ and assume that the condition

$$\sum_{i,j=1}^n a_{ij}(\mathbf{x}) \lambda_i \lambda_j \geq c \sum_{i=1}^n \lambda_i^2, \quad c > 0,$$

is satisfied; this condition imposes the requirement that the differential operator on the right-hand side of the equation is elliptic.

1°. First boundary value problem. The following conditions are prescribed:

$$w = f(\mathbf{x}) \quad \text{at} \quad t = 0 \quad \text{(initial condition)},$$
$$w = g(\mathbf{x},t) \quad \text{for} \quad \mathbf{x} \in S \quad \text{(boundary condition)}.$$

Solution:

$$w(\mathbf{x},t) = \int_0^t \int_V \Phi(\mathbf{y},\tau) G(\mathbf{x},\mathbf{y},t-\tau)\, dV_y\, d\tau + \int_V f(\mathbf{y}) G(\mathbf{x},\mathbf{y},t)\, dV_y$$
$$- \int_0^t \int_S g(\mathbf{y},\tau) \left[\frac{\partial}{\partial M_y} G(\mathbf{x},\mathbf{y},t-\tau)\right] dS_y\, d\tau. \qquad (1)$$

Here, the Green's function is given by

$$G(\mathbf{x},\mathbf{y},t) = \sum_{n=1}^\infty \frac{u_n(\mathbf{x}) u_n(\mathbf{y})}{\|u_n\|^2} \exp(-\lambda_n t), \quad \|u_n\|^2 = \int_V u_n^2(\mathbf{x})\, dV, \quad \mathbf{y} = \{y_1,\ldots,y_n\}, \qquad (2)$$

where the λ_n and $u_n(\mathbf{x})$ are the eigenvalues and corresponding eigenfunctions of the Sturm–Liouville problem for the following elliptic second-order equation with homogeneous boundary condition of the first kind:

$$\sum_{i,j=1}^n \frac{\partial}{\partial x_i}\left[a_{ij}(\mathbf{x}) \frac{\partial u}{\partial x_j}\right] - q(\mathbf{x}) u + \lambda u = 0, \qquad (3)$$
$$u = 0 \quad \text{for} \quad \mathbf{x} \in S. \qquad (4)$$

The integration in solution (1) is carried out with respect to $y_1, \ldots, y_n$; $\frac{\partial}{\partial M_y}$ is the differential operator defined as

$$\frac{\partial G}{\partial M_y} \equiv \sum_{i,j=1}^{n} a_{ij}(\mathbf{y}) N_j \frac{\partial G}{\partial y_i}, \tag{5}$$

where $\mathbf{N} = \{N_1, \ldots, N_n\}$ is the unit outward normal to the surface S. In the special case where $a_{ii}(\mathbf{x}) = 1$ and $a_{ij}(\mathbf{x}) = 0$ for $i \neq j$, the operator of (5) coincides with the usual operator of differentiation along the direction of the outward normal to the surface S.

General properties of the Sturm–Liouville problem (3)–(4):

1. There are countably many eigenvalues. All eigenvalues are real and can be ordered so that $\lambda_1 \leq \lambda_2 \leq \lambda_3 \leq \cdots$, with $\lambda_n \to \infty$ as $n \to \infty$; consequently, there can exist only finitely many negative eigenvalues.

2. For $q(\mathbf{x}) \geq 0$ all eigenvalues are positive: $\lambda_n > 0$.

3. The eigenfunctions are defined up to a constant multiplier. Any two eigenfunctions $u_n(\mathbf{x})$ and $u_m(\mathbf{x})$ corresponding to different eigenvalues λ_n and λ_m are orthogonal in the domain V:

$$\int_V u_n(\mathbf{x}) u_m(\mathbf{x}) \, dV = 0 \quad \text{for} \quad n \neq m.$$

Remark. To each eigenvalue λ_n there generally correspond finitely many linearly independent eigenfunctions $u_n^{(1)}, u_n^{(2)}, \ldots, u_n^{(m)}$. These functions can always be replaced by their linear combinations

$$\bar{u}_n^{(k)} = A_{k,1} u_n^{(1)} + \cdots + A_{k,k-1} u_n^{(k-1)} + u_n^{(k)}, \quad k = 1, 2, \ldots, m,$$

such that $\bar{u}_n^{(1)}, \bar{u}_n^{(2)}, \ldots, \bar{u}_n^{(m)}$ are now pairwise orthogonal. Thus, without loss of generality, we assume that all eigenfunctions are orthogonal.

2°. *Second boundary value problem.* The following conditions are prescribed:

$$w = f(\mathbf{x}) \quad \text{at} \quad t = 0 \quad \text{(initial condition)},$$

$$\frac{\partial w}{\partial M_x} = g(\mathbf{x}, t) \quad \text{for} \quad \mathbf{x} \in S \quad \text{(boundary condition)}.$$

Here, the left-hand side of the boundary condition is determined with the help of (5), where G, y, $\mathbf{y}$, and y_k must be replaced by w, x, $\mathbf{x}$, and x_k, respectively.

Solution:

$$w(\mathbf{x}, t) = \int_0^t \int_V \Phi(\mathbf{y}, \tau) G(\mathbf{x}, \mathbf{y}, t-\tau) \, dV_y \, d\tau + \int_V f(\mathbf{y}) G(\mathbf{x}, \mathbf{y}, t) \, dV_y$$
$$+ \int_0^t \int_S g(\mathbf{y}, \tau) G(\mathbf{x}, \mathbf{y}, t-\tau) \, dS_y \, d\tau. \tag{6}$$

Here, the Green's function is defined by (2), where the λ_n and $u_n(\mathbf{x})$ are the eigenvalues and corresponding eigenfunctions of the Sturm–Liouville problem for the elliptic second-order equation (3) with a homogeneous boundary condition of the second kind:

$$\frac{\partial u}{\partial M_x} = 0 \quad \text{for} \quad \mathbf{x} \in S. \tag{7}$$

For $q(\mathbf{x}) > 0$ the general properties of the eigenvalue problem (3), (7) are the same as for the first boundary value problem (see Item 1°). For $q(\mathbf{x}) \equiv 0$ the zero eigenvalue $\lambda_0 = 0$ arises which corresponds to the eigenfunction $u_0 = \text{const}$.

It should be noted that the Green's function of the second boundary value problem can be expressed in terms of the Green's function of the third boundary value problem (see Item 3°).

3°. *Third boundary value problem.* The following conditions are prescribed:

$$w = f(\mathbf{x}) \quad \text{at} \quad t = 0 \quad \text{(initial condition)},$$

$$\frac{\partial w}{\partial M_x} + k(\mathbf{x})w = g(\mathbf{x}, t) \quad \text{for} \quad \mathbf{x} \in S \quad \text{(boundary condition)}.$$

The solution of the third boundary value problem is given by relations (6) and (2), where the λ_n and $u_n(\mathbf{x})$ are the eigenvalues and corresponding eigenfunctions of the Sturm–Liouville problem for the second-order elliptic equation (3) with a homogeneous boundary condition of the third kind:

$$\frac{\partial u}{\partial M_x} + k(\mathbf{x})u = 0 \quad \text{for} \quad \mathbf{x} \in S. \tag{8}$$

For $q(\mathbf{x}) \geq 0$ and $k(\mathbf{x}) > 0$, the general properties of the eigenvalue problem (3), (8) are the same as for the first boundary value problem (see Item 1°).

Let $k(\mathbf{x}) = k = \text{const}$. Denote the Green's functions of the second and third boundary value problems by $G_2(\mathbf{x}, \mathbf{y}, t)$ and $G_3(\mathbf{x}, \mathbf{y}, t, k)$, respectively. Then the following relations hold:

$$G_2(\mathbf{x}, \mathbf{y}, t) = \begin{cases} \lim_{k \to 0} G_3(\mathbf{x}, \mathbf{y}, t, k), & \text{if } q(\mathbf{x}) > 0; \\ \dfrac{1}{V_0} + \lim_{k \to 0} G_3(\mathbf{x}, \mathbf{y}, t, k), & \text{if } q(\mathbf{x}) \equiv 0; \end{cases}$$

where $V_0 = \int_V dV$ is the volume of the domain in question.

⊙ *References*: V. M. Babich, M. B. Kapilevich, S. G. Mikhlin, et al. (1964), A. D. Polyanin (2000a, 2000b).

Chapter 4

Hyperbolic Equations with One Space Variable

4.1. Constant Coefficient Equations

4.1.1. Wave Equation $\dfrac{\partial^2 w}{\partial t^2} = a^2 \dfrac{\partial^2 w}{\partial x^2}$

This equation is also known as the *equation of vibration of a string*. It is often encountered in elasticity, aerodynamics, acoustics, and electrodynamics.

> 4.1.1-1. General solution. Some formulas.

1°. General solution:
$$w(x,t) = \varphi(x+at) + \psi(x-at),$$
where $\varphi(x)$ and $\psi(x)$ are arbitrary functions. *Physical interpretation*: The solution represents two traveling waves that propagate, respectively, to the left and right along the x-axis at a constant speed a.

2°. Fundamental solution:
$$\mathcal{E}(x,t) = \frac{1}{2a}\vartheta\bigl(at-|x|\bigr), \qquad \vartheta(z) = \begin{cases} 0 & \text{for } z < 0, \\ 1 & \text{for } z > 0. \end{cases}$$

3°. Infinite series solutions containing arbitrary functions of the space variable:
$$w(x,t) = f(x) + \sum_{n=1}^{\infty} \frac{(at)^{2n}}{(2n)!} f_x^{(2n)}(x), \qquad f_x^{(m)}(x) = \frac{d^m}{dx^m} f(x),$$
$$w(x,t) = tg(x) + t\sum_{n=1}^{\infty} \frac{(at)^{2n}}{(2n+1)!} g_x^{(2n)}(x),$$
where $f(x)$ and $g(x)$ are any infinitely differentiable functions. The first solution satisfies the initial conditions $w(x,0) = f(x)$ and $\partial_t w(x,0) = 0$, and the second $w(x,0) = 0$ and $\partial_t w(x,0) = g(x)$. The sums are finite if $f(x)$ and $g(x)$ are polynomials.

4°. Infinite series solutions containing arbitrary functions of time.
$$w(x,t) = f(t) + \sum_{n=1}^{\infty} \frac{1}{a^{2n}(2n)!} x^{2n} f_t^{(2n)}(t), \qquad f_t^{(m)}(t) = \frac{d^m}{dt^m} f(t),$$
$$w(x,t) = xg(t) + x\sum_{n=1}^{\infty} \frac{1}{a^{2n}(2n+1)!} x^{2n} g_t^{(2n)}(t),$$
where $f(t)$ and $g(t)$ are any infinitely differentiable functions. The sums are finite if $f(t)$ and $g(t)$ are polynomials. The first solution satisfies the boundary condition of the first kind $w(0,t) = f(t)$, and the second solution to the boundary condition of the second kind $\partial_x w(0,t) = g(t)$.

5°. If $w(x, t)$ is a solution of the wave equation, then the functions

$$w_1 = Aw(\pm\lambda x + C_1, \pm\lambda t + C_2),$$

$$w_2 = Aw\left(\frac{x - vt}{\sqrt{1 - (v/a)^2}}, \frac{t - va^{-2}x}{\sqrt{1 - (v/a)^2}}\right),$$

$$w_3 = Aw\left(\frac{x}{x^2 - a^2t^2}, \frac{t}{x^2 - a^2t^2}\right),$$

are also solutions of the equation everywhere these functions are defined (A, C_1, C_2, v, and λ are arbitrary constants). The signs at λ's in the formula for w_1 are taken arbitrarily, independently of each other. The function w_2 results from the invariance of the wave equation under the Lorentz transformations.

⊙ *References*: G. N. Polozhii (1964), A. V. Bitsadze and D. F. Kalinichenko (1985).

4.1.1-2. Domain: $-\infty < x < \infty$. Cauchy problem.

Initial conditions are prescribed:

$$w = f(x) \quad \text{at} \quad t = 0,$$
$$\partial_t w = g(x) \quad \text{at} \quad t = 0.$$

Solution (D'Alembert's formula):

$$w(x, t) = \frac{1}{2}[f(x + at) + f(x - at)] + \frac{1}{2a}\int_{x-at}^{x+at} g(\xi)\,d\xi.$$

4.1.1-3. Domain: $0 \leq x < \infty$. First boundary value problem.

1°. Problem with a homogeneous boundary condition:

$$w = f(x) \quad \text{at} \quad t = 0 \quad \text{(initial condition)},$$
$$\partial_t w = g(x) \quad \text{at} \quad t = 0 \quad \text{(initial condition)},$$
$$w = 0 \quad \text{at} \quad x = 0 \quad \text{(boundary condition)}.$$

Solution:

$$w(x, t) = \begin{cases} \dfrac{1}{2}[f(x + at) + f(x - at)] + \dfrac{1}{2a}\displaystyle\int_{x-at}^{x+at} g(\xi)\,d\xi & \text{for } t < \dfrac{x}{a}, \\ \dfrac{1}{2}[f(x + at) - f(at - x)] + \dfrac{1}{2a}\displaystyle\int_{at-x}^{x+at} g(\xi)\,d\xi & \text{for } t > \dfrac{x}{a}. \end{cases}$$

2°. Problem with a nonhomogeneous boundary condition:

$$w = f(x) \quad \text{at} \quad t = 0 \quad \text{(initial condition)},$$
$$\partial_t w = g(x) \quad \text{at} \quad t = 0 \quad \text{(initial condition)},$$
$$w = h(t) \quad \text{at} \quad x = 0 \quad \text{(boundary condition)}.$$

Solution:

$$w(x, t) = \begin{cases} \dfrac{1}{2}[f(x + at) + f(x - at)] + \dfrac{1}{2a}\displaystyle\int_{x-at}^{x+at} g(\xi)\,d\xi & \text{for } t < \dfrac{x}{a}, \\ \dfrac{1}{2}[f(x + at) - f(at - x)] + \dfrac{1}{2a}\displaystyle\int_{at-x}^{x+at} g(\xi)\,d\xi + h\!\left(t - \dfrac{x}{a}\right) & \text{for } t > \dfrac{x}{a}. \end{cases}$$

In the domain $t < x/a$ the boundary conditions have no effect on the solution and the expression of $w(x, t)$ coincides with D'Alembert's solution for an infinite line (see Paragraph 4.1.1-2).

⊙ *Reference*: A. N. Tikhonov and A. A. Samarskii (1990).

4.1.1-4. Domain: $0 \leq x < \infty$. Second boundary value problem.

1°. Problem with a homogeneous boundary condition:

$$w = f(x) \quad \text{at} \quad t = 0 \quad \text{(initial condition)},$$
$$\partial_t w = g(x) \quad \text{at} \quad t = 0 \quad \text{(initial condition)},$$
$$\partial_x w = 0 \quad \text{at} \quad x = 0 \quad \text{(boundary condition)}.$$

Solution:

$$w(x,t) = \begin{cases} \dfrac{1}{2}[f(x+at)+f(x-at)] + \dfrac{1}{2a}[G(x+at)-G(x-at)] & \text{for } t < \dfrac{x}{a}, \\ \dfrac{1}{2}[f(x+at)+f(at-x)] + \dfrac{1}{2a}[G(x+at)+G(at-x)] & \text{for } t > \dfrac{x}{a}, \end{cases}$$

where $G(z) = \displaystyle\int_0^z g(\xi)\,d\xi$.

2°. Problem with a nonhomogeneous boundary condition:

$$w = f(x) \quad \text{at} \quad t = 0 \quad \text{(initial condition)},$$
$$\partial_t w = g(x) \quad \text{at} \quad t = 0 \quad \text{(initial condition)},$$
$$\partial_x w = h(t) \quad \text{at} \quad x = 0 \quad \text{(boundary condition)}.$$

Solution:

$$w(x,t) = \begin{cases} \dfrac{1}{2}[f(x+at)+f(x-at)] + \dfrac{1}{2a}[G(x+at)-G(x-at)] & \text{for } t < \dfrac{x}{a}, \\ \dfrac{1}{2}[f(x+at)+f(at-x)] + \dfrac{1}{2a}[G(x+at)+G(at-x)] - aH\left(t-\dfrac{x}{a}\right) & \text{for } t > \dfrac{x}{a}, \end{cases}$$

where $G(z) = \displaystyle\int_0^z g(\xi)\,d\xi$ and $H(z) = \displaystyle\int_0^z h(\xi)\,d\xi$. In the domain $t < x/a$ the boundary conditions have no effect on the solution, and the expression of $w(x,t)$ coincides with D'Alembert's solution for an infinite line (see Paragraph 4.1.1-2).

⊙ *Reference*: B. M. Budak, A. N. Tikhonov, and A. A. Samarskii (1980).

4.1.1-5. Domain: $0 \leq x \leq l$. First boundary value problem.

1°. Vibration of a string with rigidly fixed ends. The following conditions are prescribed:

$$w = f(x) \quad \text{at} \quad t = 0 \quad \text{(initial condition)},$$
$$\partial_t w = g(x) \quad \text{at} \quad t = 0 \quad \text{(initial condition)},$$
$$w = 0 \quad \text{at} \quad x = 0 \quad \text{(boundary condition)},$$
$$w = 0 \quad \text{at} \quad x = l \quad \text{(boundary condition)}.$$

Solution:

$$w(x,t) = \sum_{n=1}^{\infty} \left[A_n \cos(\lambda_n a t) + B_n \sin(\lambda_n a t)\right] \sin(\lambda_n x), \qquad \lambda_n = \frac{\pi n}{l},$$

$$A_n = \frac{2}{l}\int_0^l f(x)\sin(\lambda_n x)\,dx, \qquad B_n = \frac{2}{a\pi n}\int_0^l g(x)\sin(\lambda_n x)\,dx.$$

Example 1. The initial shape of the string is a triangle with base $0 \leq x \leq l$ and height h at $x = c$, i.e.,

$$f(x) = \begin{cases} \dfrac{hx}{c} & \text{for } 0 \leq x \leq c, \\ \dfrac{h(l-x)}{l-c} & \text{for } c \leq x \leq l. \end{cases}$$

The initial velocities of the string points are zero, $g(x) = 0$.
Solution:
$$w(x,t) = \frac{2hl^2}{\pi^2 c(l-c)} \sum_{n=1}^{\infty} \frac{1}{n^2} \sin\left(\frac{n\pi c}{l}\right) \sin\left(\frac{n\pi x}{l}\right) \cos\left(\frac{n\pi a t}{l}\right).$$

Example 2. Initially, the string has the shape of a parabola symmetric about the center of the string with elevation h, so that
$$f(x) = \frac{4h}{l^2} x(l-x).$$

The initial velocities of the string points are zero, $g(x) = 0$.
Solution:
$$w(x,t) = \frac{32h}{\pi^3} \sum_{n=0}^{\infty} \frac{1}{(2n+1)^3} \sin\left[\frac{(2n+1)\pi x}{l}\right] \cos\left[\frac{(2n+1)\pi a t}{l}\right].$$

2°. For the solution of the first boundary value problem with a nonhomogeneous boundary condition, see Paragraph 4.1.2-4 with $\Phi(x,t) \equiv 0$.

⊙ *References*: B. M. Budak, A. N. Tikhonov, and A. A. Samarskii (1980), A. V. Bitsadze and D. F. Kalinichenko (1985).

4.1.1-6. Domain: $0 \le x \le l$. Second boundary value problem.

1°. Longitudinal vibration of an elastic rod with free ends. The following conditions are prescribed:

$$w = f(x) \quad \text{at} \quad t = 0 \quad \text{(initial condition)},$$
$$\partial_t w = g(x) \quad \text{at} \quad t = 0 \quad \text{(initial condition)},$$
$$\partial_x w = 0 \quad \text{at} \quad x = 0 \quad \text{(boundary condition)},$$
$$\partial_x w = 0 \quad \text{at} \quad x = l \quad \text{(boundary condition)}.$$

Solution:
$$w(x,t) = A_0 + B_0 t + \sum_{n=1}^{\infty} \left[A_n \cos(\lambda_n a t) + B_n \sin(\lambda_n a t)\right] \cos(\lambda_n x),$$

$$\lambda_n = \frac{\pi n}{l}, \qquad A_0 = \frac{1}{l} \int_0^l f(x)\,dx, \qquad B_0 = \frac{1}{l} \int_0^l g(x)\,dx,$$

$$A_n = \frac{2}{l} \int_0^l f(x) \cos(\lambda_n x)\,dx, \qquad B_n = \frac{2}{a\pi n} \int_0^l g(x) \cos(\lambda_n x)\,dx.$$

2°. For the solution of the second boundary value problem with a nonhomogeneous boundary condition, see Paragraph 4.1.2-5 with $\Phi(x,t) \equiv 0$.

⊙ *Reference*: A. V. Bitsadze and D. F. Kalinichenko (1985).

4.1.1-7. Domain: $0 \le x \le l$. Third boundary value problem.

1°. Longitudinal vibration of an elastic rod with clamped ends in the case of equal stiffness coefficients. The following conditions are prescribed:

$$w = f(x) \quad \text{at} \quad t = 0 \quad \text{(initial condition)},$$
$$\partial_t w = g(x) \quad \text{at} \quad t = 0 \quad \text{(initial condition)},$$
$$\partial_x w - kw = 0 \quad \text{at} \quad x = 0 \quad \text{(boundary condition)},$$
$$\partial_x w + kw = 0 \quad \text{at} \quad x = l \quad \text{(boundary condition)}.$$

Solution:
$$w(x,t) = \sum_{n=1}^{\infty} \left[A_n \cos(\lambda_n a t) + B_n \sin(\lambda_n a t)\right] \sin(\lambda_n x + \varphi_n),$$

where

$$A_n = \frac{1}{\|X_n\|^2} \int_0^l \sin(\lambda_n x + \varphi_n) f(x)\, dx, \quad B_n = \frac{1}{a\lambda_n \|X_n\|^2} \int_0^l \sin(\lambda_n x + \varphi_n) g(x)\, dx,$$

$$\varphi_n = \arctan \frac{\lambda_n}{k}, \quad \|X_n\|^2 = \int_0^l \sin^2(\lambda_n x + \varphi_n)\, dx = \frac{l}{2} + \frac{k}{k^2 + \lambda_n^2};$$

the λ_n are positive roots of the transcendental equation $\cot(\lambda l) = \dfrac{1}{2}\left(\dfrac{\lambda}{k} - \dfrac{k}{\lambda}\right)$.

2°. Longitudinal vibration of an elastic rod with clamped ends in the case of different stiffness coefficients. The following conditions are prescribed:

$$\begin{aligned}
w &= f(x) &\text{at}\quad t &= 0 &\text{(initial condition)}, \\
\partial_t w &= g(x) &\text{at}\quad t &= 0 &\text{(initial condition)}, \\
\partial_x w - k_1 w &= 0 &\text{at}\quad x &= 0 &\text{(boundary condition)}, \\
\partial_x w + k_2 w &= 0 &\text{at}\quad x &= l &\text{(boundary condition)}.
\end{aligned}$$

Solution:

$$w(x,t) = \sum_{n=1}^{\infty} \left[A_n \cos(\lambda_n a t) + B_n \sin(\lambda_n a t)\right] \sin(\lambda_n x + \varphi_n),$$

where

$$A_n = \frac{1}{\|X_n\|^2} \int_0^l \sin(\lambda_n x + \varphi_n) f(x)\, dx, \quad B_n = \frac{1}{a\lambda_n \|X_n\|^2} \int_0^l \sin(\lambda_n x + \varphi_n) g(x)\, dx,$$

$$\varphi_n = \arctan \frac{\lambda_n}{k_1}, \quad \|X_n\|^2 = \int_0^l \sin^2(\lambda_n x + \varphi_n)\, dx = \frac{l}{2} + \frac{(\lambda_n^2 + k_1 k_2)(k_1 + k_2)}{2(\lambda_n^2 + k_1^2)(\lambda_n^2 + k_2^2)};$$

the λ_n are positive roots of the transcendental equation $\cot(\lambda l) = \dfrac{\lambda^2 - k_1 k_2}{\lambda(k_1 + k_2)}$.

3°. For the solution of the third boundary value problem with nonhomogeneous boundary conditions, see Paragraph 4.1.2-6 with $\Phi(x,t) \equiv 0$.

⊙ *Reference*: B. M. Budak, A. N. Tikhonov, and A. A. Samarskii (1980).

4.1.1-8. Domain: $0 \leq x \leq l$. Mixed boundary value problem.

1°. Longitudinal vibration of an elastic rod with one end rigidly fixed and the other free. The following conditions are prescribed:

$$\begin{aligned}
w &= f(x) &\text{at}\quad t &= 0 &\text{(initial condition)}, \\
\partial_t w &= g(x) &\text{at}\quad t &= 0 &\text{(initial condition)}, \\
w &= 0 &\text{at}\quad x &= 0 &\text{(boundary condition)}, \\
\partial_x w &= 0 &\text{at}\quad x &= l &\text{(boundary condition)}.
\end{aligned}$$

Solution:

$$w(x,t) = \sum_{n=0}^{\infty} \left[A_n \cos(\lambda_n a t) + B_n \sin(\lambda_n a t)\right] \sin(\lambda_n x), \quad \lambda_n = \frac{\pi(2n+1)}{2l},$$

$$A_n = \frac{2}{l} \int_0^l f(x) \sin(\lambda_n x)\, dx, \quad B_n = \frac{2}{a l \lambda_n} \int_0^l g(x) \sin(\lambda_n x)\, dx.$$

2°. For the solution of the mixed boundary value problem with nonhomogeneous boundary conditions, see Paragraph 4.1.2-7 with $\Phi(x,t) \equiv 0$.

⊙ *References*: M. M. Smirnov (1975), A. V. Bitsadze and D. F. Kalinichenko (1985).

4.1.1-9. Goursat problem.

The boundary conditions are prescribed to the equation characteristics:
$$w = f(x) \quad \text{for} \quad x - at = 0 \quad (0 \le x \le b),$$
$$w = g(x) \quad \text{for} \quad x + at = 0 \quad (0 \le x \le c),$$

where $f(0) = g(0)$.

Solution:
$$w(x,t) = f\left(\frac{x+at}{2}\right) + g\left(\frac{x-at}{2}\right) - f(0).$$

The solution propagation domain is bounded by four lines:
$$x - at = 0, \quad x + at = 0, \quad x - at = 2c, \quad x + at = 2b.$$

⊙ *Reference*: A. V. Bitsadze and D. F. Kalinichenko (1985).

4.1.2. Equations of the Form $\dfrac{\partial^2 w}{\partial t^2} = a^2 \dfrac{\partial^2 w}{\partial x^2} + \Phi(x,t)$

4.1.2-1. Domain: $-\infty < x < \infty$. Cauchy problem.

Initial conditions are prescribed:
$$w = f(x) \quad \text{at} \quad t = 0,$$
$$\partial_t w = g(x) \quad \text{at} \quad t = 0.$$

Solution:
$$w(x,t) = \frac{1}{2}[f(x-at) + f(x+at)] + \frac{1}{2a}\int_{x-at}^{x+at} g(\xi)\, d\xi + \frac{1}{2a}\int_0^t \int_{x-a(t-\tau)}^{x+a(t-\tau)} \Phi(\xi,\tau)\, d\xi\, d\tau.$$

4.1.2-2. Domain: $0 \le x < \infty$. First boundary value problem.

The following conditions are prescribed:
$$w = f(x) \quad \text{at} \quad t = 0 \quad \text{(initial condition)},$$
$$\partial_t w = g(x) \quad \text{at} \quad t = 0 \quad \text{(initial condition)},$$
$$w = h(t) \quad \text{at} \quad x = 0 \quad \text{(boundary condition)}.$$

Solution:
$$w(x,t) = w_1(x,t) + \frac{1}{2a} w_2(x,t),$$

where

$$w_1(x,t) = \begin{cases} \dfrac{1}{2}[f(x+at) + f(x-at)] + \dfrac{1}{2a}\displaystyle\int_{x-at}^{x+at} g(\xi)\, d\xi & \text{for } t < \dfrac{x}{a}, \\[2mm] \dfrac{1}{2}[f(x+at) - f(at-x)] + \dfrac{1}{2a}\displaystyle\int_{at-x}^{x+at} g(\xi)\, d\xi + h\left(t - \dfrac{x}{a}\right) & \text{for } t > \dfrac{x}{a}, \end{cases}$$

$$w_2(x,t) = \begin{cases} \displaystyle\int_0^t \int_{x-a(t-\tau)}^{x+a(t-\tau)} \Phi(\xi,\tau)\, d\xi\, d\tau & \text{for } t < \dfrac{x}{a}, \\[2mm] \displaystyle\int_0^{t-x/a} \int_{a(t-\tau)-x}^{x+a(t-\tau)} \Phi(\xi,\tau)\, d\xi\, d\tau + \int_{t-x/a}^t \int_{x-a(t-\tau)}^{x+a(t-\tau)} \Phi(\xi,\tau)\, d\xi\, d\tau & \text{for } t > \dfrac{x}{a}. \end{cases}$$

⊙ *Reference*: A. V. Bitsadze and D. F. Kalinichenko (1985).

4.1.2-3. Domain: $0 \le x < \infty$. Second boundary value problem.

The following conditions are prescribed:

$$w = f(x) \quad \text{at} \quad t = 0 \quad \text{(initial condition)},$$
$$\partial_t w = g(x) \quad \text{at} \quad t = 0 \quad \text{(initial condition)},$$
$$\partial_x w = h(t) \quad \text{at} \quad x = 0 \quad \text{(boundary condition)}.$$

Solution:
$$w(x,t) = w_1(x,t) + \frac{1}{2a} w_2(x,t),$$

where

$$w_1(x,t) = \begin{cases} \dfrac{1}{2}[f(x+at) + f(x-at)] + \dfrac{1}{2a} \displaystyle\int_{x-at}^{x+at} g(\xi)\,d\xi & \text{for } t < \dfrac{x}{a}, \\[2mm] \dfrac{1}{2}[f(x+at) + f(at-x)] + \dfrac{1}{2a} \displaystyle\int_0^{x+at} g(\xi)\,d\xi \\[2mm] \qquad + \dfrac{1}{2a} \displaystyle\int_0^{at-x} g(\xi)\,d\xi - a\displaystyle\int_0^{t-x/a} h(\xi)\,d\xi & \text{for } t > \dfrac{x}{a}, \end{cases}$$

$$w_2(x,t) = \begin{cases} \displaystyle\int_0^t \int_{x-a(t-\tau)}^{x+a(t-\tau)} \Phi(\xi,\tau)\,d\xi\,d\tau & \text{for } t < \dfrac{x}{a}, \\[2mm] \displaystyle\int_0^{t-x/a}\int_0^{x+a(t-\tau)} \Phi(\xi,\tau)\,d\xi\,d\tau + \displaystyle\int_0^{t-x/a}\int_0^{a(t-\tau)-x} \Phi(\xi,\tau)\,d\xi\,d\tau \\[2mm] \qquad + \displaystyle\int_{t-x/a}^{t}\int_{x-a(t-\tau)}^{x+a(t-\tau)} \Phi(\xi,\tau)\,d\xi\,d\tau & \text{for } t > \dfrac{x}{a}. \end{cases}$$

⊙ *Reference*: A. V. Bitsadze and D. F. Kalinichenko (1985).

4.1.2-4. Domain: $0 \le x \le l$. First boundary value problem.

The following conditions are prescribed:

$$w = f_0(x) \quad \text{at} \quad t = 0 \quad \text{(initial condition)},$$
$$\partial_t w = f_1(x) \quad \text{at} \quad t = 0 \quad \text{(initial condition)},$$
$$w = g_1(t) \quad \text{at} \quad x = 0 \quad \text{(boundary condition)},$$
$$w = g_2(t) \quad \text{at} \quad x = l \quad \text{(boundary condition)}.$$

Solution:

$$w(x,t) = \frac{\partial}{\partial t}\int_0^l f_0(\xi) G(x,\xi,t)\,d\xi + \int_0^l f_1(\xi) G(x,\xi,t)\,d\xi + \int_0^t \int_0^l \Phi(\xi,\tau) G(x,\xi,t-\tau)\,d\xi\,d\tau$$
$$+ a^2 \int_0^t g_1(\tau) \left[\frac{\partial}{\partial \xi} G(x,\xi,t-\tau)\right]_{\xi=0} d\tau - a^2 \int_0^t g_2(\tau) \left[\frac{\partial}{\partial \xi} G(x,\xi,t-\tau)\right]_{\xi=l} d\tau,$$

where

$$G(x,\xi,t) = \frac{2}{a\pi} \sum_{n=1}^{\infty} \frac{1}{n} \sin\!\left(\frac{n\pi x}{l}\right) \sin\!\left(\frac{n\pi \xi}{l}\right) \sin\!\left(\frac{n\pi a t}{l}\right).$$

4.1.2-5. Domain: $0 \leq x \leq l$. Second boundary value problem.

The following conditions are prescribed:

$$w = f_0(x) \quad \text{at} \quad t = 0 \quad \text{(initial condition)},$$
$$\partial_t w = f_1(x) \quad \text{at} \quad t = 0 \quad \text{(initial condition)},$$
$$\partial_x w = g_1(t) \quad \text{at} \quad x = 0 \quad \text{(boundary condition)},$$
$$\partial_x w = g_2(t) \quad \text{at} \quad x = l \quad \text{(boundary condition)}.$$

Solution:

$$w(x,t) = \frac{\partial}{\partial t}\int_0^l f_0(\xi)G(x,\xi,t)\,d\xi + \int_0^l f_1(\xi)G(x,\xi,t)\,d\xi + \int_0^t\int_0^l \Phi(\xi,\tau)G(x,\xi,t-\tau)\,d\xi\,d\tau$$
$$- a^2\int_0^t g_1(\tau)G(x,0,t-\tau)\,d\tau + a^2\int_0^t g_2(\tau)G(x,l,t-\tau)\,d\tau,$$

where

$$G(x,\xi,t) = \frac{t}{l} + \frac{2}{a\pi}\sum_{n=1}^{\infty}\frac{1}{n}\cos\left(\frac{n\pi x}{l}\right)\cos\left(\frac{n\pi \xi}{l}\right)\sin\left(\frac{n\pi a t}{l}\right).$$

4.1.2-6. Domain: $0 \leq x \leq l$. Third boundary value problem.

The following conditions are prescribed:

$$w = f_0(x) \quad \text{at} \quad t = 0 \quad \text{(initial condition)},$$
$$\partial_t w = f_1(x) \quad \text{at} \quad t = 0 \quad \text{(initial condition)},$$
$$\partial_x w - k_1 w = g_1(t) \quad \text{at} \quad x = 0 \quad \text{(boundary condition)},$$
$$\partial_x w + k_2 w = g_2(t) \quad \text{at} \quad x = l \quad \text{(boundary condition)}.$$

The solution $w(x,t)$ is determined by the formula in Paragraph 4.1.2-5 where

$$G(x,\xi,t) = \frac{1}{a}\sum_{n=1}^{\infty}\frac{1}{\lambda_n \|u_n\|^2}\sin(\lambda_n x + \varphi_n)\sin(\lambda_n \xi + \varphi_n)\sin(\lambda_n a t),$$

$$\varphi_n = \arctan\frac{\lambda_n}{k_1}, \qquad \|u_n\|^2 = \frac{l}{2} + \frac{(\lambda_n^2 + k_1 k_2)(k_1 + k_2)}{2(\lambda_n^2 + k_1^2)(\lambda_n^2 + k_2^2)};$$

the λ_n are positive roots of the transcendental equation $\cot(\lambda l) = \dfrac{\lambda^2 - k_1 k_2}{\lambda(k_1 + k_2)}$.

4.1.2-7. Domain: $0 \leq x \leq l$. Mixed boundary value problem.

The following conditions are prescribed:

$$w = f_0(x) \quad \text{at} \quad t = 0 \quad \text{(initial condition)},$$
$$\partial_t w = f_1(x) \quad \text{at} \quad t = 0 \quad \text{(initial condition)},$$
$$w = g_1(t) \quad \text{at} \quad x = 0 \quad \text{(boundary condition)},$$
$$\partial_x w = g_2(t) \quad \text{at} \quad x = l \quad \text{(boundary condition)}.$$

Solution:

$$w(x,t) = \frac{\partial}{\partial t}\int_0^l f_0(\xi)G(x,\xi,t)\,d\xi + \int_0^l f_1(\xi)G(x,\xi,t)\,d\xi + \int_0^t\int_0^l \Phi(\xi,\tau)G(x,\xi,t-\tau)\,d\xi\,d\tau$$
$$+ a^2\int_0^t g_1(\tau)\left[\frac{\partial}{\partial \xi}G(x,\xi,t-\tau)\right]_{\xi=0}\,d\tau + a^2\int_0^t g_2(\tau)G(x,l,t-\tau)\,d\tau,$$

where

$$G(x,\xi,t) = \frac{2}{al}\sum_{n=1}^{\infty}\frac{1}{\lambda_n}\sin(\lambda_n x)\sin(\lambda_n \xi)\sin(\lambda_n a t), \qquad \lambda_n = \frac{\pi(2n+1)}{2l}.$$

4.1.3. Equation of the Form $\frac{\partial^2 w}{\partial t^2} = a^2 \frac{\partial^2 w}{\partial x^2} - bw + \Phi(x,t)$

This equation with $\Phi(x,t) \equiv 0$ and $b > 0$ is encountered in quantum field theory and a number of applications and is referred to as the *Klein–Gordon equation*.

4.1.3-1. Solutions of the homogeneous equation ($\Phi \equiv 0$).

1°. Particular solutions:

$$w(x,t) = \exp(\pm \mu t)(Ax + B), \quad b = -\mu^2,$$

$$w(x,t) = \exp(\pm \lambda x)(At + B), \quad b = a^2 \lambda^2,$$

$$w(x,t) = \cos(\lambda x)[A\cos(\mu t) + B\sin(\mu t)], \quad b = -a^2\lambda^2 + \mu^2,$$

$$w(x,t) = \sin(\lambda x)[A\cos(\mu t) + B\sin(\mu t)], \quad b = -a^2\lambda^2 + \mu^2,$$

$$w(x,t) = \exp(\pm \mu t)[A\cos(\lambda x) + B\sin(\lambda x)], \quad b = -a^2\lambda^2 - \mu^2,$$

$$w(x,t) = \exp(\pm \lambda x)[A\cos(\mu t) + B\sin(\mu t)], \quad b = a^2\lambda^2 + \mu^2,$$

$$w(x,t) = \exp(\pm \lambda x)[A\exp(\mu t) + B\exp(-\mu t)], \quad b = a^2\lambda^2 - \mu^2,$$

$$w(x,t) = AJ_0(\xi) + BY_0(\xi), \quad \xi = \frac{\sqrt{b}}{a}\sqrt{a^2(t+C_1)^2 - (x+C_2)^2}, \quad b > 0,$$

$$w(x,t) = AI_0(\xi) + BK_0(\xi), \quad \xi = \frac{\sqrt{-b}}{a}\sqrt{a^2(t+C_1)^2 - (x+C_2)^2}, \quad b < 0,$$

where A, B, C_1, and C_2 are arbitrary constants, $J_0(\xi)$ and $Y_0(\xi)$ are the Bessel functions, and $I_0(\xi)$ and $K_0(\xi)$ are the modified Bessel functions.

2°. Fundamental solutions:

$$\mathscr{E}(x,t) = \frac{\vartheta(at - |x|)}{2a} J_0\left(\frac{c}{a}\sqrt{a^2t^2 - x^2}\right) \quad \text{for} \quad b = c^2 > 0,$$

$$\mathscr{E}(x,t) = \frac{\vartheta(at - |x|)}{2a} I_0\left(\frac{c}{a}\sqrt{a^2t^2 - x^2}\right) \quad \text{for} \quad b = -c^2 < 0,$$

where $\vartheta(z)$ is the Heaviside unit step function ($\vartheta = 0$ for $z < 0$ and $\vartheta = 1$ for $z \geq 0$), $J_0(z)$ is the Bessel function, and $I_0(z)$ is the modified Bessel function.

⊙ *Reference*: V. S. Vladimirov, V. P. Mikhailov, A. A. Vasharin, et al. (1974).

4.1.3-2. Some formulas and transformations of the homogeneous equation ($\Phi \equiv 0$).

1°. Suppose $w = w(x,t)$ is a solution of the Klein–Gordon equation. Then the functions

$$w_1 = Aw(x + C_1, \pm t + C_2),$$

$$w_2 = Aw(-x + C_1, \pm t + C_2),$$

$$w_3 = Aw\left(\frac{x - vt}{\sqrt{1 - (v/a)^2}}, \frac{t - va^{-2}x}{\sqrt{1 - (v/a)^2}}\right),$$

where A, C_1, C_2, and v are arbitrary constants, are also solutions of this equation.

2°. Table 19 lists transformations of the independent variables that allow separation of variables in the Klein–Gordon equation.

Notation: $J_\sigma(z)$ and $Y_\sigma(z)$ are the Bessel functions, $I_\sigma(z)$ and $K_\sigma(z)$ are the modified Bessel functions, and $D_\lambda(z)$ is the parabolic cylinder function.

⊙ *References*: E. Kalnins (1975), W. Miller, Jr. (1977).

TABLE 19
Orthogonal coordinates $u = u(x,t)$, $v = v(x,t)$ admitting separable solutions $w = F(u)G(v)$
of the Klein–Gordon equation ($a = 1$; A_1, A_2, B_1, B_2, and λ are arbitrary constants)

No	Relation between x, t and u, v	Function $F = F(u)$ (differential equation)	Function $G = G(v)$ (differential equation)
1	$x = u$, $t = v$	$F = A_1 e^{u\sqrt{\lambda+b}} + A_2 e^{-u\sqrt{\lambda+b}}$	$G = B_1 e^{v\sqrt{\lambda}} + B_2 e^{-v\sqrt{\lambda}}$
2	$x = u\sinh v$, $t = u\cosh v$	$F = \sqrt{u}\,[A_1 J_\sigma(u\sqrt{b}) + A_2 Y_\sigma(u\sqrt{b})]$, $\sigma = \frac{1}{2}\sqrt{1+\lambda^2}$	$G = B_1 e^{\lambda v} + B_2 e^{-\lambda v}$
3	$x = uv$, $t = \frac{1}{2}(u^2+v^2)$	$F = A_1 D_\lambda(\beta u) + A_2 D_\lambda(-\beta u)$, $\beta = (-4b)^{1/4}$	$G = B_1 D_\lambda(\beta v) + B_2 D_\lambda(-\beta v)$, $\beta = (-4b)^{1/4}$
4	$x = \frac{1}{2}(u^2+v^2)$, $t = uv$	$F = A_1 D_\lambda(\beta u) + A_2 D_\lambda(-\beta u)$, $\beta = (4b)^{1/4}$	$G = B_1 D_\lambda(\beta v) + B_2 D_\lambda(-\beta v)$, $\beta = (4b)^{1/4}$
5	$x = -\frac{1}{2}(u-v)^2 + u + v$, $t = \frac{1}{2}(u-v)^2 + u + v$	$F = \sqrt{U}\,[A_1 J_{\frac{1}{3}}(\xi) + A_2 Y_{\frac{1}{3}}(\xi)]$, $U = u+\lambda$, $\xi = \frac{2}{3}\sqrt{b}\,U^{3/2}$	$G = \sqrt{V}\,[B_1 J_{\frac{1}{3}}(\eta) + B_2 Y_{\frac{1}{3}}(\eta)]$, $V = v+\lambda$, $\eta = \frac{2}{3}\sqrt{b}\,V^{3/2}$
6	$t+x = \cosh[\frac{1}{2}(u-v)]$, $t-x = \sinh[\frac{1}{2}(u+v)]$	$F'' + (\lambda + b\sinh u)F = 0$	$G'' + (\lambda + b\sinh v)G = 0$
7	$x = \sinh(u-v) - \frac{1}{2}e^{u+v}$, $t = \sinh(u-v) + \frac{1}{2}e^{u+v}$	$F = A_1 J_\lambda(\beta e^u) + A_2 Y_\lambda(\beta e^u)$, $\beta = \sqrt{b}$	$G = B_1 I_\lambda(\beta e^v) + B_1 K_\lambda(\beta e^v)$, $\beta = \sqrt{b}$
8	$x = \cosh(u-v) - \frac{1}{2}e^{u+v}$, $t = \cosh(u-v) + \frac{1}{2}e^{u+v}$	$F = A_1 J_\lambda(\beta e^u) + A_2 Y_\lambda(\beta e^u)$, $\beta = \sqrt{b}$	$G = B_1 J_\lambda(\beta e^v) + B_1 Y_\lambda(\beta e^v)$, $\beta = \sqrt{b}$
9	$x = \cosh u \sinh v$, $t = \sinh u \cosh v$	$F'' + (\lambda + \frac{1}{2}b\cosh 2u)F = 0$, modified Mathieu equation	$G'' + (\lambda - \frac{1}{2}b\cosh 2v)G = 0$, modified Mathieu equation
10	$x = \sinh u \sinh v$, $t = \cosh u \cosh v$	$F'' + (\lambda + \frac{1}{2}b\cosh 2u)F = 0$, modified Mathieu equation	$G'' + (\lambda + \frac{1}{2}b\cosh 2v)G = 0$, modified Mathieu equation
11	$x = \sin u \sin v$, $t = \cos u \cos v$	$F'' + (\lambda - \frac{1}{2}b\cos 2u)F = 0$, Mathieu equation	$G'' + (\lambda - \frac{1}{2}b\cos 2v)G = 0$, Mathieu equation

4.1.3-3. Domain: $-\infty < x < \infty$. Cauchy problem.

Initial conditions are prescribed:
$$w = f(x) \quad \text{at} \quad t = 0,$$
$$\partial_t w = g(x) \quad \text{at} \quad t = 0.$$

Solution for $b = -c^2 < 0$:
$$w(x,t) = \frac{1}{2}[f(x+at) + f(x-at)] + \frac{ct}{2a}\int_{x-at}^{x+at} \frac{I_1(c\sqrt{t^2 - (x-\xi)^2/a^2})}{\sqrt{t^2 - (x-\xi)^2/a^2}} f(\xi)\,d\xi$$
$$+ \frac{1}{2a}\int_{x-at}^{x+at} I_0(c\sqrt{t^2 - (x-\xi)^2/a^2})\,g(\xi)\,d\xi$$
$$+ \frac{1}{2a}\int_0^t \int_{x-a(t-\tau)}^{x+a(t-\tau)} I_0(c\sqrt{(t-\tau)^2 - (x-\xi)^2/a^2})\,\Phi(\xi,\tau)\,d\xi\,d\tau,$$

where $I_0(z)$ and $I_1(z)$ are the modified Bessel functions of the first kind.

Solution for $b = c^2 > 0$:

$$w(x,t) = \frac{1}{2}[f(x+at) + f(x-at)] - \frac{ct}{2a}\int_{x-at}^{x+at}\frac{J_1\left(c\sqrt{t^2-(x-\xi)^2/a^2}\right)}{\sqrt{t^2-(x-\xi)^2/a^2}}f(\xi)\,d\xi$$

$$+ \frac{1}{2a}\int_{x-at}^{x+at} J_0\left(c\sqrt{t^2-(x-\xi)^2/a^2}\right)g(\xi)\,d\xi$$

$$+ \frac{1}{2a}\int_0^t\int_{x-a(t-\tau)}^{x+a(t-\tau)} J_0\left(c\sqrt{(t-\tau)^2-(x-\xi)^2/a^2}\right)\Phi(\xi,\tau)\,d\xi\,d\tau,$$

where $J_0(z)$ and $J_1(z)$ are the Bessel functions of the first kind.

⦿ *Reference*: B. M. Budak, A. N. Tikhonov, and A. A. Samarskii (1980).

> **4.1.3-4. Domain: $0 \le x \le l$. First boundary value problem.**

The following conditions are prescribed:

$$\begin{aligned}
w &= f_0(x) & \text{at} \quad & t = 0 & \text{(initial condition)}, \\
\partial_t w &= f_1(x) & \text{at} \quad & t = 0 & \text{(initial condition)}, \\
w &= g_1(t) & \text{at} \quad & x = 0 & \text{(boundary condition)}, \\
w &= g_2(t) & \text{at} \quad & x = l & \text{(boundary condition)}.
\end{aligned}$$

Solution:

$$w(x,t) = \frac{\partial}{\partial t}\int_0^l f_0(\xi)G(x,\xi,t)\,d\xi + \int_0^l f_1(\xi)G(x,\xi,t)\,d\xi + \int_0^t\int_0^l \Phi(\xi,\tau)G(x,\xi,t-\tau)\,d\xi\,d\tau$$

$$+ a^2\int_0^t g_1(\tau)\left[\frac{\partial}{\partial\xi}G(x,\xi,t-\tau)\right]_{\xi=0}\,d\tau - a^2\int_0^t g_2(\tau)\left[\frac{\partial}{\partial\xi}G(x,\xi,t-\tau)\right]_{\xi=l}\,d\tau,$$

where

$$G(x,\xi,t) = \frac{2}{l}\sum_{n=1}^\infty \sin(\lambda_n x)\sin(\lambda_n \xi)\frac{\sin\left(t\sqrt{a^2\lambda_n^2+b}\right)}{\sqrt{a^2\lambda_n^2+b}}, \qquad \lambda_n = \frac{\pi n}{l}.$$

Remark. Let $b < 0$ and $a^2\lambda_n^2 + b < 0$ for $n = 1, \ldots, m$ and $a^2\lambda_n^2 + b > 0$ for $n = m+1, m+2, \ldots$ In this case the Green's function is modified and acquires the form

$$G(x,\xi,t) = \frac{2}{l}\sum_{n=1}^m \sin(\lambda_n x)\sin(\lambda_n \xi)\frac{\sinh\left(t\sqrt{|a^2\lambda_n^2+b|}\right)}{\sqrt{|a^2\lambda_n^2+b|}}$$

$$+ \frac{2}{l}\sum_{n=m+1}^\infty \sin(\lambda_n x)\sin(\lambda_n \xi)\frac{\sin\left(t\sqrt{a^2\lambda_n^2+b}\right)}{\sqrt{a^2\lambda_n^2+b}}, \qquad \lambda_n = \frac{\pi n}{l}.$$

Analogously, the Green's functions for the second, third, and mixed boundary value problems are modified in similar cases.

⦿ *Reference*: A. G. Butkovskiy (1979).

> **4.1.3-5. Domain: $0 \le x \le l$. Second boundary value problem.**

The following conditions are prescribed:

$$\begin{aligned}
w &= f_0(x) & \text{at} \quad & t = 0 & \text{(initial condition)}, \\
\partial_t w &= f_1(x) & \text{at} \quad & t = 0 & \text{(initial condition)}, \\
\partial_x w &= g_1(t) & \text{at} \quad & x = 0 & \text{(boundary condition)}, \\
\partial_x w &= g_2(t) & \text{at} \quad & x = l & \text{(boundary condition)}.
\end{aligned}$$

Solution:

$$w(x,t) = \frac{\partial}{\partial t}\int_0^l f_0(\xi)G(x,\xi,t)\,d\xi + \int_0^l f_1(\xi)G(x,\xi,t)\,d\xi + \int_0^t\int_0^l \Phi(\xi,\tau)G(x,\xi,t-\tau)\,d\xi\,d\tau$$
$$- a^2\int_0^t g_1(\tau)G(x,0,t-\tau)\,d\tau + a^2\int_0^t g_2(\tau)G(x,l,t-\tau)\,d\tau,$$

where

$$G(x,\xi,t) = \frac{1}{l\sqrt{b}}\sin(t\sqrt{b}) + \frac{2}{l}\sum_{n=1}^{\infty}\cos(\lambda_n x)\cos(\lambda_n\xi)\frac{\sin(t\sqrt{a^2\lambda_n^2+b})}{\sqrt{a^2\lambda_n^2+b}},\quad \lambda_n = \frac{\pi n}{l}.$$

4.1.3-6. Domain: $0 \le x \le l$. Third boundary value problem.

The following conditions are prescribed:

$$\begin{aligned}w &= f_0(x) &&\text{at}\quad t=0 &&\text{(initial condition)},\\ \partial_t w &= f_1(x) &&\text{at}\quad t=0 &&\text{(initial condition)},\\ \partial_x w - k_1 w &= g_1(t) &&\text{at}\quad x=0 &&\text{(boundary condition)},\\ \partial_x w + k_2 w &= g_2(t) &&\text{at}\quad x=l &&\text{(boundary condition)}.\end{aligned}$$

The solution $w(x,t)$ is determined by the formula in Paragraph 4.1.3-5 where

$$G(x,\xi,t) = \sum_{n=1}^{\infty}\frac{y_n(x)y_n(\xi)\sin(t\sqrt{a^2\lambda_n^2+b})}{\|y_n\|^2\sqrt{a^2\lambda_n^2+b}},$$

$$y_n(x) = \cos(\lambda_n x) + \frac{k_1}{\lambda_n}\sin(\lambda_n x),\quad \|y_n\|^2 = \frac{k_2}{2\lambda_n^2}\frac{\lambda_n^2+k_1^2}{\lambda_n^2+k_2^2} + \frac{k_1}{2\lambda_n^2} + \frac{l}{2}\left(1+\frac{k_1^2}{\lambda_n^2}\right).$$

Here, the λ_n are positive roots of the transcendental equation $\dfrac{\tan(\lambda l)}{\lambda} = \dfrac{k_1+k_2}{\lambda^2 - k_1 k_2}$.

4.1.3-7. Domain: $0 \le x \le l$. Mixed boundary value problem.

The following conditions are prescribed:

$$\begin{aligned}w &= f_0(x) &&\text{at}\quad t=0 &&\text{(initial condition)},\\ \partial_t w &= f_1(x) &&\text{at}\quad t=0 &&\text{(initial condition)},\\ w &= g_1(t) &&\text{at}\quad x=0 &&\text{(boundary condition)},\\ \partial_x w &= g_2(t) &&\text{at}\quad x=l &&\text{(boundary condition)}.\end{aligned}$$

Solution:

$$w(x,t) = \frac{\partial}{\partial t}\int_0^l f_0(\xi)G(x,\xi,t)\,d\xi + \int_0^l f_1(\xi)G(x,\xi,t)\,d\xi + \int_0^t\int_0^l \Phi(\xi,\tau)G(x,\xi,t-\tau)\,d\xi\,d\tau$$
$$+ a^2\int_0^t g_1(\tau)\left[\frac{\partial}{\partial\xi}G(x,\xi,t-\tau)\right]_{\xi=0}d\tau + a^2\int_0^t g_2(\tau)G(x,l,t-\tau)\,d\tau,$$

where

$$G(x,\xi,t) = \frac{2}{l}\sum_{n=0}^{\infty}\sin(\lambda_n x)\sin(\lambda_n\xi)\frac{\sin(t\sqrt{a^2\lambda_n^2+b})}{\sqrt{a^2\lambda_n^2+b}},\quad \lambda_n = \frac{\pi(2n+1)}{2l}.$$

4.1.4. Equation of the Form $\frac{\partial^2 w}{\partial t^2} = a^2 \frac{\partial^2 w}{\partial x^2} - b\frac{\partial w}{\partial x} + \Phi(x,t)$

4.1.4-1. Reduction to the nonhomogeneous Klein–Gordon equation.

The substitution $w(x,t) = \exp\left(\frac{1}{2}bx/a^2\right)u(x,t)$ leads the nonhomogeneous Klein–Gordon equation

$$\frac{\partial^2 u}{\partial t^2} = a^2 \frac{\partial^2 u}{\partial x^2} - \frac{b^2}{4a^2}u + \exp\left(-\frac{bx}{2a^2}\right)\Phi(x,t),$$

which is discussed in Subsection 4.1.3.

4.1.4-2. Domain: $-\infty < x < \infty$. Cauchy problem.

Initial conditions are prescribed:

$$w = f(x) \quad \text{at} \quad t = 0,$$
$$\partial_t w = g(x) \quad \text{at} \quad t = 0.$$

Solution:

$$w(x,t) = \frac{1}{2}f(x+at)\exp\left(-\frac{bt}{2a}\right) + \frac{1}{2}f(x-at)\exp\left(\frac{bt}{2a}\right)$$
$$- \frac{\sigma t}{2a}\exp\left(\frac{bx}{2a^2}\right)\int_{x-at}^{x+at}\exp\left(-\frac{b\xi}{2a^2}\right)\frac{J_1\left(\sigma\sqrt{t^2-(x-\xi)^2/a^2}\right)}{\sqrt{t^2-(x-\xi)^2/a^2}}f(\xi)\,d\xi$$
$$+ \frac{1}{2a}\exp\left(\frac{bx}{2a^2}\right)\int_{x-at}^{x+at}\exp\left(-\frac{b\xi}{2a^2}\right)J_0\left(\sigma\sqrt{t^2-(x-\xi)^2/a^2}\right)g(\xi)\,d\xi$$
$$+ \frac{1}{2a}\int_0^t\int_{x-a(t-\tau)}^{x+a(t-\tau)}\exp\left[\frac{b(x-\xi)}{2a^2}\right]J_0\left(\sigma\sqrt{(t-\tau)^2-(x-\xi)^2/a^2}\right)\Phi(\xi,\tau)\,d\xi\,d\tau,$$

where $J_0(z)$ and $J_1(z)$ are the Bessel functions of the first kind, and $\sigma = \frac{1}{2}|b|/a$.

4.1.4-3. Domain: $0 \leq x \leq l$. First boundary value problem.

The following conditions are prescribed:

$$w = f_0(x) \quad \text{at} \quad t = 0 \quad \text{(initial condition)},$$
$$\partial_t w = f_1(x) \quad \text{at} \quad t = 0 \quad \text{(initial condition)},$$
$$w = g_1(t) \quad \text{at} \quad x = 0 \quad \text{(boundary condition)},$$
$$w = g_2(t) \quad \text{at} \quad x = l \quad \text{(boundary condition)}.$$

Solution:

$$w(x,t) = \frac{\partial}{\partial t}\int_0^l f_0(\xi)G(x,\xi,t)\,d\xi + \int_0^l f_1(\xi)G(x,\xi,t)\,d\xi + \int_0^t\int_0^l \Phi(\xi,\tau)G(x,\xi,t-\tau)\,d\xi\,d\tau$$
$$+ a^2\int_0^t g_1(\tau)\left[\frac{\partial}{\partial \xi}G(x,\xi,t-\tau)\right]_{\xi=0}d\tau - a^2\int_0^t g_2(\tau)\left[\frac{\partial}{\partial \xi}G(x,\xi,t-\tau)\right]_{\xi=l}d\tau,$$

where

$$G(x,\xi,t) = \frac{2}{l}\exp\left[\frac{b}{2a^2}(x-\xi)\right]\sum_{n=1}^\infty \sin\left(\frac{\pi nx}{l}\right)\sin\left(\frac{\pi n\xi}{l}\right)\frac{\sin(\lambda_n t)}{\lambda_n}, \quad \lambda_n = \sqrt{\frac{a^2\pi^2 n^2}{l^2} + \frac{b^2}{4a^2}}.$$

⊙ *Reference*: A. G. Butkovskiy (1979).

4.1.4-4. Domain: $0 \le x \le l$. Second boundary value problem.

The following conditions are prescribed:

$$\begin{aligned}
w &= f_0(x) && \text{at} \quad t = 0 && \text{(initial condition)},\\
\partial_t w &= f_1(x) && \text{at} \quad t = 0 && \text{(initial condition)},\\
\partial_x w &= g_1(t) && \text{at} \quad x = 0 && \text{(boundary condition)},\\
\partial_x w &= g_2(t) && \text{at} \quad x = l && \text{(boundary condition)}.
\end{aligned}$$

Solution:

$$w(x,t) = \frac{\partial}{\partial t}\int_0^l f_0(\xi)G(x,\xi,t)\,d\xi + \int_0^l f_1(\xi)G(x,\xi,t)\,d\xi + \int_0^t\int_0^l \Phi(\xi,\tau)G(x,\xi,t-\tau)\,d\xi\,d\tau$$
$$- a^2\int_0^t g_1(\tau)G(x,0,t-\tau)\,d\tau + a^2\int_0^t g_2(\tau)G(x,l,t-\tau)\,d\tau,$$

where

$$G(x,\xi,t) = \frac{bt}{a^2[1-\exp(-bl/a^2)]}\exp\left(-\frac{b\xi}{a^2}\right) + \frac{2}{l}\exp\left[\frac{b}{2a^2}(x-\xi)\right]\sum_{n=1}^{\infty}\frac{y_n(x)y_n(\xi)\sin(\lambda_n t)}{\lambda_n(1+\mu_n^2)},$$

$$y_n(x) = \cos\left(\frac{\pi n x}{l}\right) - \frac{bl}{2a^2\pi n}\sin\left(\frac{\pi n x}{l}\right), \quad \lambda_n = \sqrt{\frac{a^2\pi^2 n^2}{l^2} + \frac{b^2}{4a^2}}, \quad \mu_n = \frac{bl}{2a^2\pi n}.$$

⦿ *Reference*: A. G. Butkovskiy (1979).

4.1.4-5. Domain: $0 \le x \le l$. Third boundary value problem.

The following conditions are prescribed:

$$\begin{aligned}
w &= f_0(x) && \text{at} \quad t = 0 && \text{(initial condition)},\\
\partial_t w &= f_1(x) && \text{at} \quad t = 0 && \text{(initial condition)},\\
\partial_x w - k_1 w &= g_1(t) && \text{at} \quad x = 0 && \text{(boundary condition)},\\
\partial_x w + k_2 w &= g_2(t) && \text{at} \quad x = l && \text{(boundary condition)}.
\end{aligned}$$

The solution $w(x,t)$ is determined by the formula in Paragraph 4.1.4-4 where

$$G(x,\xi,t) = \exp\left[\frac{b(x-\xi)}{2a^2}\right]\sum_{n=1}^{\infty}\frac{y_n(x)y_n(\xi)\sin(a\lambda_n t)}{a\lambda_n B_n}.$$

Here,

$$y_n(x) = \cos(\mu_n x) + \frac{2a^2 k_1 - b}{2a^2\mu_n}\sin(\mu_n x), \quad \lambda_n = \sqrt{\mu_n^2 + \frac{b^2}{4a^4}},$$

$$B_n = \frac{2a^2 k_2 + b}{4a^2\mu_n^2}\frac{4a^4\mu_n^2 + (2a^2 k_1 - b)^2}{4a^4\mu_n^2 + (2a^2 k_2 + b)^2} + \frac{2a^2 k_1 - b}{4a^2\mu_n^2} + \frac{l}{2} + \frac{l(2a^2 k_1 - b)^2}{8a^4\mu_n^2},$$

where the μ_n are positive roots of the transcendental equation

$$\frac{\tan(\mu l)}{\mu} = \frac{4a^4(k_1 + k_2)}{4a^4\mu^2 - (2a^2 k_1 - b)(2a^2 k_2 + b)}.$$

⦿ *Reference*: A. G. Butkovskiy (1979).

4.1.5. Equation of the Form $\dfrac{\partial^2 w}{\partial t^2} = a^2 \dfrac{\partial^2 w}{\partial x^2} + b\dfrac{\partial w}{\partial x} + cw + \Phi(x,t)$

4.1.5-1. Reduction to the nonhomogeneous Klein–Gordon equation.

The substitution $w(x,t) = \exp\bigl(-\tfrac{1}{2}a^{-2}bx\bigr)u(x,t)$ leads to the equation

$$\frac{\partial^2 u}{\partial t^2} = a^2 \frac{\partial^2 u}{\partial x^2} + \bigl(c - \tfrac{1}{4}a^{-2}b^2\bigr)u + \exp\bigl(\tfrac{1}{2}a^{-2}bx\bigr)\Phi(x,t),$$

which is discussed in Subsection 4.1.3.

4.1.5-2. Domain: $-\infty < x < \infty$. Cauchy problem.

Initial conditions are prescribed:

$$w = f(x) \quad \text{at} \quad t = 0,$$
$$\partial_t w = g(x) \quad \text{at} \quad t = 0.$$

Solution for $c - \tfrac{1}{4}a^{-2}b^2 = \sigma^2 > 0$:

$$w(x,t) = \frac{1}{2} f(x+at)\exp\!\left(\frac{bt}{2a}\right) + \frac{1}{2} f(x-at)\exp\!\left(-\frac{bt}{2a}\right)$$

$$+ \frac{\sigma t}{2a} \exp\!\left(-\frac{bx}{2a^2}\right) \int_{x-at}^{x+at} \exp\!\left(\frac{b\xi}{2a^2}\right) \frac{I_1\bigl(\sigma\sqrt{t^2 - (x-\xi)^2/a^2}\bigr)}{\sqrt{t^2 - (x-\xi)^2/a^2}} f(\xi)\, d\xi$$

$$+ \frac{1}{2a} \exp\!\left(-\frac{bx}{2a^2}\right) \int_{x-at}^{x+at} \exp\!\left(\frac{b\xi}{2a^2}\right) I_0\bigl(\sigma\sqrt{t^2 - (x-\xi)^2/a^2}\bigr) g(\xi)\, d\xi$$

$$+ \frac{1}{2a} \int_0^t \int_{x-a(t-\tau)}^{x+a(t-\tau)} \exp\!\left[\frac{b(\xi - x)}{2a^2}\right] I_0\bigl(\sigma\sqrt{(t-\tau)^2 - (x-\xi)^2/a^2}\bigr) \Phi(\xi,\tau)\, d\xi\, d\tau,$$

where $I_0(z)$ and $I_1(z)$ are the modified Bessel functions of the first kind.

Solution for $c - \tfrac{1}{4}a^{-2}b^2 = -\sigma^2 < 0$:

$$w(x,t) = \frac{1}{2} f(x+at)\exp\!\left(\frac{bt}{2a}\right) + \frac{1}{2} f(x-at)\exp\!\left(-\frac{bt}{2a}\right)$$

$$- \frac{\sigma t}{2a} \exp\!\left(-\frac{bx}{2a^2}\right) \int_{x-at}^{x+at} \exp\!\left(\frac{b\xi}{2a^2}\right) \frac{J_1\bigl(\sigma\sqrt{t^2 - (x-\xi)^2/a^2}\bigr)}{\sqrt{t^2 - (x-\xi)^2/a^2}} f(\xi)\, d\xi$$

$$+ \frac{1}{2a} \exp\!\left(-\frac{bx}{2a^2}\right) \int_{x-at}^{x+at} \exp\!\left(\frac{b\xi}{2a^2}\right) J_0\bigl(\sigma\sqrt{t^2 - (x-\xi)^2/a^2}\bigr) g(\xi)\, d\xi$$

$$+ \frac{1}{2a} \int_0^t \int_{x-a(t-\tau)}^{x+a(t-\tau)} \exp\!\left[\frac{b(\xi - x)}{2a^2}\right] J_0\bigl(\sigma\sqrt{(t-\tau)^2 - (x-\xi)^2/a^2}\bigr) \Phi(\xi,\tau)\, d\xi\, d\tau,$$

where $J_0(z)$ and $J_1(z)$ are the Bessel functions of the first kind.

⊙ *Reference*: A. N. Tikhonov and A. A. Samarskii (1990).

4.1.5-3. Domain: $0 \leq x \leq l$. First boundary value problem.

The following conditions are prescribed:

$$w = f_0(x) \quad \text{at} \quad t = 0 \quad \text{(initial condition)},$$
$$\partial_t w = f_1(x) \quad \text{at} \quad t = 0 \quad \text{(initial condition)},$$
$$w = g_1(t) \quad \text{at} \quad x = 0 \quad \text{(boundary condition)},$$
$$w = g_2(t) \quad \text{at} \quad x = l \quad \text{(boundary condition)}.$$

Solution:
$$w(x,t) = \int_0^t \int_0^l \Phi(\xi,\tau) G(x,\xi,t-\tau)\, d\xi\, d\tau + \frac{\partial}{\partial t} \int_0^l f_0(\xi) G(x,\xi,t)\, d\xi + \int_0^l f_1(\xi) G(x,\xi,t)\, d\xi$$
$$+ a^2 \int_0^t g_1(\tau) \left[\frac{\partial}{\partial \xi} G(x,\xi,t-\tau)\right]_{\xi=0} d\tau - a^2 \int_0^t g_2(\tau) \left[\frac{\partial}{\partial \xi} G(x,\xi,t-\tau)\right]_{\xi=l} d\tau.$$

Let $a^2\pi^2 + \frac{1}{4}a^{-2}b^2l^2 - cl^2 > 0$. Then
$$G(x,\xi,t) = \frac{2}{l} \exp\left[\frac{b(\xi-x)}{2a^2}\right] \sum_{n=1}^{\infty} \sin\left(\frac{\pi n x}{l}\right) \sin\left(\frac{\pi n \xi}{l}\right) \frac{\sin(t\sqrt{\lambda_n})}{\sqrt{\lambda_n}}, \qquad \lambda_n = \frac{a^2\pi^2 n^2}{l^2} + \frac{b^2}{4a^2} - c.$$

Let
$$a^2\pi^2 n^2 + \tfrac{1}{4}a^{-2}b^2l^2 - cl^2 \le 0 \quad \text{at} \quad n = 1, \ldots, m;$$
$$a^2\pi^2 n^2 + \tfrac{1}{4}a^{-2}b^2l^2 - cl^2 > 0 \quad \text{at} \quad n = m+1, m+2, \ldots$$

Then
$$G(x,\xi,t) = \frac{2}{l} \exp\left[\frac{b(\xi-x)}{2a^2}\right] \sum_{n=1}^{m} \sin\left(\frac{\pi n x}{l}\right) \sin\left(\frac{\pi n \xi}{l}\right) \frac{\sinh(t\sqrt{\beta_n})}{\sqrt{\beta_n}}$$
$$+ \frac{2}{l} \exp\left[\frac{b(\xi-x)}{2a^2}\right] \sum_{n=m+1}^{\infty} \sin\left(\frac{\pi n x}{l}\right) \sin\left(\frac{\pi n \xi}{l}\right) \frac{\sin(t\sqrt{\lambda_n})}{\sqrt{\lambda_n}},$$
$$\beta_n = c - \frac{a^2\pi^2 n^2}{l^2} - \frac{b^2}{4a^2}, \qquad \lambda_n = \frac{a^2\pi^2 n^2}{l^2} + \frac{b^2}{4a^2} - c.$$

For $\beta_n = 0$ the ratio $\sinh(t\sqrt{\beta_n})/\sqrt{\beta_n}$ must be replaced by t.

⊙ *Reference*: A. G. Butkovskiy (1979).

4.1.5-4. Domain: $0 \le x \le l$. Second boundary value problem.

The following conditions are prescribed:
$$w = f_0(x) \quad \text{at} \quad t = 0 \quad \text{(initial condition)},$$
$$\partial_t w = f_1(x) \quad \text{at} \quad t = 0 \quad \text{(initial condition)},$$
$$\partial_x w = g_1(t) \quad \text{at} \quad x = 0 \quad \text{(boundary condition)},$$
$$\partial_x w = g_2(t) \quad \text{at} \quad x = l \quad \text{(boundary condition)}.$$

Solution:
$$w(x,t) = \int_0^t \int_0^l \Phi(\xi,\tau) G(x,\xi,t-\tau)\, d\xi\, d\tau + \frac{\partial}{\partial t} \int_0^l f_0(\xi) G(x,\xi,t)\, d\xi + \int_0^l f_1(\xi) G(x,\xi,t)\, d\xi$$
$$- a^2 \int_0^t g_1(\tau) G(x,0,t-\tau)\, d\tau + a^2 \int_0^t g_2(\tau) G(x,l,t-\tau)\, d\tau.$$

For $c < 0$,
$$G(x,\xi,t) = \frac{b}{a^2(e^{bl/a^2}-1)} \exp\left(\frac{b\xi}{a^2}\right) \frac{\sin(t\sqrt{|c|})}{\sqrt{|c|}} + \frac{2}{l} \exp\left[\frac{b(\xi-x)}{2a^2}\right] \sum_{n=1}^{\infty} \frac{y_n(x) y_n(\xi)}{1+\mu_n^2} \frac{\sin(t\sqrt{\lambda_n})}{\sqrt{\lambda_n}},$$
$$\lambda_n = \frac{a^2\pi^2 n^2}{l^2} + \frac{b^2}{4a^2} - c, \quad y_n(x) = \cos\left(\frac{\pi n x}{l}\right) + \mu_n \sin\left(\frac{\pi n x}{l}\right), \quad \mu_n = \frac{bl}{2a^2 \pi n}.$$

For $c > 0$,
$$G(x,\xi,t) = \frac{b}{a^2(e^{bl/a^2}-1)} \exp\left(\frac{b\xi}{a^2}\right) \frac{\sinh(t\sqrt{c})}{\sqrt{c}} + \frac{2}{l} \exp\left[\frac{b(\xi-x)}{2a^2}\right] \sum_{n=1}^{\infty} \frac{y_n(x) y_n(\xi)}{1+\mu_n^2} \frac{\sin(t\sqrt{\lambda_n})}{\sqrt{\lambda_n}},$$

where the λ_n, $y_n(x)$, and μ_n were specified previously. If the inequality $\lambda_n < 0$ holds for several first values $n = 1, \ldots, m$, then the $\sqrt{\lambda_n}$ in the corresponding terms of the series should be replaced by $\sqrt{|\lambda_n|}$, and the sines by the hyperbolic sines.

4.1.5-5. Domain: $0 \leq x \leq l$. Third boundary value problem.

The following conditions are prescribed:

$$w = f_0(x) \quad \text{at} \quad t = 0 \quad \text{(initial condition)},$$
$$\partial_t w = f_1(x) \quad \text{at} \quad t = 0 \quad \text{(initial condition)},$$
$$\partial_x w - k_1 w = g_1(t) \quad \text{at} \quad x = 0 \quad \text{(boundary condition)},$$
$$\partial_x w + k_2 w = g_2(t) \quad \text{at} \quad x = l \quad \text{(boundary condition)}.$$

The solution $w(x,t)$ is determined by the formula in Paragraph 4.1.5-4 where

$$G(x, \xi, t) = \exp\left[\frac{b(\xi - x)}{2a^2}\right] \sum_{n=1}^{\infty} \frac{y_n(x) y_n(\xi) \sin(t\sqrt{\lambda_n})}{B_n \sqrt{\lambda_n}}.$$

Here,

$$y_n(x) = \cos(\mu_n x) + \frac{2a^2 k_1 + b}{2a^2 \mu_n} \sin(\mu_n x), \quad \lambda_n = a^2 \mu_n^2 + \frac{b^2}{4a^2} - c,$$

$$B_n = \frac{2a^2 k_2 - b}{4a^2 \mu_n^2} \frac{4a^4 \mu_n^2 + (2a^2 k_1 + b)^2}{4a^4 \mu_n^2 + (2a^2 k_2 - b)^2} + \frac{2a^2 k_1 + b}{4a^2 \mu_n^2} + \frac{l}{2} + \frac{l(2a^2 k_1 + b)^2}{8a^4 \mu_n^2},$$

where the μ_n are positive roots of the transcendental equation

$$\frac{\tan(\mu l)}{\mu} = \frac{4a^4(k_1 + k_2)}{4a^4 \mu^2 - (2a^2 k_1 + b)(2a^2 k_2 - b)}.$$

4.2. Wave Equation with Axial or Central Symmetry

4.2.1. Equations of the Form $\frac{\partial^2 w}{\partial t^2} = a^2 \left(\frac{\partial^2 w}{\partial r^2} + \frac{1}{r} \frac{\partial w}{\partial r} \right)$

This is the one-dimensional wave equation with axial symmetry, where $r = \sqrt{x^2 + y^2}$ is the radial coordinate. In the problems considered in Paragraphs 4.2.1-1 through 4.2.1-3, the solutions bounded at $r = 0$ are sought (this is not specially stated below).

4.2.1-1. Domain: $0 \leq r \leq R$. First boundary value problem.

The following conditions are prescribed:

$$w = f_0(r) \quad \text{at} \quad t = 0 \quad \text{(initial condition)},$$
$$\partial_t w = f_1(r) \quad \text{at} \quad t = 0 \quad \text{(initial condition)},$$
$$w = g(t) \quad \text{at} \quad r = R \quad \text{(boundary condition)}.$$

Solution:

$$w(r,t) = \frac{\partial}{\partial t} \int_0^R f_0(\xi) G(r, \xi, t)\, d\xi + \int_0^R f_1(\xi) G(r, \xi, t)\, d\xi - a^2 \int_0^t g(\tau) \left[\frac{\partial}{\partial \xi} G(r, \xi, t - \tau) \right]_{\xi = R} d\tau,$$

where

$$G(r, \xi, t) = \frac{2\xi}{aR} \sum_{n=1}^{\infty} \frac{1}{\lambda_n J_1^2(\lambda_n)} J_0\left(\frac{\lambda_n r}{R}\right) J_0\left(\frac{\lambda_n \xi}{R}\right) \sin\left(\frac{\lambda_n a t}{R}\right).$$

Here, the λ_n are positive zeros of the Bessel function, $J_0(\lambda) = 0$. The numerical values of the first ten λ_n are specified in Paragraph 1.2.1-3.

⊙ *Reference*: B. M. Budak, A. A. Samarskii, and A. N. Tikhonov (1980).

4.2.1-2. Domain: $0 \leq r \leq R$. Second boundary value problem.

The following conditions are prescribed:

$$w = f_0(r) \quad \text{at} \quad t = 0 \quad \text{(initial condition)},$$
$$\partial_t w = f_1(r) \quad \text{at} \quad t = 0 \quad \text{(initial condition)},$$
$$\partial_r w = g(t) \quad \text{at} \quad r = R \quad \text{(boundary condition)}.$$

Solution:

$$w(r,t) = \frac{\partial}{\partial t} \int_0^R f_0(\xi) G(r,\xi,t)\, d\xi + \int_0^R f_1(\xi) G(r,\xi,t)\, d\xi + a^2 \int_0^t g(\tau) G(r,R,t-\tau)\, d\tau,$$

where

$$G(r,\xi,t) = \frac{2t\xi}{R^2} + \frac{2\xi}{aR} \sum_{n=1}^{\infty} \frac{1}{\lambda_n J_0^2(\lambda_n)} J_0\!\left(\frac{\lambda_n r}{R}\right) J_0\!\left(\frac{\lambda_n \xi}{R}\right) \sin\!\left(\frac{\lambda_n a t}{R}\right).$$

Here, the λ_n are positive zeros of the first-order Bessel function, $J_1(\lambda) = 0$. The numerical values of the first ten roots λ_n are specified in Paragraph 1.2.1-4.

⊙ *References*: M. M. Smirnov (1975), B. M. Budak, A. A. Samarskii, and A. N. Tikhonov (1980).

4.2.1-3. Domain: $0 \leq r \leq R$. Third boundary value problem.

The following conditions are prescribed:

$$w = f_0(r) \quad \text{at} \quad t = 0 \quad \text{(initial condition)},$$
$$\partial_t w = f_1(r) \quad \text{at} \quad t = 0 \quad \text{(initial condition)},$$
$$\partial_r w + kw = g(t) \quad \text{at} \quad r = R \quad \text{(boundary condition)}.$$

The solution $w(r,t)$ is determined by the formula in Paragraph 4.2.1-2 where

$$G(r,\xi,t) = \frac{2\xi}{aR} \sum_{n=1}^{\infty} \frac{\lambda_n}{(k^2 R^2 + \lambda_n^2) J_0^2(\lambda_n)} J_0\!\left(\frac{\lambda_n r}{R}\right) J_0\!\left(\frac{\lambda_n \xi}{R}\right) \sin\!\left(\frac{\lambda_n a t}{R}\right).$$

Here, the λ_n are positive roots of the transcendental equation

$$\lambda J_1(\lambda) - kR J_0(\lambda) = 0.$$

The numerical values of the first six roots λ_n can be found in Carslaw and Jaeger (1984); see also Abramowitz and Stegun (1964).

4.2.1-4. Domain: $R_1 \leq r \leq R_2$. First boundary value problem.

The following conditions are prescribed:

$$w = f_0(r) \quad \text{at} \quad t = 0 \quad \text{(initial condition)},$$
$$\partial_t w = f_1(r) \quad \text{at} \quad t = 0 \quad \text{(initial condition)},$$
$$w = g_1(t) \quad \text{at} \quad r = R_1 \quad \text{(boundary condition)},$$
$$w = g_2(t) \quad \text{at} \quad r = R_2 \quad \text{(boundary condition)}.$$

Solution:

$$w(r,t) = \frac{\partial}{\partial t} \int_{R_1}^{R_2} f_0(\xi) G(r,\xi,t)\, d\xi + \int_{R_1}^{R_2} f_1(\xi) G(r,\xi,t)\, d\xi$$
$$+ a^2 \int_0^t g_1(\tau) \left[\frac{\partial}{\partial \xi} G(r,\xi,t-\tau)\right]_{\xi=R_1} d\tau - a^2 \int_0^t g_2(\tau) \left[\frac{\partial}{\partial \xi} G(r,\xi,t-\tau)\right]_{\xi=R_2} d\tau.$$

Here,

$$G(r,\xi,t) = \sum_{n=1}^{\infty} A_n \xi \Psi_n(r) \Psi_n(\xi) \sin\left(\frac{\lambda_n a t}{R_1}\right), \quad A_n = \frac{\pi^2 \lambda_n J_0^2(s\lambda_n)}{2aR_1\left[J_0^2(\lambda_n) - J_0^2(s\lambda_n)\right]},$$

$$\Psi_n(r) = Y_0(\lambda_n) J_0\left(\frac{\lambda_n r}{R_1}\right) - J_0(\lambda_n) Y_0\left(\frac{\lambda_n r}{R_1}\right), \quad s = \frac{R_2}{R_1},$$

where $J_0(z)$ and $Y_0(z)$ are the Bessel functions, the λ_n are positive roots of the transcendental equation

$$J_0(\lambda)Y_0(s\lambda) - J_0(s\lambda)Y_0(\lambda) = 0.$$

The numerical values of the first five roots $\lambda_n = \lambda_n(s)$ can be found in Abramowitz and Stegun (1964) and Carslaw and Jaeger (1984).

4.2.1-5. Domain: $R_1 \leq r \leq R_2$. Second boundary value problem.

The following conditions are prescribed:

$$\begin{array}{lll} w = f_0(r) & \text{at} \quad t=0 & \text{(initial condition)}, \\ \partial_t w = f_1(r) & \text{at} \quad t=0 & \text{(initial condition)}, \\ \partial_r w = g_1(t) & \text{at} \quad r=R_1 & \text{(boundary condition)}, \\ \partial_r w = g_2(t) & \text{at} \quad r=R_2 & \text{(boundary condition)}. \end{array}$$

Solution:

$$w(r,t) = \frac{\partial}{\partial t} \int_{R_1}^{R_2} f_0(\xi) G(r,\xi,t)\, d\xi + \int_{R_1}^{R_2} f_1(\xi) G(r,\xi,t)\, d\xi$$
$$- a^2 \int_0^t g_1(\tau) G(r,R_1,t-\tau)\, d\tau + a^2 \int_0^t g_2(\tau) G(r,R_2,t-\tau)\, d\tau.$$

Here,

$$G(r,\xi,t) = \frac{2t\xi}{R_2^2 - R_1^2} + \sum_{n=1}^{\infty} A_n \xi \Psi_n(r) \Psi_n(\xi) \sin\left(\frac{\lambda_n a t}{R_1}\right), \quad A_n = \frac{\pi^2 \lambda_n J_1^2(s\lambda_n)}{2aR_1\left[J_1^2(\lambda_n) - J_1^2(s\lambda_n)\right]},$$

$$\Psi_n(r) = Y_1(\lambda_n) J_0\left(\frac{\lambda_n r}{R_1}\right) - J_1(\lambda_n) Y_0\left(\frac{\lambda_n r}{R_1}\right), \quad s = \frac{R_2}{R_1},$$

where $J_k(z)$ and $Y_k(z)$ are the Bessel functions ($k = 0, 1$); the λ_n are positive roots of the transcendental equation

$$J_1(\lambda)Y_1(s\lambda) - J_1(s\lambda)Y_1(\lambda) = 0.$$

The numerical values of the first five roots $\lambda_n = \lambda_n(s)$ can be found in Abramowitz and Stegun (1964).

4.2.1-6. Domain: $R_1 \leq r \leq R_2$. Third boundary value problem.

The following conditions are prescribed:

$$\begin{array}{lll} w = f_0(r) & \text{at} \quad t=0 & \text{(initial condition)}, \\ \partial_t w = f_1(r) & \text{at} \quad t=0 & \text{(initial condition)}, \\ \partial_r w - k_1 w = g_1(t) & \text{at} \quad r=R_1 & \text{(boundary condition)}, \\ \partial_r w + k_2 w = g_2(t) & \text{at} \quad r=R_2 & \text{(boundary condition)}. \end{array}$$

The solution $w(r, t)$ is determined by the formula in Paragraph 4.2.1-5 where

$$G(r, \xi, t) = \frac{\pi^2}{2a} \sum_{n=1}^{\infty} \frac{\mu_n}{B_n} \left[k_2 J_0(\mu_n R_2) - \mu_n J_1(\mu_n R_2)\right]^2 \xi H_n(r) H_n(\xi) \sin(\mu_n a t).$$

Here,

$$B_n = (\mu_n^2 + k_2^2)\left[k_1 J_0(\mu_n R_1) + \mu_n J_1(\mu_n R_1)\right]^2 - (\mu_n^2 + k_1^2)\left[k_2 J_0(\mu_n R_2) - \mu_n J_1(\mu_n R_2)\right]^2,$$
$$H_n(r) = \left[k_1 Y_0(\mu_n R_1) + \mu_n Y_1(\mu_n R_1)\right] J_0(\mu_n r) - \left[k_1 J_0(\mu_n R_1) + \mu_n J_1(\mu_n R_1)\right] Y_0(\mu_n r);$$

$J_k(z)$ and $Y_k(z)$ are the Bessel functions ($k=0$, 1); and the μ_n are positive roots of the transcendental equation

$$\left[k_1 J_0(\mu R_1) + \mu J_1(\mu R_1)\right]\left[k_2 Y_0(\mu R_2) - \mu Y_1(\mu R_2)\right]$$
$$- \left[k_2 J_0(\mu R_2) - \mu J_1(\mu R_2)\right]\left[k_1 Y_0(\mu R_1) + \mu Y_1(\mu R_1)\right] = 0.$$

4.2.2. Equation of the Form $\frac{\partial^2 w}{\partial t^2} = a^2\left(\frac{\partial^2 w}{\partial r^2} + \frac{1}{r}\frac{\partial w}{\partial r}\right) + \Phi(r, t)$

4.2.2-1. Domain: $0 \leq r \leq R$. Different boundary value problems.

1°. The solution to the first boundary value problem for a circle of radius R is given by the formula from Paragraph 4.2.1-1 with the additional term

$$\int_0^t \int_0^R \Phi(\xi, \tau) G(r, \xi, t - \tau) \, d\xi \, d\tau, \tag{1}$$

which allows for the equation's nonhomogeneity.

2°. The solution to the second boundary value problem for a circle of radius R is given by the formula from Paragraph 4.2.1-2 with the additional term (1).

3°. The solution to the third boundary value problem for a circle of radius R is the sum of the solution presented in Paragraph 4.2.1-3 and expression (1).

4.2.2-2. Domain: $R_1 \leq r \leq R_2$. Different boundary value problems.

1°. The solution to the first boundary value problem for an annular domain is given by the formula from Paragraph 4.2.1-4 with the additional term

$$\int_0^t \int_{R_1}^{R_2} \Phi(\xi, \tau) G(r, \xi, t - \tau) \, d\xi \, d\tau, \tag{2}$$

which allows for the equation's nonhomogeneity.

2°. The solution to the second boundary value problem for an annular domain is given by the formula from Paragraph 4.2.1-5 with the additional term (2).

3°. The solution to the third boundary value problem for an annular domain is the sum of the solution presented in Paragraph 4.2.1-6 and expression (2).

4.2.3. Equation of the Form $\frac{\partial^2 w}{\partial t^2} = a^2\left(\frac{\partial^2 w}{\partial r^2} + \frac{2}{r}\frac{\partial w}{\partial r}\right)$

This is the equation of one-dimensional vibration of a gas with central symmetry, where $r = \sqrt{x^2 + y^2 + z^2}$ is the radial coordinate. In the problems considered in Paragraphs 4.2.3-1 through 4.2.3-3, the solutions bounded at $r = 0$ are sought; this is not specially stated below.

4.2.3-1. General solution:

$$w(t,r) = \frac{\varphi(r+at) + \psi(r-at)}{r},$$

where $\varphi(r_1)$ and $\psi(r_2)$ are arbitrary functions.

4.2.3-2. Reduction to a constant coefficient equation.

The substitution $u(r,t) = rw(r,t)$ leads to the constant coefficient equation

$$\frac{\partial^2 u}{\partial t^2} = a^2 \frac{\partial^2 u}{\partial r^2},$$

which is discussed in Subsection 4.1.1.

4.2.3-3. Domain: $0 \leq r < \infty$. Cauchy problem.

Initial conditions are prescribed:

$$w = f(r) \quad \text{at} \quad t = 0,$$
$$\partial_t w = g(r) \quad \text{at} \quad t = 0.$$

Solution:

$$w(r,t) = \frac{1}{2r}\left[(r-at)f(|r-at|) + (r+at)f(|r+at|)\right] + \frac{1}{2ar}\int_{r-at}^{r+at} \xi g(|\xi|)\,d\xi.$$

Solution at the center $r = 0$:

$$w(0,t) = at f'(at) + f(at) + t g(at).$$

⊙ *Reference*: B. M. Budak, A. A. Samarskii, and A. N. Tikhonov (1980).

4.2.3-4. Domain: $0 \leq r \leq R$. First boundary value problem.

The following conditions are prescribed:

$$w = f_0(r) \quad \text{at} \quad t = 0 \quad \text{(initial condition)},$$
$$\partial_t w = f_1(r) \quad \text{at} \quad t = 0 \quad \text{(initial condition)},$$
$$w = g(t) \quad \text{at} \quad r = R \quad \text{(boundary condition)}.$$

Solution:

$$w(r,t) = \frac{\partial}{\partial t}\int_0^R f_0(\xi)G(r,\xi,t)\,d\xi + \int_0^R f_1(\xi)G(r,\xi,t)\,d\xi - a^2\int_0^t g(\tau)\left[\frac{\partial}{\partial \xi}G(r,\xi,t-\tau)\right]_{\xi=R}\,d\tau,$$

where

$$G(r,\xi,t) = \frac{2\xi}{\pi ar}\sum_{n=1}^{\infty}\frac{1}{n}\sin\left(\frac{n\pi r}{R}\right)\sin\left(\frac{n\pi \xi}{R}\right)\sin\left(\frac{an\pi t}{R}\right).$$

4.2.3-5. Domain: $0 \leq r \leq R$. Second boundary value problem.

The following conditions are prescribed:

$$w = f_0(r) \quad \text{at} \quad t = 0 \quad \text{(initial condition)},$$
$$\partial_t w = f_1(r) \quad \text{at} \quad t = 0 \quad \text{(initial condition)},$$
$$\partial_r w = g(t) \quad \text{at} \quad r = R \quad \text{(boundary condition)}.$$

Solution:

$$w(r,t) = \frac{\partial}{\partial t}\int_0^R f_0(\xi)G(r,\xi,t)\,d\xi + \int_0^R f_1(\xi)G(r,\xi,t)\,d\xi + a^2\int_0^t g(\tau)G(r,R,t-\tau)\,d\tau,$$

where

$$G(r,\xi,t) = \frac{3t\xi^2}{R^3} + \frac{2\xi}{ar}\sum_{n=1}^{\infty}\frac{\mu_n^2+1}{\mu_n^3}\sin\left(\frac{\mu_n r}{R}\right)\sin\left(\frac{\mu_n \xi}{R}\right)\sin\left(\frac{\mu_n at}{R}\right).$$

Here, the μ_n are positive roots of the transcendental equation $\tan\mu - \mu = 0$. The numerical values of the first five roots μ_n are specified in Paragraph 1.2.3-5.

4.2.3-6. Domain: $0 \le r \le R$. Third boundary value problem.

The following conditions are prescribed:

$$\begin{aligned}
w &= f_0(r) & \text{at} \quad t &= 0 & \text{(initial condition)},\\
\partial_t w &= f_1(r) & \text{at} \quad t &= 0 & \text{(initial condition)},\\
\partial_r w + kw &= g(t) & \text{at} \quad r &= R & \text{(boundary condition)}.
\end{aligned}$$

The solution $w(r,t)$ is determined by the formula in Paragraph 4.2.3-5 where

$$G(r,\xi,t) = \frac{2\xi}{ar}\sum_{n=1}^{\infty}\frac{\mu_n^2+(kR-1)^2}{\mu_n[\mu_n^2+kR(kR-1)]}\sin\left(\frac{\mu_n r}{R}\right)\sin\left(\frac{\mu_n \xi}{R}\right)\sin\left(\frac{\mu_n at}{R}\right).$$

Here, the μ_n are positive roots of the transcendental equation

$$\mu\cot\mu + kR - 1 = 0.$$

The numerical values of the first six roots μ_n can be found in Carslaw and Jaeger (1984).

4.2.3-7. Domain: $R_1 \le r \le R_2$. First boundary value problem.

The following conditions are prescribed:

$$\begin{aligned}
w &= f_0(r) & \text{at} \quad t &= 0 & \text{(initial condition)},\\
\partial_t w &= f_1(r) & \text{at} \quad t &= 0 & \text{(initial condition)},\\
w &= g_1(t) & \text{at} \quad r &= R_1 & \text{(boundary condition)},\\
w &= g_2(t) & \text{at} \quad r &= R_2 & \text{(boundary condition)}.
\end{aligned}$$

Solution:

$$\begin{aligned}
w(r,t) &= \frac{\partial}{\partial t}\int_{R_1}^{R_2} f_0(\xi)G(r,\xi,t)\,d\xi + \int_{R_1}^{R_2} f_1(\xi)G(r,\xi,t)\,d\xi\\
&\quad + a^2\int_0^t g_1(\tau)\left[\frac{\partial}{\partial\xi}G(r,\xi,t-\tau)\right]_{\xi=R_1}d\tau - a^2\int_0^t g_2(\tau)\left[\frac{\partial}{\partial\xi}G(r,\xi,t-\tau)\right]_{\xi=R_2}d\tau,
\end{aligned}$$

where

$$G(r,\xi,t) = \frac{2\xi}{\pi ar}\sum_{n=1}^{\infty}\frac{1}{n}\sin\left[\frac{\pi n(r-R_1)}{R_2-R_1}\right]\sin\left[\frac{\pi n(\xi-R_1)}{R_2-R_1}\right]\sin\left(\frac{\pi nat}{R_2-R_1}\right).$$

4.2.3-8. Domain: $R_1 \leq r \leq R_2$. Second boundary value problem.

The following conditions are prescribed:

$$w = f_0(r) \quad \text{at} \quad t = 0 \quad \text{(initial condition)},$$
$$\partial_t w = f_1(r) \quad \text{at} \quad t = 0 \quad \text{(initial condition)},$$
$$\partial_r w = g_1(t) \quad \text{at} \quad r = R_1 \quad \text{(boundary condition)},$$
$$\partial_r w = g_2(t) \quad \text{at} \quad r = R_2 \quad \text{(boundary condition)}.$$

Solution:

$$w(r,t) = \frac{\partial}{\partial t} \int_{R_1}^{R_2} f_0(\xi) G(r,\xi,t)\, d\xi + \int_{R_1}^{R_2} f_1(\xi) G(r,\xi,t)\, d\xi$$
$$- a^2 \int_0^t g_1(\tau) G(r, R_1, t-\tau)\, d\tau + a^2 \int_0^t g_2(\tau) G(r, R_2, t-\tau)\, d\tau,$$

where

$$G(r,\xi,t) = \frac{3t\xi^2}{R_2^3 - R_1^3} + \frac{2\xi}{a(R_2 - R_1)r} \sum_{n=1}^{\infty} \frac{(1 + R_2^2 \lambda_n^2) \Psi_n(r) \Psi_n(\xi) \sin(\lambda_n a t)}{\lambda_n^3 [R_1^2 + R_2^2 + R_1 R_2 (1 + R_1 R_2 \lambda_n^2)]},$$
$$\Psi_n(r) = \sin[\lambda_n(r - R_1)] + R_1 \lambda_n \cos[\lambda_n(r - R_1)].$$

Here, the λ_n are positive roots of the transcendental equation

$$(\lambda^2 R_1 R_2 + 1) \tan[\lambda(R_2 - R_1)] - \lambda(R_2 - R_1) = 0.$$

4.2.3-9. Domain: $R_1 \leq r \leq R_2$. Third boundary value problem.

The following conditions are prescribed:

$$w = f_0(r) \quad \text{at} \quad t = 0 \quad \text{(initial condition)},$$
$$\partial_t w = f_1(r) \quad \text{at} \quad t = 0 \quad \text{(initial condition)},$$
$$\partial_r w - k_1 w = g_1(t) \quad \text{at} \quad r = R_1 \quad \text{(boundary condition)},$$
$$\partial_r w + k_2 w = g_2(t) \quad \text{at} \quad r = R_2 \quad \text{(boundary condition)}.$$

The solution $w(r,t)$ is determined by the formula in Paragraph 4.2.3-8 where

$$G(r,\xi,t) = \frac{2\xi}{a(R_2 - R_1)r} \sum_{n=1}^{\infty} \frac{1}{\lambda_n} \frac{(b_2^2 + R_2^2 \lambda_n^2) \Psi_n(r) \Psi_n(\xi) \sin(\lambda_n a t)}{(b_1^2 + R_1^2 \lambda_n^2)(b_2^2 + R_2^2 \lambda_n^2) + (b_1 R_2 + b_2 R_1)(b_1 b_2 + R_1 R_2 \lambda_n^2)},$$
$$\Psi_n(r) = b_1 \sin[\lambda_n(r - R_1)] + R_1 \lambda_n \cos[\lambda_n(r - R_1)], \quad b_1 = k_1 R_1 + 1, \quad b_2 = k_2 R_2 - 1.$$

Here, the λ_n are positive roots of the transcendental equation

$$(b_1 b_2 - R_1 R_2 \lambda^2) \sin[\lambda(R_2 - R_1)] + \lambda(R_1 b_2 + R_2 b_1) \cos[\lambda(R_2 - R_1)] = 0.$$

4.2.4. Equation of the Form $\frac{\partial^2 w}{\partial t^2} = a^2 \left(\frac{\partial^2 w}{\partial r^2} + \frac{2}{r} \frac{\partial w}{\partial r} \right) + \Phi(r,t)$

4.2.4-1. Reduction to a nonhomogeneous constant coefficient equation.

The substitution $u(r,t) = rw(r,t)$ leads to the nonhomogeneous constant coefficient equation

$$\frac{\partial^2 u}{\partial t^2} = a^2 \frac{\partial^2 u}{\partial r^2} + r\Phi(r,t),$$

which is discussed in Subsection 4.1.2.

4.2.4-2. Domain: $0 \le r < \infty$. Cauchy problem.

Initial conditions are prescribed:
$$w = f(r) \quad \text{at} \quad t = 0,$$
$$\partial_t w = g(r) \quad \text{at} \quad t = 0.$$

Solution:
$$w(r,t) = \frac{1}{2r}\left[(r-at)f(|r-at|) + (r+at)f(|r+at|)\right] + \frac{1}{2ar}\int_{r-at}^{r+at} \xi g(|\xi|)\,d\xi$$
$$+ \frac{1}{2ar}\int_0^t d\tau \int_{r-a(t-\tau)}^{r+a(t-\tau)} \xi \Phi(|\xi|,\tau)\,d\xi.$$

⊙ *Reference*: B. M. Budak, A. A. Samarskii, and A. N. Tikhonov (1980).

4.2.4-3. Domain: $0 \le r \le R$. Different boundary value problems.

1°. The solution to the first boundary value problem for a sphere of radius R is given by the formula from Paragraph 4.2.3-4 with the additional term

$$\int_0^t \int_0^R \Phi(\xi,\tau) G(r,\xi,t-\tau)\,d\xi\,d\tau, \tag{1}$$

which allows for the equation's nonhomogeneity.

2°. The solution to the second boundary value problem for a sphere of radius R is given by the formula from Paragraph 4.2.3-5 with the additional term (1).

3°. The solution to the third boundary value problem for a sphere of radius R is the sum of the solution presented in Paragraph 4.2.3-6 and expression (1).

4.2.4-4. Domain: $R_1 \le r \le R_2$. Different boundary value problems.

1°. The solution to the first boundary value problem for a spherical layer is given by the formula from Paragraph 4.2.3-7 with the additional term

$$\int_0^t \int_{R_1}^{R_2} \Phi(\xi,\tau) G(r,\xi,t-\tau)\,d\xi\,d\tau, \tag{2}$$

which allows for the equation's nonhomogeneity.

2°. The solution to the second boundary value problem for a spherical layer is given by the formula from Paragraph 4.2.3-8 with the additional term (2).

3°. The solution to the third boundary value problem for a spherical layer is the sum of the solution presented in Paragraph 4.2.3-9 and expression (2).

4.2.5. Equation of the Form $\frac{\partial^2 w}{\partial t^2} = a^2\left(\frac{\partial^2 w}{\partial r^2} + \frac{1}{r}\frac{\partial w}{\partial r}\right) - bw + \Phi(r,t)$

For $b > 0$ and $\Phi \equiv 0$, this is the Klein–Gordon equation describing one-dimensional wave phenomena with axial symmetry. In the problems considered in Paragraphs 4.2.5-1 through 4.2.5-3, the solutions bounded at $r = 0$ are sought; this is not specially stated below.

4.2.5-1. Domain: $0 \le r \le R$. First boundary value problem.

The following conditions are prescribed:

$$w = f_0(r) \quad \text{at} \quad t = 0 \quad \text{(initial condition)},$$
$$\partial_t w = f_1(r) \quad \text{at} \quad t = 0 \quad \text{(initial condition)},$$
$$w = g(t) \quad \text{at} \quad r = R \quad \text{(boundary condition)}.$$

Solution:

$$w(r,t) = \frac{\partial}{\partial t} \int_0^R f_0(\xi) G(r,\xi,t)\, d\xi + \int_0^R f_1(\xi) G(r,\xi,t)\, d\xi$$
$$- a^2 \int_0^t g(\tau) \left[\frac{\partial}{\partial \xi} G(r,\xi,t-\tau) \right]_{\xi=R} d\tau + \int_0^t \int_0^R \Phi(\xi,\tau) G(r,\xi,t-\tau)\, d\xi\, d\tau.$$

Here,

$$G(r,\xi,t) = \frac{2\xi}{R^2} \sum_{n=1}^\infty \frac{1}{J_1^2(\mu_n)} J_0\!\left(\frac{\mu_n r}{R}\right) J_0\!\left(\frac{\mu_n \xi}{R}\right) \frac{\sin(t\sqrt{\lambda_n})}{\sqrt{\lambda_n}}, \qquad \lambda_n = \frac{a^2 \mu_n^2}{R^2} + b,$$

where the μ_n are positive zeros of the Bessel function, $J_0(\mu) = 0$. The numerical values of the first ten μ_n are specified in Paragraph 1.2.1-3.

4.2.5-2. Domain: $0 \le r \le R$. Second boundary value problem.

The following conditions are prescribed:

$$w = f_0(r) \quad \text{at} \quad t = 0 \quad \text{(initial condition)},$$
$$\partial_t w = f_1(r) \quad \text{at} \quad t = 0 \quad \text{(initial condition)},$$
$$\partial_r w = g(t) \quad \text{at} \quad r = R \quad \text{(boundary condition)}.$$

Solution:

$$w(r,t) = \frac{\partial}{\partial t} \int_0^R f_0(\xi) G(r,\xi,t)\, d\xi + \int_0^R f_1(\xi) G(r,\xi,t)\, d\xi$$
$$+ a^2 \int_0^t g(\tau) G(r,R,t-\tau)\, d\tau + \int_0^t \int_0^R \Phi(\xi,\tau) G(r,\xi,t-\tau)\, d\xi\, d\tau.$$

Here,

$$G(r,\xi,t) = \frac{2\xi \sin(t\sqrt{b})}{R^2 \sqrt{b}} + \frac{2\xi}{R^2} \sum_{n=1}^\infty \frac{1}{J_0^2(\mu_n)} J_0\!\left(\frac{\mu_n r}{R}\right) J_0\!\left(\frac{\mu_n \xi}{R}\right) \frac{\sin(t\sqrt{\lambda_n})}{\sqrt{\lambda_n}}, \qquad \lambda_n = \frac{a^2 \mu_n^2}{R^2} + b,$$

where the μ_n are positive zeros of the first-order Bessel function, $J_1(\mu) = 0$. The numerical values of the first ten μ_n are specified in Paragraph 1.2.1-4.

4.2.5-3. Domain: $0 \le r \le R$. Third boundary value problem.

The following conditions are prescribed:

$$w = f_0(r) \quad \text{at} \quad t = 0 \quad \text{(initial condition)},$$
$$\partial_t w = f_1(r) \quad \text{at} \quad t = 0 \quad \text{(initial condition)},$$
$$\partial_r w + kw = g(t) \quad \text{at} \quad r = R \quad \text{(boundary condition)}.$$

The solution $w(r,t)$ is determined by the formula in Paragraph 4.2.5-2 where

$$G(r,\xi,t) = \frac{2}{R^2} \sum_{n=1}^{\infty} \frac{\mu_n^2 \xi}{(k^2 R^2 + \mu_n^2) J_0^2(\mu_n)} J_0\left(\frac{\mu_n r}{R}\right) J_0\left(\frac{\mu_n \xi}{R}\right) \frac{\sin(t\sqrt{\lambda_n})}{\sqrt{\lambda_n}}, \qquad \lambda_n = \frac{a^2 \mu_n^2}{R^2} + b.$$

Here, the μ_n are positive roots of the transcendental equation

$$\mu J_1(\mu) - kR J_0(\mu) = 0.$$

The numerical values of the first six roots μ_n can be found in Abramowitz and Stegun (1964) and Carslaw and Jaeger (1984).

4.2.5-4. Domain: $R_1 \leq r \leq R_2$. First boundary value problem.

The following conditions are prescribed:

$$\begin{aligned}
w &= f_0(r) & \text{at} \quad t &= 0 & &\text{(initial condition)}, \\
\partial_t w &= f_1(r) & \text{at} \quad t &= 0 & &\text{(initial condition)}, \\
w &= g_1(t) & \text{at} \quad r &= R_1 & &\text{(boundary condition)}, \\
w &= g_2(t) & \text{at} \quad r &= R_2 & &\text{(boundary condition)}.
\end{aligned}$$

Solution:

$$\begin{aligned}
w(r,t) &= \int_0^t \int_{R_1}^{R_2} \Phi(\xi,\tau) G(r,\xi,t-\tau)\, d\xi\, d\tau \\
&+ \frac{\partial}{\partial t} \int_{R_1}^{R_2} f_0(\xi) G(r,\xi,t)\, d\xi + \int_{R_1}^{R_2} f_1(\xi) G(r,\xi,t)\, d\xi \\
&+ a^2 \int_0^t g_1(\tau) \left[\frac{\partial}{\partial \xi} G(r,\xi,t-\tau)\right]_{\xi=R_1} d\tau - a^2 \int_0^t g_2(\tau) \left[\frac{\partial}{\partial \xi} G(r,\xi,t-\tau)\right]_{\xi=R_2} d\tau.
\end{aligned}$$

Here,

$$G(r,\xi,t) = \frac{\pi^2}{2R_1^2} \sum_{n=1}^{\infty} \frac{\mu_n^2 J_0^2(s\mu_n) \xi}{J_0^2(\mu_n) - J_0^2(s\mu_n)} \Psi_n(r) \Psi_n(\xi) \frac{\sin(t\sqrt{\lambda_n})}{\sqrt{\lambda_n}}, \qquad \lambda_n = \frac{a^2 \mu_n^2}{R_1^2} + b,$$

$$\Psi_n(r) = Y_0(\mu_n) J_0\left(\frac{\mu_n r}{R_1}\right) - J_0(\mu_n) Y_0\left(\frac{\mu_n r}{R_1}\right), \qquad s = \frac{R_2}{R_1},$$

where $J_0(z)$ and $Y_0(z)$ are the Bessel functions and the μ_n are positive roots of the transcendental equation

$$J_0(\mu) Y_0(s\mu) - J_0(s\mu) Y_0(\mu) = 0.$$

The numerical values of the first five roots $\mu_n = \mu_n(s)$ can be found in Abramowitz and Stegun (1964) and Carslaw and Jaeger (1984).

4.2.5-5. Domain: $R_1 \leq r \leq R_2$. Second boundary value problem.

The following conditions are prescribed:

$$\begin{aligned}
w &= f_0(r) & \text{at} \quad t &= 0 & &\text{(initial condition)}, \\
\partial_t w &= f_1(r) & \text{at} \quad t &= 0 & &\text{(initial condition)}, \\
\partial_r w &= g_1(t) & \text{at} \quad r &= R_1 & &\text{(boundary condition)}, \\
\partial_r w &= g_2(t) & \text{at} \quad r &= R_2 & &\text{(boundary condition)}.
\end{aligned}$$

Solution:

$$w(r,t) = \int_0^t \int_{R_1}^{R_2} \Phi(\xi,\tau) G(r,\xi,t-\tau) \, d\xi \, d\tau$$
$$+ \frac{\partial}{\partial t} \int_{R_1}^{R_2} f_0(\xi) G(r,\xi,t) \, d\xi + \int_{R_1}^{R_2} f_1(\xi) G(r,\xi,t) \, d\xi$$
$$- a^2 \int_0^t g_1(\tau) G(r,R_1,t-\tau) \, d\tau + a^2 \int_0^t g_2(\tau) G(r,R_2,t-\tau) \, d\tau.$$

Here,

$$G(r,\xi,t) = \frac{2\xi \sin(t\sqrt{b})}{(R_2^2 - R_1^2)\sqrt{b}} + \frac{\pi^2}{2R_1^2} \sum_{n=1}^{\infty} \frac{\mu_n^2 J_1^2(s\mu_n)\xi}{J_1^2(\mu_n) - J_1^2(s\mu_n)} \Psi_n(r)\Psi_n(\xi) \frac{\sin(t\sqrt{\lambda_n})}{\sqrt{\lambda_n}},$$

$$\Psi_n(r) = Y_1(\mu_n) J_0\!\left(\frac{\mu_n r}{R_1}\right) - J_1(\mu_n) Y_0\!\left(\frac{\mu_n r}{R_1}\right), \quad \lambda_n = \frac{a^2 \mu_n^2}{R_1^2} + b, \quad s = \frac{R_2}{R_1},$$

where $J_k(z)$ and $Y_k(z)$ are the Bessel functions ($k = 0, 1$); the μ_n are positive roots of the transcendental equation

$$J_1(\mu) Y_1(s\mu) - J_1(s\mu) Y_1(\mu) = 0.$$

The numerical values of the first five roots $\mu_n = \mu_n(s)$ can be found in Abramowitz and Stegun (1964).

$\boxed{\text{4.2.5-6. Domain: } R_1 \leq r \leq R_2. \text{ Third boundary value problem.}}$

The following conditions are prescribed:

$$\begin{aligned}
w &= f_0(r) & \text{at} \quad t &= 0 & &\text{(initial condition)}, \\
\partial_t w &= f_1(r) & \text{at} \quad t &= 0 & &\text{(initial condition)}, \\
\partial_r w - k_1 w &= g_1(t) & \text{at} \quad r &= R_1 & &\text{(boundary condition)}, \\
\partial_r w + k_2 w &= g_2(t) & \text{at} \quad r &= R_2 & &\text{(boundary condition)}.
\end{aligned}$$

The solution $w(r,t)$ is determined by the formula in Paragraph 4.2.5-5 where

$$G(r,\xi,t) = \frac{\pi^2}{2} \sum_{n=1}^{\infty} \frac{\beta_n^2}{B_n \sqrt{a^2 \beta_n^2 + b}} \left[k_2 J_0(\beta_n R_2) - \beta_n J_1(\beta_n R_2)\right]^2 \xi H_n(r) H_n(\xi) \sin\!\left(t\sqrt{a^2\beta_n^2 + b}\right).$$

Here,

$$B_n = (\beta_n^2 + k_2^2)\left[k_1 J_0(\beta_n R_1) + \beta_n J_1(\beta_n R_1)\right]^2 - (\beta_n^2 + k_1^2)\left[k_2 J_0(\beta_n R_2) - \beta_n J_1(\beta_n R_2)\right]^2,$$
$$H_n(r) = \left[k_1 Y_0(\beta_n R_1) + \beta_n Y_1(\beta_n R_1)\right] J_0(\beta_n r) - \left[k_1 J_0(\beta_n R_1) + \beta_n J_1(\beta_n R_1)\right] Y_0(\beta_n r),$$

where the β_n are positive roots of the transcendental equation

$$\left[k_1 J_0(\beta R_1) + \beta J_1(\beta R_1)\right]\left[k_2 Y_0(\beta R_2) - \beta Y_1(\beta R_2)\right]$$
$$- \left[k_2 J_0(\beta R_2) - \beta J_1(\beta R_2)\right]\left[k_1 Y_0(\beta R_1) + \beta Y_1(\beta R_1)\right] = 0.$$

4.2.6. Equation of the Form $\dfrac{\partial^2 w}{\partial t^2} = a^2\!\left(\dfrac{\partial^2 w}{\partial r^2} + \dfrac{2}{r}\dfrac{\partial w}{\partial r}\right) - bw + \Phi(r,t)$

For $b > 0$ and $\Phi \equiv 0$, this is the Klein–Gordon equation describing one-dimensional wave phenomena with central symmetry. In the problems considered in Paragraphs 4.2.6-1 through 4.2.6-3, the solutions bounded at $r = 0$ are sought; this is not specially stated below.

4.2.6-1. Domain: $0 \le r \le R$. First boundary value problem.

The following conditions are prescribed:

$$w = f_0(r) \quad \text{at} \quad t = 0 \quad \text{(initial condition)},$$
$$\partial_t w = f_1(r) \quad \text{at} \quad t = 0 \quad \text{(initial condition)},$$
$$w = g(t) \quad \text{at} \quad r = R \quad \text{(boundary condition)}.$$

Solution:

$$w(r,t) = \frac{\partial}{\partial t} \int_0^R f_0(\xi) G(r,\xi,t) \, d\xi + \int_0^R f_1(\xi) G(r,\xi,t) \, d\xi$$
$$- a^2 \int_0^t g(\tau) \left[\frac{\partial}{\partial \xi} G(r,\xi,t-\tau) \right]_{\xi=R} d\tau + \int_0^t \int_0^R \Phi(\xi,\tau) G(r,\xi,t-\tau) \, d\xi \, d\tau,$$

where

$$G(r,\xi,t) = \frac{2\xi}{Rr} \sum_{n=1}^{\infty} \sin\left(\frac{n\pi r}{R}\right) \sin\left(\frac{n\pi \xi}{R}\right) \frac{\sin(t\sqrt{\lambda_n})}{\sqrt{\lambda_n}}, \quad \lambda_n = \frac{a^2 \pi^2 n^2}{R^2} + b.$$

4.2.6-2. Domain: $0 \le r \le R$. Second boundary value problem.

The following conditions are prescribed:

$$w = f_0(r) \quad \text{at} \quad t = 0 \quad \text{(initial condition)},$$
$$\partial_t w = f_1(r) \quad \text{at} \quad t = 0 \quad \text{(initial condition)},$$
$$\partial_r w = g(t) \quad \text{at} \quad r = R \quad \text{(boundary condition)}.$$

Solution:

$$w(r,t) = \frac{\partial}{\partial t} \int_0^R f_0(\xi) G(r,\xi,t) \, d\xi + \int_0^R f_1(\xi) G(r,\xi,t) \, d\xi$$
$$+ a^2 \int_0^t g(\tau) G(r,R,t-\tau) \, d\tau + \int_0^t \int_0^R \Phi(\xi,\tau) G(r,\xi,t-\tau) \, d\xi \, d\tau,$$

where

$$G(r,\xi,t) = \frac{3\xi^2 \sin(t\sqrt{b})}{R^3 \sqrt{b}} + \frac{2\xi}{Rr} \sum_{n=1}^{\infty} \frac{\mu_n^2 + 1}{\mu_n^2 \sqrt{\lambda_n}} \sin\left(\frac{\mu_n r}{R}\right) \sin\left(\frac{\mu_n \xi}{R}\right) \sin(t\sqrt{\lambda_n}), \quad \lambda_n = \frac{a^2 \mu_n^2}{R^2} + b.$$

Here, the μ_n are positive roots of the transcendental equation $\tan\mu - \mu = 0$; for the numerical values of the first five roots μ_n, see Paragraph 1.2.3-5.

4.2.6-3. Domain: $0 \le r \le R$. Third boundary value problem.

The following conditions are prescribed:

$$w = f_0(r) \quad \text{at} \quad t = 0 \quad \text{(initial condition)},$$
$$\partial_t w = f_1(r) \quad \text{at} \quad t = 0 \quad \text{(initial condition)},$$
$$\partial_r w + kw = g(t) \quad \text{at} \quad r = R \quad \text{(boundary condition)}.$$

The solution $w(r,t)$ is determined by the formula in Paragraph 4.2.6-2 where

$$G(r,\xi,t) = \frac{2\xi}{Rr} \sum_{n=1}^{\infty} \frac{\mu_n^2 + (kR-1)^2}{\mu_n^2 + kR(kR-1)} \sin\left(\frac{\mu_n r}{R}\right) \sin\left(\frac{\mu_n \xi}{R}\right) \frac{\sin(t\sqrt{\lambda_n})}{\sqrt{\lambda_n}}, \quad \lambda_n = \frac{a^2 \mu_n^2}{R^2} + b.$$

Here, the μ_n are positive roots of the transcendental equation $\mu \cot\mu + kR - 1 = 0$. The numerical values of the six five roots μ_n can be found in Carslaw and Jaeger (1984).

4.2.6-4. Domain: $R_1 \leq r \leq R_2$. First boundary value problem.

The following conditions are prescribed:

$$w = f_0(r) \quad \text{at} \quad t = 0 \quad \text{(initial condition)},$$
$$\partial_t w = f_1(r) \quad \text{at} \quad t = 0 \quad \text{(initial condition)},$$
$$w = g_1(t) \quad \text{at} \quad r = R_1 \quad \text{(boundary condition)},$$
$$w = g_2(t) \quad \text{at} \quad r = R_2 \quad \text{(boundary condition)}.$$

Solution:

$$w(r,t) = \int_0^t \int_{R_1}^{R_2} \Phi(\xi,\tau) G(r,\xi,t-\tau) \, d\xi \, d\tau$$
$$+ \frac{\partial}{\partial t} \int_{R_1}^{R_2} f_0(\xi) G(r,\xi,t) \, d\xi + \int_{R_1}^{R_2} f_1(\xi) G(r,\xi,t) \, d\xi$$
$$+ a^2 \int_0^t g_1(\tau) \left[\frac{\partial}{\partial \xi} G(r,\xi,t-\tau) \right]_{\xi=R_1} d\tau - a^2 \int_0^t g_2(\tau) \left[\frac{\partial}{\partial \xi} G(r,\xi,t-\tau) \right]_{\xi=R_2} d\tau,$$

where

$$G(r,\xi,t) = \frac{2\xi}{(R_2-R_1)r} \sum_{n=1}^{\infty} \sin\left[\frac{\pi n(r-R_1)}{R_2-R_1}\right] \sin\left[\frac{\pi n(\xi-R_1)}{R_2-R_1}\right] \frac{\sin(t\sqrt{\lambda_n})}{\sqrt{\lambda_n}}, \quad \lambda_n = \frac{a^2\pi^2 n^2}{(R_2-R_1)^2} + b.$$

4.2.6-5. Domain: $R_1 \leq r \leq R_2$. Second boundary value problem.

The following conditions are prescribed:

$$w = f_0(r) \quad \text{at} \quad t = 0 \quad \text{(initial condition)},$$
$$\partial_t w = f_1(r) \quad \text{at} \quad t = 0 \quad \text{(initial condition)},$$
$$\partial_r w = g_1(t) \quad \text{at} \quad r = R_1 \quad \text{(boundary condition)},$$
$$\partial_r w = g_2(t) \quad \text{at} \quad r = R_2 \quad \text{(boundary condition)}.$$

Solution:

$$w(r,t) = \int_0^t \int_{R_1}^{R_2} \Phi(\xi,\tau) G(r,\xi,t-\tau) \, d\xi \, d\tau$$
$$+ \frac{\partial}{\partial t} \int_{R_1}^{R_2} f_0(\xi) G(r,\xi,t) \, d\xi + \int_{R_1}^{R_2} f_1(\xi) G(r,\xi,t) \, d\xi$$
$$- a^2 \int_0^t g_1(\tau) G(r,R_1,t-\tau) \, d\tau + a^2 \int_0^t g_2(\tau) G(r,R_2,t-\tau) \, d\tau.$$

Here,

$$G(r,\xi,t) = \frac{3\xi^2 \sin(t\sqrt{b})}{(R_2^3 - R_1^3)\sqrt{b}} + \frac{2\xi}{(R_2-R_1)r} \sum_{n=1}^{\infty} \frac{(1+R_2^2\lambda_n^2)\Psi_n(r)\Psi_n(\xi)\sin(t\sqrt{a^2\lambda_n^2+b})}{\lambda_n^2 [R_1^2 + R_2^2 + R_1 R_2(1+R_1 R_2 \lambda_n^2)]\sqrt{a^2\lambda_n^2+b}},$$
$$\Psi_n(r) = \sin[\lambda_n(r-R_1)] + R_1 \lambda_n \cos[\lambda_n(r-R_1)],$$

where the λ_n are positive roots of the transcendental equation

$$(\lambda^2 R_1 R_2 + 1) \tan[\lambda(R_2 - R_1)] - \lambda(R_2 - R_1) = 0.$$

4.2.6-6. Domain: $R_1 \leq r \leq R_2$. Third boundary value problem.

The following conditions are prescribed:

$$w = f_0(r) \quad \text{at} \quad t = 0 \quad \text{(initial condition)},$$
$$\partial_t w = f_1(r) \quad \text{at} \quad t = 0 \quad \text{(initial condition)},$$
$$\partial_r w - k_1 w = g_1(t) \quad \text{at} \quad r = R_1 \quad \text{(boundary condition)},$$
$$\partial_r w + k_2 w = g_2(t) \quad \text{at} \quad r = R_2 \quad \text{(boundary condition)}.$$

The solution $w(r,t)$ is determined by the formula in Paragraph 4.2.6-5 where

$$G(r,\xi,t) = \frac{2\xi}{r} \sum_{n=1}^{\infty} \frac{(b_2^2 + R_2^2 \lambda_n^2) \Psi_n(r) \Psi_n(\xi) \sin\left(t\sqrt{a^2 \lambda_n^2 + b}\right)}{\left[(R_2 - R_1)(b_1^2 + R_1^2 \lambda_n^2)(b_2^2 + R_2^2 \lambda_n^2) + (b_1 R_2 + b_2 R_1)(b_1 b_2 + R_1 R_2 \lambda_n^2)\right] \sqrt{a^2 \lambda_n^2 + b}},$$

$$\Psi_n(r) = b_1 \sin[\lambda_n(r - R_1)] + R_1 \lambda_n \cos[\lambda_n(r - R_1)], \quad b_1 = k_1 R_1 + 1, \quad b_2 = k_2 R_2 - 1.$$

Here, the λ_n are positive roots of the transcendental equation

$$(b_1 b_2 - R_1 R_2 \lambda^2) \sin[\lambda(R_2 - R_1)] + \lambda(R_1 b_2 + R_2 b_1) \cos[\lambda(R_2 - R_1)] = 0.$$

4.3. Equations Containing Power Functions and Arbitrary Parameters

4.3.1. Equations of the Form $\frac{\partial^2 w}{\partial t^2} = (ax + b)\frac{\partial^2 w}{\partial x^2} + c\frac{\partial w}{\partial x} + kw + \Phi(x,t)$

1. $\dfrac{\partial^2 w}{\partial t^2} = a^2 \dfrac{\partial}{\partial x}\left(x \dfrac{\partial w}{\partial x}\right) + \Phi(x,t).$

For $\Phi(x,t) \equiv 0$, this equation governs small-amplitude free vibration of a hanging heavy homogeneous thread (a^2 is the acceleration due to gravity, w the deflection of the thread from the vertical axis, and x the vertical coordinate).

1°. The substitution $x = \frac{1}{4}r^2$ leads to the equation

$$\frac{\partial^2 w}{\partial t^2} = a^2 \left(\frac{\partial^2 w}{\partial r^2} + \frac{1}{r}\frac{\partial w}{\partial r}\right) + \Phi\left(\tfrac{1}{4}r^2, t\right),$$

which is discussed in Subsections 4.2.1–4.2.2.

2°. Domain: $0 \leq x \leq l$. First boundary value problem.
The following conditions are prescribed:

$$w = f_0(x) \quad \text{at} \quad t = 0 \quad \text{(initial condition)},$$
$$\partial_t w = f_1(x) \quad \text{at} \quad t = 0 \quad \text{(initial condition)},$$
$$w = g(t) \quad \text{at} \quad x = l \quad \text{(boundary condition)},$$
$$w \neq \infty \quad \text{at} \quad x = 0 \quad \text{(boundedness condition)}.$$

Solution:

$$w(x,t) = \frac{\partial}{\partial t} \int_0^l f_0(\xi) G(x,\xi,t)\, d\xi + \int_0^l f_1(\xi) G(x,\xi,t)\, d\xi$$

$$- a^2 l \int_0^t g(\tau) \left[\frac{\partial}{\partial \xi} G(x,\xi,t-\tau)\right]_{\xi=l} d\tau + \int_0^t \int_0^l \Phi(\xi,\tau) G(x,\xi,t-\tau)\, d\xi\, d\tau,$$

where

$$G(x,\xi,t) = \frac{2}{a\sqrt{l}} \sum_{n=1}^{\infty} \frac{1}{\mu_n J_1^2(\mu_n)} J_0\left(\mu_n \sqrt{\frac{x}{l}}\right) J_0\left(\mu_n \sqrt{\frac{\xi}{l}}\right) \sin\left(\frac{\mu_n a t}{2\sqrt{l}}\right).$$

Here, the μ_n are positive zeros of the Bessel function, $J_0(\mu) = 0$. The numerical values of the first ten roots μ_n are specified in Paragraph 1.2.1-3.

⊙ *Reference*: M. M. Smirnov (1975).

3°. Domain: $0 \leq x \leq l$. Second boundary value problem.
The following conditions are prescribed:

$$\begin{aligned}
w &= f_0(x) &&\text{at} \quad t = 0 &&\text{(initial condition),} \\
\partial_t w &= f_1(x) &&\text{at} \quad t = 0 &&\text{(initial condition),} \\
\partial_x w &= g(t) &&\text{at} \quad x = l &&\text{(boundary condition),} \\
w &\neq \infty &&\text{at} \quad x = 0 &&\text{(boundedness condition).}
\end{aligned}$$

Solution:

$$w(x,t) = \frac{\partial}{\partial t} \int_0^l f_0(\xi) G(x,\xi,t)\, d\xi + \int_0^l f_1(\xi) G(x,\xi,t)\, d\xi$$
$$+ a^2 l \int_0^t g(\tau) G(x,l,t-\tau)\, d\tau + \int_0^t \int_0^l \Phi(\xi,\tau) G(x,\xi,t-\tau)\, d\xi\, d\tau,$$

where

$$G(x,\xi,t) = \frac{t}{l} + \frac{2}{a\sqrt{l}} \sum_{n=1}^{\infty} \frac{1}{\mu_n J_0^2(\mu_n)} J_0\!\left(\mu_n \sqrt{\frac{x}{l}}\right) J_0\!\left(\mu_n \sqrt{\frac{\xi}{l}}\right) \sin\!\left(\frac{\mu_n a t}{2\sqrt{l}}\right).$$

Here, the μ_n are positive zeros of the first-order Bessel function, $J_1(\mu) = 0$. The numerical values of the first ten roots μ_n are specified in Paragraph 1.2.1-4.

4°. Domain: $0 \leq x \leq l$. Third boundary value problem.
The following conditions are prescribed:

$$\begin{aligned}
w &= f_0(x) &&\text{at} \quad t = 0 &&\text{(initial condition),} \\
\partial_t w &= f_1(x) &&\text{at} \quad t = 0 &&\text{(initial condition),} \\
\partial_x w + kw &= g(t) &&\text{at} \quad x = l &&\text{(boundary condition),} \\
w &\neq \infty &&\text{at} \quad x = 0 &&\text{(boundedness condition).}
\end{aligned}$$

The solution $w(x,t)$ is given by the formula in Item 3° with

$$G(x,\xi,t) = \frac{2}{a\sqrt{l}} \sum_{n=1}^{\infty} \frac{\mu_n}{(4k^2 l + \mu_n^2) J_0^2(\mu_n)} J_0\!\left(\mu_n \sqrt{\frac{x}{l}}\right) J_0\!\left(\mu_n \sqrt{\frac{\xi}{l}}\right) \sin\!\left(\frac{\mu_n a t}{2\sqrt{l}}\right).$$

Here, the μ_n are positive roots of the transcendental equation

$$\mu J_1(\mu) - 2k\sqrt{l}\, J_0(\mu) = 0.$$

The numerical values of the first six roots μ_n can be found in Carslaw and Jaeger (1984).

2. $\dfrac{\partial^2 w}{\partial t^2} = a^2 \dfrac{\partial}{\partial x}\!\left(x \dfrac{\partial w}{\partial x}\right) - bw + \Phi(x,t).$

For $b < 0$ and $\Phi(x,t) \equiv 0$, this equation describes small-amplitude vibration of a heavy homogeneous thread that rotates at a constant angular velocity $\omega = \sqrt{|b|}$ about the vertical axis (a^2 is the acceleration due to gravity).

1°. The substitution $x = \tfrac{1}{4} r^2$ leads to the equation

$$\frac{\partial^2 w}{\partial t^2} = a^2 \left(\frac{\partial^2 w}{\partial r^2} + \frac{1}{r} \frac{\partial w}{\partial r}\right) - bw + \Phi\!\left(\tfrac{1}{4} r^2, t\right),$$

which is discussed in Subsection 4.2.5.

2°. Domain: $0 \le x \le l$. First boundary value problem.
The following conditions are prescribed:

$$w = f_0(x) \quad \text{at} \quad t = 0 \quad \text{(initial condition)},$$
$$\partial_t w = f_1(x) \quad \text{at} \quad t = 0 \quad \text{(initial condition)},$$
$$w = g(t) \quad \text{at} \quad x = l \quad \text{(boundary condition)},$$
$$w \ne \infty \quad \text{at} \quad x = 0 \quad \text{(boundedness condition)}.$$

Solution:
$$w(x,t) = \frac{\partial}{\partial t}\int_0^l f_0(\xi)G(x,\xi,t)\,d\xi + \int_0^l f_1(\xi)G(x,\xi,t)\,d\xi$$
$$- a^2 l \int_0^t g(\tau)\left[\frac{\partial}{\partial \xi}G(x,\xi,t-\tau)\right]_{\xi=l} d\tau + \int_0^t\int_0^l \Phi(\xi,\tau)G(x,\xi,t-\tau)\,d\xi\,d\tau.$$

Here,
$$G(x,\xi,t) = \frac{1}{l}\sum_{n=1}^{\infty} \frac{1}{J_1^2(\mu_n)} J_0\!\left(\mu_n\sqrt{\frac{x}{l}}\right) J_0\!\left(\mu_n\sqrt{\frac{\xi}{l}}\right) \frac{\sin(t\sqrt{\lambda_n})}{\sqrt{\lambda_n}}, \quad \lambda_n = \frac{a^2\mu_n^2}{4l} + b,$$

where the μ_n are positive zeros of the Bessel function, $J_0(\mu) = 0$.

⊙ *Reference*: M. M. Smirnov (1975).

3°. Domain: $0 \le x \le l$. Second boundary value problem.
The following conditions are prescribed:

$$w = f_0(x) \quad \text{at} \quad t = 0 \quad \text{(initial condition)},$$
$$\partial_t w = f_1(x) \quad \text{at} \quad t = 0 \quad \text{(initial condition)},$$
$$\partial_x w = g(t) \quad \text{at} \quad x = l \quad \text{(boundary condition)},$$
$$w \ne \infty \quad \text{at} \quad x = 0 \quad \text{(boundedness condition)}.$$

Solution:
$$w(x,t) = \frac{\partial}{\partial t}\int_0^l f_0(\xi)G(x,\xi,t)\,d\xi + \int_0^l f_1(\xi)G(x,\xi,t)\,d\xi$$
$$+ a^2 l \int_0^t g(\tau) G(x,l,t-\tau)\,d\tau + \int_0^t\int_0^l \Phi(\xi,\tau)G(x,\xi,t-\tau)\,d\xi\,d\tau.$$

Here,
$$G(r,\xi,t) = \frac{\sin(t\sqrt{b})}{l\sqrt{b}} + \frac{1}{l}\sum_{n=1}^{\infty} \frac{1}{J_0^2(\mu_n)} J_0\!\left(\mu_n\sqrt{\frac{x}{l}}\right) J_0\!\left(\mu_n\sqrt{\frac{\xi}{l}}\right) \frac{\sin(t\sqrt{\lambda_n})}{\sqrt{\lambda_n}}, \quad \lambda_n = \frac{a^2\mu_n^2}{4l} + b,$$

where the μ_n are positive zeros of the first-order Bessel function, $J_1(\mu) = 0$. The numerical values of the first ten roots μ_n are specified in Paragraph 1.2.1-4.

4°. Domain: $0 \le x \le l$. Third boundary value problem.
The following conditions are prescribed:

$$w = f_0(x) \quad \text{at} \quad t = 0 \quad \text{(initial condition)},$$
$$\partial_t w = f_1(x) \quad \text{at} \quad t = 0 \quad \text{(initial condition)},$$
$$\partial_x w + kw = g(t) \quad \text{at} \quad x = l \quad \text{(boundary condition)}.$$

The solution $w(x,t)$ is given by the formula in Item 3° with

$$G(r,\xi,t) = \frac{1}{l}\sum_{n=1}^{\infty} \frac{\mu_n^2}{(4k^2 l + \mu_n^2) J_0^2(\mu_n)} J_0\!\left(\mu_n\sqrt{\frac{x}{l}}\right) J_0\!\left(\mu_n\sqrt{\frac{\xi}{l}}\right) \frac{\sin(t\sqrt{\lambda_n})}{\sqrt{\lambda_n}}, \quad \lambda_n = \frac{a^2\mu_n^2}{4l} + b.$$

Here, the μ_n are positive roots of the transcendental equation
$$\mu J_1(\mu) - 2k\sqrt{l}\, J_0(\mu) = 0.$$

The numerical values of the first six roots μ_n can be found in Abramowitz and Stegun (1964) and Carslaw and Jaeger (1984).

3. $\dfrac{\partial^2 w}{\partial t^2} = a^2 \dfrac{\partial}{\partial x}\left[(l-x)\dfrac{\partial w}{\partial x}\right].$

This equation governs small-amplitude free vibration of a heavy homogeneous thread of length l (a^2 is the acceleration due to gravity, w the deflection of the thread from the vertical axis, and x the vertical coordinate). The change of variable $z = l - x$ leads a special case of equation 4.3.1.1 with $b = 0$ and $\Phi \equiv 0$.

4. $\dfrac{\partial^2 w}{\partial t^2} = a^2\left(\dfrac{2}{2n+1} x \dfrac{\partial^2 w}{\partial x^2} + \dfrac{\partial w}{\partial x}\right), \qquad n = 1, 2, \ldots$

General solution:
$$w(x,t) = \dfrac{\partial^{n-1}}{\partial x^{n-1}}\left[\dfrac{\Phi(\sqrt{2(2n+1)x} + at) + \Psi(\sqrt{2(2n+1)x} - at)}{\sqrt{x}}\right],$$

where Φ and Ψ are arbitrary functions.

⊙ *Reference*: M. M. Smirnov (1975).

5. $\dfrac{\partial^2 w}{\partial t^2} = (ax + b)\dfrac{\partial^2 w}{\partial x^2} + a\dfrac{\partial w}{\partial x} + cw + \Phi(x,t).$

The substitution $z = ax + b$ leads to an equation of the form 4.3.1.2:
$$\dfrac{\partial^2 w}{\partial t^2} = a^2 \dfrac{\partial}{\partial z}\left(z\dfrac{\partial w}{\partial z}\right) + cw + \Phi\left(\dfrac{z-b}{a}, t\right).$$

6. $\dfrac{\partial^2 w}{\partial t^2} = (ax + b)\dfrac{\partial^2 w}{\partial x^2} + \dfrac{1}{2}a\dfrac{\partial w}{\partial x} + cw + \Phi(x,t).$

The substitution $z = 2\sqrt{ax+b}$ leads to the equation
$$\dfrac{\partial^2 w}{\partial t^2} = a^2 \dfrac{\partial^2 w}{\partial z^2} + cw + \Phi\left(\dfrac{z^2 - 4b}{4a}, t\right),$$

which is considered in Subsection 4.1.3.

7. $\dfrac{\partial^2 w}{\partial t^2} = (a_2 x + b_2)\dfrac{\partial^2 w}{\partial x^2} + (a_1 x + b_1)\dfrac{\partial w}{\partial x} + (a_0 x + b_0)w.$

This is a special case of equation 4.5.3.4 with $f(x) = a_2 x + b_2$, $g(x) = a_1 x + b_1$, $h(x) = a_0 x + b_0$, and $\Phi \equiv 0$.

Particular solutions:
$$w(x,t) = \exp(kx) F\left(\dfrac{x+q}{p}\right)\left[A\sin(t\sqrt{\mu}) + B\cos(t\sqrt{\mu})\right] \quad \text{for} \quad \mu > 0,$$
$$w(x,t) = \exp(kx) F\left(\dfrac{x+q}{p}\right)\left[A\sinh(t\sqrt{-\mu}) + B\cosh(t\sqrt{-\mu})\right] \quad \text{for} \quad \mu < 0.$$

Here, A, B, and μ are arbitrary constants; the coefficients k, p, q and the function $F = F(\xi)$ are listed in Table 20, where
$$\mathcal{J}(\alpha, \beta; x) = C_1 \Phi(\alpha, \beta; x) + C_2 \Psi(\alpha, \beta; x), \qquad C_1, C_2 \text{ are any numbers,}$$

is an arbitrary solution of the degenerate hypergeometric equation $xy''_{xx} + (\beta - x)y'_x - \alpha y = 0$, and
$$Z_\nu(x) = C_1 J_\nu(x) + C_2 Y_\nu(x), \qquad C_1, C_2 \text{ are any numbers,}$$

is an arbitrary solution of the Bessel equation $x^2 y''_{xx} + xy'_x + (x^2 - \nu^2)y = 0$.

TABLE 20

The coefficients k, p, q and the function $F = F(\xi)$ determining the form of
particular solutions to equation 4.3.1.7. Notation: $E(k) = b_2 k^2 + b_1 k + b_0 + \mu$

Conditions	k	p	q	$F = F(\xi)$	Parameters
$a_2 \neq 0$, $D \neq 0$ $D \equiv a_1^2 - 4a_0 a_2$	$\dfrac{\sqrt{D} - a_1}{2a_2}$	$-\dfrac{a_2}{2a_2 k + a_1}$	$\dfrac{b_2}{a_2}$	$\mathcal{J}(\alpha, \beta; \xi)$	$\alpha = E(k)/(2a_2 k + a_1)$, $\beta = (a_2 b_1 - a_1 b_2) a_2^{-2}$
$a_2 = 0$, $a_1 \neq 0$	$-\dfrac{a_0}{a_1}$	1	$\dfrac{2b_2 k + b_1}{a_1}$	$\mathcal{J}(\alpha, \tfrac{1}{2}; \sigma \xi^2)$	$\alpha = E(k)/(2a_1)$, $\sigma = -a_1/(2b_2)$
$a_2 \neq 0$, $a_1^2 = 4a_0 a_2$	$-\dfrac{a_1}{2a_2}$	a_2	$\dfrac{b_2}{a_2}$	$\xi^\alpha Z_{2\alpha}(\sigma \sqrt{\xi})$	$\alpha = \dfrac{1}{2} - \dfrac{2b_2 k + b_1}{2a_2}$, $\sigma = 2\sqrt{E(k)}$
$a_2 = a_1 = 0$, $a_0 \neq 0$	$-\dfrac{b_1}{2b_2}$	1	$\dfrac{4(b_0 + \mu) b_2 - b_1^2}{4a_0 b_2}$	$\xi^{1/2} Z_{1/3}(\sigma \xi^{3/2})$	$\sigma = \dfrac{2}{3}\left(\dfrac{a_0}{b_2}\right)^{1/2}$

For the degenerate hypergeometric functions $\Phi(a,b;x)$ and $\Psi(a,b;x)$, see Supplement A.9 and the books by Abramowitz and Stegun (1964) and Bateman and Erdélyi (1953, Vol. 1). For the Bessel functions $J_\nu(x)$ and $Y_\nu(x)$, see Supplement A.6 and the books by Abramowitz and Stegun (1964) and Bateman and Erdélyi (1953, Vol. 2).

4.3.2. Equations of the Form $\dfrac{\partial^2 w}{\partial t^2} = (ax^2 + b)\dfrac{\partial^2 w}{\partial x^2} + cx\dfrac{\partial w}{\partial x} + kw + \Phi(x, t)$

1. $\dfrac{\partial^2 w}{\partial t^2} = x^2 \dfrac{\partial^2 w}{\partial x^2} + \Phi(x, t).$

This is a special case of equation 4.3.2.2 with $a = 1$ and $b = c = 0$.

1°. Domain: $1 \leq x \leq a$. First boundary value problem.
The following conditions are prescribed:

$$w = f_0(x) \quad \text{at} \quad t = 0 \quad \text{(initial condition)},$$
$$\partial_t w = f_1(x) \quad \text{at} \quad t = 0 \quad \text{(initial condition)},$$
$$w = g_1(t) \quad \text{at} \quad x = 1 \quad \text{(boundary condition)},$$
$$w = g_2(t) \quad \text{at} \quad x = a \quad \text{(boundary condition)}.$$

Solution:

$$w(x,t) = \int_0^t \int_1^a \Phi(\xi, \tau) G(x, \xi, t - \tau)\, d\xi\, d\tau$$
$$+ \frac{\partial}{\partial t} \int_1^a f_0(\xi) G(x, \xi, t)\, d\xi + \int_1^a f_1(\xi) G(x, \xi, t)\, d\xi$$
$$+ \int_0^t g_1(\tau)\left[\frac{\partial}{\partial \xi} G(x, \xi, t - \tau)\right]_{\xi=1} d\tau - a^2 \int_0^t g_2(\tau)\left[\frac{\partial}{\partial \xi} G(x, \xi, t - \tau)\right]_{\xi=a} d\tau,$$

where

$$G(x, \xi, t) = \frac{2\sqrt{x}}{\xi^{3/2} \ln a} \sum_{n=1}^{\infty} \frac{1}{\lambda_n} \sin(\mu_n \ln x) \sin(\mu_n \ln \xi) \sin(\lambda_n t), \quad \mu_n = \frac{\pi n}{\ln a}, \quad \lambda_n = \sqrt{\mu_n^2 + \tfrac{1}{4}}.$$

2°. Domain: $1 \le x \le a$. **Second boundary value problem.**
The following conditions are prescribed:

$$w = f_0(x) \quad \text{at} \quad t = 0 \quad \text{(initial condition)},$$
$$\partial_t w = f_1(x) \quad \text{at} \quad t = 0 \quad \text{(initial condition)},$$
$$\partial_x w = g_1(t) \quad \text{at} \quad x = 1 \quad \text{(boundary condition)},$$
$$\partial_x w = g_2(t) \quad \text{at} \quad x = a \quad \text{(boundary condition)}.$$

Solution:

$$w(x,t) = \int_0^t \int_1^a \Phi(\xi,\tau) G(x,\xi,t-\tau)\,d\xi\,d\tau$$
$$+ \frac{\partial}{\partial t} \int_1^a f_0(\xi) G(x,\xi,t)\,d\xi + \int_1^a f_1(\xi) G(x,\xi,t)\,d\xi$$
$$- \int_0^t g_1(\tau) G(x,1,t-\tau)\,d\tau + a^2 \int_0^t g_2(\tau) G(x,a,t-\tau)\,d\tau,$$

where

$$G(x,\xi,t) = \frac{at}{(a-1)\xi^2} + \frac{8\sqrt{x}}{\xi^{3/2} \ln a} \sum_{n=1}^{\infty} \frac{\mu_n^2}{\lambda_n(1+\mu_n^2)} \varphi_n(x) \varphi_n(\xi) \sin(\lambda_n t),$$

$$\varphi_n(x) = \cos(\mu_n \ln x) - \frac{1}{2\mu_n} \sin(\mu_n \ln x), \quad \mu_n = \frac{\pi n}{\ln a}, \quad \lambda_n = \sqrt{\mu_n^2 + \tfrac{1}{4}}.$$

⊙ *Reference*: A. G. Butkovskiy (1979).

2. $\dfrac{\partial^2 w}{\partial t^2} = ax^2 \dfrac{\partial^2 w}{\partial x^2} + bx \dfrac{\partial w}{\partial x} + cw + \Phi(x,t).$

The substitution $x = ke^z$ ($k \ne 0$) leads to the constant coefficient equation $\partial_{tt} w = a\partial_{zz} w + (b-a)\partial_z w + cw + \Phi(ke^z, t)$, which is discussed in Subsection 4.1.5.

3. $\dfrac{\partial^2 w}{\partial t^2} = (ax^2 + b) \dfrac{\partial^2 w}{\partial x^2} + ax \dfrac{\partial w}{\partial x} + cw.$

The substitution $z = \displaystyle\int \dfrac{dx}{\sqrt{ax^2 + b}}$ leads to the constant coefficient equation $\partial_{tt} w = \partial_{zz} w + cw$, which is discussed in Subsection 4.1.3.

4. $\dfrac{\partial^2 w}{\partial t^2} = a^2 \dfrac{\partial}{\partial x}\left[(l^2 - x^2) \dfrac{\partial w}{\partial x}\right] + \Phi(x,t).$

Domain: $0 \le x \le l$. First boundary value problem.
The following conditions are prescribed:

$$w = f_0(x) \quad \text{at} \quad t = 0 \quad \text{(initial condition)},$$
$$\partial_t w = f_1(x) \quad \text{at} \quad t = 0 \quad \text{(initial condition)},$$
$$w = g(t) \quad \text{at} \quad x = 0 \quad \text{(boundary condition)},$$
$$w \ne \infty \quad \text{at} \quad x = l \quad \text{(boundedness condition)}.$$

Solution:

$$w(x,t) = \frac{\partial}{\partial t} \int_0^l f_0(\xi) G(x,\xi,t)\,d\xi + \int_0^l f_1(\xi) G(x,\xi,t)\,d\xi$$
$$+ a^2 l^2 \int_0^t g(\tau) \left[\frac{\partial}{\partial \xi} G(x,\xi,t-\tau)\right]_{\xi=0} d\tau + \int_0^t \int_0^l \Phi(\xi,\tau) G(x,\xi,t-\tau)\,d\xi\,d\tau.$$

Here,
$$G(x,\xi,t) = \frac{1}{al}\sum_{n=1}^{\infty}\frac{4n-1}{\lambda_n}P_{2n-1}\left(\frac{x}{l}\right)P_{2n-1}\left(\frac{\xi}{l}\right)\sin(\lambda_n at), \quad \lambda_n = \sqrt{2n(2n-1)},$$
where $P_k(x) = \frac{1}{2^k k!}\frac{d^k}{dx^k}\left[(x^2-1)^k\right]$ are the Legendre polynomials.

⊙ *Reference*: M. M. Smirnov (1975).

4.3.3. Other Equations

1. $\dfrac{\partial^2 w}{\partial t^2} = a^2\left[\dfrac{\partial^2 w}{\partial r^2} + \dfrac{2}{r}\dfrac{\partial w}{\partial r} - \dfrac{n(n+1)}{r^2}w\right], \quad n = 1, 2, 3, \ldots$

General solution:
$$w(r,t) = r^n\left(\frac{1}{r}\frac{\partial}{\partial r}\right)^n\left[\frac{\Phi(r+at) + \Psi(r-at)}{r}\right],$$
where $\Phi(r_1)$ and $\Psi(r_2)$ are arbitrary functions.

⊙ *Reference*: M. M. Smirnov (1975).

2. $\dfrac{\partial^2 w}{\partial t^2} = \dfrac{\partial^2 w}{\partial x^2} + \dfrac{\alpha}{x}\dfrac{\partial w}{\partial x}.$

The hyperbolic Euler–Poisson–Darboux equation.

1°. For $\alpha = 1$ and $\alpha = 2$, see Subsections 4.2.1–4.2.4. For $\alpha \neq 1$, the substitution $z = x^{1-\alpha}$ leads to an equation of the form 4.5.3.1:
$$\frac{\partial^2 w}{\partial t^2} = (1-\alpha)^2 z^{\frac{2\alpha}{\alpha-1}}\frac{\partial^2 w}{\partial z^2}.$$

2°. Suppose $w_\alpha = w_\alpha(x,t)$ is a solution of the equation in question for a fixed value of the parameter α. Then the functions $\widetilde{w}_\alpha$ defined by the relations
$$\widetilde{w}_\alpha = \frac{\partial w_\alpha}{\partial t},$$
$$\widetilde{w}_\alpha = x\frac{\partial w_\alpha}{\partial x} + t\frac{\partial w_\alpha}{\partial t},$$
$$\widetilde{w}_\alpha = 2xt\frac{\partial w_\alpha}{\partial x} + (x^2+t^2)\frac{\partial w_\alpha}{\partial t} + \alpha t w_\alpha$$
are also solutions of this equation.

3°. Suppose $w_\alpha = w_\alpha(x,t)$ is a solution of the equation in question for a fixed value of the parameter α. Using this w_α, one can construct solutions of the equation with other values of the parameter by the formulas
$$w_{2-\alpha} = x^{\alpha-1}w_\alpha,$$
$$w_{\alpha-2} = x\frac{\partial w_\alpha}{\partial x} + (\alpha-1)w_\alpha,$$
$$w_{\alpha-2} = xt\frac{\partial w_\alpha}{\partial x} + x^2\frac{\partial w_\alpha}{\partial t} + (\alpha-1)tw_\alpha,$$
$$w_{\alpha-2} = x(x^2+t^2)\frac{\partial w_\alpha}{\partial x} + 2x^2 t\frac{\partial w_\alpha}{\partial t} + \left[x^2 + (\alpha-1)t^2\right]w_\alpha,$$
$$w_{\alpha+2} = \frac{1}{x}\frac{\partial w_\alpha}{\partial x},$$
$$w_{\alpha+2} = \frac{t}{x}\frac{\partial w_\alpha}{\partial x} + \frac{\partial w_\alpha}{\partial t},$$
$$w_{\alpha+2} = \frac{x^2+t^2}{x}\frac{\partial w_\alpha}{\partial x} + 2t\frac{\partial w_\alpha}{\partial t} + \alpha w_\alpha.$$

⊙ The results of Items 2° and 3° were obtained by A. V. Aksenov (2001).

3. $\dfrac{\partial^2 w}{\partial t^2} = \dfrac{\partial^2 w}{\partial x^2} + \dfrac{2a}{x}\dfrac{\partial w}{\partial x} + b^2 w, \qquad 0 < 2a < 1.$

General solution:
$$w(x,t) = \int_0^1 \frac{\Phi(t + x(2\xi - 1))}{[\xi(1-\xi)]^{1-a}} \bar{J}_{a-1}\left(2bx\sqrt{\xi(1-\xi)}\right) d\xi$$
$$+ x^{1-2a} \int_0^1 \frac{\Psi(t + x(2\xi - 1))}{[\xi(1-\xi)]^a} \bar{J}_{-a}\left(2bx\sqrt{\xi(1-\xi)}\right) d\xi,$$

where $\Phi(\xi_1)$ and $\Psi(\xi_2)$ are arbitrary functions; $\bar{J}_{-\nu}(z) = \Gamma(1-\nu)2^{-\nu}z^\nu J_{-\nu}(z)$; $J_{-\nu}(z)$ is the Bessel function.

⊙ *Reference*: M. M. Smirnov (1975).

4. $\dfrac{\partial^2 w}{\partial t^2} = ax^4 \dfrac{\partial^2 w}{\partial x^2} + \Phi(x,t).$

The transformation $z = 1/x$, $u = w/x$ leads to the equation
$$\frac{\partial^2 u}{\partial t^2} = a\frac{\partial^2 u}{\partial z^2} + z\Phi\left(\frac{1}{z}, t\right),$$
which is discussed in Subsection 4.1.2.

5. $\dfrac{\partial^2 w}{\partial t^2} = (ax+b)^4 \dfrac{\partial^2 w}{\partial x^2}.$

The transformation
$$u = \frac{w}{ax+b}, \quad z = at + \frac{1}{ax+b}, \quad y = -at + \frac{1}{ax+b}$$
leads to the equation $\partial_{zy}u = 0$. Thus, the general solution of the original equation has the form
$$w = (ax+b)[f(z) + g(y)],$$
where $f = f(z)$ and $g = g(y)$ are arbitrary functions.

⊙ *Reference*: N. H. Ibragimov (1994).

6. $\dfrac{\partial^2 w}{\partial t^2} = (a^2 - x^2)^2 \dfrac{\partial^2 w}{\partial x^2} + \Phi(x,t).$

Domain: $-l \le x \le l$. First boundary value problem.
The following conditions are prescribed:
$$\begin{aligned} w &= f(x) & \text{at} \quad & t = 0 & \text{(initial condition),} \\ \partial_t w &= g(x) & \text{at} \quad & t = 0 & \text{(initial condition),} \\ w &= 0 & \text{at} \quad & x = l & \text{(boundary condition),} \\ w &= 0 & \text{at} \quad & x = -l & \text{(boundary condition).} \end{aligned}$$

Solution for $0 < l < a$:
$$w(x,t) = \frac{\partial}{\partial t}\int_{-l}^{l} f(\xi)G(x,\xi,t)\,d\xi + \int_{-l}^{l} g(\xi)G(x,\xi,t)\,d\xi + \int_0^t \int_{-l}^{l} \Phi(\xi,\tau)G(x,\xi,t-\tau)\,d\xi\,d\tau,$$
where
$$G(x,\xi,t) = \frac{2a}{k(\xi^2 - a^2)^2} \sum_{n=1}^\infty \frac{1}{\lambda_n}\varphi_n(x)\varphi_n(\xi)\sin(\lambda_n t),$$
$$\varphi_n(x) = \sqrt{a^2 - x^2}\,\sin\left(\frac{\pi n}{2} + \frac{\pi n}{2k}\ln\frac{a+x}{a-x}\right), \quad \lambda_n = \frac{a}{k}\sqrt{\pi^2 n^2 + k^2}, \quad k = \ln\frac{a+l}{a-l}.$$

⊙ *Reference*: A. G. Butkovskiy (1979).

7. $\dfrac{\partial^2 w}{\partial t^2} = (x - a_1)^2 (x - a_2)^2 \dfrac{\partial^2 w}{\partial x^2}$, $\quad a_1 \neq a_2$.

The transformation

$$w(x, t) = (x - a_2) u(\xi, \tau), \quad \xi = \ln\left|\dfrac{x - a_1}{x - a_2}\right|, \quad \tau = |a_1 - a_2| t$$

leads to the constant coefficient equation $\partial_{\tau\tau} u = \partial_{\xi\xi} u - \partial_\xi u$, which is discussed in Subsection 4.1.4.

8. $\dfrac{\partial^2 w}{\partial t^2} = (ax^2 + bx + c)^2 \dfrac{\partial^2 w}{\partial x^2}$.

The transformation

$$w(x, t) = u(z, t) \sqrt{|ax^2 + bx + c|}, \quad z = \int \dfrac{dx}{ax^2 + bx + c}$$

leads to the constant coefficient equation $\partial_{tt} u = \partial_{zz} u + \left(ac - \tfrac{1}{4} b^2\right) u$, which is discussed in Subsection 4.1.3.

9. $\dfrac{\partial^2 w}{\partial t^2} = a^2 \dfrac{\partial}{\partial x}\left(x^m \dfrac{\partial w}{\partial x}\right) + \Phi(x, t)$.

1°. Domain: $0 \leq x \leq l$. First boundary value problem.
The following conditions are prescribed:
    1.1. Case $0 < m < 1$:

$$\begin{aligned} w &= f_0(x) & \text{at} \quad t &= 0 & &\text{(initial condition)}, \\ \partial_t w &= f_1(x) & \text{at} \quad t &= 0 & &\text{(initial condition)}, \\ w &= 0 & \text{at} \quad x &= 0 & &\text{(boundary condition)}, \\ w &= g(t) & \text{at} \quad x &= l & &\text{(boundary condition)}. \end{aligned}$$

Solution:

$$\begin{aligned} w(x, t) &= \dfrac{\partial}{\partial t} \int_0^l f_0(\xi) G(x, \xi, t)\, d\xi + \int_0^l f_1(\xi) G(x, \xi, t)\, d\xi \\ &\quad - a^2 l^m \int_0^t g(\tau) \left[\dfrac{\partial}{\partial \xi} G(x, \xi, t - \tau)\right]_{\xi = l} d\tau + \int_0^t \int_0^l \Phi(\xi, \tau) G(x, \xi, t - \tau)\, d\xi\, d\tau. \end{aligned} \quad (1)$$

Here,

$$G(x, \xi, t) = \sum_{n=1}^{\infty} \dfrac{y_n(x) y_n(\xi) \sin(\lambda_n a t)}{a \|y_n\|^2 \lambda_n}, \quad \lambda_n = \dfrac{\mu_n}{2}(2 - m) l^{\frac{m-2}{2}}, \qquad (2)$$

where

$$y_n(x) = x^{\frac{1-m}{2}} J_p\!\left(\mu_n \left(\dfrac{x}{l}\right)^{\frac{2-m}{2}}\right), \quad \|y_n\|^2 = \int_0^l y_n^2(x)\, dx, \quad p = \left|\dfrac{1 - m}{2 - m}\right|;$$

the μ_n are positive zeros of the Bessel function, $J_p(\mu) = 0$.

    1.2. Case $1 \leq m < 2$:

$$\begin{aligned} w &= f_0(x) & \text{at} \quad t &= 0 & &\text{(initial condition)}, \\ \partial_t w &= f_1(x) & \text{at} \quad t &= 0 & &\text{(initial condition)}, \\ w &\neq \infty & \text{at} \quad x &= 0 & &\text{(boundedness condition)}, \\ w &= g(t) & \text{at} \quad x &= l & &\text{(boundary condition)}. \end{aligned}$$

The solution is given by the formulas presented in Item 1.1.

⊙ *Reference*: M. M. Smirnov (1975).

2°. Domain: $0 \le x \le l$. Mixed boundary value problem.
The following conditions are prescribed:

$$w = f_0(x) \quad \text{at} \quad t = 0 \quad \text{(initial condition)},$$
$$\partial_t w = f_1(x) \quad \text{at} \quad t = 0 \quad \text{(initial condition)},$$
$$(x^m \partial_x w) = 0 \quad \text{at} \quad x = 0 \quad \text{(boundary condition)},$$
$$w = g(t) \quad \text{at} \quad x = l \quad \text{(boundary condition)}.$$

The solution for $0 < m < 1$ is given by relations (1) and (2) with

$$y_n(x) = x^{\frac{1-m}{2}} J_{-p}\left(\mu_n \left(\frac{x}{l}\right)^{\frac{2-m}{2}}\right), \quad \|y_n\|^2 = \int_0^l y_n^2(x)\, dx, \quad p = \frac{1-m}{2-m};$$

the μ_n are positive zeros of the Bessel function, $J_{-p}(\mu) = 0$.

3°. For $\Phi \equiv 0$, the change of variable $z = x^{1-m}$ leads to an equation of the form 4.3.3.10:

$$\frac{\partial^2 w}{\partial t^2} = a^2(1-m)^2 z^{\frac{m}{m-1}} \frac{\partial^2 w}{\partial z^2}.$$

10. $\dfrac{\partial^2 w}{\partial t^2} = a^2 x^m \dfrac{\partial^2 w}{\partial x^2}.$

1°. Particular solutions (A_1, A_2, B_1, B_2, and μ are arbitrary constants):

$$w(x,t) = \sqrt{x}\left[A_1 J_{\frac{1}{2q}}(\mu x^q) + A_2 Y_{\frac{1}{2q}}(\mu x^q)\right]\left[B_1 \sin(aq\mu t) + B_2 \cos(aq\mu t)\right],$$

$$w(x,t) = \sqrt{x}\left[A_1 I_{\frac{1}{2q}}(\mu x^q) + A_2 K_{\frac{1}{2q}}(\mu x^q)\right]\left[B_1 \sinh(aq\mu t) + B_2 \cosh(aq\mu t)\right],$$

where $q = \frac{1}{2}(2-m)$; $J_\nu(z)$ and $Y_\nu(z)$ are the Bessel functions; $I_\nu(z)$ and $K_\nu(z)$ are the modified Bessel functions.

2°. Below are discrete transformations that preserve the form of the original equation; what changes is the parameter n.

2.1. The point transformation

$$z = \frac{1}{x}, \quad u = \frac{w}{x} \quad \text{(transformation } \mathcal{P}\text{)}$$

leads to a similar equation

$$\frac{\partial^2 u}{\partial t^2} = a^2 z^{4-m} \frac{\partial^2 u}{\partial z^2}.$$

The transformation $\mathcal{P}$ changes the equation parameter in accordance with the rule $m \stackrel{\mathcal{P}}{\Longrightarrow} 4-m$. The double application of the transformation $\mathcal{P}$ yields the original equation.

2.2. Suppose $w = w(x,t)$ is a solution of the original equation. Then the function $v = v(\xi, \tau)$, which is related to the solution $w = w(x,t)$ by the Bäcklund transformation

$$v(\xi, \tau) = \frac{\partial}{\partial x} w(x,t), \quad x = \xi^{\frac{1}{1-m}}, \quad \tau = |1-m|t \quad \text{(transformation } \mathcal{B}\text{)},$$

is a solution of a similar equation

$$\frac{\partial^2 v}{\partial \tau^2} = a^2 \xi^{\frac{m}{m-1}} \frac{\partial^2 v}{\partial \xi^2}.$$

The transformation $\mathcal{B}$ changes the equation parameters in accordance with the rule $m \stackrel{\mathcal{B}}{\Longrightarrow} \dfrac{m}{m-1}$. The double application of the transformation $\mathcal{B}$ yields the original equation.

2.3. The composition of transformations $\mathcal{F} = \mathcal{B} \circ \mathcal{P}$ changes the equation parameter as follows:
$$m \stackrel{\mathcal{F}}{\Longrightarrow} \frac{4-m}{3-m} \stackrel{\mathcal{F}}{\Longrightarrow} \frac{8-3m}{5-2m} \stackrel{\mathcal{F}}{\Longrightarrow} \frac{12-5m}{7-3m} \stackrel{\mathcal{F}}{\Longrightarrow} \frac{16-7m}{9-4m} \stackrel{\mathcal{F}}{\Longrightarrow} \ldots$$
The n-fold application of the transformation $\mathcal{F}$ yields the equation with parameter
$$m \stackrel{\mathcal{F}^n}{\Longrightarrow} \frac{4n-(2n-1)m}{2n+1-nm}. \tag{1}$$

2.4. The composition of transformations $\mathcal{G} = \mathcal{P} \circ \mathcal{B}$ changes the equation parameter as follows:
$$m \stackrel{\mathcal{G}}{\Longrightarrow} \frac{4-3m}{1-m} \stackrel{\mathcal{G}}{\Longrightarrow} \frac{8-5m}{3-2m} \stackrel{\mathcal{G}}{\Longrightarrow} \frac{12-7m}{5-3m} \stackrel{\mathcal{G}}{\Longrightarrow} \frac{16-9m}{7-4m} \stackrel{\mathcal{G}}{\Longrightarrow} \ldots$$
The n-fold application of the transformation $\mathcal{G}$ yields the equation with parameter
$$m \stackrel{\mathcal{G}^n}{\Longrightarrow} \frac{4n-(2n+1)m}{2n-1-nm}. \tag{2}$$

2.5. Setting $m = 0$ in (1) and (2), we arrive at two families of equations
$$\frac{\partial^2 w}{\partial t^2} = a^2 x^{\frac{4n}{2n+1}} \frac{\partial^2 w}{\partial x^2} \quad \text{at} \quad n = 1, 2, \ldots;$$
$$\frac{\partial^2 w}{\partial t^2} = a^2 x^{\frac{4n}{2n-1}} \frac{\partial^2 w}{\partial x^2} \quad \text{at} \quad n = 1, 2, \ldots;$$
whose solutions can be obtained with the aid of the wave equation; for this constant coefficient wave equation, see Subsection 4.1.1.

3°. Below are some useful transformations that lead to other equations.

3.1. The substitution $\xi = x^{1-m}$ leads to an equation of the form 4.3.3.9:
$$\frac{\partial^2 w}{\partial t^2} = a^2 (1-m)^2 \frac{\partial}{\partial \xi} \left(\xi^{\frac{m}{m-1}} \frac{\partial w}{\partial \xi} \right).$$

3.2. The transformation $\tau = \frac{1}{2}a|2-m|t$, $\xi = x^{\frac{2-m}{2}}$ leads to an equation of the form 4.3.3.3:
$$\frac{\partial^2 w}{\partial \tau^2} = \frac{\partial^2 w}{\partial \xi^2} + \frac{m}{m-2} \frac{1}{\xi} \frac{\partial w}{\partial \xi}.$$

11. $\dfrac{\partial^2 w}{\partial t^2} = t^m \dfrac{\partial^2 w}{\partial x^2}.$

1°. Domain: $-\infty < x < \infty$. Cauchy problem.
Initial conditions are prescribed:
$$w = f(x) \quad \text{at} \quad t = 0,$$
$$\partial_t w = g(x) \quad \text{at} \quad t = 0.$$

Solution for $m > 0$:
$$w(x,t) = \frac{\Gamma(2\beta)}{\Gamma^2(\beta)} \int_0^1 f\left(x + \frac{2}{m+2} t^{\frac{m+2}{2}} (2\xi - 1)\right) [\xi(1-\xi)]^{\beta-1} d\xi$$
$$+ \frac{\Gamma(2-2\beta)}{\Gamma^2(1-\beta)} t \int_0^1 g\left(x + \frac{2}{m+2} t^{\frac{m+2}{2}} (2\xi - 1)\right) [\xi(1-\xi)]^{-\beta} d\xi,$$
where
$$\beta = \frac{m}{2(m+2)}, \qquad \Gamma(z) = \int_0^\infty e^{-s} s^{z-1} ds.$$

⊙ *Reference*: M. M. Smirnov (1975).

2°. Domain: $0 \le x \le l$. First boundary value problem.
The following conditions are prescribed:

$$w = f(x) \quad \text{at} \quad t = 0 \quad \text{(initial condition),}$$
$$\partial_t w = g(x) \quad \text{at} \quad t = 0 \quad \text{(initial condition),}$$
$$w = 0 \quad \text{at} \quad x = 0 \quad \text{(boundary condition),}$$
$$w = 0 \quad \text{at} \quad x = l \quad \text{(boundary condition).}$$

Solution for $m > -1$:

$$w(x,t) = \sqrt{t} \sum_{n=1}^{\infty} \left[A_n J_{-p}\left(2p\lambda_n t^{\frac{1}{2p}}\right) + B_n J_p\left(2p\lambda_n t^{\frac{1}{2p}}\right) \right] \sin(\lambda_n x),$$

$$A_n = \Gamma(1-p)(\lambda_n p)^p \frac{2}{l} \int_0^l f(x) \sin(\lambda_n x)\,dx, \quad p = \frac{1}{m+2},$$

$$B_n = \Gamma(1+p)(\lambda_n p)^{-p} \frac{2}{l} \int_0^l g(x) \sin(\lambda_n x)\,dx, \quad \lambda_n = \frac{\pi n}{l},$$

where $\Gamma(p)$ is the gamma function.
⊙ *Reference*: M. M. Smirnov (1975).

12. $\dfrac{\partial^2 w}{\partial t^2} = t^m \dfrac{\partial^2 w}{\partial x^2} + bt^{\frac{m-2}{2}} \dfrac{\partial w}{\partial x}, \quad m \ge 2.$

Domain: $-\infty < x < \infty$. Cauchy problem.
Initial conditions are prescribed:

$$w = f(x) \quad \text{at} \quad t = 0,$$
$$\partial_t w = g(x) \quad \text{at} \quad t = 0.$$

1°. Solution for $|b| < \frac{1}{2}m$:

$$w(x,t) = \frac{\Gamma(\alpha+\beta)}{\Gamma(\alpha)\Gamma(\beta)} \int_0^1 f\left(x + \frac{2}{m+2}t^{\frac{m+2}{2}}(2\xi-1)\right) \xi^{\beta-1}(1-\xi)^{\alpha-1}\,d\xi$$

$$+ \frac{\Gamma(2-\alpha-\beta)}{\Gamma(1-\alpha)\Gamma(1-\beta)} t \int_0^1 g\left(x + \frac{2}{m+2}t^{\frac{m+2}{2}}(2\xi-1)\right) \xi^{-\alpha}(1-\xi)^{-\beta}\,d\xi,$$

where

$$\alpha = \frac{m-2b}{2(m+2)}, \quad \beta = \frac{m+2b}{2(m+2)}, \quad \Gamma(z) = \int_0^{\infty} e^{-s} s^{z-1}\,ds.$$

2°. Solution for $b = \frac{1}{2}m$:

$$w(x,t) = f\left(x + \frac{2}{m+2}t^{\frac{m+2}{2}}\right) + \frac{2t}{m+2} \int_0^1 g\left(x + \frac{2}{m+2}t^{\frac{m+2}{2}}(2\xi-1)\right)(1-\xi)^{-\frac{m}{m+2}}\,d\xi.$$

3°. Solution for $b = -\frac{1}{2}m$:

$$w(x,t) = f\left(x - \frac{2}{m+2}t^{\frac{m+2}{2}}\right) + \frac{2t}{m+2} \int_0^1 g\left(x + \frac{2}{m+2}t^{\frac{m+2}{2}}(2\xi-1)\right)(1-\xi)^{-\frac{m}{m+2}}\,d\xi.$$

⊙ *Reference*: M. M. Smirnov (1975).

13. $(b+x)^2 \dfrac{\partial^2 w}{\partial t^2} = a^2 \dfrac{\partial}{\partial x}\left[(b+x)^2 \dfrac{\partial w}{\partial x}\right].$

General solution:
$$w(x,t) = \frac{f(x+at) + g(x-at)}{b+x},$$

where $f(y)$ and $g(z)$ are arbitrary functions.

4.4. Equations Containing the First Time Derivative

4.4.1. Equations of the Form $\dfrac{\partial^2 w}{\partial t^2} + k\dfrac{\partial w}{\partial t} = a^2 \dfrac{\partial^2 w}{\partial x^2} + b\dfrac{\partial w}{\partial x} + cw + \Phi(x,t)$

1. $\dfrac{\partial^2 w}{\partial t^2} + k\dfrac{\partial w}{\partial t} = a^2 \dfrac{\partial^2 w}{\partial x^2} + \Phi(x,t).$

For $\Phi(x,t) \equiv 0$, this equation governs free transverse vibration of a string, and also longitudinal vibration of a rod in a resisting medium with a velocity-proportional resistance coefficient.

1°. The substitution $w(x,t) = \exp\left(-\tfrac{1}{2}kt\right)u(x,t)$ leads to the equation

$$\frac{\partial^2 u}{\partial t^2} = a^2 \frac{\partial^2 u}{\partial x^2} + \tfrac{1}{4}k^2 u + \exp\left(\tfrac{1}{2}kt\right)\Phi(x,t),$$

which is considered in Subsection 4.1.3.

2°. Fundamental solution:

$$\mathscr{E}(x,t) = \frac{1}{2a}\vartheta(at - |x|)\exp\left(-\tfrac{1}{2}kt\right)I_0\left(\tfrac{1}{2}k\sqrt{t^2 - x^2/a^2}\right),$$

where $\vartheta(z)$ is the Heaviside unit step function and $I_0(z)$ is the modified Bessel function.

⊙ *Reference*: V. S. Vladimirov, V. P. Mikhailov, A. A. Vasharin, et al. (1974).

3°. Domain: $-\infty < x < \infty$. Cauchy problem.
Initial conditions are prescribed:

$$w = f(x) \quad \text{at} \quad t = 0,$$
$$\partial_t w = g(x) \quad \text{at} \quad t = 0.$$

Solution:

$$w(x,t) = \tfrac{1}{2}\exp\left(-\tfrac{1}{2}kt\right)\left[f(x+at) + f(x-at)\right]$$
$$+ \frac{kt}{4a}\exp\left(-\tfrac{1}{2}kt\right)\int_{x-at}^{x+at} \frac{I_1\left(\tfrac{1}{2}k\sqrt{t^2 - (x-\xi)^2/a^2}\right)}{\sqrt{t^2 - (x-\xi)^2/a^2}}f(\xi)\,d\xi$$
$$+ \frac{1}{2a}\exp\left(-\tfrac{1}{2}kt\right)\int_{x-at}^{x+at} I_0\left(\tfrac{1}{2}k\sqrt{t^2 - (x-\xi)^2/a^2}\right)\left[g(\xi) + \tfrac{1}{2}kf(\xi)\right]d\xi$$
$$+ \frac{1}{2a}\int_0^t \int_{x-a(t-\tau)}^{x+a(t-\tau)} \exp\left[-\tfrac{1}{2}k(t-\tau)\right]I_0\left(\tfrac{1}{2}k\sqrt{(t-\tau)^2 - (x-\xi)^2/a^2}\right)\Phi(\xi,\tau)\,d\xi\,d\tau,$$

where $I_0(z)$ and $I_1(z)$ are the modified Bessel functions of the first kind.

4°. Domain: $0 \le x \le l$. First boundary value problem.
The following conditions are prescribed:

$$w = f_0(x) \quad \text{at} \quad t = 0 \quad \text{(initial condition)},$$
$$\partial_t w = f_1(x) \quad \text{at} \quad t = 0 \quad \text{(initial condition)},$$
$$w = g_1(t) \quad \text{at} \quad x = 0 \quad \text{(boundary condition)},$$
$$w = g_2(t) \quad \text{at} \quad x = l \quad \text{(boundary condition)}.$$

Solution:

$$w(x,t) = \int_0^t \int_0^l \Phi(\xi,\tau)G(x,\xi,t-\tau)\,d\xi\,d\tau$$
$$+ \frac{\partial}{\partial t}\int_0^l f_0(\xi)G(x,\xi,t)\,d\xi + \int_0^l \left[f_1(\xi) + kf_0(\xi)\right]G(x,\xi,t)\,d\xi$$
$$+ a^2 \int_0^t g_1(\tau)\left[\frac{\partial}{\partial \xi}G(x,\xi,t-\tau)\right]_{\xi=0} d\tau - a^2 \int_0^t g_2(\tau)\left[\frac{\partial}{\partial \xi}G(x,\xi,t-\tau)\right]_{\xi=l} d\tau,$$

where

$$G(x,\xi,t) = \frac{2}{l}\exp\left(-\frac{kt}{2}\right)\sum_{n=1}^{\infty}\sin\left(\frac{\pi n x}{l}\right)\sin\left(\frac{\pi n \xi}{l}\right)\frac{\sin(\lambda_n t)}{\lambda_n}, \qquad \lambda_n = \sqrt{\frac{a^2\pi^2 n^2}{l^2} - \frac{k^2}{4}}.$$

Example. Consider the homogeneous equation ($\Phi \equiv 0$). The initial shape of the string is a triangle with base $0 \le x \le l$ and height h at $x = c$, that is,

$$f(x) = \begin{cases} \dfrac{hx}{c} & \text{for } 0 \le x \le c, \\ \dfrac{h(l-x)}{l-c} & \text{for } c \le x \le l. \end{cases}$$

The initial velocities of the string points are zero, $g(x) = 0$.

Solution:

$$w(x,t) = \frac{2hl^2}{\pi^2 c(l-c)}\exp(-\tfrac{1}{2}kt)\sum_{n=1}^{\infty}\frac{1}{n^2}\sin\left(\frac{n\pi c}{l}\right)\sin\left(\frac{n\pi x}{l}\right)\Theta_n(t),$$

where

$$\Theta_n(t) = \begin{cases} \cos(\lambda_n t) + \dfrac{k}{2\lambda_n}\sin(\lambda_n t) & \text{for } k < \dfrac{2\pi n a}{l}, \\ 1 + \dfrac{kt}{2} & \text{for } k = \dfrac{2\pi n a}{l}, \\ \cosh(\lambda_n t) + \dfrac{k}{2\lambda_n}\sinh(\lambda_n t) & \text{for } k > \dfrac{2\pi n a}{l}, \end{cases} \qquad \lambda_n = \sqrt{\left|\frac{a^2 n^2 \pi^2}{l^2} - \frac{k^2}{4}\right|}.$$

⊙ *References*: M. M. Smirnov (1975), B. M. Budak, A. N. Tikhonov, and A. A. Samarskii (1980).

5°. For the second and third boundary value problems on the interval $0 \le x \le l$, see equation 4.4.1.2 (Items 5° and 6° with $b = 0$).

2. $\dfrac{\partial^2 w}{\partial t^2} + k\dfrac{\partial w}{\partial t} = a^2\dfrac{\partial^2 w}{\partial x^2} + bw + \Phi(x,t).$

Telegraph equation (with $k > 0$, $b < 0$, and $\Phi(x,t) \equiv 0$).

1°. The substitution $w(x,t) = \exp\left(-\tfrac{1}{2}kt\right)u(x,t)$ leads to the equation

$$\frac{\partial^2 u}{\partial t^2} = a^2\frac{\partial^2 u}{\partial x^2} + (b + \tfrac{1}{4}k^2)u + \exp\left(\tfrac{1}{2}kt\right)\Phi(x,t),$$

which is considered in Subsection 4.1.3.

2°. Fundamental solutions:

$$\mathscr{E}(x,t) = \frac{1}{2a}\vartheta(at - |x|)\exp(-\tfrac{1}{2}kt)I_0\left(c\sqrt{t^2 - x^2/a^2}\right) \quad \text{for } b + \tfrac{1}{4}k^2 = c^2 > 0,$$

$$\mathscr{E}(x,t) = \frac{1}{2a}\vartheta(at - |x|)\exp(-\tfrac{1}{2}kt)J_0\left(c\sqrt{t^2 - x^2/a^2}\right) \quad \text{for } b + \tfrac{1}{4}k^2 = -c^2 < 0,$$

where $\vartheta(z)$ is the Heaviside unit step function, $J_0(z)$ and $J_1(z)$ are the Bessel functions, and $I_0(z)$ and $I_1(z)$ are the modified Bessel functions.

3°. Domain: $-\infty < x < \infty$. Cauchy problem.
Initial conditions are prescribed:

$$w = f(x) \quad \text{at} \quad t = 0,$$
$$\partial_t w = g(x) \quad \text{at} \quad t = 0.$$

Solution for $b + \frac{1}{4}k^2 = c^2 > 0$:

$$w(x,t) = \tfrac{1}{2}\exp(-\tfrac{1}{2}kt)\left[f(x+at)+f(x-at)\right]$$
$$+ \frac{ct}{2a}\exp(-\tfrac{1}{2}kt)\int_{x-at}^{x+at}\frac{I_1\left(c\sqrt{t^2-(x-\xi)^2/a^2}\right)}{\sqrt{t^2-(x-\xi)^2/a^2}}f(\xi)\,d\xi$$
$$+\frac{1}{2a}\exp(-\tfrac{1}{2}kt)\int_{x-at}^{x+at} I_0\left(c\sqrt{t^2-(x-\xi)^2/a^2}\right)\left[g(\xi)+\tfrac{1}{2}kf(\xi)\right]d\xi$$
$$+\frac{1}{2a}\int_0^t\int_{x-a(t-\tau)}^{x+a(t-\tau)}\exp\!\left[-\tfrac{1}{2}k(t-\tau)\right]I_0\!\left(c\sqrt{(t-\tau)^2-(x-\xi)^2/a^2}\right)\Phi(\xi,\tau)\,d\xi\,d\tau.$$

Solution for $b + \frac{1}{4}k^2 = -c^2 < 0$:

$$w(x,t) = \tfrac{1}{2}\exp(-\tfrac{1}{2}kt)\left[f(x+at)+f(x-at)\right]$$
$$- \frac{ct}{2a}\exp(-\tfrac{1}{2}kt)\int_{x-at}^{x+at}\frac{J_1\left(c\sqrt{t^2-(x-\xi)^2/a^2}\right)}{\sqrt{t^2-(x-\xi)^2/a^2}}f(\xi)\,d\xi$$
$$+\frac{1}{2a}\exp(-\tfrac{1}{2}kt)\int_{x-at}^{x+at} J_0\left(c\sqrt{t^2-(x-\xi)^2/a^2}\right)\left[g(\xi)+\tfrac{1}{2}kf(\xi)\right]d\xi$$
$$+\frac{1}{2a}\int_0^t\int_{x-a(t-\tau)}^{x+a(t-\tau)}\exp\!\left[-\tfrac{1}{2}k(t-\tau)\right]J_0\!\left(c\sqrt{(t-\tau)^2-(x-\xi)^2/a^2}\right)\Phi(\xi,\tau)\,d\xi\,d\tau.$$

4°. **Domain:** $0 \le x \le l$. **First boundary value problem.**
The following conditions are prescribed:

$$w = f_0(x) \quad \text{at} \quad t=0 \quad \text{(initial condition)},$$
$$\partial_t w = f_1(x) \quad \text{at} \quad t=0 \quad \text{(initial condition)},$$
$$w = g_1(t) \quad \text{at} \quad x=0 \quad \text{(boundary condition)},$$
$$w = g_2(t) \quad \text{at} \quad x=l \quad \text{(boundary condition)}.$$

Solution:

$$w(x,t) = \int_0^t\int_0^l \Phi(\xi,\tau)G(x,\xi,t-\tau)\,d\xi\,d\tau$$
$$+\frac{\partial}{\partial t}\int_0^l f_0(\xi)G(x,\xi,t)\,d\xi + \int_0^l\left[f_1(\xi)+kf_0(\xi)\right]G(x,\xi,t)\,d\xi$$
$$+a^2\int_0^t g_1(\tau)\left[\frac{\partial}{\partial \xi}G(x,\xi,t-\tau)\right]_{\xi=0}d\tau - a^2\int_0^t g_2(\tau)\left[\frac{\partial}{\partial \xi}G(x,\xi,t-\tau)\right]_{\xi=l}d\tau.$$

Let $a^2\pi^2 - bl^2 - \tfrac{1}{4}k^2l^2 > 0$. Then

$$G(x,\xi,t) = \frac{2}{l}\exp\!\left(-\frac{kt}{2}\right)\sum_{n=1}^{\infty}\sin\!\left(\frac{\pi n x}{l}\right)\sin\!\left(\frac{\pi n \xi}{l}\right)\frac{\sin(t\sqrt{\lambda_n})}{\sqrt{\lambda_n}}, \qquad \lambda_n = \frac{a^2\pi^2 n^2}{l^2} - b - \frac{k^2}{4}.$$

Let $a^2\pi^2 n^2 - bl^2 - \tfrac{1}{4}k^2l^2 \le 0$ for $n = 1,\ldots,m$ and $a^2\pi^2 n^2 - bl^2 - \tfrac{1}{4}k^2l^2 > 0$ for $n = m+1, m+2, \ldots$
Then

$$G(x,\xi,t) = \frac{2}{l}\exp\!\left(-\frac{kt}{2}\right)\sum_{n=1}^{m}\sin\!\left(\frac{\pi n x}{l}\right)\sin\!\left(\frac{\pi n \xi}{l}\right)\frac{\sinh(t\sqrt{\beta_n})}{\sqrt{\beta_n}}$$
$$+\frac{2}{l}\exp\!\left(-\frac{kt}{2}\right)\sum_{n=m+1}^{\infty}\sin\!\left(\frac{\pi n x}{l}\right)\sin\!\left(\frac{\pi n \xi}{l}\right)\frac{\sin(t\sqrt{\lambda_n})}{\sqrt{\lambda_n}},$$
$$\beta_n = b + \frac{k^2}{4} - \frac{a^2\pi^2 n^2}{l^2}, \qquad \lambda_n = \frac{a^2\pi^2 n^2}{l^2} - b - \frac{k^2}{4}.$$

5°. Domain: $0 \leq x \leq l$. Second boundary value problem.
The following conditions are prescribed:

$$w = f_0(x) \quad \text{at} \quad t = 0 \quad \text{(initial condition)},$$
$$\partial_t w = f_1(x) \quad \text{at} \quad t = 0 \quad \text{(initial condition)},$$
$$\partial_x w = g_1(t) \quad \text{at} \quad x = 0 \quad \text{(boundary condition)},$$
$$\partial_x w = g_2(t) \quad \text{at} \quad x = l \quad \text{(boundary condition)}.$$

Solution:

$$w(x,t) = \int_0^t \int_0^l \Phi(\xi,\tau) G(x,\xi,t-\tau) \, d\xi \, d\tau$$
$$+ \frac{\partial}{\partial t} \int_0^l f_0(\xi) G(x,\xi,t) \, d\xi + \int_0^l \left[f_1(\xi) + k f_0(\xi) \right] G(x,\xi,t) \, d\xi$$
$$- a^2 \int_0^t g_1(\tau) G(x,0,t-\tau) \, d\tau + a^2 \int_0^t g_2(\tau) G(x,l,t-\tau) \, d\tau.$$

For $p = b + \frac{1}{4} k^2 < 0$,

$$G(x,\xi,t) = \exp\left(-\tfrac{1}{2} kt\right) \left[\frac{\sin\left(t\sqrt{|p|}\right)}{l\sqrt{|p|}} + \frac{2}{l} \sum_{n=1}^{\infty} \cos(\mu_n x) \cos(\mu_n \xi) \frac{\sin\left(t\sqrt{a^2 \mu_n^2 - p}\right)}{\sqrt{a^2 \mu_n^2 - p}} \right], \quad \mu_n = \frac{\pi n}{l}.$$

For $p = b + \frac{1}{4} k^2 > 0$,

$$G(x,\xi,t) = \exp\left(-\tfrac{1}{2} kt\right) \left[\frac{\sinh\left(t\sqrt{p}\right)}{l\sqrt{p}} + \frac{2}{l} \sum_{n=1}^{\infty} \cos(\mu_n x) \cos(\mu_n \xi) \frac{\sin\left(t\sqrt{a^2 \mu_n^2 - p}\right)}{\sqrt{a^2 \mu_n^2 - p}} \right], \quad \mu_n = \frac{\pi n}{l}.$$

If the inequality $a^2 \mu_n^2 - p < 0$ holds for several first values $n = 1, \ldots, m$, then the expressions $\sqrt{a^2 \mu_n^2 - p}$ should be replaced by $\sqrt{|a^2 \mu_n^2 - p|}$ and the sines by the hyperbolic sines in the corresponding terms of the series.

6°. Domain: $0 \leq x \leq l$. Third boundary value problem.
The following conditions are prescribed:

$$w = f_0(x) \quad \text{at} \quad t = 0 \quad \text{(initial condition)},$$
$$\partial_t w = f_1(x) \quad \text{at} \quad t = 0 \quad \text{(initial condition)},$$
$$\partial_x w - s_1 w = g_1(t) \quad \text{at} \quad x = 0 \quad \text{(boundary condition)},$$
$$\partial_x w + s_2 w = g_2(t) \quad \text{at} \quad x = l \quad \text{(boundary condition)}.$$

The solution $w(x,t)$ is determined by the formula in Item 5° with

$$G(x,\xi,t) = \exp\left(-\tfrac{1}{2} kt\right) \sum_{n=1}^{\infty} \frac{y_n(x) y_n(\xi) \sin\left(t \sqrt{a^2 \mu_n^2 - p}\right)}{B_n \sqrt{a^2 \mu_n^2 - p}}, \quad p = b + \tfrac{1}{4} k^2,$$

$$y_n(x) = \cos(\mu_n x) + \frac{s_1}{\mu_n} \sin(\mu_n x), \quad B_n = \frac{s_2}{2\mu_n^2} \frac{\mu_n^2 + s_1^2}{\mu_n^2 + s_2^2} + \frac{s_1}{2\mu_n^2} + \frac{l}{2}\left(1 + \frac{s_1^2}{\mu_n^2}\right).$$

Here, the μ_n are positive roots of the transcendental equation $\dfrac{\tan(\mu l)}{\mu} = \dfrac{s_1 + s_2}{\mu^2 - s_1 s_2}$.

If the inequality $a^2 \mu_n^2 - p < 0$ holds for several first values $n = 1, \ldots, m$, then the expressions $\sqrt{a^2 \mu_n^2 - p}$ should be replaced by $\sqrt{|a^2 \mu_n^2 - p|}$ and the sines by the hyperbolic sines in the corresponding terms of the series.

3. $\dfrac{\partial^2 w}{\partial t^2} + k\dfrac{\partial w}{\partial t} = a^2 \dfrac{\partial^2 w}{\partial x^2} + b\dfrac{\partial w}{\partial x} + cw + \Phi(x,t).$

1°. The substitution $w(x,t) = \exp\left(-\tfrac{1}{2}a^{-2}bx - \tfrac{1}{2}kt\right)u(x,t)$ leads to the equation

$$\frac{\partial^2 u}{\partial t^2} = a^2 \frac{\partial^2 u}{\partial x^2} + \left(c + \tfrac{1}{4}k^2 - \tfrac{1}{4}a^{-2}b^2\right)u + \exp\left(\tfrac{1}{2}a^{-2}bx + \tfrac{1}{2}kt\right)\Phi(x,t),$$

which is discussed in Subsection 4.1.3.

2°. Fundamental solutions:

$$\mathscr{E}(x,t) = \frac{1}{2a}\vartheta(at-|x|)\exp\left(-\frac{bx}{2a^2} - \frac{kt}{2}\right) I_0\left(\sigma\sqrt{t^2 - \frac{x^2}{a^2}}\right) \quad \text{if} \quad c + \frac{k^2}{4} - \frac{b^2}{4a^2} = \sigma^2 > 0,$$

$$\mathscr{E}(x,t) = \frac{1}{2a}\vartheta(at-|x|)\exp\left(-\frac{bx}{2a^2} - \frac{kt}{2}\right) J_0\left(\sigma\sqrt{t^2 - \frac{x^2}{a^2}}\right) \quad \text{if} \quad c + \frac{k^2}{4} - \frac{b^2}{4a^2} = -\sigma^2 < 0,$$

where $\vartheta(z)$ is the Heaviside unit step function, $J_0(z)$ and $J_1(z)$ are the Bessel functions, and $I_0(z)$ and $I_1(z)$ are the modified Bessel functions.

3°. Domain: $-\infty < x < \infty$. Cauchy problem.
Initial conditions are prescribed:

$$w = f(x) \quad \text{at} \quad t = 0,$$
$$\partial_t w = g(x) \quad \text{at} \quad t = 0.$$

Solution for $c + \tfrac{1}{4}k^2 - \tfrac{1}{4}a^{-2}b^2 = \sigma^2 > 0$:

$$w(x,t) = \frac{1}{2}\exp\left(-\frac{kt}{2}\right)\left[f(x+at)\exp\left(\frac{bt}{2a}\right) + f(x-at)\exp\left(-\frac{bt}{2a}\right)\right]$$
$$+ \frac{\sigma t}{2a}\exp\left(-\frac{bx}{2a^2} - \frac{kt}{2}\right)\int_{x-at}^{x+at}\exp\left(\frac{b\xi}{2a^2}\right)\frac{I_1\left(\sigma\sqrt{t^2-(x-\xi)^2/a^2}\right)}{\sqrt{t^2-(x-\xi)^2/a^2}} f(\xi)\,d\xi$$
$$+ \frac{1}{2a}\exp\left(-\frac{bx}{2a^2} - \frac{kt}{2}\right)\int_{x-at}^{x+at}\exp\left(\frac{b\xi}{2a^2}\right) I_0\left(\sigma\sqrt{t^2-(x-\xi)^2/a^2}\right)\left[g(\xi) + \tfrac{1}{2}kf(\xi)\right] d\xi$$
$$+ \frac{1}{2a}\int_0^t\int_{x-a(t-\tau)}^{x+a(t-\tau)}\exp\left[\frac{b(\xi-x)}{2a^2} - \frac{k(t-\tau)}{2}\right] I_0\left(\sigma\sqrt{(t-\tau)^2-(x-\xi)^2/a^2}\right)\Phi(\xi,\tau)\,d\xi\,d\tau.$$

Solution for $c + \tfrac{1}{4}k^2 - \tfrac{1}{4}a^{-2}b^2 = -\sigma^2 < 0$:

$$w(x,t) = \frac{1}{2}\exp\left(-\frac{kt}{2}\right)\left[f(x+at)\exp\left(\frac{bt}{2a}\right) + f(x-at)\exp\left(-\frac{bt}{2a}\right)\right]$$
$$- \frac{\sigma t}{2a}\exp\left(-\frac{bx}{2a^2} - \frac{kt}{2}\right)\int_{x-at}^{x+at}\exp\left(\frac{b\xi}{2a^2}\right)\frac{J_1\left(\sigma\sqrt{t^2-(x-\xi)^2/a^2}\right)}{\sqrt{t^2-(x-\xi)^2/a^2}} f(\xi)\,d\xi$$
$$+ \frac{1}{2a}\exp\left(-\frac{bx}{2a^2} - \frac{kt}{2}\right)\int_{x-at}^{x+at}\exp\left(\frac{b\xi}{2a^2}\right) J_0\left(\sigma\sqrt{t^2-(x-\xi)^2/a^2}\right)\left[g(\xi) + \tfrac{1}{2}kf(\xi)\right] d\xi$$
$$+ \frac{1}{2a}\int_0^t\int_{x-a(t-\tau)}^{x+a(t-\tau)}\exp\left[\frac{b(\xi-x)}{2a^2} - \frac{k(t-\tau)}{2}\right] J_0\left(\sigma\sqrt{(t-\tau)^2-(x-\xi)^2/a^2}\right)\Phi(\xi,\tau)\,d\xi\,d\tau.$$

⊙ *Reference*: A. N. Tikhonov and A. A. Samarskii (1990).

4°. Domain: $0 \leq x \leq l$. First boundary value problem.
The following conditions are prescribed:

$$w = f_0(x) \quad \text{at} \quad t = 0 \quad \text{(initial condition)},$$
$$\partial_t w = f_1(x) \quad \text{at} \quad t = 0 \quad \text{(initial condition)},$$
$$w = g_1(t) \quad \text{at} \quad x = 0 \quad \text{(boundary condition)},$$
$$w = g_2(t) \quad \text{at} \quad x = l \quad \text{(boundary condition)}.$$

Solution:

$$w(x,t) = \int_0^t \int_0^l \Phi(\xi,\tau) G(x,\xi,t-\tau) \, d\xi \, d\tau$$
$$+ \frac{\partial}{\partial t} \int_0^l f_0(\xi) G(x,\xi,t) \, d\xi + \int_0^l \big[f_1(\xi) + k f_0(\xi) \big] G(x,\xi,t) \, d\xi$$
$$+ a^2 \int_0^t g_1(\tau) \left[\frac{\partial}{\partial \xi} G(x,\xi,t-\tau) \right]_{\xi=0} d\tau - a^2 \int_0^t g_2(\tau) \left[\frac{\partial}{\partial \xi} G(x,\xi,t-\tau) \right]_{\xi=l} d\tau.$$

Let $a^2\pi^2 + \frac{1}{4} a^{-2} b^2 l^2 - cl^2 - \frac{1}{4} k^2 l^2 > 0$. Then

$$G(x,\xi,t) = \frac{2}{l} \exp\left[\frac{b(\xi-x)}{2a^2} - \frac{kt}{2} \right] \sum_{n=1}^{\infty} \sin\left(\frac{\pi n x}{l}\right) \sin\left(\frac{\pi n \xi}{l}\right) \frac{\sin\left(t\sqrt{\lambda_n}\right)}{\sqrt{\lambda_n}},$$

$$\lambda_n = \frac{a^2 \pi^2 n^2}{l^2} + \frac{b^2}{4a^2} - c - \frac{k^2}{4}.$$

Let

$$a^2\pi^2 n^2 + \tfrac{1}{4} a^{-2} b^2 l^2 - cl^2 - \tfrac{1}{4} k^2 l^2 \leq 0 \quad \text{for} \quad n = 1, \ldots, m;$$
$$a^2\pi^2 n^2 + \tfrac{1}{4} a^{-2} b^2 l^2 - cl^2 - \tfrac{1}{4} k^2 l^2 > 0 \quad \text{for} \quad n = m+1, m+2, \ldots$$

Then

$$G(x,\xi,t) = \frac{2}{l} \exp\left[\frac{b(\xi-x)}{2a^2} - \frac{kt}{2} \right] \sum_{n=1}^{m} \sin\left(\frac{\pi n x}{l}\right) \sin\left(\frac{\pi n \xi}{l}\right) \frac{\sinh\left(t\sqrt{\beta_n}\right)}{\sqrt{\beta_n}}$$
$$+ \frac{2}{l} \exp\left[\frac{b(\xi-x)}{2a^2} - \frac{kt}{2} \right] \sum_{n=m+1}^{\infty} \sin\left(\frac{\pi n x}{l}\right) \sin\left(\frac{\pi n \xi}{l}\right) \frac{\sin\left(t\sqrt{\lambda_n}\right)}{\sqrt{\lambda_n}},$$

where $\beta_n = c + \dfrac{k^2}{4} - \dfrac{a^2 \pi^2 n^2}{l^2} - \dfrac{b^2}{4a^2}$ and $\lambda_n = \dfrac{a^2 \pi^2 n^2}{l^2} + \dfrac{b^2}{4a^2} - c - \dfrac{k^2}{4}$.

⊙ *Reference*: A. G. Butkovskiy (1979).

5°. Domain: $0 \leq x \leq l$. Second boundary value problem.
The following conditions are prescribed:

$$w = f_0(x) \quad \text{at} \quad t = 0 \quad \text{(initial condition)},$$
$$\partial_t w = f_1(x) \quad \text{at} \quad t = 0 \quad \text{(initial condition)},$$
$$\partial_x w = g_1(t) \quad \text{at} \quad x = 0 \quad \text{(boundary condition)},$$
$$\partial_x w = g_2(t) \quad \text{at} \quad x = l \quad \text{(boundary condition)}.$$

Solution:

$$w(x,t) = \int_0^t \int_0^l \Phi(\xi,\tau) G(x,\xi,t-\tau) \, d\xi \, d\tau$$
$$+ \frac{\partial}{\partial t} \int_0^l f_0(\xi) G(x,\xi,t) \, d\xi + \int_0^l \big[f_1(\xi) + k f_0(\xi) \big] G(x,\xi,t) \, d\xi$$
$$- a^2 \int_0^t g_1(\tau) G(x,0,t-\tau) \, d\tau + a^2 \int_0^t g_2(\tau) G(x,l,t-\tau) \, d\tau.$$

For $p = c + \frac{1}{4}k^2 < 0$,

$$G(x,\xi,t) = A\exp\left(\frac{b\xi}{a^2} - \frac{kt}{2}\right)\frac{\sin(t\sqrt{|p|})}{\sqrt{|p|}} + \frac{2}{l}\exp\left[\frac{b(\xi-x)}{2a^2} - \frac{kt}{2}\right]\sum_{n=1}^{\infty}\frac{y_n(x)y_n(\xi)}{1+\mu_n^2}\frac{\sin(t\sqrt{\lambda_n})}{\sqrt{\lambda_n}},$$

where

$$A = \frac{b}{a^2\left(e^{bl/a^2}-1\right)}, \quad \lambda_n = \frac{a^2\pi^2n^2}{l^2} + \frac{b^2}{4a^2} - c - \frac{k^2}{4},$$

$$y_n(x) = \cos\left(\frac{\pi n x}{l}\right) + \mu_n\sin\left(\frac{\pi n x}{l}\right), \quad \mu_n = \frac{bl}{2a^2\pi n}.$$

For $p = c + \frac{1}{4}k^2 > 0$,

$$G(x,\xi,t) = A\exp\left(\frac{b\xi}{a^2} - \frac{kt}{2}\right)\frac{\sinh(t\sqrt{p})}{\sqrt{p}} + \frac{2}{l}\exp\left[\frac{b(\xi-x)}{2a^2} - \frac{kt}{2}\right]\sum_{n=1}^{\infty}\frac{y_n(x)y_n(\xi)}{1+\mu_n^2}\frac{\sin(t\sqrt{\lambda_n})}{\sqrt{\lambda_n}},$$

where the coefficient A, λ_n, μ_n and the functions $y_n(x)$ remain as before. If the inequality $\lambda_n < 0$ holds for several first values $n = 1, \ldots, m$, then the expressions $\sqrt{\lambda_n}$ must be replaced by $\sqrt{|\lambda_n|}$ and the sines by the hyperbolic sines in the corresponding terms of the series.

6°. Domain: $0 \leq x \leq l$. Third boundary value problem.
The following conditions are prescribed:

$$\begin{aligned}
w &= f_0(x) &&\text{at}\quad t=0 &&\text{(initial condition)},\\
\partial_t w &= f_1(x) &&\text{at}\quad t=0 &&\text{(initial condition)},\\
\partial_x w - s_1 w &= g_1(t) &&\text{at}\quad x=0 &&\text{(boundary condition)},\\
\partial_x w + s_2 w &= g_2(t) &&\text{at}\quad x=l &&\text{(boundary condition)}.
\end{aligned}$$

The solution $w(x,t)$ is determined by the formula in Item 5° with

$$G(x,\xi,t) = \exp\left[\frac{b(\xi-x)}{2a^2} - \frac{kt}{2}\right]\sum_{n=1}^{\infty}\frac{y_n(x)y_n(\xi)\sin(t\sqrt{\lambda_n})}{B_n\sqrt{\lambda_n}}.$$

Here,

$$y_n(x) = \cos(\mu_n x) + \frac{2a^2 s_1 + b}{2a^2 \mu_n}\sin(\mu_n x), \quad \lambda_n = a^2\mu_n^2 + \frac{b^2}{4a^2} - c - \frac{k^2}{4},$$

$$B_n = \frac{2a^2 s_2 - b}{4a^2\mu_n^2}\frac{4a^4\mu_n^2 + (2a^2 s_1 + b)^2}{4a^4\mu_n^2 + (2a^2 s_2 - b)^2} + \frac{2a^2 s_1 + b}{4a^2\mu_n^2} + \frac{l}{2} + \frac{l(2a^2 s_1 + b)^2}{8a^4\mu_n^2},$$

where the μ_n are positive roots of the transcendental equation

$$\frac{\tan(\mu l)}{\mu} = \frac{4a^4(s_1 + s_2)}{4a^4\mu^2 - (2a^2 s_1 + b)(2a^2 s_2 - b)}.$$

4.4.2. Equations of the Form

$$\frac{\partial^2 w}{\partial t^2} + k\frac{\partial w}{\partial t} = f(x)\frac{\partial^2 w}{\partial x^2} + g(x)\frac{\partial w}{\partial x} + h(x)w + \Phi(x,t)$$

1. $\dfrac{\partial^2 w}{\partial t^2} + k\dfrac{\partial w}{\partial t} = a^2\left(\dfrac{\partial^2 w}{\partial r^2} + \dfrac{1}{r}\dfrac{\partial w}{\partial r}\right).$

This equation describes vibration of a circular membrane in a resisting medium with velocity-proportional resistance coefficient.

1°. Domain: $0 \leq r \leq R$. First boundary value problem.
The following conditions are prescribed:

$$w = f(r) \quad \text{at} \quad t = 0 \quad \text{(initial condition)},$$
$$\partial_t w = g(r) \quad \text{at} \quad t = 0 \quad \text{(initial condition)},$$
$$w = 0 \quad \text{at} \quad x = R \quad \text{(boundary condition)}.$$

Solution:

$$w(r,t) = \exp\left(-\tfrac{1}{2}kt\right) \sum_{n=1}^{\infty} \left[A_n \cos(\lambda_n t) + B_n \sin(\lambda_n t)\right] J_0\left(\frac{\mu_n r}{R}\right), \quad \lambda_n = \sqrt{\frac{a^2 \mu_n^2}{R^2} - \frac{k^2}{4}}.$$

Here,

$$A_n = \frac{2}{R^2 J_1^2(\mu_n)} \int_0^R f(r) J_0\left(\frac{\mu_n r}{R}\right) r\, dr, \quad B_n = \frac{A_n k}{2\lambda_n} + \frac{2}{\lambda_n R^2 J_1^2(\mu_n)} \int_0^R g(r) J_0\left(\frac{\mu_n r}{R}\right) r\, dr,$$

where the μ_n are positive zeros of the Bessel function, $J_0(\mu) = 0$.

2°. For the solution of the second and third boundary value problems, see equation 4.4.2.2 (Items 3° and 4° with $b = 0$).

⊙ *Reference*: B. M. Budak, A. A. Samarskii, and A. N. Tikhonov (1980).

2. $\dfrac{\partial^2 w}{\partial t^2} + k\dfrac{\partial w}{\partial t} = a^2 \left(\dfrac{\partial^2 w}{\partial r^2} + \dfrac{1}{r}\dfrac{\partial w}{\partial r}\right) - bw + \Phi(r,t).$

1°. The substitution $w(r,t) = \exp\left(-\tfrac{1}{2}kt\right) u(r,t)$ leads to the equation

$$\frac{\partial^2 u}{\partial t^2} = a^2 \left(\frac{\partial^2 u}{\partial r^2} + \frac{1}{r}\frac{\partial u}{\partial r}\right) - \left(b - \tfrac{1}{4}k^2\right) u + \exp\left(\tfrac{1}{2}kt\right) \Phi(r,t),$$

which is discussed in Subsection 4.2.5.

2°. Domain: $0 \leq r \leq R$. First boundary value problem.
The following conditions are prescribed:

$$w = f_0(r) \quad \text{at} \quad t = 0 \quad \text{(initial condition)},$$
$$\partial_t w = f_1(r) \quad \text{at} \quad t = 0 \quad \text{(initial condition)},$$
$$w = g(t) \quad \text{at} \quad r = R \quad \text{(boundary condition)}.$$

Solution:

$$w(r,t) = \frac{\partial}{\partial t} \int_0^R f_0(\xi) G(r,\xi,t)\, d\xi + \int_0^R \left[f_1(\xi) + k f_0(\xi)\right] G(r,\xi,t)\, d\xi$$
$$- a^2 \int_0^t g(\tau) \left[\frac{\partial}{\partial \xi} G(r,\xi,t-\tau)\right]_{\xi=R} d\tau + \int_0^t \int_0^R \Phi(\xi,\tau) G(r,\xi,t-\tau)\, d\xi\, d\tau.$$

Here,

$$G(r,\xi,t) = \exp\left(-\tfrac{1}{2}kt\right) \sum_{n=1}^{\infty} \frac{2\xi}{R^2 J_1^2(\mu_n)} J_0\left(\frac{\mu_n r}{R}\right) J_0\left(\frac{\mu_n \xi}{R}\right) \frac{\sin(t\sqrt{\lambda_n})}{\sqrt{\lambda_n}}, \quad \lambda_n = \frac{a^2 \mu_n^2}{R^2} + b - \frac{k^2}{4},$$

where the μ_n are positive zeros of the Bessel function, $J_0(\mu) = 0$. The numerical values of the first ten μ_n are specified in Paragraph 1.2.1-3.

3°. Domain: $0 \leq r \leq R$. Second boundary value problem.
The following conditions are prescribed:

$$w = f_0(r) \quad \text{at} \quad t = 0 \quad \text{(initial condition)},$$
$$\partial_t w = f_1(r) \quad \text{at} \quad t = 0 \quad \text{(initial condition)},$$
$$\partial_r w = g(t) \quad \text{at} \quad r = R \quad \text{(boundary condition)}.$$

Solution:

$$w(r,t) = \frac{\partial}{\partial t}\int_0^R f_0(\xi) G(r,\xi,t)\,d\xi + \int_0^R [f_1(\xi) + k f_0(\xi)] G(r,\xi,t)\,d\xi$$
$$+ a^2 \int_0^t g(\tau) G(r,R,t-\tau)\,d\tau + \int_0^t \int_0^R \Phi(\xi,\tau) G(r,\xi,t-\tau)\,d\xi\,d\tau.$$

Here,

$$G(r,\xi,t) = \exp\left(-\tfrac{1}{2}kt\right)\left[\frac{2\xi \sin(t\sqrt{\lambda_0})}{R^2 \sqrt{\lambda_0}} + \frac{2}{R^2}\sum_{n=1}^{\infty}\frac{\xi}{J_0^2(\mu_n)} J_0\!\left(\frac{\mu_n r}{R}\right) J_0\!\left(\frac{\mu_n \xi}{R}\right) \frac{\sin(t\sqrt{\lambda_n})}{\sqrt{\lambda_n}}\right],$$

where $\lambda_0 = b - \tfrac{1}{4}k^2$; $\lambda_n = a^2 \mu_n^2 R^{-2} + b - \tfrac{1}{4}k^2$; the μ_n are positive zeros of the first-order Bessel function, $J_1(\mu) = 0$. The numerical values of the first ten roots μ_n are specified in Paragraph 1.2.1-4.

4°. Domain: $0 \leq r \leq R$. Third boundary value problem.
The following conditions are prescribed:

$$w = f_0(r) \quad \text{at} \quad t = 0 \quad \text{(initial condition)},$$
$$\partial_t w = f_1(r) \quad \text{at} \quad t = 0 \quad \text{(initial condition)},$$
$$\partial_r w + sw = g(t) \quad \text{at} \quad r = R \quad \text{(boundary condition)}.$$

The solution $w(r,t)$ is given by the formula in Item 3° with

$$G(r,\xi,t) = \frac{2}{R^2} \exp\left(-\tfrac{1}{2}kt\right) \sum_{n=1}^{\infty} \frac{\mu_n^2 \xi}{(s^2 R^2 + \mu_n^2) J_0^2(\mu_n)} J_0\!\left(\frac{\mu_n r}{R}\right) J_0\!\left(\frac{\mu_n \xi}{R}\right) \frac{\sin(t\sqrt{\lambda_n})}{\sqrt{\lambda_n}}.$$

Here, $\lambda_n = a^2 \mu_n^2 R^{-2} + b - \tfrac{1}{4}k^2$ and the μ_n are positive roots of the transcendental equation

$$\mu J_1(\mu) - sR J_0(\mu) = 0.$$

The numerical values of the first six roots μ_n can be found in Abramowitz and Stegun (1964) and Carslaw and Jaeger (1984).

3. $\dfrac{\partial^2 w}{\partial t^2} + k\dfrac{\partial w}{\partial t} = a^2\left(\dfrac{\partial^2 w}{\partial r^2} + \dfrac{2}{r}\dfrac{\partial w}{\partial r}\right) - bw + \Phi(r,t).$

1°. The substitution $w(r,t) = \exp\left(-\tfrac{1}{2}kt\right) u(r,t)$ leads to the equation

$$\frac{\partial^2 u}{\partial t^2} = a^2\left(\frac{\partial^2 u}{\partial r^2} + \frac{2}{r}\frac{\partial u}{\partial r}\right) - \left(b - \tfrac{1}{4}k^2\right)u + \exp\left(\tfrac{1}{2}kt\right)\Phi(r,t),$$

which is discussed in Subsection 4.2.6.

2°. Domain: $0 \leq r \leq R$. First boundary value problem.
The following conditions are prescribed:

$$w = f_0(r) \quad \text{at} \quad t = 0 \quad \text{(initial condition)},$$
$$\partial_t w = f_1(r) \quad \text{at} \quad t = 0 \quad \text{(initial condition)},$$
$$w = g(t) \quad \text{at} \quad r = R \quad \text{(boundary condition)}.$$

Solution:

$$w(r,t) = \frac{\partial}{\partial t} \int_0^R f_0(\xi) G(r,\xi,t)\, d\xi + \int_0^R \left[f_1(\xi) + k f_0(\xi) \right] G(r,\xi,t)\, d\xi$$
$$- a^2 \int_0^t g(\tau) \left[\frac{\partial}{\partial \xi} G(r,\xi,t-\tau) \right]_{\xi=R} d\tau + \int_0^t \int_0^R \Phi(\xi,\tau) G(r,\xi,t-\tau)\, d\xi\, d\tau,$$

where

$$G(r,\xi,t) = \frac{2\xi}{Rr} \exp\left(-\tfrac{1}{2}kt\right) \sum_{n=1}^\infty \sin\left(\frac{n\pi r}{R}\right) \sin\left(\frac{n\pi \xi}{R}\right) \frac{\sin(t\sqrt{\lambda_n})}{\sqrt{\lambda_n}}, \quad \lambda_n = \frac{a^2 \pi^2 n^2}{R^2} + b - \frac{k^2}{4}.$$

3°. Domain: $0 \leq r \leq R$. Second boundary value problem.
The following conditions are prescribed:

$$w = f_0(r) \quad \text{at} \quad t = 0 \quad \text{(initial condition)},$$
$$\partial_t w = f_1(r) \quad \text{at} \quad t = 0 \quad \text{(initial condition)},$$
$$\partial_r w = g(t) \quad \text{at} \quad r = R \quad \text{(boundary condition)}.$$

Solution:

$$w(r,t) = \frac{\partial}{\partial t} \int_0^R f_0(\xi) G(r,\xi,t)\, d\xi + \int_0^R \left[f_1(\xi) + k f_0(\xi) \right] G(r,\xi,t)\, d\xi$$
$$+ a^2 \int_0^t g(\tau) G(r,R,t-\tau)\, d\tau + \int_0^t \int_0^R \Phi(\xi,\tau) G(r,\xi,t-\tau)\, d\xi\, d\tau,$$

where

$$G(r,\xi,t) = \exp\left(-\tfrac{1}{2}kt\right) \left[\frac{3\xi^2 \sin(t\sqrt{\lambda_0})}{R^3 \sqrt{\lambda_0}} + \frac{2\xi}{Rr} \sum_{n=1}^\infty \frac{\mu_n^2 + 1}{\mu_n^2 \sqrt{\lambda_n}} \sin\left(\frac{\mu_n r}{R}\right) \sin\left(\frac{\mu_n \xi}{R}\right) \sin(t\sqrt{\lambda_n}) \right].$$

Here, $\lambda_0 = b - \tfrac{1}{4}k^2$; $\lambda_n = a^2 \mu_n^2 R^{-2} + b - \tfrac{1}{4}k^2$; and the μ_n are positive roots of the transcendental equation $\tan \mu - \mu = 0$. The numerical values of the first five roots μ_n are specified in Paragraph 1.2.1-5.

4°. Domain: $0 \leq r \leq R$. Third boundary value problem.
The following conditions are prescribed:

$$w = f_0(r) \quad \text{at} \quad t = 0 \quad \text{(initial condition)},$$
$$\partial_t w = f_1(r) \quad \text{at} \quad t = 0 \quad \text{(initial condition)},$$
$$\partial_r w + sw = g(t) \quad \text{at} \quad r = R \quad \text{(boundary condition)}.$$

The solution $w(r,t)$ is given by the formula in Item 3° with

$$G(r,\xi,t) = \frac{2\xi}{Rr} \exp\left(-\tfrac{1}{2}kt\right) \sum_{n=1}^\infty \frac{\mu_n^2 + (sR-1)^2}{\mu_n^2 + sR(sR-1)} \sin\left(\frac{\mu_n r}{R}\right) \sin\left(\frac{\mu_n \xi}{R}\right) \frac{\sin(t\sqrt{\lambda_n})}{\sqrt{\lambda_n}}.$$

Here, $\lambda_n = a^2 \mu_n^2 R^{-2} + b - \tfrac{1}{4}k^2$ and the μ_n are positive roots of the transcendental equation $\mu \cot \mu + sR - 1 = 0$. The numerical values of the first six roots μ_n can be found in Carslaw and Jaeger (1984).

4. $\dfrac{\partial^2 w}{\partial t^2} + k\dfrac{\partial w}{\partial t} = a^2 \dfrac{\partial}{\partial x}\left(x\dfrac{\partial w}{\partial x}\right) - bw + \Phi(x,t).$

1°. The substitution $w(r,t) = \exp\!\left(-\tfrac{1}{2}kt\right)u(r,t)$ leads to an equation of the form 4.3.1.2:

$$\dfrac{\partial^2 u}{\partial t^2} = a^2 \dfrac{\partial}{\partial x}\left(x\dfrac{\partial u}{\partial x}\right) - \left(b - \tfrac{1}{4}k^2\right)u + \exp\!\left(\tfrac{1}{2}kt\right)\Phi(x,t).$$

2°. Domain: $0 \le x \le l$. First boundary value problem.
The following conditions are prescribed:

$$\begin{aligned}
w &= f_0(x) & \text{at} \quad t &= 0 & &\text{(initial condition)},\\
\partial_t w &= f_1(x) & \text{at} \quad t &= 0 & &\text{(initial condition)},\\
w &= g(t) & \text{at} \quad x &= l & &\text{(boundary condition)},\\
w &\ne \infty & \text{at} \quad x &= 0 & &\text{(boundedness condition)}.
\end{aligned}$$

Solution:

$$w(x,t) = \dfrac{\partial}{\partial t}\int_0^l f_0(\xi)G(x,\xi,t)\,d\xi + \int_0^l [f_1(\xi) + kf_0(\xi)]G(x,\xi,t)\,d\xi$$
$$- a^2 l \int_0^t g(\tau)\left[\dfrac{\partial}{\partial \xi}G(x,\xi,t-\tau)\right]_{\xi=l} d\tau + \int_0^t \int_0^l \Phi(\xi,\tau)G(x,\xi,t-\tau)\,d\xi\,d\tau,$$

where

$$G(x,\xi,t) = \dfrac{1}{l}\exp\!\left(-\tfrac{1}{2}kt\right)\sum_{n=1}^{\infty}\dfrac{1}{J_1^2(\mu_n)}J_0\!\left(\mu_n\sqrt{\tfrac{x}{l}}\right)J_0\!\left(\mu_n\sqrt{\tfrac{\xi}{l}}\right)\dfrac{\sin\!\left(t\sqrt{\lambda_n}\right)}{\sqrt{\lambda_n}}.$$

Here, $\lambda_n = \tfrac{1}{4}a^2\mu_n^2 l^{-1} + b - \tfrac{1}{4}k^2$; the μ_n are positive zeros of the Bessel function, $J_0(\mu) = 0$.

3°. Domain: $0 \le x \le l$. Second boundary value problem.
The following conditions are prescribed:

$$\begin{aligned}
w &= f_0(x) & \text{at} \quad t &= 0 & &\text{(initial condition)},\\
\partial_t w &= f_1(x) & \text{at} \quad t &= 0 & &\text{(initial condition)},\\
\partial_x w &= g(t) & \text{at} \quad x &= l & &\text{(boundary condition)},\\
w &\ne \infty & \text{at} \quad x &= 0 & &\text{(boundedness condition)}.
\end{aligned}$$

Solution:

$$w(x,t) = \dfrac{\partial}{\partial t}\int_0^l f_0(\xi)G(x,\xi,t)\,d\xi + \int_0^l [f_1(\xi) + kf_0(\xi)]\,d\xi$$
$$+ a^2 l \int_0^t g(\tau)G(x,l,t-\tau)\,d\tau + \int_0^t \int_0^l \Phi(\xi,\tau)G(x,\xi,t-\tau)\,d\xi\,d\tau,$$

where

$$G(r,\xi,t) = \exp\!\left(-\tfrac{1}{2}kt\right)\left[\dfrac{\sin\!\left(t\sqrt{\lambda_0}\right)}{l\sqrt{\lambda_0}} + \dfrac{1}{l}\sum_{n=1}^{\infty}\dfrac{1}{J_0^2(\mu_n)}J_0\!\left(\mu_n\sqrt{\tfrac{x}{l}}\right)J_0\!\left(\mu_n\sqrt{\tfrac{\xi}{l}}\right)\dfrac{\sin\!\left(t\sqrt{\lambda_n}\right)}{\sqrt{\lambda_n}}\right].$$

Here, $\lambda_0 = b - \tfrac{1}{4}k^2$; $\lambda_n = \tfrac{1}{4}a^2\mu_n^2 l^{-1} + b - \tfrac{1}{4}k^2$; the μ_n are positive zeros of the first-order Bessel function, $J_1(\mu) = 0$. The numerical values of the first ten roots μ_n are specified in Paragraph 1.2.1-5.

4°. Domain: $0 \le x \le l$. Third boundary value problem.
The following conditions are prescribed:
$$w = f_0(x) \quad \text{at} \quad t = 0 \quad \text{(initial condition)},$$
$$\partial_t w = f_1(x) \quad \text{at} \quad t = 0 \quad \text{(initial condition)},$$
$$\partial_x w + kw = g(t) \quad \text{at} \quad x = l \quad \text{(boundary condition)}.$$

The solution $w(x,t)$ is given by the formula in Item 3° with
$$G(r,\xi,t) = \frac{1}{l} \exp\left(-\tfrac{1}{2}kt\right) \sum_{n=1}^{\infty} \frac{\mu_n^2}{(4k^2 l + \mu_n^2) J_0^2(\mu_n)} J_0\left(\mu_n \sqrt{\frac{x}{l}}\right) J_0\left(\mu_n \sqrt{\frac{\xi}{l}}\right) \frac{\sin(t\sqrt{\lambda_n})}{\sqrt{\lambda_n}}.$$

Here, $\lambda_n = \tfrac{1}{4} a^2 \mu_n^2 l^{-1} + b - \tfrac{1}{4} k^2$, and the μ_n are positive roots of the transcendental equation
$$\mu J_1(\mu) - 2k\sqrt{l}\, J_0(\mu) = 0.$$

The numerical values of the first six roots μ_n can be found in Abramowitz and Stegun (1964) and Carslaw and Jaeger (1984).

5. $\dfrac{\partial^2 w}{\partial t^2} + k \dfrac{\partial w}{\partial t} = (ax^m + b)\dfrac{\partial^2 w}{\partial x^2} + \dfrac{1}{2} a m x^{m-1} \dfrac{\partial w}{\partial x} + cw.$

The substitution $z = \displaystyle\int \frac{dx}{\sqrt{ax^m + b}}$ leads to a constant coefficient equation of the form 4.4.1.2: $\partial_{tt} w + k \partial_t w = \partial_{zz} w + cw$.

4.4.3. Other Equations

1. $\dfrac{\partial^2 w}{\partial t^2} + \dfrac{k-1}{t} \dfrac{\partial w}{\partial t} = \dfrac{\partial^2 w}{\partial x^2}.$

Darboux equation. Domain: $-\infty < x < \infty$. Cauchy problem.
Initial conditions are prescribed:
$$w = f(x) \quad \text{at} \quad t = 0,$$
$$\partial_t w = 0 \quad \text{at} \quad t = 0.$$

Solution:
$$w(x,t) = \frac{\Gamma\left(\frac{k}{2}\right)}{\sqrt{\pi}\,\Gamma\left(\frac{k}{2} - \frac{1}{2}\right)} \int_{-1}^{1} f(x + t\xi)(1 - \xi^2)^{\frac{k-3}{2}} d\xi \qquad (k > 1).$$

⊙ *Reference*: R. Courant and D. Hilbert (1989).

2. $\dfrac{\partial^2 w}{\partial t^2} + \dfrac{2a}{t} \dfrac{\partial w}{\partial t} = \dfrac{\partial^2 w}{\partial x^2} - b^2 w.$

Domain: $-\infty < x < \infty$. Cauchy problem.
Initial conditions are prescribed:
$$w = f(x) \quad \text{at} \quad t = 0,$$
$$t^{2a} \partial_t w = g(x) \quad \text{at} \quad t = 0.$$

Solution for $0 < 2a < 1$:
$$w(x,t) = \frac{\Gamma(2a)}{\Gamma^2(a)} \int_0^1 f\big(x + t(2\xi - 1)\big) \bar{J}_{a-1}\big(2bt\sqrt{\xi(1-\xi)}\big) \xi^{a-1}(1-\xi)^{a-1} d\xi$$
$$+ \frac{\Gamma(2-2a)}{(1-2a)\Gamma^2(1-a)} t^{1-2a} \int_0^1 g\big(x + t(2\xi - 1)\big) \bar{J}_{-a}\big(2bt\sqrt{\xi(1-\xi)}\big) \xi^{-a}(1-\xi)^{-a} d\xi,$$

where
$$\bar{J}_\nu(z) = 2^\nu \Gamma(1+\nu) z^{-\nu} J_\nu(z), \qquad \Gamma(\nu) = \int_0^\infty e^{-s} s^{\nu-1} ds.$$

⊙ *Reference*: M. M. Smirnov (1975).

3. $\dfrac{\partial^2 w}{\partial t^2} + \dfrac{2a}{t}\dfrac{\partial w}{\partial t} = t^m \dfrac{\partial^2 w}{\partial x^2}$.

Domain: $-\infty < x < \infty$. Cauchy problem.
Initial conditions are prescribed:
$$w = f(x) \quad \text{at} \quad t = 0,$$
$$t^{2a}\partial_t w = g(x) \quad \text{at} \quad t = 0.$$

Solution for $0 \le 2a < 1$ and $m > 0$:
$$w(x,t) = \dfrac{\Gamma(2\beta)}{\Gamma^2(\beta)} \int_0^1 f\!\left(x + \dfrac{2}{2+m} t^{\frac{2+m}{2}}(2\xi - 1)\right) \xi^{\beta-1}(1-\xi)^{\beta-1} d\xi$$
$$+ \dfrac{\Gamma(2-2\beta)}{(1-2a)\Gamma^2(1-\beta)} t^{1-2a} \int_0^1 g\!\left(x + \dfrac{2}{2+m} t^{\frac{2+m}{2}}(2\xi - 1)\right) \xi^{-\beta}(1-\xi)^{-\beta} d\xi,$$

where
$$\beta = \dfrac{m+4a}{2(m+2)}, \quad \Gamma(z) = \int_0^\infty e^{-s} s^{z-1} ds.$$

⊙ Reference: M. M. Smirnov (1975).

4. $t^2 \dfrac{\partial^2 w}{\partial t^2} + kt\dfrac{\partial w}{\partial t} = a^2 \dfrac{\partial^2 w}{\partial x^2} + b\dfrac{\partial w}{\partial x} + cw$.

The substitution $t = Ae^\tau$ ($A \ne 0$) leads to a constant coefficient equation of the form 4.4.1.3:
$$\dfrac{\partial^2 w}{\partial \tau^2} + (k-1)\dfrac{\partial w}{\partial \tau} = a^2 \dfrac{\partial^2 w}{\partial x^2} + b\dfrac{\partial w}{\partial x} + cw.$$

5. $t^2 \dfrac{\partial^2 w}{\partial t^2} + kt\dfrac{\partial w}{\partial t} = a^2 x^2 \dfrac{\partial^2 w}{\partial x^2} + bx\dfrac{\partial w}{\partial x} + cw$.

The transformation
$$t = Ae^\tau, \quad x = Be^\xi \quad (A \ne 0, \ B \ne 0)$$
leads to a constant coefficient equation of the form 4.4.1.3:
$$\dfrac{\partial^2 w}{\partial \tau^2} + (k-1)\dfrac{\partial w}{\partial \tau} = a^2 \dfrac{\partial^2 w}{\partial \xi^2} + (b - a^2)\dfrac{\partial w}{\partial \xi} + cw.$$

6. $t^m \dfrac{\partial^2 w}{\partial t^2} + at^{m-1}\dfrac{\partial w}{\partial t} = \dfrac{\partial^2 w}{\partial x^2}$, $\quad 0 < m < 2$.

Domain: $-\infty < x < \infty$. Cauchy problem.
Initial conditions are prescribed:
$$w = f(x) \quad \text{at} \quad t = 0,$$
$$t^a \partial_t w = g(x) \quad \text{at} \quad t = 0.$$

1°. Solution for $\tfrac{1}{2}m < a < 1$:
$$w(x,t) = \dfrac{\Gamma(2\beta)}{\Gamma^2(\beta)} \int_0^1 f\!\left(x + \dfrac{2}{2-m} t^{\frac{2-m}{2}}(2\xi - 1)\right) \xi^{\beta-1}(1-\xi)^{\beta-1} d\xi$$
$$+ \dfrac{\Gamma(2-2\beta)}{(1-a)\Gamma^2(1-\beta)} t^{1-a} \int_0^1 g\!\left(x + \dfrac{2}{2-m} t^{\frac{2-m}{2}}(2\xi - 1)\right) \xi^{-\beta}(1-\xi)^{-\beta} d\xi,$$

where

$$\beta = \frac{2a-m}{2(2-m)}, \quad \Gamma(z) = \int_0^\infty e^{-s} s^{z-1} ds.$$

2°. Solution for $a = \frac{1}{2}m$:

$$w(x,t) = \frac{f(y) + f(z)}{2} + \frac{1}{2}\int_z^y g(\xi)\,d\xi,$$

$$y = x - \frac{2}{2-m} t^{\frac{2-m}{2}}, \quad z = x + \frac{2}{2-m} t^{\frac{2-m}{2}}.$$

⊙ *Reference*: M. M. Smirnov (1975).

7. $(t^m + k)\dfrac{\partial^2 w}{\partial t^2} + \dfrac{1}{2} m t^{m-1} \dfrac{\partial w}{\partial t} = a \dfrac{\partial^2 w}{\partial x^2} + b \dfrac{\partial w}{\partial x} + cw.$

The substitution $\tau = \displaystyle\int \frac{dt}{\sqrt{t^m + k}}$ leads to the equation $\partial_{\tau\tau} w = a\partial_{xx} w + b\partial_x w + cw$, which is discussed in Subsection 4.1.5.

4.5. Equations Containing Arbitrary Functions

4.5.1. Equations of the Form $s(x)\dfrac{\partial^2 w}{\partial t^2} = \dfrac{\partial}{\partial x}\left[p(x)\dfrac{\partial w}{\partial x}\right] - q(x)w + \Phi(x,t)$

It is assumed that the functions s, p, p'_x, and q are continuous and the inequalities $s > 0$, $p > 0$ hold for $x_1 \leq x \leq x_2$.

4.5.1-1. General relations to solve linear nonhomogeneous boundary value problems.

The solution of the equation in question under the general initial conditions

$$\begin{aligned} w &= f_0(x) \quad \text{at} \quad t = 0, \\ \partial_t w &= f_1(x) \quad \text{at} \quad t = 0 \end{aligned} \tag{1}$$

and the arbitrary linear nonhomogeneous boundary conditions

$$\begin{aligned} a_1 \partial_x w + b_1 w &= g_1(t) \quad \text{at} \quad x = x_1, \\ a_2 \partial_x w + b_2 w &= g_2(t) \quad \text{at} \quad x = x_2 \end{aligned} \tag{2}$$

can be represented as the sum

$$\begin{aligned} w(x,t) &= \int_0^t \int_{x_1}^{x_2} \Phi(\xi,\tau) \mathcal{G}(x,\xi,t-\tau)\,d\xi\,d\tau \\ &\quad + \frac{\partial}{\partial t} \int_{x_1}^{x_2} s(\xi) f_0(\xi) \mathcal{G}(x,\xi,t)\,d\xi + \int_{x_1}^{x_2} s(\xi) f_1(\xi) \mathcal{G}(x,\xi,t)\,d\xi \\ &\quad + p(x_1) \int_0^t g_1(\tau) \Lambda_1(x,t-\tau)\,d\tau + p(x_2) \int_0^t g_2(\tau) \Lambda_2(x,t-\tau)\,d\tau. \end{aligned} \tag{3}$$

Here, the modified Green's function is determined by

$$\mathcal{G}(x,\xi,t) = \sum_{n=1}^\infty \frac{y_n(x) y_n(\xi) \sin(t\sqrt{\lambda_n})}{\|y_n\|^2 \sqrt{\lambda_n}}, \quad \|y_n\|^2 = \int_{x_1}^{x_2} s(x) y_n^2(x)\,dx, \tag{4}$$

where the λ_n and $y_n(x)$ are the eigenvalues and corresponding eigenfunctions of the Sturm–Liouville problem for the second-order linear ordinary differential equation

$$[p(x)y'_x]'_x + [\lambda s(x) - q(x)]y = 0,$$
$$a_1 y'_x + b_1 y = 0 \quad \text{at} \quad x = x_1, \qquad (5)$$
$$a_2 y'_x + b_2 y = 0 \quad \text{at} \quad x = x_2.$$

The functions $\Lambda_1(x,t)$ and $\Lambda_2(x,t)$ that occur in the integrands of the last two terms in solution (3) are expressed in terms of the Green's function of (4). The corresponding formulas will be specified below in studying specific boundary value problems.

General properties of the Sturm–Liouville problem (5):

1°. There are finitely many eigenvalues $\lambda_1 < \lambda_2 < \lambda_3 < \cdots$, with $\lambda_n \to \infty$ as $n \to \infty$; hence the number of negative eigenvalues is finite.

2°. Any two eigenfunctions $y_n(x)$ and $y_m(x)$ for $n \neq m$ are orthogonal to each other with weight $s(x)$ on the interval $x_1 \leq x \leq x_2$; specifically,

$$\int_{x_1}^{x_2} s(x) y_n(x) y_m(x)\, dx = 0 \quad \text{at} \quad n \neq m.$$

3°. If the conditions

$$q(x) \geq 0, \quad a_1 b_1 \leq 0, \quad a_2 b_2 \geq 0 \qquad (6)$$

are satisfied, then there are no negative eigenvalues. If $q \equiv 0$ and $b_1 = b_2 = 0$, the least eigenvalue is $\lambda_1 = 0$ and the corresponding eigenfunction is $\varphi_1 = \text{const}$. In the other cases where conditions (6) are satisfied, all eigenvalues are positive.

Remark. More detailed information about the properties of the Sturm–Liouville problem (5) can be found in Subsection 1.8.9. Asymptotic and approximate formulas for eigenvalues and eigenfunctions are also presented there.

4.5.1-2. First boundary value problem (case $a_1 = a_2 = 0$, $b_1 = b_2 = 1$).

The solution of the first boundary value problem for the equation in question with the initial conditions (1) and the boundary conditions

$$w = g_1(t) \quad \text{at} \quad x = x_1,$$
$$w = g_2(t) \quad \text{at} \quad x = x_2$$

is given by relations (3) and (4) in which

$$\Lambda_1(x,t) = \frac{\partial}{\partial \xi} \mathcal{G}(x, \xi, t)\Big|_{\xi = x_1}, \qquad \Lambda_2(x,t) = -\frac{\partial}{\partial \xi} \mathcal{G}(x, \xi, t)\Big|_{\xi = x_2}.$$

4.5.1-3. Second boundary value problem (case $a_1 = a_2 = 1$, $b_1 = b_2 = 0$).

The solution of the second boundary value problem for the equation in question with the initial conditions (1) and the boundary conditions

$$\partial_x w = g_1(t) \quad \text{at} \quad x = x_1,$$
$$\partial_x w = g_2(t) \quad \text{at} \quad x = x_2$$

is given by relations (3) and (4) with

$$\Lambda_1(x,t) = -\mathcal{G}(x, x_1, t), \qquad \Lambda_2(x,t) = \mathcal{G}(x, x_2, t).$$

4.5.1-4. Third boundary value problem (case $a_1 = a_2 = 1$, $b_1 \neq 0$, $b_2 \neq 0$).

The solution of the third boundary value problem for the equation in question with the initial conditions (1) and the boundary conditions (2) with $a_1 = a_2 = 1$ is given by relations (3) and (4) in which

$$\Lambda_1(x,t) = -\mathcal{G}(x, x_1, t), \quad \Lambda_2(x,t) = \mathcal{G}(x, x_2, t).$$

4.5.1-5. Mixed boundary value problem (case $a_1 = b_2 = 0$, $a_2 = b_1 = 1$).

The solution of the mixed boundary value problem for the equation in question with the initial conditions (1) and the boundary conditions

$$w = g_1(t) \quad \text{at} \quad x = x_1,$$
$$\partial_x w = g_2(t) \quad \text{at} \quad x = x_2$$

is given by relations (3) and (4) with

$$\Lambda_1(x,t) = \frac{\partial}{\partial \xi}\mathcal{G}(x,\xi,t)\Big|_{\xi=x_1}, \quad \Lambda_2(x,t) = \mathcal{G}(x, x_2, t).$$

4.5.1-6. Mixed boundary value problem (case $a_1 = b_2 = 1$, $a_2 = b_1 = 0$).

The solution of the mixed boundary value problem with the initial conditions (1) and the boundary conditions

$$\partial_x w = g_1(t) \quad \text{at} \quad x = x_1,$$
$$w = g_2(t) \quad \text{at} \quad x = x_2$$

is given by relations (3) and (4) with

$$\Lambda_1(x,t) = -\mathcal{G}(x, x_1, t), \quad \Lambda_2(x,t) = -\frac{\partial}{\partial \xi}\mathcal{G}(x,\xi,t)\Big|_{\xi=x_2}.$$

⊙ *References for Subsection* 4.5.1: V. M. Babich, M. B. Kapilevich, S. G. Mikhlin et al. (1964), V. A. Marchenko (1986), V. S. Vladimirov (1988), A. D. Polyanin (2000a).

4.5.2. Equations of the Form
$$\frac{\partial^2 w}{\partial t^2} + a(t)\frac{\partial w}{\partial t} = b(t)\left\{\frac{\partial}{\partial x}\left[p(x)\frac{\partial w}{\partial x}\right] - q(x)w\right\} + \Phi(x,t)$$

It is assumed that the functions p, p'_x, and q are continuous and $p > 0$ for $x_1 \leq x \leq x_2$.

4.5.2-1. General relations to solve linear nonhomogeneous boundary value problems.

The solution of the equation in question under the general initial conditions

$$w = f_0(x) \quad \text{at} \quad t = 0,$$
$$\partial_t w = f_1(x) \quad \text{at} \quad t = 0 \tag{1}$$

and the arbitrary linear nonhomogeneous boundary conditions

$$s_1 \partial_x w + k_1 w = g_1(t) \quad \text{at} \quad x = x_1,$$
$$s_2 \partial_x w + k_2 w = g_2(t) \quad \text{at} \quad x = x_2 \tag{2}$$

can be represented as the sum

$$w(x,t) = \int_0^t \int_{x_1}^{x_2} \Phi(\xi,\tau) G(x,\xi,t,\tau) \, d\xi \, d\tau$$
$$- \int_{x_1}^{x_2} f_0(\xi) \left[\frac{\partial}{\partial \tau} G(x,\xi,t,\tau) \right]_{\tau=0} d\xi + \int_{x_1}^{x_2} \left[f_1(\xi) + a(0) f_0(\xi) \right] G(x,\xi,t,0) \, d\xi$$
$$+ p(x_1) \int_0^t g_1(\tau) b(\tau) \Lambda_1(x,t,\tau) \, d\tau + p(x_2) \int_0^t g_2(\tau) b(\tau) \Lambda_2(x,t,\tau) \, d\tau. \tag{3}$$

Here, the modified Green's function is determined by

$$G(x,\xi,t,\tau) = \sum_{n=1}^{\infty} \frac{y_n(x) y_n(\xi)}{\|y_n\|^2} U_n(t,\tau), \qquad \|y_n\|^2 = \int_{x_1}^{x_2} y_n^2(x) \, dx, \tag{4}$$

where the λ_n and $y_n(x)$ are the eigenvalues and corresponding eigenfunctions of the Sturm–Liouville problem for the following second-order linear ordinary differential equation with homogeneous boundary conditions:

$$\begin{aligned} [p(x) y_x']_x' + [\lambda - q(x)] y &= 0, \\ s_1 y_x' + k_1 y &= 0 \quad \text{at} \quad x = x_1, \\ s_2 y_x' + k_2 y &= 0 \quad \text{at} \quad x = x_2. \end{aligned} \tag{5}$$

The functions $U_n = U_n(t,\tau)$ are determined by solving the Cauchy problem for the linear ordinary differential equation

$$\begin{aligned} U_n'' + a(t) U_n' + \lambda_n b(t) U_n &= 0, \\ U_n\big|_{t=\tau} = 0, \quad U_n'\big|_{t=\tau} &= 1. \end{aligned} \tag{6}$$

The prime denotes the derivative with respect to t, and τ is a free parameter occurring in the initial conditions.

The functions $\Lambda_1(x,t)$ and $\Lambda_2(x,t)$ that occur in the integrands of the last two terms in solution (3) are expressed in terms of the Green's function of (4). The corresponding formulas will be specified below when studying specific boundary value problems.

The properties of the Sturm–Liouville problem (5) are detailed in Subsection 1.8.9. Asymptotic and approximate formulas for eigenvalues and eigenfunctions are also presented there.

4.5.2-2. First, second, third, and mixed boundary value problems.

1°. *First boundary value problem.* The solution of the equation in question with the initial conditions (1) and boundary conditions (2) for $s_1 = s_2 = 0$ and $k_1 = k_2 = 1$ is given by relations (3) and (4), where

$$\Lambda_1(x,t,\tau) = \frac{\partial}{\partial \xi} G(x,\xi,t,\tau) \bigg|_{\xi=x_1}, \qquad \Lambda_2(x,t,\tau) = -\frac{\partial}{\partial \xi} G(x,\xi,t,\tau) \bigg|_{\xi=x_2}.$$

2°. *Second boundary value problem.* The solution of the equation with the initial conditions (1) and boundary conditions (2) for $s_1 = s_2 = 1$ and $k_1 = k_2 = 0$ is given by relations (3) and (4) with

$$\Lambda_1(x,t,\tau) = -G(x,x_1,t,\tau), \qquad \Lambda_2(x,t,\tau) = G(x,x_2,t,\tau).$$

3°. *Third boundary value problem.* The solution of the equation with the initial conditions (1) and boundary conditions (2) for $s_1 = s_2 = 1$ and $k_1 k_2 \neq 0$ is given by relations (3) and (4) in which

$$\Lambda_1(x,t,\tau) = -G(x,x_1,t,\tau), \qquad \Lambda_2(x,t,\tau) = G(x,x_2,t,\tau).$$

4°. *Mixed boundary value problem.* The solution of the equation with the initial conditions (1) and boundary conditions (2) for $s_1 = k_2 = 0$ and $s_2 = k_1 = 1$ is given by relations (3) and (4) with

$$\Lambda_1(x,t,\tau) = \frac{\partial}{\partial \xi} G(x,\xi,t,\tau) \bigg|_{\xi=x_1}, \qquad \Lambda_2(x,t,\tau) = G(x,x_2,t,\tau).$$

5°. *Mixed boundary value problem.* The solution of the equation with the initial conditions (1) and boundary conditions (2) for $s_1 = k_2 = 1$ and $s_2 = k_1 = 0$ is given by relations (3) and (4) with

$$\Lambda_1(x,t,\tau) = -G(x,x_1,t,\tau), \quad \Lambda_2(x,t,\tau) = -\frac{\partial}{\partial \xi}G(x,\xi,t,\tau)\Big|_{\xi=x_2}.$$

⊙ *References*: V. M. Babich, M. B. Kapilevich, S. G. Mikhlin et al. (1964), A. V. Bitsadze and D. F. Kalinichenko (1985), A. D. Polyanin (2000a).

4.5.3. Other Equations

1. $\dfrac{\partial^2 w}{\partial t^2} = f(x)\dfrac{\partial^2 w}{\partial x^2}.$

This is a special case of the equation of Subsection 4.5.1 with $s(x) = 1/f(x)$, $p(x) = 1$, and $q = \Phi = 0$.

1°. Particular solutions:

$$w = C_1 xt + C_2 t + C_3 x + C_4,$$

$$w = C_1 t^2 + C_2 xt + C_3 t + C_4 x + 2C_1 \int_a^x \frac{x-\xi}{f(\xi)}\,d\xi + C_5,$$

$$w = C_1 t^3 + C_2 xt + C_3 t + C_4 x + 6C_1 t \int_a^x \frac{x-\xi}{f(\xi)}\,d\xi + C_5,$$

$$w = (C_1 x + C_2)t^2 + C_3 xt + C_4 t + C_5 x + 2\int_a^x (x-\xi)\frac{(C_1\xi+C_2)}{f(\xi)}\,d\xi + C_6,$$

where C_1, C_2, C_3, C_4, C_5, and C_6 are arbitrary constants, and a is an arbitrary real number.

2°. Separable particular solution:

$$w = (C_1 e^{\lambda t} + C_2 e^{-\lambda t})H(x),$$

where C_1, C_2, and λ are arbitrary constants, and the function $H = H(x)$ is determined by the ordinary differential equation $f(x)H''_{xx} - \lambda^2 H = 0$.

3°. Separable particular solution:

$$w = [C_1 \sin(\lambda t) + C_2 \cos(\lambda t)]Z(x),$$

where C_1, C_2, and λ are arbitrary constants, and the function $Z = Z(x)$ is determined by the ordinary differential equation $f(x)Z''_{xx} + \lambda^2 Z = 0$.

4°. Particular solutions with even powers of t:

$$w = \sum_{k=0}^{n} \varphi_k(x) t^{2k},$$

where the functions $\varphi_k = \varphi_k(x)$ are defined by the recurrence relations

$$\varphi_n(x) = A_n x + B_n,$$

$$\varphi_{k-1}(x) = A_k x + B_k + 2k(2k-1)\int_a^x (x-\xi)\frac{\varphi_k(\xi)}{f(\xi)}\,d\xi,$$

where A_k, B_k are arbitrary constants ($k = n, \ldots, 1$).

5°. Particular solutions with odd powers of t:

$$w = \sum_{k=0}^{n} \psi_k(x) t^{2k+1},$$

where the functions $\psi_k = \psi_k(x)$ are defined by the recurrence relations

$$\psi_n(x) = A_n x + B_n,$$

$$\psi_{k-1}(x) = A_k x + B_k + 2k(2k+1)\int_a^x (x-\xi)\frac{\psi_k(\xi)}{f(\xi)}\,d\xi,$$

where A_k, B_k are arbitrary constants ($k = n, \ldots, 1$).

2. $\dfrac{\partial^2 w}{\partial t^2} = \dfrac{\partial}{\partial x}\left[f(x)\dfrac{\partial w}{\partial x}\right].$

This is a special case of the equation of Subsection 4.5.1 with $s(x) = 1$, $p(x) = f(x)$, and $q = \Phi = 0$.

1°. Particular solutions:

$$w = C_1 t^2 + C_2 t + 2\int \frac{C_1 x + C_3}{f(x)}\,dx + C_4,$$

$$w = C_1 t^3 + C_2 t + 6t\int \frac{C_1 x + C_3}{f(x)}\,dx + C_4,$$

$$w = [C_1 \Phi(x) + C_2]t + C_3 \Phi(x) + C_4, \quad \Phi(x) = \int \frac{dx}{f(x)},$$

$$w = [C_1 \Phi(x) + C_2]t^2 + C_3 \Phi(x) + C_4 + 2\int\left\{\frac{1}{f(x)}\int [C_1 \Phi(x) + C_2]\,dx\right\}dx,$$

where C_1, C_2, C_3, C_4, and C_5 are arbitrary constants.

2°. Separable particular solution:

$$w = (C_1 e^{\lambda t} + C_2 e^{-\lambda t})H(x),$$

where C_1, C_2, and λ are arbitrary constants, and the function $H = H(x)$ is determined by the ordinary differential equation $[f(x)H'_x]'_x - \lambda^2 H = 0$.

3°. Separable particular solution:

$$w = [C_1 \sin(\lambda t) + C_2 \cos(\lambda t)]Z(x),$$

where C_1, C_2, and λ are arbitrary constants, and the function $Z = Z(x)$ is determined by the ordinary differential equation $[f(x)Z'_x]'_x + \lambda^2 Z = 0$.

4°. Particular solutions with even powers of t:

$$w = \sum_{k=0}^{n} \zeta_k(x) t^{2k},$$

where the functions $\zeta_k = \zeta_k(x)$ are defined by the recurrence relations

$$\zeta_n(x) = A_n \Phi(x) + B_n, \quad \Phi(x) = \int \frac{dx}{f(x)},$$

$$\zeta_{k-1}(x) = A_k \Phi(x) + B_k + 2k(2k-1)\int \frac{1}{f(x)}\left\{\int \zeta_k(x)\,dx\right\}dx,$$

where A_k, B_k are arbitrary constants ($k = n, \ldots, 1$).

5°. Particular solutions with odd powers of t:

$$w = \sum_{k=0}^{n} \eta_k(x) t^{2k+1},$$

where the functions $\eta_k = \eta_k(x)$ are defined by the recurrence relations

$$\eta_n(x) = A_n \Phi(x) + B_n, \quad \Phi(x) = \int \frac{dx}{f(x)},$$

$$\eta_{k-1}(x) = A_k \Phi(x) + B_k + 2k(2k+1)\int \frac{1}{f(x)}\left\{\int \eta_k(x)\,dx\right\}dx,$$

where A_k, B_k are arbitrary constants ($k = n, \ldots, 1$).

3. $\dfrac{\partial^2 w}{\partial t^2} = f(x)\dfrac{\partial^2 w}{\partial x^2} + g(x)\dfrac{\partial w}{\partial x} + \Phi(x, t), \qquad 0 < f(x) < \infty.$

This equation can be rewritten in the form of the equation from Subsection 4.5.1 with $q(x) \equiv 0$:

$$s(x)\dfrac{\partial^2 w}{\partial t^2} = \dfrac{\partial}{\partial x}\left[p(x)\dfrac{\partial w}{\partial x}\right] + s(x)\Phi(x, t),$$

where

$$s(x) = \dfrac{1}{f(x)}\exp\left[\int \dfrac{g(x)}{f(x)}dx\right], \quad p(x) = \exp\left[\int \dfrac{g(x)}{f(x)}dx\right].$$

4. $\dfrac{\partial^2 w}{\partial t^2} = f(x)\dfrac{\partial^2 w}{\partial x^2} + g(x)\dfrac{\partial w}{\partial x} + h(x)w + \Phi(x, t).$

This equation can be rewritten in the form of the equation from Subsection 4.5.1:

$$s(x)\dfrac{\partial^2 w}{\partial t^2} = \dfrac{\partial}{\partial x}\left[p(x)\dfrac{\partial w}{\partial x}\right] - q(x)w + s(x)\Phi(x, t),$$

where

$$s(x) = \dfrac{1}{f(x)}\exp\left[\int \dfrac{g(x)}{f(x)}dx\right], \quad p(x) = \exp\left[\int \dfrac{g(x)}{f(x)}dx\right], \quad q(x) = -\dfrac{h(x)}{f(x)}\exp\left[\int \dfrac{g(x)}{f(x)}dx\right].$$

5. $\dfrac{\partial^2 w}{\partial t^2} = f(x)\dfrac{\partial^2 w}{\partial x^2} + g(x)\dfrac{\partial w}{\partial x} + \bigl[h_1(x) + h_2(t)\bigr]w.$

1°. There are separable solutions in the product form $w(x, t) = \varphi(x)\psi(t)$, where the functions $\varphi = \varphi(x)$ and $\psi = \psi(t)$ satisfy the ordinary differential equations (λ is an arbitrary constant):

$$f(x)\varphi''_{xx} + g(x)\varphi'_x + \bigl[\lambda + h_1(x)\bigr]\varphi = 0, \qquad \psi''_{tt} + \bigl[\lambda - h_2(t)\bigr]\psi = 0.$$

2°. For the solution of various boundary value problems for the original equation, see Subsections 0.4.1 and 0.4.2.

6. $\dfrac{\partial^2 w}{\partial t^2} = f(x)\dfrac{\partial^2 w}{\partial x^2} + \dfrac{1}{2}f'(x)\dfrac{\partial w}{\partial x} + bw.$

The substitution $z = \displaystyle\int \dfrac{dx}{\sqrt{f(x)}}$ leads to the constant coefficient equation $\partial_{tt}w = \partial_{zz}w + bw$ that is discussed in Subsection 4.1.3.

7. $\dfrac{\partial^2 w}{\partial t^2} = f^2\dfrac{\partial^2 w}{\partial x^2} + f(f'_x + 2g)\dfrac{\partial w}{\partial x} + (fg'_x + g^2)w, \qquad f = f(x),\ g = g(x).$

The transformation

$$w(x, t) = u(\xi, t)\exp\left(-\int \dfrac{g}{f}dx\right), \qquad \xi = \int \dfrac{dx}{f(x)}$$

leads to the wave equation $\partial_{tt}u = \partial_{\xi\xi}u$ that is discussed in Subsection 4.1.1.

8. $\dfrac{\partial^2 w}{\partial t^2} + a\dfrac{\partial w}{\partial t} = f(x)\dfrac{\partial^2 w}{\partial x^2} + \dfrac{1}{2}f'(x)\dfrac{\partial w}{\partial x} + bw.$

The substitution $z = \displaystyle\int \dfrac{dx}{\sqrt{f(x)}}$ leads to a constant coefficient equation of the form 4.4.1.2: $\partial_{tt}w + a\partial_t w = \partial_{zz}w + bw.$

9. $f(t)\dfrac{\partial^2 w}{\partial t^2} + \dfrac{1}{2}f'(t)\dfrac{\partial w}{\partial t} = a\dfrac{\partial^2 w}{\partial x^2} + b\dfrac{\partial w}{\partial x} + cw.$

The substitution $\tau = \displaystyle\int \dfrac{dt}{\sqrt{f(t)}}$ leads to the equation $\partial_{\tau\tau}w = a\partial_{xx}w + b\partial_x w + cw$ that is discussed in Subsection 4.1.5.

10. $f(t)\dfrac{\partial^2 w}{\partial t^2} + \dfrac{1}{2}f'(t)\dfrac{\partial w}{\partial t} = g(x)\dfrac{\partial^2 w}{\partial x^2} + \dfrac{1}{2}g'(x)\dfrac{\partial w}{\partial x} + cw.$

The transformation $\tau = \displaystyle\int \dfrac{dt}{\sqrt{f(t)}}$, $z = \displaystyle\int \dfrac{dx}{\sqrt{g(x)}}$ leads to the constant coefficient equation $\partial_{\tau\tau}w = \partial_{zz}w + cw$ that is discussed in Subsection 4.1.3.

Chapter 5

Hyperbolic Equations with Two Space Variables

5.1. Wave Equation $\frac{\partial^2 w}{\partial t^2} = a^2 \Delta_2 w$

5.1.1. Problems in Cartesian Coordinates

The wave equation with two space variables in the rectangular Cartesian system of coordinates has the form

$$\frac{\partial^2 w}{\partial t^2} = a^2 \left(\frac{\partial^2 w}{\partial x^2} + \frac{\partial^2 w}{\partial y^2} \right).$$

5.1.1-1. Particular solutions and some relations.

1°. Particular solutions:

$$w(x,y,t) = A \exp\left(k_1 x + k_2 y \pm at\sqrt{k_1^2 + k_2^2}\right),$$

$$w(x,y,t) = A \sin(k_1 x + C_1) \sin(k_2 y + C_2) \sin\left(at\sqrt{k_1^2 + k_2^2}\right),$$

$$w(x,y,t) = A \sin(k_1 x + C_1) \sin(k_2 y + C_2) \cos\left(at\sqrt{k_1^2 + k_2^2}\right),$$

$$w(x,y,t) = A \sinh(k_1 x + C_1) \sinh(k_2 y + C_2) \sinh\left(at\sqrt{k_1^2 + k_2^2}\right),$$

$$w(x,y,t) = A \sinh(k_1 x + C_1) \sinh(k_2 y + C_2) \cosh\left(at\sqrt{k_1^2 + k_2^2}\right),$$

$$w(x,y,t) = \varphi(x \sin\beta + y \cos\beta + at) + \psi(x \sin\beta + y \cos\beta - at),$$

where A, C_1, C_2, k_1, k_2, and β are arbitrary constants, and $\varphi(z)$ and $\psi(z)$ are arbitrary functions.

2°. Particular solutions that are expressed in terms of solutions to simpler equations:

$$w(x,y,t) = [A\cos(ky) + B\sin(ky)]u(x,t), \quad \text{where} \quad \partial_{tt} u = a^2 \partial_{xx} u - a^2 k^2 u, \qquad (1)$$

$$w(x,y,t) = [A\cosh(ky) + B\sinh(ky)]u(x,t), \quad \text{where} \quad \partial_{tt} u = a^2 \partial_{xx} u + a^2 k^2 u, \qquad (2)$$

$$w(x,y,t) = [A\cos(kt) + B\sin(kt)]u(x,y), \quad \text{where} \quad \partial_{xx} u + \partial_{yy} u = -(k/a)^2 u, \qquad (3)$$

$$w(x,y,t) = [A\cosh(kt) + B\sinh(kt)]u(x,y), \quad \text{where} \quad \partial_{xx} u + \partial_{yy} u = (k/a)^2 u, \qquad (4)$$

$$w(x,y,t) = \exp\left(\frac{at \pm y}{2b}\right) u(x,\tau), \quad \tau = \frac{at \mp y}{2}, \quad \text{where} \quad \partial_\tau u = b \partial_{xx} u. \qquad (5)$$

For particular solutions of equations (1) and (2) for the function $u(x,t)$, see the Klein–Gordon equation 4.1.3. For particular solutions of equations (3) and (4) for the function $u(x,y)$, see Subsection 7.3.2. For particular solutions of the heat equation (5) for the function $u(x,\tau)$, see Subsection 1.1.1.

3°. Fundamental solution:

$$\mathscr{E}(x,y,t) = \frac{\vartheta(at-r)}{2\pi a\sqrt{a^2t^2-r^2}}, \qquad \vartheta(z) = \begin{cases} 1 & \text{for } z \geq 0, \\ 0 & \text{for } z < 0, \end{cases}$$

where $r = \sqrt{x^2+y^2}$.

4°. Infinite series solutions that contain arbitrary functions of the space variables:

$$w(x,y,t) = f(x,y) + \sum_{n=1}^{\infty} \frac{(at)^{2n}}{(2n)!} \Delta^n f(x,y), \qquad \Delta \equiv \frac{\partial^2}{\partial x^2} + \frac{\partial^2}{\partial y^2},$$

$$w(x,y,t) = tg(x,y) + t\sum_{n=1}^{\infty} \frac{(at)^{2n}}{(2n+1)!} \Delta^n g(x,y),$$

where $f(x,y)$ and $g(x,y)$ are any infinitely differentiable functions. The first solution satisfies the initial conditions $w(x,y,0) = f(x,y)$, $\partial_t w(x,y,0) = 0$ and the second solution to the initial conditions $w(x,y,0) = 0$, $\partial_t w(x,y,0) = g(x,y)$. The sums are finite if $f(x,y)$ and $g(x,y)$ are bivariate polynomials.

⊙ *Reference*: A. V. Bitsadze and D. F. Kalinichenko (1985).

5°. A wide class of solutions to the wave equation with two space variables are described by the formulas

$$w(x,y,t) = \operatorname{Re} F(\theta) \qquad \text{and} \qquad w(x,y,t) = \operatorname{Im} F(\theta). \tag{6}$$

Here, $F(\theta)$ is an arbitrary analytic function of the complex argument θ related to the variables (x,y,t) by the implicit relation

$$at - (x-x_0)\theta + (y-y_0)\sqrt{1-\theta^2} = G(\theta), \tag{7}$$

where $G(\theta)$ is any analytic function and x_0, y_0 are arbitrary constants. Solutions of the forms (6), (7) find wide application in the theory of diffraction. If the argument θ obtained by solving (7) with a prescribed $G(\theta)$ is real in some domain D, then one should set $\operatorname{Re} F(\theta) = F(\theta)$ in relation (6) everywhere in D.

⊙ *Reference*: V. I. Smirnov (1974, Vol. 3, Pt. 2).

6°. Suppose $w = w(x,y,t)$ is a solution of the wave equation. Then the functions

$$w_1 = Aw(\pm\lambda x + C_1, \pm\lambda y + C_2, \pm\lambda t + C_3),$$

$$w_2 = Aw\left(\frac{x-vt}{\sqrt{1-(v/a)^2}}, y, \frac{t-va^{-2}x}{\sqrt{1-(v/a)^2}}\right),$$

$$w_3 = \frac{A}{\sqrt{|r^2-a^2t^2|}} w\left(\frac{x}{r^2-a^2t^2}, \frac{y}{r^2-a^2t^2}, \frac{t}{r^2-a^2t^2}\right),$$

$$w_4 = \frac{A}{\sqrt{\Xi}} w\left(\frac{x+k_1(a^2t^2-r^2)}{\Xi}, \frac{y+k_2(a^2t^2-r^2)}{\Xi}, \frac{at+k_3(a^2t^2-r^2)}{a\Xi}\right),$$

$$r^2 = x^2+y^2, \quad \Xi = 1 - 2(k_1x + k_2y - ak_3t) + (k_1^2+k_2^2-k_3^2)(r^2-a^2t^2),$$

where A, C_1, C_2, C_3, k_1, k_2, k_3, v, and λ are arbitrary constants, are also solutions of the equation. The signs at λ in the expression of w_1 can be taken independently of one another. The function w_2 results from the invariance of the wave equation under the Lorentz transformation.

More detailed information about particular solutions and transformations of the wave equation with two space variables can be found in the references cited below.

⊙ *References*: E. Kalnins and W. Miller, Jr. (1975, 1976), W. Miller, Jr. (1977).

5.1.1-2. Domain: $-\infty < x < \infty$, $-\infty < y < \infty$. Cauchy problem.

Initial conditions are prescribed:
$$w = f(x,y) \quad \text{at} \quad t = 0,$$
$$\partial_t w = g(x,y) \quad \text{at} \quad t = 0.$$

Solution (Poisson's formula):
$$w(x,y,t) = \frac{1}{2\pi a}\frac{\partial}{\partial t}\iint_{C_{at}}\frac{f(\xi,\eta)\,d\xi\,d\eta}{\sqrt{a^2t^2 - (\xi-x)^2 - (\eta-y)^2}} + \frac{1}{2\pi a}\iint_{C_{at}}\frac{g(\xi,\eta)\,d\xi\,d\eta}{\sqrt{a^2t^2 - (\xi-x)^2 - (\eta-y)^2}},$$

where the integration is performed over the interior of the circle of radius at with center at (x,y).

⊙ *References*: N. S. Koshlyakov, E. B. Gliner, and M. M. Smirnov (1970), A. N. Tikhonov and A. A. Samarskii (1990).

5.1.1-3. Domain: $0 \le x \le l_1$, $0 \le y \le l_2$. First boundary value problem.

A rectangle is considered. The following conditions are prescribed:
$$w = f_0(x,y) \quad \text{at} \quad t = 0 \quad \text{(initial condition)},$$
$$\partial_t w = f_1(x,y) \quad \text{at} \quad t = 0 \quad \text{(initial condition)},$$
$$w = g_1(y,t) \quad \text{at} \quad x = 0 \quad \text{(boundary condition)},$$
$$w = g_2(y,t) \quad \text{at} \quad x = l_1 \quad \text{(boundary condition)},$$
$$w = g_3(x,t) \quad \text{at} \quad y = 0 \quad \text{(boundary condition)},$$
$$w = g_4(x,t) \quad \text{at} \quad y = l_2 \quad \text{(boundary condition)}.$$

Solution:
$$w(x,y,t) = \frac{\partial}{\partial t}\int_0^{l_1}\int_0^{l_2} f_0(\xi,\eta)G(x,y,\xi,\eta,t)\,d\eta\,d\xi + \int_0^{l_1}\int_0^{l_2} f_1(\xi,\eta)G(x,y,\xi,\eta,t)\,d\eta\,d\xi$$
$$+ a^2\int_0^t\int_0^{l_2} g_1(\eta,\tau)\left[\frac{\partial}{\partial \xi}G(x,y,\xi,\eta,t-\tau)\right]_{\xi=0}\,d\eta\,d\tau$$
$$- a^2\int_0^t\int_0^{l_2} g_2(\eta,\tau)\left[\frac{\partial}{\partial \xi}G(x,y,\xi,\eta,t-\tau)\right]_{\xi=l_1}\,d\eta\,d\tau$$
$$+ a^2\int_0^t\int_0^{l_1} g_3(\xi,\tau)\left[\frac{\partial}{\partial \eta}G(x,y,\xi,\eta,t-\tau)\right]_{\eta=0}\,d\xi\,d\tau$$
$$- a^2\int_0^t\int_0^{l_1} g_4(\xi,\tau)\left[\frac{\partial}{\partial \eta}G(x,y,\xi,\eta,t-\tau)\right]_{\eta=l_2}\,d\xi\,d\tau,$$

where
$$G(x,y,\xi,\eta,t) = \frac{4}{al_1l_2}\sum_{n=1}^{\infty}\sum_{m=1}^{\infty}\frac{1}{\lambda_{nm}}\sin(p_n x)\sin(q_m y)\sin(p_n \xi)\sin(q_m \eta)\sin(a\lambda_{nm}t),$$
$$p_n = \frac{n\pi}{l_1}, \quad q_m = \frac{m\pi}{l_2}, \quad \lambda_{nm} = \sqrt{p_n^2 + q_m^2}.$$

The problem of vibration of a rectangular membrane with sides l_1 and l_2 rigidly fixed in its contour is characterized by homogeneous boundary conditions, $g_s \equiv 0$ ($s = 1, 2, 3, 4$).

⊙ *References*: M. M. Smirnov (1964), B. M. Budak, A. A. Samarskii, and A. N. Tikhonov (1980).

5.1.1-4. Domain: $0 \leq x \leq l_1$, $0 \leq y \leq l_2$. Second boundary value problem.

A rectangle is considered. The following conditions are prescribed:

$$w = f_0(x, y) \quad \text{at} \quad t = 0 \quad \text{(initial condition)},$$
$$\partial_t w = f_1(x, y) \quad \text{at} \quad t = 0 \quad \text{(initial condition)},$$
$$\partial_x w = g_1(y, t) \quad \text{at} \quad x = 0 \quad \text{(boundary condition)},$$
$$\partial_x w = g_2(y, t) \quad \text{at} \quad x = l_1 \quad \text{(boundary condition)},$$
$$\partial_y w = g_3(x, t) \quad \text{at} \quad y = 0 \quad \text{(boundary condition)},$$
$$\partial_y w = g_4(x, t) \quad \text{at} \quad y = l_2 \quad \text{(boundary condition)}.$$

Solution:

$$w(x, y, t) = \frac{\partial}{\partial t} \int_0^{l_1} \int_0^{l_2} f_0(\xi, \eta) G(x, y, \xi, \eta, t) \, d\eta \, d\xi + \int_0^{l_1} \int_0^{l_2} f_1(\xi, \eta) G(x, y, \xi, \eta, t) \, d\eta \, d\xi$$
$$- a^2 \int_0^t \int_0^{l_2} g_1(\eta, \tau) G(x, y, 0, \eta, t - \tau) \, d\eta \, d\tau$$
$$+ a^2 \int_0^t \int_0^{l_2} g_2(\eta, \tau) G(x, y, l_1, \eta, t - \tau) \, d\eta \, d\tau$$
$$- a^2 \int_0^t \int_0^{l_1} g_3(\xi, \tau) G(x, y, \xi, 0, t - \tau) \, d\xi \, d\tau$$
$$+ a^2 \int_0^t \int_0^{l_1} g_4(\xi, \tau) G(x, y, \xi, l_2, t - \tau) \, d\xi \, d\tau,$$

where

$$G(x, y, \xi, \eta, t) = \frac{t}{l_1 l_2} + \frac{2}{a l_1 l_2} \sum_{n=0}^{\infty} \sum_{m=0}^{\infty} \frac{A_{nm}}{\lambda_{nm}} \cos(p_n x) \cos(q_m y) \cos(p_n \xi) \cos(q_m \eta) \sin(a \lambda_{nm} t),$$

$$p_n = \frac{n\pi}{l_1}, \quad q_m = \frac{m\pi}{l_2}, \quad \lambda_{nm} = \sqrt{p_n^2 + q_m^2}, \quad A_{nm} = \begin{cases} 0 & \text{for } n = m = 0, \\ 1 & \text{for } nm = 0 \ (n \neq m), \\ 2 & \text{for } nm \neq 0, \end{cases}$$

⊙ *References*: A. G. Butkovskiy (1979), B. M. Budak, A. A. Samarskii, and A. N. Tikhonov (1980).

5.1.1-5. Domain: $0 \leq x \leq l_1$, $0 \leq y \leq l_2$. Third boundary value problem.

A rectangle is considered. The following conditions are prescribed:

$$w = f_0(x, y) \quad \text{at} \quad t = 0 \quad \text{(initial condition)},$$
$$\partial_t w = f_1(x, y) \quad \text{at} \quad t = 0 \quad \text{(initial condition)},$$
$$\partial_x w - k_1 w = g_1(y, t) \quad \text{at} \quad x = 0 \quad \text{(boundary condition)},$$
$$\partial_x w + k_2 w = g_2(y, t) \quad \text{at} \quad x = l_1 \quad \text{(boundary condition)},$$
$$\partial_y w - k_3 w = g_3(x, t) \quad \text{at} \quad y = 0 \quad \text{(boundary condition)},$$
$$\partial_y w + k_4 w = g_4(x, t) \quad \text{at} \quad y = l_2 \quad \text{(boundary condition)}.$$

The solution $w(x, y, t)$ is determined by the formula in Paragraph 5.1.1-4 where

$$G(x, y, \xi, \eta, t) = \frac{4}{a} \sum_{n=1}^{\infty} \sum_{m=1}^{\infty} \frac{1}{E_{nm} \sqrt{\mu_n^2 + \nu_m^2}} \sin(\mu_n x + \varepsilon_n) \sin(\nu_m y + \sigma_m)$$
$$\times \sin(\mu_n \xi + \varepsilon_n) \sin(\nu_m \eta + \sigma_m) \sin\left(at\sqrt{\mu_n^2 + \nu_m^2}\right),$$

$$\varepsilon_n = \arctan \frac{\mu_n}{l_1}, \quad \sigma_m = \arctan \frac{\nu_m}{l_2}, \quad E_{nm} = \left[l_1 + \frac{(k_1 k_2 + \mu_n^2)(k_1 + k_2)}{(k_1^2 + \mu_n^2)(k_2^2 + \mu_n^2)}\right]\left[l_2 + \frac{(k_3 k_4 + \nu_m^2)(k_3 + k_4)}{(k_3^2 + \nu_m^2)(k_4^2 + \nu_m^2)}\right],$$

where the μ_n and ν_m are positive roots of the transcendental equations

$$\mu^2 - k_1 k_2 = (k_1 + k_2)\mu \cot(l_1\mu), \quad \nu^2 - k_3 k_4 = (k_3 + k_4)\nu \cot(l_2\nu).$$

⊙ *References*: B. M. Budak, A. A. Samarskii, and A. N. Tikhonov (1980).

5.1.1-6. Domain: $0 \leq x \leq l_1$, $0 \leq y \leq l_2$. Mixed boundary value problems.

1°. A rectangle is considered. The following conditions are prescribed:

$$\begin{aligned}
w &= f_0(x,y) & \text{at} \quad t &= 0 & \text{(initial condition)}, \\
\partial_t w &= f_1(x,y) & \text{at} \quad t &= 0 & \text{(initial condition)}, \\
w &= g_1(y,t) & \text{at} \quad x &= 0 & \text{(boundary condition)}, \\
w &= g_2(y,t) & \text{at} \quad x &= l_1 & \text{(boundary condition)}, \\
\partial_y w &= g_3(x,t) & \text{at} \quad y &= 0 & \text{(boundary condition)}, \\
\partial_y w &= g_4(x,t) & \text{at} \quad y &= l_2 & \text{(boundary condition)}.
\end{aligned}$$

Solution:

$$\begin{aligned}
w(x,y,t) = &\frac{\partial}{\partial t}\int_0^{l_1}\int_0^{l_2} f_0(\xi,\eta)G(x,y,\xi,\eta,t)\,d\eta\,d\xi + \int_0^{l_1}\int_0^{l_2} f_1(\xi,\eta)G(x,y,\xi,\eta,t)\,d\eta\,d\xi \\
&+ a^2\int_0^t\int_0^{l_2} g_1(\eta,\tau)\left[\frac{\partial}{\partial \xi}G(x,y,\xi,\eta,t-\tau)\right]_{\xi=0} d\eta\,d\tau \\
&- a^2\int_0^t\int_0^{l_2} g_2(\eta,\tau)\left[\frac{\partial}{\partial \xi}G(x,y,\xi,\eta,t-\tau)\right]_{\xi=l_1} d\eta\,d\tau \\
&- a^2\int_0^t\int_0^{l_1} g_3(\xi,\tau)G(x,y,\xi,0,t-\tau)\,d\xi\,d\tau \\
&+ a^2\int_0^t\int_0^{l_1} g_4(\xi,\tau)G(x,y,\xi,l_2,t-\tau)\,d\xi\,d\tau,
\end{aligned}$$

where

$$G(x,y,\xi,\eta,t) = \frac{2}{al_1l_2}\sum_{n=1}^\infty\sum_{m=0}^\infty \frac{A_m}{\lambda_{nm}}\sin(p_n x)\cos(q_m y)\sin(p_n \xi)\cos(q_m \eta)\sin(a\lambda_{nm} t),$$

$$p_n = \frac{n\pi}{l_1}, \quad q_m = \frac{m\pi}{l_2}, \quad \lambda_{nm} = \sqrt{p_n^2 + q_m^2}, \quad A_m = \begin{cases} 1 & \text{for } m = 0, \\ 2 & \text{for } m \neq 0. \end{cases}$$

2°. A rectangle is considered. The following conditions are prescribed:

$$\begin{aligned}
w &= f_0(x,y) & \text{at} \quad t &= 0 & \text{(initial condition)}, \\
\partial_t w &= f_1(x,y) & \text{at} \quad t &= 0 & \text{(initial condition)}, \\
w &= g_1(y,t) & \text{at} \quad x &= 0 & \text{(boundary condition)}, \\
\partial_x w &= g_2(y,t) & \text{at} \quad x &= l_1 & \text{(boundary condition)}, \\
w &= g_3(x,t) & \text{at} \quad y &= 0 & \text{(boundary condition)}, \\
\partial_y w &= g_4(x,t) & \text{at} \quad y &= l_2 & \text{(boundary condition)}.
\end{aligned}$$

Solution:
$$w(x,y,t) = \frac{\partial}{\partial t}\int_0^{l_1}\int_0^{l_2} f_0(\xi,\eta)G(x,y,\xi,\eta,t)\,d\eta\,d\xi + \int_0^{l_1}\int_0^{l_2} f_1(\xi,\eta)G(x,y,\xi,\eta,t)\,d\eta\,d\xi$$
$$+ a^2\int_0^t\int_0^{l_2} g_1(\eta,\tau)\left[\frac{\partial}{\partial\xi}G(x,y,\xi,\eta,t-\tau)\right]_{\xi=0} d\eta\,d\tau$$
$$+ a^2\int_0^t\int_0^{l_2} g_2(\eta,\tau)G(x,y,l_1,\eta,t-\tau)\,d\eta\,d\tau$$
$$+ a^2\int_0^t\int_0^{l_1} g_3(\xi,\tau)\left[\frac{\partial}{\partial\eta}G(x,y,\xi,\eta,t-\tau)\right]_{\eta=0} d\xi\,d\tau$$
$$+ a^2\int_0^t\int_0^{l_1} g_4(\xi,\tau)G(x,y,\xi,l_2,t-\tau)\,d\xi\,d\tau,$$

where
$$G(x,y,\xi,\eta,t) = \frac{4}{al_1l_2}\sum_{n=0}^{\infty}\sum_{m=0}^{\infty}\frac{1}{\lambda_{nm}}\sin(p_n x)\sin(q_m y)\sin(p_n\xi)\sin(q_m\eta)\sin(a\lambda_{nm}t),$$
$$p_n = \frac{\pi(2n+1)}{2l_1}, \quad q_m = \frac{\pi(2m+1)}{2l_2}, \quad \lambda_{nm} = \sqrt{p_n^2 + q_m^2}.$$

5.1.2. Problems in Polar Coordinates

The wave equation with two space variables in the polar coordinate system has the form
$$\frac{\partial^2 w}{\partial t^2} = a^2\left(\frac{\partial^2 w}{\partial r^2} + \frac{1}{r}\frac{\partial w}{\partial r} + \frac{1}{r^2}\frac{\partial^2 w}{\partial\varphi^2}\right), \quad r = \sqrt{x^2 + y^2}.$$

One-dimensional solutions $w = w(r,t)$ that are independent of the angular coordinate φ are considered in Subsection 4.2.1.

5.1.2-1. Domain: $0 \le r \le R$, $0 \le \varphi \le 2\pi$. First boundary value problem.

A circle is considered. The following conditions are prescribed:
$$w = f_0(r,\varphi) \quad \text{at} \quad t = 0 \quad \text{(initial condition)},$$
$$\partial_t w = f_1(r,\varphi) \quad \text{at} \quad t = 0 \quad \text{(initial condition)},$$
$$w = g(\varphi,t) \quad \text{at} \quad r = R \quad \text{(boundary condition)}.$$

Solution:
$$w(r,\varphi,t) = \frac{\partial}{\partial t}\int_0^{2\pi}\int_0^R f_0(\xi,\eta)G(r,\varphi,\xi,\eta,t)\xi\,d\xi\,d\eta + \int_0^{2\pi}\int_0^R f_1(\xi,\eta)G(r,\varphi,\xi,\eta,t)\xi\,d\xi\,d\eta$$
$$- a^2 R\int_0^t\int_0^{2\pi} g(\eta,\tau)\left[\frac{\partial}{\partial\xi}G(r,\varphi,\xi,\eta,t-\tau)\right]_{\xi=R} d\eta\,d\tau.$$

Here,
$$G(r,\varphi,\xi,\eta,t) = \frac{1}{\pi aR^2}\sum_{n=0}^{\infty}\sum_{m=1}^{\infty}\frac{A_n}{\mu_{nm}[J_n'(\mu_{nm}R)]^2}J_n(\mu_{nm}r)J_n(\mu_{nm}\xi)\cos[n(\varphi-\eta)]\sin(\mu_{nm}at),$$
$$A_0 = 1, \quad A_n = 2 \quad (n = 1, 2, \ldots),$$

where the $J_n(\xi)$ are the Bessel functions (the prime denotes the derivative with respect to the argument) and the μ_{nm} are positive roots of the transcendental equation $J_n(\mu R) = 0$.

The problem of vibration of a circular membrane of radius R rigidly fixed in its contour is characterized by the homogeneous boundary condition, $g(\varphi,t) \equiv 0$.

⊙ *References*: N. S. Koshlyakov, E. B. Gliner, and M. M. Smirnov (1970), A. G. Butkovskiy (1979), B. M. Budak, A. A. Samarskii, and A. N. Tikhonov (1980).

5.1.2-2. Domain: $0 \le r \le R$, $0 \le \varphi \le 2\pi$. Second boundary value problem.

A circle is considered. The following conditions are prescribed:

$$w = f_0(r, \varphi) \quad \text{at} \quad t = 0 \quad \text{(initial condition)},$$
$$\partial_t w = f_1(r, \varphi) \quad \text{at} \quad t = 0 \quad \text{(initial condition)},$$
$$\partial_r w = g(\varphi, t) \quad \text{at} \quad r = R \quad \text{(boundary condition)}.$$

Solution:

$$w(r, \varphi, t) = \frac{\partial}{\partial t} \int_0^{2\pi} \int_0^R f_0(\xi, \eta) G(r, \varphi, \xi, \eta, t) \xi \, d\xi \, d\eta + \int_0^{2\pi} \int_0^R f_1(\xi, \eta) G(r, \varphi, \xi, \eta, t) \xi \, d\xi \, d\eta$$
$$+ a^2 R \int_0^t \int_0^{2\pi} g(\eta, \tau) G(r, \varphi, R, \eta, t - \tau) \, d\eta \, d\tau.$$

Here,

$$G(r, \varphi, \xi, \eta, t) = \frac{t}{\pi R^2} + \frac{1}{\pi a} \sum_{n=0}^{\infty} \sum_{m=1}^{\infty} \frac{A_n \mu_{nm} J_n(\mu_{nm} r) J_n(\mu_{nm} \xi)}{(\mu_{nm}^2 R^2 - n^2)[J_n(\mu_{nm} R)]^2} \cos[n(\varphi - \eta)] \sin(\mu_{nm} a t),$$
$$A_0 = 1, \quad A_n = 2 \quad (n = 1, 2, \dots),$$

where the $J_n(\xi)$ are the Bessel functions and the μ_{nm} are positive roots of the transcendental equation $J'_n(\mu R) = 0$.

⊙ *References*: A. G. Butkovskiy (1979), B. M. Budak, A. A. Samarskii, and A. N. Tikhonov (1980).

5.1.2-3. Domain: $0 \le r \le R$, $0 \le \varphi \le 2\pi$. Third boundary value problem.

A circle is considered. The following conditions are prescribed:

$$w = f_0(r, \varphi) \quad \text{at} \quad t = 0 \quad \text{(initial condition)},$$
$$\partial_t w = f_1(r, \varphi) \quad \text{at} \quad t = 0 \quad \text{(initial condition)},$$
$$\partial_r w + kw = g(\varphi, t) \quad \text{at} \quad r = R \quad \text{(boundary condition)}.$$

The solution $w(r, \varphi, t)$ is determined by the formula in Paragraph 5.1.2-2 where

$$G(r, \varphi, \xi, \eta, t) = \frac{1}{\pi a} \sum_{n=0}^{\infty} \sum_{m=1}^{\infty} \frac{A_n \mu_{nm} J_n(\mu_{nm} r) J_n(\mu_{nm} \xi)}{(\mu_{nm}^2 R^2 + k^2 R^2 - n^2)[J_n(\mu_{nm} R)]^2} \cos[n(\varphi - \eta)] \sin(\mu_{nm} a t),$$
$$A_0 = 1, \quad A_n = 2 \quad (n = 1, 2, \dots).$$

Here, the $J_n(\xi)$ are the Bessel functions and the μ_{nm} are positive roots of the transcendental equation

$$\mu J'_n(\mu R) + k J_n(\mu R) = 0.$$

5.1.2-4. Domain: $R_1 \le r \le R_2$, $0 \le \varphi \le 2\pi$. First boundary value problem.

An annular domain is considered. The following conditions are prescribed:

$$w = f_0(r, \varphi) \quad \text{at} \quad t = 0 \quad \text{(initial condition)},$$
$$\partial_t w = f_1(r, \varphi) \quad \text{at} \quad t = 0 \quad \text{(initial condition)},$$
$$w = g_1(\varphi, t) \quad \text{at} \quad r = R_1 \quad \text{(boundary condition)},$$
$$w = g_2(\varphi, t) \quad \text{at} \quad r = R_2 \quad \text{(boundary condition)}.$$

Solution:

$$w(r,\varphi,t) = \frac{\partial}{\partial t}\int_0^{2\pi}\int_{R_1}^{R_2} f_0(\xi,\eta)G(r,\varphi,\xi,\eta,t)\xi\,d\xi\,d\eta + \int_0^{2\pi}\int_{R_1}^{R_2} f_1(\xi,\eta)G(r,\varphi,\xi,\eta,t)\xi\,d\xi\,d\eta$$

$$+ a^2 R_1 \int_0^t\int_0^{2\pi} g_1(\eta,\tau)\left[\frac{\partial}{\partial \xi}G(r,\varphi,\xi,\eta,t-\tau)\right]_{\xi=R_1} d\eta\,d\tau$$

$$- a^2 R_2 \int_0^t\int_0^{2\pi} g_2(\eta,\tau)\left[\frac{\partial}{\partial \xi}G(r,\varphi,\xi,\eta,t-\tau)\right]_{\xi=R_2} d\eta\,d\tau.$$

Here,

$$G(r,\varphi,\xi,\eta,t) = \frac{\pi}{2a}\sum_{n=0}^{\infty}\sum_{m=1}^{\infty} A_n B_{nm} Z_n(\mu_{nm}r)Z_n(\mu_{nm}\xi)\cos[n(\varphi-\eta)]\sin(\mu_{nm}at),$$

$$A_n = \begin{cases} 1/2 & \text{for } n=0, \\ 1 & \text{for } n\neq 0, \end{cases} \qquad B_{nm} = \frac{\mu_{nm} J_n^2(\mu_{nm}R_2)}{J_n^2(\mu_{nm}R_1) - J_n^2(\mu_{nm}R_2)},$$

$$Z_n(\mu_{nm}r) = J_n(\mu_{nm}R_1)Y_n(\mu_{nm}r) - Y_n(\mu_{nm}R_1)J_n(\mu_{nm}r),$$

where the $J_n(r)$ and $Y_n(r)$ are the Bessel functions, and the μ_{nm} are positive roots of the transcendental equation

$$J_n(\mu R_1)Y_n(\mu R_2) - Y_n(\mu R_1)J_n(\mu R_2) = 0.$$

5.1.2-5. Domain: $R_1 \leq r \leq R_2$, $0 \leq \varphi \leq 2\pi$. Second boundary value problem.

An annular domain is considered. The following conditions are prescribed:

$$\begin{aligned} w &= f_0(r,\varphi) & \text{at} \quad & t = 0 & \text{(initial condition)}, \\ \partial_t w &= f_1(r,\varphi) & \text{at} \quad & t = 0 & \text{(initial condition)}, \\ \partial_r w &= g_1(\varphi,t) & \text{at} \quad & r = R_1 & \text{(boundary condition)}, \\ \partial_r w &= g_2(\varphi,t) & \text{at} \quad & r = R_2 & \text{(boundary condition)}. \end{aligned}$$

Solution:

$$w(r,\varphi,t) = \frac{\partial}{\partial t}\int_0^{2\pi}\int_{R_1}^{R_2} f_0(\xi,\eta)G(r,\varphi,\xi,\eta,t)\xi\,d\xi\,d\eta + \int_0^{2\pi}\int_{R_1}^{R_2} f_1(\xi,\eta)G(r,\varphi,\xi,\eta,t)\xi\,d\xi\,d\eta$$

$$- a^2 R_1 \int_0^t\int_0^{2\pi} g_1(\eta,\tau)G(r,\varphi,R_1,\eta,t-\tau)\,d\eta\,d\tau$$

$$+ a^2 R_2 \int_0^t\int_0^{2\pi} g_2(\eta,\tau)G(r,\varphi,R_2,\eta,t-\tau)\,d\eta\,d\tau.$$

Here,

$$G(r,\varphi,\xi,\eta,t) = \frac{t}{\pi(R_2^2 - R_1^2)} + \frac{1}{\pi a}\sum_{n=0}^{\infty}\sum_{m=1}^{\infty} \frac{A_n \mu_{nm} Z_n(\mu_{nm}r)Z_n(\mu_{nm}\xi)\cos[n(\varphi-\eta)]\sin(\mu_{nm}at)}{(\mu_{nm}^2 R_2^2 - n^2)Z_n^2(\mu_{nm}R_2) - (\mu_{nm}^2 R_1^2 - n^2)Z_n^2(\mu_{nm}R_1)},$$

$$Z_n(\mu_{nm}r) = J_n'(\mu_{nm}R_1)Y_n(\mu_{nm}r) - Y_n'(\mu_{nm}R_1)J_n(\mu_{nm}r),$$

where $A_0 = 1$ and $A_n = 2$ for $n = 1, 2, \ldots$; the $J_n(r)$ and $Y_n(r)$ are the Bessel functions; and the μ_{nm} are positive roots of the transcendental equation

$$J_n'(\mu R_1)Y_n'(\mu R_2) - Y_n'(\mu R_1)J_n'(\mu R_2) = 0.$$

5.1.2-6. Domain: $R_1 \leq r \leq R_2$, $0 \leq \varphi \leq 2\pi$. Third boundary value problem.

An annular domain is considered. The following conditions are prescribed:

$$w = f_0(r,\varphi) \quad \text{at} \quad t = 0 \quad \text{(initial condition)},$$
$$\partial_t w = f_1(r,\varphi) \quad \text{at} \quad t = 0 \quad \text{(initial condition)},$$
$$\partial_r w - k_1 w = g_1(\varphi,t) \quad \text{at} \quad r = R_1 \quad \text{(boundary condition)},$$
$$\partial_r w + k_2 w = g_2(\varphi,t) \quad \text{at} \quad r = R_2 \quad \text{(boundary condition)}.$$

The solution $w(r,\varphi,t)$ is determined by the formula in Paragraph 5.1.2-5 where

$$G(r,\varphi,\xi,\eta,t) = \frac{1}{\pi a} \sum_{n=0}^{\infty} \sum_{m=1}^{\infty} \frac{A_n \mu_{nm} Z_n(\mu_{nm} r) Z_n(\mu_{nm}\xi) \cos[n(\varphi-\eta)] \sin(\mu_{nm} at)}{(k_2^2 R_2^2 + \mu_{nm}^2 R_2^2 - n^2) Z_n^2(\mu_{nm} R_2) - (k_1^2 R_1^2 + \mu_{nm}^2 R_1^2 - n^2) Z_n^2(\mu_{nm} R_1)},$$

$$Z_n(\mu_{nm} r) = [\mu_{nm} J_n'(\mu_{nm} R_1) - k_1 J_n(\mu_{nm} R_1)] Y_n(\mu_{nm} r)$$
$$- [\mu_{nm} Y_n'(\mu_{nm} R_1) - k_1 Y_n(\mu_{nm} R_1)] J_n(\mu_{nm} r).$$

Here, $A_0 = 1$ and $A_n = 2$ for $n = 1, 2, \ldots$; the $J_n(r)$ and $Y_n(r)$ are the Bessel functions; and the μ_{nm} are positive roots of the transcendental equation

$$[\mu J_n'(\mu R_1) - k_1 J_n(\mu R_1)][\mu Y_n'(\mu R_2) + k_2 Y_n(\mu R_2)]$$
$$= [\mu Y_n'(\mu R_1) - k_1 Y_n(\mu R_1)][\mu J_n'(\mu R_2) + k_2 J_n(\mu R_2)].$$

5.1.2-7. Domain: $0 \leq r \leq R$, $0 \leq \varphi \leq \varphi_0$. First boundary value problem.

A circular sector is considered. The following conditions are prescribed:

$$w = f_0(r,\varphi) \quad \text{at} \quad t = 0 \quad \text{(initial condition)},$$
$$\partial_t w = f_1(r,\varphi) \quad \text{at} \quad t = 0 \quad \text{(initial condition)},$$
$$w = g_1(\varphi,t) \quad \text{at} \quad r = R \quad \text{(boundary condition)},$$
$$w = g_2(r,t) \quad \text{at} \quad \varphi = 0 \quad \text{(boundary condition)},$$
$$w = g_3(r,t) \quad \text{at} \quad \varphi = \varphi_0 \quad \text{(boundary condition)}.$$

Solution:

$$w(r,\varphi,t) = \frac{\partial}{\partial t} \int_0^{\varphi_0} \int_0^R f_0(\xi,\eta) G(r,\varphi,\xi,\eta,t) \xi \, d\xi \, d\eta + \int_0^{\varphi_0} \int_0^R f_1(\xi,\eta) G(r,\varphi,\xi,\eta,t) \xi \, d\xi \, d\eta$$
$$- a^2 R \int_0^t \int_0^{\varphi_0} g_1(\eta,\tau) \left[\frac{\partial}{\partial \xi} G(r,\varphi,\xi,\eta,t-\tau)\right]_{\xi=R} d\eta \, d\tau$$
$$+ a^2 \int_0^t \int_0^R g_2(\xi,\tau) \frac{1}{\xi} \left[\frac{\partial}{\partial \eta} G(r,\varphi,\xi,\eta,t-\tau)\right]_{\eta=0} d\xi \, d\tau$$
$$- a^2 \int_0^t \int_0^R g_3(\xi,\tau) \frac{1}{\xi} \left[\frac{\partial}{\partial \eta} G(r,\varphi,\xi,\eta,t-\tau)\right]_{\eta=\varphi_0} d\xi \, d\tau.$$

Here,

$$G(r,\varphi,\xi,\eta,t) = \frac{4}{aR^2\varphi_0} \sum_{n=1}^{\infty} \sum_{m=1}^{\infty} \frac{J_{n\pi/\varphi_0}(\mu_{nm} r) J_{n\pi/\varphi_0}(\mu_{nm}\xi)}{\mu_{nm} [J'_{n\pi/\varphi_0}(\mu_{nm} R)]^2} \sin\left(\frac{n\pi\varphi}{\varphi_0}\right) \sin\left(\frac{n\pi\eta}{\varphi_0}\right) \sin(\mu_{nm} at),$$

where the $J_{n\pi/\varphi_0}(r)$ are the Bessel functions and the μ_{nm} are positive roots of the transcendental equation $J_{n\pi/\varphi_0}(\mu R) = 0$.

5.1.2-8. Domain: $0 \leq r \leq R$, $0 \leq \varphi \leq \varphi_0$. Second boundary value problem.

A circular sector is considered. The following conditions are prescribed:

$$w = f_0(r,\varphi) \quad \text{at} \quad t = 0 \quad \text{(initial condition)},$$
$$\partial_t w = f_1(r,\varphi) \quad \text{at} \quad t = 0 \quad \text{(initial condition)},$$
$$\partial_r w = g_1(\varphi,t) \quad \text{at} \quad r = R \quad \text{(boundary condition)},$$
$$r^{-1}\partial_\varphi w = g_2(r,t) \quad \text{at} \quad \varphi = 0 \quad \text{(boundary condition)},$$
$$r^{-1}\partial_\varphi w = g_3(r,t) \quad \text{at} \quad \varphi = \varphi_0 \quad \text{(boundary condition)}.$$

Solution:

$$w(r,\varphi,t) = \frac{\partial}{\partial t}\int_0^{\varphi_0}\int_0^R f_0(\xi,\eta)G(r,\varphi,\xi,\eta,t)\xi\,d\xi\,d\eta + \int_0^{\varphi_0}\int_0^R f_1(\xi,\eta)G(r,\varphi,\xi,\eta,t)\xi\,d\xi\,d\eta$$
$$+ a^2 R\int_0^t\int_0^{\varphi_0} g_1(\eta,\tau)G(r,\varphi,R,\eta,t-\tau)\,d\eta\,d\tau$$
$$- a^2\int_0^t\int_0^R g_2(\xi,\tau)G(r,\varphi,\xi,0,t-\tau)\,d\xi\,d\tau$$
$$+ a^2\int_0^t\int_0^R g_3(\xi,\tau)G(r,\varphi,\xi,\varphi_0,t-\tau)\,d\xi\,d\tau.$$

Here,

$$G(r,\varphi,\xi,\eta,t) = \frac{2t}{R^2\varphi_0} + \frac{4\varphi_0}{a}\sum_{n=0}^{\infty}\sum_{m=1}^{\infty}\frac{\mu_{nm}J_{n\pi/\varphi_0}(\mu_{nm}r)J_{n\pi/\varphi_0}(\mu_{nm}\xi)}{(R^2\varphi_0^2\mu_{nm}^2 - n^2\pi^2)\bigl[J_{n\pi/\varphi_0}(\mu_{nm}R)\bigr]^2}$$
$$\times \cos\left(\frac{n\pi\varphi}{\varphi_0}\right)\cos\left(\frac{n\pi\eta}{\varphi_0}\right)\sin(\mu_{nm}at),$$

where the $J_{n\pi/\varphi_0}(r)$ are the Bessel functions and the μ_{nm} are positive roots of the transcendental equation $J'_{n\pi/\varphi_0}(\mu R) = 0$.

5.1.2-9. Domain: $0 \leq r \leq R$, $0 \leq \varphi \leq \varphi_0$. Mixed boundary value problem.

A circular sector is considered. The following conditions are prescribed:

$$w = f_0(r,\varphi) \quad \text{at} \quad t = 0 \quad \text{(initial condition)},$$
$$\partial_t w = f_1(r,\varphi) \quad \text{at} \quad t = 0 \quad \text{(initial condition)},$$
$$\partial_r w + kw = g(\varphi,t) \quad \text{at} \quad r = R \quad \text{(boundary condition)},$$
$$\partial_\varphi w = 0 \quad \text{at} \quad \varphi = 0 \quad \text{(boundary condition)},$$
$$\partial_\varphi w = 0 \quad \text{at} \quad \varphi = \varphi_0 \quad \text{(boundary condition)}.$$

Solution:

$$w(r,\varphi,t) = \frac{\partial}{\partial t}\int_0^{\varphi_0}\int_0^R f_0(\xi,\eta)G(r,\varphi,\xi,\eta,t)\xi\,d\xi\,d\eta + \int_0^{\varphi_0}\int_0^R f_1(\xi,\eta)G(r,\varphi,\xi,\eta,t)\xi\,d\xi\,d\eta$$
$$+ a^2 R\int_0^t\int_0^{\varphi_0} g(\eta,\tau)G(r,\varphi,R,\eta,t-\tau)\,d\eta\,d\tau.$$

Here,

$$G(r,\varphi,\xi,\eta,t) = \sum_{n=0}^{\infty}\sum_{m=1}^{\infty}A_{nm}J_{s_n}(\mu_{nm}r)J_{s_n}(\mu_{nm}\xi)\cos(s_n\varphi)\cos(s_n\eta)\sin(\mu_{nm}at),$$

$$s_n = \frac{n\pi}{\varphi_0}, \qquad A_{nm} = \frac{4\mu_{nm}}{a\varphi_0(\mu_{nm}^2 R^2 + k^2 R^2 - s_n^2)\bigl[J_{s_n}(\mu_{nm}R)\bigr]^2},$$

where the $J_{s_n}(r)$ are the Bessel functions and the μ_{nm} are positive roots of the transcendental equation

$$\mu J'_{s_n}(\mu R) + k J_{s_n}(\mu R) = 0.$$

5.1.3. Axisymmetric Problems

In the axisymmetric case the wave equation in the cylindrical system of coordinates has the form

$$\frac{\partial^2 w}{\partial t^2} = a^2 \left(\frac{\partial^2 w}{\partial r^2} + \frac{1}{r}\frac{\partial w}{\partial r} + \frac{\partial^2 w}{\partial z^2} \right), \qquad r = \sqrt{x^2 + y^2}.$$

One-dimensional problems with axial symmetry that have solutions $w = w(r,t)$ are considered in Subsection 4.2.1.

In the solution of the problems considered below, the modified Green's function $\mathcal{G}(r,z,\xi,\eta,t) = 2\pi\xi G(r,z,\xi,\eta,t)$ is used for convenience.

5.1.3-1. Domain: $0 \le r \le R$, $0 \le z \le l$. First boundary value problem.

A circular cylinder of finite length is considered. The following conditions are prescribed:

$$\begin{aligned}
w &= f_0(r,z) & \text{at} \quad t &= 0 & \text{(initial condition)}, \\
\partial_t w &= f_1(r,z) & \text{at} \quad t &= 0 & \text{(initial condition)}, \\
w &= g_1(z,t) & \text{at} \quad r &= R & \text{(boundary condition)}, \\
w &= g_2(r,t) & \text{at} \quad z &= 0 & \text{(boundary condition)}, \\
w &= g_3(r,t) & \text{at} \quad z &= l & \text{(boundary condition)}.
\end{aligned}$$

Solution:

$$\begin{aligned}
w(r,z,t) =\ & \frac{\partial}{\partial t}\int_0^l\!\!\int_0^R f_0(\xi,\eta)\mathcal{G}(r,z,\xi,\eta,t)\,d\xi\,d\eta + \int_0^l\!\!\int_0^R f_1(\xi,\eta)\mathcal{G}(r,z,\xi,\eta,t)\,d\xi\,d\eta \\
& - a^2\int_0^t\!\!\int_0^l g_1(\eta,\tau)\left[\frac{\partial}{\partial\xi}\mathcal{G}(r,z,\xi,\eta,t-\tau)\right]_{\xi=R} d\eta\,d\tau \\
& + a^2\int_0^t\!\!\int_0^R g_2(\xi,\tau)\left[\frac{\partial}{\partial\eta}\mathcal{G}(r,z,\xi,\eta,t-\tau)\right]_{\eta=0} d\xi\,d\tau \\
& - a^2\int_0^t\!\!\int_0^R g_3(\xi,\tau)\left[\frac{\partial}{\partial\eta}\mathcal{G}(r,z,\xi,\eta,t-\tau)\right]_{\eta=l} d\xi\,d\tau.
\end{aligned}$$

Here,

$$\mathcal{G}(r,z,\xi,\eta,t) = \frac{4\xi}{R^2 l}\sum_{n=1}^\infty\sum_{m=1}^\infty \frac{1}{J_1^2(\mu_n)} J_0\!\left(\frac{\mu_n r}{R}\right) J_0\!\left(\frac{\mu_n \xi}{R}\right) \sin\!\left(\frac{m\pi z}{l}\right)\sin\!\left(\frac{m\pi\eta}{l}\right)\frac{\sin\!\left(at\sqrt{\lambda_{nm}}\right)}{a\sqrt{\lambda_{nm}}},$$

$$\lambda_{nm} = \frac{\mu_n^2}{R^2} + \frac{\pi^2 m^2}{l^2},$$

where the μ_n are positive zeros of the Bessel function, $J_0(\mu) = 0$.

5.1.3-2. Domain: $0 \le r \le R$, $0 \le z \le l$. Second boundary value problem.

A circular cylinder of finite length is considered. The following conditions are prescribed:

$$\begin{aligned}
w &= f_0(r,z) & \text{at} \quad t &= 0 & \text{(initial condition)}, \\
\partial_t w &= f_1(r,z) & \text{at} \quad t &= 0 & \text{(initial condition)}, \\
\partial_r w &= g_1(z,t) & \text{at} \quad r &= R & \text{(boundary condition)}, \\
\partial_z w &= g_2(r,t) & \text{at} \quad z &= 0 & \text{(boundary condition)}, \\
\partial_z w &= g_3(r,t) & \text{at} \quad z &= l & \text{(boundary condition)}.
\end{aligned}$$

Solution:

$$w(r,z,t) = \frac{\partial}{\partial t}\int_0^l\int_0^R f_0(\xi,\eta)\mathcal{G}(r,z,\xi,\eta,t)\,d\xi\,d\eta + \int_0^l\int_0^R f_1(\xi,\eta)\mathcal{G}(r,z,\xi,\eta,t)\,d\xi\,d\eta$$

$$+ a^2\int_0^t\int_0^l g_1(\eta,\tau)\mathcal{G}(r,z,R,\eta,t-\tau)\,d\eta\,d\tau$$

$$- a^2\int_0^t\int_0^R g_2(\xi,\tau)\mathcal{G}(r,z,\xi,0,t-\tau)\,d\xi\,d\tau$$

$$+ a^2\int_0^t\int_0^R g_3(\xi,\tau)\mathcal{G}(r,z,\xi,l,t-\tau)\,d\xi\,d\tau.$$

Here,

$$\mathcal{G}(r,z,\xi,\eta,t) = \frac{2t\xi}{R^2 l} + \frac{2\xi}{R^2 l}\sum_{n=0}^{\infty}\sum_{m=0}^{\infty}\frac{A_{nm}}{J_0^2(\mu_n)}J_0\!\left(\frac{\mu_n r}{R}\right)J_0\!\left(\frac{\mu_n \xi}{R}\right)$$

$$\times \cos\!\left(\frac{m\pi z}{l}\right)\cos\!\left(\frac{m\pi\eta}{l}\right)\frac{\sin(at\sqrt{\lambda_{nm}})}{a\sqrt{\lambda_{nm}}},$$

$$\lambda_{nm} = \frac{\mu_n^2}{R^2} + \frac{\pi^2 m^2}{l^2}, \quad A_{nm} = \begin{cases} 0 & \text{for } m=0, n=0, \\ 1 & \text{for } m=0, n>0, \\ 2 & \text{for } m>0, \end{cases}$$

where the μ_n are zeros of the first-order Bessel function, $J_1(\mu)=0$ ($\mu_0=0$).

5.1.3-3. Domain: $0 \le r \le R$, $0 \le z \le l$. Third boundary value problem.

A circular cylinder of finite length is considered. The following conditions are prescribed:

$$\begin{aligned}
w &= f_0(r,z) &\text{at}\quad t &= 0 &&\text{(initial condition)}, \\
\partial_t w &= f_1(r,z) &\text{at}\quad t &= 0 &&\text{(initial condition)}, \\
\partial_r w + k_1 w &= g_1(z,t) &\text{at}\quad r &= R &&\text{(boundary condition)}, \\
\partial_z w - k_2 w &= g_2(r,t) &\text{at}\quad z &= 0 &&\text{(boundary condition)}, \\
\partial_z w + k_3 w &= g_3(r,t) &\text{at}\quad z &= l &&\text{(boundary condition)}.
\end{aligned}$$

The solution $w(r,z,t)$ is determined by the formula in Paragraph 5.1.3-2 where

$$\mathcal{G}(r,z,\xi,\eta,t) = \frac{2\xi}{R^2}\sum_{n=1}^{\infty}\sum_{m=1}^{\infty}\frac{\mu_n^2}{(k_1^2 R^2 + \mu_n^2)J_0^2(\mu_n)}J_0\!\left(\frac{\mu_n r}{R}\right)J_0\!\left(\frac{\mu_n \xi}{R}\right)\frac{\varphi_m(z)\varphi_m(\eta)}{\|\varphi_m\|^2}\frac{\sin(at\sqrt{\lambda_{nm}})}{a\sqrt{\lambda_{nm}}},$$

$$\lambda_{nm} = \frac{\mu_n^2}{R^2} + \beta_m^2, \quad \varphi_m(z) = \cos(\beta_m z) + \frac{k_2}{\beta_m}\sin(\beta_m z),$$

$$\|\varphi_m\|^2 = \frac{k_3}{2\beta_m^2}\frac{\beta_m^2 + k_2^2}{\beta_m^2 + k_3^2} + \frac{k_2}{2\beta_m^2} + \frac{l}{2}\left(1 + \frac{k_2^2}{\beta_m^2}\right).$$

Here, the μ_n and β_m are positive roots of the transcendental equations

$$\mu J_1(\mu) - k_1 R J_0(\mu) = 0, \qquad \frac{\tan(\beta l)}{\beta} = \frac{k_2 + k_3}{\beta^2 - k_2 k_3}.$$

5.1.3-4. Domain: $0 \leq r \leq R$, $0 \leq z \leq l$. Mixed boundary value problems.

1°. A circular cylinder of finite length is considered. The following conditions are prescribed:

$$w = f_0(r, z) \quad \text{at} \quad t = 0 \quad \text{(initial condition),}$$
$$\partial_t w = f_1(r, z) \quad \text{at} \quad t = 0 \quad \text{(initial condition),}$$
$$w = g_1(z, t) \quad \text{at} \quad r = R \quad \text{(boundary condition),}$$
$$\partial_z w = g_2(r, t) \quad \text{at} \quad z = 0 \quad \text{(boundary condition),}$$
$$\partial_z w = g_3(r, t) \quad \text{at} \quad z = l \quad \text{(boundary condition).}$$

Solution:

$$w(r, z, t) = \frac{\partial}{\partial t} \int_0^l \int_0^R f_0(\xi, \eta) \mathcal{G}(r, z, \xi, \eta, t) \, d\xi \, d\eta + \int_0^l \int_0^R f_1(\xi, \eta) \mathcal{G}(r, z, \xi, \eta, t) \, d\xi \, d\eta$$
$$- a^2 \int_0^t \int_0^l g_1(\eta, \tau) \left[\frac{\partial}{\partial \xi} \mathcal{G}(r, z, \xi, \eta, t - \tau) \right]_{\xi = R} d\eta \, d\tau$$
$$- a^2 \int_0^t \int_0^R g_2(\xi, \tau) \mathcal{G}(r, z, \xi, 0, t - \tau) \, d\xi \, d\tau$$
$$+ a^2 \int_0^t \int_0^R g_3(\xi, \tau) \mathcal{G}(r, z, \xi, l, t - \tau) \, d\xi \, d\tau.$$

Here,

$$\mathcal{G}(r, z, \xi, \eta, t) = \frac{2\xi}{R^2 l} \sum_{n=1}^{\infty} \sum_{m=0}^{\infty} \frac{A_m}{J_1^2(\mu_n)} J_0\left(\frac{\mu_n r}{R}\right) J_0\left(\frac{\mu_n \xi}{R}\right) \cos\left(\frac{m \pi z}{l}\right) \cos\left(\frac{m \pi \eta}{l}\right) \frac{\sin\left(a t \sqrt{\lambda_{nm}}\right)}{a \sqrt{\lambda_{nm}}},$$

$$\lambda_{nm} = \frac{\mu_n^2}{R^2} + \frac{\pi^2 m^2}{l^2}, \quad A_m = \begin{cases} 1 & \text{for } m = 0, \\ 2 & \text{for } m > 0, \end{cases}$$

where the μ_n are positive zeros of the Bessel function, $J_0(\mu) = 0$.

2°. A circular cylinder of finite length is considered. The following conditions are prescribed:

$$w = f_0(r, z) \quad \text{at} \quad t = 0 \quad \text{(initial condition),}$$
$$\partial_t w = f_1(r, z) \quad \text{at} \quad t = 0 \quad \text{(initial condition),}$$
$$\partial_r w = g_1(z, t) \quad \text{at} \quad r = R \quad \text{(boundary condition),}$$
$$w = g_2(r, t) \quad \text{at} \quad z = 0 \quad \text{(boundary condition),}$$
$$w = g_3(r, t) \quad \text{at} \quad z = l \quad \text{(boundary condition).}$$

Solution:

$$w(r, z, t) = \frac{\partial}{\partial t} \int_0^l \int_0^R f_0(\xi, \eta) \mathcal{G}(r, z, \xi, \eta, t) \, d\xi \, d\eta + \int_0^l \int_0^R f_1(\xi, \eta) \mathcal{G}(r, z, \xi, \eta, t) \, d\xi \, d\eta$$
$$+ a^2 \int_0^t \int_0^l g_1(\eta, \tau) \mathcal{G}(r, z, R, \eta, t - \tau) \, d\eta \, d\tau$$
$$+ a^2 \int_0^t \int_0^R g_2(\xi, \tau) \left[\frac{\partial}{\partial \eta} \mathcal{G}(r, z, \xi, \eta, t - \tau) \right]_{\eta = 0} d\xi \, d\tau$$
$$- a^2 \int_0^t \int_0^R g_3(\xi, \tau) \left[\frac{\partial}{\partial \eta} \mathcal{G}(r, z, \xi, \eta, t - \tau) \right]_{\eta = l} d\xi \, d\tau.$$

Here,
$$\mathcal{G}(r,z,\xi,\eta,t) = \frac{4\xi}{R^2 l}\sum_{n=0}^{\infty}\sum_{m=1}^{\infty}\frac{1}{J_0^2(\mu_n)}J_0\!\left(\frac{\mu_n r}{R}\right)J_0\!\left(\frac{\mu_n \xi}{R}\right)\sin\!\left(\frac{m\pi z}{l}\right)\sin\!\left(\frac{m\pi \eta}{l}\right)\frac{\sin(at\sqrt{\lambda_{nm}})}{a\sqrt{\lambda_{nm}}},$$
$$\lambda_{nm} = \frac{\mu_n^2}{R^2} + \frac{\pi^2 m^2}{l^2},$$
where the μ_n are zeros of the first-order Bessel function, $J_1(\mu) = 0$ ($\mu_0 = 0$).

5.1.3-5. Domain: $R_1 \le r \le R_2$, $0 \le z \le l$. First boundary value problem.

A hollow circular cylinder of finite length is considered. The following conditions are prescribed:

$$\begin{aligned}
w &= f_0(r,z) & &\text{at} & t &= 0 & &\text{(initial condition),}\\
\partial_t w &= f_1(r,z) & &\text{at} & t &= 0 & &\text{(initial condition),}\\
w &= g_1(z,t) & &\text{at} & r &= R_1 & &\text{(boundary condition),}\\
w &= g_2(z,t) & &\text{at} & r &= R_2 & &\text{(boundary condition),}\\
w &= g_3(r,t) & &\text{at} & z &= 0 & &\text{(boundary condition),}\\
w &= g_4(r,t) & &\text{at} & z &= l & &\text{(boundary condition).}
\end{aligned}$$

Solution:
$$w(r,z,t) = \frac{\partial}{\partial t}\int_0^l\!\!\int_{R_1}^{R_2} f_0(\xi,\eta)\mathcal{G}(r,z,\xi,\eta,t)\,d\xi\,d\eta + \int_0^l\!\!\int_{R_1}^{R_2} f_1(\xi,\eta)\mathcal{G}(r,z,\xi,\eta,t)\,d\xi\,d\eta$$
$$+ a^2\int_0^t\!\!\int_0^l g_1(\eta,\tau)\left[\frac{\partial}{\partial \xi}\mathcal{G}(r,z,\xi,\eta,t-\tau)\right]_{\xi=R_1}d\eta\,d\tau$$
$$- a^2\int_0^t\!\!\int_0^l g_2(\eta,\tau)\left[\frac{\partial}{\partial \xi}\mathcal{G}(r,z,\xi,\eta,t-\tau)\right]_{\xi=R_2}d\eta\,d\tau$$
$$+ a^2\int_0^t\!\!\int_{R_1}^{R_2} g_3(\xi,\tau)\left[\frac{\partial}{\partial \eta}\mathcal{G}(r,z,\xi,\eta,t-\tau)\right]_{\eta=0}d\xi\,d\tau$$
$$- a^2\int_0^t\!\!\int_{R_1}^{R_2} g_4(\xi,\tau)\left[\frac{\partial}{\partial \eta}\mathcal{G}(r,z,\xi,\eta,t-\tau)\right]_{\eta=l}d\xi\,d\tau.$$

Here,
$$\mathcal{G}(r,z,\xi,\eta,t) = \frac{\pi^2\xi}{R_1^2 l}\sum_{n=1}^{\infty}\sum_{m=1}^{\infty}\frac{\mu_n^2 J_0^2(s\mu_n)}{J_0^2(\mu_n)-J_0^2(s\mu_n)}\Psi_n(r)\Psi_n(\xi)\sin\!\left(\frac{m\pi z}{l}\right)\sin\!\left(\frac{m\pi \eta}{l}\right)\frac{\sin(at\sqrt{\lambda_{nm}})}{a\sqrt{\lambda_{nm}}},$$
$$\Psi_n(r) = Y_0(\mu_n)J_0\!\left(\frac{\mu_n r}{R_1}\right) - J_0(\mu_n)Y_0\!\left(\frac{\mu_n r}{R_1}\right),\quad s = \frac{R_2}{R_1},\quad \lambda_{nm} = \frac{\mu_n^2}{R_1^2} + \frac{\pi^2 m^2}{l^2},$$
where $J_0(\mu)$ and $Y_0(\mu)$ are the Bessel functions, and the μ_n are positive roots of the transcendental equation
$$J_0(\mu)Y_0(s\mu) - J_0(s\mu)Y_0(\mu) = 0.$$

5.1.3-6. Domain: $R_1 \le r \le R_2$, $0 \le z \le l$. Second boundary value problem.

A hollow circular cylinder of finite length is considered. The following conditions are prescribed:

$$\begin{aligned}
w &= f_0(r,z) & &\text{at} & t &= 0 & &\text{(initial condition),}\\
\partial_t w &= f_1(r,z) & &\text{at} & t &= 0 & &\text{(initial condition),}\\
\partial_r w &= g_1(z,t) & &\text{at} & r &= R_1 & &\text{(boundary condition),}\\
\partial_r w &= g_2(z,t) & &\text{at} & r &= R_2 & &\text{(boundary condition),}\\
\partial_z w &= g_3(r,t) & &\text{at} & z &= 0 & &\text{(boundary condition),}\\
\partial_z w &= g_4(r,t) & &\text{at} & z &= l & &\text{(boundary condition).}
\end{aligned}$$

Solution:

$$w(r,z,t) = \frac{\partial}{\partial t}\int_0^l \int_{R_1}^{R_2} f_0(\xi,\eta)\mathcal{G}(r,z,\xi,\eta,t)\,d\xi\,d\eta + \int_0^l \int_{R_1}^{R_2} f_1(\xi,\eta)\mathcal{G}(r,z,\xi,\eta,t)\,d\xi\,d\eta$$
$$- a^2 \int_0^t \int_0^l g_1(\eta,\tau)\mathcal{G}(r,z,R_1,\eta,t-\tau)\,d\eta\,d\tau + a^2 \int_0^t \int_0^l g_2(\eta,\tau)\mathcal{G}(r,z,R_2,\eta,t-\tau)\,d\eta\,d\tau$$
$$- a^2 \int_0^t \int_{R_1}^{R_2} g_3(\xi,\tau)\mathcal{G}(r,z,\xi,0,t-\tau)\,d\xi\,d\tau + a^2 \int_0^t \int_{R_1}^{R_2} g_4(\xi,\tau)\mathcal{G}(r,z,\xi,l,t-\tau)\,d\xi\,d\tau.$$

Here,

$$\mathcal{G}(r,z,\xi,\eta,t) = \frac{2t\xi}{(R_2^2-R_1^2)l} + \frac{4\xi}{\pi a(R_2^2-R_1^2)} \sum_{m=1}^\infty \frac{1}{m} \cos\left(\frac{m\pi z}{l}\right)\cos\left(\frac{m\pi\eta}{l}\right)\sin\left(\frac{m\pi at}{l}\right)$$
$$+ \frac{\pi^2 \xi}{2R_1^2 l} \sum_{n=1}^\infty \sum_{m=0}^\infty \frac{A_m \mu_n^2 J_1^2(s\mu_n)}{J_1^2(\mu_n) - J_1^2(s\mu_n)} \Psi_n(r)\Psi_n(\xi)\cos\left(\frac{m\pi z}{l}\right)\cos\left(\frac{m\pi\eta}{l}\right)\frac{\sin(at\sqrt{\lambda_{nm}})}{a\sqrt{\lambda_{nm}}},$$

where

$$\Psi_n(r) = Y_1(\mu_n)J_0\left(\frac{\mu_n r}{R_1}\right) - J_1(\mu_n)Y_0\left(\frac{\mu_n r}{R_1}\right), \qquad s = \frac{R_2}{R_1},$$

$$A_m = \begin{cases} 1 & \text{for } m=0, \\ 2 & \text{for } m>1, \end{cases} \qquad \lambda_{nm} = \frac{\mu_n^2}{R_1^2} + \frac{\pi^2 m^2}{l^2};$$

$J_k(\mu)$ and $Y_k(\mu)$ are the Bessel functions ($k=0, 1$); and the μ_n are positive roots of the transcendental equation

$$J_1(\mu)Y_1(s\mu) - J_1(s\mu)Y_1(\mu) = 0.$$

5.2. Nonhomogeneous Wave Equation $\frac{\partial^2 w}{\partial t^2} = a^2 \Delta_2 w + \Phi(x,y,t)$

5.2.1. Problems in Cartesian Coordinates

5.2.1-1. Domain: $-\infty < x < \infty$, $-\infty < y < \infty$. Cauchy problem.

Initial conditions are prescribed:

$$w = f(x,y) \quad \text{at} \quad t=0,$$
$$\partial_t w = g(x,y) \quad \text{at} \quad t=0.$$

Solution:

$$w(x,y,t) = \frac{1}{2\pi a}\frac{\partial}{\partial t}\iint_{\rho\leq at}\frac{f(\xi,\eta)\,d\xi\,d\eta}{\sqrt{a^2t^2-\rho^2}} + \frac{1}{2\pi a}\iint_{\rho\leq at}\frac{g(\xi,\eta)\,d\xi\,d\eta}{\sqrt{a^2t^2-\rho^2}}$$
$$+ \frac{1}{2\pi a}\int_0^t \left[\iint_{\rho\leq a(t-\tau)}\frac{\Phi(\xi,\eta,\tau)\,d\xi\,d\eta}{\sqrt{a^2(t-\tau)^2-\rho^2}}\right]d\tau, \qquad \rho^2 = (\xi-x)^2 + (\eta-y)^2.$$

⊙ *Reference*: N. S. Koshlyakov, E. B. Gliner, and M. M. Smirnov (1970).

5.2.1-2. Domain: $0 \leq x \leq l_1$, $0 \leq y \leq l_2$. First boundary value problem.

A rectangle is considered. The following conditions are prescribed:

$$
\begin{aligned}
w &= f_0(x,y) & \text{at} \quad t &= 0 & \text{(initial condition)},\\
\partial_t w &= f_1(x,y) & \text{at} \quad t &= 0 & \text{(initial condition)},\\
w &= g_1(y,t) & \text{at} \quad x &= 0 & \text{(boundary condition)},\\
w &= g_2(y,t) & \text{at} \quad x &= l_1 & \text{(boundary condition)},\\
w &= g_3(x,t) & \text{at} \quad y &= 0 & \text{(boundary condition)},\\
w &= g_4(x,t) & \text{at} \quad y &= l_2 & \text{(boundary condition)}.
\end{aligned}
$$

The solution $w(x,y,t)$ is given by the formula in Paragraph 5.1.1-3 with the additional term

$$\int_0^t \int_0^{l_1} \int_0^{l_2} \Phi(\xi,\eta,\tau) G(x,y,\xi,\eta,t-\tau)\, d\eta\, d\xi\, d\tau,$$

which allows for the equation's nonhomogeneity; this term is the solution of the nonhomogeneous equation with homogeneous initial and boundary conditions.

⊙ *Reference*: B. M. Budak, A. A. Samarskii, and A. N. Tikhonov (1980).

5.2.1-3. Domain: $0 \leq x \leq l_1$, $0 \leq y \leq l_2$. Second boundary value problem.

A rectangle is considered. The following conditions are prescribed:

$$
\begin{aligned}
w &= f_1(x,y) & \text{at} \quad t &= 0 & \text{(initial condition)},\\
\partial_t w &= f_2(x,y) & \text{at} \quad t &= 0 & \text{(initial condition)},\\
\partial_x w &= g_1(y,t) & \text{at} \quad x &= 0 & \text{(boundary condition)},\\
\partial_x w &= g_2(y,t) & \text{at} \quad x &= l_1 & \text{(boundary condition)},\\
\partial_y w &= g_3(x,t) & \text{at} \quad y &= 0 & \text{(boundary condition)},\\
\partial_y w &= g_4(x,t) & \text{at} \quad y &= l_2 & \text{(boundary condition)}.
\end{aligned}
$$

The solution $w(x,y,t)$ is given by the formula in Paragraph 5.1.1-4 with the additional term specified in Paragraph 5.2.1-2 (the Green's function is taken from Paragraph 5.1.1-4).

⊙ *References*: A. G. Butkovskiy (1979), B. M. Budak, A. A. Samarskii, and A. N. Tikhonov (1980).

5.2.1-4. Domain: $0 \leq x \leq l_1$, $0 \leq y \leq l_2$. Third boundary value problem.

A rectangle is considered. The following conditions are prescribed:

$$
\begin{aligned}
w &= f_1(x,y) & \text{at} \quad t &= 0 & \text{(initial condition)},\\
\partial_t w &= f_2(x,y) & \text{at} \quad t &= 0 & \text{(initial condition)},\\
\partial_x w - k_1 w &= g_1(y,t) & \text{at} \quad x &= 0 & \text{(boundary condition)},\\
\partial_x w + k_2 w &= g_2(y,t) & \text{at} \quad x &= l_1 & \text{(boundary condition)},\\
\partial_y w - k_3 w &= g_3(x,t) & \text{at} \quad y &= 0 & \text{(boundary condition)},\\
\partial_y w + k_4 w &= g_4(x,t) & \text{at} \quad y &= l_2 & \text{(boundary condition)}.
\end{aligned}
$$

The solution $w(x,y,t)$ is the sum of the solution to the homogeneous equation with nonhomogeneous initial and boundary conditions (see Paragraph 5.1.1-5) and the solution to the nonhomogeneous equation with homogeneous initial and boundary conditions. This solution is given by the formula in Paragraph 5.2.1-2 in which one should substitute the Green's function of Paragraph 5.1.1-5).

⊙ *References*: A. G. Butkovskiy (1979), B. M. Budak, A. A. Samarskii, and A. N. Tikhonov (1980).

5.2.1-5. Domain: $0 \le x \le l_1$, $0 \le y \le l_2$. Mixed boundary value problems.

1°. A rectangle is considered. The following conditions are prescribed:

$$\begin{aligned}
w &= f_1(x,y) & \text{at} \quad t &= 0 & \text{(initial condition)},\\
\partial_t w &= f_2(x,y) & \text{at} \quad t &= 0 & \text{(initial condition)},\\
w &= g_1(y,t) & \text{at} \quad x &= 0 & \text{(boundary condition)},\\
w &= g_2(y,t) & \text{at} \quad x &= l_1 & \text{(boundary condition)},\\
\partial_y w &= g_3(x,t) & \text{at} \quad y &= 0 & \text{(boundary condition)},\\
\partial_y w &= g_4(x,t) & \text{at} \quad y &= l_2 & \text{(boundary condition)}.
\end{aligned}$$

The solution $w(x,y,t)$ is given by the formula in Paragraph 5.1.1-6, Item 1°, with the additional term specified in Paragraph 5.2.1-2.

2°. A rectangle is considered. The following conditions are prescribed:

$$\begin{aligned}
w &= f_1(x,y) & \text{at} \quad t &= 0 & \text{(initial condition)},\\
\partial_t w &= f_2(x,y) & \text{at} \quad t &= 0 & \text{(initial condition)},\\
w &= g_1(y,t) & \text{at} \quad x &= 0 & \text{(boundary condition)},\\
\partial_x w &= g_2(y,t) & \text{at} \quad x &= l_1 & \text{(boundary condition)},\\
w &= g_3(x,t) & \text{at} \quad y &= 0 & \text{(boundary condition)},\\
\partial_y w &= g_4(x,t) & \text{at} \quad y &= l_2 & \text{(boundary condition)}.
\end{aligned}$$

The solution $w(x,y,t)$ is given by the formula in Paragraph 5.1.1-6, Item 2°, with the additional term specified in Paragraph 5.2.1-2.

5.2.2. Problems in Polar Coordinates

A nonhomogeneous wave equation in the polar coordinate system has the form

$$\frac{\partial^2 w}{\partial t^2} = a^2 \left(\frac{\partial^2 w}{\partial r^2} + \frac{1}{r}\frac{\partial w}{\partial r} + \frac{1}{r^2}\frac{\partial^2 w}{\partial \varphi^2} \right) + \Phi(r,\varphi,t), \qquad r = \sqrt{x^2 + y^2}.$$

One-dimensional boundary value problems independent of the angular coordinate φ are considered in Subsection 4.2.2.

5.2.2-1. Domain: $0 \le r \le R$, $0 \le \varphi \le 2\pi$. First boundary value problem.

A circle is considered. The following conditions are prescribed:

$$\begin{aligned}
w &= f_0(r,\varphi) & \text{at} \quad t &= 0 & \text{(initial condition)},\\
\partial_t w &= f_1(r,\varphi) & \text{at} \quad t &= 0 & \text{(initial condition)},\\
w &= g(\varphi,t) & \text{at} \quad r &= R & \text{(boundary condition)}.
\end{aligned}$$

The solution $w(r,\varphi,t)$ is given by the formula in Paragraph 5.1.2-1 with the additional term

$$\int_0^t \int_0^{2\pi} \int_0^R \Phi(\xi,\eta,\tau) G(r,\varphi,\xi,\eta,t-\tau)\xi\,d\xi\,d\eta\,d\tau, \tag{1}$$

which allows for the equation's nonhomogeneity; this term is the solution of the nonhomogeneous equation with homogeneous initial and boundary conditions.

⊙ *References*: N. S. Koshlyakov, E. B. Gliner, and M. M. Smirnov (1970), B. M. Budak, A. A. Samarskii, and A. N. Tikhonov (1980).

5.2.2-2. Domain: $0 \leq r \leq R$, $0 \leq \varphi \leq 2\pi$. Second boundary value problem.

A circle is considered. The following conditions are prescribed:

$$w = f_0(r, \varphi) \quad \text{at} \quad t = 0 \quad \text{(initial condition)},$$
$$\partial_t w = f_1(r, \varphi) \quad \text{at} \quad t = 0 \quad \text{(initial condition)},$$
$$\partial_r w = g(\varphi, t) \quad \text{at} \quad r = R \quad \text{(boundary condition)}.$$

The solution $w(r, \varphi, t)$ is given by the formula in Paragraph 5.1.2-2 with the additional term (1).

⊙ *References*: A. G. Butkovskiy (1979), B. M. Budak, A. A. Samarskii, and A. N. Tikhonov (1980).

5.2.2-3. Domain: $0 \leq r \leq R$, $0 \leq \varphi \leq 2\pi$. Third boundary value problem.

A circle is considered. The following conditions are prescribed:

$$w = f_0(r, \varphi) \quad \text{at} \quad t = 0 \quad \text{(initial condition)},$$
$$\partial_t w = f_1(r, \varphi) \quad \text{at} \quad t = 0 \quad \text{(initial condition)},$$
$$\partial_r w + kw = g(\varphi, t) \quad \text{at} \quad r = R \quad \text{(boundary condition)}.$$

The solution $w(r, \varphi, t)$ is the sum of the solution to the homogeneous equation with nonhomogeneous initial and boundary conditions (see Paragraph 5.1.2-3) and the solution to the nonhomogeneous equation with homogeneous initial and boundary conditions [this solution is given by formula (1) in which one should substitute the Green's function in Paragraph 5.1.2-3].

⊙ *References*: A. G. Butkovskiy (1979), B. M. Budak, A. A. Samarskii, and A. N. Tikhonov (1980).

5.2.2-4. Domain: $R_1 \leq r \leq R_2$, $0 \leq \varphi \leq 2\pi$. First boundary value problem.

An annular domain is considered. The following conditions are prescribed:

$$w = f_0(r, \varphi) \quad \text{at} \quad t = 0 \quad \text{(initial condition)},$$
$$\partial_t w = f_1(r, \varphi) \quad \text{at} \quad t = 0 \quad \text{(initial condition)},$$
$$w = g_1(\varphi, t) \quad \text{at} \quad r = R_1 \quad \text{(boundary condition)},$$
$$w = g_2(\varphi, t) \quad \text{at} \quad r = R_2 \quad \text{(boundary condition)}.$$

The solution $w(r, \varphi, t)$ is given by the formula in Paragraph 5.1.2-4 with the additional term

$$\int_0^t \int_0^{2\pi} \int_{R_1}^{R_2} \Phi(\xi, \eta, \tau) G(r, \varphi, \xi, \eta, t - \tau) \xi \, d\xi \, d\eta \, d\tau, \tag{2}$$

which allows for the equation's nonhomogeneity; this term is the solution of the nonhomogeneous equation with homogeneous initial and boundary conditions.

5.2.2-5. Domain: $R_1 \leq r \leq R_2$, $0 \leq \varphi \leq 2\pi$. Second boundary value problem.

An annular domain is considered. The following conditions are prescribed:

$$w = f_0(r, \varphi) \quad \text{at} \quad t = 0 \quad \text{(initial condition)},$$
$$\partial_t w = f_1(r, \varphi) \quad \text{at} \quad t = 0 \quad \text{(initial condition)},$$
$$\partial_r w = g_1(\varphi, t) \quad \text{at} \quad r = R_1 \quad \text{(boundary condition)},$$
$$\partial_r w = g_2(\varphi, t) \quad \text{at} \quad r = R_2 \quad \text{(boundary condition)}.$$

The solution $w(r, \varphi, t)$ is given by the formula in Paragraph 5.1.2-5 with the additional term (2).

5.2. NONHOMOGENEOUS WAVE EQUATION $\frac{\partial^2 w}{\partial t^2} = a^2 \Delta_2 w + \Phi(x,y,t)$

5.2.2-6. Domain: $R_1 \leq r \leq R_2$, $0 \leq \varphi \leq 2\pi$. Third boundary value problem.

An annular domain is considered. The following conditions are prescribed:

$$w = f_0(r,\varphi) \quad \text{at} \quad t = 0 \quad \text{(initial condition)},$$
$$\partial_t w = f_1(r,\varphi) \quad \text{at} \quad t = 0 \quad \text{(initial condition)},$$
$$\partial_r w - k_1 w = g_1(\varphi, t) \quad \text{at} \quad r = R_1 \quad \text{(boundary condition)},$$
$$\partial_r w + k_2 w = g_2(\varphi, t) \quad \text{at} \quad r = R_2 \quad \text{(boundary condition)}.$$

The solution $w(r,\varphi,t)$ is the sum of the solution to the homogeneous equation with nonhomogeneous initial and boundary conditions (see Paragraph 5.1.2-6) and the solution to the nonhomogeneous equation with homogeneous initial and boundary conditions [this solution is given by formula (2) in which one should substitute the Green's function in Paragraph 5.1.2-6].

5.2.2-7. Domain: $0 \leq r \leq R$, $0 \leq \varphi \leq \varphi_0$. First boundary value problem.

A circular sector is considered. The following conditions are prescribed:

$$w = f_0(r,\varphi) \quad \text{at} \quad t = 0 \quad \text{(initial condition)},$$
$$\partial_t w = f_1(r,\varphi) \quad \text{at} \quad t = 0 \quad \text{(initial condition)},$$
$$w = g_1(\varphi, t) \quad \text{at} \quad r = R \quad \text{(boundary condition)},$$
$$w = g_2(r, t) \quad \text{at} \quad \varphi = 0 \quad \text{(boundary condition)},$$
$$w = g_3(r, t) \quad \text{at} \quad \varphi = \varphi_0 \quad \text{(boundary condition)}.$$

The solution $w(r,\varphi,t)$ is given by the formula in Paragraph 5.1.2-7 with the additional term

$$\int_0^t \int_0^{\varphi_0} \int_0^R \Phi(\xi,\eta,\tau) G(r,\varphi,\xi,\eta,t-\tau)\xi \, d\xi \, d\eta \, d\tau, \tag{3}$$

which allows for the equation's nonhomogeneity.

5.2.2-8. Domain: $0 \leq r \leq R$, $0 \leq \varphi \leq \varphi_0$. Second boundary value problem.

A circular sector is considered. The following conditions are prescribed:

$$w = f_0(r,\varphi) \quad \text{at} \quad t = 0 \quad \text{(initial condition)},$$
$$\partial_t w = f_1(r,\varphi) \quad \text{at} \quad t = 0 \quad \text{(initial condition)},$$
$$\partial_r w = g_1(\varphi, t) \quad \text{at} \quad r = R \quad \text{(boundary condition)},$$
$$r^{-1}\partial_\varphi w = g_2(r, t) \quad \text{at} \quad \varphi = 0 \quad \text{(boundary condition)},$$
$$r^{-1}\partial_\varphi w = g_3(r, t) \quad \text{at} \quad \varphi = \varphi_0 \quad \text{(boundary condition)}.$$

The solution $w(r,\varphi,t)$ is given by the formula in Paragraph 5.1.2-8 with the additional term (3).

5.2.2-9. Domain: $0 \leq r \leq R$, $0 \leq \varphi \leq \varphi_0$. Mixed boundary value problem.

A circular sector is considered. The following conditions are prescribed:

$$w = f_0(r,\varphi) \quad \text{at} \quad t = 0 \quad \text{(initial condition)},$$
$$\partial_t w = f_1(r,\varphi) \quad \text{at} \quad t = 0 \quad \text{(initial condition)},$$
$$\partial_r w + kw = g(\varphi, t) \quad \text{at} \quad r = R \quad \text{(boundary condition)},$$
$$\partial_\varphi w = 0 \quad \text{at} \quad \varphi = 0 \quad \text{(boundary condition)},$$
$$\partial_\varphi w = 0 \quad \text{at} \quad \varphi = \varphi_0 \quad \text{(boundary condition)}.$$

The solution $w(r,\varphi,t)$ is given by the formula in Paragraph 5.1.2-9 with the additional term (3).

5.2.3. Axisymmetric Problems

In the axisymmetric case, a nonhomogeneous wave equation in the cylindrical system of coordinates has the form

$$\frac{\partial^2 w}{\partial t^2} = a^2 \left(\frac{\partial^2 w}{\partial r^2} + \frac{1}{r} \frac{\partial w}{\partial r} + \frac{\partial^2 w}{\partial z^2} \right) + \Phi(r, z, t), \qquad r = \sqrt{x^2 + y^2}.$$

5.2.3-1. Domain: $0 \leq r \leq R$, $0 \leq z \leq l$. First boundary value problem.

A circular cylinder of finite length is considered. The following conditions are prescribed:

$$\begin{aligned}
w &= f_0(r, z) & \text{at} \quad t &= 0 & \text{(initial condition)}, \\
\partial_t w &= f_1(r, z) & \text{at} \quad t &= 0 & \text{(initial condition)}, \\
w &= g_1(z, t) & \text{at} \quad r &= R & \text{(boundary condition)}, \\
w &= g_2(r, t) & \text{at} \quad z &= 0 & \text{(boundary condition)}, \\
w &= g_3(r, t) & \text{at} \quad z &= l & \text{(boundary condition)}.
\end{aligned}$$

The solution $w(r, z, t)$ is given by the formula in Paragraph 5.1.3-1 with the additional term

$$\int_0^t \int_0^l \int_0^R \Phi(\xi, \eta, \tau) \mathcal{G}(r, z, \xi, \eta, t - \tau) \, d\xi \, d\eta \, d\tau, \tag{1}$$

which allows for the equation's nonhomogeneity; this term is the solution of the nonhomogeneous equation with homogeneous initial and boundary conditions.

5.2.3-2. Domain: $0 \leq r \leq R$, $0 \leq z \leq l$. Second boundary value problem.

A circular cylinder of finite length is considered. The following conditions are prescribed:

$$\begin{aligned}
w &= f_0(r, z) & \text{at} \quad t &= 0 & \text{(initial condition)}, \\
\partial_t w &= f_1(r, z) & \text{at} \quad t &= 0 & \text{(initial condition)}, \\
\partial_r w &= g_1(z, t) & \text{at} \quad r &= R & \text{(boundary condition)}, \\
\partial_z w &= g_2(r, t) & \text{at} \quad z &= 0 & \text{(boundary condition)}, \\
\partial_z w &= g_3(r, t) & \text{at} \quad z &= l & \text{(boundary condition)}.
\end{aligned}$$

The solution $w(r, z, t)$ is given by the formula in Paragraph 5.1.3-2 with the additional term (1).

5.2.3-3. Domain: $0 \leq r \leq R$, $0 \leq z \leq l$. Third boundary value problem.

A circular cylinder of finite length is considered. The following conditions are prescribed:

$$\begin{aligned}
w &= f_0(r, z) & \text{at} \quad t &= 0 & \text{(initial condition)}, \\
\partial_t w &= f_1(r, z) & \text{at} \quad t &= 0 & \text{(initial condition)}, \\
\partial_r w + k_1 w &= g_1(z, t) & \text{at} \quad r &= R & \text{(boundary condition)}, \\
\partial_z w - k_2 w &= g_2(r, t) & \text{at} \quad z &= 0 & \text{(boundary condition)}, \\
\partial_z w + k_3 w &= g_3(r, t) & \text{at} \quad z &= l & \text{(boundary condition)}.
\end{aligned}$$

The solution $w(r, z, t)$ is the sum of the solution to the homogeneous equation with nonhomogeneous initial and boundary conditions (see Paragraph 5.1.3-3) and the solution to the nonhomogeneous equation with homogeneous initial and boundary conditions [this solution is given by formula (1) in which one should substitute the Green's function in Paragraph 5.1.3-3].

5.2. Nonhomogeneous Wave Equation $\frac{\partial^2 w}{\partial t^2} = a^2 \Delta_2 w + \Phi(x,y,t)$

5.2.3-4. Domain: $0 \le r \le R$, $0 \le z \le l$. Mixed boundary value problems.

1°. A circular cylinder of finite length is considered. The following conditions are prescribed:

$$w = f_0(r,z) \quad \text{at} \quad t = 0 \quad \text{(initial condition)},$$
$$\partial_t w = f_1(r,z) \quad \text{at} \quad t = 0 \quad \text{(initial condition)},$$
$$w = g_1(z,t) \quad \text{at} \quad r = R \quad \text{(boundary condition)},$$
$$\partial_z w = g_2(r,t) \quad \text{at} \quad z = 0 \quad \text{(boundary condition)},$$
$$\partial_z w = g_3(r,t) \quad \text{at} \quad z = l \quad \text{(boundary condition)}.$$

The solution $w(r,z,t)$ is given by the formula in Paragraph 5.1.3-4, Item 1°, with the additional term (1).

2°. A circular cylinder of finite length is considered. The following conditions are prescribed:

$$w = f_0(r,z) \quad \text{at} \quad t = 0 \quad \text{(initial condition)},$$
$$\partial_t w = f_1(r,z) \quad \text{at} \quad t = 0 \quad \text{(initial condition)},$$
$$\partial_r w = g_1(z,t) \quad \text{at} \quad r = R \quad \text{(boundary condition)},$$
$$w = g_2(r,t) \quad \text{at} \quad z = 0 \quad \text{(boundary condition)},$$
$$w = g_3(r,t) \quad \text{at} \quad z = l \quad \text{(boundary condition)}.$$

The solution $w(r,z,t)$ is given by the formula in Paragraph 5.1.3-4, Item 2°, with the additional term (1).

5.2.3-5. Domain: $R_1 \le r \le R_2$, $0 \le z \le l$. First boundary value problem.

A hollow circular cylinder of finite length is considered. The following conditions are prescribed:

$$w = f_0(r,z) \quad \text{at} \quad t = 0 \quad \text{(initial condition)},$$
$$\partial_t w = f_1(r,z) \quad \text{at} \quad t = 0 \quad \text{(initial condition)},$$
$$w = g_1(z,t) \quad \text{at} \quad r = R_1 \quad \text{(boundary condition)},$$
$$w = g_2(z,t) \quad \text{at} \quad r = R_2 \quad \text{(boundary condition)},$$
$$w = g_3(r,t) \quad \text{at} \quad z = 0 \quad \text{(boundary condition)},$$
$$w = g_4(r,t) \quad \text{at} \quad z = l \quad \text{(boundary condition)}.$$

The solution $w(r,z,t)$ is given by the formula in Paragraph 5.1.3-5 with the additional term

$$\int_0^t \int_0^l \int_{R_1}^{R_2} \Phi(\xi,\eta,\tau) \mathcal{G}(r,z,\xi,\eta,t-\tau) \, d\xi \, d\eta \, d\tau, \tag{2}$$

which allows for the equation's nonhomogeneity; this term is the solution of the nonhomogeneous equation with homogeneous initial and boundary conditions.

5.2.3-6. Domain: $R_1 \le r \le R_2$, $0 \le z \le l$. Second boundary value problem.

A hollow circular cylinder of finite length is considered. The following conditions are prescribed:

$$w = f_0(r,z) \quad \text{at} \quad t = 0 \quad \text{(initial condition)},$$
$$\partial_t w = f_1(r,z) \quad \text{at} \quad t = 0 \quad \text{(initial condition)},$$
$$\partial_r w = g_1(z,t) \quad \text{at} \quad r = R_1 \quad \text{(boundary condition)},$$
$$\partial_r w = g_2(z,t) \quad \text{at} \quad r = R_2 \quad \text{(boundary condition)},$$
$$\partial_z w = g_3(r,t) \quad \text{at} \quad z = 0 \quad \text{(boundary condition)},$$
$$\partial_z w = g_4(r,t) \quad \text{at} \quad z = l \quad \text{(boundary condition)}.$$

The solution $w(r,z,t)$ is given by the formula in Paragraph 5.1.3-6 with the additional term (2).

5.3. Equations of the Form $\dfrac{\partial^2 w}{\partial t^2} = a^2 \Delta_2 w - bw + \Phi(x,y,t)$

5.3.1. Problems in Cartesian Coordinates

The *two-dimensional nonhomogeneous Klein–Gordon equation* with two space variables in the rectangular Cartesian coordinate system is written as

$$\frac{\partial^2 w}{\partial t^2} = a^2 \left(\frac{\partial^2 w}{\partial x^2} + \frac{\partial^2 w}{\partial y^2} \right) - bw + \Phi(x,y,t).$$

5.3.1-1. Fundamental solutions.

1°. Case $b = -\sigma^2 < 0$:

$$\mathscr{E}(x,y,t) = \frac{\vartheta(at-r)}{2\pi a^2} \frac{\cosh\left(\sigma\sqrt{t^2 - r^2/a^2}\right)}{\sqrt{t^2 - r^2/a^2}}, \qquad r = \sqrt{x^2 + y^2},$$

where $\vartheta(z)$ is the Heaviside unit step function.

2°. Case $b = \sigma^2 > 0$:

$$\mathscr{E}(x,y,t) = \frac{\vartheta(at-r)}{2\pi a^2} \frac{\cos\left(\sigma\sqrt{t^2 - r^2/a^2}\right)}{\sqrt{t^2 - r^2/a^2}}, \qquad r = \sqrt{x^2 + y^2}.$$

⊙ *References*: V. S. Vladimirov, V. P. Mikhailov, A. A. Vasharin, et al. (1974), B. M. Budak, A. A. Samarskii, and A. N. Tikhonov (1980).

5.3.1-2. Domain: $-\infty < x < \infty$, $-\infty < y < \infty$. Cauchy problem.

Initial conditions are prescribed:

$$w = f(x,y) \quad \text{at} \quad t = 0,$$
$$\partial_t w = g(x,y) \quad \text{at} \quad t = 0.$$

1°. Solution for $b = -a^2 c^2 < 0$:

$$w(x,y,t) = \frac{1}{2\pi a} \frac{\partial}{\partial t} \iint\limits_{\rho \leq at} f(\xi,\eta) \frac{\cosh\left(c\sqrt{a^2 t^2 - \rho^2}\right)}{\sqrt{a^2 t^2 - \rho^2}} d\xi\, d\eta + \frac{1}{2\pi a} \iint\limits_{\rho \leq at} g(\xi,\eta) \frac{\cosh\left(c\sqrt{a^2 t^2 - \rho^2}\right)}{\sqrt{a^2 t^2 - \rho^2}} d\xi\, d\eta$$

$$+ \frac{1}{2\pi a} \int_0^t d\tau \iint\limits_{\rho \leq a(t-\tau)} \Phi(\xi,\eta,\tau) \frac{\cosh\left(c\sqrt{a^2(t-\tau)^2 - \rho^2}\right)}{\sqrt{a^2(t-\tau)^2 - \rho^2}} d\xi\, d\eta, \qquad \rho = \sqrt{(x-\xi)^2 + (y-\eta)^2}.$$

2°. Solution for $b = a^2 c^2 > 0$:

$$w(x,y,t) = \frac{1}{2\pi a} \frac{\partial}{\partial t} \iint\limits_{\rho \leq at} f(\xi,\eta) \frac{\cos\left(c\sqrt{a^2 t^2 - \rho^2}\right)}{\sqrt{a^2 t^2 - \rho^2}} d\xi\, d\eta + \frac{1}{2\pi a} \iint\limits_{\rho \leq at} g(\xi,\eta) \frac{\cos\left(c\sqrt{a^2 t^2 - \rho^2}\right)}{\sqrt{a^2 t^2 - \rho^2}} d\xi\, d\eta$$

$$+ \frac{1}{2\pi a} \int_0^t d\tau \iint\limits_{\rho \leq a(t-\tau)} \Phi(\xi,\eta,\tau) \frac{\cos\left(c\sqrt{a^2(t-\tau)^2 - \rho^2}\right)}{\sqrt{a^2(t-\tau)^2 - \rho^2}} d\xi\, d\eta, \qquad \rho = \sqrt{(x-\xi)^2 + (y-\eta)^2}.$$

⊙ *Reference*: B. M. Budak, A. A. Samarskii, and A. N. Tikhonov (1980).

5.3. Equations of the Form $\frac{\partial^2 w}{\partial t^2} = a^2 \Delta_2 w - bw + \Phi(x, y, t)$

5.3.1-3. Domain: $0 \le x \le l_1$, $0 \le y \le l_2$. First boundary value problem.

A rectangle is considered. The following conditions are prescribed:

$$w = f_0(x, y) \quad \text{at} \quad t = 0 \quad \text{(initial condition)},$$
$$\partial_t w = f_1(x, y) \quad \text{at} \quad t = 0 \quad \text{(initial condition)},$$
$$w = g_1(y, t) \quad \text{at} \quad x = 0 \quad \text{(boundary condition)},$$
$$w = g_2(y, t) \quad \text{at} \quad x = l_1 \quad \text{(boundary condition)},$$
$$w = g_3(x, t) \quad \text{at} \quad y = 0 \quad \text{(boundary condition)},$$
$$w = g_4(x, t) \quad \text{at} \quad y = l_2 \quad \text{(boundary condition)}.$$

Solution:

$$
\begin{aligned}
w(x, y, t) = &\frac{\partial}{\partial t} \int_0^{l_1} \int_0^{l_2} f_0(\xi, \eta) G(x, y, \xi, \eta, t) \, d\eta \, d\xi + \int_0^{l_1} \int_0^{l_2} f_1(\xi, \eta) G(x, y, \xi, \eta, t) \, d\eta \, d\xi \\
&+ a^2 \int_0^t \int_0^{l_2} g_1(\eta, \tau) \left[\frac{\partial}{\partial \xi} G(x, y, \xi, \eta, t - \tau) \right]_{\xi=0} d\eta \, d\tau \\
&- a^2 \int_0^t \int_0^{l_2} g_2(\eta, \tau) \left[\frac{\partial}{\partial \xi} G(x, y, \xi, \eta, t - \tau) \right]_{\xi=l_1} d\eta \, d\tau \\
&+ a^2 \int_0^t \int_0^{l_1} g_3(\xi, \tau) \left[\frac{\partial}{\partial \eta} G(x, y, \xi, \eta, t - \tau) \right]_{\eta=0} d\xi \, d\tau \\
&- a^2 \int_0^t \int_0^{l_1} g_4(\xi, \tau) \left[\frac{\partial}{\partial \eta} G(x, y, \xi, \eta, t - \tau) \right]_{\eta=l_2} d\xi \, d\tau \\
&+ \int_0^t \int_0^{l_1} \int_0^{l_2} \Phi(\xi, \eta, \tau) G(x, y, \xi, \eta, t - \tau) \, d\eta \, d\xi \, d\tau,
\end{aligned}
$$

where

$$G(x, y, \xi, \eta, t) = \frac{4}{l_1 l_2} \sum_{n=1}^\infty \sum_{m=1}^\infty \frac{1}{\lambda_{nm}} \sin(p_n x) \sin(q_m y) \sin(p_n \xi) \sin(q_m \eta) \sin(\lambda_{nm} t),$$

$$p_n = \frac{n\pi}{l_1}, \quad q_m = \frac{m\pi}{l_2}, \quad \lambda_{nm} = \sqrt{a^2 p_n^2 + a^2 q_m^2 + b}.$$

5.3.1-4. Domain: $0 \le x \le l_1$, $0 \le y \le l_2$. Second boundary value problem.

A rectangle is considered. The following conditions are prescribed:

$$w = f_0(x, y) \quad \text{at} \quad t = 0 \quad \text{(initial condition)},$$
$$\partial_t w = f_1(x, y) \quad \text{at} \quad t = 0 \quad \text{(initial condition)},$$
$$\partial_x w = g_1(y, t) \quad \text{at} \quad x = 0 \quad \text{(boundary condition)},$$
$$\partial_x w = g_2(y, t) \quad \text{at} \quad x = l_1 \quad \text{(boundary condition)},$$
$$\partial_y w = g_3(x, t) \quad \text{at} \quad y = 0 \quad \text{(boundary condition)},$$
$$\partial_y w = g_4(x, t) \quad \text{at} \quad y = l_2 \quad \text{(boundary condition)}.$$

Solution:

$$
\begin{aligned}
w(x,y,t) = &\frac{\partial}{\partial t}\int_0^{l_1}\int_0^{l_2} f_0(\xi,\eta)G(x,y,\xi,\eta,t)\,d\eta\,d\xi + \int_0^{l_1}\int_0^{l_2} f_1(\xi,\eta)G(x,y,\xi,\eta,t)\,d\eta\,d\xi \\
&- a^2\int_0^t\int_0^{l_2} g_1(\eta,\tau)G(x,y,0,\eta,t-\tau)\,d\eta\,d\tau \\
&+ a^2\int_0^t\int_0^{l_2} g_2(\eta,\tau)G(x,y,l_1,\eta,t-\tau)\,d\eta\,d\tau \\
&- a^2\int_0^t\int_0^{l_1} g_3(\xi,\tau)G(x,y,\xi,0,t-\tau)\,d\xi\,d\tau \\
&+ a^2\int_0^t\int_0^{l_1} g_4(\xi,\tau)G(x,y,\xi,l_2,t-\tau)\,d\xi\,d\tau \\
&+ \int_0^t\int_0^{l_1}\int_0^{l_2} \Phi(\xi,\eta,\tau)G(x,y,\xi,\eta,t-\tau)\,d\eta\,d\xi\,d\tau,
\end{aligned}
$$

where

$$G(x,y,\xi,\eta,t) = \frac{\sin(t\sqrt{b})}{l_1 l_2 \sqrt{b}} + \frac{2}{l_1 l_2}\sum_{n=0}^{\infty}\sum_{m=0}^{\infty}\frac{A_{nm}}{\lambda_{nm}}\cos(p_n x)\cos(q_m y)\cos(p_n \xi)\cos(q_m \eta)\sin(\lambda_{nm} t),$$

$$p_n = \frac{n\pi}{l_1},\quad q_m = \frac{m\pi}{l_2},\quad \lambda_{nm} = \sqrt{a^2 p_n^2 + a^2 q_m^2 + b},\quad A_{nm} = \begin{cases} 0 & \text{for } n=m=0, \\ 1 & \text{for } nm=0\ (n\neq m), \\ 2 & \text{for } nm\neq 0. \end{cases}$$

5.3.1-5. Domain: $0 \le x \le l_1$, $0 \le y \le l_2$. Third boundary value problem.

A rectangle is considered. The following conditions are prescribed:

$$
\begin{aligned}
w &= f_0(x,y) & \text{at}\quad & t=0 & \text{(initial condition),} \\
\partial_t w &= f_1(x,y) & \text{at}\quad & t=0 & \text{(initial condition),} \\
\partial_x w - k_1 w &= g_1(y,t) & \text{at}\quad & x=0 & \text{(boundary condition),} \\
\partial_x w + k_2 w &= g_2(y,t) & \text{at}\quad & x=l_1 & \text{(boundary condition),} \\
\partial_y w - k_3 w &= g_3(x,t) & \text{at}\quad & y=0 & \text{(boundary condition),} \\
\partial_y w + k_4 w &= g_4(x,t) & \text{at}\quad & y=l_2 & \text{(boundary condition).}
\end{aligned}
$$

The solution $w(x,y,t)$ is determined by the formula in Paragraph 5.3.1-3 where

$$G(x,y,\xi,\eta,t) = 4\sum_{n=1}^{\infty}\sum_{m=1}^{\infty}\frac{1}{E_{nm}\sqrt{a^2\mu_n^2 + a^2\nu_m^2 + b}}\sin(\mu_n x + \varepsilon_n)\sin(\nu_m y + \sigma_m)$$

$$\times \sin(\mu_n \xi + \varepsilon_n)\sin(\nu_m \eta + \sigma_m)\sin\!\left(t\sqrt{a^2\mu_n^2 + a^2\nu_m^2 + b}\right),$$

$$\varepsilon_n = \arctan\frac{\mu_n}{l_1},\quad \sigma_m = \arctan\frac{\nu_m}{l_2},\quad E_{nm} = \left[l_1 + \frac{(k_1 k_2 + \mu_n^2)(k_1 + k_2)}{(k_1^2 + \mu_n^2)(k_2^2 + \mu_n^2)}\right]\left[l_2 + \frac{(k_3 k_4 + \nu_m^2)(k_3 + k_4)}{(k_3^2 + \nu_m^2)(k_4^2 + \nu_m^2)}\right].$$

Here, the μ_n and ν_m are positive roots of the transcendental equations

$$\mu^2 - k_1 k_2 = (k_1 + k_2)\mu \cot(l_1 \mu),$$
$$\nu^2 - k_3 k_4 = (k_3 + k_4)\nu \cot(l_2 \nu).$$

⊙ *References*: A. G. Butkovskiy (1979), B. M. Budak, A. A. Samarskii, and A. N. Tikhonov (1980).

5.3.1-6. Domain: $0 \leq x \leq l_1$, $0 \leq y \leq l_2$. **Mixed boundary value problems.**

1°. A rectangle is considered. The following conditions are prescribed:

$$\begin{aligned}
w &= f_0(x,y) &&\text{at} \quad t = 0 &&\text{(initial condition)},\\
\partial_t w &= f_1(x,y) &&\text{at} \quad t = 0 &&\text{(initial condition)},\\
w &= g_1(y,t) &&\text{at} \quad x = 0 &&\text{(boundary condition)},\\
w &= g_2(y,t) &&\text{at} \quad x = l_1 &&\text{(boundary condition)},\\
\partial_y w &= g_3(x,t) &&\text{at} \quad y = 0 &&\text{(boundary condition)},\\
\partial_y w &= g_4(x,t) &&\text{at} \quad y = l_2 &&\text{(boundary condition)}.
\end{aligned}$$

Solution:

$$\begin{aligned}
w(x,y,t) =\ & \frac{\partial}{\partial t} \int_0^{l_1}\!\!\int_0^{l_2} f_0(\xi,\eta) G(x,y,\xi,\eta,t)\,d\eta\,d\xi\\
& + \int_0^{l_1}\!\!\int_0^{l_2} f_1(\xi,\eta) G(x,y,\xi,\eta,t)\,d\eta\,d\xi\\
& + a^2 \int_0^t\!\!\int_0^{l_2} g_1(\eta,\tau)\left[\frac{\partial}{\partial \xi}G(x,y,\xi,\eta,t-\tau)\right]_{\xi=0} d\eta\,d\tau\\
& - a^2 \int_0^t\!\!\int_0^{l_2} g_2(\eta,\tau)\left[\frac{\partial}{\partial \xi}G(x,y,\xi,\eta,t-\tau)\right]_{\xi=l_1} d\eta\,d\tau\\
& - a^2 \int_0^t\!\!\int_0^{l_1} g_3(\xi,\tau) G(x,y,\xi,0,t-\tau)\,d\xi\,d\tau\\
& + a^2 \int_0^t\!\!\int_0^{l_1} g_4(\xi,\tau) G(x,y,\xi,l_2,t-\tau)\,d\xi\,d\tau\\
& + \int_0^t\!\!\int_0^{l_1}\!\!\int_0^{l_2} \Phi(\xi,\eta,\tau) G(x,y,\xi,\eta,t-\tau)\,d\eta\,d\xi\,d\tau,
\end{aligned}$$

where

$$G(x,y,\xi,\eta,t) = \frac{2}{l_1 l_2} \sum_{n=1}^{\infty} \sum_{m=0}^{\infty} \frac{A_m}{\lambda_{nm}} \sin(p_n x)\cos(q_m y)\sin(p_n \xi)\cos(q_m \eta)\sin(\lambda_{nm} t),$$

$$p_n = \frac{n\pi}{l_1},\quad q_m = \frac{m\pi}{l_2},\quad \lambda_{nm} = \sqrt{a^2 p_n^2 + a^2 q_m^2 + b},\quad A_m = \begin{cases} 1 & \text{for } m = 0,\\ 2 & \text{for } m \neq 0.\end{cases}$$

2°. A rectangle is considered. The following conditions are prescribed:

$$\begin{aligned}
w &= f_0(x,y) &&\text{at} \quad t = 0 &&\text{(initial condition)},\\
\partial_t w &= f_1(x,y) &&\text{at} \quad t = 0 &&\text{(initial condition)},\\
w &= g_1(y,t) &&\text{at} \quad x = 0 &&\text{(boundary condition)},\\
\partial_x w &= g_2(y,t) &&\text{at} \quad x = l_1 &&\text{(boundary condition)},\\
w &= g_3(x,t) &&\text{at} \quad y = 0 &&\text{(boundary condition)},\\
\partial_y w &= g_4(x,t) &&\text{at} \quad y = l_2 &&\text{(boundary condition)}.
\end{aligned}$$

Solution:

$$w(x,y,t) = \frac{\partial}{\partial t}\int_0^{l_1}\int_0^{l_2} f_0(\xi,\eta)G(x,y,\xi,\eta,t)\,d\eta\,d\xi + \int_0^{l_1}\int_0^{l_2} f_1(\xi,\eta)G(x,y,\xi,\eta,t)\,d\eta\,d\xi$$
$$+ a^2\int_0^t\int_0^{l_2} g_1(\eta,\tau)\left[\frac{\partial}{\partial \xi}G(x,y,\xi,\eta,t-\tau)\right]_{\xi=0}d\eta\,d\tau$$
$$+ a^2\int_0^t\int_0^{l_2} g_2(\eta,\tau)G(x,y,l_1,\eta,t-\tau)\,d\eta\,d\tau$$
$$+ a^2\int_0^t\int_0^{l_1} g_3(\xi,\tau)\left[\frac{\partial}{\partial \eta}G(x,y,\xi,\eta,t-\tau)\right]_{\eta=0}d\xi\,d\tau$$
$$+ a^2\int_0^t\int_0^{l_1} g_4(\xi,\tau)G(x,y,\xi,l_2,t-\tau)\,d\xi\,d\tau$$
$$+ \int_0^t\int_0^{l_1}\int_0^{l_2} \Phi(\xi,\eta,\tau)G(x,y,\xi,\eta,t-\tau)\,d\eta\,d\xi\,d\tau,$$

where

$$G(x,y,\xi,\eta,t) = \frac{4}{l_1 l_2}\sum_{n=0}^{\infty}\sum_{m=0}^{\infty}\frac{1}{\lambda_{nm}}\sin(p_n x)\sin(q_m y)\sin(p_n\xi)\sin(q_m\eta)\sin(\lambda_{nm}t),$$

$$p_n = \frac{\pi(2n+1)}{2l_1}, \qquad q_m = \frac{\pi(2m+1)}{2l_2}, \qquad \lambda_{nm} = \sqrt{a^2 p_n^2 + a^2 q_m^2 + b}.$$

5.3.2. Problems in Polar Coordinates

A nonhomogeneous Klein–Gordon equation with two space variables in the polar coordinate system has the form

$$\frac{\partial^2 w}{\partial t^2} = a^2\left(\frac{\partial^2 w}{\partial r^2} + \frac{1}{r}\frac{\partial w}{\partial r} + \frac{1}{r^2}\frac{\partial^2 w}{\partial \varphi^2}\right) - bw + \Phi(r,\varphi,t), \qquad r=\sqrt{x^2+y^2}.$$

One-dimensional solutions $w = w(r,t)$ independent of the angular coordinate φ are considered in Subsection 4.2.5.

5.3.2-1. Domain: $0 \leq r \leq R$, $0 \leq \varphi \leq 2\pi$. First boundary value problem.

A circle is considered. The following conditions are prescribed:

$$w = f_0(r,\varphi) \quad \text{at} \quad t=0 \quad \text{(initial condition)},$$
$$\partial_t w = f_1(r,\varphi) \quad \text{at} \quad t=0 \quad \text{(initial condition)},$$
$$w = g(\varphi,t) \quad \text{at} \quad r=R \quad \text{(boundary condition)}.$$

Solution:

$$w(r,\varphi,t) = \frac{\partial}{\partial t}\int_0^{2\pi}\int_0^R f_0(\xi,\eta)G(r,\varphi,\xi,\eta,t)\xi\,d\xi\,d\eta + \int_0^{2\pi}\int_0^R f_1(\xi,\eta)G(r,\varphi,\xi,\eta,t)\xi\,d\xi\,d\eta$$
$$- a^2 R\int_0^t\int_0^{2\pi} g(\eta,\tau)\left[\frac{\partial}{\partial \xi}G(r,\varphi,\xi,\eta,t-\tau)\right]_{\xi=R}d\eta\,d\tau$$
$$+ \int_0^t\int_0^{2\pi}\int_0^R \Phi(\xi,\eta,\tau)G(r,\varphi,\xi,\eta,t-\tau)\xi\,d\xi\,d\eta\,d\tau.$$

Here,*

$$G(r,\varphi,\xi,\eta,t) = \frac{1}{\pi R^2} \sum_{n=0}^{\infty} \sum_{m=1}^{\infty} \frac{A_n}{[J_n'(\mu_{nm}R)]^2} J_n(\mu_{nm}r) J_n(\mu_{nm}\xi) \cos[n(\varphi-\eta)] \frac{\sin\left(t\sqrt{a^2\mu_{nm}^2+b}\right)}{\sqrt{a^2\mu_{nm}^2+b}},$$

$$A_0 = 1, \quad A_n = 2 \quad (n = 1, 2, \ldots),$$

where the $J_n(\xi)$ are the Bessel functions (the prime denotes the derivative with respect to the argument) and the μ_{nm} are positive roots of the transcendental equation $J_n(\mu R) = 0$.

5.3.2-2. Domain: $0 \le r \le R$, $0 \le \varphi \le 2\pi$. Second boundary value problem.

A circle is considered. The following conditions are prescribed:

$$w = f_0(r,\varphi) \quad \text{at} \quad t = 0 \quad \text{(initial condition)},$$
$$\partial_t w = f_1(r,\varphi) \quad \text{at} \quad t = 0 \quad \text{(initial condition)},$$
$$\partial_r w = g(\varphi,t) \quad \text{at} \quad r = R \quad \text{(boundary condition)}.$$

Solution:

$$w(r,\varphi,t) = \frac{\partial}{\partial t} \int_0^{2\pi}\!\!\int_0^R f_0(\xi,\eta) G(r,\varphi,\xi,\eta,t)\xi\,d\xi\,d\eta + \int_0^{2\pi}\!\!\int_0^R f_1(\xi,\eta) G(r,\varphi,\xi,\eta,t)\xi\,d\xi\,d\eta$$
$$+ a^2 R \int_0^t\!\!\int_0^{2\pi} g(\eta,\tau) G(r,\varphi,R,\eta,t-\tau)\,d\eta\,d\tau$$
$$+ \int_0^t\!\!\int_0^{2\pi}\!\!\int_0^R \Phi(\xi,\eta,\tau) G(r,\varphi,\xi,\eta,t-\tau)\xi\,d\xi\,d\eta\,d\tau.$$

Here,

$$G(r,\varphi,\xi,\eta,t) = \frac{\sin(t\sqrt{b})}{\pi R^2\sqrt{b}}$$
$$+ \frac{1}{\pi}\sum_{n=0}^{\infty}\sum_{m=1}^{\infty} \frac{A_n \mu_{nm}^2 J_n(\mu_{nm}r) J_n(\mu_{nm}\xi)}{(\mu_{nm}^2 R^2 - n^2)[J_n(\mu_{nm}R)]^2} \cos[n(\varphi-\eta)] \frac{\sin\left(t\sqrt{a^2\mu_{nm}^2+b}\right)}{\sqrt{a^2\mu_{nm}^2+b}},$$

where $A_0 = 1$ and $A_n = 2$ for $n = 1, 2, \ldots$; the $J_n(\xi)$ are the Bessel functions; and the μ_m are positive roots of the transcendental equation $J_n'(\mu R) = 0$.

5.3.2-3. Domain: $0 \le r \le R$, $0 \le \varphi \le 2\pi$. Third boundary value problem.

A circle is considered. The following conditions are prescribed:

$$w = f_0(r,\varphi) \quad \text{at} \quad t = 0 \quad \text{(initial condition)},$$
$$\partial_t w = f_1(r,\varphi) \quad \text{at} \quad t = 0 \quad \text{(initial condition)},$$
$$\partial_r w + kw = g(\varphi,t) \quad \text{at} \quad r = R \quad \text{(boundary condition)}.$$

The solution $w(r,\varphi,t)$ is determined by the formula in Paragraph 5.3.2-2 where

$$G(r,\varphi,\xi,\eta,t) = \frac{1}{\pi}\sum_{n=0}^{\infty}\sum_{m=1}^{\infty} \frac{A_n \mu_{nm}^2 J_n(\mu_{nm}r) J_n(\mu_{nm}\xi)}{(\mu_{nm}^2 R^2 + k^2 R^2 - n^2)[J_n(\mu_{nm}R)]^2} \cos[n(\varphi-\eta)] \frac{\sin\left(t\sqrt{a^2\mu_{nm}^2+b}\right)}{\sqrt{a^2\mu_{nm}^2+b}},$$

$$A_0 = 1, \quad A_n = 2 \quad (n = 1, 2, \ldots).$$

Here, the $J_n(\xi)$ are the Bessel functions and the μ_m are positive roots of the transcendental equation

$$\mu J_n'(\mu R) + k J_n(\mu R) = 0.$$

---

* In the expressions of the Green's functions specified in Subsection 5.3.2, the ratios $\sin\left(t\sqrt{a^2\mu_{nm}^2+b}\right)/\sqrt{a^2\mu_{nm}^2+b}$ must be replaced by $\sinh\left(t\sqrt{|a^2\mu_{nm}^2+b|}\right)/\sqrt{|a^2\mu_{nm}^2+b|}$ if $a^2\mu_{nm}^2+b < 0$.

5.3.2-4. Domain: $R_1 \leq r \leq R_2$, $0 \leq \varphi \leq 2\pi$. First boundary value problem.

An annular domain is considered. The following conditions are prescribed:

$$w = f_0(r, \varphi) \quad \text{at} \quad t = 0 \quad \text{(initial condition)},$$
$$\partial_t w = f_1(r, \varphi) \quad \text{at} \quad t = 0 \quad \text{(initial condition)},$$
$$w = g_1(\varphi, t) \quad \text{at} \quad r = R_1 \quad \text{(boundary condition)},$$
$$w = g_2(\varphi, t) \quad \text{at} \quad r = R_2 \quad \text{(boundary condition)}.$$

Solution:

$$w(r, \varphi, t) = \frac{\partial}{\partial t} \int_0^{2\pi} \int_{R_1}^{R_2} f_0(\xi, \eta) G(r, \varphi, \xi, \eta, t) \xi \, d\xi \, d\eta + \int_0^{2\pi} \int_{R_1}^{R_2} f_1(\xi, \eta) G(r, \varphi, \xi, \eta, t) \xi \, d\xi \, d\eta$$
$$+ a^2 R_1 \int_0^t \int_0^{2\pi} g_1(\eta, \tau) \left[\frac{\partial}{\partial \xi} G(r, \varphi, \xi, \eta, t - \tau) \right]_{\xi = R_1} d\eta \, d\tau$$
$$- a^2 R_2 \int_0^t \int_0^{2\pi} g_2(\eta, \tau) \left[\frac{\partial}{\partial \xi} G(r, \varphi, \xi, \eta, t - \tau) \right]_{\xi = R_2} d\eta \, d\tau$$
$$+ \int_0^t \int_0^{2\pi} \int_{R_1}^{R_2} \Phi(\xi, \eta, \tau) G(r, \varphi, \xi, \eta, t - \tau) \xi \, d\xi \, d\eta \, d\tau.$$

Here,

$$G(r, \varphi, \xi, \eta, t) = \frac{\pi}{2} \sum_{n=0}^{\infty} \sum_{m=1}^{\infty} A_n B_{nm} Z_n(\mu_{nm} r) Z_n(\mu_{nm} \xi) \cos[n(\varphi - \eta)] \frac{\sin\left(t \sqrt{a^2 \mu_{nm}^2 + b}\right)}{\sqrt{a^2 \mu_{nm}^2 + b}},$$

$$A_n = \begin{cases} 1/2 & \text{for } n = 0, \\ 1 & \text{for } n \neq 0, \end{cases} \quad B_{nm} = \frac{\mu_{nm}^2 J_n^2(\mu_{nm} R_2)}{J_n^2(\mu_{nm} R_1) - J_n^2(\mu_{nm} R_2)},$$

$$Z_n(\mu_{nm} r) = J_n(\mu_{nm} R_1) Y_n(\mu_{nm} r) - Y_n(\mu_{nm} R_1) J_n(\mu_{nm} r),$$

where the $J_n(r)$ and $Y_n(r)$ are the Bessel functions, and the μ_{nm} are positive roots of the transcendental equation

$$J_n(\mu R_1) Y_n(\mu R_2) - Y_n(\mu R_1) J_n(\mu R_2) = 0.$$

5.3.2-5. Domain: $R_1 \leq r \leq R_2$, $0 \leq \varphi \leq 2\pi$. Second boundary value problem.

An annular domain is considered. The following conditions are prescribed:

$$w = f_0(r, \varphi) \quad \text{at} \quad t = 0 \quad \text{(initial condition)},$$
$$\partial_t w = f_1(r, \varphi) \quad \text{at} \quad t = 0 \quad \text{(initial condition)},$$
$$\partial_r w = g_1(\varphi, t) \quad \text{at} \quad r = R_1 \quad \text{(boundary condition)},$$
$$\partial_r w = g_2(\varphi, t) \quad \text{at} \quad r = R_2 \quad \text{(boundary condition)}.$$

Solution:

$$w(r, \varphi, t) = \frac{\partial}{\partial t} \int_0^{2\pi} \int_{R_1}^{R_2} f_0(\xi, \eta) G(r, \varphi, \xi, \eta, t) \xi \, d\xi \, d\eta + \int_0^{2\pi} \int_{R_1}^{R_2} f_1(\xi, \eta) G(r, \varphi, \xi, \eta, t) \xi \, d\xi \, d\eta$$
$$- a^2 R_1 \int_0^t \int_0^{2\pi} g_1(\eta, \tau) G(r, \varphi, R_1, \eta, t - \tau) \, d\eta \, d\tau$$
$$+ a^2 R_2 \int_0^t \int_0^{2\pi} g_2(\eta, \tau) G(r, \varphi, R_2, \eta, t - \tau) \, d\eta \, d\tau$$
$$+ \int_0^t \int_0^{2\pi} \int_{R_1}^{R_2} \Phi(\xi, \eta, \tau) G(r, \varphi, \xi, \eta, t - \tau) \xi \, d\xi \, d\eta \, d\tau.$$

Here,

$$G(r,\varphi,\xi,\eta,t) = \frac{\sin(t\sqrt{b})}{\pi(R_2^2-R_1^2)\sqrt{b}}$$
$$+ \frac{1}{\pi}\sum_{n=0}^{\infty}\sum_{m=1}^{\infty}\frac{A_n\mu_{nm}^2 Z_n(\mu_{nm}r)Z_n(\mu_{nm}\xi)\cos[n(\varphi-\eta)]\sin(t\sqrt{a^2\mu_{nm}^2+b})}{\left[(\mu_{nm}^2 R_2^2 - n^2)Z_n^2(\mu_{nm}R_2) - (\mu_{nm}^2 R_1^2 - n^2)Z_n^2(\mu_{nm}R_1)\right]\sqrt{a^2\mu_{nm}^2+b}},$$

where

$$Z_n(\mu_{nm}r) = J_n'(\mu_{nm}R_1)Y_n(\mu_{nm}r) - Y_n'(\mu_{nm}R_1)J_n(\mu_{nm}r), \qquad A_n = \begin{cases} 1 & \text{for } n=0, \\ 2 & \text{for } n>0, \end{cases}$$

the $J_n(r)$ and $Y_n(r)$ are the Bessel functions, and the μ_{nm} are positive roots of the transcendental equation

$$J_n'(\mu R_1)Y_n'(\mu R_2) - Y_n'(\mu R_1)J_n'(\mu R_2) = 0.$$

5.3.2-6. Domain: $R_1 \leq r \leq R_2$, $0 \leq \varphi \leq 2\pi$. Third boundary value problem.

An annular domain is considered. The following conditions are prescribed:

$$\begin{aligned}
w &= f_0(r,\varphi) & \text{at} \quad & t = 0 & &\text{(initial condition)}, \\
\partial_t w &= f_1(r,\varphi) & \text{at} \quad & t = 0 & &\text{(initial condition)}, \\
\partial_r w - k_1 w &= g_1(\varphi,t) & \text{at} \quad & r = R_1 & &\text{(boundary condition)}, \\
\partial_r w + k_2 w &= g_2(\varphi,t) & \text{at} \quad & r = R_2 & &\text{(boundary condition)}.
\end{aligned}$$

The solution $w(r,\varphi,t)$ is determined by the formula in Paragraph 5.3.2-5 where

$$G(r,\varphi,\xi,\eta,t) = \frac{1}{\pi}\sum_{n=0}^{\infty}\sum_{m=1}^{\infty}\frac{A_n\mu_{nm}^2 Z_{nm}(r)Z_{nm}(\xi)\cos[n(\varphi-\eta)]\sin(\lambda_{nm}t)}{\lambda_{nm}\left[(k_2^2 R_2^2 + \mu_{nm}^2 R_2^2 - n^2)Z_{nm}^2(R_2) - (k_1^2 R_1^2 + \mu_{nm}^2 R_1^2 - n^2)Z_{nm}^2(R_1)\right]},$$

$$Z_{nm}(r) = \left[\mu_{nm}J_n'(\mu_{nm}R_1) - k_1 J_n(\mu_{nm}R_1)\right]Y_n(\mu_{nm}r)$$
$$- \left[\mu_{nm}Y_n'(\mu_{nm}R_1) - k_1 Y_n(\mu_{nm}R_1)\right]J_n(\mu_{nm}r).$$

Here, $A_0 = 1$ and $A_n = 2$ for $n = 1, 2, \ldots$; $\lambda_{nm} = \sqrt{a^2\mu_{nm}^2 + b}$; the $J_n(r)$ and $Y_n(r)$ are the Bessel functions; and the μ_{nm} are positive roots of the transcendental equation

$$\left[\mu J_n'(\mu R_1) - k_1 J_n(\mu R_1)\right]\left[\mu Y_n'(\mu R_2) + k_2 Y_n(\mu R_2)\right]$$
$$= \left[\mu Y_n'(\mu R_1) - k_1 Y_n(\mu R_1)\right]\left[\mu J_n'(\mu R_2) + k_2 J_n(\mu R_2)\right].$$

5.3.2-7. Domain: $0 \leq r \leq R$, $0 \leq \varphi \leq \varphi_0$. First boundary value problem.

A circular sector is considered. The following conditions are prescribed:

$$\begin{aligned}
w &= f_0(r,\varphi) & \text{at} \quad & t = 0 & &\text{(initial condition)}, \\
\partial_t w &= f_1(r,\varphi) & \text{at} \quad & t = 0 & &\text{(initial condition)}, \\
w &= g_1(\varphi,t) & \text{at} \quad & r = R & &\text{(boundary condition)}, \\
w &= g_2(r,t) & \text{at} \quad & \varphi = 0 & &\text{(boundary condition)}, \\
w &= g_3(r,t) & \text{at} \quad & \varphi = \varphi_0 & &\text{(boundary condition)}.
\end{aligned}$$

Solution:
$$w(r,\varphi,t) = \frac{\partial}{\partial t}\int_0^{\varphi_0}\!\!\int_0^R f_0(\xi,\eta)G(r,\varphi,\xi,\eta,t)\xi\,d\xi\,d\eta + \int_0^{\varphi_0}\!\!\int_0^R f_1(\xi,\eta)G(r,\varphi,\xi,\eta,t)\xi\,d\xi\,d\eta$$
$$- a^2 R \int_0^t\!\!\int_0^{\varphi_0} g_1(\eta,\tau)\left[\frac{\partial}{\partial \xi}G(r,\varphi,\xi,\eta,t-\tau)\right]_{\xi=R} d\eta\,d\tau$$
$$+ a^2 \int_0^t\!\!\int_0^R g_2(\xi,\tau)\frac{1}{\xi}\left[\frac{\partial}{\partial \eta}G(r,\varphi,\xi,\eta,t-\tau)\right]_{\eta=0} d\xi\,d\tau$$
$$- a^2 \int_0^t\!\!\int_0^R g_3(\xi,\tau)\frac{1}{\xi}\left[\frac{\partial}{\partial \eta}G(r,\varphi,\xi,\eta,t-\tau)\right]_{\eta=\varphi_0} d\xi\,d\tau$$
$$+ \int_0^t\!\!\int_0^{\varphi_0}\!\!\int_0^R \Phi(\xi,\eta,\tau)G(r,\varphi,\xi,\eta,t-\tau)\xi\,d\xi\,d\eta\,d\tau.$$

Here,
$$G(r,\varphi,\xi,\eta,t) = \frac{4}{R^2\varphi_0}\sum_{n=1}^{\infty}\sum_{m=1}^{\infty}\frac{J_{n\pi/\varphi_0}(\mu_{nm}r)J_{n\pi/\varphi_0}(\mu_{nm}\xi)}{[J'_{n\pi/\varphi_0}(\mu_{nm}R)]^2}\sin\left(\frac{n\pi\varphi}{\varphi_0}\right)\sin\left(\frac{n\pi\eta}{\varphi_0}\right)\frac{\sin(\lambda_{nm}t)}{\lambda_{nm}},$$

where the $J_{n\pi/\varphi_0}(r)$ are the Bessel functions and the μ_{nm} are positive roots of the transcendental equation $J_{n\pi/\varphi_0}(\mu R) = 0$, and $\lambda_{nm} = \sqrt{a^2\mu_{nm}^2 + b}$.

5.3.2-8. Domain: $0 \leq r \leq R$, $0 \leq \varphi \leq \varphi_0$. Second boundary value problem.

A circular sector is considered. The following conditions are prescribed:

$$w = f_0(r,\varphi) \quad \text{at} \quad t = 0 \quad \text{(initial condition),}$$
$$\partial_t w = f_1(r,\varphi) \quad \text{at} \quad t = 0 \quad \text{(initial condition),}$$
$$\partial_r w = g_1(\varphi,t) \quad \text{at} \quad r = R \quad \text{(boundary condition),}$$
$$r^{-1}\partial_\varphi w = g_2(r,t) \quad \text{at} \quad \varphi = 0 \quad \text{(boundary condition),}$$
$$r^{-1}\partial_\varphi w = g_3(r,t) \quad \text{at} \quad \varphi = \varphi_0 \quad \text{(boundary condition).}$$

Solution:
$$w(r,\varphi,t) = \frac{\partial}{\partial t}\int_0^{\varphi_0}\!\!\int_0^R f_0(\xi,\eta)G(r,\varphi,\xi,\eta,t)\xi\,d\xi\,d\eta + \int_0^{\varphi_0}\!\!\int_0^R f_1(\xi,\eta)G(r,\varphi,\xi,\eta,t)\xi\,d\xi\,d\eta$$
$$+ a^2 R \int_0^t\!\!\int_0^{\varphi_0} g_1(\eta,\tau)G(r,\varphi,R,\eta,t-\tau)\,d\eta\,d\tau$$
$$- a^2 \int_0^t\!\!\int_0^R g_2(\xi,\tau)G(r,\varphi,\xi,0,t-\tau)\,d\xi\,d\tau$$
$$+ a^2 \int_0^t\!\!\int_0^R g_3(\xi,\tau)G(r,\varphi,\xi,\varphi_0,t-\tau)\,d\xi\,d\tau$$
$$+ \int_0^t\!\!\int_0^{\varphi_0}\!\!\int_0^R \Phi(\xi,\eta,\tau)G(r,\varphi,\xi,\eta,t-\tau)\xi\,d\xi\,d\eta\,d\tau.$$

Here,
$$G(r,\varphi,\xi,\eta,t) = \frac{2\sin(t\sqrt{b})}{R^2\varphi_0\sqrt{b}} + 4\varphi_0\sum_{n=0}^{\infty}\sum_{m=1}^{\infty}\frac{\mu_{nm}^2 J_{n\pi/\varphi_0}(\mu_{nm}r)J_{n\pi/\varphi_0}(\mu_{nm}\xi)}{(R^2\varphi_0^2\mu_{nm}^2 - n^2\pi^2)[J_{n\pi/\varphi_0}(\mu_{nm}R)]^2}$$
$$\times \cos\left(\frac{n\pi\varphi}{\varphi_0}\right)\cos\left(\frac{n\pi\eta}{\varphi_0}\right)\frac{\sin(t\sqrt{a^2\mu_{nm}^2 + b})}{\sqrt{a^2\mu_{nm}^2 + b}},$$

where the $J_{n\pi/\varphi_0}(r)$ are the Bessel functions and the μ_{nm} are positive roots of the transcendental equation $J'_{n\pi/\varphi_0}(\mu R) = 0$.

5.3. EQUATIONS OF THE FORM $\frac{\partial^2 w}{\partial t^2} = a^2 \Delta_2 w - bw + \Phi(x,y,t)$ 371

5.3.2-9. Domain: $0 \leq r \leq R$, $0 \leq \varphi \leq \varphi_0$. Mixed boundary value problem.

A circular sector is considered. The following conditions are prescribed:

$$w = f_0(r,\varphi) \quad \text{at} \quad t = 0 \quad \text{(initial condition)},$$
$$\partial_t w = f_1(r,\varphi) \quad \text{at} \quad t = 0 \quad \text{(initial condition)},$$
$$\partial_r w + kw = g(\varphi,t) \quad \text{at} \quad r = R \quad \text{(boundary condition)},$$
$$\partial_\varphi w = 0 \quad \text{at} \quad \varphi = 0 \quad \text{(boundary condition)},$$
$$\partial_\varphi w = 0 \quad \text{at} \quad \varphi = \varphi_0 \quad \text{(boundary condition)}.$$

Solution:

$$w(r,\varphi,t) = \frac{\partial}{\partial t} \int_0^{\varphi_0} \int_0^R f_0(\xi,\eta) G(r,\varphi,\xi,\eta,t) \xi \, d\xi \, d\eta$$
$$+ \int_0^{\varphi_0} \int_0^R f_1(\xi,\eta) G(r,\varphi,\xi,\eta,t) \xi \, d\xi \, d\eta$$
$$+ a^2 R \int_0^t \int_0^{\varphi_0} g(\eta,\tau) G(r,\varphi,R,\eta,t-\tau) \, d\eta \, d\tau$$
$$+ \int_0^t \int_0^{\varphi_0} \int_0^R \Phi(\xi,\eta,\tau) G(r,\varphi,\xi,\eta,t-\tau) \xi \, d\xi \, d\eta \, d\tau.$$

Here,

$$G(r,\varphi,\xi,\eta,t) = \sum_{n=0}^{\infty} \sum_{m=1}^{\infty} A_{nm} J_{s_n}(\mu_{nm} r) J_{s_n}(\mu_{nm} \xi) \cos(s_n \varphi) \cos(s_n \eta) \sin\left(t\sqrt{a^2 \mu_{nm}^2 + b}\right),$$

$$s_n = \frac{n\pi}{\varphi_0}, \quad A_{nm} = \frac{4\mu_{nm}^2}{\varphi_0(\mu_{nm}^2 R^2 + k^2 R^2 - s_n^2)\left[J_{s_n}(\mu_{nm} R)\right]^2 \sqrt{a^2 \mu_{nm}^2 + b}},$$

where the $J_{s_n}(r)$ are the Bessel functions and the μ_{nm} are positive roots of the transcendental equation

$$\mu J'_{s_n}(\mu R) + k J_{s_n}(\mu R) = 0.$$

5.3.3. Axisymmetric Problems

In the axisymmetric case, a nonhomogeneous Klein–Gordon equation in the cylindrical system of coordinates has the form

$$\frac{\partial^2 w}{\partial t^2} = a^2 \left(\frac{\partial^2 w}{\partial r^2} + \frac{1}{r} \frac{\partial w}{\partial r} + \frac{\partial^2 w}{\partial z^2} \right) - bw + \Phi(r,z,t), \quad r = \sqrt{x^2 + y^2}.$$

In the solutions of the problems considered below, the modified Green's function $\mathcal{G}(r,z,\xi,\eta,t) = 2\pi \xi G(r,z,\xi,\eta,t)$ is used for convenience.

5.3.3-1. Domain: $0 \leq r \leq R$, $0 \leq z \leq l$. First boundary value problem.

A circular cylinder of finite length is considered. The following conditions are prescribed:

$$w = f_0(r,z) \quad \text{at} \quad t = 0 \quad \text{(initial condition)},$$
$$\partial_t w = f_1(r,z) \quad \text{at} \quad t = 0 \quad \text{(initial condition)},$$
$$w = g_1(z,t) \quad \text{at} \quad r = R \quad \text{(boundary condition)},$$
$$w = g_2(r,t) \quad \text{at} \quad z = 0 \quad \text{(boundary condition)},$$
$$w = g_3(r,t) \quad \text{at} \quad z = l \quad \text{(boundary condition)}.$$

Solution:

$$w(r,z,t) = \frac{\partial}{\partial t}\int_0^l\int_0^R f_0(\xi,\eta)\mathcal{G}(r,z,\xi,\eta,t)\,d\xi\,d\eta$$
$$+ \int_0^l\int_0^R f_1(\xi,\eta)\mathcal{G}(r,z,\xi,\eta,t)\,d\xi\,d\eta$$
$$- a^2\int_0^t\int_0^l g_1(\eta,\tau)\left[\frac{\partial}{\partial\xi}\mathcal{G}(r,z,\xi,\eta,t-\tau)\right]_{\xi=R}d\eta\,d\tau$$
$$+ a^2\int_0^t\int_0^R g_2(\xi,\tau)\left[\frac{\partial}{\partial\eta}\mathcal{G}(r,z,\xi,\eta,t-\tau)\right]_{\eta=0}d\xi\,d\tau$$
$$- a^2\int_0^t\int_0^R g_3(\xi,\tau)\left[\frac{\partial}{\partial\eta}\mathcal{G}(r,z,\xi,\eta,t-\tau)\right]_{\eta=l}d\xi\,d\tau$$
$$+ \int_0^t\int_0^l\int_0^R \Phi(\xi,\eta,\tau)\mathcal{G}(r,z,\xi,\eta,t-\tau)\,d\xi\,d\eta\,d\tau.$$

Here,

$$\mathcal{G}(r,z,\xi,\eta,t) = \frac{4\xi}{R^2 l}\sum_{n=1}^{\infty}\sum_{m=1}^{\infty}\frac{1}{J_1^2(\mu_n)}J_0\!\left(\frac{\mu_n r}{R}\right)J_0\!\left(\frac{\mu_n \xi}{R}\right)\sin\!\left(\frac{m\pi z}{l}\right)\sin\!\left(\frac{m\pi \eta}{l}\right)\frac{\sin(t\sqrt{\lambda_{nm}})}{\sqrt{\lambda_{nm}}},$$

$$\lambda_{nm} = \frac{a^2\mu_n^2}{R^2} + \frac{a^2\pi^2 m^2}{l^2} + b,$$

where the μ_n are positive zeros of the Bessel function, $J_0(\mu) = 0$.

5.3.3-2. Domain: $0 \leq r \leq R$, $0 \leq z \leq l$. Second boundary value problem.

A circular cylinder of finite length is considered. The following conditions are prescribed:

$$\begin{aligned}
w &= f_0(r,z) \quad &\text{at}\quad& t=0 \quad &\text{(initial condition)},\\
\partial_t w &= f_1(r,z) \quad &\text{at}\quad& t=0 \quad &\text{(initial condition)},\\
\partial_r w &= g_1(z,t) \quad &\text{at}\quad& r=R \quad &\text{(boundary condition)},\\
\partial_z w &= g_2(r,t) \quad &\text{at}\quad& z=0 \quad &\text{(boundary condition)},\\
\partial_z w &= g_3(r,t) \quad &\text{at}\quad& z=l \quad &\text{(boundary condition)}.
\end{aligned}$$

Solution:

$$w(r,z,t) = \frac{\partial}{\partial t}\int_0^l\int_0^R f_0(\xi,\eta)\mathcal{G}(r,z,\xi,\eta,t)\,d\xi\,d\eta$$
$$+ \int_0^l\int_0^R f_1(\xi,\eta)\mathcal{G}(r,z,\xi,\eta,t)\,d\xi\,d\eta$$
$$+ a^2\int_0^t\int_0^l g_1(\eta,\tau)\mathcal{G}(r,z,R,\eta,t-\tau)\,d\eta\,d\tau$$
$$- a^2\int_0^t\int_0^R g_2(\xi,\tau)\mathcal{G}(r,z,\xi,0,t-\tau)\,d\xi\,d\tau$$
$$+ a^2\int_0^t\int_0^R g_3(\xi,\tau)\mathcal{G}(r,z,\xi,l,t-\tau)\,d\xi(\,d\tau$$
$$+ \int_0^t\int_0^l\int_0^R \Phi(\xi,\eta,\tau)\mathcal{G}(r,z,\xi,\eta,t-\tau)\,d\xi\,d\eta\,d\tau.$$

Here,

$$\mathcal{G}(r,z,\xi,\eta,t) = \frac{2\xi \sin(t\sqrt{b})}{R^2 l \sqrt{b}}$$

$$+ \frac{2\xi}{R^2 l} \sum_{n=0}^{\infty} \sum_{m=0}^{\infty} \frac{A_{nm}}{J_0^2(\mu_n)} J_0\left(\frac{\mu_n r}{R}\right) J_0\left(\frac{\mu_n \xi}{R}\right) \cos\left(\frac{m\pi z}{l}\right) \cos\left(\frac{m\pi \eta}{l}\right) \frac{\sin(t\sqrt{\lambda_{nm}})}{\sqrt{\lambda_{nm}}},$$

$$\lambda_{nm} = \frac{a^2 \mu_n^2}{R^2} + \frac{a^2 \pi^2 m^2}{l^2} + b, \quad A_{nm} = \begin{cases} 0 & \text{for } m=0, n=0, \\ 1 & \text{for } m=0, n>0, \\ 2 & \text{for } m>0, \end{cases}$$

where the μ_n are zeros of the first-order Bessel function, $J_1(\mu) = 0$ ($\mu_0 = 0$).

5.3.3-3. Domain: $0 \le r \le R$, $0 \le z \le l$. Third boundary value problem.

A circular cylinder of finite length is considered. The following conditions are prescribed:

$$\begin{aligned}
w &= f_0(r,z) & \text{at} \quad t &= 0 & \text{(initial condition)}, \\
\partial_t w &= f_1(r,z) & \text{at} \quad t &= 0 & \text{(initial condition)}, \\
\partial_r w + k_1 w &= g_1(z,t) & \text{at} \quad r &= R & \text{(boundary condition)}, \\
\partial_z w - k_2 w &= g_2(r,t) & \text{at} \quad z &= 0 & \text{(boundary condition)}, \\
\partial_z w + k_3 w &= g_3(r,t) & \text{at} \quad z &= l & \text{(boundary condition)}.
\end{aligned}$$

The solution $w(r,z,t)$ is determined by the formula in Paragraph 5.3.3-2 where

$$\mathcal{G}(r,z,\xi,\eta,t) = \frac{2\xi}{R^2} \sum_{n=1}^{\infty} \sum_{m=1}^{\infty} \frac{\mu_n^2}{(k_1^2 R^2 + \mu_n^2) J_0^2(\mu_n)} J_0\left(\frac{\mu_n r}{R}\right) J_0\left(\frac{\mu_n \xi}{R}\right) \frac{\varphi_m(z)\varphi_m(\eta)}{\|\varphi_m\|^2} \frac{\sin(t\sqrt{\lambda_{nm}})}{\sqrt{\lambda_{nm}}},$$

$$\lambda_{nm} = \frac{a^2 \mu_n^2}{R^2} + a^2 \beta_m^2 + b, \quad \varphi_m(z) = \cos(\beta_m z) + \frac{k_2}{\beta_m} \sin(\beta_m z),$$

$$\|\varphi_m\|^2 = \frac{k_3}{2\beta_m^2} \frac{\beta_m^2 + k_2^2}{\beta_m^2 + k_3^2} + \frac{k_2}{2\beta_m^2} + \frac{l}{2}\left(1 + \frac{k_2^2}{\beta_m^2}\right).$$

Here, the μ_n and β_m are positive roots of the transcendental equations

$$\mu J_1(\mu) - k_1 R J_0(\mu) = 0, \qquad \frac{\tan(\beta l)}{\beta} = \frac{k_2 + k_3}{\beta^2 - k_2 k_3}.$$

5.3.3-4. Domain: $0 \le r \le R$, $0 \le z \le l$. Mixed boundary value problems.

1°. A circular cylinder of finite length is considered. The following conditions are prescribed:

$$\begin{aligned}
w &= f_0(r,z) & \text{at} \quad t &= 0 & \text{(initial condition)}, \\
\partial_t w &= f_1(r,z) & \text{at} \quad t &= 0 & \text{(initial condition)}, \\
w &= g_1(z,t) & \text{at} \quad r &= R & \text{(boundary condition)}, \\
\partial_z w &= g_2(r,t) & \text{at} \quad z &= 0 & \text{(boundary condition)}, \\
\partial_z w &= g_3(r,t) & \text{at} \quad z &= l & \text{(boundary condition)}.
\end{aligned}$$

Solution:

$$w(r,z,t) = \frac{\partial}{\partial t}\int_0^l\int_0^R f_0(\xi,\eta)\mathcal{G}(r,z,\xi,\eta,t)\,d\xi\,d\eta$$

$$+ \int_0^l\int_0^R f_1(\xi,\eta)\mathcal{G}(r,z,\xi,\eta,t)\,d\xi\,d\eta$$

$$- a^2\int_0^t\int_0^l g_1(\eta,\tau)\left[\frac{\partial}{\partial\xi}\mathcal{G}(r,z,\xi,\eta,t-\tau)\right]_{\xi=R}d\eta\,d\tau$$

$$- a^2\int_0^t\int_0^R g_2(\xi,\tau)\mathcal{G}(r,z,\xi,0,t-\tau)\,d\xi\,d\tau$$

$$+ a^2\int_0^t\int_0^R g_3(\xi,\tau)\mathcal{G}(r,z,\xi,l,t-\tau)\,d\xi\,d\tau$$

$$+ \int_0^t\int_0^l\int_0^R \Phi(\xi,\eta,\tau)\mathcal{G}(r,z,\xi,\eta,t-\tau)\,d\xi\,d\eta\,d\tau.$$

Here,

$$\mathcal{G}(r,z,\xi,\eta,t) = \frac{2\xi}{R^2 l}\sum_{n=1}^{\infty}\sum_{m=0}^{\infty}\frac{A_m}{J_1^2(\mu_n)}J_0\!\left(\frac{\mu_n r}{R}\right)J_0\!\left(\frac{\mu_n \xi}{R}\right)\cos\!\left(\frac{m\pi z}{l}\right)\cos\!\left(\frac{m\pi\eta}{l}\right)\frac{\sin\!\left(t\sqrt{\lambda_{nm}}\right)}{\sqrt{\lambda_{nm}}},$$

$$\lambda_{nm} = \frac{a^2\mu_n^2}{R^2} + \frac{a^2\pi^2 m^2}{l^2} + b, \qquad A_m = \begin{cases} 1 & \text{for } m = 0, \\ 2 & \text{for } m > 0, \end{cases}$$

where the μ_n are positive zeros of the Bessel function, $J_0(\mu) = 0$.

2°. A circular cylinder of finite length is considered. The following conditions are prescribed:

$$\begin{aligned}
w &= f_0(r,z) & \text{at} \quad t &= 0 & \text{(initial condition)},\\
\partial_t w &= f_1(r,z) & \text{at} \quad t &= 0 & \text{(initial condition)},\\
\partial_r w &= g_1(z,t) & \text{at} \quad r &= R & \text{(boundary condition)},\\
w &= g_2(r,t) & \text{at} \quad z &= 0 & \text{(boundary condition)},\\
w &= g_3(r,t) & \text{at} \quad z &= l & \text{(boundary condition)}.
\end{aligned}$$

Solution:

$$w(r,z,t) = \frac{\partial}{\partial t}\int_0^l\int_0^R f_0(\xi,\eta)\mathcal{G}(r,z,\xi,\eta,t)\,d\xi\,d\eta + \int_0^l\int_0^R f_1(\xi,\eta)\mathcal{G}(r,z,\xi,\eta,t)\,d\xi\,d\eta$$

$$+ a^2\int_0^t\int_0^l g_1(\eta,\tau)\mathcal{G}(r,z,R,\eta,t-\tau)\,d\eta\,d\tau$$

$$+ a^2\int_0^t\int_0^R g_2(\xi,\tau)\left[\frac{\partial}{\partial\eta}\mathcal{G}(r,z,\xi,\eta,t-\tau)\right]_{\eta=0}d\xi\,d\tau$$

$$- a^2\int_0^t\int_0^R g_3(\xi,\tau)\left[\frac{\partial}{\partial\eta}\mathcal{G}(r,z,\xi,\eta,t-\tau)\right]_{\eta=l}d\xi\,d\tau$$

$$+ \int_0^t\int_0^l\int_0^R \Phi(\xi,\eta,\tau)\mathcal{G}(r,z,\xi,\eta,t-\tau)\,d\xi\,d\eta\,d\tau.$$

Here,

$$\mathcal{G}(r,z,\xi,\eta,t) = \frac{4\xi}{R^2 l}\sum_{n=0}^{\infty}\sum_{m=1}^{\infty}\frac{1}{J_0^2(\mu_n)}J_0\!\left(\frac{\mu_n r}{R}\right)J_0\!\left(\frac{\mu_n \xi}{R}\right)\sin\!\left(\frac{m\pi z}{l}\right)\sin\!\left(\frac{m\pi\eta}{l}\right)\frac{\sin\!\left(t\sqrt{\lambda_{nm}}\right)}{\sqrt{\lambda_{nm}}},$$

$$\lambda_{nm} = \frac{a^2\mu_n^2}{R^2} + \frac{a^2\pi^2 m^2}{l^2} + b,$$

where the μ_n are zeros of the first-order Bessel function, $J_1(\mu) = 0$ ($\mu_0 = 0$).

5.3.3-5. Domain: $R_1 \leq r \leq R_2$, $0 \leq z \leq l$. First boundary value problem.

A hollow circular cylinder of finite length is considered. The following conditions are prescribed:

$$w = f_0(r, z) \quad \text{at} \quad t = 0 \quad \text{(initial condition)},$$
$$\partial_t w = f_1(r, z) \quad \text{at} \quad t = 0 \quad \text{(initial condition)},$$
$$w = g_1(z, t) \quad \text{at} \quad r = R_1 \quad \text{(boundary condition)},$$
$$w = g_2(z, t) \quad \text{at} \quad r = R_2 \quad \text{(boundary condition)},$$
$$w = g_3(r, t) \quad \text{at} \quad z = 0 \quad \text{(boundary condition)},$$
$$w = g_4(r, t) \quad \text{at} \quad z = l \quad \text{(boundary condition)}.$$

Solution:

$$w(r,z,t) = \frac{\partial}{\partial t} \int_0^l \int_{R_1}^{R_2} f_0(\xi, \eta) \mathcal{G}(r, z, \xi, \eta, t)\, d\xi\, d\eta$$
$$+ \int_0^l \int_{R_1}^{R_2} f_1(\xi, \eta) \mathcal{G}(r, z, \xi, \eta, t)\, d\xi\, d\eta$$
$$+ a^2 \int_0^t \int_0^l g_1(\eta, \tau) \left[\frac{\partial}{\partial \xi} \mathcal{G}(r, z, \xi, \eta, t - \tau) \right]_{\xi = R_1} d\eta\, d\tau$$
$$- a^2 \int_0^t \int_0^l g_2(\eta, \tau) \left[\frac{\partial}{\partial \xi} \mathcal{G}(r, z, \xi, \eta, t - \tau) \right]_{\xi = R_2} d\eta\, d\tau$$
$$+ a^2 \int_0^t \int_{R_1}^{R_2} g_3(\xi, \tau) \left[\frac{\partial}{\partial \eta} \mathcal{G}(r, z, \xi, \eta, t - \tau) \right]_{\eta = 0} d\xi\, d\tau$$
$$- a^2 \int_0^t \int_{R_1}^{R_2} g_4(\xi, \tau) \left[\frac{\partial}{\partial \eta} \mathcal{G}(r, z, \xi, \eta, t - \tau) \right]_{\eta = l} d\xi\, d\tau$$
$$+ \int_0^t \int_0^l \int_{R_1}^{R_2} \Phi(\xi, \eta, \tau) \mathcal{G}(r, z, \xi, \eta, t - \tau)\, d\xi\, d\eta\, d\tau.$$

Here,

$$\mathcal{G}(r, z, \xi, \eta, t) = \frac{\pi^2 \xi}{R_1^2 l} \sum_{n=1}^\infty \sum_{m=1}^\infty \frac{\mu_n^2 J_0^2(s\mu_n)}{J_0^2(\mu_n) - J_0^2(s\mu_n)} \Psi_n(r) \Psi_n(\xi) \sin\left(\frac{m\pi z}{l}\right) \sin\left(\frac{m\pi \eta}{l}\right) \frac{\sin\left(t\sqrt{\lambda_{nm}}\right)}{\sqrt{\lambda_{nm}}},$$

$$\Psi_n(r) = Y_0(\mu_n) J_0\left(\frac{\mu_n r}{R_1}\right) - J_0(\mu_n) Y_0\left(\frac{\mu_n r}{R_1}\right), \quad s = \frac{R_2}{R_1}, \quad \lambda_{nm} = \frac{a^2 \mu_n^2}{R_1^2} + \frac{a^2 \pi^2 m^2}{l^2} + b,$$

where $J_0(\mu)$ and $Y_0(\mu)$ are the Bessel functions, and the μ_n are positive roots of the transcendental equation

$$J_0(\mu) Y_0(s\mu) - J_0(s\mu) Y_0(\mu) = 0.$$

5.3.3-6. Domain: $R_1 \leq r \leq R_2$, $0 \leq z \leq l$. Second boundary value problem.

A hollow circular cylinder of finite length is considered. The following conditions are prescribed:

$$w = f_0(r, z) \quad \text{at} \quad t = 0 \quad \text{(initial condition)},$$
$$\partial_t w = f_1(r, z) \quad \text{at} \quad t = 0 \quad \text{(initial condition)},$$
$$\partial_r w = g_1(z, t) \quad \text{at} \quad r = R_1 \quad \text{(boundary condition)},$$
$$\partial_r w = g_2(z, t) \quad \text{at} \quad r = R_2 \quad \text{(boundary condition)},$$
$$\partial_z w = g_3(r, t) \quad \text{at} \quad z = 0 \quad \text{(boundary condition)},$$
$$\partial_z w = g_4(r, t) \quad \text{at} \quad z = l \quad \text{(boundary condition)}.$$

Solution:

$$w(r,z,t) = \frac{\partial}{\partial t}\int_0^l \int_{R_1}^{R_2} f_0(\xi,\eta)\mathcal{G}(r,z,\xi,\eta,t)\,d\xi\,d\eta$$

$$+ \int_0^l \int_{R_1}^{R_2} f_1(\xi,\eta)\mathcal{G}(r,z,\xi,\eta,t)\,d\xi\,d\eta$$

$$- a^2 \int_0^t \int_0^l g_1(\eta,\tau)\mathcal{G}(r,z,R_1,\eta,t-\tau)\,d\eta\,d\tau$$

$$+ a^2 \int_0^t \int_0^l g_2(\eta,\tau)\mathcal{G}(r,z,R_2,\eta,t-\tau)\,d\eta\,d\tau$$

$$- a^2 \int_0^t \int_{R_1}^{R_2} g_3(\xi,\tau)\mathcal{G}(r,z,\xi,0,t-\tau)\,d\xi\,d\tau$$

$$+ a^2 \int_0^t \int_{R_1}^{R_2} g_4(\xi,\tau)\mathcal{G}(r,z,\xi,l,t-\tau)\,d\xi\,d\tau$$

$$+ \int_0^t \int_0^l \int_{R_1}^{R_2} \Phi(\xi,\eta,\tau)\mathcal{G}(r,z,\xi,\eta,t-\tau)\,d\xi\,d\eta\,d\tau.$$

Here,

$$\mathcal{G}(r,z,\xi,\eta,t) = \frac{2\xi \sin(t\sqrt{b})}{(R_2^2-R_1^2)l\sqrt{b}} + \frac{4\xi}{(R_2^2-R_1^2)l}\sum_{m=1}^{\infty}\cos\left(\frac{m\pi z}{l}\right)\cos\left(\frac{m\pi\eta}{l}\right)\frac{\sin(t\sqrt{\beta_m})}{\sqrt{\beta_m}}$$

$$+ \frac{\pi^2 \xi}{2R_1^2 l}\sum_{n=1}^{\infty}\sum_{m=0}^{\infty}\frac{A_m \mu_n^2 J_1^2(s\mu_n)}{J_1^2(\mu_n)-J_1^2(s\mu_n)}\Psi_n(r)\Psi_n(\xi)\cos\left(\frac{m\pi z}{l}\right)\cos\left(\frac{m\pi\eta}{l}\right)\frac{\sin(t\sqrt{\lambda_{nm}})}{\sqrt{\lambda_{nm}}},$$

where

$$\Psi_n(r) = Y_1(\mu_n)J_0\left(\frac{\mu_n r}{R_1}\right) - J_1(\mu_n)Y_0\left(\frac{\mu_n r}{R_1}\right), \quad s = \frac{R_2}{R_1},$$

$$A_m = \begin{cases} 1 & \text{for } m=0, \\ 2 & \text{for } m>1, \end{cases} \quad \beta_m = \frac{a^2\pi^2 m^2}{l^2} + b, \quad \lambda_{nm} = \frac{a^2\mu_n^2}{R_1^2} + \frac{a^2\pi^2 m^2}{l^2} + b;$$

$J_k(\mu)$ and $Y_k(\mu)$ are the Bessel functions ($k=0, 1$); and the μ_n are positive roots of the transcendental equation

$$J_1(\mu)Y_1(s\mu) - J_1(s\mu)Y_1(\mu) = 0.$$

5.4. Telegraph Equation

$$\frac{\partial^2 w}{\partial t^2} + k\frac{\partial w}{\partial t} = a^2 \Delta_2 w - bw + \Phi(x,y,t)$$

5.4.1. Problems in Cartesian Coordinates

A two-dimensional nonhomogeneous telegraph equation in the rectangular Cartesian coordinate system is written as

$$\frac{\partial^2 w}{\partial t^2} + k\frac{\partial w}{\partial t} = a^2\left(\frac{\partial^2 w}{\partial x^2} + \frac{\partial^2 w}{\partial y^2}\right) - bw + \Phi(x,y,t).$$

5.4.1-1. Reduction to the two-dimensional Klein–Gordon equation.

The substitution $w(x,y,t) = \exp\left(-\frac{1}{2}kt\right)u(x,y,t)$ leads to the equation

$$\frac{\partial^2 u}{\partial t^2} = a^2\left(\frac{\partial^2 u}{\partial x^2} + \frac{\partial^2 u}{\partial y^2}\right) - \left(b - \tfrac{1}{4}k^2\right)u + \exp\left(\tfrac{1}{2}kt\right)\Phi(x,y,t),$$

which is discussed in Subsection 5.3.1.

5.4. TELEGRAPH EQUATION $\frac{\partial^2 w}{\partial t^2} + k\frac{\partial w}{\partial t} = a^2 \Delta_2 w - bw + \Phi(x,y,t)$ 377

5.4.1-2. Fundamental solutions.

1°. Case $b - \frac{1}{4}k^2 = \sigma^2 > 0$:

$$\mathscr{E}(x,y,t) = \vartheta(at-r)\exp\left(-\tfrac{1}{2}kt\right)\frac{\cos\left(\sigma\sqrt{t^2 - r^2/a^2}\right)}{2\pi a^2\sqrt{t^2 - r^2/a^2}},$$

where $r = \sqrt{x^2 + y^2}$ and $\vartheta(z)$ is the Heaviside unit step function.

2°. Case $b - \frac{1}{4}k^2 = -\sigma^2 < 0$:

$$\mathscr{E}(x,y,t) = \vartheta(at-r)\exp\left(-\tfrac{1}{2}kt\right)\frac{\cosh\left(\sigma\sqrt{t^2 - r^2/a^2}\right)}{2\pi a^2\sqrt{t^2 - r^2/a^2}}.$$

⊙ *Reference*: V. S. Vladimirov, V. P. Mikhailov, A. A. Vasharin, et al. (1974).

5.4.1-3. Domain: $-\infty < x < \infty$, $-\infty < y < \infty$. Cauchy problem.

Initial conditions are prescribed:

$$w = f(x,y) \quad \text{at} \quad t = 0,$$
$$\partial_t w = g(x,y) \quad \text{at} \quad t = 0.$$

Solution:

$$w(x,y,t) = \exp\left(-\tfrac{1}{2}kt\right)\frac{\partial}{\partial t}\iint\limits_{\rho \leq at} f(\xi,\eta) H(x,y,\xi,\eta,t)\, d\xi\, d\eta$$
$$+ \exp\left(-\tfrac{1}{2}kt\right)\iint\limits_{\rho \leq at}\left[g(\xi,\eta) + \tfrac{1}{2}kf(\xi,\eta)\right] H(x,y,\xi,\eta,t)\, d\xi\, d\eta$$
$$+ \int_0^t d\tau \iint\limits_{\rho \leq a(t-\tau)} \exp\left[-\tfrac{1}{2}k(t-\tau)\right]\Phi(\xi,\eta,\tau) H(x,y,\xi,\eta,t-\tau)\, d\xi\, d\eta.$$

Here,

$$H(x,y,\xi,\eta,t) = \begin{cases} \dfrac{\cos\left(\sigma\sqrt{t^2 - \rho^2/a^2}\right)}{2\pi a^2\sqrt{t^2 - \rho^2/a^2}} & \text{for } b - \tfrac{1}{4}k^2 = \sigma^2 > 0, \\ \dfrac{\cosh\left(\sigma\sqrt{t^2 - \rho^2/a^2}\right)}{2\pi a^2\sqrt{t^2 - \rho^2/a^2}} & \text{for } b - \tfrac{1}{4}k^2 = -\sigma^2 < 0, \end{cases}$$

where $\rho = \sqrt{(x-\xi)^2 + (y-\eta)^2}$.

5.4.1-4. Domain: $0 \leq x \leq l_1$, $0 \leq y \leq l_2$. First boundary value problem.

A rectangle is considered. The following conditions are prescribed:

$$w = f_0(x,y) \quad \text{at} \quad t = 0 \quad \text{(initial condition)},$$
$$\partial_t w = f_1(x,y) \quad \text{at} \quad t = 0 \quad \text{(initial condition)},$$
$$w = g_1(y,t) \quad \text{at} \quad x = 0 \quad \text{(boundary condition)},$$
$$w = g_2(y,t) \quad \text{at} \quad x = l_1 \quad \text{(boundary condition)},$$
$$w = g_3(x,t) \quad \text{at} \quad y = 0 \quad \text{(boundary condition)},$$
$$w = g_4(x,t) \quad \text{at} \quad y = l_2 \quad \text{(boundary condition)}.$$

Solution:

$$w(x,y,t) = \frac{\partial}{\partial t}\int_0^{l_1}\int_0^{l_2} f_0(\xi,\eta) G(x,y,\xi,\eta,t)\,d\eta\,d\xi$$

$$+ \int_0^{l_1}\int_0^{l_2} \bigl[f_1(\xi,\eta) + k f_0(\xi,\eta)\bigr] G(x,y,\xi,\eta,t)\,d\eta\,d\xi$$

$$+ a^2 \int_0^t \int_0^{l_2} g_1(\eta,\tau) \left[\frac{\partial}{\partial \xi} G(x,y,\xi,\eta,t-\tau)\right]_{\xi=0} d\eta\,d\tau$$

$$- a^2 \int_0^t \int_0^{l_2} g_2(\eta,\tau) \left[\frac{\partial}{\partial \xi} G(x,y,\xi,\eta,t-\tau)\right]_{\xi=l_1} d\eta\,d\tau$$

$$+ a^2 \int_0^t \int_0^{l_1} g_3(\xi,\tau) \left[\frac{\partial}{\partial \eta} G(x,y,\xi,\eta,t-\tau)\right]_{\eta=0} d\xi\,d\tau$$

$$- a^2 \int_0^t \int_0^{l_1} g_4(\xi,\tau) \left[\frac{\partial}{\partial \eta} G(x,y,\xi,\eta,t-\tau)\right]_{\eta=l_2} d\xi\,d\tau$$

$$+ \int_0^t \int_0^{l_1} \int_0^{l_2} \Phi(\xi,\eta,\tau) G(x,y,\xi,\eta,t-\tau)\,d\eta\,d\xi\,d\tau,$$

where

$$G(x,y,\xi,\eta,t) = \frac{4}{l_1 l_2} \exp\bigl(-\tfrac{1}{2} kt\bigr) \sum_{n=1}^{\infty} \sum_{m=1}^{\infty} \frac{1}{\lambda_{nm}} \sin(p_n x)\sin(q_m y)\sin(p_n \xi)\sin(q_m \eta)\sin(\lambda_{nm} t),$$

$$p_n = \frac{n\pi}{l_1}, \quad q_m = \frac{m\pi}{l_2}, \quad \lambda_{nm} = \sqrt{a^2 p_n^2 + a^2 q_m^2 + b - \tfrac{1}{4} k^2}.$$

5.4.1-5. Domain: $0 \le x \le l_1$, $0 \le y \le l_2$. Second boundary value problem.

A rectangle is considered. The following conditions are prescribed:

$$\begin{aligned}
w &= f_0(x,y) &&\text{at}\quad t=0 &&\text{(initial condition)},\\
\partial_t w &= f_1(x,y) &&\text{at}\quad t=0 &&\text{(initial condition)},\\
\partial_x w &= g_1(y,t) &&\text{at}\quad x=0 &&\text{(boundary condition)},\\
\partial_x w &= g_2(y,t) &&\text{at}\quad x=l_1 &&\text{(boundary condition)},\\
\partial_y w &= g_3(x,t) &&\text{at}\quad y=0 &&\text{(boundary condition)},\\
\partial_y w &= g_4(x,t) &&\text{at}\quad y=l_2 &&\text{(boundary condition)}.
\end{aligned}$$

Solution:

$$w(x,y,t) = \frac{\partial}{\partial t}\int_0^{l_1}\int_0^{l_2} f_0(\xi,\eta) G(x,y,\xi,\eta,t)\,d\eta\,d\xi$$

$$+ \int_0^{l_1}\int_0^{l_2} \bigl[f_1(\xi,\eta) + k f_0(\xi,\eta)\bigr] G(x,y,\xi,\eta,t)\,d\eta\,d\xi$$

$$- a^2 \int_0^t \int_0^{l_2} g_1(\eta,\tau) G(x,y,0,\eta,t-\tau)\,d\eta\,d\tau + a^2 \int_0^t \int_0^{l_2} g_2(\eta,\tau) G(x,y,l_1,\eta,t-\tau)\,d\eta\,d\tau$$

$$- a^2 \int_0^t \int_0^{l_1} g_3(\xi,\tau) G(x,y,\xi,0,t-\tau)\,d\xi\,d\tau + a^2 \int_0^t \int_0^{l_1} g_4(\xi,\tau) G(x,y,\xi,l_2,t-\tau)\,d\xi\,d\tau$$

$$+ \int_0^t \int_0^{l_1} \int_0^{l_2} \Phi(\xi,\eta,\tau) G(x,y,\xi,\eta,t-\tau)\,d\eta\,d\xi\,d\tau.$$

Here,

$$G(x,y,\xi,\eta,t) = \exp(-\tfrac{1}{2}kt)\left[\frac{\sin(\lambda_{00}t)}{l_1 l_2 \lambda_{00}}\right.$$
$$\left. + \frac{2}{l_1 l_2}\sum_{n=0}^{\infty}\sum_{m=0}^{\infty}\frac{A_{nm}}{\lambda_{nm}}\cos(p_n x)\cos(q_m y)\cos(p_n\xi)\cos(q_m\eta)\sin(\lambda_{nm}t)\right],$$

where

$$p_n = \frac{n\pi}{l_1}, \quad q_m = \frac{m\pi}{l_2}, \quad \lambda_{nm} = \sqrt{a^2 p_n^2 + a^2 q_m^2 + b - \tfrac{1}{4}k^2}, \quad A_{nm} = \begin{cases} 0 & \text{for } n=m=0, \\ 1 & \text{for } nm=0 \ (n\neq m), \\ 2 & \text{for } nm\neq 0. \end{cases}$$

5.4.1-6. Domain: $0 \leq x \leq l_1$, $0 \leq y \leq l_2$. Third boundary value problem.

A rectangle is considered. The following conditions are prescribed:

$$\begin{aligned}
w &= f_0(x,y) & \text{at} \quad & t=0 & \text{(initial condition)}, \\
\partial_t w &= f_1(x,y) & \text{at} \quad & t=0 & \text{(initial condition)}, \\
\partial_x w - s_1 w &= g_1(y,t) & \text{at} \quad & x=0 & \text{(boundary condition)}, \\
\partial_x w + s_2 w &= g_2(y,t) & \text{at} \quad & x=l_1 & \text{(boundary condition)}, \\
\partial_y w - s_3 w &= g_3(x,t) & \text{at} \quad & y=0 & \text{(boundary condition)}, \\
\partial_y w + s_4 w &= g_4(x,t) & \text{at} \quad & y=l_2 & \text{(boundary condition)}.
\end{aligned}$$

The solution $w(x,y,t)$ is determined by the formula in Paragraph 5.4.1-5 where

$$G(x,y,\xi,\eta,t) = 4\exp(-\tfrac{1}{2}kt)\sum_{n=1}^{\infty}\sum_{m=1}^{\infty}\frac{1}{E_{nm}\sqrt{a^2\mu_n^2 + a^2\nu_m^2 + b - \tfrac{1}{4}k^2}}\sin(\mu_n x + \varepsilon_n)\sin(\nu_m y + \sigma_m)$$
$$\times \sin(\mu_n \xi + \varepsilon_n)\sin(\nu_m \eta + \sigma_m)\sin\left(t\sqrt{a^2\mu_n^2 + a^2\nu_m^2 + b - \tfrac{1}{4}k^2}\right).$$

Here,

$$\varepsilon_n = \arctan\frac{\mu_n}{l_1}, \quad \sigma_m = \arctan\frac{\nu_m}{l_2}, \quad E_{nm} = \left[l_1 + \frac{(s_1 s_2 + \mu_n^2)(s_1 + s_2)}{(s_1^2 + \mu_n^2)(s_2^2 + \mu_n^2)}\right]\left[l_2 + \frac{(s_3 s_4 + \nu_m^2)(s_3 + s_4)}{(s_3^2 + \nu_m^2)(s_4^2 + \nu_m^2)}\right];$$

the μ_n and ν_m are positive roots of the transcendental equations

$$\mu^2 - s_1 s_2 = (s_1 + s_2)\mu \cot(l_1 \mu), \quad \nu^2 - s_3 s_4 = (s_3 + s_4)\nu \cot(l_2 \nu).$$

5.4.1-7. Domain: $0 \leq x \leq l_1$, $0 \leq y \leq l_2$. Mixed boundary value problems.

1°. A rectangle is considered. The following conditions are prescribed:

$$\begin{aligned}
w &= f_0(x,y) & \text{at} \quad & t=0 & \text{(initial condition)}, \\
\partial_t w &= f_1(x,y) & \text{at} \quad & t=0 & \text{(initial condition)}, \\
w &= g_1(y,t) & \text{at} \quad & x=0 & \text{(boundary condition)}, \\
w &= g_2(y,t) & \text{at} \quad & x=l_1 & \text{(boundary condition)}, \\
\partial_y w &= g_3(x,t) & \text{at} \quad & y=0 & \text{(boundary condition)}, \\
\partial_y w &= g_4(x,t) & \text{at} \quad & y=l_2 & \text{(boundary condition)}.
\end{aligned}$$

Solution:

$$w(x,y,t) = \frac{\partial}{\partial t}\int_0^{l_1}\int_0^{l_2} f_0(\xi,\eta)G(x,y,\xi,\eta,t)\,d\eta\,d\xi$$

$$+ \int_0^{l_1}\int_0^{l_2} \big[f_1(\xi,\eta) + kf_0(\xi,\eta)\big]G(x,y,\xi,\eta,t)\,d\eta\,d\xi$$

$$+ a^2\int_0^t\int_0^{l_2} g_1(\eta,\tau)\bigg[\frac{\partial}{\partial \xi}G(x,y,\xi,\eta,t-\tau)\bigg]_{\xi=0}\,d\eta\,d\tau$$

$$- a^2\int_0^t\int_0^{l_2} g_2(\eta,\tau)\bigg[\frac{\partial}{\partial \xi}G(x,y,\xi,\eta,t-\tau)\bigg]_{\xi=l_1}\,d\eta\,d\tau$$

$$- a^2\int_0^t\int_0^{l_1} g_3(\xi,\tau)G(x,y,\xi,0,t-\tau)\,d\xi\,d\tau$$

$$+ a^2\int_0^t\int_0^{l_1} g_4(\xi,\tau)G(x,y,\xi,l_2,t-\tau)\,d\xi\,d\tau$$

$$+ \int_0^t\int_0^{l_1}\int_0^{l_2} \Phi(\xi,\eta,\tau)G(x,y,\xi,\eta,t-\tau)\,d\eta\,d\xi\,d\tau,$$

where

$$G(x,y,\xi,\eta,t) = \frac{2}{l_1 l_2}\exp\!\big(-\tfrac{1}{2}kt\big)\sum_{n=1}^\infty\sum_{m=0}^\infty \frac{A_m}{\lambda_{nm}}\sin(p_n x)\cos(q_m y)\sin(p_n\xi)\cos(q_m\eta)\sin(\lambda_{nm}t),$$

$$p_n = \frac{n\pi}{l_1},\quad q_m = \frac{m\pi}{l_2},\quad \lambda_{nm} = \sqrt{a^2 p_n^2 + a^2 q_m^2 + b - \tfrac{1}{4}k^2},\quad A_m = \begin{cases} 1 & \text{for } m=0,\\ 2 & \text{for } m\neq 0.\end{cases}$$

2°. A rectangle is considered. The following conditions are prescribed:

$$\begin{aligned}
w &= f_0(x,y) &\text{at}\quad& t=0 &\text{(initial condition)},\\
\partial_t w &= f_1(x,y) &\text{at}\quad& t=0 &\text{(initial condition)},\\
w &= g_1(y,t) &\text{at}\quad& x=0 &\text{(boundary condition)},\\
\partial_x w &= g_2(y,t) &\text{at}\quad& x=l_1 &\text{(boundary condition)},\\
w &= g_3(x,t) &\text{at}\quad& y=0 &\text{(boundary condition)},\\
\partial_y w &= g_4(x,t) &\text{at}\quad& y=l_2 &\text{(boundary condition)}.
\end{aligned}$$

Solution:

$$w(x,y,t) = \frac{\partial}{\partial t}\int_0^{l_1}\int_0^{l_2} f_0(\xi,\eta)G(x,y,\xi,\eta,t)\,d\eta\,d\xi$$

$$+ \int_0^{l_1}\int_0^{l_2} \big[f_1(\xi,\eta) + kf_0(\xi,\eta)\big]G(x,y,\xi,\eta,t)\,d\eta\,d\xi$$

$$+ a^2\int_0^t\int_0^{l_2} g_1(\eta,\tau)\bigg[\frac{\partial}{\partial \xi}G(x,y,\xi,\eta,t-\tau)\bigg]_{\xi=0}\,d\eta\,d\tau$$

$$+ a^2\int_0^t\int_0^{l_2} g_2(\eta,\tau)G(x,y,l_1,\eta,t-\tau)\,d\eta\,d\tau$$

$$+ a^2\int_0^t\int_0^{l_1} g_3(\xi,\tau)\bigg[\frac{\partial}{\partial \eta}G(x,y,\xi,\eta,t-\tau)\bigg]_{\eta=0}\,d\xi\,d\tau$$

$$+ a^2\int_0^t\int_0^{l_1} g_4(\xi,\tau)G(x,y,\xi,l_2,t-\tau)\,d\xi\,d\tau$$

$$+ \int_0^t\int_0^{l_1}\int_0^{l_2} \Phi(\xi,\eta,\tau)G(x,y,\xi,\eta,t-\tau)\,d\eta\,d\xi\,d\tau,$$

where

$$G(x, y, \xi, \eta, t) = \frac{4}{l_1 l_2} \exp(-\tfrac{1}{2}kt) \sum_{n=0}^{\infty} \sum_{m=0}^{\infty} \frac{1}{\lambda_{nm}} \sin(p_n x) \sin(q_m y) \sin(p_n \xi) \sin(q_m \eta) \sin(\lambda_{nm} t),$$

$$p_n = \frac{\pi(2n+1)}{2l_1}, \quad q_m = \frac{\pi(2m+1)}{2l_2}, \quad \lambda_{nm} = \sqrt{a^2 p_n^2 + a^2 q_m^2 + b - \tfrac{1}{4}k^2}.$$

5.4.2. Problems in Polar Coordinates

A two-dimensional nonhomogeneous telegraph equation in the polar coordinate system has the form

$$\frac{\partial^2 w}{\partial t^2} + k \frac{\partial w}{\partial t} = a^2 \left(\frac{\partial^2 w}{\partial r^2} + \frac{1}{r} \frac{\partial w}{\partial r} + \frac{1}{r^2} \frac{\partial^2 w}{\partial \varphi^2} \right) - bw + \Phi(r, \varphi, t), \qquad r = \sqrt{x^2 + y^2}.$$

For one-dimensional solutions $w = w(r, t)$, see equation 4.4.2.2.

5.4.2-1. Domain: $0 \leq r \leq R$, $0 \leq \varphi \leq 2\pi$. First boundary value problem.

A circle is considered. The following conditions are prescribed:

$$\begin{aligned} w &= f_0(r, \varphi) & \text{at} \quad t &= 0 & &\text{(initial condition)}, \\ \partial_t w &= f_1(r, \varphi) & \text{at} \quad t &= 0 & &\text{(initial condition)}, \\ w &= g(\varphi, t) & \text{at} \quad r &= R & &\text{(boundary condition)}. \end{aligned}$$

Solution:

$$\begin{aligned} w(r, \varphi, t) = {}& \frac{\partial}{\partial t} \int_0^{2\pi} \int_0^R f_0(\xi, \eta) G(r, \varphi, \xi, \eta, t) \xi \, d\xi \, d\eta \\ &+ \int_0^{2\pi} \int_0^R \left[f_1(\xi, \eta) + k f_0(\xi, \eta) \right] G(r, \varphi, \xi, \eta, t) \xi \, d\xi \, d\eta \\ &- a^2 R \int_0^t \int_0^{2\pi} g(\eta, \tau) \left[\frac{\partial}{\partial \xi} G(r, \varphi, \xi, \eta, t - \tau) \right]_{\xi=R} d\eta \, d\tau \\ &+ \int_0^t \int_0^{2\pi} \int_0^R \Phi(\xi, \eta, \tau) G(r, \varphi, \xi, \eta, t - \tau) \xi \, d\xi \, d\eta \, d\tau. \end{aligned}$$

Here,

$$G(r, \varphi, \xi, \eta, t) = \frac{1}{\pi R^2} \exp(-\tfrac{1}{2}kt) \sum_{n=0}^{\infty} \sum_{m=1}^{\infty} \frac{A_n J_n(\mu_{nm} r) J_n(\mu_{nm} \xi)}{[J_n'(\mu_{nm} R)]^2} \cos[n(\varphi - \eta)] \frac{\sin(t \sqrt{\lambda_{nm}})}{\sqrt{\lambda_{nm}}},$$

$$\lambda_{nm} = a^2 \mu_{nm}^2 + b - \tfrac{1}{4} k^2, \quad A_0 = 1, \quad A_n = 2 \quad (n = 1, 2, \ldots),$$

where the $J_n(\xi)$ are the Bessel functions (the prime denotes the derivative with respect to the argument) and the μ_{nm} are positive roots of the transcendental equation $J_n(\mu R) = 0$.

5.4.2-2. Domain: $0 \leq r \leq R$, $0 \leq \varphi \leq 2\pi$. Second boundary value problem.

A circle is considered. The following conditions are prescribed:

$$\begin{aligned} w &= f_0(r, \varphi) & \text{at} \quad t &= 0 & &\text{(initial condition)}, \\ \partial_t w &= f_1(r, \varphi) & \text{at} \quad t &= 0 & &\text{(initial condition)}, \\ \partial_r w &= g(\varphi, t) & \text{at} \quad r &= R & &\text{(boundary condition)}. \end{aligned}$$

Solution:

$$w(r,\varphi,t) = \frac{\partial}{\partial t}\int_0^{2\pi}\!\!\int_0^R f_0(\xi,\eta)G(r,\varphi,\xi,\eta,t)\xi\,d\xi\,d\eta$$
$$+ \int_0^{2\pi}\!\!\int_0^R [f_1(\xi,\eta) + kf_0(\xi,\eta)]G(r,\varphi,\xi,\eta,t)\xi\,d\xi\,d\eta$$
$$+ a^2 R \int_0^t\!\!\int_0^{2\pi} g(\eta,\tau)G(r,\varphi,R,\eta,t-\tau)\,d\eta\,d\tau$$
$$+ \int_0^t\!\!\int_0^{2\pi}\!\!\int_0^R \Phi(\xi,\eta,\tau)G(r,\varphi,\xi,\eta,t-\tau)\xi\,d\xi\,d\eta\,d\tau.$$

Here,

$$G(r,\varphi,\xi,\eta,t) = \exp\!\left(-\tfrac{1}{2}kt\right)\!\left[\frac{\sin\!\left(t\sqrt{b-k^2/4}\right)}{\pi R^2\sqrt{b-k^2/4}}\right.$$
$$\left.+ \frac{1}{\pi}\sum_{n=0}^{\infty}\sum_{m=1}^{\infty}\frac{A_n\mu_{nm}^2 J_n(\mu_{nm}r)J_n(\mu_{nm}\xi)}{(\mu_{nm}^2 R^2 - n^2)[J_n(\mu_{nm}R)]^2}\cos[n(\varphi-\eta)]\frac{\sin\!\left(t\sqrt{\lambda_{nm}}\right)}{\sqrt{\lambda_{nm}}}\right],$$
$$\lambda_{nm} = a^2\mu_{nm}^2 + b - \tfrac{1}{4}k^2, \quad A_0 = 1, \quad A_n = 2 \quad (n = 1, 2, \ldots),$$

where the $J_n(\xi)$ are the Bessel functions and the μ_m are positive roots of the transcendental equation $J_n'(\mu R) = 0$.

5.4.2-3. Domain: $0 \le r \le R$, $0 \le \varphi \le 2\pi$. Third boundary value problem.

A circle is considered. The following conditions are prescribed:

$$w = f_0(r,\varphi) \quad \text{at} \quad t = 0 \quad \text{(initial condition)},$$
$$\partial_t w = f_1(r,\varphi) \quad \text{at} \quad t = 0 \quad \text{(initial condition)},$$
$$\partial_r w + sw = g(\varphi,t) \quad \text{at} \quad r = R \quad \text{(boundary condition)}.$$

The solution $w(r,\varphi,t)$ is determined by the formula in Paragraph 5.4.2-2 where

$$G(r,\varphi,\xi,\eta,t) = \frac{1}{\pi}\exp\!\left(-\tfrac{1}{2}kt\right)\sum_{n=0}^{\infty}\sum_{m=1}^{\infty}\frac{A_n\mu_{nm}^2 J_n(\mu_{nm}r)J_n(\mu_{nm}\xi)\cos[n(\varphi-\eta)]\sin\!\left(t\sqrt{\lambda_{nm}}\right)}{(\mu_{nm}^2 R^2 + s^2 R^2 - n^2)[J_n(\mu_{nm}R)]^2\sqrt{\lambda_{nm}}},$$
$$\lambda_{nm} = a^2\mu_{nm}^2 + b - \tfrac{1}{4}k^2, \quad A_0 = 1, \quad A_n = 2 \quad (n = 1, 2, \ldots).$$

Here, the $J_n(\xi)$ are the Bessel functions and the μ_m are positive roots of the transcendental equation

$$\mu J_n'(\mu R) + s J_n(\mu R) = 0.$$

5.4.2-4. Domain: $R_1 \le r \le R_2$, $0 \le \varphi \le 2\pi$. First boundary value problem.

An annular domain is considered. The following conditions are prescribed:

$$w = f_0(r,\varphi) \quad \text{at} \quad t = 0 \quad \text{(initial condition)},$$
$$\partial_t w = f_1(r,\varphi) \quad \text{at} \quad t = 0 \quad \text{(initial condition)},$$
$$w = g_1(\varphi,t) \quad \text{at} \quad r = R_1 \quad \text{(boundary condition)},$$
$$w = g_2(\varphi,t) \quad \text{at} \quad r = R_2 \quad \text{(boundary condition)}.$$

Solution:

$$w(r,\varphi,t) = \frac{\partial}{\partial t}\int_0^{2\pi}\int_{R_1}^{R_2} f_0(\xi,\eta) G(r,\varphi,\xi,\eta,t)\xi\, d\xi\, d\eta$$

$$+ \int_0^{2\pi}\int_{R_1}^{R_2} \bigl[f_1(\xi,\eta) + kf_0(\xi,\eta)\bigr] G(r,\varphi,\xi,\eta,t)\xi\, d\xi\, d\eta$$

$$+ a^2 R_1 \int_0^t\int_0^{2\pi} g_1(\eta,\tau)\left[\frac{\partial}{\partial \xi}G(r,\varphi,\xi,\eta,t-\tau)\right]_{\xi=R_1} d\eta\, d\tau$$

$$- a^2 R_2 \int_0^t\int_0^{2\pi} g_2(\eta,\tau)\left[\frac{\partial}{\partial \xi}G(r,\varphi,\xi,\eta,t-\tau)\right]_{\xi=R_2} d\eta\, d\tau$$

$$+ \int_0^t\int_0^{2\pi}\int_{R_1}^{R_2} \Phi(\xi,\eta,\tau) G(r,\varphi,\xi,\eta,t-\tau)\xi\, d\xi\, d\eta\, d\tau.$$

Here,

$$G(r,\varphi,\xi,\eta,t) = \frac{\pi}{2}\exp\!\left(-\tfrac{1}{2}kt\right)\sum_{n=0}^{\infty}\sum_{m=1}^{\infty} A_n B_{nm} Z_n(\mu_{nm}r) Z_n(\mu_{nm}\xi)\cos[n(\varphi-\eta)]\frac{\sin\!\left(t\sqrt{\lambda_{nm}}\right)}{\sqrt{\lambda_{nm}}},$$

$$A_n = \begin{cases} 1/2 & \text{for } n=0, \\ 1 & \text{for } n\neq 0, \end{cases} \qquad B_{nm} = \frac{\mu_{nm}^2 J_n^2(\mu_{nm}R_2)}{J_n^2(\mu_{nm}R_1) - J_n^2(\mu_{nm}R_2)},$$

$$Z_n(\mu_{nm}r) = J_n(\mu_{nm}R_1)Y_n(\mu_{nm}r) - Y_n(\mu_{nm}R_1)J_n(\mu_{nm}r), \qquad \lambda_{nm} = a^2\mu_{nm}^2 + b - \tfrac{1}{4}k^2,$$

where the $J_n(r)$ and $Y_n(r)$ are the Bessel functions, and the μ_{nm} are positive roots of the transcendental equation

$$J_n(\mu R_1)Y_n(\mu R_2) - Y_n(\mu R_1)J_n(\mu R_2) = 0.$$

5.4.2-5. Domain: $R_1 \leq r \leq R_2$, $0 \leq \varphi \leq 2\pi$. **Second boundary value problem.**

An annular domain is considered. The following conditions are prescribed:

$$\begin{aligned}
w &= f_0(r,\varphi) & \text{at } & t=0 & \text{(initial condition)}, \\
\partial_t w &= f_1(r,\varphi) & \text{at } & t=0 & \text{(initial condition)}, \\
\partial_r w &= g_1(\varphi,t) & \text{at } & r=R_1 & \text{(boundary condition)}, \\
\partial_r w &= g_2(\varphi,t) & \text{at } & r=R_2 & \text{(boundary condition)}.
\end{aligned}$$

Solution:

$$w(r,\varphi,t) = \frac{\partial}{\partial t}\int_0^{2\pi}\int_{R_1}^{R_2} f_0(\xi,\eta) G(r,\varphi,\xi,\eta,t)\xi\, d\xi\, d\eta$$

$$+ \int_0^{2\pi}\int_{R_1}^{R_2} \bigl[f_1(\xi,\eta) + kf_0(\xi,\eta)\bigr] G(r,\varphi,\xi,\eta,t)\xi\, d\xi\, d\eta$$

$$- a^2 R_1 \int_0^t\int_0^{2\pi} g_1(\eta,\tau) G(r,\varphi,R_1,\eta,t-\tau)\, d\eta\, d\tau$$

$$+ a^2 R_2 \int_0^t\int_0^{2\pi} g_2(\eta,\tau) G(r,\varphi,R_2,\eta,t-\tau)\, d\eta\, d\tau$$

$$+ \int_0^t\int_0^{2\pi}\int_{R_1}^{R_2} \Phi(\xi,\eta,\tau) G(r,\varphi,\xi,\eta,t-\tau)\xi\, d\xi\, d\eta\, d\tau.$$

Here,

$$G(r,\varphi,\xi,\eta,t) = \exp\left(-\tfrac{1}{2}kt\right)\left[\frac{\sin\left(t\sqrt{b-k^2/4}\right)}{\pi(R_2^2-R_1^2)\sqrt{b-k^2/4}}\right.$$
$$\left.+\frac{1}{\pi}\sum_{n=0}^{\infty}\sum_{m=1}^{\infty}\frac{A_n\mu_{nm}^2 Z_n(\mu_{nm}r)Z_n(\mu_{nm}\xi)\cos[n(\varphi-\eta)]\sin\left(t\sqrt{a^2\mu_{nm}^2+b-k^2/4}\right)}{[(\mu_{nm}^2 R_2^2-n^2)Z_n^2(\mu_{nm}R_2)-(\mu_{nm}^2 R_1^2-n^2)Z_n^2(\mu_{nm}R_1)]\sqrt{a^2\mu_{nm}^2+b-k^2/4}}\right],$$

where

$$Z_n(\mu_{nm}r) = J_n'(\mu_{nm}R_1)Y_n(\mu_{nm}r) - Y_n'(\mu_{nm}R_1)J_n(\mu_{nm}r), \qquad A_n = \begin{cases} 1 & \text{for } n=0, \\ 2 & \text{for } n>0, \end{cases}$$

the $J_n(r)$ and $Y_n(r)$ are the Bessel functions, and the μ_{nm} are positive roots of the transcendental equation

$$J_n'(\mu R_1)Y_n'(\mu R_2) - Y_n'(\mu R_1)J_n'(\mu R_2) = 0.$$

5.4.2-6. Domain: $R_1 \leq r \leq R_2$, $0 \leq \varphi \leq 2\pi$. Third boundary value problem.

An annular domain is considered. The following conditions are prescribed:

$$\begin{aligned}
w &= f_0(r,\varphi) & \text{at} \quad & t=0 & \text{(initial condition)}, \\
\partial_t w &= f_1(r,\varphi) & \text{at} \quad & t=0 & \text{(initial condition)}, \\
\partial_r w - s_1 w &= g_1(\varphi,t) & \text{at} \quad & r=R_1 & \text{(boundary condition)}, \\
\partial_r w + s_2 w &= g_2(\varphi,t) & \text{at} \quad & r=R_2 & \text{(boundary condition)}.
\end{aligned}$$

The solution $w(r,\varphi,t)$ is determined by the formula in Paragraph 5.4.2-5 where

$$G(r,\varphi,\xi,\eta,t) = \frac{1}{\pi}\exp\left(-\tfrac{1}{2}kt\right)\sum_{n=0}^{\infty}\sum_{m=1}^{\infty}\frac{A_n\mu_{nm}^2}{B_{nm}\lambda_{nm}} Z_n(\mu_{nm}r)Z_n(\mu_{nm}\xi)\cos[n(\varphi-\eta)]\sin(\lambda_{nm}t).$$

Here,

$$A_n = \begin{cases} 1 & \text{for } n=0, \\ 2 & \text{for } n>0, \end{cases} \qquad \lambda_{nm} = \sqrt{a^2\mu_{nm}^2 + b - \tfrac{1}{4}k^2},$$

$$B_{nm} = (s_2^2 R_2^2 + \mu_{nm}^2 R_2^2 - n^2)Z_n^2(\mu_{nm}R_2) - (s_1^2 R_1^2 + \mu_{nm}^2 R_1^2 - n^2)Z_n^2(\mu_{nm}R_1),$$

$$Z_n(\mu_{nm}r) = \left[\mu_{nm}J_n'(\mu_{nm}R_1) - s_1 J_n(\mu_{nm}R_1)\right]Y_n(\mu_{nm}r)$$
$$- \left[\mu_{nm}Y_n'(\mu_{nm}R_1) - s_1 Y_n(\mu_{nm}R_1)\right]J_n(\mu_{nm}r),$$

where the $J_n(r)$ and $Y_n(r)$ are the Bessel functions, and the μ_{nm} are positive roots of the transcendental equation

$$\left[\mu J_n'(\mu R_1) - s_1 J_n(\mu R_1)\right]\left[\mu Y_n'(\mu R_2) + s_2 Y_n(\mu R_2)\right]$$
$$= \left[\mu Y_n'(\mu R_1) - s_1 Y_n(\mu R_1)\right]\left[\mu J_n'(\mu R_2) + s_2 J_n(\mu R_2)\right].$$

5.4.2-7. Domain: $0 \leq r \leq R$, $0 \leq \varphi \leq \varphi_0$. First boundary value problem.

A circular sector is considered. The following conditions are prescribed:

$$\begin{aligned}
w &= f_0(r,\varphi) & \text{at} \quad & t=0 & \text{(initial condition)}, \\
\partial_t w &= f_1(r,\varphi) & \text{at} \quad & t=0 & \text{(initial condition)}, \\
w &= g_1(\varphi,t) & \text{at} \quad & r=R & \text{(boundary condition)}, \\
w &= g_2(r,t) & \text{at} \quad & \varphi=0 & \text{(boundary condition)}, \\
w &= g_3(r,t) & \text{at} \quad & \varphi=\varphi_0 & \text{(boundary condition)}.
\end{aligned}$$

5.4. Telegraph Equation $\frac{\partial^2 w}{\partial t^2} + k\frac{\partial w}{\partial t} = a^2\Delta_2 w - bw + \Phi(x,y,t)$ 385

Solution:
$$w(r,\varphi,t) = \frac{\partial}{\partial t}\int_0^{\varphi_0}\int_0^R f_0(\xi,\eta)G(r,\varphi,\xi,\eta,t)\xi\,d\xi\,d\eta$$
$$+\int_0^{\varphi_0}\int_0^R [f_1(\xi,\eta)+kf_0(\xi,\eta)]G(r,\varphi,\xi,\eta,t)\xi\,d\xi\,d\eta$$
$$-a^2R\int_0^t\int_0^{\varphi_0} g_1(\eta,\tau)\left[\frac{\partial}{\partial\xi}G(r,\varphi,\xi,\eta,t-\tau)\right]_{\xi=R}d\eta\,d\tau$$
$$+a^2\int_0^t\int_0^R g_2(\xi,\tau)\frac{1}{\xi}\left[\frac{\partial}{\partial\eta}G(r,\varphi,\xi,\eta,t-\tau)\right]_{\eta=0}d\xi\,d\tau$$
$$-a^2\int_0^t\int_0^R g_3(\xi,\tau)\frac{1}{\xi}\left[\frac{\partial}{\partial\eta}G(r,\varphi,\xi,\eta,t-\tau)\right]_{\eta=\varphi_0}d\xi\,d\tau$$
$$+\int_0^t\int_0^{\varphi_0}\int_0^R \Phi(\xi,\eta,\tau)G(r,\varphi,\xi,\eta,t-\tau)\xi\,d\xi\,d\eta\,d\tau.$$

Here,
$$G(r,\varphi,\xi,\eta,t) = \frac{4}{R^2\varphi_0}\exp\left(-\tfrac{1}{2}kt\right)\sum_{n=1}^\infty\sum_{m=1}^\infty \frac{J_{n\pi/\varphi_0}(\mu_{nm}r)J_{n\pi/\varphi_0}(\mu_{nm}\xi)}{[J'_{n\pi/\varphi_0}(\mu_{nm}R)]^2}$$
$$\times \sin\left(\frac{n\pi\varphi}{\varphi_0}\right)\sin\left(\frac{n\pi\eta}{\varphi_0}\right)\frac{\sin\left(t\sqrt{a^2\mu_{nm}^2+b-k^2/4}\right)}{\sqrt{a^2\mu_{nm}^2+b-k^2/4}},$$

where the $J_{n\pi/\varphi_0}(r)$ are the Bessel functions and the μ_{nm} are positive roots of the transcendental equation $J_{n\pi/\varphi_0}(\mu R) = 0$.

5.4.2-8. Domain: $0 \le r \le R$, $0 \le \varphi \le \varphi_0$. Second boundary value problem.

A circular sector is considered. The following conditions are prescribed:

$$w = f_0(r,\varphi) \quad \text{at} \quad t = 0 \quad \text{(initial condition)},$$
$$\partial_t w = f_1(r,\varphi) \quad \text{at} \quad t = 0 \quad \text{(initial condition)},$$
$$\partial_r w = g_1(\varphi,t) \quad \text{at} \quad r = R \quad \text{(boundary condition)},$$
$$r^{-1}\partial_\varphi w = g_2(r,t) \quad \text{at} \quad \varphi = 0 \quad \text{(boundary condition)},$$
$$r^{-1}\partial_\varphi w = g_3(r,t) \quad \text{at} \quad \varphi = \varphi_0 \quad \text{(boundary condition)}.$$

Solution:
$$w(r,\varphi,t) = \frac{\partial}{\partial t}\int_0^{\varphi_0}\int_0^R f_0(\xi,\eta)G(r,\varphi,\xi,\eta,t)\xi\,d\xi\,d\eta$$
$$+\int_0^{\varphi_0}\int_0^R [f_1(\xi,\eta)+kf_0(\xi,\eta)]G(r,\varphi,\xi,\eta,t)\xi\,d\xi\,d\eta$$
$$+a^2R\int_0^t\int_0^{\varphi_0} g_1(\eta,\tau)G(r,\varphi,R,\eta,t-\tau)\,d\eta\,d\tau$$
$$-a^2\int_0^t\int_0^R g_2(\xi,\tau)G(r,\varphi,\xi,0,t-\tau)\,d\xi\,d\tau$$
$$+a^2\int_0^t\int_0^R g_3(\xi,\tau)G(r,\varphi,\xi,\varphi_0,t-\tau)\,d\xi\,d\tau$$
$$+\int_0^t\int_0^{\varphi_0}\int_0^R \Phi(\xi,\eta,\tau)G(r,\varphi,\xi,\eta,t-\tau)\xi\,d\xi\,d\eta\,d\tau.$$

Here,

$$G(r,\varphi,\xi,\eta,t) = \exp\left(-\tfrac{1}{2}kt\right)\left[\frac{2\sin(t\sqrt{b-k^2/4})}{R^2\varphi_0\sqrt{b-k^2/4}} + 4\varphi_0\sum_{n=0}^{\infty}\sum_{m=1}^{\infty}\frac{\mu_{nm}^2 J_{n\pi/\varphi_0}(\mu_{nm}r)J_{n\pi/\varphi_0}(\mu_{nm}\xi)}{(R^2\varphi_0^2\mu_{nm}^2 - n^2\pi^2)J_{n\pi/\varphi_0}^2(\mu_{nm}R)}\right.$$
$$\left.\times \cos\left(\frac{n\pi\varphi}{\varphi_0}\right)\cos\left(\frac{n\pi\eta}{\varphi_0}\right)\frac{\sin(t\sqrt{a^2\mu_{nm}^2 + b - k^2/4})}{\sqrt{a^2\mu_{nm}^2 + b - k^2/4}}\right],$$

where the $J_{n\pi/\varphi_0}(r)$ are the Bessel functions and the μ_{nm} are positive roots of the transcendental equation $J'_{n\pi/\varphi_0}(\mu R) = 0$.

5.4.2-9. Domain: $0 \le r \le R$, $0 \le \varphi \le \varphi_0$. Mixed boundary value problem.

A circular sector is considered. The following conditions are prescribed:

$$\begin{aligned}
w &= f_0(r,\varphi) &&\text{at}\quad t=0 &&\text{(initial condition)},\\
\partial_t w &= f_1(r,\varphi) &&\text{at}\quad t=0 &&\text{(initial condition)},\\
\partial_r w + \beta w &= g(\varphi,t) &&\text{at}\quad r=R &&\text{(boundary condition)},\\
\partial_\varphi w &= 0 &&\text{at}\quad \varphi=0 &&\text{(boundary condition)},\\
\partial_\varphi w &= 0 &&\text{at}\quad \varphi=\varphi_0 &&\text{(boundary condition)}.
\end{aligned}$$

Solution:

$$\begin{aligned}
w(r,\varphi,t) = &\frac{\partial}{\partial t}\int_0^{\varphi_0}\int_0^R f_0(\xi,\eta)G(r,\varphi,\xi,\eta,t)\xi\,d\xi\,d\eta\\
&+\int_0^{\varphi_0}\int_0^R [f_1(\xi,\eta)+kf_0(\xi,\eta)]G(r,\varphi,\xi,\eta,t)\xi\,d\xi\,d\eta\\
&+a^2R\int_0^t\int_0^{\varphi_0} g(\eta,\tau)G(r,\varphi,R,\eta,t-\tau)\,d\eta\,d\tau\\
&+\int_0^t\int_0^{\varphi_0}\int_0^R \Phi(\xi,\eta,\tau)G(r,\varphi,\xi,\eta,t-\tau)\xi\,d\xi\,d\eta\,d\tau.
\end{aligned}$$

Here,

$$G(r,\varphi,\xi,\eta,t) = \exp\left(-\tfrac{1}{2}kt\right)\sum_{n=0}^{\infty}\sum_{m=1}^{\infty} A_{nm} J_{s_n}(\mu_{nm}r)J_{s_n}(\mu_{nm}\xi)\cos(s_n\varphi)\cos(s_n\eta)\sin(\lambda_{nm}t),$$

$$s_n = \frac{n\pi}{\varphi_0},\quad A_{nm} = \frac{4\mu_{nm}^2}{\varphi_0(\mu_{nm}^2 R^2 + \beta^2 R^2 - s_n^2)[J_{s_n}(\mu_{nm}R)]^2\lambda_{nm}},\quad \lambda_{nm} = \sqrt{a^2\mu_{nm}^2 + b - \tfrac{1}{4}k^2},$$

where the $J_{s_n}(r)$ are the Bessel functions and the μ_{nm} are positive roots of the transcendental equation

$$\mu J'_{s_n}(\mu R) + \beta J_{s_n}(\mu R) = 0.$$

5.4.3. Axisymmetric Problems

In the axisymmetric case, a nonhomogeneous telegraph equation in the cylindrical coordinate system has the form

$$\frac{\partial^2 w}{\partial t^2} + k\frac{\partial w}{\partial t} = a^2\left(\frac{\partial^2 w}{\partial r^2} + \frac{1}{r}\frac{\partial w}{\partial r} + \frac{\partial^2 w}{\partial z^2}\right) - bw + \Phi(r,z,t),\qquad r = \sqrt{x^2+y^2}.$$

5.4. TELEGRAPH EQUATION $\frac{\partial^2 w}{\partial t^2} + k\frac{\partial w}{\partial t} = a^2 \Delta_2 w - bw + \Phi(x,y,t)$ 387

In the solutions of the problems considered below, the modified Green's function $\mathcal{G}(r,z,\xi,\eta,t) = 2\pi\xi G(r,z,\xi,\eta,t)$ is used for convenience.

5.4.3-1. Domain: $0 \le r \le R$, $0 \le z \le l$. First boundary value problem.

A circular cylinder of finite length is considered. The following conditions are prescribed:

$$
\begin{aligned}
w &= f_0(r,z) &&\text{at} \quad t=0 &&\text{(initial condition)}, \\
\partial_t w &= f_1(r,z) &&\text{at} \quad t=0 &&\text{(initial condition)}, \\
w &= g_1(z,t) &&\text{at} \quad r=R &&\text{(boundary condition)}, \\
w &= g_2(r,t) &&\text{at} \quad z=0 &&\text{(boundary condition)}, \\
w &= g_3(r,t) &&\text{at} \quad z=l &&\text{(boundary condition)}.
\end{aligned}
$$

Solution:

$$
\begin{aligned}
w(r,z,t) &= \frac{\partial}{\partial t} \int_0^l \int_0^R f_0(\xi,\eta)\mathcal{G}(r,z,\xi,\eta,t)\,d\xi\,d\eta \\
&\quad + \int_0^l \int_0^R \bigl[f_1(\xi,\eta) + kf_0(\xi,\eta)\bigr]\mathcal{G}(r,z,\xi,\eta,t)\,d\xi\,d\eta \\
&\quad - a^2 \int_0^t \int_0^l g_1(\eta,\tau)\left[\frac{\partial}{\partial \xi}\mathcal{G}(r,z,\xi,\eta,t-\tau)\right]_{\xi=R}\,d\eta\,d\tau \\
&\quad + a^2 \int_0^t \int_0^R g_2(\xi,\tau)\left[\frac{\partial}{\partial \eta}\mathcal{G}(r,z,\xi,\eta,t-\tau)\right]_{\eta=0}\,d\xi\,d\tau \\
&\quad - a^2 \int_0^t \int_0^R g_3(\xi,\tau)\left[\frac{\partial}{\partial \eta}\mathcal{G}(r,z,\xi,\eta,t-\tau)\right]_{\eta=l}\,d\xi\,d\tau \\
&\quad + \int_0^t \int_0^l \int_0^R \Phi(\xi,\eta,\tau)\mathcal{G}(r,z,\xi,\eta,t-\tau)\,d\xi\,d\eta\,d\tau.
\end{aligned}
$$

Here,

$$
\mathcal{G}(r,z,\xi,\eta,t) = \frac{4\xi e^{-kt/2}}{R^2 l} \sum_{n=1}^\infty \sum_{m=1}^\infty \frac{1}{J_1^2(\mu_n)} J_0\!\left(\frac{\mu_n r}{R}\right) J_0\!\left(\frac{\mu_n \xi}{R}\right) \sin\!\left(\frac{m\pi z}{l}\right) \sin\!\left(\frac{m\pi \eta}{l}\right) \frac{\sin(\lambda_{nm} t)}{\lambda_{nm}},
$$

$$
\lambda_{nm} = \sqrt{\frac{a^2 \mu_n^2}{R^2} + \frac{a^2 \pi^2 m^2}{l^2} + b - \frac{k^2}{4}},
$$

where the μ_n are positive zeros of the Bessel function, $J_0(\mu) = 0$.

5.4.3-2. Domain: $0 \le r \le R$, $0 \le z \le l$. Second boundary value problem.

A circular cylinder of finite length is considered. The following conditions are prescribed:

$$
\begin{aligned}
w &= f_0(r,z) &&\text{at} \quad t=0 &&\text{(initial condition)}, \\
\partial_t w &= f_1(r,z) &&\text{at} \quad t=0 &&\text{(initial condition)}, \\
\partial_r w &= g_1(z,t) &&\text{at} \quad r=R &&\text{(boundary condition)}, \\
\partial_z w &= g_2(r,t) &&\text{at} \quad z=0 &&\text{(boundary condition)}, \\
\partial_z w &= g_3(r,t) &&\text{at} \quad z=l &&\text{(boundary condition)}.
\end{aligned}
$$

Solution:

$$w(r,z,t) = \frac{\partial}{\partial t}\int_0^l\int_0^R f_0(\xi,\eta)\mathcal{G}(r,z,\xi,\eta,t)\,d\xi\,d\eta$$
$$+ \int_0^l\int_0^R [f_1(\xi,\eta) + kf_0(\xi,\eta)]\mathcal{G}(r,z,\xi,\eta,t)\,d\xi\,d\eta$$
$$+ a^2\int_0^t\int_0^l g_1(\eta,\tau)\mathcal{G}(r,z,R,\eta,t-\tau)\,d\eta\,d\tau$$
$$- a^2\int_0^t\int_0^R g_2(\xi,\tau)\mathcal{G}(r,z,\xi,0,t-\tau)\,d\xi\,d\tau$$
$$+ a^2\int_0^t\int_0^R g_3(\xi,\tau)\mathcal{G}(r,z,\xi,l,t-\tau)\,d\xi\,d\tau$$
$$+ \int_0^t\int_0^l\int_0^R \Phi(\xi,\eta,\tau)\mathcal{G}(r,z,\xi,\eta,t-\tau)\,d\xi\,d\eta\,d\tau.$$

Here,

$$\mathcal{G}(r,z,\xi,\eta,t) = 2\xi\exp\!\left(-\tfrac{1}{2}kt\right)\!\left[\frac{\sin(t\sqrt{c})}{R^2 l\sqrt{c}} + \frac{1}{R^2 l}\sum_{n=0}^{\infty}\sum_{m=0}^{\infty}\frac{A_{nm}}{J_0^2(\mu_n)}J_0\!\left(\frac{\mu_n r}{R}\right)J_0\!\left(\frac{\mu_n \xi}{R}\right)\right.$$
$$\left.\times\cos\!\left(\frac{m\pi z}{l}\right)\cos\!\left(\frac{m\pi\eta}{l}\right)\frac{\sin(t\sqrt{\lambda_{nm}})}{\sqrt{\lambda_{nm}}}\right],$$

where

$$c = b - \frac{k^2}{4}, \quad \lambda_{nm} = \frac{a^2\mu_n^2}{R^2} + \frac{a^2\pi^2 m^2}{l^2} + b - \frac{k^2}{4}, \quad A_{nm} = \begin{cases} 0 & \text{for } m=0, n=0, \\ 1 & \text{for } m=0, n>0, \\ 2 & \text{for } m>0, \end{cases}$$

and the μ_n are zeros of the first-order Bessel function, $J_1(\mu) = 0$ ($\mu_0 = 0$).

5.4.3-3. Domain: $0 \le r \le R$, $0 \le z \le l$. Third boundary value problem.

A circular cylinder of finite length is considered. The following conditions are prescribed:

$$w = f_0(r,z) \quad \text{at} \quad t = 0 \quad \text{(initial condition)},$$
$$\partial_t w = f_1(r,z) \quad \text{at} \quad t = 0 \quad \text{(initial condition)},$$
$$\partial_r w + s_1 w = g_1(z,t) \quad \text{at} \quad r = R \quad \text{(boundary condition)},$$
$$\partial_z w - s_2 w = g_2(r,t) \quad \text{at} \quad z = 0 \quad \text{(boundary condition)},$$
$$\partial_z w + s_3 w = g_3(r,t) \quad \text{at} \quad z = l \quad \text{(boundary condition)}.$$

The solution $w(r,z,t)$ is determined by the formula in Paragraph 5.4.3-2 where

$$\mathcal{G}(r,z,\xi,\eta,t) = \frac{2\xi}{R^2}\exp\!\left(-\tfrac{1}{2}kt\right)\sum_{n=1}^{\infty}\sum_{m=1}^{\infty} A_n J_0\!\left(\frac{\mu_n r}{R}\right)J_0\!\left(\frac{\mu_n \xi}{R}\right)\frac{\varphi_m(z)\varphi_m(\eta)}{\|\varphi_m\|^2}\frac{\sin(t\sqrt{\lambda_{nm}})}{\sqrt{\lambda_{nm}}}.$$

Here,

$$A_n = \frac{\mu_n^2}{(s_1^2 R^2 + \mu_n^2)J_0^2(\mu_n)}, \quad \lambda_{nm} = \frac{a^2\mu_n^2}{R^2} + a^2\beta_m^2 + b - \frac{k^2}{4},$$

$$\varphi_m(z) = \cos(\beta_m z) + \frac{s_2}{\beta_m}\sin(\beta_m z), \quad \|\varphi_m\|^2 = \frac{s_3}{2\beta_m^2}\frac{\beta_m^2 + s_2^2}{\beta_m^2 + s_3^2} + \frac{s_2}{2\beta_m^2} + \frac{l}{2}\!\left(1 + \frac{s_2^2}{\beta_m^2}\right);$$

the μ_n and β_m are positive roots of the transcendental equations

$$\mu J_1(\mu) - s_1 R J_0(\mu) = 0, \qquad \frac{\tan(\beta l)}{\beta} = \frac{s_2 + s_3}{\beta^2 - s_2 s_3}.$$

5.4. TELEGRAPH EQUATION $\frac{\partial^2 w}{\partial t^2} + k \frac{\partial w}{\partial t} = a^2 \Delta_2 w - bw + \Phi(x, y, t)$ 389

5.4.3-4. Domain: $0 \leq r \leq R$, $0 \leq z \leq l$. Mixed boundary value problems.

1°. A circular cylinder of finite length is considered. The following conditions are prescribed:

$$w = f_0(r, z) \quad \text{at} \quad t = 0 \quad \text{(initial condition)},$$
$$\partial_t w = f_1(r, z) \quad \text{at} \quad t = 0 \quad \text{(initial condition)},$$
$$w = g_1(z, t) \quad \text{at} \quad r = R \quad \text{(boundary condition)},$$
$$\partial_z w = g_2(r, t) \quad \text{at} \quad z = 0 \quad \text{(boundary condition)},$$
$$\partial_z w = g_3(r, t) \quad \text{at} \quad z = l \quad \text{(boundary condition)}.$$

Solution:

$$w(r, z, t) = \frac{\partial}{\partial t} \int_0^l \int_0^R f_0(\xi, \eta) \mathcal{G}(r, z, \xi, \eta, t) \, d\xi \, d\eta$$
$$+ \int_0^l \int_0^R [f_1(\xi, \eta) + k f_0(\xi, \eta)] \mathcal{G}(r, z, \xi, \eta, t) \, d\xi \, d\eta$$
$$- a^2 \int_0^t \int_0^l g_1(\eta, \tau) \left[\frac{\partial}{\partial \xi} \mathcal{G}(r, z, \xi, \eta, t - \tau) \right]_{\xi=R} d\eta \, d\tau$$
$$- a^2 \int_0^t \int_0^R g_2(\xi, \tau) \mathcal{G}(r, z, \xi, 0, t - \tau) \, d\xi \, d\tau + a^2 \int_0^t \int_0^R g_3(\xi, \tau) \mathcal{G}(r, z, \xi, l, t - \tau) \, d\xi \, d\tau$$
$$+ \int_0^t \int_0^l \int_0^R \Phi(\xi, \eta, \tau) \mathcal{G}(r, z, \xi, \eta, t - \tau) \, d\xi \, d\eta \, d\tau.$$

Here,

$$\mathcal{G}(r, z, \xi, \eta, t) = \frac{2\xi e^{-kt/2}}{R^2 l} \sum_{n=1}^{\infty} \sum_{m=0}^{\infty} \frac{A_m}{J_1^2(\mu_n)} J_0\left(\frac{\mu_n r}{R}\right) J_0\left(\frac{\mu_n \xi}{R}\right) \cos\left(\frac{m\pi z}{l}\right) \cos\left(\frac{m\pi \eta}{l}\right) \frac{\sin(\lambda_{nm} t)}{\lambda_{nm}},$$

$$\lambda_{nm} = \sqrt{\frac{a^2 \mu_n^2}{R^2} + \frac{a^2 \pi^2 m^2}{l^2} + b - \frac{k^2}{4}}, \quad A_m = \begin{cases} 1 & \text{for } m = 0, \\ 2 & \text{for } m > 0, \end{cases}$$

where the μ_n are zeros of the Bessel function, $J_0(\mu) = 0$.

2°. A circular cylinder of finite length is considered. The following conditions are prescribed:

$$w = f_0(r, z) \quad \text{at} \quad t = 0 \quad \text{(initial condition)},$$
$$\partial_t w = f_1(r, z) \quad \text{at} \quad t = 0 \quad \text{(initial condition)},$$
$$\partial_r w = g_1(z, t) \quad \text{at} \quad r = R \quad \text{(boundary condition)},$$
$$w = g_2(r, t) \quad \text{at} \quad z = 0 \quad \text{(boundary condition)},$$
$$w = g_3(r, t) \quad \text{at} \quad z = l \quad \text{(boundary condition)}.$$

Solution:

$$w(r, z, t) = \frac{\partial}{\partial t} \int_0^l \int_0^R f_0(\xi, \eta) \mathcal{G}(r, z, \xi, \eta, t) \, d\xi \, d\eta$$
$$+ \int_0^l \int_0^R [f_1(\xi, \eta) + k f_0(\xi, \eta)] \mathcal{G}(r, z, \xi, \eta, t) \, d\xi \, d\eta + a^2 \int_0^t \int_0^l g_1(\eta, \tau) \mathcal{G}(r, z, R, \eta, t - \tau) \, d\eta \, d\tau$$
$$+ a^2 \int_0^t \int_0^R g_2(\xi, \tau) \left[\frac{\partial}{\partial \eta} \mathcal{G}(r, z, \xi, \eta, t - \tau) \right]_{\eta=0} d\xi \, d\tau$$
$$- a^2 \int_0^t \int_0^R g_3(\xi, \tau) \left[\frac{\partial}{\partial \eta} \mathcal{G}(r, z, \xi, \eta, t - \tau) \right]_{\eta=l} d\xi \, d\tau$$
$$+ \int_0^t \int_0^l \int_0^R \Phi(\xi, \eta, \tau) \mathcal{G}(r, z, \xi, \eta, t - \tau) \, d\xi \, d\eta \, d\tau.$$

Here,

$$\mathcal{G}(r,z,\xi,\eta,t) = \frac{4\xi e^{-kt/2}}{R^2 l} \sum_{n=0}^{\infty} \sum_{m=1}^{\infty} \frac{1}{J_0^2(\mu_n)} J_0\!\left(\frac{\mu_n r}{R}\right) J_0\!\left(\frac{\mu_n \xi}{R}\right) \sin\!\left(\frac{m\pi z}{l}\right) \sin\!\left(\frac{m\pi \eta}{l}\right) \frac{\sin(\lambda_{nm} t)}{\lambda_{nm}},$$

$$\lambda_{nm} = \sqrt{\frac{a^2 \mu_n^2}{R^2} + \frac{a^2 \pi^2 m^2}{l^2} + b - \frac{k^2}{4}},$$

where the μ_n are zeros of the first-order Bessel function, $J_1(\mu) = 0$ ($\mu_0 = 0$).

5.4.3-5. Domain: $R_1 \leq r \leq R_2$, $0 \leq z \leq l$. First boundary value problem.

A hollow circular cylinder of finite length is considered. The following conditions are prescribed:

$$\begin{aligned}
w &= f_0(r,z) & \text{at} \quad & t = 0 & \text{(initial condition)}, \\
\partial_t w &= f_1(r,z) & \text{at} \quad & t = 0 & \text{(initial condition)}, \\
w &= g_1(z,t) & \text{at} \quad & r = R_1 & \text{(boundary condition)}, \\
w &= g_2(z,t) & \text{at} \quad & r = R_2 & \text{(boundary condition)}, \\
w &= g_3(r,t) & \text{at} \quad & z = 0 & \text{(boundary condition)}, \\
w &= g_4(r,t) & \text{at} \quad & z = l & \text{(boundary condition)}.
\end{aligned}$$

Solution:

$$\begin{aligned}
w(r,z,t) = {}& \frac{\partial}{\partial t} \int_0^l \int_{R_1}^{R_2} f_0(\xi,\eta) \mathcal{G}(r,z,\xi,\eta,t)\, d\xi\, d\eta \\
& + \int_0^l \int_{R_1}^{R_2} \bigl[f_1(\xi,\eta) + k f_0(\xi,\eta)\bigr] \mathcal{G}(r,z,\xi,\eta,t)\, d\xi\, d\eta \\
& + a^2 \int_0^t \int_0^l g_1(\eta,\tau) \left[\frac{\partial}{\partial \xi} \mathcal{G}(r,z,\xi,\eta,t-\tau)\right]_{\xi = R_1} d\eta\, d\tau \\
& - a^2 \int_0^t \int_0^l g_2(\eta,\tau) \left[\frac{\partial}{\partial \xi} \mathcal{G}(r,z,\xi,\eta,t-\tau)\right]_{\xi = R_2} d\eta\, d\tau \\
& + a^2 \int_0^t \int_{R_1}^{R_2} g_3(\xi,\tau) \left[\frac{\partial}{\partial \eta} \mathcal{G}(r,z,\xi,\eta,t-\tau)\right]_{\eta = 0} d\xi\, d\tau \\
& - a^2 \int_0^t \int_{R_1}^{R_2} g_4(\xi,\tau) \left[\frac{\partial}{\partial \eta} \mathcal{G}(r,z,\xi,\eta,t-\tau)\right]_{\eta = l} d\xi\, d\tau \\
& + \int_0^t \int_0^l \int_{R_1}^{R_2} \Phi(\xi,\eta,\tau) \mathcal{G}(r,z,\xi,\eta,t-\tau)\, d\xi\, d\eta\, d\tau.
\end{aligned}$$

Here,

$$\mathcal{G}(r,z,\xi,\eta,t) = \frac{\pi^2 \xi}{R_1^2 l} e^{-kt/2} \sum_{n=1}^{\infty} \sum_{m=1}^{\infty} \frac{\mu_n^2 J_0^2(s\mu_n) \Psi_n(r) \Psi_n(\xi)}{J_0^2(\mu_n) - J_0^2(s\mu_n)} \sin\!\left(\frac{m\pi z}{l}\right) \sin\!\left(\frac{m\pi \eta}{l}\right) \frac{\sin\!\left(t\sqrt{\lambda_{nm}}\right)}{\sqrt{\lambda_{nm}}},$$

$$\Psi_n(r) = Y_0(\mu_n) J_0\!\left(\frac{\mu_n r}{R_1}\right) - J_0(\mu_n) Y_0\!\left(\frac{\mu_n r}{R_1}\right), \quad s = \frac{R_2}{R_1}, \quad \lambda_{nm} = \frac{a^2 \mu_n^2}{R_1^2} + \frac{a^2 \pi^2 m^2}{l^2} + b - \frac{k^2}{4},$$

where $J_0(\mu)$ and $Y_0(\mu)$ are the Bessel functions, and the μ_n are positive roots of the transcendental equation

$$J_0(\mu) Y_0(s\mu) - J_0(s\mu) Y_0(\mu) = 0.$$

5.5. Other Equations with Two Space Variables

1. $\dfrac{\partial^2 w}{\partial t^2} + k\dfrac{\partial w}{\partial t} = a^2\left(\dfrac{\partial^2 w}{\partial x^2} + \dfrac{\partial^2 w}{\partial y^2}\right) + b_1\dfrac{\partial w}{\partial x} + b_2\dfrac{\partial w}{\partial y} + cw.$

The transformation

$$w(x,y,t) = u(x,y,\tau)\exp\left(-\tfrac{1}{2}kt - \dfrac{b_1 x + b_2 y}{2a^2}\right), \quad \tau = at$$

leads to the equation from Subsection 5.1.3:

$$\dfrac{\partial^2 u}{\partial \tau^2} = \dfrac{\partial^2 u}{\partial x^2} + \dfrac{\partial^2 u}{\partial y^2} + \beta u, \quad \beta = \dfrac{c}{a^2} + \dfrac{k^2}{4a^2} - \dfrac{1}{4a^4}(b_1^2 + b_2^2).$$

2. $t^m \dfrac{\partial^2 w}{\partial t^2} + \dfrac{m}{2} t^{m-1} \dfrac{\partial w}{\partial t} = \dfrac{\partial^2 w}{\partial x^2} + \dfrac{\partial^2 w}{\partial y^2}.$

Domain: $-\infty < x < \infty$, $-\infty < y < \infty$. Cauchy problem.
Initial conditions are prescribed:

$$w = f(x,y) \quad \text{at} \quad t = 0,$$
$$t^{m/2}\partial_t w = g(x,y) \quad \text{at} \quad t = 0.$$

Solution for $1 \leq m < 2$:

$$w(x,y,t) = \dfrac{1}{2\pi} t^{m/2} \dfrac{\partial}{\partial t} \iint_{C_t} \dfrac{f(\xi,\eta)\,d\xi\,d\eta}{\sqrt{k_m^2 t^{2-m} - \rho^2}} + \dfrac{1}{2\pi} \iint_{C_t} \dfrac{g(\xi,\eta)\,d\xi\,d\eta}{\sqrt{k_m^2 t^{2-m} - \rho^2}},$$

$$k_m = \dfrac{2}{2-m}, \quad \rho = \sqrt{(x-\xi)^2 + (y-\eta)^2},$$

where $C_t = \{\rho^2 \leq k_m^2 t^{2-m}\}$ is the circle with center at (x,y) and radius $k_m t^{1/k_m}$.
⊙ *Reference*: M. M. Smirnov (1975).

Chapter 6

Hyperbolic Equations with Three or More Space Variables

6.1. Wave Equation $\dfrac{\partial^2 w}{\partial t^2} = a^2 \Delta_3 w$

6.1.1. Problems in Cartesian Coordinates

The wave equation with three space variables in the rectangular Cartesian coordinate system has the form

$$\frac{\partial^2 w}{\partial t^2} = a^2 \left(\frac{\partial^2 w}{\partial x^2} + \frac{\partial^2 w}{\partial y^2} + \frac{\partial^2 w}{\partial z^2} \right).$$

This equation is of fundamental importance in sound propagation theory, the propagation of electromagnetic fields theory, and a number of other areas of physics and mechanics.

6.1.1-1. Particular solutions and their properties.

1°. Particular solutions:

$$w(x,y,z,t) = A \exp\left(k_1 x + k_2 y + k_3 z \pm at \sqrt{k_1^2 + k_2^2 + k_3^2} \right),$$

$$w(x,y,z,t) = A \sin(k_1 x + C_1) \sin(k_2 y + C_2) \sin(k_3 z + C_3) \sin\left(at \sqrt{k_1^2 + k_2^2 + k_3^2} \right),$$

$$w(x,y,z,t) = A \sin(k_1 x + C_1) \sin(k_2 y + C_2) \sin(k_3 z + C_3) \cos\left(at \sqrt{k_1^2 + k_2^2 + k_3^2} \right),$$

$$w(x,y,z,t) = A \sinh(k_1 x + C_1) \sinh(k_2 y + C_2) \sinh(k_3 z + C_3) \sinh\left(at \sqrt{k_1^2 + k_2^2 + k_3^2} \right),$$

$$w(x,y,z,t) = A \sinh(k_1 x + C_1) \sinh(k_2 y + C_2) \sinh(k_3 z + C_3) \cosh\left(at \sqrt{k_1^2 + k_2^2 + k_3^2} \right),$$

where A, C_1, C_2, C_3, k_1, k_2, and k_3 are arbitrary constants.

2°. Fundamental solution:

$$\mathscr{E}(x,y,z,t) = \frac{1}{2\pi a} \delta(a^2 t^2 - r^2), \qquad r = \sqrt{x^2 + y^2 + z^2},$$

where $\delta(\xi)$ is the Dirac delta function.

⊙ *Reference*: V. S. Vladimirov (1988).

3°. Infinite series solutions containing arbitrary functions of space variables:

$$w(x,y,z,t) = f(x,y,z) + \sum_{n=1}^{\infty} \frac{(at)^{2n}}{(2n)!} \Delta^n f(x,y,z), \qquad \Delta \equiv \frac{\partial^2}{\partial x^2} + \frac{\partial^2}{\partial y^2} + \frac{\partial^2}{\partial z^2},$$

$$w(x,y,z,t) = tg(x,y,z) + t \sum_{n=1}^{\infty} \frac{(at)^{2n}}{(2n+1)!} \Delta^n g(x,y,z),$$

where $f(x, y, z)$ and $g(x, y, z)$ are any infinitely differentiable functions. The first solution satisfies the initial conditions $w(x, y, z, 0) = f(x, y, z)$, $\partial_t w(x, y, z, 0) = 0$, and the second solution initial conditions $w(x, y, z, 0) = 0$, $\partial_t w(x, y, z, 0) = g(x, y, z)$. The sums are finite if $f(x, y, z)$ and $g(x, y, z)$ are polynomials in x, y, z.

⊙ *Reference*: A. V. Bitsadze and D. F. Kalinichenko (1985).

4°. Suppose $w = w(x, y, z, t)$ is a solution of the wave equation. Then the functions

$$w_1 = Aw(\pm \lambda x + C_1, \pm \lambda y + C_2, \pm \lambda z + C_3, \pm \lambda t + C_4),$$

$$w_2 = Aw\left(\frac{x - vt}{\sqrt{1 - (v/a)^2}}, y, z, \frac{t - va^{-2}x}{\sqrt{1 - (v/a)^2}}\right),$$

$$w_3 = \frac{A}{r^2 - a^2 t^2} w\left(\frac{x}{r^2 - a^2 t^2}, \frac{y}{r^2 - a^2 t^2}, \frac{z}{r^2 - a^2 t^2}, \frac{t}{r^2 - a^2 t^2}\right),$$

where A, C_n, v, and λ are arbitrary constants, are also solutions of the equation. The signs at λ in the expression of w_1 can be taken independently of one another. The function w_2 is a consequence of the invariance of the wave equation under the Lorentz transformation.

⊙ *References*: G. N. Polozhii (1964), W. Miller, Jr. (1977), A. V. Bitsadze and D. F. Kalinichenko (1985).

6.1.1-2. Domain: $-\infty < x < \infty$, $-\infty < y < \infty$, $-\infty < z < \infty$. Cauchy problem.

Initial conditions are prescribed:

$$w = f(x, y, z) \quad \text{at} \quad t = 0,$$
$$\partial_t w = g(x, y, z) \quad \text{at} \quad t = 0.$$

Solution (Kirchhoff's formula):

$$w(x, y, z, t) = \frac{1}{4\pi a} \frac{\partial}{\partial t} \iint_{S_{at}} \frac{f(\xi, \eta, \zeta)}{r} dS + \frac{1}{4\pi a} \iint_{S_{at}} \frac{g(\xi, \eta, \zeta)}{r} dS,$$
$$r = \sqrt{(\xi - x)^2 + (\eta - y)^2 + (\zeta - z)^2},$$

where the integration is performed over the surface of the sphere of radius at with center at (x, y, z).

⊙ *References*: N. S. Koshlyakov, E. B. Glizer, and M. M. Smirnov (1970), A. N. Tikhonov and A. A. Samarskii (1990).

6.1.1-3. Domain: $0 \leq x \leq l_1$, $0 \leq y \leq l_2$, $0 \leq z \leq l_3$. First boundary value problem.

A rectangular parallelepiped is considered. The following conditions are prescribed:

$$w = f_0(x, y, z) \quad \text{at} \quad t = 0 \quad \text{(initial condition)},$$
$$\partial_t w = f_1(x, y, z) \quad \text{at} \quad t = 0 \quad \text{(initial condition)},$$
$$w = g_1(y, z, t) \quad \text{at} \quad x = 0 \quad \text{(boundary condition)},$$
$$w = g_2(y, z, t) \quad \text{at} \quad x = l_1 \quad \text{(boundary condition)},$$
$$w = g_3(x, z, t) \quad \text{at} \quad y = 0 \quad \text{(boundary condition)},$$
$$w = g_4(x, z, t) \quad \text{at} \quad y = l_2 \quad \text{(boundary condition)},$$
$$w = g_5(x, y, t) \quad \text{at} \quad z = 0 \quad \text{(boundary condition)},$$
$$w = g_6(x, y, t) \quad \text{at} \quad z = l_3 \quad \text{(boundary condition)}.$$

Solution:

$$w(x,y,z,t) = \frac{\partial}{\partial t}\int_0^{l_3}\int_0^{l_2}\int_0^{l_1} f_0(\xi,\eta,\zeta) G(x,y,z,\xi,\eta,\zeta,t)\, d\xi\, d\eta\, d\zeta$$
$$+ \int_0^{l_3}\int_0^{l_2}\int_0^{l_1} f_1(\xi,\eta,\zeta) G(x,y,z,\xi,\eta,\zeta,t)\, d\xi\, d\eta\, d\zeta$$
$$+ a^2\int_0^t\int_0^{l_3}\int_0^{l_2} g_1(\eta,\zeta,\tau)\left[\frac{\partial}{\partial \xi}G(x,y,z,\xi,\eta,\zeta,t-\tau)\right]_{\xi=0} d\eta\, d\zeta\, d\tau$$
$$- a^2\int_0^t\int_0^{l_3}\int_0^{l_2} g_2(\eta,\zeta,\tau)\left[\frac{\partial}{\partial \xi}G(x,y,z,\xi,\eta,\zeta,t-\tau)\right]_{\xi=l_1} d\eta\, d\zeta\, d\tau$$
$$+ a^2\int_0^t\int_0^{l_3}\int_0^{l_1} g_3(\xi,\zeta,\tau)\left[\frac{\partial}{\partial \eta}G(x,y,z,\xi,\eta,\zeta,t-\tau)\right]_{\eta=0} d\xi\, d\zeta\, d\tau$$
$$- a^2\int_0^t\int_0^{l_3}\int_0^{l_1} g_4(\xi,\zeta,\tau)\left[\frac{\partial}{\partial \eta}G(x,y,z,\xi,\eta,\zeta,t-\tau)\right]_{\eta=l_2} d\xi\, d\zeta\, d\tau$$
$$+ a^2\int_0^t\int_0^{l_2}\int_0^{l_1} g_5(\xi,\eta,\tau)\left[\frac{\partial}{\partial \zeta}G(x,y,z,\xi,\eta,\zeta,t-\tau)\right]_{\zeta=0} d\xi\, d\eta\, d\tau$$
$$- a^2\int_0^t\int_0^{l_2}\int_0^{l_1} g_6(\xi,\eta,\tau)\left[\frac{\partial}{\partial \zeta}G(x,y,z,\xi,\eta,\zeta,t-\tau)\right]_{\zeta=l_3} d\xi\, d\eta\, d\tau.$$

Here,

$$G(x,y,z,\xi,\eta,\zeta,t) = \frac{8}{al_1l_2l_3}\sum_{n=1}^{\infty}\sum_{m=1}^{\infty}\sum_{k=1}^{\infty}\frac{1}{\lambda_{nmk}}\sin(\alpha_n x)\sin(\beta_m y)\sin(\gamma_k z)$$
$$\times \sin(\alpha_n \xi)\sin(\beta_m \eta)\sin(\gamma_k \zeta)\sin(a\lambda_{nmk}t),$$

where

$$\alpha_n = \frac{n\pi}{l_1}, \quad \beta_m = \frac{m\pi}{l_2}, \quad \gamma_k = \frac{k\pi}{l_3},$$
$$\lambda_{nmk} = \sqrt{\alpha_n^2 + \beta_m^2 + \gamma_k^2}.$$

6.1.1-4. Domain: $0 \le x \le l_1$, $0 \le y \le l_2$, $0 \le z \le l_3$. Second boundary value problem.

A rectangular parallelepiped is considered. The following conditions are prescribed:

$$w = f_0(x,y,z) \quad \text{at} \quad t=0 \quad \text{(initial condition)},$$
$$\partial_t w = f_1(x,y,z) \quad \text{at} \quad t=0 \quad \text{(initial condition)},$$
$$\partial_x w = g_1(y,z,t) \quad \text{at} \quad x=0 \quad \text{(boundary condition)},$$
$$\partial_x w = g_2(y,z,t) \quad \text{at} \quad x=l_1 \quad \text{(boundary condition)},$$
$$\partial_y w = g_3(x,z,t) \quad \text{at} \quad y=0 \quad \text{(boundary condition)},$$
$$\partial_y w = g_4(x,z,t) \quad \text{at} \quad y=l_2 \quad \text{(boundary condition)},$$
$$\partial_z w = g_5(x,y,t) \quad \text{at} \quad z=0 \quad \text{(boundary condition)},$$
$$\partial_z w = g_6(x,y,t) \quad \text{at} \quad z=l_3 \quad \text{(boundary condition)}.$$

Solution:

$$w(x,y,z,t) = \frac{\partial}{\partial t}\int_0^{l_3}\int_0^{l_2}\int_0^{l_1} f_0(\xi,\eta,\zeta)G(x,y,z,\xi,\eta,\zeta,t)\,d\xi\,d\eta\,d\zeta$$

$$+ \int_0^{l_3}\int_0^{l_2}\int_0^{l_1} f_1(\xi,\eta,\zeta)G(x,y,z,\xi,\eta,\zeta,t)\,d\xi\,d\eta\,d\zeta$$

$$- a^2\int_0^t\int_0^{l_3}\int_0^{l_2} g_1(\eta,\zeta,\tau)G(x,y,z,0,\eta,\zeta,t-\tau)\,d\eta\,d\zeta\,d\tau$$

$$+ a^2\int_0^t\int_0^{l_3}\int_0^{l_2} g_2(\eta,\zeta,\tau)G(x,y,z,l_1,\eta,\zeta,t-\tau)\,d\eta\,d\zeta\,d\tau$$

$$- a^2\int_0^t\int_0^{l_3}\int_0^{l_1} g_3(\xi,\zeta,\tau)G(x,y,z,\xi,0,\zeta,t-\tau)\,d\xi\,d\zeta\,d\tau$$

$$+ a^2\int_0^t\int_0^{l_3}\int_0^{l_1} g_4(\xi,\zeta,\tau)G(x,y,z,\xi,l_2,\zeta,t-\tau)\,d\xi\,d\zeta\,d\tau$$

$$- a^2\int_0^t\int_0^{l_2}\int_0^{l_1} g_5(\xi,\eta,\tau)G(x,y,z,\xi,\eta,0,t-\tau)\,d\xi\,d\eta\,d\tau$$

$$+ a^2\int_0^t\int_0^{l_2}\int_0^{l_1} g_6(\xi,\eta,\tau)G(x,y,z,\xi,\eta,l_3,t-\tau)\,d\xi\,d\eta\,d\tau,$$

where

$$G(x,y,z,\xi,\eta,\zeta,t) = \frac{t}{l_1l_2l_3} + \frac{1}{al_1l_2l_3}\sum_{n=0}^{\infty}\sum_{m=0}^{\infty}\sum_{k=0}^{\infty} \frac{A_nA_mA_k}{\lambda_{nmk}}\cos(\alpha_n x)\cos(\beta_m y)\cos(\gamma_k z)$$

$$\times \cos(\alpha_n\xi)\cos(\beta_m\eta)\cos(\gamma_k\zeta)\sin(a\lambda_{nmk}t),$$

$$\alpha_n = \frac{n\pi}{l_1},\quad \beta_m = \frac{m\pi}{l_2},\quad \gamma_k = \frac{k\pi}{l_3},\quad \lambda_{nmk} = \sqrt{\alpha_n^2+\beta_m^2+\gamma_k^2},\quad A_n = \begin{cases} 1 & \text{for } n=0, \\ 2 & \text{for } n>0. \end{cases}$$

The summation here is performed over the indices satisfying the condition $n+m+k > 0$; the term corresponding to $n = m = k = 0$ is singled out.

6.1.1-5. Domain: $0 \leq x \leq l_1$, $0 \leq y \leq l_2$, $0 \leq z \leq l_3$. Third boundary value problem.

A rectangular parallelepiped is considered. The following conditions are prescribed:

$$\begin{aligned}
w &= f_0(x,y,z) & \text{at}\quad & t = 0 & \text{(initial condition)}, \\
\partial_t w &= f_1(x,y,z) & \text{at}\quad & t = 0 & \text{(initial condition)}, \\
\partial_x w - s_1 w &= g_1(y,z,t) & \text{at}\quad & x = 0 & \text{(boundary condition)}, \\
\partial_x w + s_2 w &= g_2(y,z,t) & \text{at}\quad & x = l_1 & \text{(boundary condition)}, \\
\partial_y w - s_3 w &= g_3(x,z,t) & \text{at}\quad & y = 0 & \text{(boundary condition)}, \\
\partial_y w + s_4 w &= g_4(x,z,t) & \text{at}\quad & y = l_2 & \text{(boundary condition)}, \\
\partial_z w - s_5 w &= g_5(x,y,t) & \text{at}\quad & z = 0 & \text{(boundary condition)}, \\
\partial_z w + s_6 w &= g_6(x,y,t) & \text{at}\quad & z = l_3 & \text{(boundary condition)}.
\end{aligned}$$

The solution $w(x,y,z,t)$ is determined by the formula in Paragraph 6.1.1-4 where

$$G(x,y,\xi,\eta,t) = \frac{8}{a}\sum_{n=1}^{\infty}\sum_{m=1}^{\infty}\sum_{k=1}^{\infty}\frac{1}{E_{nmk}\sqrt{\alpha_n^2+\beta_m^2+\gamma_k^2}}\sin(\alpha_n x+\varepsilon_n)\sin(\beta_m y+\sigma_m)\sin(\gamma_k z+\nu_k)$$

$$\times \sin(\alpha_n\xi+\varepsilon_n)\sin(\beta_m\eta+\sigma_m)\sin(\gamma_k\zeta+\nu_k)\sin\!\left(at\sqrt{\alpha_n^2+\beta_m^2+\gamma_k^2}\right)$$

with
$$\varepsilon_n = \arctan\frac{\alpha_n}{l_1}, \quad \sigma_m = \arctan\frac{\beta_m}{l_2}, \quad \nu_k = \arctan\frac{\gamma_k}{l_3},$$

$$E_{nmk} = \left[l_1 + \frac{(s_1 s_2 + \alpha_n^2)(s_1 + s_2)}{(s_1^2 + \alpha_n^2)(s_2^2 + \alpha_n^2)}\right]\left[l_2 + \frac{(s_3 s_4 + \beta_m^2)(s_3 + s_4)}{(s_3^2 + \beta_m^2)(s_4^2 + \beta_m^2)}\right]\left[l_3 + \frac{(s_5 s_6 + \gamma_k^2)(s_5 + s_6)}{(s_5^2 + \gamma_k^2)(s_6^2 + \gamma_k^2)}\right].$$

Here, the α_n, β_m, and γ_k are positive roots of the transcendental equations

$$\alpha^2 - s_1 s_2 = (s_1 + s_2)\alpha \cot(l_1 \alpha), \quad \beta^2 - s_3 s_4 = (s_3 + s_4)\beta \cot(l_2 \beta), \quad \gamma^2 - s_5 s_6 = (s_5 + s_6)\gamma \cot(l_3 \gamma).$$

6.1.1-6. Domain: $0 \le x \le l_1$, $0 \le y \le l_2$, $0 \le z \le l_3$. Mixed boundary value problems.

1°. A rectangular parallelepiped is considered. The following conditions are prescribed:

$$\begin{aligned}
w &= f_0(x,y,z) & \text{at} \quad t &= 0 & \text{(initial condition)},\\
\partial_t w &= f_1(x,y,z) & \text{at} \quad t &= 0 & \text{(initial condition)},\\
w &= g_1(y,z,t) & \text{at} \quad x &= 0 & \text{(boundary condition)},\\
w &= g_2(y,z,t) & \text{at} \quad x &= l_1 & \text{(boundary condition)},\\
\partial_y w &= g_3(x,z,t) & \text{at} \quad y &= 0 & \text{(boundary condition)},\\
\partial_y w &= g_4(x,z,t) & \text{at} \quad y &= l_2 & \text{(boundary condition)},\\
\partial_z w &= g_5(x,y,t) & \text{at} \quad z &= 0 & \text{(boundary condition)},\\
\partial_z w &= g_6(x,y,t) & \text{at} \quad z &= l_3 & \text{(boundary condition)}.
\end{aligned}$$

Solution:

$$\begin{aligned}
w(x,y,z,t) = &\frac{\partial}{\partial t}\int_0^{l_3}\int_0^{l_2}\int_0^{l_1} f_0(\xi,\eta,\zeta)G(x,y,z,\xi,\eta,\zeta,t)\,d\xi\,d\eta\,d\zeta\\
&+ \int_0^{l_3}\int_0^{l_2}\int_0^{l_1} f_1(\xi,\eta,\zeta)G(x,y,z,\xi,\eta,\zeta,t)\,d\xi\,d\eta\,d\zeta\\
&+ a^2\int_0^t\int_0^{l_3}\int_0^{l_2} g_1(\eta,\zeta,\tau)\left[\frac{\partial}{\partial \xi}G(x,y,z,\xi,\eta,\zeta,t-\tau)\right]_{\xi=0} d\eta\,d\zeta\,d\tau\\
&- a^2\int_0^t\int_0^{l_3}\int_0^{l_2} g_2(\eta,\zeta,\tau)\left[\frac{\partial}{\partial \xi}G(x,y,z,\xi,\eta,\zeta,t-\tau)\right]_{\xi=l_1} d\eta\,d\zeta\,d\tau\\
&- a^2\int_0^t\int_0^{l_3}\int_0^{l_1} g_3(\xi,\zeta,\tau)G(x,y,z,\xi,0,\zeta,t-\tau)\,d\xi\,d\zeta\,d\tau\\
&+ a^2\int_0^t\int_0^{l_3}\int_0^{l_1} g_4(\xi,\zeta,\tau)G(x,y,z,\xi,l_2,\zeta,t-\tau)\,d\xi\,d\zeta\,d\tau\\
&- a^2\int_0^t\int_0^{l_2}\int_0^{l_1} g_5(\xi,\eta,\tau)G(x,y,z,\xi,\eta,0,t-\tau)\,d\xi\,d\eta\,d\tau\\
&+ a^2\int_0^t\int_0^{l_2}\int_0^{l_1} g_6(\xi,\eta,\tau)G(x,y,z,\xi,\eta,l_3,t-\tau)\,d\xi\,d\eta\,d\tau.
\end{aligned}$$

Here,

$$G(x,y,z,\xi,\eta,\zeta,t) = \frac{2}{a l_1 l_2 l_3}\sum_{n=1}^{\infty}\sum_{m=0}^{\infty}\sum_{k=0}^{\infty}\frac{A_m A_k}{\lambda_{nmk}}\sin(\alpha_n x)\cos(\beta_m y)\cos(\gamma_k z)$$
$$\times \sin(\alpha_n \xi)\cos(\beta_m \eta)\cos(\gamma_k \zeta)\sin(a\lambda_{nmk} t),$$

where
$$\alpha_n = \frac{n\pi}{l_1}, \qquad \beta_m = \frac{m\pi}{l_2}, \qquad \gamma_k = \frac{k\pi}{l_3},$$
$$\lambda_{nmk} = \sqrt{\alpha_n^2 + \beta_m^2 + \gamma_k^2}, \qquad A_m = \begin{cases} 1 & \text{for } m = 0, \\ 2 & \text{for } m > 0. \end{cases}$$

2°. A rectangular parallelepiped is considered. The following conditions are prescribed:

$$\begin{aligned}
w &= f_0(x,y,z) & \text{at} \quad t &= 0 & &\text{(initial condition)}, \\
\partial_t w &= f_1(x,y,z) & \text{at} \quad t &= 0 & &\text{(initial condition)}, \\
w &= g_1(y,z,t) & \text{at} \quad x &= 0 & &\text{(boundary condition)}, \\
\partial_x w &= g_2(y,z,t) & \text{at} \quad x &= l_1 & &\text{(boundary condition)}, \\
w &= g_3(x,z,t) & \text{at} \quad y &= 0 & &\text{(boundary condition)}, \\
\partial_y w &= g_4(x,z,t) & \text{at} \quad y &= l_2 & &\text{(boundary condition)}, \\
w &= g_5(x,y,t) & \text{at} \quad z &= 0 & &\text{(boundary condition)}, \\
\partial_z w &= g_6(x,y,t) & \text{at} \quad z &= l_3 & &\text{(boundary condition)}.
\end{aligned}$$

Solution:
$$\begin{aligned}
w(x,y,z,t) &= \frac{\partial}{\partial t} \int_0^{l_3} \int_0^{l_2} \int_0^{l_1} f_0(\xi,\eta,\zeta) G(x,y,z,\xi,\eta,\zeta,t) \, d\xi \, d\eta \, d\zeta \\
&+ \int_0^{l_3} \int_0^{l_2} \int_0^{l_1} f_1(\xi,\eta,\zeta) G(x,y,z,\xi,\eta,\zeta,t) \, d\xi \, d\eta \, d\zeta \\
&+ a^2 \int_0^t \int_0^{l_3} \int_0^{l_2} g_1(\eta,\zeta,\tau) \left[\frac{\partial}{\partial \xi} G(x,y,z,\xi,\eta,\zeta,t-\tau) \right]_{\xi=0} d\eta \, d\zeta \, d\tau \\
&+ a^2 \int_0^t \int_0^{l_3} \int_0^{l_2} g_2(\eta,\zeta,\tau) G(x,y,z,l_1,\eta,\zeta,t-\tau) \, d\eta \, d\zeta \, d\tau \\
&+ a^2 \int_0^t \int_0^{l_3} \int_0^{l_1} g_3(\xi,\zeta,\tau) \left[\frac{\partial}{\partial \eta} G(x,y,z,\xi,\eta,\zeta,t-\tau) \right]_{\eta=0} d\xi \, d\zeta \, d\tau \\
&+ a^2 \int_0^t \int_0^{l_3} \int_0^{l_1} g_4(\xi,\zeta,\tau) G(x,y,z,\xi,l_2,\zeta,t-\tau) \, d\xi \, d\zeta \, d\tau \\
&+ a^2 \int_0^t \int_0^{l_2} \int_0^{l_1} g_5(\xi,\eta,\tau) \left[\frac{\partial}{\partial \zeta} G(x,y,z,\xi,\eta,\zeta,t-\tau) \right]_{\zeta=0} d\xi \, d\eta \, d\tau \\
&+ a^2 \int_0^t \int_0^{l_2} \int_0^{l_1} g_6(\xi,\eta,\tau) G(x,y,z,\xi,\eta,l_3,t-\tau) \, d\xi \, d\eta \, d\tau.
\end{aligned}$$

Here,
$$G(x,y,z,\xi,\eta,\zeta,t) = \frac{8}{al_1 l_2 l_3} \sum_{n=1}^\infty \sum_{m=1}^\infty \sum_{k=1}^\infty \frac{1}{\lambda_{nmk}} \sin(\alpha_n x) \sin(\beta_m y) \sin(\gamma_k z)$$
$$\times \sin(\alpha_n \xi) \sin(\beta_m \eta) \sin(\gamma_k \zeta) \sin(a \lambda_{nmk} t),$$

where
$$\alpha_n = \frac{\pi(2n+1)}{2l_1}, \qquad \beta_m = \frac{\pi(2m+1)}{2l_2}, \qquad \gamma_k = \frac{\pi(2k+1)}{2l_3}, \qquad \lambda_{nmk} = \sqrt{\alpha_n^2 + \beta_m^2 + \gamma_k^2}.$$

6.1.2. Problems in Cylindrical Coordinates

The three-dimensional wave equation in the cylindrical coordinate system is written as

$$\frac{\partial^2 w}{\partial t^2} = a^2 \left[\frac{1}{r}\frac{\partial}{\partial r}\left(r\frac{\partial w}{\partial r}\right) + \frac{1}{r^2}\frac{\partial^2 w}{\partial \varphi^2} + \frac{\partial^2 w}{\partial z^2} \right], \qquad r = \sqrt{x^2 + y^2}.$$

One-dimensional problems with axial symmetry that have solutions $w = w(r, t)$ are considered in Subsection 4.2.1. Two-dimensional problems whose solutions have the form $w = w(r, \varphi, t)$ or $w = w(r, z, t)$ are discussed in Subsections 5.1.2 and 5.1.3.

6.1.2-1. Domain: $0 \leq r \leq R$, $0 \leq \varphi \leq 2\pi$, $0 \leq z \leq l$. First boundary value problem.

A circular cylinder of finite length is considered. The following conditions are prescribed:

$$\begin{aligned} w &= f_0(r,\varphi,z) & \text{at} \quad t &= 0 & \text{(initial condition)},\\ \partial_t w &= f_1(r,\varphi,z) & \text{at} \quad t &= 0 & \text{(initial condition)},\\ w &= g_1(\varphi,z,t) & \text{at} \quad r &= R & \text{(boundary condition)},\\ w &= g_2(r,\varphi,t) & \text{at} \quad z &= 0 & \text{(boundary condition)},\\ w &= g_3(r,\varphi,t) & \text{at} \quad z &= l & \text{(boundary condition)}. \end{aligned}$$

Solution:

$$\begin{aligned} w(r,\varphi,z,t) =& \frac{\partial}{\partial t}\int_0^l \int_0^{2\pi} \int_0^R \xi f_0(\xi,\eta,\zeta) G(r,\varphi,z,\xi,\eta,\zeta,t)\, d\xi\, d\eta\, d\zeta \\ &+ \int_0^l \int_0^{2\pi} \int_0^R \xi f_1(\xi,\eta,\zeta) G(r,\varphi,z,\xi,\eta,\zeta,t)\, d\xi\, d\eta\, d\zeta \\ &- a^2 R \int_0^t \int_0^l \int_0^{2\pi} g_1(\eta,\zeta,\tau)\left[\frac{\partial}{\partial \xi} G(r,\varphi,z,\xi,\eta,\zeta,t-\tau)\right]_{\xi=R} d\eta\, d\zeta\, d\tau \\ &+ a^2 \int_0^t \int_0^{2\pi} \int_0^R \xi g_2(\xi,\eta,\tau)\left[\frac{\partial}{\partial \zeta} G(r,\varphi,z,\xi,\eta,\zeta,t-\tau)\right]_{\zeta=0} d\xi\, d\eta\, d\tau \\ &- a^2 \int_0^t \int_0^{2\pi} \int_0^R \xi g_3(\xi,\eta,\tau)\left[\frac{\partial}{\partial \zeta} G(r,\varphi,z,\xi,\eta,\zeta,t-\tau)\right]_{\zeta=l} d\xi\, d\eta\, d\tau. \end{aligned}$$

Here,

$$G(r,\varphi,z,\xi,\eta,\zeta,t) = \frac{2}{\pi a R^2 l}\sum_{n=0}^{\infty}\sum_{m=1}^{\infty}\sum_{k=1}^{\infty} \frac{A_n}{[J_n'(\mu_{nm}R)]^2 \sqrt{\lambda_{nmk}}} J_n(\mu_{nm}r) J_n(\mu_{nm}\xi)$$

$$\times \cos[n(\varphi - \eta)] \sin\left(\frac{k\pi z}{l}\right) \sin\left(\frac{k\pi \zeta}{l}\right) \sin\!\left(at\sqrt{\lambda_{nmk}}\right),$$

$$\lambda_{nmk} = \mu_{nm}^2 + \frac{k^2\pi^2}{l^2}, \qquad A_n = \begin{cases} 1 & \text{for } n=0, \\ 2 & \text{for } n>0, \end{cases}$$

where the $J_n(\xi)$ are the Bessel functions (the prime denotes the derivative with respect to the argument) and the μ_{nm} are positive roots of the transcendental equation $J_n(\mu R) = 0$.

6.1.2-2. Domain: $0 \leq r \leq R$, $0 \leq \varphi \leq 2\pi$, $0 \leq z \leq l$. Second boundary value problem.

A circular cylinder of finite length is considered. The following conditions are prescribed:

$$\begin{aligned} w &= f_0(r,\varphi,z) & \text{at} \quad t &= 0 & \text{(initial condition)},\\ \partial_t w &= f_1(r,\varphi,z) & \text{at} \quad t &= 0 & \text{(initial condition)},\\ \partial_r w &= g_1(\varphi,z,t) & \text{at} \quad r &= R & \text{(boundary condition)},\\ \partial_z w &= g_2(r,\varphi,t) & \text{at} \quad z &= 0 & \text{(boundary condition)},\\ \partial_z w &= g_3(r,\varphi,t) & \text{at} \quad z &= l & \text{(boundary condition)}. \end{aligned}$$

Solution:

$$w(r,\varphi,z,t) = \frac{\partial}{\partial t}\int_0^l\int_0^{2\pi}\int_0^R \xi f_0(\xi,\eta,\zeta)G(r,\varphi,z,\xi,\eta,\zeta,t)\,d\xi\,d\eta\,d\zeta$$
$$+ \int_0^l\int_0^{2\pi}\int_0^R \xi f_1(\xi,\eta,\zeta)G(r,\varphi,z,\xi,\eta,\zeta,t)\,d\xi\,d\eta\,d\zeta$$
$$+ a^2R\int_0^t\int_0^l\int_0^{2\pi} g_1(\eta,\zeta,\tau)G(r,\varphi,z,R,\eta,\zeta,t-\tau)\,d\eta\,d\zeta\,d\tau$$
$$- a^2\int_0^t\int_0^{2\pi}\int_0^R \xi g_2(\xi,\eta,\tau)G(r,\varphi,z,\xi,\eta,0,t-\tau)\,d\xi\,d\eta\,d\tau$$
$$+ a^2\int_0^t\int_0^{2\pi}\int_0^R \xi g_3(\xi,\eta,\tau)G(r,\varphi,z,\xi,\eta,l,t-\tau)\,d\xi\,d\eta\,d\tau.$$

Here,

$$G(r,\varphi,z,\xi,\eta,\zeta,t) = \frac{t}{\pi R^2 l} + \frac{2}{\pi^2 aR^2}\sum_{k=1}^{\infty}\frac{1}{k}\cos\left(\frac{k\pi x}{l}\right)\cos\left(\frac{k\pi\xi}{l}\right)\sin\left(\frac{ak\pi t}{l}\right)$$
$$+ \frac{1}{\pi l}\sum_{n=0}^{\infty}\sum_{m=1}^{\infty}\sum_{k=0}^{\infty}\frac{A_n A_k \mu_{nm}^2 J_n(\mu_{nm}r)J_n(\mu_{nm}\xi)}{(\mu_{nm}^2 R^2 - n^2)[J_n(\mu_{nm}R)]^2}\cos[n(\varphi-\eta)]\cos\left(\frac{k\pi x}{l}\right)\cos\left(\frac{k\pi\xi}{l}\right)\frac{\sin(\lambda_{nmk}t)}{\lambda_{nmk}},$$

$$\lambda_{nmk} = a\sqrt{\mu_{nm}^2 + \frac{k^2\pi^2}{l^2}}, \qquad A_n = \begin{cases} 1 & \text{for } n=0, \\ 2 & \text{for } n>0, \end{cases}$$

where the $J_n(\xi)$ are the Bessel functions and the μ_{nm} are positive roots of the transcendental equation $J_n'(\mu R) = 0$.

6.1.2-3. Domain: $0 \le r \le R$, $0 \le \varphi \le 2\pi$, $0 \le z \le l$. Third boundary value problem.

A circular cylinder of finite length is considered. The following conditions are prescribed:

$$\begin{aligned}
w &= f_0(r,\varphi,z) & \text{at} \quad t &= 0 & &\text{(initial condition)}, \\
\partial_t w &= f_1(r,\varphi,z) & \text{at} \quad t &= 0 & &\text{(initial condition)}, \\
\partial_r w + k_1 w &= g(\varphi,z,t) & \text{at} \quad r &= R & &\text{(boundary condition)}, \\
\partial_z w - k_2 w &= g_2(r,\varphi,t) & \text{at} \quad z &= 0 & &\text{(boundary condition)}, \\
\partial_z w + k_3 w &= g_3(r,\varphi,t) & \text{at} \quad z &= l & &\text{(boundary condition)}.
\end{aligned}$$

The solution $w(r,\varphi,z,t)$ is determined by the formula in Paragraph 6.1.2-2 where

$$G(r,\varphi,z,\xi,\eta,\zeta,t) = \frac{1}{\pi}\sum_{n=0}^{\infty}\sum_{m=1}^{\infty}\sum_{s=1}^{\infty}\frac{A_n\mu_{nm}^2 J_n(\mu_{nm}r)J_n(\mu_{nm}\xi)\cos[n(\varphi-\eta)]h_s(z)h_s(\zeta)\sin(\lambda_{nms}t)}{(\mu_{nm}^2 R^2 + k_1^2 R^2 - n^2)[J_n(\mu_{nm}R)]^2\|h_s\|^2\lambda_{nms}},$$

$$\lambda_{nms} = a\sqrt{\mu_{nm}^2 + \beta_s^2}, \quad h_s(z) = \cos(\beta_s z) + \frac{k_2}{\beta_s}\sin(\beta_s z), \quad \|h_s\|^2 = \frac{k_3}{2\beta_s^2}\frac{\beta_s^2+k_2^2}{\beta_s^2+k_3^2} + \frac{k_2}{2\beta_s^2} + \frac{l}{2}\left(1+\frac{k_2^2}{\beta_s^2}\right).$$

Here, $A_0 = 1$ and $A_n = 2$ for $n = 1, 2, \ldots$; the $J_n(\xi)$ are the Bessel functions; and the μ_{nm} and β_s are positive roots of the transcendental equations

$$\mu J_n'(\mu R) + k_1 J_n(\mu R) = 0, \qquad \frac{\tan(\beta l)}{\beta} = \frac{k_2 + k_3}{\beta^2 - k_2 k_3}.$$

6.1.2-4. Domain: $0 \leq r \leq R$, $0 \leq \varphi \leq 2\pi$, $0 \leq z \leq l$. Mixed boundary value problems.

1°. A circular cylinder of finite length is considered. The following conditions are prescribed:

$$w = f_0(r, \varphi, z) \quad \text{at} \quad t = 0 \quad \text{(initial condition)},$$
$$\partial_t w = f_1(r, \varphi, z) \quad \text{at} \quad t = 0 \quad \text{(initial condition)},$$
$$w = g_1(\varphi, z, t) \quad \text{at} \quad r = R \quad \text{(boundary condition)},$$
$$\partial_z w = g_2(r, \varphi, t) \quad \text{at} \quad z = 0 \quad \text{(boundary condition)},$$
$$\partial_z w = g_3(r, \varphi, t) \quad \text{at} \quad z = l \quad \text{(boundary condition)}.$$

Solution:

$$w(r, \varphi, z, t) = \frac{\partial}{\partial t} \int_0^l \int_0^{2\pi} \int_0^R \xi f_0(\xi, \eta, \zeta) G(r, \varphi, z, \xi, \eta, \zeta, t) \, d\xi \, d\eta \, d\zeta$$
$$+ \int_0^l \int_0^{2\pi} \int_0^R \xi f_1(\xi, \eta, \zeta) G(r, \varphi, z, \xi, \eta, \zeta, t) \, d\xi \, d\eta \, d\zeta$$
$$- a^2 R \int_0^t \int_0^l \int_0^{2\pi} g_1(\eta, \zeta, \tau) \left[\frac{\partial}{\partial \xi} G(r, \varphi, z, \xi, \eta, \zeta, t-\tau)\right]_{\xi=R} d\eta \, d\zeta \, d\tau$$
$$- a^2 \int_0^t \int_0^{2\pi} \int_0^R \xi g_2(\xi, \eta, \tau) G(r, \varphi, z, \xi, \eta, 0, t-\tau) \, d\xi \, d\eta \, d\tau$$
$$+ a^2 \int_0^t \int_0^{2\pi} \int_0^R \xi g_3(\xi, \eta, \tau) G(r, \varphi, z, \xi, \eta, l, t-\tau) \, d\xi \, d\eta \, d\tau.$$

Here,

$$G(r, \varphi, z, \xi, \eta, \zeta, t) = \frac{1}{\pi a R^2 l} \sum_{n=0}^{\infty} \sum_{m=1}^{\infty} \sum_{k=0}^{\infty} \frac{A_n A_k}{[J_n'(\mu_{nm} R)]^2 \sqrt{\lambda_{nmk}}} J_n(\mu_{nm} r) J_n(\mu_{nm} \xi)$$
$$\times \cos[n(\varphi - \eta)] \cos\left(\frac{k\pi z}{l}\right) \cos\left(\frac{k\pi \zeta}{l}\right) \sin\left(at\sqrt{\lambda_{nmk}}\right),$$

$$\lambda_{nmk} = \mu_{nm}^2 + \frac{k^2 \pi^2}{l^2}, \quad A_n = \begin{cases} 1 & \text{for } n = 0, \\ 2 & \text{for } n > 0, \end{cases}$$

where the $J_n(\xi)$ are the Bessel functions (the prime denotes the derivative with respect to the argument) and the μ_{nm} are positive roots of the transcendental equation $J_n(\mu R) = 0$.

2°. A circular cylinder of finite length is considered. The following conditions are prescribed:

$$w = f_0(r, \varphi, z) \quad \text{at} \quad t = 0 \quad \text{(initial condition)},$$
$$\partial_t w = f_1(r, \varphi, z) \quad \text{at} \quad t = 0 \quad \text{(initial condition)},$$
$$\partial_r w = g_1(\varphi, z, t) \quad \text{at} \quad r = R \quad \text{(boundary condition)},$$
$$w = g_2(r, \varphi, t) \quad \text{at} \quad z = 0 \quad \text{(boundary condition)},$$
$$w = g_3(r, \varphi, t) \quad \text{at} \quad z = l \quad \text{(boundary condition)}.$$

Solution:

$$w(r, \varphi, z, t) = \frac{\partial}{\partial t} \int_0^l \int_0^{2\pi} \int_0^R \xi f_0(\xi, \eta, \zeta) G(r, \varphi, z, \xi, \eta, \zeta, t) \, d\xi \, d\eta \, d\zeta$$
$$+ \int_0^l \int_0^{2\pi} \int_0^R \xi f_1(\xi, \eta, \zeta) G(r, \varphi, z, \xi, \eta, \zeta, t) \, d\xi \, d\eta \, d\zeta$$
$$+ a^2 R \int_0^t \int_0^l \int_0^{2\pi} g_1(\eta, \zeta, \tau) G(r, \varphi, z, R, \eta, \zeta, t-\tau) \, d\eta \, d\zeta \, d\tau$$
$$+ a^2 \int_0^t \int_0^{2\pi} \int_0^R \xi g_2(\xi, \eta, \tau) \left[\frac{\partial}{\partial \zeta} G(r, \varphi, z, \xi, \eta, \zeta, t-\tau)\right]_{\zeta=0} d\xi \, d\eta \, d\tau$$
$$- a^2 \int_0^t \int_0^{2\pi} \int_0^R \xi g_3(\xi, \eta, \tau) \left[\frac{\partial}{\partial \zeta} G(r, \varphi, z, \xi, \eta, \zeta, t-\tau)\right]_{\zeta=l} d\xi \, d\eta \, d\tau.$$

Here,

$$G(r,\varphi,z,\xi,\eta,\zeta,t) = \frac{2}{\pi^2 a R^2} \sum_{k=1}^{\infty} \frac{1}{k} \sin\left(\frac{k\pi z}{l}\right) \sin\left(\frac{k\pi \zeta}{l}\right) \sin\left(\frac{k\pi a t}{l}\right)$$

$$+ \frac{2}{\pi a l} \sum_{n=0}^{\infty} \sum_{m=1}^{\infty} \sum_{k=1}^{\infty} \frac{A_n \mu_{nm}^2}{(\mu_{nm}^2 R^2 - n^2)[J_n(\mu_{nm} R)]^2 \sqrt{\lambda_{nmk}}} J_n(\mu_{nm} r) J_n(\mu_{nm} \xi)$$

$$\times \cos[n(\varphi - \eta)] \sin\left(\frac{k\pi z}{l}\right) \sin\left(\frac{k\pi \zeta}{l}\right) \sin\left(a t \sqrt{\lambda_{nmk}}\right),$$

$$\lambda_{nmk} = \mu_{nm}^2 + \frac{k^2 \pi^2}{l^2}, \quad A_n = \begin{cases} 1 & \text{for } n = 0, \\ 2 & \text{for } n > 0, \end{cases}$$

where the $J_n(\xi)$ are the Bessel functions and the μ_{nm} are positive roots of the transcendental equation $J_n'(\mu R) = 0$.

6.1.2-5. Domain: $R_1 \le r \le R_2$, $0 \le \varphi \le 2\pi$, $0 \le z \le l$. First boundary value problem.

A hollow circular cylinder of finite length is considered. The following conditions are prescribed:

$$\begin{aligned}
w &= f_0(r,\varphi,z) & \text{at} \quad t &= 0 & \text{(initial condition)}, \\
\partial_t w &= f_1(r,\varphi,z) & \text{at} \quad t &= 0 & \text{(initial condition)}, \\
w &= g_1(\varphi,z,t) & \text{at} \quad r &= R_1 & \text{(boundary condition)}, \\
w &= g_2(\varphi,z,t) & \text{at} \quad r &= R_2 & \text{(boundary condition)}, \\
w &= g_3(r,\varphi,t) & \text{at} \quad z &= 0 & \text{(boundary condition)}, \\
w &= g_4(r,\varphi,t) & \text{at} \quad z &= l & \text{(boundary condition)}.
\end{aligned}$$

Solution:

$$w(r,\varphi,z,t) = \frac{\partial}{\partial t} \int_0^l \int_0^{2\pi} \int_{R_1}^{R_2} f_0(\xi,\eta,\zeta) G(r,\varphi,z,\xi,\eta,\zeta,t) \xi\, d\xi\, d\eta\, d\zeta$$

$$+ \int_0^l \int_0^{2\pi} \int_{R_1}^{R_2} f_1(\xi,\eta,\zeta) G(r,\varphi,z,\xi,\eta,\zeta,t) \xi\, d\xi\, d\eta\, d\zeta$$

$$+ a^2 R_1 \int_0^t \int_0^l \int_0^{2\pi} g_1(\eta,\zeta,\tau) \left[\frac{\partial}{\partial \xi} G(r,\varphi,z,\xi,\eta,\zeta,t-\tau)\right]_{\xi=R_1} d\eta\, d\zeta\, d\tau$$

$$- a^2 R_2 \int_0^t \int_0^l \int_0^{2\pi} g_2(\eta,\zeta,\tau) \left[\frac{\partial}{\partial \xi} G(r,\varphi,z,\xi,\eta,\zeta,t-\tau)\right]_{\xi=R_2} d\eta\, d\zeta\, d\tau$$

$$+ a^2 \int_0^t \int_0^{2\pi} \int_{R_1}^{R_2} g_3(\xi,\eta,\tau) \left[\frac{\partial}{\partial \zeta} G(r,\varphi,z,\xi,\eta,\zeta,t-\tau)\right]_{\zeta=0} \xi\, d\xi\, d\eta\, d\tau$$

$$- a^2 \int_0^t \int_0^{2\pi} \int_{R_1}^{R_2} g_4(\xi,\eta,\tau) \left[\frac{\partial}{\partial \zeta} G(r,\varphi,z,\xi,\eta,\zeta,t-\tau)\right]_{\zeta=l} \xi\, d\xi\, d\eta\, d\tau.$$

Here,

$$G(r,\varphi,z,\xi,\eta,\zeta,t) = \frac{\pi}{2l} \sum_{n=0}^{\infty} \sum_{m=1}^{\infty} \sum_{k=1}^{\infty} \frac{A_n \mu_{nm}^2 J_n^2(\mu_{nm} R_2)}{J_n^2(\mu_{nm} R_1) - J_n^2(\mu_{nm} R_2)} Z_{nm}(r) Z_{nm}(\xi)$$

$$\times \cos[n(\varphi - \eta)] \sin\left(\frac{k\pi z}{l}\right) \sin\left(\frac{k\pi \zeta}{l}\right) \frac{\sin(a t \sqrt{\lambda_{nmk}})}{a \sqrt{\lambda_{nmk}}},$$

$$A_n = \begin{cases} 1 & \text{for } n = 0, \\ 2 & \text{for } n \ne 0, \end{cases} \quad \lambda_{nmk} = \mu_{nm}^2 + \frac{k^2 \pi^2}{l^2},$$

$$Z_{nm}(r) = J_n(\mu_{nm} R_1) Y_n(\mu_{nm} r) - Y_n(\mu_{nm} R_1) J_n(\mu_{nm} r),$$

where the $J_n(r)$ and $Y_n(r)$ are the Bessel functions, and the μ_{nm} are positive roots of the transcendental equation

$$J_n(\mu R_1)Y_n(\mu R_2) - Y_n(\mu R_1)J_n(\mu R_2) = 0.$$

6.1.2-6. Domain: $R_1 \leq r \leq R_2$, $0 \leq \varphi \leq 2\pi$, $0 \leq z \leq l$. Second boundary value problem.

A hollow circular cylinder of finite length is considered. The following conditions are prescribed:

$$\begin{aligned}
w &= f_0(r,\varphi,z) & \text{at} \quad t &= 0 & &\text{(initial condition)}, \\
\partial_t w &= f_1(r,\varphi,z) & \text{at} \quad t &= 0 & &\text{(initial condition)}, \\
\partial_r w &= g_1(\varphi,z,t) & \text{at} \quad r &= R_1 & &\text{(boundary condition)}, \\
\partial_r w &= g_2(\varphi,z,t) & \text{at} \quad r &= R_2 & &\text{(boundary condition)}, \\
\partial_z w &= g_3(r,\varphi,t) & \text{at} \quad z &= 0 & &\text{(boundary condition)}, \\
\partial_z w &= g_4(r,\varphi,t) & \text{at} \quad z &= l & &\text{(boundary condition)}.
\end{aligned}$$

Solution:

$$\begin{aligned}
w(r,\varphi,z,t) &= \frac{\partial}{\partial t} \int_0^l \int_0^{2\pi} \int_{R_1}^{R_2} f_0(\xi,\eta,\zeta) G(r,\varphi,z,\xi,\eta,\zeta,t)\xi\, d\xi\, d\eta\, d\zeta \\
&+ \int_0^l \int_0^{2\pi} \int_{R_1}^{R_2} f_1(\xi,\eta,\zeta) G(r,\varphi,z,\xi,\eta,\zeta,t)\xi\, d\xi\, d\eta\, d\zeta \\
&- a^2 R_1 \int_0^t \int_0^l \int_0^{2\pi} g_1(\eta,\zeta,\tau) G(r,\varphi,z,R_1,\eta,\zeta,t-\tau)\, d\eta\, d\zeta\, d\tau \\
&+ a^2 R_2 \int_0^t \int_0^l \int_0^{2\pi} g_2(\eta,\zeta,\tau) G(r,\varphi,z,R_2,\eta,\zeta,t-\tau)\, d\eta\, d\zeta\, d\tau \\
&- a^2 \int_0^t \int_0^{2\pi} \int_{R_1}^{R_2} g_3(\xi,\eta,\tau) G(r,\varphi,z,\xi,\eta,0,t-\tau)\xi\, d\xi\, d\eta\, d\tau \\
&+ a^2 \int_0^t \int_0^{2\pi} \int_{R_1}^{R_2} g_4(\xi,\eta,\tau) G(r,\varphi,z,\xi,\eta,l,t-\tau)\xi\, d\xi\, d\eta\, d\tau.
\end{aligned}$$

Here,

$$\begin{aligned}
G(r,\varphi,z,\xi,\eta,\zeta,t) &= \frac{t}{\pi(R_2^2-R_1^2)l} + \frac{2}{\pi^2 a(R_2^2-R_1^2)} \sum_{k=1}^{\infty} \frac{1}{k} \cos\left(\frac{k\pi z}{l}\right) \cos\left(\frac{k\pi\zeta}{l}\right) \sin\left(\frac{k\pi a t}{l}\right) \\
&+ \frac{1}{\pi l} \sum_{n=0}^{\infty} \sum_{m=1}^{\infty} \sum_{k=0}^{\infty} \frac{A_n A_k \mu_{nm}^2 Z_{nm}(r) Z_{nm}(\xi)}{(\mu_{nm}^2 R_2^2 - n^2) Z_{nm}^2(R_2) - (\mu_{nm}^2 R_1^2 - n^2) Z_{nm}^2(R_1)} \\
&\quad \times \cos[n(\varphi-\eta)] \cos\left(\frac{k\pi z}{l}\right) \cos\left(\frac{k\pi\zeta}{l}\right) \frac{\sin(at\sqrt{\lambda_{nmk}})}{a\sqrt{\lambda_{nmk}}},
\end{aligned}$$

where

$$A_n = \begin{cases} 1 & \text{for } n=0, \\ 2 & \text{for } n \neq 0, \end{cases} \qquad \lambda_{nmk} = \mu_{nm}^2 + \frac{k^2\pi^2}{l^2},$$

$$Z_{nm}(r) = J_n'(\mu_{nm}R_1)Y_n(\mu_{nm}r) - Y_n'(\mu_{nm}R_1)J_n(\mu_{nm}r);$$

the $J_n(r)$ and $Y_n(r)$ are the Bessel functions, and the μ_{nm} are positive roots of the transcendental equation

$$J_n'(\mu R_1)Y_n'(\mu R_2) - Y_n'(\mu R_1)J_n'(\mu R_2) = 0.$$

6.1.2-7. Domain: $R_1 \leq r \leq R_2$, $0 \leq \varphi \leq 2\pi$, $0 \leq z \leq l$. **Third boundary value problem.**

A hollow circular cylinder of finite length is considered. The following conditions are prescribed:

$$
\begin{aligned}
w &= f_0(r,\varphi,z) & \text{at} \quad t &= 0 & \text{(initial condition)}, \\
\partial_t w &= f_1(r,\varphi,z) & \text{at} \quad t &= 0 & \text{(initial condition)}, \\
\partial_r w - k_1 w &= g_1(\varphi,z,t) & \text{at} \quad r &= R_1 & \text{(boundary condition)}, \\
\partial_r w + k_2 w &= g_2(\varphi,z,t) & \text{at} \quad r &= R_2 & \text{(boundary condition)}, \\
\partial_z w - k_3 w &= g_3(r,\varphi,t) & \text{at} \quad z &= 0 & \text{(boundary condition)}, \\
\partial_z w + k_4 w &= g_4(r,\varphi,t) & \text{at} \quad z &= l & \text{(boundary condition)}.
\end{aligned}
$$

The solution $w(r,\varphi,z,t)$ is determined by the formula in Paragraph 6.1.2-6 where

$$
G(r,\varphi,z,\xi,\eta,\zeta,t) = \frac{1}{\pi a} \sum_{n=0}^{\infty} \sum_{m=1}^{\infty} \sum_{s=1}^{\infty} \frac{A_n \mu_{nm}^2}{\|h_s\|^2 \sqrt{\mu_{nm}^2 + \lambda_s^2}}
$$
$$
\times \frac{Z_{nm}(r) Z_{nm}(\xi) \cos[n(\varphi - \eta)] h_s(z) h_s(\zeta) \sin\left(at\sqrt{\mu_{nm}^2 + \lambda_s^2}\right)}{(k_2^2 R_2^2 + \mu_{nm}^2 R_2^2 - n^2) Z_{nm}^2(R_2) - (k_1^2 R_1^2 + \mu_{nm}^2 R_1^2 - n^2) Z_{nm}^2(R_1)}.
$$

Here,

$$
A_n = \begin{cases} 1 & \text{for } n = 0, \\ 2 & \text{for } n \neq 0, \end{cases} \quad
\begin{aligned}
Z_{nm}(r) &= \left[\mu_{nm} J_n'(\mu_{nm} R_1) - k_1 J_n(\mu_{nm} R_1)\right] Y_n(\mu_{nm} r) \\
&\quad - \left[\mu_{nm} Y_n'(\mu_{nm} R_1) - k_1 Y_n(\mu_{nm} R_1)\right] J_n(\mu_{nm} r),
\end{aligned}
$$

$$
h_s(z) = \cos(\lambda_s z) + \frac{k_3}{\lambda_s} \sin(\lambda_s z), \quad \|h_s\|^2 = \frac{k_4}{2\lambda_s^2} \frac{\lambda_s^2 + k_3^2}{\lambda_s^2 + k_4^2} + \frac{k_3}{2\lambda_s^2} + \frac{l}{2}\left(1 + \frac{k_3^2}{\lambda_s^2}\right),
$$

where the $J_n(r)$ and $Y_n(r)$ are the Bessel functions; the μ_{nm} are positive roots of the transcendental equation

$$
\left[\mu J_n'(\mu R_1) - k_1 J_n(\mu R_1)\right]\left[\mu Y_n'(\mu R_2) + k_2 Y_n(\mu R_2)\right]
$$
$$
= \left[\mu Y_n'(\mu R_1) - k_1 Y_n(\mu R_1)\right]\left[\mu J_n'(\mu R_2) + k_2 J_n(\mu R_2)\right];
$$

and the λ_s are positive roots of the transcendental equation $\dfrac{\tan(\lambda l)}{\lambda} = \dfrac{k_3 + k_4}{\lambda^2 - k_3 k_4}$.

6.1.2-8. Domain: $R_1 \leq r \leq R_2$, $0 \leq \varphi \leq 2\pi$, $0 \leq z \leq l$. **Mixed boundary value problems.**

1°. A hollow circular cylinder of finite length is considered. The following conditions are prescribed:

$$
\begin{aligned}
w &= f_0(r,\varphi,z) & \text{at} \quad t &= 0 & \text{(initial condition)}, \\
\partial_t w &= f_1(r,\varphi,z) & \text{at} \quad t &= 0 & \text{(initial condition)}, \\
w &= g_1(\varphi,z,t) & \text{at} \quad r &= R_1 & \text{(boundary condition)}, \\
w &= g_2(\varphi,z,t) & \text{at} \quad r &= R_2 & \text{(boundary condition)}, \\
\partial_z w &= g_3(r,\varphi,t) & \text{at} \quad z &= 0 & \text{(boundary condition)}, \\
\partial_z w &= g_4(r,\varphi,t) & \text{at} \quad z &= l & \text{(boundary condition)}.
\end{aligned}
$$

6.1. Wave Equation $\frac{\partial^2 w}{\partial t^2} = a^2 \Delta_3 w$

Solution:
$$w(r,\varphi,z,t) = \frac{\partial}{\partial t}\int_0^l \int_0^{2\pi}\int_{R_1}^{R_2} f_0(\xi,\eta,\zeta) G(r,\varphi,z,\xi,\eta,\zeta,t)\xi\, d\xi\, d\eta\, d\zeta$$
$$+ \int_0^l \int_0^{2\pi}\int_{R_1}^{R_2} f_1(\xi,\eta,\zeta) G(r,\varphi,z,\xi,\eta,\zeta,t)\xi\, d\xi\, d\eta\, d\zeta$$
$$+ a^2 R_1 \int_0^t \int_0^l \int_0^{2\pi} g_1(\eta,\zeta,\tau)\left[\frac{\partial}{\partial \xi} G(r,\varphi,z,\xi,\eta,\zeta,t-\tau)\right]_{\xi=R_1} d\eta\, d\zeta\, d\tau$$
$$- a^2 R_2 \int_0^t \int_0^l \int_0^{2\pi} g_2(\eta,\zeta,\tau)\left[\frac{\partial}{\partial \xi} G(r,\varphi,z,\xi,\eta,\zeta,t-\tau)\right]_{\xi=R_2} d\eta\, d\zeta\, d\tau$$
$$- a^2 \int_0^t \int_0^{2\pi}\int_{R_1}^{R_2} g_3(\xi,\eta,\tau) G(r,\varphi,z,\xi,\eta,0,t-\tau)\xi\, d\xi\, d\eta\, d\tau$$
$$+ a^2 \int_0^t \int_0^{2\pi}\int_{R_1}^{R_2} g_4(\xi,\eta,\tau) G(r,\varphi,z,\xi,\eta,l,t-\tau)\xi\, d\xi\, d\eta\, d\tau.$$

Here,
$$G(r,\varphi,z,\xi,\eta,\zeta,t) = \frac{\pi}{4l}\sum_{n=0}^\infty \sum_{m=1}^\infty \sum_{k=0}^\infty \frac{A_n A_k \mu_{nm}^2 J_n^2(\mu_{nm}R_2)}{J_n^2(\mu_{nm}R_1) - J_n^2(\mu_{nm}R_2)} Z_{nm}(r)Z_{nm}(\xi)$$
$$\times \cos[n(\varphi-\eta)]\cos\left(\frac{k\pi z}{l}\right)\cos\left(\frac{k\pi \zeta}{l}\right)\frac{\sin\left(at\sqrt{\lambda_{nmk}}\right)}{a\sqrt{\lambda_{nmk}}},$$
$$A_n = \begin{cases} 1 & \text{for } n=0, \\ 2 & \text{for } n\neq 0, \end{cases} \qquad \lambda_{nmk} = \mu_{nm}^2 + \frac{k^2\pi^2}{l^2},$$
$$Z_{nm}(r) = J_n(\mu_{nm}R_1)Y_n(\mu_{nm}r) - Y_n(\mu_{nm}R_1)J_n(\mu_{nm}r),$$

where the $J_n(r)$ and $Y_n(r)$ are the Bessel functions, and the μ_{nm} are positive roots of the transcendental equation
$$J_n(\mu R_1)Y_n(\mu R_2) - Y_n(\mu R_1)J_n(\mu R_2) = 0.$$

2°. A hollow circular cylinder of finite length is considered. The following conditions are prescribed:

$$\begin{array}{llll}
w = f_0(r,\varphi,z) & \text{at} & t=0 & \text{(initial condition)}, \\
\partial_t w = f_1(r,\varphi,z) & \text{at} & t=0 & \text{(initial condition)}, \\
\partial_r w = g_1(\varphi,z,t) & \text{at} & r=R_1 & \text{(boundary condition)}, \\
\partial_r w = g_2(\varphi,z,t) & \text{at} & r=R_2 & \text{(boundary condition)}, \\
w = g_3(r,\varphi,t) & \text{at} & z=0 & \text{(boundary condition)}, \\
w = g_4(r,\varphi,t) & \text{at} & z=l & \text{(boundary condition)}.
\end{array}$$

Solution:
$$w(r,\varphi,z,t) = \frac{\partial}{\partial t}\int_0^l \int_0^{2\pi}\int_{R_1}^{R_2} f_0(\xi,\eta,\zeta) G(r,\varphi,z,\xi,\eta,\zeta,t)\xi\, d\xi\, d\eta\, d\zeta$$
$$+ \int_0^l \int_0^{2\pi}\int_{R_1}^{R_2} f_1(\xi,\eta,\zeta) G(r,\varphi,z,\xi,\eta,\zeta,t)\xi\, d\xi\, d\eta\, d\zeta$$
$$- a^2 R_1 \int_0^t \int_0^l \int_0^{2\pi} g_1(\eta,\zeta,\tau) G(r,\varphi,z,R_1,\eta,\zeta,t-\tau)\, d\eta\, d\zeta\, d\tau$$
$$+ a^2 R_2 \int_0^t \int_0^l \int_0^{2\pi} g_2(\eta,\zeta,\tau) G(r,\varphi,z,R_2,\eta,\zeta,t-\tau)\, d\eta\, d\zeta\, d\tau$$
$$+ a^2 \int_0^t \int_0^{2\pi}\int_{R_1}^{R_2} g_3(\xi,\eta,\tau)\left[\frac{\partial}{\partial \zeta} G(r,\varphi,z,\xi,\eta,\zeta,t-\tau)\right]_{\zeta=0} \xi\, d\xi\, d\eta\, d\tau$$
$$- a^2 \int_0^t \int_0^{2\pi}\int_{R_1}^{R_2} g_4(\xi,\eta,\tau)\left[\frac{\partial}{\partial \zeta} G(r,\varphi,z,\xi,\eta,\zeta,t-\tau)\right]_{\zeta=l} \xi\, d\xi\, d\eta\, d\tau.$$

Here,

$$G(r,\varphi,z,\xi,\eta,\zeta,t) = \frac{2}{\pi^2 a(R_2^2 - R_1^2)} \sum_{k=1}^{\infty} \frac{1}{k} \sin\left(\frac{k\pi z}{l}\right) \sin\left(\frac{k\pi \zeta}{l}\right) \sin\left(\frac{k\pi a t}{l}\right)$$

$$+ \frac{2}{\pi l} \sum_{n=0}^{\infty} \sum_{m=1}^{\infty} \sum_{k=1}^{\infty} \frac{A_n \mu_{nm}^2 Z_{nm}(r) Z_{nm}(\xi)}{(\mu_{nm}^2 R_2^2 - n^2) Z_{nm}^2(R_2) - (\mu_{nm}^2 R_1^2 - n^2) Z_{nm}^2(R_1)}$$

$$\times \cos[n(\varphi - \eta)] \sin\left(\frac{k\pi z}{l}\right) \sin\left(\frac{k\pi \zeta}{l}\right) \frac{\sin(at\sqrt{\lambda_{nmk}})}{a\sqrt{\lambda_{nmk}}},$$

where

$$A_n = \begin{cases} 1 & \text{for } n = 0, \\ 2 & \text{for } n \neq 0, \end{cases} \quad \lambda_{nmk} = \mu_{nm}^2 + \frac{k^2 \pi^2}{l^2},$$

$$Z_{nm}(r) = J_n'(\mu_{nm} R_1) Y_n(\mu_{nm} r) - Y_n'(\mu_{nm} R_1) J_n(\mu_{nm} r);$$

the $J_n(r)$ and $Y_n(r)$ are the Bessel functions, and the μ_{nm} are positive roots of the transcendental equation

$$J_n'(\mu R_1) Y_n'(\mu R_2) - Y_n'(\mu R_1) J_n'(\mu R_2) = 0.$$

6.1.2-9. Domain: $0 \leq r \leq R$, $0 \leq \varphi \leq \varphi_0$, $0 \leq z \leq l$. First boundary value problem.

A cylindrical sector of finite thickness is considered. The following conditions are prescribed:

$$w = f_0(r,\varphi,z) \quad \text{at} \quad t = 0 \quad \text{(initial condition),}$$
$$\partial_t w = f_1(r,\varphi,z) \quad \text{at} \quad t = 0 \quad \text{(initial condition),}$$
$$w = g_1(\varphi,z,t) \quad \text{at} \quad r = R \quad \text{(boundary condition),}$$
$$w = g_2(r,z,t) \quad \text{at} \quad \varphi = 0 \quad \text{(boundary condition),}$$
$$w = g_3(r,z,t) \quad \text{at} \quad \varphi = \varphi_0 \quad \text{(boundary condition),}$$
$$w = g_4(r,\varphi,t) \quad \text{at} \quad z = 0 \quad \text{(boundary condition),}$$
$$w = g_5(r,\varphi,t) \quad \text{at} \quad z = l \quad \text{(boundary condition).}$$

Solution:

$$w(r,\varphi,z,t) = \frac{\partial}{\partial t} \int_0^l \int_0^{\varphi_0} \int_0^R f_0(\xi,\eta,\zeta) G(r,\varphi,z,\xi,\eta,\zeta,t) \xi \, d\xi \, d\eta \, d\zeta$$

$$+ \int_0^l \int_0^{\varphi_0} \int_0^R f_1(\xi,\eta,\zeta) G(r,\varphi,z,\xi,\eta,\zeta,t) \xi \, d\xi \, d\eta \, d\zeta$$

$$- a^2 R \int_0^t \int_0^l \int_0^{\varphi_0} g_1(\eta,\zeta,\tau) \left[\frac{\partial}{\partial \xi} G(r,\varphi,z,\xi,\eta,\zeta,t-\tau)\right]_{\xi=R} d\eta \, d\zeta \, d\tau$$

$$+ a^2 \int_0^t \int_0^l \int_0^R g_2(\xi,\zeta,\tau) \frac{1}{\xi} \left[\frac{\partial}{\partial \eta} G(r,\varphi,z,\xi,\eta,\zeta,t-\tau)\right]_{\eta=0} d\xi \, d\zeta \, d\tau$$

$$- a^2 \int_0^t \int_0^l \int_0^R g_3(\xi,\zeta,\tau) \frac{1}{\xi} \left[\frac{\partial}{\partial \eta} G(r,\varphi,z,\xi,\eta,\zeta,t-\tau)\right]_{\eta=\varphi_0} d\xi \, d\zeta \, d\tau$$

$$+ a^2 \int_0^t \int_0^{\varphi_0} \int_0^R g_4(\xi,\eta,\tau) \left[\frac{\partial}{\partial \zeta} G(r,\varphi,z,\xi,\eta,\zeta,t-\tau)\right]_{\zeta=0} \xi \, d\xi \, d\eta \, d\tau$$

$$- a^2 \int_0^t \int_0^{\varphi_0} \int_0^R g_5(\xi,\eta,\tau) \left[\frac{\partial}{\partial \zeta} G(r,\varphi,z,\xi,\eta,\zeta,t-\tau)\right]_{\zeta=l} \xi \, d\xi \, d\eta \, d\tau.$$

Here,

$$G(r,\varphi,z,\xi,\eta,\zeta,t) = \frac{8}{R^2 l \varphi_0} \sum_{n=1}^{\infty} \sum_{m=1}^{\infty} \sum_{k=1}^{\infty} \frac{J_{n\pi/\varphi_0}(\mu_{nm}r) J_{n\pi/\varphi_0}(\mu_{nm}\xi)}{[J'_{n\pi/\varphi_0}(\mu_{nm}R)]^2} \sin\left(\frac{n\pi\varphi}{\varphi_0}\right) \sin\left(\frac{n\pi\eta}{\varphi_0}\right)$$

$$\times \sin\left(\frac{k\pi z}{l}\right) \sin\left(\frac{k\pi\zeta}{l}\right) \frac{\sin\left(at\sqrt{\mu_{nm}^2 + k^2\pi^2/l^2}\right)}{a\sqrt{\mu_{nm}^2 + k^2\pi^2/l^2}},$$

where the $J_{n\pi/\varphi_0}(r)$ are the Bessel functions and the μ_{nm} are positive roots of the transcendental equation $J_{n\pi/\varphi_0}(\mu R) = 0$.

6.1.2-10. Domain: $0 \le r \le R$, $0 \le \varphi \le \varphi_0$, $0 \le z \le l$. Mixed boundary value problem.

A cylindrical sector of finite thickness is considered. The following conditions are prescribed:

$$\begin{aligned}
w &= f_0(r,\varphi,z) & \text{at} \quad t &= 0 & \text{(initial condition)},\\
\partial_t w &= f_1(r,\varphi,z) & \text{at} \quad t &= 0 & \text{(initial condition)},\\
w &= g_1(\varphi,z,t) & \text{at} \quad r &= R & \text{(boundary condition)},\\
w &= g_2(r,z,t) & \text{at} \quad \varphi &= 0 & \text{(boundary condition)},\\
w &= g_3(r,z,t) & \text{at} \quad \varphi &= \varphi_0 & \text{(boundary condition)},\\
\partial_z w &= g_4(r,\varphi,t) & \text{at} \quad z &= 0 & \text{(boundary condition)},\\
\partial_z w &= g_5(r,\varphi,t) & \text{at} \quad z &= l & \text{(boundary condition)}.
\end{aligned}$$

Solution:

$$\begin{aligned}
w(r,\varphi,z,t) &= \frac{\partial}{\partial t} \int_0^l \int_0^{\varphi_0} \int_0^R f_0(\xi,\eta,\zeta) G(r,\varphi,z,\xi,\eta,\zeta,t) \xi \, d\xi \, d\eta \, d\zeta \\
&+ \int_0^l \int_0^{\varphi_0} \int_0^R f_1(\xi,\eta,\zeta) G(r,\varphi,z,\xi,\eta,\zeta,t) \xi \, d\xi \, d\eta \, d\zeta \\
&- a^2 R \int_0^t \int_0^l \int_0^{\varphi_0} g_1(\eta,\zeta,\tau) \left[\frac{\partial}{\partial \xi} G(r,\varphi,z,\xi,\eta,\zeta,t-\tau) \right]_{\xi=R} d\eta \, d\zeta \, d\tau \\
&+ a^2 \int_0^t \int_0^l \int_0^R g_2(\xi,\zeta,\tau) \frac{1}{\xi}\left[\frac{\partial}{\partial \eta} G(r,\varphi,z,\xi,\eta,\zeta,t-\tau) \right]_{\eta=0} d\xi \, d\zeta \, d\tau \\
&- a^2 \int_0^t \int_0^l \int_0^R g_3(\xi,\zeta,\tau) \frac{1}{\xi}\left[\frac{\partial}{\partial \eta} G(r,\varphi,z,\xi,\eta,\zeta,t-\tau) \right]_{\eta=\varphi_0} d\xi \, d\zeta \, d\tau \\
&- a^2 \int_0^t \int_0^{\varphi_0} \int_0^R g_4(\xi,\eta,\tau) G(r,\varphi,z,\xi,\eta,0,t-\tau) \xi \, d\xi \, d\eta \, d\tau \\
&+ a^2 \int_0^t \int_0^{\varphi_0} \int_0^R g_5(\xi,\eta,\tau) G(r,\varphi,z,\xi,\eta,l,t-\tau) \xi \, d\xi \, d\eta \, d\tau.
\end{aligned}$$

Here,

$$G(r,\varphi,z,\xi,\eta,\zeta,t) = \frac{4}{R^2 l \varphi_0} \sum_{n=1}^{\infty} \sum_{m=1}^{\infty} \sum_{k=0}^{\infty} \frac{A_k J_{n\pi/\varphi_0}(\mu_{nm}r) J_{n\pi/\varphi_0}(\mu_{nm}\xi)}{[J'_{n\pi/\varphi_0}(\mu_{nm}R)]^2} \sin\left(\frac{n\pi\varphi}{\varphi_0}\right) \sin\left(\frac{n\pi\eta}{\varphi_0}\right)$$

$$\times \cos\left(\frac{k\pi z}{l}\right) \cos\left(\frac{k\pi\zeta}{l}\right) \frac{\sin\left(at\sqrt{\mu_{nm}^2 + k^2\pi^2/l^2}\right)}{a\sqrt{\mu_{nm}^2 + k^2\pi^2/l^2}},$$

where $A_0 = 1$ and $A_k = 2$ for $k \ge 1$; the $J_{n\pi/\varphi_0}(r)$ are the Bessel functions; and the μ_{nm} are positive roots of the transcendental equation $J_{n\pi/\varphi_0}(\mu R) = 0$.

6.1.3. Problems in Spherical Coordinates

The three-dimensional wave equation in the spherical coordinate system is represented as

$$\frac{\partial^2 w}{\partial t^2} = a^2 \left[\frac{1}{r^2}\frac{\partial}{\partial r}\left(r^2 \frac{\partial w}{\partial r}\right) + \frac{1}{r^2 \sin\theta}\frac{\partial}{\partial \theta}\left(\sin\theta \frac{\partial w}{\partial \theta}\right) + \frac{1}{r^2 \sin^2\theta}\frac{\partial^2 w}{\partial \varphi^2}\right], \qquad r = \sqrt{x^2 + y^2 + z^2}.$$

One-dimensional problems with central symmetry that have solutions $w = w(r,t)$ are considered in Subsection 4.2.3.

6.1.3-1. Domain: $0 \le r \le R$, $0 \le \theta \le \pi$, $0 \le \varphi \le 2\pi$. First boundary value problem.

A spherical domain is considered. The following conditions are prescribed:

$$w = f_0(r,\theta,\varphi) \quad \text{at} \quad t = 0 \quad \text{(initial condition)},$$
$$\partial_t w = f_1(r,\theta,\varphi) \quad \text{at} \quad t = 0 \quad \text{(initial condition)},$$
$$w = g(\theta,\varphi,t) \quad \text{at} \quad r = R \quad \text{(boundary condition)}.$$

Solution:

$$w(r,\theta,\varphi,t) = \frac{\partial}{\partial t}\int_0^{2\pi}\int_0^{\pi}\int_0^R f_0(\xi,\eta,\zeta)G(r,\theta,\varphi,\xi,\eta,\zeta,t)\xi^2 \sin\eta\, d\xi\, d\eta\, d\zeta$$
$$+ \int_0^{2\pi}\int_0^{\pi}\int_0^R f_1(\xi,\eta,\zeta)G(r,\theta,\varphi,\xi,\eta,\zeta,t)\xi^2 \sin\eta\, d\xi\, d\eta\, d\zeta$$
$$- a^2 R^2 \int_0^t\int_0^{2\pi}\int_0^{\pi} g(\eta,\zeta,\tau)\left[\frac{\partial}{\partial \xi}G(r,\theta,\varphi,\xi,\eta,\zeta,t-\tau)\right]_{\xi=R}\sin\eta\, d\eta\, d\zeta\, d\tau,$$

where

$$G(r,\theta,\varphi,\xi,\eta,\zeta,t) = \frac{1}{2\pi a R^2 \sqrt{r\xi}} \sum_{n=0}^{\infty}\sum_{m=1}^{\infty}\sum_{k=0}^{n} A_k B_{nmk} J_{n+1/2}(\lambda_{nm}r) J_{n+1/2}(\lambda_{nm}\xi)$$
$$\times P_n^k(\cos\theta)P_n^k(\cos\eta)\cos[k(\varphi-\zeta)]\sin(\lambda_{nm}at),$$

$$A_k = \begin{cases} 1 & \text{for } k = 0, \\ 2 & \text{for } k \ne 0, \end{cases} \qquad B_{nmk} = \frac{(2n+1)(n-k)!}{(n+k)!\left[J'_{n+1/2}(\lambda_{nm}R)\right]^2 \lambda_{nm}}.$$

Here, the $J_{n+1/2}(r)$ are the Bessel functions, the $P_n^k(\mu)$ are the associated Legendre functions expressed in terms of the Legendre polynomials $P_n(\mu)$ as

$$P_n^k(\mu) = (1-\mu^2)^{k/2}\frac{d^k}{d\mu^k}P_n(\mu), \qquad P_n(\mu) = \frac{1}{n!\,2^n}\frac{d^n}{d\mu^n}(\mu^2-1)^n,$$

and the λ_{nm} are positive roots of the transcendental equation $J_{n+1/2}(\lambda R) = 0$.

6.1.3-2. Domain: $0 \le r \le R$, $0 \le \theta \le \pi$, $0 \le \varphi \le 2\pi$. Second boundary value problem.

A spherical domain is considered. The following conditions are prescribed:

$$w = f_0(r,\theta,\varphi) \quad \text{at} \quad t = 0 \quad \text{(initial condition)},$$
$$\partial_t w = f_1(r,\theta,\varphi) \quad \text{at} \quad t = 0 \quad \text{(initial condition)},$$
$$\partial_r w = g(\theta,\varphi,t) \quad \text{at} \quad r = R \quad \text{(boundary condition)}.$$

Solution:

$$
\begin{aligned}
w(r,\theta,\varphi,t) = & \frac{\partial}{\partial t} \int_0^{2\pi}\!\!\int_0^{\pi}\!\!\int_0^R f_0(\xi,\eta,\zeta) G(r,\theta,\varphi,\xi,\eta,\zeta,t)\xi^2 \sin\eta\, d\xi\, d\eta\, d\zeta \\
& + \int_0^{2\pi}\!\!\int_0^{\pi}\!\!\int_0^R f_1(\xi,\eta,\zeta) G(r,\theta,\varphi,\xi,\eta,\zeta,t)\xi^2 \sin\eta\, d\xi\, d\eta\, d\zeta \\
& + a^2 R^2 \int_0^t\!\!\int_0^{2\pi}\!\!\int_0^{\pi} g(\eta,\zeta,\tau) G(r,\theta,\varphi,R,\eta,\zeta,t-\tau)\sin\eta\, d\eta\, d\zeta\, d\tau,
\end{aligned}
$$

where

$$
G(r,\theta,\varphi,\xi,\eta,\zeta,t) = \frac{3t}{4\pi R^3} + \frac{1}{2\pi a\sqrt{r\xi}} \sum_{n=0}^{\infty}\sum_{m=1}^{\infty}\sum_{k=0}^{n} A_k B_{nmk} J_{n+1/2}(\lambda_{nm}r) J_{n+1/2}(\lambda_{nm}\xi)
$$
$$
\times P_n^k(\cos\theta) P_n^k(\cos\eta) \cos[k(\varphi-\zeta)] \sin(\lambda_{nm}at),
$$

$$
A_k = \begin{cases} 1 & \text{for } k=0, \\ 2 & \text{for } k\neq 0, \end{cases} \qquad B_{nmk} = \frac{\lambda_{nm}(2n+1)(n-k)!}{(n+k)!\left[R^2\lambda_{nm}^2 - n(n+1)\right]\left[J_{n+1/2}(\lambda_{nm}R)\right]^2}.
$$

Here, the $J_{n+1/2}(r)$ are the Bessel functions, the $P_n^k(\mu)$ are the associated Legendre functions (see Paragraph 6.1.3-1), and the λ_{nm} are positive roots of the transcendental equation

$$
2\lambda R J'_{n+1/2}(\lambda R) - J_{n+1/2}(\lambda R) = 0.
$$

⊙ *Reference*: M. M. Smirnov (1975).

6.1.3-3. Domain: $0 \leq r \leq R$, $0 \leq \theta \leq \pi$, $0 \leq \varphi \leq 2\pi$. Third boundary value problem.

A spherical domain is considered. The following conditions are prescribed:

$$
\begin{aligned}
w &= f_0(r,\theta,\varphi) && \text{at } t=0 && \text{(initial condition)}, \\
\partial_t w &= f_1(r,\theta,\varphi) && \text{at } t=0 && \text{(initial condition)}, \\
\partial_r w + kw &= g(\theta,\varphi,t) && \text{at } r=R && \text{(boundary condition)}.
\end{aligned}
$$

The solution $w(r,\theta,\varphi,t)$ is determined by the formula in Paragraph 6.1.3-2 where

$$
G(r,\theta,\varphi,\xi,\eta,\zeta,t) = \frac{1}{2\pi a\sqrt{r\xi}} \sum_{n=0}^{\infty}\sum_{m=1}^{\infty}\sum_{s=0}^{n} A_s B_{nms} J_{n+1/2}(\lambda_{nm}r) J_{n+1/2}(\lambda_{nm}\xi)
$$
$$
\times P_n^s(\cos\theta) P_n^s(\cos\eta) \cos[s(\varphi-\zeta)] \sin(\lambda_{nm}at),
$$

$$
A_s = \begin{cases} 1 & \text{for } s=0, \\ 2 & \text{for } s\neq 0, \end{cases} \qquad B_{nms} = \frac{\lambda_{nm}(2n+1)(n-s)!}{(n+s)!\left[R^2\lambda_{nm}^2 + (kR+n)(kR-n-1)\right]\left[J_{n+1/2}(\lambda_{nm}R)\right]^2}.
$$

Here, the $J_{n+1/2}(r)$ are the Bessel functions, the $P_n^s(\mu)$ are the associated Legendre functions (see Paragraph 6.1.3-1), and the λ_{nm} are positive roots of the transcendental equation

$$
\lambda R J'_{n+1/2}(\lambda R) + \left(kR - \tfrac{1}{2}\right) J_{n+1/2}(\lambda R) = 0.
$$

6.1.3-4. Domain: $R_1 \leq r \leq R_2$, $0 \leq \theta \leq \pi$, $0 \leq \varphi \leq 2\pi$. First boundary value problem.

A spherical layer is considered. The following conditions are prescribed:

$$
\begin{aligned}
w &= f_0(r,\theta,\varphi) && \text{at } t=0 && \text{(initial condition)}, \\
\partial_t w &= f_1(r,\theta,\varphi) && \text{at } t=0 && \text{(initial condition)}, \\
w &= g_1(\theta,\varphi,t) && \text{at } r=R_1 && \text{(boundary condition)}, \\
w &= g_2(\theta,\varphi,t) && \text{at } r=R_2 && \text{(boundary condition)}.
\end{aligned}
$$

Solution:

$$w(r,\theta,\varphi,t) = \frac{\partial}{\partial t}\int_0^{2\pi}\int_0^{\pi}\int_{R_1}^{R_2} f_0(\xi,\eta,\zeta)G(r,\theta,\varphi,\xi,\eta,\zeta,t)\xi^2\sin\eta\,d\xi\,d\eta\,d\zeta$$

$$+ \int_0^{2\pi}\int_0^{\pi}\int_{R_1}^{R_2} f_1(\xi,\eta,\zeta)G(r,\theta,\varphi,\xi,\eta,\zeta,t)\xi^2\sin\eta\,d\xi\,d\eta\,d\zeta$$

$$+ a^2 R_1^2\int_0^t\int_0^{2\pi}\int_0^{\pi} g_1(\eta,\zeta,\tau)\left[\frac{\partial}{\partial\xi}G(r,\theta,\varphi,\xi,\eta,\zeta,t-\tau)\right]_{\xi=R_1}\sin\eta\,d\eta\,d\zeta\,d\tau$$

$$- a^2 R_2^2\int_0^t\int_0^{2\pi}\int_0^{\pi} g_2(\eta,\zeta,\tau)\left[\frac{\partial}{\partial\xi}G(r,\theta,\varphi,\xi,\eta,\zeta,t-\tau)\right]_{\xi=R_2}\sin\eta\,d\eta\,d\zeta\,d\tau,$$

where

$$G(r,\theta,\varphi,\xi,\eta,\zeta,t) = \frac{\pi}{8a\sqrt{r\xi}}\sum_{n=0}^{\infty}\sum_{m=1}^{\infty}\sum_{k=0}^{n} A_k B_{nmk} Z_{n+1/2}(\lambda_{nm}r)Z_{n+1/2}(\lambda_{nm}\xi)$$
$$\times P_n^k(\cos\theta)P_n^k(\cos\eta)\cos[k(\varphi-\zeta)]\sin(\lambda_{nm}at).$$

Here,

$$Z_{n+1/2}(\lambda_{nm}r) = J_{n+1/2}(\lambda_{nm}R_1)Y_{n+1/2}(\lambda_{nm}r) - Y_{n+1/2}(\lambda_{nm}R_1)J_{n+1/2}(\lambda_{nm}r),$$

$$A_k = \begin{cases} 1 & \text{for } k=0, \\ 2 & \text{for } k\neq 0, \end{cases} \qquad B_{nmk} = \frac{\lambda_{nm}(2n+1)(n-k)!\,J_{n+1/2}^2(\lambda_{nm}R_2)}{(n+k)!\left[J_{n+1/2}^2(\lambda_{nm}R_1) - J_{n+1/2}^2(\lambda_{nm}R_2)\right]},$$

where the $J_{n+1/2}(r)$ are the Bessel functions, the $P_n^k(\mu)$ are the associated Legendre functions expressed in terms of the Legendre polynomials $P_n(\mu)$ as

$$P_n^k(\mu) = (1-\mu^2)^{k/2}\frac{d^k}{d\mu^k}P_n(\mu), \qquad P_n(\mu) = \frac{1}{n!\,2^n}\frac{d^n}{d\mu^n}(\mu^2-1)^n,$$

and the λ_{nm} are positive roots of the transcendental equation $Z_{n+1/2}(\lambda R_2) = 0$.

6.1.3-5. Domain: $R_1 \leq r \leq R_2$, $0 \leq \theta \leq \pi$, $0 \leq \varphi \leq 2\pi$. Second boundary value problem.

A spherical layer is considered. The following conditions are prescribed:

$$w = f_0(r,\theta,\varphi) \quad \text{at} \quad t=0 \quad \text{(initial condition)},$$
$$\partial_t w = f_1(r,\theta,\varphi) \quad \text{at} \quad t=0 \quad \text{(initial condition)},$$
$$\partial_r w = g_1(\theta,\varphi,t) \quad \text{at} \quad r=R_1 \quad \text{(boundary condition)},$$
$$\partial_r w = g_2(\theta,\varphi,t) \quad \text{at} \quad r=R_2 \quad \text{(boundary condition)}.$$

Solution:

$$w(r,\theta,\varphi,t) = \frac{\partial}{\partial t}\int_0^{2\pi}\int_0^{\pi}\int_{R_1}^{R_2} f_0(\xi,\eta,\zeta)G(r,\theta,\varphi,\xi,\eta,\zeta,t)\xi^2\sin\eta\,d\xi\,d\eta\,d\zeta$$

$$+ \int_0^{2\pi}\int_0^{\pi}\int_{R_1}^{R_2} f_1(\xi,\eta,\zeta)G(r,\theta,\varphi,\xi,\eta,\zeta,t)\xi^2\sin\eta\,d\xi\,d\eta\,d\zeta$$

$$- a^2 R_1^2\int_0^t\int_0^{2\pi}\int_0^{\pi} g_1(\eta,\zeta,\tau)G(r,\theta,\varphi,R_1,\eta,\zeta,t-\tau)\sin\eta\,d\eta\,d\zeta\,d\tau$$

$$+ a^2 R_2^2\int_0^t\int_0^{2\pi}\int_0^{\pi} g_2(\eta,\zeta,\tau)G(r,\theta,\varphi,R_2,\eta,\zeta,t-\tau)\sin\eta\,d\eta\,d\zeta\,d\tau,$$

where

$$G(r,\theta,\varphi,\xi,\eta,\zeta,t) = \frac{3t}{4\pi(R_2^3 - R_1^3)} + \frac{1}{4\pi a\sqrt{r\xi}} \sum_{n=0}^{\infty}\sum_{m=1}^{\infty}\sum_{k=0}^{n} \frac{A_k}{B_{nmk}} Z_{n+1/2}(\lambda_{nm}r)Z_{n+1/2}(\lambda_{nm}\xi)$$
$$\times P_n^k(\cos\theta)P_n^k(\cos\eta)\cos[k(\varphi-\zeta)]\sin(\lambda_{nm}at),$$

$$A_k = \begin{cases} 1 & \text{for } k = 0, \\ 2 & \text{for } k \neq 0, \end{cases} \qquad B_{nmk} = \frac{\lambda_{nm}(n+k)!}{(2n+1)(n-k)!}\int_{R_1}^{R_2} rZ_{n+1/2}^2(\lambda_{nm}r)\,dr,$$

$$Z_{n+1/2}(\lambda_{nm}r) = \left[\lambda_{nm}J'_{n+1/2}(\lambda_{nm}R_1) - \frac{1}{2R_1}J_{n+1/2}(\lambda_{nm}R_1)\right]Y_{n+1/2}(\lambda_{nm}r)$$
$$- \left[\lambda_{nm}Y'_{n+1/2}(\lambda_{nm}R_1) - \frac{1}{2R_1}Y_{n+1/2}(\lambda_{nm}R_1)\right]J_{n+1/2}(\lambda_{nm}r).$$

Here, the $J_{n+1/2}(r)$ and $Y_{n+1/2}(r)$ are the Bessel functions, the $P_n^k(\mu)$ are the associated Legendre functions (see Paragraph 6.1.3-4), and the λ_{nm} are positive roots of the transcendental equation

$$\lambda Z'_{n+1/2}(\lambda R_2) - \frac{1}{2R_2}Z_{n+1/2}(\lambda R_2) = 0.$$

6.1.3-6. Domain: $R_1 \leq r \leq R_2$, $0 \leq \theta \leq \pi$, $0 \leq \varphi \leq 2\pi$. **Third boundary value problem.**

A spherical layer is considered. The following conditions are prescribed:

$$\begin{aligned} w &= f_0(r,\theta,\varphi) & \text{at} \quad t = 0 & \quad \text{(initial condition)}, \\ \partial_t w &= f_1(r,\theta,\varphi) & \text{at} \quad t = 0 & \quad \text{(initial condition)}, \\ \partial_r w - k_1 w &= g_1(\theta,\varphi,t) & \text{at} \quad r = R_1 & \quad \text{(boundary condition)}, \\ \partial_r w + k_2 w &= g_2(\theta,\varphi,t) & \text{at} \quad r = R_2 & \quad \text{(boundary condition)}. \end{aligned}$$

The solution $w(r,\theta,\varphi,t)$ is determined by the formula in Paragraph 6.1.3-5 where

$$G(r,\theta,\varphi,\xi,\eta,\zeta,t) = \frac{1}{4\pi a\sqrt{r\xi}} \sum_{n=0}^{\infty}\sum_{m=1}^{\infty}\sum_{s=0}^{n} \frac{A_s}{B_{nms}} Z_{n+1/2}(\lambda_{nm}r)Z_{n+1/2}(\lambda_{nm}\xi)$$
$$\times P_n^s(\cos\theta)P_n^s(\cos\eta)\cos[s(\varphi-\zeta)]\sin(\lambda_{nm}at).$$

Here,

$$A_s = \begin{cases} 1 & \text{for } s = 0, \\ 2 & \text{for } s \neq 0, \end{cases} \qquad B_{nms} = \frac{\lambda_{nm}(n+s)!}{(2n+1)(n-s)!}\int_{R_1}^{R_2} rZ_{n+1/2}^2(\lambda_{nm}r)\,dr,$$

$$Z_{n+1/2}(\lambda r) = \left[\lambda J'_{n+1/2}(\lambda R_1) - \left(k_1 + \frac{1}{2R_1}\right)J_{n+1/2}(\lambda R_1)\right]Y_{n+1/2}(\lambda r)$$
$$- \left[\lambda Y'_{n+1/2}(\lambda R_1) - \left(k_1 + \frac{1}{2R_1}\right)Y_{n+1/2}(\lambda R_1)\right]J_{n+1/2}(\lambda r),$$

where the $J_{n+1/2}(r)$ and $Y_{n+1/2}(r)$ are the Bessel functions, the $P_n^s(\mu)$ are the associated Legendre functions (see Paragraph 6.1.3-4), and the λ_{nm} are positive roots of the transcendental equation

$$\lambda Z'_{n+1/2}(\lambda R_2) + \left(k_2 - \frac{1}{2R_2}\right)Z_{n+1/2}(\lambda R_2) = 0.$$

6.2. Nonhomogeneous Wave Equation

$$\frac{\partial^2 w}{\partial t^2} = a^2 \Delta_3 w + \Phi(x, y, z, t)$$

6.2.1. Problems in Cartesian Coordinates

6.2.1-1. Domain: $-\infty < x < \infty$, $-\infty < y < \infty$, $-\infty < z < \infty$. Cauchy problem.

Initial conditions are prescribed:
$$w = f(x, y, z) \quad \text{at} \quad t = 0,$$
$$\partial_t w = g(x, y, z) \quad \text{at} \quad t = 0.$$

Solution:
$$w(x, y, z, t) = \frac{1}{4\pi a} \frac{\partial}{\partial t} \iint_{r=at} \frac{f(\xi, \eta, \zeta)}{r} dS + \frac{1}{4\pi a} \iint_{r=at} \frac{g(\xi, \eta, \zeta)}{r} dS$$
$$+ \frac{1}{4\pi a^2} \iiint_{r \leq at} \frac{1}{r} \Phi\left(\xi, \eta, \zeta, t - \frac{r}{a}\right) d\xi\, d\eta\, d\zeta, \qquad r = \sqrt{(\xi - x)^2 + (\eta - y)^2 + (\zeta - z)^2},$$

where the integration is performed over the surface of the sphere ($r = at$) and the volume of the sphere ($r \leq at$) with center at (x, y, z).

⊙ *Reference*: N. S. Koshlyakov, E. B. Glizer, and M. M. Smirnov (1970).

6.2.1-2. Domain: $0 \leq x \leq l_1$, $0 \leq y \leq l_2$, $0 \leq z \leq l_3$. Different boundary value problems.

1°. The solution of the first boundary value problem for a parallelepiped is given by the formula from Paragraph 6.1.1-3 with the additional term

$$\int_0^t \int_0^{l_1} \int_0^{l_2} \int_0^{l_3} \Phi(\xi, \eta, \zeta, \tau) G(x, y, z, \xi, \eta, \zeta, t - \tau)\, d\zeta\, d\eta\, d\xi\, d\tau,$$

which allows for the equation's nonhomogeneity; this term is the solution of the nonhomogeneous equation with homogeneous initial and boundary conditions.

2°. The solution of the second boundary value problem for a parallelepiped is given by the formula from Paragraph 6.1.1-4 with the additional term specified in Paragraph 6.2.1-2, Item 1°; the Green's function is taken from Paragraph 6.1.1-4.

3°. The solution of the third boundary value problem for a parallelepiped is the sum of the solution of the homogeneous equation with nonhomogeneous initial and boundary conditions (see Paragraph 6.1.1-5) and the solution of the nonhomogeneous equation with homogeneous initial and boundary conditions. The latter solution is given by the formula from Paragraph 6.2.1-2, Item 1°, in which one should substitute the Green's function from Paragraph 6.1.1-5.

4°. The solutions of mixed boundary value problems for a parallelepiped are given by the formulas from Paragraph 6.1.1-6 to which one should add the term specified in Paragraph 6.2.1-2, Item 1°.

6.2.2. Problems in Cylindrical Coordinates

A three-dimensional nonhomogeneous wave equation in the cylindrical coordinate system is written as

$$\frac{\partial^2 w}{\partial t^2} = a^2 \left[\frac{1}{r} \frac{\partial}{\partial r}\left(r \frac{\partial w}{\partial r}\right) + \frac{1}{r^2} \frac{\partial^2 w}{\partial \varphi^2} + \frac{\partial^2 w}{\partial z^2} \right] + \Phi(r, \varphi, z, t).$$

6.2.2-1. Domain: $0 \leq r \leq R$, $0 \leq \varphi \leq 2\pi$, $0 \leq z \leq l$. Different boundary value problems.

1°. The solution of the first boundary value problem for a circular cylinder of finite length is given by the formula from Paragraph 6.1.2-1 with the additional term

$$\int_0^t \int_0^l \int_0^{2\pi} \int_0^R \Phi(\xi, \eta, \zeta, \tau) G(r, \varphi, z, \xi, \eta, \zeta, t - \tau) \xi \, d\xi \, d\eta \, d\zeta \, d\tau, \tag{1}$$

which allows for the equation's nonhomogeneity.

2°. The solution of the second boundary value problem for a circular cylinder of finite length is given by the formula from Paragraph 6.1.2-2 with the additional term (1).

3°. The solution of the third boundary value problem for a circular cylinder of finite length is the sum of the solution specified in Paragraph 6.1.2-3 and expression (1).

4°. The solutions of mixed boundary value problems for a circular cylinder of finite length are given by the formulas from Paragraph 6.1.2-4 with additional terms of the form (1).

6.2.2-2. Domain: $R_1 \leq r \leq R_2$, $0 \leq \varphi \leq 2\pi$, $0 \leq z \leq l$. Different boundary value problems.

1°. The solution of the first boundary value problem for a hollow cylinder of finite dimensions is given by the formula from Paragraph 6.1.2-5 with the additional term

$$\int_0^t \int_0^l \int_0^{2\pi} \int_{R_1}^{R_2} \Phi(\xi, \eta, \zeta, \tau) G(r, \varphi, z, \xi, \eta, \zeta, t - \tau) \xi \, d\xi \, d\eta \, d\zeta \, d\tau, \tag{2}$$

which allows for the equation's nonhomogeneity.

2°. The solution of the second boundary value problem for a hollow cylinder of finite dimensions is given by the formula from Paragraph 6.1.2-6 with the additional term (2).

3°. The solution of the third boundary value problem for a hollow cylinder of finite dimensions is the sum of the solution specified in Paragraph 6.1.2-7 and expression (2).

4°. The solutions of mixed boundary value problems for a hollow cylinder of finite dimensions are given by the formulas from Paragraph 6.1.2-8 with additional terms of the form (2).

6.2.2-3. Domain: $0 \leq r \leq R$, $0 \leq \varphi \leq \varphi_0$, $0 \leq z \leq l$. Different boundary value problems.

1°. The solution of the first boundary value problem for a cylindrical sector of finite thickness is given by the formula from Paragraph 6.1.2-9 with the additional term

$$\int_0^t \int_0^l \int_0^{\varphi_0} \int_0^R \Phi(\xi, \eta, \zeta, \tau) G(r, \varphi, z, \xi, \eta, \zeta, t - \tau) \xi \, d\xi \, d\eta \, d\zeta \, d\tau, \tag{3}$$

which allows for the equation's nonhomogeneity.

2°. The solution of a mixed boundary value problem for a cylindrical sector of finite thickness is given by the formula from Paragraph 6.1.2-10 with the additional term (2).

6.2.3. Problems in Spherical Coordinates

A three-dimensional nonhomogeneous wave equation in the spherical coordinate system is represented as

$$\frac{\partial^2 w}{\partial t^2} = a^2 \left[\frac{1}{r^2} \frac{\partial}{\partial r} \left(r^2 \frac{\partial w}{\partial r} \right) + \frac{1}{r^2 \sin \theta} \frac{\partial}{\partial \theta} \left(\sin \theta \frac{\partial w}{\partial \theta} \right) + \frac{1}{r^2 \sin^2 \theta} \frac{\partial^2 w}{\partial \varphi^2} \right] + \Phi(r, \theta, \varphi, t).$$

6.2.3-1. Domain: $0 \leq r \leq R$, $0 \leq \theta \leq \pi$, $0 \leq \varphi \leq 2\pi$. Boundary value problem.

1°. The solution of the first boundary value problem for a sphere is given by the formula from Paragraph 6.1.3-1 with the additional term

$$\int_0^t \int_0^{2\pi} \int_0^\pi \int_0^R \Phi(\xi, \eta, \zeta, \tau) G(r, \theta, \varphi, \xi, \eta, \zeta, t-\tau) \xi^2 \sin\eta \, d\xi \, d\eta \, d\zeta \, d\tau, \tag{1}$$

which allows for the equation's nonhomogeneity.

2°. The solution of the second boundary value problem for a sphere is given by the formula from Paragraph 6.1.3-2 with the additional term (1).

3°. The solution of the third boundary value problem for a sphere is the sum of the solution specified in Paragraph 6.1.3-3 and expression (1).

6.2.3-2. Domain: $R_1 \leq r \leq R_2$, $0 \leq \theta \leq \pi$, $0 \leq \varphi \leq 2\pi$. Boundary value problems.

1°. The solution of the first boundary value problem for a spherical layer is given by the formula from Paragraph 6.1.3-4 with the additional term

$$\int_0^t \int_0^{2\pi} \int_0^\pi \int_{R_1}^{R_2} \Phi(\xi, \eta, \zeta, \tau) G(r, \theta, \varphi, \xi, \eta, \zeta, t-\tau) \xi^2 \sin\eta \, d\xi \, d\eta \, d\zeta \, d\tau, \tag{2}$$

which allows for the equation's nonhomogeneity.

2°. The solution of the second boundary value problem for a spherical layer is given by the formula from Paragraph 6.1.3-5 with the additional term (2).

3°. The solution of the third boundary value problem for a spherical layer is the sum of the solution specified in Paragraph 6.1.3-6 and expression (2).

6.3. Equations of the Form $\dfrac{\partial^2 w}{\partial t^2} = a^2 \Delta_3 w - bw + \Phi(x, y, z, t)$

6.3.1. Problems in Cartesian Coordinates

A *three-dimensional nonhomogeneous Klein–Gordon equation* in the rectangular Cartesian system of coordinates has the form

$$\frac{\partial^2 w}{\partial t^2} = a^2 \left(\frac{\partial^2 w}{\partial x^2} + \frac{\partial^2 w}{\partial y^2} + \frac{\partial^2 w}{\partial z^2} \right) - bw + \Phi(x, y, z, t).$$

6.3.1-1. Fundamental solutions.

1°. For $b = -c^2 < 0$,

$$\mathscr{E}(x, y, z, t) = \frac{1}{4\pi a^2} \left[\frac{\delta(t - r/a)}{r} - \frac{c}{a} \frac{I_1\left(c\sqrt{t^2 - r^2/a^2}\right)}{\sqrt{t^2 - r^2/a^2}} \vartheta(t - r/a) \right],$$

where $r = \sqrt{x^2 + y^2 + z^2}$, $\delta(\xi)$ is the Dirac delta function, $\vartheta(\xi)$ is the Heaviside unit step function, and $I_1(z)$ is the modified Bessel function.

2°. For $b = c^2 > 0$,

$$\mathscr{E}(x, y, z, t) = \frac{1}{4\pi a^2} \left[\frac{\delta(t - r/a)}{r} - \frac{c}{a} \frac{J_1\left(c\sqrt{t^2 - r^2/a^2}\right)}{\sqrt{t^2 - r^2/a^2}} \vartheta(t - r/a) \right],$$

where $J_1(z)$ is the Bessel function.

⊙ *Reference*: V. S. Vladimirov, V. P. Mikhailov, A. A. Vasharin, et al. (1974).

6.3.1-2. Domain: $-\infty < x < \infty$, $-\infty < y < \infty$, $-\infty < z < \infty$. Cauchy problem.

Initial conditions are prescribed:

$$w = f(x, y, z) \quad \text{at} \quad t = 0,$$
$$\partial_t w = g(x, y, z) \quad \text{at} \quad t = 0.$$

Let $a = 1$ and $\Phi(x, y, z, t) \equiv 0$.

1°. Solution for $b = -c^2 < 0$:

$$w(x, y, z, t) = \frac{\partial}{\partial t}\left[\frac{1}{t}\frac{\partial}{\partial t}\int_0^t r^2 I_0\left(c\sqrt{t^2 - r^2}\right) T_r\left[f(x, y, z)\right] dr\right]$$
$$+ \frac{1}{t}\frac{\partial}{\partial t}\int_0^t r^2 I_0\left(c\sqrt{t^2 - r^2}\right) T_r\left[g(x, y, z)\right] dr.$$

Here, $I_0(z)$ is the modified Bessel function and $T_r[h(x, y, z)]$ is the average of $h(x, y, z)$ over the spherical surface with center at (x, y, z) and radius r:

$$T_r[h(x, y, z)] = \frac{1}{4\pi}\int_0^{2\pi}\int_0^{\pi} h(x + r\sin\theta\cos\varphi, y + r\sin\theta\sin\varphi, z + r\cos\theta)\sin\theta\, d\theta\, d\varphi.$$

2°. Solution for $b = c^2 > 0$:

$$w(x, y, z, t) = \frac{\partial}{\partial t}\left[\frac{1}{t}\frac{\partial}{\partial t}\int_0^t r^2 J_0\left(c\sqrt{t^2 - r^2}\right) T_r\left[f(x, y, z)\right] dr\right]$$
$$+ \frac{1}{t}\frac{\partial}{\partial t}\int_0^t r^2 J_0\left(c\sqrt{t^2 - r^2}\right) T_r\left[g(x, y, z)\right] dr,$$

where $J_0(z)$ is the Bessel function.

⊙ *Reference*: V. I. Smirnov (1974, Vol. 2).

6.3.1-3. Domain: $0 \leq x \leq l_1$, $0 \leq y \leq l_2$, $0 \leq z \leq l_3$. First boundary value problem.

A rectangular parallelepiped is considered. The following conditions are prescribed:

$$w = f_0(x, y, z) \quad \text{at} \quad t = 0 \quad \text{(initial condition)},$$
$$\partial_t w = f_1(x, y, z) \quad \text{at} \quad t = 0 \quad \text{(initial condition)},$$
$$w = g_1(y, z, t) \quad \text{at} \quad x = 0 \quad \text{(boundary condition)},$$
$$w = g_2(y, z, t) \quad \text{at} \quad x = l_1 \quad \text{(boundary condition)},$$
$$w = g_3(x, z, t) \quad \text{at} \quad y = 0 \quad \text{(boundary condition)},$$
$$w = g_4(x, z, t) \quad \text{at} \quad y = l_2 \quad \text{(boundary condition)},$$
$$w = g_5(x, y, t) \quad \text{at} \quad z = 0 \quad \text{(boundary condition)},$$
$$w = g_6(x, y, t) \quad \text{at} \quad z = l_3 \quad \text{(boundary condition)}.$$

Solution:

$$w(x,y,z,t) = \frac{\partial}{\partial t}\int_0^{l_3}\int_0^{l_2}\int_0^{l_1} f_0(\xi,\eta,\zeta)G(x,y,z,\xi,\eta,\zeta,t)\,d\xi\,d\eta\,d\zeta$$

$$+ \int_0^{l_3}\int_0^{l_2}\int_0^{l_1} f_1(\xi,\eta,\zeta)G(x,y,z,\xi,\eta,\zeta,t)\,d\xi\,d\eta\,d\zeta$$

$$+ a^2\int_0^t\int_0^{l_3}\int_0^{l_2} g_1(\eta,\zeta,\tau)\left[\frac{\partial}{\partial \xi}G(x,y,z,\xi,\eta,\zeta,t-\tau)\right]_{\xi=0}\,d\eta\,d\zeta\,d\tau$$

$$- a^2\int_0^t\int_0^{l_3}\int_0^{l_2} g_2(\eta,\zeta,\tau)\left[\frac{\partial}{\partial \xi}G(x,y,z,\xi,\eta,\zeta,t-\tau)\right]_{\xi=l_1}\,d\eta\,d\zeta\,d\tau$$

$$+ a^2\int_0^t\int_0^{l_3}\int_0^{l_1} g_3(\xi,\zeta,\tau)\left[\frac{\partial}{\partial \eta}G(x,y,z,\xi,\eta,\zeta,t-\tau)\right]_{\eta=0}\,d\xi\,d\zeta\,d\tau$$

$$- a^2\int_0^t\int_0^{l_3}\int_0^{l_1} g_4(\xi,\zeta,\tau)\left[\frac{\partial}{\partial \eta}G(x,y,z,\xi,\eta,\zeta,t-\tau)\right]_{\eta=l_2}\,d\xi\,d\zeta\,d\tau$$

$$+ a^2\int_0^t\int_0^{l_2}\int_0^{l_1} g_5(\xi,\eta,\tau)\left[\frac{\partial}{\partial \zeta}G(x,y,z,\xi,\eta,\zeta,t-\tau)\right]_{\zeta=0}\,d\xi\,d\eta\,d\tau$$

$$- a^2\int_0^t\int_0^{l_2}\int_0^{l_1} g_6(\xi,\eta,\tau)\left[\frac{\partial}{\partial \zeta}G(x,y,z,\xi,\eta,\zeta,t-\tau)\right]_{\zeta=l_3}\,d\xi\,d\eta\,d\tau$$

$$+ \int_0^t\int_0^{l_3}\int_0^{l_2}\int_0^{l_1} \Phi(\xi,\eta,\zeta,\tau)G(x,y,z,\xi,\eta,\zeta,t-\tau)\,d\xi\,d\eta\,d\zeta\,d\tau.$$

Here,

$$G(x,y,z,\xi,\eta,\zeta,t) = \frac{8}{l_1 l_2 l_3}\sum_{n=1}^{\infty}\sum_{m=1}^{\infty}\sum_{k=1}^{\infty}\frac{1}{\sqrt{\lambda_{nmk}}}\sin(\alpha_n x)\sin(\beta_m y)\sin(\gamma_k z)$$
$$\times \sin(\alpha_n \xi)\sin(\beta_m \eta)\sin(\gamma_k \zeta)\sin\!\left(t\sqrt{\lambda_{nmk}}\right),$$

where

$$\alpha_n = \frac{n\pi}{l_1},\quad \beta_m = \frac{m\pi}{l_2},\quad \gamma_k = \frac{k\pi}{l_3},\quad \lambda_{nmk} = a^2(\alpha_n^2+\beta_m^2+\gamma_k^2)+b.$$

6.3.1-4. Domain: $0 \le x \le l_1$, $0 \le y \le l_2$, $0 \le z \le l_3$. Second boundary value problem.

A rectangular parallelepiped is considered. The following conditions are prescribed:

$$\begin{aligned}
w &= f_0(x,y,z) & \text{at}\quad & t = 0 & \text{(initial condition)},\\
\partial_t w &= f_1(x,y,z) & \text{at}\quad & t = 0 & \text{(initial condition)},\\
\partial_x w &= g_1(y,z,t) & \text{at}\quad & x = 0 & \text{(boundary condition)},\\
\partial_x w &= g_2(y,z,t) & \text{at}\quad & x = l_1 & \text{(boundary condition)},\\
\partial_y w &= g_3(x,z,t) & \text{at}\quad & y = 0 & \text{(boundary condition)},\\
\partial_y w &= g_4(x,z,t) & \text{at}\quad & y = l_2 & \text{(boundary condition)},\\
\partial_z w &= g_5(x,y,t) & \text{at}\quad & z = 0 & \text{(boundary condition)},\\
\partial_z w &= g_6(x,y,t) & \text{at}\quad & z = l_3 & \text{(boundary condition)}.
\end{aligned}$$

Solution:

$$w(x,y,z,t) = \frac{\partial}{\partial t}\int_0^{l_3}\int_0^{l_2}\int_0^{l_1} f_0(\xi,\eta,\zeta)G(x,y,z,\xi,\eta,\zeta,t)\,d\xi\,d\eta\,d\zeta$$

$$+ \int_0^{l_3}\int_0^{l_2}\int_0^{l_1} f_1(\xi,\eta,\zeta)G(x,y,z,\xi,\eta,\zeta,t)\,d\xi\,d\eta\,d\zeta$$

$$- a^2\int_0^t\int_0^{l_3}\int_0^{l_2} g_1(\eta,\zeta,\tau)G(x,y,z,0,\eta,\zeta,t-\tau)\,d\eta\,d\zeta\,d\tau$$

$$+ a^2\int_0^t\int_0^{l_3}\int_0^{l_2} g_2(\eta,\zeta,\tau)G(x,y,z,l_1,\eta,\zeta,t-\tau)\,d\eta\,d\zeta\,d\tau$$

$$- a^2\int_0^t\int_0^{l_3}\int_0^{l_1} g_3(\xi,\zeta,\tau)G(x,y,z,\xi,0,\zeta,t-\tau)\,d\xi\,d\zeta\,d\tau$$

$$+ a^2\int_0^t\int_0^{l_3}\int_0^{l_1} g_4(\xi,\zeta,\tau)G(x,y,z,\xi,l_2,\zeta,t-\tau)\,d\xi\,d\zeta\,d\tau$$

$$- a^2\int_0^t\int_0^{l_2}\int_0^{l_1} g_5(\xi,\eta,\tau)G(x,y,z,\xi,\eta,0,t-\tau)\,d\xi\,d\eta\,d\tau$$

$$+ a^2\int_0^t\int_0^{l_2}\int_0^{l_1} g_6(\xi,\eta,\tau)G(x,y,z,\xi,\eta,l_3,t-\tau)\,d\xi\,d\eta\,d\tau$$

$$+ \int_0^t\int_0^{l_3}\int_0^{l_2}\int_0^{l_1} \Phi(\xi,\eta,\zeta,\tau)G(x,y,z,\xi,\eta,\zeta,t-\tau)\,d\xi\,d\eta\,d\zeta\,d\tau,$$

where

$$G(x,y,z,\xi,\eta,\zeta,t) = \frac{\sin(t\sqrt{b})}{l_1 l_2 l_3 \sqrt{b}} + \frac{1}{l_1 l_2 l_3}\sum_{n=0}^{\infty}\sum_{m=0}^{\infty}\sum_{k=0}^{\infty} \frac{A_n A_m A_k}{\sqrt{\lambda_{nmk}}}\cos(\alpha_n x)\cos(\beta_m y)\cos(\gamma_k z)$$

$$\times \cos(\alpha_n\xi)\cos(\beta_m\eta)\cos(\gamma_k\zeta)\sin(t\sqrt{\lambda_{nmk}}),$$

$$\alpha_n = \frac{n\pi}{l_1},\quad \beta_m = \frac{m\pi}{l_2},\quad \gamma_k = \frac{k\pi}{l_3},\quad \lambda_{nmk} = a^2(\alpha_n^2+\beta_m^2+\gamma_k^2)+b,\quad A_n = \begin{cases} 1 & \text{for } n=0, \\ 2 & \text{for } n>0. \end{cases}$$

The summation is performed over the indices satisfying the condition $n+m+k>0$; the term corresponding to $n=m=k=0$ is singled out.

6.3.1-5. Domain: $0 \le x \le l_1$, $0 \le y \le l_2$, $0 \le z \le l_3$. Third boundary value problem.

A rectangular parallelepiped is considered. The following conditions are prescribed:

$$w = f_0(x,y,z) \quad \text{at} \quad t=0 \quad \text{(initial condition)},$$
$$\partial_t w = f_1(x,y,z) \quad \text{at} \quad t=0 \quad \text{(initial condition)},$$
$$\partial_x w - s_1 w = g_1(y,z,t) \quad \text{at} \quad x=0 \quad \text{(boundary condition)},$$
$$\partial_x w + s_2 w = g_2(y,z,t) \quad \text{at} \quad x=l_1 \quad \text{(boundary condition)},$$
$$\partial_y w - s_3 w = g_3(x,z,t) \quad \text{at} \quad y=0 \quad \text{(boundary condition)},$$
$$\partial_y w + s_4 w = g_4(x,z,t) \quad \text{at} \quad y=l_2 \quad \text{(boundary condition)},$$
$$\partial_z w - s_5 w = g_5(x,y,t) \quad \text{at} \quad z=0 \quad \text{(boundary condition)},$$
$$\partial_z w + s_6 w = g_6(x,y,t) \quad \text{at} \quad z=l_3 \quad \text{(boundary condition)}.$$

The solution $w(x,y,z,t)$ is determined by the formula in Paragraph 6.3.1-4 where

$$G(x,y,\xi,\eta,t) = 8\sum_{n=1}^{\infty}\sum_{m=1}^{\infty}\sum_{k=1}^{\infty}\frac{1}{E_{nmk}\sqrt{\lambda_{nmk}}}\sin(\alpha_n x+\varepsilon_n)\sin(\beta_m y+\sigma_m)\sin(\gamma_k z+\nu_k)$$

$$\times \sin(\alpha_n\xi+\varepsilon_n)\sin(\beta_m\eta+\sigma_m)\sin(\gamma_k\zeta+\nu_k)\sin(t\sqrt{\lambda_{nmk}}),$$

$$\varepsilon_n = \arctan\frac{\alpha_n}{l_1}, \quad \sigma_m = \arctan\frac{\beta_m}{l_2}, \quad \nu_k = \arctan\frac{\gamma_k}{l_3}, \quad \lambda_{nmk} = a^2(\alpha_n^2 + \beta_m^2 + \gamma_k^2) + b,$$

$$E_{nmk} = \left[l_1 + \frac{(s_1s_2 + \alpha_n^2)(s_1 + s_2)}{(s_1^2 + \alpha_n^2)(s_2^2 + \alpha_n^2)}\right]\left[l_2 + \frac{(s_3s_4 + \beta_m^2)(s_3 + s_4)}{(s_3^2 + \beta_m^2)(s_4^2 + \beta_m^2)}\right]\left[l_3 + \frac{(s_5s_6 + \gamma_k^2)(s_5 + s_6)}{(s_5^2 + \gamma_k^2)(s_6^2 + \gamma_k^2)}\right].$$

Here, the α_n, β_m, and γ_k are positive roots of the transcendental equations

$$\alpha^2 - s_1s_2 = (s_1 + s_2)\alpha\cot(l_1\alpha), \quad \beta^2 - s_3s_4 = (s_3 + s_4)\beta\cot(l_2\beta), \quad \gamma^2 - s_5s_6 = (s_5 + s_6)\gamma\cot(l_3\gamma).$$

6.3.1-6. Domain: $0 \le x \le l_1$, $0 \le y \le l_2$, $0 \le z \le l_3$. Mixed boundary value problems.

1°. A rectangular parallelepiped is considered. The following conditions are prescribed:

$$\begin{aligned}
w &= f_0(x,y,z) & \text{at} \quad t &= 0 & &\text{(initial condition)},\\
\partial_t w &= f_1(x,y,z) & \text{at} \quad t &= 0 & &\text{(initial condition)},\\
w &= g_1(y,z,t) & \text{at} \quad x &= 0 & &\text{(boundary condition)},\\
w &= g_2(y,z,t) & \text{at} \quad x &= l_1 & &\text{(boundary condition)},\\
\partial_y w &= g_3(x,z,t) & \text{at} \quad y &= 0 & &\text{(boundary condition)},\\
\partial_y w &= g_4(x,z,t) & \text{at} \quad y &= l_2 & &\text{(boundary condition)},\\
\partial_z w &= g_5(x,y,t) & \text{at} \quad z &= 0 & &\text{(boundary condition)},\\
\partial_z w &= g_6(x,y,t) & \text{at} \quad z &= l_3 & &\text{(boundary condition)}.
\end{aligned}$$

Solution:

$$\begin{aligned}
w(x,y,z,t) &= \frac{\partial}{\partial t}\int_0^{l_3}\int_0^{l_2}\int_0^{l_1} f_0(\xi,\eta,\zeta)G(x,y,z,\xi,\eta,\zeta,t)\,d\xi\,d\eta\,d\zeta\\
&+ \int_0^{l_3}\int_0^{l_2}\int_0^{l_1} f_1(\xi,\eta,\zeta)G(x,y,z,\xi,\eta,\zeta,t)\,d\xi\,d\eta\,d\zeta\\
&+ a^2\int_0^t\int_0^{l_3}\int_0^{l_2} g_1(\eta,\zeta,\tau)\left[\frac{\partial}{\partial\xi}G(x,y,z,\xi,\eta,\zeta,t-\tau)\right]_{\xi=0}d\eta\,d\zeta\,d\tau\\
&- a^2\int_0^t\int_0^{l_3}\int_0^{l_2} g_2(\eta,\zeta,\tau)\left[\frac{\partial}{\partial\xi}G(x,y,z,\xi,\eta,\zeta,t-\tau)\right]_{\xi=l_1}d\eta\,d\zeta\,d\tau\\
&- a^2\int_0^t\int_0^{l_3}\int_0^{l_1} g_3(\xi,\zeta,\tau)G(x,y,z,\xi,0,\zeta,t-\tau)\,d\xi\,d\zeta\,d\tau\\
&+ a^2\int_0^t\int_0^{l_3}\int_0^{l_1} g_4(\xi,\zeta,\tau)G(x,y,z,\xi,l_2,\zeta,t-\tau)\,d\xi\,d\zeta\,d\tau\\
&- a^2\int_0^t\int_0^{l_2}\int_0^{l_1} g_5(\xi,\eta,\tau)G(x,y,z,\xi,\eta,0,t-\tau)\,d\xi\,d\eta\,d\tau\\
&+ a^2\int_0^t\int_0^{l_2}\int_0^{l_1} g_6(\xi,\eta,\tau)G(x,y,z,\xi,\eta,l_3,t-\tau)\,d\xi\,d\eta\,d\tau\\
&+ \int_0^t\int_0^{l_3}\int_0^{l_2}\int_0^{l_1}\Phi(\xi,\eta,\zeta,\tau)G(x,y,z,\xi,\eta,\zeta,t-\tau)\,d\xi\,d\eta\,d\zeta\,d\tau.
\end{aligned}$$

Here,

$$G(x,y,z,\xi,\eta,\zeta,t) = \frac{2}{l_1l_2l_3}\sum_{n=1}^\infty\sum_{m=0}^\infty\sum_{k=0}^\infty \frac{A_mA_k}{\sqrt{\lambda_{nmk}}}\sin(\alpha_n x)\cos(\beta_m y)\cos(\gamma_k z)$$
$$\times\sin(\alpha_n\xi)\cos(\beta_m\eta)\cos(\gamma_k\zeta)\sin\!\left(t\sqrt{\lambda_{nmk}}\right),$$

where
$$A_m = \begin{cases} 1 & \text{for } m = 0, \\ 2 & \text{for } m > 0, \end{cases} \quad A_k = \begin{cases} 1 & \text{for } k = 0, \\ 2 & \text{for } k > 0, \end{cases}$$

$$\alpha_n = \frac{n\pi}{l_1}, \quad \beta_m = \frac{m\pi}{l_2}, \quad \gamma_k = \frac{k\pi}{l_3}, \quad \lambda_{nmk} = a^2(\alpha_n^2 + \beta_m^2 + \gamma_k^2) + b.$$

2°. A rectangular parallelepiped is considered. The following conditions are prescribed:

$$\begin{aligned}
w &= f_0(x, y, z) & \text{at} \quad t &= 0 & \text{(initial condition),} \\
\partial_t w &= f_1(x, y, z) & \text{at} \quad t &= 0 & \text{(initial condition),} \\
w &= g_1(y, z, t) & \text{at} \quad x &= 0 & \text{(boundary condition),} \\
\partial_x w &= g_2(y, z, t) & \text{at} \quad x &= l_1 & \text{(boundary condition),} \\
w &= g_3(x, z, t) & \text{at} \quad y &= 0 & \text{(boundary condition),} \\
\partial_y w &= g_4(x, z, t) & \text{at} \quad y &= l_2 & \text{(boundary condition),} \\
w &= g_5(x, y, t) & \text{at} \quad z &= 0 & \text{(boundary condition),} \\
\partial_z w &= g_6(x, y, t) & \text{at} \quad z &= l_3 & \text{(boundary condition).}
\end{aligned}$$

Solution:
$$\begin{aligned}
w(x,y,z,t) &= \frac{\partial}{\partial t} \int_0^{l_3}\!\!\int_0^{l_2}\!\!\int_0^{l_1} f_0(\xi,\eta,\zeta) G(x,y,z,\xi,\eta,\zeta,t)\, d\xi\, d\eta\, d\zeta \\
&+ \int_0^{l_3}\!\!\int_0^{l_2}\!\!\int_0^{l_1} f_1(\xi,\eta,\zeta) G(x,y,z,\xi,\eta,\zeta,t)\, d\xi\, d\eta\, d\zeta \\
&+ a^2 \int_0^t\!\!\int_0^{l_3}\!\!\int_0^{l_2} g_1(\eta,\zeta,\tau) \left[\frac{\partial}{\partial \xi} G(x,y,z,\xi,\eta,\zeta,t-\tau)\right]_{\xi=0} d\eta\, d\zeta\, d\tau \\
&+ a^2 \int_0^t\!\!\int_0^{l_3}\!\!\int_0^{l_2} g_2(\eta,\zeta,\tau) G(x,y,z,l_1,\eta,\zeta,t-\tau)\, d\eta\, d\zeta\, d\tau \\
&+ a^2 \int_0^t\!\!\int_0^{l_3}\!\!\int_0^{l_1} g_3(\xi,\zeta,\tau) \left[\frac{\partial}{\partial \eta} G(x,y,z,\xi,\eta,\zeta,t-\tau)\right]_{\eta=0} d\xi\, d\zeta\, d\tau \\
&+ a^2 \int_0^t\!\!\int_0^{l_3}\!\!\int_0^{l_1} g_4(\xi,\zeta,\tau) G(x,y,z,\xi,l_2,\zeta,t-\tau)\, d\xi\, d\zeta\, d\tau \\
&+ a^2 \int_0^t\!\!\int_0^{l_2}\!\!\int_0^{l_1} g_5(\xi,\eta,\tau) \left[\frac{\partial}{\partial \zeta} G(x,y,z,\xi,\eta,\zeta,t-\tau)\right]_{\zeta=0} d\xi\, d\eta\, d\tau \\
&+ a^2 \int_0^t\!\!\int_0^{l_2}\!\!\int_0^{l_1} g_6(\xi,\eta,\tau) G(x,y,z,\xi,\eta,l_3,t-\tau)\, d\xi\, d\eta\, d\tau \\
&+ \int_0^t\!\!\int_0^{l_3}\!\!\int_0^{l_2}\!\!\int_0^{l_1} \Phi(\xi,\eta,\zeta,\tau) G(x,y,z,\xi,\eta,\zeta,t-\tau)\, d\xi\, d\eta\, d\zeta\, d\tau.
\end{aligned}$$

Here,
$$G(x,y,z,\xi,\eta,\zeta,t) = \frac{8}{l_1 l_2 l_3} \sum_{n=1}^{\infty} \sum_{m=1}^{\infty} \sum_{k=1}^{\infty} \frac{1}{\sqrt{\lambda_{nmk}}} \sin(\alpha_n x) \sin(\beta_m y) \sin(\gamma_k z)$$
$$\times \sin(\alpha_n \xi) \sin(\beta_m \eta) \sin(\gamma_k \zeta) \sin\!\left(t\sqrt{\lambda_{nmk}}\right),$$

where
$$\alpha_n = \frac{\pi(2n+1)}{2l_1}, \quad \beta_m = \frac{\pi(2m+1)}{2l_2}, \quad \gamma_k = \frac{\pi(2k+1)}{2l_3},$$
$$\lambda_{nmk} = a^2(\alpha_n^2 + \beta_m^2 + \gamma_k^2) + b.$$

6.3.2. Problems in Cylindrical Coordinates

A *nonhomogeneous Klein–Gordon equation* in the cylindrical coordinate system is written as

$$\frac{\partial^2 w}{\partial t^2} = a^2\left[\frac{1}{r}\frac{\partial}{\partial r}\left(r\frac{\partial w}{\partial r}\right) + \frac{1}{r^2}\frac{\partial^2 w}{\partial \varphi^2} + \frac{\partial^2 w}{\partial z^2}\right] - bw + \Phi(r,\varphi,z,t), \qquad r = \sqrt{x^2+y^2}.$$

One-dimensional problems with axial symmetry that have solutions $w = w(r,t)$ are treated in Subsection 4.2.5. Two-dimensional problems whose solutions have the form $w = w(r,\varphi,t)$ or $w = w(r,z,t)$ are considered in Subsections 5.3.2 and 5.3.3.

6.3.2-1. Domain: $0 \le r \le R$, $0 \le \varphi \le 2\pi$, $0 \le z \le l$. First boundary value problem.

A circular cylinder of finite length is considered. The following conditions are prescribed:

$$\begin{aligned}
w &= f_0(r,\varphi,z) & &\text{at } t=0 & &\text{(initial condition)},\\
\partial_t w &= f_1(r,\varphi,z) & &\text{at } t=0 & &\text{(initial condition)},\\
w &= g_1(\varphi,z,t) & &\text{at } r=R & &\text{(boundary condition)},\\
w &= g_2(r,\varphi,t) & &\text{at } z=0 & &\text{(boundary condition)},\\
w &= g_3(r,\varphi,t) & &\text{at } z=l & &\text{(boundary condition)}.
\end{aligned}$$

Solution:

$$\begin{aligned}
w(r,\varphi,z,t) =\ & \frac{\partial}{\partial t}\int_0^l\int_0^{2\pi}\int_0^R \xi f_0(\xi,\eta,\zeta)G(r,\varphi,z,\xi,\eta,\zeta,t)\,d\xi\,d\eta\,d\zeta\\
& + \int_0^l\int_0^{2\pi}\int_0^R \xi f_1(\xi,\eta,\zeta)G(r,\varphi,z,\xi,\eta,\zeta,t)\,d\xi\,d\eta\,d\zeta\\
& - a^2 R\int_0^t\int_0^l\int_0^{2\pi} g_1(\eta,\zeta,\tau)\left[\frac{\partial}{\partial \xi}G(r,\varphi,z,\xi,\eta,\zeta,t-\tau)\right]_{\xi=R}\,d\eta\,d\zeta\,d\tau\\
& + a^2\int_0^t\int_0^{2\pi}\int_0^R \xi g_2(\xi,\eta,\tau)\left[\frac{\partial}{\partial \zeta}G(r,\varphi,z,\xi,\eta,\zeta,t-\tau)\right]_{\zeta=0}\,d\xi\,d\eta\,d\tau\\
& - a^2\int_0^t\int_0^{2\pi}\int_0^R \xi g_3(\xi,\eta,\tau)\left[\frac{\partial}{\partial \zeta}G(r,\varphi,z,\xi,\eta,\zeta,t-\tau)\right]_{\zeta=l}\,d\xi\,d\eta\,d\tau\\
& + \int_0^t\int_0^l\int_0^{2\pi}\int_0^R \xi\Phi(\xi,\eta,\zeta,\tau)G(r,\varphi,z,\xi,\eta,\zeta,t-\tau)\,d\xi\,d\eta\,d\zeta\,d\tau.
\end{aligned}$$

Here,

$$G(r,\varphi,z,\xi,\eta,\zeta,t) = \frac{2}{\pi R^2 l}\sum_{n=0}^{\infty}\sum_{m=1}^{\infty}\sum_{k=1}^{\infty}\frac{A_n}{[J'_n(\mu_{nm}R)]^2\sqrt{\lambda_{nmk}}}J_n(\mu_{nm}r)J_n(\mu_{nm}\xi)$$

$$\times \cos[n(\varphi-\eta)]\sin\left(\frac{k\pi z}{l}\right)\sin\left(\frac{k\pi \zeta}{l}\right)\sin(t\sqrt{\lambda_{nmk}}),$$

where

$$\lambda_{nmk} = a^2\mu_{nm}^2 + \frac{a^2 k^2 \pi^2}{l^2} + b, \qquad A_n = \begin{cases} 1 & \text{for } n=0,\\ 2 & \text{for } n>0, \end{cases}$$

the $J_n(\xi)$ are the Bessel functions (the prime denotes the derivative with respect to the argument), and the μ_{nm} are positive roots of the transcendental equation $J_n(\mu R) = 0$.

6.3. Equations of the Form $\frac{\partial^2 w}{\partial t^2} = a^2 \Delta_3 w - bw + \Phi(x,y,z,t)$

6.3.2-2. Domain: $0 \le r \le R$, $0 \le \varphi \le 2\pi$, $0 \le z \le l$. Second boundary value problem.

A circular cylinder of finite length is considered. The following conditions are prescribed:

$$w = f_0(r,\varphi,z) \quad \text{at} \quad t = 0 \quad \text{(initial condition)},$$
$$\partial_t w = f_1(r,\varphi,z) \quad \text{at} \quad t = 0 \quad \text{(initial condition)},$$
$$\partial_r w = g_1(\varphi,z,t) \quad \text{at} \quad r = R \quad \text{(boundary condition)},$$
$$\partial_z w = g_2(r,\varphi,t) \quad \text{at} \quad z = 0 \quad \text{(boundary condition)},$$
$$\partial_z w = g_3(r,\varphi,t) \quad \text{at} \quad z = l \quad \text{(boundary condition)}.$$

Solution:

$$w(r,\varphi,z,t) = \frac{\partial}{\partial t} \int_0^l \int_0^{2\pi} \int_0^R \xi f_0(\xi,\eta,\zeta) G(r,\varphi,z,\xi,\eta,\zeta,t) \, d\xi \, d\eta \, d\zeta$$
$$+ \int_0^l \int_0^{2\pi} \int_0^R \xi f_1(\xi,\eta,\zeta) G(r,\varphi,z,\xi,\eta,\zeta,t) \, d\xi \, d\eta \, d\zeta$$
$$+ a^2 R \int_0^t \int_0^l \int_0^{2\pi} g_1(\eta,\zeta,\tau) G(r,\varphi,z,R,\eta,\zeta,t-\tau) \, d\eta \, d\zeta \, d\tau$$
$$- a^2 \int_0^t \int_0^{2\pi} \int_0^R \xi g_2(\xi,\eta,\tau) G(r,\varphi,z,\xi,\eta,0,t-\tau) \, d\xi \, d\eta \, d\tau$$
$$+ a^2 \int_0^t \int_0^{2\pi} \int_0^R \xi g_3(\xi,\eta,\tau) G(r,\varphi,z,\xi,\eta,l,t-\tau) \, d\xi \, d\eta \, d\tau$$
$$+ \int_0^t \int_0^l \int_0^{2\pi} \int_0^R \xi \Phi(\xi,\eta,\zeta,\tau) G(r,\varphi,z,\xi,\eta,\zeta,t-\tau) \, d\xi \, d\eta \, d\zeta \, d\tau.$$

Here,

$$G(r,\varphi,z,\xi,\eta,\zeta,t) = \frac{\sin(t\sqrt{b})}{\pi R^2 l \sqrt{b}} + \frac{2}{\pi R^2 l} \sum_{k=1}^{\infty} \frac{1}{\sqrt{\beta_k}} \cos\left(\frac{k\pi x}{l}\right) \cos\left(\frac{k\pi \xi}{l}\right) \sin(t\sqrt{\beta_k})$$
$$+ \frac{1}{\pi l} \sum_{n=0}^{\infty} \sum_{m=1}^{\infty} \sum_{k=0}^{\infty} \frac{A_n A_k \mu_{nm}^2 J_n(\mu_{nm} r) J_n(\mu_{nm} \xi)}{(\mu_{nm}^2 R^2 - n^2)[J_n(\mu_{nm} R)]^2} \cos[n(\varphi - \eta)] \cos\left(\frac{k\pi x}{l}\right) \cos\left(\frac{k\pi \xi}{l}\right) \frac{\sin(\lambda_{nmk} t)}{\lambda_{nmk}},$$

$$\beta_k = \frac{a^2 k^2 \pi^2}{l^2} + b, \quad \lambda_{nmk} = \sqrt{a^2 \mu_{nm}^2 + \frac{a^2 k^2 \pi^2}{l^2} + b}, \quad A_n = \begin{cases} 1 & \text{for } n = 0, \\ 2 & \text{for } n > 0, \end{cases}$$

where the $J_n(\xi)$ are the Bessel functions and the μ_{nm} are positive roots of the transcendental equation $J_n'(\mu R) = 0$.

6.3.2-3. Domain: $0 \le r \le R$, $0 \le \varphi \le 2\pi$, $0 \le z \le l$. Third boundary value problem.

A circular cylinder of finite length is considered. The following conditions are prescribed:

$$w = f_0(r,\varphi,z) \quad \text{at} \quad t = 0 \quad \text{(initial condition)},$$
$$\partial_t w = f_1(r,\varphi,z) \quad \text{at} \quad t = 0 \quad \text{(initial condition)},$$
$$\partial_r w + k_1 w = g(\varphi,z,t) \quad \text{at} \quad r = R \quad \text{(boundary condition)},$$
$$\partial_z w - k_2 w = g_2(r,\varphi,t) \quad \text{at} \quad z = 0 \quad \text{(boundary condition)},$$
$$\partial_z w + k_3 w = g_3(r,\varphi,t) \quad \text{at} \quad z = l \quad \text{(boundary condition)}.$$

The solution $w(r,\varphi,z,t)$ is determined by the formula in Paragraph 6.3.2-2 where

$$G(r,\varphi,z,\xi,\eta,\zeta,t) = \frac{1}{\pi} \sum_{n=0}^{\infty} \sum_{m=1}^{\infty} \sum_{s=1}^{\infty} \frac{A_n \mu_{nm}^2 J_n(\mu_{nm} r) J_n(\mu_{nm} \xi) \cos[n(\varphi-\eta)] h_s(z) h_s(\zeta) \sin(\lambda_{nms} t)}{(\mu_{nm}^2 R^2 + k_1^2 R^2 - n^2)[J_n(\mu_{nm} R)]^2 \|h_s\|^2 \lambda_{nms}}.$$

Here,

$$A_n = \begin{cases} 1 & \text{for } n = 0, \\ 2 & \text{for } n > 0, \end{cases} \qquad \lambda_{nms} = \sqrt{a^2\mu_{nm}^2 + a^2\beta_s^2 + b},$$

$$h_s(z) = \cos(\beta_s z) + \frac{k_2}{\beta_s}\sin(\beta_s z), \qquad \|h_s\|^2 = \frac{k_3}{2\beta_s^2}\frac{\beta_s^2 + k_2^2}{\beta_s^2 + k_3^2} + \frac{k_2}{2\beta_s^2} + \frac{l}{2}\left(1 + \frac{k_2^2}{\beta_s^2}\right),$$

the $J_n(\xi)$ are the Bessel functions, and the μ_{nm} and β_s are positive roots of the transcendental equations

$$\mu J_n'(\mu R) + k_1 J_n(\mu R) = 0, \qquad \frac{\tan(\beta l)}{\beta} = \frac{k_2 + k_3}{\beta^2 - k_2 k_3}.$$

6.3.2-4. Domain: $0 \le r \le R$, $0 \le \varphi \le 2\pi$, $0 \le z \le l$. Mixed boundary value problems.

1°. A circular cylinder of finite length is considered. The following conditions are prescribed:

$$\begin{aligned}
w &= f_0(r,\varphi,z) & &\text{at} & t &= 0 & &\text{(initial condition)}, \\
\partial_t w &= f_1(r,\varphi,z) & &\text{at} & t &= 0 & &\text{(initial condition)}, \\
w &= g_1(\varphi,z,t) & &\text{at} & r &= R & &\text{(boundary condition)}, \\
\partial_z w &= g_2(r,\varphi,t) & &\text{at} & z &= 0 & &\text{(boundary condition)}, \\
\partial_z w &= g_3(r,\varphi,t) & &\text{at} & z &= l & &\text{(boundary condition)}.
\end{aligned}$$

Solution:

$$\begin{aligned}
w(r,\varphi,z,t) &= \frac{\partial}{\partial t}\int_0^l\int_0^{2\pi}\int_0^R \xi f_0(\xi,\eta,\zeta)G(r,\varphi,z,\xi,\eta,\zeta,t)\,d\xi\,d\eta\,d\zeta \\
&\quad + \int_0^l\int_0^{2\pi}\int_0^R \xi f_1(\xi,\eta,\zeta)G(r,\varphi,z,\xi,\eta,\zeta,t)\,d\xi\,d\eta\,d\zeta \\
&\quad - a^2 R\int_0^t\int_0^l\int_0^{2\pi} g_1(\eta,\zeta,\tau)\left[\frac{\partial}{\partial \xi}G(r,\varphi,z,\xi,\eta,\zeta,t-\tau)\right]_{\xi=R}\,d\eta\,d\zeta\,d\tau \\
&\quad - a^2\int_0^t\int_0^{2\pi}\int_0^R \xi g_2(\xi,\eta,\tau)G(r,\varphi,z,\xi,\eta,0,t-\tau)\,d\xi\,d\eta\,d\tau \\
&\quad + a^2\int_0^t\int_0^{2\pi}\int_0^R \xi g_3(\xi,\eta,\tau)G(r,\varphi,z,\xi,\eta,l,t-\tau)\,d\xi\,d\eta\,d\tau \\
&\quad + \int_0^t\int_0^l\int_0^{2\pi}\int_0^R \xi \Phi(\xi,\eta,\zeta,\tau)G(r,\varphi,z,\xi,\eta,\zeta,t-\tau)\,d\xi\,d\eta\,d\zeta\,d\tau.
\end{aligned}$$

Here,

$$G(r,\varphi,z,\xi,\eta,\zeta,t) = \frac{1}{\pi R^2 l}\sum_{n=0}^{\infty}\sum_{m=1}^{\infty}\sum_{k=0}^{\infty}\frac{A_n A_k}{[J_n'(\mu_{nm}R)]^2 \sqrt{\lambda_{nmk}}}J_n(\mu_{nm}r)J_n(\mu_{nm}\xi)$$

$$\times \cos[n(\varphi-\eta)]\cos\left(\frac{k\pi z}{l}\right)\cos\left(\frac{k\pi \zeta}{l}\right)\sin\left(t\sqrt{\lambda_{nmk}}\right),$$

$$\lambda_{nmk} = a^2\mu_{nm}^2 + \frac{a^2 k^2\pi^2}{l^2} + b, \qquad A_n = \begin{cases} 1 & \text{for } n = 0, \\ 2 & \text{for } n > 0, \end{cases}$$

where the $J_n(\xi)$ are the Bessel functions (the prime denotes the derivative with respect to the argument) and the μ_{nm} are positive roots of the transcendental equation $J_n(\mu R) = 0$.

2°. A circular cylinder of finite length is considered. The following conditions are prescribed:

$$w = f_0(r, \varphi, z) \quad \text{at} \quad t = 0 \quad \text{(initial condition)},$$
$$\partial_t w = f_1(r, \varphi, z) \quad \text{at} \quad t = 0 \quad \text{(initial condition)},$$
$$\partial_r w = g_1(\varphi, z, t) \quad \text{at} \quad r = R \quad \text{(boundary condition)},$$
$$w = g_2(r, \varphi, t) \quad \text{at} \quad z = 0 \quad \text{(boundary condition)},$$
$$w = g_3(r, \varphi, t) \quad \text{at} \quad z = l \quad \text{(boundary condition)}.$$

Solution:

$$\begin{aligned} w(r, \varphi, z, t) &= \frac{\partial}{\partial t} \int_0^l \int_0^{2\pi} \int_0^R \xi f_0(\xi, \eta, \zeta) G(r, \varphi, z, \xi, \eta, \zeta, t) \, d\xi \, d\eta \, d\zeta \\ &+ \int_0^l \int_0^{2\pi} \int_0^R \xi f_1(\xi, \eta, \zeta) G(r, \varphi, z, \xi, \eta, \zeta, t) \, d\xi \, d\eta \, d\zeta \\ &+ a^2 R \int_0^t \int_0^l \int_0^{2\pi} g_1(\eta, \zeta, \tau) G(r, \varphi, z, R, \eta, \zeta, t - \tau) \, d\eta \, d\zeta \, d\tau \\ &+ a^2 \int_0^t \int_0^{2\pi} \int_0^R \xi g_2(\xi, \eta, \tau) \left[\frac{\partial}{\partial \zeta} G(r, \varphi, z, \xi, \eta, \zeta, t - \tau) \right]_{\zeta=0} d\xi \, d\eta \, d\tau \\ &- a^2 \int_0^t \int_0^{2\pi} \int_0^R \xi g_3(\xi, \eta, \tau) \left[\frac{\partial}{\partial \zeta} G(r, \varphi, z, \xi, \eta, \zeta, t - \tau) \right]_{\zeta=l} d\xi \, d\eta \, d\tau \\ &+ \int_0^t \int_0^l \int_0^{2\pi} \int_0^R \xi \Phi(\xi, \eta, \zeta, \tau) G(r, \varphi, z, \xi, \eta, \zeta, t - \tau) \, d\xi \, d\eta \, d\zeta \, d\tau. \end{aligned}$$

Here,

$$\begin{aligned} G(r, \varphi, z, \xi, \eta, \zeta, t) &= \frac{2}{\pi R^2 l} \sum_{k=1}^{\infty} \frac{1}{\sqrt{\beta_k}} \sin\left(\frac{k\pi z}{l}\right) \sin\left(\frac{k\pi \zeta}{l}\right) \sin\left(t \sqrt{\beta_k}\right) \\ &+ \frac{2}{\pi l} \sum_{n=0}^{\infty} \sum_{m=1}^{\infty} \sum_{k=1}^{\infty} \frac{A_n \mu_{nm}^2}{(\mu_{nm}^2 R^2 - n^2)[J_n(\mu_{nm} R)]^2 \sqrt{\lambda_{nmk}}} J_n(\mu_{nm} r) J_n(\mu_{nm} \xi) \\ &\qquad \times \cos[n(\varphi - \eta)] \sin\left(\frac{k\pi z}{l}\right) \sin\left(\frac{k\pi \zeta}{l}\right) \sin\left(t \sqrt{\lambda_{nmk}}\right), \end{aligned}$$

$$\beta_k = \frac{a^2 k^2 \pi^2}{l^2} + b, \quad \lambda_{nmk} = a^2 \mu_{nm}^2 + \frac{a^2 k^2 \pi^2}{l^2} + b, \quad A_n = \begin{cases} 1 & \text{for } n = 0, \\ 2 & \text{for } n > 0, \end{cases}$$

where the $J_n(\xi)$ are the Bessel functions and the μ_{nm} are positive roots of the transcendental equation $J_n'(\mu R) = 0$.

6.3.2-5. Domain: $R_1 \leq r \leq R_2$, $0 \leq \varphi \leq 2\pi$, $0 \leq z \leq l$. **First boundary value problem.**

A hollow circular cylinder of finite length is considered. The following conditions are prescribed:

$$w = f_0(r, \varphi, z) \quad \text{at} \quad t = 0 \quad \text{(initial condition)},$$
$$\partial_t w = f_1(r, \varphi, z) \quad \text{at} \quad t = 0 \quad \text{(initial condition)},$$
$$w = g_1(\varphi, z, t) \quad \text{at} \quad r = R_1 \quad \text{(boundary condition)},$$
$$w = g_2(\varphi, z, t) \quad \text{at} \quad r = R_2 \quad \text{(boundary condition)},$$
$$w = g_3(r, \varphi, t) \quad \text{at} \quad z = 0 \quad \text{(boundary condition)},$$
$$w = g_4(r, \varphi, t) \quad \text{at} \quad z = l \quad \text{(boundary condition)}.$$

Solution:

$$w(r,\varphi,z,t) = \frac{\partial}{\partial t}\int_0^l\int_0^{2\pi}\int_{R_1}^{R_2} f_0(\xi,\eta,\zeta)G(r,\varphi,z,\xi,\eta,\zeta,t)\xi\,d\xi\,d\eta\,d\zeta$$

$$+ \int_0^l\int_0^{2\pi}\int_{R_1}^{R_2} f_1(\xi,\eta,\zeta)G(r,\varphi,z,\xi,\eta,\zeta,t)\xi\,d\xi\,d\eta\,d\zeta$$

$$+ a^2 R_1 \int_0^t\int_0^l\int_0^{2\pi} g_1(\eta,\zeta,\tau)\left[\frac{\partial}{\partial\xi}G(r,\varphi,z,\xi,\eta,\zeta,t-\tau)\right]_{\xi=R_1} d\eta\,d\zeta\,d\tau$$

$$- a^2 R_2 \int_0^t\int_0^l\int_0^{2\pi} g_2(\eta,\zeta,\tau)\left[\frac{\partial}{\partial\xi}G(r,\varphi,z,\xi,\eta,\zeta,t-\tau)\right]_{\xi=R_2} d\eta\,d\zeta\,d\tau$$

$$+ a^2 \int_0^t\int_0^{2\pi}\int_{R_1}^{R_2} g_3(\xi,\eta,\tau)\left[\frac{\partial}{\partial\zeta}G(r,\varphi,z,\xi,\eta,\zeta,t-\tau)\right]_{\zeta=0}\xi\,d\xi\,d\eta\,d\tau$$

$$- a^2 \int_0^t\int_0^{2\pi}\int_{R_1}^{R_2} g_4(\xi,\eta,\tau)\left[\frac{\partial}{\partial\zeta}G(r,\varphi,z,\xi,\eta,\zeta,t-\tau)\right]_{\zeta=l}\xi\,d\xi\,d\eta\,d\tau$$

$$+ \int_0^t\int_0^l\int_0^{2\pi}\int_{R_1}^{R_2} \Phi(\xi,\eta,\zeta,\tau)G(r,\varphi,z,\xi,\eta,\zeta,t-\tau)\xi\,d\xi\,d\eta\,d\zeta\,d\tau.$$

Here,

$$G(r,\varphi,z,\xi,\eta,\zeta,t) = \frac{\pi}{2l}\sum_{n=0}^{\infty}\sum_{m=1}^{\infty}\sum_{k=1}^{\infty}\frac{A_n\mu_{nm}^2 J_n^2(\mu_{nm}R_2)}{J_n^2(\mu_{nm}R_1) - J_n^2(\mu_{nm}R_2)}Z_{nm}(r)Z_{nm}(\xi)$$

$$\times \cos[n(\varphi-\eta)]\sin\left(\frac{k\pi z}{l}\right)\sin\left(\frac{k\pi\zeta}{l}\right)\frac{\sin(t\sqrt{\lambda_{nmk}})}{\sqrt{\lambda_{nmk}}},$$

where

$$A_n = \begin{cases} 1 & \text{for } n=0, \\ 2 & \text{for } n\neq 0, \end{cases} \quad \lambda_{nmk} = a^2\mu_{nm}^2 + \frac{a^2 k^2\pi^2}{l^2} + b,$$

$$Z_{nm}(r) = J_n(\mu_{nm}R_1)Y_n(\mu_{nm}r) - Y_n(\mu_{nm}R_1)J_n(\mu_{nm}r);$$

the $J_n(r)$ and $Y_n(r)$ are the Bessel functions, and the μ_{nm} are positive roots of the transcendental equation

$$J_n(\mu R_1)Y_n(\mu R_2) - Y_n(\mu R_1)J_n(\mu R_2) = 0.$$

6.3.2-6. Domain: $R_1 \leq r \leq R_2$, $0 \leq \varphi \leq 2\pi$, $0 \leq z \leq l$. Second boundary value problem.

A hollow circular cylinder of finite length is considered. The following conditions are prescribed:

$$w = f_0(r,\varphi,z) \quad \text{at} \quad t=0 \quad \text{(initial condition)},$$
$$\partial_t w = f_1(r,\varphi,z) \quad \text{at} \quad t=0 \quad \text{(initial condition)},$$
$$\partial_r w = g_1(\varphi,z,t) \quad \text{at} \quad r=R_1 \quad \text{(boundary condition)},$$
$$\partial_r w = g_2(\varphi,z,t) \quad \text{at} \quad r=R_2 \quad \text{(boundary condition)},$$
$$\partial_z w = g_3(r,\varphi,t) \quad \text{at} \quad z=0 \quad \text{(boundary condition)},$$
$$\partial_z w = g_4(r,\varphi,t) \quad \text{at} \quad z=l \quad \text{(boundary condition)}.$$

Solution:

$$w(r,\varphi,z,t) = \frac{\partial}{\partial t}\int_0^l\int_0^{2\pi}\int_{R_1}^{R_2} f_0(\xi,\eta,\zeta)G(r,\varphi,z,\xi,\eta,\zeta,t)\xi\,d\xi\,d\eta\,d\zeta$$
$$+\int_0^l\int_0^{2\pi}\int_{R_1}^{R_2} f_1(\xi,\eta,\zeta)G(r,\varphi,z,\xi,\eta,\zeta,t)\xi\,d\xi\,d\eta\,d\zeta$$
$$-a^2 R_1\int_0^t\int_0^l\int_0^{2\pi} g_1(\eta,\zeta,\tau)G(r,\varphi,z,R_1,\eta,\zeta,t-\tau)\,d\eta\,d\zeta\,d\tau$$
$$+a^2 R_2\int_0^t\int_0^l\int_0^{2\pi} g_2(\eta,\zeta,\tau)G(r,\varphi,z,R_2,\eta,\zeta,t-\tau)\,d\eta\,d\zeta\,d\tau$$
$$-a^2\int_0^t\int_0^{2\pi}\int_{R_1}^{R_2} g_3(\xi,\eta,\tau)G(r,\varphi,z,\xi,\eta,0,t-\tau)\xi\,d\xi\,d\eta\,d\tau$$
$$+a^2\int_0^t\int_0^{2\pi}\int_{R_1}^{R_2} g_4(\xi,\eta,\tau)G(r,\varphi,z,\xi,\eta,l,t-\tau)\xi\,d\xi\,d\eta\,d\tau$$
$$+\int_0^t\int_0^l\int_0^{2\pi}\int_{R_1}^{R_2} \Phi(\xi,\eta,\zeta,\tau)G(r,\varphi,z,\xi,\eta,\zeta,t-\tau)\xi\,d\xi\,d\eta\,d\zeta\,d\tau.$$

Here,

$$G(r,\varphi,z,\xi,\eta,\zeta,t) = \frac{\sin(t\sqrt{b})}{\pi(R_2^2-R_1^2)l\sqrt{b}} + \frac{2}{\pi(R_2^2-R_1^2)l}\sum_{k=1}^{\infty}\cos\left(\frac{k\pi z}{l}\right)\cos\left(\frac{k\pi\zeta}{l}\right)\frac{\sin(t\sqrt{\beta_k})}{\sqrt{\beta_k}}$$
$$+\frac{1}{\pi l}\sum_{n=0}^{\infty}\sum_{m=1}^{\infty}\sum_{k=0}^{\infty}\frac{A_n A_k \mu_{nm}^2 Z_{nm}(r)Z_{nm}(\xi)}{(\mu_{nm}^2 R_2^2 - n^2)Z_{nm}^2(R_2) - (\mu_{nm}^2 R_1^2 - n^2)Z_{nm}^2(R_1)}$$
$$\times \cos[n(\varphi-\eta)]\cos\left(\frac{k\pi z}{l}\right)\cos\left(\frac{k\pi\zeta}{l}\right)\frac{\sin(t\sqrt{\lambda_{nmk}})}{\sqrt{\lambda_{nmk}}},$$

where

$$A_n = \begin{cases} 1 & \text{for } n=0, \\ 2 & \text{for } n\neq 0, \end{cases} \quad \beta_k = \frac{a^2 k^2 \pi^2}{l^2} + b, \quad \lambda_{nmk} = a^2\mu_{nm}^2 + \frac{a^2 k^2\pi^2}{l^2} + b,$$
$$Z_{nm}(r) = J_n'(\mu_{nm}R_1)Y_n(\mu_{nm}r) - Y_n'(\mu_{nm}R_1)J_n(\mu_{nm}r);$$

the $J_n(r)$ and $Y_n(r)$ are the Bessel functions, and the μ_{nm} are positive roots of the transcendental equation

$$J_n'(\mu R_1)Y_n'(\mu R_2) - Y_n'(\mu R_1)J_n'(\mu R_2) = 0.$$

6.3.2-7. Domain: $R_1 \leq r \leq R_2$, $0 \leq \varphi \leq 2\pi$, $0 \leq z \leq l$. Third boundary value problem.

A hollow circular cylinder of finite length is considered. The following conditions are prescribed:

$$\begin{aligned}
w &= f_0(r,\varphi,z) & &\text{at} & t &= 0 & &\text{(initial condition)}, \\
\partial_t w &= f_1(r,\varphi,z) & &\text{at} & t &= 0 & &\text{(initial condition)}, \\
\partial_r w - k_1 w &= g_1(\varphi,z,t) & &\text{at} & r &= R_1 & &\text{(boundary condition)}, \\
\partial_r w + k_2 w &= g_2(\varphi,z,t) & &\text{at} & r &= R_2 & &\text{(boundary condition)}, \\
\partial_z w - k_3 w &= g_3(r,\varphi,t) & &\text{at} & z &= 0 & &\text{(boundary condition)}, \\
\partial_z w + k_4 w &= g_4(r,\varphi,t) & &\text{at} & z &= l & &\text{(boundary condition)}.
\end{aligned}$$

The solution $w(r, \varphi, z, t)$ is determined by the formula in Paragraph 6.3.2-6 where

$$G(r, \varphi, z, \xi, \eta, \zeta, t) = \frac{1}{\pi} \sum_{n=0}^{\infty} \sum_{m=1}^{\infty} \sum_{s=1}^{\infty} \frac{A_n \mu_{nm}^2}{\|h_s\|^2 \sqrt{a^2 \mu_{nm}^2 + a^2 \lambda_s^2 + b}}$$

$$\times \frac{Z_{nm}(r) Z_{nm}(\xi) \cos[n(\varphi - \eta)] h_s(z) h_s(\zeta) \sin\left(t \sqrt{a^2 \mu_{nm}^2 + a^2 \lambda_s^2 + b}\right)}{(k_2^2 R_2^2 + \mu_{nm}^2 R_2^2 - n^2) Z_{nm}^2(R_2) - (k_1^2 R_1^2 + \mu_{nm}^2 R_1^2 - n^2) Z_{nm}^2(R_1)}.$$

Here,

$$Z_{nm}(r) = \left[\mu_{nm} J_n'(\mu_{nm} R_1) - k_1 J_n(\mu_{nm} R_1)\right] Y_n(\mu_{nm} r)$$
$$- \left[\mu_{nm} Y_n'(\mu_{nm} R_1) - k_1 Y_n(\mu_{nm} R_1)\right] J_n(\mu_{nm} r),$$

$$A_n = \begin{cases} 1 & \text{for } n = 0, \\ 2 & \text{for } n \neq 0, \end{cases} \quad h_s(z) = \cos(\lambda_s z) + \frac{k_3}{\lambda_s} \sin(\lambda_s z), \quad \|h_s\|^2 = \frac{k_4}{2\lambda_s^2} \frac{\lambda_s^2 + k_3^2}{\lambda_s^2 + k_4^2} + \frac{k_3}{2\lambda_s^2} + \frac{l}{2}\left(1 + \frac{k_3^2}{\lambda_s^2}\right),$$

where the $J_n(r)$ and $Y_n(r)$ are the Bessel functions, and the μ_{nm} are positive roots of the transcendental equation

$$\left[\mu J_n'(\mu R_1) - k_1 J_n(\mu R_1)\right] \left[\mu Y_n'(\mu R_2) + k_2 Y_n(\mu R_2)\right]$$
$$= \left[\mu Y_n'(\mu R_1) - k_1 Y_n(\mu R_1)\right] \left[\mu J_n'(\mu R_2) + k_2 J_n(\mu R_2)\right],$$

and the λ_s are positive roots of the transcendental equation $\dfrac{\tan(\lambda l)}{\lambda} = \dfrac{k_3 + k_4}{\lambda^2 - k_3 k_4}$.

6.3.2-8. Domain: $R_1 \leq r \leq R_2$, $0 \leq \varphi \leq 2\pi$, $0 \leq z \leq l$. Mixed boundary value problems.

1°. A hollow circular cylinder of finite length is considered. The following conditions are prescribed:

$$\begin{aligned}
w &= f_0(r, \varphi, z) & \text{at} \quad t &= 0 & \text{(initial condition)}, \\
\partial_t w &= f_1(r, \varphi, z) & \text{at} \quad t &= 0 & \text{(initial condition)}, \\
w &= g_1(\varphi, z, t) & \text{at} \quad r &= R_1 & \text{(boundary condition)}, \\
w &= g_2(\varphi, z, t) & \text{at} \quad r &= R_2 & \text{(boundary condition)}, \\
\partial_z w &= g_3(r, \varphi, t) & \text{at} \quad z &= 0 & \text{(boundary condition)}, \\
\partial_z w &= g_4(r, \varphi, t) & \text{at} \quad z &= l & \text{(boundary condition)}.
\end{aligned}$$

Solution:

$$w(r, \varphi, z, t) = \frac{\partial}{\partial t} \int_0^l \int_0^{2\pi} \int_{R_1}^{R_2} f_0(\xi, \eta, \zeta) G(r, \varphi, z, \xi, \eta, \zeta, t) \xi \, d\xi \, d\eta \, d\zeta$$

$$+ \int_0^l \int_0^{2\pi} \int_{R_1}^{R_2} f_1(\xi, \eta, \zeta) G(r, \varphi, z, \xi, \eta, \zeta, t) \xi \, d\xi \, d\eta \, d\zeta$$

$$+ a^2 R_1 \int_0^t \int_0^l \int_0^{2\pi} g_1(\eta, \zeta, \tau) \left[\frac{\partial}{\partial \xi} G(r, \varphi, z, \xi, \eta, \zeta, t - \tau)\right]_{\xi = R_1} d\eta \, d\zeta \, d\tau$$

$$- a^2 R_2 \int_0^t \int_0^l \int_0^{2\pi} g_2(\eta, \zeta, \tau) \left[\frac{\partial}{\partial \xi} G(r, \varphi, z, \xi, \eta, \zeta, t - \tau)\right]_{\xi = R_2} d\eta \, d\zeta \, d\tau$$

$$- a^2 \int_0^t \int_0^{2\pi} \int_{R_1}^{R_2} g_3(\xi, \eta, \tau) G(r, \varphi, z, \xi, \eta, 0, t - \tau) \xi \, d\xi \, d\eta \, d\tau$$

$$+ a^2 \int_0^t \int_0^{2\pi} \int_{R_1}^{R_2} g_4(\xi, \eta, \tau) G(r, \varphi, z, \xi, \eta, l, t - \tau) \xi \, d\xi \, d\eta \, d\tau$$

$$+ \int_0^t \int_0^l \int_0^{2\pi} \int_{R_1}^{R_2} \Phi(\xi, \eta, \zeta, \tau) G(r, \varphi, z, \xi, \eta, \zeta, t - \tau) \xi \, d\xi \, d\eta \, d\zeta \, d\tau.$$

Here,
$$G(r,\varphi,z,\xi,\eta,\zeta,t) = \frac{\pi}{4l} \sum_{n=0}^{\infty} \sum_{m=1}^{\infty} \sum_{k=0}^{\infty} \frac{A_n A_k \mu_{nm}^2 J_n^2(\mu_{nm}R_2)}{J_n^2(\mu_{nm}R_1) - J_n^2(\mu_{nm}R_2)} Z_{nm}(r) Z_{nm}(\xi)$$
$$\times \cos[n(\varphi-\eta)] \cos\left(\frac{k\pi z}{l}\right) \cos\left(\frac{k\pi \zeta}{l}\right) \frac{\sin(t\sqrt{\lambda_{nmk}})}{\sqrt{\lambda_{nmk}}},$$

where
$$A_n = \begin{cases} 1 & \text{for } n = 0, \\ 2 & \text{for } n \neq 0, \end{cases} \quad \lambda_{nmk} = a^2 \mu_{nm}^2 + \frac{a^2 k^2 \pi^2}{l^2} + b,$$
$$Z_{nm}(r) = J_n(\mu_{nm}R_1) Y_n(\mu_{nm}r) - Y_n(\mu_{nm}R_1) J_n(\mu_{nm}r);$$

the $J_n(r)$ and $Y_n(r)$ are the Bessel functions, and the μ_{nm} are positive roots of the transcendental equation
$$J_n(\mu R_1) Y_n(\mu R_2) - Y_n(\mu R_1) J_n(\mu R_2) = 0.$$

2°. A hollow circular cylinder of finite length is considered. The following conditions are prescribed:

$$w = f_0(r,\varphi,z) \quad \text{at} \quad t = 0 \quad \text{(initial condition)},$$
$$\partial_t w = f_1(r,\varphi,z) \quad \text{at} \quad t = 0 \quad \text{(initial condition)},$$
$$\partial_r w = g_1(\varphi,z,t) \quad \text{at} \quad r = R_1 \quad \text{(boundary condition)},$$
$$\partial_r w = g_2(\varphi,z,t) \quad \text{at} \quad r = R_2 \quad \text{(boundary condition)},$$
$$w = g_3(r,\varphi,t) \quad \text{at} \quad z = 0 \quad \text{(boundary condition)},$$
$$w = g_4(r,\varphi,t) \quad \text{at} \quad z = l \quad \text{(boundary condition)}.$$

Solution:
$$w(r,\varphi,z,t) = \frac{\partial}{\partial t} \int_0^l \int_0^{2\pi} \int_{R_1}^{R_2} f_0(\xi,\eta,\zeta) G(r,\varphi,z,\xi,\eta,\zeta,t) \xi\, d\xi\, d\eta\, d\zeta$$
$$+ \int_0^l \int_0^{2\pi} \int_{R_1}^{R_2} f_1(\xi,\eta,\zeta) G(r,\varphi,z,\xi,\eta,\zeta,t) \xi\, d\xi\, d\eta\, d\zeta$$
$$- a^2 R_1 \int_0^t \int_0^l \int_0^{2\pi} g_1(\eta,\zeta,\tau) G(r,\varphi,z,R_1,\eta,\zeta,t-\tau)\, d\eta\, d\zeta\, d\tau$$
$$+ a^2 R_2 \int_0^t \int_0^l \int_0^{2\pi} g_2(\eta,\zeta,\tau) G(r,\varphi,z,R_2,\eta,\zeta,t-\tau)\, d\eta\, d\zeta\, d\tau$$
$$+ a^2 \int_0^t \int_0^{2\pi} \int_{R_1}^{R_2} g_3(\xi,\eta,\tau) \left[\frac{\partial}{\partial \zeta} G(r,\varphi,z,\xi,\eta,\zeta,t-\tau)\right]_{\zeta=0} \xi\, d\xi\, d\eta\, d\tau$$
$$- a^2 \int_0^t \int_0^{2\pi} \int_{R_1}^{R_2} g_4(\xi,\eta,\tau) \left[\frac{\partial}{\partial \zeta} G(r,\varphi,z,\xi,\eta,\zeta,t-\tau)\right]_{\zeta=l} \xi\, d\xi\, d\eta\, d\tau$$
$$+ \int_0^t \int_0^l \int_0^{2\pi} \int_{R_1}^{R_2} \Phi(\xi,\eta,\zeta,\tau) G(r,\varphi,z,\xi,\eta,\zeta,t-\tau) \xi\, d\xi\, d\eta\, d\zeta\, d\tau.$$

Here,
$$G(r,\varphi,z,\xi,\eta,\zeta,t) = \frac{2}{\pi(R_2^2 - R_1^2)l} \sum_{k=1}^{\infty} \sin\left(\frac{k\pi z}{l}\right) \sin\left(\frac{k\pi \zeta}{l}\right) \frac{\sin(t\sqrt{\beta_k})}{\sqrt{\beta_k}}$$
$$+ \frac{2}{\pi l} \sum_{n=0}^{\infty} \sum_{m=1}^{\infty} \sum_{k=1}^{\infty} \frac{A_n \mu_{nm}^2 Z_{nm}(r) Z_{nm}(\xi)}{(\mu_{nm}^2 R_2^2 - n^2) Z_{nm}^2(R_2) - (\mu_{nm}^2 R_1^2 - n^2) Z_{nm}^2(R_1)}$$
$$\times \cos[n(\varphi-\eta)] \sin\left(\frac{k\pi z}{l}\right) \sin\left(\frac{k\pi \zeta}{l}\right) \frac{\sin(t\sqrt{\lambda_{nmk}})}{\sqrt{\lambda_{nmk}}},$$

where

$$A_n = \begin{cases} 1 & \text{for } n = 0, \\ 2 & \text{for } n \neq 0, \end{cases} \quad \beta_k = \frac{a^2 k^2 \pi^2}{l^2} + b, \quad \lambda_{nmk} = a^2 \mu_{nm}^2 + \frac{a^2 k^2 \pi^2}{l^2} + b,$$

$$Z_{nm}(r) = J'_n(\mu_{nm} R_1) Y_n(\mu_{nm} r) - Y'_n(\mu_{nm} R_1) J_n(\mu_{nm} r);$$

the $J_n(r)$ and $Y_n(r)$ are the Bessel functions, and the μ_{nm} are positive roots of the transcendental equation

$$J'_n(\mu R_1) Y'_n(\mu R_2) - Y'_n(\mu R_1) J'_n(\mu R_2) = 0.$$

6.3.2-9. Domain: $0 \le r \le R$, $0 \le \varphi \le \varphi_0$, $0 \le z \le l$. First boundary value problem.

A cylindrical sector of finite thickness is considered. The following conditions are prescribed:

$$\begin{aligned}
w &= f_0(r, \varphi, z) & \text{at} \quad t &= 0 & \text{(initial condition)}, \\
\partial_t w &= f_1(r, \varphi, z) & \text{at} \quad t &= 0 & \text{(initial condition)}, \\
w &= g_1(\varphi, z, t) & \text{at} \quad r &= R & \text{(boundary condition)}, \\
w &= g_2(r, z, t) & \text{at} \quad \varphi &= 0 & \text{(boundary condition)}, \\
w &= g_3(r, z, t) & \text{at} \quad \varphi &= \varphi_0 & \text{(boundary condition)}, \\
w &= g_4(r, \varphi, t) & \text{at} \quad z &= 0 & \text{(boundary condition)}, \\
w &= g_5(r, \varphi, t) & \text{at} \quad z &= l & \text{(boundary condition)}.
\end{aligned}$$

Solution:

$$\begin{aligned}
w(r, \varphi, z, t) &= \frac{\partial}{\partial t} \int_0^l \int_0^{\varphi_0} \int_0^R f_0(\xi, \eta, \zeta) G(r, \varphi, z, \xi, \eta, \zeta, t) \xi \, d\xi \, d\eta \, d\zeta \\
&\quad + \int_0^l \int_0^{\varphi_0} \int_0^R f_1(\xi, \eta, \zeta) G(r, \varphi, z, \xi, \eta, \zeta, t) \xi \, d\xi \, d\eta \, d\zeta \\
&\quad - a^2 R \int_0^t \int_0^l \int_0^{\varphi_0} g_1(\eta, \zeta, \tau) \left[\frac{\partial}{\partial \xi} G(r, \varphi, z, \xi, \eta, \zeta, t - \tau) \right]_{\xi = R} d\eta \, d\zeta \, d\tau \\
&\quad + a^2 \int_0^t \int_0^l \int_0^R g_2(\xi, \zeta, \tau) \frac{1}{\xi} \left[\frac{\partial}{\partial \eta} G(r, \varphi, z, \xi, \eta, \zeta, t - \tau) \right]_{\eta = 0} d\xi \, d\zeta \, d\tau \\
&\quad - a^2 \int_0^t \int_0^l \int_0^R g_3(\xi, \zeta, \tau) \frac{1}{\xi} \left[\frac{\partial}{\partial \eta} G(r, \varphi, z, \xi, \eta, \zeta, t - \tau) \right]_{\eta = \varphi_0} d\xi \, d\zeta \, d\tau \\
&\quad + a^2 \int_0^t \int_0^{\varphi_0} \int_0^R g_4(\xi, \eta, \tau) \left[\frac{\partial}{\partial \zeta} G(r, \varphi, z, \xi, \eta, \zeta, t - \tau) \right]_{\zeta = 0} \xi \, d\xi \, d\eta \, d\tau \\
&\quad - a^2 \int_0^t \int_0^{\varphi_0} \int_0^R g_5(\xi, \eta, \tau) \left[\frac{\partial}{\partial \zeta} G(r, \varphi, z, \xi, \eta, \zeta, t - \tau) \right]_{\zeta = l} \xi \, d\xi \, d\eta \, d\tau \\
&\quad + \int_0^t \int_0^l \int_0^{\varphi_0} \int_0^R \Phi(\xi, \eta, \zeta, \tau) G(r, \varphi, z, \xi, \eta, \zeta, t - \tau) \xi \, d\xi \, d\eta \, d\zeta \, d\tau.
\end{aligned}$$

Here,

$$G(r, \varphi, z, \xi, \eta, \zeta, t) = \frac{8}{R^2 l \varphi_0} \sum_{n=1}^{\infty} \sum_{m=1}^{\infty} \sum_{k=1}^{\infty} \frac{J_{n\pi/\varphi_0}(\mu_{nm} r) J_{n\pi/\varphi_0}(\mu_{nm} \xi)}{[J'_{n\pi/\varphi_0}(\mu_{nm} R)]^2} \sin\left(\frac{n\pi \varphi}{\varphi_0}\right) \sin\left(\frac{n\pi \eta}{\varphi_0}\right)$$

$$\times \sin\left(\frac{k\pi z}{l}\right) \sin\left(\frac{k\pi \zeta}{l}\right) \frac{\sin\left(t \sqrt{a^2 \mu_{nm}^2 + a^2 k^2 \pi^2 l^{-2} + b}\right)}{\sqrt{a^2 \mu_{nm}^2 + a^2 k^2 \pi^2 l^{-2} + b}},$$

where the $J_{n\pi/\varphi_0}(r)$ are the Bessel functions and the μ_{nm} are positive roots of the transcendental equation $J_{n\pi/\varphi_0}(\mu R) = 0$.

6.3.2-10. Domain: $0 \leq r \leq R$, $0 \leq \varphi \leq \varphi_0$, $0 \leq z \leq l$. Mixed boundary value problem.

A cylindrical sector of finite thickness is considered. The following conditions are prescribed:

$$\begin{aligned}
w &= f_0(r,\varphi,z) & \text{at} \quad t &= 0 & \text{(initial condition)}, \\
\partial_t w &= f_1(r,\varphi,z) & \text{at} \quad t &= 0 & \text{(initial condition)}, \\
w &= g_1(\varphi,z,t) & \text{at} \quad r &= R & \text{(boundary condition)}, \\
w &= g_2(r,z,t) & \text{at} \quad \varphi &= 0 & \text{(boundary condition)}, \\
w &= g_3(r,z,t) & \text{at} \quad \varphi &= \varphi_0 & \text{(boundary condition)}, \\
\partial_z w &= g_4(r,\varphi,t) & \text{at} \quad z &= 0 & \text{(boundary condition)}, \\
\partial_z w &= g_5(r,\varphi,t) & \text{at} \quad z &= l & \text{(boundary condition)}.
\end{aligned}$$

Solution:

$$\begin{aligned}
w(r,\varphi,z,t) = &\frac{\partial}{\partial t} \int_0^l \int_0^{\varphi_0} \int_0^R f_0(\xi,\eta,\zeta) G(r,\varphi,z,\xi,\eta,\zeta,t) \xi\, d\xi\, d\eta\, d\zeta \\
&+ \int_0^l \int_0^{\varphi_0} \int_0^R f_1(\xi,\eta,\zeta) G(r,\varphi,z,\xi,\eta,\zeta,t) \xi\, d\xi\, d\eta\, d\zeta \\
&- a^2 R \int_0^t \int_0^l \int_0^{\varphi_0} g_1(\eta,\zeta,\tau) \left[\frac{\partial}{\partial \xi} G(r,\varphi,z,\xi,\eta,\zeta,t-\tau)\right]_{\xi=R} d\eta\, d\zeta\, d\tau \\
&+ a^2 \int_0^t \int_0^l \int_0^R g_2(\xi,\zeta,\tau) \frac{1}{\xi}\left[\frac{\partial}{\partial \eta} G(r,\varphi,z,\xi,\eta,\zeta,t-\tau)\right]_{\eta=0} d\xi\, d\zeta\, d\tau \\
&- a^2 \int_0^t \int_0^l \int_0^R g_3(\xi,\zeta,\tau) \frac{1}{\xi}\left[\frac{\partial}{\partial \eta} G(r,\varphi,z,\xi,\eta,\zeta,t-\tau)\right]_{\eta=\varphi_0} d\xi\, d\zeta\, d\tau \\
&- a^2 \int_0^t \int_0^{\varphi_0} \int_0^R g_4(\xi,\eta,\tau) G(r,\varphi,z,\xi,\eta,0,t-\tau) \xi\, d\xi\, d\eta\, d\tau \\
&+ a^2 \int_0^t \int_0^{\varphi_0} \int_0^R g_5(\xi,\eta,\tau) G(r,\varphi,z,\xi,\eta,l,t-\tau) \xi\, d\xi\, d\eta\, d\tau \\
&+ \int_0^t \int_0^l \int_0^{\varphi_0} \int_0^R \Phi(\xi,\eta,\zeta,\tau) G(r,\varphi,z,\xi,\eta,\zeta,t-\tau) \xi\, d\xi\, d\eta\, d\zeta\, d\tau.
\end{aligned}$$

Here,

$$G(r,\varphi,z,\xi,\eta,\zeta,t) = \frac{4}{R^2 l \varphi_0} \sum_{n=1}^{\infty} \sum_{m=1}^{\infty} \sum_{k=0}^{\infty} \frac{A_k J_{n\pi/\varphi_0}(\mu_{nm} r) J_{n\pi/\varphi_0}(\mu_{nm} \xi)}{[J'_{n\pi/\varphi_0}(\mu_{nm} R)]^2} \sin\left(\frac{n\pi\varphi}{\varphi_0}\right) \sin\left(\frac{n\pi\eta}{\varphi_0}\right)$$

$$\times \cos\left(\frac{k\pi z}{l}\right) \cos\left(\frac{k\pi\zeta}{l}\right) \frac{\sin\left(t\sqrt{a^2 \mu_{nm}^2 + a^2 k^2 \pi^2 l^{-2} + b}\right)}{\sqrt{a^2 \mu_{nm}^2 + a^2 k^2 \pi^2 l^{-2} + b}},$$

where $A_0 = 1$ and $A_k = 2$ for $k \geq 1$; the $J_{n\pi/\varphi_0}(r)$ are the Bessel functions; and the μ_{nm} are positive roots of the transcendental equation $J_{n\pi/\varphi_0}(\mu R) = 0$.

6.3.3. Problems in Spherical Coordinates

A *nonhomogeneous Klein–Gordon equation* in the spherical coordinate system is written as

$$\frac{\partial^2 w}{\partial t^2} = a^2 \left[\frac{1}{r^2}\frac{\partial}{\partial r}\left(r^2 \frac{\partial w}{\partial r}\right) + \frac{1}{r^2 \sin\theta} \frac{\partial}{\partial \theta}\left(\sin\theta \frac{\partial w}{\partial \theta}\right) + \frac{1}{r^2 \sin^2\theta} \frac{\partial^2 w}{\partial \varphi^2}\right] - bw + \Phi(r,\theta,\varphi,t).$$

One-dimensional problems with central symmetry that have solutions of the form $w = w(r,t)$ are treated in Subsection 4.2.6.

6.3.3-1. Domain: $0 \leq r \leq R$, $0 \leq \theta \leq \pi$, $0 \leq \varphi \leq 2\pi$. First boundary value problem.

A spherical domain is considered. The following conditions are prescribed:

$$w = f_0(r, \theta, \varphi) \quad \text{at} \quad t = 0 \quad \text{(initial condition)},$$
$$\partial_t w = f_1(r, \theta, \varphi) \quad \text{at} \quad t = 0 \quad \text{(initial condition)},$$
$$w = g(\theta, \varphi, t) \quad \text{at} \quad r = R \quad \text{(boundary condition)}.$$

Solution:

$$w(r, \theta, \varphi, t) = \frac{\partial}{\partial t} \int_0^{2\pi} \int_0^{\pi} \int_0^R f_0(\xi, \eta, \zeta) G(r, \theta, \varphi, \xi, \eta, \zeta, t) \xi^2 \sin \eta \, d\xi \, d\eta \, d\zeta$$
$$+ \int_0^{2\pi} \int_0^{\pi} \int_0^R f_1(\xi, \eta, \zeta) G(r, \theta, \varphi, \xi, \eta, \zeta, t) \xi^2 \sin \eta \, d\xi \, d\eta \, d\zeta$$
$$- a^2 R^2 \int_0^t \int_0^{2\pi} \int_0^{\pi} g(\eta, \zeta, \tau) \left[\frac{\partial}{\partial \xi} G(r, \theta, \varphi, \xi, \eta, \zeta, t - \tau) \right]_{\xi=R} \sin \eta \, d\eta \, d\zeta \, d\tau$$
$$+ \int_0^t \int_0^{2\pi} \int_0^{\pi} \int_0^R \Phi(\xi, \eta, \zeta, \tau) G(r, \theta, \varphi, \xi, \eta, \zeta, t - \tau) \xi^2 \sin \eta \, d\xi \, d\eta \, d\zeta \, d\tau.$$

Here,

$$G(r, \theta, \varphi, \xi, \eta, \zeta, t) = \frac{1}{2\pi R^2 \sqrt{r\xi}} \sum_{n=0}^{\infty} \sum_{m=1}^{\infty} \sum_{k=0}^{n} A_k B_{nmk} J_{n+1/2}(\lambda_{nm} r) J_{n+1/2}(\lambda_{nm} \xi)$$
$$\times P_n^k(\cos \theta) P_n^k(\cos \eta) \cos[k(\varphi - \zeta)] \sin\!\left(t \sqrt{a^2 \lambda_{nm}^2 + b}\right),$$

where

$$A_k = \begin{cases} 1 & \text{for } k = 0, \\ 2 & \text{for } k \neq 0, \end{cases} \quad B_{nmk} = \frac{(2n+1)(n-k)!}{(n+k)! \, [J'_{n+1/2}(\lambda_{nm} R)]^2 \sqrt{a^2 \lambda_{nm}^2 + b}};$$

the $J_{n+1/2}(r)$ are the Bessel functions, the $P_n^k(\mu)$ are the associated Legendre functions expressed in terms of the Legendre polynomials $P_n(\mu)$ as

$$P_n^k(\mu) = (1 - \mu^2)^{k/2} \frac{d^k}{d\mu^k} P_n(\mu), \qquad P_n(\mu) = \frac{1}{n! \, 2^n} \frac{d^n}{d\mu^n} (\mu^2 - 1)^n,$$

and the λ_{nm} are positive roots of the transcendental equation $J_{n+1/2}(\lambda R) = 0$.

6.3.3-2. Domain: $0 \leq r \leq R$, $0 \leq \theta \leq \pi$, $0 \leq \varphi \leq 2\pi$. Second boundary value problem.

A spherical domain is considered. The following conditions are prescribed:

$$w = f_0(r, \theta, \varphi) \quad \text{at} \quad t = 0 \quad \text{(initial condition)},$$
$$\partial_t w = f_1(r, \theta, \varphi) \quad \text{at} \quad t = 0 \quad \text{(initial condition)},$$
$$\partial_r w = g(\theta, \varphi, t) \quad \text{at} \quad r = R \quad \text{(boundary condition)}.$$

Solution:

$$w(r, \theta, \varphi, t) = \frac{\partial}{\partial t} \int_0^{2\pi} \int_0^{\pi} \int_0^R f_0(\xi, \eta, \zeta) G(r, \theta, \varphi, \xi, \eta, \zeta, t) \xi^2 \sin \eta \, d\xi \, d\eta \, d\zeta$$
$$+ \int_0^{2\pi} \int_0^{\pi} \int_0^R f_1(\xi, \eta, \zeta) G(r, \theta, \varphi, \xi, \eta, \zeta, t) \xi^2 \sin \eta \, d\xi \, d\eta \, d\zeta$$
$$+ a^2 R^2 \int_0^t \int_0^{2\pi} \int_0^{\pi} g(\eta, \zeta, \tau) G(r, \theta, \varphi, R, \eta, \zeta, t - \tau) \sin \eta \, d\eta \, d\zeta \, d\tau$$
$$+ \int_0^t \int_0^{2\pi} \int_0^{\pi} \int_0^R \Phi(\xi, \eta, \zeta, \tau) G(r, \theta, \varphi, \xi, \eta, \zeta, t - \tau) \xi^2 \sin \eta \, d\xi \, d\eta \, d\zeta \, d\tau.$$

Here,

$$G(r,\theta,\varphi,\xi,\eta,\zeta,t) = \frac{3\sin(t\sqrt{b})}{4\pi R^3\sqrt{b}} + \frac{1}{2\pi\sqrt{r\xi}}\sum_{n=0}^{\infty}\sum_{m=1}^{\infty}\sum_{k=0}^{n}\frac{A_k B_{nmk}}{\sqrt{a^2\lambda_{nm}^2+b}} J_{n+1/2}(\lambda_{nm}r)J_{n+1/2}(\lambda_{nm}\xi)$$

$$\times P_n^k(\cos\theta)P_n^k(\cos\eta)\cos[k(\varphi-\zeta)]\sin\left(t\sqrt{a^2\lambda_{nm}^2+b}\right),$$

where

$$A_k = \begin{cases} 1 & \text{for } k=0, \\ 2 & \text{for } k\neq 0, \end{cases} \quad B_{nmk} = \frac{\lambda_{nm}^2(2n+1)(n-k)!}{(n+k)!\left[R^2\lambda_{nm}^2 - n(n+1)\right]\left[J_{n+1/2}(\lambda_{nm}R)\right]^2};$$

the $J_{n+1/2}(r)$ are the Bessel functions, the $P_n^k(\mu)$ are the associated Legendre functions (see Paragraph 6.3.3-1), and the λ_{nm} are positive roots of the transcendental equation

$$2\lambda R J'_{n+1/2}(\lambda R) - J_{n+1/2}(\lambda R) = 0.$$

6.3.3-3. Domain: $0 \leq r \leq R$, $0 \leq \theta \leq \pi$, $0 \leq \varphi \leq 2\pi$. Third boundary value problem.

A spherical domain is considered. The following conditions are prescribed:

$$w = f_0(r,\theta,\varphi) \quad \text{at} \quad t=0 \quad \text{(initial condition)},$$
$$\partial_t w = f_1(r,\theta,\varphi) \quad \text{at} \quad t=0 \quad \text{(initial condition)},$$
$$\partial_r w + kw = g(\theta,\varphi,t) \quad \text{at} \quad r=R \quad \text{(boundary condition)}.$$

The solution $w(r,\theta,\varphi,t)$ is determined by the formula in Paragraph 6.3.3-2 where

$$G(r,\theta,\varphi,\xi,\eta,\zeta,t) = \frac{1}{2\pi\sqrt{r\xi}}\sum_{n=0}^{\infty}\sum_{m=1}^{\infty}\sum_{s=0}^{n}\frac{A_s B_{nms}}{\sqrt{a^2\lambda_{nm}^2+b}} J_{n+1/2}(\lambda_{nm}r)J_{n+1/2}(\lambda_{nm}\xi)$$

$$\times P_n^s(\cos\theta)P_n^s(\cos\eta)\cos[s(\varphi-\zeta)]\sin\left(t\sqrt{a^2\lambda_{nm}^2+b}\right).$$

Here,

$$A_s = \begin{cases} 1 & \text{for } s=0, \\ 2 & \text{for } s\neq 0, \end{cases} \quad B_{nms} = \frac{\lambda_{nm}^2(2n+1)(n-s)!}{(n+s)!\left[R^2\lambda_{nm}^2 + (kR+n)(kR-n-1)\right]\left[J_{n+1/2}(\lambda_{nm}R)\right]^2};$$

the $J_{n+1/2}(r)$ are the Bessel functions, the $P_n^s(\mu)$ are the associated Legendre functions (see Paragraph 6.3.3-1), and the λ_{nm} are positive roots of the transcendental equation

$$\lambda R J'_{n+1/2}(\lambda R) + \left(kR - \tfrac{1}{2}\right)J_{n+1/2}(\lambda R) = 0.$$

6.3.3-4. Domain: $R_1 \leq r \leq R_2$, $0 \leq \theta \leq \pi$, $0 \leq \varphi \leq 2\pi$. First boundary value problem.

A spherical layer is considered. The following conditions are prescribed:

$$w = f_0(r,\theta,\varphi) \quad \text{at} \quad t=0 \quad \text{(initial condition)},$$
$$\partial_t w = f_1(r,\theta,\varphi) \quad \text{at} \quad t=0 \quad \text{(initial condition)},$$
$$w = g_1(\theta,\varphi,t) \quad \text{at} \quad r=R_1 \quad \text{(boundary condition)},$$
$$w = g_2(\theta,\varphi,t) \quad \text{at} \quad r=R_2 \quad \text{(boundary condition)}.$$

Solution:
$$w(r,\theta,\varphi,t) = \frac{\partial}{\partial t}\int_0^{2\pi}\int_0^{\pi}\int_{R_1}^{R_2} f_0(\xi,\eta,\zeta)G(r,\theta,\varphi,\xi,\eta,\zeta,t)\xi^2 \sin\eta\, d\xi\, d\eta\, d\zeta$$
$$+ \int_0^{2\pi}\int_0^{\pi}\int_{R_1}^{R_2} f_1(\xi,\eta,\zeta)G(r,\theta,\varphi,\xi,\eta,\zeta,t)\xi^2 \sin\eta\, d\xi\, d\eta\, d\zeta$$
$$+ a^2 R_1^2 \int_0^t \int_0^{2\pi}\int_0^{\pi} g_1(\eta,\zeta,\tau)\left[\frac{\partial}{\partial \xi}G(r,\theta,\varphi,\xi,\eta,\zeta,t-\tau)\right]_{\xi=R_1} \sin\eta\, d\eta\, d\zeta\, d\tau$$
$$- a^2 R_2^2 \int_0^t \int_0^{2\pi}\int_0^{\pi} g_2(\eta,\zeta,\tau)\left[\frac{\partial}{\partial \xi}G(r,\theta,\varphi,\xi,\eta,\zeta,t-\tau)\right]_{\xi=R_2} \sin\eta\, d\eta\, d\zeta\, d\tau$$
$$+ \int_0^t \int_0^{2\pi}\int_0^{\pi}\int_{R_1}^{R_2} \Phi(\xi,\eta,\zeta,\tau)G(r,\theta,\varphi,\xi,\eta,\zeta,t-\tau)\xi^2 \sin\eta\, d\xi\, d\eta\, d\zeta\, d\tau,$$

where
$$G(r,\theta,\varphi,\xi,\eta,\zeta,t) = \frac{\pi}{8\sqrt{r\xi}}\sum_{n=0}^{\infty}\sum_{m=1}^{\infty}\sum_{k=0}^{n} \frac{A_k B_{nmk}}{\sqrt{a^2 \lambda_{nm}^2 + b}} Z_{n+1/2}(\lambda_{nm}r)Z_{n+1/2}(\lambda_{nm}\xi)$$
$$\times P_n^k(\cos\theta)P_n^k(\cos\eta)\cos[k(\varphi-\zeta)]\sin\!\left(t\sqrt{a^2\lambda_{nm}^2+b}\right).$$

Here,
$$Z_{n+1/2}(\lambda_{nm}r) = J_{n+1/2}(\lambda_{nm}R_1)Y_{n+1/2}(\lambda_{nm}r) - Y_{n+1/2}(\lambda_{nm}R_1)J_{n+1/2}(\lambda_{nm}r),$$
$$A_k = \begin{cases} 1 & \text{for } k=0, \\ 2 & \text{for } k\neq 0, \end{cases} \quad B_{nmk} = \frac{\lambda_{nm}(2n+1)(n-k)!\, J_{n+1/2}^2(\lambda_{nm}R_2)}{(n+k)!\,[J_{n+1/2}^2(\lambda_{nm}R_1) - J_{n+1/2}^2(\lambda_{nm}R_2)]},$$

where the $J_{n+1/2}(r)$ are the Bessel functions, the $P_n^k(\mu)$ are the associated Legendre functions expressed in terms of the Legendre polynomials $P_n(\mu)$ as
$$P_n^k(\mu) = (1-\mu^2)^{k/2}\frac{d^k}{d\mu^k}P_n(\mu), \qquad P_n(\mu) = \frac{1}{n!\,2^n}\frac{d^n}{d\mu^n}(\mu^2-1)^n,$$

and the λ_{nm} are positive roots of the transcendental equation $Z_{n+1/2}(\lambda R_2)=0$.

6.3.3-5. Domain: $R_1 \leq r \leq R_2$, $0 \leq \theta \leq \pi$, $0 \leq \varphi \leq 2\pi$. Second boundary value problem.

A spherical layer is considered. The following conditions are prescribed:

$$w = f_0(r,\theta,\varphi) \quad \text{at} \quad t=0 \quad \text{(initial condition)},$$
$$\partial_t w = f_1(r,\theta,\varphi) \quad \text{at} \quad t=0 \quad \text{(initial condition)},$$
$$\partial_r w = g_1(\theta,\varphi,t) \quad \text{at} \quad r=R_1 \quad \text{(boundary condition)},$$
$$\partial_r w = g_2(\theta,\varphi,t) \quad \text{at} \quad r=R_2 \quad \text{(boundary condition)}.$$

Solution:
$$w(r,\theta,\varphi,t) = \frac{\partial}{\partial t}\int_0^{2\pi}\int_0^{\pi}\int_{R_1}^{R_2} f_0(\xi,\eta,\zeta)G(r,\theta,\varphi,\xi,\eta,\zeta,t)\xi^2 \sin\eta\, d\xi\, d\eta\, d\zeta$$
$$+ \int_0^{2\pi}\int_0^{\pi}\int_{R_1}^{R_2} f_1(\xi,\eta,\zeta)G(r,\theta,\varphi,\xi,\eta,\zeta,t)\xi^2 \sin\eta\, d\xi\, d\eta\, d\zeta$$
$$- a^2 R_1^2 \int_0^t \int_0^{2\pi}\int_0^{\pi} g_1(\eta,\zeta,\tau)G(r,\theta,\varphi,R_1,\eta,\zeta,t-\tau)\sin\eta\, d\eta\, d\zeta\, d\tau$$
$$+ a^2 R_2^2 \int_0^t \int_0^{2\pi}\int_0^{\pi} g_2(\eta,\zeta,\tau)G(r,\theta,\varphi,R_2,\eta,\zeta,t-\tau)\sin\eta\, d\eta\, d\zeta\, d\tau$$
$$+ \int_0^t \int_0^{2\pi}\int_0^{\pi}\int_{R_1}^{R_2} \Phi(\xi,\eta,\zeta,\tau)G(r,\theta,\varphi,\xi,\eta,\zeta,t-\tau)\xi^2 \sin\eta\, d\xi\, d\eta\, d\zeta\, d\tau,$$

where

$$G(r,\theta,\varphi,\xi,\eta,\zeta,t) = \frac{3\sin(t\sqrt{b})}{4\pi(R_2^3 - R_1^3)\sqrt{b}} + \frac{1}{4\pi\sqrt{r\xi}} \sum_{n=0}^{\infty}\sum_{m=1}^{\infty}\sum_{k=0}^{n} \frac{A_k}{B_{nmk}} Z_{n+1/2}(\lambda_{nm}r) Z_{n+1/2}(\lambda_{nm}\xi)$$

$$\times P_n^k(\cos\theta) P_n^k(\cos\eta) \cos[k(\varphi - \zeta)] \frac{\sin(t\sqrt{a^2\lambda_{nm}^2 + b})}{\sqrt{a^2\lambda_{nm}^2 + b}}.$$

Here,

$$A_k = \begin{cases} 1 & \text{for } k = 0, \\ 2 & \text{for } k \neq 0, \end{cases} \qquad B_{nmk} = \frac{(n+k)!}{(2n+1)(n-k)!} \int_{R_1}^{R_2} r Z_{n+1/2}^2(\lambda_{nm}r)\, dr,$$

$$Z_{n+1/2}(\lambda_{nm}r) = \left[\lambda_{nm} J'_{n+1/2}(\lambda_{nm}R_1) - \frac{1}{2R_1} J_{n+1/2}(\lambda_{nm}R_1)\right] Y_{n+1/2}(\lambda_{nm}r)$$

$$- \left[\lambda_{nm} Y'_{n+1/2}(\lambda_{nm}R_1) - \frac{1}{2R_1} Y_{n+1/2}(\lambda_{nm}R_1)\right] J_{n+1/2}(\lambda_{nm}r),$$

where the $J_{n+1/2}(r)$ and $Y_{n+1/2}(r)$ are the Bessel functions, the $P_n^k(\mu)$ are the associated Legendre functions (see Paragraph 6.3.3-4), and the λ_{nm} are positive roots of the transcendental equation

$$\lambda Z'_{n+1/2}(\lambda R_2) - \frac{1}{2R_2} Z_{n+1/2}(\lambda R_2) = 0.$$

6.3.3-6. Domain: $R_1 \leq r \leq R_2$, $0 \leq \theta \leq \pi$, $0 \leq \varphi \leq 2\pi$. Third boundary value problem.

A spherical layer is considered. The following conditions are prescribed:

$$\begin{aligned}
w &= f_0(r,\theta,\varphi) & \text{at} \quad t &= 0 & \text{(initial condition)}, \\
\partial_t w &= f_1(r,\theta,\varphi) & \text{at} \quad t &= 0 & \text{(initial condition)}, \\
\partial_r w - k_1 w &= g_1(\theta,\varphi,t) & \text{at} \quad r &= R_1 & \text{(boundary condition)}, \\
\partial_r w + k_2 w &= g_2(\theta,\varphi,t) & \text{at} \quad r &= R_2 & \text{(boundary condition)}.
\end{aligned}$$

The solution $w(r,\theta,\varphi,t)$ is determined by the formula in Paragraph 6.3.3-5 where

$$G(r,\theta,\varphi,\xi,\eta,\zeta,t) = \frac{1}{4\pi\sqrt{r\xi}} \sum_{n=0}^{\infty}\sum_{m=1}^{\infty}\sum_{s=0}^{n} \frac{A_s}{B_{nms}} Z_{n+1/2}(\lambda_{nm}r) Z_{n+1/2}(\lambda_{nm}\xi)$$

$$\times P_n^s(\cos\theta) P_n^s(\cos\eta) \cos[s(\varphi - \zeta)] \frac{\sin(t\sqrt{a^2\lambda_{nm}^2 + b})}{\sqrt{a^2\lambda_{nm}^2 + b}}.$$

Here,

$$A_s = \begin{cases} 1 & \text{for } s = 0, \\ 2 & \text{for } s \neq 0, \end{cases} \qquad B_{nms} = \frac{(n+s)!}{(2n+1)(n-s)!} \int_{R_1}^{R_2} r Z_{n+1/2}^2(\lambda_{nm}r)\, dr,$$

$$Z_{n+1/2}(\lambda r) = \left[\lambda J'_{n+1/2}(\lambda R_1) - \left(k_1 + \frac{1}{2R_1}\right) J_{n+1/2}(\lambda R_1)\right] Y_{n+1/2}(\lambda r)$$

$$- \left[\lambda Y'_{n+1/2}(\lambda R_1) - \left(k_1 + \frac{1}{2R_1}\right) Y_{n+1/2}(\lambda R_1)\right] J_{n+1/2}(\lambda r),$$

where the $J_{n+1/2}(r)$ and $Y_{n+1/2}(r)$ are the Bessel functions, the $P_n^s(\mu)$ are the associated Legendre functions (see Paragraph 6.3.3-4), and the λ_{nm} are positive roots of the transcendental equation

$$\lambda Z'_{n+1/2}(\lambda R_2) + \left(k_2 - \frac{1}{2R_2}\right) Z_{n+1/2}(\lambda R_2) = 0.$$

6.4. Telegraph Equation

$$\frac{\partial^2 w}{\partial t^2} + k\frac{\partial w}{\partial t} = a^2 \Delta_3 w - bw + \Phi(x,y,z,t)$$

6.4.1. Problems in Cartesian Coordinates

A *three-dimensional nonhomogeneous telegraph equation* in the rectangular Cartesian system of coordinates has the form

$$\frac{\partial^2 w}{\partial t^2} + k\frac{\partial w}{\partial t} = a^2 \left(\frac{\partial^2 w}{\partial x^2} + \frac{\partial^2 w}{\partial y^2} + \frac{\partial^2 w}{\partial z^2} \right) - bw + \Phi(x,y,z,t).$$

6.4.1-1. Reduction to the three-dimensional Klein–Gordon equation.

The substitution $w(x,y,z,t) = \exp\left(-\tfrac{1}{2}kt\right) u(x,y,z,t)$ leads to the equation

$$\frac{\partial^2 u}{\partial t^2} = a^2 \left(\frac{\partial^2 u}{\partial x^2} + \frac{\partial^2 u}{\partial y^2} + \frac{\partial^2 u}{\partial z^2} \right) - \left(b - \tfrac{1}{4}k^2\right) u + \exp\left(\tfrac{1}{2}kt\right)\Phi(x,y,z,t),$$

which is discussed in Subsection 6.3.1.

6.4.1-2. Domain: $0 \le x \le l_1$, $0 \le y \le l_2$, $0 \le z \le l_3$. First boundary value problem.

A rectangular parallelepiped is considered. The following conditions are prescribed:

$$\begin{aligned}
w &= f_0(x,y,z) &&\text{at } t=0 &&\text{(initial condition)},\\
\partial_t w &= f_1(x,y,z) &&\text{at } t=0 &&\text{(initial condition)},\\
w &= g_1(y,z,t) &&\text{at } x=0 &&\text{(boundary condition)},\\
w &= g_2(y,z,t) &&\text{at } x=l_1 &&\text{(boundary condition)},\\
w &= g_3(x,z,t) &&\text{at } y=0 &&\text{(boundary condition)},\\
w &= g_4(x,z,t) &&\text{at } y=l_2 &&\text{(boundary condition)},\\
w &= g_5(x,y,t) &&\text{at } z=0 &&\text{(boundary condition)},\\
w &= g_6(x,y,t) &&\text{at } z=l_3 &&\text{(boundary condition)}.
\end{aligned}$$

Solution:

$$\begin{aligned}
w(x,y,z,t) =\ & \frac{\partial}{\partial t} \int_0^{l_3}\int_0^{l_2}\int_0^{l_1} f_0(\xi,\eta,\zeta) G(x,y,z,\xi,\eta,\zeta,t)\, d\xi\, d\eta\, d\zeta \\
& + \int_0^{l_3}\int_0^{l_2}\int_0^{l_1} \left[f_1(\xi,\eta,\zeta) + k f_0(\xi,\eta,\zeta) \right] G(x,y,z,\xi,\eta,\zeta,t)\, d\xi\, d\eta\, d\zeta \\
& + a^2 \int_0^t \int_0^{l_3}\int_0^{l_2} g_1(\eta,\zeta,\tau) \left[\frac{\partial}{\partial \xi} G(x,y,z,\xi,\eta,\zeta,t-\tau) \right]_{\xi=0} d\eta\, d\zeta\, d\tau \\
& - a^2 \int_0^t \int_0^{l_3}\int_0^{l_2} g_2(\eta,\zeta,\tau) \left[\frac{\partial}{\partial \xi} G(x,y,z,\xi,\eta,\zeta,t-\tau) \right]_{\xi=l_1} d\eta\, d\zeta\, d\tau \\
& + a^2 \int_0^t \int_0^{l_3}\int_0^{l_1} g_3(\xi,\zeta,\tau) \left[\frac{\partial}{\partial \eta} G(x,y,z,\xi,\eta,\zeta,t-\tau) \right]_{\eta=0} d\xi\, d\zeta\, d\tau \\
& - a^2 \int_0^t \int_0^{l_3}\int_0^{l_1} g_4(\xi,\zeta,\tau) \left[\frac{\partial}{\partial \eta} G(x,y,z,\xi,\eta,\zeta,t-\tau) \right]_{\eta=l_2} d\xi\, d\zeta\, d\tau
\end{aligned}$$

6.4. Telegraph Equation $\frac{\partial^2 w}{\partial t^2} + k\frac{\partial w}{\partial t} = a^2 \Delta_3 w - bw + \Phi(x,y,z,t)$

$$+ a^2 \int_0^t \int_0^{l_2} \int_0^{l_1} g_5(\xi,\eta,\tau) \left[\frac{\partial}{\partial \zeta}G(x,y,z,\xi,\eta,\zeta,t-\tau)\right]_{\zeta=0} d\xi\, d\eta\, d\tau$$

$$- a^2 \int_0^t \int_0^{l_2} \int_0^{l_1} g_6(\xi,\eta,\tau) \left[\frac{\partial}{\partial \zeta}G(x,y,z,\xi,\eta,\zeta,t-\tau)\right]_{\zeta=l_3} d\xi\, d\eta\, d\tau$$

$$+ \int_0^t \int_0^{l_3} \int_0^{l_2} \int_0^{l_1} \Phi(\xi,\eta,\zeta,\tau) G(x,y,z,\xi,\eta,\zeta,t-\tau)\, d\xi\, d\eta\, d\zeta\, d\tau.$$

Here,

$$G(x,y,z,\xi,\eta,\zeta,t) = \frac{8}{l_1 l_2 l_3} \exp\left(-\tfrac{1}{2}kt\right) \sum_{n=1}^{\infty}\sum_{m=1}^{\infty}\sum_{s=1}^{\infty} \frac{1}{\sqrt{\lambda_{nms}}} \sin(\alpha_n x)\sin(\beta_m y)\sin(\gamma_s z)$$
$$\times \sin(\alpha_n \xi)\sin(\beta_m \eta)\sin(\gamma_s \zeta)\sin\left(t\sqrt{\lambda_{nms}}\right),$$

where

$$\alpha_n = \frac{n\pi}{l_1}, \quad \beta_m = \frac{m\pi}{l_2}, \quad \gamma_s = \frac{s\pi}{l_3}, \quad \lambda_{nms} = a^2(\alpha_n^2 + \beta_m^2 + \gamma_s^2) + b - \tfrac{1}{4}k^2.$$

6.4.1-3. Domain: $0 \leq x \leq l_1,\ 0 \leq y \leq l_2,\ 0 \leq z \leq l_3$. **Second boundary value problem.**

A rectangular parallelepiped is considered. The following conditions are prescribed:

$$\begin{aligned}
w &= f_0(x,y,z) & \text{at}\quad & t=0 & \text{(initial condition)},\\
\partial_t w &= f_1(x,y,z) & \text{at}\quad & t=0 & \text{(initial condition)},\\
\partial_x w &= g_1(y,z,t) & \text{at}\quad & x=0 & \text{(boundary condition)},\\
\partial_x w &= g_2(y,z,t) & \text{at}\quad & x=l_1 & \text{(boundary condition)},\\
\partial_y w &= g_3(x,z,t) & \text{at}\quad & y=0 & \text{(boundary condition)},\\
\partial_y w &= g_4(x,z,t) & \text{at}\quad & y=l_2 & \text{(boundary condition)},\\
\partial_z w &= g_5(x,y,t) & \text{at}\quad & z=0 & \text{(boundary condition)},\\
\partial_z w &= g_6(x,y,t) & \text{at}\quad & z=l_3 & \text{(boundary condition)}.
\end{aligned}$$

Solution:

$$w(x,y,z,t) = \frac{\partial}{\partial t} \int_0^{l_3}\int_0^{l_2}\int_0^{l_1} f_0(\xi,\eta,\zeta) G(x,y,z,\xi,\eta,\zeta,t)\, d\xi\, d\eta\, d\zeta$$

$$+ \int_0^{l_3}\int_0^{l_2}\int_0^{l_1} [f_1(\xi,\eta,\zeta) + k f_0(\xi,\eta,\zeta)] G(x,y,z,\xi,\eta,\zeta,t)\, d\xi\, d\eta\, d\zeta$$

$$- a^2 \int_0^t \int_0^{l_3}\int_0^{l_2} g_1(\eta,\zeta,\tau) G(x,y,z,0,\eta,\zeta,t-\tau)\, d\eta\, d\zeta\, d\tau$$

$$+ a^2 \int_0^t \int_0^{l_3}\int_0^{l_2} g_2(\eta,\zeta,\tau) G(x,y,z,l_1,\eta,\zeta,t-\tau)\, d\eta\, d\zeta\, d\tau$$

$$- a^2 \int_0^t \int_0^{l_3}\int_0^{l_1} g_3(\xi,\zeta,\tau) G(x,y,z,\xi,0,\zeta,t-\tau)\, d\xi\, d\zeta\, d\tau$$

$$+ a^2 \int_0^t \int_0^{l_3}\int_0^{l_1} g_4(\xi,\zeta,\tau) G(x,y,z,\xi,l_2,\zeta,t-\tau)\, d\xi\, d\zeta\, d\tau$$

$$- a^2 \int_0^t \int_0^{l_2}\int_0^{l_1} g_5(\xi,\eta,\tau) G(x,y,z,\xi,\eta,0,t-\tau)\, d\xi\, d\eta\, d\tau$$

$$+ a^2 \int_0^t \int_0^{l_2}\int_0^{l_1} g_6(\xi,\eta,\tau) G(x,y,z,\xi,\eta,l_3,t-\tau)\, d\xi\, d\eta\, d\tau$$

$$+ \int_0^t \int_0^{l_3}\int_0^{l_2}\int_0^{l_1} \Phi(\xi,\eta,\zeta,\tau) G(x,y,z,\xi,\eta,\zeta,t-\tau)\, d\xi\, d\eta\, d\zeta\, d\tau,$$

where

$$G(x,y,z,\xi,\eta,\zeta,t) = \frac{e^{-kt/2}}{l_1 l_2 l_3}\left[\frac{\sin(t\sqrt{c})}{\sqrt{c}} + \sum_{n=0}^{\infty}\sum_{m=0}^{\infty}\sum_{s=0}^{\infty}\frac{A_n A_m A_s}{\sqrt{\lambda_{nms}}}\cos(\alpha_n x)\cos(\beta_m y)\cos(\gamma_s z)\right.$$
$$\left. \times \cos(\alpha_n \xi)\cos(\beta_m \eta)\cos(\gamma_s \zeta)\sin(t\sqrt{\lambda_{nms}})\right],$$

$$A_n = \begin{cases} 1 & \text{for } n = 0, \\ 2 & \text{for } n > 0, \end{cases} \quad \alpha_n = \frac{n\pi}{l_1}, \quad \beta_m = \frac{m\pi}{l_2}, \quad \gamma_s = \frac{s\pi}{l_3},$$

$$c = b - \tfrac{1}{4}k^2, \quad \lambda_{nms} = a^2(\alpha_n^2 + \beta_m^2 + \gamma_s^2) + b - \tfrac{1}{4}k^2.$$

The summation is performed over the indices satisfying the condition $n + m + s > 0$; the term corresponding to $n = m = s = 0$ is singled out.

6.4.1-4. Domain: $0 \leq x \leq l_1$, $0 \leq y \leq l_2$, $0 \leq z \leq l_3$. Third boundary value problem.

A rectangular parallelepiped is considered. The following conditions are prescribed:

$$\begin{aligned}
w &= f_0(x,y,z) & \text{at} \quad t &= 0 & \text{(initial condition)}, \\
\partial_t w &= f_1(x,y,z) & \text{at} \quad t &= 0 & \text{(initial condition)}, \\
\partial_x w - s_1 w &= g_1(y,z,t) & \text{at} \quad x &= 0 & \text{(boundary condition)}, \\
\partial_x w + s_2 w &= g_2(y,z,t) & \text{at} \quad x &= l_1 & \text{(boundary condition)}, \\
\partial_y w - s_3 w &= g_3(x,z,t) & \text{at} \quad y &= 0 & \text{(boundary condition)}, \\
\partial_y w + s_4 w &= g_4(x,z,t) & \text{at} \quad y &= l_2 & \text{(boundary condition)}, \\
\partial_z w - s_5 w &= g_5(x,y,t) & \text{at} \quad z &= 0 & \text{(boundary condition)}, \\
\partial_z w + s_6 w &= g_6(x,y,t) & \text{at} \quad z &= l_3 & \text{(boundary condition)}.
\end{aligned}$$

The solution $w(x,y,z,t)$ is determined by the formula in Paragraph 6.4.1-3 where

$$G(x,y,\xi,\eta,t) = 8\exp\left(-\tfrac{1}{2}kt\right)\sum_{n=1}^{\infty}\sum_{p=1}^{\infty}\sum_{q=1}^{\infty}\frac{1}{E_{npq}\sqrt{\lambda_{npq}}}\sin(\alpha_n x + \varepsilon_n)\sin(\beta_p y + \sigma_p)\sin(\gamma_q z + \nu_q)$$
$$\times \sin(\alpha_n \xi + \varepsilon_n)\sin(\beta_p \eta + \sigma_p)\sin(\gamma_q \zeta + \nu_q)\sin(t\sqrt{\lambda_{npq}}).$$

Here,

$$\varepsilon_n = \arctan\frac{\alpha_n}{l_1}, \quad \sigma_p = \arctan\frac{\beta_p}{l_2}, \quad \nu_q = \arctan\frac{\gamma_q}{l_3}, \quad \lambda_{npq} = a^2(\alpha_n^2 + \beta_p^2 + \gamma_q^2) + b - \tfrac{1}{4}k^2,$$

$$E_{npq} = \left[l_1 + \frac{(s_1 s_2 + \alpha_n^2)(s_1 + s_2)}{(s_1^2 + \alpha_n^2)(s_2^2 + \alpha_n^2)}\right]\left[l_2 + \frac{(s_3 s_4 + \beta_p^2)(s_3 + s_4)}{(s_3^2 + \beta_p^2)(s_4^2 + \beta_p^2)}\right]\left[l_3 + \frac{(s_5 s_6 + \gamma_q^2)(s_5 + s_6)}{(s_5^2 + \gamma_q^2)(s_6^2 + \gamma_q^2)}\right],$$

where the α_n, β_p, and γ_q are positive roots of the transcendental equations

$$\alpha^2 - s_1 s_2 = (s_1 + s_2)\alpha \cot(l_1 \alpha),$$
$$\beta^2 - s_3 s_4 = (s_3 + s_4)\beta \cot(l_2 \beta),$$
$$\gamma^2 - s_5 s_6 = (s_5 + s_6)\gamma \cot(l_3 \gamma).$$

6.4. TELEGRAPH EQUATION $\frac{\partial^2 w}{\partial t^2} + k\frac{\partial w}{\partial t} = a^2 \Delta_3 w - bw + \Phi(x,y,z,t)$

6.4.1-5. Domain: $0 \leq x \leq l_1$, $0 \leq y \leq l_2$, $0 \leq z \leq l_3$. Mixed boundary value problems.

1°. A rectangular parallelepiped is considered. The following conditions are prescribed:

$$w = f_0(x,y,z) \quad \text{at} \quad t = 0 \quad \text{(initial condition)},$$
$$\partial_t w = f_1(x,y,z) \quad \text{at} \quad t = 0 \quad \text{(initial condition)},$$
$$w = g_1(y,z,t) \quad \text{at} \quad x = 0 \quad \text{(boundary condition)},$$
$$w = g_2(y,z,t) \quad \text{at} \quad x = l_1 \quad \text{(boundary condition)},$$
$$\partial_y w = g_3(x,z,t) \quad \text{at} \quad y = 0 \quad \text{(boundary condition)},$$
$$\partial_y w = g_4(x,z,t) \quad \text{at} \quad y = l_2 \quad \text{(boundary condition)},$$
$$\partial_z w = g_5(x,y,t) \quad \text{at} \quad z = 0 \quad \text{(boundary condition)},$$
$$\partial_z w = g_6(x,y,t) \quad \text{at} \quad z = l_3 \quad \text{(boundary condition)}.$$

Solution:

$$w(x,y,z,t) = \frac{\partial}{\partial t} \int_0^{l_3} \int_0^{l_2} \int_0^{l_1} f_0(\xi,\eta,\zeta) G(x,y,z,\xi,\eta,\zeta,t) \, d\xi \, d\eta \, d\zeta$$
$$+ \int_0^{l_3} \int_0^{l_2} \int_0^{l_1} \big[f_1(\xi,\eta,\zeta) + k f_0(\xi,\eta,\zeta)\big] G(x,y,z,\xi,\eta,\zeta,t) \, d\xi \, d\eta \, d\zeta$$
$$+ a^2 \int_0^t \int_0^{l_3} \int_0^{l_2} g_1(\eta,\zeta,\tau) \left[\frac{\partial}{\partial \xi} G(x,y,z,\xi,\eta,\zeta,t-\tau)\right]_{\xi=0} d\eta \, d\zeta \, d\tau$$
$$- a^2 \int_0^t \int_0^{l_3} \int_0^{l_2} g_2(\eta,\zeta,\tau) \left[\frac{\partial}{\partial \xi} G(x,y,z,\xi,\eta,\zeta,t-\tau)\right]_{\xi=l_1} d\eta \, d\zeta \, d\tau$$
$$- a^2 \int_0^t \int_0^{l_3} \int_0^{l_1} g_3(\xi,\zeta,\tau) G(x,y,z,\xi,0,\zeta,t-\tau) \, d\xi \, d\zeta \, d\tau$$
$$+ a^2 \int_0^t \int_0^{l_3} \int_0^{l_1} g_4(\xi,\zeta,\tau) G(x,y,z,\xi,l_2,\zeta,t-\tau) \, d\xi \, d\zeta \, d\tau$$
$$- a^2 \int_0^t \int_0^{l_2} \int_0^{l_1} g_5(\xi,\eta,\tau) G(x,y,z,\xi,\eta,0,t-\tau) \, d\xi \, d\eta \, d\tau$$
$$+ a^2 \int_0^t \int_0^{l_2} \int_0^{l_1} g_6(\xi,\eta,\tau) G(x,y,z,\xi,\eta,l_3,t-\tau) \, d\xi \, d\eta \, d\tau$$
$$+ \int_0^t \int_0^{l_3} \int_0^{l_2} \int_0^{l_1} \Phi(\xi,\eta,\zeta,\tau) G(x,y,z,\xi,\eta,\zeta,t-\tau) \, d\xi \, d\eta \, d\zeta \, d\tau,$$

where

$$G(x,y,z,\xi,\eta,\zeta,t) = \frac{2}{l_1 l_2 l_3} \exp\left(-\tfrac{1}{2}kt\right) \sum_{n=1}^{\infty} \sum_{m=0}^{\infty} \sum_{s=0}^{\infty} \frac{A_m A_s}{\sqrt{\lambda_{nms}}} \sin(\alpha_n x) \cos(\beta_m y) \cos(\gamma_s z)$$
$$\times \sin(\alpha_n \xi) \cos(\beta_m \eta) \cos(\gamma_s \zeta) \sin\left(t\sqrt{\lambda_{nms}}\right),$$

$$A_m = \begin{cases} 1 & \text{for } m=0, \\ 2 & \text{for } m>0, \end{cases} \quad \alpha_n = \frac{n\pi}{l_1}, \quad \beta_m = \frac{m\pi}{l_2}, \quad \gamma_s = \frac{s\pi}{l_3},$$
$$\lambda_{nms} = a^2(\alpha_n^2 + \beta_m^2 + \gamma_s^2) + b - \tfrac{1}{4}k^2.$$

2°. A rectangular parallelepiped is considered. The following conditions are prescribed:

$$w = f_0(x,y,z) \quad \text{at} \quad t = 0 \quad \text{(initial condition)},$$
$$\partial_t w = f_1(x,y,z) \quad \text{at} \quad t = 0 \quad \text{(initial condition)},$$
$$w = g_1(y,z,t) \quad \text{at} \quad x = 0 \quad \text{(boundary condition)},$$
$$\partial_x w = g_2(y,z,t) \quad \text{at} \quad x = l_1 \quad \text{(boundary condition)},$$
$$w = g_3(x,z,t) \quad \text{at} \quad y = 0 \quad \text{(boundary condition)},$$
$$\partial_y w = g_4(x,z,t) \quad \text{at} \quad y = l_2 \quad \text{(boundary condition)},$$
$$w = g_5(x,y,t) \quad \text{at} \quad z = 0 \quad \text{(boundary condition)},$$
$$\partial_z w = g_6(x,y,t) \quad \text{at} \quad z = l_3 \quad \text{(boundary condition)}.$$

Solution:

$$\begin{aligned}
w(x,y,z,t) &= \frac{\partial}{\partial t}\int_0^{l_3}\int_0^{l_2}\int_0^{l_1} f_0(\xi,\eta,\zeta)G(x,y,z,\xi,\eta,\zeta,t)\,d\xi\,d\eta\,d\zeta \\
&+ \int_0^{l_3}\int_0^{l_2}\int_0^{l_1} \bigl[f_1(\xi,\eta,\zeta) + kf_0(\xi,\eta,\zeta)\bigr]G(x,y,z,\xi,\eta,\zeta,t)\,d\xi\,d\eta\,d\zeta \\
&+ a^2\int_0^t\int_0^{l_3}\int_0^{l_2} g_1(\eta,\zeta,\tau)\left[\frac{\partial}{\partial\xi}G(x,y,z,\xi,\eta,\zeta,t-\tau)\right]_{\xi=0} d\eta\,d\zeta\,d\tau \\
&+ a^2\int_0^t\int_0^{l_3}\int_0^{l_2} g_2(\eta,\zeta,\tau)G(x,y,z,l_1,\eta,\zeta,t-\tau)\,d\eta\,d\zeta\,d\tau \\
&+ a^2\int_0^t\int_0^{l_3}\int_0^{l_1} g_3(\xi,\zeta,\tau)\left[\frac{\partial}{\partial\eta}G(x,y,z,\xi,\eta,\zeta,t-\tau)\right]_{\eta=0} d\xi\,d\zeta\,d\tau \\
&+ a^2\int_0^t\int_0^{l_3}\int_0^{l_1} g_4(\xi,\zeta,\tau)G(x,y,z,\xi,l_2,\zeta,t-\tau)\,d\xi\,d\zeta\,d\tau \\
&+ a^2\int_0^t\int_0^{l_2}\int_0^{l_1} g_5(\xi,\eta,\tau)\left[\frac{\partial}{\partial\zeta}G(x,y,z,\xi,\eta,\zeta,t-\tau)\right]_{\zeta=0} d\xi\,d\eta\,d\tau \\
&+ a^2\int_0^t\int_0^{l_2}\int_0^{l_1} g_6(\xi,\eta,\tau)G(x,y,z,\xi,\eta,l_3,t-\tau)\,d\xi\,d\eta\,d\tau \\
&+ \int_0^t\int_0^{l_3}\int_0^{l_2}\int_0^{l_1} \Phi(\xi,\eta,\zeta,\tau)G(x,y,z,\xi,\eta,\zeta,t-\tau)\,d\xi\,d\eta\,d\zeta\,d\tau,
\end{aligned}$$

where

$$G(x,y,z,\xi,\eta,\zeta,t) = \frac{8}{l_1 l_2 l_3}\exp\!\left(-\tfrac{1}{2}kt\right)\sum_{n=1}^{\infty}\sum_{m=1}^{\infty}\sum_{s=1}^{\infty}\frac{1}{\sqrt{\lambda_{nms}}}\sin(\alpha_n x)\sin(\beta_m y)\sin(\gamma_s z)$$
$$\times \sin(\alpha_n \xi)\sin(\beta_m \eta)\sin(\gamma_s \zeta)\sin\!\left(t\sqrt{\lambda_{nms}}\right),$$

$$\alpha_n = \frac{\pi(2n+1)}{2l_1},\quad \beta_m = \frac{\pi(2m+1)}{2l_2},\quad \gamma_s = \frac{\pi(2s+1)}{2l_3},\quad \lambda_{nms} = a^2(\alpha_n^2 + \beta_m^2 + \gamma_s^2) + b - \tfrac{1}{4}k^2.$$

6.4.2. Problems in Cylindrical Coordinates

A three-dimensional nonhomogeneous telegraph equation in the cylindrical coordinate system is written as

$$\frac{\partial^2 w}{\partial t^2} + k\frac{\partial w}{\partial t} = a^2\left[\frac{1}{r}\frac{\partial}{\partial r}\!\left(r\frac{\partial w}{\partial r}\right) + \frac{1}{r^2}\frac{\partial^2 w}{\partial \varphi^2} + \frac{\partial^2 w}{\partial z^2}\right] - bw + \Phi(r,\varphi,z,t),\qquad r = \sqrt{x^2+y^2}.$$

One-dimensional problems with axial symmetry that have solutions $w = w(r,t)$ are treated in Subsection 4.4.2. Two-dimensional problems whose solutions have the form $w = w(r,\varphi,t)$ or $w = w(r,z,t)$ are considered in Subsections 5.4.2 and 5.4.3.

6.4. TELEGRAPH EQUATION $\frac{\partial^2 w}{\partial t^2} + k\frac{\partial w}{\partial t} = a^2 \Delta_3 w - bw + \Phi(x,y,z,t)$

6.4.2-1. Domain: $0 \leq r \leq R$, $0 \leq \varphi \leq 2\pi$, $0 \leq z \leq l$. First boundary value problem.

A circular cylinder of finite length is considered. The following conditions are prescribed:

$$\begin{aligned}
w &= f_0(r,\varphi,z) && \text{at} \quad t=0 && \text{(initial condition)}, \\
\partial_t w &= f_1(r,\varphi,z) && \text{at} \quad t=0 && \text{(initial condition)}, \\
w &= g_1(\varphi,z,t) && \text{at} \quad r=R && \text{(boundary condition)}, \\
w &= g_2(r,\varphi,t) && \text{at} \quad z=0 && \text{(boundary condition)}, \\
w &= g_3(r,\varphi,t) && \text{at} \quad z=l && \text{(boundary condition)}.
\end{aligned}$$

Solution:

$$\begin{aligned}
w(r,\varphi,z,t) &= \frac{\partial}{\partial t}\int_0^l\int_0^{2\pi}\int_0^R \xi f_0(\xi,\eta,\zeta)G(r,\varphi,z,\xi,\eta,\zeta,t)\,d\xi\,d\eta\,d\zeta \\
&+ \int_0^l\int_0^{2\pi}\int_0^R \xi[f_1(\xi,\eta,\zeta)+kf_0(\xi,\eta,\zeta)]G(r,\varphi,z,\xi,\eta,\zeta,t)\,d\xi\,d\eta\,d\zeta \\
&- a^2 R\int_0^t\int_0^l\int_0^{2\pi} g_1(\eta,\zeta,\tau)\left[\frac{\partial}{\partial\xi}G(r,\varphi,z,\xi,\eta,\zeta,t-\tau)\right]_{\xi=R} d\eta\,d\zeta\,d\tau \\
&+ a^2\int_0^t\int_0^{2\pi}\int_0^R \xi g_2(\xi,\eta,\tau)\left[\frac{\partial}{\partial\zeta}G(r,\varphi,z,\xi,\eta,\zeta,t-\tau)\right]_{\zeta=0} d\xi\,d\eta\,d\tau \\
&- a^2\int_0^t\int_0^{2\pi}\int_0^R \xi g_3(\xi,\eta,\tau)\left[\frac{\partial}{\partial\zeta}G(r,\varphi,z,\xi,\eta,\zeta,t-\tau)\right]_{\zeta=l} d\xi\,d\eta\,d\tau \\
&+ \int_0^t\int_0^l\int_0^{2\pi}\int_0^R \xi\Phi(\xi,\eta,\zeta,\tau)G(r,\varphi,z,\xi,\eta,\zeta,t-\tau)\,d\xi\,d\eta\,d\zeta\,d\tau.
\end{aligned}$$

Here,

$$G(r,\varphi,z,\xi,\eta,\zeta,t) = \frac{2e^{-kt/2}}{\pi R^2 l}\sum_{n=0}^{\infty}\sum_{m=1}^{\infty}\sum_{s=1}^{\infty}\frac{A_n}{[J_n'(\mu_{nm}R)]^2}J_n(\mu_{nm}r)J_n(\mu_{nm}\xi)\cos[n(\varphi-\eta)]$$

$$\times \sin\left(\frac{s\pi z}{l}\right)\sin\left(\frac{s\pi\zeta}{l}\right)\frac{\sin(t\sqrt{\lambda_{nms}})}{\sqrt{\lambda_{nms}}},$$

where

$$\lambda_{nms} = a^2\mu_{nm}^2 + \frac{a^2 s^2 \pi^2}{l^2} + b - \tfrac{1}{4}k^2, \qquad A_n = \begin{cases} 1 & \text{for } n=0, \\ 2 & \text{for } n>0, \end{cases}$$

the $J_n(\xi)$ are the Bessel functions (the prime denotes the derivative with respect to the argument), and the μ_{nm} are positive roots of the transcendental equation $J_n(\mu R) = 0$.

6.4.2-2. Domain: $0 \leq r \leq R$, $0 \leq \varphi \leq 2\pi$, $0 \leq z \leq l$. Second boundary value problem.

A circular cylinder of finite length is considered. The following conditions are prescribed:

$$\begin{aligned}
w &= f_0(r,\varphi,z) && \text{at} \quad t=0 && \text{(initial condition)}, \\
\partial_t w &= f_1(r,\varphi,z) && \text{at} \quad t=0 && \text{(initial condition)}, \\
\partial_r w &= g_1(\varphi,z,t) && \text{at} \quad r=R && \text{(boundary condition)}, \\
\partial_z w &= g_2(r,\varphi,t) && \text{at} \quad z=0 && \text{(boundary condition)}, \\
\partial_z w &= g_3(r,\varphi,t) && \text{at} \quad z=l && \text{(boundary condition)}.
\end{aligned}$$

Solution:

$$w(r,\varphi,z,t) = \frac{\partial}{\partial t}\int_0^l\int_0^{2\pi}\int_0^R \xi f_0(\xi,\eta,\zeta)G(r,\varphi,z,\xi,\eta,\zeta,t)\,d\xi\,d\eta\,d\zeta$$

$$+\int_0^l\int_0^{2\pi}\int_0^R \xi[f_1(\xi,\eta,\zeta)+kf_0(\xi,\eta,\zeta)]G(r,\varphi,z,\xi,\eta,\zeta,t)\,d\xi\,d\eta\,d\zeta$$

$$+a^2R\int_0^t\int_0^l\int_0^{2\pi} g_1(\eta,\zeta,\tau)G(r,\varphi,z,R,\eta,\zeta,t-\tau)\,d\eta\,d\zeta\,d\tau$$

$$-a^2\int_0^t\int_0^{2\pi}\int_0^R \xi g_2(\xi,\eta,\tau)G(r,\varphi,z,\xi,\eta,0,t-\tau)\,d\xi\,d\eta\,d\tau$$

$$+a^2\int_0^t\int_0^{2\pi}\int_0^R \xi g_3(\xi,\eta,\tau)G(r,\varphi,z,\xi,\eta,l,t-\tau)\,d\xi\,d\eta\,d\tau$$

$$+\int_0^t\int_0^l\int_0^{2\pi}\int_0^R \xi\Phi(\xi,\eta,\zeta,\tau)G(r,\varphi,z,\xi,\eta,\zeta,t-\tau)\,d\xi\,d\eta\,d\zeta\,d\tau.$$

Here,

$$G(r,\varphi,z,\xi,\eta,\zeta,t) = \exp\bigl(-\tfrac{1}{2}kt\bigr)\left[\frac{\sin(t\sqrt{c})}{\pi R^2 l\sqrt{c}} + \frac{2}{\pi R^2 l}\sum_{s=1}^{\infty}\frac{1}{\sqrt{\beta_s}}\cos\left(\frac{s\pi x}{l}\right)\cos\left(\frac{s\pi \xi}{l}\right)\sin(t\sqrt{\beta_s})\right.$$

$$\left.+\frac{1}{\pi l}\sum_{n=0}^{\infty}\sum_{m=1}^{\infty}\sum_{s=0}^{\infty}\frac{A_nA_s\mu_{nm}^2 J_n(\mu_{nm}r)J_n(\mu_{nm}\xi)}{(\mu_{nm}^2R^2-n^2)[J_n(\mu_{nm}R)]^2}\cos[n(\varphi-\eta)]\cos\left(\frac{s\pi x}{l}\right)\cos\left(\frac{s\pi\xi}{l}\right)\frac{\sin(\lambda_{nms}t)}{\lambda_{nms}}\right],$$

$$c = b - \frac{k^2}{4},\quad \beta_s = \frac{a^2s^2\pi^2}{l^2}+b-\frac{k^2}{4},\quad \lambda_{nms} = \sqrt{a^2\mu_{nm}^2+\frac{a^2s^2\pi^2}{l^2}+b-\frac{k^2}{4}},\quad A_n = \begin{cases}1 & \text{for } n=0,\\ 2 & \text{for } n>0,\end{cases}$$

where the $J_n(\xi)$ are the Bessel functions and the μ_{nm} are positive roots of the transcendental equation $J_n'(\mu R) = 0$.

6.4.2-3. Domain: $0 \leq r \leq R$, $0 \leq \varphi \leq 2\pi$, $0 \leq z \leq l$. Third boundary value problem.

A circular cylinder of finite length is considered. The following conditions are prescribed:

$$\begin{aligned}
w &= f_0(r,\varphi,z) & \text{at}\quad & t=0 & \text{(initial condition)},\\
\partial_t w &= f_1(r,\varphi,z) & \text{at}\quad & t=0 & \text{(initial condition)},\\
\partial_r w + s_1 w &= g(\varphi,z,t) & \text{at}\quad & r=R & \text{(boundary condition)},\\
\partial_z w - s_2 w &= g_2(r,\varphi,t) & \text{at}\quad & z=0 & \text{(boundary condition)},\\
\partial_z w + s_3 w &= g_3(r,\varphi,t) & \text{at}\quad & z=l & \text{(boundary condition)}.
\end{aligned}$$

The solution $w(r,\varphi,z,t)$ is determined by the formula in Paragraph 6.4.2-2 where

$$G(r,\varphi,z,\xi,\eta,\zeta,t) = \frac{1}{\pi}\exp\bigl(-\tfrac{1}{2}kt\bigr)\sum_{n=0}^{\infty}\sum_{m=1}^{\infty}\sum_{p=1}^{\infty}\frac{A_n\mu_{nm}^2 J_n(\mu_{nm}r)J_n(\mu_{nm}\xi)}{(\mu_{nm}^2R^2+s_1^2R^2-n^2)[J_n(\mu_{nm}R)]^2}$$

$$\times \cos[n(\varphi-\eta)]\frac{h_p(z)h_p(\zeta)}{\|h_p\|^2}\frac{\sin(t\sqrt{\lambda_{nmp}})}{\sqrt{\lambda_{nmp}}}.$$

Here, the $J_n(\xi)$ are the Bessel functions,

$$A_n = \begin{cases}1 & \text{for } n=0,\\ 2 & \text{for } n>0,\end{cases}\qquad \lambda_{nmp} = a^2\mu_{nm}^2 + a^2\beta_p^2 + b - \tfrac{1}{4}k^2,$$

$$h_p(z) = \cos(\beta_p z) + \frac{s_2}{\beta_p}\sin(\beta_p z),\qquad \|h_p\|^2 = \frac{s_3}{2\beta_p^2}\frac{\beta_p^2+s_2^2}{\beta_p^2+s_3^2} + \frac{s_2}{2\beta_p^2} + \frac{l}{2}\left(1+\frac{s_2^2}{\beta_p^2}\right);$$

the μ_{nm} and β_p are positive roots of the transcendental equations

$$\mu J_n'(\mu R) + s_1 J_n(\mu R) = 0, \qquad \frac{\tan(\beta l)}{\beta} = \frac{s_2 + s_3}{\beta^2 - s_2 s_3}.$$

6.4.2-4. Domain: $0 \leq r \leq R$, $0 \leq \varphi \leq 2\pi$, $0 \leq z \leq l$. Mixed boundary value problems.

1°. A circular cylinder of finite length is considered. The following conditions are prescribed:

$$\begin{aligned} w &= f_0(r,\varphi,z) && \text{at} \quad t=0 && \text{(initial condition)}, \\ \partial_t w &= f_1(r,\varphi,z) && \text{at} \quad t=0 && \text{(initial condition)}, \\ w &= g_1(\varphi,z,t) && \text{at} \quad r=R && \text{(boundary condition)}, \\ \partial_z w &= g_2(r,\varphi,t) && \text{at} \quad z=0 && \text{(boundary condition)}, \\ \partial_z w &= g_3(r,\varphi,t) && \text{at} \quad z=l && \text{(boundary condition)}. \end{aligned}$$

Solution:

$$\begin{aligned}
w(r,\varphi,z,t) = & \frac{\partial}{\partial t} \int_0^l \int_0^{2\pi} \int_0^R \xi f_0(\xi,\eta,\zeta) G(r,\varphi,z,\xi,\eta,\zeta,t)\, d\xi\, d\eta\, d\zeta \\
& + \int_0^l \int_0^{2\pi} \int_0^R \xi \big[f_1(\xi,\eta,\zeta) + k f_0(\xi,\eta,\zeta)\big] G(r,\varphi,z,\xi,\eta,\zeta,t)\, d\xi\, d\eta\, d\zeta \\
& - a^2 R \int_0^t \int_0^l \int_0^{2\pi} g_1(\eta,\zeta,\tau) \left[\frac{\partial}{\partial \xi} G(r,\varphi,z,\xi,\eta,\zeta,t-\tau)\right]_{\xi=R} d\eta\, d\zeta\, d\tau \\
& - a^2 \int_0^t \int_0^{2\pi} \int_0^R \xi g_2(\xi,\eta,\tau) G(r,\varphi,z,\xi,\eta,0,t-\tau)\, d\xi\, d\eta\, d\tau \\
& + a^2 \int_0^t \int_0^{2\pi} \int_0^R \xi g_3(\xi,\eta,\tau) G(r,\varphi,z,\xi,\eta,l,t-\tau)\, d\xi\, d\eta\, d\tau \\
& + \int_0^t \int_0^l \int_0^{2\pi} \int_0^R \xi \Phi(\xi,\eta,\zeta,\tau) G(r,\varphi,z,\xi,\eta,\zeta,t-\tau)\, d\xi\, d\eta\, d\zeta\, d\tau.
\end{aligned}$$

Here,

$$G(r,\varphi,z,\xi,\eta,\zeta,t) = \frac{1}{\pi R^2 l} \exp\!\left(-\tfrac{1}{2}kt\right) \sum_{n=0}^\infty \sum_{m=1}^\infty \sum_{s=0}^\infty \frac{A_n A_s}{[J_n'(\mu_{nm}R)]^2 \sqrt{\lambda_{nms}}} J_n(\mu_{nm}r) J_n(\mu_{nm}\xi)$$

$$\times \cos[n(\varphi-\eta)] \cos\!\left(\frac{s\pi z}{l}\right) \cos\!\left(\frac{s\pi \zeta}{l}\right) \sin\!\left(t\sqrt{\lambda_{nms}}\right),$$

$$\lambda_{nms} = a^2 \mu_{nm}^2 + \frac{a^2 s^2 \pi^2}{l^2} + b - \tfrac{1}{4}k^2, \qquad A_n = \begin{cases} 1 & \text{for } n=0, \\ 2 & \text{for } n>0, \end{cases}$$

where the $J_n(\xi)$ are the Bessel functions (the prime denotes the derivative with respect to the argument) and the μ_{nm} are positive roots of the transcendental equation $J_n(\mu R) = 0$.

2°. A circular cylinder of finite length is considered. The following conditions are prescribed:

$$\begin{aligned} w &= f_0(r,\varphi,z) && \text{at} \quad t=0 && \text{(initial condition)}, \\ \partial_t w &= f_1(r,\varphi,z) && \text{at} \quad t=0 && \text{(initial condition)}, \\ \partial_r w &= g_1(\varphi,z,t) && \text{at} \quad r=R && \text{(boundary condition)}, \\ w &= g_2(r,\varphi,t) && \text{at} \quad z=0 && \text{(boundary condition)}, \\ w &= g_3(r,\varphi,t) && \text{at} \quad z=l && \text{(boundary condition)}. \end{aligned}$$

Solution:

$$
\begin{aligned}
w(r,\varphi,z,t) = {} & \frac{\partial}{\partial t}\int_0^l\int_0^{2\pi}\int_0^R \xi f_0(\xi,\eta,\zeta)G(r,\varphi,z,\xi,\eta,\zeta,t)\,d\xi\,d\eta\,d\zeta \\
& + \int_0^l\int_0^{2\pi}\int_0^R \xi\bigl[f_1(\xi,\eta,\zeta)+kf_0(\xi,\eta,\zeta)\bigr]G(r,\varphi,z,\xi,\eta,\zeta,t)\,d\xi\,d\eta\,d\zeta \\
& + a^2 R \int_0^t\int_0^l\int_0^{2\pi} g_1(\eta,\zeta,\tau)G(r,\varphi,z,R,\eta,\zeta,t-\tau)\,d\eta\,d\zeta\,d\tau \\
& + a^2 \int_0^t\int_0^{2\pi}\int_0^R \xi g_2(\xi,\eta,\tau)\left[\frac{\partial}{\partial \zeta}G(r,\varphi,z,\xi,\eta,\zeta,t-\tau)\right]_{\zeta=0} d\xi\,d\eta\,d\tau \\
& - a^2 \int_0^t\int_0^{2\pi}\int_0^R \xi g_3(\xi,\eta,\tau)\left[\frac{\partial}{\partial \zeta}G(r,\varphi,z,\xi,\eta,\zeta,t-\tau)\right]_{\zeta=l} d\xi\,d\eta\,d\tau \\
& + \int_0^t\int_0^l\int_0^{2\pi}\int_0^R \xi\Phi(\xi,\eta,\zeta,\tau)G(r,\varphi,z,\xi,\eta,\zeta,t-\tau)\,d\xi\,d\eta\,d\zeta\,d\tau.
\end{aligned}
$$

Here,

$$
\begin{aligned}
G(r,\varphi,z,\xi,\eta,\zeta,t) = {} & \frac{2}{\pi R^2 l}\exp\bigl(-\tfrac{1}{2}kt\bigr)\sum_{s=1}^{\infty}\frac{1}{\sqrt{\beta_s}}\sin\!\left(\frac{s\pi z}{l}\right)\sin\!\left(\frac{s\pi\zeta}{l}\right)\sin\!\bigl(t\sqrt{\beta_s}\bigr) \\
& + \frac{2}{\pi l}\exp\bigl(-\tfrac{1}{2}kt\bigr)\sum_{n=0}^{\infty}\sum_{m=1}^{\infty}\sum_{s=1}^{\infty}\frac{A_n\mu_{nm}^2}{(\mu_{nm}^2 R^2 - n^2)[J_n(\mu_{nm}R)]^2}J_n(\mu_{nm}r)J_n(\mu_{nm}\xi) \\
& \qquad\qquad \times \cos[n(\varphi-\eta)]\sin\!\left(\frac{s\pi z}{l}\right)\sin\!\left(\frac{s\pi\zeta}{l}\right)\frac{\sin\!\bigl(t\sqrt{\lambda_{nms}}\bigr)}{\sqrt{\lambda_{nms}}},
\end{aligned}
$$

$$
\beta_s = \frac{a^2 s^2 \pi^2}{l^2} + b - \tfrac{1}{4}k^2, \qquad A_n = \begin{cases} 1 & \text{for } n=0, \\ 2 & \text{for } n>0, \end{cases}
$$

$$
\lambda_{nms} = a^2\mu_{nm}^2 + \frac{a^2 s^2 \pi^2}{l^2} + b - \tfrac{1}{4}k^2,
$$

where the $J_n(\xi)$ are the Bessel functions and the μ_{nm} are positive roots of the transcendental equation $J_n'(\mu R) = 0$.

6.4.2-5. Domain: $R_1 \leq r \leq R_2$, $0 \leq \varphi \leq 2\pi$, $0 \leq z \leq l$. First boundary value problem.

A hollow circular cylinder of finite length is considered. The following conditions are prescribed:

$$
\begin{array}{rlll}
w = f_0(r,\varphi,z) & \text{at} & t = 0 & \text{(initial condition)}, \\
\partial_t w = f_1(r,\varphi,z) & \text{at} & t = 0 & \text{(initial condition)}, \\
w = g_1(\varphi,z,t) & \text{at} & r = R_1 & \text{(boundary condition)}, \\
w = g_2(\varphi,z,t) & \text{at} & r = R_2 & \text{(boundary condition)}, \\
w = g_3(r,\varphi,t) & \text{at} & z = 0 & \text{(boundary condition)}, \\
w = g_4(r,\varphi,t) & \text{at} & z = l & \text{(boundary condition)}.
\end{array}
$$

Solution:

$$w(r,\varphi,z,t) = \frac{\partial}{\partial t}\int_0^l \int_0^{2\pi}\int_{R_1}^{R_2} f_0(\xi,\eta,\zeta) G(r,\varphi,z,\xi,\eta,\zeta,t)\xi\,d\xi\,d\eta\,d\zeta$$

$$+ \int_0^l \int_0^{2\pi}\int_{R_1}^{R_2} \bigl[f_1(\xi,\eta,\zeta) + k f_0(\xi,\eta,\zeta)\bigr] G(r,\varphi,z,\xi,\eta,\zeta,t)\xi\,d\xi\,d\eta\,d\zeta$$

$$+ a^2 R_1 \int_0^t \int_0^l \int_0^{2\pi} g_1(\eta,\zeta,\tau)\left[\frac{\partial}{\partial \xi} G(r,\varphi,z,\xi,\eta,\zeta,t-\tau)\right]_{\xi=R_1} d\eta\,d\zeta\,d\tau$$

$$- a^2 R_2 \int_0^t \int_0^l \int_0^{2\pi} g_2(\eta,\zeta,\tau)\left[\frac{\partial}{\partial \xi} G(r,\varphi,z,\xi,\eta,\zeta,t-\tau)\right]_{\xi=R_2} d\eta\,d\zeta\,d\tau$$

$$+ a^2 \int_0^t \int_0^{2\pi}\int_{R_1}^{R_2} g_3(\xi,\eta,\tau)\left[\frac{\partial}{\partial \zeta} G(r,\varphi,z,\xi,\eta,\zeta,t-\tau)\right]_{\zeta=0} \xi\,d\xi\,d\eta\,d\tau$$

$$- a^2 \int_0^t \int_0^{2\pi}\int_{R_1}^{R_2} g_4(\xi,\eta,\tau)\left[\frac{\partial}{\partial \zeta} G(r,\varphi,z,\xi,\eta,\zeta,t-\tau)\right]_{\zeta=l} \xi\,d\xi\,d\eta\,d\tau$$

$$+ \int_0^t \int_0^l \int_0^{2\pi}\int_{R_1}^{R_2} \Phi(\xi,\eta,\zeta,\tau) G(r,\varphi,z,\xi,\eta,\zeta,t-\tau)\xi\,d\xi\,d\eta\,d\zeta\,d\tau.$$

Here,

$$G(r,\varphi,z,\xi,\eta,\zeta,t) = \frac{\pi}{2l}\exp\!\left(-\tfrac{1}{2}kt\right) \sum_{n=0}^{\infty}\sum_{m=1}^{\infty}\sum_{s=1}^{\infty} \frac{A_n \mu_{nm}^2 J_n^2(\mu_{nm}R_2)}{J_n^2(\mu_{nm}R_1) - J_n^2(\mu_{nm}R_2)} Z_{nm}(r) Z_{nm}(\xi)$$

$$\times \cos[n(\varphi-\eta)] \sin\!\left(\frac{s\pi z}{l}\right)\sin\!\left(\frac{s\pi \zeta}{l}\right) \frac{\sin\!\left(t\sqrt{\lambda_{nms}}\right)}{\sqrt{\lambda_{nms}}},$$

$$A_n = \begin{cases} 1 & \text{for } n=0, \\ 2 & \text{for } n\neq 0, \end{cases} \quad \lambda_{nms} = a^2\mu_{nm}^2 + \frac{a^2 s^2 \pi^2}{l^2} + b - \tfrac{1}{4}k^2,$$

$$Z_{nm}(r) = J_n(\mu_{nm}R_1) Y_n(\mu_{nm}r) - Y_n(\mu_{nm}R_1) J_n(\mu_{nm}r),$$

where the $J_n(r)$ and $Y_n(r)$ are the Bessel functions, and the μ_{nm} are positive roots of the transcendental equation

$$J_n(\mu R_1) Y_n(\mu R_2) - Y_n(\mu R_1) J_n(\mu R_2) = 0.$$

6.4.2-6. Domain: $R_1 \leq r \leq R_2$, $0 \leq \varphi \leq 2\pi$, $0 \leq z \leq l$. Second boundary value problem.

A hollow circular cylinder of finite length is considered. The following conditions are prescribed:

$$\begin{array}{lll} w = f_0(r,\varphi,z) & \text{at } t=0 & \text{(initial condition)}, \\ \partial_t w = f_1(r,\varphi,z) & \text{at } t=0 & \text{(initial condition)}, \\ \partial_r w = g_1(\varphi,z,t) & \text{at } r=R_1 & \text{(boundary condition)}, \\ \partial_r w = g_2(\varphi,z,t) & \text{at } r=R_2 & \text{(boundary condition)}, \\ \partial_z w = g_3(r,\varphi,t) & \text{at } z=0 & \text{(boundary condition)}, \\ \partial_z w = g_4(r,\varphi,t) & \text{at } z=l & \text{(boundary condition)}. \end{array}$$

Solution:

$$w(r,\varphi,z,t) = \frac{\partial}{\partial t} \int_0^l \int_0^{2\pi} \int_{R_1}^{R_2} f_0(\xi,\eta,\zeta) G(r,\varphi,z,\xi,\eta,\zeta,t) \xi\, d\xi\, d\eta\, d\zeta$$

$$+ \int_0^l \int_0^{2\pi} \int_{R_1}^{R_2} \left[f_1(\xi,\eta,\zeta) + k f_0(\xi,\eta,\zeta) \right] G(r,\varphi,z,\xi,\eta,\zeta,t) \xi\, d\xi\, d\eta\, d\zeta$$

$$- a^2 R_1 \int_0^t \int_0^l \int_0^{2\pi} g_1(\eta,\zeta,\tau) G(r,\varphi,z,R_1,\eta,\zeta,t-\tau)\, d\eta\, d\zeta\, d\tau$$

$$+ a^2 R_2 \int_0^t \int_0^l \int_0^{2\pi} g_2(\eta,\zeta,\tau) G(r,\varphi,z,R_2,\eta,\zeta,t-\tau)\, d\eta\, d\zeta\, d\tau$$

$$- a^2 \int_0^t \int_0^{2\pi} \int_{R_1}^{R_2} g_3(\xi,\eta,\tau) G(r,\varphi,z,\xi,\eta,0,t-\tau) \xi\, d\xi\, d\eta\, d\tau$$

$$+ a^2 \int_0^t \int_0^{2\pi} \int_{R_1}^{R_2} g_4(\xi,\eta,\tau) G(r,\varphi,z,\xi,\eta,l,t-\tau) \xi\, d\xi\, d\eta\, d\tau$$

$$+ \int_0^t \int_0^l \int_0^{2\pi} \int_{R_1}^{R_2} \Phi(\xi,\eta,\zeta,\tau) G(r,\varphi,z,\xi,\eta,\zeta,t-\tau) \xi\, d\xi\, d\eta\, d\zeta\, d\tau.$$

Here,

$$G(r,\varphi,z,\xi,\eta,\zeta,t) = \frac{e^{-kt/2}}{\pi(R_2^2 - R_1^2)l} \left[\frac{\sin(t\sqrt{c})}{\sqrt{c}} + 2 \sum_{s=1}^{\infty} \cos\left(\frac{s\pi z}{l}\right) \cos\left(\frac{s\pi \zeta}{l}\right) \frac{\sin(t\sqrt{\beta_s})}{\sqrt{\beta_s}} \right]$$

$$+ \frac{e^{-kt/2}}{\pi l} \sum_{n=0}^{\infty} \sum_{m=1}^{\infty} \sum_{s=0}^{\infty} \frac{A_n A_s \mu_{nm}^2 Z_{nm}(r) Z_{nm}(\xi)}{(\mu_{nm}^2 R_2^2 - n^2) Z_{nm}^2(R_2) - (\mu_{nm}^2 R_1^2 - n^2) Z_{nm}^2(R_1)}$$

$$\times \cos[n(\varphi - \eta)] \cos\left(\frac{s\pi z}{l}\right) \cos\left(\frac{s\pi \zeta}{l}\right) \frac{\sin(t\sqrt{\lambda_{nms}})}{\sqrt{\lambda_{nms}}},$$

where

$$A_n = \begin{cases} 1 & \text{for } n = 0, \\ 2 & \text{for } n \neq 0, \end{cases} \quad c = b - \tfrac{1}{4}k^2, \quad \beta_s = \frac{a^2 s^2 \pi^2}{l^2} + b - \tfrac{1}{4}k^2, \quad \lambda_{nms} = a^2 \mu_{nm}^2 + \frac{a^2 s^2 \pi^2}{l^2} + b - \tfrac{1}{4}k^2,$$

$$Z_{nm}(r) = J_n'(\mu_{nm} R_1) Y_n(\mu_{nm} r) - Y_n'(\mu_{nm} R_1) J_n(\mu_{nm} r);$$

the $J_n(r)$ and $Y_n(r)$ are the Bessel functions, and the μ_{nm} are positive roots of the transcendental equation

$$J_n'(\mu R_1) Y_n'(\mu R_2) - Y_n'(\mu R_1) J_n'(\mu R_2) = 0.$$

6.4.2-7. Domain: $R_1 \leq r \leq R_2$, $0 \leq \varphi \leq 2\pi$, $0 \leq z \leq l$. Third boundary value problem.

A hollow circular cylinder of finite length is considered. The following conditions are prescribed:

$$\begin{aligned}
w &= f_0(r,\varphi,z) & &\text{at} \quad t = 0 & &\text{(initial condition)}, \\
\partial_t w &= f_1(r,\varphi,z) & &\text{at} \quad t = 0 & &\text{(initial condition)}, \\
\partial_r w - s_1 w &= g_1(\varphi,z,t) & &\text{at} \quad r = R_1 & &\text{(boundary condition)}, \\
\partial_r w + s_2 w &= g_2(\varphi,z,t) & &\text{at} \quad r = R_2 & &\text{(boundary condition)}, \\
\partial_z w - s_3 w &= g_3(r,\varphi,t) & &\text{at} \quad z = 0 & &\text{(boundary condition)}, \\
\partial_z w + s_4 w &= g_4(r,\varphi,t) & &\text{at} \quad z = l & &\text{(boundary condition)}.
\end{aligned}$$

6.4. TELEGRAPH EQUATION $\frac{\partial^2 w}{\partial t^2} + k \frac{\partial w}{\partial t} = a^2 \Delta_3 w - bw + \Phi(x,y,z,t)$ 445

The solution $w(r,\varphi,z,t)$ is determined by the formula in Paragraph 6.4.2-6 where

$$G(r,\varphi,z,\xi,\eta,\zeta,t) = \frac{1}{\pi} \exp(-\tfrac{1}{2}kt) \sum_{n=0}^{\infty} \sum_{m=1}^{\infty} \sum_{p=1}^{\infty} \frac{A_n \mu_{nm}^2}{\|h_p\|^2 \sqrt{a^2 \mu_{nm}^2 + a^2 \lambda_p^2 + b - k^2/4}}$$

$$\times \frac{Z_{nm}(r) Z_{nm}(\xi) \cos[n(\varphi-\eta)] h_p(z) h_p(\zeta) \sin\left(t\sqrt{a^2\mu_{nm}^2 + a^2\lambda_p^2 + b - k^2/4}\right)}{(s_2^2 R_2^2 + \mu_{nm}^2 R_2^2 - n^2) Z_{nm}^2(R_2) - (s_1^2 R_1^2 + \mu_{nm}^2 R_1^2 - n^2) Z_{nm}^2(R_1)}.$$

Here,

$$Z_{nm}(r) = [\mu_{nm} J_n'(\mu_{nm} R_1) - s_1 J_n(\mu_{nm} R_1)] Y_n(\mu_{nm} r)$$
$$- [\mu_{nm} Y_n'(\mu_{nm} R_1) - s_1 Y_n(\mu_{nm} R_1)] J_n(\mu_{nm} r),$$

$$A_n = \begin{cases} 1 & \text{for } n = 0, \\ 2 & \text{for } n \neq 0, \end{cases} \quad h_p(z) = \cos(\lambda_p z) + \frac{s_3}{\lambda_p} \sin(\lambda_p z), \quad \|h_p\|^2 = \frac{s_4}{2\lambda_p^2} \frac{\lambda_p^2 + s_3^2}{\lambda_p^2 + s_4^2} + \frac{s_3}{2\lambda_p^2} + \frac{l}{2}\left(1 + \frac{s_3^2}{\lambda_p^2}\right),$$

where the $J_n(r)$ and $Y_n(r)$ are the Bessel functions, the μ_{nm} are positive roots of the transcendental equation

$$[\mu J_n'(\mu R_1) - s_1 J_n(\mu R_1)][\mu Y_n'(\mu R_2) + s_2 Y_n(\mu R_2)]$$
$$= [\mu Y_n'(\mu R_1) - s_1 Y_n(\mu R_1)][\mu J_n'(\mu R_2) + s_2 J_n(\mu R_2)],$$

and the λ_p are positive roots of the transcendental equation $\dfrac{\tan(\lambda l)}{\lambda} = \dfrac{s_3 + s_4}{\lambda^2 - s_3 s_4}$.

6.4.2-8. Domain: $R_1 \leq r \leq R_2$, $0 \leq \varphi \leq 2\pi$, $0 \leq z \leq l$. **Mixed boundary value problems.**

1°. A hollow circular cylinder of finite length is considered. The following conditions are prescribed:

$$\begin{aligned}
w &= f_0(r,\varphi,z) &\text{at}\quad t &= 0 &\text{(initial condition)},\\
\partial_t w &= f_1(r,\varphi,z) &\text{at}\quad t &= 0 &\text{(initial condition)},\\
w &= g_1(\varphi,z,t) &\text{at}\quad r &= R_1 &\text{(boundary condition)},\\
w &= g_2(\varphi,z,t) &\text{at}\quad r &= R_2 &\text{(boundary condition)},\\
\partial_z w &= g_3(r,\varphi,t) &\text{at}\quad z &= 0 &\text{(boundary condition)},\\
\partial_z w &= g_4(r,\varphi,t) &\text{at}\quad z &= l &\text{(boundary condition)}.
\end{aligned}$$

Solution:

$$w(r,\varphi,z,t) = \frac{\partial}{\partial t} \int_0^l \int_0^{2\pi} \int_{R_1}^{R_2} f_0(\xi,\eta,\zeta) G(r,\varphi,z,\xi,\eta,\zeta,t) \xi\, d\xi\, d\eta\, d\zeta$$

$$+ \int_0^l \int_0^{2\pi} \int_{R_1}^{R_2} [f_1(\xi,\eta,\zeta) + k f_0(\xi,\eta,\zeta)] G(r,\varphi,z,\xi,\eta,\zeta,t) \xi\, d\xi\, d\eta\, d\zeta$$

$$+ a^2 R_1 \int_0^t \int_0^l \int_0^{2\pi} g_1(\eta,\zeta,\tau) \left[\frac{\partial}{\partial \xi} G(r,\varphi,z,\xi,\eta,\zeta,t-\tau)\right]_{\xi=R_1} d\eta\, d\zeta\, d\tau$$

$$- a^2 R_2 \int_0^t \int_0^l \int_0^{2\pi} g_2(\eta,\zeta,\tau) \left[\frac{\partial}{\partial \xi} G(r,\varphi,z,\xi,\eta,\zeta,t-\tau)\right]_{\xi=R_2} d\eta\, d\zeta\, d\tau$$

$$- a^2 \int_0^t \int_0^{2\pi} \int_{R_1}^{R_2} g_3(\xi,\eta,\tau) G(r,\varphi,z,\xi,\eta,0,t-\tau) \xi\, d\xi\, d\eta\, d\tau$$

$$+ a^2 \int_0^t \int_0^{2\pi} \int_{R_1}^{R_2} g_4(\xi,\eta,\tau) G(r,\varphi,z,\xi,\eta,l,t-\tau) \xi\, d\xi\, d\eta\, d\tau$$

$$+ \int_0^t \int_0^l \int_0^{2\pi} \int_{R_1}^{R_2} \Phi(\xi,\eta,\zeta,\tau) G(r,\varphi,z,\xi,\eta,\zeta,t-\tau) \xi\, d\xi\, d\eta\, d\zeta\, d\tau.$$

Here,

$$G(r,\varphi,z,\xi,\eta,\zeta,t) = \frac{\pi}{4l}\exp\left(-\tfrac{1}{2}kt\right)\sum_{n=0}^{\infty}\sum_{m=1}^{\infty}\sum_{s=0}^{\infty}\frac{A_n A_s \mu_{nm}^2 J_n^2(\mu_{nm}R_2)}{J_n^2(\mu_{nm}R_1) - J_n^2(\mu_{nm}R_2)}Z_{nm}(r)Z_{nm}(\xi)$$

$$\times \cos[n(\varphi-\eta)]\cos\left(\frac{s\pi z}{l}\right)\cos\left(\frac{s\pi\zeta}{l}\right)\frac{\sin(t\sqrt{\lambda_{nms}})}{\sqrt{\lambda_{nms}}},$$

$$A_n = \begin{cases} 1 & \text{for } n=0, \\ 2 & \text{for } n\neq 0, \end{cases} \quad \lambda_{nms} = a^2\mu_{nm}^2 + \frac{a^2 s^2 \pi^2}{l^2} + b - \tfrac{1}{4}k^2,$$

$$Z_{nm}(r) = J_n(\mu_{nm}R_1)Y_n(\mu_{nm}r) - Y_n(\mu_{nm}R_1)J_n(\mu_{nm}r),$$

where the $J_n(r)$ and $Y_n(r)$ are the Bessel functions, and the μ_{nm} are positive roots of the transcendental equation

$$J_n(\mu R_1)Y_n(\mu R_2) - Y_n(\mu R_1)J_n(\mu R_2) = 0.$$

2°. A hollow circular cylinder of finite length is considered. The following conditions are prescribed:

$$\begin{array}{llll} w = f_0(r,\varphi,z) & \text{at} & t=0 & \text{(initial condition)}, \\ \partial_t w = f_1(r,\varphi,z) & \text{at} & t=0 & \text{(initial condition)}, \\ \partial_r w = g_1(\varphi,z,t) & \text{at} & r=R_1 & \text{(boundary condition)}, \\ \partial_r w = g_2(\varphi,z,t) & \text{at} & r=R_2 & \text{(boundary condition)}, \\ w = g_3(r,\varphi,t) & \text{at} & z=0 & \text{(boundary condition)}, \\ w = g_4(r,\varphi,t) & \text{at} & z=l & \text{(boundary condition)}. \end{array}$$

Solution:

$$w(r,\varphi,z,t) = \frac{\partial}{\partial t}\int_0^l\int_0^{2\pi}\int_{R_1}^{R_2} f_0(\xi,\eta,\zeta)G(r,\varphi,z,\xi,\eta,\zeta,t)\xi\,d\xi\,d\eta\,d\zeta$$

$$+ \int_0^l\int_0^{2\pi}\int_{R_1}^{R_2}[f_1(\xi,\eta,\zeta)+kf_0(\xi,\eta,\zeta)]G(r,\varphi,z,\xi,\eta,\zeta,t)\xi\,d\xi\,d\eta\,d\zeta$$

$$- a^2 R_1 \int_0^t\int_0^l\int_0^{2\pi} g_1(\eta,\zeta,\tau)G(r,\varphi,z,R_1,\eta,\zeta,t-\tau)\,d\eta\,d\zeta\,d\tau$$

$$+ a^2 R_2 \int_0^t\int_0^l\int_0^{2\pi} g_2(\eta,\zeta,\tau)G(r,\varphi,z,R_2,\eta,\zeta,t-\tau)\,d\eta\,d\zeta\,d\tau$$

$$+ a^2 \int_0^t\int_0^{2\pi}\int_{R_1}^{R_2} g_3(\xi,\eta,\tau)\left[\frac{\partial}{\partial\zeta}G(r,\varphi,z,\xi,\eta,\zeta,t-\tau)\right]_{\zeta=0}\xi\,d\xi\,d\eta\,d\tau$$

$$- a^2 \int_0^t\int_0^{2\pi}\int_{R_1}^{R_2} g_4(\xi,\eta,\tau)\left[\frac{\partial}{\partial\zeta}G(r,\varphi,z,\xi,\eta,\zeta,t-\tau)\right]_{\zeta=l}\xi\,d\xi\,d\eta\,d\tau$$

$$+ \int_0^t\int_0^l\int_0^{2\pi}\int_{R_1}^{R_2} \Phi(\xi,\eta,\zeta,\tau)G(r,\varphi,z,\xi,\eta,\zeta,t-\tau)\xi\,d\xi\,d\eta\,d\zeta\,d\tau.$$

Here,

$$G(r,\varphi,z,\xi,\eta,\zeta,t) = \frac{2e^{-kt/2}}{\pi(R_2^2-R_1^2)l}\sum_{s=1}^{\infty}\sin\left(\frac{s\pi z}{l}\right)\sin\left(\frac{s\pi\zeta}{l}\right)\frac{\sin(t\sqrt{\beta_s})}{\sqrt{\beta_s}}$$

$$+ \frac{2e^{-kt/2}}{\pi l}\sum_{n=0}^{\infty}\sum_{m=1}^{\infty}\sum_{s=1}^{\infty}\frac{A_n\mu_{nm}^2 Z_{nm}(r)Z_{nm}(\xi)}{(\mu_{nm}^2 R_2^2 - n^2)Z_{nm}^2(R_2) - (\mu_{nm}^2 R_1^2 - n^2)Z_{nm}^2(R_1)}$$

$$\times \cos[n(\varphi-\eta)]\sin\left(\frac{s\pi z}{l}\right)\sin\left(\frac{s\pi\zeta}{l}\right)\frac{\sin(t\sqrt{\lambda_{nms}})}{\sqrt{\lambda_{nms}}},$$

where

$$A_n = \begin{cases} 1 & \text{for } n = 0, \\ 2 & \text{for } n \neq 0, \end{cases} \quad \beta_s = \frac{a^2 s^2 \pi^2}{l^2} + b - \tfrac{1}{4}k^2, \quad \lambda_{nms} = a^2 \mu_{nm}^2 + \frac{a^2 s^2 \pi^2}{l^2} + b - \tfrac{1}{4}k^2,$$

$$Z_{nm}(r) = J_n'(\mu_{nm}R_1)Y_n(\mu_{nm}r) - Y_n'(\mu_{nm}R_1)J_n(\mu_{nm}r);$$

the $J_n(r)$ and $Y_n(r)$ are the Bessel functions, and the μ_{nm} are positive roots of the transcendental equation

$$J_n'(\mu R_1)Y_n'(\mu R_2) - Y_n'(\mu R_1)J_n'(\mu R_2) = 0.$$

6.4.2-9. Domain: $0 \leq r \leq R$, $0 \leq \varphi \leq \varphi_0$, $0 \leq z \leq l$. **First boundary value problem.**

A cylindrical sector of finite thickness is considered. The following conditions are prescribed:

$$\begin{aligned}
w &= f_0(r, \varphi, z) & \text{at} \quad t &= 0 & \text{(initial condition)}, \\
\partial_t w &= f_1(r, \varphi, z) & \text{at} \quad t &= 0 & \text{(initial condition)}, \\
w &= g_1(\varphi, z, t) & \text{at} \quad r &= R & \text{(boundary condition)}, \\
w &= g_2(r, z, t) & \text{at} \quad \varphi &= 0 & \text{(boundary condition)}, \\
w &= g_3(r, z, t) & \text{at} \quad \varphi &= \varphi_0 & \text{(boundary condition)}, \\
w &= g_4(r, \varphi, t) & \text{at} \quad z &= 0 & \text{(boundary condition)}, \\
w &= g_5(r, \varphi, t) & \text{at} \quad z &= l & \text{(boundary condition)}.
\end{aligned}$$

Solution:

$$\begin{aligned}
w(r, \varphi, z, t) =& \frac{\partial}{\partial t} \int_0^l \int_0^{\varphi_0} \int_0^R f_0(\xi, \eta, \zeta) G(r, \varphi, z, \xi, \eta, \zeta, t) \xi \, d\xi \, d\eta \, d\zeta \\
&+ \int_0^l \int_0^{\varphi_0} \int_0^R \big[f_1(\xi, \eta, \zeta) + k f_0(\xi, \eta, \zeta)\big] G(r, \varphi, z, \xi, \eta, \zeta, t) \xi \, d\xi \, d\eta \, d\zeta \\
&- a^2 R \int_0^t \int_0^l \int_0^{\varphi_0} g_1(\eta, \zeta, \tau) \left[\frac{\partial}{\partial \xi} G(r, \varphi, z, \xi, \eta, \zeta, t-\tau)\right]_{\xi=R} d\eta \, d\zeta \, d\tau \\
&+ a^2 \int_0^t \int_0^l \int_0^R g_2(\xi, \zeta, \tau) \frac{1}{\xi}\left[\frac{\partial}{\partial \eta} G(r, \varphi, z, \xi, \eta, \zeta, t-\tau)\right]_{\eta=0} d\xi \, d\zeta \, d\tau \\
&- a^2 \int_0^t \int_0^l \int_0^R g_3(\xi, \zeta, \tau) \frac{1}{\xi}\left[\frac{\partial}{\partial \eta} G(r, \varphi, z, \xi, \eta, \zeta, t-\tau)\right]_{\eta=\varphi_0} d\xi \, d\zeta \, d\tau \\
&+ a^2 \int_0^t \int_0^{\varphi_0} \int_0^R g_4(\xi, \eta, \tau) \left[\frac{\partial}{\partial \zeta} G(r, \varphi, z, \xi, \eta, \zeta, t-\tau)\right]_{\zeta=0} \xi \, d\xi \, d\eta \, d\tau \\
&- a^2 \int_0^t \int_0^{\varphi_0} \int_0^R g_5(\xi, \eta, \tau) \left[\frac{\partial}{\partial \zeta} G(r, \varphi, z, \xi, \eta, \zeta, t-\tau)\right]_{\zeta=l} \xi \, d\xi \, d\eta \, d\tau \\
&+ \int_0^t \int_0^l \int_0^{\varphi_0} \int_0^R \Phi(\xi, \eta, \zeta, \tau) G(r, \varphi, z, \xi, \eta, \zeta, t-\tau) \xi \, d\xi \, d\eta \, d\zeta \, d\tau.
\end{aligned}$$

Here,

$$\begin{aligned}
G(r, \varphi, z, \xi, \eta, \zeta, t) =& \frac{8 e^{-kt/2}}{R^2 l \varphi_0} \sum_{n=1}^{\infty} \sum_{m=1}^{\infty} \sum_{s=1}^{\infty} \frac{J_{n\pi/\varphi_0}(\mu_{nm}r) J_{n\pi/\varphi_0}(\mu_{nm}\xi)}{[J_{n\pi/\varphi_0}'(\mu_{nm}R)]^2} \sin\left(\frac{n\pi\varphi}{\varphi_0}\right) \sin\left(\frac{n\pi\eta}{\varphi_0}\right) \\
&\times \sin\left(\frac{s\pi z}{l}\right) \sin\left(\frac{s\pi \zeta}{l}\right) \frac{\sin\!\big(t\sqrt{a^2\mu_{nm}^2 + a^2 s^2 \pi^2 l^{-2} + b - k^2/4}\big)}{\sqrt{a^2\mu_{nm}^2 + a^2 s^2 \pi^2 l^{-2} + b - k^2/4}},
\end{aligned}$$

where the $J_{n\pi/\varphi_0}(r)$ are the Bessel functions and the μ_{nm} are positive roots of the transcendental equation $J_{n\pi/\varphi_0}(\mu R) = 0$.

6.4.2-10. Domain: $0 \le r \le R$, $0 \le \varphi \le \varphi_0$, $0 \le z \le l$. Mixed boundary value problem.

A cylindrical sector of finite thickness is considered. The following conditions are prescribed:

$$\begin{aligned}
w &= f_0(r,\varphi,z) && \text{at } t=0 && \text{(initial condition)}, \\
\partial_t w &= f_1(r,\varphi,z) && \text{at } t=0 && \text{(initial condition)}, \\
w &= g_1(\varphi,z,t) && \text{at } r=R && \text{(boundary condition)}, \\
w &= g_2(r,z,t) && \text{at } \varphi=0 && \text{(boundary condition)}, \\
w &= g_3(r,z,t) && \text{at } \varphi=\varphi_0 && \text{(boundary condition)}, \\
\partial_z w &= g_4(r,\varphi,t) && \text{at } z=0 && \text{(boundary condition)}, \\
\partial_z w &= g_5(r,\varphi,t) && \text{at } z=l && \text{(boundary condition)}.
\end{aligned}$$

Solution:

$$\begin{aligned}
w(r,\varphi,z,t) =\;& \frac{\partial}{\partial t}\int_0^l\!\int_0^{\varphi_0}\!\int_0^R f_0(\xi,\eta,\zeta)G(r,\varphi,z,\xi,\eta,\zeta,t)\xi\,d\xi\,d\eta\,d\zeta \\
&+ \int_0^l\!\int_0^{\varphi_0}\!\int_0^R [f_1(\xi,\eta,\zeta)+kf_0(\xi,\eta,\zeta)]G(r,\varphi,z,\xi,\eta,\zeta,t)\xi\,d\xi\,d\eta\,d\zeta \\
&- a^2R\int_0^t\!\int_0^l\!\int_0^{\varphi_0} g_1(\eta,\zeta,\tau)\left[\frac{\partial}{\partial\xi}G(r,\varphi,z,\xi,\eta,\zeta,t-\tau)\right]_{\xi=R}\,d\eta\,d\zeta\,d\tau \\
&+ a^2\int_0^t\!\int_0^l\!\int_0^R g_2(\xi,\zeta,\tau)\frac{1}{\xi}\left[\frac{\partial}{\partial\eta}G(r,\varphi,z,\xi,\eta,\zeta,t-\tau)\right]_{\eta=0}\,d\xi\,d\zeta\,d\tau \\
&- a^2\int_0^t\!\int_0^l\!\int_0^R g_3(\xi,\zeta,\tau)\frac{1}{\xi}\left[\frac{\partial}{\partial\eta}G(r,\varphi,z,\xi,\eta,\zeta,t-\tau)\right]_{\eta=\varphi_0}\,d\xi\,d\zeta\,d\tau \\
&- a^2\int_0^t\!\int_0^{\varphi_0}\!\int_0^R g_4(\xi,\eta,\tau)G(r,\varphi,z,\xi,\eta,0,t-\tau)\xi\,d\xi\,d\eta\,d\tau \\
&+ a^2\int_0^t\!\int_0^{\varphi_0}\!\int_0^R g_5(\xi,\eta,\tau)G(r,\varphi,z,\xi,\eta,l,t-\tau)\xi\,d\xi\,d\eta\,d\tau \\
&+ \int_0^t\!\int_0^l\!\int_0^{\varphi_0}\!\int_0^R \Phi(\xi,\eta,\zeta,\tau)G(r,\varphi,z,\xi,\eta,\zeta,t-\tau)\xi\,d\xi\,d\eta\,d\zeta\,d\tau.
\end{aligned}$$

Here,

$$\begin{aligned}
G(r,\varphi,z,\xi,\eta,\zeta,t) =\;& \frac{4e^{-kt/2}}{R^2 l\varphi_0}\sum_{n=1}^{\infty}\sum_{m=1}^{\infty}\sum_{s=0}^{\infty}\frac{A_s J_{n\pi/\varphi_0}(\mu_{nm}r)J_{n\pi/\varphi_0}(\mu_{nm}\xi)}{[J'_{n\pi/\varphi_0}(\mu_{nm}R)]^2}\sin\!\left(\frac{n\pi\varphi}{\varphi_0}\right)\sin\!\left(\frac{n\pi\eta}{\varphi_0}\right) \\
& \times \cos\!\left(\frac{s\pi z}{l}\right)\cos\!\left(\frac{s\pi\zeta}{l}\right)\frac{\sin\!\left(t\sqrt{a^2\mu_{nm}^2+a^2s^2\pi^2 l^{-2}+b-k^2/4}\right)}{\sqrt{a^2\mu_{nm}^2+a^2s^2\pi^2 l^{-2}+b-k^2/4}},
\end{aligned}$$

where $A_0=1$ and $A_s=2$ for $s\ge 1$; the $J_{n\pi/\varphi_0}(r)$ are the Bessel functions; and the μ_{nm} are positive roots of the transcendental equation $J_{n\pi/\varphi_0}(\mu R)=0$.

6.4.3. Problems in Spherical Coordinates

A three-dimensional nonhomogeneous telegraph equation in the spherical coordinate system is written as

$$\frac{\partial^2 w}{\partial t^2}+k\frac{\partial w}{\partial t}=a^2\left[\frac{1}{r^2}\frac{\partial}{\partial r}\!\left(r^2\frac{\partial w}{\partial r}\right)+\frac{1}{r^2\sin\theta}\frac{\partial}{\partial\theta}\!\left(\sin\theta\frac{\partial w}{\partial\theta}\right)+\frac{1}{r^2\sin^2\theta}\frac{\partial^2 w}{\partial\varphi^2}\right]-bw+\Phi(r,\theta,\varphi,t).$$

6.4.3-1. Domain: $0 \leq r \leq R$, $0 \leq \theta \leq \pi$, $0 \leq \varphi \leq 2\pi$. First boundary value problem.

A spherical domain is considered. The following conditions are prescribed:

$$w = f_0(r, \theta, \varphi) \quad \text{at} \quad t = 0 \quad \text{(initial condition)},$$
$$\partial_t w = f_1(r, \theta, \varphi) \quad \text{at} \quad t = 0 \quad \text{(initial condition)},$$
$$w = g(\theta, \varphi, t) \quad \text{at} \quad r = R \quad \text{(boundary condition)}.$$

Solution:

$$w(r, \theta, \varphi, t) = \frac{\partial}{\partial t} \int_0^{2\pi} \int_0^\pi \int_0^R f_0(\xi, \eta, \zeta) G(r, \theta, \varphi, \xi, \eta, \zeta, t) \xi^2 \sin \eta \, d\xi \, d\eta \, d\zeta$$
$$+ \int_0^{2\pi} \int_0^\pi \int_0^R \big[f_1(\xi, \eta, \zeta) + k f_0(\xi, \eta, \zeta) \big] G(r, \theta, \varphi, \xi, \eta, \zeta, t) \xi^2 \sin \eta \, d\xi \, d\eta \, d\zeta$$
$$- a^2 R^2 \int_0^t \int_0^{2\pi} \int_0^\pi g(\eta, \zeta, \tau) \left[\frac{\partial}{\partial \xi} G(r, \theta, \varphi, \xi, \eta, \zeta, t - \tau) \right]_{\xi = R} \sin \eta \, d\eta \, d\zeta \, d\tau$$
$$+ \int_0^t \int_0^{2\pi} \int_0^\pi \int_0^R \Phi(\xi, \eta, \zeta, \tau) G(r, \theta, \varphi, \xi, \eta, \zeta, t - \tau) \xi^2 \sin \eta \, d\xi \, d\eta \, d\zeta \, d\tau,$$

where

$$G(r, \theta, \varphi, \xi, \eta, \zeta, t) = \frac{1}{2\pi R^2 \sqrt{r \xi}} \exp\!\left(-\tfrac{1}{2} k t\right) \sum_{n=0}^\infty \sum_{m=1}^\infty \sum_{s=0}^n A_s B_{nms} J_{n+1/2}(\lambda_{nm} r) J_{n+1/2}(\lambda_{nm} \xi)$$
$$\times P_n^s(\cos \theta) P_n^s(\cos \eta) \cos[s(\varphi - \zeta)] \frac{\sin\!\left(t \sqrt{a^2 \lambda_{nm}^2 + b - k^2/4}\right)}{\sqrt{a^2 \lambda_{nm}^2 + b - k^2/4}},$$

$$A_s = \begin{cases} 1 & \text{for } s = 0, \\ 2 & \text{for } s \neq 0, \end{cases} \qquad B_{nms} = \frac{(2n+1)(n-s)!}{(n+s)! \left[J'_{n+1/2}(\lambda_{nm} R) \right]^2}.$$

Here, the $J_{n+1/2}(r)$ are the Bessel functions, the $P_n^s(\mu)$ are the associated Legendre functions expressed in terms of the Legendre polynomials $P_n(\mu)$ as

$$P_n^s(\mu) = (1 - \mu^2)^{s/2} \frac{d^s}{d\mu^s} P_n(\mu), \qquad P_n(\mu) = \frac{1}{n! \, 2^n} \frac{d^n}{d\mu^n} (\mu^2 - 1)^n,$$

and the λ_{nm} are positive roots of the transcendental equation $J_{n+1/2}(\lambda R) = 0$.

6.4.3-2. Domain: $0 \leq r \leq R$, $0 \leq \theta \leq \pi$, $0 \leq \varphi \leq 2\pi$. Second boundary value problem.

A spherical domain is considered. The following conditions are prescribed:

$$w = f_0(r, \theta, \varphi) \quad \text{at} \quad t = 0 \quad \text{(initial condition)},$$
$$\partial_t w = f_1(r, \theta, \varphi) \quad \text{at} \quad t = 0 \quad \text{(initial condition)},$$
$$\partial_r w = g(\theta, \varphi, t) \quad \text{at} \quad r = R \quad \text{(boundary condition)}.$$

Solution:

$$w(r, \theta, \varphi, t) = \frac{\partial}{\partial t} \int_0^{2\pi} \int_0^\pi \int_0^R f_0(\xi, \eta, \zeta) G(r, \theta, \varphi, \xi, \eta, \zeta, t) \xi^2 \sin \eta \, d\xi \, d\eta \, d\zeta$$
$$+ \int_0^{2\pi} \int_0^\pi \int_0^R \big[f_1(\xi, \eta, \zeta) + k f_0(\xi, \eta, \zeta) \big] G(r, \theta, \varphi, \xi, \eta, \zeta, t) \xi^2 \sin \eta \, d\xi \, d\eta \, d\zeta$$
$$+ a^2 R^2 \int_0^t \int_0^{2\pi} \int_0^\pi g(\eta, \zeta, \tau) G(r, \theta, \varphi, R, \eta, \zeta, t - \tau) \sin \eta \, d\eta \, d\zeta \, d\tau$$
$$+ \int_0^t \int_0^{2\pi} \int_0^\pi \int_0^R \Phi(\xi, \eta, \zeta, \tau) G(r, \theta, \varphi, \xi, \eta, \zeta, t - \tau) \xi^2 \sin \eta \, d\xi \, d\eta \, d\zeta \, d\tau,$$

where

$$G(r,\theta,\varphi,\xi,\eta,\zeta,t) = \frac{3e^{-kt/2}}{4\pi R^3} \frac{\sin(t\sqrt{c})}{\sqrt{c}} + \frac{e^{-kt/2}}{2\pi\sqrt{r\xi}} \sum_{n=0}^{\infty} \sum_{m=1}^{\infty} \sum_{s=0}^{n} A_s B_{nms} J_{n+1/2}(\lambda_{nm} r)$$

$$\times J_{n+1/2}(\lambda_{nm}\xi) P_n^s(\cos\theta) P_n^s(\cos\eta) \cos[s(\varphi-\zeta)] \frac{\sin\left(t\sqrt{a^2\lambda_{nm}^2 + c}\right)}{\sqrt{a^2\lambda_{nm}^2 + c}},$$

$$A_s = \begin{cases} 1 & \text{for } s=0, \\ 2 & \text{for } s \neq 0, \end{cases} \quad B_{nms} = \frac{\lambda_{nm}^2 (2n+1)(n-s)!}{(n+s)! \left[R^2\lambda_{nm}^2 - n(n+1)\right] \left[J_{n+1/2}(\lambda_{nm} R)\right]^2}, \quad c = b - \tfrac{1}{4}k^2.$$

Here, the $J_{n+1/2}(r)$ are the Bessel functions, the $P_n^s(\mu)$ are the associated Legendre functions (see Paragraph 6.4.3-1), and the λ_{nm} are positive roots of the transcendental equation

$$2\lambda R J'_{n+1/2}(\lambda R) - J_{n+1/2}(\lambda R) = 0.$$

6.4.3-3. Domain: $0 \le r \le R$, $0 \le \theta \le \pi$, $0 \le \varphi \le 2\pi$. Third boundary value problem.

A spherical domain is considered. The following conditions are prescribed:

$$\begin{aligned}
w &= f_0(r,\theta,\varphi) && \text{at} \quad t=0 && \text{(initial condition)}, \\
\partial_t w &= f_1(r,\theta,\varphi) && \text{at} \quad t=0 && \text{(initial condition)}, \\
\partial_r w + sw &= g(\theta,\varphi,t) && \text{at} \quad r=R && \text{(boundary condition)}.
\end{aligned}$$

The solution $w(r,\theta,\varphi,t)$ is determined by the formula in Paragraph 6.4.3-2 where

$$G(r,\theta,\varphi,\xi,\eta,\zeta,t) = \frac{e^{-kt/2}}{2\pi\sqrt{r\xi}} \sum_{n=0}^{\infty} \sum_{m=1}^{\infty} \sum_{l=0}^{n} A_l B_{nml} J_{n+1/2}(\lambda_{nm} r) J_{n+1/2}(\lambda_{nm}\xi)$$

$$\times P_n^l(\cos\theta) P_n^l(\cos\eta) \cos[l(\varphi-\zeta)] \frac{\sin\left(t\sqrt{a^2\lambda_{nm}^2 + b - k^2/4}\right)}{\sqrt{a^2\lambda_{nm}^2 + b - k^2/4}},$$

$$A_l = \begin{cases} 1 & \text{for } l=0, \\ 2 & \text{for } l \neq 0, \end{cases} \quad B_{nml} = \frac{\lambda_{nm}^2 (2n+1)(n-l)!}{(n+l)! \left[R^2\lambda_{nm}^2 + (sR+n)(sR-n-1)\right] \left[J_{n+1/2}(\lambda_{nm} R)\right]^2}.$$

Here, the $J_{n+1/2}(r)$ are the Bessel functions, the $P_n^l(\mu)$ are the associated Legendre functions (see Paragraph 6.4.3-1), and the λ_{nm} are positive roots of the transcendental equation

$$\lambda R J'_{n+1/2}(\lambda R) + \left(sR - \tfrac{1}{2}\right) J_{n+1/2}(\lambda R) = 0.$$

6.4.3-4. Domain: $R_1 \le r \le R_2$, $0 \le \theta \le \pi$, $0 \le \varphi \le 2\pi$. First boundary value problem.

A spherical layer is considered. The following conditions are prescribed:

$$\begin{aligned}
w &= f_0(r,\theta,\varphi) && \text{at} \quad t=0 && \text{(initial condition)}, \\
\partial_t w &= f_1(r,\theta,\varphi) && \text{at} \quad t=0 && \text{(initial condition)}, \\
w &= g_1(\theta,\varphi,t) && \text{at} \quad r=R_1 && \text{(boundary condition)}, \\
w &= g_2(\theta,\varphi,t) && \text{at} \quad r=R_2 && \text{(boundary condition)}.
\end{aligned}$$

Solution:

$$w(r,\theta,\varphi,t) = \frac{\partial}{\partial t}\int_0^{2\pi}\int_0^{\pi}\int_{R_1}^{R_2} f_0(\xi,\eta,\zeta)G(r,\theta,\varphi,\xi,\eta,\zeta,t)\xi^2\sin\eta\,d\xi\,d\eta\,d\zeta$$

$$+ \int_0^{2\pi}\int_0^{\pi}\int_{R_1}^{R_2}\big[f_1(\xi,\eta,\zeta)+kf_0(\xi,\eta,\zeta)\big]G(r,\theta,\varphi,\xi,\eta,\zeta,t)\xi^2\sin\eta\,d\xi\,d\eta\,d\zeta$$

$$+ a^2 R_1^2 \int_0^t\int_0^{2\pi}\int_0^{\pi} g_1(\eta,\zeta,\tau)\left[\frac{\partial}{\partial \xi}G(r,\theta,\varphi,\xi,\eta,\zeta,t-\tau)\right]_{\xi=R_1}\sin\eta\,d\eta\,d\zeta\,d\tau$$

$$- a^2 R_2^2 \int_0^t\int_0^{2\pi}\int_0^{\pi} g_2(\eta,\zeta,\tau)\left[\frac{\partial}{\partial \xi}G(r,\theta,\varphi,\xi,\eta,\zeta,t-\tau)\right]_{\xi=R_2}\sin\eta\,d\eta\,d\zeta\,d\tau$$

$$+ \int_0^t\int_0^{2\pi}\int_0^{\pi}\int_{R_1}^{R_2} \Phi(\xi,\eta,\zeta,\tau)G(r,\theta,\varphi,\xi,\eta,\zeta,t-\tau)\xi^2\sin\eta\,d\xi\,d\eta\,d\zeta\,d\tau,$$

where

$$G(r,\theta,\varphi,\xi,\eta,\zeta,t) = \frac{\pi e^{-kt/2}}{8\sqrt{r\xi}}\sum_{n=0}^{\infty}\sum_{m=1}^{\infty}\sum_{s=0}^{n} A_s B_{nms} Z_{n+1/2}(\lambda_{nm}r)Z_{n+1/2}(\lambda_{nm}\xi)$$

$$\times P_n^s(\cos\theta)P_n^s(\cos\eta)\cos[s(\varphi-\zeta)]\frac{\sin\left(t\sqrt{a^2\lambda_{nm}^2+b-k^2/4}\right)}{\sqrt{a^2\lambda_{nm}^2+b-k^2/4}},$$

$$Z_{n+1/2}(\lambda_{nm}r) = J_{n+1/2}(\lambda_{nm}R_1)Y_{n+1/2}(\lambda_{nm}r) - Y_{n+1/2}(\lambda_{nm}R_1)J_{n+1/2}(\lambda_{nm}r),$$

$$A_s = \begin{cases} 1 & \text{for } s=0, \\ 2 & \text{for } s\neq 0, \end{cases} \qquad B_{nms} = \frac{\lambda_{nm}(2n+1)(n-s)!\,J_{n+1/2}^2(\lambda_{nm}R_2)}{(n+s)!\,\big[J_{n+1/2}^2(\lambda_{nm}R_1)-J_{n+1/2}^2(\lambda_{nm}R_2)\big]}.$$

Here, the $J_{n+1/2}(r)$ are the Bessel functions, the $P_n^s(\mu)$ are the associated Legendre functions expressed in terms of the Legendre polynomials $P_n(\mu)$ as

$$P_n^s(\mu) = (1-\mu^2)^{s/2}\frac{d^s}{d\mu^s}P_n(\mu),\qquad P_n(\mu) = \frac{1}{n!\,2^n}\frac{d^n}{d\mu^n}(\mu^2-1)^n,$$

and the λ_{nm} are positive roots of the transcendental equation $Z_{n+1/2}(\lambda R_2) = 0$.

6.4.3-5. Domain: $R_1 \leq r \leq R_2$, $0 \leq \theta \leq \pi$, $0 \leq \varphi \leq 2\pi$. Second boundary value problem.

A spherical layer is considered. The following conditions are prescribed:

$$\begin{aligned} w &= f_0(r,\theta,\varphi) & \text{at}\quad & t=0 & \text{(initial condition)}, \\ \partial_t w &= f_1(r,\theta,\varphi) & \text{at}\quad & t=0 & \text{(initial condition)}, \\ \partial_r w &= g_1(\theta,\varphi,t) & \text{at}\quad & r=R_1 & \text{(boundary condition)}, \\ \partial_r w &= g_2(\theta,\varphi,t) & \text{at}\quad & r=R_2 & \text{(boundary condition)}. \end{aligned}$$

Solution:

$$w(r,\theta,\varphi,t) = \frac{\partial}{\partial t}\int_0^{2\pi}\int_0^{\pi}\int_{R_1}^{R_2} f_0(\xi,\eta,\zeta)G(r,\theta,\varphi,\xi,\eta,\zeta,t)\xi^2\sin\eta\,d\xi\,d\eta\,d\zeta$$

$$+ \int_0^{2\pi}\int_0^{\pi}\int_{R_1}^{R_2}\big[f_1(\xi,\eta,\zeta)+kf_0(\xi,\eta,\zeta)\big]G(r,\theta,\varphi,\xi,\eta,\zeta,t)\xi^2\sin\eta\,d\xi\,d\eta\,d\zeta$$

$$- a^2 R_1^2 \int_0^t\int_0^{2\pi}\int_0^{\pi} g_1(\eta,\zeta,\tau)G(r,\theta,\varphi,R_1,\eta,\zeta,t-\tau)\sin\eta\,d\eta\,d\zeta\,d\tau$$

$$+ a^2 R_2^2 \int_0^t\int_0^{2\pi}\int_0^{\pi} g_2(\eta,\zeta,\tau)G(r,\theta,\varphi,R_2,\eta,\zeta,t-\tau)\sin\eta\,d\eta\,d\zeta\,d\tau$$

$$+ \int_0^t\int_0^{2\pi}\int_0^{\pi}\int_{R_1}^{R_2}\Phi(\xi,\eta,\zeta,\tau)G(r,\theta,\varphi,\xi,\eta,\zeta,t-\tau)\xi^2\sin\eta\,d\xi\,d\eta\,d\zeta\,d\tau,$$

where

$$G(r,\theta,\varphi,\xi,\eta,\zeta,t) = \frac{3e^{-kt/2}\sin(t\sqrt{c})}{4\pi(R_2^3-R_1^3)\sqrt{c}} + \frac{e^{-kt/2}}{4\pi\sqrt{r\xi}}\sum_{n=0}^{\infty}\sum_{m=1}^{\infty}\sum_{s=0}^{n}\frac{A_s}{B_{nms}}Z_{n+1/2}(\lambda_{nm}r)Z_{n+1/2}(\lambda_{nm}\xi)$$
$$\times P_n^s(\cos\theta)P_n^s(\cos\eta)\cos[s(\varphi-\zeta)]\frac{\sin(t\sqrt{a^2\lambda_{nm}^2+c})}{\sqrt{a^2\lambda_{nm}^2+c}}.$$

Here,

$$A_s = \begin{cases} 1 & \text{for } s=0, \\ 2 & \text{for } s\neq 0, \end{cases} \quad B_{nms} = \frac{(n+s)!}{(2n+1)(n-s)!}\int_{R_1}^{R_2} rZ_{n+1/2}^2(\lambda_{nm}r)\,dr, \quad c=b-\tfrac{1}{4}k^2,$$

$$Z_{n+1/2}(\lambda_{nm}r) = \left[\lambda_{nm}J'_{n+1/2}(\lambda_{nm}R_1) - \frac{1}{2R_1}J_{n+1/2}(\lambda_{nm}R_1)\right]Y_{n+1/2}(\lambda_{nm}r)$$
$$- \left[\lambda_{nm}Y'_{n+1/2}(\lambda_{nm}R_1) - \frac{1}{2R_1}Y_{n+1/2}(\lambda_{nm}R_1)\right]J_{n+1/2}(\lambda_{nm}r),$$

where the $J_{n+1/2}(r)$ and $Y_{n+1/2}(r)$ are the Bessel functions, the $P_n^s(\mu)$ are the associated Legendre functions (see Paragraph 6.4.3-4), and the λ_{nm} are positive roots of the transcendental equation

$$\lambda Z'_{n+1/2}(\lambda R_2) - \frac{1}{2R_2}Z_{n+1/2}(\lambda R_2) = 0.$$

6.4.3-6. Domain: $R_1 \leq r \leq R_2$, $0 \leq \theta \leq \pi$, $0 \leq \varphi \leq 2\pi$. Third boundary value problem.

A spherical layer is considered. The following conditions are prescribed:

$$w = f_0(r,\theta,\varphi) \quad \text{at} \quad t=0 \quad \text{(initial condition)},$$
$$\partial_t w = f_1(r,\theta,\varphi) \quad \text{at} \quad t=0 \quad \text{(initial condition)},$$
$$\partial_r w - s_1 w = g_1(\theta,\varphi,t) \quad \text{at} \quad r=R_1 \quad \text{(boundary condition)},$$
$$\partial_r w + s_2 w = g_2(\theta,\varphi,t) \quad \text{at} \quad r=R_2 \quad \text{(boundary condition)}.$$

The solution $w(r,\theta,\varphi,t)$ is determined by the formula in Paragraph 6.4.3-5 where

$$G(r,\theta,\varphi,\xi,\eta,\zeta,t) = \frac{e^{-kt/2}}{4\pi\sqrt{r\xi}}\sum_{n=0}^{\infty}\sum_{m=1}^{\infty}\sum_{l=0}^{n}\frac{A_l}{B_{nml}}Z_{n+1/2}(\lambda_{nm}r)Z_{n+1/2}(\lambda_{nm}\xi)$$
$$\times P_n^l(\cos\theta)P_n^l(\cos\eta)\cos[l(\varphi-\zeta)]\frac{\sin(t\sqrt{a^2\lambda_{nm}^2+c})}{\sqrt{a^2\lambda_{nm}^2+c}}.$$

Here,

$$A_l = \begin{cases} 1 & \text{for } l=0, \\ 2 & \text{for } l\neq 0, \end{cases} \quad B_{nml} = \frac{(n+l)!}{(2n+1)(n-l)!}\int_{R_1}^{R_2} rZ_{n+1/2}^2(\lambda_{nm}r)\,dr, \quad c=b-\tfrac{1}{4}k^2,$$

$$Z_{n+1/2}(\lambda r) = \left[\lambda J'_{n+1/2}(\lambda R_1) - \left(s_1 + \frac{1}{2R_1}\right)J_{n+1/2}(\lambda R_1)\right]Y_{n+1/2}(\lambda r)$$
$$- \left[\lambda Y'_{n+1/2}(\lambda R_1) - \left(s_1 + \frac{1}{2R_1}\right)Y_{n+1/2}(\lambda R_1)\right]J_{n+1/2}(\lambda r),$$

where the $J_{n+1/2}(r)$ and $Y_{n+1/2}(r)$ are the Bessel functions, the $P_n^l(\mu)$ are the associated Legendre functions (see Paragraph 6.4.3-4), and the λ_{nm} are positive roots of the transcendental equation

$$\lambda Z'_{n+1/2}(\lambda R_2) + \left(s_2 - \frac{1}{2R_2}\right)Z_{n+1/2}(\lambda R_2) = 0.$$

6.5. Other Equations with Three Space Variables

6.5.1. Equations Containing Arbitrary Parameters

1. $\dfrac{\partial^2 w}{\partial t^2} = \dfrac{\partial}{\partial x}\left(ax^n \dfrac{\partial w}{\partial x}\right) + \dfrac{\partial}{\partial y}\left(by^m \dfrac{\partial w}{\partial y}\right) + \dfrac{\partial}{\partial z}\left(cz^k \dfrac{\partial w}{\partial z}\right).$

This equation admits separable solutions. In addition, for $n \neq 2$, $m \neq 2$, and $k \neq 2$, there are particular solutions of the form

$$w = w(\xi, t), \qquad \xi^2 = 4\left[\dfrac{x^{2-n}}{a(2-n)^2} + \dfrac{y^{2-m}}{b(2-m)^2} + \dfrac{z^{2-k}}{c(2-k)^2}\right],$$

where $w(\xi, t)$ is determined by the one-dimensional nonstationary equation

$$\dfrac{\partial^2 w}{\partial t^2} = \dfrac{\partial^2 w}{\partial \xi^2} + \dfrac{A}{\xi}\dfrac{\partial w}{\partial \xi}, \qquad A = 2\left(\dfrac{1}{2-n} + \dfrac{1}{2-m} + \dfrac{1}{2-k}\right) - 1.$$

2. $\dfrac{\partial^2 w}{\partial t^2} + k\dfrac{\partial w}{\partial t} = a^2\left(\dfrac{\partial^2 w}{\partial x^2} + \dfrac{\partial^2 w}{\partial y^2} + \dfrac{\partial^2 w}{\partial z^2}\right) + b_1\dfrac{\partial w}{\partial x} + b_2\dfrac{\partial w}{\partial y} + b_3\dfrac{\partial w}{\partial z} + cw.$

The transformation

$$w(x, y, z, t) = u(x, y, z, \tau)\exp\left(-\dfrac{1}{2}kt - \dfrac{b_1 x + b_2 y + b_3 z}{2a^2}\right), \qquad \tau = at$$

leads to the equation in Subsection 6.3.1:

$$\dfrac{\partial^2 u}{\partial \tau^2} = \dfrac{\partial^2 u}{\partial x^2} + \dfrac{\partial^2 u}{\partial y^2} + \dfrac{\partial^2 u}{\partial z^2} + \beta u, \qquad \beta = \dfrac{c}{a^2} + \dfrac{k^2}{4a^2} - \dfrac{1}{4a^4}(b_1^2 + b_2^2 + b_3^2).$$

6.5.2. Equation of the Form

$$\rho(x, y, z)\dfrac{\partial^2 w}{\partial t^2} = \mathrm{div}[a(x, y, z)\nabla w] - q(x, y, z)w + \Phi(x, y, z, t)$$

Such equations are encountered when studying vibration of finite volumes. The equation is written using the notation

$$\mathrm{div}[a(\mathbf{r})\nabla w] = \dfrac{\partial}{\partial x}\left[a(\mathbf{r})\dfrac{\partial w}{\partial x}\right] + \dfrac{\partial}{\partial y}\left[a(\mathbf{r})\dfrac{\partial w}{\partial y}\right] + \dfrac{\partial}{\partial z}\left[a(\mathbf{r})\dfrac{\partial w}{\partial z}\right], \qquad \mathbf{r} = \{x, y, z\}.$$

The problems for the equation in question are considered below for the interior of a bounded domain V with smooth surface S. In what follows, it is assumed that $\rho(\mathbf{r}) > 0$, $a(\mathbf{r}) > 0$, and $q(\mathbf{r}) \geq 0$.

6.5.2-1. First boundary value problem.

The solution of the equation in question with the initial conditions

$$\begin{aligned} w &= f_0(\mathbf{r}) \quad \text{at} \quad t = 0, \\ \partial_t w &= f_1(\mathbf{r}) \quad \text{at} \quad t = 0 \end{aligned} \qquad (1)$$

and the nonhomogeneous boundary conditions of the first kind

$$w = g(\mathbf{r}, t) \quad \text{for} \quad \mathbf{r} \in S \qquad (2)$$

can be written as the sum

$$w(\mathbf{r}, t) = \frac{\partial}{\partial t} \int_V f_0(\boldsymbol{\xi})\rho(\boldsymbol{\xi})\mathcal{G}(\mathbf{r}, \boldsymbol{\xi}, t)\, dV_\xi + \int_V f_1(\boldsymbol{\xi})\rho(\boldsymbol{\xi})\mathcal{G}(\mathbf{r}, \boldsymbol{\xi}, t)\, dV_\xi$$
$$- \int_0^t \int_S g(\boldsymbol{\xi}, \tau)a(\boldsymbol{\xi})\left[\frac{\partial}{\partial N_\xi}\mathcal{G}(\mathbf{r}, \boldsymbol{\xi}, t - \tau)\right] dS_\xi\, d\tau + \int_0^t \int_V \Phi(\boldsymbol{\xi}, \tau)\mathcal{G}(\mathbf{r}, \boldsymbol{\xi}, t - \tau)\, dV_\xi\, d\tau. \quad (3)$$

Here, the modified Green's function is expressed as

$$\mathcal{G}(\mathbf{r}, \boldsymbol{\xi}, t) = \sum_{n=1}^\infty \frac{1}{\sqrt{\lambda_n}\|u_n\|^2} u_n(\mathbf{r})u_n(\boldsymbol{\xi}) \sin(\sqrt{\lambda_n}\, t),$$
$$\|u_n\|^2 = \int_V \rho(\mathbf{r})u_n^2(\mathbf{r})\, dV, \quad \boldsymbol{\xi} = \{\xi_1, \xi_2, \xi_3\}, \quad (4)$$

where the λ_n and $u_n(\mathbf{r})$ are the eigenvalues and corresponding eigenfunctions of the Sturm–Liouville problem for the following second-order elliptic equation with homogeneous boundary conditions of the first kind:

$$\operatorname{div}[a(\mathbf{r})\nabla u] - q(\mathbf{r})u + \lambda\rho(\mathbf{r})u = 0, \quad (5)$$
$$u = 0 \quad \text{for} \quad \mathbf{r} \in S. \quad (6)$$

The integration in solution (3) is performed with respect to ξ_1, ξ_2, and ξ_3; $\frac{\partial}{\partial N_\xi}$ is the derivative along the outward normal to the surface S with respect to ξ_1, ξ_2, and ξ_3.

General properties of the Sturm–Liouville problem (5)–(6):

1°. There are finitely many eigenvalues. All eigenvalues are real and can be ordered so that $\lambda_1 \le \lambda_2 \le \lambda_3 \le \cdots$ and $\lambda_n \to \infty$ as $n \to \infty$; therefore the number of negative eigenvalues is finite.

2°. If $\rho(\mathbf{r}) > 0$, $a(\mathbf{r}) > 0$, and $q(\mathbf{r}) \ge 0$, then all eigenvalues are positive, $\lambda_n > 0$.

3°. An eigenfunction is determined up to a constant multiplier. Two eigenfunctions $u_n(\mathbf{r})$ and $u_m(\mathbf{r})$ corresponding to different eigenvalues λ_n and λ_m are orthogonal with weight $\rho(\mathbf{r})$ in the domain V, that is,

$$\int_V \rho(\mathbf{r})u_n(\mathbf{r})u_m(\mathbf{r})\, dV = 0 \quad \text{for} \quad n \ne m.$$

4°. An arbitrary function $F(\mathbf{r})$ twice continuously differentiable and satisfying the boundary condition of the Sturm–Liouville problem ($F = 0$ for $\mathbf{r} \in S$) can be expanded into an absolutely and uniformly convergent series in the eigenfunctions:

$$F(\mathbf{r}) = \sum_{n=1}^\infty F_n u_n(\mathbf{r}), \quad F_n = \frac{1}{\|u_n\|^2} \int_V F(\mathbf{r})\rho(\mathbf{r})u_n(\mathbf{r})\, dV,$$

where $\|u_n\|^2$ is defined in (4).

Remark. In a three-dimensional problem, finitely many linearly independent eigenfunctions $u_n^{(1)}, \ldots, u_n^{(m)}$ generally correspond to each eigenvalue λ_n. These functions can always be replaced by their linear combinations

$$\bar{u}_n^{(k)} = A_{k,1}u_n^{(1)} + \cdots + A_{k,k-1}u_n^{(k-1)} + u_n^{(k)}, \quad k = 1, \ldots, m,$$

so that $\bar{u}_n^{(1)}, \ldots, \bar{u}_n^{(m)}$ are now orthogonal pairwise. For this reason, without loss of generality, all eigenfunctions can be assumed orthogonal.

6.5.2-2. Second boundary value problem.

The solution of the equation with the initial conditions (1) and nonhomogeneous boundary conditions of the second kind,

$$\frac{\partial w}{\partial N} = g(\mathbf{r}, t) \quad \text{for} \quad \mathbf{r} \in S,$$

can be represented as the sum

$$w(\mathbf{r}, t) = \frac{\partial}{\partial t} \int_V f_0(\boldsymbol{\xi}) \rho(\boldsymbol{\xi}) \mathcal{G}(\mathbf{r}, \boldsymbol{\xi}, t) \, dV_\xi + \int_V f_1(\boldsymbol{\xi}) \rho(\boldsymbol{\xi}) \mathcal{G}(\mathbf{r}, \boldsymbol{\xi}, t) \, dV_\xi$$
$$+ \int_0^t \int_S g(\boldsymbol{\xi}, \tau) a(\boldsymbol{\xi}) \mathcal{G}(\mathbf{r}, \boldsymbol{\xi}, t - \tau) \, dS_\xi \, d\tau + \int_0^t \int_V \Phi(\boldsymbol{\xi}, \tau) \mathcal{G}(\mathbf{r}, \boldsymbol{\xi}, t - \tau) \, dV_\xi \, d\tau. \qquad (7)$$

Here, the modified Green's function $\mathcal{G}$ is given by relation (4), the λ_n and $u_n(\mathbf{r})$ are the eigenvalues and corresponding eigenfunctions of the Sturm–Liouville problem for the second-order elliptic equation (5) with homogeneous boundary conditions of the second kind,

$$\frac{\partial u}{\partial N} = 0 \quad \text{for} \quad \mathbf{r} \in S. \qquad (8)$$

For $q(\mathbf{r}) > 0$, the general properties of the eigenvalue problem (5), (8) are the same as those of the first boundary value problem (all λ_n are positive).

6.5.2-3. Third boundary value problem.

The solution of the equation with the initial conditions (1) and nonhomogeneous boundary conditions of the third kind,

$$\frac{\partial w}{\partial N} + k(\mathbf{r})w = g(\mathbf{r}, t) \quad \text{for} \quad \mathbf{r} \in S,$$

is determined by relations (7) and (4), where the λ_n and $u_n(\mathbf{r})$ are the eigenvalues and eigenfunctions of the Sturm–Liouville problem for the second-order elliptic equation (5) with homogeneous boundary conditions of the third kind,

$$\frac{\partial u}{\partial N} + k(\mathbf{r})u = 0 \quad \text{for} \quad \mathbf{r} \in S. \qquad (9)$$

If $q(\mathbf{r}) \geq 0$ and $k(\mathbf{r}) > 0$, the general properties of the eigenvalue problem (5), (9) are the same as those of the first boundary value problem (see Paragraph 6.5.2-1).

Suppose $k(\mathbf{r}) = k = \text{const}$. Denote the Green's functions of the second and third boundary value problems by $G_2(\mathbf{r}, \boldsymbol{\xi}, t)$ and $G_3(\mathbf{r}, \boldsymbol{\xi}, t, k)$, respectively. If $q(\mathbf{r}) > 0$, the limit relation $G_2(\mathbf{r}, \boldsymbol{\xi}, t) = \lim_{k \to 0} G_3(\mathbf{r}, \boldsymbol{\xi}, t, k)$ holds.

⊙ *References for Subsection* 6.5.2: V. S. Vladimirov (1988), A. D. Polyanin (2000a).

6.6. Equations with n Space Variables

Throughout this section the following notation is used:

$$\Delta_n w = \sum_{k=1}^n \frac{\partial^2 w}{\partial x_k^2}, \quad \mathbf{x} = \{x_1, \ldots, x_n\}, \quad \mathbf{y} = \{y_1, \ldots, y_n\}, \quad |\mathbf{x}| = \sqrt{x_1^2 + \cdots + x_n^2}.$$

6.6.1. Wave Equation $\dfrac{\partial^2 w}{\partial t^2} = a^2 \Delta_n w$

6.6.1-1. Fundamental solution:

$$\mathscr{E}(\mathbf{x},t) = \begin{cases} \dfrac{(-1)^{\frac{n-2}{2}}}{2a\pi^{\frac{n+1}{2}}} \Gamma\left(\dfrac{n-1}{2}\right) \dfrac{\vartheta(at-|\mathbf{x}|)}{(a^2 t^2 - |\mathbf{x}|^2)^{\frac{n-1}{2}}} & \text{if } n \geq 2 \text{ is even;} \\[2ex] \dfrac{1}{2\pi a}\left(\dfrac{1}{2\pi a^2 t}\dfrac{\partial}{\partial t}\right)^{\frac{n-3}{2}} \delta(a^2 t^2 - |\mathbf{x}|^2) & \text{if } n \geq 3 \text{ is odd;} \end{cases}$$

where $\vartheta(z)$ is the Heaviside unit step function and $\delta(z)$ is the Dirac delta function.

⊙ *Reference*: V. S. Vladimirov (1988).

6.6.1-2. Properties of solutions.

Suppose $w(x_1, \ldots, x_n, t)$ is a solution of the wave equation. Then the functions

$$w_1 = Aw(\pm \lambda x_1 + C_1, \ldots, \pm \lambda x_n + C_n, \pm \lambda t + C_{n+1}),$$

$$w_2 = Aw\left(\dfrac{x_1 - vt}{\sqrt{1-(v/a)^2}}, x_2, \ldots, x_n, \dfrac{t - v a^{-2} x_1}{\sqrt{1-(v/a)^2}}\right),$$

$$w_3 = A|r^2 - a^2 t^2|^{-\frac{n-1}{2}} w\left(\dfrac{x_1}{r^2 - a^2 t^2}, \ldots, \dfrac{x_n}{r^2 - a^2 t^2}, \dfrac{t}{r^2 - a^2 t^2}\right), \quad r = |\mathbf{x}|,$$

are also solutions of this equation everywhere they are defined; A, C_1, ..., C_{n+1}, v, and λ are arbitrary constants. The signs at λ in the expression of w_1 can be taken independently of one another.

6.6.1-3. Domain: $-\infty < x_k < \infty$; $k = 1, \ldots, n$. **Cauchy problem.**

Initial conditions are prescribed:

$$w = f(\mathbf{x}) \quad \text{at} \quad t = 0,$$
$$\partial_t w = g(\mathbf{x}) \quad \text{at} \quad t = 0.$$

Solution:

$$w(\mathbf{x},t) = \dfrac{1}{a^{n-1}(n-2)!} \dfrac{\partial^{n-1}}{\partial t^{n-1}} \int_0^{at} (a^2 t^2 - r^2)^{\frac{n-3}{2}} r T_r[f(\mathbf{x})] \, dr$$
$$+ \dfrac{1}{a^{n-1}(n-2)!} \dfrac{\partial^{n-2}}{\partial t^{n-2}} \int_0^{at} (a^2 t^2 - r^2)^{\frac{n-3}{2}} r T_r[g(\mathbf{x})] \, dr.$$

Here, $T_r[f(\mathbf{x})]$ is the average of f over the surface of the sphere of radius r with center at $\mathbf{x}$:

$$T_r[f(\mathbf{x})] \equiv \dfrac{1}{\sigma_n r^{n-1}} \int_{|\mathbf{x}-\mathbf{y}|=r} f(\mathbf{y}) \, dS_y, \qquad \sigma_n = \dfrac{2\pi^{n/2}}{\Gamma(n/2)},$$

where $\sigma_n r^{n-1}$ is the area of the surface of an n-dimensional sphere of radius r, dS_y is the area element of this surface, and $|\mathbf{x}-\mathbf{y}|^2 = (x_1 - y_1)^2 + \cdots + (x_n - y_n)^2$.

For odd n, the solution can be alternatively represented as

$$w(\mathbf{x},t) = \dfrac{1}{1 \times 3 \ldots (n-2)} \dfrac{\partial}{\partial t} \left(\dfrac{1}{t}\dfrac{\partial}{\partial t}\right)^{\frac{n-3}{2}} \left(t^{n-2} T_{at}[f(\mathbf{x})]\right)$$
$$+ \dfrac{1}{1 \times 3 \ldots (n-2)} \left(\dfrac{1}{t}\dfrac{\partial}{\partial t}\right)^{\frac{n-3}{2}} \left(t^{n-2} T_{at}[g(\mathbf{x})]\right).$$

For even n, the solution can be alternatively represented as

$$w(\mathbf{x},t) = \frac{1}{2\times 4\ldots(n-2)a^{n-1}} \frac{\partial}{\partial t}\left(\frac{1}{t}\frac{\partial}{\partial t}\right)^{\frac{n-2}{2}} \int_0^{at} T_r[f(\mathbf{x})]\frac{r^{n-1}\,dr}{\sqrt{a^2t^2-r^2}}$$

$$+ \frac{1}{2\times 4\ldots(n-2)a^{n-1}}\left(\frac{1}{t}\frac{\partial}{\partial t}\right)^{\frac{n-2}{2}} \int_0^{at} T_r[g(\mathbf{x})]\frac{r^{n-1}\,dr}{\sqrt{a^2t^2-r^2}}.$$

⊙ *References*: V. M. Babich, M. B. Kapilevich, S. G. Mikhlin, et al. (1964), R. Courant and D. Hilbert (1989), D. Zwillinger (1998).

6.6.1-4. Domain: $0 \le x_k \le l_k$; $k=1,\ldots,n$. Boundary value problems.

For solutions of the first, second, third, and mixed boundary value problems with nonhomogeneous conditions of general form, see Paragraphs 6.6.2-2, 6.6.2-3, 6.6.2-4, and 6.6.2-5 for $\Phi \equiv 0$, respectively.

6.6.2. Nonhomogeneous Wave Equation

$$\frac{\partial^2 w}{\partial t^2} = a^2 \Delta_n w + \Phi(x_1,\ldots,x_n,t)$$

6.6.2-1. Domain: $-\infty < x_k < \infty$; $k=1,\ldots,n$. Cauchy problem.

Initial conditions are prescribed:

$$w = f(\mathbf{x}) \quad \text{at} \quad t = 0,$$
$$\partial_t w = g(\mathbf{x}) \quad \text{at} \quad t = 0.$$

Solution:

$$w(\mathbf{x},t) = \frac{1}{a^{n-1}(n-2)!} \frac{\partial^{n-1}}{\partial t^{n-1}} \int_0^{at} \left(a^2t^2-r^2\right)^{\frac{n-3}{2}} rT_r[f(\mathbf{x})]\,dr$$

$$+ \frac{1}{a^{n-1}(n-2)!} \frac{\partial^{n-2}}{\partial t^{n-2}} \int_0^{at} \left(a^2t^2-r^2\right)^{\frac{n-3}{2}} rT_r[g(\mathbf{x})]\,dr$$

$$+ \frac{1}{a^{n-1}(n-2)!} \frac{\partial^{n-2}}{\partial t^{n-2}} \int_0^{at} d\tau \int_0^{a\tau} \left(a^2\tau^2-r^2\right)^{\frac{n-3}{2}} rT_r[\Phi(\mathbf{x},t-\tau)]\,dr.$$

Here, $T_r[f(\mathbf{x})]$ is the average of f over the spherical surface of radius r with center at $\mathbf{x}$:

$$T_r[f(\mathbf{x})] \equiv \frac{1}{\sigma_n r^{n-1}} \int_{|\mathbf{x}-\mathbf{y}|=r} f(\mathbf{y})\,dS_y, \qquad \sigma_n = \frac{2\pi^{n/2}}{\Gamma(n/2)},$$

where $\sigma_n r^{n-1}$ is the area of the surface of an n-dimensional sphere of radius r and dS_y is the area element of this surface.

⊙ *References*: V. M. Babich, M. B. Kapilevich, S. G. Mikhlin, et al. (1964), R. Courant and D. Hilbert (1989).

6.6.2-2. Domain: $V = \{0 \le x_k \le l_k;\ k=1,\ldots,n\}$. First boundary value problem.

The following conditions are prescribed:

$$w = f_0(\mathbf{x}) \quad \text{at} \quad t = 0 \quad \text{(initial condition)},$$
$$\partial_t w = f_1(\mathbf{x}) \quad \text{at} \quad t = 0 \quad \text{(initial condition)},$$
$$w = g_k(\mathbf{x},t) \quad \text{at} \quad x_k = 0 \quad \text{(boundary conditions)},$$
$$w = h_k(\mathbf{x},t) \quad \text{at} \quad x_k = l_k \quad \text{(boundary conditions)}.$$

Solution:

$$w(\mathbf{x}, t) = \int_0^t \int_V \Phi(\mathbf{y}, \tau) G(\mathbf{x}, \mathbf{y}, t - \tau) \, d\mathbf{y} \, d\tau$$
$$+ \frac{\partial}{\partial t} \int_V f_0(\mathbf{y}) G(\mathbf{x}, \mathbf{y}, t) \, d\mathbf{y} + \int_V f_1(\mathbf{y}) G(\mathbf{x}, \mathbf{y}, t) \, d\mathbf{y}$$
$$+ a^2 \sum_{k=1}^n \int_0^t \int_{S^{(k)}} \left[g_k(\mathbf{y}, \tau) \frac{\partial}{\partial y_k} G(\mathbf{x}, \mathbf{y}, t - \tau) \right]_{y_k=0} dS_y^{(k)} \, d\tau$$
$$- a^2 \sum_{k=1}^n \int_0^t \int_{S^{(k)}} \left[h_k(\mathbf{y}, \tau) \frac{\partial}{\partial y_k} G(\mathbf{x}, \mathbf{y}, t - \tau) \right]_{y_k=l_k} dS_y^{(k)} \, d\tau,$$

where

$$\mathbf{x} = \{x_1, \ldots, x_n\}, \quad \mathbf{y} = \{y_1, \ldots, y_n\}, \quad d\mathbf{y} = dy_1 \, dy_2 \ldots dy_n, \quad dS_y^{(k)} = dy_1 \ldots dy_{k-1} \, dy_{k+1} \ldots dy_n,$$
$$S^{(k)} = \{0 \leq y_m \leq l_m \text{ for } m = 1, \ldots, k-1, k+1, \ldots, n\}.$$

Green's function:

$$G(\mathbf{x}, \mathbf{y}, t) = \frac{2^n}{l_1 l_2 \ldots l_n} \sum_{s_1=1}^\infty \sum_{s_2=1}^\infty \cdots \sum_{s_n=1}^\infty \sin(\lambda_{s_1} x_1) \sin(\lambda_{s_2} x_2) \ldots \sin(\lambda_{s_n} x_n)$$
$$\times \sin(\lambda_{s_1} y_1) \sin(\lambda_{s_2} y_2) \ldots \sin(\lambda_{s_n} y_n) \frac{\sin\left(at\sqrt{\lambda_{s_1}^2 + \cdots + \lambda_{s_n}^2}\right)}{a\sqrt{\lambda_{s_1}^2 + \cdots + \lambda_{s_n}^2}},$$

where

$$\lambda_{s_1} = \frac{s_1 \pi}{l_1}, \quad \lambda_{s_2} = \frac{s_2 \pi}{l_2}, \quad \ldots, \quad \lambda_{s_n} = \frac{s_n \pi}{l_n}.$$

6.6.2-3. Domain: $V = \{0 \leq x_k \leq l_k; \ k = 1, \ldots, n\}$. Second boundary value problem.

The following conditions are prescribed:

$$w = f_0(\mathbf{x}) \quad \text{at} \quad t = 0 \quad \text{(initial condition)},$$
$$\partial_t w = f_1(\mathbf{x}) \quad \text{at} \quad t = 0 \quad \text{(initial condition)},$$
$$\partial_{x_k} w = g_k(\mathbf{x}, t) \quad \text{at} \quad x_k = 0 \quad \text{(boundary conditions)},$$
$$\partial_{x_k} w = h_k(\mathbf{x}, t) \quad \text{at} \quad x_k = l_k \quad \text{(boundary conditions)}.$$

Solution:

$$w(\mathbf{x}, t) = \int_0^t \int_V \Phi(\mathbf{y}, \tau) G(\mathbf{x}, \mathbf{y}, t - \tau) \, d\mathbf{y} \, d\tau$$
$$+ \int_V f_0(\mathbf{y}) G(\mathbf{x}, \mathbf{y}, t) \, d\mathbf{y} + \int_V f_1(\mathbf{y}) G(\mathbf{x}, \mathbf{y}, t) \, d\mathbf{y}$$
$$- a^2 \sum_{k=1}^n \int_0^t \int_{S^{(k)}} \left[g_k(\mathbf{y}, \tau) G(\mathbf{x}, \mathbf{y}, t - \tau) \right]_{y_k=0} dS_y^{(k)} \, d\tau$$
$$+ a^2 \sum_{k=1}^n \int_0^t \int_{S^{(k)}} \left[h_k(\mathbf{y}, \tau) G(\mathbf{x}, \mathbf{y}, t - \tau) \right]_{y_k=l_k} dS_y^{(k)} \, d\tau.$$

Here,

$$G(\mathbf{x}, \mathbf{y}, t) = \frac{t}{l_1 l_2 \ldots l_n} + \frac{1}{l_1 l_2 \ldots l_n} \sum_{s_1=0}^\infty \sum_{s_2=0}^\infty \cdots \sum_{s_n=0}^\infty \frac{A_{s_1} A_{s_2} \ldots A_{s_n}}{a\sqrt{\lambda_{s_1}^2 + \cdots + \lambda_{s_n}^2}} \sin\left(at\sqrt{\lambda_{s_1}^2 + \cdots + \lambda_{s_n}^2}\right)$$
$$\times \cos(\lambda_{s_1} x_1) \cos(\lambda_{s_2} x_2) \ldots \cos(\lambda_{s_n} x_n) \cos(\lambda_{s_1} y_1) \cos(\lambda_{s_2} y_2) \ldots \cos(\lambda_{s_n} y_n),$$

where
$$\lambda_{s_1} = \frac{s_1\pi}{l_1}, \quad \lambda_{s_2} = \frac{s_2\pi}{l_2}, \quad \ldots, \quad \lambda_{s_n} = \frac{s_n\pi}{l_n}; \quad A_{s_m} = \begin{cases} 1 & \text{for } s_m = 0, \\ 2 & \text{for } s_m \neq 0, \end{cases} \quad m = 1, 2, \ldots, n.$$

The summation is performed over the indices satisfying the condition $s_1 + \cdots + s_n > 0$; the term corresponding to $s_1 = \cdots = s_n = 0$ is singled out.

6.6.2-4. Domain: $V = \{0 \leq x_k \leq l_k; \; k = 1, \ldots, n\}$. Third boundary value problem.

The following conditions are prescribed:

$$\begin{aligned}
w &= f_0(\mathbf{x}) & &\text{at} & t &= 0 & &\text{(initial condition)}, \\
\partial_t w &= f_1(\mathbf{x}) & &\text{at} & t &= 0 & &\text{(initial condition)}, \\
\partial_{x_k} w - b_k w &= g_k(\mathbf{x}, t) & &\text{at} & x_k &= 0 & &\text{(boundary conditions)}, \\
\partial_{x_k} w + c_k w &= h_k(\mathbf{x}, t) & &\text{at} & x_k &= l_k & &\text{(boundary conditions)}.
\end{aligned}$$

The solution $w(\mathbf{x}, t)$ is determined by the formula in Paragraph 6.6.2-3 where

$$G(\mathbf{x}, \mathbf{y}, t) = 2^n \sum_{s_1=1}^{\infty} \sum_{s_2=1}^{\infty} \cdots \sum_{s_n=1}^{\infty} \frac{\sin\left(at\sqrt{\lambda_{s_1}^2 + \lambda_{s_2}^2 + \cdots + \lambda_{s_n}^2}\right)}{a E_{s_1} E_{s_2} \ldots E_{s_n} \sqrt{\lambda_{s_1}^2 + \lambda_{s_2}^2 + \cdots + \lambda_{s_n}^2}}$$
$$\times \sin(\lambda_{s_1} x_1 + \varphi_{s_1}) \sin(\lambda_{s_2} x_2 + \varphi_{s_2}) \ldots \sin(\lambda_{s_n} x_n + \varphi_{s_n})$$
$$\times \sin(\lambda_{s_1} y_1 + \varphi_{s_1}) \sin(\lambda_{s_2} y_2 + \varphi_{s_2}) \ldots \sin(\lambda_{s_n} y_n + \varphi_{s_n}).$$

Here,

$$\varphi_{s_m} = \arctan\frac{\lambda_{s_m}}{l_m}, \quad E_{s_m} = l_m + \frac{(b_m c_m + \lambda_{s_m}^2)(b_m + c_m)}{(b_m^2 + \lambda_{s_m}^2)(c_m^2 + \lambda_{s_m}^2)}, \quad m = 1, 2, \ldots, n;$$

the λ_{s_m} are positive roots of the transcendental equations

$$\frac{1}{b_m + c_m}\left(\lambda - \frac{b_m c_m}{\lambda}\right) = \cot(l_m \lambda), \quad m = 1, 2, \ldots, n.$$

6.6.2-5. Domain: $V = \{0 \leq x_k \leq l_k; \; k = 1, \ldots, n\}$. Mixed boundary value problem.

The following conditions are prescribed:

$$\begin{aligned}
w &= f_0(\mathbf{x}) & &\text{at} & t &= 0 & &\text{(initial condition)}, \\
\partial_t w &= f_1(\mathbf{x}) & &\text{at} & t &= 0 & &\text{(initial condition)}, \\
w &= g_k(\mathbf{x}, t) & &\text{at} & x_k &= 0 & &\text{(boundary conditions)}, \\
\partial_{x_k} w &= h_k(\mathbf{x}, t) & &\text{at} & x_k &= l_k & &\text{(boundary conditions)}.
\end{aligned}$$

Solution:

$$w(\mathbf{x}, t) = \int_0^t \int_V \Phi(\mathbf{y}, \tau) G(\mathbf{x}, \mathbf{y}, t - \tau) \, d\mathbf{y} \, d\tau$$
$$+ \frac{\partial}{\partial t} \int_V f_0(\mathbf{y}) G(\mathbf{x}, \mathbf{y}, t) \, d\mathbf{y} + \int_V f_1(\mathbf{y}) G(\mathbf{x}, \mathbf{y}, t) \, d\mathbf{y}$$
$$+ a^2 \sum_{k=1}^n \int_0^t \int_{S^{(k)}} \left[g_k(\mathbf{y}, \tau) \frac{\partial}{\partial y_k} G(\mathbf{x}, \mathbf{y}, t - \tau) \right]_{y_k = 0} dS_y^{(k)} \, d\tau$$
$$+ a^2 \sum_{k=1}^n \int_0^t \int_{S^{(k)}} \left[h_k(\mathbf{y}, \tau) G(\mathbf{x}, \mathbf{y}, t - \tau) \right]_{y_k = l_k} dS_y^{(k)} \, d\tau,$$

where
$$G(\mathbf{x},\mathbf{y},t) = \frac{2^n}{l_1 l_2 \ldots l_n} \sum_{s_1=1}^{\infty} \sum_{s_2=1}^{\infty} \ldots \sum_{s_n=1}^{\infty} \sin(\lambda_{s_1} x_1) \sin(\lambda_{s_2} x_2) \ldots \sin(\lambda_{s_n} x_n)$$
$$\times \sin(\lambda_{s_1} y_1) \sin(\lambda_{s_2} y_2) \ldots \sin(\lambda_{s_n} y_n) \frac{\sin\left(at\sqrt{\lambda_{s_1}^2 + \cdots + \lambda_{s_n}^2}\right)}{a\sqrt{\lambda_{s_1}^2 + \cdots + \lambda_{s_n}^2}},$$
$$\lambda_{s_1} = \frac{\pi(2s_1+1)}{2l_1}, \quad \lambda_{s_2} = \frac{\pi(2s_2+1)}{2l_2}, \quad \ldots, \quad \lambda_{s_n} = \frac{\pi(2s_n+1)}{2l_n}.$$

6.6.3. Equations of the Form $\frac{\partial^2 w}{\partial t^2} = a^2 \Delta_n w - bw + \Phi(x_1, \ldots, x_n, t)$

6.6.3-1. Domain: $-\infty < x_k < \infty$; $k = 1, \ldots, n$. **Cauchy problem.**

Initial conditions are prescribed:
$$w = f(\mathbf{x}) \quad \text{at} \quad t = 0,$$
$$\partial_t w = g(\mathbf{x}) \quad \text{at} \quad t = 0,$$
where $\mathbf{x} = \{x_1, \ldots, x_n\}$.

1°. Let $b = -c^2 < 0$ and $\Phi \equiv 0$. The solution is sought by the descent method in the form
$$w(\mathbf{x}, t) = \frac{1}{\exp(cx_{n+1})} u(\mathbf{x}, x_{n+1}, t), \tag{1}$$
where u is the solution of the Cauchy problem for the auxiliary $(n+1)$-dimensional wave equation
$$\frac{\partial^2 u}{\partial t^2} = \Delta_{n+1} u \tag{2}$$
with the initial conditions
$$u = \exp(cx_{n+1}) f(\mathbf{x}) \quad \text{at} \quad t = 0,$$
$$\partial_t u = \exp(cx_{n+1}) g(\mathbf{x}) \quad \text{at} \quad t = 0. \tag{3}$$
For the solution of problem (2), (3), see Paragraph 6.6.1-3.

2°. Let $b = c^2 > 0$ and $\Phi \equiv 0$. In this case the function $\exp(cx_{n+1})$ in (1) and (3) must be replaced by $\cos(cx_{n+1})$.

⊙ *Reference*: R. Courant and D. Hilbert (1989).

6.6.3-2. Domain: $V = \{0 \leq x_k \leq l_k;\ k = 1, \ldots, n\}$. **First boundary value problem.**

The following conditions are prescribed:
$$w = f_0(\mathbf{x}) \quad \text{at} \quad t = 0 \quad \text{(initial condition)},$$
$$\partial_t w = f_1(\mathbf{x}) \quad \text{at} \quad t = 0 \quad \text{(initial condition)},$$
$$w = g_k(\mathbf{x}, t) \quad \text{at} \quad x_k = 0 \quad \text{(boundary conditions)},$$
$$w = h_k(\mathbf{x}, t) \quad \text{at} \quad x_k = l_k \quad \text{(boundary conditions)}.$$

Solution:
$$w(\mathbf{x}, t) = \int_0^t \int_V \Phi(\mathbf{y}, \tau) G(\mathbf{x}, \mathbf{y}, t-\tau) \, d\mathbf{y} \, d\tau$$
$$+ \frac{\partial}{\partial t} \int_V f_0(\mathbf{y}) G(\mathbf{x}, \mathbf{y}, t) \, d\mathbf{y} + \int_V f_1(\mathbf{y}) G(\mathbf{x}, \mathbf{y}, t) \, d\mathbf{y}$$
$$+ a^2 \sum_{k=1}^{n} \int_0^t \int_{S^{(k)}} \left[g_k(\mathbf{y}, \tau) \frac{\partial}{\partial y_k} G(\mathbf{x}, \mathbf{y}, t-\tau) \right]_{y_k=0} dS_y^{(k)} \, d\tau$$
$$- a^2 \sum_{k=1}^{n} \int_0^t \int_{S^{(k)}} \left[h_k(\mathbf{y}, \tau) \frac{\partial}{\partial y_k} G(\mathbf{x}, \mathbf{y}, t-\tau) \right]_{y_k=l_k} dS_y^{(k)} \, d\tau,$$

where

$$\mathbf{x} = \{x_1, \ldots, x_n\}, \quad \mathbf{y} = \{y_1, \ldots, y_n\}, \quad d\mathbf{y} = dy_1\, dy_2 \ldots dy_n, \quad dS_y^{(k)} = dy_1 \ldots dy_{k-1}\, dy_{k+1} \ldots dy_n,$$
$$S^{(k)} = \{0 \leq y_m \leq l_m \text{ for } m = 1, \ldots, k-1, k+1, \ldots, n\}.$$

Green's function:

$$G(\mathbf{x}, \mathbf{y}, t) = \frac{2^n}{l_1 l_2 \ldots l_n} \sum_{s_1=1}^{\infty} \sum_{s_2=1}^{\infty} \cdots \sum_{s_n=1}^{\infty} \sin(\lambda_{s_1} x_1) \sin(\lambda_{s_2} x_2) \ldots \sin(\lambda_{s_n} x_n)$$
$$\times \sin(\lambda_{s_1} y_1) \sin(\lambda_{s_2} y_2) \ldots \sin(\lambda_{s_n} y_n) \frac{\sin\left(t\sqrt{a^2(\lambda_{s_1}^2 + \cdots + \lambda_{s_n}^2) + b}\right)}{\sqrt{a^2(\lambda_{s_1}^2 + \cdots + \lambda_{s_n}^2) + b}},$$

where

$$\lambda_{s_1} = \frac{s_1 \pi}{l_1}, \quad \lambda_{s_2} = \frac{s_2 \pi}{l_2}, \quad \ldots, \quad \lambda_{s_n} = \frac{s_n \pi}{l_n}.$$

6.6.3-3. Domain: $V = \{0 \leq x_k \leq l_k; \ k = 1, \ldots, n\}$. **Second boundary value problem.**

The following conditions are prescribed:

$$\begin{aligned}
w &= f_0(\mathbf{x}) & \text{at} \quad & t = 0 & & \text{(initial condition)}, \\
\partial_t w &= f_1(\mathbf{x}) & \text{at} \quad & t = 0 & & \text{(initial condition)}, \\
\partial_{x_k} w &= g_k(\mathbf{x}, t) & \text{at} \quad & x_k = 0 & & \text{(boundary conditions)}, \\
\partial_{x_k} w &= h_k(\mathbf{x}, t) & \text{at} \quad & x_k = l_k & & \text{(boundary conditions)}.
\end{aligned}$$

Solution:

$$w(\mathbf{x}, t) = \int_0^t \int_V \Phi(\mathbf{y}, \tau) G(\mathbf{x}, \mathbf{y}, t - \tau)\, d\mathbf{y}\, d\tau$$
$$+ \int_V f_0(\mathbf{y}) G(\mathbf{x}, \mathbf{y}, t)\, d\mathbf{y} + \int_V f_1(\mathbf{y}) G(\mathbf{x}, \mathbf{y}, t)\, d\mathbf{y}$$
$$- a^2 \sum_{k=1}^n \int_0^t \int_{S^{(k)}} \left[g_k(\mathbf{y}, \tau) G(\mathbf{x}, \mathbf{y}, t - \tau)\right]_{y_k=0} dS_y^{(k)}\, d\tau$$
$$+ a^2 \sum_{k=1}^n \int_0^t \int_{S^{(k)}} \left[h_k(\mathbf{y}, \tau) G(\mathbf{x}, \mathbf{y}, t - \tau)\right]_{y_k=l_k} dS_y^{(k)}\, d\tau.$$

Here,

$$G(\mathbf{x}, \mathbf{y}, t) = \frac{1}{l_1 l_2 \ldots l_n} \sum_{s_1=0}^{\infty} \sum_{s_2=0}^{\infty} \cdots \sum_{s_n=0}^{\infty} A_{s_1} A_{s_2} \ldots A_{s_n} \cos(\lambda_{s_1} x_1) \cos(\lambda_{s_2} x_2) \ldots \cos(\lambda_{s_n} x_n)$$
$$\times \cos(\lambda_{s_1} y_1) \cos(\lambda_{s_2} y_2) \ldots \cos(\lambda_{s_n} y_n) \frac{\sin\left(t\sqrt{a^2(\lambda_{s_1}^2 + \cdots + \lambda_{s_n}^2) + b}\right)}{\sqrt{a^2(\lambda_{s_1}^2 + \cdots + \lambda_{s_n}^2) + b}},$$

where

$$\lambda_{s_1} = \frac{s_1 \pi}{l_1}, \quad \lambda_{s_2} = \frac{s_2 \pi}{l_2}, \quad \ldots, \quad \lambda_{s_n} = \frac{s_n \pi}{l_n}; \quad A_{s_m} = \begin{cases} 1 & \text{for } s_m = 0, \\ 2 & \text{for } s_m \neq 0, \end{cases} \quad m = 1, 2, \ldots, n.$$

6.6.3-4. Domain: $V = \{0 \leq x_k \leq l_k;\ k = 1, \ldots, n\}$. **Third boundary value problem.**

The following conditions are prescribed:

$$w = f_0(\mathbf{x}) \quad \text{at} \quad t = 0 \quad \text{(initial condition)},$$
$$\partial_t w = f_1(\mathbf{x}) \quad \text{at} \quad t = 0 \quad \text{(initial condition)},$$
$$\partial_{x_k} w - b_k w = g_k(\mathbf{x}, t) \quad \text{at} \quad x_k = 0 \quad \text{(boundary conditions)},$$
$$\partial_{x_k} w + c_k w = h_k(\mathbf{x}, t) \quad \text{at} \quad x_k = l_k \quad \text{(boundary conditions)}.$$

The solution $w(\mathbf{x}, t)$ is determined by the formula in Paragraph 6.6.3-3 where

$$G(\mathbf{x}, \mathbf{y}, t) = 2^n \sum_{s_1=1}^{\infty} \sum_{s_2=1}^{\infty} \cdots \sum_{s_n=1}^{\infty} \frac{\sin\!\left(t\sqrt{a^2(\lambda_{s_1}^2 + \lambda_{s_2}^2 + \cdots + \lambda_{s_n}^2) + b}\right)}{E_{s_1} E_{s_2} \ldots E_{s_n} \sqrt{a^2(\lambda_{s_1}^2 + \lambda_{s_2}^2 + \cdots + \lambda_{s_n}^2) + b}}$$
$$\times \sin(\lambda_{s_1} x_1 + \varphi_{s_1}) \sin(\lambda_{s_2} x_2 + \varphi_{s_2}) \ldots \sin(\lambda_{s_n} x_n + \varphi_{s_n})$$
$$\times \sin(\lambda_{s_1} y_1 + \varphi_{s_1}) \sin(\lambda_{s_2} y_2 + \varphi_{s_2}) \ldots \sin(\lambda_{s_n} y_n + \varphi_{s_n}).$$

Here,

$$\varphi_{s_m} = \arctan \frac{\lambda_{s_m}}{l_m}, \qquad E_{s_m} = l_m + \frac{(b_m c_m + \lambda_{s_m}^2)(b_m + c_m)}{(b_m^2 + \lambda_{s_m}^2)(c_m^2 + \lambda_{s_m}^2)}, \qquad m = 1, 2, \ldots, n;$$

the λ_{s_m} are positive roots of the transcendental equations

$$\frac{1}{b_m + c_m}\left(\lambda - \frac{b_m c_m}{\lambda}\right) = \cot(l_m \lambda), \qquad m = 1, 2, \ldots, n.$$

6.6.3-5. Domain: $V = \{0 \leq x_k \leq l_k;\ k = 1, \ldots, n\}$. **Mixed boundary value problem.**

The following conditions are prescribed:

$$w = f_0(\mathbf{x}) \quad \text{at} \quad t = 0 \quad \text{(initial condition)},$$
$$\partial_t w = f_1(\mathbf{x}) \quad \text{at} \quad t = 0 \quad \text{(initial condition)},$$
$$w = g_k(\mathbf{x}, t) \quad \text{at} \quad x_k = 0 \quad \text{(boundary conditions)},$$
$$\partial_{x_k} w = h_k(\mathbf{x}, t) \quad \text{at} \quad x_k = l_k \quad \text{(boundary conditions)}.$$

Solution:

$$w(\mathbf{x}, t) = \int_0^t \int_V \Phi(\mathbf{y}, \tau) G(\mathbf{x}, \mathbf{y}, t - \tau)\, d\mathbf{y}\, d\tau$$
$$+ \frac{\partial}{\partial t} \int_V f_0(\mathbf{y}) G(\mathbf{x}, \mathbf{y}, t)\, d\mathbf{y} + \int_V f_1(\mathbf{y}) G(\mathbf{x}, \mathbf{y}, t)\, d\mathbf{y}$$
$$+ a^2 \sum_{k=1}^{n} \int_0^t \int_{S^{(k)}} \left[g_k(\mathbf{y}, \tau) \frac{\partial}{\partial y_k} G(\mathbf{x}, \mathbf{y}, t - \tau) \right]_{y_k = 0} dS_y^{(k)}\, d\tau$$
$$+ a^2 \sum_{k=1}^{n} \int_0^t \int_{S^{(k)}} \left[h_k(\mathbf{y}, \tau) G(\mathbf{x}, \mathbf{y}, t - \tau) \right]_{y_k = l_k} dS_y^{(k)}\, d\tau.$$

Here,

$$G(\mathbf{x}, \mathbf{y}, t) = \frac{2^n}{l_1 l_2 \ldots l_n} \sum_{s_1=1}^{\infty} \sum_{s_2=1}^{\infty} \cdots \sum_{s_n=1}^{\infty} \sin(\lambda_{s_1} x_1) \sin(\lambda_{s_2} x_2) \ldots \sin(\lambda_{s_n} x_n)$$
$$\times \sin(\lambda_{s_1} y_1) \sin(\lambda_{s_2} y_2) \ldots \sin(\lambda_{s_n} y_n) \frac{\sin\!\left(t\sqrt{a^2(\lambda_{s_1}^2 + \cdots + \lambda_{s_n}^2) + b}\right)}{\sqrt{a^2(\lambda_{s_1}^2 + \cdots + \lambda_{s_n}^2) + b}},$$

where

$$\lambda_{s_1} = \frac{\pi(2s_1 + 1)}{2 l_1}, \qquad \lambda_{s_2} = \frac{\pi(2s_2 + 1)}{2 l_2}, \qquad \ldots, \qquad \lambda_{s_n} = \frac{\pi(2s_n + 1)}{2 l_n}.$$

6.6.4. Equations Containing the First Time Derivative

1. $\dfrac{\partial^2 w}{\partial t^2} + \beta \dfrac{\partial w}{\partial t} = a^2 \Delta_n w - bw + \Phi(x_1, \ldots, x_n, t).$

Nonhomogeneous telegraph equation with n space variables.

1°. The substitution $w = \exp\left(-\tfrac{1}{2}\beta t\right) u$ leads to the equation

$$\frac{\partial^2 u}{\partial t^2} = a^2 \Delta_n u - \left(b - \tfrac{1}{4}\beta^2\right) u + \exp\left(\tfrac{1}{2}\beta t\right) \Phi(x_1, \ldots, x_n, t),$$

which is considered in Subsection 6.6.3.

2°. Domain: $V = \{0 \le x_k \le l_k;\ k = 1, \ldots, n\}$. First boundary value problem. The following conditions are prescribed:

$$\begin{aligned}
w &= f_0(\mathbf{x}) & \text{at} \quad t &= 0 & \text{(initial condition)}, \\
\partial_t w &= f_1(\mathbf{x}) & \text{at} \quad t &= 0 & \text{(initial condition)}, \\
w &= g_k(\mathbf{x}, t) & \text{at} \quad x_k &= 0 & \text{(boundary conditions)}, \\
w &= h_k(\mathbf{x}, t) & \text{at} \quad x_k &= l_k & \text{(boundary conditions)}.
\end{aligned}$$

Solution:

$$\begin{aligned}
w(\mathbf{x}, t) &= \int_0^t \int_V \Phi(\mathbf{y}, \tau) G(\mathbf{x}, \mathbf{y}, t - \tau)\, d\mathbf{y}\, d\tau \\
&\quad + \frac{\partial}{\partial t} \int_V f_0(\mathbf{y}) G(\mathbf{x}, \mathbf{y}, t)\, d\mathbf{y} + \int_V \left[f_1(\mathbf{y}) + \beta f_0(\mathbf{y})\right] G(\mathbf{x}, \mathbf{y}, t)\, d\mathbf{y} \\
&\quad + a^2 \sum_{k=1}^n \int_0^t \int_{S^{(k)}} \left[g_k(\mathbf{y}, \tau) \frac{\partial}{\partial y_k} G(\mathbf{x}, \mathbf{y}, t - \tau)\right]_{y_k = 0} dS_y^{(k)}\, d\tau \\
&\quad - a^2 \sum_{k=1}^n \int_0^t \int_{S^{(k)}} \left[h_k(\mathbf{y}, \tau) \frac{\partial}{\partial y_k} G(\mathbf{x}, \mathbf{y}, t - \tau)\right]_{y_k = l_k} dS_y^{(k)}\, d\tau,
\end{aligned}$$

where

$\mathbf{x} = \{x_1, \ldots, x_n\}$, $\mathbf{y} = \{y_1, \ldots, y_n\}$, $d\mathbf{y} = dy_1\, dy_2 \ldots dy_n$, $dS_y^{(k)} = dy_1 \ldots dy_{k-1}\, dy_{k+1} \ldots dy_n$,

$S^{(k)} = \{0 \le y_m \le l_m \text{ for } m = 1, \ldots, k-1, k+1, \ldots, n\}$.

Green's function:

$$G(\mathbf{x}, \mathbf{y}, t) = \frac{2^n e^{-\beta t/2}}{l_1 l_2 \ldots l_n} \sum_{s_1=1}^{\infty} \sum_{s_2=1}^{\infty} \ldots \sum_{s_n=1}^{\infty} \sin(\lambda_{s_1} x_1) \sin(\lambda_{s_2} x_2) \ldots \sin(\lambda_{s_n} x_n)$$

$$\times \sin(\lambda_{s_1} y_1) \sin(\lambda_{s_2} y_2) \ldots \sin(\lambda_{s_n} y_n) \frac{\sin\!\left(t\sqrt{a^2(\lambda_{s_1}^2 + \cdots + \lambda_{s_n}^2) + b - \beta^2/4}\right)}{\sqrt{a^2(\lambda_{s_1}^2 + \cdots + \lambda_{s_n}^2) + b - \beta^2/4}},$$

where

$$\lambda_{s_1} = \frac{s_1 \pi}{l_1}, \quad \lambda_{s_2} = \frac{s_2 \pi}{l_2}, \quad \ldots, \quad \lambda_{s_n} = \frac{s_n \pi}{l_n}.$$

3°. Domain: $V = \{0 \le x_k \le l_k;\ k = 1, \ldots, n\}$. Second boundary value problem. The following conditions are prescribed:

$$\begin{aligned}
w &= f_0(\mathbf{x}) & \text{at} \quad t &= 0 & \text{(initial condition)}, \\
\partial_t w &= f_1(\mathbf{x}) & \text{at} \quad t &= 0 & \text{(initial condition)}, \\
\partial_{x_k} w &= g_k(\mathbf{x}, t) & \text{at} \quad x_k &= 0 & \text{(boundary conditions)}, \\
\partial_{x_k} w &= h_k(\mathbf{x}, t) & \text{at} \quad x_k &= l_k & \text{(boundary conditions)}.
\end{aligned}$$

Solution:

$$w(\mathbf{x},t) = \int_0^t \int_V \Phi(\mathbf{y},\tau) G(\mathbf{x},\mathbf{y},t-\tau)\,d\mathbf{y}\,d\tau$$
$$+ \int_V f_0(\mathbf{y}) G(\mathbf{x},\mathbf{y},t)\,d\mathbf{y} + \int_V \left[f_1(\mathbf{y}) + \beta f_0(\mathbf{y})\right] G(\mathbf{x},\mathbf{y},t)\,d\mathbf{y}$$
$$- a^2 \sum_{k=1}^n \int_0^t \int_{S^{(k)}} \left[g_k(\mathbf{y},\tau) G(\mathbf{x},\mathbf{y},t-\tau)\right]_{y_k=0} dS_y^{(k)}\,d\tau$$
$$+ a^2 \sum_{k=1}^n \int_0^t \int_{S^{(k)}} \left[h_k(\mathbf{y},\tau) G(\mathbf{x},\mathbf{y},t-\tau)\right]_{y_k=l_k} dS_y^{(k)}\,d\tau.$$

Here,

$$G(\mathbf{x},\mathbf{y},t) = \frac{e^{-\beta t/2}}{l_1 l_2 \ldots l_n} \sum_{s_1=0}^{\infty} \sum_{s_2=0}^{\infty} \ldots \sum_{s_n=0}^{\infty} A_{s_1} A_{s_2} \ldots A_{s_n} \cos(\lambda_{s_1} x_1) \cos(\lambda_{s_2} x_2) \ldots \cos(\lambda_{s_n} x_n)$$
$$\times \cos(\lambda_{s_1} y_1) \cos(\lambda_{s_2} y_2) \ldots \cos(\lambda_{s_n} y_n) \frac{\sin\left(t\sqrt{a^2(\lambda_{s_1}^2 + \cdots + \lambda_{s_n}^2) + b - \beta^2/4}\right)}{\sqrt{a^2(\lambda_{s_1}^2 + \cdots + \lambda_{s_n}^2) + b - \beta^2/4}},$$

where

$$\lambda_{s_1} = \frac{s_1 \pi}{l_1}, \quad \lambda_{s_2} = \frac{s_2 \pi}{l_2}, \quad \ldots, \quad \lambda_{s_n} = \frac{s_n \pi}{l_n}; \quad A_{s_m} = \begin{cases} 1 & \text{for } s_m = 0, \\ 2 & \text{for } s_m \neq 0, \end{cases} \quad m = 1, 2, \ldots, n.$$

4°. Domain: $V = \{0 \leq x_k \leq l_k; \; k = 1, \ldots, n\}$. Third boundary value problem.
The following conditions are prescribed:

$$\begin{aligned}
w &= f_0(\mathbf{x}) & \text{at} \quad t &= 0 & \text{(initial condition)},\\
\partial_t w &= f_1(\mathbf{x}) & \text{at} \quad t &= 0 & \text{(initial condition)},\\
\partial_{x_k} w - b_k w &= g_k(\mathbf{x},t) & \text{at} \quad x_k &= 0 & \text{(boundary conditions)},\\
\partial_{x_k} w + c_k w &= h_k(\mathbf{x},t) & \text{at} \quad x_k &= l_k & \text{(boundary conditions)}.
\end{aligned}$$

The solution $w(\mathbf{x},t)$ is given by the formula in Item 3° with

$$G(\mathbf{x},\mathbf{y},t) = 2^n e^{-\beta t/2} \sum_{s_1=1}^{\infty} \sum_{s_2=1}^{\infty} \ldots \sum_{s_n=1}^{\infty} \frac{\sin\left(t\sqrt{a^2(\lambda_{s_1}^2 + \lambda_{s_2}^2 + \cdots + \lambda_{s_n}^2) + b - \beta^2/4}\right)}{E_{s_1} E_{s_2} \ldots E_{s_n} \sqrt{a^2(\lambda_{s_1}^2 + \lambda_{s_2}^2 + \cdots + \lambda_{s_n}^2) + b - \beta^2/4}}$$
$$\times \sin(\lambda_{s_1} x_1 + \varphi_{s_1}) \sin(\lambda_{s_2} x_2 + \varphi_{s_2}) \ldots \sin(\lambda_{s_n} x_n + \varphi_{s_n})$$
$$\times \sin(\lambda_{s_1} y_1 + \varphi_{s_1}) \sin(\lambda_{s_2} y_2 + \varphi_{s_2}) \ldots \sin(\lambda_{s_n} y_n + \varphi_{s_n}).$$

Here,

$$\varphi_{s_m} = \arctan \frac{\lambda_{s_m}}{l_m}, \quad E_{s_m} = l_m + \frac{(b_m c_m + \lambda_{s_m}^2)(b_m + c_m)}{(b_m^2 + \lambda_{s_m}^2)(c_m^2 + \lambda_{s_m}^2)}, \quad m = 1, 2, \ldots, n;$$

the λ_{s_m} are positive roots of the transcendental equation

$$\frac{1}{b_m + c_m}\left(\lambda - \frac{b_m c_m}{\lambda}\right) = \cot(l_m \lambda), \quad m = 1, 2, \ldots, n.$$

5°. Domain: $V = \{0 \leq x_k \leq l_k;\ k = 1, \ldots, n\}$. Mixed boundary value problem. The following conditions are prescribed:

$$w = f_0(\mathbf{x}) \quad \text{at} \quad t = 0 \quad \text{(initial condition)},$$
$$\partial_t w = f_1(\mathbf{x}) \quad \text{at} \quad t = 0 \quad \text{(initial condition)},$$
$$w = g_k(\mathbf{x}, t) \quad \text{at} \quad x_k = 0 \quad \text{(boundary conditions)},$$
$$\partial_{x_k} w = h_k(\mathbf{x}, t) \quad \text{at} \quad x_k = l_k \quad \text{(boundary conditions)}.$$

Solution:

$$w(\mathbf{x}, t) = \int_0^t \int_V \Phi(\mathbf{y}, \tau) G(\mathbf{x}, \mathbf{y}, t - \tau)\, d\mathbf{y}\, d\tau$$
$$+ \frac{\partial}{\partial t} \int_V f_0(\mathbf{y}) G(\mathbf{x}, \mathbf{y}, t)\, d\mathbf{y} + \int_V [f_1(\mathbf{y}) + \beta f_0(\mathbf{y})] G(\mathbf{x}, \mathbf{y}, t)\, d\mathbf{y}$$
$$+ a^2 \sum_{k=1}^n \int_0^t \int_{S^{(k)}} \left[g_k(\mathbf{y}, \tau) \frac{\partial}{\partial y_k} G(\mathbf{x}, \mathbf{y}, t - \tau)\right]_{y_k = 0} dS_y^{(k)}\, d\tau$$
$$+ a^2 \sum_{k=1}^n \int_0^t \int_{S^{(k)}} \left[h_k(\mathbf{y}, \tau) G(\mathbf{x}, \mathbf{y}, t - \tau)\right]_{y_k = l_k} dS_y^{(k)}\, d\tau,$$

where

$$G(\mathbf{x}, \mathbf{y}, t) = \frac{2^n e^{-\beta t/2}}{l_1 l_2 \ldots l_n} \sum_{s_1=1}^\infty \sum_{s_2=1}^\infty \ldots \sum_{s_n=1}^\infty \sin(\lambda_{s_1} x_1) \sin(\lambda_{s_2} x_2) \ldots \sin(\lambda_{s_n} x_n)$$
$$\times \sin(\lambda_{s_1} y_1) \sin(\lambda_{s_2} y_2) \ldots \sin(\lambda_{s_n} y_n) \frac{\sin\left(t\sqrt{a^2(\lambda_{s_1}^2 + \cdots + \lambda_{s_n}^2) + b - \beta^2/4}\right)}{\sqrt{a^2(\lambda_{s_1}^2 + \cdots + \lambda_{s_n}^2) + b - \beta^2/4}},$$

$$\lambda_{s_1} = \frac{\pi(2s_1 + 1)}{2l_1}, \quad \lambda_{s_2} = \frac{\pi(2s_2 + 1)}{2l_2}, \quad \ldots, \quad \lambda_{s_n} = \frac{\pi(2s_n + 1)}{2l_n}.$$

2. $\dfrac{\partial^2 w}{\partial t^2} + \beta \dfrac{\partial w}{\partial t} = a^2 \Delta_n w + \sum\limits_{k=1}^n b_k \dfrac{\partial w}{\partial x_k} + cw.$

The transformation

$$w(x_1, \ldots, x_n, t) = u(x_1, \ldots, x_n, \tau) \exp\left(-\frac{1}{2}\beta t - \frac{1}{2a^2} \sum_{k=1}^n b_k x_k\right), \quad \tau = at$$

leads to the equation

$$\frac{\partial^2 u}{\partial \tau^2} = \Delta_n u + \lambda u, \quad \lambda = \frac{c}{a^2} + \frac{\beta^2}{4a^2} - \frac{1}{4a^4} \sum_{k=1}^n b_k^2,$$

which is considered in Subsection 6.6.3.

⊙ *Reference*: R. Courant and D. Hilbert (1989).

3. $\dfrac{\partial^2 w}{\partial t^2} + \dfrac{n-1}{t} \dfrac{\partial w}{\partial t} = \Delta_n w.$

Darboux equation. Cauchy problem.
Initial conditions are prescribed:

$$w = f(\mathbf{x}) \quad \text{at} \quad t = 0,$$
$$\partial_t w = 0 \quad \text{at} \quad t = 0.$$

Solution:
$$w(\mathbf{x}, t) = \frac{1}{\sigma_n t^{n-1}} \int_{|\mathbf{x}-\mathbf{y}|=t} f(\mathbf{y}) \, dS_y, \qquad \sigma_n = \frac{2\pi^{n/2}}{\Gamma(n/2)},$$

where $\sigma_n t^{n-1}$ is the area of the surface of an n-dimensional sphere of radius t, and dS_y is the area element of this surface (i.e., the solution w is the average of the function f over the sphere for radius t with center at $\mathbf{x}$).

⊙ *Reference*: R. Courant and D. Hilbert (1989).

Chapter 7

Elliptic Equations with Two Space Variables

7.1. Laplace Equation $\Delta_2 w = 0$

The Laplace equation is often encountered in heat and mass transfer theory, fluid mechanics, elasticity, electrostatics, and other areas of mechanics and physics. For example, in heat and mass transfer theory, this equation describes steady-state temperature distribution in the absence of heat sources and sinks in the domain under study.

A regular solution of the Laplace equation is called a harmonic function. The first boundary value problem for the Laplace equation is often referred to as the Dirichlet problem, and the second boundary value problem as the Neumann problem.

Extremum principle: Given a domain D, a harmonic function w in D that is not identically constant in D cannot attain its maximum or minimum value at any interior point of D.

7.1.1. Problems in Cartesian Coordinate System

The Laplace equation with two space variables in the rectangular Cartesian system of coordinates is written as

$$\frac{\partial^2 w}{\partial x^2} + \frac{\partial^2 w}{\partial y^2} = 0.$$

7.1.1-1. Particular solutions and a method for their construction.

1°. Particular solutions:

$$w(x,y) = Ax + By + C,$$
$$w(x,y) = A(x^2 - y^2) + Bxy,$$
$$w(x,y) = A(x^3 - 3xy^2) + B(3x^2 y - y^3),$$
$$w(x,y) = \frac{Ax + By}{x^2 + y^2} + C,$$
$$w(x,y) = \exp(\pm \mu x)(A \cos \mu y + B \sin \mu y),$$
$$w(x,y) = (A \cos \mu x + B \sin \mu x) \exp(\pm \mu y),$$
$$w(x,y) = (A \sinh \mu x + B \cosh \mu x)(C \cos \mu y + D \sin \mu y),$$
$$w(x,y) = (A \cos \mu x + B \sin \mu x)(C \sinh \mu y + D \cosh \mu y),$$
$$w(x,y) = A \ln\left[(x-x_0)^2 + (y-y_0)^2\right] + B,$$

where A, B, C, D, x_0, y_0, and μ are arbitrary constants.

2°. Fundamental solution:

$$\mathcal{E}(x,y) = \frac{1}{2\pi} \ln \frac{1}{r}, \qquad r = \sqrt{x^2 + y^2}.$$

3°. If $w(x, y)$ is a solution of the Laplace equation, then the functions

$$w_1 = Aw(\pm\lambda x + C_1, \pm\lambda y + C_2),$$
$$w_2 = Aw(x\cos\beta + y\sin\beta, -x\sin\beta + y\cos\beta),$$
$$w_3 = Aw\left(\frac{x}{x^2+y^2}, \frac{y}{x^2+y^2}\right),$$

are also solutions everywhere they are defined; A, C_1, C_2, β, and λ are arbitrary constants. The signs at λ in w_1 are taken independently of each other.

4°. A fairly general method for constructing particular solutions involves the following. Let $f(z) = u(x, y) + iv(x, y)$ be any analytic function of the complex variable $z = x + iy$ (u and v are real functions of the real variables x and y; $i^2 = -1$). Then the real and imaginary parts of f both satisfy the two-dimensional Laplace equation,

$$\Delta_2 u = 0, \qquad \Delta_2 v = 0.$$

Recall that the Cauchy–Riemann conditions

$$\frac{\partial u}{\partial x} = \frac{\partial v}{\partial y}, \qquad \frac{\partial u}{\partial y} = -\frac{\partial v}{\partial x}$$

are necessary and sufficient conditions for the function f to be analytic. Thus, by specifying analytic functions $f(z)$ and taking their real and imaginary parts, one obtains various solutions of the two-dimensional Laplace equation.

⊙ *References*: M. A. Lavrent'ev and B. V. Shabat (1973), A. G. Sveshnikov and A. N. Tikhonov (1974), A. V. Bitsadze and D. F. Kalinichenko (1985).

7.1.1-2. Specific features of stating boundary value problems for the Laplace equation.

1°. For outer boundary value problems on the plane, it is (usually) required to set the additional condition that the solution of the Laplace equation must be bounded at infinity.

2°. The solution of the second boundary value problem is determined up to an arbitrary additive term.

3°. Let the second boundary value problem in a closed bounded domain D with piecewise smooth boundary Σ be characterized by the boundary condition*

$$\frac{\partial w}{\partial N} = f(\mathbf{r}) \quad \text{for} \quad \mathbf{r} \in \Sigma,$$

where $\frac{\partial w}{\partial N}$ is the derivative along the (outward) normal to Σ. The necessary and sufficient condition of solvability of the problem has the form

$$\int_\Sigma f(\mathbf{r})\, d\Sigma = 0.$$

Remark. The same solvability condition occurs for the outer second boundary value problem if the domain is infinite but has a finite boundary.

⊙ *Reference*: V. M. Babich, M. B. Kapilevich, S. G. Mikhlin, et al. (1964).

---

* More rigorously, Σ must satisfy the Lyapunov condition [see Babich, Kapilevich, Mikhlin, et al. (1964) and Tikhonov and Samarskii (1990)].

7.1.1-3. Domain: $-\infty < x < \infty$, $0 \leq y < \infty$. **First boundary value problem.**

A half-plane is considered. A boundary condition is prescribed:
$$w = f(x) \quad \text{at} \quad y = 0.$$

Solution:
$$w(x, y) = \frac{1}{\pi} \int_{-\infty}^{\infty} \frac{y f(\xi)\, d\xi}{(x - \xi)^2 + y^2} = \frac{1}{\pi} \int_{-\pi/2}^{\pi/2} f(x + y \tan \theta)\, d\theta.$$

⊙ *References*: V. M. Babich, M. B. Kapilevich, S. G. Mikhlin, et al. (1964), H. S. Carslaw and J. C. Jaeger (1984).

7.1.1-4. Domain: $-\infty < x < \infty$, $0 \leq y < \infty$. **Second boundary value problem.**

A half-plane is considered. A boundary condition is prescribed:
$$\partial_y w = f(x) \quad \text{at} \quad y = 0.$$

Solution:
$$w(x, y) = \frac{1}{\pi} \int_{-\infty}^{\infty} f(\xi) \ln \sqrt{(x - \xi)^2 + y^2}\, d\xi + C,$$

where C is an arbitrary constant.

⊙ *Reference*: V. S. Vladimirov (1988).

7.1.1-5. Domain: $0 \leq x < \infty$, $0 \leq y < \infty$. **First boundary value problem.**

A quadrant of the plane is considered. Boundary conditions are prescribed:
$$w = f_1(y) \quad \text{at} \quad x = 0, \qquad w = f_2(x) \quad \text{at} \quad y = 0.$$

Solution:
$$w(x, y) = \frac{4}{\pi} xy \int_0^{\infty} \frac{f_1(\eta)\eta\, d\eta}{[x^2 + (y - \eta)^2][x^2 + (y + \eta)^2]} + \frac{4}{\pi} xy \int_0^{\infty} \frac{f_2(\xi)\xi\, d\xi}{[(x - \xi)^2 + y^2][(x + \xi)^2 + y^2]}.$$

⊙ *Reference*: V. S. Vladimirov, V. P. Mikhailov, A. A. Vasharin, et al. (1974).

7.1.1-6. Domain: $-\infty < x < \infty$, $0 \leq y \leq a$. **First boundary value problem.**

An infinite strip is considered. Boundary conditions are prescribed:
$$w = f_1(x) \quad \text{at} \quad y = 0, \qquad w = f_2(x) \quad \text{at} \quad y = a.$$

Solution:
$$w(x, y) = \frac{1}{2a} \sin\left(\frac{\pi y}{a}\right) \int_{-\infty}^{\infty} \frac{f_1(\xi)\, d\xi}{\cosh[\pi(x - \xi)/a] - \cos(\pi y/a)}$$
$$+ \frac{1}{2a} \sin\left(\frac{\pi y}{a}\right) \int_{-\infty}^{\infty} \frac{f_2(\xi)\, d\xi}{\cosh[\pi(x - \xi)/a] + \cos(\pi y/a)}.$$

⊙ *Reference*: H. S. Carslaw and J. C. Jaeger (1984).

7.1.1-7. Domain: $-\infty < x < \infty$, $0 \leq y \leq a$. **Second boundary value problem.**

An infinite strip is considered. Boundary conditions are prescribed:
$$\partial_y w = f_1(x) \quad \text{at} \quad y = 0, \qquad \partial_y w = f_2(x) \quad \text{at} \quad y = a.$$

Solution:
$$w(x, y) = \frac{1}{2\pi} \int_{-\infty}^{\infty} f_1(\xi) \ln\{\cosh[\pi(x - \xi)/a] - \cos(\pi y/a)\}\, d\xi$$
$$- \frac{1}{2\pi} \int_{-\infty}^{\infty} f_2(\xi) \ln\{\cosh[\pi(x - \xi)/a] + \cos(\pi y/a)\}\, d\xi + C,$$

where C is an arbitrary constant.

7.1.1-8. Domain: $0 \le x < \infty$, $0 \le y \le a$. **First boundary value problem.**

A semiinfinite strip is considered. Boundary conditions are prescribed:

$$w = f_1(y) \text{ at } x = 0, \qquad w = f_2(x) \text{ at } y = 0, \qquad w = f_3(x) \text{ at } y = a.$$

Solution:

$$w(x,y) = \frac{2}{a} \sum_{n=1}^{\infty} \exp\left(-\frac{n\pi x}{a}\right) \sin\left(\frac{n\pi y}{a}\right) \int_0^a f_1(\eta) \sin\left(\frac{n\pi \eta}{a}\right) d\eta$$

$$+ \frac{1}{2a} \sin\left(\frac{\pi y}{a}\right) \int_0^\infty \left\{ \frac{1}{\cosh[\pi(x-\xi)/a]-\cos(\pi y/a)} - \frac{1}{\cosh[\pi(x+\xi)/a]-\cos(\pi y/a)} \right\} f_2(\xi)\, d\xi$$

$$+ \frac{1}{2a} \sin\left(\frac{\pi y}{a}\right) \int_0^\infty \left\{ \frac{1}{\cosh[\pi(x-\xi)/a]+\cos(\pi y/a)} - \frac{1}{\cosh[\pi(x+\xi)/a]+\cos(\pi y/a)} \right\} f_3(\xi)\, d\xi.$$

Example. Consider the first boundary value problem for the Laplace equation in a semiinfinite strip with $f_1(y) = 1$ and $f_2(x) = f_3(x) = 0$.

Using the general formula and carrying out transformations, we obtain the solution

$$w(x,y) = \frac{2}{\pi} \arctan\left[\frac{\sin(\pi y/a)}{\sinh(\pi x/a)}\right].$$

⊙ *Reference*: H. S. Carslaw and J. C. Jaeger (1984).

7.1.1-9. Domain: $0 \le x \le a$, $0 \le y \le b$. **First boundary value problem.**

A rectangle is considered. Boundary conditions are prescribed:

$$w = f_1(y) \text{ at } x = 0, \qquad w = f_2(y) \text{ at } x = a,$$
$$w = f_3(x) \text{ at } y = 0, \qquad w = f_4(x) \text{ at } y = b.$$

Solution:

$$w(x,y) = \sum_{n=1}^{\infty} A_n \sinh\left[\frac{n\pi}{b}(a-x)\right] \sin\left(\frac{n\pi}{b}y\right) + \sum_{n=1}^{\infty} B_n \sinh\left(\frac{n\pi}{b}x\right) \sin\left(\frac{n\pi}{b}y\right)$$

$$+ \sum_{n=1}^{\infty} C_n \sin\left(\frac{n\pi}{a}x\right) \sinh\left[\frac{n\pi}{a}(b-y)\right] + \sum_{n=1}^{\infty} D_n \sin\left(\frac{n\pi}{a}x\right) \sinh\left(\frac{n\pi}{a}y\right),$$

where the coefficients A_n, B_n, C_n, and D_n are expressed as

$$A_n = \frac{2}{\lambda_n} \int_0^b f_1(\xi) \sin\left(\frac{n\pi\xi}{b}\right) d\xi, \qquad B_n = \frac{2}{\lambda_n} \int_0^b f_2(\xi) \sin\left(\frac{n\pi\xi}{b}\right) d\xi,$$

$$C_n = \frac{2}{\mu_n} \int_0^a f_3(\xi) \sin\left(\frac{n\pi\xi}{a}\right) d\xi, \qquad D_n = \frac{2}{\mu_n} \int_0^a f_4(\xi) \sin\left(\frac{n\pi\xi}{a}\right) d\xi,$$

$$\lambda_n = b \sinh\left(\frac{n\pi a}{b}\right), \qquad \mu_n = a \sinh\left(\frac{n\pi b}{a}\right).$$

⊙ *References*: M. M. Smirnov (1975), H. S. Carslaw and J. C. Jaeger (1984).

7.1.1-10. Domain: $0 \leq x \leq a$, $0 \leq y \leq b$. Second boundary value problem.

A rectangle is considered. Boundary conditions are prescribed:

$$\partial_x w = f_1(y) \quad \text{at} \quad x = 0, \qquad \partial_x w = f_2(y) \quad \text{at} \quad x = a,$$
$$\partial_y w = f_3(x) \quad \text{at} \quad y = 0, \qquad \partial_y w = f_4(x) \quad \text{at} \quad y = b.$$

Solution:

$$w(x,y) = -\frac{A_0}{4a}(x-a)^2 + \frac{B_0}{4a}x^2 - \frac{C_0}{4b}(x-b)^2 + \frac{D_0}{4b}y^2 + K$$
$$- b\sum_{n=1}^{\infty} \frac{A_n}{\lambda_n} \cosh\left[\frac{n\pi}{b}(a-x)\right] \cos\left(\frac{n\pi}{b}y\right) + b\sum_{n=1}^{\infty} \frac{B_n}{\lambda_n} \cosh\left(\frac{n\pi}{b}x\right) \cos\left(\frac{n\pi}{b}y\right)$$
$$- a\sum_{n=1}^{\infty} \frac{C_n}{\mu_n} \cos\left(\frac{n\pi}{a}x\right) \cosh\left[\frac{n\pi}{a}(b-y)\right] + a\sum_{n=1}^{\infty} \frac{D_n}{\mu_n} \cos\left(\frac{n\pi}{a}x\right) \cosh\left(\frac{n\pi}{a}y\right),$$

where K is an arbitrary constant, and the coefficients A_n, B_n, C_n, D_n, λ_n, and μ_n are expressed as

$$A_n = \frac{2}{b}\int_0^b f_1(\xi) \cos\left(\frac{n\pi\xi}{b}\right) d\xi, \qquad B_n = \frac{2}{b}\int_0^b f_2(\xi) \cos\left(\frac{n\pi\xi}{b}\right) d\xi,$$
$$C_n = \frac{2}{a}\int_0^a f_3(\xi) \cos\left(\frac{n\pi\xi}{a}\right) d\xi, \qquad D_n = \frac{2}{a}\int_0^a f_4(\xi) \cos\left(\frac{n\pi\xi}{a}\right) d\xi,$$
$$\lambda_n = n\pi \sinh\left(\frac{n\pi a}{b}\right), \qquad \mu_n = n\pi \sinh\left(\frac{n\pi b}{a}\right).$$

The solvability condition for the problem in question has the form (see Paragraph 7.1.1-2, Item 3°)

$$\int_0^b f_1(y)\,dy + \int_0^b f_2(y)\,dy - \int_0^a f_3(x)\,dx - \int_0^a f_4(x)\,dx = 0.$$

7.1.1-11. Domain: $0 \leq x \leq a$, $0 \leq y \leq b$. Third boundary value problem.

A rectangle is considered. Boundary conditions are prescribed:

$$\partial_x w - k_1 w = f_1(y) \quad \text{at} \quad x = 0, \qquad \partial_x w + k_2 w = f_2(y) \quad \text{at} \quad x = a,$$
$$\partial_y w - k_3 w = f_3(x) \quad \text{at} \quad y = 0, \qquad \partial_y w + k_4 w = f_4(x) \quad \text{at} \quad y = b.$$

For the solution, see Paragraph 7.2.2-14 with $\Phi \equiv 0$.

7.1.1-12. Domain: $0 \leq x \leq a$, $0 \leq y \leq b$. Mixed boundary value problems.

1°. A rectangle is considered. Boundary conditions are prescribed:

$$\partial_x w = f(y) \quad \text{at} \quad x = 0, \qquad \partial_x w = g(y) \quad \text{at} \quad x = a,$$
$$w = h(x) \quad \text{at} \quad y = 0, \qquad w = s(x) \quad \text{at} \quad y = b.$$

Solution:

$$w(x,y) = -\frac{b}{\pi}\sum_{n=1}^{\infty} \frac{f_n}{n\lambda_n} \cosh\left[\frac{\pi n}{b}(a-x)\right] \sin\left(\frac{\pi n y}{b}\right) + \frac{b}{\pi}\sum_{n=1}^{\infty} \frac{g_n}{n\lambda_n} \cosh\left(\frac{\pi n x}{b}\right) \sin\left(\frac{\pi n y}{b}\right)$$
$$+ \sum_{n=1}^{\infty} \frac{h_n}{\mu_n} \cos\left(\frac{\pi n x}{a}\right) \sinh\left[\frac{\pi n}{a}(b-y)\right] + \sum_{n=1}^{\infty} \frac{s_n}{\mu_n} \cos\left(\frac{\pi n x}{a}\right) \sinh\left(\frac{\pi n y}{a}\right)$$
$$+ \frac{b-y}{ab}\int_0^a h(x)\,dx + \frac{y}{ab}\int_0^a s(x)\,dx,$$

where

$$f_n = \frac{2}{b}\int_0^b f(\xi)\sin\left(\frac{\pi n\xi}{b}\right)d\xi, \quad g_n = \frac{2}{b}\int_0^b g(\xi)\sin\left(\frac{\pi n\xi}{b}\right)d\xi,$$

$$h_n = \frac{2}{a}\int_0^a h(\xi)\cos\left(\frac{\pi n\xi}{a}\right)d\xi, \quad s_n = \frac{2}{a}\int_0^a s(\xi)\cos\left(\frac{\pi n\xi}{a}\right)d\xi,$$

$$\lambda_n = \sinh\left(\frac{\pi na}{b}\right), \quad \mu_n = \sinh\left(\frac{\pi nb}{a}\right),$$

⊙ *Reference*: M. M. Smirnov (1975).

2°. A rectangle is considered. Boundary conditions are prescribed:

$$w = f(y) \quad \text{at} \quad x = 0, \qquad \partial_x w = g(y) \quad \text{at} \quad x = a,$$
$$w = h(x) \quad \text{at} \quad y = 0, \qquad \partial_y w = s(x) \quad \text{at} \quad y = b,$$

where $f(0) = h(0)$.

Solution:

$$w(x,y) = \sum_{n=0}^\infty \frac{f_n}{\cosh \lambda_n}\cosh\left(\lambda_n\frac{a-x}{a}\right)\sin\left(\lambda_n\frac{y}{a}\right) + a\sum_{n=0}^\infty \frac{g_n}{\lambda_n\cosh\lambda_n}\sinh\left(\lambda_n\frac{x}{a}\right)\sin\left(\lambda_n\frac{y}{a}\right)$$
$$+ \sum_{n=0}^\infty \frac{h_n}{\cosh\mu_n}\sin\left(\mu_n\frac{x}{b}\right)\cosh\left(\mu_n\frac{b-y}{b}\right) + b\sum_{n=0}^\infty \frac{s_n}{\mu_n\cosh\mu_n}\sin\left(\mu_n\frac{x}{b}\right)\sinh\left(\mu_n\frac{y}{b}\right),$$

where

$$f_n = \frac{2}{b}\int_0^b f(\xi)\sin\left[\frac{\pi(2n+1)}{b}\xi\right]d\xi, \quad g_n = \frac{2}{b}\int_0^b g(\xi)\sin\left[\frac{\pi(2n+1)}{b}\xi\right]d\xi,$$

$$h_n = \frac{2}{a}\int_0^a h(\xi)\sin\left[\frac{\pi(2n+1)}{a}\xi\right]d\xi, \quad s_n = \frac{2}{a}\int_0^a s(\xi)\sin\left[\frac{\pi(2n+1)}{a}\xi\right]d\xi,$$

$$\lambda_n = \frac{\pi(2n+1)a}{2b}, \quad \mu_n = \frac{\pi(2n+1)b}{2a}.$$

⊙ *Reference*: M. M. Smirnov (1975).

7.1.2. Problems in Polar Coordinate System

The two-dimensional Laplace equation in the polar coordinate system is written as

$$\frac{1}{r}\frac{\partial}{\partial r}\left(r\frac{\partial w}{\partial r}\right) + \frac{1}{r^2}\frac{\partial^2 w}{\partial \varphi^2} = 0, \quad r = \sqrt{x^2+y^2}.$$

7.1.2-1. Particular solutions:

$$w(r) = A\ln r + B,$$
$$w(r,\varphi) = \left(Ar^m + \frac{B}{r^m}\right)(C\cos m\varphi + D\sin m\varphi),$$

where $m = 1, 2, \ldots$; A, B, C, and D are arbitrary constants.

7.1.2-2. Domain: $0 \leq r \leq R$ or $R \leq r < \infty$. First boundary value problem.

The condition
$$w = f(\varphi) \quad \text{at} \quad r = R$$
is set at the boundary of the circle; $f(\varphi)$ is a given function.

1°. Solution of the inner problem ($r \leq R$):
$$w(r, \varphi) = \frac{1}{2\pi} \int_0^{2\pi} f(\psi) \frac{R^2 - r^2}{r^2 - 2Rr \cos(\varphi - \psi) + R^2} \, d\psi.$$

This formula is conventionally referred to as the Poisson integral.

Solution of the outer problem in series form:
$$w(r, \varphi) = \frac{a_0}{2} + \sum_{n=1}^{\infty} \left(\frac{r}{R}\right)^n (a_n \cos n\varphi + b_n \sin n\varphi),$$

$$a_n = \frac{1}{\pi} \int_0^{2\pi} f(\psi) \cos(n\psi) \, d\psi, \quad n = 0, 1, 2, \ldots,$$

$$b_n = \frac{1}{\pi} \int_0^{2\pi} f(\psi) \sin(n\psi) \, d\psi, \quad n = 1, 2, 3, \ldots$$

2°. Bounded solution of the outer problem ($r \geq R$):
$$w(r, \varphi) = \frac{1}{2\pi} \int_0^{2\pi} f(\psi) \frac{r^2 - R^2}{r^2 - 2Rr \cos(\varphi - \psi) + R^2} \, d\psi.$$

Bounded solution of the outer problem in series form:
$$w(r, \varphi) = \frac{a_0}{2} + \sum_{n=1}^{\infty} \left(\frac{R}{r}\right)^n (a_n \cos n\varphi + b_n \sin n\varphi),$$

where the coefficients a_0, a_n, and b_n are defined by the same relations as in the inner problem.

In hydrodynamics and other applications, outer problems are sometimes encountered in which one has to consider unbounded solutions for $r \to \infty$.

Example. The potential flow of an ideal (inviscid) incompressible fluid about a circular cylinder of radius R with a constant incident velocity U at infinity is characterized by the following boundary conditions for the stream function:
$$w = 0 \quad \text{at} \quad r = R, \qquad w \to Ur \sin \varphi \quad \text{as} \quad r \to \infty.$$

Solution:
$$w(r, \varphi) = U\left(r - \frac{R^2}{r}\right) \sin \varphi.$$

⊙ *References*: V. M. Babich, M. B. Kapilevich, S. G. Mikhlin, et al. (1964), A. N. Tikhonov and A. A. Samarskii (1990).

7.1.2-3. Domain: $0 \leq r \leq R$ or $R \leq r < \infty$. Second boundary value problem.

The condition
$$\partial_r w = f(\varphi) \quad \text{at} \quad r = R$$
is set at the boundary of the circle. The function $f(\varphi)$ must satisfy the solvability condition
$$\int_0^{2\pi} f(\varphi) \, d\varphi = 0.$$

1°. Solution of the inner problem ($r \leq R$):

$$w(r,\varphi) = \frac{R}{2\pi} \int_0^{2\pi} f(\psi) \ln \frac{r^2 - 2Rr\cos(\varphi - \psi) + R^2}{R^2} \, d\psi + C,$$

where C is an arbitrary constant; this formula is known as the Dini integral.

Series solution of the inner problem:

$$w(r,\varphi) = \sum_{n=1}^{\infty} \frac{R}{n} \left(\frac{r}{R}\right)^n (a_n \cos n\varphi + b_n \sin n\varphi) + C,$$

$$a_n = \frac{1}{\pi} \int_0^{2\pi} f(\psi) \cos(n\psi) \, d\psi, \quad b_n = \frac{1}{\pi} \int_0^{2\pi} f(\psi) \sin(n\psi) \, d\psi,$$

where C is an arbitrary constant.

2°. Solution of the outer problem ($r \geq R$):

$$w(r,\varphi) = -\frac{R}{2\pi} \int_0^{2\pi} f(\psi) \ln \frac{r^2 - 2Rr\cos(\varphi - \psi) + R^2}{r^2} \, d\psi + C,$$

where C is an arbitrary constant.

Series solution of the outer problem:

$$w(r,\varphi) = -\sum_{n=1}^{\infty} \frac{R}{n} \left(\frac{R}{r}\right)^n (a_n \cos n\varphi + b_n \sin n\varphi) + C,$$

where the coefficient a_n and b_n are defined by the same relations as in the inner problem, and C is an arbitrary constant.

⊙ *Reference*: V. M. Babich, M. B. Kapilevich, S. G. Mikhlin, et al. (1964).

7.1.2-4. Domain: $0 \leq r \leq R$ or $R \leq r < \infty$. Third boundary value problem.

The condition

$$\partial_r w + kw = f(\varphi) \quad \text{at} \quad r = R.$$

is set at the circle boundary; $f(\varphi)$ is a given function.

1°. Solution of the inner problem ($r \leq R$):

$$w(r,\varphi) = \frac{a_0}{2k} + \sum_{n=1}^{\infty} \frac{R}{kR+n} \left(\frac{r}{R}\right)^n (a_n \cos n\varphi + b_n \sin n\varphi),$$

$$a_n = \frac{1}{\pi} \int_0^{2\pi} f(\psi) \cos(n\psi) \, d\psi, \quad n = 0, 1, 2, \ldots,$$

$$b_n = \frac{1}{\pi} \int_0^{2\pi} f(\psi) \sin(n\psi) \, d\psi, \quad n = 1, 2, 3, \ldots$$

2°. Solution of the outer problem ($r \geq R$):

$$w(r,\varphi) = \frac{a_0}{2k} + \sum_{n=1}^{\infty} \frac{R}{kR-n} \left(\frac{R}{r}\right)^n (a_n \cos n\varphi + b_n \sin n\varphi),$$

where the coefficient a_0, a_n, and b_n are defined by the same relations as in the inner problem.

⊙ *Reference*: V. M. Babich, M. B. Kapilevich, S. G. Mikhlin, et al. (1964).

7.1.2-5. Domain: $R_1 \leq r \leq R_2$. First boundary value problem.

An annular domain is considered. Boundary conditions are prescribed:

$$w = f_1(\varphi) \quad \text{at} \quad r = R_1, \qquad w = f_2(\varphi) \quad \text{at} \quad r = R_2.$$

Solution:

$$w(r, \varphi) = A_0 + B_0 \ln r + \sum_{n=1}^{\infty} r^n (A_n \cos n\varphi + B_n \sin n\varphi) + \sum_{n=1}^{\infty} \frac{1}{r^n}(C_n \cos n\varphi + D_n \sin n\varphi),$$

where the coefficient A_0, B_0, A_n, B_n, C_n, and D_n are expressed as

$$A_0 = \frac{1}{2}\frac{a_0^{(1)} \ln R_2 - a_0^{(2)} \ln R_1}{\ln R_2 - \ln R_1}, \qquad B_0 = \frac{1}{2}\frac{a_0^{(2)} - a_0^{(1)}}{\ln R_2 - \ln R_1},$$

$$A_n = \frac{R_2^n a_n^{(2)} - R_1^n a_n^{(1)}}{R_2^{2n} - R_1^{2n}}, \qquad B_n = \frac{R_2^n b_n^{(2)} - R_1^n b_n^{(1)}}{R_2^{2n} - R_1^{2n}},$$

$$C_n = (R_1 R_2)^n \frac{R_2^n a_n^{(1)} - R_1^n a_n^{(2)}}{R_2^{2n} - R_1^{2n}}, \qquad D_n = (R_1 R_2)^n \frac{R_2^n b_n^{(1)} - R_1^n b_n^{(2)}}{R_2^{2n} - R_1^{2n}}.$$

Here, the $a_n^{(i)}$ and $b_n^{(i)}$ ($i = 1, 2$) are the coefficients of the Fourier series expansions of the functions $f_1(\varphi)$ and $f_2(\varphi)$:

$$a_n^{(i)} = \frac{1}{\pi} \int_0^{2\pi} f_i(\psi) \cos(n\psi)\, d\psi, \qquad n = 0, 1, 2, \ldots,$$

$$b_n^{(i)} = \frac{1}{\pi} \int_0^{2\pi} f_i(\psi) \sin(n\psi)\, d\psi, \qquad n = 1, 2, 3, \ldots$$

⊙ *Reference*: M. M. Smirnov (1975).

7.1.2-6. Domain: $R_1 \leq r \leq R_2$. Second boundary value problem.

An annular domain is considered. Boundary conditions are prescribed:

$$\partial_r w = f_1(\varphi) \quad \text{at} \quad r = R_1, \qquad \partial_r w = f_2(\varphi) \quad \text{at} \quad r = R_2.$$

Solution:

$$w(r, \varphi) = B \ln r + \sum_{n=1}^{\infty} r^n (A_n \cos n\varphi + B_n \sin n\varphi) + \sum_{n=1}^{\infty} \frac{1}{r^n}(C_n \cos n\varphi + D_n \sin n\varphi) + K.$$

Here, the coefficients B, A_n, B_n, C_n, and D_n are expressed as

$$B = \frac{1}{2} R_1 a_0^{(1)}, \quad A_n = \frac{R_2^{n+1} a_n^{(2)} - R_1^{n+1} a_n^{(1)}}{n(R_2^{2n} - R_1^{2n})}, \quad B_n = \frac{R_2^{n+1} b_n^{(2)} - R_1^{n+1} b_n^{(1)}}{n(R_2^{2n} - R_1^{2n})},$$

$$C_n = (R_1 R_2)^{n+1} \frac{R_1^{n-1} a_n^{(2)} - R_2^{n-1} a_n^{(1)}}{n(R_2^{2n} - R_1^{2n})}, \quad D_n = (R_1 R_2)^{n+1} \frac{R_1^{n-1} b_n^{(2)} - R_2^{n-1} b_n^{(1)}}{n(R_2^{2n} - R_1^{2n})},$$

where the constants $a_n^{(i)}$ and $b_n^{(i)}$ ($i = 1, 2$) are defined by the same relations as in the first boundary value problem; K is an arbitrary constant.

Remark. Note that the condition $a_0^{(1)} R_1 = a_0^{(2)} R_2$ must hold; this relation is a consequence of the solvability condition for the problem,

$$\int_{r=R_1} f_1\, dS - \int_{r=R_2} f_2\, dS = 0.$$

TABLE 21
Two-dimensional Laplace operator in some curvilinear orthogonal systems of coordinates

Coordinates	Transformation ($c > 0$)	Laplace operator, $\Delta_2 w$
Parabolic coordinates u, v	$x = cuv$, $y = \frac{1}{2}c(v^2 - u^2)$ $-\infty < u < \infty$, $0 \le v < \infty$	$\dfrac{1}{c^2(u^2+v^2)} \left(\dfrac{\partial^2 w}{\partial u^2} + \dfrac{\partial^2 w}{\partial v^2} \right)$
Elliptic coordinates ξ, η	$x = c\cosh\xi \cos\eta$, $y = c\sinh\xi \sin\eta$ $0 \le \xi < \infty$, $0 \le \eta < 2\pi$	$\dfrac{1}{c^2(\sinh^2\xi + \sin^2\eta)} \left(\dfrac{\partial^2 w}{\partial \xi^2} + \dfrac{\partial^2 w}{\partial \eta^2} \right)$
Bipolar coordinates σ, τ	$x = \dfrac{c\sinh\tau}{\cosh\tau - \cos\sigma}$, $y = \dfrac{c\sin\sigma}{\cosh\tau - \cos\sigma}$ $0 \le \sigma < 2\pi$, $-\infty < \tau < \infty$	$\dfrac{1}{c^2}(\cosh\tau - \cos\sigma)^2 \left(\dfrac{\partial^2 w}{\partial \sigma^2} + \dfrac{\partial^2 w}{\partial \tau^2} \right)$

7.1.2-7. Domain: $R_1 \le r \le R_2$. Mixed boundary value problem.

An annular domain is considered. Boundary conditions are prescribed:

$$\partial_r w = f_1(\varphi) \quad \text{at} \quad r = R_1, \qquad w = f_2(\varphi) \quad \text{at} \quad r = R_2.$$

Solution:

$$w(r, \varphi) = \tfrac{1}{2}a_0^{(2)} + \tfrac{1}{2}a_0^{(1)} R_1 \ln\frac{r}{R_2} + \sum_{n=1}^{\infty} r^n (A_n \cos n\varphi + B_n \sin n\varphi) + \sum_{n=1}^{\infty} \frac{1}{r^n}(C_n \cos n\varphi + D_n \sin n\varphi).$$

Here, the coefficients A_n, B_n, C_n, and D_n are expressed as

$$A_n = \frac{nR_2^n a_n^{(2)} + R_1^{n+1} a_n^{(1)}}{n(R_2^{2n} + R_1^{2n})}, \qquad B_n = \frac{nR_2^n b_n^{(2)} + R_1^{n+1} b_n^{(1)}}{n(R_2^{2n} + R_1^{2n})},$$

$$C_n = R_1^{n+1} R_2^n \frac{nR_1^{n-1} a_n^{(2)} - R_2^n a_n^{(1)}}{n(R_2^{2n} + R_1^{2n})}, \qquad D_n = R_1^{n+1} R_2^n \frac{nR_1^{n-1} b_n^{(2)} - R_2^n b_n^{(1)}}{n(R_2^{2n} + R_1^{2n})},$$

where the constants $a_n^{(i)}$ and $b_n^{(i)}$ ($i = 1, 2$) are defined by the same formulas as in the first boundary value problem.

⊙ *Reference*: M. M. Smirnov (1975).

7.1.3. Other Coordinate Systems. Conformal Mappings Method

7.1.3-1. Parabolic, elliptic, and bipolar coordinate systems.

In a number of applications, it is convenient to solve the Laplace equation in other orthogonal systems of coordinates. Some of those commonly encountered are displayed in Table 21. In all the coordinate systems presented, the Laplace equation $\Delta_2 w = 0$ is reduced to the equation considered in Paragraph 7.1.1-1 in detail (particular solutions and solutions to boundary value problems are given there).

The orthogonal transformations presented in Table 21 can be written in the language of complex variables as follows:

$$x + iy = -\tfrac{1}{2}ic(u + iv)^2 \qquad \text{(parabolic coordinates)},$$
$$x + iy = c\cosh(\xi + i\eta) \qquad \text{(elliptic coordinates)},$$
$$x + iy = ic\cot\left[\tfrac{1}{2}(\sigma + i\tau)\right] \qquad \text{(bipolar coordinates)}.$$

The real parts, as well as the imaginary parts, in both sides of these relations must be equated to each other ($i^2 = -1$).

Example. Plane hydrodynamic problems of potential flows of ideal (inviscid) incompressible fluid are reduced to the Laplace equation for the stream function. In particular, the motion of an elliptic cylinder with semiaxes a and b at a velocity U in the direction parallel to the major semiaxis ($a > b$) in ideal fluid is described by the stream function

$$w(\xi, \eta) = -Ub\left(\frac{a+b}{a-b}\right)^{1/2} e^{\xi} \sin\eta, \qquad c^2 = a^2 - b^2,$$

where ξ and η are the elliptic coordinates.

⊙ *References*: G. Lamb (1945), J. Happel and H. Brenner (1965), G. Korn and T. Korn (1968).

7.1.3-2. Domain of arbitrary shape. Method of conformal mappings.

1°. Let $\zeta = \zeta(z)$ be an analytic function that defines a conformal mapping from the complex plane $z = x + iy$ into a complex plane $\zeta = u + iv$, where $u = u(x, y)$ and $v = v(x, y)$ are new independent variables. With reference to the fact that the real and imaginary parts of an analytic function satisfy the Cauchy–Riemann conditions, we have $\partial_x u = \partial_y v$ and $\partial_y u = -\partial_x v$, and hence

$$\frac{\partial^2 w}{\partial x^2} + \frac{\partial^2 w}{\partial y^2} = |\zeta'(z)|^2 \left(\frac{\partial^2 w}{\partial u^2} + \frac{\partial^2 w}{\partial v^2}\right).$$

Therefore, the Laplace equation in the xy-plane transforms under a conformal mapping into the Laplace equation in the uv-plane.

2°. Any simply connected domain D in the xy-plane with a piecewise smooth boundary can be mapped, with appropriate conformal mappings, onto the upper half-plane or into a unit circle in the uv-plane. Consequently, a first and a second boundary value problem for the Laplace equation in D can be reduced, respectively, to a first and a second boundary value problem for the upper half-space or a circle; such problems are considered in Subsections 7.1.1 and 7.1.2.

Subsection 7.2.4 presents conformal mappings of some domains onto the upper half-plane or a unit circle. Moreover, examples of solving specific boundary value problems for the Poisson equation by the conformal mappings method are given there; the Green's functions for a semicircle and a quadrant of a circle are obtained.

A large number of conformal mappings of various domains can be found, for example, in the references cited below.

⊙ *References*: V. I. Lavrik and V. N. Savenkov (1970), M. A. Lavrent'ev and B. V. Shabat (1973), V. I. Ivanov and M. K. Trubetskov (1994).

7.1.3-3. Reduction of the two-dimensional Neumann problem to the Dirichlet problem.

Let the position of any point (x_*, y_*) located on the boundary Σ of a domain D be specified by a parameter s, so that $x_* = x_*(s)$ and $y_* = y_*(s)$. Then a function of two variables, $f(x, y)$, is determined on Σ by the parameter s as well, $f(x, y)|_\Sigma = f(x_*(s), y_*(s)) = f_*(s)$.

The solution of the two-dimensional Neumann problem for the Laplace equation $\Delta_2 w = 0$ in D with the boundary condition of the second kind

$$\frac{\partial w}{\partial N} = f_*(s) \quad \text{for} \quad \mathbf{r} \in \Sigma$$

can be expressed in terms of the solution of the two-dimensional Dirichlet problem for the Laplace equation $\Delta_2 u = 0$ in D with the boundary condition of the first kind

$$u = F_*(s) \quad \text{for} \quad \mathbf{r} \in \Sigma,$$

where $F_*(s) = \int f_*(s)\, ds$, as follows:

$$w(x, y) = \int_{x_0}^{x} \frac{\partial u}{\partial y}(t, y_0)\, dt - \int_{y_0}^{y} \frac{\partial u}{\partial x}(x, t)\, dt + C.$$

Here, (x_0, y_0) are the coordinates of any point in D, and C is an arbitrary constant.

⊙ *Reference*: V. M. Babich, M. B. Kapilevich, S. G. Mikhlin, et al. (1964).

7.2. Poisson Equation $\Delta_2 w = -\Phi(\mathbf{x})$

7.2.1. Preliminary Remarks. Solution Structure

Just as the Laplace equation, the Poisson equation is often encountered in heat and mass transfer theory, fluid mechanics, elasticity, electrostatics, and other areas of mechanics and physics. For example, it describes steady-state temperature distribution in the presence of heat sources or sinks in the domain under study.

The Laplace equation is a special case of the Poisson equation with $\Phi \equiv 0$.

In what follows, we consider a finite domain S with a sufficiently smooth boundary L. Let $\mathbf{r} \in S$ and $\rho \in S$, where $\mathbf{r} = \{x, y\}$, $\rho = \{\xi, \eta\}$, $|\mathbf{r} - \rho|^2 = (x - \xi)^2 + (y - \eta)^2$.

7.2.1-1. First boundary value problem.

The solution of the first boundary value problem for the Poisson equation

$$\Delta_2 w = -\Phi(\mathbf{r}) \tag{1}$$

in the domain S with the nonhomogeneous boundary condition

$$w = f(\mathbf{r}) \quad \text{for} \quad \mathbf{r} \in L$$

can be represented as

$$w(\mathbf{r}) = \int_S \Phi(\rho) G(\mathbf{r}, \rho) \, dS_\rho - \int_L f(\rho) \frac{\partial G}{\partial N_\rho} \, dL_\rho. \tag{2}$$

Here, $G(\mathbf{r}, \rho)$ is the Green's function of the first boundary value problem, $\frac{\partial G}{\partial N_\rho}$ is the derivative of the Green's function with respect to ξ, η along the outward normal $\mathbf{N}$ to the boundary L. The integration is performed with respect to ξ, η, with $dS_\rho = d\xi \, d\eta$.

The Green's function $G = G(\mathbf{r}, \rho)$ of the first boundary value problem is determined by the following conditions.

1°. The function G satisfies the Laplace equation in x, y in the domain S everywhere except for the point (ξ, η), at which G has a singularity of the form $\frac{1}{2\pi} \ln \frac{1}{|\mathbf{r}-\rho|}$.

2°. With respect to x, y, the function G satisfies the homogeneous boundary condition of the first kind at the domain boundary, i.e., the condition $G|_L = 0$.

The Green's function can be represented in the form

$$G(\mathbf{r}, \rho) = \frac{1}{2\pi} \ln \frac{1}{|\mathbf{r} - \rho|} + u, \tag{3}$$

where the auxiliary function $u = u(\mathbf{r}, \rho)$ is determined by solving the first boundary value problem for the Laplace equation $\Delta_2 u = 0$ with the boundary condition $u|_L = -\frac{1}{2\pi} \ln \frac{1}{|\mathbf{r}-\rho|}$; in this problem, ρ is treated as a two-dimensional free parameter.

The Green's function is symmetric with respect to its arguments: $G(\mathbf{r}, \rho) = G(\rho, \mathbf{r})$.

Remark 1. When using the polar coordinate system, one should set

$$\mathbf{r} = \{r, \varphi\}, \quad \rho = \{\xi, \eta\}, \quad |\mathbf{r} - \rho|^2 = r^2 + \xi^2 - 2r\xi \cos(\varphi - \eta), \quad dS_\rho = \xi \, d\xi \, d\eta$$

in relations (2) and (3).

7.2.1-2. Second boundary value problem.

The second boundary value problem for the Poisson equation (1) is characterized by the boundary condition

$$\frac{\partial w}{\partial N} = f(\mathbf{r}) \quad \text{for} \quad \mathbf{r} \in L.$$

The necessary solvability condition for this problem is

$$\int_S \Phi(\mathbf{r})\, dS + \int_L f(\mathbf{r})\, dL = 0. \tag{4}$$

The solution of the second boundary value problem, provided that condition (4) is satisfied, can be represented as

$$w(\mathbf{r}) = \int_S \Phi(\boldsymbol{\rho}) G(\mathbf{r}, \boldsymbol{\rho})\, dS_\rho + \int_L f(\boldsymbol{\rho}) G(\mathbf{r}, \boldsymbol{\rho})\, dL_\rho + C, \tag{5}$$

where C is an arbitrary constant.

The Green's function $G = G(\mathbf{r}, \boldsymbol{\rho})$ of the second boundary value problem is determined by the following conditions:

$1°$. The function G satisfies the Laplace equation in x, y in the domain S everywhere except for the point (ξ, η), at which G has a singularity of the form $\frac{1}{2\pi} \ln \frac{1}{|\mathbf{r}-\boldsymbol{\rho}|}$.

$2°$. With respect to x, y, the function G satisfies the homogeneous boundary condition of the second kind at the domain boundary:

$$\left.\frac{\partial G}{\partial N}\right|_L = \frac{1}{L_0},$$

where L_0 is the length of the boundary of S.

The Green's function is unique up to an additive constant.

Remark 2. The Green's function cannot be determined by condition $1°$ and the homogeneous boundary condition $\left.\frac{\partial G}{\partial N}\right|_L = 0$. The point is that the problem is unsolvable for G in this case, because, on representing G in the form (3), for u we obtain a problem with a nonhomogeneous boundary condition of the second kind for which the solvability condition (4) now is not satisfied.

7.2.1-3. Third boundary value problem.

The solution of the third boundary value problem for the Poisson equation (1) in the domain S with the nonhomogeneous boundary condition

$$\frac{\partial w}{\partial N} + kw = f(\mathbf{r}) \quad \text{for} \quad \mathbf{r} \in L$$

is given by formula (5) with $C = 0$, where $G = G(\mathbf{r}, \boldsymbol{\rho})$ is the Green's function of the third boundary value problem and is determined by the following conditions:

$1°$. The function G satisfies the Laplace equation in x, y in the domain S everywhere except for the point (ξ, η), at which G has a singularity of the form $\frac{1}{2\pi} \ln \frac{1}{|\mathbf{r}-\boldsymbol{\rho}|}$.

$2°$. With respect to x, y, the function G satisfies the homogeneous boundary condition of the third kind at the domain boundary, i.e., the condition $\left[\frac{\partial G}{\partial N} + kG\right]_L = 0$.

The Green's function can be represented in the form (3); the auxiliary function u is identified by solving the corresponding third boundary value problem for the Laplace equation $\Delta_2 u = 0$.

The Green's function is symmetric with respect to its arguments: $G(\mathbf{r}, \boldsymbol{\rho}) = G(\boldsymbol{\rho}, \mathbf{r})$.

⊙ *References for Subsection* 7.2.1: V. M. Babich, M. B. Kapilevich, S. G. Mikhlin, et al. (1964), N. S. Koshlyakov, E. B. Gliner, and M. M. Smirnov (1970).

7.2.2. Problems in Cartesian Coordinate System

The two-dimensional Poisson equation in the rectangular Cartesian coordinate system has the form

$$\frac{\partial^2 w}{\partial x^2} + \frac{\partial^2 w}{\partial y^2} + \Phi(x,y) = 0.$$

7.2.2-1. Particular solutions of the Poisson equation with a special right-hand side.

1°. If $\Phi(x,y) = \sum_{i=1}^{n}\sum_{j=1}^{n} a_{ij} \exp(b_i x + c_j y)$, the equation has solutions of the form

$$w(x,y) = -\sum_{i=1}^{n}\sum_{j=1}^{n} \frac{a_{ij}}{b_i^2 + c_j^2} \exp(b_i x + c_j y).$$

2°. If $\Phi(x,y) = \sum_{i=1}^{n}\sum_{j=1}^{n} a_{ij} \sin(b_i x + p_i) \sin(c_j y + q_j)$, the equation admits solutions of the form

$$w(x,y) = \sum_{i=1}^{n}\sum_{j=1}^{n} \frac{a_{ij}}{b_i^2 + c_j^2} \sin(b_i x + p_i) \sin(c_j y + q_j).$$

7.2.2-2. Domain: $-\infty < x < \infty$, $-\infty < y < \infty$.

Solution:

$$w(x,y) = \frac{1}{2\pi} \int_{-\infty}^{\infty} \int_{-\infty}^{\infty} \Phi(\xi,\eta) \ln \frac{1}{\sqrt{(x-\xi)^2 + (y-\eta)^2}} \, d\xi \, d\eta.$$

7.2.2-3. Domain: $-\infty < x < \infty$, $0 \leq y < \infty$. First boundary value problem.

A half-plane is considered. A boundary condition is prescribed:

$$w = f(x) \quad \text{at} \quad y = 0.$$

Solution:

$$w(x,y) = \frac{1}{\pi} \int_{-\infty}^{\infty} \frac{y f(\xi) \, d\xi}{(x-\xi)^2 + y^2} + \frac{1}{2\pi} \int_{0}^{\infty} \int_{-\infty}^{\infty} \Phi(\xi,\eta) \ln \frac{\sqrt{(x-\xi)^2 + (y+\eta)^2}}{\sqrt{(x-\xi)^2 + (y-\eta)^2}} \, d\xi \, d\eta.$$

⊙ *Reference*: A. G. Butkovskiy (1979).

7.2.2-4. Domain: $-\infty < x < \infty$, $0 \leq y < \infty$. Second boundary value problem.

A half-plane is considered. A boundary condition is prescribed:

$$\partial_y w = f(x) \quad \text{at} \quad y = 0.$$

Solution:

$$w(x,y) = \frac{1}{\pi} \int_{-\infty}^{\infty} f(\xi) \ln \sqrt{(x-\xi)^2 + y^2} \, d\xi$$
$$+ \frac{1}{2\pi} \int_{0}^{\infty} \int_{-\infty}^{\infty} \Phi(\xi,\eta) \left[\ln \frac{1}{\sqrt{(x-\xi)^2 + (y-\eta)^2}} + \ln \frac{1}{\sqrt{(x-\xi)^2 + (y+\eta)^2}} \right] d\xi \, d\eta + C,$$

where C is an arbitrary constant.

⊙ *Reference*: V. S. Vladimirov (1988).

7.2.2-5. Domain: $-\infty < x < \infty$, $0 \leq y \leq a$. **First boundary value problem.**

An infinite strip is considered. Boundary conditions are prescribed:

$$w = f_1(x) \quad \text{at} \quad y = 0, \qquad w = f_2(x) \quad \text{at} \quad y = a.$$

Solution:

$$\begin{aligned} w(x,y) &= \frac{1}{2a}\sin\!\left(\frac{\pi y}{a}\right)\int_{-\infty}^{\infty}\frac{f_1(\xi)\,d\xi}{\cosh[\pi(x-\xi)/a]-\cos(\pi y/a)} \\ &+ \frac{1}{2a}\sin\!\left(\frac{\pi y}{a}\right)\int_{-\infty}^{\infty}\frac{f_2(\xi)\,d\xi}{\cosh[\pi(x-\xi)/a]+\cos(\pi y/a)} \\ &+ \frac{1}{4\pi}\int_0^a\!\!\int_{-\infty}^{\infty}\Phi(\xi,\eta)\ln\frac{\cosh[\pi(x-\xi)/a]-\cos[\pi(y+\eta)/a]}{\cosh[\pi(x-\xi)/a]-\cos[\pi(y-\eta)/a]}\,d\xi\,d\eta. \end{aligned}$$

⊙ *Reference*: H. S. Carslaw and J. C. Jaeger (1984).

7.2.2-6. Domain: $-\infty < x < \infty$, $0 \leq y \leq a$. **Second boundary value problem.**

An infinite strip is considered. Boundary conditions are prescribed:

$$\partial_y w = f_1(x) \quad \text{at} \quad y = 0, \qquad \partial_y w = f_2(x) \quad \text{at} \quad y = a.$$

Solution:

$$\begin{aligned} w(x,y) &= -\int_{-\infty}^{\infty} f_1(\xi)G(x,y,\xi,0)\,d\xi + \int_{-\infty}^{\infty} f_2(\xi)G(x,y,\xi,a)\,d\xi \\ &+ \int_0^a\!\!\int_{-\infty}^{\infty} \Phi(\xi,\eta)G(x,y,\xi,\eta)\,d\xi\,d\eta + C. \end{aligned}$$

Here,

$$G(x,y,\xi,\eta)=\frac{1}{4\pi}\ln\frac{1}{\cosh[\pi(x-\xi)/a]-\cos[\pi(y-\eta)/a]} + \frac{1}{4\pi}\ln\frac{1}{\cosh[\pi(x-\xi)/a]-\cos[\pi(y+\eta)/a]},$$

where C is an arbitrary constant.

7.2.2-7. Domain: $-\infty < x < \infty$, $0 \leq y \leq a$. **Third boundary value problem.**

An infinite strip is considered. Boundary conditions are prescribed:

$$\partial_y w - k_1 w = f_1(x) \quad \text{at} \quad y = 0, \qquad \partial_y w + k_2 w = f_2(x) \quad \text{at} \quad y = a.$$

The solution $w(x,y)$ is determined by the formula in Paragraph 7.2.2-6 where

$$G(x,y,\xi,\eta) = \frac{1}{2}\sum_{n=1}^{\infty}\frac{\varphi_n(y)\varphi_n(\eta)}{\|\varphi_n\|^2\mu_n}\exp(-\mu_n|x-\xi|),$$

$$\varphi_n(y) = \mu_n\cos(\mu_n y) + k_1\sin(\mu_n y), \quad \|\varphi_n\|^2 = \frac{1}{2}(\mu_n^2 + k_1^2)\left[a + \frac{(k_1+k_2)(\mu_n^2 + k_1 k_2)}{(\mu_n^2+k_1^2)(\mu_n^2+k_2^2)}\right].$$

Here, the μ_n are positive roots of the transcendental equation $\tan(\mu a) = \dfrac{(k_1+k_2)\mu}{\mu^2 - k_1 k_2}$.

7.2.2-8. Domain: $-\infty < x < \infty$, $0 \le y \le a$. Mixed boundary value problem.

An infinite strip is considered. Boundary conditions are prescribed:

$$w = f_1(x) \quad \text{at} \quad y = 0, \qquad \partial_y w = f_2(x) \quad \text{at} \quad y = a.$$

Solution:

$$w(x,y) = \int_{-\infty}^{\infty} f_1(\xi) \left[\frac{\partial}{\partial \eta} G(x,y,\xi,\eta)\right]_{\eta=0} d\xi + \int_{-\infty}^{\infty} f_2(\xi) G(x,y,\xi,a) \, d\xi$$
$$+ \int_0^a \int_{-\infty}^{\infty} \Phi(\xi,\eta) G(x,y,\xi,\eta) \, d\xi \, d\eta,$$

where

$$G(x,y,\xi,\eta) = \frac{1}{a} \sum_{n=0}^{\infty} \frac{1}{\mu_n} \exp(-\mu_n |x-\xi|) \sin(\mu_n y) \sin(\mu_n \eta), \qquad \mu_n = \frac{\pi(2n+1)}{2a}.$$

7.2.2-9. Domain: $0 \le x < \infty$, $0 \le y \le a$. First boundary value problem.

A semiinfinite strip is considered. Boundary conditions are prescribed:

$$w = f_1(y) \quad \text{at} \quad x=0, \qquad w = f_2(x) \quad \text{at} \quad y=0, \qquad w = f_3(x) \quad \text{at} \quad y=a.$$

Solution:

$$w(x,y) = \int_0^a f_1(\eta) \left[\frac{\partial}{\partial \xi} G(x,y,\xi,\eta)\right]_{\xi=0} d\eta + \int_0^{\infty} f_2(\xi) \left[\frac{\partial}{\partial \eta} G(x,y,\xi,\eta)\right]_{\eta=0} d\xi$$
$$- \int_0^{\infty} f_3(\xi) \left[\frac{\partial}{\partial \eta} G(x,y,\xi,\eta)\right]_{\eta=a} d\xi + \int_0^a \int_0^{\infty} \Phi(\xi,\eta) G(x,y,\xi,\eta) \, d\xi \, d\eta,$$

where

$$G(x,y,\xi,\eta) = \frac{1}{4\pi} \ln \frac{\cosh[\pi(x-\xi)/a] - \cos[\pi(y+\eta)/a]}{\cosh[\pi(x-\xi)/a] - \cos[\pi(y-\eta)/a]} - \frac{1}{4\pi} \ln \frac{\cosh[\pi(x+\xi)/a] - \cos[\pi(y+\eta)/a]}{\cosh[\pi(x+\xi)/a] - \cos[\pi(y-\eta)/a]}.$$

Alternatively, the Green's function can be represented in the series form

$$G(x,y,\xi,\eta) = \frac{1}{a} \sum_{n=1}^{\infty} \frac{1}{q_n} \left[\exp(-q_n |x-\xi|) - \exp(-q_n |x+\xi|)\right] \sin(q_n y) \sin(q_n \eta), \qquad q_n = \frac{\pi n}{a}.$$

⊙ *References*: N. N. Lebedev, I. P. Skal'skaya, and Ya. S. Uflyand (1955), A. G. Butkovskiy (1979).

7.2.2-10. Domain: $0 \le x < \infty$, $0 \le y \le a$. Third boundary value problem.

A semiinfinite strip is considered. Boundary conditions are prescribed:

$$\partial_x w - k_1 w = f_1(y) \text{ at } x=0, \quad \partial_y w - k_2 w = f_2(x) \text{ at } y=0, \quad \partial_y w + k_3 w = f_3(x) \text{ at } y=a.$$

Solution:

$$w(x,y) = \int_0^a \int_0^{\infty} \Phi(\xi,\eta) G(x,y,\xi,\eta) \, d\xi \, d\eta - \int_0^a f_1(\eta) G(x,y,0,\eta) \, d\eta$$
$$- \int_0^{\infty} f_2(\xi) G(x,y,\xi,0) \, d\xi + \int_0^{\infty} f_3(\xi) G(x,y,\xi,a) \, d\xi,$$

where
$$G(x,y,\xi,\eta) = \sum_{n=1}^{\infty} \frac{\varphi_n(y)\varphi_n(\eta)}{\|\varphi_n\|^2 \mu_n(\mu_n+k_1)} H_n(x,\xi),$$

$$\varphi_n(y) = \mu_n\cos(\mu_n y) + k_2\sin(\mu_n y), \quad \|\varphi_n\|^2 = \frac{1}{2}(\mu_n^2+k_2^2)\left[a + \frac{(k_2+k_3)(\mu_n^2+k_2k_3)}{(\mu_n^2+k_2^2)(\mu_n^2+k_3^2)}\right],$$

$$H_n(x,\xi) = \begin{cases} \exp(-\mu_n x)\big[\mu_n\cosh(\mu_n\xi) + k_1\sinh(\mu_n\xi)\big] & \text{for } x > \xi, \\ \exp(-\mu_n\xi)\big[\mu_n\cosh(\mu_n x) + k_1\sinh(\mu_n x)\big] & \text{for } \xi > x. \end{cases}$$

Here, the μ_n are positive roots of the transcendental equation $\tan(\mu a) = \dfrac{(k_2+k_3)\mu}{\mu^2 - k_2 k_3}$.

7.2.2-11. Domain: $0 \le x < \infty$, $0 \le y \le a$. Mixed boundary value problems.

1°. A semiinfinite strip is considered. Boundary conditions are prescribed:
$$w = f_1(y) \text{ at } x=0, \quad \partial_y w = f_2(x) \text{ at } y=0, \quad \partial_y w = f_3(x) \text{ at } y=a.$$
Solution:
$$w(x,y) = \int_0^a f_1(\eta)\left[\frac{\partial}{\partial \xi}G(x,y,\xi,\eta)\right]_{\xi=0} d\eta - \int_0^\infty f_2(\xi)G(x,y,\xi,0)\,d\xi$$
$$+ \int_0^\infty f_3(\xi)G(x,y,\xi,a)\,d\xi + \int_0^a\int_0^\infty \Phi(\xi,\eta)G(x,y,\xi,\eta)\,d\xi\,d\eta,$$
where
$$G(x,y,\xi,\eta) = \frac{1}{2a}\sum_{n=0}^{\infty}\frac{\varepsilon_n}{q_n}\big[\exp(-q_n|x-\xi|) - \exp(-q_n|x+\xi|)\big]\cos(q_n y)\cos(q_n\eta),$$
$$q_n = \frac{\pi n}{a}, \quad \varepsilon = \begin{cases} 1 & \text{for } n=0, \\ 2 & \text{for } n\ne 0. \end{cases}$$

2°. A semiinfinite strip is considered. Boundary conditions are prescribed:
$$\partial_x w = f_1(y) \text{ at } x=0, \quad w = f_2(x) \text{ at } y=0, \quad w = f_3(x) \text{ at } y=a.$$
Solution:
$$w(x,y) = -\int_0^a f_1(\eta)G(x,y,0,\eta)\,d\eta + \int_0^\infty f_2(\xi)\left[\frac{\partial}{\partial\eta}G(x,y,\xi,\eta)\right]_{\eta=0}d\xi$$
$$-\int_0^\infty f_3(\xi)\left[\frac{\partial}{\partial\eta}G(x,y,\xi,\eta)\right]_{\eta=a}d\xi + \int_0^a\int_0^\infty \Phi(\xi,\eta)G(x,y,\xi,\eta)\,d\xi\,d\eta,$$
where
$$G(x,y,\xi,\eta) = \frac{1}{a}\sum_{n=1}^{\infty}\frac{1}{q_n}\big[\exp(-q_n|x-\xi|) + \exp(-q_n|x+\xi|)\big]\sin(q_n y)\sin(q_n\eta), \quad q_n = \frac{\pi n}{a}.$$

7.2.2-12. Domain: $0 \le x < \infty$, $0 \le y < \infty$. First boundary value problem.

A quadrant of the plane is considered. Boundary conditions are prescribed:
$$w = f_1(y) \text{ at } x=0, \quad w = f_2(x) \text{ at } y=0.$$
Solution:
$$w(x,y) = \frac{4}{\pi}xy\int_0^\infty \frac{f_1(\eta)\eta\,d\eta}{[x^2+(y-\eta)^2][x^2+(y+\eta)^2]} + \frac{4}{\pi}xy\int_0^\infty \frac{f_2(\xi)\xi\,d\xi}{[(x-\xi)^2+y^2][(x+\xi)^2+y^2]}$$
$$+ \frac{1}{2\pi}\int_0^\infty\int_0^\infty \Phi(\xi,\eta)\ln\frac{\sqrt{(x-\xi)^2+(y+\eta)^2}\sqrt{(x+\xi)^2+(y-\eta)^2}}{\sqrt{(x-\xi)^2+(y-\eta)^2}\sqrt{(x+\xi)^2+(y+\eta)^2}}\,d\xi\,d\eta.$$

⊙ *References*: V. S. Vladimirov, V. P. Mikhailov, A. A. Vasharin, et al. (1974), A. G. Butkovskiy (1979).

7.2.2-13. Domain: $0 \le x \le a$, $0 \le y \le b$. First boundary value problem.

A rectangle is considered. Boundary conditions are prescribed:

$$w = f_1(y) \quad \text{at} \quad x = 0, \qquad w = f_2(y) \quad \text{at} \quad x = a,$$
$$w = f_3(x) \quad \text{at} \quad y = 0, \qquad w = f_4(x) \quad \text{at} \quad y = b.$$

Solution:

$$w(x,y) = \int_0^a \int_0^b \Phi(\xi,\eta) G(x,y,\xi,\eta) \, d\eta \, d\xi$$
$$+ \int_0^b f_1(\eta) \left[\frac{\partial}{\partial \xi} G(x,y,\xi,\eta)\right]_{\xi=0} d\eta - \int_0^b f_2(\eta) \left[\frac{\partial}{\partial \xi} G(x,y,\xi,\eta)\right]_{\xi=a} d\eta$$
$$+ \int_0^a f_3(\xi) \left[\frac{\partial}{\partial \eta} G(x,y,\xi,\eta)\right]_{\eta=0} d\xi - \int_0^a f_4(\xi) \left[\frac{\partial}{\partial \eta} G(x,y,\xi,\eta)\right]_{\eta=b} d\xi.$$

Two forms of representation of the Green's function:

$$G(x,y,\xi,\eta) = \frac{2}{a} \sum_{n=1}^{\infty} \frac{\sin(p_n x) \sin(p_n \xi)}{p_n \sinh(p_n b)} H_n(y,\eta) = \frac{2}{b} \sum_{m=1}^{\infty} \frac{\sin(q_m y) \sin(q_m \eta)}{q_m \sinh(q_m a)} Q_m(x,\xi),$$

where

$$p_n = \frac{\pi n}{a}, \quad H_n(y,\eta) = \begin{cases} \sinh(p_n \eta) \sinh[p_n(b-y)] & \text{for } b \ge y > \eta \ge 0, \\ \sinh(p_n y) \sinh[p_n(b-\eta)] & \text{for } b \ge \eta > y \ge 0, \end{cases}$$

$$q_m = \frac{\pi m}{b}, \quad Q_m(x,\xi) = \begin{cases} \sinh(q_m \xi) \sinh[q_m(a-x)] & \text{for } a \ge x > \xi \ge 0, \\ \sinh(q_m x) \sinh[q_m(a-\xi)] & \text{for } a \ge \xi > x \ge 0. \end{cases}$$

The Green's function can be written in form of a double series:

$$G(x,y,\xi,\eta) = \frac{4}{ab} \sum_{n=1}^{\infty} \sum_{m=1}^{\infty} \frac{\sin(p_n x) \sin(q_m y) \sin(p_n \xi) \sin(q_m \eta)}{p_n^2 + q_m^2}, \quad p_n = \frac{\pi n}{a}, \quad q_m = \frac{\pi m}{b}.$$

⊙ *Reference*: A. G. Butkovskiy (1979).

7.2.2-14. Domain: $0 \le x \le a$, $0 \le y \le b$. Third boundary value problem.

A rectangle is considered. Boundary conditions are prescribed:

$$\partial_x w - k_1 w = f_1(y) \quad \text{at} \quad x = 0, \qquad \partial_x w + k_2 w = f_2(y) \quad \text{at} \quad x = a,$$
$$\partial_y w - k_3 w = f_3(x) \quad \text{at} \quad y = 0, \qquad \partial_y w + k_4 w = f_4(x) \quad \text{at} \quad y = b.$$

Solution:

$$w(x,y) = \int_0^a \int_0^b \Phi(\xi,\eta) G(x,y,\xi,\eta) \, d\eta \, d\xi$$
$$- \int_0^b f_1(\eta) G(x,y,0,\eta) \, d\eta + \int_0^b f_2(\eta) G(x,y,a,\eta) \, d\eta$$
$$- \int_0^a f_3(\xi) G(x,y,\xi,0) \, d\xi + \int_0^a f_4(\xi) G(x,y,\xi,b) \, d\xi.$$

Here,

$$G(x,y,\xi,\eta) = \sum_{n=1}^{\infty} \sum_{m=1}^{\infty} \frac{\varphi_n(x) \varphi_n(\xi) \psi_m(y) \psi_m(\eta)}{\|\varphi_n\|^2 \|\psi_m\|^2 (\mu_n^2 + \lambda_m^2)},$$

$$\varphi_n(x) = \cos(\mu_n x) + \frac{k_1}{\mu_n} \sin(\mu_n x), \quad \|\varphi_n\|^2 = \frac{k_2}{2\mu_n^2} \frac{\mu_n^2 + k_1^2}{\mu_n^2 + k_2^2} + \frac{k_1}{2\mu_n^2} + \frac{a}{2}\left(1 + \frac{k_1^2}{\mu_n^2}\right),$$

$$\psi_m(y) = \cos(\lambda_m y) + \frac{k_3}{\lambda_m} \sin(\lambda_m y), \quad \|\psi_m\|^2 = \frac{k_4}{2\lambda_m^2} \frac{\lambda_m^2 + k_3^2}{\lambda_m^2 + k_4^2} + \frac{k_3}{2\lambda_m^2} + \frac{b}{2}\left(1 + \frac{k_3^2}{\lambda_m^2}\right),$$

where the μ_n and λ_m are positive roots of the transcendental equations

$$\frac{\tan(\mu a)}{\mu} = \frac{k_1 + k_2}{\mu^2 - k_1 k_2}, \qquad \frac{\tan(\lambda b)}{\lambda} = \frac{k_3 + k_4}{\lambda^2 - k_3 k_4}.$$

7.2.2-15. Domain: $0 \le x \le a$, $0 \le y \le b$. Mixed boundary value problem.

A rectangle is considered. Boundary conditions are prescribed:

$$w = f_1(y) \quad \text{at} \quad x = 0, \qquad \partial_x w = f_2(y) \quad \text{at} \quad x = a,$$
$$w = f_3(x) \quad \text{at} \quad y = 0, \qquad \partial_y w = f_4(x) \quad \text{at} \quad y = b.$$

Solution:

$$w(x,y) = \int_0^a \int_0^b \Phi(\xi,\eta) G(x,y,\xi,\eta)\, d\eta\, d\xi$$
$$+ \int_0^b f_1(\eta) \left[\frac{\partial}{\partial \xi} G(x,y,\xi,\eta)\right]_{\xi=0} d\eta + \int_0^b f_2(\eta) G(x,y,a,\eta)\, d\eta$$
$$+ \int_0^a f_3(\xi) \left[\frac{\partial}{\partial \eta} G(x,y,\xi,\eta)\right]_{\eta=0} d\xi + \int_0^a f_4(\xi) G(x,y,\xi,b)\, d\xi.$$

Two forms of representation of the Green's function:

$$G(x,y,\xi,\eta) = \frac{2}{a} \sum_{n=0}^\infty \frac{\sin(p_n x) \sin(p_n \xi)}{p_n \cosh(p_n b)} H_n(y,\eta) = \frac{2}{b} \sum_{m=0}^\infty \frac{\sin(q_m y) \sin(q_m \eta)}{q_m \cosh(q_m a)} Q_m(x,\xi),$$

where

$$p_n = \frac{\pi(2n+1)}{a}, \quad H_n(y,\eta) = \begin{cases} \sinh(p_n \eta) \cosh[p_n(b-y)] & \text{for } b \ge y > \eta \ge 0, \\ \sinh(p_n y) \cosh[p_n(b-\eta)] & \text{for } b \ge \eta > y \ge 0, \end{cases}$$

$$q_m = \frac{\pi(2m+1)}{b}, \quad Q_m(x,\xi) = \begin{cases} \sinh(q_m \xi) \cosh[q_m(a-x)] & \text{for } a \ge x > \xi \ge 0, \\ \sinh(q_m x) \cosh[q_m(a-\xi)] & \text{for } a \ge \xi > x \ge 0. \end{cases}$$

The Green's function can be written in form of a double series:

$$G(x,y,\xi,\eta) = \frac{4}{ab} \sum_{n=0}^\infty \sum_{m=0}^\infty \frac{\sin(p_n x) \sin(q_m y) \sin(p_n \xi) \sin(q_m \eta)}{p_n^2 + q_m^2},$$

$$p_n = \frac{\pi(2n+1)}{2a}, \qquad q_m = \frac{\pi(2m+1)}{2b}.$$

7.2.3. Problems in Polar Coordinate System

The two-dimensional Poisson equation in the polar coordinate system is written as

$$\frac{1}{r} \frac{\partial}{\partial r}\left(r \frac{\partial w}{\partial r}\right) + \frac{1}{r^2} \frac{\partial^2 w}{\partial \varphi^2} + \Phi(r,\varphi) = 0, \qquad r = \sqrt{x^2 + y^2}.$$

7.2.3-1. Domain: $0 \le r \le R$, $0 \le \varphi \le 2\pi$. First boundary value problem.

A circle is considered. A boundary condition is prescribed:

$$w = f(\varphi) \quad \text{at} \quad r = R.$$

Solution:

$$w(r,\varphi) = \frac{1}{2\pi}\int_0^{2\pi} f(\eta)\frac{R^2 - r^2}{r^2 - 2Rr\cos(\varphi - \eta) + R^2}\,d\eta + \int_0^{2\pi}\int_0^R \Phi(\xi,\eta)G(r,\varphi,\xi,\eta)\xi\,d\xi\,d\eta,$$

where

$$G(r,\varphi,\xi,\eta) = \frac{1}{2\pi}\ln\frac{1}{|\mathbf{r} - \mathbf{r}_0|} - \frac{1}{2\pi}\ln\frac{R}{r_0|(R/r_0)^2\mathbf{r}_0 - \mathbf{r}|},$$

$$\mathbf{r} = \{x, y\}, \qquad x = r\cos\varphi, \qquad y = r\sin\varphi,$$

$$\mathbf{r}_0 = \{x_0, y_0\}, \qquad x_0 = \xi\cos\eta, \qquad y_0 = \xi\sin\eta.$$

The magnitude of a vector difference is calculated as $|a\mathbf{r} - b\mathbf{r}_0|^2 = a^2 r^2 - 2ab r\xi\cos(\varphi - \eta) + b^2\xi^2$ (a and b are any scalars). Thus, we obtain

$$G(r,\varphi,\xi,\eta) = \frac{1}{4\pi}\ln\frac{r^2\xi^2 - 2R^2 r\xi\cos(\varphi - \eta) + R^4}{R^2[r^2 - 2r\xi\cos(\varphi - \eta) + \xi^2]}.$$

⊙ *References*: V. M. Babich, M. B. Kapilevich, S. G. Mikhlin, et al. (1964), A. G. Butkovskiy (1979).

7.2.3-2. Domain: $0 \le r \le R$, $0 \le \varphi \le 2\pi$. Third boundary value problem.

A circle is considered. A boundary condition is prescribed:

$$\partial_r w + kw = f(\varphi) \quad \text{at} \quad r = R.$$

Solution:

$$w(r,\varphi) = R\int_0^{2\pi} f(\eta)G(r,\varphi,R,\eta)\,d\eta + \int_0^{2\pi}\int_0^R \Phi(\xi,\eta)G(r,\varphi,\xi,\eta)\xi\,d\xi\,d\eta,$$

where

$$G(r,\varphi,\xi,\eta) = \frac{1}{\pi}\sum_{n=0}^{\infty}\sum_{m=1}^{\infty}\frac{A_n J_n(\mu_{nm}r)J_n(\mu_{nm}\xi)}{(\mu_{nm}^2 R^2 + k^2 R^2 - n^2)[J_n(\mu_{nm}R)]^2}\cos[n(\varphi - \eta)],$$

$$A_0 = 1, \qquad A_n = 2 \quad (n = 1, 2, \ldots).$$

Here, the $J_n(\xi)$ are the Bessel functions and the μ_{nm} are positive roots of the transcendental equation

$$\mu J_n'(\mu R) + k J_n(\mu R) = 0.$$

7.2.3-3. Domain: $R \le r < \infty$, $0 \le \varphi \le 2\pi$. First boundary value problem.

The exterior of a circle is considered. A boundary condition is prescribed:

$$w = f(\varphi) \quad \text{at} \quad r = R.$$

Solution:

$$w(r,\varphi) = \frac{1}{2\pi}\int_0^{2\pi} f(\eta)\frac{r^2 - R^2}{r^2 - 2Rr\cos(\varphi - \eta) + R^2}\,d\eta + \int_0^{2\pi}\int_R^{\infty}\Phi(\xi,\eta)G(r,\varphi,\xi,\eta)\xi\,d\xi\,d\eta,$$

where the Green's function $G(r,\varphi,\xi,\eta)$ is defined by the formula presented in Paragraph 7.2.3-1.

⊙ *Reference*: A. G. Butkovskiy (1979).

7.2.3-4. Domain: $R_1 \leq r \leq R_2$, $0 \leq \varphi \leq 2\pi$. First boundary value problem.

An annular domain is considered. Boundary conditions are prescribed:
$$w = f_1(\varphi) \quad \text{at} \quad r = R_1, \qquad w = f_2(\varphi) \quad \text{at} \quad r = R_2.$$

Solution:
$$w(r,\varphi) = R_1 \int_0^{2\pi} f_1(\eta) \left[\frac{\partial}{\partial \xi} G(r,\varphi,\xi,\eta)\right]_{\xi=R_1} d\eta - R_2 \int_0^{2\pi} f_2(\eta) \left[\frac{\partial}{\partial \xi} G(r,\varphi,\xi,\eta)\right]_{\xi=R_2} d\eta$$
$$+ \int_0^{2\pi} \int_{R_1}^{R_2} \Phi(\xi,\eta) G(r,\varphi,\xi,\eta) \xi \, d\xi \, d\eta.$$

Here,
$$G(r,\varphi,\xi,\eta) = \frac{1}{2\pi} \sum_{n=0}^{\infty} \left(\ln \frac{1}{r_n} - \ln \frac{R_1}{\xi r_n^*}\right),$$

where
$$r_n^2 = r^2 + \rho_n^2 - 2r\rho_n \cos(\varphi - \eta), \quad (r_n^*)^2 = r^2 + (\rho_n^*)^2 - 2r\rho_n^* \cos(\varphi - \eta),$$
$$\rho_n = \begin{cases} (R_1/R_2)^{2k} \xi & \text{for } n = 2k, \\ (R_2/R_1)^{2k+2} \xi & \text{for } n = 2k+1, \end{cases} \qquad \rho_n^* = \frac{R_1^2}{\rho_n}.$$

⊙ *Reference*: B. M. Budak, A. A. Samarskii, and A. N. Tikhonov (1980).

7.2.3-5. Domain: $0 \leq r \leq R$, $0 \leq \varphi \leq \pi$. First boundary value problem.

A semicircle is considered. Boundary conditions are prescribed:
$$w = f_1(\varphi) \quad \text{at} \quad r = R, \qquad w = f_2(r) \quad \text{at} \quad \varphi = 0, \qquad w = f_3(r) \quad \text{at} \quad \varphi = \pi.$$

Solution:
$$w(r,\varphi) = -R \int_0^{\pi} f_1(\eta) \left[\frac{\partial}{\partial \xi} G(r,\varphi,\xi,\eta)\right]_{\xi=R} d\eta + \int_0^R f_2(\xi) \frac{1}{\xi} \left[\frac{\partial}{\partial \eta} G(r,\varphi,\xi,\eta)\right]_{\eta=0} d\xi$$
$$- \int_0^R f_3(\xi) \frac{1}{\xi} \left[\frac{\partial}{\partial \eta} G(r,\varphi,\xi,\eta)\right]_{\eta=\pi} d\xi + \int_0^{\pi} \int_0^R \Phi(\xi,\eta) G(r,\varphi,\xi,\eta) \xi \, d\xi \, d\eta,$$

where
$$G(r,\varphi,\xi,\eta) = \frac{1}{4\pi} \ln \frac{r^2 \xi^2 - 2R^2 r\xi \cos(\varphi - \eta) + R^4}{R^2[r^2 - 2r\xi \cos(\varphi - \eta) + \xi^2]} - \frac{1}{4\pi} \ln \frac{r^2 \xi^2 - 2R^2 r\xi \cos(\varphi + \eta) + R^4}{R^2[r^2 - 2r\xi \cos(\varphi + \eta) + \xi^2]}.$$

See also Example 2 in Paragraph 7.2.4-2.

⊙ *References*: V. S. Vladimirov, V. P. Mikhailov, A. A. Vasharin, et al. (1974), B. M. Budak, A. A. Samarskii, and A. N. Tikhonov (1980).

7.2.3-6. Domain: $0 \leq r \leq R$, $0 \leq \varphi \leq \pi/2$. First boundary value problem.

A quadrant of a circle is considered. Boundary conditions are prescribed:
$$w = f_1(\varphi) \quad \text{at} \quad r = R, \qquad w = f_2(r) \quad \text{at} \quad \varphi = 0, \qquad w = f_3(r) \quad \text{at} \quad \varphi = \pi/2.$$

Solution:
$$w(r,\varphi) = -R \int_0^{\pi/2} f_1(\eta) \left[\frac{\partial}{\partial \xi} G(r,\varphi,\xi,\eta)\right]_{\xi=R} d\eta + \int_0^R f_2(\xi) \frac{1}{\xi} \left[\frac{\partial}{\partial \eta} G(r,\varphi,\xi,\eta)\right]_{\eta=0} d\xi$$
$$- \int_0^R f_3(\xi) \frac{1}{\xi} \left[\frac{\partial}{\partial \eta} G(r,\varphi,\xi,\eta)\right]_{\eta=\pi/2} d\xi + \int_0^{\pi/2} \int_0^R \Phi(\xi,\eta) G(r,\varphi,\xi,\eta) \xi \, d\xi \, d\eta,$$

where
$$G(r,\varphi,\xi,\eta) = G_1(r,\varphi,\xi,\eta) - G_1(r,\varphi,\xi,2\pi-\eta) - G_1(r,\varphi,\xi,\pi-\eta) + G_1(r,\varphi,\xi,\pi+\eta),$$
$$G_1(r,\varphi,\xi,\eta) = \frac{1}{4\pi}\ln\frac{r^2\xi^2 - 2R^2r\xi\cos(\varphi-\eta) + R^4}{R^2[r^2 - 2r\xi\cos(\varphi-\eta) + \xi^2]}.$$

See also Example 3 in Paragraph 7.2.4-2.

⊙ *References*: V. S. Vladimirov, V. P. Mikhailov, A. A. Vasharin, et al. (1974), B. M. Budak, A. A. Samarskii, and A. N. Tikhonov (1980).

7.2.3-7. Domain: $0 \le r \le R$, $0 \le \varphi \le \beta$. First boundary value problem.

A circular sector is considered. Boundary conditions are prescribed:
$$w = f_1(\varphi) \text{ at } r = R, \qquad w = f_2(r) \text{ at } \varphi = 0, \qquad w = f_3(r) \text{ at } \varphi = \beta.$$

Solution:
$$w(r,\varphi) = -R\int_0^\beta f_1(\eta)\left[\frac{\partial}{\partial\xi}G(r,\varphi,\xi,\eta)\right]_{\xi=R} d\eta + \int_0^R f_2(\xi)\frac{1}{\xi}\left[\frac{\partial}{\partial\eta}G(r,\varphi,\xi,\eta)\right]_{\eta=0} d\xi$$
$$- \int_0^R f_3(\xi)\frac{1}{\xi}\left[\frac{\partial}{\partial\eta}G(r,\varphi,\xi,\eta)\right]_{\eta=\beta} d\xi + \int_0^\beta\int_0^R \Phi(\xi,\eta)G(r,\varphi,\xi,\eta)\xi\,d\xi\,d\eta.$$

1°. For $\beta = \pi/n$, where n is a positive integer, the Green's function is expressed as
$$G(r,\varphi,\xi,\eta) = \sum_{k=0}^{n-1}\left[G_1(r,\varphi,\xi,2k\beta+\eta) - G_1(r,\varphi,\xi,2k\beta-\eta)\right],$$
$$G_1(r,\varphi,\xi,\eta) = \frac{1}{4\pi}\ln\frac{r^2\xi^2 - 2R^2r\xi\cos(\varphi-\eta) + R^4}{R^2[r^2 - 2r\xi\cos(\varphi-\eta) + \xi^2]}.$$

⊙ *Reference*: B. M. Budak, A. A. Samarskii, and A. N. Tikhonov (1980).

2°. For arbitrary β, the Green's function is given by
$$G(r,\varphi,\xi,\eta) = \frac{1}{2\pi}\ln\frac{\left|z^{\pi/\beta} - \bar{\zeta}^{\pi/\beta}\right|\left|R^{2\pi/\beta} - (\bar{\zeta}z)^{\pi/\beta}\right|}{\left|z^{\pi/\beta} - \zeta^{\pi/\beta}\right|\left|R^{2\pi/\beta} - (\zeta z)^{\pi/\beta}\right|},$$
where $z = re^{i\varphi}$, $\zeta = \xi e^{i\eta}$, $\bar{\zeta} = \xi e^{-i\eta}$, and $i^2 = -1$.

7.2.3-8. Domain: $0 \le r < \infty$, $0 \le \varphi \le \beta$. First boundary value problem.

A wedge domain is considered. Boundary conditions are prescribed:
$$w = f_1(r) \text{ at } \varphi = 0, \qquad w = f_2(r) \text{ at } \varphi = \beta.$$

Solution:
$$w(r,\varphi) = \int_0^\infty f_1(\xi)\frac{1}{\xi}\left[\frac{\partial}{\partial\eta}G(r,\varphi,\xi,\eta)\right]_{\eta=0} d\xi - \int_0^\infty f_2(\xi)\frac{1}{\xi}\left[\frac{\partial}{\partial\eta}G(r,\varphi,\xi,\eta)\right]_{\eta=\beta} d\xi$$
$$+ \int_0^\beta\int_0^\infty \Phi(\xi,\eta)G(r,\varphi,\xi,\eta)\xi\,d\xi\,d\eta,$$

where
$$G(r,\varphi,\xi,\eta) = \frac{1}{4\pi}\ln\frac{r^{2\pi/\beta} - 2(r\xi)^{\pi/\beta}\cos[\pi(\varphi+\eta)/\beta] + \xi^{2\pi/\beta}}{r^{2\pi/\beta} - 2(r\xi)^{\pi/\beta}\cos[\pi(\varphi-\eta)/\beta] + \xi^{2\pi/\beta}}.$$

Alternatively, the Green's function can be represented in the complex form
$$G(r,\varphi,\xi,\eta) = \frac{1}{2\pi}\ln\frac{\left|z^{\pi/\beta} - \bar{\zeta}^{\pi/\beta}\right|}{\left|z^{\pi/\beta} - \zeta^{\pi/\beta}\right|}, \qquad z = re^{i\varphi}, \quad \zeta = \xi e^{i\eta}, \quad \bar{\zeta} = \xi e^{-i\eta}, \quad i^2 = -1.$$

7.2.4. Arbitrary Shape Domain. Conformal Mappings Method

7.2.4-1. Description of the method. Tables of conformal mappings.

Any simply connected domain D in the xy-plane with a piecewise smooth boundary can be mapped in a mutually unique way, with an appropriate conformal mapping, onto the upper half-plane or into a unit circle in a uv-plane. Under a conformal mapping, a Poisson equation in the xy-plane transforms into a Poisson equation in the uv-plane; what is changed is the function Φ, as well as the function f in the boundary condition. Consequently, a first and a second boundary value problem for the plane domain D can be reduced, respectively, to a first and a second boundary value problem for the upper half-plane or a unit circle. The latter problems are considered above (see Subsections 7.2.2 and 7.2.3).

A large number of conformal mappings (mappings defined by analytic functions) of various domains onto the upper half-plane or a unit circle can be found, for example, in Lavrik and Savenkov (1970), Lavrent'ev and Shabat (1973), and Ivanov and Trubetskov (1994).

Table 22 presents conformal mappings of some domains D in the complex plane z onto the upper half-plane $\operatorname{Im} \omega \geq 0$ in the complex plane ω. In the relations involving square roots, it is assumed that $\sqrt{\zeta} = \sqrt{|\zeta|}\,[\cos(\tfrac{1}{2}\varphi) + i\sin(\tfrac{1}{2}\varphi)]$, where $\varphi = \arg \zeta$ (i.e., the first branch of $\sqrt{\zeta}$ is taken).

Table 23 presents conformal mappings of some domains D in the complex plane z onto the unit circle $|\omega| \leq 1$ in the complex plane ω.

7.2.4-2. General formula for the Green's function. Example boundary value problems.

Let a function $\omega = \omega(z)$ define a conformal mapping of a domain D in the complex plane z onto the upper half-plane in the complex plane ω. Then the Green's function of the first boundary value problem in D for the Poisson (Laplace) equation is expressed as

$$G(x, y, \xi, \eta) = \frac{1}{2\pi} \ln\left|\frac{\omega(z) - \bar{\omega}(\zeta)}{\omega(z) - \omega(\zeta)}\right|, \qquad z = x + iy, \quad \zeta = \xi + i\eta, \tag{1}$$

where $\omega(z) = u(x,y) + iv(x,y)$ and $\bar{\omega}(z) = u(x,y) - iv(x,y)$.

The solution of the first boundary value problem for the Poisson equation is determined by the above Green's function in accordance with formula (2) specified in Paragraph 7.2.1-1.

Example 1. Consider the first boundary value problem for the Poisson equation in the strip $-\infty < x < \infty$, $0 \leq y \leq a$. The function that maps this strip onto the upper half-plane has the form $\omega(z) = \exp(\pi z/a)$ (see the second row of Table 22). Substituting this expression into relation (1) and performing elementary transformations, we obtain the Green's function

$$G(x, y, \xi, \eta) = \frac{1}{4\pi} \ln \frac{\cosh[\pi(x-\xi)/a] - \cos[\pi(y+\eta)/a]}{\cosh[\pi(x-\xi)/a] - \cos[\pi(y-\eta)/a]}.$$

Example 2. Consider the first boundary value problem for the Poisson equation in a semicircle of radius a such that $D = \{x^2 + y^2 \leq a^2,\ y \geq 0\}$. The domain D is conformally mapped onto the upper half-plane by the function $\omega(z) = -(z/a + a/z)$ (see the sixth row of Table 22). Substituting this expression into (1), we arrive at the Green's function

$$G(x, y, \xi, \eta) = \frac{1}{2\pi} \ln \frac{|z - \bar{\zeta}|\,|a^2 - z\bar{\zeta}|}{|z - \zeta|\,|a^2 - z\zeta|}, \qquad z = x + iy, \quad \zeta = \xi + i\eta.$$

Example 3. Consider the first boundary value problem for the Poisson equation in a quadrant of a circle of radius a, so that $D = \{x^2 + y^2 \leq a^2,\ x \geq 0,\ y \geq 0\}$. The conformal mapping of the domain D onto the upper half-plane is performed with the function $\omega(z) = -(z/a)^2 - (a/z)^2$ (see the seventh row of Table 22). Substituting this expression into (1) yields

$$G(x, y, \xi, \eta) = \frac{1}{2\pi} \ln \frac{|z^2 - \bar{\zeta}^2|\,|a^4 - z^2\bar{\zeta}^2|}{|z^2 - \zeta^2|\,|a^4 - z^2\zeta^2|}, \qquad z = x + iy, \quad \zeta = \xi + i\eta.$$

3°. Let a function $\omega = \omega(z)$ define a conformal mapping of a domain D in the complex plane z onto the unit circle $|\omega| \leq 1$ in the complex plane ω. Then the Green's function of the first boundary value problem in D for the Laplace equation is given by

$$G(x, y, \xi, \eta) = \frac{1}{2\pi} \ln\left|\frac{1 - \bar{\omega}(\zeta)\omega(z)}{\omega(z) - \omega(\zeta)}\right|, \qquad z = x + iy, \quad \zeta = \xi + i\eta. \tag{2}$$

⊙ *References for Subsection* 7.2.4: N. N. Lebedev, I. P. Skal'skaya, and Ya. S. Uflyand (1955), A. G. Sveshnikov and A. N. Tikhonov (1974).

TABLE 22
Conformal mapping of some domains D in the z-plane onto the upper half-plane $\operatorname{Im}\omega \geq 0$ in the ω-plane. Notation: $z = x + iy$ and $\omega = u + iv$

No	Domain D in the z-plane	Transformation
1	First quadrant: $0 \leq x < \infty, 0 \leq y < \infty$	$\omega = a^2 z^2 + b$, a, b are real numbers
2	Infinite strip of width a: $-\infty < x < \infty, 0 \leq y \leq a$	$\omega = \exp(\pi z / a)$
3	Semiinfinite strip of width a: $0 \leq x < \infty, 0 \leq y \leq a$	$\omega = \cosh(\pi z / a)$
4	Plane with the cut in the real axis	$\omega = \sqrt{z}$
5	Interior of an infinite sector with angle β: $0 \leq \arg z \leq \beta, 0 \leq |z| < \infty$ $(0 < \beta \leq 2\pi)$	$\omega = z^{\pi/\beta}$
6	Upper half of a circle of radius a: $x^2 + y^2 \leq a^2, y \geq 0$	$\omega = -\dfrac{z}{a} - \dfrac{a}{z}$
7	Quadrant of a circle of radius a: $x^2 + y^2 \leq a^2, x \geq 0, y \geq 0$	$\omega = -\dfrac{z^2}{a^2} - \dfrac{a^2}{z^2}$
8	Sector of a circle of radius a with angle β: $x^2 + y^2 \leq a^2, 0 \leq \arg z \leq \beta$	$\omega = -\left(\dfrac{z}{a}\right)^{\pi/\beta} - \left(\dfrac{a}{z}\right)^{\pi/\beta}$
9	Upper half-plane with a circular domain or radius a removed: $y \geq 0, x^2 + y^2 \geq a^2$	$\omega = \dfrac{z}{a} + \dfrac{a}{z}$
10	Exterior of a parabola: $y^2 - 2px \geq 0$	$\omega = \sqrt{z - \tfrac{1}{2}p} - i\sqrt{\tfrac{1}{2}p}$
11	Interior of a parabola: $y^2 - 2px \leq 0$	$\omega = i\cosh\left(\pi\sqrt{\tfrac{1}{2}z/p - \tfrac{1}{4}}\right)$

7.3. Helmholtz Equation $\Delta_2 w + \lambda w = -\Phi(\mathbf{x})$

Many problems related to steady-state oscillations (mechanical, acoustical, thermal, electromagnetic, etc.) lead to the two-dimensional Helmholtz equation. For $\lambda < 0$, this equation describes mass transfer processes with volume chemical reactions of the first order. Moreover, any elliptic equation with constant coefficients can be reduced to the Helmholtz equation.

7.3.1. General Remarks, Results, and Formulas

7.3.1-1. Some definitions.

The Helmholtz equation is called homogeneous if $\Phi = 0$ and nonhomogeneous if $\Phi \neq 0$. A homogeneous boundary value problem is a boundary value problem for the homogeneous Helmholtz equation with homogeneous boundary conditions; a particular solution of a homogeneous boundary value problem is $w = 0$.

The values λ_n of the parameter λ for which there are nontrivial solutions (solutions other

TABLE 23
Conformal mapping of some domains D in the z-plane onto the unit circle $|\omega| \leq 1$. Notation: $z = x + iy$, $\omega = u + iv$, $z_0 = x_0 + iy_0$, and $\bar{z}_0 = x_0 - iy_0$

No	Domain D in z-plane	Transformation
1	Upper half-plane: $-\infty < x < \infty, 0 \leq y < \infty$	$\omega = e^{i\lambda}\dfrac{z-z_0}{z-\bar{z}_0}$, λ is a real number
2	A circle of unit radius: $x^2 + y^2 \leq 1$	$\omega = e^{i\lambda}\dfrac{z-z_0}{1-\bar{z}_0 z}$, λ is a real number
3	Exterior of a circle of radius a: $x^2 + y^2 \geq a^2$	$\omega = \dfrac{a}{z}$
4	Infinite strip of width a: $-\infty < x < \infty, 0 \leq y \leq a$	$\omega = \dfrac{\exp(\pi z/a) - \exp(\pi z_0/a)}{\exp(\pi z/a) - \exp(\pi \bar{z}_0/a)}$
5	Semicircle of radius a: $x^2 + y^2 \leq a^2$, $x \geq 0$	$\omega = i\dfrac{z^2 + 2az - a^2}{z^2 - 2az - a^2}$
6	Sector of a unit circle with angle β: $|z| \leq 1, 0 \leq \arg z \leq \beta$	$\omega = \dfrac{(1+z^{\pi/\beta})^2 - i(1-z^{\pi/\beta})^2}{(1+z^{\pi/\beta})^2 + i(1-z^{\pi/\beta})^2}$
7	Exterior of an ellipse with semiaxes a and b: $(x/a)^2 + (y/b)^2 \geq 1$	$z = \dfrac{1}{2}\left[(a-b)\omega + \dfrac{a+b}{\omega}\right]$

than identical zero) of the homogeneous boundary value problem are called eigenvalues and the corresponding solutions, $w = w_n$, are called eigenfunctions of the boundary value problem.

In what follows, the first, second, and third boundary value problems for the two-dimensional Helmholtz equation in a finite two-dimensional domain S with boundary L are considered. For the third boundary value problem with the boundary condition

$$\frac{\partial w}{\partial N} + kw = 0 \quad \text{for} \quad \mathbf{r} \in L,$$

it is assumed that $k > 0$. Here, $\frac{\partial w}{\partial N}$ is the derivative along the outward normal to the contour L, and $\mathbf{r} = \{x, y\}$.

7.3.1-2. Properties of eigenvalues and eigenfunctions.

1°. There are infinitely many eigenvalues $\{\lambda_n\}$; the set of eigenvalues forms a discrete spectrum for the given boundary value problem.

2°. All eigenvalues are positive, except for the eigenvalue $\lambda_0 = 0$ existing in the second boundary value problem (the corresponding eigenfunction is $w_0 = \text{const}$). We number the eigenvalues in order of increasing magnitudes, $\lambda_1 < \lambda_2 < \lambda_3 < \cdots$.

3°. The eigenvalues tend to infinity as the number n increases. The following asymptotic estimate holds:

$$\lim_{n \to \infty} \frac{n}{\lambda_n} = \frac{S_2}{4\pi},$$

where S_2 is the area of the two-dimensional domain under study.

4°. The eigenfunctions $w_n = w_n(x, y)$ are defined up to a constant multiplier. Any two eigenfunctions corresponding to different eigenvalues, $\lambda_n \neq \lambda_m$, are orthogonal:

$$\int_S w_n w_m \, dS = 0.$$

5°. Any twice continuously differentiable function $f = f(\mathbf{r})$ that satisfies the boundary conditions of a boundary value problem can be expanded into a uniformly convergent series in the eigenfunctions of the boundary value problem:

$$f = \sum_{n=1}^{\infty} f_n w_n, \qquad \text{where} \quad f_n = \frac{1}{\|w_n\|^2} \int_S f w_n \, dS, \quad \|w_n\|^2 = \int_S w_n^2 \, dS.$$

If f is square summable, then the series converges in mean.

6°. The eigenvalues of the first boundary value problem do not increase if the domain is extended.

Remark 1. In a two-dimensional problem, generally correspond to each eigenvalue λ_n finitely many linearly independent eigenfunctions $w_n^{(1)}, w_n^{(2)}, \ldots, w_n^{(p)}$. These functions can always be replaced by their linear combinations

$$\bar{w}_n^{(j)} = c_{j,1} w_n^{(1)} + \cdots + c_{j,j-1} w_n^{(j-1)} + w_n^{(j)}, \qquad j = 1, 2, \ldots, p,$$

so that the new eigenfunctions $\bar{w}_n^{(1)}, \bar{w}_n^{(2)}, \ldots, \bar{w}_n^{(p)}$ now are pairwise orthogonal. Therefore, without loss of generality, we assume that all the eigenfunctions are orthogonal.

⦿ *Reference*: V. M. Babich, M. B. Kapilevich, S. G. Mikhlin, et al. (1964).

7.3.1-3. Nonhomogeneous Helmholtz equation with homogeneous boundary conditions.

Three cases are possible.

1°. If the equation parameter λ is not equal to any one of the eigenvalues, then there exists the series solution

$$w = \sum_{n=1}^{\infty} \frac{A_n}{\lambda_n - \lambda} w_n, \qquad \text{where} \quad A_n = \frac{1}{\|w_n\|^2} \int_S \Phi w_n \, dS, \quad \|w_n\|^2 = \int_S w_n^2 \, dS.$$

2°. If λ is equal to some eigenvalue, $\lambda = \lambda_m$, then the solution of the nonhomogeneous problem exists only if the function Φ is orthogonal to w_m, i.e.,

$$\int_S \Phi w_m \, dS = 0.$$

In this case the system is expressed as

$$w = \sum_{n=1}^{m-1} \frac{A_n}{\lambda_n - \lambda_m} w_n + \sum_{n=m+1}^{\infty} \frac{A_n}{\lambda_n - \lambda_m} w_n + C w_m, \qquad A_n = \frac{1}{\|w_n\|^2} \int_S \Phi w_n \, dS,$$

where $\|w_n\|^2 = \int_S w_n^2 \, dS$, and C is an arbitrary constant.

3°. If $\lambda = \lambda_m$ and $\int_S \Phi w_m \, dS \neq 0$, then the boundary value problem for the nonhomogeneous equation does not have solutions.

Remark 2. If p_n mutually orthogonal eigenfunctions $w_n^{(j)}$ ($j = 1, 2, \ldots, p_n$) correspond to each eigenvalue λ_n, then, for $\lambda \neq \lambda_n$, the solution is written as

$$w = \sum_{n=1}^{\infty} \sum_{j=1}^{p_n} \frac{A_n^{(j)}}{\lambda_n - \lambda} w_n^{(j)}, \qquad \text{where} \quad A_n^{(j)} = \frac{1}{\|w_n^{(j)}\|^2} \int_S \Phi w_n^{(j)} \, dS, \quad \|w_n^{(j)}\|^2 = \int_S \left[w_n^{(j)}\right]^2 dS.$$

⦿ *Reference*: V. M. Babich, M. B. Kapilevich, S. G. Mikhlin, et al. (1964).

7.3.1-4. Solution of nonhomogeneous boundary value problem of general form.

1°. The solution of the first boundary value problem for the Helmholtz equation with the boundary condition

$$w = f(\mathbf{r}) \quad \text{for} \quad \mathbf{r} \in L$$

can be represented in the form

$$w(\mathbf{r}) = \int_S \Phi(\boldsymbol{\rho}) G(\mathbf{r}, \boldsymbol{\rho}) \, dS_\rho - \int_L f(\boldsymbol{\rho}) \frac{\partial}{\partial N_\rho} G(\mathbf{r}, \boldsymbol{\rho}) \, dL_\rho. \tag{1}$$

Here, $\mathbf{r} = \{x, y\}$ and $\boldsymbol{\rho} = \{\xi, \eta\}$ ($\mathbf{r} \in S$, $\boldsymbol{\rho} \in S$); $\frac{\partial}{\partial N_\rho}$ denotes the derivative along the outward normal to the contour L with respect to the variables ξ and η. The Green's function is given by the series

$$G(\mathbf{r}, \boldsymbol{\rho}) = \sum_{n=1}^{\infty} \frac{w_n(\mathbf{r}) w_n(\boldsymbol{\rho})}{\|w_n\|^2 (\lambda_n - \lambda)}, \quad \lambda \ne \lambda_n, \tag{2}$$

where the w_n and λ_n are the eigenfunctions and eigenvalues of the homogeneous first boundary value problem.

2°. The solution of the second boundary value problem with the boundary condition

$$\frac{\partial w}{\partial N} = f(\mathbf{r}) \quad \text{for} \quad \mathbf{r} \in L$$

can be written as

$$w(\mathbf{r}) = \int_S \Phi(\boldsymbol{\rho}) G(\mathbf{r}, \boldsymbol{\rho}) \, dS_\rho + \int_L f(\boldsymbol{\rho}) G(\mathbf{r}, \boldsymbol{\rho}) \, dL_\rho. \tag{3}$$

Here, the Green's function is given by the series

$$G(\mathbf{r}, \boldsymbol{\rho}) = -\frac{1}{S_2 \lambda} + \sum_{n=1}^{\infty} \frac{w_n(\mathbf{r}) w_n(\boldsymbol{\rho})}{\|w_n\|^2 (\lambda_n - \lambda)}, \quad \lambda \ne \lambda_n, \tag{4}$$

where S_2 is the area of the two-dimensional domain under consideration, and the λ_n and w_n are the positive eigenvalues and the corresponding eigenfunctions of the homogeneous second boundary value problem. For clarity, the term corresponding to the zero eigenvalue $\lambda_0 = 0$ ($w_0 = $ const) is singled out in (4).

3°. The solution of the third boundary value problem for the Helmholtz equation with the boundary condition

$$\frac{\partial w}{\partial N} + kw = f(\mathbf{r}) \quad \text{for} \quad \mathbf{r} \in L$$

is given by formula (3), where the Green's function is defined by series (2), which involves the eigenfunctions w_n and eigenvalues λ_n of the homogeneous third boundary value problem.

7.3.1-5. Boundary conditions at infinity in the case of an infinite domain.

In what follows, the function Φ is assumed to be finite or sufficiently rapidly decaying as $r \to \infty$.

1°. For $\lambda < 0$, in the case of an infinite domain, the vanishing condition of the solution at infinity is set,

$$w \to 0 \quad \text{as} \quad r \to \infty.$$

2°. For $\lambda > 0$, if the domain is unbounded, the radiation conditions (Sommerfeld conditions) at infinity are used. In two-dimensional problems, these conditions are written as

$$\lim_{r \to \infty} \sqrt{r}\, w = \text{const}, \quad \lim_{r \to \infty} \sqrt{r} \left(\frac{\partial w}{\partial r} + i\sqrt{\lambda}\, w \right) = 0,$$

where $i^2 = -1$.

To identify a single solution, the principle of limit absorption and the principle of limit amplitude are also used.

⊙ *Reference*: A. N. Tikhonov and A. A. Samarskii (1990).

7.3.2. Problems in Cartesian Coordinate System

A two-dimensional nonhomogeneous Helmholtz equation in the rectangular Cartesian system of coordinates has the form

$$\frac{\partial^2 w}{\partial x^2} + \frac{\partial^2 w}{\partial y^2} + \lambda w = -\Phi(x, y).$$

7.3.2-1. Particular solutions and some relations.

1°. Particular solutions of the homogeneous equation ($\Phi \equiv 0$):

$w = (Ax + B)(C \cos \mu y + D \sin \mu y), \qquad \lambda = \mu^2,$

$w = (Ax + B)(C \cosh \mu y + D \sinh \mu y), \qquad \lambda = -\mu^2,$

$w = (A \cos \mu x + B \sin \mu x)(Cy + D), \qquad \lambda = \mu^2,$

$w = (A \cosh \mu x + B \sinh \mu x)(Cy + D), \qquad \lambda = -\mu^2,$

$w = (A \cos \mu_1 x + B \sin \mu_1 x)(C \cos \mu_2 y + D \sin \mu_2 y), \qquad \lambda = \mu_1^2 + \mu_2^2,$

$w = (A \cos \mu_1 x + B \sin \mu_1 x)(C \cosh \mu_2 y + D \sinh \mu_2 y), \qquad \lambda = \mu_1^2 - \mu_2^2,$

$w = (A \cosh \mu_1 x + B \sinh \mu_1 x)(C \cos \mu_2 y + D \sin \mu_2 y), \qquad \lambda = -\mu_1^2 + \mu_2^2,$

$w = (A \cosh \mu_1 x + B \sinh \mu_1 x)(C \cosh \mu_2 y + D \sinh \mu_2 y), \qquad \lambda = -\mu_1^2 - \mu_2^2,$

where A, B, C, and D are arbitrary constants.

2°. Fundamental solutions:

$$\mathscr{E}(x, y) = \frac{1}{2\pi} K_0(sr) \quad \text{if} \quad \lambda = -s^2 < 0,$$

$$\mathscr{E}(x, y) = \frac{i}{4} H_0^{(1)}(kr) \quad \text{if} \quad \lambda = k^2 > 0,$$

$$\mathscr{E}(x, y) = -\frac{i}{4} H_0^{(2)}(kr) \quad \text{if} \quad \lambda = k^2 > 0,$$

where $r = \sqrt{x^2 + y^2}$, $K_0(z)$ is the modified Bessel function of the second kind, $H_0^{(1)}(z)$ and $H_0^{(2)}(z)$ are the Hankel functions of the first and second kind of order 0, x_0 and y_0 are arbitrary constants, and $i^2 = -1$. The leading term of the asymptotic expansion of the fundamental solutions, as $r \to 0$, is given by $\frac{1}{2\pi} \ln \frac{1}{r}$.

3°. Suppose $w = w(x, y)$ is a solution of the homogeneous Helmholtz equation. Then the functions

$w_1 = w(x + C_1, \pm y + C_2),$

$w_2 = w(-x + C_1, \pm y + C_2),$

$w_3 = w(x \cos \theta + y \sin \theta + C_1, -x \sin \theta + y \cos \theta + C_2),$

where C_1, C_2, and θ are arbitrary constants, are also solutions of the equation.

⊙ *Reference*: A. N. Tikhonov and A. A. Samarskii (1990).

7.3.2-2. Domain: $-\infty < x < \infty$, $-\infty < y < \infty$.

1°. Solution for $\lambda = -s^2 < 0$:

$$w(x, y) = \frac{1}{2\pi} \int_{-\infty}^{\infty} \int_{-\infty}^{\infty} \Phi(\xi, \eta) K_0(s\varrho) \, d\xi \, d\eta, \qquad \varrho = \sqrt{(x - \xi)^2 + (y - \eta)^2}.$$

2°. Solution for $\lambda = k^2 > 0$:

$$w(x, y) = -\frac{i}{4} \int_{-\infty}^{\infty} \int_{-\infty}^{\infty} \Phi(\xi, \eta) H_0^{(2)}(k\varrho) \, d\xi \, d\eta, \qquad \varrho = \sqrt{(x - \xi)^2 + (y - \eta)^2}.$$

The radiation conditions (Sommerfeld conditions) at infinity were used to obtain this solution (see Paragraph 7.3.1-5, Item 2°).

⊙ *References*: B. M. Budak, A. A. Samarskii, and A. N. Tikhonov (1980), A. N. Tikhonov and A. A. Samarskii (1990).

7.3.2-3. Domain: $-\infty < x < \infty$, $0 \le y < \infty$. First boundary value problem.

A half-plane is considered. A boundary condition is prescribed:

$$w = f(x) \quad \text{at} \quad y = 0.$$

Solution:

$$w(x,y) = \int_{-\infty}^{\infty} f(\xi) \left[\frac{\partial}{\partial \eta} G(x,y,\xi,\eta) \right]_{\eta=0} d\xi + \int_{0}^{\infty} \int_{-\infty}^{\infty} \Phi(\xi,\eta) G(x,y,\xi,\eta) \, d\xi \, d\eta.$$

$1°$. The Green's function for $\lambda = -s^2 < 0$:

$$G(x,y,\xi,\eta) = \frac{1}{2\pi} \left[K_0(s\varrho_1) - K_0(s\varrho_2) \right],$$

$$\varrho_1 = \sqrt{(x-\xi)^2 + (y-\eta)^2}, \quad \varrho_2 = \sqrt{(x-\xi)^2 + (y+\eta)^2}.$$

$2°$. The Green's function for $\lambda = k^2 > 0$:

$$G(x,y,\xi,\eta) = -\frac{i}{4} \left[H_0^{(2)}(k\varrho_1) - H_0^{(2)}(k\varrho_2) \right].$$

The radiation conditions at infinity were used to obtain this relation (see Paragraph 7.3.1-5, Item $2°$).

⊙ *Reference*: B. M. Budak, A. A. Samarskii, and A. N. Tikhonov (1980).

7.3.2-4. Domain: $-\infty < x < \infty$, $0 \le y < \infty$. Second boundary value problem.

A half-plane is considered. A boundary condition is prescribed:

$$\partial_y w = f(x) \quad \text{at} \quad y = 0.$$

Solution:

$$w(x,y) = -\int_{-\infty}^{\infty} f(\xi) G(x,y,\xi,0) \, d\xi + \int_{0}^{\infty} \int_{-\infty}^{\infty} \Phi(\xi,\eta) G(x,y,\xi,\eta) \, d\xi \, d\eta.$$

$1°$. The Green's function for $\lambda = -s^2 < 0$:

$$G(x,y,\xi,\eta) = \frac{1}{2\pi} \left[K_0(s\varrho_1) + K_0(s\varrho_2) \right],$$

$$\varrho_1 = \sqrt{(x-\xi)^2 + (y-\eta)^2}, \quad \varrho_2 = \sqrt{(x-\xi)^2 + (y+\eta)^2}.$$

$2°$. The Green's function for $\lambda = k^2 > 0$:

$$G(x,y,\xi,\eta) = -\frac{i}{4} \left[H_0^{(2)}(k\varrho_1) + H_0^{(2)}(k\varrho_2) \right].$$

The radiation conditions at infinity were used to obtain this relation (see Paragraph 7.3.1-5, Item $2°$).

⊙ *Reference*: B. M. Budak, A. A. Samarskii, and A. N. Tikhonov (1980).

7.3.2-5. Domain: $0 \leq x < \infty$, $0 \leq y < \infty$. First boundary value problem.

A quadrant of the plane is considered. Boundary conditions are prescribed:

$$w = f_1(y) \quad \text{at} \quad x = 0, \qquad w = f_2(x) \quad \text{at} \quad y = 0.$$

Solution:

$$w(x,y) = \int_0^\infty f_1(\eta) \left[\frac{\partial}{\partial \xi} G(x,y,\xi,\eta) \right]_{\xi=0} d\eta + \int_0^\infty f_2(\xi) \left[\frac{\partial}{\partial \eta} G(x,y,\xi,\eta) \right]_{\eta=0} d\xi$$
$$+ \int_0^\infty \int_0^\infty \Phi(\xi,\eta) G(x,y,\xi,\eta)\, d\xi\, d\eta.$$

1°. The Green's function for $\lambda = -s^2 < 0$:

$$G(x,y,\xi,\eta) = \frac{1}{2\pi} \left[K_0(s\varrho_1) - K_0(s\varrho_2) - K_0(s\varrho_3) + K_0(s\varrho_4) \right],$$

$$\varrho_1 = \sqrt{(x-\xi)^2 + (y-\eta)^2}, \qquad \varrho_2 = \sqrt{(x-\xi)^2 + (y+\eta)^2},$$
$$\varrho_3 = \sqrt{(x+\xi)^2 + (y-\eta)^2}, \qquad \varrho_4 = \sqrt{(x+\xi)^2 + (y+\eta)^2}.$$

2°. The Green's function for $\lambda = k^2 > 0$:

$$G(x,y,\xi,\eta) = -\frac{i}{4} \left[H_0^{(2)}(k\varrho_1) - H_0^{(2)}(k\varrho_2) - H_0^{(2)}(k\varrho_3) + H_0^{(2)}(k\varrho_4) \right].$$

7.3.2-6. Domain: $0 \leq x < \infty$, $0 \leq y < \infty$. Second boundary value problem.

A quadrant of the plane is considered. Boundary conditions are prescribed:

$$\partial_x w = f_1(y) \quad \text{at} \quad x = 0, \qquad \partial_y w = f_2(x) \quad \text{at} \quad y = 0.$$

Solution:

$$w(x,y) = -\int_0^\infty f_1(\eta) G(x,y,0,\eta)\, d\eta - \int_0^\infty f_2(\xi) G(x,y,\xi,0)\, d\xi$$
$$+ \int_0^\infty \int_0^\infty \Phi(\xi,\eta) G(x,y,\xi,\eta)\, d\xi\, d\eta.$$

1°. The Green's function for $\lambda = -s^2 < 0$:

$$G(x,y,\xi,\eta) = \frac{1}{2\pi} \left[K_0(s\varrho_1) + K_0(s\varrho_2) + K_0(s\varrho_3) + K_0(s\varrho_4) \right],$$

$$\varrho_1 = \sqrt{(x-\xi)^2 + (y-\eta)^2}, \qquad \varrho_2 = \sqrt{(x-\xi)^2 + (y+\eta)^2},$$
$$\varrho_3 = \sqrt{(x+\xi)^2 + (y-\eta)^2}, \qquad \varrho_4 = \sqrt{(x+\xi)^2 + (y+\eta)^2}.$$

2°. The Green's function for $\lambda = k^2 > 0$:

$$G(x,y,\xi,\eta) = -\frac{i}{4} \left[H_0^{(2)}(k\varrho_1) + H_0^{(2)}(k\varrho_2) + H_0^{(2)}(k\varrho_3) + H_0^{(2)}(k\varrho_4) \right].$$

7.3.2-7. Domain: $-\infty < x < \infty$, $0 \leq y \leq a$. First boundary value problem.

An infinite strip is considered. Boundary conditions are prescribed:

$$w = f_1(x) \quad \text{at} \quad y = 0, \qquad w = f_2(x) \quad \text{at} \quad y = a.$$

Solution:

$$w(x,y) = \int_{-\infty}^{\infty} f_1(\xi)\left[\frac{\partial}{\partial \eta}G(x,y,\xi,\eta)\right]_{\eta=0} d\xi - \int_{-\infty}^{\infty} f_2(\xi)\left[\frac{\partial}{\partial \eta}G(x,y,\xi,\eta)\right]_{\eta=a} d\xi$$
$$+ \int_0^a \int_{-\infty}^{\infty} \Phi(\xi,\eta) G(x,y,\xi,\eta)\,d\xi\,d\eta.$$

Green's function:

$$G(x,y,\xi,\eta) = \frac{1}{a}\sum_{n=1}^{\infty} \frac{1}{\beta_n} \exp(-\beta_n|x-\xi|)\sin(q_n y)\sin(q_n \eta), \qquad q_n = \frac{\pi n}{a}, \quad \beta_n = \sqrt{q_n^2 - \lambda}.$$

Alternatively, the Green's function for $\lambda = -s^2 < 0$ can be represented as

$$G(x,y,\xi,\eta) = \frac{1}{2\pi}\sum_{n=-\infty}^{\infty}\bigl[K_0(s\varrho_{1n}) - K_0(s\varrho_{2n})\bigr],$$
$$\varrho_{n1} = \sqrt{(x-\xi)^2 + (y-\eta-2na)^2}, \qquad \varrho_{n2} = \sqrt{(x-\xi)^2 + (y+\eta+2na)^2}.$$

7.3.2-8. Domain: $-\infty < x < \infty$, $0 \leq y \leq a$. Second boundary value problem.

An infinite strip is considered. Boundary conditions are prescribed:

$$\partial_y w = f_1(x) \quad \text{at} \quad y = 0, \qquad \partial_y w = f_2(x) \quad \text{at} \quad y = a.$$

Solution:

$$w(x,y) = -\int_{-\infty}^{\infty} f_1(\xi)G(x,y,\xi,0)\,d\xi + \int_{-\infty}^{\infty} f_2(\xi)G(x,y,\xi,a)\,d\xi$$
$$+ \int_0^a \int_{-\infty}^{\infty} \Phi(\xi,\eta)G(x,y,\xi,\eta)\,d\xi\,d\eta.$$

Green's function:

$$G(x,y,\xi,\eta) = \frac{1}{2a}\sum_{n=0}^{\infty} \frac{\varepsilon_n}{\beta_n}\exp(-\beta_n|x-\xi|)\cos(q_n y)\cos(q_n \eta),$$
$$q_n = \frac{\pi n}{a}, \quad \beta_n = \sqrt{q_n^2 - \lambda}, \quad \varepsilon = \begin{cases} 1 & \text{for } n = 0, \\ 2 & \text{for } n \neq 0. \end{cases}$$

Alternatively, the Green's function for $\lambda = -s^2 < 0$ can be represented as

$$G(x,y,\xi,\eta) = \frac{1}{2\pi}\sum_{n=-\infty}^{\infty}\bigl[K_0(s\varrho_{1n}) + K_0(s\varrho_{2n})\bigr],$$
$$\varrho_{n1} = \sqrt{(x-\xi)^2 + (y-\eta_{n1})^2}, \qquad \eta_{n1} = 2na + \eta,$$
$$\varrho_{n2} = \sqrt{(x-\xi)^2 + (y-\eta_{n2})^2}, \qquad \eta_{n2} = 2na - \eta.$$

⊙ *Reference*: B. M. Budak, A. A. Samarskii, and A. N. Tikhonov (1980).

> **7.3.2-9. Domain: $-\infty < x < \infty$, $0 \leq y \leq a$. Third boundary value problem.**

An infinite strip is considered. Boundary conditions are prescribed:

$$\partial_y w - k_1 w = f_1(x) \quad \text{at} \quad y = 0, \qquad \partial_y w + k_2 w = f_2(x) \quad \text{at} \quad y = a.$$

The solution $w(x, y)$ is determined by the formula in Paragraph 7.3.2-8 where

$$G(x, y, \xi, \eta) = \frac{1}{2} \sum_{n=1}^{\infty} \frac{\varphi_n(y) \varphi_n(\eta)}{\|\varphi_n\|^2 \beta_n} \exp(-\beta_n |x - \xi|), \qquad \beta_n = \sqrt{\mu_n^2 - \lambda},$$

$$\varphi_n(y) = \mu_n \cos(\mu_n y) + k_1 \sin(\mu_n y), \quad \|\varphi_n\|^2 = \frac{1}{2}(\mu_n^2 + k_1^2)\left[a + \frac{(k_1 + k_2)(\mu_n^2 + k_1 k_2)}{(\mu_n^2 + k_1^2)(\mu_n^2 + k_2^2)}\right].$$

Here, the μ_n are positive roots of the transcendental equation $\tan(\mu a) = \dfrac{(k_1 + k_2)\mu}{\mu^2 - k_1 k_2}$.

> **7.3.2-10. Domain: $-\infty < x < \infty$, $0 \leq y \leq a$. Mixed boundary value problem.**

An infinite strip is considered. Boundary conditions are prescribed:

$$w = f_1(x) \quad \text{at} \quad y = 0, \qquad \partial_y w = f_2(x) \quad \text{at} \quad y = a.$$

Solution:

$$w(x, y) = \int_{-\infty}^{\infty} f_1(\xi) \left[\frac{\partial}{\partial \eta} G(x, y, \xi, \eta)\right]_{\eta=0} d\xi + \int_{-\infty}^{\infty} f_2(\xi) G(x, y, \xi, a) \, d\xi$$
$$+ \int_0^a \int_{-\infty}^{\infty} \Phi(\xi, \eta) G(x, y, \xi, \eta) \, d\xi \, d\eta,$$

where

$$G(x, y, \xi, \eta) = \frac{1}{a} \sum_{n=0}^{\infty} \frac{1}{\beta_n} \exp(-\beta_n |x - \xi|) \sin(q_n y) \sin(q_n \eta), \quad q_n = \frac{\pi(2n+1)}{2a}, \quad \beta_n = \sqrt{q_n^2 - \lambda}.$$

> **7.3.2-11. Domain: $0 \leq x < \infty$, $0 \leq y \leq a$. First boundary value problem.**

A semiinfinite strip is considered. Boundary conditions are prescribed:

$$w = f_1(y) \quad \text{at} \quad x = 0, \qquad w = f_2(x) \quad \text{at} \quad y = 0, \qquad w = f_3(x) \quad \text{at} \quad y = a.$$

Solution:

$$w(x, y) = \int_0^a \int_0^{\infty} \Phi(\xi, \eta) G(x, y, \xi, \eta) \, d\xi \, d\eta + \int_0^a f_1(\eta) \left[\frac{\partial}{\partial \xi} G(x, y, \xi, \eta)\right]_{\xi=0} d\eta$$
$$+ \int_0^{\infty} f_2(\xi) \left[\frac{\partial}{\partial \eta} G(x, y, \xi, \eta)\right]_{\eta=0} d\xi - \int_0^{\infty} f_3(\xi) \left[\frac{\partial}{\partial \eta} G(x, y, \xi, \eta)\right]_{\eta=a} d\xi,$$

where

$$G(x, y, \xi, \eta) = \frac{1}{a} \sum_{n=1}^{\infty} \frac{1}{\beta_n} \left[\exp(-\beta_n |x - \xi|) - \exp(-\beta_n |x + \xi|)\right] \sin(q_n y) \sin(q_n \eta),$$

$$q_n = \frac{\pi n}{a}, \quad \beta_n = \sqrt{q_n^2 - \lambda}.$$

7.3.2-12. Domain: $0 \leq x < \infty$, $0 \leq y \leq a$. Second boundary value problem.

A semiinfinite strip is considered. Boundary conditions are prescribed:

$$\partial_x w = f_1(y) \quad \text{at} \quad x = 0, \qquad \partial_y w = f_2(x) \quad \text{at} \quad y = 0, \qquad \partial_y w = f_3(x) \quad \text{at} \quad y = a.$$

Solution:

$$w(x,y) = \int_0^a \int_0^\infty \Phi(\xi,\eta) G(x,y,\xi,\eta)\, d\xi\, d\eta - \int_0^a f_1(\eta) G(x,y,0,\eta)\, d\eta$$
$$- \int_0^\infty f_2(\xi) G(x,y,\xi,0)\, d\xi + \int_0^\infty f_3(\xi) G(x,y,\xi,a)\, d\xi,$$

where

$$G(x,y,\xi,\eta) = \frac{1}{2a} \sum_{n=0}^\infty \frac{\varepsilon_n}{\beta_n} \left[\exp(-\beta_n |x-\xi|) + \exp(-\beta_n |x+\xi|) \right] \cos(q_n y) \cos(q_n \eta),$$

$$q_n = \frac{\pi n}{a}, \quad \beta_n = \sqrt{q_n^2 - \lambda}, \quad \varepsilon = \begin{cases} 1 & \text{for } n = 0, \\ 2 & \text{for } n \neq 0. \end{cases}$$

7.3.2-13. Domain: $0 \leq x < \infty$, $0 \leq y \leq a$. Third boundary value problem.

A semiinfinite strip is considered. Boundary conditions are prescribed:

$$\partial_x w - k_1 w = f_1(y) \text{ at } x=0, \quad \partial_y w - k_2 w = f_2(x) \text{ at } y=0, \quad \partial_y w + k_3 w = f_3(x) \text{ at } y=a.$$

The solution $w(x,y)$ is determined by the formula in Paragraph 7.3.2-12 where

$$G(x,y,\xi,\eta) = \sum_{n=1}^\infty \frac{\varphi_n(y)\varphi_n(\eta)}{\|\varphi_n\|^2 \beta_n(\beta_n + k_1)} H_n(x,\xi), \qquad \beta_n = \sqrt{\mu_n^2 - \lambda},$$

$$\varphi_n(y) = \mu_n \cos(\mu_n y) + k_2 \sin(\mu_n y), \quad \|\varphi_n\|^2 = \frac{1}{2}(\mu_n^2 + k_2^2)\left[a + \frac{(k_2+k_3)(\mu_n^2+k_2 k_3)}{(\mu_n^2+k_2^2)(\mu_n^2+k_3^2)} \right],$$

$$H_n(x,\xi) = \begin{cases} \exp(-\beta_n x)\left[\beta_n \cosh(\beta_n \xi) + k_1 \sinh(\beta_n \xi)\right] & \text{for } x > \xi, \\ \exp(-\beta_n \xi)\left[\beta_n \cosh(\beta_n x) + k_1 \sinh(\beta_n x)\right] & \text{for } \xi > x. \end{cases}$$

Here, the μ_n are positive roots of the transcendental equation $\tan(\mu a) = \dfrac{(k_2+k_3)\mu}{\mu^2 - k_2 k_3}$.

7.3.2-14. Domain: $0 \leq x < \infty$, $0 \leq y \leq a$. Mixed boundary value problems.

1°. A semiinfinite strip is considered. Boundary conditions are prescribed:

$$w = f_1(y) \quad \text{at} \quad x = 0, \qquad \partial_y w = f_2(x) \quad \text{at} \quad y = 0, \qquad \partial_y w = f_3(x) \quad \text{at} \quad y = a.$$

Solution:

$$w(x,y) = \int_0^a f_1(\eta)\left[\frac{\partial}{\partial \xi} G(x,y,\xi,\eta)\right]_{\xi=0} d\eta - \int_0^\infty f_2(\xi) G(x,y,\xi,0)\, d\xi$$
$$+ \int_0^\infty f_3(\xi) G(x,y,\xi,a)\, d\xi + \int_0^a \int_0^\infty \Phi(\xi,\eta) G(x,y,\xi,\eta)\, d\xi\, d\eta,$$

where

$$G(x,y,\xi,\eta) = \frac{1}{2a} \sum_{n=0}^\infty \frac{\varepsilon_n}{\beta_n} \left[\exp(-\beta_n |x-\xi|) - \exp(-\beta_n |x+\xi|) \right] \cos(q_n y) \cos(q_n \eta),$$

$$q_n = \frac{\pi n}{a}, \quad \beta_n = \sqrt{q_n^2 - \lambda}, \quad \varepsilon = \begin{cases} 1 & \text{for } n = 0, \\ 2 & \text{for } n \neq 0. \end{cases}$$

2°. A semiinfinite strip is considered. Boundary conditions are prescribed:
$$\partial_x w = f_1(y) \quad \text{at} \quad x = 0, \qquad w = f_2(x) \quad \text{at} \quad y = 0, \qquad w = f_3(x) \quad \text{at} \quad y = a.$$
Solution:
$$w(x,y) = -\int_0^a f_1(\eta) G(x,y,0,\eta)\, d\eta + \int_0^\infty f_2(\xi) \left[\frac{\partial}{\partial \eta} G(x,y,\xi,\eta)\right]_{\eta=0} d\xi$$
$$- \int_0^\infty f_3(\xi) \left[\frac{\partial}{\partial \eta} G(x,y,\xi,\eta)\right]_{\eta=a} d\xi + \int_0^a \int_0^\infty \Phi(\xi,\eta) G(x,y,\xi,\eta)\, d\xi\, d\eta,$$
where
$$G(x,y,\xi,\eta) = \frac{1}{a} \sum_{n=1}^\infty \frac{1}{\beta_n} \left[\exp(-\beta_n |x-\xi|) + \exp(-\beta_n |x+\xi|)\right] \sin(q_n y) \sin(q_n \eta),$$
$$q_n = \frac{\pi n}{a}, \qquad \beta_n = \sqrt{q_n^2 - \lambda}.$$

7.3.2-15. Domain: $0 \le x \le a$, $0 \le y \le b$. First boundary value problem.

A rectangle is considered. Boundary conditions are prescribed:
$$w = f_1(y) \quad \text{at} \quad x = 0, \qquad w = f_2(y) \quad \text{at} \quad x = a,$$
$$w = f_3(x) \quad \text{at} \quad y = 0, \qquad w = f_4(x) \quad \text{at} \quad y = b.$$

1°. Eigenvalues of the one-dimensional problem (it is convenient to label them with a double subscript):
$$\lambda_{nm} = \pi^2 \left(\frac{n^2}{a^2} + \frac{m^2}{b^2}\right); \qquad n = 1, 2, \ldots; \quad m = 1, 2, \ldots$$

Eigenfunctions and the norm squared:
$$w_{nm} = \sin\left(\frac{n\pi x}{a}\right) \sin\left(\frac{m\pi y}{b}\right), \qquad \|w_{nm}\|^2 = \frac{ab}{4}.$$

⊙ *References*: V. M. Babich, M. B. Kapilevich, S. G. Mikhlin, et al. (1964), B. M. Budak, A. A. Samarskii, and A. N. Tikhonov (1980).

2°. Solution for $\lambda \ne \lambda_{nm}$:
$$w(x,y) = \int_0^a \int_0^b \Phi(\xi,\eta) G(x,y,\xi,\eta)\, d\eta\, d\xi$$
$$+ \int_0^b f_1(\eta) \left[\frac{\partial}{\partial \xi} G(x,y,\xi,\eta)\right]_{\xi=0} d\eta - \int_0^b f_2(\eta) \left[\frac{\partial}{\partial \xi} G(x,y,\xi,\eta)\right]_{\xi=a} d\eta$$
$$+ \int_0^a f_3(\xi) \left[\frac{\partial}{\partial \eta} G(x,y,\xi,\eta)\right]_{\eta=0} d\xi - \int_0^a f_4(\xi) \left[\frac{\partial}{\partial \eta} G(x,y,\xi,\eta)\right]_{\eta=b} d\xi.$$

Two forms of representation of the Green's function:
$$G(x,y,\xi,\eta) = \frac{2}{a} \sum_{n=1}^\infty \frac{\sin(p_n x) \sin(p_n \xi)}{\beta_n \sinh(\beta_n b)} H_n(y,\eta) = \frac{2}{b} \sum_{m=1}^\infty \frac{\sin(q_m y) \sin(q_m \eta)}{\mu_m \sinh(\mu_m a)} Q_m(x,\xi),$$
where
$$p_n = \frac{\pi n}{a}, \quad \beta_n = \sqrt{p_n^2 - \lambda}, \quad H_n(y,\eta) = \begin{cases} \sinh(\beta_n \eta) \sinh[\beta_n(b-y)] & \text{for } b \ge y > \eta \ge 0, \\ \sinh(\beta_n y) \sinh[\beta_n(b-\eta)] & \text{for } b \ge \eta > y \ge 0, \end{cases}$$
$$q_m = \frac{\pi m}{b}, \quad \mu_m = \sqrt{q_m^2 - \lambda}, \quad Q_m(x,\xi) = \begin{cases} \sinh(\mu_m \xi) \sinh[\mu_m(a-x)] & \text{for } a \ge x > \xi \ge 0, \\ \sinh(\mu_m x) \sinh[\mu_m(a-\xi)] & \text{for } a \ge \xi > x \ge 0. \end{cases}$$

Alternatively, the Green's function can be written as the double series
$$G(x,y,\xi,\eta) = \frac{4}{ab} \sum_{n=1}^\infty \sum_{m=1}^\infty \frac{\sin(p_n x) \sin(q_m y) \sin(p_n \xi) \sin(q_m \eta)}{p_n^2 + q_m^2 - \lambda}, \qquad p_n = \frac{\pi n}{a}, \quad q_m = \frac{\pi m}{b}.$$

7.3.2-16. Domain: $0 \le x \le a$, $0 \le y \le b$. Second boundary value problem.

A rectangle is considered. Boundary conditions are prescribed:

$$\partial_x w = f_1(y) \quad \text{at} \quad x = 0, \qquad \partial_x w = f_2(y) \quad \text{at} \quad x = a,$$
$$\partial_y w = f_3(x) \quad \text{at} \quad y = 0, \qquad \partial_y w = f_4(x) \quad \text{at} \quad y = b.$$

1°. Eigenvalues of the homogeneous problem:

$$\lambda_{nm} = \pi^2 \left(\frac{n^2}{a^2} + \frac{m^2}{b^2} \right); \qquad n = 0, 1, 2, \ldots; \quad m = 0, 1, 2, \ldots$$

Eigenfunctions and the norm squared:

$$w_{nm} = \cos\left(\frac{n\pi x}{a}\right) \cos\left(\frac{m\pi y}{b}\right), \quad \|w_{nm}\|^2 = \frac{ab}{4}(1+\delta_{n0})(1+\delta_{m0}), \quad \delta_{n0} = \begin{cases} 1 & \text{for } n = 0, \\ 0 & \text{for } n \neq 0. \end{cases}$$

⊙ *References*: V. M. Babich, M. B. Kapilevich, S. G. Mikhlin, et al. (1964), B. M. Budak, A. A. Samarskii, and A. N. Tikhonov (1980).

2°. Solution for $\lambda \neq \lambda_{nm}$:

$$w(x,y) = \int_0^a \int_0^b \Phi(\xi,\eta) G(x,y,\xi,\eta) \, d\eta \, d\xi$$
$$- \int_0^b f_1(\eta) G(x,y,0,\eta) \, d\eta + \int_0^b f_2(\eta) G(x,y,a,\eta) \, d\eta$$
$$- \int_0^a f_3(\xi) G(x,y,\xi,0) \, d\xi + \int_0^a f_4(\xi) G(x,y,\xi,b) \, d\xi.$$

Two forms of representation of the Green's function:

$$G(x,y,\xi,\eta) = \frac{1}{a} \sum_{n=0}^{\infty} \frac{\varepsilon_n \cos(p_n x) \cos(p_n \xi)}{\beta_n \sinh(\beta_n b)} H_n(y,\eta) = \frac{1}{b} \sum_{m=0}^{\infty} \frac{\varepsilon_m \cos(q_m y) \cos(q_m \eta)}{\mu_m \sinh(\mu_m a)} Q_m(x,\xi),$$

where

$$p_n = \frac{\pi n}{a}, \quad H_n(y,\eta) = \begin{cases} \cosh(\beta_n \eta) \cosh[\beta_n(b-y)] & \text{for } y > \eta, \\ \cosh(\beta_n y) \cosh[\beta_n(b-\eta)] & \text{for } \eta > y, \end{cases}$$

$$q_m = \frac{\pi m}{b}, \quad Q_m(x,\xi) = \begin{cases} \cosh(\mu_m \xi) \cosh[\mu_m(a-x)] & \text{for } x > \xi, \\ \cosh(\mu_m x) \cosh[\mu_m(a-\xi)] & \text{for } \xi > x, \end{cases}$$

$$\beta_n = \sqrt{p_n^2 - \lambda}, \quad \mu_m = \sqrt{q_m^2 - \lambda}, \quad \varepsilon_n = \begin{cases} 1 & \text{for } n = 0, \\ 2 & \text{for } n \neq 0. \end{cases}$$

The Green's function can also be written as the double series

$$G(x,y,\xi,\eta) = \frac{1}{ab} \sum_{n=0}^{\infty} \sum_{m=0}^{\infty} \frac{\varepsilon_n \varepsilon_m \cos(p_n x) \cos(q_m y) \cos(p_n \xi) \cos(q_m \eta)}{p_n^2 + q_m^2 - \lambda}, \quad p_n = \frac{\pi n}{a}, \quad q_m = \frac{\pi m}{b}.$$

▶ *In Paragraphs 7.3.2-17 through 7.3.2-20, only the eigenvalues and eigenfunctions of homogeneous boundary value problems for the homogeneous Helmholtz equation (with $\Phi \equiv 0$) are given. The solutions of the corresponding nonhomogeneous problems can be constructed using formulas presented in Paragraphs 7.3.1-3 and 7.3.1-4.*

7.3.2-17. Domain: $0 \leq x \leq a$, $0 \leq y \leq b$. Third boundary value problem.

A rectangle is considered. Boundary conditions are prescribed:

$$\partial_x w - k_1 w = 0 \quad \text{at} \quad x = 0, \qquad \partial_x w + k_2 w = 0 \quad \text{at} \quad x = a,$$
$$\partial_y w - k_3 w = 0 \quad \text{at} \quad y = 0, \qquad \partial_y w + k_4 w = 0 \quad \text{at} \quad y = b.$$

Eigenvalues:
$$\lambda_{nm} = \mu_n^2 + \nu_m^2,$$

where the μ_n and ν_m are positive roots of the transcendental equations

$$\tan(\mu a) = \frac{(k_1 + k_2)\mu}{\mu^2 - k_1 k_2}, \qquad \tan(\nu b) = \frac{(k_3 + k_4)\nu}{\nu^2 - k_3 k_4}.$$

Eigenfunctions:

$$w_{nm} = (\mu_n \cos \mu_n x + k_1 \sin \mu_n x)(\nu_m \cos \nu_m y + k_3 \sin \nu_m y).$$

The square of the norm of an eigenfunction:

$$\|w_{nm}\|^2 = \frac{1}{4}(\mu_n^2 + k_1^2)(\nu_m^2 + k_3^2)\left[a + \frac{(k_1 + k_2)(\mu_n^2 + k_1 k_2)}{(\mu_n^2 + k_1^2)(\mu_n^2 + k_2^2)}\right]\left[b + \frac{(k_3 + k_4)(\nu_m^2 + k_3 k_4)}{(\nu_m^2 + k_3^2)(\nu_m^2 + k_4^2)}\right].$$

⊙ *Reference*: B. M. Budak, A. A. Samarskii, and A. N. Tikhonov (1980).

7.3.2-18. Domain: $0 \leq x \leq a$, $0 \leq y \leq b$. Mixed boundary value problems.

1°. A rectangle is considered. Boundary conditions are prescribed:

$$w = 0 \quad \text{at} \quad x = 0, \qquad w = 0 \quad \text{at} \quad x = a,$$
$$\partial_y w = 0 \quad \text{at} \quad y = 0, \qquad \partial_y w = 0 \quad \text{at} \quad y = b.$$

Eigenvalues:
$$\lambda_{nm} = \pi^2\left(\frac{n^2}{a^2} + \frac{m^2}{b^2}\right); \qquad n = 1, 2, 3, \ldots; \quad m = 0, 1, 2, \ldots$$

Eigenfunctions and the norm squared:

$$w_{nm} = \sin\left(\frac{n\pi x}{a}\right)\cos\left(\frac{m\pi y}{b}\right), \quad \|w_{nm}\|^2 = \frac{ab}{4}(1 + \delta_{m0}), \quad \delta_{m0} = \begin{cases} 1 & \text{for } m = 0, \\ 0 & \text{for } m \neq 0. \end{cases}$$

2°. A rectangle is considered. Boundary conditions are prescribed:

$$w = 0 \quad \text{at} \quad x = 0, \qquad \partial_x w = 0 \quad \text{at} \quad x = a,$$
$$w = 0 \quad \text{at} \quad y = 0, \qquad \partial_y w = 0 \quad \text{at} \quad y = b.$$

Eigenvalues:
$$\lambda_{nm} = \frac{\pi^2}{4}\left[\frac{(2n+1)^2}{a^2} + \frac{(2m+1)^2}{b^2}\right]; \qquad n = 0, 1, 2, \ldots; \quad m = 0, 1, 2, \ldots$$

Eigenfunctions and the norm squared:

$$w_{nm} = \sin\left[\frac{\pi(2n+1)x}{2a}\right]\sin\left[\frac{\pi(2m+1)y}{2b}\right], \qquad \|w_{nm}\|^2 = \frac{ab}{4}.$$

⊙ *References*: V. M. Babich, M. B. Kapilevich, S. G. Mikhlin, et al. (1964), B. M. Budak, A. A. Samarskii, and A. N. Tikhonov (1980).

7.3.2-19. First boundary value problem for a triangular domain.

The sides of the triangle are defined by the equations

$$x = 0, \quad y = 0, \quad y = a - x.$$

The unknown quantity is zero for these sides.

Eigenvalues:

$$\lambda_{nm} = \frac{\pi^2}{a^2}\left[(n+m)^2 + m^2\right]; \qquad n = 1, 2, \ldots; \quad m = 1, 2, \ldots$$

Eigenfunctions:

$$w_{nm} = \sin\left[\frac{\pi}{a}(n+m)x\right]\sin\left(\frac{\pi}{a}my\right) - (-1)^n \sin\left(\frac{\pi}{a}mx\right)\sin\left[\frac{\pi}{a}(n+m)y\right].$$

⊙ *Reference*: V. M. Babich, M. B. Kapilevich, S. G. Mikhlin, et al. (1964).

7.3.2-20. Second boundary value problem for a triangular domain.

The sides of the triangle are defined by the equations

$$x = 0, \quad y = 0, \quad y = a - x.$$

The normal derivative of the unknown quantity for these sides is zero.

Eigenvalues:

$$\lambda_{nm} = \frac{\pi^2}{a^2}\left[(n+m)^2 + m^2\right]; \qquad n = 0, 1, \ldots; \quad m = 0, 1, \ldots$$

Eigenfunctions:

$$w_{nm} = \cos\left[\frac{\pi}{a}(n+m)x\right]\cos\left(\frac{\pi}{a}my\right) - (-1)^n \cos\left(\frac{\pi}{a}mx\right)\cos\left[\frac{\pi}{a}(n+m)y\right].$$

7.3.3. Problems in Polar Coordinate System

A two-dimensional nonhomogeneous Helmholtz equation in the polar coordinate system is written as

$$\frac{1}{r}\frac{\partial}{\partial r}\left(r\frac{\partial w}{\partial r}\right) + \frac{1}{r^2}\frac{\partial^2 w}{\partial \varphi^2} + \lambda w = -\Phi(r, \varphi, z), \qquad r = \sqrt{x^2 + y^2}.$$

7.3.3-1. Particular solutions of the homogeneous equation ($\Phi \equiv 0$):

$$w = [AJ_0(\mu r) + BY_0(\mu r)](C\varphi + D), \quad \lambda = \mu^2,$$
$$w = [AI_0(\mu r) + BK_0(\mu r)](C\varphi + D), \quad \lambda = -\mu^2,$$
$$w = [AJ_m(\mu r) + BY_m(\mu r)](C\cos m\varphi + D\sin m\varphi), \quad \lambda = \mu^2,$$
$$w = [AI_m(\mu r) + BK_m(\mu r)](C\cos m\varphi + D\sin m\varphi), \quad \lambda = -\mu^2,$$

where $m = 1, 2, \ldots$; A, B, C, D are arbitrary constants; the $J_m(\mu)$ and $Y_m(\mu)$ are the Bessel functions; and the $I_m(\mu)$ and $K_m(\mu)$ are the modified Bessel functions.

▶ *In Paragraphs 7.3.3-2 through 7.3.3-11, only the eigenvalues and eigenfunctions of homogeneous boundary value problems for the homogeneous Helmholtz equation (with $\Phi \equiv 0$) are given. The solutions of the corresponding nonhomogeneous problems can be constructed using formulas presented in Paragraphs 7.3.1-3 and 7.3.1-4.*

7.3.3-2. Domain: $0 \leq r \leq R$. First boundary value problem.

A circle is considered. A boundary condition is prescribed:

$$w = 0 \quad \text{at} \quad r = R.$$

Eigenvalues:

$$\lambda_{nm} = \frac{\mu_{nm}^2}{R^2}; \qquad n = 0, 1, 2, \ldots; \ m = 1, 2, 3, \ldots$$

Here, the μ_{nm} are positive zeros of the Bessel functions, $J_n(\mu) = 0$.

Eigenfunctions:

$$w_{nm}^{(1)} = J_n\left(r\sqrt{\lambda_{nm}}\right)\cos n\varphi, \quad w_{nm}^{(2)} = J_n\left(r\sqrt{\lambda_{nm}}\right)\sin n\varphi.$$

Eigenfunctions possessing the axial symmetry property: $w_{0m}^{(1)} = J_0\left(r\sqrt{\lambda_{0m}}\right)$.

The square of the norm of an eigenfunction is given by

$$\|w_{nm}^{(k)}\|^2 = \tfrac{1}{2}\pi R^2(1+\delta_{n0})[J_n'(\mu_{nm})]^2, \qquad k = 1, 2; \qquad \delta_{ij} = \begin{cases} 1 & \text{for } i = j, \\ 0 & \text{for } i \neq j. \end{cases}$$

⊙ *Reference*: B. M. Budak, A. A. Samarskii, and A. N. Tikhonov (1980).

7.3.3-3. Domain: $0 \leq r \leq R$. Second boundary value problem.

A circle is considered. A boundary condition is prescribed:

$$\partial_r w = 0 \quad \text{at} \quad r = R.$$

Eigenvalues:

$$\lambda_{nm} = \frac{\mu_{nm}^2}{R^2},$$

where the μ_{nm} are roots of the transcendental equation $J_n'(\mu) = 0$.

Eigenfunctions:

$$w_{nm}^{(1)} = J_n(r\sqrt{\lambda_{nm}})\cos n\varphi, \quad w_{nm}^{(2)} = J_n(r\sqrt{\lambda_{nm}})\sin n\varphi.$$

Here, $n = 0, 1, 2, \ldots$; for $n \neq 0$, the parameter m assumes the values $m = 1, 2, 3, \ldots$; for $n = 0$, a root $\mu_{00} = 0$ (the corresponding eigenfunction is $w_{00} = 1$).

Eigenfunctions possessing the axial symmetry property: $w_{0m}^{(1)} = J_0\left(r\sqrt{\lambda_{0m}}\right)$.

The square of the norm of an eigenfunction is given by

$$\|w_{nm}^{(k)}\|^2 = \frac{\pi^2 R^2(1+\delta_{n0})}{2\mu_{nm}^2}(\mu_{nm}^2 - n^2)[J_n(\mu_{nm})]^2, \quad \|w_{00}\|^2 = \pi R^2,$$

where $k = 1, 2$; $\delta_{ij} = \begin{cases} 1 & \text{for } i = j, \\ 0 & \text{for } i \neq j. \end{cases}$

⊙ *References*: V. M. Babich, M. B. Kapilevich, S. G. Mikhlin, et al. (1964), B. M. Budak, A. A. Samarskii, and A. N. Tikhonov (1980).

7.3.3-4. Domain: $0 \leq r \leq R$. Third boundary value problem.

A circle is considered. A boundary condition is prescribed:

$$\partial_r w + kw = 0 \quad \text{at} \quad r = R.$$

Eigenvalues:

$$\lambda_{nm} = \frac{\mu_{nm}^2}{R^2}; \qquad n = 0, 1, 2, \ldots; \ m = 1, 2, 3, \ldots$$

Here, the μ_{nm} is the mth root of the transcendental equation $\mu J'_n(\mu) + kR J_n(\mu) = 0$.

Eigenfunctions:

$$w_{nm}^{(1)} = J_n\!\left(r\sqrt{\lambda_{nm}}\right)\cos n\varphi, \quad w_{nm}^{(2)} = J_n\!\left(r\sqrt{\lambda_{nm}}\right)\sin n\varphi.$$

The square of the norm of an eigenfunction is given by

$$\|w_{nm}^{(1)}\|^2 = \|w_{nm}^{(2)}\|^2 = \frac{\pi R^2 (1+\delta_{n0})}{2\mu_{nm}^2}(k^2 R^2 + \mu_{nm}^2 - n^2)[J_n(\mu_{nm})]^2, \qquad \delta_{ij} = \begin{cases} 1 & \text{for } i = j, \\ 0 & \text{for } i \neq j. \end{cases}$$

⊙ *References*: V. M. Babich, M. B. Kapilevich, S. G. Mikhlin, et al. (1964), B. M. Budak, A. A. Samarskii, and A. N. Tikhonov (1980).

7.3.3-5. Domain: $R_1 \leq r \leq R_2$. First boundary value problem.

An annular domain is considered. Boundary conditions are prescribed:

$$w = 0 \quad \text{at} \quad r = R_1, \qquad w = 0 \quad \text{at} \quad r = R_2.$$

Eigenvalues:

$$\lambda_{nm} = \mu_{nm}^2; \qquad n = 0, 1, 2, \ldots; \ m = 1, 2, 3, \ldots$$

Here, the μ_{nm} are positive roots of the transcendental equation

$$J_n(\mu R_1) Y_n(\mu R_2) - J_n(\mu R_2) Y_n(\mu R_1) = 0.$$

Eigenfunctions:

$$w_{nm}^{(1)} = [J_n(\mu_{nm} r) Y_n(\mu_{nm} R_1) - J_n(\mu_{nm} R_1) Y_n(\mu_{nm} r)]\cos n\varphi,$$
$$w_{nm}^{(2)} = [J_n(\mu_{nm} r) Y_n(\mu_{nm} R_1) - J_n(\mu_{nm} R_1) Y_n(\mu_{nm} r)]\sin n\varphi.$$

The square of the norm of an eigenfunction is given by

$$\|w_{nm}^{(1)}\|^2 = \|w_{nm}^{(2)}\|^2 = \frac{2(1+\delta_{n0})}{\pi \mu_{nm}^2}\frac{J_n^2(\mu_{nm} R_1) - J_n^2(\mu_{nm} R_2)}{J_n^2(\mu_{nm} R_2)}, \qquad \delta_{ij} = \begin{cases} 1 & \text{for } i = j, \\ 0 & \text{for } i \neq j. \end{cases}$$

⊙ *References*: V. M. Babich, M. B. Kapilevich, S. G. Mikhlin, et al. (1964), B. M. Budak, A. A. Samarskii, and A. N. Tikhonov (1980).

7.3.3-6. Domain: $R_1 \leq r \leq R_2$. Second boundary value problem.

An annular domain is considered. Boundary conditions are prescribed:

$$\partial_r w = 0 \quad \text{at} \quad r = R_1, \qquad \partial_r w = 0 \quad \text{at} \quad r = R_2.$$

Eigenvalues:

$$\lambda_{nm} = \mu_{nm}^2; \qquad n = 0, 1, 2, \ldots; \ m = 0, 1, 2, \ldots$$

Here, the μ_{nm} are roots of the transcendental equation

$$J'_n(\mu R_1)Y'_n(\mu R_2) - J'_n(\mu R_2)Y'_n(\mu R_1) = 0.$$

If $n = 0$, there is a root $\mu_{00} = 0$ and the corresponding eigenfunction is $w^{(1)}_{00} = 1$.

Eigenfunctions:

$$w^{(1)}_{nm} = [J_n(\mu_{nm}r)Y'_n(\mu_{nm}R_1) - J'_n(\mu_{nm}R_1)Y_n(\mu_{nm}r)]\cos n\varphi,$$
$$w^{(2)}_{nm} = [J_n(\mu_{nm}r)Y'_n(\mu_{nm}R_1) - J'_n(\mu_{nm}R_1)Y_n(\mu_{nm}r)]\sin n\varphi.$$

The square of the norm of an eigenfunction is given by

$$\|w^{(1)}_{nm}\|^2 = \|w^{(2)}_{nm}\|^2 = \frac{2(1+\delta_{n0})}{\pi\mu_{nm}^2}\left\{\left(1-\frac{n^2}{R_2^2\mu_{nm}^2}\right)\left[\frac{J'_n(\mu_{nm}R_1)}{J'_n(\mu_{nm}R_2)}\right]^2 - \left(1-\frac{n^2}{R_1^2\mu_{nm}^2}\right)\right\},$$

$$\|w^{(1)}_{00}\|^2 = \pi(R_2^2 - R_1^2); \quad \delta_{ij} = \begin{cases} 1 & \text{for } i = j, \\ 0 & \text{for } i \neq j. \end{cases}$$

⊙ *Reference*: B. M. Budak, A. A. Samarskii, and A. N. Tikhonov (1980).

7.3.3-7. Domain: $R_1 \leq r \leq R_2$. Third boundary value problem.

An annular domain is considered. Boundary conditions are prescribed:

$$\partial_r w - kw = 0 \quad \text{at} \quad r = R_1, \qquad \partial_r w + kw = 0 \quad \text{at} \quad r = R_2.$$

Eigenvalues:

$$\lambda_{nm} = \mu_{nm}^2; \qquad n = 0, 1, 2, \ldots; \quad m = 1, 2, 3, \ldots;$$

where the μ_{nm} are positive roots of the transcendental equation

$$A_1(\mu R_1)B_2(\mu R_2) - A_2(\mu R_2)B_1(\mu R_1) = 0.$$

Here, we use the notation

$$A_1(\mu R) = J'_n(\mu R) - \frac{k}{\mu}J_n(\mu R), \quad B_1(\mu R) = Y'_n(\mu R) - \frac{k}{\mu}Y_n(\mu R),$$
$$A_2(\mu R) = J'_n(\mu R) + \frac{k}{\mu}J_n(\mu R), \quad B_2(\mu R) = Y'_n(\mu R) + \frac{k}{\mu}Y_n(\mu R).$$

Eigenfunctions:

$$w^{(1)}_{nm} = [B_1(\mu_{nm}R_1)J_n(\mu_{nm}r) - A_1(\mu_{nm}R_1)Y_n(\mu_{nm}r)]\cos n\varphi,$$
$$w^{(2)}_{nm} = [B_1(\mu_{nm}R_1)J_n(\mu_{nm}r) - A_1(\mu_{nm}R_1)Y_n(\mu_{nm}r)]\sin n\varphi.$$

The square of the norm of an eigenfunction is given by ($s = 1, 2$)

$$\|w^{(s)}_{nm}\|^2 = \tfrac{1}{2}\pi\varepsilon_n R_2^2\left\{[F'_{nm}(R_2)]^2 + \left(1-\frac{n^2}{R_2^2\mu_{nm}^2}\right)F_{nm}^2(R_2)\right\}$$
$$-\tfrac{1}{2}\pi\varepsilon_n R_1^2\left\{[F'_{nm}(R_1)]^2 + \left(1-\frac{n^2}{R_1^2\mu_{nm}^2}\right)F_{nm}^2(R_1)\right\},$$

$$F_{nm}(r) = B_1(\mu_{nm}R_1)J_n(\mu_{nm}r) - A_1(\mu_{nm}R_1)Y_n(\mu_{nm}r), \quad \varepsilon_{ij} = \begin{cases} 2 & \text{for } i = j, \\ 1 & \text{for } i \neq j. \end{cases}$$

⊙ *Reference*: B. M. Budak, A. A. Samarskii, and A. N. Tikhonov (1980).

7.3.3-8. Domain: $0 \leq r \leq R$, $0 \leq \varphi \leq \alpha$. First boundary value problem.

A circular sector is considered. Boundary conditions are prescribed:
$$w = 0 \quad \text{at} \quad r = R, \qquad w = 0 \quad \text{at} \quad \varphi = 0, \qquad w = 0 \quad \text{at} \quad \varphi = \alpha.$$

Eigenvalues:
$$\lambda_{nm} = \frac{\mu_{nm}^2}{R^2}; \qquad n = 1, 2, 3, \ldots; \qquad m = 1, 2, 3, \ldots$$

Here, the μ_{nm} are positive zeros of the Bessel functions, $J_{\frac{n\pi}{\alpha}}(\mu) = 0$.

Eigenfunctions:
$$w_{nm} = J_{\frac{n\pi}{\alpha}}\left(\mu_{nm}\frac{r}{R}\right)\sin\left(\frac{n\pi}{\alpha}\varphi\right).$$

The square of the norm of an eigenfunction is given by
$$\|w_{nm}\|^2 = \frac{\alpha R^2}{4}\left[J'_{\frac{n\pi}{\alpha}}(\mu_{nm})\right]^2.$$

⊙ *References*: V. M. Babich, M. B. Kapilevich, S. G. Mikhlin, et al. (1964), B. M. Budak, A. A. Samarskii, and A. N. Tikhonov (1980).

7.3.3-9. Domain: $0 \leq r \leq R$, $0 \leq \varphi \leq \alpha$. Second boundary value problem.

A circular sector is considered. Boundary conditions are prescribed:
$$\partial_r w = 0 \quad \text{at} \quad r = R, \qquad \partial_\varphi w = 0 \quad \text{at} \quad \varphi = 0, \qquad \partial_\varphi w = 0 \quad \text{at} \quad \varphi = \alpha.$$

Eigenvalues:
$$\lambda_{nm} = \frac{\mu_{nm}^2}{R^2}; \qquad n = 0, 1, 2, \ldots; \qquad m = 0, 1, 2, \ldots$$

Here, the μ_{nm} are roots of the transcendental equation $J'_{\frac{n\pi}{\alpha}}(\mu) = 0$.

Eigenfunctions:
$$w_{nm} = J_{\frac{n\pi}{\alpha}}\left(\mu_{nm}\frac{r}{R}\right)\cos\left(\frac{n\pi}{\alpha}\varphi\right), \qquad w_{00} = 1.$$

The square of the norm of an eigenfunction is given by
$$\|w_{nm}\|^2 = \frac{\alpha R^2}{4}(1+\delta_{n0})\left(1 - \frac{n^2}{\mu_{nm}^2}\right)\left[J_{\frac{n\pi}{\alpha}}(\mu_{nm})\right]^2, \qquad \|w_{00}\|^2 = \frac{\alpha R^2}{2}.$$

⊙ *Reference*: B. M. Budak, A. A. Samarskii, and A. N. Tikhonov (1980).

7.3.3-10. Domain: $0 \leq r \leq R$, $0 \leq \varphi \leq \alpha$. Third boundary value problem.

A circular sector is considered. Boundary conditions are prescribed:
$$\partial_r w + k_1 w = 0 \quad \text{at} \quad r = R, \qquad \partial_\varphi w - k_2 w = 0 \quad \text{at} \quad \varphi = 0, \qquad \partial_\varphi w + k_3 w = 0 \quad \text{at} \quad \varphi = \alpha.$$

Eigenvalues:
$$\lambda_{nm} = \frac{\mu_{nm}^2}{R^2}, \qquad n = 1, 2, 3, \ldots; \qquad m = 1, 2, 3, \ldots$$

Here, the μ_{nm} are positive roots of the transcendental equation $\mu J'_{\nu_n}(\mu) + k_1 R J_{\nu_n}(\mu) = 0$; the ν_n are positive roots of the transcendental equation $\tan(\alpha\nu) = \dfrac{(k_2+k_3)\nu}{\nu^2 - k_2 k_3}$.

Eigenfunctions:
$$w_{nm} = J_{\nu_n}\left(\mu_{nm}\frac{r}{R}\right)\frac{\nu_n \cos(\nu_n\varphi) + k_2 \sin(\nu_n\varphi)}{\sqrt{\nu_n^2 + k_2^2}}.$$

The square of the norm of an eigenfunction is given by
$$\|w_{nm}\|^2 = \frac{R^2}{4}\left[\alpha + \frac{(k_2+k_3)(\nu_n^2 + k_2 k_3)}{(\nu_n^2 + k_2^2)(\nu_n^2 + k_3^2)}\right]\left(1 + \frac{k_1^2 R^2 - \nu_n^2}{\mu_{nm}^2}\right)J_{\nu_n}^2(\mu_{nm}).$$

⊙ *Reference*: B. M. Budak, A. A. Samarskii, and A. N. Tikhonov (1980).

7.3.3-11. Domain: $R_1 \leq r \leq R_2$, $0 \leq \varphi \leq \alpha$. **First boundary value problem.**

Boundary conditions are prescribed:

$$w = 0 \quad \text{at} \quad r = R_1, \qquad w = 0 \quad \text{at} \quad r = R_2,$$
$$w = 0 \quad \text{at} \quad \varphi = 0, \qquad w = 0 \quad \text{at} \quad \varphi = \alpha.$$

Eigenvalues:

$$\lambda_{nm} = \mu_{nm}^2,$$

where the μ_{nm} are positive roots of the transcendental equation

$$J_{\nu_n}(\mu R_1)Y_{\nu_n}(\mu R_2) - J_{\nu_n}(\mu R_2)Y_{\nu_n}(\mu R_1) = 0, \qquad \nu_n = \frac{n\pi}{\alpha}.$$

Eigenfunctions:

$$w_{nm} = \left[J_{\nu_n}(\mu_{nm}r)Y_{\nu_n}(\mu_{nm}R_1) - J_{\nu_n}(\mu_{nm}R_1)Y_{\nu_n}(\mu_{nm}r)\right]\sin(\nu_n\varphi).$$

The square of the norm of an eigenfunction is given by

$$\|w_{nm}\|^2 = \frac{\alpha}{\pi^2\mu_{nm}^2} \frac{\left[J_{\nu_n}(\mu_{nm}R_1)\right]^2 - \left[J_{\nu_n}(\mu_{nm}R_2)\right]^2}{\left[J_{\nu_n}(\mu_{nm}R_2)\right]^2}.$$

⊙ *References*: V. M. Babich, M. B. Kapilevich, S. G. Mikhlin, et al. (1964), B. M. Budak, A. A. Samarskii, and A. N. Tikhonov (1980).

7.3.4. Other Orthogonal Coordinate Systems. Elliptic Domain

In Paragraphs 7.3.4-1 and 7.3.4-2, two other orthogonal systems of coordinates are described in which the homogeneous Helmholtz equation admits separation of variables.

7.3.4-1. Parabolic coordinate system.

In the parabolic coordinates that are introduced by the relations

$$x = \tfrac{1}{2}(\xi^2 - \eta^2), \quad y = \xi\eta \qquad (0 \leq \xi < \infty, \ -\infty < \eta < \infty),$$

the Helmholtz equation has the form

$$\frac{\partial^2 w}{\partial \xi^2} + \frac{\partial^2 w}{\partial \eta^2} + \lambda(\xi^2 + \eta^2)w = 0.$$

Setting $w = f(\xi)g(\eta)$, we arrive at the following linear ordinary differential equations for $f = f(\xi)$ and $g = g(\eta)$:

$$f'' + (\lambda\xi^2 + k)f = 0, \qquad g'' + (\lambda\eta^2 - k)g = 0,$$

where k is the separation constant. The general solutions of these equations are given by

$$f(\xi) = A_1 D_{\mu-1/2}(\sigma\xi) + A_2 D_{\mu-1/2}(-\sigma\xi), \qquad g(\eta) = B_1 D_{-\mu-1/2}(\sigma\eta) + B_2 D_{-\mu-1/2}(-\sigma\eta),$$
$$\mu = \tfrac{1}{2}k(-\lambda)^{-1/2}, \qquad \sigma = (-4\lambda)^{1/4}.$$

Here, A_1, B_1, A_2, and B_2 are arbitrary constants, and $D_\nu(z)$ is the parabolic cylinder function,

$$D_\nu(z) = 2^{1/2}\exp\!\left(-\tfrac{1}{4}z^2\right)\left[\frac{\Gamma(\tfrac{1}{2})}{\Gamma(\tfrac{1}{2}-\tfrac{\nu}{2})}\Phi\!\left(-\tfrac{\nu}{2},\tfrac{1}{2};\tfrac{1}{2}z^2\right) + 2^{-1/2}\frac{\Gamma(-\tfrac{1}{2})}{\Gamma(-\tfrac{\nu}{2})}z\Phi\!\left(\tfrac{1}{2}-\tfrac{\nu}{2},\tfrac{3}{2};\tfrac{1}{2}z^2\right)\right].$$

For $\nu = n = 0, 1, 2, \ldots$, we have

$$D_n(z) = 2^{-n/2}\exp\!\left(-\tfrac{1}{4}z^2\right)H_n\!\left(2^{-1/2}z\right), \quad \text{where} \quad H_n(z) = (-1)^n\exp\!\left(z^2\right)\frac{d^n}{z^n}\exp\!\left(-z^2\right).$$

⊙ *References*: M. Abramowitz and I. Stegun (1964), W. Miller, Jr. (1977).

7.3.4-2. Elliptic coordinate system.

In the elliptic coordinates that are introduced by the relations

$$x = a \cosh u \cos v, \quad y = a \sinh u \sin v \qquad (0 \le u < \infty,\ 0 \le v < 2\pi,\ a > 0),$$

the Helmholtz equation is expressed as

$$\frac{\partial^2 w}{\partial \xi^2} + \frac{\partial^2 w}{\partial \eta^2} + a^2 \lambda (\cosh^2 u - \cos^2 v) w = 0.$$

Setting $w = F(u)G(v)$, we arrive at the following linear ordinary differential equations for $F = F(u)$ and $G = G(v)$:

$$F'' + \left(\tfrac{1}{2} a^2 \lambda \cosh 2u - k\right) F = 0, \quad G'' - \left(\tfrac{1}{2} a^2 \lambda \cos 2v - k\right) G = 0,$$

where k is the separation constant. The solutions of these equations periodic in v are given by

$$F(u) = \begin{cases} \mathrm{Ce}_n(u, q), \\ \mathrm{Se}_n(u, q), \end{cases} \quad G(v) = \begin{cases} \mathrm{ce}_n(v, q), \\ \mathrm{se}_n(v, q), \end{cases} \quad q = \tfrac{1}{4} a^2 \lambda,$$

where $\mathrm{Ce}_n(u, q)$ and $\mathrm{Se}_n(u, q)$ are the modified Mathieu functions, and $\mathrm{ce}_n(v, q)$ and $\mathrm{se}_n(v, q)$ are the Mathieu functions; to each value of q there is a corresponding $k = k_n(q)$.

⊙ *References*: M. Abramowitz and I. Stegun (1964), W. Miller, Jr. (1977).

7.3.4-3. Domain: $(x/a)^2 + (y/b)^2 \le 1$. First boundary value problem.

The unknown quantity is zero at the boundary of the elliptic domain:

$$w = 0 \quad \text{if} \quad (x/a)^2 + (y/b)^2 = 1 \qquad (a \ge b).$$

The first three eigenvalues and eigenfunctions are given by the approximate relations

$$\lambda_1 = \frac{\gamma_{10}^2}{2} \left(\frac{1}{a^2} + \frac{1}{b^2} \right), \qquad w_1(\mathcal{R}) = J_0(\gamma_{10} \mathcal{R}),$$

$$\lambda_2^{(c)} = \frac{\gamma_{11}^2}{4} \left(\frac{3}{a^2} + \frac{1}{b^2} \right), \qquad w_2^{(c)}(\mathcal{R}, \varphi) = J_1(\gamma_{11} \mathcal{R}) \cos \varphi,$$

$$\lambda_2^{(s)} = \frac{\gamma_{11}^2}{4} \left(\frac{1}{a^2} + \frac{3}{b^2} \right), \qquad w_2^{(s)}(\mathcal{R}, \varphi) = J_1(\gamma_{11} \mathcal{R}) \sin \varphi,$$

where $\gamma_{10} = 2.4048$ and $\gamma_{11} = 3.8317$ are the first roots of the Bessel functions J_0 and J_1, i.e., $J_0(\gamma_{10}) = 0$ and $J_1(\gamma_{11}) = 0$; $\mathcal{R} = \sqrt{(x/a)^2 + (y/b)^2}$.

The above relations were obtained using the generalized (nonorthogonal) polar coordinates $\mathcal{R}, \varphi$ defined by

$$x = a\mathcal{R} \cos \varphi, \quad y = b\mathcal{R} \sin \varphi \qquad (0 \le \mathcal{R} \le 1,\ 0 \le \varphi \le 2\pi)$$

and the variational method.

For $\varepsilon = \sqrt{1 - (b/a)^2} \le 0.9$, the above formulas provide an accuracy of 1% for λ_1 and 2% for $\lambda_2^{(c)}$ and $\lambda_2^{(s)}$. For $\varepsilon \le 0.5$, the errors in calculating λ_1 and $\lambda_2^{(c)}$ do not exceed 0.01%, and the maximum error in determining $\lambda_2^{(s)}$ is 0.12%. In the limit case $\varepsilon = 0$ that corresponds to a circular domain, the above formulas are exact.

⊙ *Reference*: L. D. Akulenko and S. V. Nesterov (2000).

TABLE 24
Transformations reducing equation 7.4.1.3 to the Helmholtz equation $\frac{\partial^2 w}{\partial \xi^2} + \frac{\partial^2 w}{\partial \eta^2} = bw$

No	Exponent k	Transformation	Factor b
1	$k = 1$	$\xi = \frac{1}{2}(x^2 - y^2)$, $\eta = xy$	$b = a$
2	$k = 2$	$\xi = \frac{1}{3}x^3 - xy^2$, $\eta = x^2 y - \frac{1}{3}y^3$	$b = a$
3	$k = -1$	$\xi = \frac{1}{2}\ln(x^2 + y^2)$, $\eta = \arctan\frac{y}{x}$	$b = a$
4	$k = -2$	$\xi = -\dfrac{x}{x^2 + y^2}$, $\eta = \dfrac{y}{x^2 + y^2}$	$b = a$
5	$k = -\frac{1}{2}$	$x = \frac{1}{2}(\xi^2 - \eta^2)$, $y = \xi\eta$	$b = 2a$
6	$k = \pm 3, \pm 4, \ldots$	$\xi = \dfrac{(x+iy)^{k+1} + (x-iy)^{k+1}}{2(k+1)}$, $\eta = \dfrac{(x+iy)^{k+1} - (x-iy)^{k+1}}{2(k+1)i}$	$b = a$
7	k is any $(k \neq -1)$	$\xi = \dfrac{\rho^{k+1}\cos[(k+1)\varphi]}{k+1}$, $\eta = \dfrac{\rho^{k+1}\sin[(k+1)\varphi]}{k+1}$ $x = \rho\cos\varphi$, $y = \rho\sin\varphi$	$b = a$

7.4. Other Equations

7.4.1. Stationary Schrödinger Equation $\Delta_2 w = f(x,y)w$

1. $\dfrac{\partial^2 w}{\partial x^2} + \dfrac{\partial^2 w}{\partial y^2} = a(x^2 + y^2)w.$

The transformation
$$z = \tfrac{1}{2}(x^2 - y^2), \quad \zeta = xy$$
leads to the Helmholtz equation
$$\frac{\partial^2 w}{\partial z^2} + \frac{\partial^2 w}{\partial \zeta^2} - aw = 0,$$
which is discussed in Subsection 7.3.2.

2. $\dfrac{\partial^2 w}{\partial x^2} + \dfrac{\partial^2 w}{\partial y^2} = a(x^2 + y^2)^2 w.$

The transformation
$$z = \tfrac{1}{3}x^3 - xy^2, \quad \zeta = x^2 y - \tfrac{1}{3}y^3$$
leads to the Helmholtz equation
$$\frac{\partial^2 w}{\partial z^2} + \frac{\partial^2 w}{\partial \zeta^2} - aw = 0,$$
which is discussed in Subsection 7.3.2.

3. $\dfrac{\partial^2 w}{\partial x^2} + \dfrac{\partial^2 w}{\partial y^2} = a(x^2 + y^2)^k w.$

This is a special case of equation 7.4.1.7 for $f(u) = au^k$. Table 24 presents transformations that reduce this equation to the Helmholtz equation that is discussed in Subsection 7.3.2; the sixth row involves the imaginary unit, $i^2 = -1$.

4. $\dfrac{\partial^2 w}{\partial x^2} + \dfrac{\partial^2 w}{\partial y^2} = ae^{\beta x} w.$

The transformation
$$u(x, y) = \exp\left(\tfrac{1}{2}\beta x\right) \cos\left(\tfrac{1}{2}\beta y\right), \quad v(x, y) = \exp\left(\tfrac{1}{2}\beta x\right) \sin\left(\tfrac{1}{2}\beta y\right)$$
leads to the Helmholtz equation
$$\dfrac{\partial^2 w}{\partial u^2} + \dfrac{\partial^2 w}{\partial v^2} = 4a\beta^{-2} w,$$
which is discussed in Subsection 7.3.2.

5. $\dfrac{\partial^2 w}{\partial x^2} + \dfrac{\partial^2 w}{\partial y^2} = ke^{ax+by} w.$

The transformation
$$\xi = ax + by, \quad \eta = bx - ay$$
leads to an equation of the form 7.4.1.4:
$$\dfrac{\partial^2 w}{\partial \xi^2} + \dfrac{\partial^2 w}{\partial \eta^2} = \dfrac{k}{a^2 + b^2} e^{\xi} w.$$

6. $\dfrac{\partial^2 w}{\partial x^2} + \dfrac{\partial^2 w}{\partial y^2} = f(ax + by) w.$

This is a special case of equation 7.4.1.9 for $g(u) = 0$. Particular solutions:
$$w(x, y) = \{C_1 \cos[k(bx - ay)] + C_2 \sin[k(bx - ay)]\} \varphi(ax + by),$$
where C_1, C_2, and k are arbitrary constants, and the function $\varphi = \varphi(\xi)$ is determined by the ordinary differential equation
$$\varphi''_{\xi\xi} - \left[\dfrac{1}{a^2 + b^2} f(\xi) + k^2\right] \varphi = 0.$$

7. $\dfrac{\partial^2 w}{\partial x^2} + \dfrac{\partial^2 w}{\partial y^2} = f(x^2 + y^2) w.$

1°. This equation admits separation of variables in the polar coordinates ρ, φ ($x = \rho \cos \varphi$, $y = \rho \sin \varphi$). Particular solution:
$$w(x, y) = [C_1 \cos(k\varphi) + C_2 \sin(k\varphi)] U(\rho),$$
where C_1, C_2, and k are arbitrary constants, and the function $U = U(\rho)$ is determined by the ordinary differential equation
$$\rho(\rho U'_\rho)'_\rho - [k^2 + \rho^2 f(\rho^2)] U = 0.$$

2°. The transformation
$$z = \tfrac{1}{2}(x^2 - y^2), \quad \zeta = xy$$
leads to a similar equation
$$\dfrac{\partial^2 w}{\partial z^2} + \dfrac{\partial^2 w}{\partial \zeta^2} = F(z^2 + \zeta^2) w, \quad F(u) = \dfrac{f(2\sqrt{u})}{2\sqrt{u}}.$$

In the special case $f(u) = 2a$, we have $F(u) = a/\sqrt{u}$. For $f(u) = bu^3$, we obtain an equation of the form 7.4.1.1 with $F(u) = 4bu$.

8. $\dfrac{\partial^2 w}{\partial x^2} + \dfrac{\partial^2 w}{\partial y^2} = [f(x) + g(y)]w.$

A particular separable solution:
$$w(x,y) = \varphi(x)\psi(y),$$
where the functions $\varphi(x)$ and $\psi(y)$ are determined by the second-order ordinary differential equations
$$\varphi''_{xx} - [f(x) - C]\varphi = 0, \quad \psi''_{yy} - [g(y) + C]\psi = 0,$$
where C is an arbitrary constant.

9. $\dfrac{\partial^2 w}{\partial x^2} + \dfrac{\partial^2 w}{\partial y^2} = [f(ax + by) + g(bx - ay)]w.$

The transformation
$$\xi = ax + by, \quad \eta = bx - ay$$
leads to an equation of the form 7.4.1.8:
$$\dfrac{\partial^2 w}{\partial \xi^2} + \dfrac{\partial^2 w}{\partial \eta^2} = \left[\dfrac{f(\xi)}{a^2 + b^2} + \dfrac{g(\eta)}{a^2 + b^2}\right] w.$$

10. $\dfrac{\partial^2 w}{\partial x^2} + \dfrac{\partial^2 w}{\partial y^2} = (x^2 + y^2)[f(x^2 - y^2) + g(xy)]w.$

The transformation
$$z = \tfrac{1}{2}(x^2 - y^2), \quad \zeta = xy$$
leads to an equation of the form 7.4.1.8:
$$\dfrac{\partial^2 w}{\partial z^2} + \dfrac{\partial^2 w}{\partial \zeta^2} = [f(2z) + g(\zeta)]w.$$

7.4.2. Convective Heat and Mass Transfer Equations

1. $\dfrac{\partial^2 w}{\partial x^2} + \dfrac{\partial^2 w}{\partial y^2} = \alpha \dfrac{\partial w}{\partial x}.$

This is a convective heat and mass transfer equation. It describes a stationary temperature (concentration) field in a continuous medium moving with a constant velocity along the x-axis. In particular, it models convective-molecular heat transfer from a heated flat plate in a flow of a thermal-transfer ideal fluid moving along the plate. This occurs, for example, if a liquid-metal coolant flows past a flat plate or if a plate is in a seepage flow through a granular medium.

In the sequel, it is assumed that the equation is written in dimensionless variables x, y related to the characteristic length (for a flat plate of length $2h$, the characteristic length is taken to be h).

1°. The substitution $w(x,y) = \exp(\tfrac{1}{2}\alpha x) U(x,y)$ brings the original equation to the Helmholtz equation
$$\dfrac{\partial^2 U}{\partial x^2} + \dfrac{\partial^2 U}{\partial y^2} = \dfrac{1}{4}\alpha^2 U.$$

Particular solutions of this equation in Cartesian and polar coordinates can be found in Subsections 7.3.2 and 7.3.3.

2°. In the elliptic coordinates
$$x = \cosh\zeta\cos\eta, \qquad y = \sinh\zeta\sin\eta$$
a wide class of particular solutions (vanishing as $\zeta \to \infty$) can be indicated; this class of solutions of the original equation is represented in series form as
$$w = \exp\left(\tfrac{1}{2}\alpha x\right)\sum_{m=0}^{\infty} A_m\,\text{ce}_m(\eta,-q)\,\text{Fek}_m(\zeta,-q), \qquad q = -\tfrac{1}{16}\alpha^2,$$
where the A_m are arbitrary constants, the $\text{ce}_m(\eta,-q)$ are the Mathieu functions, and the $\text{Fek}_m(\zeta,-q)$ are the modified Mathieu functions [e.g., see McLachlan (1947) and Bateman and Erdélyi (1955)].

3°. Consider the first boundary value problem in the upper half-plane ($-\infty < x < \infty$, $0 \le y < \infty$). We assume that the surface of a plate of finite length is maintained at a constant temperature w_0 and the medium has a temperature $w_\infty = \text{const}$ far away from the plate:
$$w = w_0 \quad \text{for} \quad y = 0,\ |x| < 1,$$
$$\partial_y w = 0 \quad \text{for} \quad y = 0,\ |x| > 1,$$
$$w \to w_\infty \quad \text{for} \quad x^2 + y^2 \to \infty.$$
The solution of this problem in the elliptic coordinates ζ,η (see Item 2°) has the form
$$w(\eta,\zeta) = w_\infty + (w_0 - w_\infty)\exp\left(\tfrac{1}{2}\alpha\cos\eta\cosh\zeta\right)\sum_{m=0}^{\infty} D_m\,\text{ce}_m(\eta,-q)\,\frac{\text{Fek}_m(\zeta,-q)}{\text{Fer}_m(0,-q)},$$
where
$$D_{2n} = 2\,\frac{\text{ce}_{2n}(0,-q)}{\text{ce}_{2n}(0,q)}\,A_0^{(2n)}, \qquad D_{2n+1} = -\frac{1}{2}\,\frac{\text{ce}_{2n+1}(0,-q)}{\text{ce}_{2n+1}(0,q)}\,\alpha B_1^{(2n+1)}, \qquad q = -\frac{1}{16}\alpha^2.$$
Here, the $A_0^{(2n)}$ and $B_1^{(2n+1)}$ are the coefficients in the series expansions of the Mathieu functions; these can be found in McLachlan (1947).

4°. Consider the second boundary value problem in the upper half-plane ($-\infty < x < \infty$, $0 \le y < \infty$). We assume that a thermal flux is prescribed on the surface of a plate of finite length and the medium has a constant temperature far away from the plate:
$$\partial_y w = f(x) \quad \text{for} \quad y = 0,\ |x| < 1,$$
$$\partial_y w = 0 \quad \text{for} \quad y = 0,\ |x| > 1,$$
$$w \to w_\infty \quad \text{as} \quad x^2 + y^2 \to \infty.$$
The solution of this problem in the Cartesian coordinates has the form
$$w(x,y) = w_\infty - \frac{1}{\pi}\int_{-1}^{1} f(\xi)\exp\left[\tfrac{1}{2}\alpha(x-\xi)\right] K_0\left(\tfrac{1}{2}\alpha\sqrt{(x-\xi)^2 + y^2}\right) d\xi,$$
where $K_0(z)$ is the modified Bessel function of the second kind.

⊙ *References*. P. V. Cherpakov (1975), A. A. Borzykh and G. P. Cherepanov (1978).

2. $\dfrac{\partial^2 w}{\partial x^2} + \dfrac{\partial^2 w}{\partial y^2} = \alpha\dfrac{\partial w}{\partial x} + \beta\dfrac{\partial w}{\partial y} + \gamma w.$

This equation describes a stationary temperature field in a medium moving with a constant velocity, provided there is volume release heat (or absorption) proportional to temperature.

The substitution
$$w(x,y) = \exp\left[\tfrac{1}{2}(\alpha x + \beta y)\right] U(x,y)$$
brings the original equation to the Helmholtz equation
$$\frac{\partial^2 U}{\partial x^2} + \frac{\partial^2 U}{\partial y^2} = \left(\gamma + \tfrac{1}{4}\alpha^2 + \tfrac{1}{4}\beta^2\right)U,$$
which is discussed in Subsections 7.3.1 through 7.3.3.

3. $\dfrac{\partial^2 w}{\partial x^2} + \dfrac{\partial^2 w}{\partial y^2} = \text{Pe}\,(1 - y^2)\dfrac{\partial w}{\partial x}.$

The Graetz–Nusselt equation. It governs steady-state heat exchange in a laminar fluid flow with a parabolic velocity profile in a plane channel. The equation is written in terms of the dimensionless Cartesian coordinates x, y related to the channel half-width h; $\text{Pe} = Uh/a$ is the Peclet number and U is the fluid velocity at the channel axis ($y = 0$). The walls of the channel correspond to $y = \pm 1$.

1°. Particular solutions:
$$w(y) = A + By,$$
$$w(x, y) = 12Ax + A\,\text{Pe}\,(6y^2 - y^4) + B,$$
$$w(x, y) = \sum_{n=1}^{m} A_n \exp\!\left(-\dfrac{\lambda_n^2}{\text{Pe}} x\right) f_n(y).$$

Here, A, B, A_n, and λ_n are arbitrary constants, and the functions f_n are defined by
$$f_n(y) = \exp\!\left(-\tfrac{1}{2}\lambda_n y^2\right)\Phi\!\left(\alpha_n, \tfrac{1}{2}; \lambda_n y^2\right), \qquad \alpha_n = \tfrac{1}{4} - \tfrac{1}{4}\lambda_n - \tfrac{1}{4}\lambda_n^3 \text{Pe}^{-2}, \tag{1}$$
where $\Phi(\alpha, \beta; \xi) = 1 + \sum\limits_{k=1}^{\infty} \dfrac{\alpha(\alpha+1)\ldots(\alpha+k-1)}{\beta(\beta+1)\ldots(\beta+k-1)}\dfrac{\xi^k}{k!}$ is the degenerate hypergeometric function.

2°. Let the walls of the channel be maintained at a constant temperature, $w = 0$ for $x < 0$ and $w = w_0$ for $x > 0$. Due to the symmetry of the problem about the x-axis, it suffices to consider only half of the domain, $0 \le y \le 1$. The boundary conditions are written as
$$y = 0, \quad \dfrac{\partial w}{\partial y} = 0; \qquad y = 1, \quad w = \begin{cases} 0 & \text{for } x < 0, \\ w_0 & \text{for } x > 0; \end{cases}$$
$$x \to -\infty, \quad w \to 0; \qquad x \to \infty, \quad w \to w_0.$$

The solution of the original equation under these boundary conditions is sought in the form
$$w(x, y) = w_0 \sum_{n=1}^{\infty} B_n \exp\!\left(\dfrac{\mu_n^2}{\text{Pe}} x\right) g_n(y) \qquad \text{for } x < 0,$$
$$w(x, y) = w_0 \left[1 - \sum_{n=1}^{\infty} A_n \exp\!\left(-\dfrac{\lambda_n^2}{\text{Pe}} x\right) f_n(y)\right] \qquad \text{for } x > 0.$$

The series coefficients must satisfy the matching conditions at the boundary:
$$w(x, y)\big|_{x\to 0,\, x<0} - w(x, y)\big|_{x\to 0,\, x>0} = 0,$$
$$\partial_x w(x, y)\big|_{x\to 0,\, x<0} - \partial_x w(x, y)\big|_{x\to 0,\, x>0} = 0.$$

For $x > 0$, the function $f_n(y)$ is defined by relation (1), where the eigenvalues λ_n are roots of the transcendental equation
$$\Phi\!\left(\alpha_n, \tfrac{1}{2}; \lambda_n\right) = 0, \quad \text{where} \quad \alpha_n = \tfrac{1}{4} - \tfrac{1}{4}\lambda_n - \tfrac{1}{4}\lambda_n^3 \text{Pe}^{-2}.$$

For $\text{Pe} \to \infty$, it is convenient to use the following approximate relation to identify the λ_n:
$$\lambda_n = 4(n - 1) + 1.68 \qquad (n = 1, 2, 3, \ldots). \tag{2}$$

The error of this formula does not exceed 0.2%. The corresponding numerical values of the coefficients A_n are rather well approximated by the relations
$$A_1 = 1.2, \qquad A_n = 2.27\,(-1)^{n-1}\lambda_n^{-7/6} \qquad \text{for } n = 2, 3, 4, \ldots,$$
whose maximum error is less than 0.1%, provided that the λ_n are calculated by (2).

For $\text{Pe} \to 0$, the following asymptotic relations hold:
$$\lambda_n = \sqrt{\pi\!\left(n - \tfrac{1}{2}\right)\text{Pe}}, \qquad A_n = \dfrac{4(-1)^{n-1}}{\pi^2(2n-1)^2}, \qquad f_n(y) = \cos\!\left[\pi\!\left(n - \tfrac{1}{2}\right)y\right] \qquad (n = 1, 2, 3, \ldots).$$

No results for $x < 0$ are given here, because they are of secondary importance in applications.

3°. Let a constant thermal flux be prescribed at the walls for $x > 0$ and let, for $x < 0$, the walls be insulated from heat and the temperature vanishes as $x \to -\infty$. Then the boundary conditions have the form

$$y = 0, \quad \frac{\partial w}{\partial y} = 0; \quad y = 1, \quad \frac{\partial w}{\partial y} = \begin{cases} 0 & \text{for } x < 0, \\ q & \text{for } x > 0; \end{cases} \quad x \to -\infty, \quad w \to 0.$$

In the domain of thermal stabilization, the asymptotic behavior of the solution (as $x \to \infty$) is as follows:

$$w(x, y) = q\left(\frac{3}{2}\frac{x}{\text{Pe}} + \frac{3}{4}y^2 - \frac{1}{8}y^4 + \frac{9}{4\,\text{Pe}^2} - \frac{39}{280}\right).$$

⊙ *References*: L. Graetz (1883), W. Nusselt (1910), C. A. Deavours (1974), A. D. Polyanin, A. M. Kutepov, A. V. Vyazmin, and D. A. Kazenin (2001).

4. $\dfrac{\partial^2 w}{\partial r^2} + \dfrac{1}{r}\dfrac{\partial w}{\partial r} + \dfrac{\partial^2 w}{\partial z^2} = \text{Pe}\,(1 - r^2)\dfrac{\partial w}{\partial z}.$

This equation governs steady-state heat exchange in a laminar fluid flow with parabolic (Poiseuille's) velocity profile in a circular tube. The equation is written in terms of the dimensionless cylindrical coordinates x, y related to the tube radius R; $\text{Pe} = UR/a$ is the Peclet number and U is the fluid velocity at the tube axis (at $r = 0$). The walls of the tube correspond to $r = 1$.

1°. Particular solutions:
$$w(r) = A + B \ln r,$$
$$w(r, z) = 16Az + A\,\text{Pe}\,(4r^2 - r^4) + B,$$
$$w(r, z) = \sum_{n=1}^{m} A_n \exp\left(-\frac{\lambda_n^2}{\text{Pe}}z\right) f_n(r).$$

Here, A, B, A_n, and λ_n are arbitrary constants, and the functions f_n are defined by

$$f_n(r) = \exp\left(-\tfrac{1}{2}\lambda_n r^2\right)\Phi(\alpha_n, 1; \lambda_n r^2), \quad \alpha_n = \tfrac{1}{2} - \tfrac{1}{4}\lambda_n - \tfrac{1}{4}\lambda_n^3\,\text{Pe}^{-2}, \qquad (1)$$

where $\Phi(\alpha, \beta; \xi)$ is the degenerate hypergeometric function (see equation 7.4.2.3, Item 1°).

2°. Let the tube wall be maintained at a constant temperature such that $w = 0$ for $z < 0$ and $w = w_0$ for $z > 0$. The boundary conditions are written as

$$r = 0, \quad \frac{\partial w}{\partial r} = 0; \quad r = 1, \quad w = \begin{cases} 0 & \text{for } z < 0, \\ w_0 & \text{for } z > 0; \end{cases}$$
$$z \to -\infty, \quad w \to 0; \quad z \to \infty, \quad w \to w_0.$$

The solution of the original equation under these boundary conditions is sought in the form

$$w(r, z) = w_0 \sum_{n=1}^{\infty} B_n \exp\left(\frac{\mu_n^2}{\text{Pe}}z\right) g_n(r) \quad \text{for} \quad z < 0,$$

$$w(r, z) = w_0 \left[1 - \sum_{n=1}^{\infty} A_n \exp\left(-\frac{\lambda_n^2}{\text{Pe}}z\right) f_n(r)\right] \quad \text{for} \quad z > 0.$$

The series coefficients must satisfy the matching conditions at the boundary,

$$w(r, z)\big|_{z\to 0,\, z<0} - w(r, z)\big|_{z\to 0,\, z>0} = 0,$$
$$\partial_z w(r, z)\big|_{z\to 0,\, z<0} - \partial_z w(r, z)\big|_{z\to 0,\, z>0} = 0.$$

For $z > 0$, the functions $f_n(r)$ are defined by relations (1), where the eigenvalues λ_n are roots of the transcendental equation

$$\Phi(\alpha_n, 1; \lambda_n) = 0, \quad \text{where} \quad \alpha_n = \tfrac{1}{2} - \tfrac{1}{4}\lambda_n - \tfrac{1}{4}\lambda_n^3\,\text{Pe}^{-2}.$$

For Pe $\to \infty$, it is convenient to use the following approximate relation to identify the λ_n:

$$\lambda_n = 4(n-1) + 2.7 \qquad (n = 1, 2, 3, \ldots). \tag{2}$$

The error of this formula does not exceed 0.3%. The corresponding numerical values of the coefficients A_n are rather well approximated by the relations

$$A_n = 2.85(-1)^{n-1}\lambda_n^{-2/3} \qquad \text{for} \quad n = 1, 2, 3, \ldots,$$

whose maximum error is 0.5%.

No results for $z < 0$ are given here, since they are of secondary importance in applications.

3°. Let a constant thermal flux be prescribed at the wall for $z > 0$ and let, for $z < 0$, the tube surface be insulated from heat and the temperature vanishes as $z \to -\infty$. Then the boundary conditions have the form

$$r = 0, \quad \frac{\partial w}{\partial r} = 0; \qquad r = 1, \quad \frac{\partial w}{\partial r} = \begin{cases} 0 & \text{for } z < 0, \\ q & \text{for } z > 0; \end{cases} \qquad z \to -\infty, \quad w \to 0.$$

In the domain of thermal stabilization, the asymptotic behavior of the solution (as $z \to \infty$) is as follows:

$$w(r, z) = q\left(4\frac{z}{\text{Pe}} + r^2 - \frac{1}{4}r^4 + \frac{8}{\text{Pe}^2} - \frac{7}{24}\right).$$

⊙ *References*: C. A. Deavours (1974), A. D. Polyanin, A. M. Kutepov, A. V. Vyazmin, and D. A. Kazenin (2001).

5. $\dfrac{\partial^2 w}{\partial x^2} + \dfrac{\partial^2 w}{\partial y^2} = f(y)\dfrac{\partial w}{\partial x}.$

This equation describes steady-state heat exchange in a laminar fluid flow with an arbitrary velocity profile $f = f(y)$ in a plane channel.

1°. Particular solutions:

$$w(x, y) = Ax + A \int_{y_0}^{y} (y - \xi)f(\xi)\, d\xi + By + C, \tag{1}$$

$$w(x, y) = B + \sum_{n=1}^{m} A_n \exp(-\beta_n x) u_n(y). \tag{2}$$

Here, A, B, C, y_0, A_n, and β_n are arbitrary constants, and the functions $u_n = u_n(y)$ are determined by the second-order linear ordinary differential equation

$$\frac{d^2 u_n}{dy^2} + \left[\beta_n f(y) + \beta_n^2\right] u_n = 0.$$

2°. Solution (1) describes the temperature distribution far away from the inlet section of the tube, in the domain of thermal stabilization, provided that a constant thermal flux is prescribed at the channel walls.

6. $a\left(\dfrac{\partial^2 w}{\partial x^2} + \dfrac{\partial^2 w}{\partial y^2}\right) = v_1(x, y)\dfrac{\partial w}{\partial x} + v_2(x, y)\dfrac{\partial w}{\partial y}.$

This is an equation of steady-state convective heat and mass transfer in the Cartesian coordinate system. Here, $v_1 = v_1(x, y)$ and $v_2 = v_2(x, y)$ are the components of the fluid velocity that are assumed to be known from the solution of the hydrodynamic problem.

1°. In plane problems of convective heat exchange in liquid metals modeled by an ideal fluid, as well as in describing seepage (filtration) streams employing the model of potential flows, the fluid velocity components $v_1(x,y)$ and $v_2(x,y)$ can be expressed in terms of the potential $\varphi = \varphi(x,y)$ and stream function $\psi = \psi(x,y)$ as follows:

$$v_1 = \frac{\partial \varphi}{\partial x} = -\frac{\partial \psi}{\partial y}, \qquad v_2 = \frac{\partial \varphi}{\partial y} = \frac{\partial \psi}{\partial x}. \tag{1}$$

The function φ is determined by solving the Laplace equation $\Delta \varphi = 0$. In specific problems, the potential φ and stream function ψ may be identified by invoking the complex variable theory [e.g., see Lavrent'ev and Shabat (1973) and Sedov (1980)].

By passing in the convective heat exchange equation from x, y to the new variables φ, ψ (Boussinesq transformation) and taking into account (1), we arrive a simpler equation with constant coefficients of the form 7.4.2.1:

$$\frac{\partial^2 w}{\partial \varphi^2} + \frac{\partial^2 w}{\partial \psi^2} = \frac{1}{a}\frac{\partial w}{\partial \varphi}. \tag{2}$$

The Boussinesq transformation brings any plane contour in a potential flow to a cut in the φ-axis, simultaneously with the reduction of the original equation to the form (2). Consequently, the heat transfer problem of a potential flow about this contour is reduced to the heat exchange problem of a longitudinal flow of an ideal fluid past a flat plate (see equation 7.4.2.1, Items 3° and 4°).

2°. Asymptotic analyses of plane problems on heat/mass exchange of bodies of various shape with laminar translational and shear flows of a viscous (and ideal) incompressible fluid for large and small Peclet numbers were carried out in the references cited below. In the thermal boundary layer approximation, the solution of the heat exchange problem for a flat plate in a longitudinal translational flow of a viscous incompressible fluid at large Reynolds numbers is presented in 1.9.1.4, Item 3°.

⊙ *References*: V. G. Levich (1962), P. V. Cherpakov (1975), A. A. Borzykh and G. P. Cherepanov (1978), Yu. P. Gupalo, A. D. Polyanin, and Yu. S. Ryazantsev (1985), A. D. Polyanin, A. M. Kutepov, A. V. Vyazmin, and D. A. Kazenin (2001).

7. $\dfrac{1}{r^2}\dfrac{\partial}{\partial r}\left(r^2\dfrac{\partial w}{\partial r}\right) + \dfrac{1}{r^2\sin\theta}\dfrac{\partial}{\partial \theta}\left(\sin\theta\dfrac{\partial w}{\partial \theta}\right) = \cos\theta\,\dfrac{\partial w}{\partial r} - \dfrac{\sin\theta}{r}\dfrac{\partial w}{\partial \theta}.$

This is a special case of equation 7.4.1.8 with $a = 1$, $v_r = \cos\theta$, and $v_\theta = -\sin\theta$. This equation is obtained from the equation $\partial_{xx}w + \partial_{yy}w + \partial_{zz}w = \partial_x w$ by the passage to the spherical coordinate system in the axisymmetric case.

The general solution satisfying the decay condition ($w \to 0$ as $r \to \infty$) is expressed as

$$w(r,\theta) = \left(\frac{\pi}{r}\right)^{1/2}\exp\left(\frac{r\cos\theta}{2}\right)\sum_{n=0}^{\infty} A_n K_{n+\frac{1}{2}}\left(\frac{r}{2}\right)P_n(\cos\theta),$$

where the A_n are arbitrary constants. The Legendre polynomials $P_n(\xi)$ and the modified Bessel functions $K_{n+\frac{1}{2}}(z)$ are given by

$$P_n(\xi) = \frac{1}{n!\,2^n}\frac{d^n}{d\xi^n}(\xi^2-1)^n, \qquad K_{n+\frac{1}{2}}\left(\frac{r}{2}\right) = \left(\frac{\pi}{r}\right)^{1/2}\exp\left(-\frac{r}{2}\right)\sum_{m=0}^{n}\frac{(n+m)!}{(n-m)!\,m!\,r^m}.$$

⊙ *Reference*: P. L. Rimmer (1968).

8. $a\left[\dfrac{1}{r^2}\dfrac{\partial}{\partial r}\left(r^2\dfrac{\partial w}{\partial r}\right) + \dfrac{1}{r^2\sin\theta}\dfrac{\partial}{\partial \theta}\left(\sin\theta\dfrac{\partial w}{\partial \theta}\right)\right] = v_r\dfrac{\partial w}{\partial r} + \dfrac{v_\theta}{r}\dfrac{\partial w}{\partial \theta}.$

This equation is often encountered in axisymmetric problems of convective heat and mass exchange of solid particles, drops, and bubbles with a flow of a viscous incompressible fluid. The fluid velocity components $v_r = v_r(r,\theta)$ and $v_\theta = v_\theta(r,\theta)$ can be expressed in terms of the stream function $\psi = \psi(r,\theta)$ as

$$v_r = \frac{1}{r^2\sin\theta}\frac{\partial \psi}{\partial \theta}, \qquad v_\theta = -\frac{1}{r\sin\theta}\frac{\partial \psi}{\partial r}. \tag{1}$$

Asymptotic analyses for a wide class of axisymmetric problems on heat/mass exchange of solid particles, drops, and bubbles of various shape with a laminar translational or straining flow of a viscous incompressible fluid at large and small Peclet numbers $\text{Pe} = UR/a$ are performed in the books cited below. The Peclet number is written in terms of the characteristic velocity U (e.g., the unperturbed fluid velocity far away from the particle in the case of translation flow), the characteristic size of the particle R (e.g., the radius for a spherical particle), and the thermal conductivity or diffusion coefficient a.

The following boundary conditions are usually specified:

$$w = w_0 \quad \text{at} \quad r = R, \qquad w \to w_\infty \quad \text{as} \quad r \to \infty, \tag{2}$$

where R is the particle radius, w_0 the temperature at the particle surface, and w_∞ the temperature far away from the particle (w_0 and w_∞ are constant).

Convective mass transfer problems are characterized by large Peclet numbers. To solve such problems, the diffusion boundary layer approximation is often used; in this case, the left-hand side of the equation takes into account only the diffusion mass transfer in the normal direction to the particle surface (the tangential mass transfer is neglected). The convective terms on the right-hand side are partially preserved—the fluid velocity components are approximated by their leading terms of the asymptotic expansion near the phase surface. Presented below are some important results obtained by solving the original equation under the boundary conditions (2) in the diffusion boundary layer approximation.

Example 1. For the translational Stokes flow of a viscous incompressible fluid about a spherical bubble, the stream function is expressed as

$$\psi(r, \theta) = \tfrac{1}{2} U r (r - R) \sin^2 \theta.$$

Here, U is the unperturbed fluid velocity in the incident flow, R the bubble radius (the value $\theta = \pi$ corresponds to the front critical point at the bubble surface).

In this case, the solution of the convective heat/mass transfer equation with the boundary conditions (2) for $\text{Pe} = UR/a \gg 1$ in the diffusion boundary layer approximation is given by

$$w(r, \theta) = w_0 + (w_\infty - w_0) \operatorname{erf} \xi, \qquad \xi = \sqrt{\tfrac{3}{8} \text{Pe}} \left(\frac{r}{R} - 1 \right) \frac{1 - \cos \theta}{\sqrt{2 - \cos \theta}},$$

where $\operatorname{erf} \xi$ is the error function.

Example 2. For the translational Stokes flow of a viscous incompressible fluid about a solid spherical particle, the stream function is expressed as

$$\psi(r, \theta) = \frac{1}{4} U (r - R)^2 \left(2 + \frac{R}{r} \right) \sin^2 \theta.$$

Here, the notation is the same as in the case of a bubble above.

For a solid particle, the solution of the convective heat/mass transfer equation with the boundary conditions (2) for $\text{Pe} = UR/a \gg 1$ in the diffusion boundary layer approximation is given by

$$w(r, \theta) = w_0 + (w_\infty - w_0) \left[\Gamma\left(\tfrac{1}{3}\right) \right]^{-1} \gamma\left(\tfrac{1}{3}, \xi\right), \qquad \xi = \frac{\text{Pe}\,(r - R)^3 \sin^3 \theta}{3 R^3 \left(\pi - \theta + \tfrac{1}{2} \sin 2\theta \right)},$$

where $\Gamma(\beta)$ is the gamma function and $\gamma(\beta, \xi) = \int_0^\xi e^{-z} z^{\beta - 1}\, dz$ is the incomplete gamma function.

⊙ *References*: V. G. Levich (1962), Yu. P. Gupalo, A. D. Polyanin, and Yu. S. Ryazantsev (1985), A. D. Polyanin, A. M. Kutepov, A. V. Vyazmin, and D. A. Kazenin (2001).

7.4.3. Equations of Heat and Mass Transfer in Anisotropic Media

1. $\dfrac{\partial}{\partial x}\left(a x^n \dfrac{\partial w}{\partial x} \right) + \dfrac{\partial}{\partial y}\left(b y^m \dfrac{\partial w}{\partial y} \right) = 0.$

This is a two-dimensional equation of the heat and mass transfer theory in a inhomogeneous anisotropic medium. Here, $a_1(x) = a x^n$ and $a_2(y) = b y^m$ are the principal thermal diffusivities.

$1°$. Particular solutions (A, B, C are arbitrary constants):

$$w(x,y) = Ax^{1-n} + By^{1-m} + C,$$

$$w(x,y) = A\left[\frac{x^{2-n}}{a(2-n)} - \frac{y^{2-m}}{b(2-m)}\right] + B,$$

$$w(x,y) = Ax^{1-n}y^{1-m} + B.$$

$2°$. For $n \neq 2$ and $m \neq 2$, there are particular solutions of the form

$$w = w(\xi), \qquad \xi = \left[b(2-m)^2 x^{2-n} + a(2-n)^2 y^{2-m}\right]^{1/2}.$$

The function $w = w(\xi)$ is determined by the ordinary differential equation

$$w''_{\xi\xi} + \frac{A}{\xi} w'_\xi = 0, \qquad A = \frac{4-nm}{(2-n)(2-m)}. \tag{1}$$

The general solution of equation (1) is given by

$$w(\xi) = \begin{cases} C_1 \xi^{1-A} + C_2 & \text{for } A \neq 1, \\ C_1 \ln \xi + C_2 & \text{for } A = 1, \end{cases}$$

where C_1 and C_2 are arbitrary constants.

$3°$. There are multiplicatively separable particular solutions in the form

$$w(x,y) = \varphi(x)\psi(y), \tag{2}$$

where $\varphi(x)$ and $\psi(y)$ are determined by the following second-order linear ordinary differential equations (A_1 is an arbitrary constant):

$$(ax^n \varphi'_x)'_x = -A_1 \varphi, \tag{3}$$

$$(by^m \psi'_y)'_y = A_1 \psi. \tag{4}$$

The solution of equation (3) is given by

$$\varphi(x) = \begin{cases} x^{\frac{1-n}{2}} \left[C_1 J_\nu\left(\beta x^{\frac{2-n}{2}}\right) + C_2 Y_\nu\left(\beta x^{\frac{2-n}{2}}\right)\right] & \text{for } A_1 > 0, \\ x^{\frac{1-n}{2}} \left[C_1 I_\nu\left(\beta x^{\frac{2-n}{2}}\right) + C_2 K_\nu\left(\beta x^{\frac{2-n}{2}}\right)\right] & \text{for } A_1 < 0, \end{cases}$$

$$\nu = \frac{|1-n|}{2-n}, \qquad \beta = \frac{2}{2-n}\sqrt{\frac{|A_1|}{a}},$$

where C_1 and C_2 are arbitrary constants, $J_\nu(z)$ and $Y_\nu(z)$ are the Bessel functions, and $I_\nu(z)$ and $K_\nu(z)$ are the modified Bessel functions.

The solution of equation (4) is expressed as

$$\psi(y) = \begin{cases} y^{\frac{1-m}{2}} \left[C_1 J_\sigma\left(\mu y^{\frac{2-m}{2}}\right) + C_2 Y_\sigma\left(\mu y^{\frac{2-m}{2}}\right)\right] & \text{for } A_1 < 0, \\ y^{\frac{1-m}{2}} \left[C_1 I_\sigma\left(\mu y^{\frac{2-m}{2}}\right) + C_2 K_\sigma\left(\mu y^{\frac{2-m}{2}}\right)\right] & \text{for } A_1 > 0, \end{cases}$$

$$\sigma = \frac{|1-m|}{2-m}, \qquad \mu = \frac{2}{2-m}\sqrt{\frac{|A_1|}{b}},$$

where C_1 and C_2 are arbitrary constants.

The sum of solutions of the form (2) corresponding to different values of the parameter A_1 is also a solution of the original equation; the solutions of some boundary value problems may be obtained by separation of variables.

$4°$. See equation 7.4.3.3, Item $4°$, for $c = 0$.

2. $\dfrac{\partial}{\partial x}\left(ax^n \dfrac{\partial w}{\partial x}\right) + \dfrac{\partial}{\partial y}\left(by^m \dfrac{\partial w}{\partial y}\right) = c.$

This is a two-dimensional equation of the heat and mass transfer theory with constant volume release of heat in an inhomogeneous anisotropic medium. Here, $a_1(x) = ax^n$ and $a_2(y) = by^m$ are the principal thermal diffusivities.

1°. For $n \neq 2$ and $m \neq 2$, there are particular solutions of the form

$$w = w(\xi), \qquad \xi = \left[b(2-m)^2 x^{2-n} + a(2-n)^2 y^{2-m}\right]^{1/2}. \tag{1}$$

The function $w = w(\xi)$ is determined by the ordinary differential equation

$$w''_{\xi\xi} + \dfrac{A}{\xi} w'_\xi = B, \tag{2}$$

where

$$A = \dfrac{4 - nm}{(2-n)(2-m)}, \qquad B = \dfrac{4c}{ab(2-n)^2(2-m)^2}. \tag{3}$$

The general solution of equation (2) is given by

$$w(\xi) = \begin{cases} C_1 \xi^{1-A} + C_2 + \dfrac{B}{2(A+1)}\xi^2 & \text{for } A \neq \pm 1, \\ C_1 \ln \xi + C_2 + \tfrac{1}{4} B \xi^2 & \text{for } A = 1, \\ C_1 \xi^2 + C_2 + \tfrac{1}{2} B \xi^2 \ln \xi & \text{for } A = -1, \end{cases}$$

where C_1 and C_2 are arbitrary constants.

2°. The substitution

$$w(x,y) = U(x,y) + \dfrac{c}{a(2-n)} x^{2-n}$$

leads to a homogeneous equation of the form 7.4.3.1:

$$\dfrac{\partial}{\partial x}\left(ax^n \dfrac{\partial U}{\partial x}\right) + \dfrac{\partial}{\partial y}\left(by^m \dfrac{\partial U}{\partial y}\right) = 0.$$

3. $\dfrac{\partial}{\partial x}\left(ax^n \dfrac{\partial w}{\partial x}\right) + \dfrac{\partial}{\partial y}\left(by^m \dfrac{\partial w}{\partial y}\right) = cw.$

This is a two-dimensional equation of the heat and mass transfer theory with a linear source in an inhomogeneous anisotropic medium.

1°. For $n \neq 2$ and $m \neq 2$, there are particular solutions of the form

$$w = w(\xi), \qquad \xi = \left[b(2-m)^2 x^{2-n} + a(2-n)^2 y^{2-m}\right]^{1/2}.$$

The function $w = w(\xi)$ is determined by the ordinary differential equation

$$w''_{\xi\xi} + \dfrac{A}{\xi} w'_\xi = Bw, \tag{1}$$

where

$$A = \dfrac{4 - nm}{(2-n)(2-m)}, \qquad B = \dfrac{4c}{ab(2-n)^2(2-m)^2}.$$

The general solution of equation (1) is given by

$$w(\xi) = \xi^{\frac{1-A}{2}}\left[C_1 J_\nu\left(\xi\sqrt{|B|}\right) + C_2 Y_\nu\left(\xi\sqrt{|B|}\right)\right] \quad \text{for } B < 0,$$

$$w(\xi) = \xi^{\frac{1-A}{2}}\left[C_1 I_\nu\left(\xi\sqrt{B}\right) + C_2 K_\nu\left(\xi\sqrt{B}\right)\right] \quad \text{for } B > 0,$$

where $\nu = \tfrac{1}{2}|1-A|$; C_1 and C_2 are arbitrary constants; $J_\nu(z)$ and $Y_\nu(z)$ are the Bessel functions; and $I_\nu(z)$ and $K_\nu(z)$ are the modified Bessel functions.

2°. There are multiplicatively separable particular solutions of the form

$$w(x,y) = \varphi(x)\psi(y),$$

where $\varphi(x)$ and $\psi(y)$ are determined by the following second-order linear ordinary differential equations (A_1 is an arbitrary constant):

$$(ax^n \varphi'_x)'_x = A_1 \varphi, \qquad (by^m \psi'_y)'_y = (c - A_1)\psi. \qquad (2)$$

The solutions of equations (2) are expressed in terms of the Bessel functions (or modified Bessel functions); see equation 7.4.3.1, Item 3°.

3°. There are additively separable particular solutions of the form

$$w(x,y) = f(x) + g(y),$$

where $f(x)$ and $g(y)$ are determined by the following second-order linear ordinary differential equations (A_2 is an arbitrary constant):

$$(ax^n f'_x)'_x - cf = A_2, \qquad (by^m g'_y)'_y - cg = -A_2. \qquad (3)$$

The solutions of equations (3) are expressed in terms of the Bessel functions (or modified Bessel functions).

4°. The transformation (specified by A. I. Zhurov, 2001)

$$x^{\frac{2-n}{2}} = Ar\cos\theta, \qquad y^{\frac{2-m}{2}} = Br\sin\theta,$$

where $A^2 = a(2-n)^2$ and $B^2 = b(2-m)^2$, leads to the equation

$$\frac{\partial^2 w}{\partial r^2} + \frac{4-nm}{(2-n)(2-m)} \frac{1}{r} \frac{\partial w}{\partial r} + \frac{1}{r^2} \frac{\partial^2 w}{\partial \theta^2} - \frac{2}{r^2} \frac{(nm-n-m)\cos 2\theta + (n-m)}{(2-n)(2-m)\sin 2\theta} \frac{\partial w}{\partial \theta} = 4cw,$$

which admits separable solutions of the form $w(r,\theta) = F_1(r)F_2(\theta)$.

4. $\dfrac{\partial}{\partial x}\left[a(x+k)^n \dfrac{\partial w}{\partial x}\right] + \dfrac{\partial}{\partial y}\left[b(y+s)^m \dfrac{\partial w}{\partial y}\right] = c.$

The transformation $\zeta = x + k$, $\eta = y + s$ leads to an equation of the form 7.4.3.2:

$$\frac{\partial}{\partial \zeta}\left(a\zeta^n \frac{\partial w}{\partial \zeta}\right) + \frac{\partial}{\partial \eta}\left(b\eta^m \frac{\partial w}{\partial \eta}\right) = c.$$

5. $\dfrac{\partial}{\partial x}\left[a(x+k)^n \dfrac{\partial w}{\partial x}\right] + \dfrac{\partial}{\partial y}\left[b(y+s)^m \dfrac{\partial w}{\partial y}\right] = cw.$

The transformation $\zeta = x + k$, $\eta = y + s$ leads to an equation of the form 7.4.3.3:

$$\frac{\partial}{\partial \zeta}\left(a\zeta^n \frac{\partial w}{\partial \zeta}\right) + \frac{\partial}{\partial \eta}\left(b\eta^m \frac{\partial w}{\partial \eta}\right) = cw.$$

6. $\dfrac{\partial}{\partial x}\left(ae^{\beta x} \dfrac{\partial w}{\partial x}\right) + \dfrac{\partial}{\partial y}\left(be^{\mu y} \dfrac{\partial w}{\partial y}\right) = 0.$

This is a two-dimensional equation of the heat and mass transfer theory in an inhomogeneous anisotropic medium. Here, $a_1(x) = ae^{\beta x}$ and $a_2(y) = be^{\mu y}$ are the principal thermal diffusivities.

1°. Particular solutions (A, B, C are arbitrary constants):

$$w(x,y) = Ae^{-\beta x} + Be^{-\mu y} + C,$$

$$w(x,y) = \frac{A}{a\beta^2}(\beta x + 1)e^{-\beta x} - \frac{A}{b\mu^2}(\mu y + 1)e^{-\mu y} + B,$$

$$w(x,y) = Ae^{-\beta x - \mu y} + B.$$

2°. There are multiplicatively separable particular solutions of the form

$$w(x,y) = \varphi(x)\psi(y), \tag{1}$$

where $\varphi(x)$ and $\psi(y)$ are determined by the following second-order linear ordinary differential equations (A_1 is an arbitrary constant):

$$(ae^{\beta x}\varphi'_x)'_x = -A_1\varphi, \tag{2}$$

$$(be^{\mu y}\psi'_y)'_y = A_1\psi. \tag{3}$$

The solution of equation (2) is given by

$$\varphi(x) = \begin{cases} e^{-\beta x/2}\left[C_1 J_1\left(ke^{-\beta x/2}\right) + C_2 Y_1\left(ke^{-\beta x/2}\right)\right] & \text{for } A_1 > 0, \\ e^{-\beta x/2}\left[C_1 I_1\left(ke^{-\beta x/2}\right) + C_2 K_1\left(ke^{-\beta x/2}\right)\right] & \text{for } A_1 < 0, \end{cases}$$

where $k = -(2/\beta)\sqrt{|A_1|/a}$; C_1 and C_2 are arbitrary constants; $J_1(z)$ and $Y_1(z)$ are the Bessel functions; and $I_1(z)$ and $K_1(z)$ are the modified Bessel functions.

The solution of equation (3) is given by

$$\psi(y) = \begin{cases} e^{-\mu y/2}\left[C_1 J_1\left(se^{-\mu y/2}\right) + C_2 Y_1\left(se^{-\mu y/2}\right)\right] & \text{for } A_1 < 0, \\ e^{-\mu y/2}\left[C_1 I_1\left(se^{-\mu y/2}\right) + C_2 K_1\left(se^{-\mu y/2}\right)\right] & \text{for } A_1 > 0, \end{cases}$$

where $s = -(2/\mu)\sqrt{|A_1|/b}$; C_1 and C_2 are arbitrary constants.

The sum of solutions of the form (1) corresponding to different values of the parameter A_1 is also a solution of the original equation.

3°. See equation 7.4.3.8, Item 3°, for $c = 0$.

7. $\dfrac{\partial}{\partial x}\left(ae^{\beta x}\dfrac{\partial w}{\partial x}\right) + \dfrac{\partial}{\partial y}\left(be^{\mu y}\dfrac{\partial w}{\partial y}\right) = c.$

This is a two-dimensional equation of the heat and mass transfer theory with constant volume release of heat in an inhomogeneous anisotropic medium. Here, $a_1(x) = ae^{\beta x}$ and $a_2(y) = be^{\mu y}$ are the principal thermal diffusivities.

The substitution

$$w(x,y) = U(x,y) - \frac{c}{a\beta^2}(\beta x + 1)e^{-\beta x}$$

leads to a homogeneous equation of the form 7.4.3.6:

$$\frac{\partial}{\partial x}\left(ae^{\beta x}\frac{\partial U}{\partial x}\right) + \frac{\partial}{\partial y}\left(be^{\mu y}\frac{\partial U}{\partial y}\right) = 0.$$

8. $\dfrac{\partial}{\partial x}\left(ae^{\beta x}\dfrac{\partial w}{\partial x}\right) + \dfrac{\partial}{\partial y}\left(be^{\mu y}\dfrac{\partial w}{\partial y}\right) = cw.$

This is a two-dimensional equation of the heat and mass transfer theory with a linear source in an inhomogeneous anisotropic medium.

1°. For $\beta\mu \neq 0$, there are particular solutions of the form
$$w = w(\xi), \qquad \xi = \big(b\mu^2 e^{-\beta x} + a\beta^2 e^{-\mu y}\big)^{1/2}.$$
The function $w = w(\xi)$ is determined by the ordinary differential equation
$$w''_{\xi\xi} - \frac{1}{\xi} w'_\xi = Bw, \qquad B = \frac{4c}{ab\beta^2\mu^2}.$$
For the solution of this equation, see 7.4.3.3 (Item 1° for $A = -1$).

2°. The original equation admits multiplicatively (and additively) separable solutions. See equation 7.4.3.12 with $f(x) = ae^{\beta x}$ and $g(y) = be^{\mu y}$.

3°. The transformation (specified by A. I. Zhurov, 2001)
$$e^{-\beta x/2} = Ar\cos\theta, \qquad e^{-\mu y/2} = Br\sin\theta,$$
where $A^2 = a\beta^2$ and $B^2 = b\mu^2$, leads to the equation
$$\frac{\partial^2 w}{\partial r^2} - \frac{1}{r}\frac{\partial w}{\partial r} + \frac{1}{r^2}\frac{\partial^2 w}{\partial \theta^2} - \frac{2}{r^2}\cot 2\theta \frac{\partial w}{\partial \theta} = 4cw,$$
which admits separable solutions of the form $w(r,\theta) = F_1(r)F_2(\theta)$.

9. $\dfrac{\partial}{\partial x}\left(ax^n \dfrac{\partial w}{\partial x}\right) + \dfrac{\partial}{\partial y}\left(be^{\beta y}\dfrac{\partial w}{\partial x}\right) = cw.$

1°. For $n \neq 2$ and $\beta \neq 0$, there are particular solutions of the form
$$w = w(r), \qquad r^2 = \frac{x^{2-n}}{a(2-n)^2} + \frac{e^{-\beta y}}{b\beta^2}.$$
The function $w = w(r)$ is determined by the ordinary differential equation
$$\frac{\partial^2 w}{\partial r^2} + \frac{n}{2-n}\frac{1}{r}\frac{\partial w}{\partial r} = 4cw.$$
For the solution of this equation, see 7.4.3.3 (Item 1°).

2°. The original equation admits multiplicatively (and additively) separable solutions. See equation 7.4.3.12 with $f(x) = ax^n$ and $g(y) = be^{\beta y}$.

3°. The transformation (specified by A. I. Zhurov, 2001)
$$x^{1-\frac{1}{2}n} = Ar\cos\theta, \qquad e^{-\frac{1}{2}\beta y} = Br\sin\theta,$$
where $A^2 = a(2-n)^2$ and $B^2 = b\beta^2$, leads to the equation
$$\frac{\partial^2 w}{\partial r^2} + \frac{n}{2-n}\frac{1}{r}\frac{\partial w}{\partial r} + \frac{1}{r^2}\frac{\partial^2 w}{\partial \theta^2} - \frac{2}{r^2}\frac{(1-n)\cos 2\theta + 1}{(2-n)\sin 2\theta}\frac{\partial w}{\partial \theta} = 4cw,$$
which admits separable solutions of the form $w(r,\theta) = F_1(r)F_2(\theta)$.

10. $\dfrac{\partial}{\partial x}\left[f(x)\dfrac{\partial w}{\partial x}\right] + \dfrac{\partial^2 w}{\partial y^2} = 0.$

1°. Particular solutions:
$$w = C_1 y^2 + C_2 y - 2\int \frac{C_1 x + C_3}{f(x)}\,dx + C_4,$$
$$w = C_1 y^3 + C_2 y - 6y\int \frac{C_1 x + C_3}{f(x)}\,dx + C_4,$$
$$w = [C_1\Phi(x) + C_2]y + C_3\Phi(x) + C_4, \qquad \Phi(x) = \int \frac{dx}{f(x)},$$
$$w = [C_1\Phi(x) + C_2]y^2 + C_3\Phi(x) + C_4 - 2\int \left\{\frac{1}{f(x)}\int [C_1\Phi(x) + C_2]\,dx\right\}dx,$$
where C_1, C_2, C_3, C_4, and C_5 are arbitrary constants.

2°. Separable particular solution:
$$w = (C_1 e^{\lambda y} + C_2 e^{-\lambda y})H(x),$$

where C_1, C_2, and λ are arbitrary constants, and the function $H = H(x)$ is determined by the ordinary differential equation $[f(x)H'_x]'_x + \lambda^2 H = 0$.

3°. Separable particular solution:
$$w = [C_1 \sin(\lambda y) + C_2 \cos(\lambda y)]Z(x),$$

where C_1, C_2, and λ are arbitrary constants, and the function $Z = Z(x)$ is determined by the ordinary differential equation $[f(x)Z'_x]'_x - \lambda^2 Z = 0$.

4°. Particular solutions with even powers of y:
$$w = \sum_{k=0}^{n} \zeta_k(x) y^{2k},$$

where the functions $\zeta_k = \zeta_k(x)$ are defined by the recurrence relations
$$\zeta_n(x) = A_n \Phi(x) + B_n, \qquad \Phi(x) = \int \frac{dx}{f(x)},$$
$$\zeta_{k-1}(x) = A_k \Phi(x) + B_k - 2k(2k-1) \int \frac{1}{f(x)} \left\{ \int \zeta_k(x)\, dx \right\} dx,$$

where A_k and B_k are arbitrary constants ($k = n, \ldots, 1$).

5°. Particular solutions with odd powers of y:
$$w = \sum_{k=0}^{n} \eta_k(x) y^{2k+1},$$

where the functions $\eta_k = \eta_k(x)$ are defined by the recurrence relations
$$\eta_n(x) = A_n \Phi(x) + B_n, \qquad \Phi(x) = \int \frac{dx}{f(x)},$$
$$\eta_{k-1}(x) = A_k \Phi(x) + B_k - 2k(2k+1) \int \frac{1}{f(x)} \left\{ \int \eta_k(x)\, dx \right\} dx,$$

where A_k and B_k are arbitrary constants ($k = n, \ldots, 1$).

11. $\dfrac{\partial}{\partial x}\left[f(x) \dfrac{\partial w}{\partial x} \right] + \dfrac{\partial}{\partial y}\left[g(y) \dfrac{\partial w}{\partial y} \right] = 0.$

This is a two-dimensional sourceless equation of the heat and mass transfer theory in an inhomogeneous anisotropic medium. The functions $f = f(x)$ and $g = g(y)$ are the principal thermal diffusivities.

1°. Particular solutions:
$$w(x,y) = A_1 \int \frac{dx}{f(x)} + B_1 \int \frac{dy}{g(y)} + C_1,$$
$$w(x,y) = A_2 \int \frac{x\, dx}{f(x)} - A_2 \int \frac{y\, dy}{g(y)} + B_2,$$
$$w(x,y) = A_3 \int \frac{dx}{f(x)} \int \frac{dy}{g(y)} + B_3,$$

where the A_k, B_k, and C_1 are arbitrary constants. A linear combination of these solutions is also a solution of the original equation.

2°. There are multiplicatively separable particular solutions of the form

$$w(x,y) = \varphi(x)\psi(y), \tag{1}$$

where $\varphi(x)$ and $\psi(y)$ are determined by the following second-order linear ordinary differential equations (A is an arbitrary constant):

$$\begin{aligned} (f\varphi'_x)'_x &= A\varphi, & f &= f(x), \\ (g\psi'_y)'_y &= -A\psi, & g &= g(y). \end{aligned} \tag{2}$$

The sum of solutions of the form (1) corresponding to different values of the parameter A in (2) is also a solution of the original equation (the solutions of some boundary value problems may be obtained by separation of variables).

12. $\dfrac{\partial}{\partial x}\left[f(x)\dfrac{\partial w}{\partial x}\right] + \dfrac{\partial}{\partial y}\left[g(y)\dfrac{\partial w}{\partial y}\right] = \beta w.$

This is a two-dimensional equation of the heat and mass transfer theory with a linear source in an inhomogeneous anisotropic medium. The functions $f = f(x)$ and $g = g(y)$ are the principal thermal diffusivities.

1°. There are multiplicatively separable particular solutions of the form

$$w(x,y) = \varphi(x)\psi(y), \tag{1}$$

where $\varphi(x)$ and $\psi(y)$ are determined by the following second-order linear ordinary differential equations (A is an arbitrary constant):

$$\begin{aligned} (f\varphi'_x)'_x &= A\varphi, & f &= f(x), \\ (g\psi'_y)'_y &= (\beta - A)\psi, & g &= g(y). \end{aligned} \tag{2}$$

The sum of solutions of the form (1) corresponding to different values of the parameter A in (2) is also a solution of the original equation; the solutions of some boundary value problems may be obtained by separation of variables.

2°. There are additively separable particular solutions of the form

$$w(x,y) = \Phi(x) + \Psi(y),$$

where $\Phi(x)$ and $\Psi(y)$ are determined by the following second-order linear ordinary differential equations (C is an arbitrary constant):

$$\begin{aligned} (f\Phi'_x)'_x - \beta\Phi &= C, & f &= f(x), \\ (g\Psi'_y)'_y - \beta\Psi &= -C, & g &= g(y). \end{aligned}$$

In the special case $\beta = 0$, the solutions of these equations can be represented as

$$\Phi(x) = C\int \frac{x\,dx}{f(x)} + A_1\int \frac{dx}{f(x)} + B_1,$$

$$\Psi(y) = -C\int \frac{y\,dy}{g(y)} + A_2\int \frac{dy}{g(y)} + B_2,$$

where A_1, A_2, B_1, and B_2 are arbitrary constants.

7.4.4. Other Equations Arising in Applications

1. $y\dfrac{\partial^2 w}{\partial x^2} + \dfrac{\partial^2 w}{\partial y^2} = 0.$

Tricomi equation. It is used to describe near-sonic flows of gas.

1°. Particular solutions:
$$w = Axy + Bx + Cy + D,$$
$$w = A(3x^2 - y^3) + B(x^3 - xy^3) + C(6yx^2 - y^4),$$
where A, B, C, and D are arbitrary constants.

2°. Particular solutions with even powers of x:
$$w = \sum_{k=0}^{n} \varphi_k(y) x^{2k},$$
where the functions $\varphi_k = \varphi_k(y)$ are defined by the recurrence relations
$$\varphi_n(y) = A_n y + B_n, \quad \varphi_{k-1}(y) = A_k y + B_k - 2k(2k-1) \int_0^y (y-t) t \varphi_k(t)\, dt,$$
where A_k and B_k are arbitrary constants ($k = n, \dots, 1$).

3°. Particular solutions with odd powers of x:
$$w = \sum_{k=0}^{n} \psi_k(y) x^{2k+1},$$
where the functions $\psi_k = \psi_k(y)$ are defined by the recurrence relations
$$\psi_n(y) = A_n y + B_n, \quad \psi_{k-1}(y) = A_k y + B_k - 2k(2k+1) \int_0^y (y-t) t \psi_k(t)\, dt,$$
where A_k and B_k are arbitrary constants ($k = n, \dots, 1$).

4°. Separable particular solutions:
$$w(x,y) = \left[A \sinh(3\lambda x) + B \cosh(3\lambda x)\right] \sqrt{y} \left[C J_{1/3}(2\lambda y^{3/2}) + D Y_{1/3}(2\lambda y^{3/2})\right],$$
$$w(x,y) = \left[A \sin(3\lambda x) + B \cos(3\lambda x)\right] \sqrt{y} \left[C I_{1/3}(2\lambda y^{3/2}) + D K_{1/3}(2\lambda y^{3/2})\right],$$
where A, B, C, D, and λ are arbitrary constants, $J_{1/3}(z)$ and $Y_{1/3}(z)$ are the Bessel functions, and $I_{1/3}(z)$ and $K_{1/3}(z)$ are the modified Bessel functions.

5°. For $y > 0$, see also equation 7.4.4.2 with $n = 1$. For $y < 0$, the change of variable $y = -t$ leads to an equation of the form 4.3.3.11 with $n = 1$.

⊙ *Reference*: V. M. Babich, M. B. Kapilevich, S. G. Mikhlin, et al. (1964).

2. $y^n \dfrac{\partial^2 w}{\partial x^2} + \dfrac{\partial^2 w}{\partial y^2} = 0.$

1°. Particular solutions:
$$w = Axy + Bx + Cy + D,$$
$$w = Ax^2 - \frac{2A}{(n+1)(n+2)} y^{n+2},$$
$$w = Ax^3 - \frac{6A}{(n+1)(n+2)} xy^{n+2},$$
$$w = Ayx^2 - \frac{2A}{(n+2)(n+3)} y^{n+3},$$
where A, B, C, and D are arbitrary constants.

2°. Particular solutions with even powers of x:

$$w = \sum_{k=0}^{m} \varphi_k(y) x^{2k},$$

where the functions $\varphi_k = \varphi_k(y)$ are defined by the recurrence relations

$$\varphi_m(y) = A_m y + B_m, \quad \varphi_{k-1}(y) = A_k y + B_k - 2k(2k-1) \int_a^y (y-t) t^n \varphi_k(t)\, dt,$$

where A_k and B_k are arbitrary constants ($k = m, \ldots, 1$), a is any number.

3°. Particular solutions with odd powers of x:

$$w = \sum_{k=0}^{m} \psi_k(y) x^{2k+1},$$

where the functions $\psi_k = \psi_k(y)$ are defined by the recurrence relations

$$\psi_m(y) = A_m y + B_m, \quad \psi_{k-1}(y) = A_k y + B_k - 2k(2k+1) \int_a^y (y-t) t^n \psi_k(t)\, dt,$$

where A_k and B_k are arbitrary constants ($k = m, \ldots, 1$), a is any number.

4°. Separable particular solutions:

$$w(x,y) = \left[A \sinh(\lambda q x) + B \cosh(\lambda q x)\right] \sqrt{y} \left[C J_{\frac{1}{2q}}(\lambda y^q) + D Y_{\frac{1}{2q}}(\lambda y^q)\right], \quad q = \tfrac{1}{2}(n+2),$$

$$w(x,y) = \left[A \sin(\lambda q x) + B \cos(\lambda q x)\right] \sqrt{y} \left[C I_{\frac{1}{2q}}(\lambda y^q) + D K_{\frac{1}{2q}}(\lambda y^q)\right],$$

where A, B, C, D, and λ are arbitrary constants, $J_\nu(z)$ and $Y_\nu(z)$ are the Bessel functions, and $I_\nu(z)$ and $K_\nu(z)$ are the modified Bessel functions.

5°. Fundamental solutions (for $y > 0$):

$$w_1(x,y,x_0,y_0) = k_1 (r_1^2)^{-\beta} F(\beta, \beta, 2\beta; 1-\xi), \quad \beta = \frac{n}{2(n+2)}, \quad \xi = \frac{r_2^2}{r_1^2},$$

$$w_2(x,y,x_0,y_0) = k_2 (r_1^2)^{-\beta} (1-\xi)^{1-2\beta} F(1-\beta, 1-\beta, 2-2\beta; 1-\xi).$$

Here, $F(a,b,c;\xi)$ is the hypergeometric function and

$$r_1^2 = (x-x_0)^2 + \frac{4}{(n+2)^2}\left(y^{\frac{n+2}{2}} + y_0^{\frac{n+2}{2}}\right), \quad k_1 = \frac{1}{4\pi}\left(\frac{4}{n+2}\right)^{2\beta} \frac{\Gamma^2(\beta)}{\Gamma(2\beta)},$$

$$r_2^2 = (x-x_0)^2 + \frac{4}{(n+2)^2}\left(y^{\frac{n+2}{2}} - y_0^{\frac{n+2}{2}}\right), \quad k_2 = \frac{1}{4\pi}\left(\frac{4}{n+2}\right)^{2\beta} \frac{\Gamma^2(1-\beta)}{\Gamma(2-2\beta)},$$

where $\Gamma(\beta)$ is the gamma function; x_0 and y_0 are arbitrary constants.

The fundamental solutions satisfy the conditions

$$\partial_y w_1\big|_{y=0} = 0, \quad w_2\big|_{y=0} = 0 \quad (x \text{ and } x_0 \text{ are any, } y_0 > 0).$$

The solutions of some boundary value problems can be found in the first book cited below.

⊙ *Reference*: V. M. Babich, M. B. Kapilevich, S. G. Mikhlin, et al. (1964), A. D. Polyanin (2001a).

3. $\dfrac{\partial^2 w}{\partial r^2} + \dfrac{\alpha}{r}\dfrac{\partial w}{\partial r} + \dfrac{\partial^2 w}{\partial z^2} = 0.$

Elliptic analogue of the Euler–Poisson–Darboux equation.

1°. For $\alpha = 1$, see Subsections 8.1.2 and 8.2.3 with $w = w(r, z)$. For $\alpha \neq 1$, the transformation $x = (1 - \alpha)z$, $y = r^{1-\alpha}$ leads to an equation of the form 7.4.4.1:

$$y^{\frac{2\alpha}{1-\alpha}} \dfrac{\partial^2 w}{\partial x^2} + \dfrac{\partial^2 w}{\partial y^2} = 0.$$

2°. Suppose $w_\alpha = w_\alpha(r, z)$ is a solution of the equation in question for a fixed value of the parameter α. Then the functions $\widetilde{w}_\alpha$ defined by the relations

$$\widetilde{w}_\alpha = \dfrac{\partial w_\alpha}{\partial z},$$

$$\widetilde{w}_\alpha = r\dfrac{\partial w_\alpha}{\partial r} + z\dfrac{\partial w_\alpha}{\partial z},$$

$$\widetilde{w}_\alpha = 2rz\dfrac{\partial w_\alpha}{\partial r} + (z^2 - r^2)\dfrac{\partial w_\alpha}{\partial z} + \alpha z w_\alpha$$

are also solutions of this equation.

3°. Suppose $w_\alpha = w_\alpha(r, z)$ is a solution of the equation in question for a fixed value of the parameter α. Using this w_α, one can construct solutions of the equation with other values of the parameter by the formulas

$$w_{2-\alpha} = r^{\alpha-1} w_\alpha,$$

$$w_{\alpha-2} = r\dfrac{\partial w_\alpha}{\partial r} + (\alpha - 1)w_\alpha,$$

$$w_{\alpha-2} = rz\dfrac{\partial w_\alpha}{\partial r} - r^2\dfrac{\partial w_\alpha}{\partial z} + (\alpha - 1)z w_\alpha,$$

$$w_{\alpha-2} = r(r^2 - z^2)\dfrac{\partial w_\alpha}{\partial r} + 2r^2 z\dfrac{\partial w_\alpha}{\partial z} + \left[r^2 - (\alpha - 1)z^2\right] w_\alpha,$$

$$w_{\alpha+2} = \dfrac{1}{r}\dfrac{\partial w_\alpha}{\partial r},$$

$$w_{\alpha+2} = \dfrac{z}{r}\dfrac{\partial w_\alpha}{\partial r} - \dfrac{\partial w_\alpha}{\partial z},$$

$$w_{\alpha+2} = \dfrac{r^2 - z^2}{r}\dfrac{\partial w_\alpha}{\partial r} + 2z\dfrac{\partial w_\alpha}{\partial z} + \alpha w_\alpha.$$

⊙ *Reference*: A. V. Aksenov (2001).

4. $\dfrac{\partial^2 w}{\partial x^2} + f(x)\dfrac{\partial^2 w}{\partial y^2} = 0.$

1°. Particular solutions:

$$w = C_1 xy + C_2 y + C_3 x + C_4,$$

$$w = C_1 y^2 + C_2 xy + C_3 y + C_4 x - 2C_1 \int_a^x (x-t)f(t)\,dt + C_5,$$

$$w = C_1 y^3 + C_2 xy + C_3 y + C_4 x - 6C_1 y \int_a^x (x-t)f(t)\,dt + C_5,$$

$$w = (C_1 x + C_2)y^2 + C_3 xy + C_4 y + C_5 x - 2\int_a^x (x-t)(C_1 t + C_2)f(t)\,dt + C_6,$$

where C_1, C_2, C_3, C_4, C_5, and C_6 are arbitrary constants, a is any number.

2°. Separable particular solution:
$$w = (C_1 e^{\lambda y} + C_2 e^{-\lambda y})H(x),$$
where C_1, C_2, and λ are arbitrary constants, and the function $H = H(x)$ is determined by the ordinary differential equation $H''_{xx} + \lambda^2 f(x)H = 0$.

3°. Separable particular solution:
$$w = [C_1 \sin(\lambda y) + C_2 \cos(\lambda y)]Z(x),$$
where C_1, C_2, and λ are arbitrary constants, and the function $Z = Z(x)$ is determined by the ordinary differential equation $Z''_{xx} - \lambda^2 f(x)Z = 0$.

4°. Particular solutions with even powers of y:
$$w = \sum_{k=0}^{n} \varphi_k(x) y^{2k},$$
where the functions $\varphi_k = \varphi_k(x)$ are defined by the recurrence relations
$$\varphi_n(x) = A_n x + B_n, \quad \varphi_{k-1}(x) = A_k x + B_k - 2k(2k-1)\int_a^x (x-t)f(t)\varphi_k(t)\,dt,$$
where A_k and B_k are arbitrary constants ($k = n, \ldots, 1$), a is any number.

5°. Particular solutions with odd powers of y:
$$w = \sum_{k=0}^{n} \psi_k(x) y^{2k+1},$$
where the functions $\psi_k = \psi_k(x)$ are defined by the recurrence relations
$$\psi_n(x) = A_n x + B_n, \quad \psi_{k-1}(x) = A_k x + B_k - 2k(2k+1)\int_a^x (x-t)f(t)\psi_k(t)\,dt,$$
where A_k and B_k are arbitrary constants ($k = n, \ldots, 1$), and a is any number.

5. $\dfrac{\partial}{\partial x}\left[f_1(x)\dfrac{\partial w}{\partial x}\right] + \dfrac{\partial}{\partial y}\left[f_2(y)\dfrac{\partial w}{\partial y}\right] + \lambda\bigl[g_1(x) + g_2(y)\bigr]w = 0.$

This equation is encountered in the theory of vibration of inhomogeneous membranes. Its separable solutions are sought in the form $w(x, y) = \varphi(x)\psi(y)$.

The article cited below presents an algorithm for accelerated convergence of solutions to eigenvalue boundary value problems for this equation.

⊙ *Reference*: L. D. Akulenko and S. V. Nesterov (1999).

7.4.5. Equations of the Form
$$a(x)\frac{\partial^2 w}{\partial x^2} + \frac{\partial^2 w}{\partial y^2} + b(x)\frac{\partial w}{\partial x} + c(x)w = -\Phi(x, y)$$

7.4.5-1. Statements of boundary value problems. Relations for the Green's function.

Consider two-dimensional boundary value problems for the equation
$$a(x)\frac{\partial^2 w}{\partial x^2} + \frac{\partial^2 w}{\partial y^2} + b(x)\frac{\partial w}{\partial x} + c(x)w = -\Phi(x, y) \tag{1}$$

with general boundary conditions in x,

$$s_1 \partial_x w - k_1 w = f_1(y) \quad \text{at} \quad x = x_1,$$
$$s_2 \partial_x w + k_2 w = f_2(y) \quad \text{at} \quad x = x_2, \qquad (2)$$

and different boundary conditions in y. We assume that the coefficients of equation (1) and the boundary conditions (2) meet the requirement

$a(x)$, $b(x)$, $c(x)$ are continuous functions ($x_1 \leq x \leq x_2$); $\quad a > 0, \quad |s_1| + |k_1| > 0, \quad |s_2| + |k_2| > 0.$

In the general case, the Green's function can be represented as

$$G(x, y, \xi, \eta) = \rho(\xi) \sum_{n=1}^{\infty} \frac{u_n(x) u_n(\xi)}{\|u_n\|^2} \Psi_n(y, \eta; \lambda_n). \qquad (3)$$

Here,

$$\rho(x) = \frac{1}{a(x)} \exp\left[\int \frac{b(x)}{a(x)} dx\right], \quad \|u_n\|^2 = \int_{x_1}^{x_2} \rho(x) u_n^2(x) dx, \qquad (4)$$

and the λ_n and $u_n(x)$ are the eigenvalues and eigenfunctions of the homogeneous boundary value problem for the ordinary differential equation

$$a(x) u''_{xx} + b(x) u'_x + [\lambda + c(x)] u = 0, \qquad (5)$$
$$s_1 u'_x - k_1 u = 0 \quad \text{at} \quad x = x_1, \qquad (6)$$
$$s_2 u'_x + k_2 u = 0 \quad \text{at} \quad x = x_2. \qquad (7)$$

The functions Ψ_n for various boundary conditions in y are specified in Table 25.

Equation (5) can be rewritten in self-adjoint form as

$$[p(x) u'_x]'_x + [\lambda \rho(x) - q(x)] u = 0, \qquad (8)$$

where the functions $p(x)$ and $q(x)$ are given by

$$p(x) = \exp\left[\int \frac{b(x)}{a(x)} dx\right], \quad q(x) = -\frac{c(x)}{a(x)} \exp\left[\int \frac{b(x)}{a(x)} dx\right],$$

and $\rho(x)$ is defined in (4).

The eigenvalue problem (8), (6), (7) possesses the following properties:

1°. All eigenvalues $\lambda_1, \lambda_2, \ldots$ are real and $\lambda_n \to \infty$ as $n \to \infty$.

2°. The system of eigenfunctions $\{u_1(x), u_2(x), \ldots\}$ is orthogonal on the interval $x_1 \leq x \leq x_2$ with weight $\rho(x)$, that is,

$$\int_{x_1}^{x_2} \rho(x) u_n(x) u_m(x) dx = 0 \quad \text{for} \quad n \neq m.$$

3°. If the conditions

$$q(x) \geq 0, \quad s_1 k_1 \geq 0, \quad s_2 k_2 \geq 0 \qquad (9)$$

are satisfied, there are no negative eigenvalues. If $q \equiv 0$ and $k_1 = k_2 = 0$, then the least eigenvalue is $\lambda_0 = 0$ and the corresponding eigenfunction is $u_0 = \text{const}$; in this case, the summation in (3) must start with $n = 0$. In the other cases, if conditions (9) are satisfied, all eigenvalues are positive; for example, the first inequality in (9) holds if $c(x) \leq 0$.

TABLE 25
The functions Ψ_n in (3) for various boundary conditions.* Notation: $\beta_n = \sqrt{\lambda_n}$

Domain	Boundary conditions	Function $\Psi_n(y, \eta; \lambda_n)$
$-\infty < y < \infty$	$\|w\| < \infty$ for $y \to \pm\infty$	$\dfrac{1}{2\beta_n} e^{-\beta_n\|y-\eta\|}$
$0 \leq y < \infty$	$w = 0$ for $y = 0$	$\dfrac{1}{\beta_n} \begin{cases} e^{-\beta_n y} \sinh(\beta_n \eta) & \text{for } y > \eta, \\ e^{-\beta_n \eta} \sinh(\beta_n y) & \text{for } \eta > y \end{cases}$
$0 \leq y < \infty$	$\partial_y w = 0$ for $y = 0$	$\dfrac{1}{\beta_n} \begin{cases} e^{-\beta_n y} \cosh(\beta_n \eta) & \text{for } y > \eta, \\ e^{-\beta_n \eta} \cosh(\beta_n y) & \text{for } \eta > y \end{cases}$
$0 \leq y < \infty$	$\partial_y w - k_3 w = 0$ for $y = 0$	$\dfrac{1}{\beta_n(\beta_n + k_3)} \begin{cases} e^{-\beta_n y}[\beta_n \cosh(\beta_n \eta) + k_3 \sinh(\beta_n \eta)] & \text{for } y > \eta, \\ e^{-\beta_n \eta}[\beta_n \cosh(\beta_n y) + k_3 \sinh(\beta_n y)] & \text{for } \eta > y \end{cases}$
$0 \leq y \leq h$	$w = 0$ at $y = 0$, $w = 0$ at $y = h$	$\dfrac{1}{\beta_n \sinh(\beta_n h)} \begin{cases} \sinh(\beta_n \eta) \sinh[\beta_n(h-y)] & \text{for } y > \eta, \\ \sinh(\beta_n y) \sinh[\beta_n(h-\eta)] & \text{for } \eta > y \end{cases}$
$0 \leq y \leq h$	$\partial_y w = 0$ at $y = 0$, $\partial_y w = 0$ at $y = h$	$\dfrac{1}{\beta_n \sinh(\beta_n h)} \begin{cases} \cosh(\beta_n \eta) \cosh[\beta_n(h-y)] & \text{for } y > \eta, \\ \cosh(\beta_n y) \cosh[\beta_n(h-\eta)] & \text{for } \eta > y \end{cases}$
$0 \leq y \leq h$	$w = 0$ at $y = 0$, $\partial_y w = 0$ at $y = h$	$\dfrac{1}{\beta_n \cosh(\beta_n h)} \begin{cases} \sinh(\beta_n \eta) \cosh[\beta_n(h-y)] & \text{for } y > \eta, \\ \sinh(\beta_n y) \cosh[\beta_n(h-\eta)] & \text{for } \eta > y \end{cases}$

Subsection 1.8.9 presents some relations for estimating the eigenvalues λ_n and eigenfunctions $u_n(x)$.

The Green's function of the two-dimensional third boundary value problem (1)–(2) augmented by the boundary conditions

$$\frac{\partial w}{\partial y} - k_3 w = 0 \quad \text{at} \quad y = 0, \qquad \frac{\partial w}{\partial y} + k_4 w = 0 \quad \text{at} \quad y = h$$

is given by relation (3) with

$$\Psi_n(y, \eta; \lambda_n) = \begin{cases} \dfrac{[\beta_n \cosh(\beta_n \eta) + k_3 \sinh(\beta_n \eta)]\{\beta_n \cosh[\beta_n(h-y)] + k_4 \sinh[\beta_n(h-y)]\}}{\beta_n[\beta_n(k_3 + k_4) \cosh(\beta_n h) + (\beta_n^2 + k_3 k_4) \sinh(\beta_n h)]} & \text{for } y > \eta, \\[2mm] \dfrac{[\beta_n \cosh(\beta_n y) + k_3 \sinh(\beta_n y)]\{\beta_n \cosh[\beta_n(h-\eta)] + k_4 \sinh[\beta_n(h-\eta)]\}}{\beta_n[\beta_n(k_3 + k_4) \cosh(\beta_n h) + (\beta_n^2 + k_3 k_4) \sinh(\beta_n h)]} & \text{for } y < \eta. \end{cases}$$

7.4.5-2. Representation of solutions to boundary value problems using the Green's function.

1°. The solution of the first boundary value problem for equation (1) with the boundary conditions

$$w = f_1(y) \quad \text{at} \quad x = x_1, \qquad w = f_2(y) \quad \text{at} \quad x = x_2,$$
$$w = f_3(x) \quad \text{at} \quad y = 0, \qquad w = f_4(x) \quad \text{at} \quad y = h$$

---

* For unbounded domains, the condition of boundedness of the solution as $y \to \pm\infty$ is set; in Table 25, this condition is omitted.

is expressed in terms of the Green's function as

$$w(x,y) = a(x_1)\int_0^h f_1(\eta)\left[\frac{\partial}{\partial \xi}G(x,y,\xi,\eta)\right]_{\xi=x_1} d\eta - a(x_2)\int_0^h f_2(\eta)\left[\frac{\partial}{\partial \xi}G(x,y,\xi,\eta)\right]_{\xi=x_2} d\eta$$

$$+ \int_{x_1}^{x_2} f_3(\xi)\left[\frac{\partial}{\partial \eta}G(x,y,\xi,\eta)\right]_{\eta=0} d\xi - \int_{x_1}^{x_2} f_4(\xi)\left[\frac{\partial}{\partial \eta}G(x,y,\xi,\eta)\right]_{\eta=h} d\xi$$

$$+ \int_{x_1}^{x_2}\int_0^h \Phi(\xi,\eta)G(x,y,\xi,\eta)\, d\eta\, d\xi.$$

2°. The solution of the second boundary value problem for equation (1) with boundary conditions

$$\partial_x w = f_1(y) \quad \text{at} \quad x = x_1, \qquad \partial_x w = f_2(y) \quad \text{at} \quad x = x_2,$$
$$\partial_y w = f_3(x) \quad \text{at} \quad y = 0, \qquad \partial_y w = f_4(x) \quad \text{at} \quad y = h$$

is expressed in terms of the Green's function as

$$w(x,y) = -a(x_1)\int_0^h f_1(\eta)G(x,y,x_1,\eta)\, d\eta + a(x_2)\int_0^h f_2(\eta)G(x,y,x_2,\eta)\, d\eta$$

$$- \int_{x_1}^{x_2} f_3(\xi)G(x,y,\xi,0)\, d\xi + \int_{x_1}^{x_2} f_4(\xi)G(x,y,\xi,h)\, d\xi$$

$$+ \int_{x_1}^{x_2}\int_0^h \Phi(\xi,\eta)G(x,y,\xi,\eta)\, d\eta\, d\xi.$$

3°. The solution of the third boundary value problem for equation (1) in terms of the Green's function is represented in the same way as the solution of the second boundary value problem (the Green's function is now different).

Chapter 8

Elliptic Equations with Three or More Space Variables

8.1. Laplace Equation $\Delta_3 w = 0$

The three-dimensional Laplace equation is often encountered in heat and mass transfer theory, fluid mechanics, elasticity, electrostatics, and other areas of mechanics and physics. For example, in heat and mass transfer theory, this equation describes stationary temperature distribution in the absence of heat sources and sinks in the domain under study.

A regular solution of the Laplace equation is called a harmonic function. The first boundary value problem for the Laplace equation is often referred to as the Dirichlet problem, and the second boundary value problem, as the Neumann problem.

Extremum principle: Given a domain D, a harmonic function w in D that is not identically constant in D cannot attain its maximum or minimum value at any interior point of D.

8.1.1. Problems in Cartesian Coordinates

The three-dimensional Laplace equation in the rectangular Cartesian system of coordinates is written as

$$\frac{\partial^2 w}{\partial x^2} + \frac{\partial^2 w}{\partial y^2} + \frac{\partial^2 w}{\partial z^2} = 0.$$

8.1.1-1. Particular solutions and some relations.

1°. Particular solutions:

$$w(x, y, z) = Ax + By + Cz + D,$$
$$w(x, y, z) = Ax^2 + By^2 - (A+B)z^2 + Cxy + Dxz + Eyz,$$
$$w(x, y, z) = \cos(\mu_1 x + \mu_2 y) \exp(\pm \mu z),$$
$$w(x, y, z) = \sin(\mu_1 x + \mu_2 y) \exp(\pm \mu z),$$
$$w(x, y, z) = \exp(\mu_1 x + \mu_2 y) \cos(\mu z + A),$$
$$w(x, y, z) = \exp(\pm \mu x) \cos(\mu_1 y + A) \cos(\mu_2 z + B),$$
$$w(x, y, z) = \cosh(\mu_1 x) \cosh(\mu_2 y) \cos(\mu z + B),$$
$$w(x, y, z) = \cosh(\mu_1 x) \sinh(\mu_2 y) \cos(\mu z + B),$$
$$w(x, y, z) = \cosh(\mu x) \cos(\mu_1 y + A) \cos(\mu_2 z + B),$$
$$w(x, y, z) = \sinh(\mu_1 x) \sinh(\mu_2 y) \sin(\mu z + B),$$
$$w(x, y, z) = \sinh(\mu x) \sin(\mu_1 y + A) \sin(\mu_2 z + B),$$

where A, B, C, D, E, μ_1, and μ_2 are arbitrary constants, and $\mu = \sqrt{\mu_1^2 + \mu_2^2}$.

2°. Fundamental solution:

$$\mathscr{E}(x, y, z) = \frac{1}{4\pi\sqrt{x^2 + y^2 + z^2}}.$$

3°. Suppose $w = w(x, y, z)$ is a solution of the Laplace equation. Then the functions

$$w_1 = Aw(\pm\lambda x + C_1, \pm\lambda y + C_2, \pm\lambda x + C_3),$$

$$w_2 = \frac{A}{r}w\left(\frac{x}{r^2}, \frac{y}{r^2}, \frac{z}{r^2}\right), \quad r = \sqrt{x^2 + y^2 + z^2},$$

$$w_3 = \frac{A}{\sqrt{\Xi}}w\left(\frac{x - ar^2}{\Xi}, \frac{y - br^2}{\Xi}, \frac{z - cr^2}{\Xi}\right), \quad \Xi = 1 - 2(ax + by + cz) + (a^2 + b^2 + c^2)r^2,$$

where A, C_n, a, b, c, and λ are arbitrary constants, are also solutions of this equation. The signs at λ in the expression of w_1 can be taken independently of one another.

⦿ *References*: W. Miller, Jr. (1977), R. Courant and D. Hilbert (1989).

8.1.1-2. Domain: $-\infty < x < \infty$, $-\infty < y < \infty$, $0 \le z < \infty$. First boundary value problem.

A half-space is considered. A boundary condition is prescribed:

$$w = f(x, y) \quad \text{at} \quad z = 0.$$

Solution:

$$w(x, y, z) = \frac{1}{2\pi}\int_{-\infty}^{\infty}\int_{-\infty}^{\infty}\frac{zf(\xi, \eta)\,d\xi\,d\eta}{\left[(x - \xi)^2 + (y - \eta)^2 + z^2\right]^{3/2}}.$$

⦿ *Reference*: V. M. Babich, M. B. Kapilevich, S. G. Mikhlin, et al. (1964).

8.1.1-3. Domain: $-\infty < x < \infty$, $-\infty < y < \infty$, $0 \le z < \infty$. Second boundary value problem.

A half-space is considered. A boundary condition is prescribed:

$$\partial_z w = f(x, y) \quad \text{at} \quad z = 0.$$

Solution:

$$w(x, y, z) = -\frac{1}{2\pi}\int_{-\infty}^{\infty}\int_{-\infty}^{\infty}\frac{f(\xi, \eta)\,d\xi\,d\eta}{\sqrt{(x - \xi)^2 + (y - \eta)^2 + z^2}} + C,$$

where C is an arbitrary constant.

⦿ *Reference*: V. S. Vladimirov, V. P. Mikhailov, A. A. Vasharin, et al. (1974).

8.1.1-4. Domain: $0 \le x \le a$, $0 \le y \le b$, $0 \le z \le c$. First boundary value problem.

A rectangular parallelepiped is considered. Boundary conditions are prescribed:

$$w = f_1(y, z) \quad \text{at} \quad x = 0, \quad w = f_2(y, z) \quad \text{at} \quad x = a,$$
$$w = f_3(x, z) \quad \text{at} \quad y = 0, \quad w = f_4(x, z) \quad \text{at} \quad y = b,$$
$$w = f_5(x, y) \quad \text{at} \quad z = 0, \quad w = f_6(x, y) \quad \text{at} \quad z = c.$$

Solution:

$$w(x, y, z) = \sum_{n=1}^{\infty}\sum_{m=1}^{\infty}\frac{f_{nm}^2\sinh(\lambda_{nm}^1 x) + f_{nm}^1\sinh[\lambda_{nm}^1(a - x)]}{\sinh(\lambda_{nm}^1 a)}\sin\left(\frac{\pi n y}{b}\right)\sin\left(\frac{\pi m z}{c}\right)$$

$$+ \sum_{n=1}^{\infty}\sum_{m=1}^{\infty}\frac{f_{nm}^4\sinh(\lambda_{nm}^2 y) + f_{nm}^3\sinh[\lambda_{nm}^2(b - y)]}{\sinh(\lambda_{nm}^2 b)}\sin\left(\frac{\pi n x}{a}\right)\sin\left(\frac{\pi m z}{c}\right)$$

$$+ \sum_{n=1}^{\infty}\sum_{m=1}^{\infty}\frac{f_{nm}^6\sinh(\lambda_{nm}^3 z) + f_{nm}^5\sinh[\lambda_{nm}^3(c - z)]}{\sinh(\lambda_{nm}^3 c)}\sin\left(\frac{\pi n x}{a}\right)\sin\left(\frac{\pi m y}{b}\right),$$

where the constant coefficients are given by

$$\lambda^1_{nm} = \pi\sqrt{\frac{n^2}{b^2} + \frac{m^2}{c^2}}, \quad \lambda^2_{nm} = \pi\sqrt{\frac{n^2}{a^2} + \frac{m^2}{c^2}}, \quad \lambda^3_{nm} = \pi\sqrt{\frac{n^2}{a^2} + \frac{m^2}{b^2}},$$

$$f^i_{nm} = \begin{cases} \dfrac{4}{bc}\displaystyle\int_0^b\int_0^c f_i(y,z)\sin\left(\frac{\pi n y}{b}\right)\sin\left(\frac{\pi m z}{c}\right) dy\, dz & \text{for } i = 1, 2; \\[2pt] \dfrac{4}{ac}\displaystyle\int_0^a\int_0^c f_i(x,z)\sin\left(\frac{\pi n x}{a}\right)\sin\left(\frac{\pi m z}{c}\right) dx\, dz & \text{for } i = 3, 4; \\[2pt] \dfrac{4}{ab}\displaystyle\int_0^a\int_0^b f_i(x,y)\sin\left(\frac{\pi n x}{a}\right)\sin\left(\frac{\pi m y}{b}\right) dx\, dy & \text{for } i = 5, 6. \end{cases}$$

Example. The planes $x = 0$ and $x = a$ have constant temperatures w_1 and w_2, respectively. The other planes are maintained at zero temperature ($f_3 = f_4 = f_5 = f_6 = 0$).

Solution:

$$w = \frac{16}{\pi^2}\sum_{n=1}^\infty\sum_{m=1}^\infty \frac{w_2\sinh(\mu_{nm}x) + w_1\sinh[\mu_{nm}(a-x)]}{(2n+1)(2m+1)\sinh(\mu_{nm}a)}\sin(p_n y)\sin(q_m z),$$

$$p_n = \frac{\pi(2n+1)}{b}, \quad q_m = \frac{\pi(2m+1)}{c}, \quad \mu_{nm} = \sqrt{p_n^2 + q_m^2}.$$

⊙ *References*: V. M. Babich, M. B. Kapilevich, S. G. Mikhlin, et al. (1964), B. M. Budak, A. A. Samarskii, and A. N. Tikhonov (1980), H. S. Carslaw and J. C. Jaeger (1984).

▶ *For the solution of other boundary value problems for the three-dimensional Laplace equation in the Cartesian coordinate system, see Subsection 8.2.2 for $\Phi \equiv 0$.*

8.1.2. Problems in Cylindrical Coordinates

The three-dimensional Laplace equation in the cylindrical coordinate system is written as

$$\Delta_3 w \equiv \frac{1}{r}\frac{\partial}{\partial r}\left(r\frac{\partial w}{\partial r}\right) + \frac{1}{r^2}\frac{\partial^2 w}{\partial \varphi^2} + \frac{\partial^2 w}{\partial z^2} = 0, \qquad r = \sqrt{x^2 + y^2}.$$

8.1.2-1. Particular solutions:

$$w(r,\varphi,z) = \left(Ar^m + \frac{B}{r^m}\right)(C\cos m\varphi + D\sin m\varphi)(\alpha + \beta z),$$

$$w(r,\varphi,z) = J_m(\mu r)(A\cos m\varphi + B\sin m\varphi)(C\cosh\mu z + D\sinh\mu z),$$

$$w(r,\varphi,z) = Y_m(\mu r)(A\cos m\varphi + B\sin m\varphi)(C\cosh\mu z + D\sinh\mu z),$$

$$w(r,\varphi,z) = I_m(\mu r)(A\cos m\varphi + B\sin m\varphi)(C\cos\mu z + D\sin\mu z),$$

$$w(r,\varphi,z) = K_m(\mu r)(A\cos m\varphi + B\sin m\varphi)(C\cos\mu z + D\sin\mu z),$$

where $m = 0, 1, 2, \ldots$; A, B, C, D, α, β, and μ are arbitrary constants; the $J_m(\xi)$ and $Y_m(\xi)$ are the Bessel functions; and the $I_m(\xi)$ and $K_m(\xi)$ are the modified Bessel functions.

8.1.2-2. Domain: $0 \le r \le a$, $0 \le \varphi \le 2\pi$, $-\infty < z < \infty$. First boundary value problem.

An infinite circular cylinder is considered. A boundary condition is prescribed:

$$w = f(\varphi, z) \quad \text{at} \quad r = a.$$

Solution:

$$w(r,\varphi,z) = -\frac{1}{a}\sum_{n=0}^\infty\sum_{m=1}^\infty \frac{J_n(\lambda_{nm}r)}{J_n'(\lambda_{nm}a)}\int_{-\infty}^\infty \left[A_n(\xi)\cos n\varphi + B_n(\xi)\sin n\varphi\right]\exp(-\lambda_{nm}|\xi - z|)\, d\xi,$$

where the $J_n(r)$ are the Bessel functions and λ_{mn} are positive roots of the transcendental equation $J_n(a\lambda) = 0$. The functions $A_n(z)$ and $B_n(z)$ are the coefficients of the Fourier series expansion of $f(\varphi, z)$,

$$A_n(z) = \frac{\varepsilon_n}{\pi} \int_0^{2\pi} f(\varphi, z) \cos(n\varphi)\, d\varphi, \qquad B_n(z) = \frac{1}{\pi} \int_0^{2\pi} f(\varphi, z) \sin(n\varphi)\, d\varphi,$$

where $\varepsilon_0 = 1/2$ and $\varepsilon_n = 1$ for $n = 1, 2, \ldots$

If the surface temperature is independent of φ, i.e., $f(z, \varphi) = f(z)$, then the solution takes the form

$$w(r, \varphi, z) = \frac{1}{a} \sum_{m=1}^{\infty} \frac{J_0(\lambda_m r)}{J_1(\lambda_m a)} \int_0^{\infty} [f(z+\zeta) + f(z-\zeta)] \exp(-\lambda_m \zeta)\, d\zeta,$$

where the λ_m are positive roots of the transcendental equation $J_0(a\lambda) = 0$.

⊙ *Reference*: H. S. Carslaw and J. C. Jaeger (1984).

8.1.2-3. Domain: $0 \le r \le a$, $0 \le \varphi \le 2\pi$, $0 \le z \le b$. First boundary value problem.

A circular cylinder of finite length is considered. Boundary conditions are prescribed:

$$w = f(\varphi, z) \quad \text{at} \quad r = a, \qquad w = g_1(r, \varphi) \quad \text{at} \quad z = 0, \qquad w = g_2(r, \varphi) \quad \text{at} \quad z = b.$$

Solution:

$$w(r, \varphi, z) = \sum_{n=0}^{\infty} \sum_{m=1}^{\infty} \frac{I_n\left(\frac{\pi m r}{b}\right)}{I_n\left(\frac{\pi m a}{b}\right)} (A_{nm} \cos n\varphi + B_{nm} \sin n\varphi) \sin\left(\frac{\pi m z}{b}\right)$$

$$+ \sum_{n=0}^{\infty} \sum_{m=1}^{\infty} J_n\left(\frac{\mu_{mn} r}{a}\right) \left(C_{nm}^{(1)} \cos n\varphi + D_{nm}^{(1)} \sin n\varphi\right) \frac{\sinh\left(\frac{\mu_{mn}(b-z)}{a}\right)}{\sinh\left(\frac{\mu_{mn} b}{a}\right)}$$

$$+ \sum_{n=0}^{\infty} \sum_{m=1}^{\infty} J_n\left(\frac{\mu_{mn} r}{a}\right) \left(C_{nm}^{(2)} \cos n\varphi + D_{nm}^{(2)} \sin n\varphi\right) \frac{\sinh\left(\frac{\mu_{mn} z}{a}\right)}{\sinh\left(\frac{\mu_{mn} b}{a}\right)},$$

where the $J_n(r)$ are the Bessel functions, the $I_n(r)$ are the modified Bessel functions, and μ_{mn} is the mth root of the equation $J_n(\mu) = 0$. The coefficients A_{nm}, B_{nm}, $C_{nm}^{(i)}$, and $D_{nm}^{(i)}$ are defined by

$$A_{nm} = \frac{\varepsilon_n}{\pi b} \int_0^{2\pi} \int_0^b f(\varphi, z) \cos(n\varphi) \sin\left(\frac{\pi m z}{b}\right) d\varphi\, dz,$$

$$B_{nm} = \frac{2}{\pi b} \int_0^{2\pi} \int_0^b f(\varphi, z) \sin(n\varphi) \sin\left(\frac{\pi m z}{b}\right) d\varphi\, dz,$$

$$C_{nm}^{(i)} = \frac{\varepsilon_n}{\pi a^2 [J_n'(\mu_{mn})]^2} \int_0^{2\pi} \int_0^a g_i(r, \varphi) \cos(n\varphi) J_n\left(\frac{\mu_{mn} r}{a}\right) r\, dr\, d\varphi,$$

$$D_{nm}^{(i)} = \frac{2}{\pi a^2 [J_n'(\mu_{mn})]^2} \int_0^{2\pi} \int_0^a g_i(r, \varphi) \sin(n\varphi) J_n\left(\frac{\mu_{mn} r}{a}\right) r\, dr\, d\varphi,$$

$$\varepsilon_n = \begin{cases} 1 & \text{for } n = 0, \\ 2 & \text{for } n \ne 0, \end{cases} \qquad i = 1, 2.$$

⊙ *References*: V. M. Babich, M. B. Kapilevich, S. G. Mikhlin, et al. (1964), B. M. Budak, A. A. Samarskii, and A. N. Tikhonov (1980).

▶ *For the solution of other boundary value problems for the three-dimensional Laplace equation in the cylindrical coordinate system, see Subsection 8.2.3 for $\Phi \equiv 0$.*

8.1.3. Problems in Spherical Coordinates

The three-dimensional Laplace equation in the spherical coordinate system is written as

$$\frac{1}{r^2}\frac{\partial}{\partial r}\left(r^2\frac{\partial w}{\partial r}\right) + \frac{1}{r^2\sin\theta}\frac{\partial}{\partial \theta}\left(\sin\theta\frac{\partial w}{\partial \theta}\right) + \frac{1}{r^2\sin^2\theta}\frac{\partial^2 w}{\partial \varphi^2} = 0, \qquad r = \sqrt{x^2+y^2+z^2}.$$

8.1.3-1. Particular solutions:

$$w(r) = A + \frac{B}{r},$$

$$w(r,\theta) = \left(Ar^n + \frac{B}{r^{n+1}}\right)P_n(\cos\theta),$$

$$w(r,\theta,\varphi) = \left(Ar^n + \frac{B}{r^{n+1}}\right)P_n^m(\cos\theta)(C\cos m\varphi + D\sin m\varphi),$$

where $n = 0, 1, 2, \ldots$; $m = 0, 1, 2, \ldots, n$; A, B, C, D are arbitrary constants; the $P_n(\xi)$ are the Legendre polynomials; and the $P_n^m(\xi)$ are the associated Legendre functions that are expressed as

$$P_n(x) = \frac{1}{n!\,2^n}\frac{d^n}{dx^n}(x^2-1)^n, \qquad P_n^m(x) = (1-x^2)^{m/2}\frac{d^m}{dx^m}P_n(x).$$

8.1.3-2. Domain: $0 \leq r \leq R$ or $R \leq r < \infty$. First boundary value problem.

A boundary condition at the sphere surface is prescribed:

$$w = f(\theta,\varphi) \quad \text{at} \quad r = R.$$

1°. Solution of the inner problem (for $r \leq R$):

$$w(r,\theta,\varphi) = \frac{R}{4\pi}\int_0^{2\pi}\!\!\int_0^{\pi} f(\theta_0,\varphi_0)\frac{R^2-r^2}{(r^2-2Rr\cos\gamma+R^2)^{3/2}}\sin\theta_0\,d\theta_0\,d\varphi_0,$$
$$\cos\gamma = \cos\theta\cos\theta_0 + \sin\theta\sin\theta_0\cos(\varphi-\varphi_0).$$

This formula is conventionally called the Poisson integral for a sphere.

Series solution:

$$w(r,\theta,\varphi) = \sum_{n=0}^{\infty}\left(\frac{r}{R}\right)^n Y_n(\theta,\varphi), \qquad Y_n(\theta,\varphi) = \sum_{m=0}^{n}(A_{nm}\cos m\varphi + B_{nm}\sin m\varphi)P_n^m(\cos\theta),$$

where

$$A_{00} = \frac{1}{4\pi}\int_0^{2\pi}\!\!\int_0^{\pi} f(\theta,\varphi)\sin\theta\,d\theta\,d\varphi,$$

$$A_{nm} = \frac{(2n+1)(n-m)!}{2\pi(n+m)!}\int_0^{2\pi}\!\!\int_0^{\pi} f(\theta,\varphi)P_n^m(\cos\theta)\cos m\varphi\sin\theta\,d\theta\,d\varphi,$$

$$B_{nm} = \frac{(2n+1)(n-m)!}{2\pi(n+m)!}\int_0^{2\pi}\!\!\int_0^{\pi} f(\theta,\varphi)P_n^m(\cos\theta)\sin m\varphi\sin\theta\,d\theta\,d\varphi.$$

2°. Solution of the outer problem (for $r \geq R$):

$$w(r,\theta,\varphi) = \frac{R}{4\pi}\int_0^{2\pi}\!\!\int_0^{\pi} f(\theta_0,\varphi_0)\frac{r^2-R^2}{(r^2-2Rr\cos\gamma+R^2)^{3/2}}\sin\theta_0\,d\theta_0\,d\varphi_0,$$

where $\cos\gamma$ is expressed in the same way as in the inner problem.

Series solution:

$$w(r,\theta,\varphi) = \sum_{n=0}^{\infty} \left(\frac{R}{r}\right)^{n+1} Y_n(\theta,\varphi),$$

$$Y_n(\theta,\varphi) = \sum_{m=0}^{n} (A_{nm} \cos m\varphi + B_{nm} \sin m\varphi) P_n^m(\cos\theta),$$

where the coefficients A_{nm} and B_{nm} are defined by the same relations as in the inner problem.

⊙ *References*: G. N. Polozhii (1964), V. M. Babich, M. B. Kapilevich, S. G. Mikhlin, et al. (1964), B. M. Budak, A. A. Samarskii, and A. N. Tikhonov (1980), A. N. Tikhonov and A. A. Samarskii (1990).

8.1.3-3. Domain: $0 \leq r \leq R$ or $R \leq r < \infty$. Second boundary value problem.

A boundary condition at the sphere surface is prescribed:

$$\frac{\partial w}{\partial r} = f(\theta,\varphi) \quad \text{at} \quad r = R.$$

The function $f(\theta,\varphi)$ must satisfy the solvability condition

$$\int_0^{2\pi} \int_0^{\pi} f(\theta,\varphi) \sin\theta \, d\theta \, d\varphi = 0.$$

1°. Solution of the inner problem (for $r \leq R$):

$$w(r,\theta,\varphi) = \frac{R}{4\pi} \int_0^{2\pi} \int_0^{\pi} f(\theta_0,\varphi_0) \left[\frac{1}{R}\ln(R + r_1 - r\cos\gamma) - \frac{2}{r_1}\right] \sin\theta_0 \, d\theta_0 \, d\varphi_0,$$

$$r_1 = \sqrt{r^2 - 2Rr\cos\gamma + R^2}, \quad \cos\gamma = \cos\theta \cos\theta_0 + \sin\theta \sin\theta_0 \cos(\varphi - \varphi_0).$$

Series solution:

$$w(r,\theta,\varphi) = \sum_{n=1}^{\infty} \sum_{m=0}^{n} \frac{R}{n} \left(\frac{r}{R}\right)^n (A_{nm} \cos m\varphi + B_{nm} \sin m\varphi) P_n^m(\cos\theta) + C,$$

where the coefficients A_{nm} and B_{nm} are expressed in the same way as in the inner first boundary value problem (see Paragraph 8.1.3-2), and C is an arbitrary constant.

2°. Solution of the outer problem (for $r \geq R$):

$$w(r,\theta,\varphi) = -\frac{R}{4\pi} \int_0^{2\pi} \int_0^{\pi} f(\theta_0,\varphi_0) \left[\frac{1}{R}\ln\frac{R + r_1 - r\cos\gamma}{r(1-\cos\gamma)} - \frac{2}{r_1}\right] \sin\theta_0 \, d\theta_0 \, d\varphi_0,$$

$$r_1 = \sqrt{r^2 - 2Rr\cos\gamma + R^2}, \quad \cos\gamma = \cos\theta \cos\theta_0 + \sin\theta \sin\theta_0 \cos(\varphi - \varphi_0).$$

Series solution:

$$w(r,\theta,\varphi) = -\sum_{n=0}^{\infty} \sum_{m=0}^{n} \frac{R}{n+1} \left(\frac{R}{r}\right)^{n+1} (A_{nm} \cos m\varphi + B_{nm} \sin m\varphi) P_n^m(\cos\theta) + C,$$

where the coefficients A_{nm} and B_{nm} are expressed in the same way as in the inner first boundary value problem, and C is an arbitrary constant.

$3°$. Outer boundary value problems where unbounded solutions as $r \to \infty$ are sought are also encountered in applications.

Example. A potential translational flow of an ideal incompressible fluid about a sphere of radius R is governed by the Laplace equation with the boundary conditions:

$$\partial_r w = 0 \quad \text{at} \quad r = R, \quad |w - Ur\cos\theta| \to 0 \quad \text{as} \quad r \to \infty,$$

where w is the potential, U the unperturbed flow velocity at infinity; the fluid velocity is expressed in terms of the potential as $\mathbf{v} = \nabla\varphi$.

Solution:

$$w = Ur\left(1 + \frac{R^3}{2r^3}\right)\cos\theta.$$

This solution is a special case of the second formula from Paragraph for 8.1.3-1 for $n = 1$.

⊙ *References*: G. N. Polozhii (1964), V. M. Babich, M. B. Kapilevich, S. G. Mikhlin, et al. (1964), B. M. Budak, A. A. Samarskii, and A. N. Tikhonov (1980), L. G. Loitsyanskii (1996).

▶ *For the solution of other boundary value problems for the three-dimensional Laplace equation in the spherical coordinate system, see Subsection 8.2.4 for $\Phi \equiv 0$.*

8.1.4. Other Orthogonal Curvilinear Systems of Coordinates

The three-dimensional Laplace equation admits separation of variables in the eleven orthogonal coordinate systems that are listed in Table 26.

For the general ellipsoidal and conical coordinate systems, the functions f, g, and h are determined by Lamé equations that involve the Jacobian elliptic function $\operatorname{sn} z = \operatorname{sn}(z, k)$. The solutions of these equations under some conditions can be represented in the form of finite series called Lamé polynomials. For details about the Lamé equation and its solutions, see Whittaker and Watson (1963), Bateman and Erdélyi (1955), Arscott (1964), and Miller, Jr. (1977).

There are also coordinate systems that allow the so-called $\mathcal{R}$-separation of variables of the three-dimensional Laplace equation. Such solutions in the new coordinate system, μ, ν, ρ, can be represented in the form $w = \sqrt{\mathcal{R}(\mu,\nu,\rho)}\, f(\mu)g(\nu)h(\rho)$. Coordinates that allow the $\mathcal{R}$-separation of variables are listed in Table 27.

Only the bicylindrical and toroidal coordinate systems are fairly widely used in applications. In three subsequent coordinate systems, the functions $f = f(\mu)$ and $g = g(\rho)$ are determined by identical equations. With the change of variables $\mu = \operatorname{sn}^2(\alpha, k)$, $\rho = \operatorname{sn}^2(\beta, k)$, where $k = a^{-1/2}$, these equations are reduced to Lamé equations (α and β are the new independent variables).

⊙ *References for Subsection* 8.1.4: M. Bôcher (1894), F. M. Morse and H. Feshbach (1953, Vols. 1–2), N. N. Lebedev, I. P. Skal'skaya, and Ya. S. Uflyand (1955), P. Moon and D. Spencer (1961), A. Makarov, J. Smorodinsky, K. Valiev, and P. Winternitz (1967), W. Miller, Jr. (1977).

8.2. Poisson Equation $\Delta_3 w + \Phi(\mathbf{x}) = 0$

8.2.1. Preliminary Remarks. Solution Structure

Like the three-dimensional Laplace equation, the three-dimensional Poisson equation is often encountered in heat and mass transfer theory, fluid mechanics, elasticity, electrostatics, and other areas of mechanics and physics. In particular, the Poisson equation describes stationary temperature distribution in the presence of thermal sources or sinks in the domain under consideration.

The Laplace equation is a special case of the Poisson equation with $\Phi \equiv 0$.

Throughout this section, we consider a three-dimensional bounded domain V with a sufficiently smooth boundary S. We assume that $\mathbf{r} \in V$ and $\boldsymbol{\rho} \in V$, where $\mathbf{r} = \{x, y, z\}$ and $\boldsymbol{\rho} = \{\xi, \eta, \zeta\}$.

TABLE 26
Orthogonal coordinates $\bar{x}, \bar{y}, \bar{z}$ that allow separable solutions of the form $w = f(\bar{x})g(\bar{y})h(\bar{z})$ for the three-dimensional Laplace equation $\Delta_3 w = 0$

Coordinates	Transformations	Particular solutions (or equations for f, g, h)
Cartesian x, y, z	$x = x,$ $y = y,$ $z = z$	$w = \cos(k_1 x + s_1)\cos(k_2 y + s_2)\cosh(k_3 z + s_3),$ where $k_1^2 + k_2^2 = k_3^2$; see also Paragraph 8.1.1-1
Cylindrical r, φ, z	$x = r\cos\varphi,$ $y = r\sin\varphi,$ $z = z$	$w = [AJ_n(kr) + BY_n(kr)]\cos(n\varphi + c)\exp(\pm kz),$ $J_n(z)$ and $Y_n(z)$ are the Bessel functions; see also Paragraph 8.1.2-1
Parabolic cylindrical ξ, η, z	$x = \frac{1}{2}(\xi^2 - \eta^2),$ $y = \xi\eta,$ $z = z$	$w = D_{\mu-1/2}(\pm\sigma\xi)D_{-\mu-1/2}(\pm\sigma\eta)\cos(kz + s),$ where $\sigma = \sqrt{2k}$, $D_\mu(z)$ is the parabolic cylinder function
Elliptic cylindrical u, v, z	$x = a\cosh u \cos v,$ $y = a\sinh u \sin v,$ $z = z$	$w = \begin{cases} \text{Ce}_n(u,-q)\,\text{ce}_n(v,-q)\cos(kz+s), \\ \text{Se}_n(u,-q)\,\text{se}_n(v,-q)\cos(kz+s), \end{cases}$ Ce_n and Se_n are the modified Mathieu functions, ce_n and se_n are the Mathieu functions, $q = \frac{1}{4}a^2 k^2$
Spherical r, θ, φ	$x = r\sin\theta\cos\varphi,$ $y = r\sin\theta\sin\varphi,$ $z = r\cos\theta$	$w = (Ar^n + Br^{-n-1})P_n^k(\cos\theta)\cos(k\varphi + s),$ $P_n^k(\xi)$ are the associated Legendre functions, see also Paragraph 8.1.3-1
Prolate spheroidal u, v, φ	$x = a\sinh u \sin v \cos\varphi,$ $y = a\sinh u \sin v \sin\varphi,$ $z = a\cosh u \cos v$	$w = P_n^k(\cosh u)P_n^k(\cos v)\cos(k\varphi + s),$ $P_n^k(\xi)$ are the associated Legendre functions
Oblate spheroidal u, v, φ	$x = a\cosh u \sin v \cos\varphi,$ $y = a\cosh u \sin v \sin\varphi,$ $z = a\sinh u \cos v$	$w = P_n^k(-i\sinh u)P_n^k(\cos v)\cos(k\varphi + s),$ $P_n^k(\xi)$ are the associated Legendre functions
Parabolic ξ, η, φ	$x = a\xi\eta\cos\varphi,$ $y = a\xi\eta\sin\varphi,$ $z = \frac{1}{2}a(\xi^2 - \eta^2)$	$w = I_{\pm k}(\beta\xi)J_{\pm k}(\beta\eta)\cos(k\varphi + s),$ $J_k(z)$ are the Bessel functions, $I_k(z)$ are the modified Bessel functions
Paraboloidal u, v, φ	$x = 2a\cosh u \cos v \sinh\varphi,$ $y = 2a\sinh u \sin v \cosh\varphi,$ $z = \frac{1}{2}a(\cosh 2u + \cos 2v - \cosh 2\varphi)$	$w = \begin{cases} \text{Ce}_n(u,-b)\,\text{ce}_n(v,-b)\,\text{Ce}_n(\varphi + i\pi/2,-b), \\ \text{Se}_n(u,-b)\,\text{se}_n(v,-b)\,\text{Se}_n(\varphi + i\pi/2,-b), \end{cases}$ $b = \frac{1}{2}a\beta$; ce_n and se_n are the Mathieu functions, Ce_n and Se_n are the modified Mathieu functions
General ellipsoidal μ, ν, ρ	$x = \sqrt{\frac{(\mu-a)(\nu-a)(\rho-a)}{a(a-1)}},$ $y = \sqrt{\frac{(\mu-1)(\nu-1)(\rho-1)}{1-a}},$ $z = \sqrt{\frac{\mu\nu\rho}{a}}$	$f''_{\xi\xi} + (\beta_2 + \beta_1 \,\text{sn}^2 \xi)f = 0$ (Lamé equation), $g''_{\eta\eta} + (\beta_2 + \beta_1 \,\text{sn}^2 \eta)g = 0$ (Lamé equation), $h''_{\zeta\zeta} + (\beta_2 + \beta_1 \,\text{sn}^2 \zeta)g = 0$ (Lamé equation), $\mu = \text{sn}^2(\xi,k),\ \nu = \text{sn}^2(\eta,k),\ \rho = \text{sn}^2(\zeta,k),\ k = a^{-1/2}$
Conical ρ, μ, ν	$x = \rho\sqrt{\frac{(a\mu-1)(a\nu-1)}{1-a}},$ $y = \rho\sqrt{\frac{a(\mu-1)(\nu-1)}{a-1}},$ $z = \rho\sqrt{a\mu\nu}$	$f(\rho) = A\rho^n + B\rho^{-n-1},\ n = 0, 1, \ldots,$ $g''_{\xi\xi} + [\beta - n(n+1)k^2 \,\text{sn}^2 \xi]g = 0$ (Lamé equation), $h''_{\eta\eta} + [\beta - n(n+1)k^2 \,\text{sn}^2 \eta]h = 0$ (Lamé equation), where $\mu = \text{sn}^2(\xi,k),\ \nu = \text{sn}^2(\eta,k),\ k = a^{1/2},$ g and h are expressed in terms of the Lamé polynomials

TABLE 27

Coordinates $\bar{x}, \bar{y}, \bar{z}$ that allow $\mathcal{R}$-separated solutions of the form
$w = \sqrt{\mathcal{R}(\bar{x},\bar{y},\bar{z})}\, f(\bar{x})g(\bar{y})h(\bar{z})$ for the three-dimensional Laplace equation $\Delta_3 w = 0$

New coordinates, function $\mathcal{R}$	Transformations of coordinates	Functions f, g, h (equations for f, g, h)
Bicylindrical coordinates α, β, φ, $\mathcal{R} = \cosh\beta - \cos\alpha$	$x = c\mathcal{R}^{-1}\sin\alpha\cos\varphi$, $y = c\mathcal{R}^{-1}\sin\alpha\sin\varphi$, $z = c\mathcal{R}^{-1}\sinh\beta$; $0 \leq \alpha \leq \pi$, β is any, $0 \leq \varphi < 2\pi$	$f(\alpha) = A_1 P_n^m(\cos\alpha) + A_2 Q_n^m(\cos\alpha)$, $g(\beta) = B_1\cosh\left[(n+\tfrac{1}{2})\beta\right] + B_2\sinh\left[(n+\tfrac{1}{2})\beta\right]$, $h(\varphi) = C_1\cos(m\varphi) + C_2\sin(m\varphi)$, $n = 0, 1, 2, \ldots$; $m = 0, 1, 2, \ldots$
Toroidal coordinates α, β, φ, $\mathcal{R} = \cosh\alpha - \cos\beta$	$x = c\mathcal{R}^{-1}\sinh\alpha\cos\varphi$, $y = c\mathcal{R}^{-1}\sinh\alpha\sin\varphi$, $z = c\mathcal{R}^{-1}\sin\beta$; $\alpha \geq 0$, $-\pi \leq \beta \leq \pi$, $0 \leq \varphi < 2\pi$	$f(\alpha) = A_1 P_{n-1/2}^m(\cosh\alpha) + A_2 Q_{n-1/2}^m(\cosh\alpha)$, $g(\beta) = B_1\cos(n\beta) + B_2\sin(n\beta)$, $h(\varphi) = C_1\cos(m\varphi) + C_2\sin(m\varphi)$, $n = 0, 1, 2, \ldots$; $m = 0, 1, 2, \ldots$
Coordinates μ, ρ, φ, $\mathcal{R} = \sqrt{\frac{(\mu-a)(a-\rho)}{a(a-1)}} - \sqrt{\frac{(\mu-1)(1-\rho)}{a-1}}$	$x = \mathcal{R}^{-1}\cos\varphi$, $y = \mathcal{R}^{-1}\sin\varphi$, $z = \mathcal{R}^{-1}\sqrt{-\mu\rho/a}$; $\mu > a > 1$, $\rho < 0$, $0 \leq \varphi < 2\pi$	$\sqrt{U(\mu)}\left[\sqrt{U(\mu)}\,f'\right]' + \left[(\tfrac{1}{4} - n^2)\mu - \lambda\right]f = 0$, $\sqrt{U(\rho)}\left[\sqrt{U(\rho)}\,g'\right]' + \left[(\tfrac{1}{4} - n^2)\rho - \lambda\right]g = 0$, $h(\varphi) = C_1\cos(n\varphi) + C_2\sin(n\varphi)$, $U(t) = 4t(t-1)(t-a)$
Coordinates μ, ρ, φ, $\mathcal{R} = \sqrt{\frac{\mu\rho}{a}} + \sqrt{\frac{(\mu-1)(\rho-1)}{a-1}}$	$x = \mathcal{R}^{-1}\cos\varphi$, $y = \mathcal{R}^{-1}\sin\varphi$, $z = \mathcal{R}^{-1}\sqrt{\frac{(\mu-a)(a-\rho)}{a(a-1)}}$; $1 < \rho < a < \mu$, $0 \leq \varphi < 2\pi$	$\sqrt{U(\mu)}\left[\sqrt{U(\mu)}\,f'\right]' + \left[(\tfrac{1}{4} - n^2)\mu - \lambda\right]f = 0$, $\sqrt{U(\rho)}\left[\sqrt{U(\rho)}\,g'\right]' + \left[(\tfrac{1}{4} - n^2)\rho - \lambda\right]g = 0$, $h(\varphi) = C_1\cos(n\varphi) + C_2\sin(n\varphi)$, $U(t) = 4t(t-1)(t-a)$
Coordinates μ, ρ, φ, $\mathcal{R} = 2\,\mathrm{Re}\sqrt{\frac{i(\mu-a)(\rho-a)}{a(a-b)}}$, $a = \bar{b} = \alpha + i\beta$, α, β are real numbers	$x = \mathcal{R}^{-1}\cos\varphi$, $y = \mathcal{R}^{-1}\sin\varphi$, $z = \mathcal{R}^{-1}\sqrt{-\mu\rho/(ab)}$; $\mu > 0$, $\rho < 0$, $0 \leq \varphi < 2\pi$	$\sqrt{U(\mu)}\left[\sqrt{U(\mu)}\,f'\right]' + \left[(\tfrac{1}{4} - n^2)\mu - \lambda\right]f = 0$, $\sqrt{U(\rho)}\left[\sqrt{U(\rho)}\,g'\right]' + \left[(\tfrac{1}{4} - n^2)\rho - \lambda\right]g = 0$, $h(\varphi) = C_1\cos(n\varphi) + C_2\sin(n\varphi)$, $U(t) = 4t(t-a)(t-b)$
Coordinates μ, ν, ρ, $\mathcal{R} = 1 + \sqrt{\frac{\mu\nu\rho}{ab}}$	$x = \mathcal{R}^{-1}\sqrt{\frac{(\mu-a)(\nu-a)(\rho-a)}{(b-a)(a-1)a}}$, $y = \mathcal{R}^{-1}\sqrt{\frac{(\mu-b)(\nu-b)(\rho-b)}{(a-b)(b-1)b}}$, $z = \mathcal{R}^{-1}\sqrt{\frac{(\mu-1)(\nu-1)(\rho-1)}{(a-1)(b-1)}}$; $0 < \rho < 1 < \nu < b < \mu < a$	$\sqrt{U(\mu)}\left[\sqrt{U(\mu)}\,f'\right]' - (3\mu^2 + \lambda_1\mu + \lambda_2)f = 0$, $\sqrt{U(\nu)}\left[\sqrt{U(\nu)}\,g'\right]' - (3\nu^2 + \lambda_1\nu + \lambda_2)g = 0$, $\sqrt{U(\rho)}\left[\sqrt{U(\rho)}\,h'\right]' - (3\rho^2 + \lambda_1\rho + \lambda_2)h = 0$, $U(t) = 16t(t-1)(t-a)(t-b)$
Coordinates μ, ν, ρ, $\mathcal{R} = 2\,\mathrm{Re}\sqrt{\frac{(\mu-a)(\nu-a)(\rho-a)}{ia(a-1)(a-b)}}$, $a = \bar{b} = \alpha + i\beta$, α, β are real numbers	$x = \mathcal{R}^{-1}\sqrt{\frac{(\mu-1)(\nu-1)(\rho-1)}{(a-1)(b-1)}}$, $y = \mathcal{R}^{-1}\sqrt{-\frac{\mu\nu\rho}{ab}}$, $z = \mathcal{R}^{-1}$; $\rho < 0 < \mu < 1 < \nu$	$\sqrt{U(\mu)}\left[\sqrt{U(\mu)}\,f'\right]' - (3\mu^2 + \lambda_1\mu + \lambda_2)f = 0$, $\sqrt{U(\nu)}\left[\sqrt{U(\nu)}\,g'\right]' - (3\nu^2 + \lambda_1\nu + \lambda_2)g = 0$, $\sqrt{U(\rho)}\left[\sqrt{U(\rho)}\,h'\right]' - (3\rho^2 + \lambda_1\rho + \lambda_2)h = 0$, $U(t) = 16t(t-1)(t-a)(t-b)$

8.2.1-1. First boundary value problem.

The solution of the first boundary value problem for the Poisson equation

$$\Delta_3 w + \Phi(\mathbf{r}) = 0 \tag{1}$$

in a domain V with the nonhomogeneous boundary condition

$$w = f(\mathbf{r}) \quad \text{for} \quad \mathbf{r} \in S$$

TABLE 28

The volume elements and distances occurring in relations (2) and (5) in some coordinate systems. In all cases, $\boldsymbol{\rho} = \{\xi, \eta, \zeta\}$

Coordinate system	Volume element, dV_ρ	Gradient, $\nabla_\rho u$ ($|\mathbf{i}_\xi| = |\mathbf{i}_\eta| = |\mathbf{i}_\zeta| = 1$)	Distance, $d = |\mathbf{r} - \boldsymbol{\rho}|$
Cartesian $\mathbf{r} = \{x, y, z\}$	$d\xi\, d\eta\, d\zeta$	$\mathbf{i}_\xi \frac{\partial u}{\partial \xi} + \mathbf{i}_\eta \frac{\partial u}{\partial \eta} + \mathbf{i}_\zeta \frac{\partial u}{\partial \zeta}$	$d = \sqrt{(x-\xi)^2 + (y-\eta)^2 + (z-\zeta)^2}$
Cylindrical $\mathbf{r} = \{r, \varphi, z\}$	$\xi\, d\xi\, d\eta\, d\zeta$	$\mathbf{i}_\xi \frac{\partial u}{\partial \xi} + \mathbf{i}_\eta \frac{1}{\xi}\frac{\partial u}{\partial \eta} + \mathbf{i}_\zeta \frac{\partial u}{\partial \zeta}$	$d = \sqrt{r^2 + \xi^2 - 2r\xi\cos(\varphi - \eta) + (z - \zeta)^2}$
Spherical $\mathbf{r} = \{r, \theta, \varphi\}$	$\xi^2 \sin\eta\, d\xi\, d\eta\, d\zeta$	$\mathbf{i}_\xi \frac{\partial u}{\partial \xi} + \mathbf{i}_\eta \frac{1}{\xi}\frac{\partial u}{\partial \eta} + \mathbf{i}_\zeta \frac{1}{\xi \sin\eta}\frac{\partial u}{\partial \zeta}$	$d = \sqrt{r^2 + \xi^2 - 2r\xi\cos\gamma}$, where $\cos\gamma = \cos\theta\cos\eta + \sin\theta\sin\eta\cos(\varphi - \zeta)$

can be represented in the form

$$w(\mathbf{r}) = \int_V \Phi(\boldsymbol{\rho}) G(\mathbf{r}, \boldsymbol{\rho})\, dV_\rho - \int_S f(\boldsymbol{\rho}) \frac{\partial G}{\partial N_\rho}\, dS_\rho. \tag{2}$$

Here, $G(\mathbf{r}, \boldsymbol{\rho})$ is the Green's function of the first boundary value problem, $\frac{\partial G}{\partial N_\rho}$ is the derivative of the Green's function with respect to ξ, η, ζ along the outward normal $\mathbf{N}$ the boundary S of the domain V. Integration is everywhere with respect to ξ, η, ζ.

The volume elements in solution (2) for basic coordinate systems are presented in Table 28. In addition, the expressions of the gradients are given, which enable one to find the derivative along the normal in accordance with the formula $\frac{\partial G}{\partial N_\rho} = (\mathbf{N} \cdot \nabla_\rho G)$.

The Green's function $G = G(\mathbf{r}, \boldsymbol{\rho})$ of the first boundary value problem is determined by the following conditions:

1°. The function G satisfies the Laplace equation with respect to x, y, z in the domain V everywhere except for the point (ξ, η, ζ), at which it can have a singularity of the form $\frac{1}{4\pi}\frac{1}{|\mathbf{r}-\boldsymbol{\rho}|}$.

2°. The function G, with respect to x, y, z, satisfies the homogeneous boundary condition of the first kind at the boundary, i.e., the condition $G|_S = 0$.

The Green's function can be represented as

$$G(\mathbf{r}, \boldsymbol{\rho}) = \frac{1}{4\pi}\frac{1}{|\mathbf{r} - \boldsymbol{\rho}|} + u, \tag{3}$$

where the auxiliary function $u = u(\mathbf{r}, \boldsymbol{\rho})$ is determined by solving the first boundary value problem for the Laplace equation $\Delta_3 u = 0$ with the boundary condition $u|_S = -\frac{1}{4\pi}\frac{1}{|\mathbf{r}-\boldsymbol{\rho}|}$; the vector quantity $\boldsymbol{\rho}$ in this problem is treated as a three-dimensional free parameter.

The Green's function possesses the symmetry property with respect to their arguments: $G(\mathbf{r}, \boldsymbol{\rho}) = G(\boldsymbol{\rho}, \mathbf{r})$.

The construction of Green's functions is discussed in Paragraphs 8.3.1-4 and 8.3.1-6 through 8.3.1-8 for $\lambda = 0$.

Remark 1. For outer first boundary value problems for the Laplace equation, the following condition is usually set at infinity: $|w| < A/|\mathbf{r}|$ ($|\mathbf{r}| \to \infty$, $A = \text{const}$).

8.2.1-2. Second boundary value problem.

The second boundary value problem for the Poisson equation (1) is characterized by the boundary condition

$$\frac{\partial w}{\partial N} = f(\mathbf{r}) \quad \text{for} \quad \mathbf{r} \in S.$$

Necessary condition solvability of the inner problem:

$$\int_V \Phi(\mathbf{r})\,d\mathbf{r} + \int_S f(\mathbf{r})\,dS = 0. \tag{4}$$

The solution of the second boundary value problem can be written as

$$w(\mathbf{r}) = \int_V \Phi(\rho) G(\mathbf{r},\rho)\,dV_\rho + \int_S f(\rho) G(\mathbf{r},\rho)\,dS_\rho + C, \tag{5}$$

where C is an arbitrary constant, provided that the solvability condition is met.

The Green's function $G = G(\mathbf{r},\rho)$ of the second boundary value problem is determined by the following conditions:

1°. The function G satisfies the Laplace equation with respect to x, y, z in the domain V everywhere except for the point (ξ, η, ζ) at which it has a singularity of the form $\frac{1}{4\pi}\frac{1}{|\mathbf{r}-\rho|}$.

2°. The function G, with respect to x, y, z, satisfies the homogeneous condition of the second kind at the boundary, i.e., the condition

$$\left.\frac{\partial G}{\partial N}\right|_S = \frac{1}{S_0},$$

where S_0 is the area of the surface S.

The Green's function is unique up to an additive constant.

Remark 2. The Green's function cannot be identified with condition 1° and the homogeneous boundary condition $\left.\frac{\partial G}{\partial N}\right|_S = 0$; this problem for G has no solution, because, on representing G in the form (3), for u we obtain a problem with a nonhomogeneous boundary condition of the second kind, for which the solvability condition (2) is not met.

Remark 3. Condition (4) is not extended to the outer second boundary value problem (for infinite domain).

8.2.1-3. Third boundary value problem.

The solution of the third boundary value problem for the Poisson equation (1) in a bounded domain V with the nonhomogeneous boundary condition

$$\frac{\partial w}{\partial N} + kw = f(\mathbf{r}) \quad \text{for} \quad \mathbf{r} \in S$$

is given by relation (5) with $C = 0$, where $G = G(\mathbf{r},\rho)$ is the Green's function of the third boundary value problem; the Green's function is determined by the following conditions:

1°. The function G satisfies the Laplace equation with respect to x, y, z in V everywhere except for the point (ξ, η, ζ) at which it has a singularity of the form $\frac{1}{4\pi}\frac{1}{|\mathbf{r}-\rho|}$.

2°. The function G, with respect to x, y, z, satisfies the homogeneous boundary condition of the third kind at the boundary, i.e., the condition $\left[\frac{\partial G}{\partial N} + kG\right]_S = 0$.

The Green's function can be represented in the form (3), where the auxiliary function u is determined by solving the corresponding third boundary value problem for the Laplace equation $\Delta_3 u = 0$.

The construction of Green's functions is discussed in Paragraphs 8.3.1-4 and 8.3.1-6 through 8.3.1-8 for $\lambda = 0$.

⊙ *References for Subsection* 8.2.1: V. M. Babich, M. B. Kapilevich, S. G. Mikhlin, et al. (1964), N. S. Koshlyakov, E. B. Gliner, and M. M. Smirnov (1970).

8.2.2. Problems in Cartesian Coordinates

The three-dimensional Poisson equation in the rectangular Cartesian system of coordinates has the form

$$\frac{\partial^2 w}{\partial x^2} + \frac{\partial^2 w}{\partial y^2} + \frac{\partial^2 w}{\partial z^2} + \Phi(x, y, z) = 0.$$

8.2.2-1. Domain: $-\infty < x < \infty$, $-\infty < y < \infty$, $-\infty < z < \infty$.

Solution:

$$w(x, y, z) = \frac{1}{4\pi} \int_{-\infty}^{\infty} \int_{-\infty}^{\infty} \int_{-\infty}^{\infty} \frac{\Phi(\xi, \eta, \zeta) \, d\xi \, d\eta \, d\zeta}{\sqrt{(x-\xi)^2 + (y-\eta)^2 + (z-\zeta)^2}}.$$

⊙ *Reference*: R. Courant and D. Hilbert (1989).

8.2.2-2. Domain: $-\infty < x < \infty$, $-\infty < y < \infty$, $0 \leq z < \infty$. First boundary value problem.

A half-space is considered. A boundary condition is prescribed:

$$w = f(x, y) \quad \text{at} \quad z = 0.$$

Solution:

$$w(x, y, z) = \frac{1}{2\pi} \int_{-\infty}^{\infty} \int_{-\infty}^{\infty} \frac{z f(\xi, \eta) \, d\xi \, d\eta}{\left[(x-\xi)^2 + (y-\eta)^2 + z^2\right]^{3/2}}$$

$$+ \frac{1}{4\pi} \int_{0}^{\infty} \int_{-\infty}^{\infty} \int_{-\infty}^{\infty} \left(\frac{1}{R_-} - \frac{1}{R_+}\right) \Phi(\xi, \eta, \zeta) \, d\xi \, d\eta \, d\zeta,$$

where

$$R_- = \sqrt{(x-\xi)^2 + (y-\eta)^2 + (z-\zeta)^2}, \quad R_+ = \sqrt{(x-\xi)^2 + (y-\eta)^2 + (z+\zeta)^2}.$$

⊙ *References*: A. G. Butkovskiy (1979), B. M. Budak, A. A. Samarskii, and A. N. Tikhonov (1980).

8.2.2-3. Domain: $-\infty < x < \infty$, $-\infty < y < \infty$, $0 \leq z < \infty$. Third boundary value problem.

A half-space is considered. A boundary condition is prescribed:

$$\partial_z w - kw = f(x, y) \quad \text{at} \quad z = 0.$$

Solution:

$$w(x, y, z) = -\int_{-\infty}^{\infty} \int_{-\infty}^{\infty} f(\xi, \eta) G(x, y, z, \xi, \eta, 0) \, d\xi \, d\eta$$

$$+ \int_{0}^{\infty} \int_{-\infty}^{\infty} \int_{-\infty}^{\infty} \Phi(\xi, \eta, \zeta) G(x, y, z, \xi, \eta, \zeta) \, d\xi \, d\eta \, d\zeta,$$

where

$$G(x, y, z, \xi, \eta, \zeta) = \frac{1}{4\pi} \left[\frac{1}{\sqrt{(x-\xi)^2 + (y-\eta)^2 + (z-\zeta)^2}} + \frac{1}{\sqrt{(x-\xi)^2 + (y-\eta)^2 + (z+\zeta)^2}} \right.$$

$$\left. - 2k \int_{0}^{\infty} \frac{\exp(-ks) \, ds}{\sqrt{(x-\xi)^2 + (y-\eta)^2 + (z+\zeta+s)^2}} \right].$$

8.2.2-4. Domain: $-\infty < x < \infty$, $0 \leq y < \infty$, $0 \leq z < \infty$. First boundary value problem.

A dihedral angle is considered. Boundary conditions are prescribed:

$$w = f_1(x,z) \quad \text{at} \quad y = 0, \qquad w = f_2(x,y) \quad \text{at} \quad z = 0.$$

Solution:

$$w(x,y,z) = \int_0^\infty \int_{-\infty}^\infty f_1(\xi,\zeta) \left[\frac{\partial}{\partial \eta} G(x,y,z,\xi,\eta,\zeta) \right]_{\eta=0} d\xi\, d\zeta$$
$$+ \int_0^\infty \int_{-\infty}^\infty f_2(\xi,\eta) \left[\frac{\partial}{\partial \zeta} G(x,y,z,\xi,\eta,\zeta) \right]_{\zeta=0} d\xi\, d\eta$$
$$+ \int_0^\infty \int_0^\infty \int_{-\infty}^\infty \Phi(\xi,\eta,\zeta) G(x,y,z,\xi,\eta,\zeta)\, d\xi\, d\eta\, d\zeta,$$

where

$$G(x,y,z,\xi,\eta,\zeta) = \frac{1}{4\pi} \Bigg[\frac{1}{\sqrt{(x-\xi)^2+(y-\eta)^2+(z-\zeta)^2}} - \frac{1}{\sqrt{(x-\xi)^2+(y-\eta)^2+(z+\zeta)^2}}$$
$$- \frac{1}{\sqrt{(x-\xi)^2+(y+\eta)^2+(z-\zeta)^2}} + \frac{1}{\sqrt{(x-\xi)^2+(y+\eta)^2+(z+\zeta)^2}} \Bigg].$$

⊙ *References*: V. S. Vladimirov, V. P. Mikhailov, A. A. Vasharin, et al. (1974), A. G. Butkovskiy (1979).

8.2.2-5. Domain: $0 \leq x < \infty$, $0 \leq y < \infty$, $0 \leq z < \infty$. First boundary value problem.

An octant is considered. Boundary conditions are prescribed:

$$w = f_1(y,z) \quad \text{at} \quad x = 0, \qquad w = f_2(x,z) \quad \text{at} \quad y = 0, \qquad w = f_3(x,y) \quad \text{at} \quad z = 0.$$

Solution:

$$w(x,y,z) = \int_0^\infty \int_0^\infty f_1(\eta,\zeta) \left[\frac{\partial}{\partial \xi} G(x,y,z,\xi,\eta,\zeta) \right]_{\xi=0} d\eta\, d\zeta$$
$$+ \int_0^\infty \int_0^\infty f_2(\xi,\zeta) \left[\frac{\partial}{\partial \eta} G(x,y,z,\xi,\eta,\zeta) \right]_{\eta=0} d\xi\, d\zeta$$
$$+ \int_0^\infty \int_0^\infty f_3(\xi,\eta) \left[\frac{\partial}{\partial \zeta} G(x,y,z,\xi,\eta,\zeta) \right]_{\zeta=0} d\xi\, d\eta$$
$$+ \int_0^\infty \int_0^\infty \int_0^\infty \Phi(\xi,\eta,\zeta) G(x,y,z,\xi,\eta,\zeta)\, d\xi\, d\eta\, d\zeta,$$

where

$$G(x,y,z,\xi,\eta,\zeta) = \frac{1}{4\pi} \Bigg[\frac{1}{\sqrt{(x-\xi)^2+(y-\eta)^2+(z-\zeta)^2}} - \frac{1}{\sqrt{(x-\xi)^2+(y-\eta)^2+(z+\zeta)^2}}$$
$$- \frac{1}{\sqrt{(x-\xi)^2+(y+\eta)^2+(z-\zeta)^2}} + \frac{1}{\sqrt{(x-\xi)^2+(y+\eta)^2+(z+\zeta)^2}}$$
$$- \frac{1}{\sqrt{(x+\xi)^2+(y-\eta)^2+(z-\zeta)^2}} + \frac{1}{\sqrt{(x+\xi)^2+(y-\eta)^2+(z+\zeta)^2}}$$
$$+ \frac{1}{\sqrt{(x+\xi)^2+(y+\eta)^2+(z-\zeta)^2}} - \frac{1}{\sqrt{(x+\xi)^2+(y+\eta)^2+(z+\zeta)^2}} \Bigg].$$

⊙ *References*: V. S. Vladimirov, V. P. Mikhailov, A. A. Vasharin, et al. (1974), A. G. Butkovskiy (1979).

8.2.2-6. Domain: $-\infty < x < \infty$, $-\infty < y < \infty$, $0 \le z \le a$. First boundary value problem.

An infinite layer is considered. Boundary conditions are prescribed:

$$w = f_1(x,y) \quad \text{at} \quad z = 0, \qquad w = f_2(x,y) \quad \text{at} \quad z = a.$$

Solution:

$$w(x,y,z) = \int_{-\infty}^{\infty}\int_{-\infty}^{\infty} f_1(\xi,\eta)\left[\frac{\partial}{\partial \zeta}G(x,y,z,\xi,\eta,\zeta)\right]_{\zeta=0} d\xi\, d\eta$$
$$- \int_{-\infty}^{\infty}\int_{-\infty}^{\infty} f_2(\xi,\eta)\left[\frac{\partial}{\partial \zeta}G(x,y,z,\xi,\eta,\zeta)\right]_{\zeta=a} d\xi\, d\eta$$
$$+ \int_{0}^{a}\int_{-\infty}^{\infty}\int_{-\infty}^{\infty} \Phi(\xi,\eta,\zeta)G(x,y,z,\xi,\eta,\zeta)\, d\xi\, d\eta\, d\zeta.$$

Green's function:

$$G(x,y,z,\xi,\eta,\zeta) = \frac{1}{4\pi}\sum_{n=-\infty}^{\infty}\left(\frac{1}{r_{n1}} - \frac{1}{r_{n2}}\right),$$

where

$$r_{n1} = \sqrt{(x-\xi)^2 + (y-\eta)^2 + (z-\zeta-2na)^2},$$
$$r_{n2} = \sqrt{(x-\xi)^2 + (y-\eta)^2 + (z+\zeta-2na)^2}.$$

⊙ *References*: A. G. Butkovskiy (1979), B. M. Budak, A. A. Samarskii, and A. N. Tikhonov (1980).

8.2.2-7. Domain: $-\infty < x < \infty$, $-\infty < y < \infty$, $0 \le z \le a$. Mixed boundary value problem.

An infinite layer is considered. Boundary conditions are prescribed:

$$w = f_1(x,y) \quad \text{at} \quad z = 0, \qquad \partial_z w = f_2(x,y) \quad \text{at} \quad z = a.$$

Solution:

$$w(x,y,z) = \int_{-\infty}^{\infty}\int_{-\infty}^{\infty} f_1(\xi,\eta)\left[\frac{\partial}{\partial \zeta}G(x,y,z,\xi,\eta,\zeta)\right]_{\zeta=0} d\xi\, d\eta$$
$$+ \int_{-\infty}^{\infty}\int_{-\infty}^{\infty} f_2(\xi,\eta)G(x,y,z,\xi,\eta,a)\, d\xi\, d\eta$$
$$+ \int_{0}^{a}\int_{-\infty}^{\infty}\int_{-\infty}^{\infty} \Phi(\xi,\eta,\zeta)G(x,y,z,\xi,\eta,\zeta)\, d\xi\, d\eta\, d\zeta.$$

Green's function:

$$G(x,y,z,\xi,\eta,\zeta) = \frac{1}{4\pi}\sum_{n=-\infty}^{\infty}\left(\frac{1}{r_{n1}} - \frac{1}{r_{n2}}\right),$$

where

$$r_{n1} = \sqrt{(x-\xi)^2 + (y-\eta)^2 + [z-(-1)^n\zeta - 2na]^2},$$
$$r_{n2} = \sqrt{(x-\xi)^2 + (y-\eta)^2 + [z+(-1)^n\zeta - 2na]^2}.$$

⊙ *Reference*: B. M. Budak, A. A. Samarskii, and A. N. Tikhonov (1980).

8.2.2-8. Domain: $0 \leq x < \infty$, $-\infty < y < \infty$, $0 \leq z \leq a$. First boundary value problem.

A semiinfinite layer is considered. Boundary conditions are prescribed:
$$w = f_1(y,z) \quad \text{at} \quad x = 0, \qquad w = f_2(x,y) \quad \text{at} \quad z = 0, \qquad w = f_3(x,y) \quad \text{at} \quad z = a.$$

Solution:
$$\begin{aligned} w(x,y,z) = &\int_0^a \int_{-\infty}^{\infty} f_1(\eta,\zeta) \left[\frac{\partial}{\partial \xi} G(x,y,z,\xi,\eta,\zeta)\right]_{\xi=0} d\eta\, d\zeta \\ &+ \int_{-\infty}^{\infty} \int_0^{\infty} f_2(\xi,\eta) \left[\frac{\partial}{\partial \zeta} G(x,y,z,\xi,\eta,\zeta)\right]_{\zeta=0} d\xi\, d\eta \\ &- \int_{-\infty}^{\infty} \int_0^{\infty} f_3(\xi,\eta) \left[\frac{\partial}{\partial \zeta} G(x,y,z,\xi,\eta,\zeta)\right]_{\zeta=a} d\xi\, d\eta \\ &+ \int_0^a \int_{-\infty}^{\infty} \int_0^{\infty} \Phi(\xi,\eta,\zeta) G(x,y,z,\xi,\eta,\zeta)\, d\xi\, d\eta\, d\zeta. \end{aligned}$$

Green's function:
$$G(x,y,z,\xi,\eta,\zeta) = \frac{1}{4\pi} \sum_{n=-\infty}^{\infty} \left(\frac{1}{r_{n1}} - \frac{1}{r_{n2}} - \frac{1}{r_{n3}} + \frac{1}{r_{n4}}\right),$$

where
$$\begin{aligned} r_{n1} &= \sqrt{(x-\xi)^2 + (y-\eta)^2 + (z-\zeta-2na)^2}, \\ r_{n2} &= \sqrt{(x-\xi)^2 + (y-\eta)^2 + (z+\zeta-2na)^2}, \\ r_{n3} &= \sqrt{(x+\xi)^2 + (y-\eta)^2 + (z-\zeta-2na)^2}, \\ r_{n4} &= \sqrt{(x+\xi)^2 + (y-\eta)^2 + (z+\zeta-2na)^2}. \end{aligned}$$

⊙ *Reference*: B. M. Budak, A. A. Samarskii, and A. N. Tikhonov (1980).

8.2.2-9. Domain: $0 \leq x \leq a$, $0 \leq y \leq b$, $-\infty < z < \infty$. First boundary value problem.

An infinite cylindrical domain of a rectangular cross-section is considered. Boundary conditions are prescribed:
$$\begin{aligned} w &= f_1(y,z) \quad \text{at} \quad x = 0, & w &= f_2(y,z) \quad \text{at} \quad x = a, \\ w &= f_3(x,z) \quad \text{at} \quad y = 0, & w &= f_4(x,z) \quad \text{at} \quad y = b. \end{aligned}$$

Solution:
$$\begin{aligned} w(x,y,z) = &\int_0^b \int_{-\infty}^{\infty} f_1(\eta,\zeta) \left[\frac{\partial}{\partial \xi} G(x,y,z,\xi,\eta,\zeta)\right]_{\xi=0} d\zeta\, d\eta \\ &- \int_0^b \int_{-\infty}^{\infty} f_2(\eta,\zeta) \left[\frac{\partial}{\partial \xi} G(x,y,z,\xi,\eta,\zeta)\right]_{\xi=a} d\zeta\, d\eta \\ &+ \int_0^a \int_{-\infty}^{\infty} f_3(\xi,\zeta) \left[\frac{\partial}{\partial \eta} G(x,y,z,\xi,\eta,\zeta)\right]_{\eta=0} d\zeta\, d\xi \\ &- \int_0^a \int_{-\infty}^{\infty} f_4(\xi,\zeta) \left[\frac{\partial}{\partial \eta} G(x,y,z,\xi,\eta,\zeta)\right]_{\eta=b} d\zeta\, d\xi \\ &+ \int_0^a \int_0^b \int_{-\infty}^{\infty} \Phi(\xi,\eta,\zeta) G(x,y,z,\xi,\eta,\zeta)\, d\zeta\, d\eta\, d\xi. \end{aligned}$$

Green's function:
$$G(x,y,z,\xi,\eta,\zeta) = \frac{2}{ab} \sum_{n=1}^{\infty} \sum_{m=1}^{\infty} \frac{1}{\beta_{nm}} \sin(p_n x) \sin(q_m y) \sin(p_n \xi) \sin(q_m \eta) \exp(-\beta_{nm}|z-\zeta|),$$
$$p_n = \frac{n\pi}{a}, \qquad q_m = \frac{m\pi}{b}, \qquad \beta_{nm} = \sqrt{p_n^2 + q_m^2}.$$

Alternatively, the Green's function can be represented as

$$G(x, y, z, \xi, \eta, \zeta) = \frac{1}{4\pi} \sum_{n=-\infty}^{\infty} \sum_{m=-\infty}^{\infty} \left(\frac{1}{r_{nm}^{(1)}} - \frac{1}{r_{nm}^{(2)}} - \frac{1}{r_{nm}^{(3)}} + \frac{1}{r_{nm}^{(4)}} \right),$$

where

$$r_{nm}^{(1)} = \sqrt{(x - \xi - 2na)^2 + (y - \eta - 2mb)^2 + (z - \zeta)^2},$$
$$r_{nm}^{(2)} = \sqrt{(x + \xi - 2na)^2 + (y - \eta - 2mb)^2 + (z - \zeta)^2},$$
$$r_{nm}^{(3)} = \sqrt{(x - \xi - 2na)^2 + (y + \eta - 2mb)^2 + (z - \zeta)^2},$$
$$r_{nm}^{(4)} = \sqrt{(x + \xi - 2na)^2 + (y + \eta - 2mb)^2 + (z - \zeta)^2}.$$

8.2.2-10. Domain: $0 \le x \le a$, $0 \le y \le b$, $-\infty < z < \infty$. Third boundary value problem.

An infinite cylindrical domain of a rectangular cross-section is considered. Boundary conditions are prescribed:

$$\partial_x w - k_1 w = f_1(y, z) \quad \text{at} \quad x = 0, \qquad \partial_x w + k_2 w = f_2(y, z) \quad \text{at} \quad x = a,$$
$$\partial_y w - k_3 w = f_3(x, z) \quad \text{at} \quad y = 0, \qquad \partial_y w + k_4 w = f_4(x, z) \quad \text{at} \quad y = b.$$

Solution:

$$w(x, y, z) = -\int_0^b \int_{-\infty}^{\infty} f_1(\eta, \zeta) G(x, y, z, 0, \eta, \zeta) \, d\zeta \, d\eta + \int_0^b \int_{-\infty}^{\infty} f_2(\eta, \zeta) G(x, y, z, a, \eta, \zeta) \, d\zeta \, d\eta$$
$$- \int_0^a \int_{-\infty}^{\infty} f_3(\xi, \zeta) G(x, y, z, \xi, 0, \zeta) \, d\zeta \, d\xi + \int_0^a \int_{-\infty}^{\infty} f_4(\xi, \zeta) G(x, y, z, \xi, b, \zeta) \, d\zeta \, d\xi$$
$$+ \int_0^a \int_0^b \int_{-\infty}^{\infty} \Phi(\xi, \eta, \zeta) G(x, y, z, \xi, \eta, \zeta) \, d\zeta \, d\eta \, d\xi.$$

Green's function:

$$G(x, y, z, \xi, \eta, \zeta) = \frac{1}{2} \sum_{n=1}^{\infty} \sum_{m=1}^{\infty} \frac{u_{nm}(x, y) u_{nm}(\xi, \eta)}{\|u_{nm}\|^2 \beta_{nm}} \exp(-\beta_{nm}|z - \zeta|),$$

where

$$w_{nm}(x, y) = (\mu_n \cos \mu_n x + k_1 \sin \mu_n x)(\nu_m \cos \nu_m y + k_3 \sin \nu_m y), \qquad \beta_{nm} = \sqrt{\mu_n^2 + \nu_m^2},$$
$$\|w_{nm}\|^2 = \frac{1}{4}(\mu_n^2 + k_1^2)(\nu_m^2 + k_3^2)\left[a + \frac{(k_1 + k_2)(\mu_n^2 + k_1 k_2)}{(\mu_n^2 + k_1^2)(\mu_n^2 + k_2^2)}\right]\left[b + \frac{(k_3 + k_4)(\nu_m^2 + k_3 k_4)}{(\nu_m^2 + k_3^2)(\nu_m^2 + k_4^2)}\right].$$

Here, the μ_n and ν_m are positive roots of the transcendental equations

$$\tan(\mu a) = \frac{(k_1 + k_2)\mu}{\mu^2 - k_1 k_2}, \qquad \tan(\nu b) = \frac{(k_3 + k_4)\nu}{\nu^2 - k_3 k_4}.$$

8.2.2-11. Domain: $0 \le x \le a$, $0 \le y \le b$, $-\infty < z < \infty$. Mixed boundary value problems.

1°. An infinite cylindrical domain of a rectangular cross-section is considered. Boundary conditions are prescribed:

$$w = f_1(y, z) \quad \text{at} \quad x = 0, \qquad \partial_x w = f_2(y, z) \quad \text{at} \quad x = a,$$
$$w = f_3(x, z) \quad \text{at} \quad y = 0, \qquad \partial_y w = f_4(x, z) \quad \text{at} \quad y = b.$$

Solution:
$$w(x,y,z) = \int_0^b \int_{-\infty}^{\infty} f_1(\eta,\zeta) \left[\frac{\partial}{\partial \xi} G(x,y,z,\xi,\eta,\zeta)\right]_{\xi=0} d\zeta\, d\eta$$
$$+ \int_0^b \int_{-\infty}^{\infty} f_2(\eta,\zeta) G(x,y,z,a,\eta,\zeta)\, d\zeta\, d\eta$$
$$+ \int_0^a \int_{-\infty}^{\infty} f_3(\xi,\zeta) \left[\frac{\partial}{\partial \eta} G(x,y,z,\xi,\eta,\zeta)\right]_{\eta=0} d\zeta\, d\xi$$
$$+ \int_0^a \int_{-\infty}^{\infty} f_4(\xi,\zeta) G(x,y,z,\xi,b,\zeta)\, d\zeta\, d\xi$$
$$+ \int_0^a \int_0^b \int_{-\infty}^{\infty} \Phi(\xi,\eta,\zeta) G(x,y,z,\xi,\eta,\zeta)\, d\zeta\, d\eta\, d\xi.$$

Green's function:
$$G(x,y,z,\xi,\eta,\zeta) = \frac{2}{ab} \sum_{n=0}^{\infty} \sum_{m=0}^{\infty} \frac{1}{\beta_{nm}} \sin(p_n x) \sin(q_m y) \sin(p_n \xi) \sin(q_m \eta) \exp(-\beta_{nm}|z-\zeta|),$$

where
$$p_n = \frac{(2n+1)\pi}{2a}, \quad q_m = \frac{(2m+1)\pi}{2b}, \quad \beta_{nm} = \sqrt{p_n^2 + q_m^2}.$$

2°. An infinite cylindrical domain of a rectangular cross-section is considered. Boundary conditions are prescribed:
$$w = f_1(y,z) \quad \text{at} \quad x=0, \qquad w = f_2(y,z) \quad \text{at} \quad x=a,$$
$$\partial_y w = f_3(x,z) \quad \text{at} \quad y=0, \qquad \partial_y w = f_4(x,z) \quad \text{at} \quad y=b.$$

Solution:
$$w(x,y,z) = \int_0^b \int_{-\infty}^{\infty} f_1(\eta,\zeta) \left[\frac{\partial}{\partial \xi} G(x,y,z,\xi,\eta,\zeta)\right]_{\xi=0} d\zeta\, d\eta$$
$$- \int_0^b \int_{-\infty}^{\infty} f_2(\eta,\zeta) \left[\frac{\partial}{\partial \xi} G(x,y,z,\xi,\eta,\zeta)\right]_{\xi=a} d\zeta\, d\eta$$
$$- \int_0^a \int_{-\infty}^{\infty} f_3(\xi,\zeta) G(x,y,z,\xi,0,\zeta)\, d\zeta\, d\xi$$
$$+ \int_0^a \int_{-\infty}^{\infty} f_4(\xi,\zeta) G(x,y,z,\xi,b,\zeta)\, d\zeta\, d\xi$$
$$+ \int_0^a \int_0^b \int_{-\infty}^{\infty} \Phi(\xi,\eta,\zeta) G(x,y,z,\xi,\eta,\zeta)\, d\zeta\, d\eta\, d\xi.$$

Green's function:
$$G(x,y,z,\xi,\eta,\zeta) = \frac{1}{ab} \sum_{n=1}^{\infty} \sum_{m=0}^{\infty} \frac{A_m}{\beta_{nm}} \sin(p_n x) \cos(q_m y) \sin(p_n \xi) \cos(q_m \eta) \exp(-\beta_{nm}|z-\zeta|),$$

where
$$p_n = \frac{n\pi}{a}, \quad q_m = \frac{m\pi}{b}, \quad \beta_{nm} = \sqrt{p_n^2 + q_m^2}, \quad A_m = \begin{cases} 1 & \text{for } m=0, \\ 2 & \text{for } m \neq 0. \end{cases}$$

▶ *Paragraphs 8.2.2-12 through 8.2.2-17 present only Green's functions; the complete solution is constructed with the formulas given in Paragraphs 8.2.1-1 through 8.2.1-3.*

8.2.2-12. Domain: $0 \leq x \leq a$, $0 \leq y \leq x$, $-\infty < z < \infty$. First boundary value problem.

An infinite cylindrical domain of triangular cross-section is considered. Boundary conditions are prescribed:
$$w = f_1(y,z) \quad \text{at} \quad x=0, \qquad w = f_2(x,z) \quad \text{at} \quad y=0, \qquad w = f_3(x,z) \quad \text{at} \quad y=x.$$

Green's function:
$$G(x,y,z,\xi,\eta,\zeta) = H(x,y,z,\xi,\eta,\zeta) - H(x,y,z,\eta,\xi,\zeta),$$

where
$$H(x,y,z,\xi,\eta,\zeta) = \frac{2}{a^2} \sum_{n=1}^{\infty} \sum_{m=1}^{\infty} \frac{1}{\beta_{nm}} \sin(p_n x) \sin(p_m y) \sin(p_n \xi) \sin(p_m \eta) \exp(-\beta_{nm}|z-\zeta|),$$

$$p_n = \frac{n\pi}{a}, \quad p_m = \frac{m\pi}{a}, \quad \beta_{nm} = \sqrt{p_n^2 + p_m^2}.$$

An alternative representation of the Green's function can be obtained by setting
$$H(x,y,z,\xi,\eta,\zeta) = \frac{1}{4\pi} \sum_{n=-\infty}^{\infty} \sum_{m=-\infty}^{\infty} \left(\frac{1}{r_{nm}^{(1)}} - \frac{1}{r_{nm}^{(2)}} - \frac{1}{r_{nm}^{(3)}} + \frac{1}{r_{nm}^{(4)}} \right),$$

where the functions $r_{nm}^{(k)}$ ($k = 1, 2, 3, 4$) are specified in Paragraph 8.2.2-9 for $a = b$.

⊙ *Reference*: B. M. Budak, A. A. Samarskii, and A. N. Tikhonov (1980).

8.2.2-13. Domain: $0 \le x \le a$, $0 \le y \le b$, $0 \le z < \infty$. First boundary value problem.

A semiinfinite cylindrical domain of a rectangular cross-section is considered. Boundary conditions are prescribed:
$$w = f_1(y,z) \quad \text{at} \quad x = 0, \qquad w = f_2(y,z) \quad \text{at} \quad x = a,$$
$$w = f_3(x,z) \quad \text{at} \quad y = 0, \qquad w = f_4(x,z) \quad \text{at} \quad y = b,$$
$$w = f_5(x,y) \quad \text{at} \quad z = 0.$$

Green's function:
$$G(x,y,z,\xi,\eta,\zeta) = \frac{4}{ab} \sum_{n=1}^{\infty} \sum_{m=1}^{\infty} \frac{1}{\beta_{nm}} \sin(p_n x) \sin(q_m y) \sin(p_n \xi) \sin(q_m \eta) H_{nm}(z,\zeta),$$

$$p_n = \frac{n\pi}{a}, \quad q_m = \frac{m\pi}{b}, \quad \beta_{nm} = \sqrt{p_n^2 + q_m^2},$$

$$H_{nm}(z,\zeta) = \begin{cases} \exp(-\beta_{nm} z) \sinh(\beta_{nm} \zeta) & \text{for } z > \zeta \ge 0, \\ \exp(-\beta_{nm} \zeta) \sinh(\beta_{nm} z) & \text{for } \zeta > z \ge 0. \end{cases}$$

An alternative representation of the Green's function:
$$G(x,y,z,\xi,\eta,\zeta) = \frac{1}{4\pi} \sum_{n=-\infty}^{\infty} \sum_{m=-\infty}^{\infty} \left(\frac{1}{r_{nm}^{(1)}} - \frac{1}{r_{nm}^{(2)}} - \frac{1}{r_{nm}^{(3)}} + \frac{1}{r_{nm}^{(4)}} - \frac{1}{r_{nm}^{(5)}} + \frac{1}{r_{nm}^{(6)}} + \frac{1}{r_{nm}^{(7)}} - \frac{1}{r_{nm}^{(8)}} \right),$$

where
$$r_{nm}^{(1)} = \sqrt{(x-\xi-2na)^2 + (y-\eta-2mb)^2 + (z-\zeta)^2},$$
$$r_{nm}^{(2)} = \sqrt{(x+\xi-2na)^2 + (y-\eta-2mb)^2 + (z-\zeta)^2},$$
$$r_{nm}^{(3)} = \sqrt{(x-\xi-2na)^2 + (y+\eta-2mb)^2 + (z-\zeta)^2},$$
$$r_{nm}^{(4)} = \sqrt{(x+\xi-2na)^2 + (y+\eta-2mb)^2 + (z-\zeta)^2},$$
$$r_{nm}^{(5)} = \sqrt{(x-\xi-2na)^2 + (y-\eta-2mb)^2 + (z+\zeta)^2},$$
$$r_{nm}^{(6)} = \sqrt{(x+\xi-2na)^2 + (y-\eta-2mb)^2 + (z+\zeta)^2},$$
$$r_{nm}^{(7)} = \sqrt{(x-\xi-2na)^2 + (y+\eta-2mb)^2 + (z+\zeta)^2},$$
$$r_{nm}^{(8)} = \sqrt{(x+\xi-2na)^2 + (y+\eta-2mb)^2 + (z+\zeta)^2}.$$

8.2.2-14. Domain: $0 \leq x \leq a$, $0 \leq y \leq b$, $0 \leq z < \infty$. Third boundary value problem.

A semiinfinite cylindrical domain of a rectangular cross-section is considered. Boundary conditions are prescribed:

$$\partial_x w - k_1 w = f_1(y,z) \quad \text{at} \quad x = 0, \qquad \partial_x w + k_2 w = f_2(y,z) \quad \text{at} \quad x = a,$$
$$\partial_y w - k_3 w = f_3(x,z) \quad \text{at} \quad y = 0, \qquad \partial_y w + k_4 w = f_4(x,z) \quad \text{at} \quad y = b,$$
$$\partial_z w - k_5 w = f_5(x,y) \quad \text{at} \quad z = 0.$$

Green's function:

$$G(x,y,z,\xi,\eta,\zeta) = \sum_{n=1}^{\infty}\sum_{m=1}^{\infty} \frac{u_{nm}(x,y)u_{nm}(\xi,\eta)}{\|u_{nm}\|^2} H_{nm}(z,\zeta),$$

where

$$w_{nm}(x,y) = (\mu_n \cos\mu_n x + k_1 \sin\mu_n x)(\nu_m \cos\nu_m y + k_3 \sin\nu_m y),$$

$$\|w_{nm}\|^2 = \frac{1}{4}(\mu_n^2 + k_1^2)(\nu_m^2 + k_3^2)\left[a + \frac{(k_1+k_2)(\mu_n^2 + k_1 k_2)}{(\mu_n^2 + k_1^2)(\mu_n^2 + k_2^2)}\right]\left[b + \frac{(k_3+k_4)(\nu_m^2 + k_3 k_4)}{(\nu_m^2 + k_3^2)(\nu_m^2 + k_4^2)}\right],$$

$$H_{nm}(z,\zeta) = \begin{cases} \dfrac{\exp(-\beta_{nm}z)[\beta_{nm}\cosh(\beta_{nm}\zeta) + k_5 \sinh(\beta_{nm}\zeta)]}{\beta_{nm}(\beta_{nm} + k_5)} & \text{for } z > \zeta, \\[6pt] \dfrac{\exp(-\beta_{nm}\zeta)[\beta_{nm}\cosh(\beta_{nm}z) + k_5 \sinh(\beta_{nm}z)]}{\beta_{nm}(\beta_{nm} + k_5)} & \text{for } \zeta > z, \end{cases} \qquad \beta_{nm} = \sqrt{\mu_n^2 + \nu_m^2}.$$

Here, the μ_n and ν_m are positive roots of the transcendental equations

$$\tan(\mu a) = \frac{(k_1+k_2)\mu}{\mu^2 - k_1 k_2}, \qquad \tan(\nu b) = \frac{(k_3+k_4)\nu}{\nu^2 - k_3 k_4}.$$

8.2.2-15. Domain: $0 \leq x \leq a$, $0 \leq y \leq b$, $0 \leq z < \infty$. Mixed boundary value problems.

1°. A semiinfinite cylindrical domain of a rectangular cross-section is considered. Boundary conditions are prescribed:

$$w = f_1(y,z) \quad \text{at} \quad x = 0, \qquad w = f_2(y,z) \quad \text{at} \quad x = a,$$
$$w = f_3(x,z) \quad \text{at} \quad y = 0, \qquad w = f_4(x,z) \quad \text{at} \quad y = b,$$
$$\partial_z w = f_5(x,y) \quad \text{at} \quad z = 0.$$

Green's function:

$$G(x,y,z,\xi,\eta,\zeta) = \frac{4}{ab}\sum_{n=1}^{\infty}\sum_{m=1}^{\infty} \frac{1}{\beta_{nm}} \sin(p_n x)\sin(q_m y)\sin(p_n \xi)\sin(q_m \eta) H_{nm}(z,\zeta),$$

$$p_n = \frac{n\pi}{a}, \qquad q_m = \frac{m\pi}{b}, \qquad \beta_{nm} = \sqrt{p_n^2 + q_m^2},$$

$$H_{nm}(z,\zeta) = \begin{cases} \exp(-\beta_{nm}z)\cosh(\beta_{nm}\zeta) & \text{for } z > \zeta \geq 0, \\ \exp(-\beta_{nm}\zeta)\cosh(\beta_{nm}z) & \text{for } \zeta > z \geq 0. \end{cases}$$

2°. A semiinfinite cylindrical domain of a rectangular cross-section is considered. Boundary conditions are prescribed:

$$\partial_x w = f_1(y,z) \quad \text{at} \quad x = 0, \qquad \partial_x w = f_2(y,z) \quad \text{at} \quad x = a,$$
$$\partial_y w = f_3(x,z) \quad \text{at} \quad y = 0, \qquad \partial_y w = f_4(x,z) \quad \text{at} \quad y = b,$$
$$w = f_5(x,y) \quad \text{at} \quad z = 0.$$

Green's function:

$$G(x,y,z,\xi,\eta,\zeta) = \frac{1}{ab}\sum_{n=0}^{\infty}\sum_{m=0}^{\infty}\frac{A_n A_m}{\beta_{nm}}\cos(p_n x)\cos(q_m y)\cos(p_n \xi)\cos(q_m \eta)H_{nm}(z,\zeta),$$

$$p_n = \frac{n\pi}{a}, \quad q_m = \frac{m\pi}{b}, \quad \beta_{nm} = \sqrt{p_n^2+q_m^2}, \quad A_n = \begin{cases} 1 & \text{for } n=0, \\ 2 & \text{for } n\neq 0, \end{cases}$$

$$H_{nm}(z,\zeta) = \begin{cases} \exp(-\beta_{nm}z)\sinh(\beta_{nm}\zeta) & \text{for } z>\zeta\geq 0, \\ \exp(-\beta_{nm}\zeta)\sinh(\beta_{nm}z) & \text{for } \zeta>z\geq 0. \end{cases}$$

8.2.2-16. Domain: $0\leq x\leq a$, $0\leq y\leq b$, $0\leq z\leq c$. First boundary value problem.

A rectangular parallelepiped is considered. Boundary conditions are prescribed:

$$\begin{array}{llll} w=f_1(y,z) & \text{at} \ x=0, & w=f_2(y,z) & \text{at} \ x=a, \\ w=f_3(x,z) & \text{at} \ y=0, & w=f_4(x,z) & \text{at} \ y=b, \\ w=f_5(x,y) & \text{at} \ z=0, & w=f_6(x,y) & \text{at} \ z=c. \end{array}$$

1°. Representation of the Green's function in the form of a double series:

$$G(x,y,z,\xi,\eta,\zeta) = \frac{4}{ab}\sum_{n=1}^{\infty}\sum_{m=1}^{\infty}\sin(p_n x)\sin(q_m y)\sin(p_n \xi)\sin(q_m \eta)F_{nm}(z,\zeta),$$

$$F_{nm}(z,\zeta) = \begin{cases} \dfrac{\sinh(\beta_{nm}\zeta)\sinh[\beta_{nm}(c-z)]}{\beta_{nm}\sinh(\beta_{nm}c)} & \text{for } c\geq z>\zeta\geq 0, \\ \dfrac{\sinh(\beta_{nm}z)\sinh[\beta_{nm}(c-\zeta)]}{\beta_{nm}\sinh(\beta_{nm}c)} & \text{for } c\geq \zeta>z\geq 0, \end{cases}$$

$$p_n = \frac{\pi n}{a}, \quad q_m = \frac{\pi m}{b}, \quad \beta_{nm} = \sqrt{p_n^2+q_m^2}.$$

This relation can be used to obtain two other representations of the Green's function by means of the following cyclic permutations:

$$\begin{array}{c}(x,\xi,a)\\ \nearrow\quad\searrow\\ (z,\zeta,c)\longleftarrow(y,\eta,b)\end{array}$$

2°. Representation of the Green's function in the form of a triple series:

$$G(x,y,z,\xi,\eta,\zeta) = \frac{8}{abc}\sum_{n=1}^{\infty}\sum_{m=1}^{\infty}\sum_{k=1}^{\infty}\frac{\sin(p_n x)\sin(q_m y)\sin(s_k z)\sin(p_n \xi)\sin(q_m \eta)\sin(s_k \zeta)}{p_n^2+q_m^2+s_k^2},$$

$$p_n = \frac{\pi n}{a}, \quad q_m = \frac{\pi m}{b}, \quad s_k = \frac{\pi k}{c}.$$

3°. An alternative representation of the Green's function in the form of a triple series:

$$G(x,y,z,\xi,\eta,\zeta) = \frac{1}{4\pi}\sum_{n=-\infty}^{\infty}\sum_{m=-\infty}^{\infty}\sum_{k=-\infty}^{\infty}\left(\frac{1}{r_{nmk}^{(1)}}-\frac{1}{r_{nmk}^{(2)}}-\frac{1}{r_{nmk}^{(3)}}+\frac{1}{r_{nmk}^{(4)}}\right.$$
$$\left.-\frac{1}{r_{nmk}^{(5)}}+\frac{1}{r_{nmk}^{(6)}}+\frac{1}{r_{nmk}^{(7)}}-\frac{1}{r_{nmk}^{(8)}}\right),$$

where

$$r_{nmk}^{(1)} = \sqrt{(x-\xi-2na)^2 + (y-\eta-2mb)^2 + (z-\zeta-2kc)^2},$$
$$r_{nmk}^{(2)} = \sqrt{(x+\xi-2na)^2 + (y-\eta-2mb)^2 + (z-\zeta-2kc)^2},$$
$$r_{nmk}^{(3)} = \sqrt{(x-\xi-2na)^2 + (y+\eta-2mb)^2 + (z-\zeta-2kc)^2},$$
$$r_{nmk}^{(4)} = \sqrt{(x+\xi-2na)^2 + (y+\eta-2mb)^2 + (z-\zeta-2kc)^2},$$
$$r_{nmk}^{(5)} = \sqrt{(x-\xi-2na)^2 + (y-\eta-2mb)^2 + (z+\zeta-2kc)^2},$$
$$r_{nmk}^{(6)} = \sqrt{(x+\xi-2na)^2 + (y-\eta-2mb)^2 + (z+\zeta-2kc)^2},$$
$$r_{nmk}^{(7)} = \sqrt{(x-\xi-2na)^2 + (y+\eta-2mb)^2 + (z+\zeta-2kc)^2},$$
$$r_{nmk}^{(8)} = \sqrt{(x+\xi-2na)^2 + (y+\eta-2mb)^2 + (z+\zeta-2kc)^2}.$$

8.2.2-17. Domain: $0 \le x \le a$, $0 \le y \le b$, $0 \le z \le c$. Third boundary value problem.

A rectangular parallelepiped is considered. Boundary conditions are prescribed:

$$\partial_x w - k_1 w = f_1(y,z) \quad \text{at} \quad x = 0, \qquad \partial_x w + k_2 w = f_2(y,z) \quad \text{at} \quad x = a,$$
$$\partial_y w - k_3 w = f_3(x,z) \quad \text{at} \quad y = 0, \qquad \partial_y w + k_4 w = f_4(x,z) \quad \text{at} \quad y = b,$$
$$\partial_z w - k_5 w = f_5(x,y) \quad \text{at} \quad z = 0, \qquad \partial_z w + k_6 w = f_6(x,y) \quad \text{at} \quad z = c.$$

Green's function:

$$G(x,y,z,\xi,\eta,\zeta) = \sum_{n=1}^{\infty}\sum_{m=1}^{\infty}\sum_{s=1}^{\infty} \frac{\varphi_n(x)\varphi_n(\xi)\psi_m(y)\psi_m(\eta)\chi_s(z)\chi_s(\zeta)}{\|\varphi_n\|^2 \|\psi_m\|^2 \|\chi_s(\zeta)\|^2 (\mu_n^2 + \lambda_m^2 + \nu_s^2)},$$

$$\varphi_n(x) = \cos(\mu_n x) + \frac{k_1}{\mu_n}\sin(\mu_n x), \quad \|\varphi_n\|^2 = \frac{k_2}{2\mu_n^2}\frac{\mu_n^2 + k_1^2}{\mu_n^2 + k_2^2} + \frac{k_1}{2\mu_n^2} + \frac{a}{2}\left(1 + \frac{k_1^2}{\mu_n^2}\right),$$

$$\psi_m(y) = \cos(\lambda_m y) + \frac{k_3}{\lambda_m}\sin(\lambda_m y), \quad \|\psi_m\|^2 = \frac{k_4}{2\lambda_m^2}\frac{\lambda_m^2 + k_3^2}{\lambda_m^2 + k_4^2} + \frac{k_3}{2\lambda_m^2} + \frac{b}{2}\left(1 + \frac{k_3^2}{\lambda_m^2}\right),$$

$$\chi_s(z) = \cos(\nu_s z) + \frac{k_5}{\nu_s}\sin(\nu_s z), \quad \|\chi_s\|^2 = \frac{k_6}{2\nu_s^2}\frac{\nu_s^2 + k_5^2}{\nu_s^2 + k_6^2} + \frac{k_5}{2\nu_s^2} + \frac{c}{2}\left(1 + \frac{k_5^2}{\nu_s^2}\right),$$

where the μ_n, λ_m, and ν_s are positive roots of the transcendental equations

$$\frac{\tan(\mu a)}{\mu} = \frac{k_1 + k_2}{\mu^2 - k_1 k_2}, \qquad \frac{\tan(\lambda b)}{\lambda} = \frac{k_3 + k_4}{\lambda^2 - k_3 k_4}, \qquad \frac{\tan(\nu c)}{\nu} = \frac{k_5 + k_6}{\nu^2 - k_5 k_6}.$$

8.2.2-18. Domain: $0 \le x \le a$, $0 \le y \le b$, $0 \le z \le c$. Mixed boundary value problem.

A rectangular parallelepiped is considered. Boundary conditions are prescribed:

$$w = f_1(y,z) \quad \text{at} \quad x = 0, \qquad \partial_x w = f_2(y,z) \quad \text{at} \quad x = a,$$
$$w = f_3(x,z) \quad \text{at} \quad y = 0, \qquad \partial_y w = f_4(x,z) \quad \text{at} \quad y = b,$$
$$w = f_5(x,y) \quad \text{at} \quad z = 0, \qquad \partial_z w = f_6(x,y) \quad \text{at} \quad z = c.$$

Solution:

$$w(x,y,z) = \int_0^a \int_0^b \int_0^c \Phi(\xi,\eta,\zeta) G(x,y,z,\xi,\eta,\zeta) \, d\zeta \, d\eta \, d\xi$$

$$+ \int_0^b \int_0^c f_1(\eta,\zeta) \left[\frac{\partial}{\partial \xi} G(x,y,z,\xi,\eta,\zeta)\right]_{\xi=0} d\zeta \, d\eta + \int_0^b \int_0^c f_2(\eta,\zeta) G(x,y,z,a,\eta,\zeta) \, d\zeta \, d\eta$$

$$+ \int_0^a \int_0^c f_3(\xi,\zeta) \left[\frac{\partial}{\partial \eta} G(x,y,z,\xi,\eta,\zeta)\right]_{\eta=0} d\zeta \, d\xi + \int_0^a \int_0^c f_4(\xi,\zeta) G(x,y,z,\xi,b,\zeta) \, d\zeta \, d\xi$$

$$+ \int_0^a \int_0^b f_5(\xi,\eta) \left[\frac{\partial}{\partial \zeta} G(x,y,z,\xi,\eta,\zeta)\right]_{\zeta=0} d\eta \, d\xi + \int_0^a \int_0^b f_6(\xi,\eta) G(x,y,z,\xi,\eta,c) \, d\eta \, d\xi.$$

1°. A double-series representation of the Green's function:

$$G(x,y,z,\xi,\eta,\zeta) = \frac{4}{ab} \sum_{n=0}^{\infty} \sum_{m=0}^{\infty} \sin(p_n x) \sin(q_m y) \sin(p_n \xi) \sin(q_m \eta) F_{nm}(z,\zeta),$$

$$F_{nm}(z,\zeta) = \begin{cases} \dfrac{\sinh(\beta_{nm}\zeta) \cosh[\beta_{nm}(c-z)]}{\beta_{nm} \cosh(\beta_{nm}c)} & \text{for } c \geq z > \zeta \geq 0, \\ \dfrac{\sinh(\beta_{nm}z) \cosh[\beta_{nm}(c-\zeta)]}{\beta_{nm} \cosh(\beta_{nm}c)} & \text{for } c \geq \zeta > z \geq 0, \end{cases}$$

$$p_n = \frac{\pi(2n+1)}{2a}, \quad q_m = \frac{\pi(2m+1)}{2b}, \quad \beta_{nm} = \sqrt{p_n^2 + q_m^2}.$$

This relation can be used to obtain two other representations of the Green's function by means of the following cyclic permutations:

$$\begin{array}{c} (x,\xi,a) \\ \nearrow \quad \searrow \\ (z,\zeta,c) \longleftarrow (y,\eta,b) \end{array}$$

2°. A triple series representation of the Green's function:

$$G(x,y,z,\xi,\eta,\zeta) = \frac{8}{abc} \sum_{n=0}^{\infty} \sum_{m=0}^{\infty} \sum_{k=0}^{\infty} \frac{\sin(p_n x) \sin(q_m y) \sin(s_k z) \sin(p_n \xi) \sin(q_m \eta) \sin(s_k \zeta)}{p_n^2 + q_m^2 + s_k^2},$$

$$p_n = \frac{\pi(2n+1)}{2a}, \quad q_m = \frac{\pi(2m+1)}{2b}, \quad s_k = \frac{\pi(2k+1)}{2c}.$$

8.2.3. Problems in Cylindrical Coordinates

The three-dimensional Poisson equation in the cylindrical coordinate system is written as

$$\frac{1}{r}\frac{\partial}{\partial r}\left(r\frac{\partial w}{\partial r}\right) + \frac{1}{r^2}\frac{\partial^2 w}{\partial \varphi^2} + \frac{\partial^2 w}{\partial z^2} = -\Phi(r,\varphi,z), \quad r = \sqrt{x^2+y^2}.$$

8.2.3-1. Domain: $0 \leq r \leq R$, $0 \leq \varphi \leq 2\pi$, $-\infty < z < \infty$. First boundary value problem.

An infinite circular cylinder is considered. A boundary condition is prescribed:

$$w = f(\varphi, z) \quad \text{at} \quad r = R.$$

Solution:

$$w(r,\varphi,z) = -R \int_0^{2\pi} \int_{-\infty}^{\infty} f(\eta,\zeta) \left[\frac{\partial}{\partial \xi} G(r,\varphi,z,\xi,\eta,\zeta)\right]_{\xi=R} d\zeta \, d\eta$$

$$+ \int_0^R \int_0^{2\pi} \int_{-\infty}^{\infty} \Phi(\xi,\eta,\zeta) G(r,\varphi,z,\xi,\eta,\zeta) \xi \, d\zeta \, d\eta \, d\xi.$$

Green's function:

$$G(r,\varphi,z,\xi,\eta,\zeta) = \frac{1}{2\pi R^2} \sum_{n=0}^{\infty} \sum_{m=1}^{\infty} \frac{A_n J_n(\mu_{nm}r) J_n(\mu_{nm}\xi)}{\left[J_n'(\mu_{nm}R)\right]^2 \mu_{nm}} \cos[n(\varphi-\eta)] \exp\bigl(-\mu_{nm}|z-\zeta|\bigr),$$

where $A_0 = 1$ and $A_n = 2$ for $n \neq 0$; the $J_n(\xi)$ are the Bessel functions; and the μ_{nm} are positive roots of the transcendental equation $J_n(\mu R) = 0$.

8.2.3-2. Domain: $0 \leq r \leq R$, $0 \leq \varphi \leq 2\pi$, $-\infty < z < \infty$. Third boundary value problem.

An infinite circular cylinder is considered. A boundary condition is prescribed:

$$\partial_r w + kw = f(\varphi, z) \quad \text{at} \quad r = R.$$

Solution:

$$w(r,\varphi,z) = R \int_0^{2\pi} \int_{-\infty}^{\infty} f(\eta,\zeta) G(r,\varphi,z,R,\eta,\zeta) \, d\zeta \, d\eta$$

$$+ \int_0^R \int_0^{2\pi} \int_{-\infty}^{\infty} \Phi(\xi,\eta,\zeta) G(r,\varphi,z,\xi,\eta,\zeta) \xi \, d\zeta \, d\eta \, d\xi.$$

Green's function:

$$G(r,\varphi,z,\xi,\eta,\zeta) = \frac{1}{2\pi} \sum_{n=0}^{\infty} \sum_{m=1}^{\infty} \frac{A_n \mu_{nm} J_n(\mu_{nm}r) J_n(\mu_{nm}\xi) \cos[n(\varphi-\eta)]}{(\mu_{nm}^2 R^2 + k^2 R^2 - n^2) J_n^2(\mu_{nm}R)} \exp\bigl(-\mu_{nm}|z-\zeta|\bigr),$$

where $A_0 = 1$ and $A_n = 2$ for $n \neq 0$; the $J_n(\xi)$ are the Bessel functions; and the μ_{nm} are positive roots of the transcendental equation

$$\mu J_n'(\mu R) + k J_n(\mu R) = 0.$$

8.2.3-3. Domain: $0 \leq r \leq R$, $0 \leq \varphi \leq 2\pi$, $0 \leq z < \infty$. First boundary value problem.

A semiinfinite circular cylinder is considered. Boundary conditions are prescribed:

$$w = f_1(\varphi, z) \quad \text{at} \quad r = R, \qquad w = f_2(r, \varphi) \quad \text{at} \quad z = 0.$$

Solution:

$$w(r,\varphi,z) = -R \int_0^{2\pi} \int_0^{\infty} f_1(\eta,\zeta) \left[\frac{\partial}{\partial \xi} G(r,\varphi,z,\xi,\eta,\zeta)\right]_{\xi=R} d\zeta \, d\eta$$

$$+ \int_0^{2\pi} \int_0^R f_2(\xi,\eta) \left[\frac{\partial}{\partial \zeta} G(r,\varphi,z,\xi,\eta,\zeta)\right]_{\zeta=0} \xi \, d\xi \, d\eta$$

$$+ \int_0^R \int_0^{2\pi} \int_0^{\infty} \Phi(\xi,\eta,\zeta) G(r,\varphi,z,\xi,\eta,\zeta) \xi \, d\zeta \, d\eta \, d\xi.$$

Green's function:

$$G(r,\varphi,z,\xi,\eta,\zeta) = \frac{1}{2\pi R^2} \sum_{n=0}^{\infty} \sum_{m=1}^{\infty} \frac{A_n J_n(\mu_{nm}r) J_n(\mu_{nm}\xi)}{\left[J_n'(\mu_{nm}R)\right]^2 \mu_{nm}} \cos[n(\varphi-\eta)] F_{nm}(z,\zeta),$$

$$F_{nm}(z,\zeta) = \exp(-\mu_{nm}|z-\zeta|) - \exp(-\mu_{nm}|z+\zeta|), \qquad A_n = \begin{cases} 1 & \text{for } n = 0, \\ 2 & \text{for } n \neq 0, \end{cases}$$

where the μ_{nm} are positive roots of the transcendental equation $J_n(\mu R) = 0$.

8.2.3-4. Domain: $0 \le r \le R$, $0 \le \varphi \le 2\pi$, $0 \le z < \infty$. Third boundary value problem.

A semiinfinite circular cylinder is considered. Boundary conditions are prescribed:

$$\partial_r w + k_1 w = f_1(\varphi, z) \quad \text{at} \quad r = R, \qquad \partial_z w - k_2 w = f_2(r, \varphi) \quad \text{at} \quad z = 0.$$

Solution:

$$w(r, \varphi, z) = R \int_0^{2\pi} \int_0^\infty f_1(\eta, \zeta) G(r, \varphi, z, R, \eta, \zeta)\, d\zeta\, d\eta$$
$$- \int_0^{2\pi} \int_0^R f_2(\xi, \eta) G(r, \varphi, z, R, \eta, 0) \xi\, d\xi\, d\eta$$
$$+ \int_0^R \int_0^{2\pi} \int_0^\infty \Phi(\xi, \eta, \zeta) G(r, \varphi, z, \xi, \eta, \zeta) \xi\, d\zeta\, d\eta\, d\xi.$$

Green's function:

$$G(r, \varphi, z, \xi, \eta, \zeta) = \frac{1}{\pi} \sum_{n=0}^\infty \sum_{m=1}^\infty \frac{A_n \mu_{nm}^2 J_n(\mu_{nm} r) J_n(\mu_{nm} \xi) \cos[n(\varphi-\eta)]}{(\mu_{nm}^2 R^2 + k_1^2 R^2 - n^2) J_n^2(\mu_{nm} R)} F_{nm}(z, \zeta),$$

$$A_n = \begin{cases} 1 & \text{for } n = 0, \\ 2 & \text{for } n \ne 0, \end{cases} \quad F_{nm}(z, \zeta) = \begin{cases} \dfrac{\exp(-\mu_{nm} z)[\mu_{nm} \cosh(\mu_{nm} \zeta) + k_2 \sinh(\mu_{nm} \zeta)]}{\mu_{nm}(\mu_{nm} + k_2)} & \text{for } z > \zeta, \\ \dfrac{\exp(-\mu_{nm} \zeta)[\mu_{nm} \cosh(\mu_{nm} z) + k_2 \sinh(\mu_{nm} z)]}{\mu_{nm}(\mu_{nm} + k_2)} & \text{for } \zeta > z, \end{cases}$$

where the $J_n(\xi)$ are the Bessel functions and the μ_{nm} are positive roots of the transcendental equation

$$\mu J_n'(\mu R) + k_1 J_n(\mu R) = 0.$$

8.2.3-5. Domain: $0 \le r \le R$, $0 \le \varphi \le 2\pi$, $0 \le z < \infty$. Mixed boundary value problem.

A semiinfinite circular cylinder is considered. Boundary conditions are prescribed:

$$w = f_1(\varphi, z) \quad \text{at} \quad r = R, \qquad \partial_z w = f_2(r, \varphi) \quad \text{at} \quad z = 0.$$

Solution:

$$w(r, \varphi, z) = -R \int_0^{2\pi} \int_0^\infty f_1(\eta, \zeta) \left[\frac{\partial}{\partial \xi} G(r, \varphi, z, \xi, \eta, \zeta) \right]_{\xi=R} d\zeta\, d\eta$$
$$- \int_0^{2\pi} \int_0^R f_2(\xi, \eta) G(r, \varphi, z, \xi, \eta, 0) \xi\, d\xi\, d\eta$$
$$+ \int_0^R \int_0^{2\pi} \int_0^\infty \Phi(\xi, \eta, \zeta) G(r, \varphi, z, \xi, \eta, \zeta) \xi\, d\zeta\, d\eta\, d\xi.$$

Green's function:

$$G(r, \varphi, z, \xi, \eta, \zeta) = \frac{1}{2\pi R^2} \sum_{n=0}^\infty \sum_{m=1}^\infty \frac{A_n J_n(\mu_{nm} r) J_n(\mu_{nm} \xi)}{[J_n'(\mu_{nm} R)]^2 \mu_{nm}} \cos[n(\varphi - \eta)] F_{nm}(z, \zeta),$$

$$F_{nm}(z, \zeta) = \exp(-\mu_{nm}|z - \zeta|) + \exp(-\mu_{nm}|z + \zeta|), \qquad A_n = \begin{cases} 1 & \text{for } n = 0, \\ 2 & \text{for } n \ne 0, \end{cases}$$

where the $J_n(\xi)$ are the Bessel functions and the μ_{nm} are roots of the transcendental equation $J_n(\mu R) = 0$.

▶ *Paragraphs 8.2.3-6 through 8.3.3-10 present only Green's functions; the complete solution is constructed with the formulas given in Subsection 8.2.1. See also Paragraphs 8.3.1-4 and 8.3.1-8 for $\lambda = 0$.*

8.2.3-6. Domain: $0 \leq r \leq R$, $0 \leq \varphi \leq 2\pi$, $0 \leq z \leq a$. **First boundary value problem.**

A circular cylinder of finite length is considered. Boundary conditions are prescribed:

$$w = f_1(\varphi, z) \quad \text{at} \quad r = R, \qquad w = f_2(r, \varphi) \quad \text{at} \quad z = 0, \qquad w = f_3(r, \varphi) \quad \text{at} \quad z = a.$$

A double series representation of the Green's function:

$$G(r, \varphi, z, \xi, \eta, \zeta) = \frac{1}{\pi R^2} \sum_{n=0}^{\infty} \sum_{m=1}^{\infty} \frac{A_n J_n(\mu_{nm} r) J_n(\mu_{nm} \xi)}{[J_n'(\mu_{nm} R)]^2 \mu_{nm} \sinh(\mu_{nm} a)} \cos[n(\varphi - \eta)] F_{nm}(z, \zeta),$$

$$F_{nm}(z, \zeta) = \begin{cases} \sinh(\mu_{nm} \zeta) \sinh[\mu_{nm}(a - z)] & \text{for } a \geq z > \zeta \geq 0, \\ \sinh(\mu_{nm} z) \sinh[\mu_{nm}(a - \zeta)] & \text{for } a \geq \zeta > z \geq 0, \end{cases} \qquad A_n = \begin{cases} 1 & \text{for } n = 0, \\ 2 & \text{for } n \neq 0, \end{cases}$$

where the $J_n(\xi)$ are the Bessel functions (the prime denotes the derivative with respect to the argument) and the μ_{nm} are positive roots of the transcendental equation $J_n(\mu R) = 0$.

A triple series representation of the Green's function:

$$G(r, \varphi, z, \xi, \eta, \zeta) = \frac{2a}{\pi R^2} \sum_{n=0}^{\infty} \sum_{m=1}^{\infty} \sum_{k=1}^{\infty} \frac{A_n}{[J_n'(\mu_{nm} R)]^2 [(a\mu_{nm})^2 + (\pi k)^2]} J_n(\mu_{nm} r) J_n(\mu_{nm} \xi)$$
$$\times \cos[n(\varphi - \eta)] \sin\left(\frac{k\pi z}{a}\right) \sin\left(\frac{k\pi \zeta}{a}\right).$$

8.2.3-7. Domain: $0 \leq r \leq R$, $0 \leq \varphi \leq 2\pi$, $0 \leq z \leq a$. **Third boundary value problem.**

A circular cylinder of finite length is considered. Boundary conditions are prescribed:

$$\partial_r w + k_1 w = f_1(\varphi, z) \quad \text{at} \quad r = R,$$
$$\partial_z w - k_2 w = f_2(r, \varphi) \quad \text{at} \quad z = 0,$$
$$\partial_z w + k_3 w = f_3(r, \varphi) \quad \text{at} \quad z = a.$$

Green's function:

$$G(r, \varphi, z, \xi, \eta, \zeta) = \frac{1}{\pi} \sum_{n=0}^{\infty} \sum_{m=1}^{\infty} \sum_{s=1}^{\infty} \frac{A_n \mu_{nm}^2 J_n(\mu_{nm} r) J_n(\mu_{nm} \xi) \cos[n(\varphi - \eta)] h_s(z) h_s(\zeta)}{(\mu_{nm}^2 R^2 + k_1^2 R^2 - n^2)(\mu_{nm}^2 + \lambda_s^2)[J_n(\mu_{nm} R)]^2 \|h_s\|^2},$$

$$h_s(z) = \cos(\lambda_s z) + \frac{k_2}{\lambda_s} \sin(\lambda_s z), \qquad \|h_s\|^2 = \frac{k_3}{2\lambda_s^2} \frac{\lambda_s^2 + k_2^2}{\lambda_s^2 + k_3^2} + \frac{k_2}{2\lambda_s^2} + \frac{a}{2}\left(1 + \frac{k_2^2}{\lambda_s^2}\right).$$

Here, $A_0 = 1$ and $A_n = 2$ for $n \neq 0$; the $J_n(\xi)$ are the Bessel functions; and the μ_{nm} and λ_s are positive roots of the transcendental equations

$$\mu J_n'(\mu R) + k_1 J_n(\mu R) = 0, \qquad \frac{\tan(\lambda a)}{\lambda} = \frac{k_2 + k_3}{\lambda^2 - k_2 k_3}.$$

8.2.3-8. Domain: $0 \leq r \leq R$, $0 \leq \varphi \leq 2\pi$, $0 \leq z \leq a$. **Mixed boundary value problem.**

A circular cylinder of finite length is considered. Boundary conditions are prescribed:

$$w = f_1(\varphi, z) \quad \text{at} \quad r = R, \qquad \partial_z w = f_2(r, \varphi) \quad \text{at} \quad z = 0, \qquad \partial_z w = f_3(r, \varphi) \quad \text{at} \quad z = a.$$

Green's function:

$$G(r, \varphi, z, \xi, \eta, \zeta) = \frac{a}{\pi R^2} \sum_{n=0}^{\infty} \sum_{m=1}^{\infty} \sum_{k=1}^{\infty} \frac{A_n A_k}{[J_n'(\mu_{nm} R)]^2 [(a\mu_{nm})^2 + (\pi k)^2]} J_n(\mu_{nm} r) J_n(\mu_{nm} \xi)$$
$$\times \cos[n(\varphi - \eta)] \cos\left(\frac{k\pi z}{a}\right) \cos\left(\frac{k\pi \zeta}{a}\right),$$

where $A_0 = 1$ and $A_n = 2$ for $n \neq 0$; the $J_n(\xi)$ are the Bessel functions (the prime denotes the derivative with respect to the argument); and the μ_{nm} are positive roots of the transcendental equation $J_n(\mu R) = 0$.

8.2.3-9. Domain: $0 \leq r \leq R$, $0 \leq \varphi \leq \varphi_0$, $0 \leq z \leq a$. First boundary value problem.

A cylindrical sector of finite thickness is considered. Boundary conditions are prescribed:

$w = f_1(r, z)$ at $\varphi = 0$, $\quad w = f_2(r, z)$ at $\varphi = \varphi_0$, $\quad w = f_3(\varphi, z)$ at $r = R$,
$w = f_4(r, \varphi)$ at $z = 0$, $\quad w = f_5(r, \varphi)$ at $z = a$.

Green's function:

$$G(r, \varphi, z, \xi, \eta, \zeta) = \frac{8a}{R^2 \varphi_0} \sum_{n=1}^{\infty} \sum_{m=1}^{\infty} \sum_{k=1}^{\infty} \frac{J_{n\pi/\varphi_0}(\mu_{nm} r) J_{n\pi/\varphi_0}(\mu_{nm} \xi)}{[J'_{n\pi/\varphi_0}(\mu_{nm} R)]^2 [(a\mu_{nm})^2 + (\pi k)^2]}$$
$$\times \sin\left(\frac{n\pi\varphi}{\varphi_0}\right) \sin\left(\frac{n\pi\eta}{\varphi_0}\right) \sin\left(\frac{k\pi z}{a}\right) \sin\left(\frac{k\pi \zeta}{a}\right),$$

where the $J_{n\pi/\varphi_0}(r)$ are the Bessel functions and the μ_{nm} are positive roots of the transcendental equation $J_{n\pi/\varphi_0}(\mu R) = 0$.

8.2.3-10. Domain: $0 \leq r \leq R$, $0 \leq \varphi \leq \varphi_0$, $0 \leq z \leq a$. Mixed boundary value problem.

A cylindrical sector of finite thickness is considered. Boundary conditions are prescribed:

$w = f_1(r, z)$ at $\varphi = 0$, $\quad w = f_2(r, z)$ at $\varphi = \varphi_0$, $\quad w = f_3(\varphi, z)$ at $r = R$,
$\partial_z w = f_4(r, \varphi)$ at $z = 0$, $\quad \partial_z w = f_5(r, \varphi)$ at $z = a$.

Green's function:

$$G(r, \varphi, z, \xi, \eta, \zeta) = \frac{4a}{R^2 \varphi_0} \sum_{n=1}^{\infty} \sum_{m=1}^{\infty} \sum_{k=0}^{\infty} \frac{A_k J_{n\pi/\varphi_0}(\mu_{nm} r) J_{n\pi/\varphi_0}(\mu_{nm} \xi)}{[J'_{n\pi/\varphi_0}(\mu_{nm} R)]^2 [(a\mu_{nm})^2 + (\pi k)^2]}$$
$$\times \sin\left(\frac{n\pi\varphi}{\varphi_0}\right) \sin\left(\frac{n\pi\eta}{\varphi_0}\right) \cos\left(\frac{k\pi z}{a}\right) \cos\left(\frac{k\pi \zeta}{a}\right),$$

where $A_0 = 1$ and $A_k = 2$ for $k \neq 0$; the $J_{n\pi/\varphi_0}(r)$ are the Bessel functions; and the μ_{nm} are positive roots of the transcendental equation $J_{n\pi/\varphi_0}(\mu R) = 0$.

8.2.4. Problems in Spherical Coordinates

The three-dimensional Poisson equation in the spherical coordinate system is written as

$$\frac{1}{r^2} \frac{\partial}{\partial r}\left(r^2 \frac{\partial w}{\partial r}\right) + \frac{1}{r^2 \sin\theta} \frac{\partial}{\partial \theta}\left(\sin\theta \frac{\partial w}{\partial \theta}\right) + \frac{1}{r^2 \sin^2\theta} \frac{\partial^2 w}{\partial \varphi^2} = -\Phi(r, \theta, \varphi), \qquad r = \sqrt{x^2 + y^2 + z^2}.$$

▶ Only Green's functions are presented below; the complete solutions can be constructed with the formulas given in Subsection 8.2.1.

8.2.4-1. Domain: $0 \leq r \leq R$, $0 \leq \theta \leq \pi$, $0 \leq \varphi \leq 2\pi$. First boundary value problem.

A spherical domain is considered. A boundary condition is prescribed:

$$w = f(\varphi, \theta) \quad \text{at} \quad r = R.$$

Green's function:

$$G(r, \theta, \varphi, \xi, \eta, \zeta) = \frac{1}{4\pi \sqrt{r^2 - 2r\xi \cos\gamma + \xi^2}} - \frac{1}{4\pi \sqrt{r^2 \xi^2 - 2R^2 r\xi \cos\gamma + R^4}},$$
$$\cos\gamma = \cos\theta \cos\eta + \sin\theta \sin\eta \cos(\varphi - \zeta).$$

An alternative representation of the Green's function:

$$G(\mathbf{r}, \mathbf{r}_0) = \frac{1}{4\pi} \frac{1}{|\mathbf{r} - \mathbf{r}_0|} - \frac{1}{4\pi} \frac{R}{r_0 |(R/r_0)^2 \mathbf{r}_0 - \mathbf{r}|}, \qquad r_0 = |\mathbf{r}_0|,$$

where

$$\mathbf{r} = \{x, y, z\}, \qquad x = r \sin\theta \cos\varphi, \qquad y = r \sin\theta \sin\varphi, \qquad z = r \cos\theta$$

$$\mathbf{r}_0 = \{x_0, y_0, z_0\}, \qquad x_0 = \xi \sin\eta \cos\zeta, \qquad y_0 = \xi \sin\eta \sin\zeta, \qquad z_0 = \xi \cos\eta.$$

⊙ *References*: V. M. Babich, M. B. Kapilevich, S. G. Mikhlin, et al. (1964), B. M. Budak, A. A. Samarskii, and A. N. Tikhonov (1980).

8.2.4-2. Domain: $0 \le r \le R$, $0 \le \theta \le \pi$, $0 \le \varphi \le 2\pi$. Second boundary value problem.

A spherical domain is considered. A boundary condition is prescribed:

$$\partial_r w = f(\varphi, \theta) \quad \text{at} \quad r = R.$$

Green's function:

$$G(r, \theta, \varphi, \xi, \eta, \zeta) = \frac{1}{4\pi} \left\{ \frac{1}{|\mathbf{r} - \mathbf{r}_0|} + \frac{R}{|\mathbf{r}_0||\mathbf{r}_1|} + \frac{1}{R} \ln \frac{2R^2}{R^2 + |\mathbf{r}_0||\mathbf{r}_1| - (\mathbf{r} \cdot \mathbf{r}_0)} \right\},$$

where

$$|\mathbf{r} - \mathbf{r}_0| = \sqrt{r^2 - 2r\xi \cos\gamma + \xi^2}, \qquad |\mathbf{r}_0||\mathbf{r}_1| = \sqrt{r^2 \xi^2 - 2R^2 r\xi \cos\gamma + R^4},$$

$$|\mathbf{r}_0| = \xi, \qquad (\mathbf{r} \cdot \mathbf{r}_0) = r\xi \cos\gamma, \qquad \cos\gamma = \cos\theta \cos\eta + \sin\theta \sin\eta \cos(\varphi - \zeta).$$

For a solution of the second boundary value problem to exist the solvability condition must be satisfied (see Paragraph 8.2.1-2).

⊙ *Reference*: N. S. Koshlyakov, E. B. Gliner, and M. M. Smirnov (1970).

8.2.4-3. Domain: $0 \le r \le R$, $0 \le \theta \le \pi$, $0 \le \varphi \le 2\pi$. Third boundary value problem.

A spherical domain is considered. A boundary condition is prescribed:

$$\partial_r w + kw = f(\theta, \varphi) \quad \text{at} \quad r = R.$$

Green's function:

$$G(r, \theta, \varphi, \xi, \eta, \zeta) = \frac{1}{2\pi \sqrt{r\xi}} \sum_{n=0}^{\infty} \sum_{m=1}^{\infty} \sum_{s=0}^{n} A_s B_{nms} J_{n+1/2}(\lambda_{nm} r) J_{n+1/2}(\lambda_{nm} \xi)$$

$$\times \bar{P}_n^s(\cos\theta) \bar{P}_n^s(\cos\eta) \cos[s(\varphi - \zeta)],$$

$$A_s = \begin{cases} 1 & \text{for } s = 0, \\ 2 & \text{for } s \ne 0, \end{cases} \qquad B_{nms} = \frac{(2n+1)(n-s)!}{(n+s)! \left[R^2 \lambda_{nm}^2 + (kR+n)(kR-n-1) \right] \left[J_{n+1/2}(\lambda_{nm} R) \right]^2}.$$

Here, the $J_{n+1/2}(r)$ are the Bessel functions, the $P_n^s(\mu)$ are the associated Legendre functions that are expressed in terms of the Legendre polynomials $P_n(\mu)$ as

$$P_n^s(\mu) = (1 - \mu^2)^{s/2} \frac{d^s}{d\mu^s} P_n(\mu), \qquad P_n(\mu) = \frac{1}{n! \, 2^n} \frac{d^n}{d\mu^n} (\mu^2 - 1)^n,$$

and the λ_{nm} are positive roots of the transcendental equation

$$\lambda R J'_{n+1/2}(\lambda R) + \left(kR - \tfrac{1}{2} \right) J_{n+1/2}(\lambda R) = 0.$$

8.2.4-4. Domain: $R \leq r < \infty$, $0 \leq \theta \leq \pi$, $0 \leq \varphi \leq 2\pi$. First boundary value problem.

Three-dimensional space with a spherical cavity is considered. A boundary condition is prescribed:
$$w = f(\varphi, \theta) \quad \text{at} \quad r = R.$$

The Green's function of the outer first boundary value problem is given by the same relation as that for the inner first boundary value problem (see Paragraph 8.2.4-1), except that $r \geq R$ and $\xi \geq R$.

8.2.4-5. Domain: $R \leq r < \infty$, $0 \leq \theta \leq \pi$, $0 \leq \varphi \leq 2\pi$. Second boundary value problem.

Three-dimensional space with a spherical cavity is considered. A boundary condition is prescribed:
$$\partial_r w = f(\varphi, \theta) \quad \text{at} \quad r = R.$$

Green's function:
$$G(r, \theta, \varphi, \xi, \eta, \zeta) = \frac{1}{4\pi} \left\{ \frac{1}{|\mathbf{r} - \mathbf{r}_0|} + \frac{R}{|\mathbf{r}_0||\mathbf{r}_1|} + \frac{1}{R} \ln \frac{(1 - \cos\gamma)|\mathbf{r}||\mathbf{r}_0|}{R^2 + |\mathbf{r}_0||\mathbf{r}_1| - (\mathbf{r} \cdot \mathbf{r}_0)} \right\},$$

where
$$|\mathbf{r}| = r, \quad |\mathbf{r}_0| = \xi, \quad |\mathbf{r} - \mathbf{r}_0| = \sqrt{r^2 - 2r\xi\cos\gamma + \xi^2}, \quad |\mathbf{r}_0||\mathbf{r}_1| = \sqrt{r^2\xi^2 - 2R^2 r\xi \cos\gamma + R^4},$$
$$(\mathbf{r} \cdot \mathbf{r}_0) = r\xi\cos\gamma, \quad \cos\gamma = \cos\theta\cos\eta + \sin\theta\sin\eta\cos(\varphi - \zeta).$$

⊙ *Reference*: N. S. Koshlyakov, E. B. Gliner, and M. M. Smirnov (1970).

8.2.4-6. Domain: $R_1 \leq r \leq R_2$, $0 \leq \theta \leq \pi$, $0 \leq \varphi \leq 2\pi$. First boundary value problem.

A spherical layer is considered. Boundary conditions are prescribed:
$$w = f_1(\theta, \varphi) \quad \text{at} \quad r = R_1, \qquad w = f_2(\theta, \varphi) \quad \text{at} \quad r = R_2.$$

Green's function:
$$G(r, \theta, \varphi, \xi, \eta, \zeta) = \frac{\pi}{8\sqrt{r\xi}} \sum_{n=0}^{\infty} \sum_{m=1}^{\infty} \sum_{k=0}^{n} A_k B_{nmk} Z_{n+1/2}(\lambda_{nm} r) Z_{n+1/2}(\lambda_{nm} \xi)$$
$$\times P_n^k(\cos\theta) P_n^k(\cos\eta) \cos[k(\varphi - \zeta)],$$

where
$$Z_{n+1/2}(\lambda_{nm} r) = J_{n+1/2}(\lambda_{nm} R_1) Y_{n+1/2}(\lambda_{nm} r) - Y_{n+1/2}(\lambda_{nm} R_1) J_{n+1/2}(\lambda_{nm} r),$$
$$A_k = \begin{cases} 1 & \text{for } k = 0, \\ 2 & \text{for } k \neq 0, \end{cases} \qquad B_{nmk} = \frac{(2n+1)(n-k)! J_{n+1/2}^2(\lambda_{nm} R_2)}{(n+k)! [J_{n+1/2}^2(\lambda_{nm} R_1) - J_{n+1/2}^2(\lambda_{nm} R_2)]};$$

the $J_{n+1/2}(r)$ are the Bessel functions, the $P_n^k(\mu)$ are the associated Legendre functions (see Paragraph 8.2.4-3), and the λ_{nm} are positive roots of the transcendental equation $Z_{n+1/2}(\lambda R_2) = 0$.

8.2.4-7. Domain: $0 \leq r \leq R$, $0 \leq \theta \leq \pi/2$, $0 \leq \varphi \leq 2\pi$. First boundary value problem.

A hemisphere is considered. Boundary conditions are prescribed:
$$w = f_1(\varphi, \theta) \quad \text{at} \quad r = R, \qquad w = f_2(r, \varphi) \quad \text{at} \quad \theta = \pi/2.$$

Green's function in the spherical coordinate system:
$$G(r, \theta, \varphi, \xi, \eta, \zeta) = G_s(r, \theta, \varphi, \xi, \eta, \zeta) - G_s(r, \theta, \varphi, \xi, \pi - \eta, \zeta),$$
where $G_s(r, \theta, \varphi, \xi, \eta, \zeta)$ is the Green's functions for a sphere; see Paragraph 8.2.4-1, where G must be replaced by G_s.

Green's function in the Cartesian coordinate system:
$$G(x, y, z, x_0, y_0, z_0) = \frac{1}{4\pi}\left(\frac{1}{|\mathbf{r} - \mathbf{r}_0|} - \frac{R}{|\mathbf{r}_0||\mathbf{r} - \mathbf{r}_0^*|} \right) - \frac{1}{4\pi}\left(\frac{1}{|\mathbf{r} - \mathbf{r}_1|} - \frac{R}{|\mathbf{r}_0||\mathbf{r} - \mathbf{r}_1^*|} \right),$$
$$\mathbf{r} = \{x, y, z\}, \quad \mathbf{r}_0 = \{x_0, y_0, z_0\}, \quad \mathbf{r}_1 = \{x_0, y_0, -z_0\}, \quad \mathbf{r}_k^* = (R/r_0)^2 \mathbf{r}_k, \quad k = 0, 1.$$

⊙ *References*: V. S. Vladimirov, V. P. Mikhailov, A. A. Vasharin, et al. (1974), B. M. Budak, A. A. Samarskii, and A. N. Tikhonov (1980).

8.2.4-8. Domain: $0 \leq r \leq R$, $0 \leq \theta \leq \pi/2$, $0 \leq \varphi \leq \pi$. **First boundary value problem.**

A quarter of a sphere is considered. Boundary conditions are prescribed:

$$w = f_1(\varphi, \theta) \quad \text{at} \quad r = R, \qquad w = f_2(r, \varphi) \quad \text{at} \quad \theta = \pi/2,$$
$$w = f_3(r, \theta) \quad \text{at} \quad \varphi = 0, \qquad w = f_4(r, \theta) \quad \text{at} \quad \varphi = \pi.$$

Green's function in the spherical coordinate system:

$$G(r, \theta, \varphi, \xi, \eta, \zeta) = G_s(r, \theta, \varphi, \xi, \eta, \zeta) - G_s(r, \theta, \varphi, \xi, \pi - \eta, \zeta)$$
$$+ G_s(r, \theta, \varphi, \xi, \pi - \eta, 2\pi - \zeta) - G_s(r, \theta, \varphi, \xi, \eta, 2\pi - \zeta),$$

where $G_s(r, \theta, \varphi, \xi, \eta, \zeta)$ is the Green's function for a sphere; see Paragraph 8.2.4-1, where G must be replaced by G_s.

Green's function in the Cartesian coordinate system:

$$G(x, y, z, x_0, y_0, z_0) = \frac{1}{4\pi} \sum_{n,k=0}^{1} (-1)^{n+k} \left(\frac{1}{|\mathbf{r} - \mathbf{r}_{nk}|} - \frac{R}{|\mathbf{r}_0| |\mathbf{r} - \mathbf{r}^*_{nk}|} \right),$$

$$\mathbf{r} = \{x, y, z\}, \quad \mathbf{r}_0 = \{x_0, y_0, z_0\}, \quad \mathbf{r}_{nk} = \{x_0, (-1)^n y_0, (-1)^k z_0\}, \quad \mathbf{r}^*_{nk} = (R/r_0)^2 \mathbf{r}_{nk},$$

where $r_0 = |\mathbf{r}_0|$; $n = 0, 1$; $k = 0, 1$.

⊙ *References*: V. S. Vladimirov, V. P. Mikhailov, A. A. Vasharin, et al. (1974), B. M. Budak, A. A. Samarskii, and A. N. Tikhonov (1980).

8.3. Helmholtz Equation $\Delta_3 w + \lambda w = -\Phi(\mathbf{x})$

A variety of problems related to steady-state oscillations (mechanical, acoustic, thermal, electromagnetic, etc.) lead to the three-dimensional Helmholtz equation with $\lambda > 0$. This equation governs mass transfer phenomena with volume chemical reaction of the first order for $\lambda < 0$. Any elliptic equation with constant coefficients can be reduced to the Helmholtz equation.

8.3.1. General Remarks, Results, and Formulas

8.3.1-1. Some definitions.

The Helmholtz equation is called homogeneous if $\Phi = 0$ and nonhomogeneous if $\Phi \neq 0$. A homogeneous boundary value problem is a boundary value problem for a homogeneous equation with homogeneous boundary conditions; $w = 0$ is a particular solution of a homogeneous boundary value problem.

The values λ_n of the parameter λ for which there are nontrivial solutions (i.e., not identically zero solutions) of a homogeneous boundary value problem are called eigenvalues. The corresponding solutions, $w = w_n$, are called eigenfunctions of this boundary value problem.

In what follows, we consider simultaneously the first, second, and third boundary value problems for the three-dimensional Helmholtz equation in a finite three-dimensional domain V with a sufficiently smooth surface S. It is assumed that $k > 0$ for the third boundary value problem with the boundary condition

$$\frac{\partial w}{\partial N} + kw = 0 \quad \text{for} \quad \mathbf{r} \in S,$$

where $\frac{\partial w}{\partial N}$ is the derivative along the outward normal to the surface S, and $\mathbf{r} = \{x, y, z\}$.

8.3.1-2. Properties of eigenvalues and eigenfunctions.

$1°$. There are infinitely many eigenvalues $\{\lambda_n\}$; they form a discrete spectrum of the boundary value problem.

$2°$. All eigenvalues are positive, except for one eigenvalue $\lambda_0 = 0$ of the second boundary value problem (the corresponding eigenfunction is $w_0 = \text{const}$). The eigenvalues are assumed to be ordered so that $\lambda_1 < \lambda_2 < \lambda_3 < \cdots$.

$3°$. The eigenvalues tend to infinity as the number n increases. The following asymptotic estimate holds:

$$\lim_{n\to\infty} \frac{n}{\lambda_n^{3/2}} = \frac{V_3}{6\pi^2},$$

where V_3 is the volume of the domain under consideration.

$4°$. The eigenfunctions are defined up to a constant multiplier. Any two eigenfunctions, w_n and w_m, that correspond to different eigenvalues $\lambda_n \neq \lambda_m$ are orthogonal, that is,

$$\int_V w_n w_m \, dV = 0.$$

$5°$. Any twice continuously differentiable function $f = f(\mathbf{r})$ that satisfies the boundary conditions of a boundary value problem can be expanded into a uniformly convergent series in the eigenfunctions of this boundary value problem, specifically,

$$f = \sum_{n=1}^{\infty} a_n w_n, \qquad \text{where} \quad a_n = \frac{1}{\|w_n\|^2} \int_V f w_n \, dV, \quad \|w_n\|^2 = \int_V w_n^2 \, dV.$$

If f is square summable, then the series is convergent in mean.

$6°$. The eigenvalues of the first boundary value problem do not increase if the domain is extended.

Remark 1. In a three-dimensional problem, to each eigenvalue λ_n finitely many linearly independent eigenfunctions $w_n^{(1)}, \ldots, w_n^{(p)}$ generally correspond. These functions can always be replaced by their linear combinations

$$\bar{w}_n^{(s)} = c_{s,1} w_n^{(1)} + \cdots + c_{s,s-1} w_n^{(s-1)} + w_n^{(s)}, \qquad s = 1, 2, \ldots, p,$$

such that $\bar{w}_n^{(1)}, \ldots, \bar{w}_n^{(p)}$ are now pairwise orthogonal. Therefore, without loss of generality, we assume that all eigenfunctions are orthogonal.

⊙ *Reference*: V. M. Babich, M. B. Kapilevich, S. G. Mikhlin, et al. (1984).

8.3.1-3. Nonhomogeneous Helmholtz equation with homogeneous boundary conditions.

Three cases are possible.

$1°$. If λ is not equal to any one of the eigenvalues, then the solution of the problem is given by

$$w = \sum_{n=1}^{\infty} \frac{A_n}{\lambda_n - \lambda} w_n, \qquad \text{where} \quad A_n = \frac{1}{\|w_n\|^2} \int_V \Phi w_n \, dV, \quad \|w_n\|^2 = \int_V w_n^2 \, dV.$$

$2°$. If λ coincides with one of the eigenvalues, $\lambda = \lambda_m$, then the condition of the orthogonality of the function Φ to the eigenfunction w_m,

$$\int_V \Phi w_m \, dV = 0,$$

is a necessary condition for a solution of the nonhomogeneous problem to exist. The solution is then given by

$$w = \sum_{n=1}^{m-1} \frac{A_n}{\lambda_n - \lambda_m} w_n + \sum_{n=m+1}^{\infty} \frac{A_n}{\lambda_n - \lambda_m} w_n + C w_m, \qquad A_n = \frac{1}{\|w_n\|^2} \int_V \Phi w_n \, dV,$$

where C is an arbitrary constant and $\|w_n\|^2 = \int_V w_n^2 \, dV$.

3°. If $\lambda = \lambda_m$ and $\int_V \Phi w_m \, dV \neq 0$, then the boundary value problem for the nonhomogeneous equation has no solution.

Remark 2. If to each eigenvalue λ_n there are corresponding p_n mutually orthogonal eigenfunctions $w_n^{(s)}$ ($s = 1, \ldots, p_n$), then the solution is written as

$$w = \sum_{n=1}^{\infty} \sum_{s=1}^{p_n} \frac{A_n^{(s)}}{\lambda_n - \lambda} w_n^{(s)}, \quad \text{where} \quad A_n^{(s)} = \frac{1}{\|w_n^{(s)}\|^2} \int_V \Phi w_n^{(s)} \, dV, \quad \|w_n^{(s)}\|^2 = \int_V \left[w_n^{(s)}\right]^2 dV,$$

provided that $\lambda \neq \lambda_n$.

⊙ *Reference*: V. M. Babich, M. B. Kapilevich, S. G. Mikhlin, et al. (1984).

8.3.1-4. Solution of nonhomogeneous boundary value problems of general form.

1°. The solution of the first boundary value problem for the Helmholtz equation with the boundary condition

$$w = f(\mathbf{r}) \quad \text{for} \quad \mathbf{r} \in S$$

can be represented in the form

$$w(\mathbf{r}) = \int_V \Phi(\boldsymbol{\rho}) G(\mathbf{r}, \boldsymbol{\rho}) \, dV_\rho - \int_S f(\boldsymbol{\rho}) \frac{\partial}{\partial N_\rho} G(\mathbf{r}, \boldsymbol{\rho}) \, dS_\rho. \tag{1}$$

Here, $\mathbf{r} = \{x, y, z\}$, $\boldsymbol{\rho} = \{\xi, \eta, \zeta\}$ ($\mathbf{r} \in V$, $\boldsymbol{\rho} \in V$); $\frac{\partial}{\partial N_\rho}$ denotes the derivative along the outward normal to the surface S with respect to ξ, η, ζ. The Green's function is given by the series

$$G(\mathbf{r}, \boldsymbol{\rho}) = \sum_{n=1}^{\infty} \frac{w_n(\mathbf{r}) w_n(\boldsymbol{\rho})}{\|w_n\|^2 (\lambda_n - \lambda)}, \quad \lambda \neq \lambda_n, \tag{2}$$

where the w_n and λ_n are the eigenfunctions and eigenvalues of the homogeneous first boundary value problem.

2°. The solution of the second boundary value problem with the boundary condition

$$\frac{\partial w}{\partial N} = f(\mathbf{r}) \quad \text{for} \quad \mathbf{r} \in S$$

can be represented in the form

$$w(\mathbf{r}) = \int_V \Phi(\boldsymbol{\rho}) G(\mathbf{r}, \boldsymbol{\rho}) \, dV_\rho + \int_S f(\boldsymbol{\rho}) G(\mathbf{r}, \boldsymbol{\rho}) \, dS_\rho. \tag{3}$$

Here, the Green's function is given by the series

$$G(\mathbf{r}, \boldsymbol{\rho}) = -\frac{1}{V_3 \lambda} + \sum_{n=1}^{\infty} \frac{w_n(\mathbf{r}) w_n(\boldsymbol{\rho})}{\|w_n\|^2 (\lambda_n - \lambda)}, \tag{4}$$

where V_3 is the volume of the three-dimensional domain under consideration, and the λ_n and w_n are the positive eigenvalues and corresponding eigenfunctions of the homogeneous second boundary value problem. For clarity, the term corresponding to the zero eigenvalue $\lambda_0 = 0$ ($w_0 = \text{const}$) is singled out in (4). It is assumed that $\lambda \neq 0$ and $\lambda \neq \lambda_n$.

3°. The solution of the third boundary value problem for the Helmholtz equation with the boundary condition

$$\frac{\partial w}{\partial N} + kw = f(\mathbf{r}) \quad \text{for} \quad \mathbf{r} \in S$$

is given by relation (3) in which the Green's function is defined by series (2) with the eigenfunctions w_n and eigenvalues λ_n of the homogeneous third boundary value problem.

4°. Let nonhomogeneous boundary conditions of various types be set on different portions S_i of the surface $S = \sum_{i=1}^{m} S_i$,
$$\Gamma_i[w] = f_i(\mathbf{r}) \quad \text{for} \quad \mathbf{r} \in S_i.$$
Then the solution of the corresponding mixed boundary value problem can be written as
$$w(\mathbf{r}) = \int_V \Phi(\rho)G(\mathbf{r}, \rho)\, dV_\rho + \sum_{i=1}^{m} \int_{S_i} f_i(\rho)\Lambda_i(\mathbf{r}, \rho)\, dS_\rho^{(i)},$$
where
$$\Lambda_i(\mathbf{r}, \rho) = \begin{cases} -\dfrac{\partial}{\partial N_\rho} G(\mathbf{r}, \rho) & \text{if a first-kind boundary condition is set on } S_i, \\ G(\mathbf{r}, \rho) & \text{if a second- or third-kind boundary condition is set on } S_i. \end{cases}$$
The Green's function is expressed by series (2) that involves the eigenfunctions w_n and eigenvalues λ_n of the homogeneous mixed boundary value problem.

8.3.1-5. Boundary conditions at infinity in the case of an unbounded domain.

Below it is assumed that the function Φ is finite or sufficiently rapidly decaying as $r \to \infty$.

1°. If $\lambda < 0$ and the domain is unbounded, the additional condition that the solution must vanish at infinity is set:
$$w \to 0 \quad \text{as} \quad r \to \infty.$$

2°. If $\lambda > 0$, the radiation conditions (Sommerfeld conditions) are often used at infinity. In three-dimensional problems, these conditions are expressed as
$$\lim_{r \to \infty} rw = \text{const}, \quad \lim_{r \to \infty} r\left(\frac{\partial w}{\partial r} + i\sqrt{\lambda}\, w\right) = 0,$$
where $i^2 = -1$.

The principle of limit absorption and the principle of limit amplitude are also employed to separate a single solution.

⊙ *Reference*: A. N. Tikhonov and A. A. Samarskii (1990).

8.3.1-6. Green's function for an infinite cylindrical domain of arbitrary cross-section.

Consider the three-dimensional Helmholtz equation
$$\Delta_3 w + \lambda w = -\Phi(\mathbf{r}) \tag{5}$$
inside an infinite cylindrical domain $V = \{(x,y) \in D,\ -\infty < z < \infty\}$ with arbitrary cross-section D. On the surface of this domain, let $S = \{(x,y) \in L,\ -\infty < z < \infty\}$, where L is the boundary of D, the homogeneous boundary condition of general form
$$s\frac{\partial w}{\partial N} + kw = 0 \quad \text{for} \quad \mathbf{r} \in S \tag{6}$$
be set, with $sk \geq 0$. By appropriately choosing the constants s and k in (6), one can obtain boundary conditions of the first ($s=0$, $k=1$), second ($s=1$, $k=0$), and third ($sk \neq 0$) kind.

The Green's function of the first or third boundary value problem can be represented in the form*
$$G(x,y,z,\xi,\eta,\zeta) = \frac{1}{2} \sum_{n=1}^{\infty} \frac{u_n(x,y)u_n(\xi,\eta)}{\|u_n\|^2 \sqrt{\mu_n - \lambda}} e^{-\sqrt{\mu_n - \lambda}\,|z-\zeta|}, \quad \|u_n\|^2 = \int_D u_n^2(x,y)\, dx\, dy, \tag{7}$$

---

* In Paragraphs 8.3.1-6 through 8.3.1-8, the cross-section D is assumed to have finite dimensions.

where the μ_n and u_n are the eigenvalues and eigenfunctions of the corresponding two-dimensional boundary value problem in D,

$$\Delta_2 u + \mu u = 0 \quad \text{for } (x, y) \in D,$$
$$s\frac{\partial u}{\partial N} + ku = 0 \quad \text{for } (x, y) \in L. \tag{8}$$

Recall that all μ_n are positive.

In the second boundary value problem, the zero eigenvalue $\mu_0 = 0$ appears, and hence the summation in (7) must start with $n = 0$. In this case, $u_0 = 1$ and $\|u_0\|^2 = D_2$, where D_2 is the area of the cross-section D.

⊙ *References*: B. M. Budak, A. A. Samarskii, and A. N. Tikhonov (1980), A. N. Tikhonov and A. A. Samarskii (1990).

8.3.1-7. Green's function for a semiinfinite cylindrical domain.

1°. The Green's function of the three-dimensional first boundary value problem for equation (5) in a semiinfinite cylindrical domain $V = \{(x, y) \in D, \ 0 \le z < \infty\}$ with arbitrary cross-section D is given by

$$G(x, y, z, \xi, \eta, \zeta) = \sum_{n=1}^{\infty} \frac{u_n(x, y) u_n(\xi, \eta)}{\|u_n\|^2} H_n(z, \zeta), \tag{9}$$

where

$$H_n(z, \zeta) = \frac{1}{2\beta_n}\left[\exp(-\beta_n|z - \zeta|) - \exp(-\beta_n|z + \zeta|)\right]$$
$$= \begin{cases} \dfrac{1}{\beta_n}\exp(-\beta_n z)\sinh(\beta_n \zeta) & \text{for } z > \zeta \ge 0, \\ \dfrac{1}{\beta_n}\exp(-\beta_n \zeta)\sinh(\beta_n z) & \text{for } \zeta > z \ge 0, \end{cases} \quad \beta_n = \sqrt{\mu_n - \lambda}. \tag{10}$$

Relations (9) and (10) involve the eigenfunctions u_n and eigenvalues μ_n of the two-dimensional first boundary value problem (8) with $s = 0$ and $k = 1$.

2°. The Green's function of the three-dimensional second boundary value problem for equation (5) in a semiinfinite cylindrical domain $V = \{(x, y) \in D, \ 0 \le z < \infty\}$ with arbitrary cross-section D is given by

$$G(x, y, z, \xi, \eta, \zeta) = \frac{1}{D_2} H_0(z, \zeta) + \sum_{n=1}^{\infty} \frac{u_n(x, y) u_n(\xi, \eta)}{\|u_n\|^2} H_n(z, \zeta), \tag{11}$$

where

$$H_n(z, \zeta) = \frac{1}{2\beta_n}\left[\exp(-\beta_n|z - \zeta|) + \exp(-\beta_n|z + \zeta|)\right]$$
$$= \begin{cases} \dfrac{1}{\beta_n}\exp(-\beta_n z)\cosh(\beta_n \zeta) & \text{for } z > \zeta \ge 0, \\ \dfrac{1}{\beta_n}\exp(-\beta_n \zeta)\cosh(\beta_n z) & \text{for } \zeta > z \ge 0, \end{cases} \quad \beta_n = \sqrt{\mu_n - \lambda}. \tag{12}$$

Relations (11) and (12) involve the eigenfunctions u_n and eigenvalues μ_n of the two-dimensional second boundary value problem (8) with $s = 1$ and $k = 0$. Note that in (11) the term corresponding to the zero eigenvalue $\mu_0 = 0$ is specially singled out; D_2 is the area of the cross-section D.

3°. The Green's function of the three-dimensional third boundary value problem for equation (5) with the boundary conditions

$$\frac{\partial w}{\partial z} - k_1 w = 0 \quad \text{for} \quad z = 0, \qquad \frac{\partial w}{\partial N} + k_2 w = 0 \quad \text{for} \quad \mathbf{r} \in S$$

in a semiinfinite cylindrical domain $V = \{(x,y) \in D,\ 0 \leq z < \infty\}$ with arbitrary cross-section D and lateral surface S is given by relation (9) with

$$H_n(z,\zeta) = \begin{cases} \dfrac{\exp(-\beta_n z)\left[\beta_n \cosh(\beta_n \zeta) + k_1 \sinh(\beta_n \zeta)\right]}{\beta_n(\beta_n + k_1)} & \text{for } z > \zeta \geq 0, \\ \dfrac{\exp(-\beta_n \zeta)\left[\beta_n \cosh(\beta_n z) + k_1 \sinh(\beta_n z)\right]}{\beta_n(\beta_n + k_1)} & \text{for } \zeta > z \geq 0, \end{cases} \qquad \beta_n = \sqrt{\mu_n - \lambda}. \tag{13}$$

Relations (9) and (13) involve the eigenfunctions u_n and eigenvalues μ_n of the two-dimensional third boundary value problem (8) with $s = 1$ and $k = k_2$.

4°. The Green's function of the three-dimensional mixed boundary value problem for equation (5) with a second-kind boundary condition at the end face and a first-kind boundary condition at the lateral surface is given by relations (9) and (12), where the μ_n and u_n are the eigenvalues and eigenfunctions of the two-dimensional first boundary value problem (8) with $s = 0$ and $k = 1$.

The Green's functions of other mixed boundary value problems can be constructed likewise.

8.3.1-8. Green's function for a cylindrical domain of finite dimensions.

1°. The Green's function of the three-dimensional first boundary value problem for equation (5) in a cylindrical domain of finite dimensions $V = \{(x,y) \in D,\ 0 \leq z \leq a\}$ with arbitrary cross-section D is given by relation (9) with

$$H_n(z,\zeta) = \begin{cases} \dfrac{\sinh(\beta_n \zeta) \sinh[\beta_n(a-z)]}{\beta_n \sinh(\beta_n a)} & \text{for } a \geq z > \zeta \geq 0, \\ \dfrac{\sinh(\beta_n z) \sinh[\beta_n(a-\zeta)]}{\beta_n \sinh(\beta_n a)} & \text{for } a \geq \zeta > z \geq 0, \end{cases} \qquad \beta_n = \sqrt{\mu_n - \lambda}. \tag{14}$$

Relations (9) and (14) involve the eigenfunctions u_n and eigenvalues μ_n of the two-dimensional first boundary value problem (8) with $s = 0$ and $k = 1$.

Another representation of the Green's function:

$$G(x,y,z,\xi,\eta,\zeta) = \frac{2}{a} \sum_{n=1}^{\infty} \sum_{m=1}^{\infty} \frac{u_n(x,y) u_n(\xi,\eta) \sin(q_m z) \sin(q_m \zeta)}{\|u_n\|^2 (\mu_n + q_m^2 - \lambda)}, \qquad q_m = \frac{\pi m}{a}.$$

It is a consequence of formula (2).

2°. The Green's function of the three-dimensional second boundary value problem for equation (5) in a cylindrical domain of finite dimensions $V = \{(x,y) \in D,\ 0 \leq z \leq a\}$ with arbitrary cross-section D is given by relation (11) with

$$H_n(z,\zeta) = \begin{cases} \dfrac{\cosh(\beta_n \zeta) \cosh[\beta_n(a-z)]}{\beta_n \sinh(\beta_n a)} & \text{for } a \geq z > \zeta \geq 0, \\ \dfrac{\cosh(\beta_n z) \cosh[\beta_n(a-\zeta)]}{\beta_n \sinh(\beta_n a)} & \text{for } a \geq \zeta > z \geq 0, \end{cases} \qquad \beta_n = \sqrt{\mu_n - \lambda}. \tag{15}$$

Relations (11) and (15) involve the eigenfunctions u_n and eigenvalues μ_n of the two-dimensional second boundary value problem (8) with $s = 1$ and $k = 0$.

Another representation of the Green's function:

$$G(x,y,z,\xi,\eta,\zeta) = \frac{1}{a}\sum_{n=0}^{\infty}\sum_{m=0}^{\infty}\frac{\varepsilon_m u_n(x,y)u_n(\xi,\eta)\cos(q_m z)\cos(q_m \zeta)}{\|u_n\|^2(\mu_n + q_m^2 - \lambda)},$$

$$q_m = \frac{\pi m}{a}, \qquad \varepsilon_m = \begin{cases} 1 & \text{for } m=0, \\ 2 & \text{for } m\neq 0, \end{cases} \qquad \mu_0 = 0, \quad u_0 = 1.$$

It is a consequence of formula (4).

⊙ *Reference*: B. M. Budak, A. A. Samarskii, and A. N. Tikhonov (1980).

3°. The Green's function of the three-dimensional third boundary value problem for equation (5) with the boundary conditions

$$\frac{\partial w}{\partial z} - k_1 w = 0 \quad \text{at} \quad z = 0, \qquad \frac{\partial w}{\partial z} + k_2 w = 0 \quad \text{at} \quad z = a, \qquad \frac{\partial w}{\partial N} + k_3 w = 0 \quad \text{for} \quad \mathbf{r} \in S$$

in a cylindrical domain of finite dimensions $V = \{(x,y)\in D,\ 0\leq z\leq a\}$ with arbitrary cross-section D and lateral surface S is given by relation (9) with

$$H_n(z,\zeta) = \begin{cases} \dfrac{[\beta_n\cosh(\beta_n\zeta)+k_1\sinh(\beta_n\zeta)]\{\beta_n\cosh[\beta_n(a-z)]+k_2\sinh[\beta_n(a-z)]\}}{\beta_n[\beta_n(k_1+k_2)\cosh(\beta_n a)+(\beta_n^2+k_1 k_2)\sinh(\beta_n a)]} & \text{for } z>\zeta, \\[2mm] \dfrac{[\beta_n\cosh(\beta_n z)+k_1\sinh(\beta_n z)]\{\beta_n\cosh[\beta_n(a-\zeta)]+k_2\sinh[\beta_n(a-\zeta)]\}}{\beta_n[\beta_n(k_1+k_2)\cosh(\beta_n a)+(\beta_n^2+k_1 k_2)\sinh(\beta_n a)]} & \text{for } z<\zeta, \end{cases} \quad (16)$$

$$\beta_n = \sqrt{\mu_n - \lambda} \qquad (0\leq z\leq a,\ 0\leq \zeta\leq a).$$

Relations (9) and (16) involve the eigenfunctions u_n and eigenvalues μ_n of the two-dimensional third boundary value problem (8) with $s=1$ and $k=k_3$.

4°. The Green's function of the three-dimensional mixed boundary value problem for equation (5) with second-kind boundary conditions at the end faces and a first-kind boundary condition at the lateral surface is given by relations (9) and (15), where the μ_n and u_n are the eigenvalues and eigenfunctions of the two-dimensional first boundary value problem (8) with $s=0$ and $k=1$.

The Green's function of the three-dimensional mixed boundary value problem for equation (5) with the boundary conditions

$$w = 0 \quad \text{for} \quad z = 0, \qquad \partial_z w = 0 \quad \text{for} \quad z = a, \qquad w = 0 \quad \text{for} \quad \mathbf{r} \in S$$

in a cylindrical domain of finite dimensions $V = \{(x,y)\in D,\ 0\leq z\leq a\}$ with arbitrary cross-section D and lateral surface S is given by relation (9) with

$$H_n(z,\zeta) = \begin{cases} \dfrac{\sinh(\beta_n\zeta)\cosh[\beta_n(a-z)]}{\beta_n\cosh(\beta_n a)} & \text{for } a\geq z>\zeta\geq 0, \\[2mm] \dfrac{\sinh(\beta_n z)\cosh[\beta_n(a-\zeta)]}{\beta_n\cosh(\beta_n a)} & \text{for } a\geq \zeta>z\geq 0, \end{cases} \qquad \beta_n = \sqrt{\mu_n - \lambda}. \quad (17)$$

Relations (9) and (17) involve the eigenfunctions u_n and eigenvalues μ_n of the two-dimensional first boundary value problem (8) with $s=0$ and $k=1$.

The Green's functions of other mixed boundary value problems can be constructed likewise.

8.3.2. Problems in Cartesian Coordinates

The three-dimensional nonhomogeneous Helmholtz equation in the rectangular Cartesian system of coordinates has the form

$$\frac{\partial^2 w}{\partial x^2} + \frac{\partial^2 w}{\partial y^2} + \frac{\partial^2 w}{\partial z^2} + \lambda w = -\Phi(x,y,z).$$

8.3.2-1. Particular solutions of the homogeneous equation ($\Phi \equiv 0$):

$w = (A_1 \cos kx + A_2 \sin kx)(B_1 \cos my + B_2 \sin my)(C_1 z + C_2)$, $\quad \lambda = k^2 + m^2$;

$w = (A_1 \cos kx + A_2 \sin kx)(B_1 \cosh my + B_2 \sinh my)(C_1 z + C_2)$, $\quad \lambda = k^2 - m^2$;

$w = (A_1 \cos kx + A_2 \sin kx)(B_1 \cos my + B_2 \sin my)(C_1 \cos nz + C_2 \sin nz)$, $\quad \lambda = k^2 + m^2 + n^2$;

$w = (A_1 \cosh kx + A_2 \sinh kx)(B_1 \cos my + B_2 \sin my)(C_1 \cos nz + C_2 \sin nz)$, $\quad \lambda = -k^2 + m^2 + n^2$;

$w = (A_1 \cosh kx + A_2 \sinh kx)(B_1 \cosh my + B_2 \sinh my)(C_1 \cos nz + C_2 \sin nz)$, $\quad \lambda = -k^2 - m^2 + n^2$;

$w = (A_1 \cosh kx + A_2 \sinh kx)(B_1 \cosh my + B_2 \sinh my)(C_1 \cosh nz + C_2 \sinh nz)$, $\quad \lambda = -k^2 - m^2 - n^2$,

where A_1, A_2, B_1, B_2, C_1, and C_2 are arbitrary constants.

Fundamental solutions:

$$\mathscr{E}(x, y, z) = \frac{1}{4\pi r} \exp(-kr), \qquad \lambda = -k^2 < 0,$$

$$\mathscr{E}(x, y, z) = \frac{1}{4\pi r} \exp(\mp ikr), \qquad \lambda = k^2 > 0,$$

where $r = \sqrt{x^2 + y^2 + z^2}$, $k > 0$, $i^2 = -1$.

8.3.2-2. Domain: $-\infty < x < \infty$, $-\infty < y < \infty$, $-\infty < z < \infty$.

1°. Solution for $\lambda = -k^2 < 0$:

$$w(x, y, z) = \frac{1}{4\pi} \int_{-\infty}^{\infty} \int_{-\infty}^{\infty} \int_{-\infty}^{\infty} \Phi(\xi, \eta, \zeta) \frac{\exp\left[-k\sqrt{(x-\xi)^2 + (y-\eta)^2 + (z-\zeta)^2}\right]}{\sqrt{(x-\xi)^2 + (y-\eta)^2 + (z-\zeta)^2}} \, d\xi \, d\eta \, d\zeta.$$

2°. Solution for $\lambda = k^2 > 0$:

$$w(x, y, z) = \frac{1}{4\pi} \int_{-\infty}^{\infty} \int_{-\infty}^{\infty} \int_{-\infty}^{\infty} \Phi(\xi, \eta, \zeta) \frac{\exp\left[-ik\sqrt{(x-\xi)^2 + (y-\eta)^2 + (z-\zeta)^2}\right]}{\sqrt{(x-\xi)^2 + (y-\eta)^2 + (z-\zeta)^2}} \, d\xi \, d\eta \, d\zeta.$$

This solution was obtained taking into account the radiation condition at infinity (see Paragraph 8.3.1-5, Item 2°).

⊙ *Reference*: A. N. Tikhonov and A. A. Samarskii (1990).

8.3.2-3. Domain: $-\infty < x < \infty$, $-\infty < y < \infty$, $0 \leq z < \infty$. **First boundary value problem.**

A half-space is considered. A boundary condition is prescribed:

$$w = f(x, y) \quad \text{at} \quad z = 0.$$

Solution:

$$w(x, y, z) = \int_{-\infty}^{\infty} \int_{-\infty}^{\infty} f(\xi, \eta) \left[\frac{\partial}{\partial \zeta} G(x, y, z, \xi, \eta, \zeta) \right]_{\zeta = 0} d\xi \, d\eta$$
$$+ \int_{0}^{\infty} \int_{-\infty}^{\infty} \int_{-\infty}^{\infty} \Phi(\xi, \eta, \zeta) G(x, y, z, \xi, \eta, \zeta) \, d\xi \, d\eta \, d\zeta.$$

Green's function for $\lambda = -k^2 < 0$:

$$G(x, y, z, \xi, \eta, \zeta) = \frac{\exp(-k\mathcal{R}_1)}{4\pi \mathcal{R}_1} - \frac{\exp(-k\mathcal{R}_2)}{4\pi \mathcal{R}_2},$$

$$\mathcal{R}_1 = \sqrt{(x-\xi)^2 + (y-\eta)^2 + (z-\zeta)^2}, \quad \mathcal{R}_2 = \sqrt{(x-\xi)^2 + (y-\eta)^2 + (z+\zeta)^2}.$$

⊙ *Reference*: A. N. Tikhonov and A. A. Samarskii (1990).

8.3.2-4. Domain: $-\infty < x < \infty$, $-\infty < y < \infty$, $0 \leq z < \infty$. Second boundary value problem.

A half-space is considered. A boundary condition is prescribed:
$$\partial_z w = f(x,y) \quad \text{at} \quad z = 0.$$

Solution:
$$w(x,y,z) = -\int_{-\infty}^{\infty}\int_{-\infty}^{\infty} f(\xi,\eta) G(x,y,z,\xi,\eta,0)\, d\xi\, d\eta$$
$$+ \int_{0}^{\infty}\int_{-\infty}^{\infty}\int_{-\infty}^{\infty} \Phi(\xi,\eta,\zeta) G(x,y,z,\xi,\eta,\zeta)\, d\xi\, d\eta\, d\zeta.$$

Green's function for $\lambda = -k^2 < 0$:
$$G(x,y,z,\xi,\eta,\zeta) = \frac{\exp(-k\mathcal{R}_1)}{4\pi\mathcal{R}_1} + \frac{\exp(-k\mathcal{R}_2)}{4\pi\mathcal{R}_2},$$
$$\mathcal{R}_1 = \sqrt{(x-\xi)^2 + (y-\eta)^2 + (z-\zeta)^2}, \quad \mathcal{R}_2 = \sqrt{(x-\xi)^2 + (y-\eta)^2 + (z+\zeta)^2}.$$

⊙ *Reference*: B. M. Budak, A. A. Samarskii, and A. N. Tikhonov (1980).

8.3.2-5. Domain: $-\infty < x < \infty$, $0 \leq y < \infty$, $0 \leq z < \infty$. First boundary value problem.

A dihedral angle is considered. Boundary conditions are prescribed:
$$w = f_1(x,z) \quad \text{at} \quad y = 0, \qquad w = f_2(x,y) \quad \text{at} \quad z = 0.$$

Solution:
$$w(x,y,z) = \int_0^{\infty}\int_{-\infty}^{\infty} f_1(\xi,\zeta)\left[\frac{\partial}{\partial \eta} G(x,y,z,\xi,\eta,\zeta)\right]_{\eta=0} d\xi\, d\zeta$$
$$+ \int_0^{\infty}\int_{-\infty}^{\infty} f_2(\xi,\eta)\left[\frac{\partial}{\partial \zeta} G(x,y,z,\xi,\eta,\zeta)\right]_{\zeta=0} d\xi\, d\eta$$
$$+ \int_0^{\infty}\int_0^{\infty}\int_{-\infty}^{\infty} \Phi(\xi,\eta,\zeta) G(x,y,z,\xi,\eta,\zeta)\, d\xi\, d\eta\, d\zeta.$$

Green's function for $\lambda = -k^2 < 0$:
$$G(x,y,z,\xi,\eta,\zeta) = \frac{\exp(-k\mathcal{R}_1)}{4\pi\mathcal{R}_1} - \frac{\exp(-k\mathcal{R}_2)}{4\pi\mathcal{R}_2} - \frac{\exp(-k\mathcal{R}_3)}{4\pi\mathcal{R}_3} + \frac{\exp(-k\mathcal{R}_4)}{4\pi\mathcal{R}_4},$$
$$\mathcal{R}_1 = \sqrt{(x-\xi)^2 + (y-\eta)^2 + (z-\zeta)^2}, \quad \mathcal{R}_2 = \sqrt{(x-\xi)^2 + (y-\eta)^2 + (z+\zeta)^2},$$
$$\mathcal{R}_3 = \sqrt{(x-\xi)^2 + (y+\eta)^2 + (z-\zeta)^2}, \quad \mathcal{R}_4 = \sqrt{(x-\xi)^2 + (y+\eta)^2 + (z+\zeta)^2}.$$

8.3.2-6. Domain: $-\infty < x < \infty$, $0 \leq y < \infty$, $0 \leq z < \infty$. Second boundary value problem.

A dihedral angle is considered. Boundary conditions are prescribed:
$$\partial_y w = f_1(x,z) \quad \text{at} \quad y = 0, \qquad \partial_z w = f_2(x,y) \quad \text{at} \quad z = 0.$$

Solution:
$$w(x,y,z) = -\int_0^{\infty}\int_{-\infty}^{\infty} f_1(\xi,\zeta) G(x,y,z,\xi,0,\zeta)\, d\xi\, d\zeta$$
$$- \int_0^{\infty}\int_{-\infty}^{\infty} f_2(\xi,\eta) G(x,y,z,\xi,\eta,0)\, d\xi\, d\eta$$
$$+ \int_0^{\infty}\int_0^{\infty}\int_{-\infty}^{\infty} \Phi(\xi,\eta,\zeta) G(x,y,z,\xi,\eta,\zeta)\, d\xi\, d\eta\, d\zeta.$$

Green's function for $\lambda = -k^2 < 0$:
$$G(x,y,z,\xi,\eta,\zeta) = \frac{\exp(-k\mathcal{R}_1)}{4\pi\mathcal{R}_1} + \frac{\exp(-k\mathcal{R}_2)}{4\pi\mathcal{R}_2} + \frac{\exp(-k\mathcal{R}_3)}{4\pi\mathcal{R}_3} + \frac{\exp(-k\mathcal{R}_4)}{4\pi\mathcal{R}_4},$$
$$\mathcal{R}_1 = \sqrt{(x-\xi)^2 + (y-\eta)^2 + (z-\zeta)^2}, \quad \mathcal{R}_2 = \sqrt{(x-\xi)^2 + (y-\eta)^2 + (z+\zeta)^2},$$
$$\mathcal{R}_3 = \sqrt{(x-\xi)^2 + (y+\eta)^2 + (z-\zeta)^2}, \quad \mathcal{R}_4 = \sqrt{(x-\xi)^2 + (y+\eta)^2 + (z+\zeta)^2}.$$

8.3.2-7. Domain: $-\infty < x < \infty$, $-\infty < y < \infty$, $0 \leq z \leq a$. **First boundary value problem.**

An infinite layer is considered. Boundary conditions are prescribed:

$$w = f_1(x,y) \quad \text{at} \quad z = 0, \qquad w = f_2(x,y) \quad \text{at} \quad z = a.$$

Solution:

$$w(x,y,z) = \int_{-\infty}^{\infty}\int_{-\infty}^{\infty} f_1(\xi,\eta) \left[\frac{\partial}{\partial \zeta} G(x,y,z,\xi,\eta,\zeta)\right]_{\zeta=0} d\xi\, d\eta$$

$$- \int_{-\infty}^{\infty}\int_{-\infty}^{\infty} f_2(\xi,\eta) \left[\frac{\partial}{\partial \zeta} G(x,y,z,\xi,\eta,\zeta)\right]_{\zeta=a} d\xi\, d\eta$$

$$+ \int_0^a \int_{-\infty}^{\infty}\int_{-\infty}^{\infty} \Phi(\xi,\eta,\zeta) G(x,y,z,\xi,\eta,\zeta)\, d\xi\, d\eta\, d\zeta.$$

Green's function for $\lambda = -k^2 < 0$:

$$G(x,y,z,\xi,\eta,\zeta) = \sum_{n=-\infty}^{\infty} \left[\frac{\exp(-k\mathcal{R}_{n1})}{4\pi\mathcal{R}_{n1}} - \frac{\exp(-k\mathcal{R}_{2n})}{4\pi\mathcal{R}_{2n}}\right],$$

$$\mathcal{R}_{1n} = \sqrt{(x-\xi)^2 + (y-\eta)^2 + (z-\zeta-2na)^2},$$

$$\mathcal{R}_{2n} = \sqrt{(x-\xi)^2 + (y-\eta)^2 + (z+\zeta-2na)^2}.$$

8.3.2-8. Domain: $-\infty < x < \infty$, $-\infty < y < \infty$, $0 \leq z \leq a$. **Second boundary value problem.**

An infinite layer is considered. Boundary conditions are prescribed:

$$\partial_z w = f_1(x,y) \quad \text{at} \quad z = 0, \qquad \partial_z w = f_2(x,y) \quad \text{at} \quad z = a.$$

Solution:

$$w(x,y,z) = -\int_{-\infty}^{\infty}\int_{-\infty}^{\infty} f_1(\xi,\eta) G(x,y,z,\xi,\eta,0)\, d\xi\, d\eta$$

$$+ \int_{-\infty}^{\infty}\int_{-\infty}^{\infty} f_2(\xi,\eta) G(x,y,z,\xi,\eta,a)\, d\xi\, d\eta$$

$$+ \int_0^a \int_{-\infty}^{\infty}\int_{-\infty}^{\infty} \Phi(\xi,\eta,\zeta) G(x,y,z,\xi,\eta,\zeta)\, d\xi\, d\eta\, d\zeta.$$

Green's function for $\lambda = -k^2 < 0$:

$$G(x,y,z,\xi,\eta,\zeta) = \sum_{n=-\infty}^{\infty} \left[\frac{\exp(-k\mathcal{R}_{n1})}{4\pi\mathcal{R}_{n1}} + \frac{\exp(-k\mathcal{R}_{2n})}{4\pi\mathcal{R}_{2n}}\right],$$

$$\mathcal{R}_{1n} = \sqrt{(x-\xi)^2 + (y-\eta)^2 + (z-\zeta-2na)^2},$$

$$\mathcal{R}_{2n} = \sqrt{(x-\xi)^2 + (y-\eta)^2 + (z+\zeta-2na)^2}.$$

⊙ *Reference*: B. M. Budak, A. A. Samarskii, and A. N. Tikhonov (1980).

8.3.2-9. Domain: $0 \leq x \leq a$, $0 \leq y \leq b$, $-\infty < z < \infty$. **First boundary value problem.**

An infinite cylindrical domain of a rectangular cross-section is considered. Boundary conditions are prescribed:

$$w = f_1(y,z) \quad \text{at} \quad x = 0, \qquad w = f_2(y,z) \quad \text{at} \quad x = a,$$
$$w = f_3(x,z) \quad \text{at} \quad y = 0, \qquad w = f_4(x,z) \quad \text{at} \quad y = b.$$

8.3. HELMHOLTZ EQUATION $\Delta_3 w + \lambda w = -\Phi(x)$ 571

Solution:
$$w(x,y,z) = \int_0^b \int_{-\infty}^{\infty} f_1(\eta,\zeta) \left[\frac{\partial}{\partial \xi} G(x,y,z,\xi,\eta,\zeta)\right]_{\xi=0} d\zeta\, d\eta$$
$$- \int_0^b \int_{-\infty}^{\infty} f_2(\eta,\zeta) \left[\frac{\partial}{\partial \xi} G(x,y,z,\xi,\eta,\zeta)\right]_{\xi=a} d\zeta\, d\eta$$
$$+ \int_0^a \int_{-\infty}^{\infty} f_3(\xi,\zeta) \left[\frac{\partial}{\partial \eta} G(x,y,z,\xi,\eta,\zeta)\right]_{\eta=0} d\zeta\, d\xi$$
$$- \int_0^a \int_{-\infty}^{\infty} f_4(\xi,\zeta) \left[\frac{\partial}{\partial \eta} G(x,y,z,\xi,\eta,\zeta)\right]_{\eta=b} d\zeta\, d\xi$$
$$+ \int_0^a \int_0^b \int_{-\infty}^{\infty} \Phi(\xi,\eta,\zeta) G(x,y,z,\xi,\eta,\zeta)\, d\zeta\, d\eta\, d\xi.$$

Green's function:
$$G(x,y,z,\xi,\eta,\zeta) = \frac{2}{ab} \sum_{n=1}^{\infty} \sum_{m=1}^{\infty} \frac{1}{\beta_{nm}} \sin(p_n x) \sin(q_m y) \sin(p_n \xi) \sin(q_m \eta) \exp(-\beta_{nm}|z-\zeta|),$$
$$p_n = \frac{n\pi}{a}, \quad q_m = \frac{m\pi}{b}, \quad \beta_{nm} = \sqrt{p_n^2 + q_m^2 - \lambda}.$$

⊙ *Reference*: A. N. Tikhonov and A. A. Samarskii (1990).

8.3.2-10. Domain: $0 \le x \le a$, $0 \le y \le b$, $-\infty < z < \infty$. Second boundary value problem.

An infinite cylindrical domain of a rectangular cross-section is considered. Boundary conditions are prescribed:
$$\partial_x w = f_1(y,z) \quad \text{at} \quad x=0, \qquad \partial_x w = f_2(y,z) \quad \text{at} \quad x=a,$$
$$\partial_y w = f_3(x,z) \quad \text{at} \quad y=0, \qquad \partial_y w = f_4(x,z) \quad \text{at} \quad y=b.$$

Solution:
$$w(x,y,z) = -\int_0^b \int_{-\infty}^{\infty} f_1(\eta,\zeta) G(x,y,z,0,\eta,\zeta)\, d\zeta\, d\eta + \int_0^b \int_{-\infty}^{\infty} f_2(\eta,\zeta) G(x,y,z,a,\eta,\zeta)\, d\zeta\, d\eta$$
$$- \int_0^a \int_{-\infty}^{\infty} f_3(\xi,\zeta) G(x,y,z,\xi,0,\zeta)\, d\zeta\, d\xi + \int_0^a \int_{-\infty}^{\infty} f_4(\xi,\zeta) G(x,y,z,\xi,b,\zeta)\, d\zeta\, d\xi$$
$$+ \int_0^a \int_0^b \int_{-\infty}^{\infty} \Phi(\xi,\eta,\zeta) G(x,y,z,\xi,\eta,\zeta)\, d\zeta\, d\eta\, d\xi.$$

Green's function:
$$G(x,y,z,\xi,\eta,\zeta) = \frac{1}{2ab} \sum_{n=0}^{\infty} \sum_{m=0}^{\infty} \frac{A_n A_m}{\beta_{nm}} \cos(p_n x) \cos(q_m y) \cos(p_n \xi) \cos(q_m \eta) \exp(-\beta_{nm}|z-\zeta|),$$
$$p_n = \frac{n\pi}{a}, \quad q_m = \frac{m\pi}{b}, \quad \beta_{nm} = \sqrt{p_n^2 + q_m^2 - \lambda}, \quad A_n = \begin{cases} 1 & \text{for } n=0, \\ 2 & \text{for } n \ne 0. \end{cases}$$

⊙ *Reference*: A. N. Tikhonov and A. A. Samarskii (1990).

8.3.2-11. Domain: $0 \le x \le a$, $0 \le y \le b$, $-\infty < z < \infty$. Third boundary value problem.

An infinite cylindrical domain of a rectangular cross-section is considered. Boundary conditions are prescribed:
$$\partial_x w - k_1 w = f_1(y,z) \quad \text{at} \quad x=0, \qquad \partial_x w + k_2 w = f_2(y,z) \quad \text{at} \quad x=a,$$
$$\partial_y w - k_3 w = f_3(x,z) \quad \text{at} \quad y=0, \qquad \partial_y w + k_4 w = f_4(x,z) \quad \text{at} \quad y=b.$$

The solution $w(x, y, z)$ is determined by the formula in Paragraph 8.3.2-10 where

$$G(x, y, z, \xi, \eta, \zeta) = \frac{1}{2} \sum_{n=1}^{\infty} \sum_{m=1}^{\infty} \frac{u_{nm}(x, y) u_{nm}(\xi, \eta)}{\|u_{nm}\|^2 \beta_{nm}} \exp(-\beta_{nm}|z - \zeta|).$$

Here,

$$w_{nm}(x, y) = (\mu_n \cos \mu_n x + k_1 \sin \mu_n x)(\nu_m \cos \nu_m y + k_3 \sin \nu_m y), \quad \beta_{nm} = \sqrt{\mu_n^2 + \nu_m^2 - \lambda},$$

$$\|w_{nm}\|^2 = \frac{1}{4}(\mu_n^2 + k_1^2)(\nu_m^2 + k_3^2)\left[a + \frac{(k_1 + k_2)(\mu_n^2 + k_1 k_2)}{(\mu_n^2 + k_1^2)(\mu_n^2 + k_2^2)}\right]\left[b + \frac{(k_3 + k_4)(\nu_m^2 + k_3 k_4)}{(\nu_m^2 + k_3^2)(\nu_m^2 + k_4^2)}\right],$$

where the μ_n and ν_m are positive roots of the transcendental equations

$$\tan(\mu a) = \frac{(k_1 + k_2)\mu}{\mu^2 - k_1 k_2}, \quad \tan(\nu b) = \frac{(k_3 + k_4)\nu}{\nu^2 - k_3 k_4}.$$

8.3.2-12. Domain: $0 \le x \le a$, $0 \le y \le b$, $-\infty < z < \infty$. Mixed boundary value problems.

1°. An infinite cylindrical domain of a rectangular cross-section is considered. Boundary conditions are prescribed:

$$w = f_1(y, z) \quad \text{at} \quad x = 0, \qquad \partial_x w = f_2(y, z) \quad \text{at} \quad x = a,$$
$$w = f_3(x, z) \quad \text{at} \quad y = 0, \qquad \partial_y w = f_4(x, z) \quad \text{at} \quad y = b.$$

Solution:

$$w(x, y, z) = \int_0^b \int_{-\infty}^{\infty} f_1(\eta, \zeta) \left[\frac{\partial}{\partial \xi} G(x, y, z, \xi, \eta, \zeta)\right]_{\xi=0} d\zeta\, d\eta$$
$$+ \int_0^b \int_{-\infty}^{\infty} f_2(\eta, \zeta) G(x, y, z, a, \eta, \zeta)\, d\zeta\, d\eta$$
$$+ \int_0^a \int_{-\infty}^{\infty} f_3(\xi, \zeta) \left[\frac{\partial}{\partial \eta} G(x, y, z, \xi, \eta, \zeta)\right]_{\eta=0} d\zeta\, d\xi$$
$$+ \int_0^a \int_{-\infty}^{\infty} f_4(\xi, \zeta) G(x, y, z, \xi, b, \zeta)\, d\zeta\, d\xi$$
$$+ \int_0^a \int_0^b \int_{-\infty}^{\infty} \Phi(\xi, \eta, \zeta) G(x, y, z, \xi, \eta, \zeta)\, d\zeta\, d\eta\, d\xi.$$

Green's function:

$$G(x, y, z, \xi, \eta, \zeta) = \frac{2}{ab} \sum_{n=0}^{\infty} \sum_{m=0}^{\infty} \frac{1}{\beta_{nm}} \sin(p_n x) \sin(q_m y) \sin(p_n \xi) \sin(q_m \eta) \exp(-\beta_{nm}|z - \zeta|),$$

$$p_n = \frac{(2n + 1)\pi}{2a}, \quad q_m = \frac{(2m + 1)\pi}{2b}, \quad \beta_{nm} = \sqrt{p_n^2 + q_m^2 - \lambda}.$$

2°. An infinite cylindrical domain of a rectangular cross-section is considered. Boundary conditions are prescribed:

$$w = f_1(y, z) \quad \text{at} \quad x = 0, \qquad w = f_2(y, z) \quad \text{at} \quad x = a,$$
$$\partial_y w = f_3(x, z) \quad \text{at} \quad y = 0, \qquad \partial_y w = f_4(x, z) \quad \text{at} \quad y = b.$$

8.3. HELMHOLTZ EQUATION $\Delta_3 w + \lambda w = -\Phi(x)$ 573

Solution:
$$w(x,y,z) = \int_0^b \int_{-\infty}^\infty f_1(\eta,\zeta)\left[\frac{\partial}{\partial \xi}G(x,y,z,\xi,\eta,\zeta)\right]_{\xi=0} d\zeta\, d\eta$$
$$- \int_0^b \int_{-\infty}^\infty f_2(\eta,\zeta)\left[\frac{\partial}{\partial \xi}G(x,y,z,\xi,\eta,\zeta)\right]_{\xi=a} d\zeta\, d\eta$$
$$- \int_0^a \int_{-\infty}^\infty f_3(\xi,\zeta)G(x,y,z,\xi,0,\zeta)\, d\zeta\, d\xi$$
$$+ \int_0^a \int_{-\infty}^\infty f_4(\xi,\zeta)G(x,y,z,\xi,b,\zeta)\, d\zeta\, d\xi$$
$$+ \int_0^a \int_0^b \int_{-\infty}^\infty \Phi(\xi,\eta,\zeta)G(x,y,z,\xi,\eta,\zeta)\, d\zeta\, d\eta\, d\xi.$$

Green's function:
$$G(x,y,z,\xi,\eta,\zeta) = \frac{1}{ab}\sum_{n=1}^\infty \sum_{m=0}^\infty \frac{A_m}{\beta_{nm}}\sin(p_n x)\cos(q_m y)\sin(p_n \xi)\cos(q_m \eta)\exp(-\beta_{nm}|z-\zeta|),$$
$$p_n = \frac{n\pi}{a}, \quad q_m = \frac{m\pi}{b}, \quad \beta_{nm} = \sqrt{p_n^2 + q_m^2 - \lambda}, \quad A_m = \begin{cases} 1 & \text{for } m=0, \\ 2 & \text{for } m \neq 0. \end{cases}$$

8.3.2-13. Domain: $0 \leq x \leq a$, $0 \leq y \leq b$, $0 \leq z < \infty$. First boundary value problem.

A semiinfinite cylindrical domain of a rectangular cross-section is considered. Boundary conditions are prescribed:
$$w = f_1(y,z) \quad \text{at} \quad x = 0, \qquad w = f_2(y,z) \quad \text{at} \quad x = a,$$
$$w = f_3(x,z) \quad \text{at} \quad y = 0, \qquad w = f_4(x,z) \quad \text{at} \quad y = b,$$
$$w = f_5(x,y) \quad \text{at} \quad z = 0.$$

Solution:
$$w(x,y,z) = \int_0^b \int_0^\infty f_1(\eta,\zeta)\left[\frac{\partial}{\partial \xi}G(x,y,z,\xi,\eta,\zeta)\right]_{\xi=0} d\zeta\, d\eta$$
$$- \int_0^b \int_0^\infty f_2(\eta,\zeta)\left[\frac{\partial}{\partial \xi}G(x,y,z,\xi,\eta,\zeta)\right]_{\xi=a} d\zeta\, d\eta$$
$$+ \int_0^a \int_0^\infty f_3(\xi,\zeta)\left[\frac{\partial}{\partial \eta}G(x,y,z,\xi,\eta,\zeta)\right]_{\eta=0} d\zeta\, d\xi$$
$$- \int_0^a \int_0^\infty f_4(\xi,\zeta)\left[\frac{\partial}{\partial \eta}G(x,y,z,\xi,\eta,\zeta)\right]_{\eta=b} d\zeta\, d\xi$$
$$+ \int_0^a \int_0^b f_5(\xi,\eta)\left[\frac{\partial}{\partial \zeta}G(x,y,z,\xi,\eta,\zeta)\right]_{\zeta=0} d\eta\, d\xi$$
$$+ \int_0^a \int_0^b \int_0^\infty \Phi(\xi,\eta,\zeta)G(x,y,z,\xi,\eta,\zeta)\, d\zeta\, d\eta\, d\xi.$$

Green's function:
$$G(x,y,z,\xi,\eta,\zeta) = \frac{4}{ab}\sum_{n=1}^\infty \sum_{m=1}^\infty \frac{1}{\beta_{nm}}\sin(p_n x)\sin(q_m y)\sin(p_n \xi)\sin(q_m \eta)H_{nm}(z,\zeta),$$
$$p_n = \frac{n\pi}{a}, \quad q_m = \frac{m\pi}{b}, \quad \beta_{nm} = \sqrt{p_n^2 + q_m^2 - \lambda},$$
$$H_{nm}(z,\zeta) = \begin{cases} \exp(-\beta_{nm}z)\sinh(\beta_{nm}\zeta) & \text{for } z > \zeta \geq 0, \\ \exp(-\beta_{nm}\zeta)\sinh(\beta_{nm}z) & \text{for } \zeta > z \geq 0. \end{cases}$$

8.3.2-14. Domain: $0 \leq x \leq a$, $0 \leq y \leq b$, $0 \leq z < \infty$. Second boundary value problem.

A semiinfinite cylindrical domain of a rectangular cross-section is considered. Boundary conditions are prescribed:

$$\partial_x w = f_1(y, z) \quad \text{at} \quad x = 0, \qquad \partial_x w = f_2(y, z) \quad \text{at} \quad x = a,$$
$$\partial_y w = f_3(x, z) \quad \text{at} \quad y = 0, \qquad \partial_y w = f_4(x, z) \quad \text{at} \quad y = b,$$
$$\partial_z w = f_5(x, y) \quad \text{at} \quad z = 0.$$

Solution:

$$w(x, y, z) = \int_0^a \int_0^b \int_0^\infty \Phi(\xi, \eta, \zeta) G(x, y, z, \xi, \eta, \zeta) \, d\zeta \, d\eta \, d\xi$$
$$- \int_0^b \int_0^\infty f_1(\eta, \zeta) G(x, y, z, 0, \eta, \zeta) \, d\zeta \, d\eta + \int_0^b \int_0^\infty f_2(\eta, \zeta) G(x, y, z, a, \eta, \zeta) \, d\zeta \, d\eta$$
$$- \int_0^a \int_0^\infty f_3(\xi, \zeta) G(x, y, z, \xi, 0, \zeta) \, d\zeta \, d\xi + \int_0^a \int_0^\infty f_4(\xi, \zeta) G(x, y, z, \xi, b, \zeta) \, d\zeta \, d\xi$$
$$- \int_0^a \int_0^b f_5(\xi, \eta) G(x, y, z, \xi, \eta, 0) \, d\eta \, d\xi.$$

Green's function:

$$G(x, y, z, \xi, \eta, \zeta) = \frac{1}{ab} \sum_{n=0}^\infty \sum_{m=0}^\infty \frac{A_n A_m}{\beta_{nm}} \cos(p_n x) \cos(q_m y) \cos(p_n \xi) \cos(q_m \eta) H_{nm}(z, \zeta),$$

$$p_n = \frac{n\pi}{a}, \quad q_m = \frac{m\pi}{b}, \quad \beta_{nm} = \sqrt{p_n^2 + q_m^2 - \lambda}, \quad A_n = \begin{cases} 1 & \text{for } n = 0, \\ 2 & \text{for } n \neq 0, \end{cases}$$

$$H_{nm}(z, \zeta) = \begin{cases} \exp(-\beta_{nm} z) \cosh(\beta_{nm} \zeta) & \text{for } z > \zeta \geq 0, \\ \exp(-\beta_{nm} \zeta) \cosh(\beta_{nm} z) & \text{for } \zeta > z \geq 0. \end{cases}$$

8.3.2-15. Domain: $0 \leq x \leq a$, $0 \leq y \leq b$, $0 \leq z < \infty$. Third boundary value problem.

A semiinfinite cylindrical domain of a rectangular cross-section is considered. Boundary conditions are prescribed:

$$\partial_x w - k_1 w = f_1(y, z) \quad \text{at} \quad x = 0, \qquad \partial_x w + k_2 w = f_2(y, z) \quad \text{at} \quad x = a,$$
$$\partial_y w - k_3 w = f_3(x, z) \quad \text{at} \quad y = 0, \qquad \partial_y w + k_4 w = f_4(x, z) \quad \text{at} \quad y = b,$$
$$\partial_z w - k_5 w = f_5(x, y) \quad \text{at} \quad z = 0.$$

The solution $w(x, y, z)$ is determined by the formula in Paragraph 8.3.2-14 where

$$G(x, y, z, \xi, \eta, \zeta) = \sum_{n=1}^\infty \sum_{m=1}^\infty \frac{u_{nm}(x, y) u_{nm}(\xi, \eta)}{\|u_{nm}\|^2} H_{nm}(z, \zeta).$$

Here,

$$w_{nm}(x, y) = (\mu_n \cos \mu_n x + k_1 \sin \mu_n x)(\nu_m \cos \nu_m y + k_3 \sin \nu_m y),$$

$$\|w_{nm}\|^2 = \frac{1}{4}(\mu_n^2 + k_1^2)(\nu_m^2 + k_3^2) \left[a + \frac{(k_1 + k_2)(\mu_n^2 + k_1 k_2)}{(\mu_n^2 + k_1^2)(\mu_n^2 + k_2^2)} \right] \left[b + \frac{(k_3 + k_4)(\nu_m^2 + k_3 k_4)}{(\nu_m^2 + k_3^2)(\nu_m^2 + k_4^2)} \right],$$

$$H_{nm}(z, \zeta) = \begin{cases} \dfrac{\exp(-\beta_{nm} z)[\beta_{nm} \cosh(\beta_{nm} \zeta) + k_5 \sinh(\beta_{nm} \zeta)]}{\beta_{nm}(\beta_{nm} + k_5)} & \text{for } z > \zeta, \\ \dfrac{\exp(-\beta_{nm} \zeta)[\beta_{nm} \cosh(\beta_{nm} z) + k_5 \sinh(\beta_{nm} z)]}{\beta_{nm}(\beta_{nm} + k_5)} & \text{for } \zeta > z, \end{cases} \quad \beta_{nm} = \sqrt{\mu_n^2 + \nu_m^2 - \lambda},$$

where the μ_n and ν_m are positive roots of the transcendental equations

$$\tan(\mu a) = \frac{(k_1 + k_2)\mu}{\mu^2 - k_1 k_2}, \qquad \tan(\nu b) = \frac{(k_3 + k_4)\nu}{\nu^2 - k_3 k_4}.$$

8.3.2-16. Domain: $0 \le x \le a$, $0 \le y \le b$, $0 \le z < \infty$. Mixed boundary value problems.

1°. A semiinfinite cylindrical domain of a rectangular cross-section is considered. Boundary conditions are prescribed:

$$w = f_1(y,z) \quad \text{at} \quad x = 0, \qquad w = f_2(y,z) \quad \text{at} \quad x = a,$$
$$w = f_3(x,z) \quad \text{at} \quad y = 0, \qquad w = f_4(x,z) \quad \text{at} \quad y = b,$$
$$\partial_z w = f_5(x,y) \quad \text{at} \quad z = 0.$$

Solution:

$$w(x,y,z) = \int_0^b \int_0^\infty f_1(\eta,\zeta) \left[\frac{\partial}{\partial \xi} G(x,y,z,\xi,\eta,\zeta)\right]_{\xi=0} d\zeta\, d\eta$$
$$- \int_0^b \int_0^\infty f_2(\eta,\zeta) \left[\frac{\partial}{\partial \xi} G(x,y,z,\xi,\eta,\zeta)\right]_{\xi=a} d\zeta\, d\eta$$
$$+ \int_0^a \int_0^\infty f_3(\xi,\zeta) \left[\frac{\partial}{\partial \eta} G(x,y,z,\xi,\eta,\zeta)\right]_{\eta=0} d\zeta\, d\xi$$
$$- \int_0^a \int_0^\infty f_4(\xi,\zeta) \left[\frac{\partial}{\partial \eta} G(x,y,z,\xi,\eta,\zeta)\right]_{\eta=b} d\zeta\, d\xi$$
$$- \int_0^a \int_0^b f_5(\xi,\eta) G(x,y,z,\xi,\eta,0)\, d\eta\, d\xi$$
$$+ \int_0^a \int_0^b \int_0^\infty \Phi(\xi,\eta,\zeta) G(x,y,z,\xi,\eta,\zeta)\, d\zeta\, d\eta\, d\xi.$$

Green's function:

$$G(x,y,z,\xi,\eta,\zeta) = \frac{4}{ab} \sum_{n=1}^\infty \sum_{m=1}^\infty \frac{1}{\beta_{nm}} \sin(p_n x) \sin(q_m y) \sin(p_n \xi) \sin(q_m \eta) H_{nm}(z,\zeta),$$

$$p_n = \frac{n\pi}{a}, \qquad q_m = \frac{m\pi}{b}, \qquad \beta_{nm} = \sqrt{p_n^2 + q_m^2 - \lambda},$$

$$H_{nm}(z,\zeta) = \begin{cases} \exp(-\beta_{nm} z) \cosh(\beta_{nm}\zeta) & \text{for } z > \zeta \ge 0, \\ \exp(-\beta_{nm}\zeta) \cosh(\beta_{nm} z) & \text{for } \zeta > z \ge 0. \end{cases}$$

2°. A semiinfinite cylindrical domain of a rectangular cross-section is considered. Boundary conditions are prescribed:

$$\partial_x w = f_1(y,z) \quad \text{at} \quad x = 0, \qquad \partial_x w = f_2(y,z) \quad \text{at} \quad x = a,$$
$$\partial_y w = f_3(x,z) \quad \text{at} \quad y = 0, \qquad \partial_y w = f_4(x,z) \quad \text{at} \quad y = b,$$
$$w = f_5(x,y) \quad \text{at} \quad z = 0.$$

Solution:

$$w(x,y,z) = \int_0^a \int_0^b \int_0^\infty \Phi(\xi,\eta,\zeta) G(x,y,z,\xi,\eta,\zeta)\, d\zeta\, d\eta\, d\xi$$
$$- \int_0^b \int_0^\infty f_1(\eta,\zeta) G(x,y,z,0,\eta,\zeta)\, d\zeta\, d\eta + \int_0^b \int_0^\infty f_2(\eta,\zeta) G(x,y,z,a,\eta,\zeta)\, d\zeta\, d\eta$$
$$- \int_0^a \int_0^\infty f_3(\xi,\zeta) G(x,y,z,\xi,0,\zeta)\, d\zeta\, d\xi + \int_0^a \int_0^\infty f_4(\xi,\zeta) G(x,y,z,\xi,b,\zeta)\, d\zeta\, d\xi$$
$$+ \int_0^a \int_0^b f_5(\xi,\eta) \left[\frac{\partial}{\partial \zeta} G(x,y,z,\xi,\eta,\xi)\right]_{\zeta=0} d\eta\, d\xi.$$

Green's function:
$$G(x,y,z,\xi,\eta,\zeta) = \frac{1}{ab}\sum_{n=0}^{\infty}\sum_{m=0}^{\infty} \frac{A_n A_m}{\beta_{nm}} \cos(p_n x)\cos(q_m y)\cos(p_n \xi)\cos(q_m \eta) H_{nm}(z,\zeta).$$

Here,
$$p_n = \frac{n\pi}{a}, \quad q_m = \frac{m\pi}{b}, \quad \beta_{nm} = \sqrt{p_n^2 + q_m^2 - \lambda}, \quad A_n = \begin{cases} 1 & \text{for } n=0, \\ 2 & \text{for } n \neq 0, \end{cases}$$

$$H_{nm}(z,\zeta) = \begin{cases} \exp(-\beta_{nm}z)\sinh(\beta_{nm}\zeta) & \text{for } z > \zeta \geq 0, \\ \exp(-\beta_{nm}\zeta)\sinh(\beta_{nm}z) & \text{for } \zeta > z \geq 0. \end{cases}$$

▶ *Paragraphs 8.3.2-17 through 8.3.2-23 present only the eigenvalues and eigenfunctions of homogeneous boundary value problems for the homogeneous Helmholtz equation (with $\Phi \equiv 0$). The solutions of the corresponding nonhomogeneous boundary value problems (with $\Phi \not\equiv 0$) can be constructed by the relations specified in Paragraphs 8.3.1-4 and 8.3.1-8.*

8.3.2-17. Domain: $0 \leq x \leq a$, $0 \leq y \leq b$, $0 \leq z \leq c$. First boundary value problem.

A rectangular parallelepiped is considered. Boundary conditions are prescribed:

$$w = f_1(y,z) \quad \text{at} \quad x = 0, \qquad w = f_2(y,z) \quad \text{at} \quad x = a,$$
$$w = f_3(x,z) \quad \text{at} \quad y = 0, \qquad w = f_4(x,z) \quad \text{at} \quad y = b,$$
$$w = f_5(x,y) \quad \text{at} \quad z = 0, \qquad w = f_6(x,y) \quad \text{at} \quad z = c.$$

1°. Eigenvalues of the homogeneous problem:
$$\lambda_{nmk} = \pi^2\left(\frac{n^2}{a^2} + \frac{m^2}{b^2} + \frac{k^2}{c^2}\right); \qquad n, m, k = 1, 2, 3, \ldots$$

Eigenfunctions and the norm squared:
$$w_{nmk} = \sin\left(\frac{\pi n x}{a}\right)\sin\left(\frac{\pi m y}{b}\right)\sin\left(\frac{\pi k z}{c}\right), \qquad \|w_{nmk}\|^2 = \frac{abc}{8}.$$

2°. A double-series representation of the Green's function:
$$G(x,y,z,\xi,\eta,\zeta) = \frac{4}{ab}\sum_{n=1}^{\infty}\sum_{m=1}^{\infty} \sin(p_n x)\sin(q_m y)\sin(p_n \xi)\sin(q_m \eta) H_{nm}(z,\zeta),$$

$$H_{nm}(z,\zeta) = \begin{cases} \dfrac{\sinh(\beta_{nm}\zeta)\sinh[\beta_{nm}(c-z)]}{\beta_{nm}\sinh(\beta_{nm}c)} & \text{for } c \geq z > \zeta \geq 0, \\ \dfrac{\sinh(\beta_{nm}z)\sinh[\beta_{nm}(c-\zeta)]}{\beta_{nm}\sinh(\beta_{nm}c)} & \text{for } c \geq \zeta > z \geq 0, \end{cases}$$

$$p_n = \frac{\pi n}{a}, \quad q_m = \frac{\pi m}{b}, \quad \beta_{nm} = \sqrt{p_n^2 + q_m^2 - \lambda}.$$

This relation can be used to obtain two other representations of the Green's function with the aid of the cyclic permutations of triples:

$$\begin{array}{c} (x,\xi,a) \\ \nearrow \quad \searrow \\ (z,\zeta,c) \longleftarrow (y,\eta,b) \end{array}$$

A triple series representation of the Green's function:
$$G(x,y,z,\xi,\eta,\zeta) = \frac{8}{abc}\sum_{n=1}^{\infty}\sum_{m=1}^{\infty}\sum_{k=1}^{\infty} \frac{\sin(p_n x)\sin(q_m y)\sin(s_k z)\sin(p_n \xi)\sin(q_m \eta)\sin(s_k \zeta)}{p_n^2 + q_m^2 + s_k^2 - \lambda},$$

$$p_n = \frac{\pi n}{a}, \quad q_m = \frac{\pi m}{b}, \quad s_k = \frac{\pi k}{c}.$$

⊙ *References*: V. M. Babich, M. B. Kapilevich, S. G. Mikhlin, et al. (1964), B. M. Budak, A. A. Samarskii, and A. N. Tikhonov (1980).

8.3.2-18. Domain: $0 \leq x \leq a$, $0 \leq y \leq b$, $0 \leq z \leq c$. Second boundary value problem.

A rectangular parallelepiped is considered. Boundary conditions are prescribed:

$$\partial_x w = f_1(y,z) \quad \text{at} \quad x = 0, \qquad \partial_x w = f_2(y,z) \quad \text{at} \quad x = a,$$
$$\partial_y w = f_3(x,z) \quad \text{at} \quad y = 0, \qquad \partial_y w = f_4(x,z) \quad \text{at} \quad y = b,$$
$$\partial_z w = f_5(x,y) \quad \text{at} \quad z = 0, \qquad \partial_z w = f_6(x,y) \quad \text{at} \quad z = c.$$

1°. Eigenvalues of the homogeneous problem:

$$\lambda_{nmk} = \pi^2 \left(\frac{n^2}{a^2} + \frac{m^2}{b^2} + \frac{k^2}{c^2} \right); \qquad n, m, k = 0, 1, 2, \dots$$

Eigenfunctions:

$$w_{nmk} = \cos\left(\frac{\pi n x}{a}\right) \cos\left(\frac{\pi m y}{b}\right) \cos\left(\frac{\pi k z}{c}\right).$$

The square of the norm of an eigenfunction is defined as

$$\|w_{nmk}\|^2 = \frac{abc}{8}(1+\delta_{n0})(1+\delta_{m0})(1+\delta_{k0}), \qquad \delta_{n0} = \begin{cases} 1 & \text{for } n = 0, \\ 0 & \text{for } n \neq 0. \end{cases}$$

2°. A double series representation of the Green's function:

$$G(x,y,z,\xi,\eta,\zeta) = \frac{1}{ab} \sum_{n=0}^{\infty} \sum_{m=0}^{\infty} \varepsilon_n \varepsilon_m \cos(p_n x) \cos(q_m y) \cos(p_n \xi) \cos(q_m \eta) H_{nm}(z,\zeta),$$

$$H_{nm}(z,\zeta) = \begin{cases} \dfrac{\cosh(\beta_{nm}\zeta) \cosh[\beta_{nm}(c-z)]}{\beta_{nm} \sinh(\beta_{nm}c)} & \text{for } c \geq z > \zeta \geq 0, \\ \dfrac{\cosh(\beta_{nm}z) \cosh[\beta_{nm}(c-\zeta)]}{\beta_{nm} \sinh(\beta_{nm}c)} & \text{for } c \geq \zeta > z \geq 0, \end{cases}$$

$$p_n = \frac{\pi n}{a}, \quad q_m = \frac{\pi m}{b}, \quad \beta_{nm} = \sqrt{p_n^2 + q_m^2 - \lambda}, \quad \varepsilon_n = \begin{cases} 1 & \text{for } n = 0, \\ 2 & \text{for } n \neq 0. \end{cases}$$

This relation can be used to obtain two other representations of the Green's function with the aid of the cyclic permutations:

$$\begin{array}{c} (x, \xi, a) \\ \nearrow \qquad \searrow \\ (z, \zeta, c) \longleftarrow (y, \eta, b) \end{array}$$

A triple series representation of the Green's function:

$$G(x,y,z,\xi,\eta,\zeta) = \frac{1}{abc} \sum_{n=0}^{\infty} \sum_{m=0}^{\infty} \sum_{k=0}^{\infty} \frac{\varepsilon_n \varepsilon_m \varepsilon_k \cos(p_n x) \cos(q_m y) \cos(s_k z) \cos(p_n \xi) \cos(q_m \eta) \cos(s_k \zeta)}{p_n^2 + q_m^2 + s_k^2 - \lambda},$$

$$p_n = \frac{\pi n}{a}, \quad q_m = \frac{\pi m}{b}, \quad s_k = \frac{\pi k}{c}.$$

⊙ *References*: V. M. Babich, M. B. Kapilevich, S. G. Mikhlin, et al. (1964), B. M. Budak, A. A. Samarskii, and A. N. Tikhonov (1980).

8.3.2-19. Domain: $0 \leq x \leq a$, $0 \leq y \leq b$, $0 \leq z \leq c$. Third boundary value problem.

A rectangular parallelepiped is considered. Boundary conditions are prescribed:

$$\partial_x w - k_1 w = f_1(y,z) \quad \text{at} \quad x = 0, \qquad \partial_x w + k_2 w = f_2(y,z) \quad \text{at} \quad x = a,$$
$$\partial_y w - k_3 w = f_3(x,z) \quad \text{at} \quad y = 0, \qquad \partial_y w + k_4 w = f_4(x,z) \quad \text{at} \quad y = b,$$
$$\partial_z w - k_5 w = f_5(x,y) \quad \text{at} \quad z = 0, \qquad \partial_z w + k_6 w = f_6(x,y) \quad \text{at} \quad z = c.$$

Eigenvalues of the homogeneous problem:
$$\lambda_{nml} = \mu_n^2 + \nu_m^2 + \sigma_l^2; \qquad n, m, l = 1, 2, 3, \ldots$$

Here, the μ_n, ν_m, and σ_l are positive roots of the transcendental equations
$$\tan(\mu a) = \frac{(k_1+k_2)\mu}{\mu^2 - k_1 k_2}, \qquad \tan(\nu b) = \frac{(k_3+k_4)\nu}{\nu^2 - k_3 k_4}, \qquad \tan(\sigma c) = \frac{(k_5+k_6)\sigma}{\sigma^2 - k_5 k_6}.$$

Eigenfunctions:
$$w_{nml} = \frac{1}{A_n B_m C_l}(\mu_n \cos \mu_n x + k_1 \sin \mu_n x)(\nu_m \cos \nu_m y + k_3 \sin \nu_m y)(\sigma_l \cos \sigma_l z + k_5 \sin \sigma_l z),$$
$$A_n = \sqrt{\mu_n^2 + k_1^2}, \qquad B_m = \sqrt{\nu_m^2 + k_3^2}, \qquad C_l = \sqrt{\sigma_l^2 + k_5^2}.$$

The square of the norm of an eigenfunction is defined as
$$\|w_{nml}\|^2 = \frac{1}{8}\left[a + \frac{(k_1+k_2)(\mu_n^2 + k_1 k_2)}{(\mu_n^2 + k_1^2)(\mu_n^2 + k_2^2)}\right]\left[b + \frac{(k_3+k_4)(\nu_m^2 + k_3 k_4)}{(\nu_m^2 + k_3^2)(\nu_m^2 + k_4^2)}\right]\left[c + \frac{(k_5+k_6)(\sigma_l^2 + k_5 k_6)}{(\sigma_l^2 + k_5^2)(\sigma_l^2 + k_6^2)}\right].$$

⊙ *Reference*: B. M. Budak, A. A. Samarskii, and A. N. Tikhonov (1980).

8.3.2-20. Domain: $0 \le x \le a$, $0 \le y \le b$, $0 \le z \le c$. Mixed boundary value problems.

1°. A rectangular parallelepiped is considered. Boundary conditions are prescribed:
$$w = f_1(y,z) \ \text{at} \ x = 0, \qquad w = f_2(y,z) \ \text{at} \ x = a,$$
$$w = f_3(x,z) \ \text{at} \ y = 0, \qquad w = f_4(x,z) \ \text{at} \ y = b,$$
$$\partial_z w = f_5(x,y) \ \text{at} \ z = 0, \qquad \partial_z w = f_6(x,y) \ \text{at} \ z = c.$$

Eigenvalues of the homogeneous problem:
$$\lambda_{nmk} = \pi^2 \left(\frac{n^2}{a^2} + \frac{m^2}{b^2} + \frac{k^2}{c^2}\right); \qquad n, m = 1, 2, 3, \ldots; \quad k = 0, 1, 2, \ldots$$

Eigenfunctions:
$$w_{nmk} = \sin\left(\frac{\pi n x}{a}\right) \sin\left(\frac{\pi m y}{b}\right) \cos\left(\frac{\pi k z}{c}\right).$$

The square of the norm of an eigenfunction is defined as
$$\|w_{nmk}\|^2 = \frac{abc}{8}(1 + \delta_{k0}), \qquad \delta_{k0} = \begin{cases} 1 & \text{for } k = 0, \\ 0 & \text{for } k \ne 0. \end{cases}$$

2°. A rectangular parallelepiped is considered. Boundary conditions are prescribed:
$$w = f_1(y,z) \ \text{at} \ x = 0, \qquad w = f_2(y,z) \ \text{at} \ x = a,$$
$$\partial_y w = f_3(x,z) \ \text{at} \ y = 0, \qquad \partial_y w = f_4(x,z) \ \text{at} \ y = b,$$
$$\partial_z w = f_5(x,y) \ \text{at} \ z = 0, \qquad \partial_z w = f_6(x,y) \ \text{at} \ z = c.$$

Eigenvalues of the homogeneous problem:
$$\lambda_{nmk} = \pi^2 \left(\frac{n^2}{a^2} + \frac{m^2}{b^2} + \frac{k^2}{c^2}\right); \qquad n = 1, 2, 3, \ldots; \quad m, k = 0, 1, 2, \ldots$$

Eigenfunctions:
$$w_{nmk} = \sin\left(\frac{\pi n x}{a}\right) \cos\left(\frac{\pi m y}{b}\right) \cos\left(\frac{\pi k z}{c}\right).$$

The square of the norm of an eigenfunction is defined as
$$\|w_{nmk}\|^2 = \frac{abc}{8}(1 + \delta_{m0})(1 + \delta_{k0}), \qquad \delta_{m0} = \begin{cases} 1 & \text{for } m = 0, \\ 0 & \text{for } m \ne 0. \end{cases}$$

8.3. HELMHOLTZ EQUATION $\Delta_3 w + \lambda w = -\Phi(\mathbf{x})$

8.3.2-21. Domain: $0 \leq x \leq a$, $0 \leq y \leq x$, $0 \leq z \leq c$. **First boundary value problem.**

A right prism whose base is an isosceles right-angled triangle is considered. Boundary conditions are prescribed:

$$w = f_1(y,z) \quad \text{at} \quad x = 0, \quad w = f_2(x,z) \quad \text{at} \quad y = 0, \quad w = f_3(x,z) \quad \text{at} \quad y = x,$$
$$w = f_4(x,y) \quad \text{at} \quad z = 0, \quad w = f_5(x,y) \quad \text{at} \quad z = c.$$

Eigenvalues of the homogeneous problem:

$$\lambda_{nmk} = \frac{\pi^2}{a^2}\left[(n+m)^2 + m^2\right] + \frac{\pi^2 k^2}{c^2}; \quad n, m, k = 1, 2, 3, \ldots$$

Eigenfunctions:

$$w_{nmk} = \left\{\sin\left[\frac{\pi}{a}(n+m)x\right]\sin\left(\frac{\pi}{a}my\right) - (-1)^n \sin\left(\frac{\pi}{a}mx\right)\sin\left[\frac{\pi}{a}(n+m)y\right]\right\}\sin\left(\frac{\pi k z}{c}\right).$$

8.3.2-22. Domain: $0 \leq x \leq a$, $0 \leq y \leq x$, $0 \leq z \leq c$. **Second boundary value problem.**

A right prism whose base is an isosceles right-angled triangle is considered. Boundary conditions are prescribed:

$$\partial_x w = f_1(y,z) \quad \text{at} \quad x = 0, \quad \partial_y w = f_2(x,z) \quad \text{at} \quad y = 0, \quad \partial_N w = f_3(x,z) \quad \text{at} \quad y = x,$$
$$\partial_z w = f_4(x,y) \quad \text{at} \quad z = 0, \quad \partial_z w = f_5(x,y) \quad \text{at} \quad z = c,$$

where $\partial_N w = \mathbf{N} \cdot \nabla w = \frac{1}{\sqrt{2}}(\partial_x w + \partial_y w)$.

Eigenvalues of the homogeneous problem:

$$\lambda_{nmk} = \frac{\pi^2}{a^2}\left[(n+m)^2 + m^2\right] + \frac{\pi^2 k^2}{c^2}; \quad n, m, k = 0, 1, 2, \ldots$$

Eigenfunctions:

$$w_{nmk} = \left\{\cos\left[\frac{\pi}{a}(n+m)x\right]\cos\left(\frac{\pi}{a}my\right) - (-1)^n \cos\left(\frac{\pi}{a}mx\right)\cos\left[\frac{\pi}{a}(n+m)y\right]\right\}\cos\left(\frac{\pi k z}{c}\right).$$

8.3.2-23. Domain: $0 \leq x \leq a$, $0 \leq y \leq x$, $0 \leq z \leq c$. **Mixed boundary value problems.**

1°. A right prism whose base is an isosceles right-angled triangle is considered. Boundary conditions are prescribed:

$$w = f_1(y,z) \quad \text{at} \quad x = 0, \quad w = f_2(x,z) \quad \text{at} \quad y = 0, \quad w = f_3(x,z) \quad \text{at} \quad y = x,$$
$$\partial_z w = f_4(x,y) \quad \text{at} \quad z = 0, \quad \partial_z w = f_5(x,y) \quad \text{at} \quad z = c.$$

Eigenvalues of the homogeneous problem:

$$\lambda_{nmk} = \frac{\pi^2}{a^2}\left[(n+m)^2 + m^2\right] + \frac{\pi^2 k^2}{c^2}; \quad n, m = 1, 2, 3, \ldots; \quad k = 0, 1, 2, \ldots$$

Eigenfunctions:

$$w_{nmk} = \left\{\sin\left[\frac{\pi}{a}(n+m)x\right]\sin\left(\frac{\pi}{a}my\right) - (-1)^n \sin\left(\frac{\pi}{a}mx\right)\sin\left[\frac{\pi}{a}(n+m)y\right]\right\}\cos\left(\frac{\pi k z}{c}\right).$$

2°. A right prism whose base is an isosceles right-angled triangle is considered. Boundary conditions are prescribed:

$$\partial_x w = f_1(y,z) \quad \text{at} \quad x = 0, \quad \partial_y w = f_2(x,z) \quad \text{at} \quad y = 0, \quad \partial_N w = f_3(x,z) \quad \text{at} \quad y = x,$$
$$w = f_4(x,y) \quad \text{at} \quad z = 0, \quad w = f_5(x,y) \quad \text{at} \quad z = c.$$

Eigenvalues of the homogeneous problem:

$$\lambda_{nmk} = \frac{\pi^2}{a^2}\left[(n+m)^2 + m^2\right] + \frac{\pi^2 k^2}{c^2}; \quad n, m = 0, 1, 2, \ldots; \quad k = 1, 2, 3, \ldots$$

Eigenfunctions:

$$w_{nmk} = \left\{\cos\left[\frac{\pi}{a}(n+m)x\right]\cos\left(\frac{\pi}{a}my\right) - (-1)^n \cos\left(\frac{\pi}{a}mx\right)\cos\left[\frac{\pi}{a}(n+m)y\right]\right\}\sin\left(\frac{\pi k z}{c}\right).$$

8.3.3. Problems in Cylindrical Coordinates

The three-dimensional nonhomogeneous Helmholtz equation in the cylindrical coordinate system is written as

$$\frac{1}{r}\frac{\partial}{\partial r}\left(r\frac{\partial w}{\partial r}\right) + \frac{1}{r^2}\frac{\partial^2 w}{\partial \varphi^2} + \frac{\partial^2 w}{\partial z^2} + \lambda w = -\Phi(r, \varphi, z), \qquad r = \sqrt{x^2 + y^2}.$$

8.3.3-1. Particular solutions of the homogeneous equation ($\Phi \equiv 0$):

$$w = [A J_0(r\sqrt{\lambda}) + B Y_0(r\sqrt{\lambda})](C_1\varphi + D_1)(C_2 z + D_2),$$
$$w = J_m(r\sqrt{\lambda - \mu^2})(A\cos m\varphi + B\sin m\varphi)(C\cos \mu z + D\sin \mu z), \qquad \lambda > \mu^2,$$
$$w = Y_m(r\sqrt{\lambda - \mu^2})(A\cos m\varphi + B\sin m\varphi)(C\cos \mu z + D\sin \mu z), \qquad \lambda > \mu^2,$$
$$w = J_m(r\sqrt{\lambda + \mu^2})(A\cos m\varphi + B\sin m\varphi)(C\cosh \mu z + D\sinh \mu z), \qquad \lambda > -\mu^2,$$
$$w = Y_m(r\sqrt{\lambda + \mu^2})(A\cos m\varphi + B\sin m\varphi)(C\cosh \mu z + D\sinh \mu z), \qquad \lambda > -\mu^2,$$
$$w = I_m(r\sqrt{\mu^2 - \lambda})(A\cos m\varphi + B\sin m\varphi)(C\cos \mu z + D\sin \mu z), \qquad \lambda < \mu^2,$$
$$w = Y_m(r\sqrt{\mu^2 - \lambda})(A\cos m\varphi + B\sin m\varphi)(C\cos \mu z + D\sin \mu z), \qquad \lambda < \mu^2,$$

where $m = 0, 1, 2, \ldots$; A, B, C, D, C_1, C_2, D_1, D_2, and μ are arbitrary constants; the $J_m(\xi)$ and $Y_m(\xi)$ are the Bessel functions; and the $I_m(\xi)$ and $K_m(\xi)$ are the modified Bessel functions.

8.3.3-2. Domain: $0 \leq r \leq R$, $0 \leq \varphi \leq 2\pi$, $-\infty < z < \infty$. First boundary value problem.

An infinite circular cylinder is considered. A boundary condition is prescribed:

$$w = f(\varphi, z) \quad \text{at} \quad r = R.$$

Solution:

$$w(r, \varphi, z) = -R \int_0^{2\pi}\!\!\int_{-\infty}^{\infty} f(\eta, \zeta)\left[\frac{\partial}{\partial \xi} G(r, \varphi, z, \xi, \eta, \zeta)\right]_{\xi=R} d\zeta\, d\eta$$
$$+ \int_0^R\!\!\int_0^{2\pi}\!\!\int_{-\infty}^{\infty} \Phi(\xi, \eta, \zeta) G(r, \varphi, z, \xi, \eta, \zeta)\xi\, d\zeta\, d\eta\, d\xi.$$

Here,

$$G(r, \varphi, z, \xi, \eta, \zeta) = \frac{1}{2\pi R^2} \sum_{n=0}^{\infty}\sum_{m=1}^{\infty} \frac{A_n J_n(\mu_{nm}r) J_n(\mu_{nm}\xi)}{[J_n'(\mu_{nm}R)]^2 \beta_{nm}} \cos[n(\varphi - \eta)]\exp(-\beta_{nm}|z - \zeta|),$$

$$\beta_{nm} = \sqrt{\mu_{nm}^2 - \lambda}, \qquad A_n = \begin{cases} 1 & \text{for } n = 0, \\ 2 & \text{for } n \neq 0, \end{cases}$$

where the $J_n(\xi)$ are the Bessel functions and the μ_{nm} are positive roots of the transcendental equation $J_n(\mu R) = 0$.

⊙ *Reference*: A. N. Tikhonov and A. A. Samarskii (1990).

8.3.3-3. Domain: $0 \leq r \leq R$, $0 \leq \varphi \leq 2\pi$, $-\infty < z < \infty$. Second boundary value problem.

An infinite circular cylinder is considered. A boundary condition is prescribed:

$$\partial_r w = f(\varphi, z) \quad \text{at} \quad r = R.$$

Solution:

$$w(r,\varphi,z) = R\int_0^{2\pi}\int_{-\infty}^{\infty} f(\eta,\zeta)G(r,\varphi,z,R,\eta,\zeta)\,d\zeta\,d\eta$$
$$+ \int_0^R\int_0^{2\pi}\int_{-\infty}^{\infty} \Phi(\xi,\eta,\zeta)G(r,\varphi,z,\xi,\eta,\zeta)\xi\,d\zeta\,d\eta\,d\xi.$$

Here,

$$G(r,\varphi,z,\xi,\eta,\zeta) = \frac{\exp(-\sqrt{-\lambda}|z-\zeta|)}{2\pi R^2\sqrt{-\lambda}}$$
$$+ \frac{1}{2\pi}\sum_{n=0}^{\infty}\sum_{m=1}^{\infty}\frac{A_n\mu_{nm}^2 J_n(\mu_{nm}r)J_n(\mu_{nm}\xi)\cos[n(\varphi-\eta)]}{(\mu_{nm}^2 R^2 - n^2)J_n^2(\mu_{nm}R)\beta_{nm}}\exp(-\beta_{nm}|z-\zeta|),$$

$$\beta_{nm} = \sqrt{\mu_{nm}^2 - \lambda},\quad A_n = \begin{cases} 1 & \text{for } n = 0, \\ 2 & \text{for } n \neq 0, \end{cases}$$

where the $J_n(\xi)$ are the Bessel functions and the μ_{nm} are positive roots of the transcendental equation $J_n'(\mu R) = 0$.

⊙ *Reference*: A. N. Tikhonov and A. A. Samarskii (1990).

8.3.3-4. Domain: $0 \leq r \leq R$, $0 \leq \varphi \leq 2\pi$, $-\infty < z < \infty$. Third boundary value problem.

An infinite circular cylinder is considered. A boundary condition is prescribed:

$$\partial_r w + kw = f(\varphi,z) \quad \text{at} \quad r = R.$$

Solution:

$$w(r,\varphi,z) = R\int_0^{2\pi}\int_{-\infty}^{\infty} f(\eta,\zeta)G(r,\varphi,z,R,\eta,\zeta)\,d\zeta\,d\eta$$
$$+ \int_0^R\int_0^{2\pi}\int_{-\infty}^{\infty} \Phi(\xi,\eta,\zeta)G(r,\varphi,z,\xi,\eta,\zeta)\xi\,d\zeta\,d\eta\,d\xi.$$

Here,

$$G(r,\varphi,z,\xi,\eta,\zeta) = \frac{1}{2\pi}\sum_{n=0}^{\infty}\sum_{m=1}^{\infty}\frac{A_n\mu_{nm}^2 J_n(\mu_{nm}r)J_n(\mu_{nm}\xi)\cos[n(\varphi-\eta)]}{(\mu_{nm}^2 R^2 + k^2 R^2 - n^2)J_n^2(\mu_{nm}R)\beta_{nm}}\exp(-\beta_{nm}|z-\zeta|),$$

$$\beta_{nm} = \sqrt{\mu_{nm}^2 - \lambda},\quad A_n = \begin{cases} 1 & \text{for } n = 0, \\ 2 & \text{for } n \neq 0, \end{cases}$$

where the $J_n(\xi)$ are the Bessel functions and the μ_{nm} are positive roots of the transcendental equation

$$\mu J_n'(\mu R) + k J_n(\mu R) = 0.$$

8.3.3-5. Domain: $0 \leq r \leq R$, $0 \leq \varphi \leq 2\pi$, $0 \leq z < \infty$. First boundary value problem.

A semiinfinite circular cylinder is considered. Boundary conditions are prescribed:

$$w = f_1(\varphi,z) \quad \text{at} \quad r = R, \qquad w = f_2(r,\varphi) \quad \text{at} \quad z = 0.$$

Solution:

$$w(r,\varphi,z) = -R\int_0^{2\pi}\int_0^\infty f_1(\eta,\zeta)\left[\frac{\partial}{\partial\xi}G(r,\varphi,z,\xi,\eta,\zeta)\right]_{\xi=R} d\zeta\, d\eta$$

$$+ \int_0^{2\pi}\int_0^R f_2(\xi,\eta)\left[\frac{\partial}{\partial\zeta}G(r,\varphi,z,\xi,\eta,\zeta)\right]_{\zeta=0} \xi\, d\xi\, d\eta$$

$$+ \int_0^R\int_0^{2\pi}\int_0^\infty \Phi(\xi,\eta,\zeta)G(r,\varphi,z,\xi,\eta,\zeta)\xi\, d\zeta\, d\eta\, d\xi.$$

Here,

$$G(r,\varphi,z,\xi,\eta,\zeta) = \frac{1}{2\pi R^2}\sum_{n=0}^\infty\sum_{m=1}^\infty \frac{A_n J_n(\mu_{nm}r)J_n(\mu_{nm}\xi)}{[J_n'(\mu_{nm}R)]^2 \beta_{nm}} \cos[n(\varphi-\eta)]F_{nm}(z,\zeta),$$

$$F_{nm}(z,\zeta) = \exp(-\beta_{nm}|z-\zeta|) - \exp(-\beta_{nm}|z+\zeta|), \quad \beta_{nm} = \sqrt{\mu_{nm}^2 - \lambda}, \quad A_n = \begin{cases} 1 & \text{for } n=0, \\ 2 & \text{for } n\neq 0, \end{cases}$$

where the $J_n(\xi)$ are the Bessel functions and the μ_{nm} are positive roots of the transcendental equation $J_n(\mu R) = 0$.

8.3.3-6. Domain: $0 \leq r \leq R$, $0 \leq \varphi \leq 2\pi$, $0 \leq z < \infty$. Second boundary value problem.

A semiinfinite circular cylinder is considered. Boundary conditions are prescribed:

$$\partial_r w = f_1(\varphi,z) \quad \text{at} \quad r = R, \qquad \partial_z w = f_2(r,\varphi) \quad \text{at} \quad z = 0.$$

Solution:

$$w(r,\varphi,z) = R\int_0^{2\pi}\int_0^\infty f_1(\eta,\zeta)G(r,\varphi,z,R,\eta,\zeta)\, d\zeta\, d\eta$$

$$- \int_0^{2\pi}\int_0^R f_2(\xi,\eta)G(r,\varphi,z,R,\eta,0)\xi\, d\xi\, d\eta$$

$$+ \int_0^R\int_0^{2\pi}\int_0^\infty \Phi(\xi,\eta,\zeta)G(r,\varphi,z,\xi,\eta,\zeta)\xi\, d\zeta\, d\eta\, d\xi.$$

Here,

$$G(r,\varphi,z,\xi,\eta,\zeta) = \frac{\exp(-\sqrt{-\lambda}|z-\zeta|) + \exp(-\sqrt{-\lambda}|z+\zeta|)}{2\pi R^2 \sqrt{-\lambda}}$$

$$+ \frac{1}{2\pi}\sum_{n=0}^\infty\sum_{m=1}^\infty \frac{A_n \mu_{nm}^2 J_n(\mu_{nm}r)J_n(\mu_{nm}\xi)\cos[n(\varphi-\eta)]}{(\mu_{nm}^2 R^2 - n^2)J_n^2(\mu_{nm}R)\beta_{nm}} F_{nm}(z,\zeta),$$

$$F_{nm}(z,\zeta) = \exp(-\beta_{nm}|z-\zeta|) + \exp(-\beta_{nm}|z+\zeta|), \quad \beta_{nm} = \sqrt{\mu_{nm}^2 - \lambda}, \quad A_n = \begin{cases} 1 & \text{for } n=0, \\ 2 & \text{for } n\neq 0, \end{cases}$$

where the $J_n(\xi)$ are the Bessel functions and the μ_{nm} are positive roots of the transcendental equation $J_n'(\mu R) = 0$.

8.3.3-7. Domain: $0 \leq r \leq R$, $0 \leq \varphi \leq 2\pi$, $0 \leq z < \infty$. Third boundary value problem.

A semiinfinite circular cylinder is considered. Boundary conditions are prescribed:

$$\partial_r w + k_1 w = f(\varphi,z) \quad \text{at} \quad r = R, \qquad \partial_z w - k_2 w = f_2(r,\varphi) \quad \text{at} \quad z = 0.$$

Solution:

$$w(r,\varphi,z) = R \int_0^{2\pi} \int_0^\infty f_1(\eta,\zeta) G(r,\varphi,z,R,\eta,\zeta)\, d\zeta\, d\eta$$

$$- \int_0^{2\pi} \int_0^R f_2(\xi,\eta) G(r,\varphi,z,\xi,\eta,0) \xi\, d\xi\, d\eta$$

$$+ \int_0^R \int_0^{2\pi} \int_0^\infty \Phi(\xi,\eta,\zeta) G(r,\varphi,z,\xi,\eta,\zeta) \xi\, d\zeta\, d\eta\, d\xi.$$

Here,

$$G(r,\varphi,z,\xi,\eta,\zeta) = \frac{1}{\pi} \sum_{n=0}^\infty \sum_{m=1}^\infty \frac{A_n \mu_{nm}^2 J_n(\mu_{nm} r) J_n(\mu_{nm}\xi) \cos[n(\varphi-\eta)]}{(\mu_{nm}^2 R^2 + k_1^2 R^2 - n^2) J_n^2(\mu_{nm} R)} F_{nm}(z,\zeta),$$

$$A_n = \begin{cases} 1 & \text{for } n = 0, \\ 2 & \text{for } n \neq 0, \end{cases} \quad F_{nm}(z,\zeta) = \begin{cases} \dfrac{\exp(-\beta_{nm} z)[\beta_{nm}\cosh(\beta_{nm}\zeta) + k_2 \sinh(\beta_{nm}\zeta)]}{\beta_{nm}(\beta_{nm}+k_2)} & \text{for } z > \zeta, \\ \dfrac{\exp(-\beta_{nm}\zeta)[\beta_{nm}\cosh(\beta_{nm} z) + k_2 \sinh(\beta_{nm} z)]}{\beta_{nm}(\beta_{nm}+k_2)} & \text{for } \zeta > z, \end{cases}$$

where the $J_n(\xi)$ are the Bessel functions, $\beta_{nm} = \sqrt{\mu_{nm}^2 - \lambda}$, and the μ_{nm} are positive roots of the transcendental equation

$$\mu J_n'(\mu R) + k_1 J_n(\mu R) = 0.$$

8.3.3-8. Domain: $0 \leq r \leq R$, $0 \leq \varphi \leq 2\pi$, $0 \leq z < \infty$. Mixed boundary value problem.

A semiinfinite circular cylinder is considered. Boundary conditions are prescribed:

$$w = f_1(\varphi, z) \quad \text{at} \quad r = R, \qquad \partial_z w = f_2(r,\varphi) \quad \text{at} \quad z = 0.$$

Solution:

$$w(r,\varphi,z) = -R \int_0^{2\pi} \int_0^\infty f_1(\eta,\zeta) \left[\frac{\partial}{\partial \xi} G(r,\varphi,z,\xi,\eta,\zeta) \right]_{\xi=R} d\zeta\, d\eta$$

$$- \int_0^{2\pi} \int_0^R f_2(\xi,\eta) G(r,\varphi,z,\xi,\eta,0) \xi\, d\xi\, d\eta$$

$$+ \int_0^R \int_0^{2\pi} \int_0^\infty \Phi(\xi,\eta,\zeta) G(r,\varphi,z,\xi,\eta,\zeta) \xi\, d\zeta\, d\eta\, d\xi.$$

Here,

$$G(r,\varphi,z,\xi,\eta,\zeta) = \frac{1}{2\pi R^2} \sum_{n=0}^\infty \sum_{m=1}^\infty \frac{A_n J_n(\mu_{nm} r) J_n(\mu_{nm}\xi)}{[J_n'(\mu_{nm} R)]^2 \beta_{nm}} \cos[n(\varphi-\eta)] F_{nm}(z,\zeta),$$

$$F_{nm}(z,\zeta) = \exp(-\beta_{nm}|z-\zeta|) + \exp(-\beta_{nm}|z+\zeta|), \quad \beta_{nm} = \sqrt{\mu_{nm}^2 - \lambda}, \quad A_n = \begin{cases} 1 & \text{for } n = 0, \\ 2 & \text{for } n \neq 0, \end{cases}$$

where the $J_n(\xi)$ are the Bessel functions and the μ_{nm} are positive roots of the transcendental equation $J_n(\mu R) = 0$.

▶ *Paragraphs 8.3.3-9 through 8.3.3-16 present only the eigenvalues and eigenfunctions of homogeneous boundary value problems for the homogeneous Helmholtz equation (with $\Phi \equiv 0$). The solutions of the corresponding nonhomogeneous boundary value problems ($\Phi \not\equiv 0$) can be constructed by the relations specified in Paragraphs 8.3.1-4 and 8.3.1-8.*

8.3.3-9. Domain: $0 \leq r \leq R$, $0 \leq \varphi \leq 2\pi$, $0 \leq z \leq a$. First boundary value problem.

A circular cylinder of finite length is considered. Boundary conditions are prescribed:

$$w = 0 \text{ at } r = R, \qquad w = 0 \text{ at } z = 0, \qquad w = 0 \text{ at } z = a.$$

Eigenvalues:

$$\lambda_{nmk} = \frac{\pi^2 k^2}{a^2} + \frac{\mu_{nm}^2}{R^2}; \qquad n = 0, 1, \ldots; \quad m, k = 1, 2, \ldots$$

Here, the μ_{nm} are positive zeros of the Bessel functions, $J_n(\mu) = 0$.

Eigenfunctions:

$$w_{nmk}^{(1)} = J_n\left(\mu_{nm}\frac{r}{R}\right)\cos(n\varphi)\sin\left(\frac{\pi k z}{a}\right),$$

$$w_{nmk}^{(2)} = J_n\left(\mu_{nm}\frac{r}{R}\right)\sin(n\varphi)\sin\left(\frac{\pi k z}{a}\right).$$

Eigenfunctions possessing the axial symmetry property:

$$w_{0mk}^{(1)} = J_0\left(\mu_{0m}\frac{r}{R}\right)\sin\left(\frac{\pi k z}{a}\right).$$

The square of the norm of an eigenfunction is defined as

$$\|w_{nmk}^{(1)}\|^2 = \|w_{nmk}^{(2)}\|^2 = \frac{\pi R^2 a}{4}(1 + \delta_{n0})[J_n'(\mu_{nm})]^2, \qquad \delta_{nm} = \begin{cases} 1 & \text{for } n = m, \\ 0 & \text{for } n \neq m. \end{cases}$$

⦿ *References*: V. M. Babich, M. B. Kapilevich, S. G. Mikhlin, et al. (1964), B. M. Budak, A. A. Samarskii, and A. N. Tikhonov (1980).

8.3.3-10. Domain: $0 \leq r \leq R$, $0 \leq \varphi \leq 2\pi$, $0 \leq z \leq a$. Second boundary value problem.

A circular cylinder of finite length is considered. Boundary conditions are prescribed:

$$\partial_r w = 0 \text{ at } r = R, \qquad \partial_z w = 0 \text{ at } z = 0, \qquad \partial_z w = 0 \text{ at } z = a.$$

Eigenvalues:

$$\lambda_{000} = 0, \qquad \lambda_{nmk} = \frac{\pi^2 k^2}{a^2} + \frac{\mu_{nm}^2}{R^2}; \qquad n = 0, 1, \ldots; \quad k, m = 0, 1, \ldots$$

Here, the μ_{nm} are roots of the transcendental equation $J_n'(\mu) = 0$.

Eigenfunctions:

$$w_{nmk}^{(1)} = J_n\left(\mu_{nm}\frac{r}{R}\right)\cos(n\varphi)\cos\left(\frac{\pi k z}{a}\right), \qquad w_{000}^{(1)} = 1,$$

$$w_{nmk}^{(2)} = J_n\left(\mu_{nm}\frac{r}{R}\right)\sin(n\varphi)\cos\left(\frac{\pi k z}{a}\right).$$

The square of the norm of an eigenfunction is defined as

$$\|w_{nmk}^{(1)}\|^2 = \|w_{nmk}^{(2)}\|^2 = \frac{\pi R^2 a}{4\mu_{nm}^2}(1 + \delta_{n0})(\mu_{nm}^2 - n^2)[J_n(\mu_{nm})]^2, \qquad \|w_{000}^{(1)}\|^2 = \pi R^2 a,$$

where δ_{n0} is the Kronecker delta.

⦿ *References*: V. M. Babich, M. B. Kapilevich, S. G. Mikhlin, et al. (1964), B. M. Budak, A. A. Samarskii, and A. N. Tikhonov (1980).

8.3.3-11. Domain: $0 \leq r \leq R$, $0 \leq \varphi \leq 2\pi$, $0 \leq z \leq a$. **Third boundary value problem.**

A circular cylinder of finite length is considered. Boundary conditions are prescribed:

$$\partial_r w + k_1 w = 0 \quad \text{at} \quad r = R, \qquad \partial_z w - k_2 w = 0 \quad \text{at} \quad z = 0, \qquad \partial_z w + k_3 w = 0 \quad \text{at} \quad z = a.$$

Eigenvalues:
$$\lambda_{nml} = \nu_l^2 + \frac{\mu_{nm}^2}{R^2},$$

where the ν_l and μ_{nm} are positive roots of the transcendental equations

$$\tan(\nu a) = \frac{(k_2 + k_3)\nu}{\nu^2 - k_2 k_3}, \qquad \mu J_n'(\mu) + R k_1 J_n(\mu) = 0.$$

Eigenfunctions:

$$w_{nml}^{(1)} = J_n\left(\mu_{nm}\frac{r}{R}\right) \cos(n\varphi) \frac{\nu_l \cos \nu_l z + k_2 \sin \nu_l z}{\sqrt{\nu_l^2 + k_2^2}},$$

$$w_{nml}^{(2)} = J_n\left(\mu_{nm}\frac{r}{R}\right) \sin(n\varphi) \frac{\nu_l \cos \nu_l z + k_2 \sin \nu_l z}{\sqrt{\nu_l^2 + k_2^2}}.$$

The square of the norm of an eigenfunction is defined as

$$\|w_{nml}^{(i)}\|^2 = \frac{\pi R^2}{4\mu_{nm}^2}(1 + \delta_{n0})(R^2 k_1^2 + \mu_{nm}^2 - n^2)\left[J_n(\mu_{nm})\right]^2 \left[a + \frac{(k_2 + k_3)(\nu_l^2 + k_2 k_3)}{(\nu_l^2 + k_2^2)(\nu_l^2 + k_3^2)}\right],$$

where δ_{n0} is the Kronecker delta.

8.3.3-12. Domain: $R_1 \leq r \leq R_2$, $0 \leq \varphi \leq 2\pi$, $0 \leq z \leq a$. **First boundary value problem.**

A hollow circular cylinder of finite length is considered. Boundary conditions are prescribed:

$$w = 0 \quad \text{at} \quad r = R_1, \qquad w = 0 \quad \text{at} \quad r = R_2,$$
$$w = 0 \quad \text{at} \quad z = 0, \qquad w = 0 \quad \text{at} \quad z = a.$$

Eigenvalues:
$$\lambda_{nmk} = \frac{\pi^2 k^2}{a^2} + \mu_{nm}^2; \qquad n = 0, 1, 2, \ldots; \quad m, k = 1, 2, 3, \ldots$$

Here, the μ_{nm} are positive roots of the transcendental equation

$$J_n(\mu R_1) Y_n(\mu R_2) - J_n(\mu R_2) Y_n(\mu R_1) = 0.$$

Eigenfunctions:

$$w_{nmk}^{(1)} = [J_n(\mu_{nm} r) Y_n(\mu_{nm} R_1) - J_n(\mu_{nm} R_1) Y_n(\mu_{nm} r)] \cos(n\varphi) \sin\left(\frac{\pi k z}{a}\right),$$

$$w_{nmk}^{(2)} = [J_n(\mu_{nm} r) Y_n(\mu_{nm} R_1) - J_n(\mu_{nm} R_1) Y_n(\mu_{nm} r)] \sin(n\varphi) \sin\left(\frac{\pi k z}{a}\right).$$

The square of the norm of an eigenfunction is defined as

$$\|w_{nmk}^{(1)}\|^2 = \|w_{nmk}^{(2)}\|^2 = \frac{a}{\pi \mu_{nm}^2}(1 + \delta_{n0}) \frac{[J_n(\mu_{nm} R_1)]^2 - [J_n(\mu_{nm} R_2)]^2}{[J_n(\mu_{nm} R_2)]^2}, \qquad \delta_{ij} = \begin{cases} 1 & \text{for } i = j, \\ 0 & \text{for } i \neq j. \end{cases}$$

⊙ *References*: V. M. Babich, M. B. Kapilevich, S. G. Mikhlin, et al. (1964), B. M. Budak, A. A. Samarskii, and A. N. Tikhonov (1980).

8.3.3-13. Domain: $R_1 \leq r \leq R_2$, $0 \leq \varphi \leq 2\pi$, $0 \leq z \leq a$. Second boundary value problem.

A hollow circular cylinder of finite length is considered. Boundary conditions are prescribed:

$$\partial_r w = 0 \quad \text{at} \quad r = R_1, \qquad \partial_r w = 0 \quad \text{at} \quad r = R_2,$$
$$\partial_z w = 0 \quad \text{at} \quad z = 0, \qquad \partial_z w = 0 \quad \text{at} \quad z = a.$$

Eigenvalues:

$$\lambda_{nmk} = \frac{\pi^2 k^2}{a^2} + \mu_{nm}^2; \qquad n, m, k = 0, 1, 2, \ldots$$

Here, the μ_{nm} are roots of the transcendental equation

$$J_n'(\mu R_1) Y_n'(\mu R_2) - J_n'(\mu R_2) Y_n'(\mu R_1) = 0.$$

Eigenfunctions:

$$w_{nmk}^{(1)} = [J_n(\mu_{nm} r) Y_n'(\mu_{nm} R_1) - J_n'(\mu_{nm} R_1) Y_n(\mu_{nm} r)] \cos(n\varphi) \cos\left(\frac{\pi k z}{a}\right),$$
$$w_{nmk}^{(2)} = [J_n(\mu_{nm} r) Y_n'(\mu_{nm} R_1) - J_n'(\mu_{nm} R_1) Y_n(\mu_{nm} r)] \sin(n\varphi) \cos\left(\frac{\pi k z}{a}\right).$$

To the zero eigenvalue $\lambda_{000} = 0$ there is a corresponding eigenfunction $w_{000}^{(1)} = 1$.

The square of the norm of an eigenfunction is defined as

$$\|w_{nmk}^{(1)}\|^2 = \|w_{nmk}^{(2)}\|^2 = \frac{a(1+\delta_{n0})(1+\delta_{k0})}{\pi \mu_{nm}^2} \left\{ \left(1 - \frac{n^2}{R_2^2 \mu_{nm}^2}\right) \left[\frac{J_n'(\mu_{nm} R_1)}{J_n'(\mu_{nm} R_2)}\right]^2 - \left(1 - \frac{n^2}{R_1^2 \mu_{nm}^2}\right) \right\},$$

where δ_{n0} is the Kronecker delta.

⊙ *References*: V. M. Babich, M. B. Kapilevich, S. G. Mikhlin, et al. (1964), B. M. Budak, A. A. Samarskii, and A. N. Tikhonov (1980).

8.3.3-14. Domain: $R_1 \leq r \leq R_2$, $0 \leq \varphi \leq 2\pi$, $0 \leq z \leq a$. Mixed boundary value problems.

1°. A hollow circular cylinder of finite length is considered. Boundary conditions are prescribed:

$$w = 0 \quad \text{at} \quad r = R_1, \qquad w = 0 \quad \text{at} \quad r = R_2,$$
$$\partial_z w = 0 \quad \text{at} \quad z = 0, \qquad \partial_z w = 0 \quad \text{at} \quad z = a.$$

Eigenvalues:

$$\lambda_{nmk} = \frac{\pi^2 k^2}{a^2} + \mu_{nm}^2; \qquad n, k = 0, 1, 2, \ldots; \quad m = 1, 2, 3, \ldots$$

Here, the μ_{nm} are roots of the transcendental equation

$$J_n(\mu R_1) Y_n(\mu R_2) - J_n(\mu R_2) Y_n(\mu R_1) = 0.$$

Eigenfunctions:

$$w_{nmk}^{(1)} = [J_n(\mu_{nm} r) Y_n(\mu_{nm} R_1) - J_n(\mu_{nm} R_1) Y_n(\mu_{nm} r)] \cos(n\varphi) \cos\left(\frac{\pi k z}{a}\right),$$
$$w_{nmk}^{(2)} = [J_n(\mu_{nm} r) Y_n(\mu_{nm} R_1) - J_n(\mu_{nm} R_1) Y_n(\mu_{nm} r)] \sin(n\varphi) \cos\left(\frac{\pi k z}{a}\right).$$

The square of the norm of an eigenfunction is defined as

$$\|w_{nmk}^{(1)}\|^2 = \|w_{nmk}^{(2)}\|^2 = \frac{a\varepsilon_n \varepsilon_k}{\pi \mu_{nm}^2} \frac{[J_n(\mu_{nm} R_1)]^2 - [J_n(\mu_{nm} R_2)]^2}{[J_n(\mu_{nm} R_2)]^2}, \qquad \varepsilon_n = \begin{cases} 2 & \text{for } n = 0, \\ 1 & \text{for } n \neq 0. \end{cases}$$

2°. A hollow circular cylinder of finite length is considered. Boundary conditions are prescribed:
$$\partial_r w = 0 \quad \text{at} \quad r = R_1, \qquad \partial_r w = 0 \quad \text{at} \quad r = R_2,$$
$$w = 0 \quad \text{at} \quad z = 0, \qquad w = 0 \quad \text{at} \quad z = a.$$

Eigenvalues:
$$\lambda_{nmk} = \frac{\pi^2 k^2}{a^2} + \mu_{nm}^2; \qquad n = 0, 1, 2, \dots; \quad m, k = 1, 2, 3, \dots$$

Here, the μ_{nm} are roots of the transcendental equation
$$J_n'(\mu R_1) Y_n'(\mu R_2) - J_n'(\mu R_2) Y_n'(\mu R_1) = 0.$$

Eigenfunctions:
$$w_{nmk}^{(1)} = [J_n(\mu_{nm} r) Y_n'(\mu_{nm} R_1) - J_n'(\mu_{nm} R_1) Y_n(\mu_{nm} r)] \cos(n\varphi) \sin\left(\frac{\pi k z}{a}\right),$$
$$w_{nmk}^{(2)} = [J_n(\mu_{nm} r) Y_n'(\mu_{nm} R_1) - J_n'(\mu_{nm} R_1) Y_n(\mu_{nm} r)] \sin(n\varphi) \sin\left(\frac{\pi k z}{a}\right).$$

The square of the norm of an eigenfunction is defined as
$$\|w_{nmk}^{(1)}\|^2 = \|w_{nmk}^{(2)}\|^2 = \frac{a\varepsilon_n}{\pi \mu_{nm}^2} \left\{ \left(1 - \frac{n^2}{R_2^2 \mu_{nm}^2}\right) \left[\frac{J_n'(\mu_{nm} R_1)}{J_n'(\mu_{nm} R_2)}\right]^2 - \left(1 - \frac{n^2}{R_1^2 \mu_{nm}^2}\right) \right\},$$

where ε_n is defined in Item 1°.

8.3.3-15. Domain: $0 \le r \le R$, $0 \le \varphi \le \varphi_0$, $0 \le z \le a$. First boundary value problem.

A cylindrical sector of finite thickness is considered. Boundary conditions are prescribed:
$$w = 0 \quad \text{at} \quad \varphi = 0, \qquad w = 0 \quad \text{at} \quad \varphi = \varphi_0, \qquad w = 0 \quad \text{at} \quad r = R,$$
$$w = 0 \quad \text{at} \quad z = 0, \qquad w = 0 \quad \text{at} \quad z = a.$$

Eigenvalues:
$$\lambda_{nmk} = \frac{\pi^2 k^2}{a^2} + \frac{\mu_{nm}^2}{R^2}; \qquad n, m, k = 1, 2, 3, \dots$$

Here, the μ_{nm} are positive roots of the transcendental equation $J_{n\pi/\varphi_0}(\mu) = 0$.

Eigenfunctions:
$$w_{nmk} = J_{n\pi/\varphi_0}\left(\frac{\mu_{nm} r}{R}\right) \sin\left(\frac{n\pi\varphi}{\varphi_0}\right) \sin\left(\frac{k\pi z}{a}\right).$$

The square of the norm of an eigenfunction is defined as
$$\|w_{nmk}\|^2 = \tfrac{1}{8} a R^2 \varphi_0 \left[J_{n\pi/\varphi_0}'(\mu_{nm})\right]^2.$$

8.3.3-16. Domain: $0 \le r \le R$, $0 \le \varphi \le \varphi_0$, $0 \le z \le a$. Mixed boundary value problem.

A cylindrical sector of finite thickness is considered. Boundary conditions are prescribed:
$$w = 0 \quad \text{at} \quad \varphi = 0, \qquad w = 0 \quad \text{at} \quad \varphi = \varphi_0, \qquad w = 0 \quad \text{at} \quad r = R,$$
$$\partial_z w = 0 \quad \text{at} \quad z = 0, \qquad \partial_z w = 0 \quad \text{at} \quad z = a.$$

Eigenvalues:
$$\lambda_{nmk} = \frac{\pi^2 k^2}{a^2} + \frac{\mu_{nm}^2}{R^2}; \qquad n, m = 1, 2, 3, \dots; \quad k = 0, 1, 2, \dots$$

Here, the μ_{nm} are positive roots of the transcendental equation $J_{n\pi/\varphi_0}(\mu) = 0$.

Eigenfunctions:
$$w_{nmk} = J_{n\pi/\varphi_0}\left(\frac{\mu_{nm} r}{R}\right) \sin\left(\frac{n\pi\varphi}{\varphi_0}\right) \cos\left(\frac{k\pi z}{a}\right).$$

The square of the norm of an eigenfunction is defined as
$$\|w_{nmk}\|^2 = \tfrac{1}{8} a R^2 \varphi_0 (1 + \delta_{k0}) \left[J_{n\pi/\varphi_0}'(\mu_{nm})\right]^2, \qquad \delta_{k0} = \begin{cases} 1 & \text{for } k = 0, \\ 0 & \text{for } k \ne 0. \end{cases}$$

8.3.4. Problems in Spherical Coordinates

The three-dimensional homogeneous Helmholtz equation in the spherical coordinate system is written as

$$\frac{1}{r^2}\frac{\partial}{\partial r}\left(r^2\frac{\partial w}{\partial r}\right) + \frac{1}{r^2\sin\theta}\frac{\partial}{\partial \theta}\left(\sin\theta\frac{\partial w}{\partial \theta}\right) + \frac{1}{r^2\sin^2\theta}\frac{\partial^2 w}{\partial \varphi^2} + \lambda w = 0, \qquad r = \sqrt{x^2+y^2+z^2}.$$

8.3.4-1. Particular solutions:

$$w = \frac{1}{r}(A\sin\mu r + B\cos\mu r), \qquad \lambda = \mu^2,$$

$$w = \frac{1}{r}(A\sinh\mu r + B\cosh\mu r), \qquad \lambda = -\mu^2,$$

$$w = \frac{1}{\sqrt{r}}J_{n+1/2}(\mu r)P_n^m(\cos\theta)(A\cos m\varphi + B\sin m\varphi), \qquad \lambda = \mu^2,$$

$$w = \frac{1}{\sqrt{r}}Y_{n+1/2}(\mu r)P_n^m(\cos\theta)(A\cos m\varphi + B\sin m\varphi), \qquad \lambda = \mu^2,$$

$$w = \frac{1}{\sqrt{r}}I_{n+1/2}(\mu r)P_n^m(\cos\theta)(A\cos m\varphi + B\sin m\varphi), \qquad \lambda = -\mu^2,$$

$$w = \frac{1}{\sqrt{r}}K_{n+1/2}(\mu r)P_n^m(\cos\theta)(A\cos m\varphi + B\sin m\varphi), \qquad \lambda = -\mu^2,$$

where $n, m = 0, 1, 2, \ldots$; A and B are arbitrary constants; $J_\nu(\xi)$ and $Y_\nu(\xi)$ are the Bessel functions; $I_\nu(\xi)$ and $K_\nu(\xi)$ are the modified Bessel functions; and the $P_n^m(\xi)$ are the associated Legendre functions that are expressed in terms of the Legendre polynomials $P_n(\xi)$ as

$$P_n^m(\xi) = (1-\xi^2)^{m/2}\frac{d^m}{d\xi^m}P_n(\xi), \qquad P_n(\xi) = \frac{1}{n!\,2^n}\frac{d^n}{d\xi^n}(\xi^2-1)^n.$$

8.3.4-2. Domain: $0 \le r \le R$. First boundary value problem.

1°. A spherical domain is considered. A homogeneous boundary condition is prescribed,

$$w = 0 \quad \text{at} \quad r = R.$$

Eigenvalues:

$$\lambda_{nk} = \frac{\mu_{nk}^2}{R^2}; \qquad n = 0, 1, 2, \ldots; \quad k = 1, 2, 3, \ldots$$

Here, the μ_{nk} are positive zeros of the Bessel functions, $J_{n+1/2}(\mu) = 0$. Note that the $J_{n+1/2}(\mu)$ can be expressed in terms of elementary functions, see Bateman and Erdélyi (1953, Vol. 2).

Eigenfunctions:

$$w_{nmk}^{(1)} = \frac{1}{\sqrt{r}}J_{n+1/2}\left(\mu_{nk}\frac{r}{R}\right)P_n^m(\cos\theta)\cos m\varphi, \qquad m = 0, 1, 2, \ldots;$$

$$w_{nmk}^{(2)} = \frac{1}{\sqrt{r}}J_{n+1/2}\left(\mu_{nk}\frac{r}{R}\right)P_n^m(\cos\theta)\sin m\varphi, \qquad m = 1, 2, 3, \ldots$$

Here, the $P_n^m(\xi)$ are the associated Legendre functions.

Eigenfunctions possessing central symmetry (i.e., independent of θ and φ):

$$w_{00k}^{(1)} = J_{1/2}\left(\mu_{0k}\frac{r}{R}\right).$$

Eigenfunctions possessing axial symmetry (i.e., independent of φ):

$$w_{n0k}^{(1)} = J_{n+1/2}\left(\mu_{nk}\frac{r}{R}\right)P_n(\cos\theta).$$

The square of the norm of an eigenfunction:

$$\|w_{nmk}^{(1)}\|^2 = \frac{\pi R^2(1+\delta_{m0})(n+m)!}{(2n+1)(n-m)!}\left[J'_{n+1/2}(\mu_{nk})\right]^2, \qquad \delta_{m0} = \begin{cases} 1 & \text{for } m=0, \\ 0 & \text{for } m\neq 0, \end{cases}$$

$$\|w_{nmk}^{(1)}\|^2 = \|w_{nmk}^{(2)}\|^2, \qquad m=1,2,3,\ldots$$

2°. A spherical domain is considered. A nonhomogeneous boundary condition is prescribed,

$$w = f(\theta,\varphi) \quad \text{at} \quad r=R.$$

Solution:

$$w(r,\theta,\varphi) = \sum_{n=0}^{\infty}\sum_{m=-n}^{n} f_{nm}\frac{\Psi_n(r\sqrt{\lambda})}{\Psi_n(R\sqrt{\lambda})}Y_n^m(\theta,\varphi), \qquad \Psi_n(x) = \frac{1}{\sqrt{x}}J_{n+1/2}(x),$$

where

$$f_{nm} = \frac{1}{\|Y_n^m\|}\int_0^{2\pi}\int_0^{\pi} f(\theta,\varphi)Y_n^m(\theta,\varphi)\sin\theta\,d\theta\,d\varphi, \qquad \|Y_n^m\| = \frac{2\pi\varepsilon_m}{2n+1}\frac{(n+m)!}{(n-m)!},$$

$$Y_n^m(\theta,\varphi) = \begin{cases} P_n(\cos\theta) & \text{for } m=0, \\ P_n^m(\cos\theta)\sin m\varphi & \text{for } m=1,2,\ldots, \\ P_n^{|m|}(\cos\theta)\cos m\varphi & \text{for } m=-1,-2,\ldots, \end{cases} \qquad \varepsilon_m = \begin{cases} 2 & \text{for } m=0, \\ 1 & \text{for } m\neq 0. \end{cases}$$

The solution was written out under the assumption that $J_{n+1/2}(R\sqrt{\lambda})\neq 0$ for $n=0,1,2,\ldots$

⊙ *References*: M. M. Smirnov (1975), A. N. Tikhonov and A. A. Samarskii (1990).

▶ *Paragraphs 8.3.4-3 through 8.3.4-6 present only the eigenvalues and eigenfunctions of homogeneous boundary value problems for the homogeneous Helmholtz equation (with $\Phi\equiv 0$). The solutions of the corresponding nonhomogeneous boundary value problems ($\Phi\not\equiv 0$) can be constructed by the relations specified in Paragraph 8.3.1-4.*

8.3.4-3. Domain: $0 \leq r \leq R$. Second boundary value problem.

A spherical domain is considered. A boundary condition is prescribed:

$$\partial_r w = 0 \quad \text{at} \quad r=R.$$

Eigenvalues:

$$\lambda_{00} = 0, \qquad \lambda_{nk} = \frac{\mu_{nk}^2}{R^2}; \qquad n=0,1,2,\ldots; \quad k=1,2,3,\ldots$$

Here, the μ_{nk} are roots of the transcendental equation

$$2\mu J'_{n+1/2}(\mu) - J_{n+1/2}(\mu) = 0.$$

Eigenfunctions:

$$w_{000}^{(1)} = 1, \qquad w_{nmk}^{(1)} = \frac{1}{\sqrt{r}}J_{n+1/2}\left(\mu_{nk}\frac{r}{R}\right)P_n^m(\cos\theta)\cos m\varphi, \qquad m=0,1,2,\ldots;$$

$$w_{nmk}^{(2)} = \frac{1}{\sqrt{r}}J_{n+1/2}\left(\mu_{nk}\frac{r}{R}\right)P_n^m(\cos\theta)\sin m\varphi, \qquad m=1,2,3,\ldots$$

The square of the norm of an eigenfunction:

$$\|w_{000}^{(1)}\|^2 = \tfrac{4}{3}\pi R^3, \qquad \|w_{nmk}^{(1)}\|^2 = \frac{\pi R^2 \varepsilon_m(n+m)!}{(2n+1)(n-m)!}\left[1 - \frac{n(n+1)}{\mu_{nk}^2}\right]J_{n+1/2}^2(\mu_{nk}),$$

$$\|w_{nmk}^{(1)}\|^2 = \|w_{nmk}^{(2)}\|^2, \qquad m=1,2,3,\ldots,$$

where $\varepsilon_m = \begin{cases} 2 & \text{for } m=0, \\ 1 & \text{for } m\neq 0. \end{cases}$

⊙ *Reference*: V. M. Babich, M. B. Kapilevich, S. G. Mikhlin, et al. (1964).

8.3.4-4. Domain: $0 \leq r \leq R$. Third boundary value problem.

A spherical domain is considered. A boundary condition is prescribed:

$$\partial_r w + sw = 0 \quad \text{at} \quad r = R.$$

Eigenvalues:

$$\lambda_{nk} = \frac{\mu_{nk}^2}{R^2}; \qquad n = 0, 1, 2, \ldots; \quad k = 1, 2, 3, \ldots$$

Here, the μ_{nk} are positive roots of the transcendental equation

$$2\mu J'_{n+1/2}(\mu) - (1 - 2Rs) J_{n+1/2}(\mu) = 0.$$

Eigenfunctions:

$$w^{(1)}_{nmk} = \frac{1}{\sqrt{r}} J_{n+1/2}\left(\mu_{nk} \frac{r}{R}\right) P_n^m(\cos\theta) \cos m\varphi, \qquad m = 0, 1, 2, \ldots;$$

$$w^{(2)}_{nmk} = \frac{1}{\sqrt{r}} J_{n+1/2}\left(\mu_{nk} \frac{r}{R}\right) P_n^m(\cos\theta) \sin m\varphi, \qquad m = 1, 2, 3, \ldots$$

Here, the $P_n^m(\xi)$ are the associated Legendre functions.

The square of the norm of an eigenfunction:

$$\|w^{(1)}_{nmk}\|^2 = \frac{\pi R^2 \varepsilon_m (n+m)!}{(2n+1)(n-m)!} \left[1 + \frac{(Rs+n)(Rs-n-1)}{\mu_{nk}^2}\right] J_{n+1/2}^2(\mu_{nk}), \quad \varepsilon_m = \begin{cases} 2 & \text{for } m = 0, \\ 1 & \text{for } m \neq 0, \end{cases}$$

$$\|w^{(1)}_{nmk}\|^2 = \|w^{(2)}_{nmk}\|^2, \qquad m = 1, 2, 3, \ldots$$

⊙ *Reference*: V. M. Babich, M. B. Kapilevich, S. G. Mikhlin, et al. (1964).

8.3.4-5. Domain: $R \leq r < \infty$. First boundary value problem.

A spherical cavity is considered and the dependent variable is prescribed at its surface:

$$w = f(\theta, \varphi) \quad \text{at} \quad r = R,$$

and the radiation conditions are prescribed at infinity (see Paragraph 8.3.1-5, Item 2°).

Solution for $\lambda = k^2 > 0$:

$$w(r, \theta, \varphi) = \sum_{n=0}^{\infty} \sum_{m=-n}^{n} f_{nm} \frac{\Xi_n(kr)}{\Xi_n(kR)} Y_n^m(\theta, \varphi), \qquad \Xi_n(\rho) = \frac{1}{\sqrt{\rho}} H_{n+1/2}^{(2)}(\rho),$$

where $H_{n+1/2}^{(2)}(\rho)$ is the Hankel function of the second kind and the other quantities are defined just as in Paragraph 8.3.4-2, Item 2°.

⊙ *Reference*: A. N. Tikhonov and A. A. Samarskii (1990).

8.3.4-6. Domain: $R_1 \leq r \leq R_2$. First boundary value problem.

A spherical layer is considered. Boundary conditions are prescribed:

$$w = 0 \quad \text{at} \quad r = R_1, \qquad w = 0 \quad \text{at} \quad r = R_2.$$

Eigenvalues:

$$\lambda_{nk} = \mu_{nk}^2; \qquad n = 0, 1, 2, \ldots; \quad k = 1, 2, 3, \ldots$$

Here, the μ_{nk} are positive roots of the transcendental equation

$$J_{n+1/2}(\mu R_1) Y_{n+1/2}(\mu R_2) - J_{n+1/2}(\mu R_2) Y_{n+1/2}(\mu R_1) = 0.$$

Eigenfunctions:

$$w_{nmk}^{(1)} = \frac{1}{\sqrt{r}} Z_{n+1/2}(\mu_{nk} r) P_n^m(\cos\theta) \cos m\varphi, \qquad m = 0, 1, 2, \ldots;$$

$$w_{nmk}^{(2)} = \frac{1}{\sqrt{r}} Z_{n+1/2}(\mu_{nk} r) P_n^m(\cos\theta) \sin m\varphi, \qquad m = 1, 2, 3, \ldots$$

Here, the $P_n^m(\xi)$ are the associated Legendre functions and

$$Z_{n+1/2}(\mu r) = J_{n+1/2}(\mu R_1) Y_{n+1/2}(\mu r) - Y_{n+1/2}(\mu R_1) J_{n+1/2}(\mu r).$$

The square of the norm of an eigenfunction:

$$\|w_{nmk}^{(1)}\|^2 = \frac{4\varepsilon_m (n+m)!}{\pi (2n+1)(n-m)!} \frac{J_{n+1/2}^2(\mu_{nk} R_1) - J_{n+1/2}^2(\mu_{nk} R_2)}{\mu_{nk}^2 J_{n+1/2}^2(\mu_{nk} R_2)}, \qquad \varepsilon_m = \begin{cases} 2 & \text{for } m = 0, \\ 1 & \text{for } m \neq 0, \end{cases}$$

$$\|w_{nmk}^{(1)}\|^2 = \|w_{nmk}^{(2)}\|^2, \qquad m = 1, 2, 3, \ldots$$

8.3.5. Other Orthogonal Curvilinear Coordinates

The homogenous three-dimensional Helmholtz equation admits separation of variables in the eleven orthogonal systems of coordinates listed in Table 29.

For the parabolic cylindrical system of coordinates, the multipliers f and g are expressed in terms of the parabolic cylinder functions as

$$f(\xi) = A_1 D_{\mu-1/2}(\sigma\xi) + A_2 D_{\mu-1/2}(-\sigma\xi), \qquad g(\eta) = B_1 D_{-\mu-1/2}(\sigma\eta) + B_2 D_{-\mu-1/2}(-\sigma\eta),$$

$$\mu = \tfrac{1}{2}\beta(k^2 - \lambda)^{-1/2}, \qquad \sigma = \left[4(k^2 - \lambda)\right]^{1/4},$$

where A_1, B_1, A_2, and B_2 are arbitrary constants.

For the elliptic cylindrical system of coordinates, the functions f and g are determined by the modified Mathieu equation and Mathieu equation, respectively, so that

$$f(u) = \begin{cases} \text{Ce}_n(u,q), \\ \text{Se}_n(u,q), \end{cases} \quad g(v) = \begin{cases} \text{ce}_n(v,q), \\ \text{se}_n(v,q), \end{cases} \quad q = \tfrac{1}{4} a^2 (\lambda - k^2),$$

where $\text{Ce}_n(u,q)$ and $\text{Se}_n(u,q)$ are the modified Mathieu functions, and $\text{ce}_n(v,q)$ and $\text{se}_n(v,q)$ are the Mathieu functions; to each value of the parameter q there are certain corresponding eigenvalues $\beta = \beta_n(q)$ [see Abramowitz and Stegun (1964)].

In the prolate and oblate spheroidal systems of coordinates, the equations for f and g are different forms of the spheroidal wave equation, whose bounded solutions are given by

$$f(u) = \text{Ps}_n^{|k|}(\cosh u, a^2 \lambda), \qquad g(u) = \text{Ps}_n^{|k|}(\cos v, a^2 \lambda) \qquad \text{for prolate spheroid,}$$

$$f(u) = \text{Ps}_n^{|k|}(-i \sinh u, a^2 \lambda), \qquad g(u) = \text{Ps}_n^{|k|}(\cos v, -a^2 \lambda) \qquad \text{for oblate spheroid,}$$

$$k \text{ is an integer}, \quad n = 0, 1, 2, \ldots, \quad -n \leq k \leq n,$$

where $\text{Ps}_n^k(z, a)$ are the spheroidal wave functions; see Bateman and Erdélyi (1955, Vol. 3), Arscott (1964), and Meixner and Schäfke (1965). The separation of variables for the Helmholtz equation in modified prolate and oblate spheroidal systems of coordinates, as well as the spheroidal wave functions, are discussed in Abramowitz and Stegun (1964).

In the parabolic coordinate system, the solutions of the equations for f and g are expressed in terms of the degenerate hypergeometric functions [see Miller, Jr. (1977)] as follows:

$$f(\xi) = \xi^k \exp\left(\pm \tfrac{1}{2}\omega\xi^2\right) \Phi\left(-\frac{\beta}{4\omega} + \frac{k+1}{2}, k+1; \mp\omega\xi^2\right), \qquad \omega = \sqrt{-\lambda},$$

$$g(\eta) = \eta^k \exp\left(\pm \tfrac{1}{2}\omega\eta^2\right) \Phi\left(\frac{\beta}{4\omega} + \frac{k+1}{2}, k+1; \mp\omega\eta^2\right).$$

TABLE 29
Orthogonal coordinates $\bar{x}, \bar{y}, \bar{z}$ that allow separable solutions of the form $w = f(\bar{x})g(\bar{y})h(\bar{z})$ for the three-dimensional Helmholtz equation $\Delta_3 w + \lambda w = 0$

Coordinates	Transformations	Particular solutions (or equations for f, g, h)
Cartesian x, y, z	$x = x$, $y = y$, $z = z$	$w = \cos(k_1 x + s_1)\cos(k_2 y + s_2)\cos(k_3 z + s_3)$, where $k_1^2 + k_2^2 + k_3^2 = \lambda$; see also Paragraph 8.3.2-1
Cylindrical r, φ, z	$x = r\cos\varphi$, $y = r\sin\varphi$, $z = z$	$w = [AJ_n(\beta r) + BY_n(\beta r)]\cos(n\varphi + c)\cos(kz + s)$, where $k^2 + \beta^2 = \lambda$, see also Paragraph 8.3.3-1 (J_n and Y_n are the Bessel functions)
Parabolic cylindrical ξ, η, z	$x = \tfrac{1}{2}(\xi^2 - \eta^2)$, $y = \xi\eta$, $z = z$	$w = f(\xi)g(\eta)\cos(kz + s)$, $f'' + [(\lambda - k^2)\xi^2 + \beta]f = 0$, $g'' + [(\lambda - k^2)\eta^2 - \beta]g = 0$
Elliptic cylindrical u, v, z	$x = a\cosh u\cos v$, $y = a\sinh u\sin v$, $z = z$	$w = f(u)g(v)\cos(kz + s)$, $f'' + [\tfrac{1}{2}a^2(\lambda - k^2)\cosh 2u - \beta]f = 0$, $g'' - [\tfrac{1}{2}a^2(\lambda - k^2)\cos 2v - \beta]g = 0$
Spherical r, θ, φ	$x = r\sin\theta\cos\varphi$, $y = r\sin\theta\sin\varphi$, $z = r\cos\theta$	$w = r^{-1/2}J_{n+1/2}(\beta r)P_n^m(\cos\theta)\cos(m\varphi + s)$, $w = r^{-1/2}Y_{n+1/2}(\beta r)P_n^m(\cos\theta)\cos(m\varphi + s)$, where $\lambda = \beta^2$; see also Paragraph 8.3.4-1
Prolate spheroidal u, v, φ	$x = a\sinh u\sin v\cos\varphi$, $y = a\sinh u\sin v\sin\varphi$, $z = a\cosh u\cos v$	$w = f(u)g(v)\cos(k\varphi + s)$, $f'' + f'\coth u + (-\beta + a^2\lambda\sinh^2 u - k^2/\sinh^2 u)f = 0$, $g'' + g'\cot v + (\beta + a^2\lambda\sin^2 v - k^2/\sin^2 v)g = 0$
Oblate spheroidal u, v, φ	$x = a\cosh u\sin v\cos\varphi$, $y = a\cosh u\sin v\sin\varphi$, $z = a\sinh u\cos v$	$w = f(u)g(v)\cos(k\varphi + s)$, $f'' + f'\tanh u + (-\beta + a^2\lambda\cosh^2 u + k^2/\cosh^2 u)f = 0$, $g'' + g'\cot v + (\beta - a^2\lambda\sin^2 v - k^2/\sin^2 v)g = 0$
Parabolic ξ, η, φ	$x = \xi\eta\cos\varphi$, $y = \xi\eta\sin\varphi$, $z = \tfrac{1}{2}(\xi^2 - \eta^2)$	$w = f(\xi)g(\eta)\cos(k\varphi + s)$, $\xi^2 f'' + \xi f' + (\lambda\xi^4 - \beta\xi^2 - k^2)f = 0$, $\eta^2 g'' + \eta g' + (\lambda\eta^4 + \beta\eta^2 - k^2)g = 0$
Paraboloidal u, v, φ	$x = 2a\cosh u\cos v\sinh\varphi$, $y = 2a\sinh u\sin v\cosh\varphi$, $z = \tfrac{1}{2}a(\cosh 2u + \cos 2v - \cosh 2\varphi)$	$f'' + (-k - a\beta\cosh 2u + \tfrac{1}{2}a^2\lambda\cosh 4u)f = 0$, $g'' + (k + a\beta\cos 2v - \tfrac{1}{2}a^2\lambda\cos 4v)g = 0$, $h'' + (-k + a\beta\cosh 2\varphi - \tfrac{1}{2}a^2\lambda\cosh 4\varphi)h = 0$
General ellipsoidal μ, ν, ρ	$x = \sqrt{\frac{(\mu-a)(\nu-a)(\rho-a)}{a(a-1)}}$, $y = \sqrt{\frac{(\mu-1)(\nu-1)(\rho-1)}{1-a}}$, $z = \sqrt{\frac{\mu\nu\rho}{a}}$	$4\sqrt{\varphi(\mu)}\,[\sqrt{\varphi(\mu)}\,f']' + (\lambda\mu^2 + \beta_1\mu + \beta_2)f = 0$, $4\sqrt{\varphi(\nu)}\,[\sqrt{\varphi(\nu)}\,g']' + (\lambda\nu^2 + \beta_1\nu + \beta_2)g = 0$, $4\sqrt{\varphi(\rho)}\,[\sqrt{\varphi(\rho)}\,h']' + (\lambda\rho^2 + \beta_1\rho + \beta_2)h = 0$, $\varphi(t) = t(t-1)(t-a)$
Conical ρ, μ, ν	$x = \rho\sqrt{\frac{(a\mu-1)(a\nu-1)}{1-a}}$, $y = \rho\sqrt{\frac{a(\mu-1)(\nu-1)}{a-1}}$, $z = \rho\sqrt{a\mu\nu}$	$w = \rho^{-1/2}J_{\pm(n+1/2)}(\rho\sqrt{\lambda})g(\xi)h(\eta)$, $g'' + [\beta - n(n+1)k^2\operatorname{sn}^2\xi]g = 0$, $h'' + [\beta - n(n+1)k^2\operatorname{sn}^2\eta]h = 0$, where $\mu = \operatorname{sn}^2(\xi, k)$, $\nu = \operatorname{sn}^2(\eta, k)$, $k = \sqrt{a}$

In the case of the paraboloidal coordinate system, the equations for f, g, and h are reduced to the Whittaker–Hill equation

$$G''_{\theta\theta} + \left(\mu + \tfrac{1}{8}b^2 + bc\cos 2\theta - \tfrac{1}{8}b^2\cos 4\theta\right)G = 0.$$

Denote by $\mathrm{gc}_n(\theta; b, c)$ and $\mathrm{gs}_n(\theta; b, c)$, respectively, the even and odd 2π-periodic solutions of the Whittaker–Hill equation, which is a generalization of the Mathieu equation. The subscript $n = 0, 1, 2, \ldots$ labels the discrete eigenvalues $\mu = \mu_n$. Each of the solutions gc_n and gs_n can be represented in the form of an infinite convergent trigonometric series in $\cos n\theta$ and $\sin n\theta$, respectively; see Urvin and Arscott (1970). The functions f, g, and h can be expressed in terms of the periodic solutions of the Whittaker–Hill equation as follows [Miller, Jr. (1977)]:

$$f(u) = \begin{cases} \mathrm{gc}_n(iu; 2a\omega, \tfrac{1}{2}\beta/\omega), \\ \mathrm{gs}_n(iu; 2a\omega, \tfrac{1}{2}\beta/\omega), \end{cases} \quad g(v) = \begin{cases} \mathrm{gc}_n(v; 2a\omega, \tfrac{1}{2}\beta/\omega), \\ \mathrm{gs}_n(v; 2a\omega, \tfrac{1}{2}\beta/\omega), \end{cases} \quad h(\varphi) = \begin{cases} \mathrm{gc}_n(i\varphi + \tfrac{\pi}{2}; 2a\omega, \tfrac{1}{2}\beta/\omega), \\ \mathrm{gs}_n(i\varphi + \tfrac{\pi}{2}; 2a\omega, \tfrac{1}{2}\beta/\omega), \end{cases}$$

where $\omega = \sqrt{\lambda}$ and $k = \mu_n - \tfrac{1}{2}a^2\lambda$.

For the general ellipsoidal coordinates, the functions f, g, and h are expressed in terms of the ellipsoidal wave functions; for details, see Arscott (1964) and Miller, Jr. (1977).

For the conical coordinate system, the functions g and h are determined by the Lamé equations that involve the Jacobian elliptic function $\mathrm{sn}\, z = \mathrm{sn}(z, k)$.

The unambiguity conditions for the transformation yield $n = 0, 1, 2, \ldots$ It is known that, for any positive integer n, there exist exactly $2n+1$ solutions corresponding to $2n+1$ different eigenvalues β. These solutions can be represented the form of finite series known as Lamé polynomials. For more details about the Lamé equation and its solutions, see Whittaker and Watson (1963), Arscott (1964), Bateman and Erdélyi (1955), and Miller, Jr. (1977).

Unlike the Laplace equation, there are no nontrivial transformations for the three-dimensional Helmholtz equation that allow the $\mathcal{R}$-separation of variables.

⊙ *References for Subsection* 8.3.5: F. M. Morse and H. Feshbach (1953, Vols. 1–2), P. Moon and D. Spencer (1961), A. Makarov, J. Smorodinsky, K. Valiev, and P. Winternitz (1967), W. Miller, Jr. (1977).

8.4. Other Equations with Three Space Variables

8.4.1. Equations Containing Arbitrary Functions

1. $\dfrac{\partial^2 w}{\partial x^2} + \dfrac{\partial^2 w}{\partial y^2} + \dfrac{\partial^2 w}{\partial z^2} + \left(\lambda + \dfrac{a}{r}\right)w = 0, \qquad r^2 = x^2 + y^2 + z^2.$

Schrödinger's equation. It governs the motion of an electron in the Coulomb field of a nucleus ($a > 0$).

The desired solutions must satisfy the normalizing condition

$$\int_{-\infty}^{\infty}\int_{-\infty}^{\infty}\int_{-\infty}^{\infty} |w(x, y, z)|^2 \, dx\, dy\, dz = 1.$$

Eigenvalues:

$$\lambda_n = -\dfrac{a^2}{4n^2}; \qquad n = 1, 2, 3, \ldots$$

Normalized eigenfunctions (in the spherical coordinate system r, θ, φ):

$$w_{nmk} = \left(\dfrac{2}{n}\right)^{3/2}\sqrt{\dfrac{(2k+1)(k-m)!\,(n-k-1)!}{4\pi\varepsilon_m n(n+k)!\,(m+k)!}}\left(\dfrac{ar}{n}\right)^k \exp\left(-\dfrac{ar}{2n}\right) L_{n-k-1}^{2k+1}\left(\dfrac{ar}{n}\right) Y_k^{(m)}(\theta, \varphi),$$

$$n = 1, 2, 3, \ldots\,; \quad m = 0, \pm 1, \pm 2, \ldots, \pm k; \quad k = 0, 1, 2, \ldots, n-1;$$

where

$$\varepsilon_m = \begin{cases} 2 & \text{for } m = 0, \\ 1 & \text{for } m \neq 0, \end{cases} \quad Y_k^{(m)}(\theta, \varphi) = \begin{cases} P_k(\cos\theta) & \text{for } m = 0, \\ P_k^m(\cos\theta)\sin m\varphi & \text{for } m = 1, 2, \ldots, \\ P_k^{|m|}(\cos\theta)\cos m\varphi & \text{for } m = -1, -2, \ldots, \end{cases}$$

$$L_k^s(x) = \frac{1}{k!}x^{-s}e^x \frac{d^k}{dx^k}\left(x^{k+s}e^{-x}\right), \quad P_k^m(x) = (1-x^2)^{m/2}\frac{d^m}{dx^m}P_k(x), \quad P_k(x) = \frac{1}{k!\,2^k}\frac{d^k}{dx^k}(x^2-1)^k.$$

These relations involve the generalized Laguerre polynomials $L_k^s(x)$ and the associated Legendre functions $P_n^m(\xi)$; the $P_n(\xi)$ are the Legendre polynomials.

⊙ *References*: G. Korn and T. Korn (1968), A. N. Tikhonov and A. A. Samarskii (1990).

2. $\dfrac{\partial^2 w}{\partial x^2} + \dfrac{\partial^2 w}{\partial y^2} + \dfrac{\partial^2 w}{\partial z^2} = ay\dfrac{\partial w}{\partial x}.$

This equation is encountered in problems of convective heat and mass transfer in a simple shear flow.

Fundamental solution:

$$\mathcal{E}(x,y,z,\xi,\eta,\zeta) = \frac{1}{(4\pi)^{3/2}}\int_0^\infty \exp\left\{-\frac{[x-\xi-\frac{1}{2}at(y+\eta)]^2}{4t(1+\frac{1}{12}a^2t^2)} - \frac{(y-\eta)^2+(z-\zeta)^2}{4t}\right\}\frac{dt}{\sqrt{t^3(1+\frac{1}{12}a^2t^2)}}.$$

⊙ *References*: E. A. Novikov (1958), D. E. Elrick (1962).

3. $\dfrac{\partial^2 w}{\partial x^2} + \dfrac{\partial^2 w}{\partial y^2} + \dfrac{\partial^2 w}{\partial z^2} + a_1 x\dfrac{\partial w}{\partial x} + a_2 y\dfrac{\partial w}{\partial y} + a_3 z\dfrac{\partial w}{\partial z} = 0.$

This equation is encountered in problems of convective heat and mass transfer in a straining flow.

Fundamental solution:

$$\mathcal{E}(x,y,z,\xi,\eta,\zeta) = \int_0^\infty F(x,\xi,t;a_1)F(y,\eta,t;a_2)F(z,\zeta,t;a_3)\,dt,$$

$$F(x,\xi,t;a) = \left[\frac{2\pi}{a}\left(e^{2at}-1\right)\right]^{-1/2}\exp\left[-\frac{a(xe^{at}-\xi)^2}{2(e^{2at}-1)}\right].$$

4. $\dfrac{\partial^2 w}{\partial x_1^2} + \dfrac{\partial^2 w}{\partial x_2^2} + \dfrac{\partial^2 w}{\partial x_3^2} = \displaystyle\sum_{n,k=1}^{3} a_{nk}x_n\dfrac{\partial w}{\partial x_k}.$

This equation is encountered in problems of convective heat and mass transfer in an arbitrary linear shear flow.

The solution that corresponds to a source of unit power at the origin of coordinates is given by

$$w(x_1, x_2, x_3) = \frac{1}{(4\pi)^{3/2}}\int_0^\infty \exp\left[-\sum_{n,k=1}^{3}\frac{b_{nk}(t)x_n x_k}{4D(t)}\right]\frac{dt}{\sqrt{D(t)}}.$$

Here, $D = D(t)$ is the determinant of the matrix $\mathbf{B} = \{B_{nk}\}$; the $b_{nk} = b_{nk}(t)$ are the cofactors of the entries $B_{nk} = B_{nk}(t)$; the B_{nk} are determined by solving the following system of ordinary differential equations with constant coefficients:

$$\frac{dB_{nk}}{dt} = \delta_{nk} + \sum_{m=1}^{3} a_{nm}B_{km} + \sum_{m=1}^{3} a_{km}B_{nm},$$

$$B_{nk} \to \delta_{nk}t \quad \text{as} \quad t \to 0 \quad \text{(initial conditions)},$$

where $\delta_{nn} = 1$ and $\delta_{nk} = 0$ if $n \neq k$.

⊙ *Reference*: G. K. Batchelor (1979).

5. $\dfrac{\partial}{\partial x}\left[f_1(x)\dfrac{\partial w}{\partial x}\right] + \dfrac{\partial}{\partial y}\left[f_2(y)\dfrac{\partial w}{\partial y}\right] + \dfrac{\partial}{\partial z}\left[f_3(z)\dfrac{\partial w}{\partial z}\right] = \beta w.$

This is a three-dimensional linear equation of heat and mass transfer theory with a source in an inhomogeneous anisotropic medium. Here, $f_1 = f_1(x)$, $f_2 = f_2(y)$, and $f_3 = f_3(z)$ are the principal thermal diffusivities.

1°. The equation admits multiplicatively separable solutions, $w(x, y, z) = \varphi_1(x)\varphi_2(y)\varphi_3(z)$.

2°. There are also additively separable solutions, $w(x, y, z) = \psi_1(x) + \psi_2(y) + \psi_3(z)$.

3°. If $f_1 = ax^n$, $f_2 = by^m$, and $f_3 = cz^k$ ($n \neq 2$, $m \neq 2$, $k \neq 2$), there are particular solutions of the form

$$w = w(\xi), \qquad \xi^2 = 4\left[\dfrac{x^{2-n}}{a(2-n)^2} + \dfrac{y^{2-m}}{b(2-m)^2} + \dfrac{z^{2-k}}{c(2-k)^2}\right],$$

where the function $w(\xi)$ is determined by the ordinary differential equation

$$\dfrac{d^2 w}{d\xi^2} + \dfrac{A}{\xi}\dfrac{dw}{d\xi} = \beta w, \qquad A = 2\left(\dfrac{1}{2-n} + \dfrac{1}{2-m} + \dfrac{1}{2-k}\right) - 1,$$

whose solutions are expressed in terms of the Bessel functions.

8.4.2. Equations of the Form $\operatorname{div}[a(x,y,z)\nabla w] - q(x,y,z)w = -\Phi(x,y,z)$

Equations of this sort are often encountered in heat and mass transfer theory. For brevity, the equation is written using the notation

$$\operatorname{div}[a(\mathbf{r})\nabla w] = \dfrac{\partial}{\partial x}\left[a(\mathbf{r})\dfrac{\partial w}{\partial x}\right] + \dfrac{\partial}{\partial y}\left[a(\mathbf{r})\dfrac{\partial w}{\partial y}\right] + \dfrac{\partial}{\partial z}\left[a(\mathbf{r})\dfrac{\partial w}{\partial z}\right], \qquad \mathbf{r} = \{x, y, z\}.$$

In what follows, the problems for the equation in question will be considered in a bounded domain V with a sufficiently smooth surface S. It is assumed that $a(\mathbf{r}) > 0$ and $q(\mathbf{r}) \geq 0$.

8.4.2-1. First boundary value problem.

The following boundary condition of the first kind is imposed:

$$w = f(\mathbf{r}) \quad \text{for} \quad \mathbf{r} \in S.$$

Solution:

$$w(\mathbf{r}) = \int_V \Phi(\rho)G(\mathbf{r},\rho)\,dV_\rho - \int_S f(\rho)a(\rho)\dfrac{\partial}{\partial N_\rho}G(\mathbf{r},\rho)\,dS_\rho. \tag{1}$$

Here, the Green's function is given by

$$G(\mathbf{r},\rho) = \sum_{n=1}^{\infty}\dfrac{u_n(\mathbf{r})u_n(\rho)}{\|u_n\|^2 \lambda_n}, \qquad \|u_n\|^2 = \int_V u_n^2(\mathbf{r})\,dV, \qquad \rho = \{\xi, \eta, \zeta\}, \tag{2}$$

where the λ_n and $u_n(\mathbf{r})$ are the eigenvalues and eigenfunctions of the Sturm–Liouville problem for the following second-order elliptic equation with a homogeneous boundary condition of the first kind:

$$\operatorname{div}\left[a(\mathbf{r})\nabla u\right] - q(\mathbf{r})u + \lambda u = 0, \tag{3}$$

$$u = 0 \quad \text{for} \quad \mathbf{r} \in S. \tag{4}$$

The integration in (1) is performed with respect to ξ, η, ζ; $\dfrac{\partial}{\partial N_\rho}$ denotes the derivative along the outward normal to the surface S with respect to ξ, η, ζ.

General properties of the Sturm–Liouville problem (3)–(4):

1°. There are countably many eigenvalues. All eigenvalues are real and can be ordered so that $\lambda_1 \le \lambda_2 \le \lambda_3 \le \cdots$, with $\lambda_n \to \infty$ as $n \to \infty$; therefore the number of negative eigenvalues is finite.

2°. If $a(\mathbf{r}) > 0$ and $q(\mathbf{r}) \ge 0$, all eigenvalues are positive, $\lambda_n > 0$.

3°. The eigenfunctions are defined up to a constant multiplier. Any two eigenfunctions, $u_n(\mathbf{r})$ and $u_m(\mathbf{r})$, corresponding to different eigenvalues, λ_n and λ_m, are orthogonal to each other in V:

$$\int_V u_n(\mathbf{r}) u_m(\mathbf{r}) \, dV = 0 \quad \text{for} \quad n \ne m.$$

4°. An arbitrary function $F(\mathbf{r})$ that is twice continuously differentiable and satisfies the boundary condition of the Sturm–Liouville problem ($F = 0$ for $\mathbf{r} \in S$) can be expanded into an absolutely and uniformly convergent series in the eigenfunctions; specifically,

$$F(\mathbf{r}) = \sum_{n=1}^{\infty} F_n u_n(\mathbf{r}), \quad F_n = \frac{1}{\|u_n\|^2} \int_V F(\mathbf{r}) u_n(\mathbf{r}) \, dV,$$

where the norm squared $\|u_n\|^2$ is defined in (2).

Remark. In a three-dimensional problem, to each eigenvalue λ_n finitely many linearly independent eigenfunctions $u_n^{(1)}, \ldots, u_n^{(m)}$ generally correspond. These functions can always be replaced by their linear combinations

$$\bar{u}_n^{(k)} = A_{k,1} u_n^{(1)} + \cdots + A_{k,k-1} u_n^{(k-1)} + u_n^{(k)}, \quad k = 1, 2, \ldots, m,$$

such that $\bar{u}_n^{(1)}, \ldots, \bar{u}_n^{(m)}$ are now pairwise orthogonal. Therefore, without loss of generality, we can assume that all eigenfunctions are orthogonal.

8.4.2-2. Second boundary value problem.

A boundary condition of the second kind is imposed,

$$\frac{\partial w}{\partial N} = f(\mathbf{r}) \quad \text{for} \quad \mathbf{r} \in S.$$

It is assumed that $q(\mathbf{r}) > 0$.

Solution:

$$w(\mathbf{r}) = \int_V \Phi(\boldsymbol{\rho}) G(\mathbf{r}, \boldsymbol{\rho}) \, dV_\rho + \int_S f(\boldsymbol{\rho}) a(\boldsymbol{\rho}) G(\mathbf{r}, \boldsymbol{\rho}) \, dS_\rho. \qquad (5)$$

Here, the Green's function is defined by relation (2), where the λ_n and $u_n(\mathbf{r})$ are the eigenvalues and eigenfunctions of the Sturm–Liouville problem for the second-order elliptic equation (3) with the following homogeneous boundary condition of the second kind:

$$\frac{\partial u}{\partial N} = 0 \quad \text{for} \quad \mathbf{r} \in S. \qquad (6)$$

If $q(\mathbf{r}) > 0$, the general properties of the eigenvalue problem (3), (6) are the same as those of the first boundary value problem (see Paragraph 8.4.2-1).

8.4.2-3. Third boundary value problem.

The following boundary condition of the third kind is set:

$$\frac{\partial w}{\partial N} + k(\mathbf{r}) w = f(\mathbf{r}) \quad \text{for} \quad \mathbf{r} \in S.$$

The solution of the third boundary value problem is given by relations (5) and (2), where the λ_n and $u_n(\mathbf{r})$ are the eigenvalues and eigenfunctions of the Sturm–Liouville problem for the

second-order elliptic equation (3) with the following homogeneous boundary condition of the third kind:
$$\frac{\partial u}{\partial N} + k(\mathbf{r})u = 0 \quad \text{for} \quad \mathbf{r} \in S. \tag{7}$$

If $q(\mathbf{r}) \geq 0$ and $k(\mathbf{r}) > 0$, the general properties of the eigenvalue problem (3), (7) are the same as those of the first boundary value problem (see Paragraph 8.4.2-1).

Let $k(\mathbf{r}) = k = \text{const}$. Denote the Green's functions of the second and third boundary value problems by $G_2(\mathbf{r}, \boldsymbol{\rho})$ and $G_3(\mathbf{r}, \boldsymbol{\rho}, k)$, respectively. For $q(\mathbf{r}) > 0$, the following limit relation holds:
$$G_2(\mathbf{r}, \boldsymbol{\rho}) = \lim_{k \to 0} G_3(\mathbf{r}, \boldsymbol{\rho}, k).$$

8.5. Equations with n Space Variables

8.5.1. Laplace Equation $\Delta_n w = 0$

The n-dimensional Laplace equation in the rectangular Cartesian system of coordinates $x_1, \ldots, x_n$ has the form
$$\frac{\partial^2 w}{\partial x_1^2} + \frac{\partial^2 w}{\partial x_2^2} + \cdots + \frac{\partial^2 w}{\partial x_n^2} = 0.$$

For $n = 2$ and $n = 3$, see Subsections 7.1.1 and 8.1.1.

A regular solution of the Laplace equation is called a harmonic function.

In what follows we use the notation: $\mathbf{x} = \{x_1, \ldots, x_n\}$ and $|\mathbf{x}| = \sqrt{x_1^2 + \cdots + x_n^2}$.

8.5.1-1. Particular solutions.

1°. Fundamental solution:
$$\mathscr{E}(\mathbf{x}) = -\frac{1}{(n-2)\sigma_n |\mathbf{x}|^{n-2}}, \quad \sigma_n = \frac{2\pi^{n/2}}{\Gamma(n/2)} \quad (n \geq 3).$$

2°. Solution containing arbitrary functions of $n-1$ variables:
$$w(x_1, \ldots, x_n) = \sum_{k=0}^{\infty} (-1)^k \left[\frac{x_n^{2k}}{(2k)!} \Delta^k f(x_1, \ldots, x_{n-1}) + \frac{x_n^{2k+1}}{(2k+1)!} \Delta^k g(x_1, \ldots, x_{n-1}) \right],$$

where $f(x_1, \ldots, x_{n-1})$ and $g(x_1, \ldots, x_{n-1})$ are arbitrary infinitely differentiable functions.

3°. Let $w(x_1, \ldots, x_n)$ be a harmonic function. Then the functions
$$w_1 = A w(\pm \lambda x_1 + C_1, \ldots, \pm \lambda x_n + C_n),$$
$$w_2 = \frac{A}{|\mathbf{x}|^{n-2}} w\left(\frac{x_1}{|\mathbf{x}|^2}, \ldots, \frac{x_n}{|\mathbf{x}|^2} \right),$$

are also harmonic functions everywhere they are defined; A, C_1, ..., C_n, and λ are arbitrary constants. The signs at λ in the expression of w_1 can be taken independently of one another.

⊙ *References*: A. V. Bitsadze and D. F. Kalinichenko (1985), R. Courant and D. Hilbert (1989).

8.5.1-2. Domain: $-\infty < x_1 < \infty, \ldots, -\infty < x_{n-1} < \infty, 0 \leq x_n < \infty$.

The first boundary value problem for an n-dimensional half-space is considered. A boundary condition is prescribed:
$$w = f(x_1, \ldots, x_{n-1}) \quad \text{at} \quad x_n = 0.$$

Solution:
$$w(x_1, \ldots, x_n) = \frac{\Gamma(n/2)}{\pi^{n/2}} \int_{-\infty}^{\infty} \cdots \int_{-\infty}^{\infty} \left[\sum_{k=1}^{n-1} (y_k - x_k)^2 + x_n^2 \right]^{-n/2} x_n f(y_1, \ldots, y_{n-1}) \, dy_1 \ldots dy_{n-1},$$

where $\Gamma(z)$ is the gamma function.

⊙ *Reference*: A. V. Bitsadze and D. F. Kalinichenko (1985).

8.5.1-3. Domain: $|\mathbf{x}| \leq 1$. First boundary value problem.

A sphere of unit radius in the n-dimensional space is considered. A boundary condition is prescribed:
$$w = f(\mathbf{x}) \quad \text{for} \quad |\mathbf{x}| = 1.$$

Solution (Poisson integral):
$$w(\mathbf{x}) = \frac{\Gamma(n/2)}{2\pi^{n/2}} \int_{|\mathbf{y}|=1} \frac{1 - |\mathbf{x}|^2}{|\mathbf{y} - \mathbf{x}|^n} f(\mathbf{y}) \, dS_{\mathbf{y}}.$$

⊙ *Reference*: A. V. Bitsadze and D. F. Kalinichenko (1985).

8.5.2. Other Equations

1. $\Delta_n w = -\Phi(x_1, \ldots, x_n)$.

This is the *Poisson equation* in n independent variables. For $n = 2$ and $n = 3$, see Sections 7.2 and 8.2.

1°. Solution:
$$w(x_1, \ldots, x_n) = \frac{\Gamma(n/2)}{2(n-2)\pi^{n/2}} \int_{\mathbb{R}^n} \frac{\Phi(y_1, \ldots, y_n) \, dy_1 \ldots dy_n}{\left[(x_1 - y_1)^2 + \cdots + (x_n - y_n)^2\right]^{\frac{n-2}{2}}}.$$

⊙ *Reference*: S. G. Krein (1972).

2°. Domain: $0 \leq x_k \leq a_k$; $k = 1, \ldots, n$. First boundary value problem.
A rectangular parallelepiped is considered. Boundary conditions are prescribed:
$$w = f_k(x_1, \ldots, x_{k-1}, x_{k+1}, \ldots, x_n) \quad \text{at} \quad x_k = 0,$$
$$w = g_k(x_1, \ldots, x_{k-1}, x_{k+1}, \ldots, x_n) \quad \text{at} \quad x_k = a_k.$$

Green's function:
$$G(x_1, \ldots, x_n, y_1, \ldots, y_n) = \frac{2^n}{a_1 \ldots a_n} \sum_{k_1=1}^{\infty} \cdots \sum_{k_n=1}^{\infty} \frac{\sin(p_{k_1} x_1) \sin(p_{k_1} y_1) \ldots \sin(p_{k_n} x_n) \sin(p_{k_n} y_n)}{p_{k_1}^2 + \cdots + p_{k_n}^2},$$
$$p_{k_1} = \frac{\pi k_1}{a_1}, \quad p_{k_2} = \frac{\pi k_2}{a_2}, \quad \ldots, \quad p_{k_n} = \frac{\pi k_n}{a_n}.$$

3°. Domain: $0 \leq x_k \leq a_k$; $k = 1, \ldots, n$. Mixed boundary value problem.
A rectangular parallelepiped is considered. Boundary conditions are prescribed:
$$w = f_k(x_1, \ldots, x_{k-1}, x_{k+1}, \ldots, x_n) \quad \text{at} \quad x_k = 0,$$
$$\partial_{x_k} w = g_k(x_1, \ldots, x_{k-1}, x_{k+1}, \ldots, x_n) \quad \text{at} \quad x_k = a_k.$$

Green's function:
$$G(x_1, \ldots, x_n, y_1, \ldots, y_n) = \frac{2^n}{a_1 \ldots a_n} \sum_{k_1=0}^{\infty} \cdots \sum_{k_n=0}^{\infty} \frac{\sin(p_{k_1} x_1) \sin(p_{k_1} y_1) \ldots \sin(p_{k_n} x_n) \sin(p_{k_n} y_n)}{p_{k_1}^2 + \cdots + p_{k_n}^2},$$
$$p_{k_1} = \frac{\pi(2k_1 + 1)}{2a_1}, \quad p_{k_2} = \frac{\pi(2k_2 + 1)}{2a_2}, \quad \ldots, \quad p_{k_n} = \frac{\pi(2k_n + 1)}{2a_n}.$$

2. $\Delta_n w + \lambda w = 0$.

This is the *Helmholtz equation* in n independent variables. For $n = 2$ and $n = 3$, see Sections 7.3 and 8.3.

1°. Fundamental solution for $\lambda = k^2 > 0$:

$$\mathscr{E}(\mathbf{x},\mathbf{y}) = \frac{k^{\frac{n-2}{2}}}{4(2\pi)^{\frac{n-2}{2}}} r^{-\frac{n-2}{2}} Y_{\frac{n-2}{2}}(kr), \quad r = |\mathbf{x}-\mathbf{y}| \qquad \text{for even } n,$$

$$\mathscr{E}(\mathbf{x},\mathbf{y}) = \frac{k^{\frac{n-2}{2}}}{4(2\pi)^{\frac{n-2}{2}} \sin(\frac{1}{2}\pi n)} r^{-\frac{n-2}{2}} J_{-\frac{n-2}{2}}(kr) \qquad \text{for odd } n,$$

where $J_\nu(z)$ and $Y_\nu(z)$ are the Bessel functions.

2°. Domain: $0 \le x_k \le a_k$; $k = 1, \ldots, n$. First boundary value problem.
A rectangular parallelepiped is considered. Boundary conditions are prescribed:

$$w = f_k(x_1, \ldots, x_{k-1}, x_{k+1}, \ldots, x_n) \quad \text{at} \quad x_k = 0,$$
$$w = g_k(x_1, \ldots, x_{k-1}, x_{k+1}, \ldots, x_n) \quad \text{at} \quad x_k = a_k.$$

Green's function:

$$G(x_1, \ldots, x_n, y_1, \ldots, y_n) = \frac{2^n}{a_1 a_2 \ldots a_n} \sum_{k_1=1}^{\infty} \cdots \sum_{k_n=1}^{\infty} \frac{\sin(p_{k_1} x_1)\sin(p_{k_1} y_1) \ldots \sin(p_{k_n} x_n)\sin(p_{k_n} y_n)}{p_{k_1}^2 + \cdots + p_{k_n}^2 - \lambda},$$

$$p_{k_1} = \frac{\pi k_1}{a_1}, \quad p_{k_2} = \frac{\pi k_2}{a_2}, \quad \ldots, \quad p_{k_n} = \frac{\pi k_n}{a_n}.$$

3°. Domain: $0 \le x_k \le a_k$; $k = 1, \ldots, n$. Second boundary value problem.
A rectangular parallelepiped is considered. Boundary conditions are prescribed:

$$\partial_{x_k} w = f_k(x_1, \ldots, x_{k-1}, x_{k+1}, \ldots, x_n) \quad \text{at} \quad x_k = 0,$$
$$\partial_{x_k} w = g_k(x_1, \ldots, x_{k-1}, x_{k+1}, \ldots, x_n) \quad \text{at} \quad x_k = a_k.$$

Green's function:

$$G(x_1, \ldots, x_n, y_1, \ldots, y_n) = \sum_{k_1=0}^{\infty} \cdots \sum_{k_n=0}^{\infty} A_{k_1 k_2 \ldots k_n} \frac{\cos(p_{k_1} x_1)\cos(p_{k_1} y_1) \ldots \cos(p_{k_n} x_n)\cos(p_{k_n} y_n)}{p_{k_1}^2 + \cdots + p_{k_n}^2 - \lambda},$$

$$A_{k_1 k_2 \ldots k_n} = \frac{\varepsilon_{k_1} \varepsilon_{k_2} \ldots \varepsilon_{k_n}}{a_1 a_2 \ldots a_n}, \quad p_{k_1} = \frac{\pi k_1}{a_1}, \quad p_{k_2} = \frac{\pi k_2}{a_2}, \quad \ldots, \quad p_{k_n} = \frac{\pi k_n}{a_n}, \quad \varepsilon_m = \begin{cases} 1 & \text{for } m = 0, \\ 2 & \text{for } m \ne 0. \end{cases}$$

4°. Domain: $0 \le x_k \le a_k$; $k = 1, \ldots, n$. Mixed boundary value problem.
A rectangular parallelepiped is considered. Boundary conditions are prescribed:

$$w = f_k(x_1, \ldots, x_{k-1}, x_{k+1}, \ldots, x_n) \quad \text{at} \quad x_k = 0,$$
$$\partial_{x_k} w = g_k(x_1, \ldots, x_{k-1}, x_{k+1}, \ldots, x_n) \quad \text{at} \quad x_k = a_k.$$

Green's function:

$$G(x_1, \ldots, x_n, y_1, \ldots, y_n) = \frac{2^n}{a_1 a_2 \ldots a_n} \sum_{k_1=0}^{\infty} \cdots \sum_{k_n=0}^{\infty} \frac{\sin(p_{k_1} x_1)\sin(p_{k_1} y_1) \ldots \sin(p_{k_n} x_n)\sin(p_{k_n} y_n)}{p_{k_1}^2 + \cdots + p_{k_n}^2 - \lambda},$$

$$p_{k_1} = \frac{\pi(2k_1+1)}{2a_1}, \quad p_{k_2} = \frac{\pi(2k_2+1)}{2a_2}, \quad \ldots, \quad p_{k_n} = \frac{\pi(2k_n+1)}{2a_n}.$$

⊙ *Reference*: V. M. Babich, M. B. Kapilevich, S. G. Mikhlin, et al. (1964).

3. $\displaystyle\sum_{i,j=1}^{n} a_{ij}\frac{\partial^2 w}{\partial x_i \partial x_j} = 0.$

It is assumed that for any real numbers $y_1, \ldots, y_n$ the relation $\left|\sum_{i,j=1}^{n} a_{ij} y_i y_j\right| \geq k \sum_{i=1}^{n} y_i^2$ holds, where k is some positive constant.

Fundamental solution:

$$\mathscr{E}(x_1,\ldots,x_n,y_1,\ldots,y_n) = \begin{cases} \dfrac{\Gamma(n/2)}{2(n-2)\pi^{n/2}\sqrt{A}}\left[\displaystyle\sum_{i,j=1}^{n} b_{ij}(x_i-y_i)(x_j-y_j)\right]^{-\frac{n-2}{2}} & \text{for } n \geq 3, \\[2ex] \dfrac{1}{2\pi\sqrt{A}}\ln\left[\displaystyle\sum_{i,j=1}^{2} b_{ij}(x_i-y_i)(x_j-y_j)\right]^{-1/2} & \text{for } n = 2, \end{cases}$$

where A is the determinant of the matrix $\mathbf{A} = \{a_{ij}\}$ and the b_{ij} are the entries of the inverse of $\mathbf{A}$.

⊙ *Reference*: V. M. Babich, M. B. Kapilevich, S. G. Mikhlin, et al. (1964).

4. $\displaystyle\sum_{i=1}^{n-1}\frac{\partial^2 w}{\partial x_i^2} + \frac{\partial}{\partial x_n}\left(x_n^\beta \frac{\partial w}{\partial x_n}\right) + \lambda w = 0.$

Domain: $a_i \leq x_i \leq b_i$ ($i = 1, \ldots, n-1$), $0 \leq x_n \leq c$.

1°. Case $0 < \beta < 1$. First boundary value problem. The condition $w = 0$ is set on the entire boundary of the domain.

Eigenvalues and eigenfunctions:

$$\lambda_{k_1,\ldots,k_{n-1},m} = \sum_{i=1}^{n-1}\frac{k_i^2 \pi^2}{(b_i-a_i)^2} + \frac{(2-\beta)^2 \gamma_{\nu m}}{4c^{2-\beta}},$$

$$w_{k_1,\ldots,k_{n-1},m} = x_n^{\frac{1-\beta}{2}} J_\nu\left(\gamma_{\nu m}\left(\frac{x_n}{a}\right)^{\frac{2-\beta}{2}}\right)\prod_{i=1}^{n-1}\sin\frac{k_i\pi(x_i-a_i)}{b_i-a_i},$$

where $\gamma_{\nu m}$ is the mth positive root of the equation $J_\nu(\gamma) = 0$,

$$k_1, \ldots, k_{n-1} = 1, 2, \ldots; \qquad m = 1, 2, \ldots; \qquad \nu = \frac{1-\beta}{2-\beta}.$$

2°. Case $1 \leq \beta < 2$. Boundary conditions: the solution must be bounded at $x_n = 0$, and the condition $w = 0$ must hold on the rest of the boundary of the domain.

The eigenvalues and eigenfunctions of this problem are given by the relations of Item 1° with $\nu = (\beta-1)/(2-\beta)$.

⊙ *Reference*: M. M. Smirnov (1975).

Chapter 9
Higher-Order Partial Differential Equations

9.1. Third-Order Partial Differential Equations

1. $\dfrac{\partial w}{\partial t} + \dfrac{\partial^3 w}{\partial x^3} = 0.$

Linearized Corteveg–de Vries equation.

1°. Particular solutions:

$$w(x,t) = a(x^3 - 6t) + bx^2 + cx + k,$$
$$w(x,t) = a(x^5 - 60x^2 t) + b(x^4 - 24xt),$$
$$w(x,t) = a\sin(\lambda x + \lambda^3 t) + b\cos(\lambda x + \lambda^3 t) + c,$$
$$w(x,t) = a\sinh(\lambda x - \lambda^3 t) + b\cosh(\lambda x - \lambda^3 t) + c,$$
$$w(x,t) = \exp(-\lambda^3 t)\left[a\exp(\lambda x) + b\exp(-\tfrac{1}{2}\lambda x)\sin\left(\tfrac{\sqrt{3}}{2}\lambda x + c\right)\right],$$

where a, b, c, k, and λ are arbitrary constants.

2°. Domain: $-\infty < x \le 0$. Boundary value problem.
Initial and boundary conditions are prescribed:

$$w = 0 \quad \text{at} \quad t = 0, \qquad w = f(t) \quad \text{at} \quad x = 0, \qquad w \to 0 \quad \text{as} \quad x \to -\infty.$$

Solution:

$$w(x,t) = -\frac{3}{2}\int_0^t \operatorname{Ai}''\left(\frac{x}{(t-\tau)^{1/3}}\right)\frac{f(\tau)}{t-\tau}\,d\tau,$$

where $\operatorname{Ai}''(z)$ is the second derivative of the Airy function.

3°. Domain: $0 \le x < \infty$. The function

$$w(x,t) = 3\int_0^t \operatorname{Ai}''\left(\frac{x}{(t-\tau)^{1/3}}\right)\frac{f(\tau)}{t-\tau}\,d\tau,$$

satisfies the equation and the first two conditions specified in Item 2°.

⊙ *Reference*: A. V. Faminskii (1999).

2. $\dfrac{\partial w}{\partial t} = ax^6 \dfrac{\partial^3 w}{\partial x^3}.$

The transformation

$$u(z,\tau) = wx^{-2}, \quad z = 1/x, \quad \tau = at$$

leads to a constant coefficient equation of the form 9.1.1:

$$\frac{\partial u}{\partial \tau} = -\frac{\partial^3 u}{\partial z^3}.$$

3. $\dfrac{\partial w}{\partial t} = k(t)\dfrac{\partial^3 w}{\partial x^3} + [xf(t) + g(t)]\dfrac{\partial w}{\partial x} + h(t)w.$

The transformation

$$w(x,t) = u(z,\tau)\exp\left[\int h(t)\,dt\right], \quad z = xF(t) + \int g(t)F(t)\,dt, \quad \tau = \int k(t)F^3(t)\,dt,$$

where $F(t) = \exp\left[\int f(t)\,dt\right]$, leads to a constant coefficient equation of the form 9.1.1:

$$\dfrac{\partial u}{\partial \tau} = \dfrac{\partial^3 u}{\partial z^3}.$$

4. $\dfrac{\partial w}{\partial t} = (ax^2 + bx + c)^3\dfrac{\partial^3 w}{\partial x^3}.$

This is a special case of equation 9.6.4.4 with $k = 1$ and $n = 3$. The transformation

$$w(x,t) = u(z,t)(ax^2 + bx + c), \quad z = \int \dfrac{dx}{ax^2 + bx + c}$$

leads to the constant coefficient equation

$$\dfrac{\partial u}{\partial t} = \dfrac{\partial^3 u}{\partial z^3} + (4ac - b^2)\dfrac{\partial u}{\partial z}.$$

5. $\dfrac{\partial^2 w}{\partial t^2} = ax^6\dfrac{\partial^3 w}{\partial x^3}.$

The transformation $z = 1/x$, $u = wx^{-2}$ leads to the constant coefficient equation

$$\dfrac{\partial^2 u}{\partial t^2} = -a\dfrac{\partial^3 u}{\partial z^3}.$$

6. $\dfrac{\partial^2 w}{\partial t^2} = (ax^2 + bx + c)^3\dfrac{\partial^3 w}{\partial x^3}.$

This is a special case of equation 9.6.4.4 with $k = 2$ and $n = 3$. The transformation

$$w(x,t) = u(z,t)(ax^2 + bx + c), \quad z = \int \dfrac{dx}{ax^2 + bx + c}$$

leads to the constant coefficient equation

$$\dfrac{\partial^2 u}{\partial t^2} = \dfrac{\partial^3 u}{\partial z^3} + (4ac - b^2)\dfrac{\partial u}{\partial z}.$$

9.2. Fourth-Order One-Dimensional Nonstationary Equations

9.2.1. Equations of the Form $\dfrac{\partial w}{\partial t} + a^2\dfrac{\partial^4 w}{\partial x^4} = \Phi(x,t)$

9.2.1-1. Particular solutions of the homogeneous equation ($\Phi \equiv 0$):

$$w(x) = Ax^3 + Bx^2 + Cx + D,$$

$$w(x,t) = A(x^5 - 120a^2xt) + B(x^4 - 24a^2t),$$

$$w(x,t) = \left[A\sin(\lambda x) + B\cos(\lambda x) + C\sinh(\lambda x) + D\cosh(\lambda x)\right]\exp(-\lambda^4 a^2 t),$$

where A, B, C, D, and λ are arbitrary constants.

9.2.1-2. Domain: $0 \leq x \leq l$. Solution in terms of the Green's function.

1°. We consider problems on an interval $0 \leq x \leq l$ with the general initial condition

$$w = f(x) \quad \text{at} \quad t = 0$$

and various homogeneous boundary conditions. The solution can be represented in terms of the Green's function as

$$w(x,t) = \int_0^l f(\xi) G(x,\xi,t) \, d\xi + \int_0^t \int_0^l \Phi(\xi,\tau) G(x,\xi,t-\tau) \, d\xi \, d\tau.$$

2°. Paragraphs 9.2.1-3 through 9.2.1-10 present the Green's functions for various types of boundary conditions. The Green's functions can be evaluated from the formula

$$G(x,\xi,t) = \sum_{n=1}^{\infty} \frac{\varphi_n(x)\varphi_n(\xi)}{\|\varphi_n\|^2} \exp(-\lambda_n^4 a^2 t), \tag{1}$$

where the λ_n and $\varphi_n(x)$ are determined by solving the self-adjoint eigenvalue problem for the fourth-order ordinary differential equation

$$\varphi'''' - \lambda^4 \varphi = 0$$

subject to appropriate boundary conditions; the prime denotes differentiation with respect to x. The norms of eigenfunctions can be calculated by the formula

$$\|\varphi_n\|^2 = \int_0^l \varphi_n^2(x) \, dx = \frac{l}{4} \varphi_n^2(l) + \frac{l}{4\lambda_n^4} \left[\varphi_n''(l)\right]^2 - \frac{l}{2\lambda_n^4} \varphi_n'(l) \varphi_n'''(l). \tag{2}$$

Relations (1) and (2) are written under the assumption that $\lambda = 0$ is not an eigenvalue.

9.2.1-3. The function and its first derivative are prescribed at the boundaries:

$$w = \partial_x w = 0 \quad \text{at} \quad x = 0, \qquad w = \partial_x w = 0 \quad \text{at} \quad x = l.$$

Green's function:

$$G(x,\xi,t) = \frac{4}{l} \sum_{n=1}^{\infty} \frac{\lambda_n^4}{\left[\varphi_n''(l)\right]^2} \varphi_n(x) \varphi_n(\xi) \exp(-\lambda_n^4 a^2 t),$$

where

$$\varphi_n(x) = \left[\sinh(\lambda_n l) - \sin(\lambda_n l)\right] \left[\cosh(\lambda_n x) - \cos(\lambda_n x)\right] - \left[\cosh(\lambda_n l) - \cos(\lambda_n l)\right] \left[\sinh(\lambda_n x) - \sin(\lambda_n x)\right];$$

the λ_n are positive roots of the transcendental equation $\cosh(\lambda l)\cos(\lambda l) = 1$. The numerical values of the roots can be calculated from the formulas given in Paragraph 9.2.3-2.

9.2.1-4. The function and its second derivative are prescribed at the boundaries:

$$w = \partial_{xx} w = 0 \quad \text{at} \quad x = 0, \qquad w = \partial_{xx} w = 0 \quad \text{at} \quad x = l.$$

Green's function:

$$G(x,\xi,t) = \frac{2}{l} \sum_{n=1}^{\infty} \sin(\lambda_n x) \sin(\lambda_n \xi) \exp(-\lambda_n^4 a^2 t), \qquad \lambda_n = \frac{\pi n}{l}.$$

9.2.1-5. The first and third derivatives are prescribed at the boundaries:

$$\partial_x w = \partial_{xxx} w = 0 \quad \text{at} \quad x = 0, \qquad \partial_x w = \partial_{xxx} w = 0 \quad \text{at} \quad x = l.$$

Green's function:

$$G(x, \xi, t) = \frac{1}{l} + \frac{2}{l} \sum_{n=1}^{\infty} \cos(\lambda_n x) \cos(\lambda_n \xi) \exp(-\lambda_n^4 a^2 t), \qquad \lambda_n = \frac{\pi n}{l}.$$

9.2.1-6. The second and third derivatives are prescribed at the boundaries:

$$w_{xx} = \partial_{xxx} w = 0 \quad \text{at} \quad x = 0, \qquad w_{xx} = \partial_{xxx} w = 0 \quad \text{at} \quad x = l.$$

Green's function:

$$G(x, \xi, t) = \frac{1}{l} + \frac{3}{l^3}(2x - l)(2\xi - l) + \frac{4}{l} \sum_{n=1}^{\infty} \frac{\varphi_n(x)\varphi_n(\xi)}{\varphi_n^2(l)} \exp(-\lambda_n^4 a^2 t),$$

where

$$\varphi_n(x) = \big[\sinh(\lambda_n l) - \sin(\lambda_n l)\big]\big[\cosh(\lambda_n x) + \cos(\lambda_n x)\big] - \big[\cosh(\lambda_n l) - \cos(\lambda_n l)\big]\big[\sinh(\lambda_n x) + \sin(\lambda_n x)\big];$$

the λ_n are positive roots of the transcendental equation $\cosh(\lambda l)\cos(\lambda l) = 1$. The numerical values of the roots can be calculated from the formulas given in Paragraph 9.2.3-2.

9.2.1-7. Mixed conditions are prescribed at the boundaries (case 1):

$$w = \partial_x w = 0 \quad \text{at} \quad x = 0, \qquad w = \partial_{xx} w = 0 \quad \text{at} \quad x = l.$$

Green's function:

$$G(x, \xi, t) = \frac{2}{l} \sum_{n=1}^{\infty} \lambda_n^4 \frac{\varphi_n(x)\varphi_n(\xi)}{|\varphi_n'(l)\varphi_n'''(l)|} \exp(-\lambda_n^4 a^2 t),$$

where

$$\varphi_n(x) = \big[\sinh(\lambda_n l) - \sin(\lambda_n l)\big]\big[\cosh(\lambda_n x) - \cos(\lambda_n x)\big] - \big[\cosh(\lambda_n l) - \cos(\lambda_n l)\big]\big[\sinh(\lambda_n x) - \sin(\lambda_n x)\big];$$

the λ_n are positive roots of the transcendental equation $\tan(\lambda l) - \tanh(\lambda l) = 0$.

9.2.1-8. Mixed conditions are prescribed at the boundaries (case 2):

$$w = \partial_x w = 0 \quad \text{at} \quad x = 0, \qquad \partial_{xx} w = \partial_{xxx} w = 0 \quad \text{at} \quad x = l.$$

Green's function:

$$G(x, \xi, t) = \frac{4}{l} \sum_{n=1}^{\infty} \frac{\varphi_n(x)\varphi_n(\xi)}{\varphi_n^2(l)} \exp(-\lambda_n^4 a^2 t),$$

where

$$\varphi_n(x) = \big[\sinh(\lambda_n l) + \sin(\lambda_n l)\big]\big[\cosh(\lambda_n x) - \cos(\lambda_n x)\big] - \big[\cosh(\lambda_n l) + \cos(\lambda_n l)\big]\big[\sinh(\lambda_n x) - \sin(\lambda_n x)\big];$$

the λ_n are positive roots of the transcendental equation $\cosh(\lambda l)\cos(\lambda l) = -1$.

9.2.1-9. Mixed conditions are prescribed at the boundaries (case 3):

$$w = \partial_{xx}w = 0 \quad \text{at} \quad x = 0, \qquad \partial_x w = \partial_{xxx}w = 0 \quad \text{at} \quad x = l.$$

Green's function:

$$G(x, \xi, t) = \frac{2}{l} \sum_{n=0}^{\infty} \sin(\lambda_n x) \sin(\lambda_n \xi) \exp(-\lambda_n^4 a^2 t), \qquad \lambda_n = \frac{\pi(2n+1)}{2l}.$$

9.2.1-10. Mixed conditions are prescribed at the boundaries (case 4):

$$w = \partial_{xx}w = 0 \quad \text{at} \quad x = 0, \qquad \partial_{xx}w = \partial_{xxx}w = 0 \quad \text{at} \quad x = l.$$

Green's function:

$$G(x, \xi, t) = \frac{4}{l} \sum_{n=1}^{\infty} \frac{\varphi_n(x)\varphi_n(\xi)}{\varphi_n^2(l)} \exp(-\lambda_n^4 a^2 t),$$

where

$$\varphi_n(x) = \sin(\lambda_n l) \sinh(\lambda_n x) + \sinh(\lambda_n l) \sin(\lambda_n x);$$

the λ_n are positive roots of the transcendental equation $\tan(\lambda l) - \tanh(\lambda l) = 0$.

9.2.2. Equations of the Form $\dfrac{\partial^2 w}{\partial t^2} + a^2 \dfrac{\partial^4 w}{\partial x^4} = 0$

This equation is encountered in studying transverse vibration of elastic rods.

9.2.2-1. Particular solutions:

$$w(x,t) = (Ax^3 + Bx^2 + Cx + D)t + A_1 x^3 + B_1 x^2 + C_1 x + D_1,$$
$$w(x,t) = \big[A \sin(\lambda x) + B \cos(\lambda x) + C \sinh(\lambda x) + D \cos(\lambda x)\big] \sin(\lambda^2 a t),$$
$$w(x,t) = \big[A \sin(\lambda x) + B \cos(\lambda x) + C \sinh(\lambda x) + D \cos(\lambda x)\big] \cos(\lambda^2 a t),$$

where $A, B, C, D, A_1, B_1, C_1, D_1$, and λ are arbitrary constants.

9.2.2-2. Domain: $-\infty < x < \infty$. Cauchy problem.

Initial conditions are prescribed:

$$w = f(x) \quad \text{at} \quad t = 0, \qquad \partial_t w = ag''(x) \quad \text{at} \quad t = 0.$$

Boussinesq solution:

$$w(x,t) = \frac{1}{\sqrt{2\pi}} \int_{-\infty}^{\infty} f\big(x - 2\xi\sqrt{at}\,\big) \big(\cos \xi^2 + \sin \xi^2\big) d\xi$$
$$+ \frac{1}{a\sqrt{2\pi}} \int_{-\infty}^{\infty} g\big(x - 2\xi\sqrt{at}\,\big) \big(\cos \xi^2 - \sin \xi^2\big) d\xi.$$

⊙ *Reference*: I. Sneddon (1951).

9.2.2-3. Domain: $0 \leq x < \infty$. Free vibration of a semiinfinite rod.

The following conditions are prescribed:

$$w = 0 \quad \text{at} \quad t = 0, \qquad \partial_t w = 0 \quad \text{at} \quad t = 0 \qquad \text{(initial conditions)},$$
$$w = f(t) \quad \text{at} \quad x = 0, \qquad \partial_{xx}w = 0 \quad \text{at} \quad x = 0 \qquad \text{(boundary conditions)}.$$

Boussinesq solution:

$$w(x,t) = \frac{1}{\sqrt{\pi}} \int_{x/\sqrt{2at}}^{\infty} f\left(t - \frac{x^2}{2a\xi^2}\right) \left(\sin \frac{\xi^2}{2} + \cos \frac{\xi^2}{2}\right) d\xi.$$

⊙ *Reference*: I. Sneddon (1951).

9.2.2-4. Domain: $0 \le x \le l$. Boundary value problems.

For solutions of various boundary value problems, see Subsection 9.2.3 for $\Phi \equiv 0$.

9.2.3. Equations of the Form $\dfrac{\partial^2 w}{\partial t^2} + a^2 \dfrac{\partial^4 w}{\partial x^4} = \Phi(x,t)$

This equation is encountered in studying forced (transverse) vibration of elastic rods.

9.2.3-1. Domain: $0 \le x \le l$. Solution in terms of the Green's function.

1°. We consider boundary value problems on an interval $0 \le x \le l$ with the general initial condition

$$w = f(x) \quad \text{at} \quad t = 0, \qquad \partial_t w = g(x) \quad \text{at} \quad t = 0$$

and various homogeneous boundary conditions. The solution can be represented in terms of the Green's function as

$$w(x,t) = \frac{\partial}{\partial t}\int_0^l f(\xi)G(x,\xi,t)\,d\xi + \int_0^l g(\xi)G(x,\xi,t)\,d\xi + \int_0^t\int_0^l \Phi(\xi,\tau)G(x,\xi,t-\tau)\,d\xi\,d\tau.$$

2°. Paragraphs 9.2.3-2 through 9.2.3-9 present the Green's functions for various types of boundary conditions. The Green's functions can be evaluated from the formula

$$G(x,\xi,t) = \frac{1}{a}\sum_{n=1}^{\infty} \frac{\varphi_n(x)\varphi_n(\xi)}{\lambda_n^2 \|\varphi_n\|^2} \sin(\lambda_n^2 a t), \qquad (1)$$

where the λ_n and $\varphi_n(x)$ are determined by solving the self-adjoint eigenvalue problem for the fourth-order ordinary differential equation

$$\varphi'''' - \lambda^4 \varphi = 0$$

subject to appropriate boundary conditions; the prime denotes differentiation with respect to x. The norms of eigenfunctions can be calculated by Krylov's formula [see Krylov (1949)]:

$$\|\varphi_n\|^2 = \int_0^l \varphi_n^2(x)\,dx = \frac{l}{4}\varphi_n^2(l) + \frac{l}{4\lambda_n^4}\left[\varphi_n''(l)\right]^2 - \frac{l}{2\lambda_n^4}\varphi_n'(l)\varphi_n'''(l). \qquad (2)$$

Relations (1) and (2) are written under the assumption that $\lambda = 0$ is not an eigenvalue.

9.2.3-2. Both ends of the rod are clamped.

Boundary conditions are prescribed:

$$w = \partial_x w = 0 \quad \text{at} \quad x = 0, \qquad w = \partial_x w = 0 \quad \text{at} \quad x = l.$$

Green's function:

$$G(x,\xi,t) = \frac{4}{al}\sum_{n=1}^{\infty} \frac{\lambda_n^2}{\left[\varphi_n''(l)\right]^2}\varphi_n(x)\varphi_n(\xi)\sin(\lambda_n^2 a t),$$

where

$$\varphi_n(x) = \left[\sinh(\lambda_n l) - \sin(\lambda_n l)\right]\left[\cosh(\lambda_n x) - \cos(\lambda_n x)\right] - \left[\cosh(\lambda_n l) - \cos(\lambda_n l)\right]\left[\sinh(\lambda_n x) - \sin(\lambda_n x)\right];$$

the λ_n are positive roots of the transcendental equation $\cosh(\lambda l)\cos(\lambda l) = 1$. The numerical values of the roots can be calculated from the formulas

$$\lambda_n = \frac{\mu_n}{l}, \quad \text{where} \quad \mu_1 = 1.875, \quad \mu_2 = 4.694, \quad \mu_n = \frac{\pi}{2}(2n-1) \quad \text{for} \quad n \ge 3.$$

⊙ *Reference*: B. M. Budak, A. A. Samarskii, and A. N. Tikhonov (1980).

9.2.3-3. Both ends of the rod are hinged.

Boundary conditions are prescribed:

$$w = \partial_{xx} w = 0 \quad \text{at} \quad x = 0, \qquad w = \partial_{xx} w = 0 \quad \text{at} \quad x = l.$$

Green's function:

$$G(x, \xi, t) = \frac{2l}{a\pi^2} \sum_{n=1}^{\infty} \frac{1}{n^2} \sin(\lambda_n x) \sin(\lambda_n \xi) \sin(\lambda_n^2 at), \qquad \lambda_n = \frac{\pi n}{l}.$$

⊙ *References*: A. N. Krylov (1949), B. M. Budak, A. A. Samarskii, and A. N. Tikhonov (1980).

9.2.3-4. Both ends of the rod are free.

Boundary conditions are prescribed:

$$w_{xx} = \partial_{xxx} w = 0 \quad \text{at} \quad x = 0, \qquad w_{xx} = \partial_{xxx} w = 0 \quad \text{at} \quad x = l.$$

Green's function:

$$G(x, \xi, t) = \frac{t}{l} + \frac{3t}{l^3}(2x - l)(2\xi - l) + \frac{4}{al} \sum_{n=1}^{\infty} \frac{\varphi_n(x)\varphi_n(\xi)}{\lambda_n^2 \varphi_n^2(l)} \sin(\lambda_n^2 at),$$

where

$$\varphi_n(x) = \big[\sinh(\lambda_n l) - \sin(\lambda_n l)\big]\big[\cosh(\lambda_n x) + \cos(\lambda_n x)\big] - \big[\cosh(\lambda_n l) - \cos(\lambda_n l)\big]\big[\sinh(\lambda_n x) + \sin(\lambda_n x)\big];$$

the λ_n are positive roots of the transcendental equation $\cosh(\lambda l) \cos(\lambda l) = 1$. For the numerical values of the roots, see Paragraph 9.2.3-2.

The first two terms in the expression of the Green's function correspond to the zero eigenvalue $\lambda_0 = 0$, to which two orthogonal eigenfunctions $w_0^{(1)} = 1$ and $w_0^{(2)} = 2x - l$ correspond with $\|w_0^{(1)}\|^2 = l$ and $\|w_0^{(2)}\|^2 = \frac{1}{3}l^3$.

⊙ *Reference*: A. N. Krylov (1949).

9.2.3-5. One end of the rod is clamped and the other is hinged.

Boundary conditions are prescribed:

$$w = \partial_x w = 0 \quad \text{at} \quad x = 0, \qquad w = \partial_{xx} w = 0 \quad \text{at} \quad x = l.$$

Green's function:

$$G(x, \xi, t) = \frac{2}{al} \sum_{n=1}^{\infty} \lambda_n^2 \frac{\varphi_n(x)\varphi_n(\xi)}{|\varphi_n'(l)\varphi_n'''(l)|} \sin(\lambda_n^2 at),$$

where

$$\varphi_n(x) = \big[\sinh(\lambda_n l) - \sin(\lambda_n l)\big]\big[\cosh(\lambda_n x) - \cos(\lambda_n x)\big] - \big[\cosh(\lambda_n l) - \cos(\lambda_n l)\big]\big[\sinh(\lambda_n x) - \sin(\lambda_n x)\big];$$

the λ_n are positive roots of the transcendental equation $\tan(\lambda l) - \tanh(\lambda l) = 0$.

9.2.3-6. One end of the rod is clamped and the other is free.

Boundary conditions are prescribed:
$$w = \partial_x w = 0 \quad \text{at} \quad x = 0, \qquad \partial_{xx} w = \partial_{xxx} w = 0 \quad \text{at} \quad x = l.$$

Green's function:
$$G(x, \xi, t) = \frac{4}{al} \sum_{n=1}^{\infty} \frac{\varphi_n(x) \varphi_n(\xi)}{\lambda_n^2 \varphi_n^2(l)} \sin(\lambda_n^2 a t),$$

where
$$\varphi_n(x) = \big[\sinh(\lambda_n l) + \sin(\lambda_n l)\big]\big[\cosh(\lambda_n x) - \cos(\lambda_n x)\big] - \big[\cosh(\lambda_n l) + \cos(\lambda_n l)\big]\big[\sinh(\lambda_n x) - \sin(\lambda_n x)\big];$$
the λ_n are positive roots of the transcendental equation $\cosh(\lambda l) \cos(\lambda l) = -1$.

9.2.3-7. One end of the rod is hinged and the other is free.

Boundary conditions are prescribed:
$$w = \partial_{xx} w = 0 \quad \text{at} \quad x = 0, \qquad \partial_{xx} w = \partial_{xxx} w = 0 \quad \text{at} \quad x = l.$$

Green's function:
$$G(x, \xi, t) = \frac{4}{al} \sum_{n=1}^{\infty} \frac{\varphi_n(x) \varphi_n(\xi)}{\lambda_n^2 \varphi_n^2(l)} \sin(\lambda_n^2 a t),$$

where
$$\varphi_n(x) = \sin(\lambda_n l) \sinh(\lambda_n x) + \sinh(\lambda_n l) \sin(\lambda_n x);$$
the λ_n are positive roots of the transcendental equation $\tan(\lambda l) - \tanh(\lambda l) = 0$.

9.2.3-8. The first and third derivatives are prescribed at the ends:
$$\partial_x w = \partial_{xxx} w = 0 \quad \text{at} \quad x = 0, \qquad \partial_x w = \partial_{xxx} w = 0 \quad \text{at} \quad x = l.$$

Green's function:
$$G(x, \xi, t) = \frac{t}{l} + \frac{2}{al} \sum_{n=1}^{\infty} \frac{1}{\lambda_n^2} \cos(\lambda_n x) \cos(\lambda_n \xi) \sin(\lambda_n^2 a t), \qquad \lambda_n = \frac{\pi n}{l}.$$

9.2.3-9. Mixed boundary conditions are prescribed at the ends:
$$w = \partial_{xx} w = 0 \quad \text{at} \quad x = 0, \qquad \partial_x w = \partial_{xxx} w = 0 \quad \text{at} \quad x = l.$$

Green's function:
$$G(x, \xi, t) = \frac{2}{al} \sum_{n=0}^{\infty} \frac{1}{\lambda_n^2} \sin(\lambda_n x) \sin(\lambda_n \xi) \sin(\lambda_n^2 a t), \qquad \lambda_n = \frac{\pi(2n+1)}{2l}.$$

9.2.4. Equations of the Form $\dfrac{\partial^2 w}{\partial t^2} + a^2 \dfrac{\partial^4 w}{\partial x^4} + kw = \Phi(x, t)$

9.2.4-1. Particular solutions of the homogeneous equation ($\Phi \equiv 0$):

$$w(x, t) = (Ax^3 + Bx^2 + Cx + D) \sin(t\sqrt{k}),$$
$$w(x, t) = (Ax^3 + Bx^2 + Cx + D) \cos(t\sqrt{k}),$$
$$w(x, t) = \big[A \sin(\lambda x) + B \cos(\lambda x) + C \sinh(\lambda x) + D \cos(\lambda x)\big] \sin\!\big(t\sqrt{a^2 \lambda^4 + k}\,\big),$$
$$w(x, t) = \big[A \sin(\lambda x) + B \cos(\lambda x) + C \sinh(\lambda x) + D \cos(\lambda x)\big] \cos\!\big(t\sqrt{a^2 \lambda^4 + k}\,\big),$$

where A, B, C, D, and λ are arbitrary constants.

9.2.4-2. Domain: $0 \leq x \leq l$. Solution in terms of the Green's function.

1°. We consider boundary value problems on an interval $0 \leq x \leq l$ with the general initial condition

$$w = f(x) \quad \text{at} \quad t = 0, \qquad \partial_t w = g(x) \quad \text{at} \quad t = 0$$

and various homogeneous boundary conditions. The solution can be represented in terms of the Green's function as

$$w(x,t) = \frac{\partial}{\partial t} \int_0^l f(\xi) G(x,\xi,t)\, d\xi + \int_0^l g(\xi) G(x,\xi,t)\, d\xi + \int_0^t \int_0^l \Phi(\xi,\tau) G(x,\xi,t-\tau)\, d\xi\, d\tau.$$

2°. Paragraphs 9.2.4-3 through 9.2.4-10 present the Green's functions for various types of boundary conditions. The Green's functions can be evaluated from the formula

$$G(x,\xi,t) = \sum_{n=1}^{\infty} \frac{\varphi_n(x)\varphi_n(\xi)}{\|\varphi_n\|^2} \frac{\sin\left(t\sqrt{a^2\lambda_n^4 + k}\right)}{\sqrt{a^2\lambda_n^4 + k}},$$

where the λ_n and $\varphi_n(x)$ are determined by solving the self-adjoint eigenvalue problem for the fourth-order ordinary differential equation $\varphi'''' - \lambda^4 \varphi = 0$ subject to appropriate boundary conditions. The norms of eigenfunctions can be calculated by formula (2) from Paragraph 9.2.3-1.

9.2.4-3. The function and its first derivative are prescribed at the ends:

$$w = \partial_x w = 0 \quad \text{at} \quad x = 0, \qquad w = \partial_x w = 0 \quad \text{at} \quad x = l.$$

Green's function:

$$G(x,\xi,t) = \frac{4}{l}\sum_{n=1}^{\infty} \lambda_n^4 \frac{\varphi_n(x)\varphi_n(\xi)}{[\varphi_n''(l)]^2} \frac{\sin\left(t\sqrt{a^2\lambda_n^4+k}\right)}{\sqrt{a^2\lambda_n^4+k}}, \qquad \varphi_n''(x) = \frac{d^2\varphi_n}{dx^2},$$

where

$$\varphi_n(x) = [\sinh(\lambda_n l) - \sin(\lambda_n l)][\cosh(\lambda_n x) - \cos(\lambda_n x)] - [\cosh(\lambda_n l) - \cos(\lambda_n l)][\sinh(\lambda_n x) - \sin(\lambda_n x)];$$

the λ_n are positive roots of the transcendental equation $\cosh(\lambda l)\cos(\lambda l) = 1$.

9.2.4-4. The function and its second derivative are prescribed at the ends:

$$w = \partial_{xx} w = 0 \quad \text{at} \quad x = 0, \qquad w = \partial_{xx} w = 0 \quad \text{at} \quad x = l.$$

Green's function:

$$G(x,\xi,t) = \frac{2}{l}\sum_{n=1}^{\infty} \sin(\lambda_n x)\sin(\lambda_n \xi) \frac{\sin\left(t\sqrt{a^2\lambda_n^4+k}\right)}{\sqrt{a^2\lambda_n^4+k}}, \qquad \lambda_n = \frac{\pi n}{l}.$$

9.2.4-5. The first and third derivatives are prescribed at the ends:

$$\partial_x w = \partial_{xxx} w = 0 \quad \text{at} \quad x = 0, \qquad \partial_x w = \partial_{xxx} w = 0 \quad \text{at} \quad x = l.$$

Green's function:

$$G(x,\xi,t) = \frac{\sin(t\sqrt{k})}{l\sqrt{k}} + \frac{2}{l}\sum_{n=1}^{\infty} \cos(\lambda_n x)\cos(\lambda_n \xi)\frac{\sin\left(t\sqrt{a^2\lambda_n^4+k}\right)}{\sqrt{a^2\lambda_n^4+k}}, \qquad \lambda_n = \frac{\pi n}{l}.$$

9.2.4-6. The second and third derivatives are prescribed at the ends:

$$w_{xx} = \partial_{xxx}w = 0 \quad \text{at} \quad x = 0, \qquad w_{xx} = \partial_{xxx}w = 0 \quad \text{at} \quad x = l.$$

Green's function:

$$G(x,\xi,t) = \left[1 + \frac{3}{l^2}(2x-l)(2\xi-l)\right]\frac{\sin(t\sqrt{k})}{l\sqrt{k}} + \frac{4}{l}\sum_{n=1}^{\infty}\frac{\varphi_n(x)\varphi_n(\xi)}{\varphi_n^2(l)}\frac{\sin(t\sqrt{a^2\lambda_n^4 + k})}{\sqrt{a^2\lambda_n^4 + k}},$$

where

$$\varphi_n(x) = \left[\sinh(\lambda_n l) - \sin(\lambda_n l)\right]\left[\cosh(\lambda_n x) + \cos(\lambda_n x)\right] - \left[\cosh(\lambda_n l) - \cos(\lambda_n l)\right]\left[\sinh(\lambda_n x) + \sin(\lambda_n x)\right];$$

the λ_n are positive roots of the transcendental equation $\cosh(\lambda l)\cos(\lambda l) = 1$. For the numerical values of the roots, see Paragraph 9.2.3-2.

9.2.4-7. Mixed boundary conditions are prescribed at the ends (case 1):

$$w = \partial_x w = 0 \quad \text{at} \quad x = 0, \qquad w = \partial_{xx}w = 0 \quad \text{at} \quad x = l.$$

Green's function:

$$G(x,\xi,t) = \frac{2}{l}\sum_{n=1}^{\infty}\lambda_n^4 \frac{\varphi_n(x)\varphi_n(\xi)}{|\varphi_n'(l)\varphi_n'''(l)|}\frac{\sin(t\sqrt{a^2\lambda_n^4 + k})}{\sqrt{a^2\lambda_n^4 + k}},$$

where

$$\varphi_n(x) = \left[\sinh(\lambda_n l) - \sin(\lambda_n l)\right]\left[\cosh(\lambda_n x) - \cos(\lambda_n x)\right] - \left[\cosh(\lambda_n l) - \cos(\lambda_n l)\right]\left[\sinh(\lambda_n x) - \sin(\lambda_n x)\right];$$

the λ_n are positive roots of the transcendental equation $\tan(\lambda l) - \tanh(\lambda l) = 0$.

9.2.4-8. Mixed boundary conditions are prescribed at the ends (case 2):

$$w = \partial_x w = 0 \quad \text{at} \quad x = 0, \qquad \partial_{xx}w = \partial_{xxx}w = 0 \quad \text{at} \quad x = l.$$

Green's function:

$$G(x,\xi,t) = \frac{4}{l}\sum_{n=1}^{\infty}\frac{\varphi_n(x)\varphi_n(\xi)}{\varphi_n^2(l)}\frac{\sin(t\sqrt{a^2\lambda_n^4 + k})}{\sqrt{a^2\lambda_n^4 + k}},$$

where

$$\varphi_n(x) = \left[\sinh(\lambda_n l) + \sin(\lambda_n l)\right]\left[\cosh(\lambda_n x) - \cos(\lambda_n x)\right] - \left[\cosh(\lambda_n l) + \cos(\lambda_n l)\right]\left[\sinh(\lambda_n x) - \sin(\lambda_n x)\right];$$

the λ_n are positive roots of the transcendental equation $\cosh(\lambda l)\cos(\lambda l) = -1$.

9.2.4-9. Mixed boundary conditions are prescribed at the ends (case 3):

$$w = \partial_{xx}w = 0 \quad \text{at} \quad x = 0, \qquad \partial_x w = \partial_{xxx}w = 0 \quad \text{at} \quad x = l.$$

Green's function:

$$G(x,\xi,t) = \frac{2}{l}\sum_{n=0}^{\infty}\sin(\lambda_n x)\sin(\lambda_n \xi)\frac{\sin(t\sqrt{a^2\lambda_n^4 + k})}{\sqrt{a^2\lambda_n^4 + k}}, \qquad \lambda_n = \frac{\pi(2n+1)}{2l}.$$

9.2.4-10. Mixed boundary conditions are prescribed at the ends (case 4):

$$w = \partial_{xx}w = 0 \quad \text{at} \quad x = 0, \qquad \partial_{xx}w = \partial_{xxx}w = 0 \quad \text{at} \quad x = l.$$

Green's function:

$$G(x, \xi, t) = \frac{4}{l} \sum_{n=1}^{\infty} \frac{\varphi_n(x)\varphi_n(\xi)}{\varphi_n^2(l)} \frac{\sin\left(t\sqrt{a^2\lambda_n^4 + k}\right)}{\sqrt{a^2\lambda_n^4 + k}},$$

where

$$\varphi_n(x) = \sin(\lambda_n l) \sinh(\lambda_n x) + \sinh(\lambda_n l) \sin(\lambda_n x);$$

the λ_n are positive roots of the transcendental equation $\tan(\lambda l) - \tanh(\lambda l) = 0$.

9.2.5. Other Equations

9.2.5-1. Equations containing the first derivative with respect to t.

1. $\dfrac{\partial w}{\partial t} + a^2 \dfrac{\partial^4 w}{\partial x^4} + kw = \Phi(x, t).$

The change of variable $w(x, t) = e^{-kt}u(x, t)$ leads to the equation

$$\frac{\partial u}{\partial t} + a^2 \frac{\partial^4 u}{\partial x^4} = e^{kt}\Phi(x, t),$$

which is discussed in Subsection 9.2.1.

2. $\dfrac{\partial w}{\partial t} = ax^8 \dfrac{\partial^4 w}{\partial x^4}.$

This is a special case of equation 9.6.4.2 with $k = 1$ and $n = 4$.

3. $\dfrac{\partial w}{\partial t} = k(t)\dfrac{\partial^4 w}{\partial x^4} + [xf(t) + g(t)]\dfrac{\partial w}{\partial x} + h(t)w.$

This is a special case of equation 9.6.4.1 with $n = 4$. The transformation

$$w(x,t) = u(z,\tau)\exp\left[\int h(t)\,dt\right], \quad z = xF(t) + \int g(t)F(t)\,dt, \quad \tau = \int k(t)F^4(t)\,dt,$$

where $F(t) = \exp\left[\int f(t)\,dt\right]$, leads to the constant coefficient equation

$$\frac{\partial u}{\partial \tau} = \frac{\partial^4 u}{\partial z^4},$$

which is discussed in Subsection 9.2.1.

4. $\dfrac{\partial w}{\partial t} = (ax^2 + bx + c)^4 \dfrac{\partial^4 w}{\partial x^4}.$

This is a special case of equation 9.6.4.4 with $k = 1$ and $n = 4$.

9.2.5-2. Equations containing the second derivative with respect to t.

5. $\dfrac{\partial^2 w}{\partial t^2} + k\dfrac{\partial w}{\partial t} + a^2 \dfrac{\partial^4 w}{\partial x^4} = \Phi(x,t).$

With $\Phi(x,t) \equiv 0$ this equation governs transverse vibration of an elastic rod in a resisting medium with velocity-proportional resistance coefficient.

The change of variable $w(x,t) = \exp\left(-\tfrac{1}{2}kt\right) u(x,t)$ leads to the equation

$$\dfrac{\partial^2 u}{\partial t^2} + a^2 \dfrac{\partial^4 u}{\partial x^4} - \tfrac{1}{4}k^2 u = \exp\left(\tfrac{1}{2}kt\right)\Phi(x,t),$$

which is discussed in Subsection 9.2.4.

6. $\dfrac{\partial^2 w}{\partial t^2} = ax^8 \dfrac{\partial^4 w}{\partial x^4}.$

This is a special case of equation 9.6.4.2 with $k = 2$ and $n = 4$.

7. $\dfrac{\partial^2 w}{\partial t^2} = (ax^2 + bx + c)^4 \dfrac{\partial^4 w}{\partial x^4}.$

This is a special case of equation 9.6.4.4 with $k = 2$ and $n = 4$.

8. $\left(\dfrac{\partial}{\partial t} - \dfrac{\partial^2}{\partial x^2}\right)^2 w = 0.$

1°. General solution (two representations):
$$w(x,t) = t u_1(x,t) + u_0(x,t),$$
$$w(x,t) = x u_1(x,t) + u_0(x,t),$$

where $u_k = u_k(x,t)$ is an arbitrary function satisfying the heat equation $\partial_t u_k - \partial_{xx} u_k = 0$; $k = 1, 2$.

2°. Fundamental solution:
$$\mathscr{E}(x,t) = \dfrac{\sqrt{t}}{2\sqrt{\pi}} \exp\left(-\dfrac{x^2}{4t}\right).$$

3°. Domain: $-\infty < x < \infty$. Cauchy problem.
Initial conditions are prescribed:
$$w = 0 \quad \text{at} \quad t = 0, \qquad \partial_t w = f(x) \quad \text{at} \quad t = 0.$$

Solution:
$$w(x,t) = \dfrac{\sqrt{t}}{2\sqrt{\pi}} \int_{-\infty}^{\infty} \exp\left[-\dfrac{(x-\xi)^2}{4t}\right] f(\xi)\, d\xi.$$

⦿ *Reference*: G. E. Shilov (1965).

9. $\dfrac{\partial^4 w}{\partial t^4} - \dfrac{\partial^4 w}{\partial x^4} = 0.$

1°. Fundamental solution:
$$\mathscr{E}(x,t) = \dfrac{1}{2\pi}\left\{ t \ln\sqrt{x^2 + t^2} - x \arctan\dfrac{x}{t} - \dfrac{1}{2}(t+x)\ln|t+x| \right.$$
$$\left. - \dfrac{1}{2}(t-x)\ln|t-x| + \dfrac{1}{8}|t+x| + \dfrac{1}{8}|t-x| \right\}.$$

2°. Domain: $-\infty < x < \infty$. Cauchy problem.
Initial conditions are prescribed:
$$w = 0 \quad \text{at} \quad t = 0, \qquad \partial_t w = 0 \quad \text{at} \quad t = 0, \qquad \partial_{tt} w = f(x) \quad \text{at} \quad t = 0.$$

Solution:
$$w(x,t) = \int_{-\infty}^{\infty} \mathscr{E}(x - \xi, t) f(\xi)\, d\xi.$$

⦿ *Reference*: G. E. Shilov (1965).

10. $\dfrac{\partial^4 w}{\partial t^4} - 2\dfrac{\partial^4 w}{\partial t^2 \partial x^2} + \dfrac{\partial^4 w}{\partial x^4} = 0.$

General solution (three representations):

$$w(x,t) = f_1(t-x) + f_2(t+x) + t\big[g_1(t-x) + g_2(t+x)\big],$$
$$w(x,t) = f_1(t-x) + f_2(t+x) + x\big[g_1(t-x) + g_2(t+x)\big],$$
$$w(x,t) = f_1(t-x) + f_2(t+x) + (t+x)g_1(t-x) + (t-x)g_2(t+x),$$

where $f_1(y)$, $f_2(z)$, $g_1(y)$, and $g_2(z)$ are arbitrary functions.

⊙ *Reference*: A. V. Bitsadze and D. F. Kalinichenko (1985).

9.3. Two-Dimensional Nonstationary Fourth-Order Equations

9.3.1. Equations of the Form $\dfrac{\partial w}{\partial t} + a^2\left(\dfrac{\partial^4 w}{\partial x^4} + \dfrac{\partial^4 w}{\partial y^4}\right) = \Phi(x,y,t)$

9.3.1-1. Domain: $0 \le x \le l_1$, $0 \le y \le l_2$. Solution in terms of the Green's function.

We consider boundary value problems in a rectangular domain $0 \le x \le l_1$, $0 \le y \le l_2$ with the general initial condition

$$w = f(x,y) \quad \text{at} \quad t = 0$$

and various homogeneous boundary conditions. The solution can be represented in terms of the Green's function as

$$w(x,y,t) = \int_0^{l_1}\int_0^{l_2} f(\xi,\eta) G(x,y,\xi,\eta,t)\, d\eta\, d\xi + \int_0^t \int_0^{l_1}\int_0^{l_2} \Phi(\xi,\eta,\tau) G(x,y,\xi,\eta,t-\tau)\, d\eta\, d\xi\, d\tau.$$

Below are the Green's functions for various types of boundary conditions.

9.3.1-2. The function and its first derivatives are prescribed at the sides of a rectangle:

$$w = \partial_x w = 0 \quad \text{at} \quad x = 0, \qquad w = \partial_x w = 0 \quad \text{at} \quad x = l_1,$$
$$w = \partial_y w = 0 \quad \text{at} \quad y = 0, \qquad w = \partial_y w = 0 \quad \text{at} \quad y = l_2.$$

Green's function:

$$G(x,y,\xi,\eta,t) = \frac{16}{l_1 l_2} \sum_{n=1}^{\infty}\sum_{m=1}^{\infty} \frac{p_n^4 q_m^4}{\big[\varphi_n''(l_1)\psi_m''(l_2)\big]^2}\varphi_n(x)\psi_m(y)\varphi_n(\xi)\psi_m(\eta) \exp\big[-(p_n^4 + q_m^4)a^2 t\big],$$

$$\varphi_n''(x) = \frac{d^2\varphi_n}{dx^2}, \qquad \psi_m''(y) = \frac{d^2\psi_m}{dy^2}.$$

Here,

$$\varphi_n(x) = \big[\sinh(p_n l_1) - \sin(p_n l_1)\big]\big[\cosh(p_n x) - \cos(p_n x)\big]$$
$$\qquad - \big[\cosh(p_n l_1) - \cos(p_n l_1)\big]\big[\sinh(p_n x) - \sin(p_n x)\big],$$
$$\psi_m(y) = \big[\sinh(q_m l_2) - \sin(q_m l_2)\big]\big[\cosh(q_m y) - \cos(q_m y)\big]$$
$$\qquad - \big[\cosh(q_m l_2) - \cos(q_m l_2)\big]\big[\sinh(q_m y) - \sin(q_m y)\big],$$

where the p_n and q_m are positive roots of the transcendental equations

$$\cosh(pl_1)\cos(pl_1) = 1, \quad \cosh(ql_2)\cos(ql_2) = 1 \qquad (q_m = p_m l_1/l_2).$$

9.3.1-3. The function and its second derivatives are prescribed at the sides of a rectangle:

$$w = \partial_{xx}w = 0 \quad \text{at} \quad x = 0, \qquad w = \partial_{xx}w = 0 \quad \text{at} \quad x = l_1,$$
$$w = \partial_{yy}w = 0 \quad \text{at} \quad y = 0, \qquad w = \partial_{yy}w = 0 \quad \text{at} \quad y = l_2.$$

Green's function:

$$G(x,y,\xi,\eta,t) = \frac{4}{l_1 l_2} \sum_{n=1}^{\infty}\sum_{m=1}^{\infty} \sin(p_n x)\sin(q_m y)\sin(p_n \xi)\sin(q_m \eta)\exp\bigl[-(p_n^4 + q_m^4)a^2 t\bigr],$$

$$p_n = \frac{\pi n}{l_1}, \qquad q_m = \frac{\pi m}{l_2}.$$

9.3.1-4. The first and third derivatives are prescribed at the sides of a rectangle:

$$w_x = \partial_{xxx}w = 0 \quad \text{at} \quad x = 0, \qquad w_x = \partial_{xxx}w = 0 \quad \text{at} \quad x = l_1,$$
$$w_y = \partial_{yyy}w = 0 \quad \text{at} \quad y = 0, \qquad w_y = \partial_{yyy}w = 0 \quad \text{at} \quad y = l_2.$$

Green's function:

$$G(x,y,\xi,\eta,t) = \frac{1}{l_1 l_2} \sum_{n=0}^{\infty}\sum_{m=0}^{\infty} \varepsilon_n \varepsilon_m \cos(p_n x)\sin(q_m y)\cos(p_n \xi)\cos(q_m \eta)\exp\bigl[-(p_n^4 + q_m^4)a^2 t\bigr],$$

$$p_n = \frac{\pi n}{l_1}, \qquad q_m = \frac{\pi m}{l_2}, \qquad \varepsilon_n = \begin{cases} 1 & \text{for } n = 0, \\ 2 & \text{for } n \ne 0. \end{cases}$$

9.3.1-5. The second and third derivatives are prescribed at the sides of a rectangle:

$$w_{xx} = \partial_{xxx}w = 0 \quad \text{at} \quad x = 0, \qquad w_{xx} = \partial_{xxx}w = 0 \quad \text{at} \quad x = l_1,$$
$$w_{yy} = \partial_{yyy}w = 0 \quad \text{at} \quad y = 0, \qquad w_{yy} = \partial_{yyy}w = 0 \quad \text{at} \quad y = l_2.$$

Green's function:

$$G(x,y,\xi,\eta,t) = G_1(x,\xi,t)G_2(y,\eta,t),$$

$$G_1(x,\xi,t) = \frac{1}{l_1} + \frac{3}{l_1^3}(2x - l_1)(2\xi - l_1) + \frac{4}{l_1}\sum_{n=1}^{\infty}\frac{\varphi_n(x)\varphi_n(\xi)}{\varphi_n^2(l_1)}\exp(-p_n^4 a^2 t),$$

$$G_2(y,\eta,t) = \frac{1}{l_2} + \frac{3}{l_2^3}(2y - l_2)(2\eta - l_2) + \frac{4}{l_2}\sum_{m=1}^{\infty}\frac{\psi_m(y)\psi_m(\eta)}{\psi_m^2(l_2)}\exp(-q_m^4 a^2 t).$$

Here,

$$\varphi_n(x) = \bigl[\sinh(p_n l_1) - \sin(p_n l_1)\bigr]\bigl[\cosh(p_n x) + \cos(p_n x)\bigr]$$
$$- \bigl[\cosh(p_n l_1) - \cos(p_n l_1)\bigr]\bigl[\sinh(p_n x) + \sin(p_n x)\bigr],$$
$$\psi_m(y) = \bigl[\sinh(q_m l_2) - \sin(q_m l_2)\bigr]\bigl[\cosh(q_m y) + \cos(q_m y)\bigr]$$
$$- \bigl[\cosh(q_m l_2) - \cos(q_m l_2)\bigr]\bigl[\sinh(q_m y) + \sin(q_m y)\bigr],$$

where the p_n and q_m are positive roots of the transcendental equations

$$\cosh(p l_1)\cos(p l_1) = 1, \qquad \cosh(q l_2)\cos(q l_2) = 1.$$

9.3.1-6. Mixed boundary conditions are prescribed at the sides of a rectangle:

$$w = \partial_{xx}w = 0 \quad \text{at} \quad x = 0, \qquad \partial_x w = \partial_{xxx}w = 0 \quad \text{at} \quad x = l_1,$$
$$w = \partial_{yy}w = 0 \quad \text{at} \quad y = 0, \qquad \partial_y w = \partial_{yyy}w = 0 \quad \text{at} \quad y = l_2.$$

Green's function:

$$G(x, y, \xi, \eta, t) = \frac{4}{l_1 l_2} \sum_{n=0}^{\infty} \sum_{m=0}^{\infty} \sin(p_n x) \sin(q_m y) \sin(p_n \xi) \sin(q_m \eta) \exp[-(p_n^4 + q_m^4)a^2 t],$$

$$p_n = \frac{\pi(2n+1)}{2l_1}, \quad q_m = \frac{\pi(2m+1)}{2l_2}.$$

9.3.2. Two-Dimensional Equations of the Form $\frac{\partial^2 w}{\partial t^2} + a^2 \Delta \Delta w = 0$

This equation governs two-dimensional free transverse vibration of a thin elastic plate; the unknown w is the deflection (transverse displacement) of the plate's midplane points relative to the original plane position. Here, $\Delta\Delta = \Delta^2$ and Δ is the Laplace operator that is defined as

$$\Delta = \begin{cases} \frac{\partial^2}{\partial x^2} + \frac{\partial^2}{\partial y^2} & \text{in the Cartesian coordinate system,} \\ \frac{\partial^2}{\partial r^2} + \frac{1}{r}\frac{\partial}{\partial r} + \frac{1}{r^2}\frac{\partial^2}{\partial \varphi^2} & \text{in the polar coordinate system.} \end{cases}$$

9.3.2-1. Particular solutions:

$$w(x, y, t) = \left[A_1 \sin(k_1 x) + B_1 \cos(k_1 x)\right]\left[A_2 \sin(k_2 y) + B_2 \cos(k_2 y)\right] \sin\left[(k_1^2 + k_2^2)at\right],$$
$$w(x, y, t) = \left[A_1 \sin(k_1 x) + B_1 \cos(k_1 x)\right]\left[A_2 \sin(k_2 y) + B_2 \cos(k_2 y)\right] \cos\left[(k_1^2 + k_2^2)at\right],$$
$$w(x, y, t) = \left[A_1 \sinh(k_1 x) + B_1 \cosh(k_1 x)\right]\left[A_2 \sinh(k_2 y) + B_2 \cosh(k_2 y)\right] \sin\left[(k_1^2 + k_2^2)at\right],$$
$$w(x, y, t) = \left[A_1 \sinh(k_1 x) + B_1 \cosh(k_1 x)\right]\left[A_2 \sinh(k_2 y) + B_2 \cosh(k_2 y)\right] \cos\left[(k_1^2 + k_2^2)at\right],$$
$$w(r, \varphi, t) = \left[A_1 J_n(kr) + A_2 Y_n(kr) + A_3 I_n(kr) + A_4 K_n(kr)\right] \cos(n\varphi) \sin(k^2 at),$$
$$w(r, \varphi, t) = \left[A_1 J_n(kr) + A_2 Y_n(kr) + A_3 I_n(kr) + A_4 K_n(kr)\right] \sin(n\varphi) \cos(k^2 at),$$

where $A_1, A_2, A_3, A_4, B_1, B_2, k, k_1, k_2$ are arbitrary constants, the $J_n(\xi)$ and $Y_n(\xi)$ are the Bessel functions of the first and second kind, the $I_n(\xi)$ and $K_n(\xi)$ are the modified Bessel functions of the first and second kind, $r = \sqrt{x^2 + y^2}$, and $n = 0, 1, 2, \ldots$

9.3.2-2. Domain: $-\infty < x < \infty$, $-\infty < y < \infty$. Cauchy problem.

Initial conditions are prescribed:

$$w = f(x, y) \quad \text{at} \quad t = 0, \qquad \partial_t w = g(x, y) \quad \text{at} \quad t = 0.$$

Poisson solution:

$$w(x, y, t) = \frac{1}{\pi} \int_{-\infty}^{\infty} \int_{-\infty}^{\infty} f\left(x + 2\xi\sqrt{at},\, y + 2\eta\sqrt{at}\right) \sin\left(\xi^2 + \eta^2\right) d\xi\, d\eta$$
$$+ \frac{1}{\pi} \int_0^t d\tau \int_{-\infty}^{\infty} \int_{-\infty}^{\infty} g\left(x + 2\xi\sqrt{a\tau},\, y + 2\eta\sqrt{a\tau}\right) \sin\left(\xi^2 + \eta^2\right) d\xi\, d\eta.$$

Green's function:

$$G(x, y, \xi, \eta, t) = \frac{1}{4\pi a} \int_0^t \sin\left[\frac{(x-\xi)^2 + (y-\eta)^2}{4a\tau}\right] \frac{d\tau}{\tau}.$$

⊙ *References*: A. N. Krylov (1949), I. Sneddon (1951), B. M. Budak, A. A. Samarskii, and A. N. Tikhonov (1980).

9.3.2-3. Domain: $0 \leq x \leq l_1$, $0 \leq y \leq l_2$. Solution in terms of the Green's function.

We consider boundary value problems in a rectangular domain $0 \leq x \leq l_1$, $0 \leq y \leq l_2$ with the general initial conditions

$$w = f(x,y) \quad \text{at} \quad t = 0, \qquad \partial_t w = g(x,y) \quad \text{at} \quad t = 0$$

and various homogeneous boundary conditions. The solution can be represented in terms of the Green's function as

$$w(x,y,t) = \frac{\partial}{\partial t} \int_0^{l_1} \int_0^{l_2} f(\xi,\eta) G(x,y,\xi,\eta,t) \, d\eta \, d\xi + \int_0^{l_1} \int_0^{l_2} g(\xi,\eta) G(x,y,\xi,\eta,t) \, d\eta \, d\xi.$$

Paragraphs 9.3.2-4 through 9.3.2-6 present the Green's functions for three types of boundary conditions.

9.3.2-4. Domain: $0 \leq x \leq l_1$, $0 \leq y \leq l_2$. All sides of the plate are hinged.

Boundary conditions are prescribed:

$$w = \partial_{xx} w = 0 \quad \text{at} \quad x = 0, \qquad w = \partial_{xx} w = 0 \quad \text{at} \quad x = l_1,$$
$$w = \partial_{yy} w = 0 \quad \text{at} \quad y = 0, \qquad w = \partial_{yy} w = 0 \quad \text{at} \quad y = l_2.$$

Green's function:

$$G(x,y,\xi,\eta,t) = \frac{4}{al_1 l_2} \sum_{n=1}^{\infty} \sum_{m=1}^{\infty} \sin(p_n x) \sin(q_m y) \sin(p_n \xi) \sin(q_m \eta) \frac{\sin(\lambda_{nm} a t)}{\lambda_{nm}},$$

$$p_n = \frac{\pi n}{l_1}, \qquad q_m = \frac{\pi m}{l_2}, \qquad \lambda_{nm} = p_n^2 + q_m^2.$$

9.3.2-5. Domain: $0 \leq x \leq l_1$, $0 \leq y \leq l_2$. The 1st and 3rd derivatives are prescribed at the sides:

$$\partial_x w = \partial_{xxx} w = 0 \quad \text{at} \quad x = 0, \qquad \partial_x w = \partial_{xxx} w = 0 \quad \text{at} \quad x = l_1,$$
$$\partial_y w = \partial_{yyy} w = 0 \quad \text{at} \quad y = 0, \qquad \partial_y w = \partial_{yyy} w = 0 \quad \text{at} \quad y = l_2.$$

Green's function:

$$G(x,y,\xi,\eta,t) = \frac{1}{al_1 l_2} \sum_{n=0}^{\infty} \sum_{m=0}^{\infty} \varepsilon_n \varepsilon_m \cos(p_n x) \cos(q_m y) \cos(p_n \xi) \cos(q_m \eta) \frac{\sin(\lambda_{nm} a t)}{\lambda_{nm}},$$

$$p_n = \frac{\pi n}{l_1}, \quad q_m = \frac{\pi m}{l_2}, \quad \lambda_{nm} = p_n^2 + q_m^2, \quad \varepsilon_n = \begin{cases} 1 & \text{for } n = 0, \\ 2 & \text{for } n \neq 0. \end{cases}$$

If $n = m = 0$, the ratio $\sin(\lambda_{nm} a t)/\lambda_{nm}$ must be replaced by at.

9.3.2-6. Domain: $0 \leq x \leq l_1$, $0 \leq y \leq l_2$. Mixed boundary conditions are set at the sides:

$$w = \partial_{xx} w = 0 \quad \text{at} \quad x = 0, \qquad w = \partial_{xx} w = 0 \quad \text{at} \quad x = l_1,$$
$$\partial_y w = \partial_{yyy} w = 0 \quad \text{at} \quad y = 0, \qquad \partial_y w = \partial_{yyy} w = 0 \quad \text{at} \quad y = l_2.$$

Green's function:

$$G(x,y,\xi,\eta,t) = \frac{2}{al_1 l_2} \sum_{n=1}^{\infty} \sum_{m=0}^{\infty} \varepsilon_m \sin(p_n x) \cos(q_m y) \sin(p_n \xi) \cos(q_m \eta) \frac{\sin(\lambda_{nm} a t)}{\lambda_{nm}},$$

$$p_n = \frac{\pi n}{l_1}, \quad q_m = \frac{\pi m}{l_2}, \quad \lambda_{nm} = p_n^2 + q_m^2, \quad \varepsilon_m = \begin{cases} 1 & \text{for } m = 0, \\ 2 & \text{for } m \neq 0. \end{cases}$$

9.3.2-7. Domain: $0 \le r < \infty$, $0 \le \varphi \le 2\pi$. Cauchy problem.

Initial conditions for the symmetric case in the polar coordinate system:
$$w = f(r) \quad \text{at} \quad t = 0, \qquad \partial_t w = 0 \quad \text{at} \quad t = 0.$$

Solution:
$$w(r,t) = \frac{1}{2at} \int_0^\infty \xi f(\xi) J_0\left(\frac{\xi r}{2at}\right) \sin\left(\frac{\xi^2 + r^2}{4at}\right) d\xi,$$

where $J_0(z)$ is the zeroth Bessel function.

⊙ *References*: I. Sneddon (1951), B. M. Budak, A. A. Samarskii, and A. N. Tikhonov (1980).

9.3.2-8. Domain: $0 \le r \le R$, $0 \le \varphi \le 2\pi$. Transverse vibration of a circular plate.

Initial and boundary conditions for symmetric transverse vibrations of a circular plate of radius R with clamped contour in the polar coordinate system:
$$w = f(r) \quad \text{at} \quad t = 0, \qquad \partial_t w = g(t) \quad \text{at} \quad t = 0;$$
$$w = 0 \quad \text{at} \quad r = R, \qquad \partial_r w = 0 \quad \text{at} \quad r = R.$$

Solution:
$$w(r,t) = \sum_{n=1}^\infty \left[A_n \cos(a k_n^2 t) + B_n \sin(a k_n^2 t)\right] \Psi_n(r),$$
$$\Psi_n(r) = I_0(k_n R) J_0(k_n r) - J_0(k_n R) I_0(k_n r),$$

where the k_n are positive roots of the transcendental equation (the prime denotes the derivative)
$$J_0(kR) I_0'(kR) - I_0(kR) J_0'(kR) = 0,$$

and the coefficients A_n and B_n are given by
$$A_n = \frac{1}{\|\Psi_n\|^2} \int_0^R f(r) \Psi_n(r) r\, dr, \qquad B_n = \frac{1}{a k_n^2 \|\Psi_n\|^2} \int_0^R g(r) \Psi_n(r) r\, dr,$$
$$\|\Psi_n\|^2 = \tfrac{1}{4} R^6 \left[\Psi_n''(R)\right]^2 = R^2 J_0^2(k_n R) I_0^2(k_n R).$$

⊙ *Reference*: B. M. Budak, A. A. Samarskii, and A. N. Tikhonov (1980).

9.3.3. Three- and n-Dimensional Equations of the Form $\dfrac{\partial^2 w}{\partial t^2} + a^2 \Delta \Delta w = 0$

9.3.3-1. Three-dimensional case. Cauchy problem.

Domain: $-\infty < x < \infty$, $-\infty < y < \infty$, $-\infty < z < \infty$. Initial conditions are prescribed:
$$w = f(x,y,z) \quad \text{at} \quad t = 0, \qquad \partial_t w = 0 \quad \text{at} \quad t = 0.$$

Solution:
$$w(x,y,z,t) = \frac{1}{(2\sqrt{\pi at})^3} \int_{-\infty}^\infty \int_{-\infty}^\infty \int_{-\infty}^\infty f(x+\xi, y+\eta, z+\zeta) \cos\left(\frac{\xi^2 + \eta^2 + \zeta^2}{4at} - \frac{3\pi}{4}\right) d\xi\, d\eta\, d\zeta.$$

⊙ *Reference*: V. S. Vladimirov, V. P. Mikhailov, A. A. Vasharin, et al. (1974).

9.3.3-2. Three-dimensional case. Boundary value problem.

Domain: $0 \leq x \leq l_1$, $0 \leq y \leq l_2$, $0 \leq z \leq l_3$ (rectangular parallelepiped).
Initial conditions:

$$w = f(x,y,z) \quad \text{at} \quad t = 0, \qquad \partial_t w = g(x,y,z) \quad \text{at} \quad t = 0.$$

Boundary conditions:

$$\begin{aligned}
w = \partial_{xx} w = 0 &\quad \text{at} \quad x = 0, & w = \partial_{xx} w = 0 &\quad \text{at} \quad x = l_1, \\
w = \partial_{yy} w = 0 &\quad \text{at} \quad y = 0, & w = \partial_{yy} w = 0 &\quad \text{at} \quad y = l_2, \\
w = \partial_{zz} w = 0 &\quad \text{at} \quad z = 0, & w = \partial_{zz} w = 0 &\quad \text{at} \quad z = l_3.
\end{aligned}$$

Solution:

$$w(x,y,z,t) = \frac{\partial}{\partial t} \int_0^{l_1} \int_0^{l_2} \int_0^{l_3} f(\xi,\eta,\zeta) G(x,y,z,\xi,\eta,\zeta,t) \, d\zeta \, d\eta \, d\xi$$
$$+ \int_0^{l_1} \int_0^{l_2} \int_0^{l_3} g(\xi,\eta,\zeta) G(x,y,z,\xi,\eta,\zeta,t) \, d\zeta \, d\eta \, d\xi,$$

where

$$G(x,y,z,\xi,\eta,\zeta,t) = \frac{8}{al_1 l_2 l_3} \sum_{n=1}^{\infty} \sum_{m=1}^{\infty} \sum_{k=1}^{\infty} \frac{1}{\lambda_{nmk}} \sin(p_n x) \sin(q_m y) \sin(s_k z)$$
$$\times \sin(p_n \xi) \sin(q_m \eta) \sin(s_k \zeta) \sin(\lambda_{nmk} a t),$$
$$p_n = \frac{\pi n}{l_1}, \quad q_m = \frac{\pi m}{l_2}, \quad s_k = \frac{\pi k}{l_3}, \quad \lambda_{nmk} = p_n^2 + q_m^2 + s_k^2.$$

9.3.3-3. n-dimensional case. Cauchy problem.

Domain: $\mathbb{R}^n = \{-\infty < x_k < \infty; \; k = 1, \ldots, n\}$. Initial conditions are prescribed:

$$w = f(\mathbf{x}) \quad \text{at} \quad t = 0, \qquad \partial_t w = 0 \quad \text{at} \quad t = 0,$$

where $\mathbf{x} = \{x_1, \ldots, x_n\}$.
Solution:

$$w(\mathbf{x}, t) = \frac{1}{(2\sqrt{\pi a t})^n} \int_{\mathbb{R}^n} f(\mathbf{y}) \cos\left(\frac{|\mathbf{x}-\mathbf{y}|}{4at} - \frac{\pi n}{4}\right) d\mathbf{y},$$

where $\mathbf{y} = \{y_1, \ldots, y_n\}$ and $d\mathbf{y} = dy_1 \ldots dy_n$.

⊙ *Reference*: V. S. Vladimirov, V. P. Mikhailov, A. A. Vasharin, et al. (1974).

9.3.3-4. n-dimensional case. Boundary value problem.

Domain: $V = \{0 \leq x_k \leq l_k; \; k = 1, 2, \ldots n\}$ (n-dimensional rectangular parallelepiped).
Initial conditions:

$$w = f(\mathbf{x}) \quad \text{at} \quad t = 0, \qquad \partial_t w = g(\mathbf{x}) \quad \text{at} \quad t = 0.$$

Boundary conditions:

$$w = \partial_{x_k x_k} w = 0 \quad \text{at} \quad x_k = 0, \qquad w = \partial_{x_k x_k} w = 0 \quad \text{at} \quad x_k = l_k.$$

Solution:

$$w(\mathbf{x}, t) = \frac{\partial}{\partial t} \int_V f(\mathbf{y}) G(\mathbf{x}, \mathbf{y}, t) \, d\mathbf{y} + \int_V g(\mathbf{y}) G(\mathbf{x}, \mathbf{y}, t) \, d\mathbf{y},$$

where

$$G(\mathbf{x},\mathbf{y},t) = \frac{2^n}{a l_1 l_2 \dots l_n} \sum_{k_1=1}^{\infty} \sum_{k_2=1}^{\infty} \dots \sum_{k_n=1}^{\infty} \frac{1}{\lambda_{k_1,k_2,\dots,k_n}} \sin(p_{k_1}x_1)\sin(p_{k_2}x_2)\dots\sin(p_{k_n}x_n)$$
$$\times \sin(p_{k_1}y_1)\sin(p_{k_2}y_2)\dots\sin(p_{k_n}y_n)\sin(\lambda_{k_1,k_2,\dots,k_n}at),$$
$$p_{k_1} = \frac{\pi k_1}{l_1}, \quad p_{k_2} = \frac{\pi k_2}{l_2}, \quad \dots, \quad p_{k_n} = \frac{\pi k_n}{l_n}, \quad \lambda_{k_1,k_2,\dots,k_n} = \sqrt{p_{k_1}^2 + p_{k_2}^2 + \dots + p_{k_n}^2}.$$

9.3.4. Equations of the Form $\frac{\partial^2 w}{\partial t^2} + a^2 \Delta\Delta w + kw = \Phi(x,y,t)$

9.3.4-1. Domain: $0 \le x \le l_1$, $0 \le y \le l_2$. Solution in terms of the Green's function.

We consider boundary value problems in a rectangular domain $0 \le x \le l_1$, $0 \le y \le l_2$ with the general initial conditions

$$w = f(x,y) \quad \text{at} \quad t = 0, \qquad \partial_t w = g(x,y) \quad \text{at} \quad t = 0$$

and various homogeneous boundary conditions. The solution can be represented in terms of the Green's function as

$$w(x,y,t) = \frac{\partial}{\partial t}\int_0^{l_1}\int_0^{l_2} f(\xi,\eta) G(x,y,\xi,\eta,t)\,d\eta\,d\xi + \int_0^{l_1}\int_0^{l_2} g(\xi,\eta) G(x,y,\xi,\eta,t)\,d\eta\,d\xi$$
$$+ \int_0^t \int_0^{l_1}\int_0^{l_2} \Phi(\xi,\eta,\tau) G(x,y,z,\xi,\eta,\zeta,t-\tau)\,d\eta\,d\xi\,d\tau.$$

Paragraphs 9.3.4-2 through 9.3.4-4 present the Green's functions for three types of boundary conditions.

9.3.4-2. The function and its second derivatives are prescribed at the sides of a rectangle:

$$w = \partial_{xx}w = 0 \quad \text{at} \quad x = 0, \qquad w = \partial_{xx}w = 0 \quad \text{at} \quad x = l_1,$$
$$w = \partial_{yy}w = 0 \quad \text{at} \quad y = 0, \qquad w = \partial_{yy}w = 0 \quad \text{at} \quad y = l_2.$$

Green's function:

$$G(x,y,\xi,\eta,t) = \frac{4}{l_1 l_2} \sum_{n=1}^{\infty}\sum_{m=1}^{\infty} \sin(p_n x)\sin(q_m y)\sin(p_n \xi)\sin(q_m \eta)\frac{\sin(\lambda_{nm}t)}{\lambda_{nm}},$$
$$p_n = \frac{\pi n}{l_1}, \quad q_m = \frac{\pi m}{l_2}, \quad \lambda_{nm} = \sqrt{a^2(p_n^2 + q_m^2)^2 + k}.$$

9.3.4-3. The first and third derivatives are prescribed at the sides of a rectangle:

$$\partial_x w = \partial_{xxx}w = 0 \quad \text{at} \quad x = 0, \qquad \partial_x w = \partial_{xxx}w = 0 \quad \text{at} \quad x = l_1,$$
$$\partial_y w = \partial_{yyy}w = 0 \quad \text{at} \quad y = 0, \qquad \partial_y w = \partial_{yyy}w = 0 \quad \text{at} \quad y = l_2.$$

Green's function:

$$G(x,y,\xi,\eta,t) = \frac{1}{l_1 l_2} \sum_{n=0}^{\infty}\sum_{m=0}^{\infty} \varepsilon_n \varepsilon_m \cos(p_n x)\cos(q_m y)\cos(p_n \xi)\cos(q_m \eta)\frac{\sin(\lambda_{nm}t)}{\lambda_{nm}},$$
$$p_n = \frac{\pi n}{l_1}, \quad q_m = \frac{\pi m}{l_2}, \quad \lambda_{nm} = \sqrt{a^2(p_n^2 + q_m^2)^2 + k}, \quad \varepsilon_n = \begin{cases} 1 & \text{for } n = 0, \\ 2 & \text{for } n \ne 0. \end{cases}$$

9.3.4-4. Mixed boundary conditions are prescribed at the sides of a rectangle:

$$w = \partial_{xx}w = 0 \quad \text{at} \quad x = 0, \qquad w = \partial_{xx}w = 0 \quad \text{at} \quad x = l_1,$$
$$\partial_y w = \partial_{yyy}w = 0 \quad \text{at} \quad y = 0, \qquad \partial_y w = \partial_{yyy}w = 0 \quad \text{at} \quad y = l_2.$$

Green's function:

$$G(x, y, \xi, \eta, t) = \frac{2}{l_1 l_2} \sum_{n=1}^{\infty} \sum_{m=0}^{\infty} \varepsilon_m \sin(p_n x) \cos(q_m y) \sin(p_n \xi) \cos(q_m \eta) \frac{\sin(\lambda_{nm} t)}{\lambda_{nm}},$$

$$p_n = \frac{\pi n}{l_1}, \quad q_m = \frac{\pi m}{l_2}, \quad \lambda_{nm} = \sqrt{a^2(p_n^2 + q_m^2)^2 + k}, \quad \varepsilon_m = \begin{cases} 1 & \text{for } m = 0, \\ 2 & \text{for } m \neq 0. \end{cases}$$

9.3.5. Equations of the Form $\dfrac{\partial^2 w}{\partial t^2} + a^2 \left(\dfrac{\partial^4 w}{\partial x^4} + \dfrac{\partial^4 w}{\partial y^4} \right) + kw = \Phi(x, y, t)$

9.3.5-1. Domain: $0 \leq x \leq l_1$, $0 \leq y \leq l_2$. Solution in terms of the Green's function.

We consider boundary value problems in a rectangular domain $0 \leq x \leq l_1$, $0 \leq y \leq l_2$ with the general initial conditions

$$w = f(x, y) \quad \text{at} \quad t = 0, \qquad \partial_t w = g(x, y) \quad \text{at} \quad t = 0$$

and various homogeneous boundary conditions. The solution can be represented in terms of the Green's function as

$$w(x, y, t) = \frac{\partial}{\partial t} \int_0^{l_1} \int_0^{l_2} f(\xi, \eta) G(x, y, \xi, \eta, t) \, d\eta \, d\xi + \int_0^{l_1} \int_0^{l_2} g(\xi, \eta) G(x, y, \xi, \eta, t) \, d\eta \, d\xi$$
$$+ \int_0^t \int_0^{l_1} \int_0^{l_2} \Phi(\xi, \eta, \tau) G(x, y, z, \xi, \eta, \zeta, t - \tau) \, d\eta \, d\xi \, d\tau.$$

Paragraphs 9.3.5-2 through 9.3.5-4 present the Green's functions for three types of boundary conditions.

9.3.5-2. The function and its first derivatives are prescribed at the sides of a rectangle:

$$w = \partial_x w = 0 \quad \text{at} \quad x = 0, \qquad w = \partial_x w = 0 \quad \text{at} \quad x = l_1,$$
$$w = \partial_y w = 0 \quad \text{at} \quad y = 0, \qquad w = \partial_y w = 0 \quad \text{at} \quad y = l_2.$$

Green's function:

$$G(x, y, \xi, \eta, t) = \frac{16}{l_1 l_2} \sum_{n=1}^{\infty} \sum_{m=1}^{\infty} \frac{p_n^4 q_m^4}{[\varphi_n''(l_1)\psi_m''(l_2)]^2} \varphi_n(x)\psi_m(y)\varphi_n(\xi)\psi_m(\eta) \frac{\sin(\lambda_{nm} t)}{\lambda_{nm}},$$

$$\lambda_{nm} = \sqrt{a^2(p_n^4 + q_m^4) + k}, \quad \varphi_n''(x) = \frac{d^2 \varphi_n}{dx^2}, \quad \psi_m''(y) = \frac{d^2 \psi_m}{dy^2}.$$

Here,

$$\varphi_n(x) = \big[\sinh(p_n l_1) - \sin(p_n l_1)\big]\big[\cosh(p_n x) - \cos(p_n x)\big]$$
$$\qquad - \big[\cosh(p_n l_1) - \cos(p_n l_1)\big]\big[\sinh(p_n x) - \sin(p_n x)\big],$$
$$\psi_m(y) = \big[\sinh(q_m l_2) - \sin(q_m l_2)\big]\big[\cosh(q_m y) - \cos(q_m y)\big]$$
$$\qquad - \big[\cosh(q_m l_2) - \cos(q_m l_2)\big]\big[\sinh(q_m y) - \sin(q_m y)\big],$$

where the p_n and q_m are positive roots of the transcendental equations

$$\cosh(p l_1) \cos(p l_1) = 1, \quad \cosh(q l_2) \cos(q l_2) = 1.$$

9.3.5-3. The function and its second derivatives are prescribed at the sides of a rectangle:

$$w = \partial_{xx}w = 0 \text{ at } x = 0, \qquad w = \partial_{xx}w = 0 \text{ at } x = l_1,$$
$$w = \partial_{yy}w = 0 \text{ at } y = 0, \qquad w = \partial_{yy}w = 0 \text{ at } y = l_2.$$

Green's function:

$$G(x, y, \xi, \eta, t) = \frac{4}{l_1 l_2} \sum_{n=1}^{\infty} \sum_{m=1}^{\infty} \sin(p_n x) \sin(q_m y) \sin(p_n \xi) \sin(q_m \eta) \frac{\sin(\lambda_{nm} t)}{\lambda_{nm}},$$

$$p_n = \frac{\pi n}{l_1}, \quad q_m = \frac{\pi m}{l_2}, \quad \lambda_{nm} = \sqrt{a^2(p_n^4 + q_m^4) + k}.$$

9.3.5-4. The first and third derivatives are prescribed at the sides of a rectangle:

$$\partial_x w = \partial_{xxx}w = 0 \text{ at } x = 0, \qquad \partial_x w = \partial_{xxx}w = 0 \text{ at } x = l_1,$$
$$\partial_y w = \partial_{yyy}w = 0 \text{ at } y = 0, \qquad \partial_y w = \partial_{yyy}w = 0 \text{ at } y = l_2.$$

Green's function:

$$G(x, y, \xi, \eta, t) = \frac{1}{l_1 l_2} \sum_{n=0}^{\infty} \sum_{m=0}^{\infty} \varepsilon_n \varepsilon_m \cos(p_n x) \cos(q_m y) \cos(p_n \xi) \cos(q_m \eta) \frac{\sin(\lambda_{nm} t)}{\lambda_{nm}},$$

$$p_n = \frac{\pi n}{l_1}, \quad q_m = \frac{\pi m}{l_2}, \quad \lambda_{nm} = \sqrt{a^2(p_n^4 + q_m^4) + k}, \quad \varepsilon_n = \begin{cases} 1 & \text{for } n = 0, \\ 2 & \text{for } n \neq 0. \end{cases}$$

9.4. Fourth-Order Stationary Equations

9.4.1. Biharmonic Equation $\Delta\Delta w = 0$

The biharmonic equation is encountered in plane problems of elasticity (w is the Airy stress function). It is also used to describe slow flows of viscous incompressible fluids (w is the stream function).

All solutions of the Laplace equation $\Delta w = 0$ (see Sections 7.1 and 8.1) are also solutions of the biharmonic equation.

9.4.1-1. Two-dimensional equation. Particular solutions.

In the rectangular Cartesian system of coordinates, the biharmonic operator has the form

$$\Delta\Delta = \Delta^2 = \frac{\partial^4}{\partial x^4} + 2\frac{\partial^4}{\partial x^2 \partial y^2} + \frac{\partial^4}{\partial y^4}.$$

1°. Particular solutions.

$$w(x,y) = Ax^3 + Bx^2 y + Cxy^2 + Dy^3 + ax^2 + bxy + cy^2 + \alpha x + \beta y + \gamma,$$
$$w(x,y) = (A \cosh \beta x + B \sinh \beta x + Cx \cosh \beta x + Dx \sinh \beta x)(a \cos \beta y + b \sin \beta y),$$
$$w(x,y) = (A \cos \beta x + B \sin \beta x + Cx \cos \beta x + Dx \sin \beta x)(a \cosh \beta y + b \sinh \beta y),$$
$$w(x,y) = Ar^2 \ln r + Br^2 + C \ln r + D, \quad r = \sqrt{(x-a)^2 + (y-b)^2},$$
$$w(x,y) = (Ax + By + C)(D \cosh \beta x + E \sinh \beta x)(a \cos \beta y + b \sin \beta y),$$
$$w(x,y) = (Ax + By + C)(D \cosh \beta y + E \sinh \beta y)(a \cos \beta x + b \sin \beta x),$$
$$w(x,y) = (x^2 + y^2)(D \cosh \beta x + E \sinh \beta x)(a \cos \beta y + b \sin \beta y),$$
$$w(x,y) = (x^2 + y^2)(D \cosh \beta y + E \sinh \beta y)(a \cos \beta x + b \sin \beta x),$$

where A, B, C, D, E, a, b, c, α, β, and γ are arbitrary constants.

TABLE 30
Particular solutions of the biharmonic equation in some orthogonal curvilinear
coordinate systems; A, B, C, D, a, b, and λ are arbitrary constants

Transformation	Particular solutions
Polar coordinates r, φ: $x = r\cos\varphi$, $y = r\sin\varphi$	$w = (Ar^{2+\lambda} + Br^{2-\lambda} + Cr^{\lambda} + Dr^{-\lambda})(a\cos\lambda\varphi + b\sin\lambda\varphi)$, $w = Ar^2\ln r + Br^2 + C\ln r + D$ (at $\lambda = 0$)
Bipolar coordinates ξ, η: $x = \dfrac{c\sinh\xi}{\cosh\xi - \cos\eta}$, $y = \dfrac{c\sin\eta}{\cosh\xi - \cos\eta}$	$w = \dfrac{1}{\cosh\xi - \cos\eta}\big[A\cosh(\lambda+1)\xi + B\sinh(\lambda+1)\xi + C\cosh(\lambda-1)\xi + D\sinh(\lambda-1)\xi\big](a\cos\lambda\eta + b\sin\lambda\eta)$
Degenerate bipolar coordinates u, v: $x = \dfrac{u}{u^2+v^2}$, $y = -\dfrac{v}{u^2+v^2}$	$w = \dfrac{1}{(u^2+v^2)^2}\big[A\cosh(\lambda u) + B\sinh(\lambda u) + Cu\cosh(\lambda u) + Du\sinh(\lambda u)\big]\big[a\cos(\lambda v) + b\sin(\lambda v)\big]$

2°. Fundamental solution:

$$\mathscr{E}(x, y) = \frac{1}{8\pi}r^2 \ln r, \qquad r = \sqrt{x^2 + y^2}.$$

3°. Particular solutions of the biharmonic equation in some orthogonal curvilinear coordinate systems are listed in Table 30.

⊙ *Reference*: N. N. Lebedev, I. P. Skal'skaya, and Ya. S. Uflyand (1972).

9.4.1-2. Two-dimensional equation. Various representations of the general solution.

1°. Various representations of the general solution in terms of harmonic functions:

$$w(x, y) = xu_1(x, y) + u_2(x, y),$$
$$w(x, y) = yu_1(x, y) + u_2(x, y),$$
$$w(x, y) = (x^2 + y^2)u_1(x, y) + u_2(x, y),$$

where u_1 and u_2 are arbitrary functions satisfying the Laplace equation $\Delta u_k = 0$ ($k = 1, 2$).

⊙ *Reference*: A. N. Tikhonov and A. A. Samarskii (1990).

2°. Complex form of representation of the general solution:

$$w(x, y) = \operatorname{Re}\big[\bar{z}f(z) + g(z)\big],$$

where $f(z)$ and $g(z)$ are arbitrary analytic functions of the complex variable $z = x + iy$; $\bar{z} = x - iy$, $i^2 = -1$. The symbol $\operatorname{Re}[A]$ stands for the real part of the complex quantity A.

⊙ *Reference*: A. V. Bitsadze and D. F. Kalinichenko (1985).

9.4.1-3. Two-dimensional boundary value problems for the upper half-plane.

1°. Domain: $-\infty < x < \infty$, $0 \leq y < \infty$. The desired function and its derivative along the normal are prescribed at the boundary:

$$w = 0 \quad \text{at} \quad y = 0, \qquad \partial_y w = f(x) \quad \text{at} \quad y = 0.$$

Solution:

$$w(x, y) = \int_{-\infty}^{\infty} f(\xi)G(x - \xi, y)\,d\xi, \qquad G(x, y) = \frac{1}{\pi}\frac{y^2}{x^2 + y^2}.$$

⊙ *Reference*: G. E. Shilov (1965).

2°. Domain: $-\infty < x < \infty$, $0 \le y < \infty$. The derivatives of the desired function are prescribed at the boundary:
$$\partial_x w = f(x) \quad \text{at} \quad y = 0, \qquad \partial_y w = g(x) \quad \text{at} \quad y = 0.$$

Solution:
$$w(x,y) = \frac{1}{\pi} \int_{-\infty}^{\infty} f(\xi) \left[\arctan\left(\frac{x-\xi}{y}\right) + \frac{y(x-\xi)}{(x-\xi)^2 + y^2} \right] d\xi + \frac{y^2}{\pi} \int_{-\infty}^{\infty} \frac{g(\xi)\, d\xi}{(x-\xi)^2 + y^2} + C,$$

where C is an arbitrary constant.

Example. Let us consider the problem of a slow (Stokes) inflow of a viscous fluid into the half-plane through a slit of width $2a$ with a constant velocity U that makes an angle β with the normal to the boundary (the angle is reckoned from the normal counterclockwise).

With the stream function w introduced by the relations $v_x = -\frac{\partial w}{\partial y}$ and $v_y = \frac{\partial w}{\partial x}$ (v_x and v_y are the fluid velocity components), the problem is reduced to the special case of the previous problem with

$$f(x) = \begin{cases} U\cos\beta & \text{for } |x| < a, \\ 0 & \text{for } |x| > a, \end{cases} \qquad g(x) = \begin{cases} U\sin\beta & \text{for } |x| < a, \\ 0 & \text{for } |x| > a. \end{cases}$$

Dean's solution:
$$w(x,y) = \frac{U}{\pi}\left[(x-a)\cos\beta + y\sin\beta\right] \arctan\left(\frac{y}{x-a}\right) - \frac{U}{\pi}\left[(x+a)\cos\beta + y\sin\beta\right] \arctan\left(\frac{y}{x+a}\right) + C.$$

⊙ *Reference*: I. Sneddon (1951).

9.4.1-4. Two-dimensional boundary value problem for a circle.

Domain: $0 \le r \le a$, $0 \le \varphi \le 2\pi$. Boundary conditions in the polar coordinate system:
$$w = f(\varphi) \quad \text{at} \quad r = a, \qquad \partial_r w = g(\varphi) \quad \text{at} \quad r = a.$$

Solution:
$$w(r,\varphi) = \frac{1}{2\pi a}(r^2 - a^2)^2 \left[\int_0^{2\pi} \frac{[a - r\cos(\eta - \varphi)]f(\eta)\, d\eta}{[r^2 + a^2 - 2ar\cos(\eta - \varphi)]^2} - \frac{1}{2} \int_0^{2\pi} \frac{g(\eta)\, d\eta}{r^2 + a^2 - 2ar\cos(\eta - \varphi)} \right].$$

⊙ *Reference*: A. N. Tikhonov and A. A. Samarskii (1990).

9.4.1-5. Three-dimensional equation.

In the rectangular Cartesian coordinate system, the three-dimensional biharmonic operator is expressed as
$$\Delta\Delta \equiv \Delta^2 = \frac{\partial^4}{\partial x^4} + \frac{\partial^4}{\partial y^4} + \frac{\partial^4}{\partial z^4} + 2\frac{\partial^4}{\partial x^2 \partial y^2} + 2\frac{\partial^4}{\partial x^2 \partial z^2} + 2\frac{\partial^4}{\partial y^2 \partial z^2}.$$

1° Particular solutions in the Cartesian coordinate system:
$$w(x,y,z) = Ar^2 + Br + C + \frac{D}{r}, \qquad r = \sqrt{(x-a)^2 + (y-b)^2 + (z-c)^2},$$
$$w(x,y,z) = \left[Ax\sin(\beta x) + B\sin(\beta x) + Cx\cos(\beta x) + D\cos(\beta x)\right] \sin(\mu y) \exp\!\left(\pm z\sqrt{\beta^2 + \mu^2}\right),$$
$$w(x,y,z) = \left[Ax\sin(\beta x) + B\sin(\beta x) + Cx\cos(\beta x) + D\cos(\beta x)\right] \cos(\mu y) \exp\!\left(\pm z\sqrt{\beta^2 + \mu^2}\right),$$
$$w(x,y,z) = \left[Ax\sin(\beta x) + B\sin(\beta x) + Cx\cos(\beta x) + D\cos(\beta x)\right] \sinh(\mu y) \exp\!\left(\pm z\sqrt{\beta^2 - \mu^2}\right),$$
$$w(x,y,z) = \left[Ax\sin(\beta x) + B\sin(\beta x) + Cx\cos(\beta x) + D\cos(\beta x)\right] \cosh(\mu y) \exp\!\left(\pm z\sqrt{\beta^2 - \mu^2}\right),$$
$$w(x,y,z) = \left[Ax\sinh(\beta x) + B\sinh(\beta x) + Cx\cosh(\beta x) + D\cosh(\beta x)\right] \sinh(\mu y) \sin\!\left(z\sqrt{\beta^2 + \mu^2}\right),$$
$$w(x,y,z) = \left[Ax\sinh(\beta x) + B\sinh(\beta x) + Cx\cosh(\beta x) + D\cosh(\beta x)\right] \cosh(\mu y) \cos\!\left(z\sqrt{\beta^2 + \mu^2}\right),$$

where A, B, C, D, β, and μ are arbitrary constants.

2°. Particular solutions in the cylindrical coordinate system $\left(r = \sqrt{x^2 + y^2}\right)$:

$w(r, \varphi, z) = J_n(\mu r)(Ar \cos \varphi + Br \sin \varphi + C)(a_1 \cos n\varphi + b_1 \sin n\varphi)(a_2 \cosh \mu z + b_2 \sinh \mu z),$

$w(r, \varphi, z) = Y_n(\mu r)(Ar \cos \varphi + Br \sin \varphi + C)(a_1 \cos n\varphi + b_1 \sin n\varphi)(a_2 \cosh \mu z + b_2 \sinh \mu z),$

$w(r, \varphi, z) = I_n(\mu r)(Ar \cos \varphi + Br \sin \varphi + C)(a_1 \cos n\varphi + b_1 \sin n\varphi)(a_2 \cos \mu z + b_2 \sin \mu z),$

$w(r, \varphi, z) = K_n(\mu r)(Ar \cos \varphi + Br \sin \varphi + C)(a_1 \cos n\varphi + b_1 \sin n\varphi)(a_2 \cos \mu z + b_2 \sin \mu z),$

$w(r, \varphi, z) = J_n(\mu r)(A \cos n\varphi + B \sin n\varphi)(a_1 \cosh \mu z + b_1 \sinh \mu z + a_2 z \cosh \mu z + b_2 z \sinh \mu z),$

$w(r, \varphi, z) = Y_n(\mu r)(A \cos n\varphi + B \sin n\varphi)(a_1 \cosh \mu z + b_1 \sinh \mu z + a_2 z \cosh \mu z + b_2 z \sinh \mu z),$

$w(r, \varphi, z) = I_n(\mu r)(A \cos n\varphi + B \sin n\varphi)(a_1 \cos \mu z + b_1 \sin \mu z + a_2 z \cos \mu z + b_2 z \sin \mu z),$

$w(r, \varphi, z) = K_n(\mu r)(A \cos n\varphi + B \sin n\varphi)(a_1 \cos \mu z + b_1 \sin \mu z + a_2 z \cos \mu z + b_2 z \sin \mu z),$

where $n = 0, 1, 2, \ldots$; $A, B, C, a_1, a_2, b_1, b_2,$ and μ are arbitrary constants; the $J_n(\xi)$ and $Y_n(\xi)$ are the Bessel functions; and the $I_n(\xi)$ and $K_n(\xi)$ are the modified Bessel functions.

3°. Particular solutions in the spherical coordinate system $\left(r = \sqrt{x^2 + y^2 + z^2}\right)$:

$$w(r) = Ar^2 + Br + C + Dr^{-1},$$

$$w(r, \theta) = \left(Ar^{n+2} + Br^n + Cr^{1-n} + Dr^{-1-n}\right) P_n(\cos \theta),$$

$$w(r, \theta, \varphi) = \left(Ar^{n+2} + Br^n + Cr^{1-n} + Dr^{-1-n}\right) P_n^m(\cos \theta)(a \cos m\varphi + b \sin m\varphi),$$

where $n = 0, 1, 2, \ldots$; $m = 0, 1, 2, \ldots, n$; $A, B, C, D, a,$ and b are arbitrary constants; the $P_n(\xi)$ are the Legendre polynomials; and the $P_n^m(\xi)$ are the associated Legendre functions defined by

$$P_n(x) = \frac{1}{n!\,2^n} \frac{d^n}{dx^n}(x^2 - 1)^n, \quad P_n^m(x) = (1 - x^2)^{m/2} \frac{d^m}{dx^m} P_n(x).$$

4°. Fundamental solution:

$$\mathscr{E}(x, y, z) = -\frac{1}{8\pi}\sqrt{x^2 + y^2 + z^2}.$$

5°. Representations of solutions to the biharmonic equation in terms of harmonic functions:

$$w(x, y, z) = x u_1(x, y, z) + u_2(x, y, z),$$

$$w(x, y, z) = (x^2 + y^2 + z^2) u_1(x, y, z) + u_2(x, y, z),$$

where u_1 and u_2 are arbitrary functions satisfying the three-dimensional Laplace equation $\Delta_3 u_k = 0$ ($k = 1, 2$). The coefficient x of u_1 in the first formula can be replaced by y or z.

⊙ *Reference*: A. V. Bitsadze and D. F. Kalinichenko (1985).

9.4.1-6. *n*-dimensional equation.

1°. Particular solutions:

$$w(\mathbf{x}) = \sum_{i,j,k=1}^n A_{ijk} x_i x_j x_k + \sum_{i,j=1}^n B_{ij} x_i x_j + \sum_{i=1}^n C_i x_i + D,$$

$$w(\mathbf{x}) = Ar^2 + B + Cr^{4-n} + Dr^{2-n}, \quad r^2 = \sum_{k=1}^n (x_k - \alpha_k)^2,$$

$$w(\mathbf{x}) = (A + Br^{2-n})\left(\sum_{i=1}^n C_i x_i + D\right), \quad r^2 = \sum_{k=1}^n (x_k - \alpha_k)^2,$$

$$w(\mathbf{x}) = \exp\left(\pm x_n \sqrt{\lambda_n}\right) \left(\sum_{i=1}^n A_i x_i + B\right) \prod_{k=1}^{n-1} \sin(\alpha_k x_k + \beta_k), \quad \lambda_n = \sum_{k=1}^{n-1} \alpha_k^2,$$

$$w(\mathbf{x}) = \left(\sum_{i=1}^n A_i x_i + B\right) \left[\prod_{k=1}^{m-1} \sin(\alpha_k x_k + \beta_k)\right] \left[\prod_{k=m}^n \sinh(\gamma_k x_k)\right], \quad \sum_{k=1}^{m-1} \alpha_k^2 - \sum_{k=m}^n \gamma_k^2 = 0,$$

where the $A_{ijk}, B_{ij}, A_i, C_i, A, B, C, D, \alpha_k, \beta_k,$ and γ_k are arbitrary constants.

2°. Fundamental solution:

$$\mathscr{E}(\mathbf{x}) = \begin{cases} \dfrac{\Gamma(n/2)|\mathbf{x}|^{4-n}}{4\pi^{n/2}(n-2)(n-4)} & \text{for } n = 3, 5, 6, 7, \ldots; \\ -\dfrac{1}{8\pi^2}\ln|\mathbf{x}| & \text{for } n = 4. \end{cases}$$

For $n = 2$, see Paragraph 9.4.1-1, Item 2°.

⊙ *Reference*: G. E. Shilov (1965).

3°. Various representations of solutions to the biharmonic equation in terms of harmonic functions:

$$w(\mathbf{x}) = x_s u_1(\mathbf{x}) + u_2(\mathbf{x}), \qquad s = 1, 2, \ldots, n;$$

$$w(\mathbf{x}) = |\mathbf{x}|^2 u_1(\mathbf{x}) + u_2(\mathbf{x}), \qquad |\mathbf{x}|^2 = \sum_{k=1}^{n} x_k^2,$$

where u_1 and u_2 are arbitrary functions satisfying the n-dimensional Laplace equation $\Delta_n u_m = 0$ ($m = 1, 2$).

⊙ *Reference*: A. V. Bitsadze and D. F. Kalinichenko (1985).

9.4.2. Equations of the Form $\Delta\Delta w = \Phi(x, y)$

Nonhomogeneous biharmonic equation. It is encountered in plane problems of elasticity and hydrodynamics.

9.4.2-1. Domain: $-\infty < x < \infty$, $-\infty < y < \infty$.

Solution:

$$w(x, y) = \int_{-\infty}^{\infty} \int_{-\infty}^{\infty} \Phi(\xi, \eta) \mathscr{E}(x - \xi, y - \eta)\, d\xi\, d\eta, \qquad \mathscr{E}(x, y) = \frac{1}{8\pi}(x^2 + y^2)\ln\sqrt{x^2 + y^2}.$$

⊙ *Reference*: A. V. Bitsadze and D. F. Kalinichenko (1985).

9.4.2-2. Domain: $-\infty < x < \infty$, $0 \le y < \infty$. Boundary value problem.

The upper half-plane is considered. The derivatives are prescribed at the boundary:

$$\partial_x w = f(x) \quad \text{at} \quad y = 0, \qquad \partial_y w = g(x) \quad \text{at} \quad y = 0.$$

Solution:

$$w(x, y) = \frac{1}{\pi}\int_{-\infty}^{\infty} f(\xi)\left[\arctan\left(\frac{x-\xi}{y}\right) + \frac{y(x-\xi)}{(x-\xi)^2 + y^2}\right] d\xi + \frac{y^2}{\pi}\int_{-\infty}^{\infty} \frac{g(\xi)\, d\xi}{(x-\xi)^2 + y^2}$$
$$+ \frac{1}{8\pi}\int_{-\infty}^{\infty} d\xi \int_{0}^{\infty} \left[\frac{1}{2}(R_+^2 - R_-^2) - R_-^2 \ln\frac{R_+}{R_-}\right] \Phi(\xi, \eta)\, d\eta + C,$$

where C is an arbitrary constant,

$$R_+^2 = (x - \xi)^2 + (y + \eta)^2, \qquad R_-^2 = (x - \xi)^2 + (y - \eta)^2.$$

⊙ *Reference*: I. Sneddon (1951).

9.4.2-3. Domain: $0 \leq x \leq l_1$, $0 \leq y \leq l_2$. The sides of the plate are hinged.

A rectangle is considered. Boundary conditions are prescribed:

$$w = \partial_{xx} w = 0 \quad \text{at} \quad x = 0, \qquad w = \partial_{xx} w = 0 \quad \text{at} \quad x = l_1,$$
$$w = \partial_{yy} w = 0 \quad \text{at} \quad y = 0, \qquad w = \partial_{yy} w = 0 \quad \text{at} \quad y = l_2.$$

Solution:

$$w(x, y) = \int_0^{l_1} \int_0^{l_2} \Phi(\xi, \eta) G(x, y, \xi, \eta) \, d\eta \, d\xi,$$

where

$$G(x, y, \xi, \eta) = \frac{4}{l_1 l_2} \sum_{n=1}^{\infty} \sum_{m=1}^{\infty} \frac{1}{(p_n^2 + q_m^2)^2} \sin(p_n x) \sin(q_m y) \sin(p_n \xi) \sin(q_m \eta),$$

$$p_n = \frac{\pi n}{l_1}, \qquad q_m = \frac{\pi m}{l_2}.$$

9.4.3. Equations of the Form $\Delta\Delta w - \lambda w = \Phi(x, y)$

9.4.3-1. Homogeneous equation ($\Phi \equiv 0$).

This equation describes the shapes of two-dimensional free transverse vibrations of a thin elastic plate; the function w defines the deflection (transverse displacement) of the plate's midplane points relative to the original plane position and $k = \lambda^{1/4}$ is the frequency parameter. Here, $\Delta\Delta = \Delta^2$ is the biharmonic operator and Δ is the Laplace operator defined as

$$\Delta = \begin{cases} \frac{\partial^2}{\partial x^2} + \frac{\partial^2}{\partial y^2} & \text{in the Cartesian coordinate system,} \\ \frac{\partial^2}{\partial r^2} + \frac{1}{r}\frac{\partial}{\partial r} + \frac{1}{r^2}\frac{\partial^2}{\partial \varphi^2} & \text{in the polar coordinate system.} \end{cases}$$

1°. Particular solutions (A_1, A_2, A_3, A_4, B_1, and B_2 are arbitrary constants):

$$w(x, y) = \big[A_1 \sin(k_1 x) + B_1 \cos(k_1 x)\big] \big[A_2 \sin(k_2 y) + B_2 \cos(k_2 y)\big], \qquad \lambda = (k_1^2 + k_2^2)^2,$$
$$w(x, y) = \big[A_1 \sin(k_1 x) + B_1 \cos(k_1 x)\big] \big[A_2 \sinh(k_2 y) + B_2 \cosh(k_2 y)\big], \qquad \lambda = (k_1^2 - k_2^2)^2,$$
$$w(x, y) = \big[A_1 \sinh(k_1 x) + B_1 \cosh(k_1 x)\big] \big[A_2 \sin(k_2 y) + B_2 \cos(k_2 y)\big], \qquad \lambda = (k_1^2 - k_2^2)^2,$$
$$w(x, y) = \big[A_1 \sinh(k_1 x) + B_1 \cosh(k_1 x)\big] \big[A_2 \sinh(k_2 y) + B_2 \cosh(k_2 y)\big], \qquad \lambda = (k_1^2 + k_2^2)^2,$$
$$w(r, \varphi) = \big[A_1 J_n(kr) + A_2 Y_n(kr) + A_3 I_n(kr) + A_4 K_n(kr)\big] \cos(n\varphi), \qquad \lambda = k^4 > 0,$$
$$w(r, \varphi) = \big[A_1 J_n(kr) + A_2 Y_n(kr) + A_3 I_n(kr) + A_4 K_n(kr)\big] \sin(n\varphi), \qquad \lambda = k^4 > 0,$$

where the $J_n(\xi)$ and $Y_n(\xi)$ are the Bessel functions of the first and second kind, the $I_n(\xi)$ and $K_n(\xi)$ are the modified Bessel functions of the first and second kind, $r = \sqrt{x^2 + y^2}$, and $n = 0, 1, 2, \ldots$

2°. General solution:

$$w(x, y) = u_1(x, y) + u_2(x, y),$$

where u_1 and u_2 are arbitrary functions satisfying the Helmholtz equations

$$\Delta u_1 + \sqrt{\lambda}\, u_1 = 0, \qquad \Delta u_2 - \sqrt{\lambda}\, u_2 = 0.$$

For solutions to these equations, see Section 7.3.

9.4.3-2. Domain: $0 \leq x \leq l_1$, $0 \leq y \leq l_2$. Boundary value problem.

A rectangle is considered. Boundary conditions are prescribed:

$$w = \partial_{xx}w = 0 \quad \text{at} \quad x = 0, \qquad w = \partial_{xx}w = 0 \quad \text{at} \quad x = l_1,$$
$$w = \partial_{yy}w = 0 \quad \text{at} \quad y = 0, \qquad w = \partial_{yy}w = 0 \quad \text{at} \quad y = l_2.$$

Solution:

$$w(x,y) = \int_0^{l_1}\int_0^{l_2} \Phi(\xi,\eta)G(x,y,\xi,\eta)\,d\eta\,d\xi,$$

where

$$G(x,y,\xi,\eta,t) = \frac{4}{l_1 l_2}\sum_{n=1}^{\infty}\sum_{m=1}^{\infty}\frac{\sin(p_n x)\sin(q_m y)\sin(p_n \xi)\sin(q_m \eta)}{(p_n^2 + q_m^2)^2 - \lambda}, \qquad p_n = \frac{\pi n}{l_1}, \quad q_m = \frac{\pi m}{l_2}.$$

9.4.3-3. Domain: $0 \leq r \leq R$, $0 \leq \varphi \leq 2\pi$. Eigenvalue problem with $\Phi \equiv 0$.

The unknown and its normal derivative are zero on the boundary of a circular domain:

$$w = \frac{\partial w}{\partial r} = 0 \quad \text{at} \quad r = R.$$

Eigenvalues:

$$\lambda_{nm} = \frac{\beta_{nm}^4}{R^4}, \qquad n = 0, 1, 2, \ldots, \quad m = 1, 2, 3, \ldots,$$

where the β_{nm} are positive roots of the transcendental equation

$$J_n(\beta)I_n'(\beta) - I_n(\beta)J_n'(\beta) = 0.$$

Numerical values of some roots:

$$\beta_{01} = 3.196, \quad \beta_{02} = 6.306, \quad \beta_{03} = 9.439, \quad \beta_{04} = 12.58;$$
$$\beta_{11} = 4.611, \quad \beta_{12} = 7.799, \quad \beta_{13} = 10.96, \quad \beta_{14} = 14.11;$$
$$\beta_{21} = 5.906, \quad \beta_{22} = 9.197, \quad \beta_{23} = 12.40, \quad \beta_{24} = 15.58,$$
$$\beta_{31} = 7.144, \quad \beta_{32} = 10.54, \quad \beta_{33} = 13.79, \quad \beta_{34} = 17.01.$$

Eigenvalues:

$$w_{nm}^{(c)}(r,\varphi) = \left[I_n(\beta_{nm})J_n\!\left(\beta_{nm}\frac{r}{R}\right) - J_n(\beta_{nm})I_n\!\left(\beta_{nm}\frac{r}{R}\right)\right]\cos(n\varphi),$$
$$w_{nm}^{(s)}(r,\varphi) = \left[I_n(\beta_{nm})J_n\!\left(\beta_{nm}\frac{r}{R}\right) - J_n(\beta_{nm})I_n\!\left(\beta_{nm}\frac{r}{R}\right)\right]\sin(n\varphi).$$

⊙ *Reference*: V. V. Bolotin (1978).

9.4.3-4. Domain: $(x/a)^2 + (y/b)^2 \leq 1$. Eigenvalue problem with $\Phi \equiv 0$.

The unknown and its normal derivative are zero on the boundary of an elliptic domain:

$$w = \frac{\partial w}{\partial N} = 0 \quad \text{on} \quad (x/a)^2 + (y/b)^2 = 1 \qquad (a \geq b).$$

Eigenvalues and eigenfunctions (approximate formulas):

$$\lambda_{01} = \frac{\beta_{01}^4}{8}\left(\frac{3}{a^4} + \frac{3}{b^4} + \frac{2}{a^2 b^2}\right), \qquad w_{01}(\mathcal{R}) = I_0(\beta_{01})J_0(\beta_{01}\mathcal{R}) - J_0(\beta_{01})I_0(\beta_{01}\mathcal{R}),$$

$$\lambda_{11}^{(c)} = \frac{\beta_{11}^4}{8}\left(\frac{5}{a^4} + \frac{1}{b^4} + \frac{2}{a^2 b^2}\right), \qquad w_{11}^{(c)}(\mathcal{R},\varphi) = \left[I_1(\beta_{11})J_1(\beta_{11}\mathcal{R}) - J_1(\beta_{11})I_1(\beta_{11}\mathcal{R})\right]\cos\varphi,$$

$$\lambda_{11}^{(s)} = \frac{\beta_{11}^4}{8}\left(\frac{1}{a^4} + \frac{5}{b^4} + \frac{2}{a^2 b^2}\right), \qquad w_{11}^{(s)}(\mathcal{R},\varphi) = \left[I_1(\beta_{11})J_1(\beta_{11}\mathcal{R}) - J_1(\beta_{11})I_1(\beta_{11}\mathcal{R})\right]\sin\varphi,$$

where $\mathcal{R} = \sqrt{(x/a)^2 + (y/b)^2}$, and $\beta_{01} = 3.196$ and $\beta_{11} = 4.611$ are the least roots of the transcendental equations

$$J_0(\beta)I_1(\beta) + J_1(\beta)I_0(\beta) = 0,$$
$$J_1(\beta)I_1'(\beta) - J_1'(\beta)I_1(\beta) = 0.$$

The above formulas were obtained with the aid of generalized (nonorthogonal) polar coordinates $\mathcal{R}, \varphi$ defined by

$$x = a\mathcal{R}\cos\varphi, \quad y = b\mathcal{R}\sin\varphi \quad (0 \leq \mathcal{R} \leq 1,\ 0 \leq \varphi \leq 2\pi)$$

and the variational method.

The maximum error in the eigenvalue λ_1 for $\varepsilon = \sqrt{1-(b/a)^2} \leq 0{,}86$ is less than 1%. The errors in $\lambda_{11}^{(c)}$ and $\lambda_{11}^{(s)}$ for $\varepsilon \leq 0{,}6$ do not exceed 2%. In the limit case $\varepsilon = 0$ that corresponds to a circular domain, the formulas provide exact results.

⦿ *Reference*: L. D. Akulenko, S. V. Nesterov, and A. L. Popov (2001).

9.4.4. Equations of the Form $\dfrac{\partial^4 w}{\partial x^4} + \dfrac{\partial^4 w}{\partial y^4} = \Phi(x,y)$

9.4.4-1. Homogeneous equation ($\Phi \equiv 0$).

1°. Particular solutions:

$$w(x,y) = \left[A\sin(\lambda x) + B\cos(\lambda x) + C\sinh(\lambda x) + D\cosh(\lambda x)\right]\exp\left(\tfrac{1}{\sqrt{2}}\lambda y\right)\sin\left(\tfrac{1}{\sqrt{2}}\lambda y\right),$$
$$w(x,y) = \left[A\sin(\lambda x) + B\cos(\lambda x) + C\sinh(\lambda x) + D\cosh(\lambda x)\right]\exp\left(\tfrac{1}{\sqrt{2}}\lambda y\right)\cos\left(\tfrac{1}{\sqrt{2}}\lambda y\right),$$
$$w(x,y) = \left[A\sin(\lambda x) + B\cos(\lambda x) + C\sinh(\lambda x) + D\cosh(\lambda x)\right]\exp\left(-\tfrac{1}{\sqrt{2}}\lambda y\right)\sin\left(\tfrac{1}{\sqrt{2}}\lambda y\right),$$
$$w(x,y) = \left[A\sin(\lambda x) + B\cos(\lambda x) + C\sinh(\lambda x) + D\cosh(\lambda x)\right]\exp\left(-\tfrac{1}{\sqrt{2}}\lambda y\right)\cos\left(\tfrac{1}{\sqrt{2}}\lambda y\right),$$

where A, B, C, D, and λ are arbitrary constants.

2°. General solution:

$$w(x,y) = \operatorname{Re}\left[f(z_1) + g(z_2)\right].$$

Here, $f(z_1)$ and $g(z_2)$ are arbitrary analytic functions of the complex variables $z_1 = x - \tfrac{1}{\sqrt{2}}(1+i)y$ and $z_2 = x + \tfrac{1}{\sqrt{2}}(1+i)y$. The symbol $\operatorname{Re}[A]$ stands for the real part of the complex quantity A.

⦿ *Reference*: A. V. Bitsadze end D. F. Kalinichenko (1985).

3°. Domain: $-\infty < x < \infty$, $0 \leq y < \infty$. Boundary value problem.
The upper half-space is considered. Boundary conditions are prescribed:

$$w = 0 \quad \text{at} \quad y = 0, \qquad \partial_y w = f(x) \quad \text{at} \quad y = 0.$$

Solution:

$$w(x,y) = \int_{-\infty}^{\infty} f(\xi)G(x-\xi, y)\,d\xi,$$

where

$$G(x,y) = \frac{1}{\pi\sqrt{2}}\left[\arctan\left(1 - \frac{x\sqrt{2}}{y}\right) + \arctan\left(1 + \frac{x\sqrt{2}}{y}\right)\right].$$

⦿ *Reference*: G. E. Shilov (1965).

9.4.4-2. Nonhomogeneous equation. Boundary value problems in a rectangle.

We consider problems in a rectangular domain $0 \leq x \leq l_1$, $0 \leq y \leq l_2$ with different homogeneous boundary conditions. The solution can be expressed in terms of the Green's function as

$$w(x,y) = \int_0^{l_1} \int_0^{l_2} \Phi(\xi,\eta) G(x,y,\xi,\eta)\, d\eta\, d\xi.$$

Below are the Green's functions for two types of boundary conditions.

1°. The function and its first derivatives are prescribed at the sides of the rectangle:

$$w = \partial_x w = 0 \quad \text{at} \quad x = 0, \qquad w = \partial_x w = 0 \quad \text{at} \quad x = l_1,$$
$$w = \partial_y w = 0 \quad \text{at} \quad y = 0, \qquad w = \partial_y w = 0 \quad \text{at} \quad y = l_2.$$

Green's function:

$$G(x,y,\xi,\eta) = \frac{16}{l_1 l_2} \sum_{n=1}^{\infty} \sum_{m=1}^{\infty} \frac{p_n^4 q_m^4 \varphi_n(x)\psi_m(y)\varphi_n(\xi)\psi_m(\eta)}{(p_n^4 + q_m^4)\left[\varphi_n''(l_1)\psi_m''(l_2)\right]^2}.$$

Here,

$$\varphi_n(x) = \bigl[\sinh(p_n l_1) - \sin(p_n l_1)\bigr]\bigl[\cosh(p_n x) - \cos(p_n x)\bigr]$$
$$\qquad - \bigl[\cosh(p_n l_1) - \cos(p_n l_1)\bigr]\bigl[\sinh(p_n x) - \sin(p_n x)\bigr],$$
$$\psi_m(y) = \bigl[\sinh(q_m l_2) - \sin(q_m l_2)\bigr]\bigl[\cosh(q_m y) - \cos(q_m y)\bigr]$$
$$\qquad - \bigl[\cosh(q_m l_2) - \cos(q_m l_2)\bigr]\bigl[\sinh(q_m y) - \sin(q_m y)\bigr],$$

where the p_n and q_m are positive roots of the transcendental equations

$$\cosh(pl_1)\cos(pl_1) = 1, \qquad \cosh(ql_2)\cos(ql_2) = 1.$$

2°. The function and its second derivatives are prescribed at the sides of the rectangle:

$$w = \partial_{xx} w = 0 \quad \text{at} \quad x = 0, \qquad w = \partial_{xx} w = 0 \quad \text{at} \quad x = l_1,$$
$$w = \partial_{yy} w = 0 \quad \text{at} \quad y = 0, \qquad w = \partial_{yy} w = 0 \quad \text{at} \quad y = l_2.$$

Green's function:

$$G(x,y,\xi,\eta) = \frac{4}{l_1 l_2} \sum_{n=1}^{\infty} \sum_{m=1}^{\infty} \frac{1}{p_n^4 + q_m^4} \sin(p_n x)\sin(q_m y)\sin(p_n \xi)\sin(q_m \eta),$$
$$p_n = \frac{\pi n}{l_1}, \qquad q_m = \frac{\pi m}{l_2}.$$

9.4.5. Equations of the Form $\dfrac{\partial^4 w}{\partial x^4} + \dfrac{\partial^4 w}{\partial y^4} + kw = \Phi(x,y)$

9.4.5-1. Particular solutions of the homogeneous equation ($\Phi \equiv 0$):

$$w(x,y) = \bigl[A\sin(\lambda x) + B\cos(\lambda x) + C\sinh(\lambda x) + D\cosh(\lambda x)\bigr] \exp(\beta y)\sin(\beta y),$$
$$w(x,y) = \bigl[A\sin(\lambda x) + B\cos(\lambda x) + C\sinh(\lambda x) + D\cosh(\lambda x)\bigr] \exp(\beta y)\cos(\beta y),$$
$$w(x,y) = \bigl[A\sin(\lambda x) + B\cos(\lambda x) + C\sinh(\lambda x) + D\cosh(\lambda x)\bigr] \exp(-\beta y)\sin(\beta y),$$
$$w(x,y) = \bigl[A\sin(\lambda x) + B\cos(\lambda x) + C\sinh(\lambda x) + D\cosh(\lambda x)\bigr] \exp(-\beta y)\cos(\beta y),$$

where $\beta = \frac{1}{\sqrt{2}}(\lambda^4 + k)^{1/4}$; A, B, C, D, and λ are arbitrary constants.

9.4.5-2. Domain: $0 \le x \le l_1$, $0 \le y \le l_2$. **Boundary value problems.**

1°. We consider problems in a rectangular domain with different homogeneous boundary conditions. The solution can be expressed in terms of the Green's function as

$$w(x,y) = \int_0^{l_1} \int_0^{l_2} \Phi(\xi,\eta) G(x,y,\xi,\eta) \, d\eta \, d\xi.$$

Below are the Green's functions for two types of boundary conditions.

2°. The function and its first derivatives are prescribed at the sides of the rectangle:

$$w = \partial_x w = 0 \quad \text{at} \quad x = 0, \qquad w = \partial_x w = 0 \quad \text{at} \quad x = l_1,$$
$$w = \partial_y w = 0 \quad \text{at} \quad y = 0, \qquad w = \partial_y w = 0 \quad \text{at} \quad y = l_2.$$

Green's function:

$$G(x,y,\xi,\eta) = \frac{16}{l_1 l_2} \sum_{n=1}^{\infty} \sum_{m=1}^{\infty} \frac{p_n^4 q_m^4 \varphi_n(x) \psi_m(y) \varphi_n(\xi) \psi_m(\eta)}{(p_n^4 + q_m^4 + k)\left[\varphi_n''(l_1) \psi_m''(l_2)\right]^2}.$$

Here,

$$\varphi_n(x) = \left[\sinh(p_n l_1) - \sin(p_n l_1)\right]\left[\cosh(p_n x) - \cos(p_n x)\right]$$
$$- \left[\cosh(p_n l_1) - \cos(p_n l_1)\right]\left[\sinh(p_n x) - \sin(p_n x)\right],$$
$$\psi_m(y) = \left[\sinh(q_m l_2) - \sin(q_m l_2)\right]\left[\cosh(q_m y) - \cos(q_m y)\right]$$
$$- \left[\cosh(q_m l_2) - \cos(q_m l_2)\right]\left[\sinh(q_m y) - \sin(q_m y)\right],$$

where the p_n and q_m are positive roots of the transcendental equations

$$\cosh(p l_1)\cos(p l_1) = 1, \quad \cosh(q l_2)\cos(q l_2) = 1.$$

3°. The function and its second derivatives are prescribed at the sides of the rectangle:

$$w = \partial_{xx} w = 0 \quad \text{at} \quad x = 0, \qquad w = \partial_{xx} w = 0 \quad \text{at} \quad x = l_1,$$
$$w = \partial_{yy} w = 0 \quad \text{at} \quad y = 0, \qquad w = \partial_{yy} w = 0 \quad \text{at} \quad y = l_2.$$

Green's function:

$$G(x,y,\xi,\eta) = \frac{4}{l_1 l_2} \sum_{n=1}^{\infty} \sum_{m=1}^{\infty} \frac{1}{p_n^4 + q_m^4 + k} \sin(p_n x)\sin(q_m y)\sin(p_n \xi)\sin(q_m \eta),$$

$$p_n = \frac{\pi n}{l_1}, \quad q_m = \frac{\pi m}{l_2}.$$

9.4.6. Stokes Equation (Axisymmetric Flows of Viscous Fluids)

9.4.6-1. Stokes equation for the stream function in the spherical coordinate system.

The Stokes equation for the stream function in the axisymmetric case is written as

$$E^2(E^2 w) = 0, \qquad E^2 \equiv \frac{\partial^2}{\partial r^2} + \frac{\sin\theta}{r^2}\frac{\partial}{\partial \theta}\left(\frac{1}{\sin\theta}\frac{\partial}{\partial \theta}\right).$$

It governs slow axisymmetric flows of viscous incompressible fluids, with w being the stream function, r and θ the spherical coordinates. The components of the fluid velocity are related to the stream function by $v_r = \dfrac{1}{r^2 \sin\theta}\dfrac{\partial w}{\partial \theta}$ and $v_\theta = -\dfrac{1}{r \sin\theta}\dfrac{\partial w}{\partial r}$.

General solution (A_n, B_n, C_n, D_n, $\widetilde{A}_n$, $\widetilde{B}_n$, $\widetilde{C}_n$, and $\widetilde{D}_n$ are arbitrary constants):

$$w(r,\theta) = \sum_{n=0}^{\infty}\left(A_n r^n + B_n r^{1-n} + C_n r^{n+2} + D_n r^{3-n}\right)\mathcal{J}_n(\cos\theta) \qquad (1)$$
$$+ \sum_{n=2}^{\infty}(\widetilde{A}_n r^n + \widetilde{B}_n r^{1-n} + \widetilde{C}_n r^{n+2} + \widetilde{D}_n r^{3-n})\mathcal{H}_n(\cos\theta),$$

where the $\mathcal{J}_n(\zeta)$ and $\mathcal{H}_n(\zeta)$ are the Gegenbauer functions of the first and second kind, respectively. These are linearly related to the Legendre functions $P_n(\zeta)$ and $Q_n(\zeta)$ by

$$\mathcal{J}_n(\zeta) = \frac{P_{n-2}(\zeta) - P_n(\zeta)}{2n-1}, \quad \mathcal{H}_n(\zeta) = \frac{Q_{n-2}(\zeta) - Q_n(\zeta)}{2n-1} \quad (n \geq 2).$$

The Gegenbauer functions of the first kind are represented in the form of a finite power series as

$$\mathcal{J}_n(\zeta) = -\frac{1}{(n-1)!}\left(\frac{d}{d\zeta}\right)^{n-2}\left(\frac{\zeta^2-1}{2}\right)^{n-1}$$
$$= \frac{1\cdot 3 \ldots (2n-3)}{1\cdot 2 \ldots n}\left[\zeta^n - \frac{n(n-1)}{2(2n-3)}\zeta^{n-2} + \frac{n(n-1)(n-2)(n-3)}{2\cdot 4(2n-3)(2n-5)}\zeta^{n-4} - \cdots\right].$$

In particular,

$$\mathcal{J}_0(\zeta) = 1, \quad \mathcal{J}_1(\zeta) = -\zeta, \quad \mathcal{J}_2(\zeta) = \tfrac{1}{2}(1-\zeta^2), \quad \mathcal{J}_3(\zeta) = \tfrac{1}{2}\zeta(1-\zeta^2),$$
$$\mathcal{J}_4(\zeta) = \tfrac{1}{8}(1-\zeta^2)(5\zeta^2 - 1), \quad \mathcal{J}_5(\zeta) = \tfrac{1}{8}\zeta(1-\zeta^2)(7\zeta^2 - 3).$$

The Gegenbauer functions of the second kind are defined as

$$\mathcal{H}_0(\zeta) = -\zeta, \quad \mathcal{H}_1(\zeta) = -1, \quad \mathcal{H}_n(\zeta) = \frac{1}{2}\mathcal{J}_n(\zeta)\ln\frac{1+\zeta}{1-\zeta} + \mathcal{K}_n(\zeta) \quad \text{at} \quad n \geq 2,$$

where the functions $\mathcal{K}_n(\zeta)$ are expressed in terms of the Gegenbauer functions of the first kind as

$$\mathcal{K}_n(\zeta) = -\sum_{k}^{\frac{1}{2}n \leq k \leq \frac{1}{2}n+\frac{1}{2}} \frac{(2n-4k+1)}{(2k-1)(n-k)}\left[1 - \frac{(2k-1)(n-k)}{n(n-1)}\right]\mathcal{J}_{n-2k+1}(\zeta);$$

the series start with $\mathcal{J}_0$ or $\mathcal{J}_1$, depending on whether n is odd or even. In particular,

$$\mathcal{K}_2(\zeta) = \tfrac{1}{2}\zeta, \quad \mathcal{K}_3(\zeta) = \tfrac{1}{6}(3\zeta^2 - 2), \quad \mathcal{K}_4(\zeta) = \tfrac{1}{24}\zeta(15\zeta^2 - 13), \quad \mathcal{K}_5(\zeta) = \tfrac{1}{120}(105\zeta^4 - 115\zeta^2 + 16).$$

For $n \geq 2$, the Gegenbauer functions of the second kind assume infinite values at the points $\zeta = \pm 1$, which correspond to $\theta = 0$ and $\theta = \pi$. Therefore, if physically there are no singularities in the problem, then the quantities in (1) labeled with a tilde must be set equal to zero. In the overwhelming majority of problems on the flow about particles, drops, or bubbles, the stream function in the spherical coordinates is given by formula (1) with

$$A_1 = A_0 = B_1 = B_0 = C_1 = C_0 = D_1 = D_0 = 0; \quad \widetilde{A}_n = \widetilde{B}_n = \widetilde{C}_n = \widetilde{D}_n = 0 \quad \text{for} \quad n = 2, 3, \ldots$$

Example 1. In the problem on the translational Stokes flow about a solid spherical particle, the following boundary conditions are imposed on the stream function w:

$$w = 0 \quad \text{at} \quad r = R, \qquad \partial_r w = 0 \quad \text{at} \quad r = R, \qquad w \to \tfrac{1}{2}Ur^2\sin^2\theta \quad \text{as} \quad r \to \infty,$$

where R is the radius of the particle and U is the unperturbed fluid velocity at infinity.
Stokes solution:

$$w(r,\theta) = \frac{1}{4}U(r-R)^2\left(2 + \frac{R}{r}\right)\sin^2\theta.$$

Here, only the terms for $n = 2$ in the first sum of (1) remain.

Example 2. In the problem on the axisymmetric straining Stokes flow about a solid spherical particle, the following boundary conditions are imposed on the stream function w:

$$w = 0 \quad \text{at} \quad r = R, \qquad \partial_r w = 0 \quad \text{at} \quad r = R, \qquad w \to \tfrac{1}{2} E r^3 \sin^2 \theta \cos \theta \quad \text{as} \quad r \to \infty,$$

where R is the radius of the particle and E is the shear coefficient.

Solution:

$$w(r, \theta) = \frac{1}{2} E R^3 \left(\frac{r^3}{R^3} - \frac{5}{2} + \frac{3}{2} \frac{R^2}{r^2} \right) \sin^2 \theta \cos \theta.$$

Here, only the terms for $n = 3$ in the first sum of (1) remain.

Example 3. Solving the problem of the translational Stokes flow about a spherical drop (or bubble) is reduced to solving the Stokes equation outside and inside the drop. The boundary condition at infinity is specified in example 1. Conjugate boundary conditions are set at the drop surface; these conditions can be found in the references cited below and are not written out here.

Hadamard–Rybczynski solution:

$$w(r, \theta) = \frac{1}{4} U r^2 \left(2 - \frac{3\beta + 2}{\beta + 1} \frac{R}{r} + \frac{\beta}{\beta + 1} \frac{R^3}{r^3} \right) \sin^2 \theta \quad \text{for} \quad r > R,$$

$$w(r, \theta) = \frac{U}{4(\beta + 1)} r^2 \left(\frac{r^2}{R^2} - 1 \right) \sin^2 \theta \quad \text{for} \quad r < R,$$

where R is the radius of the drop, U the unperturbed fluid velocity at infinity, β the ratio of the dynamic viscosities of the fluids inside and outside the drop (the value $\beta = 0$ corresponds to a gas bubble and $\beta = \infty$ to a solid particle).

Example 4. Solving the problem of the axisymmetric straining Stokes flow about a spherical drop (or bubble) is reduced to solving the Stokes equation outside and inside the drop. The boundary condition at infinity is specified in example 2. Conjugate boundary conditions are set at the drop surface; these conditions can be found in the references cited below and are not written out here.

Taylor solution:

$$w(r, \theta) = \frac{1}{2} E R^3 \left(\frac{r^3}{R^3} - \frac{1}{2} \frac{5\beta + 2}{\beta + 1} + \frac{3}{2} \frac{\beta}{\beta + 1} \frac{R^2}{r^2} \right) \sin^2 \theta \cos \theta \quad \text{for} \quad r > R,$$

$$w(r, \theta) = \frac{3}{4} \frac{E R^3}{\beta + 1} \frac{r^3}{R^3} \left(\frac{r^2}{R^2} - 1 \right) \sin^2 \theta \cos \theta \quad \text{for} \quad r < R,$$

where R is the drop radius, E the shear coefficient, β the ratio of the dynamic viscosities of the fluids inside and outside the drop (the value $\beta = 0$ corresponds to a gas bubble and $\beta = \infty$ to a solid particle).

⊙ *References*: G. I. Taylor (1932), V. G. Levich (1962), J. Happel and H. Brenner (1965), A. D. Polyanin, A. M. Kutepov, A. V. Vyazmin, and D. A. Kazenin (2001).

9.4.6-2. Stokes equation in the bipolar coordinate system.

When studying axisymmetric problems of a flow about two spherical particles (drops, bubbles), one uses the bipolar coordinates ξ, η; these are related to the cylindrical coordinates $\rho = r \cos \theta$, $z = r \sin \theta$ by

$$\rho = \frac{a \sin \xi}{\cosh \eta - \cos \xi}, \qquad z = \frac{a \sinh \eta}{\cosh \eta - \cos \xi}.$$

The general solution of the equation $E^2(E^2 w) = 0$ in the bipolar coordinate system has the form

$$w(\xi, \eta) = \frac{1}{(\cosh \eta - \cos \xi)^{3/2}} \left[\sum_{n=0}^{\infty} \mathcal{J}_{n+1}(\cos \xi) f_n(\eta) + \sum_{n=0}^{\infty} \mathcal{H}_{n+1}(\cos \xi) g_n(\eta) \right],$$

$$f_n(\eta) = A_n \cosh\left[(n - \tfrac{1}{2})\eta\right] + B_n \sinh\left[(n - \tfrac{1}{2})\eta\right] + C_n \cosh\left[(n + \tfrac{3}{2})\eta\right] + D_n \sinh\left[(n + \tfrac{3}{2})\eta\right],$$

$$g_n(\eta) = \widetilde{A}_n \cosh\left[(n - \tfrac{1}{2})\eta\right] + \widetilde{B}_n \sinh\left[(n - \tfrac{1}{2})\eta\right] + \widetilde{C}_n \cosh\left[(n + \tfrac{3}{2})\eta\right] + \widetilde{D}_n \sinh\left[(n + \tfrac{3}{2})\eta\right],$$

where the $A_n, B_n, C_n, D_n, \widetilde{A}_n, \widetilde{B}_n, \widetilde{C}_n,$ and $\widetilde{D}_n$ are arbitrary constants and the $\mathcal{J}_n(\zeta)$ and $\mathcal{H}_n(\zeta)$ are the Gegenbauer functions.

⊙ *Reference*: J. Happel and H. Brenner (1965).

9.4.6-3. Stokes equation in the oblate spheroidal coordinate system.

When studying axisymmetric problems of flows about spheroidal particles, one uses the oblate spheroidal coordinates ξ, η; these are related to the cylindrical coordinates $\rho = r\cos\theta$, $z = r\sin\theta$ by

$$\rho = c\cosh\xi \sin\eta, \quad z = c\sinh\xi \cos\eta.$$

The solution of the equation $E^2(E^2w) = 0$ that describes the flow of a fluid about a prolate spheroid in the direction parallel to the spheroid axis is expressed as

$$w = \frac{1}{2}Uc^2 \cosh^2\xi \sin^2\eta \left\{1 - \frac{[\lambda/(\lambda^2+1)] - [(\lambda_0^2-1)/(\lambda_0^2+1)]\operatorname{arccot}\lambda}{[\lambda_0/(\lambda_0^2+1)] - [(\lambda_0^2-1)/(\lambda_0^2+1)]\operatorname{arccot}\lambda_0}\right\}, \quad \lambda = \sinh\xi, \ \lambda_0 = \sinh\xi_0.$$

Here, w is the stream function, U is the fluid velocity at infinity, c and λ_0 are the constants related to the spheroid semiaxes a and b ($a > b$) by $c = \sqrt{a^2 - b^2}$ and $\lambda_0 = b/c$.

⊙ *Reference*: J. Happel and H. Brenner (1965).

9.5. Higher-Order Linear Equations with Constant Coefficients

▶ Throughout Section 9.5 the following notation is used:

$$\mathbf{x} = \{x_1, \ldots, x_n\}, \quad \mathbf{y} = \{y_1, \ldots, y_n\}, \quad \boldsymbol{\omega} = \{\omega_1, \ldots, \omega_n\}, \quad \boldsymbol{\xi} = \{\xi_1, \ldots, \xi_n\},$$
$$|\mathbf{x}| = \sqrt{x_1^2 + \cdots + x_n^2}, \quad |\boldsymbol{\omega}| = \sqrt{\omega_1^2 + \cdots + \omega_n^2}, \quad \boldsymbol{\omega}\cdot\mathbf{x} = \omega_1 x_1 + \cdots + \omega_n x_n.$$

9.5.1. Fundamental Solutions. Cauchy Problem

9.5.1-1. Domain: $\mathbb{R}^n = \{-\infty < x_k < \infty; \ k = 1, \ldots, n\}$.

Let P be a constant coefficient linear differential operator such that

$$P\left(\frac{\partial}{\partial x_1}, \ldots, \frac{\partial}{\partial x_n}\right) \equiv \sum_{s=0}^{M} a_{s_1,\ldots,s_n} \frac{\partial^s}{\partial x_1^{s_1} \ldots \partial x_n^{s_n}}, \quad s = s_1 + \cdots + s_n,$$

where $s_1, \ldots, s_n$ are nonnegative integers, $a_{s_1,\ldots,s_n}$ are some constants, and M is the order of the operator. A generalized function (distribution) $\mathscr{E}(\mathbf{x}) = \mathscr{E}(x_1, \ldots, x_n)$ that satisfies the equation

$$P\left(\frac{\partial}{\partial x_1}, \ldots, \frac{\partial}{\partial x_n}\right) \mathscr{E}(\mathbf{x}) = \delta(\mathbf{x}),$$

where $\delta(\mathbf{x}) = \delta(x_1)\ldots\delta(x_n)$ is the Dirac delta function in the n-dimensional Euclidian space, is called the fundamental solution corresponding to the operator P.

Any constant coefficient linear differential operator has a fundamental solution $\mathscr{E}(\mathbf{x})$. The fundamental solution is not unique — it is defined up to an additive term $w_0(\mathbf{x})$ that is an arbitrary solution of the homogeneous equation $P\left(\frac{\partial}{\partial x_1}, \ldots, \frac{\partial}{\partial x_n}\right) w_0(\mathbf{x}) = 0$.

The solution of the nonhomogeneous equation

$$P\left(\frac{\partial}{\partial x_1}, \ldots, \frac{\partial}{\partial x_n}\right) w = \Phi(\mathbf{x})$$

with an arbitrary right-hand side has the form

$$w(\mathbf{x}) = \mathscr{E}(\mathbf{x}) * \Phi(\mathbf{x}), \quad \mathscr{E}(\mathbf{x}) * \Phi(\mathbf{x}) = \int_{\mathbb{R}^n} \mathscr{E}(\mathbf{x} - \mathbf{y}) \Phi(\mathbf{y})\, d\mathbf{y}.$$

Here, $d\mathbf{y} = dy_1 \ldots dy_n$ and the convolution $\mathscr{E} * \Phi$ is assumed to be meaningful.

⊙ *References*: G. E. Shilov (1965), S. G. Krein (1972), L. Hörmander (1983), V. S. Vladimirov (1988).

9.5.1-2. Domain: $0 \le t < \infty$, $-\infty < x_k < \infty$; $k = 1, \ldots, n$. **Cauchy problem.**

Now let $P\left(\frac{\partial}{\partial t}, \frac{\partial}{\partial x_1}, \ldots, \frac{\partial}{\partial x_n}\right)$ be a constant coefficient linear differential operator of order m with respect to t. Then a distribution $\mathscr{E}(t, \mathbf{x}) = \mathscr{E}(t, x_1, \ldots, x_n)$, which is a solution of the equation

$$P\left(\frac{\partial}{\partial t}, \frac{\partial}{\partial x_1}, \ldots, \frac{\partial}{\partial x_n}\right) \mathscr{E}(t, \mathbf{x}) = 0$$

and satisfies the initial conditions*

$$\mathscr{E}\big|_{t=0} = 0, \quad \frac{\partial \mathscr{E}}{\partial t}\bigg|_{t=0} = 0, \quad \ldots, \quad \frac{\partial^{m-2} \mathscr{E}}{\partial t^{m-2}}\bigg|_{t=0} = 0, \quad \frac{\partial^{m-1} \mathscr{E}}{\partial t^{m-1}}\bigg|_{t=0} = \delta(\mathbf{x}), \tag{1}$$

is called a *fundamental solution of the Cauchy problem* corresponding to the operators P.

The solution of the Cauchy problem for the linear differential equation

$$P\left(\frac{\partial}{\partial t}, \frac{\partial}{\partial x_1}, \ldots, \frac{\partial}{\partial x_n}\right) w = 0 \tag{2}$$

with the special initial conditions

$$w\big|_{t=0} = 0, \quad \frac{\partial w}{\partial t}\bigg|_{t=0} = 0, \quad \ldots, \quad \frac{\partial^{m-2} w}{\partial t^{m-2}}\bigg|_{t=0} = 0, \quad \frac{\partial^{m-1} w}{\partial t^{m-1}}\bigg|_{t=0} = f(\mathbf{x})$$

is given by

$$w(t, \mathbf{x}) = \mathscr{E}(t, \mathbf{x}) * f(\mathbf{x}), \qquad \mathscr{E}(t, \mathbf{x}) * f(\mathbf{x}) \equiv \int_{\mathbb{R}^n} \mathscr{E}(t, \mathbf{x} - \mathbf{y}) f(\mathbf{y}) \, d\mathbf{y}.$$

⊙ *Reference*: S. G. Krein (1972).

9.5.1-3. Solution of the Cauchy problem for general initial conditions.

If the general initial conditions

$$w\big|_{t=0} = f_0(\mathbf{x}), \quad \frac{\partial w}{\partial t}\bigg|_{t=0} = f_1(\mathbf{x}), \quad \ldots, \quad \frac{\partial^{m-2} w}{\partial t^{m-2}}\bigg|_{t=0} = f_{m-2}(\mathbf{x}), \quad \frac{\partial^{m-1} w}{\partial t^{m-1}}\bigg|_{t=0} = f_{m-1}(\mathbf{x}) \tag{3}$$

are set, the solution of equation (2) is sought in the form

$$w(t, \mathbf{x}) = \mathscr{E}(t, \mathbf{x}) * \varphi_0(\mathbf{x}) + \frac{\partial \mathscr{E}(t, \mathbf{x})}{\partial t} * \varphi_1(\mathbf{x}) + \cdots + \frac{\partial^{m-1} \mathscr{E}(t, \mathbf{x})}{\partial t^{m-1}} * \varphi_{m-1}(\mathbf{x}). \tag{4}$$

Each term in (4) satisfies equation (2), and the functions $\varphi_{m-1}, \varphi_{m-2}, \ldots, \varphi_0$ are determined successively from the linear system

$$f_0(\mathbf{x}) = \varphi_{m-1}(\mathbf{x}),$$
$$f_1(\mathbf{x}) = \varphi_{m-2}(\mathbf{x}) + \frac{\partial^m \mathscr{E}(0, \mathbf{x})}{\partial t^m} * \varphi_{m-1}(\mathbf{x}),$$
$$\ldots\ldots\ldots\ldots\ldots\ldots\ldots\ldots\ldots\ldots\ldots\ldots\ldots\ldots\ldots$$
$$f_k(\mathbf{x}) = \varphi_{m-k-1}(\mathbf{x}) + \frac{\partial^m \mathscr{E}(0, \mathbf{x})}{\partial t^m} * \varphi_{m-k}(\mathbf{x}) + \cdots + \frac{\partial^{m+k-1} \mathscr{E}(0, \mathbf{x})}{\partial t^{m+k-1}} * \varphi_{m-1}(\mathbf{x}), \quad k = 2, \ldots, m-1.$$

This system of equations is obtained by successively differentiating relation (4) followed by substituting $t = 0$ and taking into account the initial conditions (1) and (3).

⊙ *Reference*: G. E. Shilov (1965).

---

* The number of initial conditions can be less than m (see Paragraph 9.5.4-1).

9.5.2. Elliptic Equations

9.5.2-1. Homogeneous elliptic differential operator.

A constant coefficient linear homogeneous differential operator of order k has the form

$$P_k\left(\frac{\partial}{\partial x_1}, \ldots, \frac{\partial}{\partial x_n}\right) \equiv \sum a_{s_1,\ldots,s_n}\left(\frac{\partial}{\partial x_1}\right)^{s_1} \ldots \left(\frac{\partial}{\partial x_n}\right)^{s_n}, \qquad \sum_{i=1}^n s_i = k,$$

where $s_1, \ldots, s_n$ are nonnegative integers. From now on, we adopt the notation

$$\left(\frac{\partial}{\partial x_1}\right)^{s_1} \ldots \left(\frac{\partial}{\partial x_n}\right)^{s_n} \equiv \frac{\partial^{s_1+\cdots+s_n}}{\partial x_1^{s_1} \ldots \partial x_n^{s_n}}.$$

A linear homogeneous differential operator of order k possesses the property

$$P_k\left(b\frac{\partial}{\partial x_1}, \ldots, b\frac{\partial}{\partial x_n}\right) = b^k P_k\left(\frac{\partial}{\partial x_1}, \ldots, \frac{\partial}{\partial x_n}\right), \qquad b \neq 0 \text{ is an arbitrary constant.}$$

A linear homogeneous differential operator P_k is called elliptic if, on replacing in P_k the symbols $\frac{\partial}{\partial x_1}, \ldots, \frac{\partial}{\partial x_n}$ by variables $\omega_1, \ldots, \omega_n$, one obtains a polynomial $P_k(\omega_1, \ldots, \omega_n)$ that does not vanish if $\boldsymbol{\omega} \neq 0$, i.e.,

$$P_k(\omega_1, \ldots, \omega_n) \equiv \sum a_{s_1,\ldots,s_n} \omega_1^{s_1} \ldots \omega_n^{s_n} \neq 0 \qquad \text{if} \quad |\boldsymbol{\omega}| \neq 0.$$

A linear differential equation

$$P_k\left(\frac{\partial}{\partial x_1}, \ldots, \frac{\partial}{\partial x_n}\right)w \equiv \sum a_{s_1,\ldots,s_n} \frac{\partial^{s_1+\cdots+s_n} w}{\partial x_1^{s_1} \ldots \partial x_n^{s_n}} = 0, \qquad \sum_{i=1}^n s_i = k \qquad (1)$$

is called elliptic if the linear homogeneous differential operator P_k is elliptic.

9.5.2-2. Elliptic differential operator of general form.

In general, a constant coefficient linear differential operator of order k has the form

$$\mathscr{P}_k\left(\frac{\partial}{\partial x_1}, \ldots, \frac{\partial}{\partial x_n}\right) = P_k\left(\frac{\partial}{\partial x_1}, \ldots, \frac{\partial}{\partial x_n}\right) + \sum_{i=0}^{k-1} P_i\left(\frac{\partial}{\partial x_1}, \ldots, \frac{\partial}{\partial x_n}\right),$$

where P_k is the leading part of the operator and P_j ($j = 0, 1 \ldots, k$) is a linear homogeneous differential operator of order j. The operator $\mathscr{P}_k$ is said to be elliptic if its leading part P_k is elliptic.

A linear differential equation

$$\mathscr{P}_k\left(\frac{\partial}{\partial x_1}, \ldots, \frac{\partial}{\partial x_n}\right)w = 0 \qquad (2)$$

is called elliptic if the linear differential operator $\mathscr{P}_k$ is elliptic.

Remark. A linear elliptic operator and a linear elliptic differential equation can only be of even order $k = 2m$, where m is a positive integer.

9.5.2-3. Fundamental solution of a homogeneous elliptic equation.

The fundamental solution of the homogeneous elliptic equation (1) with $k = 2m$ is given by

$$\mathscr{E}(\mathbf{x}) = \frac{(-1)^{\frac{n-1}{2}}}{4(2\pi)^{n-1}(2m-n)!} \int_{\Omega_n} |\boldsymbol{\omega} \cdot \mathbf{x}|^{2m-n} \frac{d\Omega_n}{P_{2m}(\boldsymbol{\omega})} \qquad \text{if } n \text{ is odd and } 2m \geq n;$$

$$\mathscr{E}(\mathbf{x}) = \frac{(-1)^{\frac{n-2}{2}}}{(2\pi)^n (2m-n)!} \int_{\Omega_n} |\boldsymbol{\omega} \cdot \mathbf{x}|^{2m-n} \ln|\boldsymbol{\omega} \cdot \mathbf{x}| \frac{d\Omega_n}{P_{2m}(\boldsymbol{\omega})} \qquad \text{if } n \text{ is even and } 2m \geq n;$$

$$\mathscr{E}(\mathbf{x}) = \frac{(-1)^{\frac{n-1}{2}}}{2(2\pi)^{\frac{n-1}{2}}} \int_{\Omega_n} \delta^{(n-2m-1)}(\boldsymbol{\omega} \cdot \mathbf{x}) \frac{d\Omega_n}{P_{2m}(\boldsymbol{\omega})} \qquad \text{if } n \text{ is odd and } 2m < n;$$

$$\mathscr{E}(\mathbf{x}) = \frac{(-1)^{\frac{n}{2}}(n-2m-1)!}{(2\pi)^n} \int_{\Omega_n} |\boldsymbol{\omega} \cdot \mathbf{x}|^{2m-n} \frac{d\Omega_n}{P_{2m}(\boldsymbol{\omega})} \qquad \text{if } n \text{ is even and } 2m < n.$$

Here, the integration is performed over the surface of the n-dimensional sphere Ω_n of unit radius defined by the equation $|\boldsymbol{\omega}| = 1$; $\boldsymbol{\omega} \cdot \mathbf{x} = \omega_1 x_1 + \cdots + \omega_n x_n$ and $P_{2m}(\boldsymbol{\omega}) = P_{2m}(\omega_1, \ldots, \omega_n)$.

A fundamental solution is an ordinary function, analytic at any point $\mathbf{x} \neq 0$; this function is described, in a neighborhood of the origin of coordinates (as $|\mathbf{x}| \to 0$), by the relations

$$\mathscr{E}(\mathbf{x}) = \begin{cases} b_{n,m} |\mathbf{x}|^{2m-n} & \text{if } n \text{ is odd or } n \text{ is even and } n > 2m; \\ c_{n,m} |\mathbf{x}|^{2m-n} \ln|\mathbf{x}| & \text{if } n \text{ is even and } n \leq 2m. \end{cases}$$

Here, $b_{n,m}$ and $c_{n,m}$ are some nonzero constants. If $2m > n$, the fundamental solution has continuous derivatives up to order $2m - n - 1$ inclusive at the origin.

9.5.2-4. Fundamental solution of a general elliptic equation.

The fundamental solution of the general elliptic equation (2) with $k = 2m$ is determined from the relation

$$\mathscr{E}(\mathbf{x}) = \int_{\Omega_n} Z_{\boldsymbol{\omega}}(\boldsymbol{\omega} \cdot \mathbf{x}, -n) \, d\Omega_n, \qquad (3)$$

where

$$Z_{\boldsymbol{\omega}}(\xi, \lambda) = \frac{1}{\sigma_n \pi^{\frac{n-1}{2}} \Gamma\left(\frac{\lambda+1}{2}\right)} \int_{-\infty}^{\infty} G(\xi - \eta, \boldsymbol{\omega}) |\eta|^{\lambda} \, d\eta, \qquad \sigma_n = \frac{2\pi^{n/2}}{\Gamma(n/2)}.$$

Here, the function $G(\xi, \boldsymbol{\omega})$ is a fundamental solution of the constant coefficient linear ordinary differential equation

$$\mathscr{P}_{2m}\left(\omega_1 \frac{d}{d\xi}, \ldots, \omega_n \frac{d}{d\xi}\right) G(\xi, \boldsymbol{\omega}) = \delta(\xi).$$

If n is odd, the fundamental solution (3) can be represented as

$$\mathscr{E}(\mathbf{x}) = A_n \int_{\Omega_n} \left[\frac{\partial^{n-1}}{\partial \xi^{n-1}} G(\xi, \boldsymbol{\omega})\right] d\Omega_n, \qquad A_n = \frac{(-1)^{\frac{n-1}{2}}}{1 \times 3 \ldots (n-2) \sigma_n (2\pi)^{\frac{n-1}{2}}}.$$

⊙ *References*: I. M. Gel'fand, G. E. Shilov (1959), S. G. Krein (1972).

9.5.3. Hyperbolic Equations

Let $P\left(\frac{\partial}{\partial t}, \frac{\partial}{\partial x_1}, \ldots, \frac{\partial}{\partial x_n}\right)$ be a constant coefficient linear homogeneous differential operator of order m with respect to t. The operator P is called hyperbolic if for any numbers $\omega_1, \ldots, \omega_n$ such that $\sum_{s=1}^{n} \omega_s^2 = 1$, the mth-order algebraic equation

$$P(\lambda, \omega_1, \ldots, \omega_n) = 0$$

with respect to λ has m different real roots.

Fundamental solution of the Cauchy problem for $m \geq n - 1$:

$$\mathscr{E}(t, \mathbf{x}) = \frac{(-1)^{\frac{n+1}{2}}}{2(2\pi)^{n-1}(m-n-1)!} \int_{H=0} (\boldsymbol{\xi} \cdot \mathbf{x} + t)^{m-n-1} \frac{[\mathrm{sign}(\boldsymbol{\xi} \cdot \mathbf{x} + t)]^{m-1}}{|\nabla H|\, \mathrm{sign}(\boldsymbol{\xi} \cdot \nabla H)} \, d\sigma_H \qquad \text{if } n \text{ is odd;}$$

$$\mathscr{E}(t, \mathbf{x}) = \frac{2(-1)^{\frac{n}{2}}}{(2\pi)^n (m-n-1)!} \int_{H=0} \frac{(\boldsymbol{\xi} \cdot \mathbf{x} + t)^{m-n-1}}{|\nabla H|\, \mathrm{sign}(\boldsymbol{\xi} \cdot \nabla H)} \ln\left|\frac{\boldsymbol{\xi} \cdot \mathbf{x} + t}{\boldsymbol{\xi} \cdot \mathbf{x}}\right| d\sigma_H \qquad \text{if } n \text{ is even,}$$

where

$$H = P(1, \xi_1, \ldots, \xi_n), \quad |\nabla H| = \sqrt{\left(\frac{\partial H}{\partial \xi_1}\right)^2 + \cdots + \left(\frac{\partial H}{\partial \xi_n}\right)^2}, \quad \boldsymbol{\xi} \cdot \nabla H = \xi_1 \frac{\partial H}{\partial \xi_1} + \cdots + \xi_n \frac{\partial H}{\partial \xi_n},$$

and $d\sigma_H$ is the element of the surface $H = 0$.

Fundamental solution of the Cauchy problem for $m < n - 1$:

$$\mathscr{E}(t, \mathbf{x}) = \frac{(-1)^{\frac{n+1}{2}}}{(2\pi)^{n-1}} \int_{H=0} \frac{\delta^{(n-m)}(\boldsymbol{\xi} \cdot \mathbf{x} + t)}{|\nabla H|\, \mathrm{sign}(\boldsymbol{\xi} \cdot \nabla H)} \, d\sigma_H \qquad \text{if } n \text{ is odd;}$$

$$\mathscr{E}(t, \mathbf{x}) = \frac{(-1)^{\frac{n}{2}}(n-m)!}{(2\pi)^n} \int_{H=0} \frac{(\boldsymbol{\xi} \cdot \mathbf{x} + t)^{m-n-1}}{|\nabla H|\, \mathrm{sign}(\boldsymbol{\xi} \cdot \nabla H)} \, d\sigma_H \qquad \text{if } n \text{ is even.}$$

⊙ *References*: I. M. Gel'fand, G. E. Shilov (1959), S. G. Krein (1972).

9.5.4. Regular Equations. Number of Initial Conditions in the Cauchy Problem

9.5.4-1. Equations with two independent variables $(0 \leq t < \infty, -\infty < x < \infty)$.

1°. Consider the constant coefficient linear differential equation

$$\frac{\partial^m w}{\partial t^m} = \sum_{k=0}^{m-1} p_k\left(i\frac{\partial}{\partial x}\right) \frac{\partial^k w}{\partial t^k}, \tag{1}$$

where $p_k(z)$ is a polynomial of degree k, $i^2 = -1$. Let $r = r(\sigma)$ be the number of roots (taking into account their multiplicities) of the characteristic equation

$$\lambda^m - \sum_{k=0}^{m-1} p_k(\sigma)\lambda^k = 0 \tag{2}$$

whose real parts are nonpositive (or bounded above) for given a σ. If r is the same (up to a set of measure zero) for all $\sigma \in (-\infty, \infty)$, the equation (1) will be called regular with regularity index r.

Classical equations such as the heat, wave, and Laplace equations are regular.

2°. In the Cauchy problem for the regular equation (1), one should set r initial conditions of the form

$$w\big|_{t=0} = f_0(x), \quad \frac{\partial w}{\partial t}\bigg|_{t=0} = f_1(x), \quad \ldots, \quad \frac{\partial^{r-2} w}{\partial t^{r-2}}\bigg|_{t=0} = f_{r-2}(x), \quad \frac{\partial^{r-1} w}{\partial t^{r-1}}\bigg|_{t=0} = f_{r-1}(x). \qquad (3)$$

It should be emphasized that the regularity index r can, in general, differ from the equation order m with respect to t. In particular, for the two-dimensional Laplace equation $\partial_{tt} w = -\partial_{xx} w$, we have $r = 1$ and $m = 2$; here, y is replaced by t and the first boundary value problem in the upper half-plane $t \geq 0$ is considered. For the heat equation $\partial_t w = \partial_{xx} w$ and the wave equation $\partial_{tt} w = \partial_{xx} w$, we have $r = m = 1$ and $r = m = 2$, respectively.

Example 1. Below are the regularity indices for some fourth-order equations:

$$\frac{\partial^2 w}{\partial t^2} - a^2 \frac{\partial^4 w}{\partial x^4} = 0 \quad (r = 1, \ m = 2); \qquad \frac{\partial^2 w}{\partial t^2} + a^2 \frac{\partial^4 w}{\partial x^4} = 0 \quad (r = 2, \ m = 2);$$

$$\left(\frac{\partial^2}{\partial t^2} + \frac{\partial^2}{\partial x^2}\right)^2 w = 0 \quad (r = 2, \ m = 4); \qquad \frac{\partial^4 w}{\partial t^4} - a^2 \frac{\partial^4 w}{\partial x^4} = 0 \quad (r = 3, \ m = 4).$$

3°. The special solution $\mathscr{E} = \mathscr{E}(t, x)$ that satisfies the initial conditions

$$\mathscr{E}\big|_{t=0} = 0, \quad \frac{\partial \mathscr{E}}{\partial t}\bigg|_{t=0} = 0, \quad \ldots, \quad \frac{\partial^{r-2} \mathscr{E}}{\partial t^{r-2}}\bigg|_{t=0} = 0, \quad \frac{\partial^{r-1} \mathscr{E}}{\partial t^{r-1}}\bigg|_{t=0} = \delta(x) \qquad (4)$$

is called fundamental.

The fundamental solution can be found by applying the Fourier transform in the space variable to equation (1) (with $w = \mathscr{E}$) and the initial conditions (4).

Example 2. Consider the polyharmonic equation:

$$\left(\frac{\partial^2}{\partial t^2} + \frac{\partial^2}{\partial x^2}\right)^n w = 0. \qquad (5)$$

Taking into account the representation $\frac{\partial^2}{\partial x^2} = -\left(i\frac{\partial}{\partial x}\right)^2$, we rewrite the characteristic equation (2) in the form

$$(\lambda^2 - \sigma^2)^n = 0.$$

It has only one solution whose real part is nonpositive, specifically, $\lambda = -|\sigma|$. Considering the multiplicity of the root, we find that the regularity index r is equal to n.

In equation (5) with $w = \mathscr{E}$ and the initial conditions (4) with $r = n$, we perform the Fourier transform with respect to the space variable,

$$U(t, \sigma) = \int_{-\infty}^{\infty} e^{i\sigma x} \mathscr{E}(t, x)\, dx.$$

As a result, we arrive at the ordinary differential equation

$$\left(\frac{d^2}{dt^2} - \sigma^2\right)^n U = 0 \qquad (6)$$

and the initial conditions

$$U\big|_{t=0} = 0, \quad U'_t\big|_{t=0} = 0, \quad \ldots, \quad U_t^{(n-2)}\big|_{t=0} = 0, \quad U_t^{(n-1)}\big|_{t=0} = 1. \qquad (7)$$

The bounded solution of problem (6), (7) is given by

$$U(t, \sigma) = \frac{t^{n-1}}{(n-1)!} e^{-|\sigma|t}.$$

By applying the inverse Fourier transform, we obtain the fundamental solution of the polyharmonic equation in the form

$$\mathscr{E}(t, x) = \frac{1}{2\pi} \int_{-\infty}^{\infty} e^{-i\sigma x} U(t, \sigma)\, d\sigma = \frac{1}{2\pi} \frac{t^{n-1}}{(n-1)!} \int_{-\infty}^{\infty} e^{-i\sigma x - |\sigma| t}\, d\sigma$$

$$= \frac{1}{2\pi} \frac{t^{n-1}}{(n-1)!} \left(\int_0^{\infty} e^{-i\sigma x - \sigma t}\, d\sigma + \int_{-\infty}^{0} e^{-i\sigma x + \sigma t}\, d\sigma\right)$$

$$= \frac{1}{2\pi} \frac{t^{n-1}}{(n-1)!} \left(\frac{1}{t + ix} + \frac{1}{t - ix}\right) = \frac{1}{\pi} \frac{t^{n-1}}{(n-1)!} \frac{t}{t^2 + x^2}.$$

4°. For general initial conditions of the form (3), the solution of equation (1) is determined on the basis of the fundamental solution from the relation

$$w(t,x) = \mathscr{E}(t,x) * \varphi_0(x) + \frac{\partial \mathscr{E}(t,x)}{\partial t} * \varphi_1(x) + \cdots + \frac{\partial^{r-1}\mathscr{E}(t,x)}{\partial t^{r-1}} * \varphi_{r-1}(x). \tag{8}$$

Each term in (8) satisfies equation (1), and the functions $\varphi_{r-1}, \varphi_{r-2}, \ldots, \varphi_0$ are calculated successively by solving the linear system

$$f_0(x) = \varphi_{r-1}(x),$$
$$f_1(x) = \varphi_{r-2}(x) + \frac{\partial^r \mathscr{E}(0,x)}{\partial t^r} * \varphi_{r-1}(x),$$
$$\cdots\cdots\cdots\cdots\cdots\cdots\cdots\cdots\cdots\cdots\cdots\cdots\cdots\cdots\cdots\cdots\cdots\cdots\cdots\cdots\cdots\cdots$$
$$f_k(x) = \varphi_{r-k-1}(x) + \frac{\partial^r \mathscr{E}(0,x)}{\partial t^r} * \varphi_{r-k}(x) + \cdots + \frac{\partial^{r+k-1}\mathscr{E}(0,x)}{\partial t^{r+k-1}} * \varphi_{r-1}(x), \quad k = 2, \ldots, r-1.$$

This system of equations is obtained by successively differentiating relation (8) followed by substituting $t = 0$ and taking into account the initial conditions (3) and (4).

In the special case $f_0(x) = f_1(x) = \cdots = f_{r-2}(x) = 0$, one should set $\varphi_0(x) = f_{r-1}(x)$ and $\varphi_1(x) = \cdots = \varphi_{r-1}(x) = 0$ in (8).

⊙ *Reference*: G. E. Shilov (1965).

9.5.4-2. **Equations with many independent variables** ($0 \leq t < \infty$, $\mathbf{x} \in \mathbb{R}^n$).

Solving the Cauchy problem for the constant coefficient linear differential equation

$$P\left(\frac{\partial}{\partial t}, \frac{\partial}{\partial x_1}, \ldots, \frac{\partial}{\partial x_n}\right) w = 0 \tag{9}$$

with arbitrarily many space variables $x_1, \ldots, x_n$ can be reduced to solving the Cauchy problem for an equation with one space variable ξ. We take an auxiliary linear differential operator

$$P_\omega\left(\frac{\partial}{\partial t}, \frac{\partial}{\partial \xi}\right) \equiv P\left(\frac{\partial}{\partial t}, \omega_1 \frac{\partial}{\partial \xi}, \ldots, \omega_n \frac{\partial}{\partial \xi}\right)$$

that depends on two independent variables t and ξ so that the Cauchy problem for the equation

$$P_\omega\left(\frac{\partial}{\partial t}, \frac{\partial}{\partial \xi}\right) v = 0 \tag{10}$$

is well posed. Then the fundamental solution of the Cauchy problem for the original equation (9) is given by

$$\mathscr{E}(t,\mathbf{x}) = \int_{\Omega_n} v_\omega(t, \boldsymbol{\omega} \cdot \mathbf{x}, -n)\, d\Omega_n.$$

Here,

$$v_\omega(t, \xi, \lambda) = \frac{1}{\sigma_n \pi^{\frac{n-1}{2}} \Gamma\left(\frac{\lambda+1}{2}\right)} \int_{-\infty}^\infty G_\omega(t, \xi - \eta)|\eta|^\lambda\, d\eta, \qquad \sigma_n = \frac{2\pi^{n/2}}{\Gamma(n/2)},$$

where $G_\omega(t,\xi)$ is the fundamental solution of the Cauchy problem for the auxiliary equation (10).

If the number of space variables is odd, one can use the simpler formula

$$\mathscr{E}(t,\mathbf{x}) = \frac{(-1)^{\frac{n-1}{2}} \left(\frac{n-1}{2}\right)!}{\sigma_n \pi^{\frac{n-1}{2}} (n-1)!} \int_{\Omega_n} \left[\frac{d^{n-1}}{d\xi^{n-1}} G_\omega(t,\xi)\right] d\Omega_n, \qquad \xi = \boldsymbol{\omega} \cdot \mathbf{x}.$$

Remark. The above relations hold for all equations for which the Cauchy problem is well posed.

⊙ *References*: I. M. Gel'fand, G. E. Shilov (1959), S. G. Krein (1972).

9.5.4-3. Stationary homogeneous regular equations ($\mathbf{x} \in \mathbb{R}^n$).

A linear differential operator $P_k\left(\frac{\partial}{\partial x_1}, \ldots, \frac{\partial}{\partial x_n}\right)$ is called regular if it is homogeneous and if the gradient of the function $P_k(\omega_1, \ldots, \omega_n)$ on the set defined by the equation $P_k(\omega_1, \ldots, \omega_n) = 0$ is everywhere nonzero whenever $|\boldsymbol{\omega}| \neq 0$.

The fundamental solution of the linear regular differential equation $P_k\left(\frac{\partial}{\partial x_1}, \ldots, \frac{\partial}{\partial x_n}\right)w = 0$ generated by the linear regular differential operator P_k is expressed as

$$\mathscr{E}(\mathbf{x}) = \int_{\Omega_n} \frac{\varphi_{nk}(\boldsymbol{\omega} \cdot \mathbf{x})}{P_k(\boldsymbol{\omega})} \, d\Omega_n, \tag{11}$$

where the function $\varphi_{nk}(z)$ is defined by

$$\varphi_{nk}(z) = \frac{(-1)^{\frac{n-2}{2}}}{(2\pi)^n (k-n)!} z^{k-n} \ln|z| \qquad \text{if } n \text{ is even and } k \geq n;$$

$$\varphi_{nk}(z) = \frac{(-1)^{\frac{n+2k}{2}}(n-k-1)!}{(2\pi)^n} z^{k-n} \qquad \text{if } n \text{ is even and } k < n;$$

$$\varphi_{nk}(z) = \frac{(-1)^{\frac{n-1}{2}}}{4(2\pi)^{n-1}(k-n)!} z^{k-n} \operatorname{sign} z \qquad \text{if } n \text{ is odd and } k \geq n;$$

$$\varphi_{nk}(z) = \frac{(-1)^{\frac{n-1}{2}}}{2(2\pi)^{n-1}} \delta^{(n-k-1)}(z) \qquad \text{if } n \text{ is odd and } k < n.$$

The integral in (11) is understood in the sense of its regularized value, i.e.,

$$\mathscr{E}(\mathbf{x}) = \lim_{\varepsilon \to 0} \mathscr{E}_\varepsilon(\mathbf{x}), \qquad \mathscr{E}_\varepsilon(\mathbf{x}) = \int_{\Omega_n^{(\varepsilon)}} \frac{\varphi_{nk}(\boldsymbol{\omega} \cdot \mathbf{x})}{P_k(\boldsymbol{\omega})} \, d\Omega_n^{(\varepsilon)},$$

where $\Omega_n^{(\varepsilon)}$ is the set of points on a sphere of unit radius for which $|P_k(\boldsymbol{\omega})| > \varepsilon$.

⊙ *References*: I. M. Gel'fand, G. E. Shilov (1959), S. G. Krein (1972).

9.5.5. Some Special-Type Equations

1. $\dfrac{\partial w}{\partial t} = P_n\left(i\dfrac{\partial}{\partial x}\right) w, \qquad P_n(z) = a_n z^n + \cdots + a_1 z + a_0, \qquad i^2 = -1.$

The condition $\operatorname{Re} P_n(z) \leq C < \infty$ is assumed to be met for all real z.

1°. Domain: $-\infty < x < \infty$. Cauchy problem.
An initial condition is prescribed:

$$w = f(x) \quad \text{at} \quad t = 0. \tag{1}$$

Solution:

$$w(x,t) = \int_{-\infty}^{\infty} G(x-\xi, t) f(\xi) \, d\xi, \qquad G(x,t) = \frac{1}{2\pi} \int_{-\infty}^{\infty} \exp\left[t P_n(\lambda) - i x \lambda\right] d\lambda.$$

2°. The solution of the Cauchy problem with the initial condition (1) for the nonhomogeneous equation

$$\frac{\partial w}{\partial t} = P_n\left(i \frac{\partial}{\partial x}\right) w + \Phi(x,t)$$

is given by

$$w(x,t) = \int_{-\infty}^{\infty} G(x-\xi, t) f(\xi) \, d\xi + \int_0^t \int_{-\infty}^{\infty} G(x-\xi, t-\tau) \Phi(\xi, \tau) \, d\xi \, d\tau,$$

where the function $G(x,t)$ is defined in Item 1°.

⊙ *References*: S. G. Krein (1972), V. S. Vladimirov, V. P. Mikhailov, A. A. Vasharin, et al. (1974).

2. $\left(\dfrac{\partial}{\partial t} - a\dfrac{\partial}{\partial x}\right)^n w = 0$, $n = 1, 2, \ldots$

1°. General solution (two representations):
$$w(x,t) = \sum_{k=0}^{n-1} t^k f_k(x + at),$$
$$w(x,t) = \sum_{k=0}^{n-1} x^k f_k(x + at),$$

where the $f_k = f_k(z)$ are arbitrary functions.

2°. Fundamental solution:
$$\mathscr{E}(x,t) = \dfrac{t^{n-1}}{(n-1)!}\delta(x + at).$$

3. $\left(\dfrac{\partial}{\partial t} - \dfrac{\partial^2}{\partial x^2}\right)^n w = 0$, $n = 1, 2, \ldots$

1°. General solution (two representations):
$$w(x,t) = \sum_{k=0}^{n-1} t^k u_k(x,t),$$
$$w(x,t) = \sum_{k=0}^{n-1} x^k u_k(x,t),$$

where the $u_k = u_k(x,t)$ are arbitrary functions that satisfy the heat equations $\partial_t u_k - \partial_{xx} u_k = 0$.

2°. Fundamental solution:
$$\mathscr{E}(x,t) = \dfrac{1}{2\sqrt{\pi}\,(n-1)!} t^{n-3/2} \exp\!\left(-\dfrac{x^2}{4t}\right).$$

3°. Domain: $-\infty < x < \infty$. Cauchy problem.
Initial conditions are prescribed:
$$w\big|_{t=0} = 0, \quad \dfrac{\partial w}{\partial t}\bigg|_{t=0} = 0, \quad \ldots, \quad \dfrac{\partial^{n-2} w}{\partial t^{n-2}}\bigg|_{t=0} = 0, \quad \dfrac{\partial^{n-1} w}{\partial t^{n-1}}\bigg|_{t=0} = f(x).$$

Solution:
$$w(x,t) = \int_{-\infty}^{\infty} f(\xi)\mathscr{E}(x - \xi, t)\, d\xi.$$

⊙ *Reference*: G. E. Shilov (1965).

4. $\left(\dfrac{\partial^2}{\partial t^2} - \dfrac{\partial^2}{\partial x^2}\right)^n w = 0$, $n = 1, 2, \ldots$

1°. General solution (two representations):
$$w(x,t) = \sum_{k=0}^{n-1} t^k \big[f_k(x+t) + g_k(x-t)\big],$$
$$w(x,t) = \sum_{k=0}^{n-1} x^k \big[f_k(x+t) + g_k(x-t)\big],$$

where the $f_k = f_k(y)$ and $g_k = g_k(z)$ are arbitrary functions.

2°. Fundamental solution:

$$\mathscr{E}(x,t) = \frac{(-1)^{n-1}}{4^n(n-1)!}\left[\text{sign}(x-t)\sum_{k=0}^{n-1}\frac{(2t)^k(x-t)^{2n-k-2}}{k!(n-k-1)!}\right.$$
$$\left. + (-1)^n\,\text{sign}(x+t)\sum_{k=0}^{n-1}\frac{(-2t)^k(x+t)^{2n-k-2}}{k!(n-k-1)!}\right].$$

⊙ *Reference*: G. E. Shilov (1965).

5. $\left(\dfrac{\partial^2}{\partial x^2}+\dfrac{\partial^2}{\partial y^2}\right)^n w = 0,\qquad n = 1, 2, \ldots$

This is the *polyharmonic equation* of order n with two independent variables.

1°. General solution (two representations):

$$w(x,y) = \sum_{k=0}^{n-1} r^{2k} u_k(x,y),\qquad r = \sqrt{x^2+y^2},$$

$$w(x,y) = \sum_{k=0}^{n-1} x^k u_k(x,y),$$

where the $u_k(x,y)$ are arbitrary harmonic functions ($\Delta u_k = 0$). In the second relation, x^k can be replaced by y^k.

2°. Domain: $-\infty < x < \infty$, $-\infty < y < \infty$. Fundamental solution:

$$\mathscr{E}(x,y) = \frac{1}{\pi 2^{2n-1}[(n-1)!]^2} r^{2n-2}\ln r,\qquad r = \sqrt{x^2+y^2}.$$

3°. Domain: $-\infty < x < \infty$, $0 \le y < \infty$. Boundary value problem.
Boundary conditions are prescribed:

$$w\big|_{y=0} = 0,\quad \frac{\partial w}{\partial y}\bigg|_{y=0} = 0,\quad \ldots,\quad \frac{\partial^{n-2}w}{\partial y^{n-2}}\bigg|_{y=0} = 0,\quad \frac{\partial^{n-1}w}{\partial y^{n-1}}\bigg|_{y=0} = f(x).$$

Solution:

$$w(x,y) = \int_{-\infty}^{\infty} f(\xi)G(x-\xi,y)\,d\xi,\qquad G(x,y) = \frac{1}{\pi(n-1)!}\frac{y^n}{x^2+y^2}.$$

See also Example 2 in Paragraph 9.5.4-1.

⊙ *References*: G. E. Shilov (1965), L. D. Faddeev (1998).

6. $\left(\dfrac{\partial^2}{\partial x^2}+\dfrac{\partial^2}{\partial y^2}\right)^n w = \Phi(x,y),\qquad n = 1, 2, \ldots$

This is a *nonhomogeneous polyharmonic equation* of order n with two independent variables.
  Particular solution:

$$w(x,y) = \frac{1}{\pi 2^{2n}[(n-1)!]^2}\int_{-\infty}^{\infty}\int_{-\infty}^{\infty}\Phi(\xi,\eta)[(x-\xi)^2+(y-\eta)^2]^{n-1}\ln[(x-\xi)^2+(y-\eta)^2]\,d\xi\,d\eta.$$

The general solution is given by the sum of any particular solution of the nonhomogeneous equation and the general solution of the homogeneous equation (see equation 9.5.5.5, Item 1°).

7. $\Delta_n^m w = 0$, $\quad \Delta_n = \sum_{k=1}^{n} \dfrac{\partial^2}{\partial x_k^2}$.

This is the *polyharmonic equation* of order m with n independent variables. For $m = 1$, see Sections 7.1 and 8.1. For $m = 2$, see Subsection 9.4.1. For $n = 2$, see equations 9.5.5.5 and 9.5.5.6.

1°. Particular solutions:
$$w(\mathbf{x}) = \sum_{s=0}^{m-1} x_j^s u_s(\mathbf{x}), \qquad j = 1, 2, \ldots, n,$$
where the $u_s(\mathbf{x})$ are arbitrary harmonic functions ($\Delta_n u_s = 0$).

2°. Fundamental solution for $m \geq 1$ and $n \geq 3$:
$$\mathscr{E}(\mathbf{x}) = \begin{cases} b_{n,m}|\mathbf{x}|^{2m-n} & \text{if } n \text{ is odd or } n \text{ is even and } n > 2m; \\ c_{n,m}|\mathbf{x}|^{2m-n} \ln|\mathbf{x}| & \text{if } n \text{ is even and } n \leq 2m. \end{cases}$$

Here,
$$b_{n,m} = \frac{\Gamma(n/2)}{2^m(m-1)!\,\pi^{n/2}(2-n)(4-n)\ldots(2m-n)},$$
$$c_{n,m} = \frac{\Gamma(n/2)}{2^m(m-1)!\,\pi^{n/2}(2-n)(4-n)\ldots(2m_0-2-n)(2m_0+2-n)(2m_0+4-n)\ldots(2m-n)},$$
where $m_0 = n/2$. The expression of the coefficient $c_{n,m}$ can be obtained formally from the expression of $b_{n,m}$ by removing the multiplier $(2m_0 - n)$ equal to zero from the denominator.

⊙ *Reference*: G. E. Shilov (1965).

8. $\sum_{k=0}^{m} a_k \Delta^k w = 0$, $\quad \Delta \equiv \dfrac{\partial^2}{\partial x^2} + \dfrac{\partial^2}{\partial y^2}$.

Particular solutions:
$$w(x,y) = \sum_{n=1}^{m} u_n(x,y),$$
where the u_n are solutions of the Helmholtz equations $\Delta u_n - \lambda_n u_n = 0$ and the λ_n are roots of the characteristic equation $\sum_{k=0}^{m} a_k \lambda^k = 0$.

⊙ *Reference*: A. V. Bitsadze and D. F. Kalinichenko (1985).

9. $\sum_{k=0}^{m} a_k L^k[w] = 0$.

Here, L is any constant coefficient linear differential operator with arbitrarily many independent variables $x_1, \ldots, x_n$.

Particular solutions:
$$w(x_1, \ldots, x_n) = \sum_{s=1}^{m} C_s u_s(x_1, \ldots, x_n),$$
where the u_s are solution of the equations $L[u_s] - \lambda_s u_s = 0$ the λ_s are roots of the characteristic equation $\sum_{k=0}^{m} a_k \lambda^k = 0$, and the C_s are arbitrary constants.

9.6. Higher-Order Linear Equations with Variable Coefficients

9.6.1. Equations Containing the First Time Derivative

9.6.1-1. Statement of the problem for an equation with two independent variables.

Consider the linear nonhomogeneous partial differential equation

$$\frac{\partial w}{\partial t} - L_{x,t}[w] = \Phi(x,t), \tag{1}$$

where $L_{x,t}$ is a general linear differential operator of order n with respect to the space variable x,

$$L_{x,t}[w] \equiv \sum_{k=0}^{n} a_k(x,t) \frac{\partial^k w}{\partial x^k}, \tag{2}$$

whose coefficients $a_k = a_k(x,t)$ are sufficiently smooth functions of both arguments for $t \geq 0$ and $x_1 \leq x \leq x_2$. The subscripts x and t indicate that the operator $L_{x,t}$ is dependent on the variables x and t.

We set the initial condition

$$w = f(x) \quad \text{at} \quad t = 0 \tag{3}$$

and the general nonhomogeneous boundary conditions

$$\begin{aligned}
\Gamma_m^{(1)}[w] &\equiv \sum_{k=0}^{n-1} b_{mk}^{(1)}(t) \frac{\partial^k w}{\partial x^k} = g_m^{(1)}(t) \quad \text{at} \quad x = x_1 \quad (m = 1, \ldots, s), \\
\Gamma_m^{(2)}[w] &\equiv \sum_{k=0}^{n-1} b_{mk}^{(2)}(t) \frac{\partial^k w}{\partial x^k} = g_m^{(2)}(t) \quad \text{at} \quad x = x_2 \quad (m = s+1, \ldots, n),
\end{aligned} \tag{4}$$

where $s \geq 1$ and $n \geq s+1$. We assume that both sets of the boundary forms $\Gamma_m^{(1)}[w]$ ($m = 1, \ldots, s$) and $\Gamma_m^{(2)}[w]$ ($m = s+1, \ldots, n$) are linearly independent, which means that for any nonzero $\psi_m = \psi_m(t)$ the following relations hold:

$$\sum_{m=1}^{s} \psi_m(t) \Gamma_m^{(1)}[w] \not\equiv 0, \quad \sum_{m=s+1}^{n} \psi_m(t) \Gamma_m^{(2)}[w] \not\equiv 0.$$

In what follows, we deal with the nonstationary boundary value problem (1), (3), (4).

9.6.1-2. The case of general homogeneous boundary conditions. The Green's function.

The solution of equation (1) with the initial condition (3) and the homogeneous boundary conditions

$$\begin{aligned}
\Gamma_m^{(1)}[w] &= 0 \quad \text{at} \quad x = x_1 \quad (m = 1, \ldots, s), \\
\Gamma_m^{(2)}[w] &= 0 \quad \text{at} \quad x = x_2 \quad (m = s+1, \ldots, n)
\end{aligned} \tag{5}$$

can be written as

$$w(x,t) = \int_{x_1}^{x_2} f(y) G(x,y,t,0) \, dy + \int_0^t \int_{x_1}^{x_2} \Phi(y,\tau) G(x,y,t,\tau) \, dy \, d\tau. \tag{6}$$

Here, $G(x,y,t,\tau)$ is the Green's function that satisfies, for $t > \tau \geq 0$, the homogeneous equation

$$\frac{\partial G}{\partial t} - L_{x,t}[G] = 0 \tag{7}$$

with the special nonhomogeneous initial condition
$$G = \delta(x - y) \quad \text{at} \quad t = \tau \tag{8}$$
and the homogeneous boundary conditions
$$\begin{aligned}\Gamma_m^{(1)}[G] &= 0 \quad \text{at} \quad x = x_1 \quad (m = 1, \ldots, s), \\ \Gamma_m^{(2)}[G] &= 0 \quad \text{at} \quad x = x_2 \quad (m = s+1, \ldots, n).\end{aligned} \tag{9}$$

The quantities y and τ appear in problem (7)–(9) as free parameters ($x_1 \leq y \leq x_2$), and $\delta(x)$ is the Dirac delta function.

It should be emphasized that the Green's function G is independent of the functions $\Phi(x,t)$, $f(x)$, $g_m^{(1)}(t)$, and $g_m^{(2)}(t)$ that characterize various nonhomogeneities of the boundary value problem. If the coefficients a_k, $b_{mk}^{(1)}$, and $b_{mk}^{(2)}$ determining the differential operator (2) and boundary conditions (4) are independent of time t, then the Green's function depends only on three arguments, $G(x,y,t,\tau) = G(x,y,t-\tau)$.

⊙ *Reference*: Mathematical Encyclopedia (1977, Vol. 1).

9.6.1-3. The case of nonhomogeneous boundary conditions. Preliminary transformations.

To solve the problem with nonhomogeneous boundary conditions (1), (3), (4), we choose a sufficiently smooth "test function" $\varphi = \varphi(x,t)$ that satisfies the same boundary conditions as the unknown function; thus,
$$\begin{aligned}\Gamma_m^{(1)}[\varphi] &= g_m^{(1)}(t) \quad \text{at} \quad x = x_1 \quad (m = 1, \ldots, s), \\ \Gamma_m^{(2)}[\varphi] &= g_m^{(2)}(t) \quad \text{at} \quad x = x_2 \quad (m = s+1, \ldots, n).\end{aligned} \tag{10}$$

Otherwise the choice of the "test function" φ is arbitrary and is not linked to the solution of the equation in question; there are infinitely many such functions.

Let us pass from $w = w(x,t)$ to the new unknown $u = u(x,t)$ by the relation
$$w(x,t) = u(x,t) + \varphi(x,t). \tag{11}$$

Substituting (11) into (1), (3), and (4), we arrive at the problem for an equation with a modified right-hand side,
$$\frac{\partial u}{\partial t} - L_{x,t}[u] = \overline{\Phi}(x,t), \qquad \overline{\Phi}(x,t) = \Phi(x,t) - \frac{\partial \varphi}{\partial t} + L_{x,t}[\varphi], \tag{12}$$

subject to the nonhomogeneous initial condition
$$u = f(x) - \varphi(x,0) \quad \text{at} \quad t = 0 \tag{13}$$
and the homogeneous boundary conditions
$$\begin{aligned}\Gamma_m^{(1)}[u] &= 0 \quad \text{at} \quad x = x_1 \quad (m = 1, \ldots, s), \\ \Gamma_m^{(2)}[u] &= 0 \quad \text{at} \quad x = x_2 \quad (m = s+1, \ldots, n).\end{aligned} \tag{14}$$

The solution of problem (12)–(14) can be found using the Green's function by formula (6) in which one should replace w by u, $\Phi(x,t)$ by $\overline{\Phi}(x,t)$, and $f(x)$ by $\overline{f}(x) = f(x) - \varphi(x,0)$. Taking into account relation (11), for w we obtain the representation
$$\begin{aligned}w(x,t) = &\int_{x_1}^{x_2} f(y)G(x,y,t,0)\,dy + \int_0^t \int_{x_1}^{x_2} \Phi(y,\tau)G(x,y,t,\tau)\,dy\,d\tau + \varphi(x,t) \\ &- \int_{x_1}^{x_2} \varphi(y,0)G(x,y,t,0)\,dy - \int_0^t \int_{x_1}^{x_2} \frac{\partial \varphi}{\partial \tau}(y,\tau) G(x,y,t,\tau)\,dy\,d\tau \\ &+ \int_0^t \int_{x_1}^{x_2} G(x,y,t,\tau) L_{y,\tau}[\varphi(y,\tau)]\,dy\,d\tau.\end{aligned} \tag{15}$$

Changing the order of integration and integrating by parts with respect to τ, we find, with reference to the initial condition (8) for the Green's function,
$$\int_0^t \frac{\partial \varphi}{\partial \tau} G\,d\tau = \varphi(y,t)\delta(x-y) - \varphi(y,0)G(x,y,t,0) - \int_0^t \varphi(y,\tau)\frac{\partial G}{\partial \tau}(x,y,t,\tau)\,d\tau. \tag{16}$$

We transform the inner integral of the last term in (15) using the Lagrange–Green formula [see Kamke (1977)] to obtain

$$\int_{x_1}^{x_2} G L_{y,\tau}[\varphi]\, dy = \int_{x_1}^{x_2} \varphi L_{y,\tau}^*[G]\, dy + \mathscr{L}[\varphi, G]\big|_{y=x_1}^{y=x_2}, \tag{17}$$

$$L_{y,\tau}^*[G] \equiv \sum_{k=0}^{n}(-1)^k \frac{\partial^k}{\partial y^k}[a_k(y,\tau) G], \quad \mathscr{L}[\varphi, G] \equiv \sum_{r=0}^{n-1}\sum_{p+q=r}(-1)^p \frac{\partial^q \varphi}{\partial y^q}\frac{\partial^p}{\partial y^p}[a_{r+1}(y,\tau) G],$$

where $L_{x,t}^*[w]$ is the differential form adjoint with $L_{x,t}[w]$ of (2); $\varphi = \varphi(y,\tau)$; and p and q are nonnegative integers.

Using relations (16) and (17), we rewrite solution (15) in the form

$$w(x,t) = \int_{x_1}^{x_2} f(y) G(x,y,t,0)\, dy + \int_0^t \int_{x_1}^{x_2} \Phi(y,\tau) G(x,y,t,\tau)\, dy\, d\tau + \int_0^t \mathscr{L}[\varphi, G]\big|_{y=x_1}^{y=x_2} d\tau. \tag{18}$$

This formula was derived taking into account the fact that the Green's function with respect to y and τ satisfies the adjoint equation*

$$\frac{\partial G}{\partial \tau} + L_{y,\tau}^*[G] = 0.$$

For subsequent analysis, it is convenient to represent the bilinear differential form $\mathscr{L}[\varphi, G]$ as

$$\mathscr{L}[\varphi, G] = \sum_{k=0}^{n-1} \frac{\partial^k \varphi}{\partial y^k} \Psi_k[G], \quad \Psi_k[G] = \sum_{s=0}^{n-k-1}(-1)^s \frac{\partial^s}{\partial y^s}\big[a_{s+k+1}(y,\tau) G\big]. \tag{19}$$

Note that in the special case where operator (2) is binomial,

$$L_{x,t}[w] = a_n \frac{\partial^n w}{\partial x^n} + a_0(x,t) w, \quad a_n = \text{const},$$

the differential forms in (19) are written as

$$\mathscr{L}[\varphi, G] = a_n \sum_{k=0}^{n-1}(-1)^{n-k-1}\frac{\partial^k \varphi}{\partial y^k}\frac{\partial^{n-k-1}G}{\partial y^{n-k-1}}, \quad \Psi_k[G] = a_n(-1)^{n-k-1}\frac{\partial^{n-k-1}G}{\partial y^{n-k-1}}.$$

9.6.1-4. The case of special nonhomogeneous boundary conditions.

Consider the following nonhomogeneous boundary conditions of special form that are often encountered in applications:

$$\begin{aligned}\frac{\partial^{k_m} w}{\partial x^{k_m}} &= g_{k_m}^{(1)}(t) \quad \text{at} \quad x = x_1 \quad (m = 1, \ldots, s), \\ \frac{\partial^{k_m} w}{\partial x^{k_m}} &= g_{k_m}^{(2)}(t) \quad \text{at} \quad x = x_2 \quad (m = s+1, \ldots, n).\end{aligned} \tag{20}$$

Without loss of generality, we assume that the following inequalities hold:

$$n - 1 \geq k_1 > k_2 > \cdots > k_s, \quad n - 1 \geq k_{s+1} > k_{s+2} > \cdots > k_n.$$

The Green's function satisfies the corresponding homogeneous boundary conditions that can be obtained from (20) by replacing w by G and setting $g_{k_m}^{(1)}(t) = g_{k_m}^{(2)}(t) = 0$.

---

* This equation can be derived by considering the case of homogeneous initial and boundary conditions and using arbitrariness in the choice of the test function $\varphi = \varphi(x,t)$; it should be taken into account that the solution itself must be independent of the specific form of φ, because φ does not occur in the original statement of the problem. By appropriately selecting the test function, one can also derive the boundary conditions (21).

The adjoint homogeneous boundary conditions, with respect to (20), which must be met by the Green's function with respect to y and τ have the form

$$\begin{aligned}\Psi_{k_\beta}[G] = 0 \quad \text{at} \quad x = x_1 \quad & (k_\beta \neq k_m,\ \beta = s+1,\ldots,n;\ m = 1,\ldots,s), \\ \Psi_{k_\beta}[G] = 0 \quad \text{at} \quad x = x_2 \quad & (k_\beta \neq k_m,\ \beta = 1,\ldots,s;\ m = s+1,\ldots,n). \end{aligned} \quad (21)$$

These conditions involve the linear differential forms $\Psi_k[G]$ defined in (19). For each endpoint of the interval in question, the set $\{k_\beta\}$ of the indices in the boundary operators (21) together with the set $\{k_m\}$ of the orders of derivatives in the boundary conditions (20) make up a complete set of nonnegative integers from 0 to $n - 1$.

Taking into account the fact that the test function φ must satisfy the boundary conditions (20) and the Green's function G to conditions (21), we rewrite solution (18) to obtain

$$\begin{aligned} w(x,t) = &\int_{x_1}^{x_2} f(y) G(x,y,t,0)\, dy + \int_0^t \int_{x_1}^{x_2} \Phi(y,\tau) G(x,y,t,\tau)\, dy\, d\tau \\ & - \sum_{m=1}^{s} \int_0^t g_{k_m}^{(1)}(\tau) \Psi_{k_m}[G]\big|_{y=x_1}\, d\tau + \sum_{m=s+1}^{n} \int_0^t g_{k_m}^{(2)}(\tau) \Psi_{k_m}[G]\big|_{y=x_2}\, d\tau, \end{aligned} \quad (22)$$

where the $\Psi_{k_m}[G]$ are differential operators with respect to y, which are defined in (19).

If the Green's function is known, formula (22) can be used to immediately obtain the solution of the nonhomogeneous boundary value problem (1), (3), (20) for arbitrary $\Phi(x,t)$, $f(x)$, $g_{k_m}^{(1)}(t)$ ($m = 1,\ldots,s$), and $g_{k_m}^{(2)}(t)$ ($m = s+1,\ldots,n$).

9.6.1-5. The case of general nonhomogeneous boundary conditions.

On solving (4) for the highest derivatives, we reduce the boundary conditions (4) to the canonical form

$$\begin{aligned} \frac{\partial^{k_m} w}{\partial x^{k_m}} + \sum_{i=0}^{k_m - 1} c_{mi}^{(1)}(t) \frac{\partial^i w}{\partial x^i} = h_{k_m}^{(1)}(t) \quad \text{at} \quad x = x_1 \quad (m = 1,\ldots,s), \\ \frac{\partial^{k_m} w}{\partial x^{k_m}} + \sum_{i=0}^{k_m - 1} c_{mi}^{(2)}(t) \frac{\partial^i w}{\partial x^i} = h_{k_m}^{(2)}(t) \quad \text{at} \quad x = x_2 \quad (m = s+1,\ldots,n), \end{aligned} \quad (23)$$

where the leading terms in different boundary conditions are different,

$$n - 1 \geq k_1 > k_2 > \cdots > k_s, \quad n - 1 \geq k_{s+1} > k_{s+2} > \cdots > k_n.$$

The sums in (23) do not contain the derivatives of orders $k_1,\ldots,k_s$ (for $x = x_1$) and $k_{s+1},\ldots,k_n$ (for $x = x_2$); thus,

$$\begin{aligned} c_{mi}^{(1)}(t) = 0 \quad \text{at} \quad i = k_j \quad (j = 1,\ldots,s), \\ c_{mi}^{(2)}(t) = 0 \quad \text{at} \quad i = k_j \quad (j = s+1,\ldots,n). \end{aligned}$$

It can be shown that the solution of problem (1), (3), (23) is given by

$$\begin{aligned} w(x,t) = &\int_{x_1}^{x_2} f(y) G(x,y,t,0)\, dy + \int_0^t \int_{x_1}^{x_2} \Phi(y,\tau) G(x,y,t,\tau)\, dy\, d\tau \\ & - \sum_{m=1}^{s} \int_0^t h_{k_m}^{(1)}(\tau) \Psi_{k_m}[G]\big|_{y=x_1}\, d\tau + \sum_{m=s+1}^{n} \int_0^t h_{k_m}^{(2)}(\tau) \Psi_{k_m}[G]\big|_{y=x_2}\, d\tau, \end{aligned} \quad (24)$$

where the $\Psi_{k_m}[G]$ are differential operators with respect to y, which are defined in (19). Relation (24) is similar to (22) but contains the Green's function satisfying the more complicated boundary conditions that can be obtained from (23) by substituting G for w and setting $h_{k_m}^{(1)}(t) = h_{k_m}^{(2)}(t) = 0$.

9.6.2. Equations Containing the Second Time Derivative

9.6.2-1. The case of homogeneous initial and boundary conditions.

Consider the linear nonhomogeneous differential equation

$$\frac{\partial^2 w}{\partial t^2} + \psi(x,t)\frac{\partial w}{\partial t} - \sum_{k=0}^{n} a_k(x,t)\frac{\partial^k w}{\partial x^k} = \Phi(x,t). \tag{1}$$

We set the homogeneous initial conditions

$$\begin{aligned} w &= 0 \quad \text{at} \quad t = 0, \\ \partial_t w &= 0 \quad \text{at} \quad t = 0 \end{aligned} \tag{2}$$

and the homogeneous boundary conditions

$$\begin{aligned} \Gamma_m^{(1)}[w] &= 0 \quad \text{at} \quad x = x_1 \quad (m = 1, \ldots, s), \\ \Gamma_m^{(2)}[w] &= 0 \quad \text{at} \quad x = x_2 \quad (m = s+1, \ldots, n), \end{aligned} \tag{3}$$

where the boundary operators $\Gamma_m^{(1)}[w]$ and $\Gamma_m^{(2)}[w]$ are defined in Paragraph 9.6.1-1.

The solution of problem (1)–(3) can be represented in the form*

$$w(x,t) = \int_0^t \int_{x_1}^{x_2} \Phi(y,\tau) G(x,y,t,\tau)\, dy\, d\tau. \tag{4}$$

Here, $G = G(x,y,t,\tau)$ is the Green's function; for $t > \tau \geq 0$, it satisfies the homogeneous equation

$$\frac{\partial^2 G}{\partial t^2} + \psi(x,t)\frac{\partial G}{\partial t} - \sum_{k=0}^{n} a_k(x,t)\frac{\partial^k G}{\partial x^k} = 0 \tag{5}$$

with the special semihomogeneous initial conditions

$$\begin{aligned} G &= 0 \quad \text{at} \quad t = \tau, \\ \partial_t G &= \delta(x-y) \quad \text{at} \quad t = \tau \end{aligned} \tag{6}$$

and the corresponding homogeneous boundary conditions

$$\begin{aligned} \Gamma_m^{(1)}[G] &= 0 \quad \text{at} \quad x = x_1 \quad (m = 1, \ldots, s), \\ \Gamma_m^{(2)}[G] &= 0 \quad \text{at} \quad x = x_2 \quad (m = s+1, \ldots, n). \end{aligned} \tag{7}$$

The quantities y and τ appear in problem (5)–(7) as free parameters ($x_1 \leq y \leq x_2$), and $\delta(x)$ is the Dirac delta function.

One can verify by direct substitution into the equation and the initial and boundary conditions (1)–(3) that formula (4) is correct, taking into account the properties (5)–(7) of the Green's function.

9.6.2-2. The case of nonhomogeneous initial and boundary conditions.

Consider the linear nonhomogeneous differential equation (1) with the general nonhomogeneous initial conditions

$$\begin{aligned} w &= f_0(x) \quad \text{at} \quad t = 0, \\ \partial_t w &= f_1(x) \quad \text{at} \quad t = 0 \end{aligned} \tag{8}$$

---

* Problem (1)–(3) is assumed to be well posed.

and the nonhomogeneous boundary conditions, reduced to the canonical form (see Paragraph 9.6.1-5):

$$\frac{\partial^{k_m} w}{\partial x^{k_m}} + \sum_{i=0}^{k_m-1} c_{mi}^{(1)}(t) \frac{\partial^i w}{\partial x^i} = h_{k_m}^{(1)}(t) \quad \text{at} \quad x = x_1 \quad (m = 1, \dots, s),$$

$$\frac{\partial^{k_m} w}{\partial x^{k_m}} + \sum_{i=0}^{k_m-1} c_{mi}^{(2)}(t) \frac{\partial^i w}{\partial x^i} = h_{k_m}^{(2)}(t) \quad \text{at} \quad x = x_2 \quad (m = s+1, \dots, n).$$
(9)

Introducing a test function $\varphi = \varphi(x,t)$ that satisfies the nonhomogeneous initial and boundary conditions (8), (9) and using the same line of reasoning as in Paragraph 9.6.1-3 for a simpler equation, we arrive at the solution of problem (1), (8), (9) in the form

$$w(x,t) = \int_0^t \int_{x_1}^{x_2} \Phi(y,\tau) G(x,y,t,\tau) \, dy \, d\tau$$

$$- \int_{x_1}^{x_2} f_0(y) \frac{\partial}{\partial \tau} \Big[G(x,y,t,\tau)\Big]_{\tau=0} dy + \int_{x_1}^{x_2} \big[f_1(y) + f_0(y)\psi(y,0)\big] G(x,y,t,0) \, dy$$

$$- \sum_{m=1}^{s} \int_0^t h_{k_m}^{(1)}(\tau) \Psi_{k_m}[G]\Big|_{y=x_1} d\tau + \sum_{m=s+1}^{n} \int_0^t h_{k_m}^{(2)}(\tau) \Psi_{k_m}[G]\Big|_{y=x_2} d\tau, \qquad (10)$$

where the $\Psi_{k_m}[G]$ are differential operators with respect to y, which are defined in relations (19), Paragraph 9.6.1-3.

Remark. If the coefficients of equation (1) and those of the boundary conditions (9) are time independent, i.e.,

$$\psi = \psi(x), \quad a_k = a_k(x), \quad c_{mi}^{(1)} = \text{const}, \quad c_{mi}^{(2)} = \text{const},$$

then in solution (10) one should set

$$G(x,y,t,\tau) = \widetilde{G}(x,y,t-\tau), \quad \frac{\partial}{\partial \tau} G(x,y,t,\tau)\Big|_{\tau=0} = -\frac{\partial}{\partial t} \widetilde{G}(x,y,t).$$

9.6.3. Nonstationary Problems with Many Space Variables

9.6.3-1. Equations with the first-order partial derivative with respect to t.

Consider the following linear differential operator with respect to variables $x_1, \dots, x_n$:

$$\mathfrak{L}_{\mathbf{x},t}[w] \equiv \sum A_{k_1,\dots,k_n}(x_1,\dots,x_n,t) \frac{\partial^{k_1+\dots+k_n} w}{\partial x_1^{k_1} \dots \partial x_n^{k_n}}. \qquad (1)$$

The coefficients $A_{k_1,\dots,k_n}$ of the operator are assumed to be sufficiently smooth functions of $x_1,\dots,x_n$ and t (and also bounded if necessary). The coefficients of the highest derivatives are assumed to be everywhere nonzero.

1°. *Cauchy problem* ($t \geq 0$, $\mathbf{x} \in \mathbb{R}^n$). The solution of the Cauchy problem for the linear nonhomogeneous parabolic differential equation with variable coefficients

$$\frac{\partial w}{\partial t} - \mathfrak{L}_{\mathbf{x},t}[w] = \Phi(\mathbf{x},t) \qquad (2)$$

under the initial conditions

$$w = f(\mathbf{x}) \quad \text{at} \quad t = 0 \qquad (3)$$

is given by

$$w(\mathbf{x}, t) = \int_0^t \int_{\mathbb{R}^n} \Phi(\mathbf{y}, \tau) \mathscr{E}(\mathbf{x}, \mathbf{y}, t, \tau) \, d\mathbf{y} \, d\tau + \int_{\mathbb{R}^n} f(\mathbf{y}) \mathscr{E}(\mathbf{x}, \mathbf{y}, t, 0) \, d\mathbf{y}, \quad d\mathbf{y} = dy_1 \ldots dy_n. \tag{4}$$

Here, $\mathscr{E} = \mathscr{E}(\mathbf{x}, \mathbf{y}, t, \tau)$ is the fundamental solution of the Cauchy problem, which satisfies for $t > \tau \geq 0$ the equation

$$\frac{\partial \mathscr{E}}{\partial t} - \mathfrak{L}_{\mathbf{x},t}[\mathscr{E}] = 0 \tag{5}$$

and the special initial condition

$$\mathscr{E}\big|_{t=\tau} = \delta(\mathbf{x} - \mathbf{y}). \tag{6}$$

The quantities $\mathbf{y}$ and τ appear in problem (5), (6) as free parameters ($\mathbf{y} \in \mathbb{R}^n$), and $\delta(\mathbf{x})$ is the n-dimensional Dirac delta function.

If the coefficients $A_{k_1,\ldots,k_n}$ of operator (1) are independent of time t, then the fundamental solution depends on only three arguments, $\mathscr{E}(\mathbf{x}, \mathbf{y}, t, \tau) = \mathscr{E}(\mathbf{x}, \mathbf{y}, t - \tau)$. If the coefficients of operator (1) are constants, then $\mathscr{E}(\mathbf{x}, \mathbf{y}, t, \tau) = \mathscr{E}(\mathbf{x} - \mathbf{y}, t - \tau)$.

2°. *Boundary value problems* ($t \geq 0$, $\mathbf{x} \in \mathcal{D}$). The solutions of linear boundary value problems in a spatial domain $\mathcal{D}$ for equation (2) with initial condition (3) and homogeneous boundary conditions for $\mathbf{x} \in \partial \mathcal{D}$ (these conditions are not written out here) are given by formula (4) in which the domain of integration $\mathbb{R}^n$ should be replaced by $\mathcal{D}$. Here, by $\mathscr{E}$ we mean the Green's function that must satisfy, apart from equation (5) and the boundary condition (6), the same homogeneous boundary conditions for $\mathbf{x} \in \partial \mathcal{D}$ as the original equation (2). For boundary value problems, the parameter $\mathbf{y}$ belongs to the same domain as $\mathbf{x}$, i.e., $\mathbf{y} \in \mathcal{D}$.

⊙ *Reference*: Mathematical Encyclopedia (1977, Vol. 1).

9.6.3-2. Equations with the second-order partial derivative with respect to t.

1°. *Cauchy problem* ($t \geq 0$, $\mathbf{x} \in \mathbb{R}^n$). The solution of the Cauchy problem for the linear nonhomogeneous differential equation with variable coefficients

$$\frac{\partial^2 w}{\partial t^2} - \mathfrak{L}_{\mathbf{x},t}[w] = \Phi(\mathbf{x}, t) \tag{7}$$

under the initial conditions

$$\begin{aligned} w &= f(\mathbf{x}) \quad \text{at} \quad t = 0, \\ \partial_t w &= g(\mathbf{x}) \quad \text{at} \quad t = 0 \end{aligned} \tag{8}$$

is given by

$$w(\mathbf{x}, t) = \int_0^t \int_{\mathbb{R}^n} \Phi(\mathbf{y}, \tau) \mathscr{E}(\mathbf{x}, \mathbf{y}, t, \tau) \, d\mathbf{y} \, d\tau$$
$$- \int_{\mathbb{R}^n} f(\mathbf{y}) \left[\frac{\partial}{\partial \tau} \mathscr{E}(\mathbf{x}, \mathbf{y}, t, \tau) \right]_{\tau=0} d\mathbf{y} + \int_{\mathbb{R}^n} g(\mathbf{y}) \mathscr{E}(\mathbf{x}, \mathbf{y}, t, 0) \, d\mathbf{y}.$$

Here, $\mathscr{E} = \mathscr{E}(\mathbf{x}, \mathbf{y}, t, \tau)$ is the fundamental solution of the Cauchy problem

$$\frac{\partial^2 \mathscr{E}}{\partial t^2} - \mathfrak{L}_{\mathbf{x},t}[\mathscr{E}] = 0,$$

$$\mathscr{E}\big|_{t=\tau} = 0, \quad \frac{\partial \mathscr{E}}{\partial t}\bigg|_{t=\tau} = \delta(\mathbf{x} - \mathbf{y}),$$

where $\mathbf{y}$ and τ play the role of parameters.

If the coefficients $A_{k_1,\ldots,k_n}$ of operator (1) are independent of time t, then the fundamental solution depends on only three arguments, $\mathscr{E}(\mathbf{x}, \mathbf{y}, t, \tau) = \mathscr{E}(\mathbf{x}, \mathbf{y}, t - \tau)$, and the relation $\frac{\partial}{\partial \tau} \mathscr{E}(\mathbf{x}, \mathbf{y}, t, \tau)\big|_{\tau=0} = -\frac{\partial}{\partial t} \mathscr{E}(\mathbf{x}, \mathbf{y}, t)$ holds. If the coefficients of operator (1) are constants, then $\mathscr{E}(\mathbf{x}, \mathbf{y}, t, \tau) = \mathscr{E}(\mathbf{x} - \mathbf{y}, t - \tau)$.

2°. The solution of the Cauchy problem for the more complicated linear nonhomogeneous differential equation with variable coefficients

$$\frac{\partial^2 w}{\partial t^2} + \psi(\mathbf{x}, t)\frac{\partial w}{\partial t} - \mathfrak{L}_{\mathbf{x},t}[w] = \Phi(\mathbf{x}, t)$$

with initial conditions (8) is expressed as

$$w(\mathbf{x}, t) = \int_0^t \int_{\mathbb{R}^n} \Phi(\mathbf{y}, \tau)\mathscr{E}(\mathbf{x}, \mathbf{y}, t, \tau)\, d\mathbf{y}\, d\tau \\
- \int_{\mathbb{R}^n} f(\mathbf{y})\left[\frac{\partial}{\partial \tau}\mathscr{E}(\mathbf{x}, \mathbf{y}, t, \tau)\right]_{\tau=0} d\mathbf{y} + \int_{\mathbb{R}^n} [g(\mathbf{y}) + \psi(\mathbf{y}, 0)f(\mathbf{y})]\mathscr{E}(\mathbf{x}, \mathbf{y}, t, 0)\, d\mathbf{y}. \qquad (9)$$

Here, $\mathscr{E}(\mathbf{x}, \mathbf{y}, t, \tau)$ is the corresponding fundamental solution of the Cauchy problem,

$$\frac{\partial^2 \mathscr{E}}{\partial t^2} + \psi(\mathbf{x}, t)\frac{\partial \mathscr{E}}{\partial t} - \mathfrak{L}_{\mathbf{x},t}[\mathscr{E}] = 0,$$

$$\mathscr{E}\big|_{t=\tau} = 0, \quad \frac{\partial \mathscr{E}}{\partial t}\bigg|_{t=\tau} = \delta(\mathbf{x} - \mathbf{y}).$$

3°. *Boundary value problems* ($t \geq 0$, $\mathbf{x} \in \mathcal{D}$). The solutions of linear boundary value problems in a spatial domain $\mathcal{D}$ for equation (7) with initial condition (8) and homogeneous boundary conditions for $\mathbf{x} \in \partial \mathcal{D}$ (these conditions are not written out here) are given by formula (9) in which the domain of integration $\mathbb{R}^n$ should be replaced by $\mathcal{D}$. Here, by $\mathscr{E}$ we mean the Green's function that must satisfy, apart from equation (7) and the initial conditions (8), the same homogeneous boundary conditions as the original equation (7).

9.6.4. Some Special-Type Equations

1. $\dfrac{\partial w}{\partial t} = k(t)\dfrac{\partial^n w}{\partial x^n} + [xf(t) + g(t)]\dfrac{\partial w}{\partial x} + h(t)w.$

The transformation

$$w(x, t) = u(z, \tau)\exp\left[\int h(t)\, dt\right], \quad z = xF(t) + \int g(t)F(t)\, dt, \quad \tau = \int k(t)F^n(t)\, dt,$$

where $F(t) = \exp\left[\int f(t)\, dt\right]$, leads to the simpler constant coefficient equation

$$\frac{\partial u}{\partial \tau} = \frac{\partial^n u}{\partial z^n}.$$

2. $\dfrac{\partial^k w}{\partial t^k} = ax^{2n}\dfrac{\partial^n w}{\partial x^n}.$

The transformation $z = 1/x$, $u = wx^{1-n}$ leads to the constant coefficient equation

$$\frac{\partial^k u}{\partial t^k} = a(-1)^n \frac{\partial^n u}{\partial z^n}.$$

3. $\dfrac{\partial^k w}{\partial t^k} = \displaystyle\sum_{m=0}^{n} a_m x^m \dfrac{\partial^m w}{\partial x^m}.$

The change of variable $z = \ln|x|$ leads to a constant coefficient equation.

4. $\dfrac{\partial^k w}{\partial t^k} = (ax^2 + bx + c)^n \dfrac{\partial^n w}{\partial x^n}.$

The transformation

$$w(x,t) = u(z,t)|ax^2 + bx + c|^{\frac{n-1}{2}}, \qquad z = \int \dfrac{dx}{ax^2 + bx + c}$$

leads to a constant coefficient equation.

5. $\left(\dfrac{\partial}{\partial t} - L_x\right)^n w = 0, \qquad n = 1, 2, \ldots$

Here, L_x is a linear differential operator of any order with respect to the space variable x whose coefficients can depend on x.

1°. General solution:

$$w(x,t) = \sum_{k=0}^{n-1} t^k u_k(x,t),$$

where the $u_k = u_k(x,t)$ are arbitrary functions that satisfy the original equation with $n = 1$: $(\partial_t - L_x)u_k = 0$.

2°. Fundamental solution:

$$\mathscr{E}_n(x,t) = \dfrac{t^{n-1}}{(n-1)!} \mathscr{E}_1(x,t),$$

where $\mathscr{E}_1(x,t)$ is the fundamental solution of the equation with $n = 1$.

Remark. The linear differential operator L_x can involve arbitrarily many space variables.

6. $\left(\dfrac{\partial^2}{\partial t^2} - L_x\right)^n w = 0, \qquad n = 1, 2, \ldots$

Here, L_x is a linear differential operator of any order with respect to the space variable x whose coefficients can depend on x.

1°. General solution:

$$w(x,t) = \sum_{k=0}^{n-1} t^k u_k(x,t),$$

where the $u_k = u_k(x,t)$ are arbitrary functions that satisfy the original equation with $n = 1$: $(\partial_{tt} - L_x)u_k = 0$.

2°. Suppose that the Cauchy problem for the special case of the equation with $n = 1$ is well posed if only one initial condition is set at $t = 0$; this means that the constant coefficient differential operator L_x is such that the equation with $n = 1$ is regular with regularity index $r = 1$. Then the fundamental solution of the original equation can be found by the formula

$$\mathscr{E}_n(x,t) = \dfrac{t^{n-1}}{(n-1)!} \mathscr{E}_1(x,t),$$

where $\mathscr{E}_1(x,t)$ is the fundamental solution for $n = 1$.

Remark. The linear differential operator L_x can involve arbitrarily many space variables.

7. $\sum_{k=0}^{m} a_k L^k[w] = 0.$

Here, L is any linear differential operator with arbitrarily many independent variables $x_1, \ldots, x_n$.

Particular solutions:
$$w(x_1, \ldots, x_n) = \sum_{s=1}^{m} C_s u_s(x_1, \ldots, x_n),$$

where the u_s are solutions of the equations $L[u_s] - \lambda_s u_s = 0$, the λ_s are roots of the characteristic equation $\sum_{k=0}^{m} a_k \lambda^k = 0$, and the C_s are arbitrary constants.

Supplement A

Special Functions and Their Properties

Throughout Supplement A it is assumed that n is a positive integer, unless otherwise specified.

A.1. Some Symbols and Coefficients

A.1.1. Factorials

Definitions and some properties:

$$0! = 1! = 1, \quad n! = 1 \cdot 2 \cdot 3 \ldots (n-1)n, \quad n = 2, 3, \ldots,$$
$$(2n)!! = 2 \cdot 4 \cdot 6 \ldots (2n-2)(2n) = 2^n n!,$$
$$(2n+1)!! = 1 \cdot 3 \cdot 5 \ldots (2n-1)(2n+1) = \frac{2^{n+1}}{\sqrt{\pi}} \Gamma\left(n + \frac{3}{2}\right),$$
$$n!! = \begin{cases} (2k)!! & \text{if } n = 2k, \\ (2k+1)!! & \text{if } n = 2k+1, \end{cases} \quad 0!! = 1.$$

A.1.2. Binomial Coefficients

Definition:

$$C_n^k = \frac{n!}{k!(n-k)!}, \quad \text{where} \quad k = 1, \ldots, n,$$
$$C_a^k = (-1)^k \frac{(-a)_k}{k!} = \frac{a(a-1)\ldots(a-k+1)}{k!}, \quad \text{where} \quad k = 1, 2, \ldots$$

General case:

$$C_a^b = \frac{\Gamma(a+1)}{\Gamma(b+1)\Gamma(a-b+1)}, \quad \text{where} \quad \Gamma(x) \text{ is the gamma function.}$$

Properties:

$$C_a^0 = 1, \quad C_n^k = 0 \quad \text{for} \quad k = -1, -2, \ldots \text{ or } k > n,$$
$$C_a^{b+1} - \frac{a}{b+1} C_{a-1}^b = \frac{a\ b}{b+1} C_a^b, \quad C_u^b + C_u^{b+1} = C_{u+1}^{b+1},$$
$$C_{-1/2}^n = \frac{(-1)^n}{2^{2n}} C_{2n}^n = (-1)^n \frac{(2n-1)!!}{(2n)!!},$$
$$C_{1/2}^n = \frac{(-1)^{n-1}}{n 2^{2n-1}} C_{2n-2}^{n-1} = \frac{(-1)^{n-1}}{n} \frac{(2n-3)!!}{(2n-2)!!},$$
$$C_{n+1/2}^{2n+1} = (-1)^n 2^{-4n-1} C_{2n}^n, \quad C_{2n+1/2}^n = 2^{-2n} C_{4n+1}^{2n},$$
$$C_n^{1/2} = \frac{2^{2n+1}}{\pi C_{2n}^n}, \quad C_n^{n/2} = \frac{2^{2n}}{\pi} C_n^{(n-1)/2}.$$

A.1.3. Pochhammer Symbol

Definition and some properties ($k = 1, 2, \ldots$):

$$(a)_n = a(a+1)\ldots(a+n-1) = \frac{\Gamma(a+n)}{\Gamma(a)} = (-1)^n \frac{\Gamma(1-a)}{\Gamma(1-a-n)},$$

$$(a)_0 = 1, \quad (a)_{n+k} = (a)_n (a+n)_k, \quad (n)_k = \frac{(n+k-1)!}{(n-1)!},$$

$$(a)_{-n} = \frac{\Gamma(a-n)}{\Gamma(a)} = \frac{(-1)^n}{(1-a)_n}, \quad \text{where } a \neq 1, \ldots, n;$$

$$(1)_n = n!, \quad (1/2)_n = 2^{-2n} \frac{(2n)!}{n!}, \quad (3/2)_n = 2^{-2n} \frac{(2n+1)!}{n!},$$

$$(a+mk)_{nk} = \frac{(a)_{mk+nk}}{(a)_{mk}}, \quad (a+n)_n = \frac{(a)_{2n}}{(a)_n}, \quad (a+n)_k = \frac{(a)_k (a+k)_n}{(a)_n}.$$

A.1.4. Bernoulli Numbers

Definition:

$$\frac{x}{e^x - 1} = \sum_{n=0}^{\infty} B_n \frac{x^n}{n!}.$$

The numbers:

$B_0 = 1$, $B_1 = -\frac{1}{2}$, $B_2 = \frac{1}{6}$, $B_4 = -\frac{1}{30}$, $B_6 = \frac{1}{42}$, $B_8 = -\frac{1}{30}$, $B_{10} = \frac{5}{66}$, ...,
$B_{2m+1} = 0$ for $m = 1, 2, \ldots$

A.2. Error Functions and Exponential Integral

A.2.1. Error Function and Complementary Error Function

Definitions:

$$\operatorname{erf} x = \frac{2}{\sqrt{\pi}} \int_0^x \exp(-t^2)\, dt, \qquad \operatorname{erfc} x = 1 - \operatorname{erf} x = \frac{2}{\sqrt{\pi}} \int_x^\infty \exp(-t^2)\, dt.$$

Expansion of $\operatorname{erf} x$ into series in powers of x as $x \to 0$:

$$\operatorname{erf} x = \frac{2}{\sqrt{\pi}} \sum_{k=0}^{\infty} (-1)^k \frac{x^{2k+1}}{(k)!(2k+1)} = \frac{2}{\sqrt{\pi}} \exp(-x^2) \sum_{k=0}^{\infty} \frac{2^k x^{2k+1}}{2k+1)!!}.$$

Asymptotic expansion of $\operatorname{erfc} x$ as $x \to \infty$:

$$\operatorname{erfc} x = \frac{1}{\sqrt{\pi}} \exp(-x^2) \left[\sum_{m=0}^{M-1} (-1)^m \frac{\left(\frac{1}{2}\right)_m}{x^{2m+1}} + O\left(|x|^{-2M-1}\right) \right], \qquad M = 1, 2, \ldots$$

A.2.2. Exponential Integral

Definition:

$$\operatorname{Ei}(x) = \int_{-\infty}^x \frac{e^t}{t}\, dt \qquad \text{for} \quad x < 0,$$

$$\operatorname{Ei}(x) = \lim_{\varepsilon \to +0} \left(\int_{-\infty}^{-\varepsilon} \frac{e^t}{t}\, dt + \int_\varepsilon^x \frac{e^t}{t}\, dt \right) \qquad \text{for} \quad x > 0.$$

Other integral representations:

$$\text{Ei}(-x) = -e^{-x} \int_0^\infty \frac{x \sin t + t \cos t}{x^2 + t^2}\, dt \quad \text{for} \quad x > 0,$$

$$\text{Ei}(-x) = e^{-x} \int_0^\infty \frac{x \sin t - t \cos t}{x^2 + t^2}\, dt \quad \text{for} \quad x < 0,$$

$$\text{Ei}(-x) = -x \int_1^\infty e^{-xt} \ln t\, dt \quad \text{for} \quad x > 0.$$

Expansion into series in powers of x as $x \to 0$:

$$\text{Ei}(x) = \begin{cases} \mathcal{C} + \ln(-x) + \sum_{k=1}^\infty \dfrac{x^k}{k \cdot k!} & \text{if } x < 0, \\ \mathcal{C} + \ln x + \sum_{k=1}^\infty \dfrac{x^k}{k \cdot k!} & \text{if } x > 0, \end{cases}$$

where $\mathcal{C} = 0.5772\ldots$ is the Euler constant.

Asymptotic expansion as $x \to \infty$:

$$\text{Ei}(-x) = e^{-x} \sum_{k=1}^n (-1)^k \frac{(k-1)!}{x^k} + R_n, \qquad R_n < \frac{n!}{x^n}.$$

A.2.3. Logarithmic Integral

Definition:

$$\text{li}(x) = \begin{cases} \displaystyle\int_0^x \frac{dt}{\ln t} = \text{Ei}(\ln x) & \text{if } 0 < x < 1, \\ \displaystyle\lim_{\varepsilon \to +0} \left(\int_0^{1-\varepsilon} \frac{dt}{\ln t} + \int_{1+\varepsilon}^x \frac{dt}{\ln t} \right) & \text{if } x > 1. \end{cases}$$

For small x,

$$\text{li}(x) \approx \frac{x}{\ln(1/x)}.$$

Asymptotic expansion as $x \to 1$:

$$\text{li}(x) = \mathcal{C} + \ln|\ln x| + \sum_{k=1}^\infty \frac{\ln^k x}{k \cdot k!}.$$

A.3. Sine Integral and Cosine Integral. Fresnel Integrals

A.3.1. Sine Integral

Definition:

$$\text{Si}(x) = \int_0^x \frac{\sin t}{t}\, dt, \qquad \text{si}(x) = -\int_x^\infty \frac{\sin t}{t}\, dt = \text{Si}(x) - \frac{\pi}{2}.$$

Specific values:

$$\text{Si}(0) = 0, \quad \text{Si}(\infty) = \frac{\pi}{2}, \quad \text{si}(\infty) = 0.$$

Properties:

$$\text{Si}(-x) = -\text{Si}(x), \quad \text{si}(x) + \text{si}(-x) = -\pi, \quad \lim_{x \to -\infty} \text{si}(x) = -\pi.$$

Expansion into series in powers of x as $x \to 0$:

$$\mathrm{Si}(x) = \sum_{k=1}^{\infty} \frac{(-1)^{k+1} x^{2k-1}}{(2k-1)(2k-1)!}.$$

Asymptotic expansion as $x \to \infty$:

$$\mathrm{si}(x) = -\cos x \left[\sum_{m=0}^{M-1} \frac{(-1)^m (2m)!}{x^{2m+1}} + O\left(|x|^{-2M-1}\right) \right] + \sin x \left[\sum_{m=1}^{N-1} \frac{(-1)^m (2m-1)!}{x^{2m}} + O\left(|x|^{-2N}\right) \right],$$

where $M, N = 1, 2, \ldots$

A.3.2. Cosine Integral

Definition:

$$\mathrm{Ci}(x) = -\int_x^{\infty} \frac{\cos t}{t}\, dt = \mathcal{C} + \ln x + \int_0^x \frac{\cos t - 1}{t}\, dt, \qquad \mathcal{C} = 0.5772\ldots$$

Expansion into series in powers of x as $x \to 0$:

$$\mathrm{Ci}(x) = \mathcal{C} + \ln x + \sum_{k=1}^{\infty} \frac{(-1)^k x^{2k}}{2k\,(2k)!}.$$

Asymptotic expansion as $x \to \infty$:

$$\mathrm{Ci}(x) = \cos x \left[\sum_{m=1}^{M-1} \frac{(-1)^m (2m-1)!}{x^{2m}} + O\left(|x|^{-2M}\right) \right] + \sin x \left[\sum_{m=0}^{N-1} \frac{(-1)^m (2m)!}{x^{2m+1}} + O\left(|x|^{-2N-1}\right) \right],$$

where $M, N = 1, 2, \ldots$

A.3.3. Fresnel Integrals

Definitions:

$$S(x) = \frac{1}{\sqrt{2\pi}} \int_0^x \frac{\sin t}{\sqrt{t}}\, dt = \sqrt{\frac{2}{\pi}} \int_0^{\sqrt{x}} \sin t^2\, dt,$$

$$C(x) = \frac{1}{\sqrt{2\pi}} \int_0^x \frac{\cos t}{\sqrt{t}}\, dt = \sqrt{\frac{2}{\pi}} \int_0^{\sqrt{x}} \cos t^2\, dt.$$

Expansion into series in powers of x as $x \to 0$:

$$S(x) = \sqrt{\frac{2}{\pi}}\, x \sum_{k=0}^{\infty} \frac{(-1)^k x^{2k+1}}{(4k+3)(2k+1)!},$$

$$C(x) = \sqrt{\frac{2}{\pi}}\, x \sum_{k=0}^{\infty} \frac{(-1)^k x^{2k}}{(4k+1)(2k)!}.$$

Asymptotic expansion as $x \to \infty$:

$$S(x) = \frac{1}{2} - \frac{\cos x}{\sqrt{2\pi x}} P(x) - \frac{\sin x}{\sqrt{2\pi x}} Q(x),$$

$$C(x) = \frac{1}{2} + \frac{\sin x}{\sqrt{2\pi x}} P(x) - \frac{\cos x}{\sqrt{2\pi x}} Q(x),$$

$$P(x) = 1 - \frac{1 \cdot 3}{(2x)^2} + \frac{1 \cdot 3 \cdot 5 \cdot 7}{(2x)^4} - \cdots, \qquad Q(x) = \frac{1}{2x} - \frac{1 \cdot 3 \cdot 5}{(2x)^3} + \cdots.$$

A.4. Gamma and Beta Functions

A.4.1. Gamma Function

A.4.1-1. Definition. Integral representations.

The gamma function, $\Gamma(z)$, is an analytic function of the complex argument z everywhere, except for the points $z = 0, -1, -2, \ldots$

For Re $z > 0$,
$$\Gamma(z) = \int_0^\infty t^{z-1} e^{-t}\, dt.$$

For $-(n+1) < \operatorname{Re} z < -n$, where $n = 0, 1, 2, \ldots,$
$$\Gamma(z) = \int_0^\infty \left[e^{-t} - \sum_{m=0}^{n} \frac{(-1)^m}{m!} \right] t^{z-1}\, dt.$$

A.4.1-2. Some formulas.

Euler formula
$$\Gamma(z) = \lim_{n\to\infty} \frac{n!\, n^z}{z(z+1)\ldots(z+n)} \qquad (z \neq 0, -1, -2, \ldots).$$

Simplest properties:
$$\Gamma(z+1) = z\Gamma(z), \quad \Gamma(n+1) = n!, \quad \Gamma(1) = \Gamma(2) = 1.$$

Symmetry formulas:
$$\Gamma(z)\Gamma(-z) = -\frac{\pi}{z \sin(\pi z)}, \quad \Gamma(z)\Gamma(1-z) = \frac{\pi}{\sin(\pi z)},$$
$$\Gamma\!\left(\frac{1}{2}+z\right)\Gamma\!\left(\frac{1}{2}-z\right) = \frac{\pi}{\cos(\pi z)}.$$

Multiple argument formulas:
$$\Gamma(2z) = \frac{2^{2z-1}}{\sqrt{\pi}} \Gamma(z)\Gamma\!\left(z+\frac{1}{2}\right),$$
$$\Gamma(3z) = \frac{3^{3z-1/2}}{2\pi} \Gamma(z)\Gamma\!\left(z+\frac{1}{3}\right)\Gamma\!\left(z+\frac{2}{3}\right),$$
$$\Gamma(nz) = (2\pi)^{(1-n)/2} n^{nz-1/2} \prod_{k=0}^{n-1} \Gamma\!\left(z+\frac{k}{n}\right).$$

Fractional values of the argument:
$$\Gamma\!\left(\frac{1}{2}\right) = \sqrt{\pi}, \qquad \Gamma\!\left(n+\frac{1}{2}\right) = \frac{\sqrt{\pi}}{2^n}(2n-1)!!,$$
$$\Gamma\!\left(-\frac{1}{2}\right) = -2\sqrt{\pi}, \qquad \Gamma\!\left(\frac{1}{2}-n\right) = (-1)^n \frac{2^n \sqrt{\pi}}{(2n-1)!!}.$$

Asymptotic expansion (Stirling formula):
$$\Gamma(z) = \sqrt{2\pi}\, e^{-z} z^{z-1/2} \left[1 + \tfrac{1}{12} z^{-1} + \tfrac{1}{288} z^{-2} + O(z^{-3})\right] \qquad (|\arg|z < \pi).$$

A.4.1-3. Logarithmic derivative of the gamma function.

Definition:
$$\psi(z) = \frac{d \ln \Gamma(z)}{dz} = \frac{\Gamma'_z(z)}{\Gamma(z)}.$$

Functional relations:
$$\psi(z) - \psi(1+z) = -\frac{1}{z},$$
$$\psi(z) - \psi(1-z) = -\pi \cot(\pi z),$$
$$\psi(z) - \psi(-z) = -\pi \cot(\pi z) - \frac{1}{z},$$
$$\psi\left(\tfrac{1}{2}+z\right) - \psi\left(\tfrac{1}{2}-z\right) = \pi \tan(\pi z),$$
$$\psi(mz) = \ln m + \frac{1}{m} \sum_{k=0}^{m-1} \psi\left(z + \frac{k}{m}\right).$$

Integral representations (Re $z > 0$):
$$\psi(z) = \int_0^\infty \left[e^{-t} - (1+t)^{-z}\right] t^{-1} \, dt,$$
$$\psi(z) = \ln z + \int_0^\infty \left[t^{-1} - (1-e^{-t})^{-1}\right] e^{-tz} \, dt,$$
$$\psi(z) = -\mathcal{C} + \int_0^1 \frac{1 - t^{z-1}}{1-t} \, dt,$$

where $\mathcal{C} = -\psi(1) = 0.5772\ldots$ is the Euler constant.

Values for integer argument:
$$\psi(1) = -\mathcal{C}, \qquad \psi(n) = -\mathcal{C} + \sum_{k=1}^{n-1} k^{-1} \quad (n = 2, 3, \ldots)$$

A.4.2. Beta Function

Definition:
$$B(x, y) = \int_0^1 t^{x-1}(1-t)^{y-1} \, dt,$$

where Re $x > 0$ and Re $y > 0$.

Relationship with the gamma function:
$$B(x, y) = \frac{\Gamma(x)\Gamma(y)}{\Gamma(x+y)}.$$

A.5. Incomplete Gamma and Beta Functions

A.5.1. Incomplete Gamma Function

Definitions (integral representations):
$$\gamma(\alpha, x) = \int_0^x e^{-t} t^{\alpha-1} \, dt, \qquad \text{Re } \alpha > 0,$$
$$\Gamma(\alpha, x) = \int_x^\infty e^{-t} t^{\alpha-1} \, dt = \Gamma(\alpha) - \gamma(\alpha, x).$$

Recurrent formulas:
$$\gamma(\alpha+1, x) = \alpha\gamma(\alpha, x) - x^\alpha e^{-x},$$
$$\Gamma(\alpha+1, x) = \alpha\Gamma(\alpha, x) + x^\alpha e^{-x}.$$

Asymptotic expansions as $x \to 0$:
$$\gamma(\alpha, x) = \sum_{n=0}^{\infty} \frac{(-1)^n x^{\alpha+n}}{n!\,(\alpha+n)},$$
$$\Gamma(\alpha, x) = \Gamma(\alpha) - \sum_{n=0}^{\infty} \frac{(-1)^n x^{\alpha+n}}{n!\,(\alpha+n)}.$$

Asymptotic expansions as $x \to \infty$:
$$\gamma(\alpha, x) = \Gamma(\alpha) - x^{\alpha-1} e^{-x} \left[\sum_{m=0}^{M-1} \frac{(1-\alpha)_m}{(-x)^m} + O(|x|^{-M}) \right],$$
$$\Gamma(\alpha, x) = x^{\alpha-1} e^{-x} \left[\sum_{m=0}^{M-1} \frac{(1-\alpha)_m}{(-x)^m} + O(|x|^{-M}) \right] \quad (-\tfrac{3}{2}\pi < \arg x < \tfrac{3}{2}).$$

Integral functions related to the gamma function:
$$\operatorname{erf} x = \frac{1}{\sqrt{\pi}} \gamma\!\left(\tfrac{1}{2}, x^2\right), \quad \operatorname{erfc} x = \frac{1}{\sqrt{\pi}} \Gamma\!\left(\tfrac{1}{2}, x^2\right), \quad \operatorname{Ei}(-x) = -\Gamma(0, x).$$

A.5.2. Incomplete Beta Function

Definition:
$$B_x(p, q) = \int_0^1 t^{p-1}(1-t)^{q-1}\, dt,$$

where $\operatorname{Re} x > 0$ and $\operatorname{Re} y > 0$.

A.6. Bessel Functions

A.6.1. Definitions and Basic Formulas

A.6.1-1. The Bessel functions of the first and the second kinds.

The Bessel function of the first kind, $J_\nu(x)$, and the Bessel function of the second kind, $Y_\nu(x)$ (also called the Neumann function), are solutions of the Bessel equation
$$x^2 y''_{xx} + x y'_x + (x^2 - \nu^2) y = 0$$

and are defined by the formulas
$$J_\nu(x) = \sum_{k=0}^{\infty} \frac{(-1)^k (x/2)^{\nu+2k}}{k!\,\Gamma(\nu+k+1)}, \quad Y_\nu(x) = \frac{J_\nu(x) \cos \pi\nu - J_{-\nu}(x)}{\sin \pi\nu}. \tag{1}$$

The formula for $Y_\nu(x)$ is valid for $\nu \neq 0, \pm 1, \pm 2, \ldots$ (the cases $\nu \neq 0, \pm 1, \pm 2, \ldots$ are discussed in what follows).

The general solution of the Bessel equation has the form $Z_\nu(x) = C_1 J_\nu(x) + C_2 Y_\nu(x)$ and is called the cylinder function.

A.6.1-2. Some formulas.

$$2\nu Z_\nu(x) = x[Z_{\nu-1}(x) + Z_{\nu+1}(x)],$$

$$\frac{d}{dx} Z_\nu(x) = \frac{1}{2}[Z_{\nu-1}(x) - Z_{\nu+1}(x)] = \pm\left[\frac{\nu}{x} Z_\nu(x) - Z_{\nu\pm1}(x)\right],$$

$$\frac{d}{dx}[x^\nu Z_\nu(x)] = x^\nu Z_{\nu-1}(x), \qquad \frac{d}{dx}[x^{-\nu} Z_\nu(x)] = -x^{-\nu} Z_{\nu+1}(x),$$

$$\left(\frac{1}{x}\frac{d}{dx}\right)^n [x^\nu J_\nu(x)] = x^{\nu-n} J_{\nu-n}(x), \qquad \left(\frac{1}{x}\frac{d}{dx}\right)^n [x^{-\nu} J_\nu(x)] = (-1)^n x^{-\nu-n} J_{\nu+n}(x),$$

$$J_{-n}(x) = (-1)^n J_n(x), \quad Y_{-n}(x) = (-1)^n Y_n(x), \qquad n = 0, 1, 2, \ldots$$

A.6.1-3. The Bessel functions for $\nu = \pm n \pm \frac{1}{2}$, where $n = 0, 1, 2, \ldots$:

$$J_{1/2}(x) = \sqrt{\frac{2}{\pi x}} \sin x, \qquad\qquad J_{-1/2}(x) = \sqrt{\frac{2}{\pi x}} \cos x,$$

$$J_{3/2}(x) = \sqrt{\frac{2}{\pi x}}\left(\frac{1}{x}\sin x - \cos x\right), \quad J_{-3/2}(x) = \sqrt{\frac{2}{\pi x}}\left(-\frac{1}{x}\cos x - \sin x\right),$$

$$J_{n+1/2}(x) = \sqrt{\frac{2}{\pi x}}\left[\sin\left(x - \frac{n\pi}{2}\right) \sum_{k=0}^{[n/2]} \frac{(-1)^k (n+2k)!}{(2k)!\,(n-2k)!\,(2x)^{2k}} \right.$$

$$\left. + \cos\left(x - \frac{n\pi}{2}\right) \sum_{k=0}^{[(n-1)/2]} \frac{(-1)^k (n+2k+1)!}{(2k+1)!\,(n-2k-1)!\,(2x)^{2k+1}}\right],$$

$$J_{-n-1/2}(x) = \sqrt{\frac{2}{\pi x}}\left[\cos\left(x + \frac{n\pi}{2}\right) \sum_{k=0}^{[n/2]} \frac{(-1)^k (n+2k)!}{(2k)!\,(n-2k)!\,(2x)^{2k}} \right.$$

$$\left. - \sin\left(x + \frac{n\pi}{2}\right) \sum_{k=0}^{[(n-1)/2]} \frac{(-1)^k (n+2k+1)!}{(2k+1)!\,(n-2k-1)!\,(2x)^{2k+1}}\right],$$

$$Y_{1/2}(x) = -\sqrt{\frac{2}{\pi x}} \cos x, \qquad Y_{-1/2}(x) = \sqrt{\frac{2}{\pi x}} \sin x,$$

$$Y_{n+1/2}(x) = (-1)^{n+1} J_{-n-1/2}(x), \qquad Y_{-n-1/2}(x) = (-1)^n J_{n+1/2}(x).$$

A.6.1-4. The Bessel functions for $\nu = \pm n$, where $n = 0, 1, 2, \ldots$

Let $\nu = n$ be an arbitrary integer. The relations

$$J_{-n}(x) = (-1)^n J_n(x), \quad Y_{-n}(x) = (-1)^n Y_n(x)$$

are valid. The function $J_n(x)$ is given by the first formula in (1) with $\nu = n$, and $Y_n(x)$ can be obtained from the second formula in (1) by proceeding to the limit $\nu \to n$. For nonnegative n, $Y_n(x)$ can be represented in the form

$$Y_n(x) = \frac{2}{\pi} J_n(x) \ln\frac{x}{2} - \frac{1}{\pi}\sum_{k=0}^{n-1} \frac{(n-k-1)!}{k!}\left(\frac{2}{x}\right)^{n-2k} - \frac{1}{\pi}\sum_{k=0}^{\infty} (-1)^k \left(\frac{x}{2}\right)^{n+2k} \frac{\psi(k+1) + \psi(n+k+1)}{k!\,(n+k)!},$$

where $\psi(1) = -\mathcal{C}$, $\psi(n) = -\mathcal{C} + \sum_{k=1}^{n-1} k^{-1}$, $\mathcal{C} = 0.5772\ldots$ is the Euler constant, $\psi(x) = [\ln \Gamma(x)]'_x$ is the logarithmic derivative of the gamma function.

A.6.1-5. Wronskians and similar formulas:

$$W(J_\nu, J_{-\nu}) = -\frac{2}{\pi x}\sin(\pi\nu), \qquad W(J_\nu, Y_\nu) = \frac{2}{\pi x},$$

$$J_\nu(x)J_{-\nu+1}(x) + J_{-\nu}(x)J_{\nu-1}(x) = \frac{2\sin(\pi\nu)}{\pi x}, \qquad J_\nu(x)Y_{\nu+1}(x) - J_{\nu+1}(x)Y_\nu(x) = -\frac{2}{\pi x}.$$

Here, the notation $W(f, g) = fg'_x - f'_x g$ is used.

A.6.2. Integral Representations and Asymptotic Expansions

A.6.2-1. Integral representations.

The functions J_ν and Y_ν can be represented in the form of definite integrals (for $x > 0$):

$$\pi J_\nu(x) = \int_0^\pi \cos(x\sin\theta - \nu\theta)\,d\theta - \sin\pi\nu \int_0^\infty \exp(-x\sinh t - \nu t)\,dt,$$

$$\pi Y_\nu(x) = \int_0^\pi \sin(x\sin\theta - \nu\theta)\,d\theta - \int_0^\infty (e^{\nu t} + e^{-\nu t}\cos\pi\nu)e^{-x\sinh t}\,dt.$$

For $|\nu| < \frac{1}{2}$, $x > 0$,

$$J_\nu(x) = \frac{2^{1+\nu}x^{-\nu}}{\pi^{1/2}\Gamma(\frac{1}{2}-\nu)} \int_1^\infty \frac{\sin(xt)\,dt}{(t^2-1)^{\nu+1/2}},$$

$$Y_\nu(x) = -\frac{2^{1+\nu}x^{-\nu}}{\pi^{1/2}\Gamma(\frac{1}{2}-\nu)} \int_1^\infty \frac{\cos(xt)\,dt}{(t^2-1)^{\nu+1/2}}.$$

For $\nu > -\frac{1}{2}$,

$$J_\nu(x) = \frac{2(x/2)^\nu}{\pi^{1/2}\Gamma(\frac{1}{2}+\nu)} \int_0^{\pi/2} \cos(x\cos t)\sin^{2\nu} t\,dt \quad \text{(Poisson's formula)}.$$

For $\nu = 0$, $x > 0$,

$$J_0(x) = \frac{2}{\pi}\int_0^\infty \sin(x\cosh t)\,dt, \qquad Y_0(x) = -\frac{2}{\pi}\int_0^\infty \cos(x\cosh t)\,dt.$$

For integer $\nu = n = 0, 1, 2, \ldots$,

$$J_n(x) = \frac{1}{\pi}\int_0^\pi \cos(nt - x\sin t)\,dt \quad \text{(Bessel's formula)},$$

$$J_{2n}(x) = \frac{2}{\pi}\int_0^{\pi/2} \cos(x\sin t)\cos(2nt)\,dt,$$

$$J_{2n+1}(x) = \frac{2}{\pi}\int_0^{\pi/2} \sin(x\sin t)\sin[(2n+1)t]\,dt.$$

A.6.2-2. Integrals with Bessel functions:

$$\int_0^x x^\lambda J_\nu(x)\,dx = \frac{x^{\lambda+\nu+1}}{2^\nu(\lambda+\nu+1)\Gamma(\nu+1)} F\left(\frac{\lambda+\nu+1}{2}, \frac{\lambda+\nu+3}{2}, \nu+1; -\frac{x^2}{4}\right), \qquad \text{Re}(\lambda+\nu) > -1,$$

where $F(a, b, c; x)$ is the hypergeometric series (see Section 10.9 of this supplement),

$$\int_0^x x^\lambda Y_\nu(x)\,dx = -\frac{\cos(\nu\pi)\Gamma(-\nu)}{2^\nu\pi(\lambda+\nu+1)} x^{\lambda+\nu+1} F\left(\frac{\lambda+\nu+1}{2}, \nu+1, \frac{\lambda+\nu+3}{2}, -\frac{x^2}{4}\right)$$

$$- \frac{2^\nu\Gamma(\nu)}{\lambda-\nu+1} x^{\lambda-\nu+1} F\left(\frac{\lambda-\nu+1}{2}, 1-\nu, \frac{\lambda-\nu+3}{2}, -\frac{x^2}{4}\right), \qquad \text{Re}\,\lambda > |\text{Re}\,\nu| - 1.$$

A.6.2-3. Asymptotic expansions as $|x| \to \infty$:

$$J_\nu(x) = \sqrt{\frac{2}{\pi x}} \left\{ \cos\left(\frac{4x - 2\nu\pi - \pi}{4}\right) \left[\sum_{m=0}^{M-1} (-1)^m (\nu, 2m)(2x)^{-2m} + O(|x|^{-2M}) \right] \right.$$

$$\left. - \sin\left(\frac{4x - 2\nu\pi - \pi}{4}\right) \left[\sum_{m=0}^{M-1} (-1)^m (\nu, 2m+1)(2x)^{-2m-1} + O(|x|^{-2M-1}) \right] \right\},$$

$$Y_\nu(x) = \sqrt{\frac{2}{\pi x}} \left\{ \sin\left(\frac{4x - 2\nu\pi - \pi}{4}\right) \left[\sum_{m=0}^{M-1} (-1)^m (\nu, 2m)(2x)^{-2m} + O(|x|^{-2M}) \right] \right.$$

$$\left. + \cos\left(\frac{4x - 2\nu\pi - \pi}{4}\right) \left[\sum_{m=0}^{M-1} (-1)^m (\nu, 2m+1)(2x)^{-2m-1} + O(|x|^{-2M-1}) \right] \right\},$$

where $(\nu, m) = \dfrac{1}{2^{2m} m!} (4\nu^2 - 1)(4\nu^2 - 3^2) \ldots [4\nu^2 - (2m-1)^2] = \dfrac{\Gamma(\frac{1}{2} + \nu + m)}{m!\, \Gamma(\frac{1}{2} + \nu - m)}.$

For nonnegative integer n and large x,

$$\sqrt{\pi x}\, J_{2n}(x) = (-1)^n (\cos x + \sin x) + O(x^{-2}),$$
$$\sqrt{\pi x}\, J_{2n+1}(x) = (-1)^{n+1} (\cos x - \sin x) + O(x^{-2}).$$

A.6.2-4. Asymptotic for large ν ($\nu \to \infty$):

$$J_\nu(x) \simeq \frac{1}{\sqrt{2\pi\nu}} \left(\frac{ex}{2\nu}\right)^\nu, \quad Y_\nu(x) \simeq -\sqrt{\frac{2}{\pi\nu}} \left(\frac{ex}{2\nu}\right)^{-\nu},$$

where x is fixed,

$$J_\nu(\nu) \simeq \frac{2^{1/3}}{3^{2/3}\Gamma(2/3)} \frac{1}{\nu^{1/3}}, \quad Y_\nu(\nu) \simeq -\frac{2^{1/3}}{3^{1/6}\Gamma(2/3)} \frac{1}{\nu^{1/3}}.$$

A.6.3. Zeros and Orthogonality Properties of Bessel Functions

A.6.3-1. Zeros of Bessel functions.

Each of the functions $J_\nu(x)$ and $Y_\nu(x)$ has infinitely many real zeros (for real ν). All zeros are simple, except possibly for the point $x = 0$.

The zeros γ_m of $J_0(x)$, i.e., the roots of the equation $J_0(\gamma_m) = 0$, are approximately given by

$$\gamma_m = 2.4 + 3.13\,(m-1) \qquad (m = 1, 2, \ldots),$$

with maximum error 0.2%.

A.6.3-2. Orthogonality properties of Bessel functions.

1°. Let $\mu = \mu_m$ be positive roots of the Bessel function $J_\nu(\mu)$, where $\nu > -1$ and $m = 1, 2, 3, \ldots$ Then the set of functions $J_\nu(\mu_m r/a)$ is orthogonal on the interval $0 \le r \le a$ with weight r:

$$\int_0^a J_\nu\!\left(\frac{\mu_m r}{a}\right) J_\nu\!\left(\frac{\mu_k r}{a}\right) r\, dr = \begin{cases} 0 & \text{if } m \ne k, \\ \tfrac{1}{2} a^2 \left[J_\nu'(\mu_m)\right]^2 = \tfrac{1}{2} a^2 J_{\nu+1}^2(\mu_m) & \text{if } m = k. \end{cases}$$

2°. Let $\mu = \mu_m$ be positive zeros of the Bessel function derivative $J'_\nu(\mu)$, where $\nu > -1$ and $m = 1, 2, 3, \ldots$ Then the set of functions $J_\nu(\mu_m r/a)$ is orthogonal on the interval $0 \leq r \leq a$ with weight r:

$$\int_0^a J_\nu\left(\frac{\mu_m r}{a}\right) J_\nu\left(\frac{\mu_k r}{a}\right) r\, dr = \begin{cases} 0 & \text{if } m \neq k, \\ \frac{1}{2} a^2 \left(1 - \frac{\nu^2}{\mu_m^2}\right) J_\nu^2(\mu_m) & \text{if } m = k. \end{cases}$$

3°. Let $\mu = \mu_m$ be positive roots of the transcendental equation $\mu J'_\nu(\mu) + s J_\nu(\mu) = 0$, where $\nu > -1$ and $m = 1, 2, 3, \ldots$ Then the set of functions $J_\nu(\mu_m r/a)$ is orthogonal on the interval $0 \leq r \leq a$ with weight r:

$$\int_0^a J_\nu\left(\frac{\mu_m r}{a}\right) J_\nu\left(\frac{\mu_k r}{a}\right) r\, dr = \begin{cases} 0 & \text{if } m \neq k, \\ \frac{1}{2} a^2 \left(1 + \frac{s^2 - \nu^2}{\mu_m^2}\right) J_\nu^2(\mu_m) & \text{if } m = k. \end{cases}$$

4°. Let $\mu = \mu_m$ be positive roots of the transcendental equation
$$J_\nu(\lambda_m b) Y_\nu(\lambda_m a) - J_\nu(\lambda_m a) Y_\nu(\lambda_m b) = 0 \qquad (\nu > -1,\ m = 1, 2, 3, \ldots).$$
Then the set of functions
$$Z_\nu(\lambda_m r) = J_\nu(\lambda_m r) Y_\nu(\lambda_m a) - J_\nu(\lambda_m a) Y_\nu(\lambda_m r), \qquad m = 1, 2, 3, \ldots,$$
satisfying the conditions $Z_\nu(\lambda_m a) = Z_\nu(\lambda_m b) = 0$ is orthogonal on the interval $a \leq r \leq b$ with weight r:

$$\int_a^b Z_\nu(\lambda_m r) Z_\nu(\lambda_k r) r\, dr = \begin{cases} 0 & \text{if } m \neq k, \\ \dfrac{2}{\pi^2 \lambda_m^2} \dfrac{J_\nu^2(\lambda_m a) - J_\nu^2(\lambda_m b)}{J_\nu^2(\lambda_m b)} & \text{if } m = k. \end{cases}$$

5°. Let $\mu = \mu_m$ be positive roots of the transcendental equation
$$J'_\nu(\lambda_m b) Y'_\nu(\lambda_m a) - J'_\nu(\lambda_m a) Y'_\nu(\lambda_m b) = 0 \qquad (\nu > -1,\ m = 1, 2, 3, \ldots)$$
Then the set of functions
$$Z_\nu(\lambda_m r) = J_\nu(\lambda_m r) Y'_\nu(\lambda_m a) - J'_\nu(\lambda_m a) Y_\nu(\lambda_m r), \qquad m = 1, 2, 3, \ldots,$$
satisfying the conditions $Z'_\nu(\lambda_m a) = Z'_\nu(\lambda_m b) = 0$ is orthogonal on the interval $a \leq r \leq b$ with weight r:

$$\int_a^b Z_\nu(\lambda_m r) Z_\nu(\lambda_k r) r\, dr = \begin{cases} 0 & \text{if } m \neq k, \\ \dfrac{2}{\pi^2 \lambda_m^2} \left[\left(1 - \dfrac{\nu^2}{b^2 \lambda_m^2}\right) \dfrac{[J'_\nu(\lambda_m a)]^2}{[J'_\nu(\lambda_m b)]^2} - \left(1 - \dfrac{\nu^2}{a^2 \lambda_m^2}\right)\right] & \text{if } m = k. \end{cases}$$

A.6.4. Hankel Functions (Bessel Functions of the Third Kind)

The Hankel functions of the first kind and the second kind are related to Bessel functions by
$$H_\nu^{(1)}(z) = J_\nu(z) + i Y_\nu(z), \qquad H_\nu^{(2)}(z) = J_\nu(z) - i Y_\nu(z), \qquad i^2 = -1.$$

Asymptotics for $z \to 0$:
$$H_0^{(1)}(z) \simeq \frac{2i}{\pi} \ln z, \qquad H_\nu^{(1)}(z) \simeq -\frac{i}{\pi} \frac{\Gamma(\nu)}{(z/2)^\nu} \qquad (\operatorname{Re} \nu > 0),$$
$$H_0^{(2)}(z) \simeq -\frac{2i}{\pi} \ln z, \qquad H_\nu^{(2)}(z) \simeq \frac{i}{\pi} \frac{\Gamma(\nu)}{(z/2)^\nu} \qquad (\operatorname{Re} \nu > 0).$$

Asymptotics for $|z| \to \infty$:
$$H_\nu^{(1)}(z) \simeq \sqrt{\frac{2}{\pi z}} \exp\left[i\left(z - \tfrac{1}{2}\pi\nu - \tfrac{1}{4}\pi\right)\right] \qquad (-\pi < \arg z < 2\pi),$$
$$H_\nu^{(2)}(z) \simeq \sqrt{\frac{2}{\pi z}} \exp\left[-i\left(z - \tfrac{1}{2}\pi\nu - \tfrac{1}{4}\pi\right)\right] \qquad (-2\pi < \arg z < \pi).$$

A.7. Modified Bessel Functions

A.7.1. Definitions. Basic Formulas

A.7.1-1. The modified Bessel functions of the first and the second kinds.

The modified Bessel functions of the first kind, $I_\nu(x)$, and the second kind, $K_\nu(x)$ (also called the Macdonald function), of order ν are solutions of the modified Bessel equation

$$x^2 y''_{xx} + x y'_x - (x^2 + \nu^2) y = 0$$

and are defined by the formulas

$$I_\nu(x) = \sum_{k=0}^\infty \frac{(x/2)^{2k+\nu}}{k!\,\Gamma(\nu+k+1)}, \qquad K_\nu(x) = \frac{\pi}{2} \frac{I_{-\nu} - I_\nu}{\sin \pi \nu},$$

(see below for $K_\nu(x)$ with $\nu = 0, 1, 2, \dots$).

A.7.1-2. Some formulas.

The modified Bessel functions possess the properties

$$K_{-\nu}(x) = K_\nu(x); \qquad I_{-n}(x) = (-1)^n I_n(x), \quad n = 0, 1, 2, \dots$$
$$2\nu I_\nu(x) = x[I_{\nu-1}(x) - I_{\nu+1}(x)], \qquad 2\nu K_\nu(x) = -x[K_{\nu-1}(x) - K_{\nu+1}(x)],$$
$$\frac{d}{dx} I_\nu(x) = \frac{1}{2}[I_{\nu-1}(x) + I_{\nu+1}(x)], \qquad \frac{d}{dx} K_\nu(x) = -\frac{1}{2}[K_{\nu-1}(x) + K_{\nu+1}(x)].$$

A.7.1-3. Modified Bessel functions for $\nu = \pm n \pm \frac{1}{2}$, where $n = 0, 1, 2, \dots$:

$$I_{1/2}(x) = \sqrt{\frac{2}{\pi x}} \sinh x, \qquad I_{-1/2}(x) = \sqrt{\frac{2}{\pi x}} \cosh x,$$

$$I_{3/2}(x) = \sqrt{\frac{2}{\pi x}} \left(-\frac{1}{x} \sinh x + \cosh x\right), \qquad I_{-3/2}(x) = \sqrt{\frac{2}{\pi x}} \left(-\frac{1}{x} \cosh x + \sinh x\right),$$

$$I_{n+1/2}(x) = \frac{1}{\sqrt{2\pi x}} \left[e^x \sum_{k=0}^n \frac{(-1)^k (n+k)!}{k!\,(n-k)!\,(2x)^k} - (-1)^n e^{-x} \sum_{k=0}^n \frac{(n+k)!}{k!\,(n-k)!\,(2x)^k}\right],$$

$$I_{-n-1/2}(x) = \frac{1}{\sqrt{2\pi x}} \left[e^x \sum_{k=0}^n \frac{(-1)^k (n+k)!}{k!\,(n-k)!\,(2x)^k} + (-1)^n e^{-x} \sum_{k=0}^n \frac{(n+k)!}{k!\,(n-k)!\,(2x)^k}\right],$$

$$K_{\pm 1/2}(x) = \sqrt{\frac{\pi}{2x}} e^{-x}, \qquad K_{\pm 3/2}(x) = \sqrt{\frac{\pi}{2x}} \left(1 + \frac{1}{x}\right) e^{-x},$$

$$K_{n+1/2}(x) = K_{-n-1/2}(x) = \sqrt{\frac{\pi}{2x}} e^{-x} \sum_{k=0}^n \frac{(n+k)!}{k!\,(n-k)!\,(2x)^k}.$$

A.7.1-4. Modified Bessel functions for $\nu = n$, where $n = 0, 1, 2, \dots$

If $\nu = n$ is a nonnegative integer, then

$$K_n(x) = (-1)^{n+1} I_n(x) \ln \frac{x}{2} + \frac{1}{2} \sum_{m=0}^{n-1} (-1)^m \left(\frac{x}{2}\right)^{2m-n} \frac{(n-m-1)!}{m!}$$
$$+ \frac{1}{2}(-1)^n \sum_{m=0}^\infty \left(\frac{x}{2}\right)^{n+2m} \frac{\psi(n+m+1) + \psi(m+1)}{m!\,(n+m)!}; \quad n = 0, 1, 2, \dots,$$

where $\psi(z)$ is the logarithmic derivative of the gamma function; for $n = 0$, the first sum is dropped.

A.7.1-5. Wronskians and similar formulas:

$$W(I_\nu, I_{-\nu}) = -\frac{2}{\pi x}\sin(\pi\nu), \quad W(I_\nu, K_\nu) = -\frac{1}{x},$$

$$I_\nu(x)I_{-\nu+1}(x) - I_{-\nu}(x)I_{\nu-1}(x) = -\frac{2\sin(\pi\nu)}{\pi x}, \quad I_\nu(x)K_{\nu+1}(x) + I_{\nu+1}(x)K_\nu(x) = \frac{1}{x},$$

where $W(f, g) = fg'_x - f'_x g$.

A.7.2. Integral Representations and Asymptotic Expansions

A.7.2-1. Integral representations.

The functions $I_\nu(x)$ and $K_\nu(x)$ can be represented in terms of definite integrals:

$$I_\nu(x) = \frac{x^\nu}{\pi^{1/2} 2^\nu \Gamma(\nu + \tfrac{1}{2})} \int_{-1}^{1} \exp(-xt)(1-t^2)^{\nu-1/2}\,dt \qquad (x > 0,\ \nu > -\tfrac{1}{2}),$$

$$K_\nu(x) = \int_0^\infty \exp(-x\cosh t)\cosh(\nu t)\,dt \qquad (x > 0),$$

$$K_\nu(x) = \frac{1}{\cos(\tfrac{1}{2}\pi\nu)} \int_0^\infty \cos(x\sinh t)\cosh(\nu t)\,dt \qquad (x > 0,\ -1 < \nu < 1),$$

$$K_\nu(x) = \frac{1}{\sin(\tfrac{1}{2}\pi\nu)} \int_0^\infty \sin(x\sinh t)\sinh(\nu t)\,dt \qquad (x > 0,\ -1 < \nu < 1).$$

For integer $\nu = n$,

$$I_n(x) = \frac{1}{\pi} \int_0^\pi \exp(x\cos t)\cos(nt)\,dt \qquad (n = 0, 1, 2, \dots),$$

$$K_0(x) = \int_0^\infty \cos(x\sinh t)\,dt = \int_0^\infty \frac{\cos(xt)}{\sqrt{t^2+1}}\,dt \qquad (x > 0).$$

A.7.2-2. Integrals with modified Bessel functions:

$$\int_0^x x^\lambda I_\nu(x)\,dx = \frac{x^{\lambda+\nu+1}}{2^\nu(\lambda+\nu+1)\Gamma(\nu+1)} F\left(\frac{\lambda+\nu+1}{2}, \frac{\lambda+\nu+3}{2}, \nu+1;\, \frac{x^2}{4}\right), \qquad \mathrm{Re}(\lambda+\nu) > -1,$$

where $F(a, b, c; x)$ is the hypergeometric series (see Section 10.9 of this supplement),

$$\int_0^x x^\lambda K_\nu(x)\,dx = \frac{2^{\nu-1}\Gamma(\nu)}{\lambda-\nu+1} x^{\lambda-\nu+1} F\left(\frac{\lambda-\nu+1}{2}, 1-\nu, \frac{\lambda-\nu+3}{2},\, \frac{x^2}{4}\right)$$

$$+ \frac{2^{-\nu-1}\Gamma(-\nu)}{\lambda+\nu+1} x^{\lambda+\nu+1} F\left(\frac{\lambda+\nu+1}{2}, 1+\nu, \frac{\lambda+\nu+3}{2},\, \frac{x^2}{4}\right), \qquad \mathrm{Re}\,\lambda > |\mathrm{Re}\,\nu| - 1.$$

A.7.2-3. Asymptotic expansions as $x \to \infty$:

$$I_\nu(x) = \frac{e^x}{\sqrt{2\pi x}}\left\{1 + \sum_{m=1}^M (-1)^m \frac{(4\nu^2-1)(4\nu^2-3^2)\dots[4\nu^2-(2m-1)^2]}{m!\,(8x)^m}\right\},$$

$$K_\nu(x) = \sqrt{\frac{\pi}{2x}}\,e^{-x}\left\{1 + \sum_{m=1}^M \frac{(4\nu^2-1)(4\nu^2-3^2)\dots[4\nu^2-(2m-1)^2]}{m!\,(8x)^m}\right\}.$$

The terms of the order of $O(x^{-M-1})$ are omitted in the braces.

A.8. Airy Functions

A.8.1. Definition and Basic Formulas

A.8.1-1. The Airy functions of the first and the second kinds.

The Airy function of the first kind, $\operatorname{Ai}(x)$, and the Airy function of the second kind, $\operatorname{Bi}(x)$, are solutions of the Airy equation

$$y''_{xx} - xy = 0$$

and are defined by the formulas

$$\operatorname{Ai}(x) = \frac{1}{\pi}\int_0^\infty \cos\left(\tfrac{1}{3}t^3 + xt\right)dt,$$

$$\operatorname{Bi}(x) = \frac{1}{\pi}\int_0^\infty \left[\exp\left(-\tfrac{1}{3}t^3 + xt\right) + \sin\left(\tfrac{1}{3}t^3 + xt\right)\right]dt.$$

Wronskian: $W\{\operatorname{Ai}(x), \operatorname{Bi}(x)\} = 1/\pi$.

A.8.1-2. Connection with the Bessel functions and the modified Bessel functions:

$$\operatorname{Ai}(x) = \tfrac{1}{3}\sqrt{x}\left[I_{-1/3}(z) - I_{1/3}(z)\right] = \pi^{-1}\sqrt{\tfrac{1}{3}x}\, K_{1/3}(z), \quad z = \tfrac{2}{3}x^{3/2},$$

$$\operatorname{Ai}(-x) = \tfrac{1}{3}\sqrt{x}\left[J_{-1/3}(z) + J_{1/3}(z)\right],$$

$$\operatorname{Bi}(x) = \sqrt{\tfrac{1}{3}x}\left[I_{-1/3}(z) + I_{1/3}(z)\right],$$

$$\operatorname{Bi}(-x) = \sqrt{\tfrac{1}{3}x}\left[J_{-1/3}(z) - J_{1/3}(z)\right].$$

A.8.2. Power Series and Asymptotic Expansions

A.8.2-1. Power series expansions as $x \to 0$:

$$\operatorname{Ai}(x) = c_1 f(x) - c_2 g(x),$$

$$\operatorname{Bi}(x) = \sqrt{3}\left[c_1 f(x) + c_2 g(x)\right],$$

$$f(x) = 1 + \frac{1}{3!}x^3 + \frac{1\cdot 4}{6!}x^6 + \frac{1\cdot 4\cdot 7}{9!}x^9 + \ldots = \sum_{k=0}^\infty 3^k \left(\tfrac{1}{3}\right)_k \frac{x^{3k}}{(3k)!},$$

$$g(x) = x + \frac{2}{4!}x^4 + \frac{2\cdot 5}{7!}x^7 + \frac{2\cdot 5\cdot 8}{10!}x^{10} + \ldots = \sum_{k=0}^\infty 3^k \left(\tfrac{2}{3}\right)_k \frac{x^{3k+1}}{(3k+1)!},$$

where $c_1 = 3^{-2/3}/\Gamma(2/3) \approx 0.3550$ and $c_2 = 3^{-1/3}/\Gamma(1/3) \approx 0.2588$.

A.8.2-2. Asymptotic expansions as $x \to \infty$.

For large values of x, the leading terms of asymptotic expansions of the Airy functions are

$$\operatorname{Ai}(x) \simeq \tfrac{1}{2}\pi^{-1/2}x^{-1/4}\exp(-z), \quad z = \tfrac{2}{3}x^{3/2},$$

$$\operatorname{Ai}(-x) \simeq \pi^{-1/2}x^{-1/4}\sin\left(z + \tfrac{\pi}{4}\right),$$

$$\operatorname{Bi}(x) \simeq \pi^{-1/2}x^{-1/4}\exp(z),$$

$$\operatorname{Bi}(-x) \simeq \pi^{-1/2}x^{-1/4}\cos\left(z + \tfrac{\pi}{4}\right).$$

⊙ *Reference*: M. Abramowitz and I. Stegun (1964).

TABLE A1
Special cases of the Kummer function $\Phi(a,b;z)$

a	b	z	Φ	Conventional notation
a	a	x	e^x	
1	2	$2x$	$\dfrac{1}{x}e^x \sinh x$	
a	$a+1$	$-x$	$ax^{-a}\gamma(a,x)$	Incomplete gamma function $\gamma(a,x) = \displaystyle\int_0^x e^{-t} t^{a-1}\, dt$
$\dfrac{1}{2}$	$\dfrac{3}{2}$	$-x^2$	$\dfrac{\sqrt{\pi}}{2}\operatorname{erf} x$	Error function $\operatorname{erf} x = \dfrac{2}{\sqrt{\pi}}\displaystyle\int_0^x \exp(-t^2)\, dt$
$-n$	$\dfrac{1}{2}$	$\dfrac{x^2}{2}$	$\dfrac{n!}{(2n)!}\left(-\dfrac{1}{2}\right)^{-n} H_{2n}(x)$	Hermite polynomials $H_n = (-1)^n e^{x^2} \dfrac{d^n}{dx^n}\left(e^{-x^2}\right),$
$-n$	$\dfrac{3}{2}$	$\dfrac{x^2}{2}$	$\dfrac{n!}{(2n+1)!}\left(-\dfrac{1}{2}\right)^{-n} H_{2n+1}(x)$	$n = 0, 1, 2, \ldots$
$-n$	b	x	$\dfrac{n!}{(b)_n} L_n^{(b-1)}(x)$	Laguerre polynomials $L_n^{(\alpha)}(x) = \dfrac{e^x x^{-\alpha}}{n!}\dfrac{d^n}{dx^n}\left(e^{-x} x^{n+\alpha}\right),$ $\alpha = b-1,$ $(b)_n = b(b+1)\ldots(b+n-1)$
$\nu+\dfrac{1}{2}$	$2\nu+1$	$2x$	$\Gamma(1+\nu) e^x \left(\dfrac{x}{2}\right)^{-\nu} I_\nu(x)$	Modified Bessel functions $I_\nu(x)$
$n+1$	$2n+2$	$2x$	$\Gamma\left(n+\dfrac{3}{2}\right) e^x \left(\dfrac{x}{2}\right)^{-n-\frac{1}{2}} I_{n+\frac{1}{2}}(x)$	

A.9. Degenerate Hypergeometric Functions

A.9.1. Definitions and Basic Formulas

A.9.1-1. The degenerate hypergeometric functions $\Phi(a,b;x)$ and $\Psi(a,b;x)$.

The degenerate hypergeometric functions $\Phi(a,b;x)$ and $\Psi(a,b;x)$ are solutions of the degenerate hypergeometric equation

$$xy''_{xx} + (b-x)y'_x - ay = 0.$$

In the case $b \neq 0, -1, -2, -3, \ldots$, the function $\Phi(a,b;x)$ can be represented as Kummer's series:

$$\Phi(a,b;x) = 1 + \sum_{k=1}^{\infty} \frac{(a)_k}{(b)_k} \frac{x^k}{k!},$$

where $(a)_k = a(a+1)\ldots(a+k-1)$, $(a)_0 = 1$.

Table A1 presents some special cases where Φ can be expressed in terms of simpler functions. The function $\Psi(a,b;x)$ is defined as follows:

$$\Psi(a,b;x) = \frac{\Gamma(1-b)}{\Gamma(a-b+1)}\Phi(a,b;x) + \frac{\Gamma(b-1)}{\Gamma(a)} x^{1-b}\Phi(a-b+1,\, 2-b;\, x).$$

A.9.1-2. Kummer transformation and linear relations.

Kummer transformation:

$$\Phi(a,b;x) = e^x \Phi(b-a,b;-x), \qquad \Psi(a,b;x) = x^{1-b}\Psi(1+a-b,2-b;x).$$

Linear relations for Φ:

$$(b-a)\Phi(a-1,b;x) + (2a-b+x)\Phi(a,b;x) - a\Phi(a+1,b;x) = 0,$$
$$b(b-1)\Phi(a,b-1;x) - b(b-1+x)\Phi(a,b;x) + (b-a)x\Phi(a,b+1;x) = 0,$$
$$(a-b+1)\Phi(a,b;x) - a\Phi(a+1,b;x) + (b-1)\Phi(a,b-1;x) = 0,$$
$$b\Phi(a,b;x) - b\Phi(a-1,b;x) - x\Phi(a,b+1;x) = 0,$$
$$b(a+x)\Phi(a,b;x) - (b-a)x\Phi(a,b+1;x) - ab\Phi(a+1,b;x) = 0,$$
$$(a-1+x)\Phi(a,b;x) + (b-a)\Phi(a-1,b;x) - (b-1)\Phi(a,b-1;x) = 0.$$

Linear relations for Ψ:

$$\Psi(a-1,b;x) - (2a-b+x)\Psi(a,b;x) + a(a-b+1)\Psi(a+1,b;x) = 0,$$
$$(b-a-1)\Psi(a,b-1;x) - (b-1+x)\Psi(a,b;x) + x\Psi(a,b+1;x) = 0,$$
$$\Psi(a,b;x) - a\Psi(a+1,b;x) - \Psi(a,b-1;x) = 0,$$
$$(b-a)\Psi(a,b;x) - x\Psi(a,b+1;x) + \Psi(a-1,b;x) = 0,$$
$$(a+x)\Psi(a,b;x) + a(b-a-1)\Psi(a+1,b;x) - x\Psi(a,b+1;x) = 0,$$
$$(a-1+x)\Psi(a,b;x) - \Psi(a-1,b;x) + (a-c+1)\Psi(a,b-1;x) = 0.$$

A.9.1-3. Differentiation formulas and Wronskian.

Differentiation formulas:

$$\frac{d}{dx}\Phi(a,b;x) = \frac{a}{b}\Phi(a+1,b+1;x), \qquad \frac{d^n}{dx^n}\Phi(a,b;x) = \frac{(a)_n}{(b)_n}\Phi(a+n,b+n;x),$$
$$\frac{d}{dx}\Psi(a,b;x) = -a\Psi(a+1,b+1;x), \qquad \frac{d^n}{dx^n}\Psi(a,b;x) = (-1)^n(a)_n\Psi(a+n,b+n;x).$$

Wronskian:

$$W(\Phi,\Psi) = \Phi\Psi'_x - \Phi'_x\Psi = -\frac{\Gamma(b)}{\Gamma(a)}x^{-b}e^x.$$

A.9.1-4. Degenerate hypergeometric functions for $n = 0, 1, 2, \ldots$:

$$\Psi(a,n+1;x) = \frac{(-1)^{n-1}}{n!\,\Gamma(a-n)} \bigg\{ \Phi(a,n+1;x)\ln x$$
$$+ \sum_{r=0}^{\infty} \frac{(a)_r}{(n+1)_r}\big[\psi(a+r) - \psi(1+r) - \psi(1+n+r)\big]\frac{x^r}{r!} \bigg\} + \frac{(n-1)!}{\Gamma(a)}\sum_{r=0}^{n-1}\frac{(a-n)_r}{(1-n)_r}\frac{x^{r-n}}{r!},$$

where $n = 0, 1, 2, \ldots$ (the last sum is dropped for $n = 0$), $\psi(z) = [\ln \Gamma(z)]'_z$ is the logarithmic derivative of the gamma function,

$$\psi(1) = -\mathcal{C}, \qquad \psi(n) = -\mathcal{C} + \sum_{k=1}^{n-1} k^{-1},$$

where $\mathcal{C} = 0.5772\ldots$ is the Euler constant.

If $b < 0$, then the formula

$$\Psi(a, b; x) = x^{1-b}\Psi(a - b + 1, 2 - b; x)$$

is valid for any x.

For $b \neq 0, -1, -2, -3, \ldots$, the general solution of the degenerate hypergeometric equation can be represented in the form

$$y = C_1\Phi(a, b; x) + C_2\Psi(a, b; x),$$

and for $b = 0, -1, -2, -3, \ldots$, in the form

$$y = x^{1-b}\left[C_1\Phi(a - b + 1, 2 - b; x) + C_2\Psi(a - b + 1, 2 - b; x)\right].$$

A.9.2. Integral Representations and Asymptotic Expansions

A.9.2-1. Integral representations:

$$\Phi(a, b; x) = \frac{\Gamma(b)}{\Gamma(a)\Gamma(b-a)}\int_0^1 e^{xt}t^{a-1}(1-t)^{b-a-1}\,dt \qquad (\text{for } b > a > 0),$$

$$\Psi(a, b; x) = \frac{1}{\Gamma(a)}\int_0^\infty e^{-xt}t^{a-1}(1+t)^{b-a-1}\,dt \qquad (\text{for } a > 0,\ x > 0),$$

where $\Gamma(a)$ is the gamma function.

A.9.2-2. Integrals with degenerate hypergeometric functions:

$$\int \Phi(a, b; x)\,dx = \frac{b-1}{a-1}\Psi(a-1, b-1; x) + C,$$

$$\int \Psi(a, b; x)\,dx = \frac{1}{1-a}\Psi(a-1, b-1; x) + C,$$

$$\int x^n\Phi(a, b; x)\,dx = n!\sum_{k=1}^{n+1}\frac{(-1)^{k+1}(1-b)_k x^{n-k+1}}{(1-a)_k(n-k+1)!}\Phi(a-k, b-k; x) + C,$$

$$\int x^n\Psi(a, b; x)\,dx = n!\sum_{k=1}^{n+1}\frac{(-1)^{k+1}x^{n-k+1}}{(1-a)_k(n-k+1)!}\Psi(a-k, b-k; x) + C.$$

A.9.2-3. Asymptotic expansion as $|x| \to \infty$:

$$\Phi(a, b; x) = \frac{\Gamma(b)}{\Gamma(a)}e^x x^{a-b}\left[\sum_{n=0}^{N}\frac{(b-a)_n(1-a)_n}{n!}x^{-n} + \varepsilon\right], \quad x > 0,$$

$$\Phi(a, b; x) = \frac{\Gamma(b)}{\Gamma(b-a)}(-x)^{-a}\left[\sum_{n=0}^{N}\frac{(a)_n(a-b+1)_n}{n!}(-x)^{-n} + \varepsilon\right], \quad x < 0,$$

$$\Psi(a, b; x) = x^{-a}\left[\sum_{n=0}^{N}(-1)^n\frac{(a)_n(a-b+1)_n}{n!}x^{-n} + \varepsilon\right], \quad -\infty < x < \infty,$$

where $\varepsilon = O(x^{-N-1})$.

A.10. Hypergeometric Functions

A.10.1. Definition and Some Formulas

The hypergeometric function $F(\alpha, \beta, \gamma; x)$ is a solution of the Gaussian hypergeometric equation

$$x(x-1)y''_{xx} + [(\alpha + \beta + 1)x - \gamma]y'_x + \alpha\beta y = 0.$$

For $\gamma \neq 0, -1, -2, -3, \ldots$, the function $F(\alpha, \beta, \gamma; x)$ can be expressed in terms of the hypergeometric series:

$$F(\alpha, \beta, \gamma; x) = 1 + \sum_{k=1}^{\infty} \frac{(\alpha)_k (\beta)_k}{(\gamma)_k} \frac{x^k}{k!}, \quad (\alpha)_k = \alpha(\alpha+1)\ldots(\alpha+k-1),$$

which certainly converges for $|x| < 1$.

Table A2 shows some special cases where F can be expressed in term of elementary functions.

A.10.2. Basic Properties and Integral Representations

A.10.2-1. Some properties.

The function F possesses the following properties:

$$F(\alpha, \beta, \gamma; x) = F(\beta, \alpha, \gamma; x),$$
$$F(\alpha, \beta, \gamma; x) = (1-x)^{\gamma-\alpha-\beta} F(\gamma-\alpha, \gamma-\beta, \gamma; x),$$
$$F(\alpha, \beta, \gamma; x) = (1-x)^{-\alpha} F\left(\alpha, \gamma-\beta, \gamma; \frac{x}{x-1}\right),$$
$$\frac{d^n}{dx^n} F(\alpha, \beta, \gamma; x) = \frac{(\alpha)_n (\beta)_n}{(\gamma)_n} F(\alpha+n, \beta+n, \gamma+n; x).$$

If γ is not an integer, then the general solution of the hypergeometric equation can be written in the form

$$y = C_1 F(\alpha, \beta, \gamma; x) + C_2 x^{1-\gamma} F(\alpha-\gamma+1, \beta-\gamma+1, 2-\gamma; x).$$

A.10.2-2. Integral representations.

For $\gamma > \beta > 0$, the hypergeometric function can be expressed in terms of a definite integral:

$$F(\alpha, \beta, \gamma; x) = \frac{\Gamma(\gamma)}{\Gamma(\beta)\Gamma(\gamma-\beta)} \int_0^1 t^{\beta-1}(1-t)^{\gamma-\beta-1}(1-tx)^{-\alpha} dt,$$

where $\Gamma(\beta)$ is the gamma function.

See M. Abramowitz and I. Stegun (1964) and H. Bateman and A. Erdélyi (1953, Vol. 1) for more detailed information about hypergeometric functions.

A.11. Whittaker Functions

The Whittaker functions $M_{k,\mu}(x)$ and $W_{k,\mu}(x)$ are linearly independent solutions of the Whittaker equation:

$$y''_{xx} + \left[-\tfrac{1}{4} + \tfrac{1}{2}k + \left(\tfrac{1}{4} - \mu^2\right)x^{-2}\right]y = 0.$$

The Whittaker functions are expressed in terms of degenerate hypergeometric functions as

$$M_{k,\mu}(x) = x^{\mu+1/2} e^{-x/2} \Phi\left(\tfrac{1}{2} + \mu - k, 1 + 2\mu, x\right),$$
$$W_{k,\mu}(x) = x^{\mu+1/2} e^{-x/2} \Psi\left(\tfrac{1}{2} + \mu - k, 1 + 2\mu, x\right).$$

TABLE A2
Some special cases where the hypergeometric function $F(\alpha, \beta, \gamma; z)$ can be expressed in terms of elementary functions

α	β	γ	z	F
$-n$	β	γ	x	$\sum_{k=0}^{n} \frac{(-n)_k (\beta)_k}{(\gamma)_k} \frac{x^k}{k!}$, where $n = 1, 2, \ldots$
$-n$	β	$-n-m$	x	$\sum_{k=0}^{n} \frac{(-n)_k (\beta)_k}{(-n-m)_k} \frac{x^k}{k!}$, where $n = 1, 2, \ldots$
α	β	β	x	$(1-x)^{-\alpha}$
α	$\alpha + \frac{1}{2}$	$\frac{1}{2}$	x^2	$\frac{1}{2}\left[(1+x)^{-2\alpha} + (1-x)^{-2\alpha}\right]$
α	$\alpha + \frac{1}{2}$	$\frac{3}{2}$	x^2	$\dfrac{(1+x)^{1-2\alpha} - (1-x)^{1-2\alpha}}{2x(1-2\alpha)}$
α	$-\alpha$	$\frac{1}{2}$	$-x^2$	$\frac{1}{2}\left[\left(\sqrt{1+x^2}+x\right)^{2\alpha} + \left(\sqrt{1+x^2}-x\right)^{2\alpha}\right]$
α	$1-\alpha$	$\frac{1}{2}$	$-x^2$	$\dfrac{\left(\sqrt{1+x^2}+x\right)^{2\alpha-1} + \left(\sqrt{1+x^2}-x\right)^{2\alpha-1}}{2\sqrt{1+x^2}}$
α	$\alpha - \frac{1}{2}$	$2\alpha - 1$	x	$2^{2\alpha-2}\left(1+\sqrt{1-x}\right)^{2-2\alpha}$
α	$1-\alpha$	$\frac{3}{2}$	$\sin^2 x$	$\dfrac{\sin[(2\alpha-1)x]}{(\alpha-1)\sin(2x)}$
α	$2-\alpha$	$\frac{3}{2}$	$\sin^2 x$	$\dfrac{\sin[(2\alpha-2)x]}{(\alpha-1)\sin(2x)}$
α	$1-\alpha$	$\frac{1}{2}$	$\sin^2 x$	$\dfrac{\cos[(2\alpha-1)x]}{\cos x}$
α	$\alpha+1$	$\frac{1}{2}\alpha$	x	$(1+x)(1-x)^{-\alpha-1}$
α	$\alpha+\frac{1}{2}$	$2\alpha+1$	x	$\left(\dfrac{1+\sqrt{1-x}}{2}\right)^{-2\alpha}$
α	$\alpha+\frac{1}{2}$	2α	x	$\dfrac{1}{\sqrt{1-x}}\left(\dfrac{1+\sqrt{1-x}}{2}\right)^{1-2\alpha}$
$\frac{1}{2}$	$\frac{1}{2}$	$\frac{3}{2}$	x^2	$\dfrac{1}{x}\arcsin x$
$\frac{1}{2}$	1	$\frac{3}{2}$	$-x^2$	$\dfrac{1}{x}\arctan x$
1	1	2	$-x$	$\dfrac{1}{x}\ln(x+1)$
$\frac{1}{2}$	1	$\frac{3}{2}$	x^2	$\dfrac{1}{2x}\ln\dfrac{1+x}{1-x}$
$n+1$	$n+m+1$	$n+m+l+2$	x	$\dfrac{(-1)^m (n+m+l+1)!}{n!\, l!\, (n+m)!\, (m+l)!} \dfrac{d^{n+m}}{dx^{n+m}}\left\{(1-x)^{m+l}\dfrac{d^l F}{dx^l}\right\}$, $F = -\dfrac{\ln(1-x)}{x}$, $n, m, l = 0, 1, 2, \ldots$

A.12. Legendre Polynomials and Legendre Functions

A.12.1. Definitions. Basic Formulas

The Legendre polynomials $P_n = P_n(x)$ and the Legendre functions $Q_n(x)$ are solutions of the equation

$$(1-x^2)y''_{xx} - 2xy'_x + n(n+1)y = 0.$$

The Legendre polynomials $P_n(x)$ and the Legendre functions $Q_n(x)$ are defined by the formulas

$$P_n(x) = \frac{1}{n!\,2^n} \frac{d^n}{dx^n}(x^2-1)^n, \quad Q_n(x) = \frac{1}{2}P_n(x)\ln\frac{1+x}{1-x} - \sum_{m=1}^{n}\frac{1}{m}P_{m-1}(x)P_{n-m}(x).$$

The polynomials $P_n = P_n(x)$ can be calculated recursively using the relations

$$P_0(x) = 1, \quad P_1(x) = x, \quad P_2(x) = \frac{1}{2}(3x^2-1), \quad \ldots, \quad P_{n+1}(x) = \frac{2n+1}{n+1}xP_n(x) - \frac{n}{n+1}P_{n-1}(x).$$

The first three functions $Q_n = Q_n(x)$ have the form

$$Q_0(x) = \frac{1}{2}\ln\frac{1+x}{1-x}, \quad Q_1(x) = \frac{x}{2}\ln\frac{1+x}{1-x} - 1, \quad Q_2(x) = \frac{3x^2-1}{4}\ln\frac{1+x}{1-x} - \frac{3}{2}x.$$

The polynomials $P_n(x)$ have the implicit representation

$$P_n(x) = 2^{-n}\sum_{m=0}^{[n/2]}(-1)^m C_n^m C_{2n-2m}^n x^{n-2m},$$

where $[A]$ is the integer part of a number A.

A.12.2. Zeros of Legendre Polynomials and the Generating Function

All zeros of $P_n(x)$ are real and lie on the interval $-1 < x < +1$; the functions $P_n(x)$ form an orthogonal system on the interval $-1 \le x \le +1$, with

$$\int_{-1}^{+1} P_n(x)P_m(x)\,dx = \begin{cases} 0 & \text{if } n \ne m, \\ \dfrac{2}{2n+1} & \text{if } n = m. \end{cases}$$

The generating function is

$$\frac{1}{\sqrt{1-2sx+s^2}} = \sum_{n=0}^{\infty} P_n(x)s^n \quad (|s|<1).$$

A.12.3. Associated Legendre Functions

The associated Legendre functions $P_n^m(x)$ of order m are defined by the formulas

$$P_n^m(x) = (1-x^2)^{m/2}\frac{d^m}{dx^m}P_n(x), \quad n = 1, 2, 3, \ldots, \quad m = 0, 1, 2, \ldots$$

It is assumed by definition that $P_n^0(x) = P_n(x)$.

The functions $P_n^m(x)$ form an orthogonal system on the interval $-1 \le x \le +1$, with

$$\int_{-1}^{+1} P_n^m(x)P_k^m(x)\,dx = \begin{cases} 0 & \text{if } n \ne k, \\ \dfrac{2}{2n+1}\dfrac{(n+m)!}{(n-m)!} & \text{if } n = k. \end{cases}$$

The functions $P_n^m(x)$ (with $m \ne 0$) are orthogonal on the interval $-1 \le x \le +1$ with weight $(1-x^2)^{-1}$, that is,

$$\int_{-1}^{+1} \frac{P_n^m(x)P_k^m(x)}{(1-x^2)}\,dx = \begin{cases} 0 & \text{if } n \ne k, \\ \dfrac{(n+m)!}{m(n-m)!} & \text{if } n = k. \end{cases}$$

A.13. Parabolic Cylinder Functions

A.13.1. Definitions. Basic Formulas

The Weber parabolic cylinder function $D_\nu(z)$ is a solution of the linear differential equation:

$$y''_{zz} + \left(-\tfrac{1}{4}z^2 + \nu + \tfrac{1}{2}\right)y = 0,$$

where the parameter ν and the variable z can assume arbitrary real or complex values. Another linearly independent solution of this equation is the function $D_{-\nu-1}(iz)$; if ν is noninteger, then $D_\nu(-z)$ can also be taken as a linearly independent solution.

The parabolic cylinder functions can be expressed in terms of degenerate hypergeometric functions as

$$D_\nu(z) = 2^{1/2} \exp\left(-\tfrac{1}{4}z^2\right) \left[\frac{\Gamma\left(\tfrac{1}{2}\right)}{\Gamma\left(\tfrac{1}{2} - \tfrac{\nu}{2}\right)} \Phi\left(-\tfrac{\nu}{2}, \tfrac{1}{2}, \tfrac{1}{2}z^2\right) + 2^{-1/2} \frac{\Gamma\left(-\tfrac{1}{2}\right)}{\Gamma\left(-\tfrac{\nu}{2}\right)} z \Phi\left(\tfrac{1}{2} - \tfrac{\nu}{2}, \tfrac{3}{2}, \tfrac{1}{2}z^2\right) \right].$$

For nonnegative integer $\nu = n$, we have

$$D_n(z) = 2^{-n/2} \exp\left(-\tfrac{1}{4}z^2\right) H_n\left(2^{-1/2}z\right), \quad n = 0, 1, 2, \ldots;$$

$$H_n(z) = (-1)^n \exp\left(z^2\right) \frac{d^n}{dz^n} \exp\left(-z^2\right),$$

where $H_n(z)$ is the Hermitean polynomial of order n.

A.13.2. Integral Representations and Asymptotic Expansions

Integral representations:

$$D_\nu(z) = \sqrt{2/\pi} \, \exp\left(\tfrac{1}{4}z^2\right) \int_0^\infty t^\nu \exp\left(-\tfrac{1}{2}t^2\right) \cos\left(zt - \tfrac{1}{2}\pi\nu\right) dt \quad \text{for} \quad \operatorname{Re}\nu > -1,$$

$$D_\nu(z) = \frac{1}{\Gamma(-\nu)} \exp\left(-\tfrac{1}{4}z^2\right) \int_0^\infty t^{-\nu-1} \exp\left(-zt - \tfrac{1}{2}t^2\right) dt \quad \text{for} \quad \operatorname{Re}\nu < 0.$$

Asymptotic expansion as $|z| \to \infty$:

$$D_\nu(z) \sim z^\nu \exp\left(-\tfrac{1}{4}z^2\right) \left[\sum_{n=0}^N \frac{(-2)^n \left(-\tfrac{\nu}{2}\right)_n \left(\tfrac{1}{2} - \tfrac{\nu}{2}\right)_n}{n!} \frac{1}{z^{2n}} + O\left(|z|^{-2N-2}\right) \right] \quad \text{for} \quad |\arg z| < \frac{3\pi}{4},$$

where $(a)_0 = 1$, $(a)_n = a(a+1)\ldots(a+n-1)$ for $n = 1, 2, 3, \ldots$

A.14. Mathieu Functions

A.14.1. Definitions and Basic Formulas

A.14.1-1. Mathieu equation and Mathieu functions.

The Mathieu functions $ce_n(x,q)$ and $se_n(x,q)$ are periodical solutions of the Mathieu equation

$$y''_{xx} + (a - 2q\cos 2x)y = 0.$$

Such solutions exist for definite values of parameters a and q (those values of a are referred to as eigenvalues). The Mathieu functions are listed in Table A3.

TABLE A3

The Mathieu functions $\mathrm{ce}_n = \mathrm{ce}_n(x,q)$ and $\mathrm{se}_n = \mathrm{se}_n(x,q)$ (for odd n, functions ce_n and se_n are 2π-periodical, and for even n, they are π-periodical); definite eigenvalues $a = a_n(q)$ and $a = b_n(q)$ correspond to each value of parameter q.

Mathieu functions	Recurrence relations for coefficients	Normalization conditions
$\mathrm{ce}_{2n} = \sum_{m=0}^{\infty} A_{2m}^{2n} \cos 2mx$	$qA_2^{2n} = a_{2n} A_0^{2n};$ $qA_4^{2n} = (a_{2n}-4)A_2^{2n} - 2qA_0^{2n};$ $qA_{2m+2}^{2n} = (a_{2n}-4m^2)A_{2m}^{2n}$ $\quad - qA_{2m-2}^{2n}, \quad m \geq 2$	$(A_0^{2n})^2 + \sum_{m=0}^{\infty}(A_{2m}^{2n})^2$ $= \begin{cases} 2 & \text{if } n=0 \\ 1 & \text{if } n \geq 1 \end{cases}$
$\mathrm{ce}_{2n+1} = \sum_{m=0}^{\infty} A_{2m+1}^{2n+1} \cos(2m+1)x$	$qA_3^{2n+1} = (a_{2n+1}-1-q)A_1^{2n+1};$ $qA_{2m+3}^{2n+1} = [a_{2n+1}-(2m+1)^2]A_{2m+1}^{2n+1}$ $\quad - qA_{2m-1}^{2n+1}, \quad m \geq 1$	$\sum_{m=0}^{\infty}(A_{2m+1}^{2n+1})^2 = 1$
$\mathrm{se}_{2n} = \sum_{m=0}^{\infty} B_{2m}^{2n} \sin 2mx,$ $\mathrm{se}_0 = 0$	$qB_4^{2n} = (b_{2n}-4)B_2^{2n};$ $qB_{2m+2}^{2n} = (b_{2n}-4m^2)B_{2m}^{2n}$ $\quad - qB_{2m-2}^{2n}, \quad m \geq 2$	$\sum_{m=0}^{\infty}(B_{2m}^{2n})^2 = 1$
$\mathrm{se}_{2n+1} = \sum_{m=0}^{\infty} B_{2m+1}^{2n+1} \sin(2m+1)x$	$qB_3^{2n+1} = (b_{2n+1}-1-q)B_1^{2n+1};$ $qB_{2m+3}^{2n+1} = [b_{2n+1}-(2m+1)^2]B_{2m+1}^{2n+1}$ $\quad - qB_{2m-1}^{2n+1}, \quad m \geq 1$	$\sum_{m=0}^{\infty}(B_{2m+1}^{2n+1})^2 = 1$

A.14.1-2. Properties of the Mathieu functions.

The Mathieu functions possess the following properties:

$$\mathrm{ce}_{2n}(x, -q) = (-1)^n \mathrm{ce}_{2n}\left(\frac{\pi}{2}-x, q\right), \qquad \mathrm{ce}_{2n+1}(x, -q) = (-1)^n \mathrm{se}_{2n+1}\left(\frac{\pi}{2}-x, q\right),$$

$$\mathrm{se}_{2n}(x, -q) = (-1)^{n-1} \mathrm{se}_{2n}\left(\frac{\pi}{2}-x, q\right), \qquad \mathrm{se}_{2n+1}(x, -q) = (-1)^n \mathrm{ce}_{2n+1}\left(\frac{\pi}{2}-x, q\right).$$

Selecting sufficiently large number m and omitting the term with the maximum number in the recurrence relations (indicated in Table A3), we can obtain approximate relations for eigenvalues a_n (or b_n) with respect to parameter q. Then, equating the determinant of the corresponding homogeneous linear system of equations for coefficients A_m^n (or B_m^n) to zero, we obtain an algebraic equation for finding $a_n(q)$ (or $b_n(q)$).

For fixed real $q \neq 0$, eigenvalues a_n and b_n are all real and different, while

$$\text{if} \quad q > 0 \quad \text{then} \quad a_0 < b_1 < a_1 < b_2 < a_2 < \cdots$$
$$\text{if} \quad q < 0 \quad \text{then} \quad a_0 < a_1 < b_1 < b_2 < a_2 < a_3 < b_3 < b_4 < \cdots$$

The eigenvalues possess the properties

$$a_{2n}(-q) = a_{2n}(q), \quad b_{2n}(-q) = b_{2n}(q), \quad a_{2n+1}(-q) = b_{2n+1}(q).$$

Tables of the eigenvalues $a_n = a_n(q)$ and $b_n = b_n(q)$ can be found in Abramowitz and Stegun (1964, Chapter 20).

The solution of the Mathieu equation corresponding to eigenvalue a_n (or b_n) has n zeros on the interval $0 \leq x < \pi$ (q is a real number).

A.14.1-3. Asymptotic expansions at $q \to 0$ and $q \to \infty$.

Listed below are two leading terms of asymptotic expansions of the Mathieu functions $\mathrm{ce}_n(x, q)$ and $\mathrm{se}_n(x, q)$, as well as of the corresponding eigenvalues $a_n(q)$ and $b_n(q)$, as $q \to 0$:

$$\mathrm{ce}_0(x,q) = \frac{1}{\sqrt{2}}\left(1 - \frac{q}{2}\cos 2x\right), \quad a_0(q) = -\frac{q^2}{2} + \frac{7q^4}{128};$$

$$\mathrm{ce}_1(x,q) = \cos x - \frac{q}{8}\cos 3x, \quad a_1(q) = 1 + q;$$

$$\mathrm{ce}_2(x,q) = \cos 2x + \frac{q}{4}\left(1 - \frac{\cos 4x}{3}\right), \quad a_2(q) = 4 + \frac{5q^2}{12};$$

$$\mathrm{ce}_n(x,q) = \cos nx + \frac{q}{4}\left[\frac{\cos(n+2)x}{n+1} - \frac{\cos(n-2)x}{n-1}\right], \quad a_n(q) = n^2 + \frac{q^2}{2(n^2-1)} \quad (n \geq 3);$$

$$\mathrm{se}_1(x,q) = \sin x - \frac{q}{8}\sin 3x, \quad b_1(q) = 1 - q;$$

$$\mathrm{se}_2(x,q) = \sin 2x - q\frac{\sin 4x}{12}, \quad b_2(q) = 4 - \frac{q^2}{12};$$

$$\mathrm{se}_n(x,q) = \sin nx - \frac{q}{4}\left[\frac{\sin(n+2)x}{n+1} - \frac{\sin(n-2)x}{n-1}\right], \quad b_n(q) = n^2 + \frac{q^2}{2(n^2-1)} \quad (n \geq 3).$$

Asymptotic results as $q \to \infty$ ($-\pi/2 < x < \pi/2$):

$$a_n(q) \approx -2q + 2(2n+1)\sqrt{q} + \tfrac{1}{4}(2n^2 + 2n + 1),$$

$$b_{n+1}(q) \approx -2q + 2(2n+1)\sqrt{q} + \tfrac{1}{4}(2n^2 + 2n + 1),$$

$$\mathrm{ce}_n(x,q) \approx \lambda_n q^{-1/4}\cos^{-n-1}x\left[\cos^{2n+1}\xi \exp(2\sqrt{q}\sin x) + \sin^{2n+1}\xi \exp(-2\sqrt{q}\sin x)\right], \quad \xi = \tfrac{1}{2}x + \tfrac{\pi}{4},$$

$$\mathrm{se}_{n+1}(x,q) \approx \mu_{n+1} q^{-1/4}\cos^{-n-1}x\left[\cos^{2n+1}\xi \exp(2\sqrt{q}\sin x) - \sin^{2n+1}\xi \exp(-2\sqrt{q}\sin x)\right],$$

where the λ_n and μ_n some constants independent of the parameter q.

⊙ *References*: H. Bateman and A. Erdélyi (1955, Vol. 3), M. Abramowitz and I. Stegun (1964).

A.15. Modified Mathieu Functions

The modified Mathieu functions $\mathrm{Ce}_n(x, q)$ and $\mathrm{Se}_n(x, q)$ are solutions of the modified Mathieu equation

$$y''_{xx} - (a - 2q\cosh 2x)y = 0,$$

with $a = a_n(q)$ and $a = b_n(q)$ being the eigenvalues of the Mathieu equation (see Section A.12).

The modified Mathieu functions are defined as

$$\mathrm{Ce}_{2n+p}(x, q) = \mathrm{ce}_{2n+p}(ix, q) = \sum_{k=0}^{\infty} A_{2k+p}^{2n+p} \cosh[(2k+p)x],$$

$$\mathrm{Se}_{2n+p}(x, q) = -i\,\mathrm{se}_{2n+p}(ix, q) = \sum_{k=0}^{\infty} B_{2k+p}^{2n+p} \sinh[(2k+p)x],$$

where p may be equal to 0 and 1, and coefficients A_{2k+p}^{2n+p} and B_{2k+p}^{2n+p} are indicated in Subsection A.12.

⊙ *References*: H. Bateman and A. Erdélyi (1955, Vol. 3), M. Abramowitz and I. Stegun (1964).

A.16. Orthogonal Polynomials

All zeros of each of the orthogonal polynomials $\mathcal{P}_n(x)$ considered in this section are real and simple. The zeros of the polynomials $\mathcal{P}_n(x)$ and $\mathcal{P}_{n+1}(x)$ are alternating.

For Legendre polynomials see Section A.12.

A.16.1. Laguerre Polynomials and Generalized Laguerre Polynomials

A.16.1-1. Laguerre polynomials.

The Laguerre polynomials $L_n = L_n(x)$ satisfy the equation

$$xy''_{xx} + (1-x)y'_x + ny = 0$$

and are defined by the formulas

$$L_n(x) = \frac{1}{n!}e^x \frac{d^n}{dx^n}\left(x^n e^{-x}\right) = \frac{(-1)^n}{n!}\left[x^n - n^2 x^{n-1} + \frac{n^2(n-1)^2}{2!}x^{n-2} + \cdots\right].$$

The first four polynomials have the form

$$L_0 = 1, \quad L_1 = -x+1, \quad L_2 = \tfrac{1}{2}(x^2 - 4x + 2), \quad L_3 = \tfrac{1}{6}(-x^3 + 9x^2 - 18x + 6).$$

To calculate $L_n(x)$ for $n \geq 2$, one can use the recurrent formulas

$$L_{n+1}(x) = \frac{1}{n+1}\left[(2n+1-x)L_n(x) - nL_{n-1}(x)\right].$$

The functions $L_n(x)$ form an orthonormal system on the interval $0 < x < \infty$ with weight e^{-x}:

$$\int_0^\infty e^{-x} L_n(x) L_m(x)\, dx = \begin{cases} 0 & \text{if } n \neq m, \\ 1 & \text{if } n = m. \end{cases}$$

The generating function is

$$\frac{1}{1-s}\exp\left(-\frac{sx}{1-s}\right) = \sum_{n=0}^\infty L_n(x)s^n, \qquad |s| < 1.$$

A.16.1-2. Generalized Laguerre polynomials.

The generalized Laguerre polynomials $L_n^\alpha = L_n^\alpha(x)$ ($\alpha > -1$) satisfy the equation

$$xy''_{xx} + (\alpha + 1 - x)y'_x + ny = 0$$

and are defined by the formulas

$$L_n^\alpha(x) = \frac{1}{n!}x^{-\alpha}e^x \frac{d^n}{dx^n}\left(x^{n+\alpha}e^{-x}\right) = \sum_{m=0}^n C_{n+\alpha}^{n-m}\frac{(-x)^m}{m!}.$$

The first two polynomials have the form

$$L_0^\alpha = 1, \quad L_1^\alpha = \alpha + 1 - x.$$

To calculate $L_n^\alpha(x)$ for $n \geq 2$, one can use the recurrent formulas

$$L_{n+1}^\alpha(x) = \frac{1}{n+1}\left[(2n+\alpha+1-x)L_n^\alpha(x) - (n+\alpha)L_{n-1}^\alpha(x)\right].$$

The functions $L_n^\alpha(x)$ form an orthogonal system on the interval $0 < x < \infty$ with weight $x^\alpha e^{-x}$:

$$\int_0^\infty x^\alpha e^{-x} L_n^\alpha(x) L_m^\alpha(x)\, dx = \begin{cases} 0 & \text{if } n \neq m, \\ \frac{\Gamma(\alpha+n+1)}{n!} & \text{if } n = m. \end{cases}$$

The generating function is

$$(1-s)^{-\alpha-1}\exp\left(-\frac{sx}{1-s}\right) = \sum_{n=0}^\infty L_n^\alpha(x)s^n, \qquad |s| < 1.$$

A.16.2. Chebyshev Polynomials and Functions

A.16.2-1. Chebyshev polynomials.

The Chebyshev polynomials $T_n = T_n(x)$ satisfy the equation

$$(1-x^2)y''_{xx} - xy'_x + n^2 y = 0 \tag{1}$$

and are defined by the formulas

$$T_n(x) = \cos(n \arccos x) = \frac{(-2)^n n!}{(2n)!} \sqrt{1-x^2} \frac{d^n}{dx^n}\left[(1-x^2)^{n-\frac{1}{2}}\right]$$

$$= \frac{n}{2} \sum_{m=0}^{[n/2]} (-1)^m \frac{(n-m-1)!}{m!(n-2m)!} (2x)^{n-2m} \quad (n = 0, 1, 2, \ldots),$$

where $[A]$ stands for the integer part of a number A.

The first four polynomials are

$$T_0 = 1, \quad T_1 = x, \quad T_2 = 2x^2 - 1, \quad T_3 = 4x^3 - 3x.$$

The recurrent formulas:

$$T_{n+1}(x) = 2xT_n(x) - T_{n-1}(x), \quad n \geq 2.$$

The functions $T_n(x)$ form an orthogonal system on the interval $-1 < x < +1$, with

$$\int_{-1}^{+1} \frac{T_n(x)T_m(x)}{\sqrt{1-x^2}} \, dx = \begin{cases} 0 & \text{if } n \neq m, \\ \frac{1}{2}\pi & \text{if } n = m \neq 0, \\ \pi & \text{if } n = m = 0. \end{cases}$$

A.16.2-2. Chebyshev functions of the second kind.

The Chebyshev functions of the second kind,

$$U_0(x) = \arcsin x,$$

$$U_n(x) = \sin(n \arcsin x) = \frac{\sqrt{1-x^2}}{n} \frac{dT_n(x)}{dx} \quad (n = 1, 2, \ldots),$$

just as the Chebyshev polynomials, also satisfy the differential equation (1).

The generating function is

$$\frac{1-sx}{1-2sx+s^2} = \sum_{n=0}^{\infty} T_n(x)s^n \quad (|s| < 1).$$

A.16.3. Hermite Polynomial

The Hermite polynomial $H_n = H_n(x)$ satisfies the equation

$$y''_{xx} - 2xy'_x + 2ny = 0$$

and is defined by the formulas

$$H_n(x) = (-1)^n \exp(x^2) \frac{d^n}{dx^n} \exp(-x^2).$$

The first four polynomials are

$$H_0 = 1, \quad H_1 = x, \quad H_2 = 4x^2 - 2, \quad H_3 = 8x^3 - 12x.$$

The recurrent formulas:

$$H_{n+1}(x) = 2xH_n(x) - 2nH_{n-1}(x), \qquad n \geq 2.$$

The functions $H_n(x)$ form an orthogonal system on the interval $-\infty < x < \infty$ with weight e^{-x^2}:

$$\int_{-\infty}^{\infty} \exp(-x^2) H_n(x) H_m(x)\, dx = \begin{cases} 0 & \text{if } n \neq m, \\ \sqrt{\pi}\, 2^n n! & \text{if } n = m. \end{cases}$$

The Hermite functions $\psi_n(x)$ are introduced by the formula $\psi_n(x) = \exp\left(-\tfrac{1}{2}x^2\right) H_n(x)$, where $n = 0, 1, 2, \ldots$

The generating function:

$$\exp(-s^2 + 2sx) = \sum_{n=0}^{\infty} H_n(x) \frac{s^n}{n!}.$$

A.16.4. Jacobi Polynomials

The Jacobi polynomials $P_n^{\alpha,\beta} = P_n^{\alpha,\beta}(x)$ satisfy the equation

$$(1-x^2) y''_{xx} + \left[\beta - \alpha - (\alpha + \beta + 2)x\right] y'_x + n(n + \alpha + \beta + 1) y = 0$$

and are defined by the formulas

$$P_n^{\alpha,\beta} = \frac{(-1)^n}{2^n n!} (1-x)^{-\alpha} (1+x)^{-\beta} \frac{d^n}{dx^n} \left[(1-x)^{\alpha+n} (1+x)^{\beta+n}\right] = 2^{-n} \sum_{m=0}^{n} C_{n+\alpha}^m C_{n+\beta}^{n-m} (x-1)^{n-m} (x+1)^m,$$

where C_b^a are binomial coefficients.

⊙ References for Supplement: H. Bateman and A. Erdélyi (1953, 1955), M. Abramowitz and I. A. Stegun (1964).

Supplement B

Methods of Generalized and Functional Separation of Variables in Nonlinear Equations of Mathematical Physics*

B.1. Introduction

B.1.1. Preliminary Remarks

Separation of variables is the most common approach to solve linear equations of mathematical physics. This approach involves searching for exact solutions in the form of the product of functions depending on different arguments (see Section 0.4).

As far as nonlinear equations with two independent variables x, y and a dependent variable w are concerned, some of these equations also have solutions with the form

$$w(x,y) = \varphi(x)\psi(y) \quad \text{or} \quad w(x,y) = \varphi(x) + \psi(y)$$

that are called *multiplicatively* and *additively separable*, respectively. We call such solutions *ordinary separable solutions*. In particular, integrating a few classes of first-order nonlinear partial differential equations is based on searching for additively separable solutions [e.g., see Appell (1953), Kamke (1965), Markeev (1990), Zwillinger (1998), Polyanin, Zaitsev, and Moussiaux (2001)].

Over the last decade, more sophisticated, *generalized* and *functional separable solutions* have been obtained for a number of second-order nonlinear equations of mathematical physics. For example, Galaktionov and Posashkov (1989) and Galaktionov, Posashkov, and Svirshchevskii (1995) obtained generalized separable solutions with the forms $w(x,t) = \varphi(t)\psi(x) + \chi(t)$ and $w(x,t) = \varphi(t)\psi(x) + \chi(x)$ for some classes of parabolic and hyperbolic equations with quadratic nonlinearities. In Galaktionov and Posashkov (1994), Galaktionov (1995), and Svirshchevskii (1995), more complicated generalized separable solutions are presented. The results of Galaktionov and Posashkov (1994) and Galaktionov (1995) are based on finding finite-dimensional subspaces that are invariant under appropriate nonlinear differential operators (in practice, the authors had to find a system of coordinate functions in one of the variables by the method of undetermined coefficients).

In Grundland and Infeld (1992), Miller and Rubel (1993), Zhdanov (1994), and Andreev, Kaptsov, Pukhnachev, and Rodionov (1994), all nonlinear heat (diffusion) and wave equations of the form $\partial_{xx}w \pm \partial_{yy}w = f(w)$ which admit functional separable solutions having the form $w(x,y) = F(z)$, where $z = \varphi(x) + \psi(y)$, are described. Doyle and Vassiliou (1998) indicated all one-dimensional nonstationary heat equations $\partial_t w = \partial_x[f(w)\partial_x w]$ which admit solutions of the form $w(x,t) = F(z)$, $z = \varphi(x) + \psi(t)$. In Zaitsev and Polyanin (1996), Polyanin and Zhurov (1998), Polyanin, Vyazmin, Zhurov, and Kazenin (1998), and Polyanin, Zhurov, and Vyazmin (2000), many nonlinear mathematical physics equations of various types that admit generalized and functional separable solutions are described (special attention was paid to equations of general form which depend on arbitrary functions).

---

* Sections B.1–B.5 were written with A. I. Zhurov.

Functional differential equations that involve unknown functions (and their derivatives) with different arguments arise when searching for ordinary, generalized, and functional separable solutions. The current supplement presents direct methods for and examples of constructing such solutions and reviews application of these methods to solving various classes of the second-, third-, fourth-, and higher-order partial differential equations (in total, about 150 nonlinear equations with solutions are described). Special attention is paid to equations of heat and mass transfer theory, wave theory, and hydrodynamics, as well as mathematical physics equations of general form that involve arbitrary functions.

It should be noted that often exact generalized and functional separable solutions cannot be obtained by group theoretic methods or other well-known methods.

B.1.2. Simple Cases of Variable Separation in Nonlinear Equations

In isolated cases, the separation of variables in nonlinear equations is carried out following the same technique as in linear equations. Specifically, an exact solution is sought in the form of the product or sum of functions depending on different arguments. On substituting it into the equation and performing elementary algebraic manipulations, one obtains an equation with the two sides dependent on different variables (for equations with two variables). Then one concludes that the expressions on each side must be equal to the same constant quantity, called a separation constant. Below we consider specific examples.

Example 1. The heat equation with a power nonlinearity

$$\frac{\partial w}{\partial t} = a \frac{\partial}{\partial x}\left(w^k \frac{\partial w}{\partial x}\right) \tag{1}$$

has an exact solution in the product form

$$w = \varphi(x)\psi(t). \tag{2}$$

Substituting (2) into (1) yields

$$\varphi \psi'_t = a\psi^{k+1}(\varphi^k \varphi'_x)'_x.$$

Separating the variables by dividing both sides by $\varphi \psi^{k+1}$, we obtain

$$\frac{\psi'_t}{\psi^{k+1}} = \frac{a(\varphi^k \varphi'_x)'_x}{\varphi}.$$

The left-hand side depends of t alone and the right-hand side on x alone. This is possible only if

$$\frac{\psi'_t}{\psi^{k+1}} = C, \quad \frac{a(\varphi^k \varphi'_x)'_x}{\varphi} = C, \tag{3}$$

where C is an arbitrary constant (separation constant). On solving the ordinary differential equations (3), we obtain a solution of equation (1) with the form (2).

The procedure for constructing a separable solution (2) of the nonlinear equation (1) is identical to that used in solving linear equations [in particular, equation (1) with $k = 0$]. We refer to the cases of similar separation of variables as *simple separable* cases.

Example 2. The wave equation with an exponential nonlinearity

$$\frac{\partial^2 w}{\partial t^2} = a \frac{\partial}{\partial x}\left(e^{\lambda w} \frac{\partial w}{\partial x}\right) \tag{4}$$

has an additively separable solution

$$w = \varphi(x) + \psi(t). \tag{5}$$

On substituting (5) into (4) and dividing by $e^{\lambda \psi}$, we arrive at the equation

$$e^{-\lambda \psi} \psi''_{tt} = a(e^{\lambda \varphi} \varphi'_x)'_x,$$

whose left-hand side depends on t alone and the right-hand side on x alone. This is possible only if

$$e^{-\lambda \psi} \psi''_{tt} = C, \quad a(e^{\lambda \varphi} \varphi'_x)'_x = C, \tag{6}$$

where C is an arbitrary constant. Solving the ordinary differential equations (6) yields a solution of equation (4) with the form (5).

Example 3. The heat equation in an anisotropic medium with a logarithmic source

$$\frac{\partial}{\partial x}\left[f(x)\frac{\partial w}{\partial x}\right] + \frac{\partial}{\partial y}\left[g(y)\frac{\partial w}{\partial y}\right] = aw\ln w \tag{7}$$

has a multiplicatively separable solution

$$w = \varphi(x)\psi(y). \tag{8}$$

On substituting (8) into (7), dividing by $\varphi\psi$, and transposing individual terms of the resulting equation, we obtain

$$\frac{1}{\varphi}[f(x)\varphi'_x]'_x - a\ln\varphi = -\frac{1}{\psi}[g(y)\psi'_y]'_y + a\ln\psi.$$

The left-hand side of this equation depends only on x and the right-hand only on y. By equating both sides to a constant quantity, one obtains ordinary differential equations for $\varphi(x)$ and $\psi(y)$.

B.1.3. Examples of Nontrivial Variable Separation in Nonlinear Equations

Unlike linear equations, the variables in nonlinear equations often separate differently. We exemplify this below.

Example 4. Consider the equation with a cubic nonlinearity

$$\frac{\partial w}{\partial t} = f(t)\frac{\partial^2 w}{\partial x^2} + w\left(\frac{\partial w}{\partial x}\right)^2 - aw^3, \tag{9}$$

where $f(t)$ is an arbitrary function. We look for exact solutions in the product form. We substitute (2) into (9) and divide the resulting equation by $f(t)\varphi(x)\psi(t)$ to obtain

$$\frac{\psi'_t}{f\psi} = \frac{\varphi''_{xx}}{\varphi} + \frac{\psi^2}{f}[(\varphi'_x)^2 - a\varphi^2]. \tag{10}$$

In the general case, this expression cannot be represented as the sum of two functions depending on different arguments. This however does not mean that equation (9) has no solutions of the form (2).

$1°$. One can make sure by direct check that, for $a > 0$, the functional differential equation (10) has solutions

$$\varphi(x) = C\exp(\pm x\sqrt{a}), \quad \psi(t) = \exp\left[a\int f(t)\,dt\right], \tag{11}$$

where C is an arbitrary constant. Solution (11) for φ makes the expression in square brackets in (10) vanish, which allows separation of variables.

$2°$. There is a more general solution of the functional differential equation (10) for $a > 0$:

$$\varphi(x) = C_1\exp(x\sqrt{a}) + C_2\exp(-x\sqrt{a}),$$
$$\psi(t) = e^F\left(C_3 + 8aC_1C_2\int e^{2F}\,dt\right)^{-1/2}, \quad F = a\int f(t)\,dt,$$

where C_1, C_2, and C_3 are arbitrary constants. The function $\varphi = \varphi(x)$ makes each of the terms in (10) that depend on x constant, namely,

$$\varphi''_{xx}/\varphi = \text{const}, \quad (\varphi'_x)^2 - a\varphi^2 = \text{const}.$$

It is this circumstance that makes it possible to separate the variables.
Note that the function $\psi = \psi(t)$ satisfies the Bernoulli equation $\psi'_t = af(t)\psi - 4aC_1C_2\psi^3$.

$3°$. There is another solution of the functional differential equation (10) for $a < 0$:

$$\varphi(x) = C_1\sin(x\sqrt{-a}) + C_2\cos(x\sqrt{-a}),$$
$$\psi(t) = e^F\left[C_3 + 2a(C_1^2 + C_2^2)\int e^{2F}\,dt\right]^{-1/2}, \quad F = a\int f(t)\,dt,$$

where C_1, C_2, and C_3 are arbitrary constants. The function $\varphi = \varphi(x)$ makes both terms in (10) that depend on x constant. Note that the function $\psi = \psi(t)$ is determined by the Bernoulli equation $\psi'_t = af(t)\psi - a(C_1^2 + C_2^2)\psi^3$.

Example 5. Consider the third-order equation

$$\frac{\partial w}{\partial y}\frac{\partial^2 w}{\partial x^2} + a\frac{\partial w}{\partial x}\frac{\partial^2 w}{\partial y^2} = b\frac{\partial^3 w}{\partial x^3} + c\frac{\partial^3 w}{\partial y^3}. \tag{12}$$

We look for additive separable solutions

$$w = f(x) + g(y). \tag{13}$$

Substituting (13) into (12) yields

$$g'_y f''_{xx} + a f'_x g''_{yy} = b f'''_{xxx} + c g'''_{yyy}. \tag{14}$$

This expression cannot be rewritten as the sum of two functions depending on different arguments.

It is not difficult to see that the functional differential equation (14) is satisfied if

$$g'_y = C_1 \implies g(y) = C_1 y + C_2, \quad f(x) = C_3 \exp(C_1 x/b) + C_4 x \quad \text{(case 1)},$$
$$f'_x = C_1 \implies f(x) = C_1 x + C_2, \quad g(y) = C_3 \exp(aC_1 y/c) + C_4 y \quad \text{(case 2)},$$

where C_1, C_2, C_3, and C_4 are arbitrary constants. In both cases, two terms of the four in (14) vanish, which makes it possible to separate the variables.

In addition, equation (12) has a more complicated solution of the form (13):

$$w = C_1 e^{-a\lambda x} + \frac{c\lambda}{a} x + C_2 e^{\lambda y} - ab\lambda y + C_3,$$

where C_1, C_2, C_3, and λ are arbitrary constants. The mechanism of separation of variables is different here: both nonlinear terms on the left-hand side in (14) contain terms which cannot be rewritten in additive form but are equal in magnitude and have unlike signs. In adding, the two terms cancel, thus resulting in separation of variables:

$$\begin{aligned}
g'_y f''_{xx} &= C_1 C_2 a^2 \lambda^3 e^{\lambda y - a\lambda x} - C_1 b(a\lambda)^3 e^{-a\lambda x} \\
+\quad af'_x g''_{yy} &= -C_1 C_2 a^2 \lambda^3 e^{\lambda y - a\lambda x} + C_2 c\lambda^3 e^{\lambda y} \\
\hline
g'_y f''_{xx} + af'_x g''_{yy} &= -C_1 b(a\lambda)^3 e^{-a\lambda x} + C_2 c\lambda^3 e^{\lambda y} = b f'''_{xxx} + c g'''_{yyy}.
\end{aligned}$$

Example 6. Consider the second-order equation with a cubic nonlinearity

$$(1+w^2)\left(\frac{\partial^2 w}{\partial x^2} + \frac{\partial^2 w}{\partial y^2}\right) - 2w\left(\frac{\partial w}{\partial x}\right)^2 - 2w\left(\frac{\partial w}{\partial y}\right)^2 = aw(1-w^2). \tag{15}$$

We seek an exact solution of this equation in the product form

$$w = f(x)g(y). \tag{16}$$

Substituting (16) into (15) yields

$$(1+f^2g^2)(gf''_{xx} + fg''_{yy}) - 2fg[g^2(f'_x)^2 + f^2(g'_y)^2] = afg(1-f^2g^2). \tag{17}$$

This expression cannot be rewritten as the sum of two functions with different arguments. Nevertheless, equation (15) has solutions of the form (16). One can make sure by direct check that the functions $f = f(x)$ and $g = g(y)$ satisfying the nonlinear ordinary differential equations

$$\begin{aligned}
(f'_x)^2 &= Af^4 + Bf^2 + C, \\
(g'_y)^2 &= Cg^4 + (a-B)g^2 + A,
\end{aligned} \tag{18}$$

where A, B, and C are arbitrary constants, reduce equation (17) to an identity; to verify this, one should use the relations $f''_{xx} = 2Af^3 + Bf$ and $g''_{yy} = 2Cg^3 + (a-B)g$ that follow from (18).

Remark. By the variable change $u = 4\arctan w$ equation (15) can be reduced to a nonlinear heat equation with a sinusoidal source, $\Delta u = a\sin u$.

The examples considered above illustrate some specific features of separable solutions to nonlinear equations. Sections B.2 and B.3 outline fairly general methods for constructing similar and more complicated solutions to nonlinear partial differential equations.

B.2. Methods of Generalized Separation of Variables

B.2.1. Structure of Generalized Separable Solutions

B.2.1-1. General form of solutions. The classes of nonlinear equations considered.

To simplify the presentation, we confine ourselves to the case of mathematical physics equations with two independent variables x, y and a dependent variable w (one of the independent variables can play the role of time).

Linear separable equations of mathematical physics admit exact solutions in the form

$$w(x,y) = \varphi_1(x)\psi_1(y) + \varphi_2(x)\psi_2(y) + \cdots + \varphi_n(x)\psi_n(y), \tag{1}$$

where the $w_i = \varphi_i(x)\psi_i(y)$ are particular solutions; the functions $\varphi_i(x)$, as well as the functions $\psi_i(y)$, with different numbers i are not related to one another.

Also having exact solutions of the form (1) are many nonlinear partial differential equations with quadratic or power nonlinearities

$$f_1(x)g_1(y)\Pi_1[w] + f_2(x)g_2(y)\Pi_2[w] + \cdots + f_m(x)g_m(y)\Pi_m[w] = 0, \tag{2}$$

where the $\Pi_i[w]$ are differential forms that are the products of nonnegative integer powers of the function w and its partial derivatives $\partial_x w$, $\partial_y w$, $\partial_{xx} w$, $\partial_{xy} w$, $\partial_{yy} w$, $\partial_{xxx} w$, etc. We will refer to solutions (1) of nonlinear equations (2) as *generalized separable solutions*. Unlike linear equations, in nonlinear equations the functions $\varphi_i(x)$ with different subscripts i are usually related to one another [and to the functions $\psi_j(y)$]. Subsections B.1.2 and B.1.3 give examples of exact solutions (1) to nonlinear equations (2) for some simple cases with $n = 1$ or $n = 2$ (for $\psi_1 = \varphi_2 = 1$).

Remark. If the $f_s(x)$ and $g_s(y)$ in (2) are all constant, then one can seek solutions in the more general form

$$w(x,y) = \sum_{m=1}^{n} \varphi_m(\xi)\psi_m(\eta), \quad \xi = a_1 x + a_2 y, \quad \eta = b_1 x + b_2 y,$$

where a_1, a_2, b_1, and b_2 are constants. Some solutions of this sort are discussed in Subsections B.7.1 and B.8.1.

B.2.1-2. General form of functional differential equations.

In general, on substituting expression (1) into the differential equation (2), one arrives at a functional differential equation

$$\Phi_1(X)\Psi_1(Y) + \Phi_2(X)\Psi_2(Y) + \cdots + \Phi_k(X)\Psi_k(Y) = 0 \tag{3}$$

for the $\varphi_i(x)$ and $\psi_i(y)$. The functionals $\Phi_j(X)$ and $\Psi_j(Y)$ depend only on x and y, respectively,

$$\begin{aligned}\Phi_j(X) &\equiv \Phi_j\big(x, \varphi_1, \varphi_1', \varphi_1'', \ldots, \varphi_n, \varphi_n', \varphi_n''\big), \\ \Psi_j(Y) &\equiv \Psi_j\big(y, \psi_1, \psi_1', \psi_1'', \ldots, \psi_n, \psi_n', \psi_n''\big).\end{aligned} \tag{4}$$

Here, for simplicity, the formulas are written out for the case of a second-order equation (2); for higher-order equations, the right-hand sides of relations (4) will contain higher-order derivatives of φ_i and ψ_j.

B.2.2. Solution of Functional Differential Equations by Differentiation

B.2.2-1. Description of the method.

1°. Assume that $\Psi_k \not\equiv 0$. We divide equation (3) by Ψ_k and differentiate with respect to y. This results in a similar equation but with fewer terms:

$$\widetilde{\Phi}_1(X)\widetilde{\Psi}_1(Y) + \widetilde{\Phi}_2(X)\widetilde{\Psi}_2(Y) + \cdots + \widetilde{\Phi}_{k-1}(X)\widetilde{\Psi}_{k-1}(Y) = 0,$$
$$\widetilde{\Phi}_j(X) = \Phi_j(X), \quad \widetilde{\Psi}_j(Y) = [\Psi_j(Y)/\Psi_k(Y)]'_y.$$

We continue the above procedure until we obtain a separable two-term equation

$$\widehat{\Phi}_1(X)\widehat{\Psi}_1(Y) + \widehat{\Phi}_2(X)\widehat{\Psi}_2(Y) = 0. \tag{5}$$

Three cases must be considered.

Nondegenerate case: $|\widehat{\Phi}_1(X)| + |\widehat{\Phi}_2(X)| \not\equiv 0$ and $|\widehat{\Psi}_1(Y)| + |\widehat{\Psi}_2(Y)| \not\equiv 0$. Then equation (5) is equivalent to the ordinary differential equations

$$\widehat{\Phi}_1(X) + C\widehat{\Phi}_2(X) = 0, \quad C\widehat{\Psi}_1(Y) - \widehat{\Psi}_2(Y) = 0,$$

where C is an arbitrary constant. The equations $\widehat{\Phi}_2 = 0$ and $\widehat{\Psi}_1 = 0$ correspond to the limit case $C = \infty$.

Two degenerate cases:

$$\widehat{\Phi}_1(X) \equiv 0, \quad \widehat{\Phi}_2(X) \equiv 0 \implies \widehat{\Psi}_{1,2}(Y) \text{ are any};$$
$$\widehat{\Psi}_1(Y) \equiv 0, \quad \widehat{\Psi}_2(Y) \equiv 0 \implies \widehat{\Phi}_{1,2}(X) \text{ are any}.$$

2°. The solutions of the two-term equation (5) should be substituted into the original functional differential equation (3) to "remove" redundant constants of integration [these arise because equation (5) is obtained from (3) by differentiation].

3°. The case $\Psi_k \equiv 0$ should be treated separately (since we divided the equation by Ψ_k at the first stage). Likewise, we have to study all other cases where the functionals by which the intermediate functional differential equations were divided vanish.

Remark 1. The functional differential equation (3) can happen to have no solutions.

Remark 2. At each subsequent stage, the number of terms in the functional differential equation can be reduced by differentiation with respect to either y or x. For example, we can assume at the first stage that $\Phi_k \not\equiv 0$. On dividing equation (3) by Φ_k and differentiating with respect to x, we again obtain a similar equation that has fewer terms.

B.2.2-2. Examples of constructing exact generalized separable solutions.

Below we consider specific examples illustrating the application of the above method to constructing exact generalized separable solutions of nonlinear equations.

Example 1. The two-dimensional stationary equations of motion of a viscous incompressible fluid are reduced to a single fourth-order nonlinear equation for the stream function (see equation 1 in Subsection B.7.1), specifically,

$$\frac{\partial w}{\partial y}\frac{\partial}{\partial x}(\Delta w) - \frac{\partial w}{\partial x}\frac{\partial}{\partial y}(\Delta w) = \nu \Delta \Delta w, \quad \Delta w = \frac{\partial^2 w}{\partial x^2} + \frac{\partial^2 w}{\partial y^2}. \tag{6}$$

We seek exact separable solutions of equation (6) in the form

$$w = f(x) + g(y). \tag{7}$$

Substituting (7) into (6) yields

$$g'_y f'''_{xxx} - f'_x g'''_{yyy} = \nu f''''_{xxxx} + \nu g''''_{yyyy}. \tag{8}$$

Differentiating (8) with respect to x and y, we obtain

$$g''_{yy} f''''_{xxxx} - f''_{xx} g''''_{yyyy} = 0. \tag{9}$$

Nondegenerate case. If $f''_{xx} \not\equiv 0$ and $g''_{yy} \not\equiv 0$, we separate the variables in (9) to obtain the ordinary differential equations

$$f''''_{xxxx} = C f''_{xx}, \tag{10}$$
$$g''''_{yyyy} = C g''_{yy}, \tag{11}$$

which have different solutions depending on the value of the integration constant C.

$1°$. Solutions of equations (10) and (11) for $C = 0$:

$$f(x) = A_1 + A_2 x + A_3 x^2 + A_4 x^3,$$
$$g(y) = B_1 + B_2 y + B_3 y^2 + B_4 y^3, \qquad (12)$$

where A_k and B_k are arbitrary constants ($k = 1, 2, 3, 4$). On substituting (12) into (8), we evaluate the integration constants. Three cases are possible:

$$A_4 = B_4 = 0, \quad A_n, B_n \text{ are any numbers} \quad (n = 1, 2, 3);$$
$$A_k = 0, \quad B_k \text{ are any numbers} \quad (k = 1, 2, 3, 4);$$
$$B_k = 0, \quad A_k \text{ are any numbers} \quad (k = 1, 2, 3, 4).$$

The first two sets of constants determine two simple solutions (7) of equation (6):

$$w = C_1 x^2 + C_2 x + C_3 y^2 + C_4 y + C_5,$$
$$w = C_1 y^3 + C_2 y^2 + C_3 y + C_4,$$

where $C_1, \ldots, C_5$ are arbitrary constants.

$2°$. Solutions of equations (10) and (11) for $C = \lambda^2 > 0$:

$$f(x) = A_1 + A_2 x + A_3 e^{\lambda x} + A_4 e^{-\lambda x},$$
$$g(y) = B_1 + B_2 y + B_3 e^{\lambda y} + B_4 e^{-\lambda y}. \qquad (13)$$

Substituting (13) into (8), dividing by λ^3, and collecting terms, we obtain

$$A_3(\nu\lambda - B_2)e^{\lambda x} + A_4(\nu\lambda + B_2)e^{-\lambda x} + B_3(\nu\lambda + A_2)e^{\lambda y} + B_4(\nu\lambda - A_2)e^{-\lambda y} = 0.$$

Equating the coefficients of the exponentials to zero, we find

$$A_3 = A_4 = B_3 = 0, \quad A_2 = \nu\lambda \qquad \text{(case 1)},$$
$$A_3 = B_3 = 0, \quad A_2 = \nu\lambda, \quad B_2 = -\nu\lambda \qquad \text{(case 2)},$$
$$A_3 = B_4 = 0, \quad A_2 = -\nu\lambda, \quad B_2 = -\nu\lambda \qquad \text{(case 3)}.$$

(The other constants are arbitrary.) These sets of constants determine three solutions (7) of equation (6):

$$w = C_1 e^{-\lambda y} + C_2 y + C_3 + \nu\lambda x,$$
$$w = C_1 e^{-\lambda x} + \nu\lambda x + C_2 e^{-\lambda y} - \nu\lambda y + C_3,$$
$$w = C_1 e^{-\lambda x} - \nu\lambda x + C_2 e^{\lambda y} - \nu\lambda y + C_3,$$

where C_1, C_2, C_3, and λ are arbitrary constants.

$3°$. Solution of equations (10) and (11) for $C = -\lambda^2 < 0$:

$$f(x) = A_1 + A_2 x + A_3 \cos(\lambda x) + A_4 \sin(\lambda x),$$
$$g(y) = B_1 + B_2 y + B_3 \cos(\lambda y) + B_4 \sin(\lambda y). \qquad (14)$$

Substituting (14) into (8) does not yield new real solutions.

Degenerate cases. If $f''_{xx} \equiv 0$ or $g''_{yy} \equiv 0$, equation (9) becomes an identity for any $y = y(y)$ or $f = f(x)$, respectively. These cases should be treated separately from the nondegenerate case. For example, if $f''_{xx} \equiv 0$, we have $f(x) = Ax + B$, where A and B are arbitrary numbers. Substituting this f into (8), we arrive at the equation $-A g'''_{yyy} = \nu g''''_{yyyy}$. Its general solution is given by $g(y) = C_1 \exp(-Ay/\nu) + C_2 y^2 + C_3 y + C_4$. Thus, we obtain another solution (7) of equation (6):

$$w = C_1 e^{-\lambda y} + C_2 y^2 + C_3 y + C_4 + \nu\lambda x \qquad (A = \nu\lambda, \ B = 0).$$

Example 2. Consider the second-order nonlinear parabolic equation

$$\frac{\partial w}{\partial t} = aw \frac{\partial^2 w}{\partial x^2} + b\left(\frac{\partial w}{\partial x}\right)^2 + c. \qquad (15)$$

We look for exact separable solutions of equation (15) in the form

$$w = \varphi(t) + \psi(t)\theta(x). \qquad (16)$$

Substituting (15) into (16) and collecting terms yields

$$\varphi'_t - c + \psi'_t \theta = a\varphi\psi\theta''_{xx} + \psi^2\left[a\theta\theta''_{xx} + b(\theta'_x)^2\right]. \qquad (17)$$

On dividing this relation by ψ^2 and differentiating with respect to t and x, we obtain

$$(\psi'_t/\psi^2)'_t \theta'_x = a(\varphi/\psi)'_t \theta'''_{xxx}.$$

Separating the variables, we arrive at the ordinary differential equations

$$\theta'''_{xxx} = K\theta'_x, \tag{18}$$

$$(\psi'_t/\psi^2)'_t = aK(\varphi/\psi)'_t, \tag{19}$$

where K is an arbitrary constant. The general solution of equation (18) is given by

$$\theta = \begin{cases} A_1 x^2 + A_2 x + A_3 & \text{if } K = 0, \\ A_1 e^{\lambda x} + A_2 e^{-\lambda x} + A_3 & \text{if } K = \lambda^2 > 0, \\ A_1 \sin(\lambda x) + A_2 \cos(\lambda x) + A_3 & \text{if } K = -\lambda^2 < 0, \end{cases} \tag{20}$$

where A_1, A_2, and A_3 are arbitrary constants. Integrating (19) yields

$$\begin{aligned} \psi &= \frac{B}{t+C_1}, \quad \varphi(t) \text{ is any} & \text{if } K = 0, \\ \varphi &= B\psi + \frac{1}{aK}\frac{\psi'_t}{\psi}, \quad \psi(t) \text{ is any} & \text{if } K \neq 0, \end{aligned} \tag{21}$$

where B is an arbitrary constant. On substituting solutions (20) and (21) into (17), one can "remove" the redundant constants and define the functions φ and ψ. Below we summarize the results.

1°. Solution for $a \neq -b$ and $a \neq -2b$:

$$w = \frac{c(a+2b)}{2(a+b)}(t+C_1) + C_2(t+C_1)^{-\frac{a}{a+2b}} - \frac{(x+C_3)^2}{2(a+2b)(t+C_1)} \quad \text{(corresponds to } K = 0\text{)},$$

where C_1, C_2, and C_3 are arbitrary constants.

2°. Solution for $b = -a$:

$$w = \frac{1}{a\lambda^2}\frac{\psi'_t}{\psi} + \psi(A_1 e^{\lambda x} + A_2 e^{-\lambda x}) \quad \text{(corresponds to } K = \lambda^2 > 0\text{)},$$

where the function $\psi = \psi(t)$ is determined from the autonomous ordinary differential equation

$$Z''_{tt} = ac\lambda^2 + 4a^2\lambda^4 A_1 A_2 e^{2Z}, \quad \psi = e^Z,$$

whose solution can be found in implicit form. In the special case $A_1 = 0$ or $A_2 = 0$, we have $\psi = C_1 \exp\left(\frac{1}{2}ac\lambda^2 t^2 + C_2 t\right)$.

3°. Solution for $b = -a$:

$$w = -\frac{1}{a\lambda^2}\frac{\psi'_t}{\psi} + \psi[A_1 \sin(\lambda x) + A_2 \cos(\lambda x)] \quad \text{(corresponds to } K = -\lambda^2 < 0\text{)}.$$

where the function $\psi = \psi(t)$ is determined from the autonomous ordinary differential equation

$$Z''_{tt} = -ac\lambda^2 + a^2\lambda^4(A_1^2 + A_2^2)e^{2Z}, \quad \psi = e^Z,$$

whose solution can be found in implicit form.

B.2.3. Solution of Functional Differential Equations by Splitting

B.2.3-1. Preliminary remarks. Description of the method.

As one reduces the number of terms in the functional differential equation (3) by differentiation, redundant constants of integration arise. These constants must be "removed" at the final stage. Furthermore, the resulting equation can be of a higher-order than the original equation. To avoid these difficulties, it is convenient to reduce the solution of the functional differential equation to the solution of a linear functional equation of a standard form and solution of a system of ordinary differential equations. Thus, the original problem splits into two simpler problems. Below we outline the basic stages of the splitting method.

The case of even number of terms in equation (3), k = 2s.

1°. At the first stage, we treat equation (3) as a purely functional equation that depends on two variables X and Y, where $\Phi_1(X), \ldots, \Phi_s(X), \Psi_{s+1}(Y), \ldots, \Psi_{2s}(Y)$ are unknown quantities and the functions $\Phi_{s+1}(X), \ldots, \Phi_{2s}(X), \Psi_1(Y), \ldots, \Psi_s(Y)$ are assumed to be known.

It can be shown (by induction and differentiation) that the functional equation (3) has a solution depending on s^2 arbitrary constants

$$\begin{aligned}\Phi_i(X) &= C_{i1}\Phi_{s+1}(X) + C_{i2}\Phi_{s+2}(X) + \cdots + C_{is}\Phi_{2s}(X) & (i=1,\ldots,s),\\ \Psi_{s+i}(Y) &= -C_{1i}\Psi_1(Y) - C_{2i}\Psi_2(Y) - \cdots - C_{si}\Psi_s(Y) & (i=1,\ldots,s),\end{aligned} \quad (22)$$

where the C_{ij} are arbitrary constants. Note that there are also "degenerate" solutions depending on fewer arbitrary constants (see Item 2° in Paragraph B.2.3-2).

2°. At the second stage, we substitute the $\Phi_i(X)$ and $\Psi_j(Y)$ of (4) into (22). This results in an overdetermined system of ordinary differential equations for the unknown functions $\varphi_p(x)$ and $\psi_q(y)$.

The case of odd number of terms in equation (3), k = 2s − 1.

1°. If the number or terms is odd ($k = 2s-1$), the functional equation (3) has two different solutions with $s(s-1)$ arbitrary constants. One of them can be obtained from formulas (22) by setting $\Phi_{2s} \equiv 0$ and discarding the last term with Ψ_{2s}. The other solution can be obtained from the first one by renaming $\Phi_i(X) \rightleftarrows \Psi_i(Y)$.

2°. Further analysis for each solution should be performed following the same scheme as in the case of even number of terms in (3).

B.2.3-2. Solutions of simple functional equations and their application.

Below we give solutions of two simple functional equations of the form (3) that will be used subsequently for solving specific nonlinear partial differential equations.

1°. The functional equation

$$\Phi_1\Psi_1 + \Phi_2\Psi_2 + \Phi_3\Psi_3 = 0 \quad (23)$$

where the Φ_i are all functions of the same argument and the Ψ_i are all functions of another argument, has two solutions:

$$\begin{aligned}\Phi_1 &= A_1\Phi_3, & \Phi_2 &= A_2\Phi_3, & \Psi_3 &= -A_1\Psi_1 - A_2\Psi_2,\\ \Psi_1 &= A_1\Psi_3, & \Psi_2 &= A_2\Psi_3, & \Phi_3 &= -A_1\Phi_1 - A_2\Phi_2,\end{aligned} \quad (24)$$

where A_1 and A_2 are arbitrary constants.

2°. The functional equation

$$\Phi_1\Psi_1 + \Phi_2\Psi_2 + \Phi_3\Psi_3 + \Phi_4\Psi_4 = 0, \quad (25)$$

where the Φ_i are all functions of the same argument and the Ψ_i are all functions of another argument, has a solution

$$\begin{aligned}\Phi_1 &= A_1\Phi_3 + A_2\Phi_4, & \Phi_2 &= A_3\Phi_3 + A_4\Phi_4,\\ \Psi_3 &= -A_1\Psi_1 - A_3\Psi_2, & \Psi_4 &= -A_2\Psi_1 - A_4\Psi_2\end{aligned} \quad (26a)$$

depending on four arbitrary constants A_m [see solution (22) with $s=2$, $C_{11}=A_1$, $C_{12}=A_2$, $C_{21}=A_3$, and $C_{22}=A_4$].

Equation (25) has also two "degenerate" solutions

$$\begin{aligned}\Phi_1 &= A_1\Phi_4, & \Phi_2 &= A_2\Phi_4, & \Phi_3 &= A_3\Phi_4, & \Psi_4 &= -A_1\Psi_1 - A_2\Psi_2 - A_3\Psi_3,\\ \Psi_1 &= A_1\Psi_4, & \Psi_2 &= A_2\Psi_4, & \Psi_3 &= A_3\Psi_4, & \Phi_4 &= -A_1\Phi_1 - A_2\Phi_2 - A_3\Phi_3\end{aligned} \quad (26b)$$

involving three arbitrary constants.

Example 3. Consider the nonlinear hyperbolic equation

$$\frac{\partial^2 w}{\partial t^2} = a\frac{\partial}{\partial x}\left(w\frac{\partial w}{\partial x}\right) + f(t)w + g(t), \tag{27}$$

where $f(t)$ and $g(t)$ are arbitrary functions. We look for generalized separable solutions with the form

$$w(x,t) = \varphi(x)\psi(t) + \chi(t). \tag{28}$$

Substituting (28) into (27) and collecting terms yield

$$a\psi^2(\varphi\varphi_x')_x' + a\psi\chi\varphi_{xx}'' + (f\psi - \psi_{tt}'')\varphi + f\chi + g - \chi_{tt}'' = 0.$$

This equation can be represented as a functional equation (25) in which

$$\begin{aligned}
&\Phi_1 = (\varphi\varphi_x')_x', \quad \Phi_2 = \varphi_{xx}'', \quad \Phi_3 = \varphi, \quad \Phi_4 = 1, \\
&\Psi_1 = a\psi^2, \quad \Psi_2 = a\psi\chi, \quad \Psi_3 = f\psi - \psi_{tt}'', \quad \Psi_4 = f\chi + g - \chi_{tt}''.
\end{aligned} \tag{29}$$

On substituting (29) into (26a), we obtain the following overdetermined system of ordinary differential equations for the functions $\varphi = \varphi(x)$, $\psi = \psi(t)$, and $\chi = \chi(t)$:

$$\begin{aligned}
&(\varphi\varphi_x')_x' = A_1\varphi + A_2, \quad &\varphi_{xx}'' = A_3\varphi + A_4, \\
&f\psi - \psi_{tt}'' = -A_1 a\psi^2 - A_3 a\psi\chi, \quad &f\chi + g - \chi_{tt}'' = -A_2 a\psi^2 - A_4 a\psi\chi.
\end{aligned} \tag{30}$$

The first two equations in (30) are consistent only if

$$A_1 = 6B_2, \quad A_2 = B_1^2 - 4B_0 B_2, \quad A_3 = 0, \quad A_4 = 2B_2, \tag{31}$$

where B_0, B_1, and B_2 are arbitrary constants, and the solution is given by

$$\varphi(x) = B_2 x^2 + B_1 x + B_0. \tag{32}$$

On substituting the expressions (31) into the last two equations in (30), we obtain the following system of equations for $\psi(t)$ and $\chi(t)$:

$$\begin{aligned}
&\psi_{tt}'' = 6aB_2\psi^2 + f(t)\psi, \\
&\chi_{tt}'' = [2aB_2\psi + f(t)]\chi + a(B_1^2 - 4B_0 B_2)\psi^2 + g(t),
\end{aligned} \tag{33}$$

Relations (28), (32) and system (33) determine a generalized separable solution of equation (27). The first equation in (33) can be solved independently; it is linear if $B_2 = 0$ and is integrable in quadrature for $f(t) = $ const. The second equation in (33) is linear in χ (for ψ known).

Equation (27) does not have other solutions with the form (28) if f and g are arbitrary function and $\varphi \not\equiv 0$, $\psi \not\equiv 0$, and $\chi \not\equiv 0$.

Remark. It can be shown that equation (27) has a more general solution with the form

$$w(x,y) = \varphi_1(x)\psi_1(t) + \varphi_2(x)\psi_2(t) + \psi_3(t), \qquad \varphi_1(x) = x^2, \quad \varphi_2(x) = x, \tag{34}$$

where the functions $\psi_i = \psi_i(t)$ are determined by the ordinary differential equations

$$\begin{aligned}
&\psi_1'' = 6a\psi_1^2 + f(t)\psi_1, \\
&\psi_2'' = [6a\psi_1 + f(t)]\psi_2, \\
&\psi_3'' = [2a\psi_1 + f(t)]\psi_3 + a\psi_2^2 + g(t).
\end{aligned} \tag{35}$$

(The prime denotes the derivative with respect to t.) The second equation in (35) has a particular solution $\psi_2 = \psi_1$. Hence, its general solution can be represented as (see Polyanin and Zaitsev, 1995)

$$\psi_2 = C_1\psi_1 + C_2\psi_1 \int \frac{dt}{\psi_1^2}.$$

The solution obtained in Example 3 corresponds to the special case $C_2 = 0$.

Example 4. Consider the nonlinear equation

$$\frac{\partial^2 w}{\partial x \partial t} + \left(\frac{\partial w}{\partial x}\right)^2 - w\frac{\partial^2 w}{\partial x^2} = \nu\frac{\partial^3 w}{\partial x^3}, \tag{36}$$

which arises in hydrodynamics (see equations B.6.2.1, Item 3° and B.7.2.1, Item 2°).

We look for exact solutions of the form

$$w = \varphi(t)\theta(x) + \psi(t). \tag{37}$$

Substituting (37) into (36) yields

$$\varphi'_t \theta'_x - \varphi\psi\theta''_{xx} + \varphi^2\left[(\theta'_x)^2 - \theta\theta''_{xx}\right] - \nu\varphi\theta'''_{xxx} = 0.$$

This functional differential equation can be reduced to the functional equation (25) by setting

$$\Phi_1 = \varphi'_t, \quad \Phi_2 = \varphi\psi, \quad \Phi_3 = \varphi^2, \quad \Phi_4 = \nu\varphi,$$
$$\Psi_1 = \theta'_x, \quad \Psi_2 = -\theta''_{xx}, \quad \Psi_3 = (\theta'_x)^2 - \theta\theta''_{xx}, \quad \Psi_4 = -\theta'''_{xxx}.$$

On substituting these expressions into (26a), we obtain the system of equations

$$\varphi'_t = A_1\varphi^2 + A_2\nu\varphi, \qquad \varphi\psi = A_3\varphi^2 + A_4\nu\varphi,$$
$$(\theta'_x)^2 - \theta\theta''_{xx} = -A_1\theta'_x + A_3\theta''_{xx}, \quad \theta'''_{xxx} = A_2\theta'_x - A_4\theta''_{xx}. \tag{38}$$

It can be shown that the last two equations in (38) are consistent only if the function θ and its derivative are linearly dependent,

$$\theta'_x = B_1\theta + B_2. \tag{39}$$

The six constants B_1, B_2, A_1, A_2, A_3, and A_4 must satisfy the three conditions

$$B_1(A_1 + B_2 - A_3B_1) = 0,$$
$$B_2(A_1 + B_2 - A_3B_1) = 0, \tag{40}$$
$$B_1^2 + A_4B_1 - A_2 = 0.$$

Integrating (39) yields

$$\theta = \begin{cases} B_3\exp(B_1 x) - \dfrac{B_2}{B_1} & \text{if } B_1 \neq 0, \\ B_2 x + B_3 & \text{if } B_1 = 0, \end{cases} \tag{41}$$

where B_3 is an arbitrary constant.

The first two equations in (38) lead to the following expressions for φ and ψ:

$$\varphi = \begin{cases} \dfrac{A_2\nu}{C\exp(-A_2\nu t) - A_1} & \text{if } A_2 \neq 0, \\ -\dfrac{1}{A_1 t + C} & \text{if } A_2 = 0, \end{cases} \qquad \psi = A_3\varphi + A_4\nu, \tag{42}$$

where C is an arbitrary constant.

Formulas (41), (42) and relations (40) allow us to find the following solutions of equation (36) with the form (37):

$$w = \frac{x + C_1}{t + C_2} + C_3 \qquad \text{if} \quad A_2 = B_1 = 0, \ B_2 = -A_1;$$

$$w = \frac{C_1 e^{-\lambda x} + 1}{\lambda t + C_2} + \nu\lambda \qquad \text{if} \quad A_2 = 0, \ B_1 = -A_4, \ B_2 = -A_1 - A_3A_4;$$

$$w = C_1 e^{-\lambda(x+\beta\nu t)} + \nu(\lambda + \beta) \qquad \text{if} \quad A_1 = A_3 = B_2 = 0, \ A_2 = B_1^2 + A_4B_1;$$

$$w = \frac{\nu\beta + C_1 e^{-\lambda x}}{1 + C_2 e^{-\nu\lambda\beta t}} + \nu(\lambda - \beta) \qquad \text{if} \quad A_1 = A_3B_1 - B_2, \ A_2 = B_1^2 + A_4B_1,$$

where C_1, C_2, C_3, β, and λ are arbitrary constants (these can be expressed in terms of the A_k and B_k).

The analysis of the second degenerate solution (26b) of the functional equation (25) leads to the following two more general solutions of the differential equation (36):

$$w = \frac{x}{t + C_1} + \psi(t),$$

$$w = \varphi(t)e^{-\lambda x} - \frac{\varphi'_t(t)}{\lambda\varphi(t)} + \nu\lambda,$$

where $\varphi(t)$ and $\psi(t)$ are arbitrary functions, and C_1 and λ are arbitrary constants.

B.2.4. Simplified Scheme for Constructing Exact Solutions of Equations with Quadratic Nonlinearities

B.2.4-1. Description of the simplified scheme.

To construct exact solutions of equations (2) with quadratic or power nonlinearities that do not depend explicitly on x (all f_i constant), it is reasonable to use the following simplified approach. As before, we seek solutions in the form of finite sums (1). We assume that the system of coordinate functions $\{\varphi_i(x)\}$ is governed by linear differential equations with constant coefficients. The most common solutions of such equations are of the form

$$\varphi_i(x) = x^i, \quad \varphi_i(x) = e^{\lambda_i x}, \quad \varphi_i(x) = \sin(\alpha_i x), \quad \varphi_i(x) = \cos(\beta_i x). \tag{43}$$

Finite chains of these functions (in various combinations) can be used to search for separable solutions (1), where the quantities λ_i, α_i, and β_i are regarded as free parameters. The other system of functions $\{\psi_i(y)\}$ is determined by solving the nonlinear equations resulting from substituting (1) into the equation under consideration.

This simplified approach lacks the generality of the methods outlined in Subsections B.2.2 and B.2.3. However, specifying one of the systems of coordinate functions, $\{\varphi_i(x)\}$, simplifies the procedure of finding exact solutions substantially. The drawback of this approach is that some solutions of the form (1) can be overlooked. It is significant that the overwhelming majority of generalized separable solutions known to date, for partial differential equations with quadratic nonlinearities, are determined by coordinate functions (36) (usually with $n = 2$).

B.2.4-2. Examples of constructing exact solutions of higher-order equations.

Below we consider specific examples that illustrate the application of the above simplified scheme to constructing generalized separable solutions of higher-order nonlinear equations.

Example 5. The equations of laminar boundary layer on a flat plate are reduced to a single third-order nonlinear equation for the stream function (see Schlichting 1981, Loitsyanskiy 1996):

$$\frac{\partial w}{\partial y}\frac{\partial^2 w}{\partial x \partial y} - \frac{\partial w}{\partial x}\frac{\partial^2 w}{\partial y^2} = \nu \frac{\partial^3 w}{\partial y^3}. \tag{44}$$

We look for generalized separable solutions with the form

$$w(x, y) = x\psi(y) + \theta(y), \tag{45}$$

which corresponds to the simplest set of functions $\varphi_1(x) = x$, $\varphi_2(x) = 1$ with $n = 2$ in formula (1). On substituting (45) into (44) and collecting terms, we obtain

$$x[(\psi')^2 - \psi\psi'' - \nu\psi'''] + [\psi'\theta' - \psi\theta'' - \nu\theta'''] = 0.$$

(The prime denotes the derivative with respect to y.) To meet this equation for any x, one should equate both expressions in square brackets to zero. This results in a system of ordinary differential equations for $\psi = \psi(y)$ and $\theta = \theta(y)$:

$$(\psi')^2 - \psi\psi'' - \nu\psi''' = 0,$$

$$\psi'\theta' - \psi\theta'' - \nu\theta''' = 0.$$

For example, this system has an exact solution

$$\psi = \frac{6\nu}{y + C_1}, \quad \theta = \frac{C_2}{y + C_1} + \frac{C_3}{(y + C_1)^2} + C_4,$$

where C_1, C_2, C_3, and C_4 are arbitrary constants.

Other generalized separable solutions of equation (44) can be found in Polyanin (2001b, 2001c) and Subsection B.6.1.

Example 6. Consider the nth-order nonlinear equation

$$\frac{\partial w}{\partial y}\frac{\partial^2 w}{\partial x \partial y} - \frac{\partial w}{\partial x}\frac{\partial^2 w}{\partial y^2} = f(x)\frac{\partial^n w}{\partial y^n}, \tag{46}$$

where $f(x)$ is an arbitrary function. In the special case $n = 3$ with $f(x) = \nu = \text{const}$, this equation coincides with the boundary layer equation (44).

We look for generalized separable solutions of the form

$$w(x, y) = \varphi(x)e^{\lambda y} + \theta(x), \tag{47}$$

which correspond to the set of functions $\psi_1(y) = e^{\lambda y}$, $\psi_2(y) = 1$ in (1). On substituting (47) into (46) and rearranging terms, we obtain

$$\lambda^2 e^{\lambda y}\varphi[\theta'_x + \lambda^{n-2}f(x)] = 0.$$

This equation is met if

$$\theta(x) = -\lambda^{n-2}\int f(x)\, dx + C, \quad \varphi(x) \text{ is any}, \tag{48}$$

where C is an arbitrary constant. (The other case $\varphi = 0$ and θ is any is of little interest.) Formulas (47) and (48) define an exact solution of equation (46),

$$w(x, y) = \varphi(x)e^{\lambda y} - \lambda^{n-2}\int f(x)\, dx + C, \tag{49}$$

which involves an arbitrary function $\varphi(x)$ and two arbitrary constants C and λ.

Note that solution (49) with $n = 3$ and $f(x) = \text{const}$ was obtained by Ignatovich (1993) by a more complicated approach.

B.3. Methods of Functional Separation of Variables

B.3.1. Structure of Functional Separable Solutions

B.3.1-1. Functional separable solutions.

Suppose a nonlinear equation for $w = w(x,y)$ is obtained from a separable linear mathematical physics equation for $z = z(x,y)$ by a nonlinear change of variable $w = F(z)$. Then, obviously, the former has exact solutions of the form

$$w(x,y) = F(z), \quad \text{where} \quad z = \sum_{m=1}^{n} \varphi_m(x)\psi_m(y). \tag{1}$$

It is noteworthy that many nonlinear partial differential equations that are not reduced to linear equation have exact solutions of the form (1) as well. We will call such solutions *functional separable solutions*. In general, the functions $\varphi_m(x)$, $\psi_m(y)$, and $F(z)$ in (1) are not known in advance and are to be identified.

Remark 1. In functional separation of variables, searching for solutions in the forms $w = F\bigl(\varphi(x) + \psi(y)\bigr)$ and $w = F\bigl(\varphi(x)\psi(y)\bigr)$ leads to equivalent results, because the two forms are functionally equivalent. Indeed, we have $F\bigl(\varphi(x)\psi(y)\bigr) = F_1\bigl(\varphi_1(x) + \psi_1(y)\bigr)$, where $F_1(z) = F(e^z)$, $\varphi_1(x) = \ln \varphi(x)$, and $\psi_1(y) = \ln \psi(y)$.

Remark 2. In constructing functional separable solutions with the form $w = F\bigl(\varphi(x) + \psi(y)\bigr)$, it is assumed that $\varphi \not\equiv \text{const}$ and $\psi \not\equiv \text{const}$.

B.3.1-2. Various modifications.

Below we give three more general modifications of solution structure (1):

$$w(x,y) = F(z), \quad z = \sum_{m=1}^{n} \varphi_m(\xi)\psi_m(\eta), \quad \xi = a_1 x + a_2 y, \quad \eta = b_1 x + b_2 y; \tag{2}$$

$$w(x,y) = \theta_1(x)F(z) + \theta_2(x), \quad z = \sum_{m=1}^{n} \varphi_m(x)\psi_m(y); \tag{3}$$

$$w(x,y) = \theta_1(y)F(z) + \theta_2(y), \quad z = \sum_{m=1}^{n} \varphi_m(x)\psi_m(y). \tag{4}$$

These can also be used to find exact solutions of nonlinear mathematical physics equations.

The solution structures of (1)–(4) cover all most common types of solutions—traveling wave, self-similar, and additively and multiplicatively separable solutions (as well as many invariant solutions). In general, the functions $\varphi_m(\xi)$, $\psi_m(\eta)$, $\varphi_m(x)$, $\psi_m(y)$, $F(z)$, $\theta_m(x)$, and $\theta_m(y)$ are not known in advance and are to be determined in the analysis.

Note that Miller and Rubel (1993) studied functional separable solutions of a different form (for a stationary heat equation with a nonlinear source); see also Clarkson and Kruskal (1989) and Burde (1994).

B.3.2. Special Functional Separable Solutions

To simplify the analysis, some of the functions in (1) can be specified a priori and the other functions will be defined in the analysis. We call such solutions *special functional separable solutions*.

B.3.2-1. Solutions of the form (1) with z linear in one of the independent variables.

Consider functional separable solutions of the form (1) in the special case where the composite argument z is linear in one of the independent variables (e.g., in x). We substitute (1) into the equation under study and eliminate x using the expression of z to obtain a functional differential equation with two arguments. In many cases, this equation can be solved by the methods outlined in Section B.2.

Example 1. Consider the nonstationary heat equation with a nonlinear source

$$\frac{\partial w}{\partial t} = \frac{\partial^2 w}{\partial x^2} + \mathcal{F}(w). \tag{5}$$

We look for functional separable solutions of the special form

$$w = w(z), \quad z = \varphi(t)x + \psi(t). \tag{6}$$

The functions $w(z)$, $\varphi(t)$, $\psi(t)$, and $\mathcal{F}(w)$ are to be determined.

On substituting (6) into (5) and on dividing by w'_z, we have

$$\varphi'_t x + \psi'_t = \varphi^2 \frac{w''_{zz}}{w'_z} + \frac{\mathcal{F}(w)}{w'_z}. \tag{7}$$

We express x from (6) in terms of z and substitute into (7) to obtain a functional differential equation with two variables t and z,

$$-\psi'_t + \frac{\psi}{\varphi}\varphi'_t - \frac{\varphi'_t}{\varphi}z + \varphi^2 \frac{w''_{zz}}{w'_z} + \frac{\mathcal{F}(w)}{w'_z} = 0,$$

which can be treated as the functional equation (25) in Section B.2 where

$$\Phi_1 = -\psi'_t + \frac{\psi}{\varphi}\varphi'_t, \quad \Phi_2 = -\frac{\varphi'_t}{\varphi}, \quad \Phi_3 = \varphi^2, \quad \Phi_4 = 1,$$

$$\Psi_1 = 1, \quad \Psi_2 = z, \quad \Psi_3 = \frac{w''_{zz}}{w'_z}, \quad \Psi_4 = \frac{\mathcal{F}(w)}{w'_z}.$$

Substituting these expressions into relations (26a) of Section B.2 yields the system of ordinary differential equations

$$-\psi'_t + \frac{\psi}{\varphi}\varphi'_t = A_1\varphi^2 + A_2, \quad -\frac{\varphi'_t}{\varphi} = A_3\varphi^2 + A_4,$$

$$\frac{w''_{zz}}{w'_z} = -A_1 - A_3 z, \quad \frac{\mathcal{F}(w)}{w'_z} = -A_2 - A_4 z, \tag{8}$$

where A_1, A_2, A_3, and A_4 are arbitrary constants.

The solution of system (8) is given by

$$\varphi(t) = \pm\left(C_1 e^{-2A_4 t} - \frac{A_3}{A_4}\right)^{-1/2},$$

$$\psi(t) = -\varphi(t)\left[A_1 \int \varphi(t)\,dt + A_2 \int \frac{dt}{\varphi(t)} + C_2\right], \tag{9}$$

$$w(z) = C_3 \int \exp\!\left(-\tfrac{1}{2}A_3 z^2 - A_1 z\right) dz + C_4,$$

$$\mathcal{F}(w) = -C_3(A_4 z + A_2)\exp\!\left(-\tfrac{1}{2}A_3 z^2 - A_1 z\right),$$

where C_1, C_2, C_3, and C_4 are arbitrary constants. The dependence $\mathcal{F} = \mathcal{F}(w)$ is defined by the last two relations in parametric form (z is considered the parameter).

In the special case $A_3 = C_4 = 0$, $A_1 = -1$, and $C_3 = 1$, the source function can be represented in explicit form as

$$\mathcal{F}(w) = -w(A_4 \ln w + A_2). \tag{10}$$

If $A_3 \neq 0$ in (9), the source function is expressed in terms of elementary functions and the inverse of the error function.

Example 2. Consider the more general equation

$$\frac{\partial w}{\partial t} = a(t)\frac{\partial^2 w}{\partial x^2} + b(t)\frac{\partial w}{\partial x} + c(t)\mathcal{F}(w).$$

We look for solutions in the form (6). In this case, only the first two equations in system (8) will change, and the functions $w(z)$ and $\mathcal{F}(w)$ will be given by (9).

Example 3. The nonlinear heat equation

$$\frac{\partial w}{\partial t} = \frac{\partial}{\partial x}\left[\mathcal{G}(w)\frac{\partial w}{\partial x}\right] + \mathcal{F}(w)$$

has also solutions of the form (6). The unknown quantities are governed by system (8) in which w''_{zz} must be replaced by $[\mathcal{G}(w)w'_z]'_z$. The functions $\varphi(t)$ and $\psi(t)$ are determined by the first two formulas in (9). One of the two functions $\mathcal{G}(w)$ and $\mathcal{F}(w)$ can be assumed arbitrary and the other is identified in the course of the solution. The special case $\mathcal{F}(w) =$ const yields $\mathcal{G}(w) = C_1 e^{2ke} + (C_2 w + C_3) e^{kw}$.

Example 4. Likewise, we can treat the nth-order nonlinear equation

$$\frac{\partial w}{\partial t} = \frac{\partial^n w}{\partial x^n} + \mathcal{F}(w).$$

As before, we look for solutions in the form (6). In this case, the quantities φ^2 and w''_{zz} in (8) must be replaced by φ^n and $w_z^{(n)}$, respectively. In particular, for $A_3 = 0$, apart from equations with logarithmic nonlinearities of the form (10), we obtain other equations.

Example 5. For the nth-order nonlinear equation

$$\frac{\partial w}{\partial t} = \frac{\partial^n w}{\partial x^n} + \mathcal{F}(w)\frac{\partial w}{\partial x},$$

the search for exact solutions of the form (6) leads to the following system of equations for $\varphi(t)$, $\psi(t)$, $w(z)$, and $\mathcal{F}(w)$:

$$-\psi'_t + \frac{\psi}{\varphi}\varphi'_t = A_1\varphi^n + A_2\varphi, \qquad -\frac{\varphi'_t}{\varphi} = A_3\varphi^n + A_4\varphi,$$

$$\frac{w_z^{(n)}}{w'_z} = -A_1 - A_3 z, \qquad \mathcal{F}(w) = -A_2 - A_4 z,$$

where A_1, A_2, A_3, and A_4 are arbitrary constants.

In the case $n = 3$, we assume $A_3 = 0$ and $A_1 > 0$ to find in particular that $\mathcal{F}(w) = -A_2 - A_4 \arcsin(kw)$.

Example 6. In addition, searching for solutions of equation (5) with z quadratically dependent on x,

$$w = w(z), \quad z = \varphi(t)x^2 + \psi(t), \tag{11}$$

also makes sense here. Indeed, on substituting (11) into (5), we arrive at an equation that contains terms with x^2 and does not contain terms linear in x. Eliminating x^2 from the resulting equation with the aid of (11), we obtain

$$-\psi'_t + \frac{\psi}{\varphi}\varphi'_t + 2\varphi - \frac{\varphi'_t}{\varphi}z + 4\varphi z\frac{w''_{zz}}{w'_z} - 4\varphi\psi\frac{w''_{zz}}{w'_z} + \frac{\mathcal{F}(w)}{w'_z} = 0.$$

To solve this functional differential equation with two arguments, we apply the splitting method outlined in Subsection B.2.3. It can be shown that, for equations (5), this equation has a solution with a logarithmic nonlinearity of the form (10).

B.3.2-2. Solution by reduction to equations with quadratic nonlinearities.

In some cases, solutions of the form (1) can be searched for in two stages. First, one looks for a transformation that would reduce the original equation to an equation with a quadratic (or power) nonlinearity. Then the methods outlined in Section B.2 are used to find solutions of the resulting equation.

Sometimes, quadratically nonlinear equations can be obtained using the substitutions

$$w(z) = z^\lambda \quad \text{(for equations with power nonlinearities)},$$
$$w(z) = \lambda \ln z \quad \text{(for equations with exponential nonlinearities)},$$
$$w(z) = e^{\lambda z} \quad \text{(for equations with logarithmic nonlinearities)},$$

where λ is a constant to be determined. This approach is equivalent to specifying the form of the function $F(z)$ in (1) a priori.

Example 6. The nonlinear heat equation with a logarithmic source

$$\frac{\partial w}{\partial t} = a\frac{\partial^2 w}{\partial x^2} + f(t)w \ln w + g(t)w$$

can be reduced by the change of variable $w = e^z$ to the quadratically nonlinear equation

$$\frac{\partial z}{\partial t} = a\frac{\partial^2 z}{\partial x^2} + a\left(\frac{\partial z}{\partial x}\right)^2 + f(t)z + g(t),$$

which admits separable solutions with the form

$$z = \varphi_1(x)\psi_1(t) + \varphi_2(x)\psi_2(t) + \psi_3(t),$$

where $\varphi_1(x) = x^2$, $\varphi_2(x) = x$, and the functions $\psi_k(t)$ are determined by an appropriate system of ordinary differential equations.

B.3.3. Differentiation Method

B.3.3-1. Basic ideas of the method. Reduction to a standard equation.

In general, the substitution of expression (1) into the nonlinear partial differential equation under study leads to a functional differential equation with three arguments—two arguments are usual, x and y, and the third is composite, z. In many cases, the resulting equation can be reduced by differentiation to a standard functional differential equation with two arguments (either x or y is eliminated). Two solve the two-argument equation, one can use the methods outlined in Section B.2.

B.3.3-2. Examples of constructing functional separable solutions.

Below we consider specific examples illustrating the application of the differentiation method for constructing functional separable solutions of nonlinear equations.

Example 7. Consider the nonlinear heat equation

$$\frac{\partial w}{\partial t} = \frac{\partial}{\partial x}\left[f(w)\frac{\partial w}{\partial x}\right]. \tag{12}$$

We look for exact solutions with the form

$$w = w(z), \quad z = \varphi(x) + \psi(t). \tag{13}$$

On substituting (13) into (12) and dividing by w'_z, we obtain the functional differential equation

$$\psi'_t = \varphi''_{xx}f(w) + (\varphi'_x)^2 Q(z), \tag{14}$$

where

$$Q(z) = f(w)\frac{w''_{zz}}{w'_z} + f'_z(w), \quad w = w(z). \tag{15}$$

Differentiating (14) with respect to x yields

$$\varphi'''_{xxx}f(w) + \varphi'_x\varphi''_{xx}[f'_z(w) + 2Q(z)] + (\varphi'_x)^3 Q'_z = 0. \tag{16}$$

This functional differential equation with two variables can be treated as the functional equation (23) of Section B.2. This three-term functional equation has two different solutions. Accordingly, we consider two cases.

Case 1. The solutions of the functional differential equation (16) are determined from the system of ordinary differential equations

$$\begin{aligned} f'_z + 2Q &= 2A_1 f, \quad Q'_z = A_2 f, \\ \varphi'''_{xxx} + 2A_1\varphi'_x\varphi''_{xx} + A_2(\varphi'_x)^3 &= 0, \end{aligned} \tag{17}$$

where A_1 and A_2 are arbitrary constants.

The first two equations (17) are linear and independent of the third equation. Their general solution is given by

$$f = \begin{cases} e^{A_1 z}(B_1 e^{kz} + B_2 e^{-kz}) & \text{if } A_1^2 > 2A_2, \\ e^{A_1 z}(B_1 + B_2 z) & \text{if } A_1^2 = 2A_2, \quad Q = A_1 f - \tfrac{1}{2}f'_z, \quad k = \sqrt{|A_1^2 - 2A_2|}. \\ e^{A_1 z}[B_1 \sin(kz) + B_2 \cos(kz)] & \text{if } A_1^2 < 2A_2, \end{cases} \tag{18}$$

Substituting Q of (18) into (15) yields a differential equation for $w = w(z)$. On integrating this equation, we obtain

$$w = C_1 \int e^{A_1 z} |f(z)|^{-3/2} dz + C_2, \tag{19}$$

where C_1 and C_2 are arbitrary constants. The expression of f in (18) together with expression (19) define the function $f = f(w)$ in parametric form.

Without full analysis, we will study the case $A_2 = 0$ ($k = A_1$) and $A_1 \neq 0$ in more detail. It follows from (18) and (19) that

$$f(z) = B_1 e^{2A_1 z} + B_2, \quad Q = A_1 B_2, \quad w(z) = C_3 (B_1 + B_2 e^{-2A_1 z})^{-1/2} + C_2 \quad (C_1 = A_1 B_2 C_3). \tag{20}$$

Eliminating z yields

$$f(w) = \frac{B_2 C_3^2}{C_3^2 - B_1 w^2}. \tag{21}$$

The last equation in (17) with $A_2 = 0$ has the first integral $\varphi''_{xx} + A_1 (\varphi'_x)^2 = \text{const}$. The corresponding general solution is given by

$$\varphi(x) = -\frac{1}{2A_1} \ln \left[\frac{D_2}{D_1} \frac{1}{\sinh^2(A_1 \sqrt{D_2}\, x + D_3)} \right] \quad \text{for} \quad D_1 > 0 \text{ and } D_2 > 0;$$

$$\varphi(x) = -\frac{1}{2A_1} \ln \left[-\frac{D_2}{D_1} \frac{1}{\cos^2(A_1 \sqrt{-D_2}\, x + D_3)} \right] \quad \text{for} \quad D_1 > 0 \text{ and } D_2 < 0; \tag{22}$$

$$\varphi(x) = -\frac{1}{2A_1} \ln \left[-\frac{D_2}{D_1} \frac{1}{\cosh^2(A_1 \sqrt{D_2}\, x + D_3)} \right] \quad \text{for} \quad D_1 < 0 \text{ and } D_2 > 0;$$

where D_1, D_2, and D_3 are constants of integration. In all three cases, the following relations hold:

$$(\varphi'_x) = D_1 e^{-2A_1 \varphi} + D_2, \quad \varphi''_{xx} = -A_1 D_1 e^{-2A_1 \varphi}. \tag{23}$$

We substitute (20) and (23) into the original functional differential equation (14). With reference to the expression of z in (13), we obtain the following equation for $\psi = \psi(t)$:

$$\psi'_t = -A_1 B_1 D_1 e^{2A_1 \psi} + A_1 B_2 D_2.$$

Its general solution is given by

$$\psi(t) = \frac{1}{2A_1} \ln \frac{B_2 D_2}{D_4 \exp(-2A_1^2 B_2 D_2 t) + B_1 D_1}, \tag{24}$$

where D_4 is an arbitrary constant.

Formulas (13), (20) for w, (22), and (24) define three solutions of the nonlinear equation (12) with $f(w)$ of the form (21) [recall that these solutions correspond to the special case $A_2 = 0$ in (18) and (19)].

Case 2. The solutions of the functional differential equation (16) are determined from the system of ordinary differential equations

$$\varphi'''_{xxx} = A_1 (\varphi'_x)^3, \quad \varphi'_x \varphi''_{xx} = A_2 (\varphi'_x)^3,$$
$$A_1 f + A_2 (f'_z + 2Q) + Q'_z = 0. \tag{25}$$

The first two equations in (25) are consistent in the two cases

$$\begin{aligned} A_1 = A_2 = 0 &\implies \varphi(x) = B_1 x + B_2, \\ A_1 = 2A_2^2 &\implies \varphi(x) = -\frac{1}{A_2} \ln|B_1 x + B_2|. \end{aligned} \tag{26}$$

The first solution in (26) eventually leads to the traveling wave solution $w = w(B_1 x + B_2 t)$ of equation (12) and the second solution to the self-similar solution of the form $w = \widetilde{w}(x^2/t)$. In both cases, the function $f(w)$ in (12) is arbitrary.

A more detailed analysis of functional separable solutions (13) of equation (12) can be found in the reference cited below.

⊙ *Reference*: P. W. Doyle and P. J. Vassiliou (1998).

Example 8. One can look for more complicated functional separable solutions of equation (12) with the form

$$w = w(z), \quad z = \varphi(\xi) + \psi(t), \quad \xi = x + at \quad (a = \text{const}).$$

We substitute this into (12), divide the resulting functional differential equation by w'_z, and differentiate with respect to x to obtain

$$-a\varphi''_{\xi\xi} + \varphi'''_{\xi\xi\xi} f(w) + \varphi'_\xi \varphi''_{\xi\xi} [f'_z(w) + 2Q(z)] + (\varphi'_\xi)^3 Q'_z = 0,$$

where the function $Q = Q(z)$ is defined by (15). This functional differential equation with two variables ξ and z can be treated as the functional equation (25) of Section B.2. The solution of (25) is given by relations (26), thus representing a system of ordinary differential equations for f, Q, and φ.

Example 9. Consider the nonlinear Klein–Gordon equation

$$\frac{\partial^2 w}{\partial t^2} - \frac{\partial^2 w}{\partial x^2} = \mathcal{F}(w). \tag{27}$$

We look for functional separable solutions in additive form:

$$w = w(z), \qquad z = \varphi(x) + \psi(t). \tag{28}$$

Substituting (28) into (27) yields

$$\psi''_{tt} - \varphi''_{xx} + \left[(\psi'_t)^2 - (\varphi'_x)^2\right] g(z) = h(z), \tag{29}$$

where

$$g(z) = w''_{zz}/w'_z, \qquad h(z) = \mathcal{F}(w(z))/w'_z. \tag{30}$$

On differentiating (29) first with respect to t and then with respect to x and on dividing by $\psi'_t \varphi'_x$, we have

$$2(\psi''_{tt} - \varphi''_{xx})g'_z + \left[(\psi'_t)^2 - (\varphi'_x)^2\right] g''_{zz} = h''_{zz}.$$

Eliminating $\psi''_{tt} - \varphi''_{xx}$ from this equation with the aid of (29), we obtain

$$\left[(\psi'_t)^2 - (\varphi'_x)^2\right](g''_{zz} - 2gg'_z) = h''_{zz} - 2g'_z h. \tag{31}$$

This relation holds in the following cases:

$$\begin{aligned} g''_{zz} - 2gg'_z &= 0, \quad h''_{zz} - 2g'_z h = 0 & \text{(case 1)}, \\ (\psi'_t)^2 &= A\psi + B, \quad (\varphi'_x)^2 = -A\varphi + B - C, \quad h''_{zz} - 2g'_z h = (Az + C)(g''_{zz} - 2gg'_z) & \text{(case 2*)}, \end{aligned} \tag{32}$$

where A, B, and C are arbitrary constants. We consider both cases.

Case 1. The first two equations in (32) enable one to determine $g(z)$ and $h(z)$. Integrating the first equation once yields $g'_z = g^2 +$ const. Further, the following cases are possible:

$$g = k, \tag{33a}$$
$$g = -1/(z + C_1), \tag{33b}$$
$$g = -k \tanh(kz + C_1), \tag{33c}$$
$$g = -k \coth(kz + C_1), \tag{33d}$$
$$g = k \tan(kz + C_1), \tag{33e}$$

where C_1 and k are arbitrary constants.

The second equation in (32) has the particular solution $h = g(z)$. Hence, its general solution in expressed by (e.g., see Polyanin and Zaitsev 1995)

$$h = C_2 g(z) + C_3 g(z) \int \frac{dz}{g^2(z)}, \tag{34}$$

where C_2 and C_3 are arbitrary constants.

The functions $w(z)$ and $\mathcal{F}(w)$ are found from (30) as

$$w(z) = B_1 \int G(z)\,dz + B_2, \quad \mathcal{F}(w) = B_1 h(z) G(z), \quad \text{where} \quad G(z) = \exp\left[\int g(z)\,dz\right], \tag{35}$$

and B_1 and B_2 are arbitrary constants ($\mathcal{F}$ is defined parametrically).

Let us dwell on the case (33b). According to (34),

$$h = A_1(z + C_1)^2 + \frac{A_2}{z + C_1}, \tag{36}$$

where $A_1 = -C_3/3$ and $A_2 = -C_2$ are any numbers. Substituting (33b) and (36) into (35) yields

$$w = B_1 \ln|z + C_1| + B_2, \quad \mathcal{F} = A_1 B_1 (z + C_1) + \frac{A_2 B_1}{(z + C_1)^2}.$$

Eliminating z, we arrive at the explicit form of the right-hand side of equation (27):

$$\mathcal{F}(w) = A_1 B_1 e^u + A_2 B_1 e^{-2u}, \quad \text{where} \quad u = \frac{w - B_2}{B_1}. \tag{37}$$

For simplicity, we set $C_1 = 0$, $B_1 = 1$, and $B_2 = 0$ and denote $A_1 = a$ and $A_2 = b$. Thus, we have

$$w(z) = \ln|z|, \quad \mathcal{F}(w) = ae^w + be^{-2w}, \quad g(z) = -1/z, \quad h(z) = az^2 + b/z. \tag{38}$$

---

* In case 2, equation (31) can be represented as the functional equation considered in Paragraph B.3.5-1.

TABLE B1
Nonlinear Klein–Gordon equations $\partial_{tt}w - \partial_{xx}w = \mathcal{F}(w)$ admitting functional separable solutions of the form $w = w(z)$, $z = \varphi(x) + \psi(t)$. Notation: A, C_1, and C_2 are arbitrary constants; $\sigma = 1$ for $z > 0$ and $\sigma = -1$ for $z < 0$

No	Right-hand side $\mathcal{F}(w)$	Solution $w(z)$	Equations for $\psi(t)$ and $\varphi(x)$
1	$aw \ln w + bw$	e^z	$(\psi'_t)^2 = C_1 e^{-2\psi} + a\psi - \tfrac{1}{2}a + b + A$, $(\varphi'_x)^2 = C_2 e^{-2\varphi} - a\varphi + \tfrac{1}{2}a + A$
2	$ae^w + be^{-2w}$	$\ln\|z\|$	$(\psi'_t)^2 = 2a\psi^3 + A\psi^2 + C_1\psi + C_2$, $(\varphi'_x)^2 = -2a\varphi^3 + A\varphi^2 - C_1\varphi + C_2 + b$
3	$a\sin w + b\left(\sin w \ln\tan\dfrac{w}{4} + 2\sin\dfrac{w}{4}\right)$	$4\arctan e^z$	$(\psi'_t)^2 = C_1 e^{2\psi} + C_2 e^{-2\psi} + b\psi + a + A$, $(\varphi'_x)^2 = -C_2 e^{2\varphi} - C_1 e^{-2\varphi} - b\varphi + A$
4	$a\sinh w + b\left(\sinh w \ln\tanh\dfrac{w}{4} + 2\sinh\dfrac{w}{2}\right)$	$2\ln\left\|\coth\dfrac{z}{2}\right\|$	$(\psi'_t)^2 = C_1 e^{2\psi} + C_2 e^{-2\psi} - \sigma b\psi + a + A$, $(\varphi'_x)^2 = C_2 e^{2\varphi} + C_1 e^{-2\varphi} + \sigma b\varphi + A$
5	$a\sinh w + 2b\left(\sinh w \arctan e^{w/2} + \cosh\dfrac{w}{2}\right)$	$2\ln\left\|\tan\dfrac{z}{2}\right\|$	$(\psi'_t)^2 = C_1 \sin 2\psi + C_2 \cos 2\psi + \sigma b\psi + a + A$, $(\varphi'_x)^2 = -C_1 \sin 2\varphi + C_2 \cos 2\varphi - \sigma b\varphi + A$

It remains to determine $\psi(t)$ and $\varphi(x)$. We substitute (38) into the functional differential equation (29). Taking into account (28), we find

$$[\psi''_{tt}\psi - (\psi'_t)^2 - a\psi^3 - b] - [\varphi''_{xx}\varphi - (\varphi'_x)^2 + a\varphi^3] + (\psi''_{tt} - 3a\psi^2)\varphi - \psi(\varphi''_{xx} + 3a\varphi^2) = 0. \tag{39}$$

Differentiating (39) with respect to t and x yields the separable equation*

$$(\psi'''_{ttt} - 6a\psi\psi'_t)\varphi'_x - (\varphi'''_{xxx} + 6a\varphi\varphi'_x)\psi'_t = 0,$$

whose solution is determined by the ordinary differential equations

$$\psi'''_{ttt} - 6a\psi\psi'_t = A\psi'_t,$$
$$\varphi'''_{xxx} + 6a\varphi\varphi'_x = A\varphi'_x,$$

where A is the separation constant. Each equation can be integrated twice, thus resulting in

$$(\psi'_t)^2 = 2a\psi^3 + A\psi^2 + C_1\psi + C_2,$$
$$(\varphi'_x)^2 = -2a\varphi^3 + A\varphi^2 + C_3\varphi + C_4, \tag{40}$$

where C_1, C_2, C_3, and C_4 are arbitrary constants. Eliminating the derivatives from (39) with the aid of (40), we find that the arbitrary constants are related by $C_3 = -C_1$ and $C_4 = C_2 + b$. So, the functions $\psi(t)$ and $\varphi(x)$ are determined by the first-order nonlinear autonomous equations

$$(\psi'_t)^2 = 2a\psi^3 + A\psi^2 + C_1\psi + C_2,$$
$$(\varphi'_x)^2 = -2a\varphi^3 + A\varphi^2 - C_1\varphi + C_2 + b.$$

The solutions of these equations are expressed in terms of elliptic functions.

For the other cases in (33), the analysis is performed in a similar way. Table B1 presents the final results for the cases (33a)–(33e).

Case 2. Integrating the third and fourth equations in (32) yields

$$\psi = \pm\sqrt{B}\, t + D_1, \qquad \varphi = \pm\sqrt{B-C}\, t + D_2 \qquad \text{if}\quad A = 0;$$
$$\psi = \frac{1}{4A}(At + D_1)^2 - \frac{B}{A}, \qquad \varphi = \frac{1}{4A}(Ax + D_2)^2 + \frac{B-C}{A} \qquad \text{if}\quad A \neq 0; \tag{41}$$

where D_1 and D_2 are arbitrary constants. In both cases, the function $\mathcal{F}(w)$ in equation (27) is arbitrary. The first row in (41) corresponds to the traveling wave solution $w = w(kx + \lambda t)$. The second row leads to a solution of the form $w = w(x^2 - t^2)$.

⊙ *References*: A. M. Grundland and E. Infeld (1992), J. Miller and L. A. Rubel (1993), R. Z. Zhdanov (1994), V. K. Andreev, O. V. Kaptsov, V. V. Pukhnachev, and A. A. Rodionov (1994).

Example 10. The nonlinear stationary heat (diffusion) equation

$$\frac{\partial^2 w}{\partial x^2} + \frac{\partial^2 w}{\partial y^2} = \mathcal{F}(w)$$

is analyzed just as the nonlinear Klein–Gordon equation considered in Example 9. The final results are listed in Table B2; the traveling wave solutions $w = w(kx + \lambda t)$ and solutions of the form $w = w(x^2 + y^2)$, existing for any $\mathcal{F}(w)$, are omitted.

⊙ *References*: A. M. Grundland and E. Infeld (1992), J. Miller and L. A. Rubel (1993), R. Z. Zhdanov (1994), V. K. Andreev, O. V. Kaptsov, V. V. Pukhnachev, and A. A. Rodionov (1994).

---

* To solve equation (39), one can use the solution of equation (25) in Section B.2 [see (26a)].

TABLE B2
Nonlinear equations $\partial_{xx}w + \partial_{yy}w = \mathcal{F}(w)$ admitting functional separable solutions of the form $w = w(z)$, $z = \varphi(x) + \psi(y)$. Notation: A, C_1, and C_2 are arbitrary constants; $\sigma = 1$ for $z > 0$, $\sigma = -1$ for $z < 0$

No	Right-hand side $\Phi(w)$	Solution $w(z)$	Equations for $\varphi(x)$ and $\psi(y)$
1	$aw \ln w + bw$	e^z	$(\varphi'_x)^2 = C_1 e^{-2\varphi} + a\varphi - \tfrac{1}{2}a + b + A$, $(\psi'_y)^2 = C_2 e^{-2\psi} + a\psi - \tfrac{1}{2}a - A$
2	$ae^w + be^{-2w}$	$\ln\|z\|$	$(\varphi'_x)^2 = 2a\varphi^3 + A\varphi^2 + C_1\varphi + C_2$, $(\psi'_y)^2 = 2a\psi^3 - A\psi^2 + C_1\psi - C_2 - b$
3	$a \sin w + b\left(\sin w \ln \tan \dfrac{w}{4} + 2 \sin \dfrac{w}{4}\right)$	$4 \arctan e^z$	$(\varphi'_x)^2 = C_1 e^{2\varphi} + C_2 e^{-2\varphi} + b\varphi + a + A$, $(\psi'_y)^2 = C_2 e^{2\psi} + C_1 e^{-2\psi} + b\psi - A$
4	$a \sinh w + b\left(\sinh w \ln \tanh \dfrac{w}{4} + 2 \sinh \dfrac{w}{2}\right)$	$2 \ln\left\|\coth \dfrac{z}{2}\right\|$	$(\varphi'_x)^2 = C_1 e^{2\varphi} + C_2 e^{-2\varphi} - \sigma b\varphi + a + A$, $(\psi'_y)^2 = -C_2 e^{2\psi} - C_1 e^{-2\psi} - \sigma b\psi - A$
5	$a \sinh w + 2b\left(\sinh w \arctan e^{w/2} + \cosh \dfrac{w}{2}\right)$	$2 \ln\left\|\tan \dfrac{z}{2}\right\|$	$(\varphi'_x)^2 = C_1 \sin 2\varphi + C_2 \cos 2\varphi + \sigma b\varphi + a + A$, $(\psi'_y)^2 = C_1 \sin 2\psi - C_2 \cos 2\psi + \sigma b\psi - A$

B.3.4. Splitting Method. Reduction to a Functional Equation with Two Variables

B.3.4-1. Splitting method. Reduction to a standard functional equation.

The general procedure for constructing functional separable solutions, which is based on the splitting method, involves several stages outlined below.

$1°$. Substitute expression (1) into the nonlinear partial differential equation under study. This results in a functional differential equation with three arguments—the first two are usual, x and y, and the third is composite, z.

$2°$. Reduce the functional differential equation to a purely functional equation with three arguments x, y, and z with the aid of elementary differential substitutions (by selecting and renaming terms with derivatives).

$3°$. Reduce the three-argument functional differential equation by the differentiation method to the standard functional equation with two arguments (either x or y is eliminated) considered in Section B.2.

$4°$. Construct the solution of the two-argument functional equation using the formulas given in Subsection B.2.3.

$5°$. Solve the (overdetermined) system formed by the solution of Item $4°$ and the differential substitutions of Item $2°$.

$6°$. Substitute the solution of Item $5°$ into the original functional differential equation of Item $1°$ to establish the relations for the constants of integration and determine all unknown quantities.

$7°$. Consider all degenerate cases possibly arising due to violation of assumptions adopted in the previous analysis.

The splitting method reduces solving the three-argument functional differential equation to (i) solving a purely functional equation with three arguments (by reducing it to a standard functional equation with two arguments) and (ii) solving a system of ordinary differential equations. Thus, the initial problem splits into several simpler problems. Examples of constructing functional separable solutions by the splitting method are given in Subsection B.3.5.

B.3.4-2. Three-argument functional equations of special form.

The substitution of expression (1) with $n = 2$ into nonlinear partial differential equation often leads to functional differential equations of the form

$$\Phi_1(x)\Psi_1(y,z) + \Phi_2(x)\Psi_2(y,z) + \cdots + \Phi_k(x)\Psi_k(y,z) \\ + \Psi_{k+1}(y,z) + \Psi_{k+2}(y,z) + \cdots + \Psi_n(y,z) = 0, \quad (42)$$

where $\Phi_j(x)$ and $\Psi_j(y,z)$ are functionals dependent on the variables x and y, z, respectively,

$$\Phi_j(x) \equiv \Phi_j(x, \varphi, \varphi'_x, \varphi''_{xx}), \quad \Psi_j(y,z) \equiv \Psi_j(y, \psi, \psi'_y, \psi''_{yy}, F, F'_z, F''_{zz}). \quad (43)$$

(These expressions correspond to a second-order equation.)

It is reasonable to solve equation (42) by the splitting method. To this end, we treat (42) at the first stage as a purely functional equation, thus disregarding (43). Assuming that $\Psi_1 \not\equiv 0$, we divide (42) by Ψ_1 and differentiate with respect to y to obtain a similar equation but with fewer terms:

$$\Phi_2(x)\Psi_2^{(2)}(y,z) + \cdots + \Phi_k(x)\Psi_k^{(2)}(y,z) + \Psi_{k+1}^{(2)}(y,z) + \cdots + \Psi_n^{(2)}(y,z) = 0, \quad (44)$$

where $\Psi_m^{(2)} = \frac{\partial}{\partial y}(\Psi_m/\Psi_1) + \psi'_y \frac{\partial}{\partial z}(\Psi_m/\Psi_1)$. We continue this procedure until we arrive at an equation independent of x explicitly:

$$\Psi_{k+1}^{(k+1)}(y,z) + \cdots + \Psi_n^{(k+1)}(y,z) = 0, \quad (45)$$

where $\Psi_m^{(k+1)} = \frac{\partial}{\partial y}\big(\Psi_m^{(k)}/\Psi_k^{(k)}\big) + \psi'_y \frac{\partial}{\partial z}\big(\Psi_m^{(k)}/\Psi_k^{(k)}\big)$.

Relation (45) can be regarded as an equation with two independent variables y and z. If $\Psi_m^{(k+1)}(y,z) = Q_m(y)R_m(z)$ for all $m = k+1, \ldots, n$, then equation (45) can be solved using the results of Section B.2.

B.3.5. Some Functional Equations and Their Solutions. Exact Solutions of Heat and Wave Equations

In this subsection, we discuss several types of three-argument functional equations that arise most frequently in functional separation of variables in nonlinear equations of mathematical physics. The results are used to construct exact solutions for some classes of nonlinear heat and wave equations.

B.3.5-1. The functional equation $f(x) + g(y) = Q(z)$, where $z = \varphi(x) + \psi(y)$.

Here, one of the two functions $f(x)$ and $\varphi(x)$ is prescribed and the other is assumed unknown, also one of the functions $g(y)$ and $\psi(y)$ is prescribed and the other is unknown, and the function $Q(z)$ is assumed unknown.*

Differentiating the equation with respect to x and y yields $Q''_{zz} = 0$. Consequently, the solution is given by

$$f(x) = A\varphi(x) + B, \quad g(y) = A\psi(y) - B + C, \quad Q(z) = Az + C, \quad (46)$$

where A, B, and C are arbitrary constants.

B.3.5-2. The functional equation $f(t) + g(x) + h(x)Q(z) + R(z) = 0$, where $z = \varphi(x) + \psi(t)$.

Differentiating the equation with respect to x yields the two-argument equation

$$g'_x + h'_x Q + h\varphi'_x Q'_z + \varphi'_x R'_z = 0. \quad (47)$$

---

* In similar equations with a composite argument, it is assumed that $\varphi(x) \not\equiv \text{const}$ and $\psi(y) \not\equiv \text{const}$.

Such equations were discussed in Section B.2. Hence, the following relations hold [see formulas (25) and (26a) in Section B.2]:

$$\begin{aligned} g'_x &= A_1 h \varphi'_x + A_2 \varphi'_x, \\ h'_x &= A_3 h \varphi'_x + A_4 \varphi'_x, \\ Q'_z &= -A_1 - A_3 Q, \\ R'_z &= -A_2 - A_4 Q, \end{aligned} \qquad (48)$$

where A_1, A_2, A_3, and A_4 are arbitrary constants. By integrating system (48) and substituting the resulting solutions into the original functional equation, one obtains the results given below.

Case 1. If $A_3 = 0$ in (48), the corresponding solution of the functional equation is given by

$$\begin{aligned} f &= -\tfrac{1}{2} A_1 A_4 \psi^2 + (A_1 B_1 + A_2 + A_4 B_3) \psi - B_2 - B_1 B_3 - B_4, \\ g &= \tfrac{1}{2} A_1 A_4 \varphi^2 + (A_1 B_1 + A_2) \varphi + B_2, \\ h &= A_4 \varphi + B_1, \\ Q &= -A_1 z + B_3, \\ R &= \tfrac{1}{2} A_1 A_4 z^2 - (A_2 + A_4 B_3) z + B_4, \end{aligned} \qquad (49)$$

where the A_k and B_k are arbitrary constants and $\varphi = \varphi(x)$ and $\psi = \psi(t)$ are arbitrary functions.

Case 2. If $A_3 \neq 0$ in (48), the corresponding solution of the functional equation is

$$\begin{aligned} f &= -B_1 B_3 e^{-A_3 \psi} + \left(A_2 - \frac{A_1 A_4}{A_3} \right) \psi - B_2 - B_4 - \frac{A_1 A_4}{A_3^2}, \\ g &= \frac{A_1 B_1}{A_3} e^{A_3 \varphi} + \left(A_2 - \frac{A_1 A_4}{A_3} \right) \varphi + B_2, \\ h &= B_1 e^{A_3 \varphi} - \frac{A_4}{A_3}, \\ Q &= B_3 e^{-A_3 z} - \frac{A_1}{A_3}, \\ R &= \frac{A_4 B_3}{A_3} e^{-A_3 z} + \left(\frac{A_1 A_4}{A_3} - A_2 \right) z + B_4, \end{aligned} \qquad (50)$$

where the A_k and B_k are arbitrary constants and $\varphi = \varphi(x)$ and $\psi = \psi(t)$ are arbitrary functions.

Case 3. In addition, the functional equation has the two degenerate solutions:

$$f = A_1 \psi + B_1, \quad g = A_1 \varphi + B_2, \quad h = A_2, \quad R = -A_1 z - A_2 Q - B_1 - B_2, \qquad (51a)$$

where $\varphi = \varphi(x)$, $\psi = \psi(t)$, and $Q = Q(z)$ are arbitrary functions, A_1, A_2, B_1, and B_2 are arbitrary constants; and

$$f = A_1 \psi + B_1, \quad g = A_1 \varphi + A_2 h + B_2, \quad Q = -A_2, \quad R = -A_1 z - B_1 - B_2, \qquad (51b)$$

where $\varphi = \varphi(x)$, $\psi = \psi(t)$, and $h = h(x)$ are arbitrary functions, A_1, A_2, B_1, and B_2 are arbitrary constants. The degenerate solutions (51a) and (51b) can be obtained directly from the original equation or its consequence (47) using formulas (26b) in Section B.2.

Example 11. Consider the nonstationary heat equation with a nonlinear source

$$\frac{\partial w}{\partial t} = \frac{\partial^2 w}{\partial x^2} + \mathcal{F}(w). \qquad (52)$$

We look for exact solutions of the form

$$w = w(z), \quad z = \varphi(x) + \psi(t). \qquad (53)$$

Substituting (53) into (52) and dividing by w'_z yields the functional differential equation
$$\psi'_t = \varphi''_{xx} + (\varphi'_x)^2 \frac{w''_{zz}}{w'_z} + \frac{\mathcal{F}(w(z))}{w'_z}.$$

We rewrite it as the functional equation B.3.5-2 in which
$$f(t) = -\psi'_t, \quad g(x) = \varphi''_{xx}, \quad h(x) = (\varphi'_x)^2, \quad Q(z) = w''_{zz}/w'_z, \quad R(z) = f(w(z))/w'_z. \tag{54}$$

We now use the solutions of equation B.3.5-2. On substituting the expressions of g and h of (54) into (49)–(51), we arrive at overdetermined systems of equations for $\varphi = \varphi(x)$.

Case 1. The system
$$\varphi''_{xx} = \tfrac{1}{2}A_1 A_4 \varphi^2 + (A_1 B_1 + A_2)\varphi + B_2,$$
$$(\varphi'_x)^2 = A_4 \varphi + B_1$$
following from (49) and corresponding to $A_3 = 0$ in (48) is consistent in the cases
$$\begin{array}{ll}
\varphi = C_1 x + C_2 & \text{for} \quad A_2 = -A_1 C_1^2,\ A_4 = B_2 = 0,\ B_1 = C_1^2, \\
\varphi = \tfrac{1}{4} A_4 x^2 + C_1 x + C_2 & \text{for} \quad A_1 = A_2 = 0,\ B_1 = C_1^2 - A_4 C_2,\ B_2 = \tfrac{1}{2} A_4,
\end{array} \tag{55}$$
where C_1 and C_2 are arbitrary constants.

The first solution in (55) with $A_1 \neq 0$ leads to a right-hand side of equation (52) containing the inverse of the error function [the form of the right-hand side is identified from the last two relations in (49) and (54)]. The second solution in (55) corresponds to the right-hand side $\mathcal{F}(w) = k_1 w \ln w + k_2 w$ in (52). In both cases, the first relation in (49) is, taking into account that $f = -\psi'_t$, a first-order linear solution with constant coefficients, whose solution is an exponential plus a constant.

Case 2. The system
$$\varphi''_{xx} = \frac{A_1 B_1}{A_3} e^{A_3 \varphi} + \left(A_2 - \frac{A_1 A_4}{A_3}\right)\varphi + B_2,$$
$$(\varphi'_x)^2 = B_1 e^{A_3 \varphi} - \frac{A_4}{A_3},$$
following from (50) and corresponding to $A_3 \neq 0$ in (48) is consistent in the following cases:

$$\begin{array}{ll}
\varphi = \pm\sqrt{-A_4/A_3}\, x + C_1 & \text{for } A_2 = A_1 A_4 / A_3,\ B_1 = B_2 = 0, \\
\varphi = -\dfrac{2}{A_3} \ln|x| + C_1 & \text{for } A_1 = \tfrac{1}{2} A_3^2,\ A_2 = A_4 = B_2 = 0,\ B_1 = 4 A_3^{-2} e^{-A_3 C_1}, \\
\varphi = -\dfrac{2}{A_3} \ln\left|\cos\left(\tfrac{1}{2}\sqrt{A_3 A_4}\, x + C_1\right)\right| + C_2 & \text{for } A_1 = \tfrac{1}{2} A_3^2,\ A_2 = \tfrac{1}{2} A_3 A_4,\ B_2 = 0,\ A_3 A_4 > 0, \\
\varphi = -\dfrac{2}{A_3} \ln\left|\sinh\left(\tfrac{1}{2}\sqrt{-A_3 A_4}\, x + C_1\right)\right| + C_2 & \text{for } A_1 = \tfrac{1}{2} A_3^2,\ A_2 = \tfrac{1}{2} A_3 A_4,\ B_2 = 0,\ A_3 A_4 < 0, \\
\varphi = -\dfrac{2}{A_3} \ln\left|\cosh\left(\tfrac{1}{2}\sqrt{-A_3 A_4}\, x + C_1\right)\right| + C_2 & \text{for } A_1 = \tfrac{1}{2} A_3^2,\ A_2 = \tfrac{1}{2} A_3 A_4,\ B_2 = 0,\ A_3 A_4 < 0,
\end{array}$$
where C_1 and C_2 are arbitrary constants. The right-hand sides of equation (52) corresponding to these solutions are represented in parametric form.

Case 3. Traveling wave solutions of the nonlinear heat equation (52) and solutions of the linear equation (52) with $\mathcal{F}'_w = \text{const}$ correspond to the degenerate solutions of the functional equation (51).

Example 12. Likewise, one can analyze the more general equation
$$\frac{\partial w}{\partial t} = a(x)\frac{\partial^2 w}{\partial x^2} + b(x)\frac{\partial w}{\partial x} + \mathcal{F}(w). \tag{56}$$
It arises in convective heat/mass exchange problems ($a = \text{const}$ and $b = \text{const}$), problems of heat transfer in inhomogeneous media ($b = a'_x \neq \text{const}$), and spatial heat transfer problems with axial or central symmetry ($a = \text{const}$ and $b = \text{const}/x$).

Searching for exact solutions of equation (56) in the form (53) leads to the functional equation B.3.5-2 in which
$$f(t) = -\psi'_t, \quad g(x) = a(x)\varphi''_{xx} + b(x)\varphi'(x), \quad h(x) = a(x)(\varphi'_x)^2, \quad Q(z) = w''_{zz}/w'_z, \quad R(z) = f(w(z))/w'_z.$$
Substituting these expressions into (49)–(51) yields a system of ordinary differential equations for the unknowns.

Example 13. Equation (52) also admits more complicated functional separable solutions with the form
$$w = w(z), \quad z = \varphi(\xi) + \psi(t), \quad \xi = x + at.$$
Substituting these expressions into equation (52) yields the functional equation B.3.5-2 again, in which (x must be replaced by ξ)
$$f(t) = -\psi'_t, \quad g(\xi) = \varphi''_{\xi\xi} - a\varphi'_\xi, \quad h(\xi) = (\varphi'_\xi)^2, \quad Q(z) = w''_{zz}/w'_z, \quad R(z) = f(w(z))/w'_z.$$
Further, one should follow the same procedure of constructing the solution as in Example 11.

Remark. In Examples 11–13, different equations were all reduced to the same functional equation. This demonstrates the utility of isolation and independent analysis of individual types of functional equations, as well as the expedience of developing methods for solving functional equations with a composite argument.

B.3.5-3. The functional equation $f(t) + g(x)Q(z) + h(x)R(z) = 0$, where $z = \varphi(x) + \psi(t)$.
Differentiating with respect to x yields the two-argument functional differential equation

$$g'_x Q + g\varphi'_x Q'_z + h'_x R + h\varphi'_x R'_z = 0, \tag{57}$$

which coincides with equation (25) in Section B.2, up to notation.

Nondegenerate case. Equation (57) can be solved using formulas (49) in Section B.2, just as was the case for equation (25). In this way, we arrive at the system of ordinary differential equations

$$\begin{aligned}
g'_x &= (A_1 g + A_2 h)\varphi'_x, \\
h'_x &= (A_3 g + A_4 h)\varphi'_x, \\
Q'_z &= -A_1 Q - A_3 R, \\
R'_z &= -A_2 Q - A_4 R,
\end{aligned} \tag{58}$$

where A_1, A_2, A_3, and A_4 are arbitrary constants.

The solution of equation (58) is given by

$$\begin{aligned}
g(x) &= A_2 B_1 e^{k_1 \varphi} + A_2 B_2 e^{k_2 \varphi}, \\
h(x) &= (k_1 - A_1) B_1 e^{k_1 \varphi} + (k_2 - A_1) B_2 e^{k_2 \varphi}, \\
Q(z) &= A_3 B_3 e^{-k_1 z} + A_3 B_4 e^{-k_2 z}, \\
R(z) &= (k_1 - A_1) B_3 e^{-k_1 z} + (k_2 - A_1) B_4 e^{-k_2 z},
\end{aligned} \tag{59}$$

where B_1, B_2, B_3, and B_4 are arbitrary constants and k_1 and k_2 are roots of the quadratic equation

$$(k - A_1)(k - A_4) - A_2 A_3 = 0. \tag{60}$$

In the degenerate case $k_1 = k_2$ the terms $e^{k_2 \varphi}$ and $e^{-k_2 z}$ in (59) must be replaced by $\varphi e^{k_1 \varphi}$ and $z e^{-k_1 z}$, respectively. In the case of purely imaginary or complex roots, one should extract the real (or imaginary) part of the roots in solution (59).

On substituting (59) into the original functional equation, one obtains conditions that must be met by the free coefficients and identifies the function $f(t)$, specifically,

$$\begin{aligned}
B_2 = B_4 = 0 &\implies f(t) = [A_2 A_3 + (k_1 - A_1)^2] B_1 B_3 e^{-k_1 \psi}, \\
B_1 = B_3 = 0 &\implies f(t) = [A_2 A_3 + (k_2 - A_1)^2] B_2 B_4 e^{-k_2 \psi}, \\
A_1 = 0 &\implies f(t) = (A_2 A_3 + k_1^2) B_1 B_3 e^{-k_1 \psi} + (A_2 A_3 + k_2^2) B_2 B_4 e^{-k_2 \psi}.
\end{aligned} \tag{61}$$

Solution (59), (61) involves arbitrary functions $\varphi = \varphi(x)$ and $\psi = \psi(t)$.

Degenerate case. In addition, the functional equation has the two degenerate solutions:

$$f = B_1 B_2 e^{A_1 \psi}, \quad g = A_2 B_1 e^{-A_1 \varphi}, \quad h = B_1 e^{-A_1 \varphi}, \quad R = -B_2 e^{A_1 z} - A_2 Q,$$

where $\varphi = \varphi(x)$, $\psi = \psi(t)$, and $Q = Q(z)$ are arbitrary functions, A_1, A_2, B_1, and B_2 are arbitrary constants; and

$$f = B_1 B_2 e^{A_1 \psi}, \quad h = -B_1 e^{-A_1 \varphi} - A_2 g, \quad Q = A_2 B_2 e^{A_1 z}, \quad R = B_2 e^{A_1 z},$$

where $\varphi = \varphi(x)$, $\psi = \psi(t)$, and $g = g(x)$ are arbitrary functions, A_1, A_2, B_1, and B_2 are arbitrary constants. The degenerate solutions can be obtained immediately from the original equation or its consequence (57) using formulas (26b) in Section B.2.

Example 14. For the first-order nonlinear equation

$$\frac{\partial w}{\partial t} = \mathcal{F}(w)\left(\frac{\partial w}{\partial x}\right)^2 + \mathcal{G}(x),$$

the search for exact solutions in the form (53) leads to the functional equation B.3.5-3 in which

$$f(t) = -\psi'_t, \quad g(x) = (\varphi'_x)^2, \quad h(x) = \mathcal{G}(x), \quad Q(z) = \mathcal{F}(w) w'_z, \quad R(z) = 1/w'_z, \quad w = w(z).$$

B.3.5-4. Equation $f_1(x) + f_2(y) + g_1(x)P(z) + g_2(y)Q(z) + R(z) = 0$, $z = \varphi(x) + \psi(y)$.

Differentiating with respect to y and dividing the resulting relation by $\psi'_y P'_z$ and differentiating with respect to y, one arrives at the functional equation with two arguments y and z that is discussed in Section B.2 [see equation (3) and its solution (22)].

Example 15. Consider the following equation of steady-state heat transfer in an anisotropic inhomogeneous medium with a nonlinear source:

$$\frac{\partial}{\partial x}\left[a(x)\frac{\partial w}{\partial x}\right] + \frac{\partial}{\partial y}\left[b(y)\frac{\partial w}{\partial y}\right] = \mathcal{F}(w). \tag{62}$$

The search for exact solutions in the form $w = w(z)$, $z = \varphi(x) + \psi(y)$, leads to the functional equation B.3.5-4 in which

$$f_1(x) = a(x)\varphi''_{xx} + a'_x(x)\varphi'_x, \quad f_2(y) = b(y)\psi''_{yy} + b'_y(y)\psi'_y, \quad g_1(x) = a(x)(\varphi'_x)^2, \quad g_2(y) = b(y)(\psi'_y)^2,$$

$$P(z) = Q(z) = w''_{zz}/w'_z, \quad R(z) = -\mathcal{F}(w)/w'_z, \quad w = w(z).$$

Here we confine ourselves to studying functional separable solutions existing for arbitrary right-hand side $\mathcal{F}(w)$. With the change of variable $z = \zeta^2$, we look for solutions of equation (62) in the form

$$w = w(\zeta), \quad \zeta^2 = \varphi(x) + \psi(y). \tag{63}$$

Taking into account that $\frac{\partial \zeta}{\partial x} = \frac{\varphi'_x}{2\zeta}$ and $\frac{\partial \zeta}{\partial y} = \frac{\psi'_y}{2\zeta}$, we find from (62)

$$\left[(a\varphi'_x)'_x + (b\psi'_y)'_y\right]\frac{w'_\zeta}{2\zeta} + \left[a(\varphi'_x)^2 + b(\psi'_y)^2\right]\frac{\zeta w''_{\zeta\zeta} - w'_\zeta}{4\zeta^3} = \mathcal{F}(w), \quad \mathcal{F}(w) = \mathcal{F}(w(\zeta)). \tag{64}$$

For this functional differential equation to be solvable we require that the expressions in square brackets be functions of ζ:

$$(a\varphi'_x)'_x + (b\psi'_y)'_y = M(\zeta), \quad a(\varphi'_x)^2 + b(\psi'_y)^2 = N(\zeta).$$

Differentiating the first relation with respect to x and y yields the equation $(M'_\zeta/\zeta)'_\zeta = 0$, whose general solution is $M(\zeta) = C_1\zeta^2 + C_2$. Likewise, we find $N(\zeta) = C_3\zeta^2 + C_4$. Here, C_1, C_2, C_3, and C_4 are arbitrary constants. As a result, we have

$$(a\varphi'_x)'_x + (b\psi'_y)'_y = C_1(\varphi + \psi) + C_2, \quad a(\varphi'_x)^2 + b(\psi'_y)^2 = C_3(\varphi + \psi) + C_4.$$

The separation of variables results in a system of ordinary differential equations for $\varphi(x)$, $a(x)$, $\psi(y)$, and $b(y)$:

$$(a\varphi'_x)'_x - C_1\varphi - C_2 = k_1, \quad (b\psi'_y)'_y - C_1\psi = -k_1,$$

$$a(\varphi'_x)^2 - C_3\varphi - C_4 = k_2, \quad b(\psi'_y)^2 - C_3\psi = -k_2.$$

This system is always integrable in quadrature and can be rewritten as

$$(C_3\varphi + C_4 + k_2)\varphi''_{xx} + (C_1\varphi + C_2 + k_1 - C_3)(\varphi'_x)^2 = 0, \quad a = (C_3\varphi + C_4 + k_2)(\varphi'_x)^{-2};$$

$$(C_3\psi - k_2)\psi''_{yy} + (C_1\psi - k_1 - C_3)(\psi'_y)^2 = 0, \quad b = (C_3\psi - k_2)(\psi'_y)^{-2}. \tag{65}$$

Here, the equations for φ and ψ do not involve a and b and, hence, can be solved independently. Without full analysis of system (65), we note a special case where the system can be solved in explicit form.

For $C_1 = C_2 = C_4 = k_1 = k_2 = 0$ and $C_3 = C \neq 0$, we find

$$a(x) = \alpha e^{\mu x}, \quad b(y) = \beta e^{\nu y}, \quad \varphi(x) = \frac{Ce^{-\mu x}}{\alpha\mu^2}, \quad \psi(y) = \frac{Ce^{-\nu y}}{\beta\nu^2},$$

where α, β, μ, and ν are arbitrary constants. Substituting these expressions into (64) and taking into account (63), we obtain the ordinary differential equation for $w(\zeta)$

$$w''_{\zeta\zeta} - \frac{1}{\zeta}w'_\zeta = \frac{4}{C}\mathcal{F}(w).$$

System (65) has other solutions as well; these lead to various expressions of $a(x)$ and $b(y)$. Table B3 lists the cases where these functions can be written in explicit form (the traveling wave solution, which corresponds to $a = \text{const}$ and $b = \text{const}$, is omitted). In general, the solution of system (64) enables one to represent $a(x)$ and $b(y)$ in parametric form.

⊙ *Reference*: V. F. Zaitsev and A. D. Polyanin (1996), A. D. Polyanin and A. I. Zhurov (1998).

TABLE B3
Functional separable solutions of the form $w = w(\zeta)$, $\zeta^2 = \varphi(x) + \psi(y)$, for heat equations in an anisotropic inhomogeneous medium with an arbitrary nonlinear source. Notation: C, α, β, μ, ν, n, and k are free parameters ($C \neq 0$, $\mu \neq 0$, $\nu \neq 0$, $n \neq 2$, and $k \neq 2$).

Heat equation	Functions $\varphi(x)$ and $\psi(y)$	Equation for $w = w(\zeta)$
$\frac{\partial}{\partial x}\left(\alpha x^m \frac{\partial w}{\partial x}\right) + \frac{\partial}{\partial y}\left(\beta y^n \frac{\partial w}{\partial y}\right) = \mathcal{F}(w)$	$\varphi = \frac{Cx^{2-m}}{\alpha(2-m)^2}$, $\psi = \frac{Cy^{2-n}}{\beta(2-n)^2}$	$w''_{\zeta\zeta} + \frac{4-mn}{(2-m)(2-n)} \frac{1}{\zeta} w'_\zeta = \frac{4}{C}\mathcal{F}(w)$
$\frac{\partial}{\partial x}\left(\alpha e^{\mu x} \frac{\partial w}{\partial x}\right) + \frac{\partial}{\partial y}\left(\beta e^{\nu y} \frac{\partial w}{\partial y}\right) = \mathcal{F}(w)$	$\varphi = \frac{C}{\alpha \mu^2} e^{-\mu x}$, $\psi = \frac{C}{\beta \nu^2} e^{-\nu y}$	$w''_{\zeta\zeta} - \frac{1}{\zeta} w'_\zeta = \frac{4}{C}\mathcal{F}(w)$
$\frac{\partial}{\partial x}\left(\alpha e^{\mu x} \frac{\partial w}{\partial x}\right) + \frac{\partial}{\partial y}\left(\beta y^n \frac{\partial w}{\partial y}\right) = \mathcal{F}(w)$	$\varphi = \frac{C}{\alpha \mu^2} e^{-\mu x}$, $\psi = \frac{Cy^{2-n}}{\beta(2-n)^2}$	$w''_{\zeta\zeta} + \frac{n}{2-n} \frac{1}{\zeta} w'_\zeta = \frac{4}{C}\mathcal{F}(w)$
$\frac{\partial}{\partial x}\left(\alpha x^2 \frac{\partial w}{\partial x}\right) + \frac{\partial}{\partial y}\left(\beta y^2 \frac{\partial w}{\partial y}\right) = \mathcal{F}(w)$	$\varphi = \mu \ln|x|$, $\psi = \nu \ln|y|$	Equation (64); both expressions in square brackets are constant
$\alpha \frac{\partial^2 w}{\partial x^2} + \frac{\partial}{\partial y}\left(\beta y^2 \frac{\partial w}{\partial y}\right) = \mathcal{F}(w)$	$\varphi = \mu x$, $\psi = \nu \ln|y|$	Equation (64); both expressions in square brackets are constant

B.4. First-Order Nonlinear Equations

B.4.1. Preliminary Remarks

For first-order partial differential equations with two independent variables, an exact solution

$$w = \Phi(x, y, C_1, C_2) \tag{1}$$

that depends on two arbitrary constants C_1 and C_2 is called a complete integral. The general integral (general solution) can be represented in parametric form by using the complete integral (1) and the two equations

$$\begin{aligned} C_2 &= f(C_1), \\ \frac{\partial \Phi}{\partial C_1} &+ \frac{\partial \Phi}{\partial C_2} f'(C_1) = 0, \end{aligned} \tag{2}$$

where f is an arbitrary function and the prime stands for the derivative. For details, see Kamke (1965), Courant and Hilbert (1989), and Polyanin, Zaitsev, and Moussiaux (2001).

The first-order equations with two independent variables considered below are purely illustrative. The book by Polyanin, Zaitsev, and Moussiaux (2001) presents many more first-order nonlinear equations that admit generalized separable solutions (without specifying the method for obtaining them).

B.4.2. Individual Equations

1. $\dfrac{\partial w}{\partial x} - f(y) w \dfrac{\partial w}{\partial y} = g(x) w + h(x).$

Exact solution:

$$w = \varphi(x) \int \frac{dy}{f(y)} + \psi(x),$$

where

$$\varphi(x) = G(x) \left[C_1 - \int G(x)\, dx \right]^{-1}, \qquad G(x) = \exp\left[\int g(x)\, dx\right],$$

$$\psi(x) = S(x) \left[C_2 + \int \frac{h(x)}{S(x)}\, dx \right], \qquad S(x) = G(x) \exp\left[\int \varphi(x)\, dx\right].$$

2. $\dfrac{\partial w}{\partial x} + f(y)w\dfrac{\partial w}{\partial y} = aw^2 + g(x)w + h(x).$

Exact solution:

$$w = \varphi(x) + \psi(x)\exp\left[a\int\dfrac{dy}{f(y)}\right], \qquad \psi(x) = C_1\exp\left\{\int\left[a\varphi(x) + g(x)\right]dx\right\},$$

where C_1 is an arbitrary constant and the function $\varphi(x)$ is determined by the Riccati equation

$$\varphi'_x = a\varphi^2 + g(x)\varphi + h(x).$$

This equation is integrable in quadrature for a lot of specific functions $g(x)$ and $h(x)$ [e.g., for $h(x)\equiv 0$ and any $g(x)$]. For details, see the books by Kamke (1977) and Polyanin and Zaitsev (1995).

3. $\dfrac{\partial w}{\partial x}\dfrac{\partial w}{\partial y} = f(x)y^k + g(x)y^{2k+1}.$

Exact solutions:

$$w = \varphi(x)y^{k+1} + \dfrac{1}{k+1}\int\dfrac{f(x)}{\varphi(x)}\,dx + C_1, \qquad \varphi(x) = \pm\left[\dfrac{2}{k+1}\int g(x)\,dx + C_2\right]^{1/2}.$$

4. $\dfrac{\partial w}{\partial x}\dfrac{\partial w}{\partial y} = f(x)e^{\lambda y} + g(x)e^{2\lambda y}.$

Exact solutions:

$$w = \varphi(x)e^{\lambda y} + \dfrac{1}{\lambda}\int\dfrac{f(x)}{\varphi(x)}\,dx + C_1, \qquad \varphi(x) = \pm\left[\dfrac{2}{\lambda}\int g(x)\,dx + C_2\right]^{1/2}.$$

5. $\dfrac{\partial w}{\partial x} + a\left(\dfrac{\partial w}{\partial y}\right)^2 = f(x)y + g(x).$

Exact solution:

$$w = \varphi(x)y + \int\left[g(x) - a\varphi^2(x)\right]dx + C_1, \qquad \varphi(x) = \int f(x)\,dx + C_2.$$

6. $\dfrac{\partial w}{\partial x} + a\left(\dfrac{\partial w}{\partial y}\right)^2 = f(x)y^2 + g(x)y + h(x).$

Exact solution:

$$w = \varphi(x)y^2 + \psi(x)y + \chi(x),$$

where the functions $\varphi(x)$, $\psi(x)$, and $\chi(x)$ are determined by solving the following system of ordinary differential equations:

$$\varphi'_x = -4a\varphi^2 + f(x), \tag{1}$$
$$\psi'_x = -4a\varphi\psi + g(x), \tag{2}$$
$$\chi'_x = -a\psi^2 + h(x). \tag{3}$$

The Riccati equation (1) can be integrated in quadrature for numerous $f(x)$. For details, see Kamke (1977) and Polyanin and Zaitsev (1995). Given a solution of equation (1), equations (2) and (3) are easy to integrate, because they are linear in the unknowns ψ and χ.

7. $\dfrac{\partial w}{\partial x} + a\left(\dfrac{\partial w}{\partial y}\right)^2 = bw^2 + f(x)w + g(x).$

1°. Exact solution for $b = 0$:
$$w = F(x)(C_1 + C_2 y) + F(x)\int \left[g(x) - aC_2^2 F^2(x)\right]\dfrac{dx}{F(x)}, \qquad F(x) = \exp\left[\int f(x)\,dx\right].$$

2°. Exact solution for $b \ne 0$:
$$w = \varphi(x) + \psi(x)\exp\!\left(\pm y\sqrt{b/a}\right), \qquad \psi(x) = C_1 \exp\!\left\{\int \left[2b\varphi(x) + f(x)\right]dx\right\}.$$

The function $\varphi = \varphi(x)$ is determined by the Riccati equation
$$\varphi' = b\varphi^2 + f(x)\varphi + g(x).$$

This equation can be integrated in quadrature for various f and g, in particular, for $g(x) \equiv 0$ and arbitrary $f(x)$ and for $f(x) \equiv \text{const}$ and $g(x) \equiv \text{const}$. For details, see the books by Kamke (1977) and Polyanin and Zaitsev (1995).

8. $f_1(x)\left(\dfrac{\partial w}{\partial x}\right)^2 + f_2(y)\left(\dfrac{\partial w}{\partial y}\right)^2 = g_1(x) + g_2(y).$

This equation is encountered in differential geometry in studying geodesic lines of Liouville surfaces. Exact solutions:
$$w = \pm\int\sqrt{\dfrac{g_1(x) + C_1}{f_1(x)}}\,dx \pm \int\sqrt{\dfrac{g_2(y) - C_1}{f_2(y)}}\,dy + C_2.$$

The signs before each of the integrals can be chosen independently of each other.

⊙ *References*: P. Appell (1953), E. Kamke (1965).

9. $\dfrac{\partial w}{\partial x} + f\!\left(\dfrac{\partial w}{\partial y}\right) = g(x)y + h(x).$

Exact solution:
$$w = \varphi(x)y + \int\left[h(x) - f(\varphi(x))\right]dx + C_1, \qquad \varphi(x) = \int g(x)\,dx + C_2.$$

10. $\dfrac{\partial w}{\partial x} + f\!\left(\dfrac{\partial w}{\partial y}\right) = g(x)w + h(x).$

Exact solution:
$$w = (C_1 y + C_2)\varphi(x) + \varphi(x)\int\left[h(x) - f(C_1\varphi(x))\right]\dfrac{dx}{\varphi(x)}, \qquad \varphi(x) = \exp\left[\int g(x)\,dx\right].$$

11. $F_1\!\left(x, \dfrac{\partial w}{\partial x}\right) + e^{\lambda w} F_2\!\left(y, \dfrac{\partial w}{\partial y}\right) = 0.$

Exact solution:
$$w = \varphi(x) + \psi(y).$$

The functions $\varphi = \varphi(x)$ and $\psi = \psi(y)$ are determined by solving the ordinary differential equations
$$e^{-\lambda\varphi} F_1(x, \varphi'_x) = C, \qquad e^{\lambda\psi} F_2(y, \psi'_y) = -C,$$

where C is an arbitrary constant.

B.5. Second-Order Nonlinear Equations
B.5.1. Parabolic Equations

B.5.1-1. Equations of the form $\frac{\partial w}{\partial t} = a\frac{\partial^2 w}{\partial x^2} + F(x, t, w)$.

1. $\dfrac{\partial w}{\partial t} = a\dfrac{\partial^2 w}{\partial x^2} + bw \ln w + [f(x) + g(t)]w.$

Exact solution with multiplicative form:
$$w(x,t) = \exp\left[Ce^{bt} + e^{bt}\int e^{-bt}g(t)\,dt\right]\varphi(x),$$

where C is an arbitrary constant and the function $\varphi(t)$ is determined by solving the ordinary differential equation
$$a\varphi''_{xx} + b\varphi \ln \varphi + f(x)\varphi = 0.$$

2. $\dfrac{\partial w}{\partial t} = a\dfrac{\partial^2 w}{\partial x^2} + f(t)w \ln w + g(t)w.$

1°. Exact solution:
$$w(x,t) = \exp[\Phi(t)x + \Psi(t)],$$

where the functions $\Phi(t)$ and $\Psi(t)$ are given by
$$\Phi(t) = Ae^F, \quad \Psi(t) = Be^F + e^F\int e^{-F}(aA^2e^{2F} + g)\,dt, \quad F = \int f\,dt,$$

and A and B are arbitrary constants.

2°. Exact solution:
$$w(x,t) = \exp[\varphi(t)x^2 + \psi(t)],$$

where $\varphi(t)$ and $\psi(t)$ are given by
$$\varphi(t) = e^F\left(A - 4a\int e^F\,dt\right)^{-1}, \quad \psi(t) = Be^F + e^F\int e^{-F}(2a\varphi + g)\,dt, \quad F = \int f\,dt,$$

and A and B are arbitrary constants.

3°. There are also exact solutions of the more general form
$$w(x,t) = \exp[\varphi_2(t)x^2 + \varphi_1(t)x + \varphi_0(t)],$$

where the functions $\varphi_2(t)$, $\varphi_1(t)$, and $\varphi_0(t)$ are determined by a system of ordinary differential equations that can be integrated.

⊙ *Reference*: V. F. Zaitsev, A. D. Polyanin (1996).

3. $\dfrac{\partial w}{\partial t} = a\dfrac{\partial^2 w}{\partial x^2} + f(t)w \ln w + [g(t)x^2 + h(t)x + s(t)]w.$

Exact solution:
$$w(x,t) = \exp[\varphi_2(t)x^2 + \varphi_1(t)x + \varphi_0(t)],$$

where the functions $\varphi_n(t)$ ($n = 1, 2, 3$) are determined by solving the following system of first-order ordinary differential equations with variable coefficients:
$$\varphi'_2 = 4a\varphi_2^2 + f\varphi_2 + g,$$
$$\varphi'_1 = 4a\varphi_2\varphi_1 + f\varphi_1 + h,$$
$$\varphi'_0 = f\varphi_0 + a\varphi_1^2 + 2a\varphi_2 + s$$

(the arguments of f, g, h, and s are not specified and the prime denotes the derivative with respect to t).

4. $\dfrac{\partial w}{\partial t} = a\dfrac{\partial^2 w}{\partial x^2} + [k\ln^2 w + f(t)\ln w + g(t)]w.$

The change of variable $w = \exp u$ leads to an equation of the form B.5.1.14,

$$\dfrac{\partial u}{\partial t} = a\dfrac{\partial^2 u}{\partial x^2} + a\left(\dfrac{\partial u}{\partial x}\right)^2 + ku^2 + f(t)u + g(t),$$

which has exponential and sinusoidal solutions in x.

B.5.1-2. Equations of the form $\dfrac{\partial w}{\partial t} = a\dfrac{\partial^2 w}{\partial x^2} + F\left(x, t, w, \dfrac{\partial w}{\partial x}\right).$

5. $\dfrac{\partial w}{\partial t} = \dfrac{a}{x^n}\dfrac{\partial}{\partial x}\left(x^n \dfrac{\partial w}{\partial x}\right) + f(t)w \ln w.$

Exact solution:
$$w(x, t) = \exp[\varphi(t)x^2 + \psi(t)],$$

where the functions $\varphi(t)$ and $\psi(t)$ are determined by solving the following system of first-order ordinary differential equations with variable coefficients (the arguments of f and g are not specified):

$$\varphi'_t = 4a\varphi^2 + f\varphi,$$
$$\psi'_t = 2a(n+1)\varphi + f\psi.$$

Integrating successively yields

$$\varphi(t) = e^F\left(A - 4a\int e^F\, dt\right)^{-1}, \quad \psi(t) = Be^F + 2a(n+1)e^F\int \varphi e^{-F}\, dt, \quad F = \int f\, dt,$$

where A and B are arbitrary constants.

6. $\dfrac{\partial w}{\partial t} = a\dfrac{\partial^2 w}{\partial x^2} + [xf(t) + g(t)]\dfrac{\partial w}{\partial x} + h(t)w\ln w + [xp(t) + s(t)]w.$

Exact solution:
$$w(x, t) = \exp[x\varphi(t) + \psi(t)],$$

where the functions $\varphi(t)$ and $\psi(t)$ are determined by solving the following system of first-order ordinary differential equations with variable coefficients:

$$\varphi'_t = [f(t) + h(t)]\varphi + p(t), \tag{1}$$
$$\psi'_t = h(t)\psi + a\varphi^2 + g(t)\varphi + s(t). \tag{2}$$

Integrating first (1) and then (2), we obtain (C_1 and C_2 are arbitrary constants)

$$\varphi(t) = C_1 E(t) + E(t)\int \dfrac{p(t)}{E(t)}\, dt, \quad E(t) = \exp\left[\int f(t)\, dt + \int h(t)\, dt\right],$$
$$\psi(t) = C_2 H(t) + H(t)\int \dfrac{a\varphi^2(t) + g(t)\varphi(t) + s(t)}{H(t)}\, dt, \quad H(t) = \exp\left[\int h(t)\, dt\right].$$

7. $\dfrac{\partial w}{\partial t} = a\dfrac{\partial^2 w}{\partial x^2} + [xf(t) + g(t)]\dfrac{\partial w}{\partial x} + h(t)w\ln w + [x^2 r(t) + xp(t) + s(t)]w.$

Exact solution:
$$w(x, t) = \exp[x^2\varphi(t) + x\psi(t) + \chi(t)],$$

where the functions $\varphi(t)$, $\psi(t)$, and $\chi(t)$ are determined by solving the following system of first-order ordinary differential equations with variable coefficients:

$$\varphi'_t = 4a\varphi^2 + (2f + h)\varphi + r,$$
$$\psi'_t = (4a\varphi + f + h)\psi + 2g\varphi + p,$$
$$\chi'_t = h\chi + 2a\varphi + a\psi^2 + g\psi + s.$$

8. $\dfrac{\partial w}{\partial t} = a\dfrac{\partial^2 w}{\partial x^2} + \left[xf(t) + \dfrac{g(t)}{x}\right]\dfrac{\partial w}{\partial x} + h(t)w \ln w + [x^2 p(t) + s(t)]w.$

Exact solution:
$$w(x,t) = \exp\left[\varphi(t)x^2 + \psi(t)\right],$$

where the functions $\varphi(t)$ and $\psi(t)$ are determined by solving the following system of first-order ordinary differential equations with variable coefficients:

$$\varphi'_t = 4a\varphi^2 + (2f + h)\varphi + p,$$
$$\psi'_t = h\psi + 2(a+g)\varphi + s.$$

9. $\dfrac{\partial w}{\partial t} = a\dfrac{\partial^2 w}{\partial x^2} + b\left(\dfrac{\partial w}{\partial x}\right)^2 + f(t)x^2 + g(t)x + h(t).$

Exact solution:
$$w(x,t) = \varphi(t)x^2 + \psi(t)x + \chi(t),$$

where the functions $\varphi(t)$, $\psi(t)$, and $\chi(t)$ are determined by solving the following system of first-order ordinary differential equations with variable coefficients:

$$\varphi'_t = 4b\varphi^2 + f,$$
$$\psi'_t = 4b\varphi\psi + g,$$
$$\chi'_t = 2a\varphi + b\psi^2 + h.$$

10. $\dfrac{\partial w}{\partial t} = a\dfrac{\partial^2 w}{\partial x^2} + b\left(\dfrac{\partial w}{\partial x}\right)^2 + cw + f(x) + g(t).$

Exact solution in additive form:
$$w(x,t) = \varphi(x) + Ae^{ct} + e^{ct}\int e^{-ct}g(t)\,dt.$$

Here, A is an arbitrary constant and the function $\varphi(x)$ is determined by the nonlinear ordinary differential equation
$$a\varphi''_{xx} + b(\varphi'_x)^2 + c\varphi + f(x) = 0.$$

By changing variable $\varphi'_x = \dfrac{a}{b}\dfrac{\psi'_x}{\psi}$ this equation is reduced to the second-order linear equation
$$a^2\psi''_{xx} + ac\psi'_x + bf(x)\psi = 0.$$

11. $\dfrac{\partial w}{\partial t} = a\dfrac{\partial^2 w}{\partial x^2} + b\left(\dfrac{\partial w}{\partial x}\right)^2 + f(x)\dfrac{\partial w}{\partial x} + kw + g(x) + h(t).$

Exact solution in additive form:
$$w(x,t) = \varphi(x) + Ce^{kt} + e^{kt}\int e^{-kt}h(t)\,dt,$$

where C is an arbitrary constant and the function $\varphi(x)$ is determined by the second-order ordinary differential equation with variable coefficients

$$a\varphi''_{xx} + b(\varphi'_x)^2 + f(x)\varphi'_x + k\varphi + g(x) = 0.$$

12. $\dfrac{\partial w}{\partial t} = a\dfrac{\partial^2 w}{\partial x^2} + b\left(\dfrac{\partial w}{\partial x}\right)^2 + cw\dfrac{\partial w}{\partial x} + kw^2 + f(t)w + g(t).$

The equation has exact solutions of the form

$$w(x,t) = \varphi(t) + \psi(t)\exp(\lambda x),$$

where λ is a root of the quadratic equation $b\lambda^2 + c\lambda + k = 0$.

13. $\dfrac{\partial w}{\partial t} = a\dfrac{\partial^2 w}{\partial x^2} + f(t)\left(\dfrac{\partial w}{\partial x}\right)^2 + g(t)w + h(t).$

Exact solution:

$$w(x,t) = \varphi(t)x^2 + \psi(t)x + \chi(t),$$

where the functions $\varphi(t)$, $\psi(t)$, and $\chi(t)$ are determined by solving the following system of first-order ordinary differential equations with variable coefficients:

$$\varphi'_t = 4f\varphi^2 + g\varphi, \tag{1}$$
$$\psi'_t = (4f\varphi + g)\psi, \tag{2}$$
$$\chi'_t = g\chi + 2a\varphi + f\psi^2 + h. \tag{3}$$

Equation (1) for φ is a Bernoulli equation, which is easy to integrate. After that, equations (2) and (3), which are linear in ψ and χ, are integrated successively. As a result, we find

$$\varphi = e^G \left(A_1 - 4\int e^G f\, dt\right)^{-1}, \quad G = \int g\, dt,$$

$$\psi = A_2 \exp\left[\int (4f\varphi + g)\, dt\right],$$

$$\chi = A_3 e^G + e^G \int e^{-G}(2a\varphi + f\psi^2 + h)\, dt,$$

where A_1, A_2, and A_3 are arbitrary constants. A degenerate solution with $\varphi \equiv 0$ corresponds to the limit case $A_1 \to \infty$.

⊙ *Reference*: V. F. Zaitsev, A. D. Polyanin (1996).

14. $\dfrac{\partial w}{\partial t} = a\dfrac{\partial^2 w}{\partial x^2} + f(t)\left(\dfrac{\partial w}{\partial x}\right)^2 + bf(t)w^2 + g(t)w + h(t).$

1°. Exact solution:

$$w(x,t) = \varphi(t) + \psi(t)\exp\bigl(\pm x\sqrt{-b}\,\bigr), \quad b < 0, \tag{1}$$

where the functions $\varphi(t)$ and $\psi(t)$ are determined by solving the following first-order ordinary differential equations with variable coefficients (the arguments of f, g, and h are not specified):

$$\varphi'_t = bf\varphi^2 + g\varphi + h, \tag{2}$$
$$\psi'_t = (2bf\varphi + g - ab)\psi. \tag{3}$$

Equation (2) for $\varphi(t)$ is a Riccati equation; it can be reduced to a second-order linear equation. Many solutions of equation (2) for various f, g, and h can be found in Kamke (1977) and Polyanin and Zaitsev (1995).

Whenever a solution of equation (2) is known, the solution of equation (3) for $\psi(t)$ can be evaluated from

$$\psi(t) = C\exp\left[-abt + \int(2bf\varphi + g)\, dt\right],$$

where C is an arbitrary constant.

2°. Exact solution of a more general form:
$$w(x,t) = \varphi(t) + \psi(t)\left[A\exp(x\sqrt{-b}) + B\exp(-x\sqrt{-b})\right], \qquad b < 0,$$
where the functions $\varphi(t)$ and $\psi(t)$ are determined by solving the following system of first-order ordinary differential equations with variable coefficients:
$$\varphi'_t = bf(\varphi^2 + 4AB\psi^2) + g\varphi + h, \tag{4}$$
$$\psi'_t = 2bf\varphi\psi + g\psi - ab\psi. \tag{5}$$

One can express φ in terms of ψ from (5) and then substitute into (4). As a result, one obtains a second-order nonlinear equation for ψ; if f, g, h = const, this equation is autonomous and, hence, admits reduction of order.

3°. Exact solution (c is an arbitrary constant):
$$w(x,t) = \varphi(t) + \psi(t)\cos(x\sqrt{b} + c), \qquad b > 0, \tag{6}$$
where the functions $\varphi(t)$ and $\psi(t)$ are determined by solving the following system of first-order ordinary differential equations with variable coefficients:
$$\varphi'_t = bf(\varphi^2 + \psi^2) + g\varphi + h, \tag{7}$$
$$\psi'_t = 2bf\varphi\psi + g\psi - ab\psi. \tag{8}$$

One can express φ in terms of ψ from (8) and then substitute into (7). As a result, one obtains a second-order nonlinear equation for ψ; if f, g, h = const, this equation is autonomous and, hence, admits reduction of order.

⊙ *Reference*: V. F. Zaitsev, A. D. Polyanin (1996).

15. $\dfrac{\partial w}{\partial t} = a\dfrac{\partial^2 w}{\partial x^2} + f(t)\left(\dfrac{\partial w}{\partial x}\right)^2 + bf(t)w\dfrac{\partial w}{\partial x} + cf(t)w^2 + g(t)w + h(t).$

The equation has exact solutions of the form
$$w(x,t) = \varphi(t) + \psi(t)\exp(\lambda x),$$
where λ is a root of the quadratic equation $\lambda^2 + b\lambda + c = 0$.

16. $\dfrac{\partial w}{\partial t} = a\dfrac{\partial^2 w}{\partial x^2} + f(x)\left(\dfrac{\partial w}{\partial x}\right)^2 + g(x)\dfrac{\partial w}{\partial x} + bw + h(x) + p(t).$

Exact solution in additive form:
$$w(x,t) = \varphi(x) + Ce^{bt} + e^{bt}\int e^{-bt}p(t)\,dt,$$
where C is an arbitrary constant and the function $\varphi(x)$ is determined by the following second-order ordinary differential equation with variable coefficients:
$$a\varphi''_{xx} + f(x)(\varphi'_x)^2 + g(x)\varphi'_x + b\varphi + h(x) = 0.$$

17. $\dfrac{\partial w}{\partial t} = a\dfrac{\partial^2 w}{\partial x^2} + f(t)\left(\dfrac{\partial w}{\partial x}\right)^2 + [g_1(t)x + g_0(t)]\dfrac{\partial w}{\partial x} + h(t)w + p(t)x^2 + q(t)x + s(t).$

Exact solution:
$$w(x,t) = \varphi(t)x^2 + \psi(t)x + \chi(t),$$
where the functions $\varphi(t)$, $\psi(t)$, and $\chi(t)$ are determined by solving the following system of first-order ordinary differential equations with variable coefficients:
$$\varphi'_t = 4f\varphi^2 + (2g_1 + h)\varphi + p, \tag{1}$$
$$\psi'_t = (4f\varphi + g_1 + h)\psi + 2g_0\varphi + q, \tag{2}$$
$$\chi'_t = h\chi + 2a\varphi + f\psi^2 + g_0\psi + s. \tag{3}$$

Equation (1) for $\varphi(t)$ is a Riccati equation; it can be reduced to a second-order linear equation. For solutions of Riccati equations, see Kamke (1977) and Polyanin and Zaitsev (1995). Whenever a solution of equation (1) is known, the solutions of equations (2) and (3) can be obtained successively (the equations are linear in ψ and χ).

18. $\dfrac{\partial w}{\partial t} = a\dfrac{\partial^2 w}{\partial x^2} + f\!\left(x, \dfrac{\partial w}{\partial x}\right) + bw + g(t).$

Exact solution in additive form:
$$w(x,t) = \varphi(x) + Ce^{bt} + e^{bt}\int e^{-bt} g(t)\, dt,$$

where C is an arbitrary constant and the function $\varphi(x)$ is described by the second-order ordinary differential equation
$$a\varphi''_{xx} + f(x, \varphi'_x) + b\varphi = 0.$$

B.5.1-3. Equations of the form $\dfrac{\partial w}{\partial t} = f(x,t)\dfrac{\partial^2 w}{\partial x^2} + g\!\left(x,t,w,\dfrac{\partial w}{\partial x}\right).$

19. $\dfrac{\partial w}{\partial t} = \dfrac{f(t)}{x^n}\dfrac{\partial}{\partial x}\!\left(x^n\dfrac{\partial w}{\partial x}\right) + g(t)w\ln w.$

Exact solution:
$$w(x,t) = \exp\!\bigl[\varphi(t)x^2 + \psi(t)\bigr],$$

where the functions $\varphi(t)$ and $\psi(t)$ are determined by solving the following system of first-order ordinary differential equations with variable coefficients (the arguments of f and g are not specified):
$$\varphi'_t = 4f\varphi^2 + g\varphi,$$
$$\psi'_t = 2(n+1)f\varphi + g\psi.$$

Integrating successively yields
$$\varphi(t) = e^G\!\left(A - 4\int fe^G\, dt\right)^{-1},\quad \psi(t) = Be^G + 2(n+1)e^G\int f\varphi e^{-G}\, dt,\quad G = \int g\, dt,$$

where A and B are arbitrary constants.

20. $\dfrac{\partial w}{\partial t} = f(t)\dfrac{\partial^2 w}{\partial x^2} + \left[xg(t) + \dfrac{h(t)}{x}\right]\dfrac{\partial w}{\partial x} + s(t)w\ln w + [x^2 p(t) + q(t)]w.$

Exact solution:
$$w(x,t) = \exp\!\bigl[\varphi(t)x^2 + \psi(t)\bigr],$$

where the functions $\varphi(t)$ and $\psi(t)$ are determined by solving the following system of first-order ordinary differential equations with variable coefficients:
$$\varphi'_t = 4f\varphi^2 + (2g+s)\varphi + p, \tag{1}$$
$$\psi'_t = s\psi + 2(f+h)\varphi + q. \tag{2}$$

The Riccati equation (1) for the function $\varphi(t)$ can be reduced to a second-order linear equation. For solutions of the Riccati equation, see Kamke (1977) and Polyanin and Zaitsev (1995). On solving (1), one can determine the solution of the linear equation (2) for $\psi(t)$.

21. $\dfrac{\partial w}{\partial t} = f(t)\dfrac{\partial}{\partial x}\!\left(e^{\lambda x}\dfrac{\partial w}{\partial x}\right) + g(t)w\ln w + h(t)w.$

Exact solution:
$$w(x,t) = \exp\!\bigl[\varphi(t)e^{-\lambda x} + \psi(t)\bigr],$$

where the functions $\varphi(t)$ and $\psi(t)$ are determined by the ordinary differential equations
$$\varphi'_t = \lambda^2 f(t)\varphi^2 + g(t)\varphi,$$
$$\psi'_t = g(t)\psi + h(t).$$

Integrating yields

$$\varphi(t) = G(t)\left[A - \lambda^2 \int f(t)G(t)\,dt\right]^{-1}, \quad G(t) = \exp\left[\int g(t)\,dt\right],$$

$$\psi(t) = BG(t) + G(t)\int \frac{h(t)}{G(t)}\,dt,$$

where A and B are arbitrary constants.

22. $\quad \dfrac{\partial w}{\partial t} = \dfrac{\partial}{\partial x}\left[f(x)\dfrac{\partial w}{\partial x}\right] + aw \ln w.$

Exact solution:

$$w(x,t) = \exp\left[Ae^{at} + \varphi(x)\right],$$

where A is an arbitrary constant and the function $\varphi(x)$ is determined by the ordinary differential equation

$$(f\varphi'_x)'_x + f(\varphi'_x)^2 + a\varphi = 0.$$

23. $\quad \dfrac{\partial w}{\partial t} = \dfrac{\partial}{\partial x}\left[f(x)\dfrac{\partial w}{\partial x}\right] + aw \ln w + [g(x) + h(t)]w.$

Exact solution in multiplicative form:

$$w(x,t) = \exp\left[Ce^{at} + e^{at}\int e^{-at}h(t)\,dt\right]\varphi(x),$$

where C is an arbitrary constant and the function $\varphi(x)$ is determined by the ordinary differential equation

$$(f\varphi'_x)'_x + a\varphi \ln \varphi + g(x)\varphi = 0.$$

24. $\quad \dfrac{\partial w}{\partial t} = f(x)\dfrac{\partial^2 w}{\partial x^2} + g(x)\dfrac{\partial w}{\partial x} + aw \ln w + [h(x) + s(t)]w.$

Exact solution in multiplicative form:

$$w(x,t) = \exp\left[Ce^{at} + e^{at}\int e^{-at}s(t)\,dt\right]\varphi(x),$$

where C is an arbitrary constant and the function $\varphi(x)$ is determined by the ordinary differential equation

$$f(x)\varphi''_{xx} + g(x)\varphi'_x + a\varphi \ln \varphi + h(x)\varphi = 0.$$

25. $\quad \dfrac{\partial w}{\partial t} = f(x)\dfrac{\partial^2 w}{\partial x^2} + g\!\left(x, \dfrac{\partial w}{\partial x}\right) + aw + h(t).$

Exact solution in additive form:

$$w(x,t) = \varphi(x) + Ce^{at} + e^{at}\int e^{-at}h(t)\,dt,$$

where C is an arbitrary constant and the function $\varphi(x)$ is determined by the following second-order ordinary differential equation:

$$f(x)\varphi''_{xx} + g(x, \varphi'_x) + a\varphi = 0.$$

B.5.1-4. Equations of the form $\frac{\partial w}{\partial t} = aw\frac{\partial^2 w}{\partial x^2} + f\left(x, t, w, \frac{\partial w}{\partial x}\right)$.

26. $\dfrac{\partial w}{\partial t} = aw\dfrac{\partial^2 w}{\partial x^2} + f(x)w + bx + c.$

Exact solution:
$$w(x,t) = (bx+c)t + Ax + B - \frac{1}{a}\int_{x_0}^{x}(x-\xi)f(\xi)\,d\xi,$$

where A, B, and x_0 are arbitrary constants.

27. $\dfrac{\partial w}{\partial t} = aw\dfrac{\partial^2 w}{\partial x^2} + f(t)w + g(t).$

1°. Exact solution:
$$w(x,t) = F(t)(Ax+B) + F(t)\int \frac{g(t)}{F(t)}\,dt, \quad F(t) = \exp\left[\int f(t)\,dt\right],$$

where A and B are arbitrary constants.

2°. Exact solution:
$$w(x,t) = \varphi(t)(x^2 + Ax + B) + \varphi(t)\int \frac{g(t)}{\varphi(t)}\,dt,$$
$$\varphi(t) = F(t)\left[C - 2a\int F(t)\,dt\right]^{-1}, \quad F(t) = \exp\left[\int f(t)\,dt\right],$$

where A, B, and C are arbitrary constants.

28. $\dfrac{\partial w}{\partial t} = aw\dfrac{\partial^2 w}{\partial x^2} + f(x)w\dfrac{\partial w}{\partial x} + g(t)w + h(t).$

Exact solution:
$$w(x,t) = \varphi(t)\Theta(x) + \psi(t),$$

where the functions $\varphi(t)$, $\psi(t)$, and $\Theta(x)$ are described by ordinary differential equations
$$\varphi'_t = C\varphi^2 + g(t)\varphi,$$
$$\psi'_t = [C\varphi + g(t)]\psi + h(t),$$
$$a\Theta''_{xx} + f(x)\Theta'_x = C,$$

where C is an arbitrary constant. Integrating successively yields
$$\varphi(t) = G(t)\left[A_1 - C\int G(t)\,dt\right]^{-1}, \quad G(t) = \exp\left[\int g(t)\,dt\right],$$
$$\psi(t) = A_2\varphi(t) + \varphi(t)\int \frac{h(t)}{\varphi(t)}\,dt,$$
$$\Theta(x) = B_1\int \frac{dx}{F(x)} + B_2 + \frac{C}{a}\int\left[\int F(x)\,dx\right]\frac{dx}{F(x)}, \quad F(x) = \exp\left[\frac{1}{a}\int f(x)\,dx\right],$$

where A_1, A_2, B_1, and B_2 are arbitrary constants.

29. $\dfrac{\partial w}{\partial t} = aw\dfrac{\partial^2 w}{\partial x^2} + f(t)\left(\dfrac{\partial w}{\partial x}\right)^2 + g(t)\dfrac{\partial w}{\partial x} + h(t)w + s(t).$

Exact solution:
$$w(x,t) = \varphi(t)x^2 + \psi(t)x + \chi(t),$$

where the functions $\varphi(t)$, $\psi(t)$, $\chi(t)$ are determined by solving the following system of first-order ordinary differential equations with variable coefficients (the arguments of f, g, h, and s are not specified):

$$\varphi'_t = 2(2f+a)\varphi^2 + h\varphi, \tag{1}$$
$$\psi'_t = (4f\varphi + 2a\varphi + h)\psi + 2g\varphi, \tag{2}$$
$$\chi'_t = (2a\varphi + h)\chi + f\psi^2 + g\psi + s. \tag{3}$$

Equation (1) for $\varphi = \varphi(t)$ is a Bernoulli equation; it is easy to integrate. After that, one can successively construct the solutions of equations (2) and (3); each equation is linear in the unknown function.

⊙ *Reference*: V. F. Zaitsev, A. D. Polyanin (1996).

30. $\dfrac{\partial w}{\partial t} = aw\dfrac{\partial^2 w}{\partial x^2} + b\left(\dfrac{\partial w}{\partial x}\right)^2 + cw^2 + f(t)w + g(t).$

1°. Exact solution:
$$w(x,t) = \varphi(t) + \psi(t)\exp(\pm\lambda x), \quad \lambda = \left(\dfrac{-c}{a+b}\right)^{1/2}, \tag{1}$$

where the functions $\varphi(t)$ and $\psi(t)$ are determined by solving the following first-order ordinary differential equations with variable coefficients (the arguments of f and g are not specified):

$$\varphi'_t = c\varphi^2 + f\varphi + g, \tag{2}$$
$$\psi'_t = (a\lambda^2\varphi + 2c\varphi + f)\psi. \tag{3}$$

Equation (2) for $\varphi = \varphi(t)$ is a Riccati equation; it can be reduced to a second-order linear equation. The books by Kamke (1977) and Zaitsev and Polyanin (1995) present many solutions of equation (2) for various f and g.

Given a solution of equation (2), the solution of equation (3) for $\psi = \psi(t)$ is evaluated by

$$\psi(t) = C\exp\left[\int(a\lambda^2\varphi + 2c\varphi + f)\,dt\right], \tag{4}$$

where C is an arbitrary constant.

2°. Exact solution (A is an arbitrary constant):
$$w(x,t) = \varphi(t) + \psi(t)\cosh(\lambda x + A), \quad \lambda = \left(\dfrac{-c}{a+b}\right)^{1/2}, \tag{5}$$

where the functions $\varphi(t)$ and $\psi(t)$ are determined by solving the following first-order ordinary differential equations with variable coefficients (the arguments of f and g are not specified):

$$\varphi'_t = c\varphi^2 - b\lambda^2\psi^2 + f\varphi + g, \tag{6}$$
$$\psi'_t = (a\lambda^2\varphi + 2c\varphi + f)\psi. \tag{7}$$

One can express φ from (7) in terms of ψ and substitute the resulting φ into (6). As a result, one arrives at a second-order nonlinear equation for ψ (if f, g = const, this equation is autonomous and, hence, admits reduction of order).

3°. Exact solution (A is an arbitrary constant):
$$w(x,t) = \varphi(t) + \psi(t)\sinh(\lambda x + A), \quad \lambda = \left(\frac{-c}{a+b}\right)^{1/2},$$

where the functions $\varphi(t)$ and $\psi(t)$ are determined by solving the system of first-order ordinary differential equations
$$\varphi'_t = c\varphi^2 + b\lambda^2\psi^2 + f\varphi + g,$$
$$\psi'_t = (a\lambda^2\varphi + 2c\varphi + f)\psi.$$

4°. Exact solution (A is an arbitrary constant):
$$w(x,t) = \varphi(t) + \psi(t)\cos(\lambda x + A), \quad \lambda = \left(\frac{c}{a+b}\right)^{1/2}, \tag{8}$$

where the functions $\varphi(t)$ and $\psi(t)$ are determined by solving the system of first-order ordinary differential equations
$$\varphi'_t = c\varphi^2 + b\lambda^2\psi^2 + f\varphi + g, \tag{9}$$
$$\psi'_t = (-a\lambda^2\varphi + 2c\varphi + f)\psi. \tag{10}$$

One can express φ from (10) in terms of ψ and substitute the resulting φ into (9). Thus, one arrives at a second-order nonlinear equation for ψ (if f, g = const, this equation is autonomous and, hence, admits reduction of order).

⊙ *Reference*: V. F. Zaitsev, A. D. Polyanin (1996).

31. $\dfrac{\partial w}{\partial t} = aw\dfrac{\partial^2 w}{\partial x^2} + f(t)\left(\dfrac{\partial w}{\partial x}\right)^2 + [g_1(t)x + g_0(x)]\dfrac{\partial w}{\partial x} + h(t)w + p_2(t)x^2 + p_1(t)x + p_0(t).$

The equation has an exact solution of the form
$$w(x,t) = \varphi(t)x^2 + \psi(t)x + \chi(t),$$

where the functions $\varphi(t)$, $\psi(t)$, and $\chi(t)$ are determined by a system of first-order ordinary differential equations with variable coefficients (the system is not specified here).

B.5.1-5. Equations of the form $\frac{\partial w}{\partial t} = a\frac{\partial}{\partial x}\left[f(w)\frac{\partial w}{\partial x}\right] + f\left(x, t, w, \frac{\partial w}{\partial x}\right).$

---

32. $\dfrac{\partial w}{\partial t} = a\dfrac{\partial}{\partial x}\left(w^m \dfrac{\partial w}{\partial x}\right) + f(t)w^{1-m}.$

The change of variable $u = w^m$ leads to an equation of the form B.5.1.29,
$$\frac{\partial u}{\partial t} = au\frac{\partial^2 u}{\partial x^2} + \frac{a}{m}\left(\frac{\partial u}{\partial x}\right)^2 + mf(t),$$

which admits solutions with the form $u = \varphi(t)x^2 + \psi(t)x + \chi(t)$.

33. $\dfrac{\partial w}{\partial t} = a\dfrac{\partial}{\partial x}\left(w^m \dfrac{\partial w}{\partial x}\right) + f(t)w + g(t)w^{1-m}.$

The change of variable $u = w^m$ leads to an equation of the form B.5.1.29,
$$\frac{\partial u}{\partial t} = au\frac{\partial^2 u}{\partial x^2} + \frac{a}{m}\left(\frac{\partial u}{\partial x}\right)^2 + mf(t)u + mg(t),$$

which admits solutions with the form $u = \varphi(t)x^2 + \psi(t)x + \chi(t)$.

⊙ *Reference*: V. F. Zaitsev, A. D. Polyanin (1996).

34. $\dfrac{\partial w}{\partial t} = a\dfrac{\partial}{\partial x}\left(w^m \dfrac{\partial w}{\partial x}\right) + bw^{1+m} + f(t)w + g(t)w^{1-m}.$

For $b = 0$, see equation B.5.1.33.

The change of variable $u = w^m$ leads to an equation of the form B.5.1.30,

$$\dfrac{\partial u}{\partial t} = au\dfrac{\partial^2 u}{\partial x^2} + \dfrac{a}{m}\left(\dfrac{\partial u}{\partial x}\right)^2 + bmu^2 + mf(t)u + mg(t),$$

which admits solutions with the forms

$$u(x,t) = \varphi(t) + \psi(t)\exp(\pm\lambda x),$$
$$u(x,t) = \varphi(t) + \psi(t)\cosh(\lambda x + C),$$
$$u(x,t) = \varphi(t) + \psi(t)\sinh(\lambda x + C),$$
$$u(x,t) = \varphi(t) + \psi(t)\cos(\lambda x + C),$$

where the functions $\varphi(t)$ and $\psi(t)$ are determined by a system of first-order ordinary differential equations; the parameter λ is a root of a quadratic equation and C is an arbitrary constant.

⊙ *Reference*: V. F. Zaitsev, A. D. Polyanin (1996).

35. $\dfrac{\partial w}{\partial t} = a\dfrac{\partial}{\partial x}\left(e^{\lambda w}\dfrac{\partial w}{\partial x}\right) + f(t) + g(t)e^{-\lambda w}.$

The change of variable $u = e^{\lambda w}$ leads to an equation of the form B.5.1.27,

$$\dfrac{\partial u}{\partial t} = au\dfrac{\partial^2 u}{\partial x^2} + \lambda f(t)u + \lambda g(t),$$

which admits solutions with the form $u = \varphi(t)x^2 + \psi(t)x + \chi(t)$.

36. $\dfrac{\partial w}{\partial t} = a\dfrac{\partial}{\partial x}\left(e^{\lambda w}\dfrac{\partial w}{\partial x}\right) + f(x) + (bx + c)e^{-\lambda w}.$

The change of variable $u = e^{\lambda w}$ leads to an equation of the form B.5.1.26,

$$\dfrac{\partial u}{\partial t} = au\dfrac{\partial^2 u}{\partial x^2} + \lambda f(x)u + \lambda(bx + c),$$

which admits solutions with the form $u = \lambda(bx + c)t + \varphi(x)$.

37. $\dfrac{\partial w}{\partial t} = a\dfrac{\partial}{\partial x}\left(e^{\lambda w}\dfrac{\partial w}{\partial x}\right) + be^{\lambda w} + f(t) + g(t)e^{-\lambda w}.$

For $b = 0$, see equation B.5.1.35.

The change of variable $u = e^{\lambda w}$ leads to an equation of the form B.5.1.30,

$$\dfrac{\partial u}{\partial t} = au\dfrac{\partial^2 u}{\partial x^2} + bu^2 + \lambda f(t)u + \lambda g(t),$$

which admits solutions with the forms

$$u(x,t) = \varphi(t) + \psi(t)\exp(\pm\mu x),$$
$$u(x,t) = \varphi(t) + \psi(t)\cosh(\mu x + C),$$
$$u(x,t) = \varphi(t) + \psi(t)\sinh(\mu x + C),$$
$$u(x,t) = \varphi(t) + \psi(t)\cos(\mu x + C),$$

where the functions $\varphi(t)$ and $\psi(t)$ are determined by a system of first-order ordinary differential equations; the parameter μ is a root of a quadratic equation and C is an arbitrary constant.

⊙ *Reference*: V. F. Zaitsev, A. D. Polyanin (1996).

B.5.1-6. Equations of the form $\frac{\partial w}{\partial t} = f(x,t,w)\frac{\partial^2 w}{\partial x^2} + g\left(x,t,w,\frac{\partial w}{\partial x}\right)$.

38. $\dfrac{\partial w}{\partial t} = \dfrac{f(x)}{aw+b}\dfrac{\partial^2 w}{\partial x^2}$.

Exact solution:
$$w(x,t) = \frac{1}{a}\bigl[\varphi(x)t + \psi(x) - b\bigr],$$
where the functions $\varphi(x)$ and $\psi(x)$ are determined by the ordinary differential equations
$$f(x)\varphi''_{xx} - \varphi^2 = 0, \qquad f(x)\psi''_{xx} - \varphi\psi = 0.$$
The first equation can be treated independently. The second equation has a particular solution $\psi(x) = \varphi(x)$, and hence, its general solution is given by
$$\psi(x) = C_1\varphi(x) + C_2\varphi(x)\int\frac{dx}{\varphi^2(x)},$$
where C_1 and C_2 are arbitrary constants.

39. $\dfrac{\partial w}{\partial t} = f(t)\dfrac{\partial}{\partial x}\left(w^m\dfrac{\partial w}{\partial x}\right) + g(t)w^{1-m}$.

Exact solution:
$$w(x,t) = \bigl[\varphi(t)x^2 + \psi(t)\bigr]^{1/m},$$
where the functions $\varphi = \varphi(x)$ and $\psi = \psi(x)$ are determined by the first-order ordinary differential equations
$$\varphi'_t = \frac{2(m+2)}{m}f\varphi^2, \qquad \psi'_t = 2f\varphi\psi + mg.$$
Integrating yields
$$\varphi = \frac{1}{F}, \quad \psi = F^{-\frac{m}{m+2}}\left(A + m\int gF^{\frac{m}{m+2}}\,dt\right), \quad F = B - \frac{2(m+2)}{m}\int f\,dt,$$
where A and B are arbitrary constants.

⊙ *Reference*: V. F. Zaitsev, A. D. Polyanin (1996).

40. $\dfrac{\partial w}{\partial t} = \dfrac{\partial}{\partial x}\left[f(x)e^{\beta w}\dfrac{\partial w}{\partial x}\right]$.

Exact solution in additive form:
$$w(x,t) = -\frac{1}{\beta}\ln(\beta t + C) + \frac{1}{\beta}\ln\left[\int\frac{A - \beta x}{f(x)}\,dx + B\right],$$
where A, B, and C are arbitrary constants.

B.5.1-7. Equations with three independent variables.

41. $\dfrac{\partial w}{\partial t} = \dfrac{\partial}{\partial x}\left[f(x,y)\dfrac{\partial w}{\partial x}\right] + \dfrac{\partial}{\partial y}\left[g(x,y)\dfrac{\partial w}{\partial y}\right] + kw\ln w$.

Exact solution in multiplicative form:
$$w(x,y,t) = \exp\bigl(Ae^{kt}\bigr)\Theta(x,y),$$
where A is an arbitrary constant and the function $\Theta(x,y)$ satisfies the stationary equation
$$\frac{\partial}{\partial x}\left[f(x,y)\frac{\partial\Theta}{\partial x}\right] + \frac{\partial}{\partial y}\left[g(x,y)\frac{\partial\Theta}{\partial y}\right] + k\Theta\ln\Theta = 0.$$

42. $\dfrac{\partial w}{\partial t} = \dfrac{\partial}{\partial x}\left[f(x,t)\dfrac{\partial w}{\partial x}\right] + \dfrac{\partial}{\partial y}\left[g(y,t)\dfrac{\partial w}{\partial y}\right] + h(t)w \ln w.$

Incomplete separable exact solution (the solution is separable in the space coordinates x and y but not in time t):

$$w(x, y, t) = \varphi(x, t)\psi(y, t).$$

The functions $\varphi(x, t)$ and $\psi(y, t)$ are determined from the one-dimensional nonlinear parabolic differential equations

$$\dfrac{\partial \varphi}{\partial t} = \dfrac{\partial}{\partial x}\left[f(x,t)\dfrac{\partial \varphi}{\partial x}\right] + h(t)\varphi \ln \varphi + C(t)\varphi,$$

$$\dfrac{\partial \psi}{\partial t} = \dfrac{\partial}{\partial y}\left[g(y,t)\dfrac{\partial \psi}{\partial y}\right] + h(t)\psi \ln \psi - C(t)\psi,$$

where $C(t)$ is an arbitrary function.

B.5.2. Hyperbolic Equations

B.5.2-1. Equations of the form $\dfrac{\partial^2 w}{\partial t^2} = a\dfrac{\partial^2 w}{\partial x^2} + f\!\left(x, t, w, \dfrac{\partial w}{\partial x}\right)$.

1. $\dfrac{\partial^2 w}{\partial t^2} = a\dfrac{\partial^2 w}{\partial x^2} + bw \ln w + [f(x) + g(t)]w.$

Exact solution in multiplicative form:

$$w(x, t) = \varphi(t)\psi(x),$$

where the functions $\varphi(t)$ and $\psi(x)$ are determined by the second-order ordinary differential equations

$$\varphi''_{tt} - \left[b \ln \varphi + g(t) + C\right]\varphi = 0,$$

$$a\psi''_{xx} + \left[b \ln \psi + f(x) - C\right]\psi = 0,$$

where C is an arbitrary constant.

2. $\dfrac{\partial^2 w}{\partial t^2} = a\dfrac{\partial^2 w}{\partial x^2} + b\!\left(\dfrac{\partial w}{\partial x}\right)^{\!2} + cw + f(t).$

Exact solution in additive form:

$$w(x, t) = \varphi(t) + \theta(\xi), \quad \xi = x + \lambda t,$$

where λ is an arbitrary constant and the functions $\varphi = \varphi(t)$ and $\theta = \theta(\xi)$ are determined by solving the second-order ordinary differential equations

$$\varphi''_{tt} - c\varphi - f(t) = 0, \tag{1}$$

$$(a - \lambda^2)\theta''_{\xi\xi} + b\left(\theta'_\xi\right)^2 + c\theta = 0. \tag{2}$$

The general solution of equation (1) is given by

$$\varphi(t) = C_1 \cosh(kt) + C_2 \sinh(kt) + \dfrac{1}{k}\int_0^t f(\tau)\sinh[k(t-\tau)]\,d\tau \quad \text{if} \quad c = k^2 > 0,$$

$$\varphi(t) = C_1 \cos(kt) + C_2 \sin(kt) + \dfrac{1}{k}\int_0^t f(\tau)\sin[k(t-\tau)]\,d\tau \quad \text{if} \quad c = -k^2 < 0,$$

where C_1 and C_2 are arbitrary constants.

Equation (2) can be solved with the change of variable $z(\theta) = \left(\theta'_\xi\right)^2$, which leads to a first-order linear equation.

3. $\dfrac{\partial^2 w}{\partial t^2} = a\dfrac{\partial^2 w}{\partial x^2} + b\left(\dfrac{\partial w}{\partial x}\right)^2 + cw + f(x).$

Exact solution in additive form:
$$w(x,t) = \varphi(x) + \psi(t).$$

Here,
$$\psi(t) = C_1 \cosh(kt) + C_2 \sinh(kt) \quad \text{if} \quad c = k^2 > 0,$$
$$\psi(t) = C_1 \cos(kt) + C_2 \sin(kt) \quad \text{if} \quad c = -k^2 < 0,$$

where C_1 and C_2 are arbitrary constants, and the function $\varphi(x)$ is determined by the ordinary differential equation
$$a\varphi''_{xx} + b(\varphi'_x)^2 + c\varphi + f(x) = 0.$$

4. $\dfrac{\partial^2 w}{\partial t^2} = a\dfrac{\partial^2 w}{\partial x^2} + b\left(\dfrac{\partial w}{\partial x}\right)^2 + cw\dfrac{\partial w}{\partial x} + kw^2 + f(t)w + g(t).$

Exact solution:
$$w(x,t) = \varphi(t) + \psi(t)\exp(\lambda x),$$

where λ is a root of the quadratic equation $b\lambda^2 + c\lambda + k = 0$ and the functions $\varphi(t)$ and $\psi(t)$ are determined by the following system of second-order ordinary differential equations:

$$\varphi''_{tt} = k\varphi^2 + f(t)\varphi + g(t), \tag{1}$$
$$\psi''_{tt} = \left[(c\lambda + 2k)\varphi + f(t) + a\lambda^2\right]\psi. \tag{2}$$

In the special case $f(t) = \text{const}$ and $g(t) = \text{const}$, equation (1) is autonomous and has particular solutions of the form $\varphi = \text{const}$ and, hence, can be integrated in quadrature. Equation (2) is linear in ψ, and consequently, with $\varphi = \text{const}$, its general solution is expressed in terms of exponentials or sine and cosine.

5. $\dfrac{\partial^2 w}{\partial t^2} = a\dfrac{\partial^2 w}{\partial x^2} + f(t)\left(\dfrac{\partial w}{\partial x}\right)^2 + g(t)w + h(t).$

Exact solution:
$$w(x,t) = \varphi(t)x^2 + \psi(t)x + \chi(t), \tag{1}$$

where the functions $\varphi(t)$, $\psi(t)$, and $\chi(t)$ are determined by solving the following system of second-order ordinary differential equations with variable coefficients (the arguments of f, g, and h are not specified):

$$\varphi''_{tt} = 4f\varphi^2 + g\varphi, \tag{2}$$
$$\psi''_{tt} = (4f\varphi + g)\psi, \tag{3}$$
$$\chi''_{tt} = g\chi + f\psi^2 + h + 2a\varphi. \tag{4}$$

Equation (2) has the trivial particular solution $\varphi(t) \equiv 0$; the corresponding solution (1) is linear in the coordinate x.

Equation (3) has a particular solution $\psi = \bar{\varphi}(t)$, where $\bar{\varphi}(t)$ is any nontrivial particular solution of equation (2). Hence, the general solution of equation (3) is given by

$$\psi(t) = C_1\bar{\varphi}(t) + C_2\bar{\varphi}(t)\int \dfrac{dt}{\bar{\varphi}^2(t)},$$

where C_1 and C_2 are arbitrary constants. If the functions f and g proportional, then $\varphi = -\tfrac{1}{4}g/f$ ($\varphi = \text{const}$) is a particular solution of equation (2).

Equation (4) linear in $\chi = \chi(t)$.

6. $\dfrac{\partial^2 w}{\partial t^2} = a\dfrac{\partial^2 w}{\partial x^2} + f(x)\left(\dfrac{\partial w}{\partial x}\right)^2 + g(x) + h(t).$

Exact solution in additive form:
$$w(x,t) = \tfrac{1}{2}At^2 + Bt + C + \int_0^t (t-\tau)h(\tau)\,d\tau + \varphi(x).$$

Here, A, B, and C are arbitrary constants, and the function $\varphi(x)$ is determined by solving the second-order nonlinear ordinary differential equation
$$a\varphi''_{xx} + f(x)(\varphi'_x)^2 + g(x) - A = 0.$$

7. $\dfrac{\partial^2 w}{\partial t^2} = a\dfrac{\partial^2 w}{\partial x^2} + f(x)\left(\dfrac{\partial w}{\partial x}\right)^2 + bw + g(x) + h(t).$

Exact solution in additive form:
$$w(x,t) = \varphi(t) + \psi(x).$$

Here, the functions $\varphi(t)$ and $\psi(x)$ are determined by solving the second-order ordinary differential equations
$$\varphi''_{tt} - b\varphi - h(t) = 0,$$
$$a\psi''_{xx} + f(x)(\psi'_x)^2 + b\psi + g(x) = 0.$$

The general solution of the first equation is given by

$$\varphi(t) = C_1 \cosh(kt) + C_2 \sinh(kt) + \frac{1}{k}\int_0^t h(\tau)\sinh[k(t-\tau)]\,d\tau \quad \text{if} \quad b = k^2 > 0,$$

$$\varphi(t) = C_1 \cos(kt) + C_2 \sin(kt) + \frac{1}{k}\int_0^t h(\tau)\sin[k(t-\tau)]\,d\tau \quad \text{if} \quad b = -k^2 < 0,$$

where C_1 and C_2 are arbitrary constants.

8. $\dfrac{\partial^2 w}{\partial t^2} = a\dfrac{\partial^2 w}{\partial x^2} + f(t)\left(\dfrac{\partial w}{\partial x}\right)^2 + bf(t)w^2 + g(t)w + h(t).$

1°. Exact solution:
$$w(x,t) = \varphi(t) + \psi(t)\exp(\pm x\sqrt{-b}), \qquad b < 0,$$

where the functions $\varphi(t)$ and $\psi(t)$ are determined by solving the following second-order ordinary differential equations with variable coefficients (the arguments of f, g, and h are not specified):
$$\varphi''_{tt} = bf\varphi^2 + g\varphi + h,$$
$$\psi''_{tt} = (2bf\varphi + g - ab)\psi.$$

2°. Exact solution of a more general form:
$$w(x,t) = \varphi(t) + \psi(t)\bigl[A\exp(x\sqrt{-b}) + B\exp(-x\sqrt{-b})\bigr], \qquad b < 0,$$

where A and B are arbitrary constants and the functions $\varphi(t)$ and $\psi(t)$ are determined by the following system of second-order ordinary differential equations with variable coefficients (the arguments of f, g, and h are not specified):
$$\varphi''_{tt} = bf(\varphi^2 + 4AB\psi^2) + g\varphi + h,$$
$$\psi''_{tt} = (2bf\varphi + g - ab)\psi.$$

3°. Exact solution:
$$w(x,t) = \varphi(t) + \psi(t)\cos(x\sqrt{b} + C), \qquad b > 0, \tag{9}$$

where C is an arbitrary constant and the functions $\varphi(t)$ and $\psi(t)$ are determined by the following system of second-order ordinary differential equations with variable coefficients:

$$\varphi''_{tt} = bf(\varphi^2 + \psi^2) + g\varphi + h,$$
$$\psi''_{tt} = (2bf\varphi + g - ab)\psi.$$

9. $\dfrac{\partial^2 w}{\partial t^2} = a\dfrac{\partial^2 w}{\partial x^2} + f\!\left(x, \dfrac{\partial w}{\partial x}\right) + g(t).$

Exact solution in additive form:

$$w(x,t) = \tfrac{1}{2}At^2 + Bt + C + \int_0^t (t-\tau)g(\tau)\,d\tau + \varphi(x).$$

Here, A, B, and C are arbitrary constants and the function $\varphi(x)$ is determined by solving the second-order nonlinear ordinary differential equation

$$a\varphi''_{xx} + f(x, \varphi'_x) - A = 0.$$

10. $\dfrac{\partial^2 w}{\partial t^2} = a\dfrac{\partial^2 w}{\partial x^2} + f\!\left(x, \dfrac{\partial w}{\partial x}\right) + bw + g(t).$

Exact solution in additive form:

$$w(x,t) = \varphi(t) + \psi(x).$$

Here, the functions $\varphi(t)$ and $\psi(x)$ are determined by the second-order ordinary differential equations

$$\varphi''_{tt} - b\varphi - g(t) = 0,$$
$$a\psi''_{xx} + f(x, \psi'_x) + b\psi = 0.$$

The general solution of the first equation is given by

$$\varphi(t) = C_1 \cosh(kt) + C_2 \sinh(kt) + \frac{1}{k}\int_0^t g(\tau)\sinh[k(t-\tau)]\,d\tau \quad \text{if} \quad b = k^2 > 0,$$

$$\varphi(t) = C_1 \cos(kt) + C_2 \sin(kt) + \frac{1}{k}\int_0^t g(\tau)\sin[k(t-\tau)]\,d\tau \quad \text{if} \quad b = -k^2 < 0,$$

where C_1 and C_2 are arbitrary constants.

11. $\dfrac{\partial^2 w}{\partial t^2} = a\dfrac{\partial^2 w}{\partial x^2} + wf\!\left(t, \dfrac{1}{w}\dfrac{\partial w}{\partial x}\right).$

Exact solution in multiplicative form:

$$w(x,t) = e^{\lambda x}\varphi(t),$$

where λ is an arbitrary constant and the function $\varphi(t)$ is determined by the second-order linear ordinary differential equation

$$\varphi''_{tt} = [a\lambda^2 + f(t, \lambda)]\varphi.$$

B.5.2-2. Equations of the form $\frac{\partial^2 w}{\partial t^2} = f(x)\frac{\partial^2 w}{\partial x^2} + g\left(x, t, w, \frac{\partial w}{\partial x}\right)$.

12. $\dfrac{\partial^2 w}{\partial t^2} = a(x+\beta)^n \dfrac{\partial^2 w}{\partial x^2} + f(w), \qquad a > 0.$

Exact solution for $n \neq 2$:
$$w = w(z), \qquad z = \left[\tfrac{1}{4}a(2-n)^2(t+C)^2 - (x+\beta)^{2-n}\right]^{\frac{n}{2(2-n)}},$$

where C is an arbitrary constant and the function $w = w(z)$ is determined by the generalized Emden–Fowler equation
$$w''_{zz} - \frac{4}{an^2} z^{\frac{4(1-n)}{n}} f(w) = 0. \tag{1}$$

A number of exact solutions to equation (1) for some specific $f = f(w)$ can be found in Polyanin and Zaitsev (1995). In the special case $n = 1$, the general solution of equation (1) is given by
$$\int \left[C_1 + \frac{8}{a} F(w)\right]^{-1/2} dw = \pm z + C_2, \qquad F(w) = \int f(w)\, dw,$$

where C_1 and C_2 are arbitrary constants.

13. $\dfrac{\partial^2 w}{\partial t^2} = ax^n \dfrac{\partial^2 w}{\partial x^2} + bx^{n-1}\dfrac{\partial w}{\partial x} + f(w), \qquad a > 0.$

Exact solution for $n \neq 2$:
$$w = w(\xi), \qquad \xi = \tfrac{1}{4}a(2-n)^2(t+C)^2 - x^{2-n},$$

where C is an arbitrary constant and the function $w = w(\xi)$ is determined by the ordinary differential equation
$$\xi w''_{\xi\xi} + A w'_\xi - B f(w) = 0, \quad \text{where} \quad A = \frac{a(4-3n) + 2b}{2a(2-n)}, \quad B = \frac{1}{a(2-n)^2}.$$

For $A \neq 1$, the change of variable $\xi = kz^{\frac{1}{1-A}}$ ($k = \pm 1$) brings this equation to the generalized Emden–Fowler equation
$$w''_{zz} - \frac{kB}{(1-A)^2} z^{\frac{2A-1}{1-A}} f(w) = 0,$$

whose solvable cases are presented in Polyanin and Zaitsev (1995).

14. $\dfrac{\partial^2 w}{\partial t^2} = ax^n \dfrac{\partial^2 w}{\partial x^2} + x^{n-1} f(w) \dfrac{\partial w}{\partial x}.$

Exact solution for $n \neq 2$:
$$w = w(z), \qquad z = \left[ka(2-n)^2(t+C)^2 - 4kx^{2-n}\right]^{1/2}, \qquad k = \pm 1,$$

where C is an arbitrary constant and the function $w = w(z)$ is determined by the ordinary differential equation
$$w''_{zz} + \frac{2}{a(2-n)}\left[a(1-n) + f(w)\right]\frac{1}{z} w'_z = 0.$$

The change of variable $u(w) = zw'_z$ leads to a first-order separable equation. Integrating this equation yields the general solution in implicit form:
$$\int \frac{dw}{anw - 2F(w) + C_1} = \frac{1}{a(2-n)} \ln|z| + C_2, \qquad F(w) = \int f(w)\, dw,$$

where C_1 and C_2 are arbitrary constants.

15. $\dfrac{\partial^2 w}{\partial t^2} = ax^n \dfrac{\partial^2 w}{\partial x^2} + x^{n-1} f(w) \dfrac{\partial w}{\partial x} + g(w).$

Exact solution for $n \ne 2$:
$$w = w(z), \quad z = \left[ka(2-n)^2(t+C)^2 - 4kx^{2-n}\right]^{1/2}, \quad k = \pm 1,$$
where C is an arbitrary constant and the function $w = w(z)$ is determined by the ordinary differential equation
$$w''_{zz} + \dfrac{2}{a(2-n)}\left[a(1-n) + f(w)\right]\dfrac{1}{z} w'_z - \dfrac{1}{ak(2-n)^2} g(w) = 0.$$

16. $\dfrac{\partial^2 w}{\partial t^2} = ae^{\lambda x} \dfrac{\partial^2 w}{\partial x^2} + be^{\lambda x} \dfrac{\partial w}{\partial x} + f(w), \quad a > 0.$

Exact solution for $\lambda \ne 0$:
$$w = w(z), \quad z = \left[4ke^{-\lambda x} - ak\lambda^2(t+C)^2\right]^{1/2}, \quad k = \pm 1,$$
where C is an arbitrary constant and the function $w = w(z)$ is determined by the ordinary differential equation
$$w''_{zz} + \dfrac{2(a\lambda - b)}{a\lambda}\dfrac{1}{z}w'_z + \dfrac{1}{ak\lambda^2} f(w) = 0. \qquad (1)$$

For $b = a\lambda$, the solution of equation (1) is given by
$$\int\left[C_1 - \dfrac{2}{ak\lambda^2} F(w)\right]^{-1/2} dw = \pm z + C_2, \quad F(w) = \int f(w)\, dw,$$
where C_1 and C_2 are arbitrary constants.

For $b \ne \tfrac{1}{2}a\lambda$, the change of variable $\xi = z^{\frac{2b-a\lambda}{a\lambda}}$ brings (1) to the generalized Emden–Fowler equation
$$w''_{\xi\xi} + \dfrac{a}{k(2b - a\lambda)^2} \xi^{\frac{4(a\lambda - b)}{2b - a\lambda}} f(w) = 0,$$
whose solvable cases are presented in Polyanin and Zaitsev (1995).

17. $\dfrac{\partial^2 w}{\partial t^2} = ae^{\lambda x} \dfrac{\partial^2 w}{\partial x^2} + e^{\lambda x} f(w) \dfrac{\partial w}{\partial x}.$

Exact solution for $\lambda \ne 0$:
$$w = w(z), \quad z = \left[4ke^{-\lambda x} - ak\lambda^2(t+A)^2\right]^{1/2}, \quad k = \pm 1.$$
Here, A is an arbitrary constant and the function $w = w(z)$ is determined by the ordinary differential equation
$$w''_{zz} + \dfrac{2}{z}\left[1 - \dfrac{1}{a\lambda} f(w)\right] w'_z = 0,$$
which by the change of variable $u(w) = zw'_z$ is reduced to a separable first-order equation. Integrating this equation yields the general solution in implicit form:
$$\int \dfrac{dw}{2F(w) - a\lambda w + C_1} = \dfrac{1}{a\lambda} \ln|z| + C_2, \quad F(w) = \int f(w)\, dw,$$
where C_1 and C_2 are arbitrary constants.

18. $\dfrac{\partial^2 w}{\partial t^2} = ae^{\lambda x} \dfrac{\partial^2 w}{\partial x^2} + e^{\lambda x} f(w) \dfrac{\partial w}{\partial x} + g(w).$

Exact solution for $\lambda \ne 0$:
$$w = w(z), \quad z = \left[4ke^{-\lambda x} - ak\lambda^2(t+C)^2\right]^{1/2}, \quad k = \pm 1,$$
where C is an arbitrary constant and the function $w = w(z)$ is determined by the ordinary differential equation
$$w''_{zz} + \dfrac{2}{z}\left[1 - \dfrac{1}{a\lambda} f(w)\right] w'_z + \dfrac{1}{ak\lambda^2} g(w) = 0.$$

B.5.2-3. Other equations.

19. $\dfrac{\partial^2 w}{\partial t^2} = f(t)\dfrac{\partial}{\partial x}\left(w\dfrac{\partial w}{\partial x}\right).$

1°. Exact solutions:
$$w(x,t) = (C_1 t + C_2)(C_3 x + C_4)^{1/2},$$
$$w(x,t) = (C_1 t + C_2)x + \int_a^t (t-\tau)(C_1\tau + C_2)^2 f(\tau)\, d\tau + C_3 t + C_4,$$
where C_1, C_2, C_3, C_4, and a are arbitrary constants.

2°. Exact solution:
$$w(x,t) = \varphi(t)x^2 + \psi(t)x + \chi(t),$$
where the functions $\varphi = \varphi(t)$, $\psi = \psi(t)$, and $\chi = \chi(t)$ are determined by the system of ordinary differential equations
$$\varphi''_{tt} = 6f(t)\varphi^2,$$
$$\psi''_{tt} = 6f(t)\varphi\psi,$$
$$\chi''_{tt} = 2f(t)\varphi\chi + f(t)\psi^2.$$

3°. Exact solution in multiplicative form:
$$w(x,t) = \Phi(t)\Psi(x),$$
where the functions $\Phi = \Phi(t)$ and $\Psi = \Psi(x)$ are determined by the ordinary differential equations (C is an arbitrary constant)
$$\Phi''_{tt} = Cf(t)\Phi^2,$$
$$(\Psi\Psi'_x)'_x = C\Psi.$$

The latter equation is autonomous and has a particular solution $\Psi = \tfrac{1}{6}Cx^2$ and, hence, is integrable in quadrature.

20. $\dfrac{\partial^2 w}{\partial t^2} = f(t)\dfrac{\partial}{\partial x}\left(w\dfrac{\partial w}{\partial x}\right) + g_2(t)x^2 + g_1(t)x + g_0(t).$

Exact solution:
$$w(x,t) = \varphi(t)x^2 + \psi(t)x + \chi(t),$$
where the functions $\varphi = \varphi(t)$, $\psi = \psi(t)$, and $\chi = \chi(t)$ are determined by the system of ordinary differential equations
$$\varphi''_{tt} = 6f(t)\varphi^2 + g_2(t),$$
$$\psi''_{tt} = 6f(t)\varphi\psi + g_1(t),$$
$$\chi''_{tt} = 2f(t)\varphi\chi + f(t)\psi^2 + g_0(t).$$

B.5.3. Elliptic Equations

B.5.3-1. Equations of the form $a\dfrac{\partial^2 w}{\partial x^2} + b\dfrac{\partial^2 w}{\partial y^2} = f\!\left(x, y, w, \dfrac{\partial w}{\partial x}, \dfrac{\partial w}{\partial y}\right)$, $ab > 0$.

1. $\dfrac{\partial^2 w}{\partial x^2} + \dfrac{\partial^2 w}{\partial y^2} = aw\ln w + [f(x) + g(y)]w.$

Exact solution in multiplicative form:
$$w(x,y) = \varphi(x)\psi(y),$$
where the functions $\varphi(x)$ and $\psi(y)$ are determined by the ordinary differential equations
$$\varphi''_{xx} - \big[a\ln\varphi + f(x) + C\big]\varphi = 0,$$
$$\psi''_{yy} - \big[a\ln\psi + g(y) - C\big]\psi = 0,$$
where C is an arbitrary constant.

2. $\dfrac{\partial^2 w}{\partial x^2} + \dfrac{\partial^2 w}{\partial y^2} = f(x)w \ln w + [af(x)y + g(x)]w.$

Exact solution in multiplicative form:
$$w(x,y) = e^{-ay}\varphi(x),$$
where the function $\varphi(x)$ is determined by the ordinary differential equation
$$\varphi''_{xx} = f(x)\varphi \ln\varphi + \left[g(x) - a^2\right]\varphi.$$

3. $\dfrac{\partial^2 w}{\partial x^2} + a\dfrac{\partial^2 w}{\partial y^2} = f(x)\left(\dfrac{\partial w}{\partial y}\right)^2 + g(x)w + h(x).$

Exact solution:
$$w(x,y) = \varphi(x)y^2 + \psi(x)y + \chi(x). \tag{1}$$

Here, the functions $\varphi(x)$, $\psi(x)$, and $\chi(x)$ are determined by the following second-order ordinary differential equations with variable coefficients (the arguments of f, g, and h are not specified):
$$\varphi''_{xx} = 4f\varphi^2 + g\varphi, \tag{2}$$
$$\psi''_{xx} = (4f\varphi + g)\psi, \tag{3}$$
$$\chi''_{xx} = g\chi + f\psi^2 + h - 2a\varphi. \tag{4}$$

If a solution $\varphi = \varphi(x)$ of the nonlinear equation (2) is found, then the functions $\psi = \psi(x)$ and $\chi = \chi(x)$ can be determined successively from equations (3) and (4), which are linear in ψ and χ.

By comparing equations (2) and (3), one can see that equation (3) has a particular solution $\psi = \varphi(x)$. Hence, the general solution of (3) is given by (see Polyanin and Zaitsev 1995)
$$\psi(x) = C_1\varphi(x) + C_2\varphi(x) \int \dfrac{dx}{\varphi^2(x)}, \qquad \varphi \not\equiv 0.$$

Note that equation (2) has the trivial particular solution $\varphi(x) \equiv 0$, to which there is a corresponding solution (1) linear in the coordinate y. If the functions f and g are proportional, then $\varphi = -\tfrac{1}{4}g/f$ ($\varphi = $ const) is a particular solution of equation (2).

4. $\dfrac{\partial^2 w}{\partial x^2} + a\dfrac{\partial^2 w}{\partial y^2} = f(x)\left(\dfrac{\partial w}{\partial y}\right)^2 + bf(x)w^2 + g(x)w + h(x).$

1°. Exact solution:
$$w(x,y) = \varphi(x) + \psi(x)\exp\left(\pm y\sqrt{-b}\right), \qquad b < 0,$$
where the functions $\varphi(x)$ and $\psi(x)$ are determined by the following second-order ordinary differential equations with variable coefficients (the arguments of f, g, and h are not specified):
$$\varphi''_{xx} = bf\varphi^2 + g\varphi + h,$$
$$\psi''_{xx} = (2bf\varphi + g + ab)\psi.$$

2°. Exact solution of a more general form:
$$w(x,y) = \varphi(x) + \psi(x)\left[A\exp\left(y\sqrt{-b}\right) + B\exp\left(-y\sqrt{-b}\right)\right], \qquad b < 0,$$
where the functions $\varphi(x)$ and $\psi(x)$ are determined by the following system of second-order ordinary differential equations with variable coefficients:
$$\varphi''_{xx} = bf\left(\varphi^2 + 4AB\psi^2\right) + g\varphi + h,$$
$$\psi''_{xx} = 2bf\varphi\psi + g\psi + ab\psi.$$

3°. Exact solution:
$$w(x,y) = \varphi(x) + \psi(x)\cos\left(y\sqrt{b} + C\right), \qquad b > 0,$$
where C is an arbitrary constant and the functions $\varphi(x)$ and $\psi(x)$ are determined by the following system of second-order ordinary differential equations with variable coefficients:
$$\varphi''_{xx} = bf\left(\varphi^2 + \psi^2\right) + g\varphi + h,$$
$$\psi''_{xx} = 2bf\varphi\psi + g\psi + ab\psi.$$

⊙ *Reference*: V. F. Zaitsev, A. D. Polyanin (1996).

5. $a\dfrac{\partial^2 w}{\partial x^2} + b\dfrac{\partial^2 w}{\partial y^2} = f_1\!\left(x, \dfrac{\partial w}{\partial x}\right) + f_2\!\left(y, \dfrac{\partial w}{\partial y}\right) + kw.$

Exact solution in additive form:
$$w(x,y) = \varphi(x) + \psi(y).$$
Here, the functions $\varphi(x)$ and $\psi(x)$ are determined by solving the second-order ordinary differential equations
$$a\varphi''_{xx} - f_1(x, \varphi'_x) - k\varphi = C,$$
$$b\psi''_{yy} - f_2(y, \psi'_y) - k\psi = -C,$$
where C is an arbitrary constant.

B.5.3-2. Equations of the form $\frac{\partial}{\partial x}\!\left[f(x)\frac{\partial w}{\partial x}\right] + \frac{\partial}{\partial y}\!\left[g(y)\frac{\partial w}{\partial y}\right] = h(w).$

---

6. $\dfrac{\partial}{\partial x}\!\left(ax^n \dfrac{\partial w}{\partial x}\right) + \dfrac{\partial}{\partial y}\!\left(by^m \dfrac{\partial w}{\partial y}\right) = f(w).$

1°. For $n \neq 2$ and $m \neq 2$, there are exact solutions of the form
$$w = w(\xi), \qquad \xi = \left[b(2-m)^2 x^{2-n} + a(2-n)^2 y^{2-m}\right]^{1/2}.$$
Here, the function $w = w(\xi)$ is determined by the ordinary differential equation
$$w''_{\xi\xi} + \dfrac{A}{\xi} w'_\xi = Bf(w), \tag{1}$$
where
$$A = \dfrac{4 - nm}{(2-n)(2-m)}, \qquad B = \dfrac{4}{ab(2-n)^2(2-m)^2}.$$

For $m = 4/n$, one obtains from (1) the following exact solution to the original equation with arbitrary $f = f(w)$:
$$\int \left[C_1 + \dfrac{2n^2}{ab(2-n)^4} F(w)\right]^{-1/2} dw = C_2 \pm \xi, \qquad F(w) = \int f(w)\, dw,$$
where C_1 and C_2 are arbitrary constants.

2°. The change of variable $\zeta = \xi^{1-A}$ brings (1) to the generalized Emden–Fowler equation
$$w''_{\zeta\zeta} = \dfrac{B}{(1-A)^2} \zeta^{\frac{2A}{1-A}} f(w). \tag{2}$$
A lot of exact solutions to equation (2) with various $f = f(w)$ can be found in Polyanin and Zaitsev (1995).

⊙ *Reference*: V. F. Zaitsev and A. D. Polyanin (1996).

7. $\dfrac{\partial}{\partial x}\!\left(ae^{\beta x} \dfrac{\partial w}{\partial x}\right) + \dfrac{\partial}{\partial y}\!\left(be^{\mu y} \dfrac{\partial w}{\partial y}\right) = f(w).$

For $\beta\mu \neq 0$, there are exact solutions of the form
$$w = w(\xi), \qquad \xi = \left(b\mu^2 e^{-\beta x} + a\beta^2 e^{-\mu y}\right)^{1/2},$$
where the function $w = w(\xi)$ is determined by the ordinary differential equation
$$w''_{\xi\xi} - \dfrac{1}{\xi} w'_\xi = Af(w), \qquad A = \dfrac{4}{ab\beta^2 \mu^2}. \tag{1}$$
The change of variable $\zeta = \xi^2$ brings (1) to the generalized Emden–Fowler equation
$$w''_{\zeta\zeta} = \tfrac{1}{4} A \zeta^{-1} f(w),$$
whose solutions with $f(w) = (kw + s)^{-1}$ and $f(w) = (kw + s)^{-2}$ ($k, s =$ const) can be found in Polyanin and Zaitsev (1995).

⊙ *Reference*: V. F. Zaitsev and A. D. Polyanin (1996).

8. $\dfrac{\partial}{\partial x}\left(ax^n \dfrac{\partial w}{\partial x}\right) + \dfrac{\partial}{\partial y}\left(be^{\mu y}\dfrac{\partial w}{\partial y}\right) = f(w).$

For $n \neq 2$ and $\mu \neq 0$, there are exact solutions of the form

$$w = w(\xi), \qquad \xi = \left[b\mu^2 x^{2-n} + a(2-n)^2 e^{-\mu y}\right]^{1/2},$$

where the function $w = w(\xi)$ is determined by the ordinary differential equation

$$w''_{\xi\xi} + \dfrac{n}{n-2}\dfrac{1}{\xi}w'_\xi = \dfrac{4}{ab\mu^2(2-n)^2}f(w).$$

9. $\dfrac{\partial}{\partial x}\left[f(x)\dfrac{\partial w}{\partial x}\right] + \dfrac{\partial}{\partial y}\left[g(y)\dfrac{\partial w}{\partial y}\right] = aw \ln w + [h_1(x) + h_2(y)]w.$

Exact solution in multiplicative form:

$$w(x, y) = \varphi(x)\psi(y),$$

where $\varphi = \varphi(x)$ and $\psi = \psi(y)$ are determined by the ordinary differential equations

$$[f(x)\varphi'_x]'_x = [a \ln \varphi + h_1(x) + C]\varphi,$$
$$[g(y)\psi'_y]'_y = [a \ln \psi + h_2(y) - C]\psi,$$

where C is an arbitrary constant.

B.5.3-3. Other equations with two independent variables.

10. $\dfrac{\partial^2 w}{\partial x^2} + \dfrac{\partial}{\partial y}\left\{[f(x)w + g(x)]\dfrac{\partial w}{\partial y}\right\} = 0.$

1°. Exact solution:

$$w(x, y) = (Ax + B)y - \int_{x_0}^{x}(x-t)(At+B)^2 f(t)\, dt + C_1 x + C_2,$$

where A, B, C_1, C_2, and x_0 are arbitrary constants.

2°. Exact solution:

$$w(x, y) = \varphi(x)y^2 + \psi(x)y + \chi(x),$$

where the functions $\varphi = \varphi(x)$, $\psi = \psi(x)$, and $\chi = \chi(x)$ are determined by the ordinary differential equations

$$\varphi''_{xx} + 6f\varphi^2 = 0, \qquad (1)$$
$$\psi''_{xx} + 6f\varphi\psi = 0, \qquad (2)$$
$$\chi''_{xx} + 2f\varphi\chi + 2\varphi g + f\psi^2 = 0. \qquad (3)$$

The nonlinear equation (1) can be treated independently. For $f \equiv \text{const}$, its solution can be expressed in terms of elliptic integrals. For $f = ae^{\lambda x}$, a particular solution of (1) is $\varphi = -\dfrac{\lambda^2}{6a}e^{-\lambda x}$. Equations (2) and (3) can be solved successively (these are linear in the unknowns). Because $\psi = \varphi(x)$ is a particular solution of equation (2), the general solution of (2) is given by (see Polyanin and Zaitsev 1995)

$$\psi(x) = C_1 \varphi(x) + C_2 \varphi(x)\int \dfrac{dx}{\varphi^2(x)},$$

where C_1 and C_2 are arbitrary constants.

11. $ax^n \dfrac{\partial^2 w}{\partial x^2} + by^m \dfrac{\partial^2 w}{\partial y^2} + kx^{n-1}\dfrac{\partial w}{\partial x} + sy^{m-1}\dfrac{\partial w}{\partial y} = f(w)$.

For $n \neq 2$ and $m \neq 2$, there is an exact solution of the form
$$w = w(\xi), \qquad \xi = \left[b(2-m)^2 x^{2-n} + a(2-n)^2 y^{2-m}\right]^{1/2}.$$
Here, the function $w = w(\xi)$ is determined by the ordinary differential equation
$$Aw''_{\xi\xi} + \dfrac{B}{\xi} w'_\xi = f(w),$$
where
$$A = \tfrac{1}{4}ab(2-n)^2(2-m)^2,$$
$$B = \tfrac{1}{4}(2-n)(2-m)\left[ab(3nm - 4n - 4m + 4) + 2bk(2-m) + 2as(2-n)\right].$$
⊙ *Reference*: V. F. Zaitsev and A. D. Polyanin (1996).

12. $ax^n \dfrac{\partial^2 w}{\partial x^2} + by^m \dfrac{\partial^2 w}{\partial y^2} + kx^{n-1} f(w)\dfrac{\partial w}{\partial x} + sy^{m-1} f(w)\dfrac{\partial w}{\partial y} = g(w)$.

For $n \neq 2$ and $m \neq 2$, there is an exact solution of the form
$$w = w(\xi), \qquad \xi = \left[b(2-m)^2 x^{2-n} + a(2-n)^2 y^{2-m}\right]^{1/2}.$$

13. $ae^{\beta x} \dfrac{\partial^2 w}{\partial x^2} + be^{\mu y} \dfrac{\partial^2 w}{\partial y^2} + ke^{\beta x}\dfrac{\partial w}{\partial x} + se^{\mu y}\dfrac{\partial w}{\partial y} = f(w)$.

For $\beta\mu \neq 0$, there is an exact solution of the form
$$w = w(\xi), \qquad \xi = \left(b\mu^2 e^{-\beta x} + a\beta^2 e^{-\mu y}\right)^{1/2}.$$
Here, the function $w = w(\xi)$ is determined by the ordinary differential equation
$$Aw''_{\xi\xi} + \dfrac{B}{\xi} w'_\xi = f(w),$$
where
$$A = \tfrac{1}{4}ab\beta^2\mu^2, \qquad B = \tfrac{1}{4}\beta\mu(3ab\beta\mu - 2bk\mu - 2as\beta).$$

14. $ae^{\beta x} \dfrac{\partial^2 w}{\partial x^2} + be^{\mu y} \dfrac{\partial^2 w}{\partial y^2} + ke^{\beta x} f(w)\dfrac{\partial w}{\partial x} + se^{\mu y} f(w)\dfrac{\partial w}{\partial y} = g(w)$.

For $\beta\mu \neq 0$, there is an exact solution of the form
$$w = w(\xi), \qquad \xi = \left(b\mu^2 e^{-\beta x} + a\beta^2 e^{-\mu y}\right)^{1/2}.$$

15. $ax^n \dfrac{\partial^2 w}{\partial x^2} + be^{\beta y} \dfrac{\partial^2 w}{\partial y^2} + kx^{n-1}\dfrac{\partial w}{\partial x} + se^{\beta y}\dfrac{\partial w}{\partial y} = f(w)$.

For $\beta \neq 0$ and $n \neq 2$, there is an exact solution of the form
$$w = w(\xi), \qquad \xi = \left[b\beta^2 x^{2-n} + a(2-n)^2 e^{-\beta y}\right]^{1/2}.$$
Here, the function $w = w(\xi)$ is determined by the ordinary differential equation
$$Aw''_{\xi\xi} + \dfrac{B}{\xi} w'_\xi = f(w),$$
where
$$A = \tfrac{1}{4}ab\beta^2(2-n)^2, \qquad B = \tfrac{1}{4}\beta(2-n)\left[ab\beta(4-3n) + 2bk\beta - 2as(2-n)\right].$$
⊙ *Reference*: V. F. Zaitsev and A. D. Polyanin (1996).

16. $ax^n \dfrac{\partial^2 w}{\partial x^2} + be^{\beta y} \dfrac{\partial^2 w}{\partial y^2} + kx^{n-1} f(w)\dfrac{\partial w}{\partial x} + se^{\beta y} f(w)\dfrac{\partial w}{\partial y} = g(w)$.

For $\beta \neq 0$ and $n \neq 2$, there is an exact solution of the form
$$w = w(\xi), \qquad \xi = \left[b\beta^2 x^{2-n} + a(2-n)^2 e^{-\beta y}\right]^{1/2}.$$

B.5.3-4. Equations with three independent variables.

17. $\dfrac{\partial}{\partial x}\left(ax^n \dfrac{\partial w}{\partial x}\right) + \dfrac{\partial}{\partial y}\left(by^m \dfrac{\partial w}{\partial y}\right) + \dfrac{\partial}{\partial z}\left(cz^l \dfrac{\partial w}{\partial z}\right) = f(w).$

For $n \neq 2$, $m \neq 2$, and $l \neq 2$, there is an exact solution of the form

$$w = w(\xi), \qquad \xi^2 = 4\left[\dfrac{x^{2-n}}{a(2-n)^2} + \dfrac{y^{2-m}}{b(2-m)^2} + \dfrac{z^{2-l}}{c(2-l)^2}\right],$$

where the function $w(\xi)$ is determined by the ordinary differential equation

$$w''_{\xi\xi} + \dfrac{A}{\xi} w'_\xi = f(w), \qquad A = 2\left(\dfrac{1}{2-n} + \dfrac{1}{2-m} + \dfrac{1}{2-l}\right) - 1.$$

⊙ *Reference*: A. D. Polyanin and A. I. Zhurov (1998).

18. $\dfrac{\partial}{\partial x}\left(ae^{\lambda x} \dfrac{\partial w}{\partial x}\right) + \dfrac{\partial}{\partial y}\left(be^{\mu y} \dfrac{\partial w}{\partial y}\right) + \dfrac{\partial}{\partial z}\left(ce^{\nu z} \dfrac{\partial w}{\partial z}\right) = f(w).$

For $\lambda \neq 0$, $\mu \neq 0$, and $\nu \neq 0$, there is an exact solution of the form

$$w = w(\xi), \qquad \xi^2 = 4\left(\dfrac{e^{-\lambda x}}{a\lambda^2} + \dfrac{e^{-\mu y}}{b\mu^2} + \dfrac{e^{-\nu z}}{c\nu^2}\right),$$

where the function $w(\xi)$ is determined by the ordinary differential equation

$$w''_{\xi\xi} - \dfrac{1}{\xi} w'_\xi = f(w).$$

19. $\dfrac{\partial}{\partial x}\left(ax^n \dfrac{\partial w}{\partial x}\right) + \dfrac{\partial}{\partial y}\left(by^m \dfrac{\partial w}{\partial y}\right) + \dfrac{\partial}{\partial z}\left(ce^{\nu z} \dfrac{\partial w}{\partial z}\right) = f(w).$

For $n \neq 2$, $m \neq 2$, and $\nu \neq 0$, there is an exact solution of the form

$$w = w(\xi), \qquad \xi^2 = 4\left[\dfrac{x^{2-n}}{a(2-n)^2} + \dfrac{y^{2-m}}{b(2-m)^2} + \dfrac{e^{-\nu z}}{c\nu^2}\right],$$

where the function $w(\xi)$ is determined by the ordinary differential equation

$$w''_{\xi\xi} + \dfrac{A}{\xi} w'_\xi = f(w), \qquad A = 2\left(\dfrac{1}{2-n} + \dfrac{1}{2-m}\right) - 1.$$

20. $\dfrac{\partial}{\partial x}\left(ax^n \dfrac{\partial w}{\partial x}\right) + \dfrac{\partial}{\partial y}\left(be^{\mu y} \dfrac{\partial w}{\partial y}\right) + \dfrac{\partial}{\partial z}\left(ce^{\nu z} \dfrac{\partial w}{\partial z}\right) = f(w).$

For $n \neq 2$, $\mu \neq 0$, and $\nu \neq 0$, there is an exact solution of the form

$$w = w(\xi), \qquad \xi^2 = 4\left[\dfrac{x^{2-n}}{a(2-n)^2} + \dfrac{e^{-\mu y}}{b\mu^2} + \dfrac{e^{-\nu z}}{c\nu^2}\right],$$

where the function $w(\xi)$ is determined by the ordinary differential equation

$$w''_{\xi\xi} + \dfrac{n}{2-n}\dfrac{1}{\xi} w'_\xi = f(w).$$

⊙ *Reference*: A. D. Polyanin and A. I. Zhurov (1998).

B.5.4. Equations Containing Mixed Derivatives

B.5.4-1. Monge–Ampère equations.

1. $\left(\dfrac{\partial^2 w}{\partial x \partial y}\right)^2 = \dfrac{\partial^2 w}{\partial x^2} \dfrac{\partial^2 w}{\partial y^2} + f(x).$

1°. Exact solutions:

$$w(x,y) = C_1 y^2 + C_2 xy + \dfrac{C_2^2}{4C_1} x^2 - \dfrac{1}{2C_1} \int_0^x (x-t) f(t)\, dt + C_3 y + C_4 x + C_5,$$

$$w(x,y) = \dfrac{1}{x + C_1}\left(C_2 y^2 + C_3 y + \dfrac{C_3^2}{4C_2}\right) - \dfrac{1}{2C_2} \int_0^x (x-t)(t + C_1) f(t)\, dt + C_4 y + C_5 x + C_6,$$

where C_1, C_2, C_3, C_4, C_5, and C_6 are arbitrary constants.

2°. Exact solutions for $f(x) > 0$:

$$w(x,y) = \pm y \int \sqrt{f(x)}\, dx + \varphi(x) + C_1 y,$$

where $\varphi(x)$ is an arbitrary function.

2. $\left(\dfrac{\partial^2 w}{\partial x \partial y}\right)^2 = \dfrac{\partial^2 w}{\partial x^2} \dfrac{\partial^2 w}{\partial y^2} + f(x) y.$

1°. Exact solution:

$$w(x,y) = C_1 y^2 - y \int F(x)\, dx + \dfrac{1}{2C_1} \int_a^x (x-t) F^2(t)\, dt + C_2 x + C_3 y + C_4,$$

$$F(x) = \dfrac{1}{2C_1} \int f(x)\, dx + C_5,$$

where $C_1, \ldots, C_5$, and a are arbitrary constants.

2°. Exact solution:

$$w(x,y) = \varphi(x) y^2 + \psi(x) y + \chi(x),$$

where

$$\varphi(x) = \dfrac{1}{C_1 x + C_2},\quad \psi(x) = C_3 \varphi(x) + C_4 + \dfrac{\varphi(x)}{2C_1} \int \dfrac{f(x)\, dx}{[\varphi(x)]^3} - \dfrac{1}{2C_1} \int \dfrac{f(x)\, dx}{[\varphi(x)]^2},$$

$$\chi(x) = \dfrac{1}{2} \int_a^x (x-t) \dfrac{[\psi_t'(t)]^2}{\varphi(t)}\, dt + C_5 x + C_6,$$

3°. Exact solutions cubic in y:

$$w(x,y) = C_1 y^3 - \dfrac{1}{6C_1} \int_a^x (x-t) f(t)\, dt + C_2 x + C_3 y + C_4,$$

$$w(x,y) = \dfrac{y^3}{(C_1 x + C_2)^2} - \dfrac{1}{6} \int_a^x (x-t)(C_1 t + C_2)^2 f(t)\, dt + C_3 x + C_4 y + C_5,$$

where $C_1, \ldots, C_5$, and a are arbitrary constants.

4°. See the solution of equation B.5.4.5 in Item 2° with $k = 1$.

3. $\left(\dfrac{\partial^2 w}{\partial x \partial y}\right)^2 = \dfrac{\partial^2 w}{\partial x^2}\dfrac{\partial^2 w}{\partial y^2} + f(x)y^2.$

1°. Exact solution quadratic in y:
$$w(x,y) = \varphi(x)y^2 + \left[C_1 \int \varphi^2(x)\,dx + C_2\right]y + \frac{1}{2}C_1^2 \int_a^x (x-t)\varphi^3(t)\,dt + C_3 x + C_4.$$

The function $\varphi = \varphi(x)$ is determined by the ordinary differential equation
$$\varphi \varphi''_{xx} = 2(\varphi'_x)^2 - \tfrac{1}{2}f(x).$$

2°. Exact solutions quartic in y:
$$w(x,y) = C_1 y^4 - \dfrac{1}{12 C_1}\int_a^x (x-t)f(t)\,dt + C_2 x + C_3 y + C_4,$$
$$w(x,y) = \dfrac{y^4}{(C_1 x + C_2)^3} - \dfrac{1}{12}\int_a^x (x-t)(C_1 t + C_2)^3 f(t)\,dt + C_3 x + C_4 y + C_5,$$

where $C_1, \ldots, C_5$, and a are arbitrary constants.

3°. See the solution of equation B.5.4.5 in Item 2° with $k = 2$.

4. $\left(\dfrac{\partial^2 w}{\partial x \partial y}\right)^2 = \dfrac{\partial^2 w}{\partial x^2}\dfrac{\partial^2 w}{\partial y^2} + f(x)y^2 + g(x)y + h(x).$

Exact solution:
$$w(x,y) = \varphi(x)y^2 + \psi(x)y + \chi(x),$$

where the functions $\varphi = \varphi(x)$, $\psi = \psi(x)$, and $\chi = \chi(x)$ are determined by the system of ordinary differential equations
$$\varphi \varphi''_{xx} = 2(\varphi'_x)^2 - \tfrac{1}{2}f(x),$$
$$\varphi \psi''_{xx} = 2\varphi'_x \psi'_x - \tfrac{1}{2}g(x),$$
$$\varphi \chi''_{xx} = \tfrac{1}{2}(\psi'_x)^2 - \tfrac{1}{2}h(x).$$

5. $\left(\dfrac{\partial^2 w}{\partial x \partial y}\right)^2 = \dfrac{\partial^2 w}{\partial x^2}\dfrac{\partial^2 w}{\partial y^2} + f(x)y^k.$

1°. Exact solutions:
$$w(x,y) = \dfrac{C_1 y^{k+2}}{(k+1)(k+2)} - \dfrac{1}{C_1}\int_a^x (x-t)f(t)\,dt + C_2 x + C_3 y + C_4,$$
$$w(x,y) = \dfrac{y^{k+2}}{(C_1 x + C_2)^{k+1}} - \dfrac{1}{(k+1)(k+2)}\int_a^x (x-t)(C_1 t + C_2)^{k+1}f(t)\,dt + C_3 x + C_4 y + C_5,$$

where $C_1, \ldots, C_5$, and a are arbitrary constants.

2°. Exact solution:
$$w(x,y) = \varphi(x) y^{\frac{k+2}{2}}.$$

The function $\varphi = \varphi(x)$ is determined by the ordinary differential equation
$$k(k+2)\varphi \varphi''_{xx} - (k+2)^2 (\varphi'_x)^2 + 4f(x) = 0.$$

6. $\left(\dfrac{\partial^2 w}{\partial x \partial y}\right)^2 = \dfrac{\partial^2 w}{\partial x^2}\dfrac{\partial^2 w}{\partial y^2} + f(x)y^{2k+2} + g(x)y^k.$

Exact solution:
$$w(x,y) = \varphi(x) y^{k+2} - \dfrac{1}{(k+1)(k+2)}\int_a^x (x-t)\dfrac{g(t)}{\varphi(t)}\,dt + C_1 x + C_2 y + C_3,$$

where $\varphi = \varphi(x)$ is determined by the ordinary differential equation
$$(k+1)(k+2)\varphi\varphi''_{xx} - (k+2)^2(\varphi'_x)^2 + f(x) = 0.$$

7. $\left(\dfrac{\partial^2 w}{\partial x \partial y}\right)^2 = \dfrac{\partial^2 w}{\partial x^2}\dfrac{\partial^2 w}{\partial y^2} + f(x)e^{\lambda y}.$

1°. Exact solutions:
$$w(x,y) = C_1 \int_a^x (x-t)f(t)\,dt + C_2 x - \frac{1}{C_1 \lambda^2} e^{\lambda y} + C_3 y + C_4,$$
$$w(x,y) = C_1 e^{\beta x + \lambda y} - \frac{1}{C_1 \lambda^2}\int_a^x (x-t)e^{-\beta t}f(t)\,dt + C_2 x + C_3 y + C_4,$$
where C_1, C_2, C_3, C_4, a, and β are arbitrary constants.

2°. Exact solution:
$$w(x,y) = \varphi(x)\exp\!\left(\tfrac{1}{2}\lambda y\right),$$
where $\varphi = \varphi(x)$ is determined by the ordinary differential equation
$$\varphi\varphi''_{xx} - (\varphi'_x)^2 + 4\lambda^{-2} f(x) = 0.$$

8. $\left(\dfrac{\partial^2 w}{\partial x \partial y}\right)^2 = \dfrac{\partial^2 w}{\partial x^2}\dfrac{\partial^2 w}{\partial y^2} + f(x)e^{2\lambda y} + g(x)e^{\lambda y}.$

Exact solution:
$$w(x,y) = \varphi(x)e^{\lambda y} - \frac{1}{\lambda^2}\int_a^x (x-t)\frac{g(t)}{\varphi(t)}\,dt + C_1 x + C_2 y + C_3,$$
where the function $\varphi = \varphi(x)$ is determined by the ordinary differential equation
$$\varphi\varphi''_{xx} - (\varphi'_x)^2 + \lambda^{-2} f(x) = 0.$$

9. $\left(\dfrac{\partial^2 w}{\partial x \partial y}\right)^2 = \dfrac{\partial^2 w}{\partial x^2}\dfrac{\partial^2 w}{\partial y^2} + f(x)g(y).$

Exact solution:
$$w(x,y) = C_1 \int_a^x (x-t)f(t)\,dt - \frac{1}{C_1}\int_b^y (y-\xi)g(\xi)\,d\xi + C_2 x + C_3 y + C_4,$$
where C_1, C_2, C_3, C_4, a, and b are arbitrary constants.

B.5.4-2. Other equations with quadratic nonlinearities.

10. $\dfrac{\partial w}{\partial y}\dfrac{\partial^2 w}{\partial x \partial y} - \dfrac{\partial w}{\partial x}\dfrac{\partial^2 w}{\partial y^2} = f(x).$

1°. Suppose $w(x,y)$ is a solution of the equation. Then the functions
$$w_1 = \pm w\bigl(x, \pm y + \varphi(x)\bigr) + C,$$
where $\varphi(x)$ is an arbitrary function and C is an arbitrary constant, are also solutions of the equation.

2°. Exact solutions:
$$w(x,y) = \pm y\left[2\int f(x)\,dx + C_1\right]^{1/2} + \varphi(x),$$
$$w(x,y) = C_1 y^2 + \varphi(x) y + \frac{1}{4C_1}\left[\varphi^2(x) - 2\int f(x)\,dx\right] + C_2,$$
where $\varphi(x)$ is an arbitrary function and C_1 and C_2 are arbitrary constants.

3°. Exact solutions in implicit form:
$$\int \frac{dw}{\sqrt{2F(x) + \psi(w)}} = \pm y + \varphi(x),$$
where $\varphi(x)$ and $\psi(w)$ are arbitrary functions and $F(x) = \int f(x)\,dx$.

11. $\dfrac{\partial w}{\partial y}\dfrac{\partial^2 w}{\partial x \partial y} + f(y)\dfrac{\partial w}{\partial x}\dfrac{\partial^2 w}{\partial y^2} = g(y)w + h(y)x + s(y).$

Exact solution:
$$w = \varphi(y)x + \psi(y),$$

where the functions $\varphi(y)$ and $\psi(y)$ are determined by the system of ordinary differential equations
$$f\varphi\varphi''_{yy} + (\varphi'_y)^2 = g\varphi + h,$$
$$f\varphi\psi''_{yy} + \varphi'_y\psi'_y = g\psi + s.$$

B.5.5. General Form Equations

B.5.5-1. Equations of the form $\dfrac{\partial w}{\partial t} = F\!\left(x, t, w, \dfrac{\partial w}{\partial x}, \dfrac{\partial^2 w}{\partial x^2}\right).$

1. $\dfrac{\partial w}{\partial t} = F\!\left(x, \dfrac{\partial^2 w}{\partial x^2}\right).$

Exact solution:
$$w(x, t) = (Ax + B)t + \varphi(x),$$

where A and B are arbitrary constants and the function $\varphi(x)$ is determined by the ordinary differential equation
$$F(x, \varphi''_{xx}) = Ax + B.$$

2. $\dfrac{\partial w}{\partial t} = F\!\left(\dfrac{\partial w}{\partial x}, \dfrac{\partial^2 w}{\partial x^2}\right).$

Exact solution:
$$w(x, t) = At + B + \varphi(kx + \lambda t),$$

where A, B, k, and λ are arbitrary constants and the function $\varphi(z)$ is determined by the ordinary differential equation
$$F(k\varphi'_z, k^2\varphi''_{zz}) - \lambda\varphi'_z - A = 0, \qquad z = kx + \lambda t.$$

3. $\dfrac{\partial w}{\partial t} = wF\!\left(t, \dfrac{1}{w}\dfrac{\partial^2 w}{\partial x^2}\right).$

Exact solution in multiplicative form:
$$w(x, t) = (Ae^{\lambda x} + Be^{-\lambda x})E_1(t), \qquad E_1(t) = \exp\!\left[\int F(t, \lambda^2)\,dt\right],$$
$$w(x, t) = [A\cos(\lambda x) + B\sin(\lambda x)]E_2(t), \qquad E_2(t) = \exp\!\left[\int F(t, -\lambda^2)\,dt\right],$$

where A, B, and λ are arbitrary constants.

4. $\dfrac{\partial w}{\partial t} = wF\!\left(t, \dfrac{1}{w}\dfrac{\partial^2 w}{\partial x^2}\right) + f(t)e^{\lambda x} + g(t)e^{-\lambda x}.$

Exact solution:
$$w(x, t) = e^{\lambda x}E(t)\!\left[A + \int \dfrac{f(t)}{E(t)}\,dt\right] + e^{-\lambda x}E(t)\!\left[B + \int \dfrac{g(t)}{E(t)}\,dt\right],$$
$$E(t) = \exp\!\left[\int F(t, \lambda^2)\,dt\right],$$

where A, B, and λ are arbitrary constants.

5. $\dfrac{\partial w}{\partial t} = w F_1\!\left(t, \dfrac{1}{w}\dfrac{\partial^2 w}{\partial x^2}\right) + e^{\lambda x} F_2\!\left(t, \dfrac{1}{w}\dfrac{\partial^2 w}{\partial x^2}\right) + e^{-\lambda x} F_3\!\left(t, \dfrac{1}{w}\dfrac{\partial^2 w}{\partial x^2}\right).$

There are solutions of the form
$$w(x,t) = e^{\lambda x}\varphi(t) + e^{-\lambda x}\psi(t).$$

6. $\dfrac{\partial w}{\partial t} = w F\!\left(t, \dfrac{1}{w}\dfrac{\partial^2 w}{\partial x^2}\right) + f(t)\cos(\lambda x) + g(t)\sin(\lambda x).$

Exact solution:
$$w(x,t) = \cos(\lambda x)E(t)\!\left[A + \int \dfrac{f(t)}{E(t)}\,dt\right] + \sin(\lambda x)E(t)\!\left[B + \int \dfrac{g(t)}{E(t)}\,dt\right],$$
$$E(t) = \exp\!\left[\int F(t,-\lambda^2)\,dt\right],$$

where A, B, and λ are arbitrary constants.

7. $\dfrac{\partial w}{\partial t} = w F_1\!\left(t, \dfrac{1}{w}\dfrac{\partial^2 w}{\partial x^2}\right) + \cos(\lambda x)F_2\!\left(t, \dfrac{1}{w}\dfrac{\partial^2 w}{\partial x^2}\right) + \sin(\lambda x)F_3\!\left(t, \dfrac{1}{w}\dfrac{\partial^2 w}{\partial x^2}\right).$

There are solutions of the form
$$w(x,t) = \cos(\lambda x)\varphi(t) + \sin(\lambda x)\psi(t).$$

8. $\dfrac{\partial w}{\partial t} = w F\!\left(t, \dfrac{f(x)}{w}\dfrac{\partial^2 w}{\partial x^2}\right).$

Exact solution in multiplicative form:
$$w(x,t) = \varphi(x)\exp\!\left[\int F(t,\lambda)\,dt\right],$$
where the function $\varphi = \varphi(x)$ satisfies the linear ordinary differential equation $f(x)\varphi''_{xx} = \lambda\varphi$.

9. $\dfrac{\partial w}{\partial t} = w F\!\left(t, \dfrac{1}{w}\dfrac{\partial^2 w}{\partial x^2},\, w\dfrac{\partial^2 w}{\partial x^2} - \left(\dfrac{\partial w}{\partial x}\right)^{\!2}\right).$

1°. Exact solution in multiplicative form:
$$w(x,t) = C\exp\!\left[\lambda x + \int F(t,\lambda^2,0)\,dt\right],$$
where C is an arbitrary constant.

2°. Exact solution in multiplicative form:
$$w(x,t) = (Ae^{\lambda x} + Be^{-\lambda x})\varphi(t),$$
where A and B are arbitrary constants, and the function $\varphi = \varphi(t)$ satisfies the ordinary differential equation $\varphi'_t = \varphi F\!\left(t,\lambda^2,4AB\lambda^2\varphi^2\right)$.

3°. Exact solution in multiplicative form:
$$w(x,t) = [A\sin(\lambda x) + B\cos(\lambda x)]\varphi(t),$$
where A and B are arbitrary constants, and the function $\varphi = \varphi(t)$ satisfies the ordinary differential equation $\varphi'_t = \varphi F\!\left(t,-\lambda^2,-\lambda^2(A^2+B^2)\varphi^2\right)$.

⊙ *Reference*: Ph. W. Doyle (1996), the case $\partial_t F \equiv 0$ was considered.

10. $\dfrac{\partial w}{\partial t} = w F\!\left(t, \dfrac{\partial^2 w}{\partial x^2},\, \dfrac{\partial w}{\partial x} - x\dfrac{\partial^2 w}{\partial x^2},\, 2w - 2x\dfrac{\partial w}{\partial x} + x^2\dfrac{\partial^2 w}{\partial x^2}\right).$

Exact solution in multiplicative form:
$$w(x,t) = (C_2 x^2 + C_1 x + C_0)\varphi(t),$$
where C_0, C_1, and C_2 are arbitrary constants, and the function $\varphi = \varphi(t)$ satisfies the ordinary differential equation $\varphi'_t = \varphi F\!\left(t, 2C_2\varphi, C_1\varphi, 2C_0\varphi\right)$.

⊙ *Reference*: Ph. W. Doyle (1996), the case $\partial_t F \equiv 0$ was considered.

B.5.5-2. Equations of the form $\frac{\partial^2 w}{\partial t^2} = F\left(x, t, w, \frac{\partial w}{\partial t}, \frac{\partial w}{\partial x}, \frac{\partial^2 w}{\partial x^2}\right)$.

11. $\quad \dfrac{\partial^2 w}{\partial t^2} = F\left(x, \dfrac{\partial w}{\partial x}, \dfrac{\partial^2 w}{\partial x^2}\right) + G\left(t, \dfrac{\partial w}{\partial t}\right) + bw.$

Exact solution in additive form:
$$w(x,t) = \varphi(x) + \psi(t),$$
where the functions $\varphi(x)$ and $\psi(t)$ are determined by solving the second-order nonlinear ordinary differential equations (C is an arbitrary constant)
$$F(x, \varphi'_x, \varphi''_{xx}) + b\varphi = C,$$
$$\psi''_{tt} - G(t, \psi'_t) - b\psi = C.$$

12. $\quad \dfrac{\partial^2 w}{\partial t^2} = wF\left(t, \dfrac{1}{w}\dfrac{\partial w}{\partial t}, \dfrac{1}{w}\dfrac{\partial^2 w}{\partial x^2}, w\dfrac{\partial^2 w}{\partial x^2} - \left(\dfrac{\partial w}{\partial x}\right)^2\right).$

1°. Exact solution in multiplicative form:
$$w(x,t) = (Ae^{\lambda x} + Be^{-\lambda x})\varphi(t),$$
where A and B are arbitrary constants, and the function $\varphi = \varphi(t)$ satisfies the ordinary differential equation $\varphi''_{tt} = \varphi F\left(t, \varphi'_t/\varphi, \lambda^2, 4AB\lambda^2\varphi^2\right)$.

2°. Exact solution in multiplicative form:
$$w(x,t) = [A\sin(\lambda x) + B\cos(\lambda x)]\varphi(t),$$
where A and B are arbitrary constants, and the function $\varphi = \varphi(t)$ satisfies the ordinary differential equation $\varphi''_{tt} = \varphi F\left(t, \varphi'_t/\varphi, -\lambda^2, -\lambda^2(A^2 + B^2)\varphi^2\right)$.

13. $\quad \dfrac{\partial^2 w}{\partial t^2} = wF\left(t, \dfrac{\partial^2 w}{\partial x^2}, \dfrac{\partial w}{\partial x} - x\dfrac{\partial^2 w}{\partial x^2}, 2w - 2x\dfrac{\partial w}{\partial x} + x^2\dfrac{\partial^2 w}{\partial x^2}\right).$

Exact solution in multiplicative form:
$$w(x,t) = (C_2 x^2 + C_1 x + C_0)\varphi(t),$$
where C_0, C_1, and C_2 are arbitrary constants, and the function $\varphi = \varphi(t)$ satisfies the ordinary differential equation $\varphi''_{tt} = \varphi F\left(t, 2C_2\varphi, C_1\varphi, 2C_0\varphi\right)$.

14. $\quad \dfrac{\partial^2 w}{\partial t^2} = wF_1\left(t, \dfrac{1}{w}\dfrac{\partial^2 w}{\partial x^2}\right) + e^{\lambda x}F_2\left(t, \dfrac{1}{w}\dfrac{\partial^2 w}{\partial x^2}\right) + e^{-\lambda x}F_3\left(t, \dfrac{1}{w}\dfrac{\partial^2 w}{\partial x^2}\right).$

There are solutions of the form
$$w(x,t) = e^{\lambda x}\varphi(t) + e^{-\lambda x}\psi(t).$$

15. $\quad \dfrac{\partial^2 w}{\partial t^2} = wF_1\left(t, \dfrac{1}{w}\dfrac{\partial^2 w}{\partial x^2}\right) + \cos(\lambda x)F_2\left(t, \dfrac{1}{w}\dfrac{\partial^2 w}{\partial x^2}\right) + \sin(\lambda x)F_3\left(t, \dfrac{1}{w}\dfrac{\partial^2 w}{\partial x^2}\right).$

There are solutions of the form
$$w(x,t) = \cos(\lambda x)\varphi(t) + \sin(\lambda x)\psi(t).$$

B.6. Third-Order Nonlinear Equations

B.6.1. Stationary Hydrodynamic Boundary Layer Equations

1. $\dfrac{\partial w}{\partial y}\dfrac{\partial^2 w}{\partial x \partial y} - \dfrac{\partial w}{\partial x}\dfrac{\partial^2 w}{\partial y^2} = \nu \dfrac{\partial^3 w}{\partial y^3}.$

The system of equations of stationary laminar boundary layer on a flat plate (Schlichting 1981, Loitsyanskiy 1996),

$$u_1 \frac{\partial u_1}{\partial x} + u_2 \frac{\partial u_1}{\partial y} = \nu \frac{\partial^2 u_1}{\partial y^2},$$

$$\frac{\partial u_1}{\partial x} + \frac{\partial u_2}{\partial y} = 0,$$

can be reduced to this equation by introducing the stream function w in accordance with the relations $u_1 = \frac{\partial w}{\partial y}$ and $u_2 = -\frac{\partial w}{\partial x}$ (x and y are the longitudinal and transverse coordinates, u_1 and u_2 are the longitudinal and transverse components of the fluid velocity, and ν is the kinematic fluid viscosity).

1°. Suppose $w = w(x, y)$ is a solution of the stationary hydrodynamic boundary layer equation. Then the function

$$w_1 = C_1 w\big(C_2 x + C_3, C_1 C_2 y + \varphi(x)\big) + C_4,$$

where $\varphi(x)$ is an arbitrary function and C_1, C_2, C_3, and C_4 are arbitrary constants, is also a solution of the equation.

⊙ *Reference*: Yu. N. Pavlovskii (1961), L. V. Ovsyannikov (1978).

2°. Exact solutions involving arbitrary functions:

$$w(x,y) = C_1 y + \varphi(x),$$

$$w(x,y) = C_1 y^2 + \varphi(x)y + \frac{1}{4C_1}\varphi^2(x) + C_2,$$

$$w(x,y) = \frac{6\nu x + C_1}{y + \varphi(x)} + \frac{C_2}{[y+\varphi(x)]^2} + C_3,$$

$$w(x,y) = \varphi(x)\exp(-C_1 y) + \nu C_1 x + C_2,$$

$$w(x,y) = C_1 \exp\big[-C_2 y - C_2 \varphi(x)\big] + C_3 y + C_3 \varphi(x) + \nu C_2 x + C_4,$$

$$w(x,y) = 6\nu C_1 x^{1/3}\tanh\xi + C_2, \quad \xi = C_1 y x^{-2/3} + \varphi(x),$$

$$w(x,y) = -6\nu C_1 x^{1/3}\tan\xi + C_2, \quad \xi = C_1 y x^{-2/3} + \varphi(x),$$

where C_1, C_2, C_3, and C_4 are arbitrary constants and $\varphi(x)$ is an arbitrary function. The second solution is specified in Zwillinger (1998) and the fourth and fifth were obtained by Ignatovich (1993).

3°. Exact solution:

$$w(x,y) = xf(y) + g(y), \tag{1}$$

where the functions $f = f(y)$ and $g = g(y)$ are determined by the system of ordinary differential equations

$$(f'_y)^2 - f f''_{yy} = \nu f'''_{yyy}, \tag{2}$$

$$f'_y g'_y - f g''_{yy} = \nu g'''_{yyy}. \tag{3}$$

The order of equation (2) can be reduced by two. Assume that a solution $f = f(y)$ of equation (2) is known. Then equation (3), which is linear in g, has two linearly independent particular solutions

$$g_1 = 1, \quad g_2 = f(y)$$

The second particular solution is apparent from comparing equations (2) and (3). The general solution of equation (2) can be represented in the form (see Zaitsev and Polyanin 1995):

$$g(y) = C_1 + C_2 f + C_3 \left(f \int \psi \, dy - \int f \psi \, dy \right),$$
$$f = f(y), \quad \psi = \frac{1}{(f'_y)^2} \exp\left(-\frac{1}{\nu} \int f \, dy\right). \tag{4}$$

It is not difficult to check that equation (2) has the particular solutions

$$f(y) = 6\nu(y + C)^{-1},$$
$$f(y) = Ce^{\lambda y} - \lambda \nu, \tag{5}$$

where C and λ are arbitrary constants. With reference to (1) and (4), one can see that the first solution in (5) leads to the third solution in Item 2° with $\varphi(x) = \text{const}$. Substituting the second expression in (5) into (1) and (4) yields another solution.

⊙ *Reference*: A. D. Polyanin (2001b, 2001c).

2. $\dfrac{\partial w}{\partial y} \dfrac{\partial^2 w}{\partial x \partial y} - \dfrac{\partial w}{\partial x} \dfrac{\partial^2 w}{\partial y^2} = \nu \dfrac{\partial^3 w}{\partial y^3} + f(x).$

The equation of laminar boundary layer with pressure gradient.

1°. Suppose $w = w(x, y)$ is a solution of the equation in question. Then the functions (Pavlovskii 1961)

$$w_1 = \pm w(x, \pm y + \varphi(x)) + C,$$

where $\varphi(x)$ is an arbitrary function and C is an arbitrary constant, are also solutions of the equation.

2°. Nonviscous solutions (independent of the viscosity ν):

$$w(x, y) = \pm y \left[2 \int f(x) \, dx + C_1 \right]^{1/2} + \varphi(x),$$
$$w(x, y) = C_1 y^2 + \varphi(x) y + \frac{1}{4C_1} \left[\varphi^2(x) - 2 \int f(x) \, dx \right] + C_2,$$

where $\varphi(x)$ is an arbitrary function and C_1 and C_2 are arbitrary constants.

3°. Exact solution for $f(x) = ax + b$:

$$w(x, y) = xF(y) + G(y),$$

where the functions $F = F(y)$ and $G = G(y)$ are determined by the system of ordinary differential equations

$$(F'_y)^2 - FF''_{yy} = \nu F'''_{yyy} + a, \tag{1}$$
$$F'_y G'_y - FG''_{yy} = \nu G'''_{yyy} + b. \tag{2}$$

The order of the autonomous equation (1) can be reduced by one. If a particular solution $F(y)$ of equation (1) is known, the corresponding equation (2) can be reduced to a second-order linear equation by the change of variable $H(y) = G'_y$. If $F(y) = \pm\sqrt{a}\, y + C$, equation (2) can be integrated in quadrature, because its two particular solutions are known if $b = 0$, namely, $G_1 = 1$ and $G_2 = \pm\frac{1}{2}\sqrt{a}\, y^2 + Cy$.

4°. Exact solution for $f(x) = ae^{\beta x}$:

$$w(x, y) = \varphi(x) e^{\lambda y} - \frac{a}{2\beta \lambda^2 \varphi(x)} e^{\beta x - \lambda y} - \nu \lambda x + \frac{2\nu \lambda^2}{\beta} y + \frac{2\nu \lambda}{\beta} \ln |\varphi(x)|,$$

where $\varphi(x)$ is an arbitrary function and λ is an arbitrary constant.

⊙ *Reference*: A. D. Polyanin (2001b, 2001c).

3. $\dfrac{\partial w}{\partial y}\dfrac{\partial^2 w}{\partial x \partial y} - \dfrac{\partial w}{\partial x}\dfrac{\partial^2 w}{\partial y^2} = \dfrac{\partial}{\partial y}\left[f(y)\dfrac{\partial^2 w}{\partial y^2}\right] + g(y)x + h(y).$

This equation can be used to model turbulent boundary layer.

Exact solution:
$$w = \varphi(y)x + \psi(y),$$
where the functions $\varphi(y)$ and $\psi(y)$ are determined by the system of ordinary differential equations
$$(f\varphi''_{yy})'_y + \varphi\varphi''_{yy} - (\varphi'_y)^2 + g = 0,$$
$$(f\psi''_{yy})'_y + \varphi\psi''_{yy} - \varphi'_y\psi'_y + h = 0.$$

B.6.2. Nonstationary Hydrodynamic Boundary Layer Equations

1. $\dfrac{\partial^2 w}{\partial t \partial y} + \dfrac{\partial w}{\partial y}\dfrac{\partial^2 w}{\partial x \partial y} - \dfrac{\partial w}{\partial x}\dfrac{\partial^2 w}{\partial y^2} = \nu\dfrac{\partial^3 w}{\partial y^3}.$

This is the equation of nonstationary laminar boundary layer on a flat plate; x and y are the longitudinal and transverse coordinates and w is the stream function (Schlichting 1981, Loitsyanskiy 1996).

$1°$. Suppose $w = w(x, y, t)$ is a solution of the equation is question. Then the function (see Vereshchagina 1973)
$$w_1 = w\bigl(x, y + \varphi(x, t), t\bigr) + \dfrac{\partial}{\partial t}\int \varphi(x, t)\, dx + \psi(t),$$
where $\varphi(x, t)$ and $\psi(t)$ are arbitrary functions, is also a solution of the equation.

$2°$. Exact solutions:
$$w = C_1 y + \varphi(x, t),$$
$$w = C_1 y^2 + \varphi(x, t)y + \dfrac{1}{4C_1}\varphi^2(x, t) + \dfrac{\partial}{\partial t}\int \varphi(x, t)\, dx,$$
$$w = \dfrac{6\nu x + C_1}{y + \varphi(x, t)} + \dfrac{C_2}{[y + \varphi(x, t)]^2} + \dfrac{\partial}{\partial t}\int \varphi(x, t)\, dx,$$
$$w = C_1 \exp\bigl[-C_2 y - C_2\varphi(x, t)\bigr] + C_3 y + C_3\varphi(x, t) + \nu C_2 x + \dfrac{\partial}{\partial t}\int \varphi(x, t)\, dx,$$
$$w = 6\nu C_1 x^{1/3} \tanh\xi + \dfrac{\partial}{\partial t}\int \varphi(x, t)\, dx, \qquad \xi = C_1\dfrac{y + \varphi(x, t)}{x^{2/3}},$$
$$w = -6\nu C_1 x^{1/3} \tan\xi + \dfrac{\partial}{\partial t}\int \varphi(x, t)\, dx, \qquad \xi = C_1\dfrac{y + \varphi(x, t)}{x^{2/3}},$$
where $\varphi(x, t)$ is an arbitrary function of two arguments and $C_1, C_2, C_3,$ and C_4 are arbitrary constants.

$3°$. Exact solution:
$$w(x, y, t) = xF(y, t) + G(y, t), \qquad (3)$$
where the functions $F = F(y, t)$ and $G = G(y, t)$ are determined by the simpler equations with two variables
$$\dfrac{\partial^2 F}{\partial t \partial y} + \left(\dfrac{\partial F}{\partial y}\right)^2 - F\dfrac{\partial^2 F}{\partial y^2} = \nu\dfrac{\partial^3 F}{\partial y^3}, \qquad (4)$$
$$\dfrac{\partial^2 G}{\partial t \partial y} + \dfrac{\partial F}{\partial y}\dfrac{\partial G}{\partial y} - F\dfrac{\partial^2 G}{\partial y^2} = \nu\dfrac{\partial^3 G}{\partial y^3}. \qquad (5)$$

TABLE B4
Exact solutions of equation (4)

No	Function $F = F(y, t)$ (or general form of solution)	Remarks (or determining equation)
1	$F = \psi(t)$	$\psi(t)$ is an arbitrary function
2	$F = \dfrac{y}{t + C_1} + \psi(t)$	$\psi(t)$ is an arbitrary function, C_1 is any number
3	$F = \dfrac{6\nu}{y + \psi(t)} + \psi'_t(t)$	$\psi(t)$ is an arbitrary function
4	$F = C_1 \exp[-\lambda y + \lambda \psi(t)] - \psi'_t(t) + \nu\lambda$	$\psi(t)$ is an arbitrary function, C_1, λ are any numbers
5	$F = \dfrac{C_1 \exp[-\lambda y + \lambda \psi(t)] + 1}{\lambda t + C_2} - \psi'_t(t) + \nu\lambda$	$\psi(t)$ is an arbitrary function, C_1, C_2, λ are any numbers
6	$F = \dfrac{\beta + C_1 \exp[-\lambda y + \lambda \psi(t)]}{1 + C_2 \exp(-\lambda\beta t)} - \psi'_t(t) + \nu\lambda - \beta$	$\psi(t)$ is an arbitrary function, C_1, C_2, β, λ are any numbers
7	$F = F(\xi), \ \xi = y + \lambda t$	$\lambda F''_{\xi\xi} + (F'_\xi)^2 - F F''_{\xi\xi} = \nu F'''_{\xi\xi\xi}$
8	$F = t^{-1/2}\left[H(\xi) - \tfrac{1}{2}\xi\right], \ \xi = yt^{-1/2}$	$\tfrac{3}{4} - 2H'_\xi + (H'_\xi)^2 - H H''_{\xi\xi} = \nu H'''_{\xi\xi\xi}$

Equation (4) is independent of (5). If a particular solution $F = F(y, t)$ of equation (4) is known, then the corresponding equation (5) can be reduced by the change of variable $U = \frac{\partial G}{\partial y}$ to the second-order linear equation

$$\frac{\partial U}{\partial t} - F \frac{\partial U}{\partial y} = \nu \frac{\partial^2 U}{\partial y^2} - \frac{\partial F}{\partial y} U. \tag{6}$$

Exact solutions of equation (4) are listed in Table B4. The ordinary differential equations in the last two rows are autonomous and, therefore, admit reduction of order.

Table B5 presents solutions of equation (6) that correspond to the solutions of equation (4) specified in Table B4. One can see that in the first three cases the solutions of equation (6) are expressed in terms of solutions to the classical heat equation with constant coefficients. There are other three cases where equation (6) is reduced to a separable equation.

4°. Exact solution:

$$w(x, y, t) = \left[A(t)e^{k_1 x} + B(t)e^{k_2 x}\right]e^{\lambda y} + \varphi(t)x + ay,$$

$$A(t) = C_1 \exp\left[(\nu\lambda^2 - ak_1)t + \lambda \int \varphi(t)\,dt\right],$$

$$B(t) = C_2 \exp\left[(\nu\lambda^2 - ak_2)t + \lambda \int \varphi(t)\,dt\right],$$

where $\varphi(t)$ is an arbitrary function and $C_1, C_2, a, k_1, k_2,$ and λ are arbitrary parameters.

TABLE B5
Transformations of equation (6) for the corresponding exact solutions of equation (4)
[the number in the first column corresponds to the
number of the exact solution $F = F(y, t)$ in Table B4]

No	Transformations of equation (6)	Resulting equation
1	$U = u(\zeta, t)$, $\zeta = y + \int \psi(t)\,dt$	$\frac{\partial u}{\partial t} = \nu \frac{\partial^2 u}{\partial \zeta^2}$
2	$U = \frac{1}{t+C_1} u(z, \tau)$, $\tau = \frac{1}{3}(t+C_1)^3 + C_2$, $z = (t+C_1)y + \int \psi(t)(t+C_1)\,dt + C_3$	$\frac{\partial u}{\partial \tau} = \nu \frac{\partial^2 u}{\partial z^2}$
3	$U = \zeta^{-3} u(\zeta, t)$, $\zeta = y + \psi(t)$	$\frac{\partial u}{\partial t} = \nu \frac{\partial^2 u}{\partial \zeta^2}$
4	$U = e^\eta Z(\eta, t)$, $\eta = -\lambda y + \lambda \psi(t)$	$\frac{\partial Z}{\partial t} = \nu \lambda^2 \frac{\partial^2 Z}{\partial \eta^2} + (\nu\lambda^2 - C_1 \lambda e^\eta)\frac{\partial Z}{\partial \eta}$
7	$U = u(\xi, t)$, $\xi = y + \lambda t$	$\frac{\partial u}{\partial t} = \nu \frac{\partial^2 u}{\partial \xi^2} + [F(\xi) - \lambda]\frac{\partial u}{\partial \xi} - F'_\xi(\xi) u$
8	$U = t^{-1/2} u(\xi, \tau)$, $\xi = y t^{-1/2}$, $\tau = \ln t$	$\frac{\partial u}{\partial \tau} = \nu \frac{\partial^2 u}{\partial \xi^2} + H(\xi)\frac{\partial u}{\partial \xi} + [1 - H'_\xi(\xi)] u$

$5°$. Exact solution:
$$w(x, y, t) = A(t) \exp(kx + \lambda y) + B(t) \exp(\beta k x + \beta \lambda y) + \varphi(t) x + a y,$$
$$A(t) = C_1 \exp\left[(\nu \lambda^2 - ak)t + \lambda \int \varphi(t)\,dt\right],$$
$$B(t) = C_2 \exp\left[(\nu \beta^2 \lambda^2 - ak\beta)t + \beta\lambda \int \varphi(t)\,dt\right],$$
where $\varphi(t)$ is an arbitrary function and C_1, C_2, a, k, β, and λ are arbitrary parameters.

$6°$. Exact solution:
$$w(x, y, t) = \int u(z, t)\,dz + \varphi(t) y + \psi(t) x, \qquad z = kx + \lambda y,$$
where $\varphi(t)$ and $\psi(t)$ are arbitrary functions, k and λ are arbitrary parameters, and $u(z, t)$ is a function satisfying the second-order linear parabolic equation
$$\frac{\partial u}{\partial t} + [k\varphi(t) - \lambda\psi(t)]\frac{\partial u}{\partial z} = \nu\lambda^2 \frac{\partial^2 u}{\partial z^2} - \frac{1}{\lambda}\varphi'_t(t).$$
The transformation
$$u = U(\xi, t) - \frac{1}{\lambda}\varphi(t), \qquad \xi = z - \int [k\varphi(t) - \lambda\psi(t)]\,dt$$
takes the last equation to the customary heat equation
$$\frac{\partial U}{\partial t} = \nu\lambda^2 \frac{\partial^2 U}{\partial \xi^2}.$$

$7°$. Exact solutions:
$$w = e^{\nu\lambda^2 t}(C_1 e^{\lambda z} + C_2 e^{-\lambda z}) + \frac{\partial}{\partial t}\int \varphi(x, t)\,dx, \quad z = y + \varphi(x, t),$$
$$w = e^{-\nu\lambda^2 t}[C_1 \sin(\lambda z) + C_2 \cos(\lambda z)] + \frac{\partial}{\partial t}\int \varphi(x, t)\,dx, \quad z = y + \varphi(x, t),$$
$$w = C_1 e^{-\nu\lambda^2 z}\sin(\lambda z - 2\nu\lambda^2 t + C_2) + \frac{\partial}{\partial t}\int \varphi(x, t)\,dx, \quad z = y + \varphi(x, t),$$

where $\varphi(x, t)$ is an arbitrary function of two arguments and C_1, C_2, and λ are arbitrary constants.
⊙ *Reference*: A. D. Polyanin (2001b).

2. $$\frac{\partial^2 w}{\partial t \partial y} + \frac{\partial w}{\partial y}\frac{\partial^2 w}{\partial x \partial y} - \frac{\partial w}{\partial x}\frac{\partial^2 w}{\partial y^2} = \nu \frac{\partial^3 w}{\partial y^3} + f(x, t).$$

The equation of nonstationary laminar boundary layer with pressure gradient.

1°. Suppose $w(x, y, t)$ is a solution of the equation in question. Then the functions (see Vereshchagina 1973)

$$w_1 = \pm w\big(x, \pm y + \varphi(x, t), t\big) + \frac{\partial}{\partial t}\int \varphi(x, t)\, dx + \psi(t),$$

where $\varphi(x, t)$ and $\psi(t)$ are arbitrary functions, are also solutions of the equation.

2°. Nonviscous solution for any $f(x, t)$ (independent of the viscosity ν):

$$w(x, y, t) = ay^2 + \varphi(x, t)y + \frac{1}{4a}\varphi^2(x, t) + \frac{1}{2a}\int\left[\frac{\partial \varphi}{\partial t} - f(x, t)\right]dx + \psi(t).$$

where $\varphi(x, t)$ and $\psi(t)$ are arbitrary functions and a is an arbitrary constant.

Another nonviscous solution for any $f(x, t)$:

$$w(x, y, t) = \psi(x, t)y + \varphi(x, t),$$

where $\varphi(x, t)$ is an arbitrary function, and the function $\psi = \psi(x, t)$ is determined by the first-order equation

$$\frac{\partial \psi}{\partial t} + \psi\frac{\partial \psi}{\partial x} = f(x, t).$$

Nonviscous solutions for $f(x, t) = f(x)$:

$$w(x, y, t) = \pm y\left[2\int f(x)\, dx + C_1\right]^{1/2} + \varphi(x, t).$$

3°. Exact solutions for $f(x, t) = f_1(t)x + f_2(t)$:

$$w(x, y, t) = xF(y, t) + G(y, t),$$

where the functions $F = F(y, t)$ and $G = G(y, t)$ are determined from the simpler equations with two variables

$$\frac{\partial^2 F}{\partial t \partial y} + \left(\frac{\partial F}{\partial y}\right)^2 - F\frac{\partial^2 F}{\partial y^2} = \nu\frac{\partial^3 F}{\partial y^3} + f_1(t), \tag{1}$$

$$\frac{\partial^2 G}{\partial t \partial y} + \frac{\partial F}{\partial y}\frac{\partial G}{\partial y} - F\frac{\partial^2 G}{\partial y^2} = \nu\frac{\partial^3 G}{\partial y^3} + f_2(t). \tag{2}$$

Equation (1) is independent of (2). If $F = F(y, t)$ is a solution of equation (1), then the function $F_1 = F\big(y + \psi(t), t\big) + \psi'_t(t)$ with arbitrary $\psi(t)$ is also a solution of equation (1). Table B6 presents exact solutions of equation (1) for various $f_1 = f_1(t)$.

The change of variable $U = \frac{\partial G}{\partial y}$ brings equation (2) to the second-order linear equation

$$\frac{\partial U}{\partial t} - F\frac{\partial U}{\partial y} = \nu\frac{\partial^2 U}{\partial y^2} - \frac{\partial F}{\partial y}U + f_2(t). \tag{3}$$

Let us dwell on the first solution of equation (1) in Table B6:

$$F(y, t) = a(t)y + \psi(t), \qquad \text{where} \qquad a'_t + a^2 = f_1(t). \tag{4}$$

TABLE B6
Exact solutions of equation (1) for various $f_1(t)$; $\psi(t)$ is an arbitrary function

Function $f_1 = f_1(t)$	Function $F = F(y,t)$ (or general form of solution)	Determining equation (or determining coefficients)
Any	$F = a(t)y + \psi(t)$	$a'_t + a^2 = f_1(t)$
$f_1(t) = Ae^{-\beta t}$, $A > 0$, $\beta > 0$	$F = Be^{-\frac{1}{2}\beta t}\sin[\lambda y + \lambda\psi(t)] + \psi'_t(t)$, $F = Be^{-\frac{1}{2}\beta t}\cos[\lambda y + \lambda\psi(t)] + \psi'_t(t)$	$B = \pm\sqrt{\frac{2A\nu}{\beta}}$, $\lambda = \sqrt{\frac{\beta}{2\nu}}$
$f_1(t) = Ae^{\beta t}$, $A > 0$, $\beta > 0$	$F = Be^{\frac{1}{2}\beta t}\sinh[\lambda y + \lambda\psi(t)] + \psi'_t(t)$	$B = \pm\sqrt{\frac{2A\nu}{\beta}}$, $\lambda = \sqrt{\frac{\beta}{2\nu}}$
$f_1(t) = Ae^{\beta t}$, $A < 0$, $\beta > 0$	$F = Be^{\frac{1}{2}\beta t}\cosh[\lambda y + \lambda\psi(t)] + \psi'_t(t)$	$B = \pm\sqrt{\frac{2|A|\nu}{\beta}}$, $\lambda = \sqrt{\frac{\beta}{2\nu}}$
$f_1(t) = Ae^{\beta t}$, A is any, $\beta > 0$	$F = \psi(t)e^{\lambda y} - \dfrac{Ae^{\beta t - \lambda y}}{4\lambda^2\psi(t)} + \dfrac{\psi'_t(t)}{\lambda\psi(t)} - \nu\lambda$	$\lambda = \pm\sqrt{\frac{\beta}{2\nu}}$
$f_1(t) = At^{-2}$	$F = t^{-1/2}[H(\xi) - \frac{1}{2}\xi]$, $\xi = yt^{-1/2}$	$\frac{3}{4} - A - 2H'_\xi + (H'_\xi)^2 - HH''_{\xi\xi} = \nu H'''_{\xi\xi\xi}$
$f_1(t) = A$	$F = F(\xi)$, $\xi = y + \lambda t$	$-A + \lambda F''_{\xi\xi} + (F'_\xi)^2 - FF''_{\xi\xi} = \nu F'''_{\xi\xi\xi}$

Exact solutions of the Riccati equation for $a = a(t)$ with various $f_1(t)$ can be found in Polyanin and Zaitsev (1995). The substitution $a = h'_t/h$ brings this equation to a second-order linear equation for $h(t)$: $h''_{tt} - f_1(t)h = 0$. In particular, if $f_1(t) = \text{const}$, we have

$$a(t) = k\frac{C_1\cos(kt) - C_2\sin(kt)}{C_1\sin(kt) + C_2\cos(kt)} \quad \text{for} \quad f_1 = -k^2 < 0,$$

$$a(t) = k\frac{C_1\cosh(kt) + C_2\sinh(kt)}{C_1\sinh(kt) + C_2\cosh(kt)} \quad \text{for} \quad f_1 = k^2 > 0.$$

On substituting solution (4) with arbitrary $f_1(t)$ into equation (3), we obtain

$$\frac{\partial U}{\partial t} = \nu\frac{\partial^2 U}{\partial y^2} + [a(t)y + \psi(t)]\frac{\partial U}{\partial y} - a(t)U + f_2(t). \tag{5}$$

The transformation

$$U = \frac{1}{\Phi(t)}\left[u(z,\tau) + \int f_2(t)\Phi(t)\,dt\right], \quad \tau = \int \Phi^2(t)\,dt + C_1,$$

$$z = y\Phi(t) + \int \psi(t)\Phi(t)\,dt + C_2, \quad \Phi(t) = \exp\left[\int a(t)\,dt\right],$$

takes (5) to the classical constant coefficient heat equation

$$\frac{\partial u}{\partial \tau} = \nu\frac{\partial^2 u}{\partial z^2}.$$

Remark 1. The ordinary differential equations in the last two rows in Table B6 (see the last column) are autonomous and, hence, can be reduced in order.

Remark 2. Suppose $w(x,y,t)$ is a solution of the nonstationary hydrodynamic boundary layer equation with $f(x,t) = f_1(t)x + f_2(t)$. Then the function

$$w_1 = w(x + h(t), y, t) - h'_t(t)y, \quad \text{where} \quad h''_{tt} + f_1(t)h = 0,$$

is also a solution of this equation.

4°. Exact solution for $f(x,t) = g(x)e^{\beta t}$, $\beta > 0$:

$$w(x,y,t) = \varphi(x,t)e^{\lambda y} + \psi(x,t)e^{-\lambda y} + \frac{1}{\lambda}\frac{\partial}{\partial t}\int \ln|\varphi(x,t)|\,dx - \nu\lambda x,$$

$$\psi(x,t) = -\frac{e^{\beta t}}{2\lambda^2\varphi(x,t)}\int g(x)\,dx, \qquad \lambda = \pm\sqrt{\frac{\beta}{2\nu}},$$

where $\varphi(x,t)$ is an arbitrary function of two arguments.

5°. Exact solutions for $f(x,t) = g(x)e^{\beta t}$, $\beta > 0$:

$$w(x,y,t) = \pm\frac{1}{\lambda}\exp\!\left(\tfrac{1}{2}\beta t\right)\sqrt{\psi(x)}\,\sinh[\lambda y + \varphi(x,t)] + \frac{\partial}{\partial t}\int \varphi(x,t)\,dx,$$

$$w(x,y,t) = \pm\frac{1}{\lambda}\exp\!\left(\tfrac{1}{2}\beta t\right)\sqrt{\psi(x)}\,\cosh[\lambda y + \varphi(x,t)] + \frac{\partial}{\partial t}\int \varphi(x,t)\,dx,$$

$$\psi(x) = 2\int g(x)\,dx + C_1, \qquad \lambda = \sqrt{\frac{\beta}{2\nu}},$$

where $\varphi(x,t)$ is an arbitrary function of two arguments.

6°. Exact solution for $f(x,t) = g(x)e^{-\beta t}$, $\beta < 0$:

$$w(x,y,t) = \pm\frac{1}{\lambda}\exp\!\left(-\tfrac{1}{2}\beta t\right)\sqrt{\psi(x)}\,\sin[\lambda y + \varphi(x,t)] + \frac{\partial}{\partial t}\int \varphi(x,t)\,dx,$$

$$w(x,y,t) = \pm\frac{1}{\lambda}\exp\!\left(-\tfrac{1}{2}\beta t\right)\sqrt{\psi(x)}\,\cos[\lambda y + \varphi(x,t)] + \frac{\partial}{\partial t}\int \varphi(x,t)\,dx,$$

$$\psi(x) = 2\int g(x)\,dx + C_1, \qquad \lambda = \sqrt{\frac{\beta}{2\nu}}.$$

where $\varphi(x,t)$ is an arbitrary function of two arguments.

7°. Exact solution for $f(x,t) = ae^{\beta x - \gamma t}$:

$$w(x,y,t) = \varphi(x,t)e^{\lambda y} - \frac{a}{2\beta\lambda^2\varphi(x,t)}e^{\beta x - \lambda y - \gamma t}$$

$$+ \frac{1}{\lambda}\frac{\partial}{\partial t}\int \ln|\varphi(x,t)|\,dx - \nu\lambda x + \frac{2\nu\lambda^2 + \gamma}{\beta}\left(y + \frac{1}{\lambda}\ln|\varphi(x,t)|\right),$$

where $\varphi(x,t)$ is an arbitrary function of two arguments and λ is an arbitrary constant.

8°. Exact solution for $f(x,t) = f(t)$:

$$w(x,y,t) = \int u(z,t)\,dz + \varphi(t)y + \psi(t)x, \qquad z = kx + \lambda y,$$

where $\varphi(t)$ and $\psi(t)$ are arbitrary functions, k and λ are arbitrary parameters, and $u(z,t)$ is a function satisfying the second-order linear parabolic equation

$$\frac{\partial u}{\partial t} + [k\varphi(t) - \lambda\psi(t)]\frac{\partial u}{\partial z} = \nu\lambda^2\frac{\partial^2 u}{\partial z^2} - \frac{1}{\lambda}\varphi'_t(t) + \frac{1}{\lambda}f(t).$$

The transformation

$$u = U(\xi,t) - \frac{1}{\lambda}\varphi(t) + \frac{1}{\lambda}\int f(t)\,dt, \qquad \xi = z - \int [k\varphi(t) - \lambda\psi(t)]\,dt$$

brings it to the customary heat equation

$$\frac{\partial U}{\partial t} = \nu\lambda^2\frac{\partial^2 U}{\partial \xi^2}.$$

9°. Exact solution for $f(x,t) = f(t)$:

$$w(x,y,t) = C_1 e^{-\lambda y + \lambda\varphi(x,t)} - a(t)\varphi(x,t) - \frac{\partial}{\partial t}\int \varphi(x,t)\,dx + a(t)y + \nu\lambda x, \qquad a(t) = \int f(t)\,dt + C_2,$$

where $\varphi(x,t)$ is an arbitrary function of two arguments and C_1, C_2, and λ are arbitrary constants.

10°. Exact solution for $f(x,t) = f(t)$:

$$w(x,y,t) = \varphi(x,t)e^{\lambda y} + \psi(x,t)e^{-\lambda y} + \chi(x,t) + a(t)y,$$

where λ is any number, $\varphi(x,t)$ is an arbitrary function of two arguments, and the other functions are defined by

$$\psi(x,t) = \frac{C\nu e^{2\nu\lambda^2 t}}{\varphi(x,t)}\left[x - \int a(t)\,dt\right], \quad a(t) = \int f(t)\,dt + Ce^{2\nu\lambda^2 t};$$

$$\chi(x,t) = \frac{1}{\lambda}a(t)\ln|\varphi(x,t)| + \frac{1}{\lambda}\frac{\partial}{\partial t}\int \ln|\varphi(x,t)|\,dx - \nu\lambda x.$$

11°. Exact solutions for $f(x,t) = f(t)$:

$$w = e^{\nu\lambda^2 t}(C_1 e^{\lambda z} + C_2 e^{-\lambda z}) + \frac{\partial}{\partial t}\int \varphi(x,t)\,dx + z\int f(t)\,dt, \quad z = y + \varphi(x,t),$$

$$w = e^{-\nu\lambda^2 t}\left[C_1 \sin(\lambda z) + C_2 \cos(\lambda z)\right] + \frac{\partial}{\partial t}\int \varphi(x,t)\,dx + z\int f(t)\,dt, \quad z = y + \varphi(x,t),$$

$$w = C_1 e^{-\lambda z}\sin(\lambda z - 2\nu\lambda^2 t + C_2) + \frac{\partial}{\partial t}\int \varphi(x,t)\,dx + z\int f(t)\,dt, \quad z = y + \varphi(x,t),$$

where $\varphi(x,t)$ is an arbitrary function of two arguments and C_1, C_2, and λ are arbitrary constants.

12°. Exact solutions for $f(x,t) = A$:

$$w = -\frac{A}{6\nu}z^3 + C_2 z^2 + C_1 z + \frac{\partial}{\partial t}\int \varphi(x,t)\,dx, \quad z = y + \varphi(x,t),$$

$$w = kx + C_1 \exp\left(-\frac{k}{\nu}z\right) - \frac{A}{2k}z^2 + C_2 z + \frac{\partial}{\partial t}\int \varphi(x,t)\,dx, \quad z = y + \varphi(x,t),$$

where $\varphi(x,t)$ is an arbitrary function of two arguments and C_1, C_2, and k are arbitrary constants.

⊙ *Reference*: A. D. Polyanin (2001b).

3. $\dfrac{\partial^2 w}{\partial t \partial y} + \dfrac{\partial w}{\partial y}\dfrac{\partial^2 w}{\partial x \partial y} - \dfrac{\partial w}{\partial x}\dfrac{\partial^2 w}{\partial y^2} = \dfrac{\partial}{\partial y}\left[f\left(\dfrac{\partial^2 w}{\partial y^2}\right)\right] + g(x,t).$

This equation describes the flow of a non-Newtonian fluid in a two-dimensional nonstationary boundary layer with a pressure gradient. Here, w is the stream function and the function $f = f(u)$ depends on the rheological properties of the fluid. For power-law fluids, $f = k|u|^{n-1}u$.

1°. The assertion of Item 1°, equation 1 in Subsection B.6.2, remains valid for this equation.

2°. The equation admits the nonviscous solutions presented in Item 2°, equation 2 in Subsection B.6.2, where $f(x,t)$ must be replaced by $g(x,t)$.

3°. Exact solution for $g(x,t) = g(t)$:

$$w(x,y,t) = a(t)x + \int U(y,t)\,dy,$$

where $U = U(y,t)$ is a function satisfying the second-order equation

$$\frac{\partial U}{\partial t} - a(t)\frac{\partial U}{\partial y} = \frac{\partial}{\partial y}\left[f\left(\frac{\partial U}{\partial y}\right)\right] + g(t). \qquad (1)$$

The transformation

$$U = u(z,t) + \int g(t)\,dt, \quad z = y + \int a(t)\,dt$$

brings equation (1) to the simpler equation

$$\frac{\partial u}{\partial t} = \frac{\partial}{\partial z}\left[f\left(\frac{\partial u}{\partial z}\right)\right]. \tag{2}$$

Equation (2) admits exact solutions with the forms [for any $f = f(v)$]

$$u(z,t) = H(\zeta), \qquad \zeta = kz + \lambda t \quad \Longrightarrow \quad \text{equation} \quad \lambda H = kf(kH'_\zeta) + C;$$
$$u(z,t) = az + H(\zeta), \quad \zeta = kz + \lambda t \quad \Longrightarrow \quad \text{equation} \quad \lambda H = kf(kH'_\zeta + a) + C;$$
$$u(z,t) = \sqrt{t}\, H(\zeta), \quad \zeta = z/\sqrt{t} \quad \Longrightarrow \quad \text{equation} \quad \tfrac{1}{2}H - \tfrac{1}{2}\zeta H'_\zeta = [f(H'_\zeta)]'_\zeta,$$

where a, k, C, and λ are arbitrary constants. The first two equations for $H = H(\zeta)$ can be solved in parametric form.

4°. Exact solution for $g(x,t) = g(t)$:

$$w(x,y,t) = \int v(\eta,t)\, d\eta + \varphi(t)y + \psi(t)x, \qquad \eta = y + kx,$$

where $\varphi(t)$ and $\psi(t)$ are arbitrary functions, k is an arbitrary parameter, and $v(\eta,t)$ is a function satisfying the second-order nonlinear parabolic equation

$$\frac{\partial v}{\partial t} + [k\varphi(t) - \psi(t)]\frac{\partial v}{\partial \eta} = \frac{\partial}{\partial \eta}\left[f\left(\frac{\partial v}{\partial \eta}\right)\right] - \varphi'_t(t) + g(t).$$

The transformation

$$v = R(\zeta,t) - \varphi(t) + \int g(t)\, dt, \qquad \zeta = \eta - \int [k\varphi(t) - \psi(t)]\, dt$$

leads to the simpler equation

$$\frac{\partial R}{\partial t} = \frac{\partial}{\partial \zeta}\left[f\left(\frac{\partial R}{\partial \zeta}\right)\right].$$

For exact solutions of this equation, see Item 3°.

5°. Exact solution for $g(x,t) = s(t)x + h(t)$:

$$w(x,y,t) = [a(t)y + \psi(t)]x + \int Q(y,t)\, dy,$$

where $\psi(t)$ is an arbitrary function, $a = a(t)$ is determined by the Riccati equation

$$a'_t + a^2 = s(t),$$

and $Q = Q(y,t)$ satisfies the second-order equation

$$\frac{\partial Q}{\partial t} = \frac{\partial}{\partial y}\left[f\left(\frac{\partial Q}{\partial y}\right)\right] + [a(t)y + \psi(t)]\frac{\partial Q}{\partial y} - a(t)Q + h(t).$$

The transformation

$$Q = \frac{1}{\Phi(t)}\left[Z(\xi,\tau) + \int h(t)\Phi(t)\, dt\right], \quad \tau = \int \Phi^2(t)\, dt + A, \quad \xi = y\Phi(t) + \int \psi(t)\Phi(t)\, dt + B,$$

where $\Phi(t) = \exp\left[\int a(t)\, dt\right]$, leads to the simpler equation

$$\frac{\partial Z}{\partial \tau} = \frac{\partial}{\partial \xi}\left[f\left(\frac{\partial Z}{\partial \xi}\right)\right].$$

For exact solutions of this equation, see Item 3°.

B.7. Fourth-Order Nonlinear Equations

B.7.1. Stationary Hydrodynamic Equations (Navier–Stokes Equations)

1. $\dfrac{\partial w}{\partial y}\dfrac{\partial}{\partial x}(\Delta w) - \dfrac{\partial w}{\partial x}\dfrac{\partial}{\partial y}(\Delta w) = \nu \Delta \Delta w, \qquad \Delta w = \dfrac{\partial^2 w}{\partial x^2} + \dfrac{\partial^2 w}{\partial y^2}.$

The two-dimensional equations of steady-state motion of a viscous incompressible fluid (stationary Navier–Stokes equations)

$$u_1 \frac{\partial u_1}{\partial x} + u_2 \frac{\partial u_1}{\partial y} = -\frac{1}{\rho}\frac{\partial p}{\partial x} + \nu \Delta u_1,$$
$$u_1 \frac{\partial u_2}{\partial x} + u_2 \frac{\partial u_2}{\partial y} = -\frac{1}{\rho}\frac{\partial p}{\partial y} + \nu \Delta u_2,$$
$$\frac{\partial u_1}{\partial x} + \frac{\partial u_2}{\partial y} = 0$$

are reduced to the equation under consideration. To this end, one introduces the stream function w by the formulas $u_1 = \frac{\partial w}{\partial y}$ and $u_2 = -\frac{\partial w}{\partial x}$ and eliminates the pressure p, using cross differentiation, from the first two equations.

1°. Exact solutions in additive form:

$$w(x, y) = C_1 y^3 + C_2 y^2 + C_3 y + C_4,$$
$$w(x, y) = C_1 x^2 + C_2 x + C_3 y^2 + C_4 y + C_5,$$
$$w(x, y) = C_1 \exp(-\lambda y) + C_2 y^2 + C_3 y + C_4 + \nu \lambda x,$$
$$w(x, y) = C_1 \exp(\lambda x) - \nu \lambda x + C_2 \exp(\lambda y) + \nu \lambda y + C_3,$$
$$w(x, y) = C_1 \exp(\lambda x) + \nu \lambda x + C_2 \exp(-\lambda y) + \nu \lambda y + C_3,$$

where $C_1, \dots, C_5$, and λ are arbitrary constants.

2°. Exact solutions:

$$w(x, y) = (Ax + B)e^{-\lambda y} + \nu \lambda x + C,$$
$$w(x, y) = A e^{-\lambda(y+kx)} + B(y + kx)^2 + C(y + kx) + \nu \lambda (k^2 + 1) x + D,$$
$$w(x, y) = \left[A \sinh(\beta x) + B \cosh(\beta x)\right] e^{-\lambda y} + \frac{\nu}{\lambda}(\beta^2 + \lambda^2) x + C,$$
$$w(x, y) = \left[A \sin(\beta x) + B \cos(\beta x)\right] e^{-\lambda y} + \frac{\nu}{\lambda}(\lambda^2 - \beta^2) x + C,$$
$$w(x, y) = A e^{\lambda y + \beta x} + B e^{\gamma x} + \nu \gamma y + \frac{\nu}{\lambda} \gamma (\beta - \gamma) x + C, \qquad \gamma = \pm \sqrt{\lambda^2 + \beta^2},$$

where A, B, C, D, k, β, and λ are arbitrary constants.

3°. Exact solution:

$$w(x, y) = F(y) x + G(y),$$

where the functions $F = F(y)$ and $G = G(y)$ are determined by the system of fourth-order ordinary differential equations

$$F'_y F''_{yy} - F F'''_{yyy} = \nu F''''_{yyyy}, \tag{1}$$
$$G'_y F''_{yy} - F G'''_{yyy} = \nu G''''_{yyyy}. \tag{2}$$

Integrating yields the system of third-order equations

$$(F'_y)^2 - F F''_{yy} = \nu F'''_{yyy} + A, \tag{3}$$
$$G'_y F'_y - F G''_{yy} = \nu G'''_{yyy} + B, \tag{4}$$

where A and B are arbitrary constants. The order of the autonomous equation (3) can be reduced by one.

It is not difficult to verify that equation (1) has the particular solutions
$$F(y) = ay + b, \tag{5}$$
$$F(y) = 6\nu(y + a)^{-1}, \tag{6}$$
$$F(y) = ae^{-\lambda y} + \lambda\nu, \tag{7}$$
where a, b, and λ are arbitrary constants.

In general, equation (4) can be reduced by the change of variable $U = G'_y$ to the second-order nonhomogeneous linear equation
$$\nu U''_{yy} + FU'_y - F'_y U + B = 0, \quad \text{where} \quad U = G'_y. \tag{8}$$
The corresponding homogeneous equation (with $B = 0$) has two linearly independent particular solutions
$$U_1 = \begin{cases} F''_{yy} & \text{if } F''_{yy} \neq 0, \\ F & \text{if } F''_{yy} = 0, \end{cases} \quad U_2 = U_1 \int \frac{\Phi\, dy}{U_1^2}, \quad \text{where} \quad \Phi = \exp\left(-\frac{1}{\nu}\int F\, dy\right). \tag{9}$$
(The first solution is apparent from comparing equations (1) and (8) with $B = 0$.) The general solutions of equations (8) and (2) are given by
$$U = C_1 U_1 + C_2 U_2 + C_3 \left(U_2 \int \frac{U_1}{\Phi}\, dy - U_1 \int \frac{U_2}{\Phi}\, dy \right), \quad G = \int U\, dy + C_4, \quad C_3 = -\frac{B}{\nu}. \tag{10}$$

The general solution of equation (2) corresponding to the particular solution (6) is represented as
$$G(y) = \widetilde{C}_1(y+a)^3 + \widetilde{C}_2 + \widetilde{C}_3(y+a)^{-1} + \widetilde{C}_4(y+a)^{-2},$$
where $\widetilde{C}_1$, $\widetilde{C}_2$, $\widetilde{C}_3$, and $\widetilde{C}_4$ are arbitrary constants (these are expressed in terms of C_1, C_2, C_3, and C_4).

The general solutions of (2) corresponding to the particular solutions (5) and (7) are determined from (9) and (10).

4°. Exact solution of a more general form:
$$w(x, y) = F(z)x + G(z), \quad z = y + kx,$$
where the functions $F = F(z)$ and $G = G(z)$ are determined by the system of fourth-order ordinary differential equations
$$F'_z F''_{zz} - FF'''_{zzz} = \nu(k^2 + 1)F''''_{zzzz}, \tag{11}$$
$$G'_z F''_{zz} - FG'''_{zzz} = \nu(k^2 + 1)G''''_{zzzz} + 4k\nu F'''_{zzz} + \frac{2k}{(k^2+1)} FF''_{zz}. \tag{12}$$
Integrating yields the system of third-order equations
$$(F'_z)^2 - FF''_{zz} = \nu(k^2 + 1)F'''_{zzz} + A, \tag{13}$$
$$G'_z F'_z - FG''_{zz} = \nu(k^2 + 1)G'''_{zzz} + 4k\nu F''_{zz} + \frac{2k}{k^2+1}\int FF''_{zz}\, dz + B, \tag{14}$$
where A and B are arbitrary constants. The order of the autonomous equation (13) can be reduced by one.

It is not difficult to verify that equation (11) has the particular solutions
$$F(z) = az + b, \quad z = y + kx,$$
$$F(z) = 6\nu(k^2 + 1)(z + a)^{-1},$$
$$F(z) = ae^{-\lambda z} + \lambda\nu(k^2 + 1),$$
where a, b, and λ are arbitrary constants.

In general, equation (14) can be reduced by the change of variable $U = G'_z$ to a second-order nonhomogeneous linear equation.

⊙ *Reference*: A. D. Polyanin (2001d).

2. $\dfrac{\partial w}{\partial y}\dfrac{\partial}{\partial x}(\Delta w) - \dfrac{\partial w}{\partial x}\dfrac{\partial}{\partial y}(\Delta w) = \nu\Delta\Delta w + f(y)$, $\Delta w = \dfrac{\partial^2 w}{\partial x^2} + \dfrac{\partial^2 w}{\partial y^2}$.

This equation describes plane flow of a viscous incompressible fluid under the action of a transverse force (w is the stream function). The case $F(y) = a\sin(\lambda y)$ corresponds to A. N. Kolmogorov's model which is used to describe subcritical and transcritical (laminar-turbulent) modes of flow.

1°. Exact solution in additive form for arbitrary $f(y)$:

$$w(x,y) = -\dfrac{1}{2\nu}\int_0^y (y-z)^2\Phi(z)\,dz + C_1 e^{-\lambda y} + C_2 y^2 + C_3 y + C_4 + \nu\lambda x, \quad \Phi(z) = e^{-\lambda z}\int e^{\lambda z}f(z)\,dz,$$

where C_1, C_2, C_3, C_4, and λ are arbitrary constants.

Example. In the case $f(y) = a\beta\cos(\beta y)$, which corresponds to $F(y) = a\sin(\beta y)$, it follows from the previous formula with $C_1 = C_2 = C_4 = 0$ and $B = -\nu\lambda$ that

$$w(x,y) = -\dfrac{a}{\beta^2(B^2 + \nu^2\beta^2)}\left[B\sin(\beta y) + \nu\beta\cos(\beta y)\right] + Cy - Bx,$$

where B and C are arbitrary constants. This solution was indicated by Belotserkovskii and Oparin (2000); it describes the flow with a periodic structure.

2°. Exact solution in additive form for $f(y) = Ae^{\lambda y} + Be^{-\lambda y}$:

$$w(x,y) = C_1 e^{-\lambda x} + C_2 x - \dfrac{A}{\lambda^3(C_2 + \nu\lambda)}e^{\lambda y} + \dfrac{B}{\lambda^3(C_2 - \nu\lambda)}e^{-\lambda y} - \nu\lambda y,$$

where C_1 and C_2 are arbitrary constants.

3°. Generalized separable solution for arbitrary $f(y)$:

$$w(x,y) = \varphi(y)x + \psi(y),$$

where the functions $\varphi = \varphi(y)$ and $\psi = \psi(y)$ are determined by the system of fourth-order ordinary differential equations

$$\varphi'_y\varphi''_{yy} - \varphi\varphi'''_{yyy} = \nu\varphi''''_{yyyy}, \tag{1}$$
$$\psi'_y\varphi''_{yy} - \varphi\psi'''_{yyy} = \nu\psi''''_{yyyy} + f(y). \tag{2}$$

Integrating yields the system of third-order equations

$$(\varphi'_y)^2 - \varphi\varphi''_{yy} = \nu\varphi'''_{yyy} + A, \tag{3}$$
$$\psi'_y\varphi'_y - \varphi\psi''_{yy} = \nu\psi'''_{yyy} + \int f(y)\,dy + B, \tag{4}$$

where A and B are arbitrary constants. The order of the autonomous equation (3) can be reduced by one.

It is not difficult to verify that equation (1) has the particular solutions

$$\varphi(y) = ay + b, \tag{5}$$
$$\varphi(y) = 6\nu(y+a)^{-1}, \tag{6}$$
$$\varphi(y) = ae^{-\lambda y} + \lambda\nu, \tag{7}$$

where a, b, and λ are arbitrary constants.

In general, equation (4) can be reduced by the change of variable $U = \psi'_y$ to the second-order nonhomogeneous linear equation

$$\nu U''_{yy} + \varphi U'_y - \varphi'_y U + F = 0, \quad \text{where} \quad U = \psi'_y, \quad F = \int f(y)\,dy + B. \tag{8}$$

The corresponding homogeneous equation (with $F = 0$) has two linearly independent particular solutions:

$$U_1 = \begin{cases} \varphi''_{yy} & \text{for } \varphi \neq ay + b, \\ \varphi & \text{for } \varphi = ay + b, \end{cases} \quad U_2 = U_1 \int \frac{\Phi\, dy}{U_1^2}, \quad \text{where} \quad \Phi = \exp\left(-\frac{1}{\nu}\int \varphi\, dy\right).$$

(The first solution is apparent from comparing equations (1) and (8) with $F = 0$.) The general solutions of equations (8) and (2) are given by

$$U = C_1 U_1 + C_2 U_2 + \frac{1}{\nu} U_1 \int U_2 \frac{F}{\Phi}\, dy - \frac{1}{\nu} U_2 \int U_1 \frac{F}{\Phi}\, dy, \quad \psi = \int U\, dy + C_4.$$

3. $\dfrac{1}{r}\dfrac{\partial w}{\partial \theta}\dfrac{\partial}{\partial r}(\Delta w) - \dfrac{1}{r}\dfrac{\partial w}{\partial r}\dfrac{\partial}{\partial \theta}(\Delta w) = \nu \Delta \Delta w, \qquad \Delta w = \dfrac{1}{r}\dfrac{\partial}{\partial r}\left(r\dfrac{\partial w}{\partial r}\right) + \dfrac{1}{r^2}\dfrac{\partial^2 w}{\partial \theta^2}.$

The equations of steady-state flow of a viscous incompressible fluid (stationary Navier–Stokes equations) written in polar coordinates ($x = r\cos\theta$, $y = r\sin\theta$) reduce to this equation. The radial and tangential components of the fluid velocity are expressed in terms of the stream function w by the formulas $u_r = \frac{1}{r}\frac{\partial w}{\partial \theta}$ and $u_\theta = -\frac{\partial w}{\partial r}$.

1°. Exact solution in additive form:

$$w(r, \theta) = \nu C_1 \theta + C_2 r^{C_1+2} + C_3 r^2 + C_4 \ln r + C_5,$$

where $C_1, \ldots, C_5$ are arbitrary constants.

2°. Exact solution:

$$w(r, \theta) = f(r)\theta + g(r).$$

The functions $f = f(r)$ and $g = g(r)$ are determined by the system of ordinary differential equations

$$-f'_r \mathbf{L}(f) + f[\mathbf{L}(f)]'_r = \nu r \mathbf{L}^2(f), \tag{1}$$

$$-g'_r \mathbf{L}(f) + f[\mathbf{L}(g)]'_r = \nu r \mathbf{L}^2(g), \tag{2}$$

where $\mathbf{L}(f) = r^{-1}(r f'_r)'_r$.

Exact solution of system (1)–(2):

$$f(r) = C_1 \ln r + C_2, \quad g(r) = C_3 r^2 + C_4 \ln r + C_5 \int \left[\int rQ(r)\, dr\right] \frac{dr}{r} + C_6,$$

$$Q(r) = \int r^{(C_2/\nu)-1} \exp\left(\frac{C_1}{2\nu}\ln^2 r\right) dr,$$

where $C_1, \ldots, C_6$ are arbitrary constants.

B.7.2. Nonstationary Hydrodynamic Equations

1. $\dfrac{\partial}{\partial t}(\Delta w) + \dfrac{\partial w}{\partial y}\dfrac{\partial}{\partial x}(\Delta w) - \dfrac{\partial w}{\partial x}\dfrac{\partial}{\partial y}(\Delta w) = \nu \Delta \Delta w, \qquad \Delta w = \dfrac{\partial^2 w}{\partial x^2} + \dfrac{\partial^2 w}{\partial y^2}.$

The two-dimensional equations of steady-state flow of a viscous incompressible fluid (nonstationary Navier–Stokes equations) can be reduced to this equation by introducing a stream function w.

1°. Exact solution:

$$w(x, y, t) = F(y, t)x + G(y, t), \tag{1}$$

where the functions $F(y, t)$ and $G = G(y, t)$ are determined from the system of fourth-order one-dimensional equations

$$\frac{\partial^3 F}{\partial t \partial y^2} + \frac{\partial F}{\partial y}\frac{\partial^2 F}{\partial y^2} - F\frac{\partial^3 F}{\partial y^3} = \nu \frac{\partial^4 F}{\partial y^4}, \tag{2}$$

$$\frac{\partial^3 G}{\partial t \partial y^2} + \frac{\partial G}{\partial y}\frac{\partial^2 F}{\partial y^2} - F\frac{\partial^3 G}{\partial y^3} = \nu \frac{\partial^4 G}{\partial y^4}. \tag{3}$$

Equation (2) is independent of (3). Integrating (2) and (3) with respect to y yields

$$\frac{\partial^2 F}{\partial t \partial y} + \left(\frac{\partial F}{\partial y}\right)^2 - F\frac{\partial^2 F}{\partial y^2} = \nu\frac{\partial^3 F}{\partial y^3} + f_1(t), \tag{4}$$

$$\frac{\partial^2 G}{\partial t \partial y} + \frac{\partial F}{\partial y}\frac{\partial G}{\partial y} - F\frac{\partial^2 G}{\partial y^2} = \nu\frac{\partial^3 G}{\partial y^3} + f_2(t), \tag{5}$$

where $f_1(t)$ and $f_2(t)$ are arbitrary functions. Equation (5) is linear in G. The change of variable

$$G = \int U\, dy - hF + h'_t y, \qquad \text{where} \quad U = U(y,t),\ F = F(y,t), \tag{6}$$

with $h = h(t)$ satisfying the linear ordinary differential equation

$$h''_{tt} - f_1(t)h = f_2(t), \tag{7}$$

brings (5) to the second-order homogenous linear equation

$$\frac{\partial U}{\partial t} = \nu\frac{\partial^2 U}{\partial y^2} + F\frac{\partial U}{\partial y} - \frac{\partial F}{\partial y}U. \tag{8}$$

So, if a particular solution of equation (2) or (4) is known, then determining the function G reduces to solving the linear equations (7)–(8) followed by integrating in accordance with (6).

Table B7 lists exact solutions of equation (2). The ordinary differential equations in the last two rows, which determine a traveling wave solution and a self-similar solution, are autonomous and hence admit reduction of order.

The general solution of the nonhomogeneous equation (7) can be found with the aid of the fundamental system of solutions for the corresponding homogeneous equation (with $f_2 \equiv 0$). The necessary formulas and fundamental solutions of the homogeneous equation (7) that correspond to all exact solutions of equation (2) listed in Table B7 can be found in the handbooks by Kamke (1977) and Polyanin and Zaitsev (1995).

Equation (8) for any function $F = F(y,t)$ has the trivial solution, $U = 0$. The expressions in Table B7 and relation (6) with $U = 0$ define some exact solutions of the form (1). By analyzing nontrivial solutions of equation (8), one can obtain a wider class of exact solutions.

Table B8 lists transformations that simplify equation (8) for some of the solutions of equation (2) [or (4)] given in Table B7. One can see that in the first two cases, solutions to equation (8) are expressed in terms of solutions to the classical constant coefficient heat equation. In the remaining three cases, the equation reduces to a separable equation.

2°. Exact solution of a more general form:

$$w(x,y,t) = F(\xi,t)x + G(\xi,t), \qquad \xi = y + kx,$$

where the functions $F(\xi,t)$ and $G = G(\xi,t)$ are determined from the system of fourth-order one-dimensional equations

$$\frac{\partial^3 F}{\partial t \partial \xi^2} + \frac{\partial F}{\partial \xi}\frac{\partial^2 F}{\partial \xi^2} - F\frac{\partial^3 F}{\partial \xi^3} = \nu(k^2+1)\frac{\partial^4 F}{\partial \xi^4}, \tag{9}$$

$$\frac{\partial^3 G}{\partial t \partial \xi^2} + \frac{\partial G}{\partial \xi}\frac{\partial^2 F}{\partial \xi^2} - F\frac{\partial^3 G}{\partial \xi^3} = \nu(k^2+1)\frac{\partial^4 G}{\partial \xi^4} + 4\nu k\frac{\partial^3 F}{\partial \xi^3} + \frac{2k}{k^2+1}\left(F\frac{\partial^2 F}{\partial \xi^2} - \frac{\partial^2 F}{\partial t \partial \xi}\right). \tag{10}$$

Integrating equations (9) and (10) with respect to ξ yields

$$\frac{\partial^2 F}{\partial t \partial \xi} + \left(\frac{\partial F}{\partial \xi}\right)^2 - F\frac{\partial^2 F}{\partial \xi^2} = \nu(k^2+1)\frac{\partial^3 F}{\partial \xi^3} + f_1(t), \tag{11}$$

$$\frac{\partial^2 G}{\partial t \partial \xi} + \frac{\partial F}{\partial \xi}\frac{\partial G}{\partial \xi} - F\frac{\partial^2 G}{\partial \xi^2} = \nu(k^2+1)\frac{\partial^3 G}{\partial \xi^3} + Q(\xi,t), \tag{12}$$

TABLE B7
Exact solution of equations (2) and (4); $\varphi(t)$, $\psi(t)$ are arbitrary functions and A, B, λ are arbitrary constants

No	Function $F = F(y,t)$ (or general form of solution)	Function $f_1(t)$ in equation (4)	Determining coefficients (or determining equation)
1	$F = \varphi(t)y + \psi(t)$	$f_1(t) = \varphi'_t + \varphi^2$	N/A
2	$F = \dfrac{6\nu}{y + \psi(t)} + \psi'_t(t)$	$f_1(t) = 0$	N/A
3	$F = A\exp[-\lambda y - \lambda\psi(t)] + \psi'_t(t) + \nu\lambda$	$f_1(t) = 0$	N/A
4	$F = \dfrac{A\exp[-\lambda y + \lambda\psi(t)] + 1}{\lambda t + B} - \psi'_t(t) + \nu\lambda$	$f_1(t) = 0$	N/A
5	$F = \dfrac{\beta + A\exp[-\lambda y + \lambda\psi(t)]}{1 + B\exp(-\lambda\beta t)} - \psi'_t(t) + \nu\lambda - \beta$	$f_1(t) = 0$	β is an arbitrary constant
6	$F = Ae^{-\beta t}\sin[\lambda y + \lambda\psi(t)] + \psi'_t(t)$	$f_1(t) = Be^{-2\beta t}$	$\beta = \nu\lambda^2$, $B = A^2\lambda^2 > 0$
7	$F = Ae^{-\beta t}\cos[\lambda y + \lambda\psi(t)] + \psi'_t(t)$	$f_1(t) = Be^{-2\beta t}$	$\beta = \nu\lambda^2$, $B = A^2\lambda^2 > 0$
8	$F = Ae^{\beta t}\sinh[\lambda y + \lambda\psi(t)] + \psi'_t(t)$	$f_1(t) = Be^{2\beta t}$	$\beta = \nu\lambda^2$, $B = A^2\lambda^2 > 0$
9	$F = Ae^{\beta t}\cosh[\lambda y + \lambda\psi(t)] + \psi'_t(t)$	$f_1(t) = Be^{2\beta t}$	$\beta = \nu\lambda^2$, $B = -A^2\lambda^2 < 0$
10	$F = \psi(t)e^{\lambda y} - \dfrac{Ae^{\beta t - \lambda y}}{4\lambda^2\psi(t)} + \dfrac{\psi'_t(t)}{\lambda\psi(t)} - \nu\lambda$	$f_1(t) = Ae^{\beta t}$	$\beta = 2\nu\lambda^2$
11	$F = F(\xi)$, $\xi = y + \lambda t$	$f_1(t) = A$	$-A + \lambda F''_{\xi\xi} + (F'_\xi)^2 - FF''_{\xi\xi} = \nu F'''_{\xi\xi\xi}$
12	$F = t^{-1/2}[H(\xi) - \tfrac{1}{2}\xi]$, $\xi = yt^{-1/2}$	$f_1(t) = At^{-2}$	$\tfrac{3}{4} - A - 2H'_\xi + (H'_\xi)^2 - HH''_{\xi\xi} = \nu H'''_{\xi\xi\xi}$

where $f_1(t)$ is an arbitrary function and

$$Q(\xi, t) = 4\nu k\frac{\partial^2 F}{\partial\xi^2} - \frac{2k}{k^2+1}\frac{\partial F}{\partial t} + \frac{2k}{k^2+1}\int F\frac{\partial^2 F}{\partial\xi^2}\,d\xi + f_2(t) \qquad [f_2(t) \text{ is any}].$$

Equation (12) is linear in G. The change of variable $U = \dfrac{\partial G}{\partial \xi}$ takes it to the second-order linear equation

$$\frac{\partial U}{\partial t} = \nu(k^2+1)\frac{\partial^2 U}{\partial\xi^2} + F\frac{\partial U}{\partial\xi} - \frac{\partial F}{\partial\xi}U + Q(\xi, t). \qquad (13)$$

So, if a particular solution of equation (9) or (11) is known, then determining the function G reduces to solving the linear equation (13). Scaling the independent variables by the formulas $\xi = (k^2+1)\zeta$ and $t = (k^2+1)\tau$, one can reduce equation (9) to equation (2) in which y and t must be replaced by ζ and τ, respectively. Exact solutions of equation (2) are listed in Table B7.

TABLE B8
Transformations of equation (8) for the corresponding exact solutions of equation (4)
[the number in the first column corresponds to the
number of the exact solution $F = F(y,t)$ in Table B7]

No	Transformations of equation (8)	Resulting equation
1	$U = \frac{1}{\Phi(t)} u(z,\tau),\ \tau = \int \Phi^2(t)\,dt,$ $z = y\Phi(t) + \int \psi(t)\Phi(t)\,dt,\ \Phi(t) = \exp\left[\int \varphi(t)\,dt\right]$	$\frac{\partial u}{\partial \tau} = \nu \frac{\partial^2 u}{\partial z^2}$
2	$U = \zeta^{-3} u(\zeta, t),\ \zeta = y + \psi(t)$	$\frac{\partial u}{\partial t} = \nu \frac{\partial^2 u}{\partial \zeta^2}$
3	$U = e^\eta Z(\eta, t),\ \eta = -\lambda y - \lambda \psi(t)$	$\frac{\partial Z}{\partial t} = \nu\lambda^2 \frac{\partial^2 Z}{\partial \eta^2} + (\nu\lambda^2 - A\lambda e^\eta)\frac{\partial Z}{\partial \eta}$
11	$U = u(\xi, t),\ \xi = y + \lambda t$	$\frac{\partial u}{\partial t} = \nu \frac{\partial^2 u}{\partial \xi^2} + [F(\xi) - \lambda]\frac{\partial u}{\partial \xi} - F'_\xi(\xi) u$
12	$U = t^{-1/2} u(\xi, \tau),\ \xi = y t^{-1/2},\ \tau = \ln t$	$\frac{\partial u}{\partial \tau} = \nu \frac{\partial^2 u}{\partial \xi^2} + H(\xi)\frac{\partial u}{\partial \xi} + [1 - H'_\xi(\xi)] u$

$3°$. Exact solution [special case of (1)]:
$$w(x,y,t) = e^{-\lambda y}\left[f(t)x + g(t)\right] + \varphi(t)x + \psi(t)y + \chi(t),$$
$$f(t) = C_1 E(t),\quad E(t) = \exp\left[\nu\lambda^2 t - \lambda \int \varphi(t)\,dt\right],$$
$$g(t) = C_2 E(t) - C_1 E(t) \int \psi(t)\,dt,$$

where $\varphi(t)$, $\psi(t)$, and $\chi(t)$ are arbitrary functions and C_1, C_2, and λ are arbitrary parameters.

$4°$. Exact solution:
$$w(x,y,t) = e^{-\lambda y}\left[A(t)e^{\beta x} + B(t)e^{-\beta x}\right] + \varphi(t)x + \psi(t)y + \chi(t),$$
$$A(t) = C_1 \exp\left[\nu(\lambda^2 + \beta^2)t - \beta \int \psi(t)\,dt - \lambda \int \varphi(t)\,dt\right],$$
$$B(t) = C_2 \exp\left[\nu(\lambda^2 + \beta^2)t + \beta \int \psi(t)\,dt - \lambda \int \varphi(t)\,dt\right],$$

where $\varphi(t)$, $\psi(t)$, and $\chi(t)$ are arbitrary functions and C_1, C_2, λ, and β are arbitrary parameters.

$5°$. Exact solution:
$$w(x,y,t) = e^{-\lambda y}\left[A(t)\sin(\beta x) + B(t)\cos(\beta x)\right] + \varphi(t)x + \psi(t)y + \chi(t),$$

where $\varphi(t)$, $\psi(t)$, and $\chi(t)$ are arbitrary functions, λ and β are arbitrary parameters, and the functions $A(t)$ and $B(t)$ are determined by the nonautonomous system of linear ordinary differential equations
$$A'_t = [\nu(\lambda^2 - \beta^2) - \lambda\varphi(t)]A + \beta\psi(t)B,$$
$$B'_t = [\nu(\lambda^2 - \beta^2) - \lambda\varphi(t)]B - \beta\psi(t)A. \tag{14}$$

The general solution of system (14) is given by
$$A(t) = \exp\left[\nu(\lambda^2 - \beta^2)t - \lambda\int \varphi\,dt\right]\left[C_1 \sin\left(\beta \int \psi\,dt\right) + C_2 \cos\left(\beta \int \psi\,dt\right)\right],$$
$$B(t) = \exp\left[\nu(\lambda^2 - \beta^2)t - \lambda\int \varphi\,dt\right]\left[C_1 \cos\left(\beta \int \psi\,dt\right) - C_2 \sin\left(\beta \int \psi\,dt\right)\right],$$

where $\varphi = \varphi(t)$ and $\psi = \psi(t)$; C_1 and C_2 are arbitrary constants. In particular, if $\varphi = \frac{\nu}{\lambda}(\lambda^2 - \beta^2)$ and $\psi = a$, we obtain a periodic solution
$$A(t) = C_1 \sin(a\beta t) + C_2 \cos(a\beta t),$$
$$B(t) = C_1 \cos(a\beta t) - C_2 \sin(a\beta t).$$

6°. Exact solutions:
$$w(x, y, t) = A(t) \exp(k_1 x + \lambda_1 y) + B(t) \exp(k_2 x + \lambda_2 y) + \varphi(t)x + \psi(t)y + \chi(t),$$
where $\varphi(t)$, $\psi(t)$, and $\chi(t)$ are arbitrary functions and k_1, λ_1, k_2, and λ_2 are arbitrary parameters that satisfy one of the two relations
$$k_1^2 + \lambda_1^2 = k_2^2 + \lambda_2^2 \quad \text{(first family of solutions)},$$
$$k_1 \lambda_2 = k_2 \lambda_1 \quad \text{(second family of solutions)},$$
and the functions $A(t)$ and $B(t)$ are determined by the ordinary differential equations
$$A'_t = \left[\nu(k_1^2 + \lambda_1^2) + \lambda_1 \varphi(t) - k_1 \psi(t)\right] A,$$
$$B'_t = \left[\nu(k_2^2 + \lambda_2^2) + \lambda_2 \varphi(t) - k_2 \psi(t)\right] B.$$
These equations are easy to integrate:
$$A(t) = C_1 \exp\left[\nu(k_1^2 + \lambda_1^2)t + \lambda_1 \int \varphi(t)\, dt - k_1 \int \psi(t)\, dt\right],$$
$$B(t) = C_2 \exp\left[\nu(k_2^2 + \lambda_2^2)t + \lambda_2 \int \varphi(t)\, dt - k_2 \int \psi(t)\, dt\right].$$

7°. Exact solution:
$$w(x, y, t) = \left[C_1 \sin(\lambda x) + C_2 \cos(\lambda x)\right]\left[A(t) \sin(\beta y) + B(t) \cos(\beta y)\right] + \varphi(t)x + \chi(t),$$
where $\varphi(t)$ and $\chi(t)$ are arbitrary functions, C_1, C_2, λ, and β are arbitrary parameters, and the functions $A(t)$ and $B(t)$ are determined by the nonautonomous system of linear ordinary differential equations
$$A'_t = -\nu(\lambda^2 + \beta^2)A - \beta\varphi(t)B,$$
$$B'_t = -\nu(\lambda^2 + \beta^2)B + \beta\varphi(t)A. \tag{15}$$
The general solution of system (15) is given by
$$A(t) = \exp\left[-\nu(\lambda^2 + \beta^2)t\right]\left[C_3 \sin\left(\beta \int \varphi\, dt\right) + C_4 \cos\left(\beta \int \varphi\, dt\right)\right], \quad \varphi = \varphi(t),$$
$$B(t) = \exp\left[-\nu(\lambda^2 + \beta^2)t\right]\left[-C_3 \cos\left(\beta \int \varphi\, dt\right) + C_4 \sin\left(\beta \int \varphi\, dt\right)\right],$$
where C_3 and C_4 are arbitrary constants.

8°. Exact solution:
$$w(x, y, t) = \left[C_1 \sinh(\lambda x) + C_2 \cosh(\lambda x)\right]\left[A(t) \sin(\beta y) + B(t) \cos(\beta y)\right] + \varphi(t)x + \chi(t),$$
where $\varphi(t)$ and $\chi(t)$ are arbitrary functions, C_1, C_2, λ, and β are arbitrary parameters, and the functions $A(t)$ and $B(t)$ are determined by the nonautonomous system of linear ordinary differential equations
$$A'_t = \nu(\lambda^2 - \beta^2)A - \beta\varphi(t)B,$$
$$B'_t = \nu(\lambda^2 - \beta^2)B + \beta\varphi(t)A. \tag{16}$$
The general solution of system (16) is given by
$$A(t) = \exp\left[\nu(\lambda^2 - \beta^2)t\right]\left[C_3 \sin\left(\beta \int \varphi\, dt\right) + C_4 \cos\left(\beta \int \varphi\, dt\right)\right], \quad \varphi = \varphi(t),$$
$$B(t) = \exp\left[\nu(\lambda^2 - \beta^2)t\right]\left[-C_3 \cos\left(\beta \int \varphi\, dt\right) + C_4 \sin\left(\beta \int \varphi\, dt\right)\right],$$
where C_3 and C_4 are arbitrary constants.

9°. Exact solution:
$$w(x, y, t) = u(z, t) + \varphi(t)x + \psi(t)y, \qquad z = kx + \lambda y,$$
where $\varphi(t)$ and $\psi(t)$ are arbitrary functions, k and λ are arbitrary parameters, and the function $u(z, t)$ is determined by the fourth-order linear equation
$$\frac{\partial^3 u}{\partial t \partial z^2} + \left[k\psi(t) - \lambda\varphi(t)\right]\frac{\partial^3 u}{\partial z^3} = \nu(k^2 + \lambda^2)\frac{\partial^4 u}{\partial z^4}.$$

The transformation
$$U(\xi, t) = \frac{\partial^2 u}{\partial z^2}, \qquad \xi = z - \int \left[k\psi(t) - \lambda\varphi(t)\right] dt$$
brings it to the customary heat equation
$$\frac{\partial U}{\partial t} = \nu(k^2 + \lambda^2)\frac{\partial^2 U}{\partial \xi^2}.$$

⊙ *Reference*: A. D. Polyanin (2001d).

2. $\quad \dfrac{\partial Q}{\partial t} + \dfrac{1}{r}\dfrac{\partial w}{\partial \theta}\dfrac{\partial Q}{\partial r} - \dfrac{1}{r}\dfrac{\partial w}{\partial r}\dfrac{\partial Q}{\partial \theta} = \nu\Delta Q, \qquad Q = \Delta w = \dfrac{1}{r}\dfrac{\partial}{\partial r}\left(r\dfrac{\partial w}{\partial r}\right) + \dfrac{1}{r^2}\dfrac{\partial^2 w}{\partial \theta^2}.$

The two-dimensional equations of steady-state flow of a viscous incompressible fluid written in polar coordinates are reduced to this equation (w is the stream function).

Exact solution:
$$w(r, \theta, t) = f(r, t)\theta + g(r, t).$$
The functions $f = f(r, t)$ and $g = g(r, t)$ satisfy the system of equations
$$\mathbf{L}(f_t) - r^{-1}f_r\mathbf{L}(f) + r^{-1}f[\mathbf{L}(f)]_r = \nu\mathbf{L}^2(f), \tag{1}$$
$$\mathbf{L}(g_t) - r^{-1}g_r\mathbf{L}(f) + r^{-1}f[\mathbf{L}(g)]_r = \nu\mathbf{L}^2(g), \tag{2}$$
where the subscripts r and t denote partial derivatives; $\mathbf{L}(f) = r^{-1}(rf_r)_r$ and $\mathbf{L}^2(f) = \mathbf{L}\mathbf{L}(f)$.

For the particular solution $f = \varphi(t)\ln r + \psi(t)$ of equation (1), with φ and ψ arbitrary, equation (2) can be reduced by the change of variable $U = \mathbf{L}(g)$ to a second-order linear equation.

B.8. Higher-Order Nonlinear Equations

B.8.1. Equations of the Form $\dfrac{\partial w}{\partial t} = F\left(x, t, w, \dfrac{\partial w}{\partial x}, \ldots, \dfrac{\partial^n w}{\partial x^n}\right)$

1. $\quad \dfrac{\partial w}{\partial t} = a\dfrac{\partial^n w}{\partial x^n} + bw \ln w + f(t)w.$

1°. Exact solution:
$$w(x, t) = \exp\left[Ae^{bt}x + Be^{bt} + \frac{aA^n}{b(n-1)}e^{nbt} + e^{bt}\int e^{-bt}f(t)\, dt\right],$$
where A and B are arbitrary constants.

2°. Exact solution:
$$w(x, t) = \exp\left[Ae^{bt} + e^{bt}\int e^{-bt}f(t)\, dt\right]\varphi(z), \qquad z = x + \lambda t,$$
where A and λ are arbitrary constants and the function $\varphi = \varphi(z)$ is determined from the autonomous ordinary differential equation
$$a\varphi_z^{(n)} - \lambda\varphi_z' + b\varphi \ln \varphi = 0,$$
whose order can be reduced by one.

2. $\dfrac{\partial w}{\partial t} = a\dfrac{\partial^n w}{\partial x^n} + bw\ln w + [f(x) + g(t)]w.$

Exact solution in multiplicative form:
$$w(x,t) = \exp\left[Ce^{bt} + e^{bt}\int e^{-bt} g(t)\,dt\right]\varphi(x),$$

where C is an arbitrary constant and the function $\varphi(t)$ is determined by the ordinary differential equation
$$a\varphi_x^{(n)} + b\varphi\ln\varphi + f(x)\varphi = 0.$$

3. $\dfrac{\partial w}{\partial t} = a\dfrac{\partial^n w}{\partial x^n} + f(t)w\ln w + g(t)w.$

Exact solution:
$$w(x,t) = \exp\bigl[\varphi(t)x + \psi(t)\bigr].$$

The functions $\varphi(t)$ and $\psi(t)$ are determined from
$$\varphi(t) = Ae^F, \quad \psi(t) = Be^F + e^F\int e^{-F}(aA^n e^{nF} + g)\,dt, \quad F = \int f\,dt,$$

where A and B are arbitrary constants.

4. $\dfrac{\partial w}{\partial t} = a\dfrac{\partial^n w}{\partial x^n} + f(t)w\ln w + [g(t)x + h(t)]w.$

Exact solution:
$$w(x,t) = \exp\bigl[\varphi(t)x + \psi(t)\bigr].$$

The functions $\varphi(t)$ and $\psi(t)$ are determined from
$$\varphi(t) = Ae^F + e^F\int e^{-F} g\,dt, \quad F = \int f\,dt,$$
$$\psi(t) = Be^F + e^F\int e^{-F}(a\varphi^n + h)\,dt,$$

where A and B are arbitrary constants.

5. $\dfrac{\partial w}{\partial t} = a\dfrac{\partial^n w}{\partial x^n} + b\left(\dfrac{\partial w}{\partial x}\right)^2 + cw + f(t).$

1°. Exact solution:
$$w(x,t) = \varphi_2(t)x^2 + \varphi_1(t)x + \varphi_0(t),$$

where the functions $\varphi_k(t)$ satisfy an appropriate system of ordinary differential equations.

2°. Exact solution:
$$w(x,t) = Ae^{ct} + e^{ct}\int e^{-ct} f(t)\,dt + \Theta(\xi), \quad \xi = x + \lambda t,$$

where A and λ are arbitrary constants, and the function $\Theta(\xi)$ is determined by solving the autonomous ordinary differential equation
$$a\Theta_\xi^{(n)} + b\bigl(\Theta_\xi'\bigr)^2 - \lambda\Theta_\xi' + c\Theta = 0.$$

6. $\dfrac{\partial w}{\partial t} = a\dfrac{\partial^n w}{\partial x^n} + b\left(\dfrac{\partial w}{\partial x}\right)^2 + cw\dfrac{\partial w}{\partial x} + kw^2 + f(t)w + g(t).$

Exact solution:
$$w(x,t) = \varphi(t) + \psi(t)\exp(\lambda x),$$

where λ is a root of the quadratic equation $b\lambda^2 + c\lambda + k = 0$, and the functions $\varphi(t)$ and $\psi(t)$ are determined from the system of first-order ordinary differential equations

$$\varphi'_t = k\varphi^2 + f(t)\varphi + g(t), \qquad (1)$$
$$\psi'_t = \left[(c\lambda + 2k)\varphi + f(t) + a\lambda^n\right]\psi. \qquad (2)$$

For the Riccati equation (1), see Kamke (1977) and Polyanin and Zaitsev (1995). It is integrable in quadrature if, for example,

(a) $k = 0$, (b) $g(t) \equiv 0$, (c) $f(t) = \text{const}$, $g(t) = \text{const}$.

On solving equation (1), one can readily solve equation (2), which is linear in ψ.

7. $\dfrac{\partial w}{\partial t} = a\dfrac{\partial^n w}{\partial x^n} + f(t)\left(\dfrac{\partial w}{\partial x}\right)^2 + bf(t)w^2 + g(t)w + h(t).$

1°. Exact solution:
$$w(x,t) = \varphi(t) + \psi(t)\exp\!\left(\pm x\sqrt{-b}\right), \qquad b < 0,$$

where the functions $\varphi(t)$ and $\psi(t)$ are determined by solving the following first-order ordinary differential equations with variable coefficients (the arguments of f, g, and h are not specified):

$$\varphi'_t = bf\varphi^2 + g\varphi + h, \qquad (1)$$
$$\psi'_t = \left[2bf\varphi + g + a\!\left(\pm\sqrt{-b}\right)^n\right]\psi. \qquad (2)$$

Equation (1) is a Riccati equation for $\varphi = \varphi(t)$; it can be reduced to a second-order linear equation. A lot of exact solutions to equation (1) with various f, g, and h can be found in Kamke (1977) and Polyanin and Zaitsev (1995).

Given a solution of equation (1), the corresponding solution of (2) is calculated by

$$\psi(t) = C\exp\!\left[a\!\left(\pm\sqrt{-b}\right)^n t + \int (2bf\varphi + g)\,dt\right],$$

where C is an arbitrary constant.

2°. Exact solution of a more general form:
$$w(x,t) = \varphi(t) + \psi(t)\exp\!\left(x\sqrt{-b}\right) + \chi(t)\exp\!\left(-x\sqrt{-b}\right), \qquad b < 0, \qquad (3)$$

where the functions $\varphi(t)$, $\psi(t)$, and $\chi(t)$ are determined by the following system of first-order ordinary differential equations with variable coefficients:

$$\varphi'_t = bf\varphi^2 + g\varphi + h + 4bf\psi\chi, \qquad (4)$$
$$\psi'_t = \left[2bf\varphi + g + a\!\left(\sqrt{-b}\right)^n\right]\psi, \qquad (5)$$
$$\chi'_t = \left[2bf\varphi + g + a\!\left(-\sqrt{-b}\right)^n\right]\chi. \qquad (6)$$

For equations of even order with $n = 2m$ ($m = 1, 2, \ldots$), it follows from (5) and (6) that the functions $\psi(t)$ and $\chi(t)$ are proportional. Setting $\psi(t) = A\theta(t)$ and $\chi(t) = B\theta(t)$, one can rewrite solution (3) as

$$w(x,t) = \varphi(t) + \theta(t)\!\left[A\exp\!\left(x\sqrt{-b}\right) + B\exp\!\left(-x\sqrt{-b}\right)\right], \qquad b < 0,$$

where the functions $\varphi(t)$ and $\theta(t)$ are determined by the system of ordinary differential equations

$$\varphi'_t = bf\!\left(\varphi^2 + 4AB\theta^2\right) + g\varphi + h, \qquad (7)$$
$$\theta'_t = \left[2bf\varphi + g + (-1)^m ab^m\right]\theta. \qquad (8)$$

On expressing φ from (8) in terms of θ and substituting the result into (7), one arrives at a second-order nonlinear equation for θ; if f, g, $h = \text{const}$, this equation is autonomous and hence admits reduction of order.

3°. Exact solution:
$$w(x,t) = \varphi(t) + \psi(t)\cos(x\sqrt{b}) + \chi(t)\sin(x\sqrt{b}), \qquad b > 0,$$
where the functions $\varphi(t)$, $\psi(t)$, and $\chi(t)$ are determined by a system of ordinary differential equations (not specified here).

For equations of even order with $n = 2m$ ($m = 1, 2, \ldots$), there are exact solutions with the following form (c is any number):
$$w(x,t) = \varphi(t) + \theta(t)\cos(x\sqrt{b} + c), \qquad b > 0.$$
The functions $\varphi(t)$ and $\theta(t)$ are determined by the following system of first-order ordinary differential equations with variable coefficients:
$$\varphi'_t = bf(\varphi^2 + \theta^2) + g\varphi + h, \tag{9}$$
$$\theta'_t = [2bf\varphi + g + (-1)^m ab^m]\theta. \tag{10}$$

On expressing φ from (10) in terms of θ and substituting the result into (9), one arrives at a second-order nonlinear equation for θ; if f, g, h = const, this equation is autonomous and hence admits reduction of order.

8. $\dfrac{\partial w}{\partial t} = a\dfrac{\partial^n w}{\partial x^n} + f\!\left(x, \dfrac{\partial w}{\partial x}\right) + g(t).$

Exact solution in additive form:
$$w(x,t) = At + B + \int g(t)\,dt + \varphi(x).$$
Here, A and B are arbitrary constants and the function $\varphi(x)$ is determined from the nonlinear ordinary differential equation
$$a\varphi_x^{(n)} + f(x, \varphi'_x) - A = 0.$$

9. $\dfrac{\partial w}{\partial t} = a\dfrac{\partial^n w}{\partial x^n} + f\!\left(x, \dfrac{\partial w}{\partial x}\right) + bw + g(t).$

Exact solution in additive form:
$$w(x,t) = \varphi(x) + Ae^{bt} + e^{bt}\int e^{-bt} g(t)\,dt.$$
Here, A is an arbitrary constant and the function $\varphi(x)$ is determined from the nonlinear ordinary differential equation
$$a\varphi_x^{(n)} + f(x, \varphi'_x) + b\varphi = 0.$$

10. $\dfrac{\partial w}{\partial t} = aw\dfrac{\partial^n w}{\partial x^n} + f(t)w + g(t).$

1°. Exact solution:
$$w(x,t) = F(t)\bigl(A_{n-1}x^{n-1} + \cdots + A_1 x + A_0\bigr) + F(t)\int \frac{g(t)}{F(t)}\,dt, \qquad F(t) = \exp\!\left[\int f(t)\,dt\right],$$
where $A_0, A_1, \ldots, A_{n-1}$ are arbitrary constants.

2°. Exact solution:
$$w(x,t) = \varphi(t)\bigl(x^n + A_{n-1}x^{n-1} + \cdots + A_1 x + A_0\bigr) + \varphi(t)\int \frac{g(t)}{\varphi(t)}\,dt,$$
$$\varphi(t) = F(t)\!\left[C - an!\int F(t)\,dt\right]^{-1}, \qquad F(t) = \exp\!\left[\int f(t)\,dt\right],$$
where $A_0, A_1, \ldots, A_{n-1}$, and C are arbitrary constants.

11. $\dfrac{\partial w}{\partial t} = aw\dfrac{\partial^n w}{\partial x^n} + bw^2 + f(t)w + g(t).$

Exact solution:
$$w(x,t) = \varphi(t)\Theta(x) + \psi(t),$$

where the functions $\varphi(t)$ and $\psi(t)$ are determined from the following system of first-order ordinary differential equations (C is any number):

$$\varphi'_t = C\varphi^2 + b\varphi\psi + f(t)\varphi,$$
$$\psi'_t = C\varphi\psi + b\psi^2 + f(t)\psi + g(t),$$

and the function $\Theta(x)$ is determined by the nth-order linear ordinary differential equation

$$a\Theta_x^{(n)} + b\Theta = C.$$

12. $\dfrac{\partial w}{\partial t} = aw\dfrac{\partial^n w}{\partial x^n} + f(x)w\dfrac{\partial w}{\partial x} + g(t)w + h(t).$

Exact solution:
$$w(x,t) = \varphi(t)\Theta(x) + \psi(t),$$

where the functions $\varphi(t)$, $\psi(t)$, and $\Theta(x)$ are determined by the ordinary differential equations

$$\varphi'_t = C\varphi^2 + g(t)\varphi,$$
$$\psi'_t = \big[C\varphi + g(t)\big]\psi + h(t),$$
$$a\Theta_x^{(n)} + f(x)\Theta'_x = C,$$

where C is an arbitrary constant. Integrating successively, for $\varphi(t)$ and $\psi(t)$ we obtain

$$\varphi(t) = G(t)\left[A - C\int G(t)\,dt\right]^{-1}, \quad G(t) = \exp\left[\int g(t)\,dt\right],$$
$$\psi(t) = B\varphi(t) + \varphi(t)\int \dfrac{h(t)}{\varphi(t)}\,dt,$$

where A and B are arbitrary constants.

B.8.2. Equations of the Form $\dfrac{\partial^2 w}{\partial t^2} = F\left(x, t, w, \dfrac{\partial w}{\partial x}, \ldots, \dfrac{\partial^n w}{\partial x^n}\right)$

1. $\dfrac{\partial^2 w}{\partial t^2} = a\dfrac{\partial^n w}{\partial x^n} + f(x)\dfrac{\partial w}{\partial x} + bw\ln w + [g(x) + h(t)]w.$

Exact solution in multiplicative form:
$$w(x,t) = \varphi(t)\psi(x).$$

The functions $\varphi(t)$ and $\psi(x)$ are determined by the ordinary differential equations

$$\varphi''_{tt} - \big[b\ln\varphi + h(t) + C\big]\varphi = 0,$$
$$a\psi_x^{(n)} + f(x)\psi'_x + \big[b\ln\psi + g(x) - C\big]\psi = 0,$$

where C is an arbitrary constant.

2. $\dfrac{\partial^2 w}{\partial t^2} = a\dfrac{\partial^n w}{\partial x^n} + b\left(\dfrac{\partial w}{\partial x}\right)^2 + cw + f(t).$

1°. Exact solution:
$$w(x,t) = \varphi_2(t)x^2 + \varphi_1(t)x + \varphi_0(t),$$
where the functions $\varphi_k(t)$ satisfy an appropriate system of ordinary differential equations.

2°. Exact solution:
$$w(x,t) = \psi(t) + \Theta(\xi), \quad \xi = x + \lambda t.$$
The functions $\psi(t)$ and $\Theta(\xi)$ are determined from the ordinary differential equations
$$\psi''_{tt} - c\psi - f(t) = A,$$
$$a\Theta^{(n)}_\xi - \lambda^2 \Theta''_{\xi\xi} + b(\Theta'_\xi)^2 + c\Theta = A,$$
where λ and A are arbitrary constants.

3. $\dfrac{\partial^2 w}{\partial t^2} = a\dfrac{\partial^n w}{\partial x^n} + b\left(\dfrac{\partial w}{\partial x}\right)^2 + cw\dfrac{\partial w}{\partial x} + kw^2 + f(t)w + g(t).$

Exact solution:
$$w(x,t) = \varphi(t) + \psi(t)\exp(\lambda x),$$
where λ is a root of the quadratic equation $b\lambda^2 + c\lambda + k = 0$, and the functions $\varphi(t)$ and $\psi(t)$ are determined from the system of second-order ordinary differential equations
$$\varphi''_{tt} = k\varphi^2 + f(t)\varphi + g(t), \tag{1}$$
$$\psi''_{tt} = \big[(c\lambda + 2k)\varphi + f(t) + a\lambda^n\big]\psi. \tag{2}$$

In the special case $f(t) = \text{const}$ and $g(t) = \text{const}$, equation (1) is autonomous and has particular solutions of the form $\varphi = \text{const}$ and, hence, can be integrated in quadrature. Equation (2) is linear in ψ; therefore, for $\varphi = \text{const}$, its general solution is expressed in terms of exponentials or sine and cosine.

4. $\dfrac{\partial^2 w}{\partial t^2} = a\dfrac{\partial^n w}{\partial x^n} + f(x)\left(\dfrac{\partial w}{\partial x}\right)^2 + g(x) + h(t).$

Exact solution in additive form:
$$w(x,t) = \tfrac{1}{2}At^2 + Bt + C + \int_0^t (t-\tau)h(\tau)\,d\tau + \varphi(x).$$
Here, A, B, and C are arbitrary constants, and the function $\varphi(x)$ is determined by solving the nonlinear ordinary differential equation
$$a\varphi^{(n)}_x + f(x)(\varphi'_x)^2 + g(x) - A = 0.$$

5. $\dfrac{\partial^2 w}{\partial t^2} = a\dfrac{\partial^n w}{\partial x^n} + f(x)\left(\dfrac{\partial w}{\partial x}\right)^2 + bw + g(x) + h(t).$

Exact solution in additive form:
$$w(x,t) = \varphi(t) + \psi(x).$$
The functions $\varphi(t)$ and $\psi(x)$ are determined by solving the nonlinear ordinary differential equations
$$\varphi''_{tt} - b\varphi - h(t) = 0,$$
$$a\psi^{(n)}_x + f(x)(\psi'_x)^2 + b\psi + g(x) = 0.$$
The general solution of the first equation is given by
$$\varphi(t) = C_1 \cosh(kx) + C_2 \sinh(kx) + \dfrac{1}{k}\int_0^t h(\tau)\sinh[k(t-\tau)]\,d\tau \quad \text{for} \quad b = k^2 > 0,$$
$$\varphi(t) = C_1 \cos(kx) + C_2 \sin(kx) + \dfrac{1}{k}\int_0^t h(\tau)\sin[k(t-\tau)]\,d\tau \quad \text{for} \quad b = -k^2 < 0,$$
where C_1 and C_2 are arbitrary constants.

B.8. HIGHER-ORDER NONLINEAR EQUATIONS

6. $\dfrac{\partial^2 w}{\partial t^2} = a\dfrac{\partial^{2n} w}{\partial x^{2n}} + f(t)\left(\dfrac{\partial w}{\partial x}\right)^2 + bf(t)w^2 + g(t)w + h(t).$

1°. Exact solution:
$$w(x,t) = \varphi(t) + \psi(t)\exp(\pm x\sqrt{-b}), \qquad b < 0,$$

where the functions $\varphi(t)$ and $\psi(t)$ are determined by solving the following first-order ordinary differential equations with variable coefficients (the arguments of f, g, and h are not specified):

$$\varphi''_{tt} = bf\varphi^2 + g\varphi + h, \qquad (1)$$
$$\psi''_{tt} = [2bf\varphi + g + (-1)^n ab^n]\psi. \qquad (2)$$

In the special case where f, g, h are constant, equation (1) has particular solutions of the form $\varphi = \text{const}$. Here, the general solution of equation (2) is expressed in terms of exponentials or sine and cosine.

2°. Exact solution of a more general form:
$$w(x,t) = \varphi(t) + \psi(t)\big[A\exp(x\sqrt{-b}) + B\exp(-x\sqrt{-b})\big], \qquad b < 0,$$

where the functions $\varphi(t)$ and $\psi(t)$ are determined by solving the following system of second-order ordinary differential equations with variable coefficients:

$$\varphi''_{tt} = bf(\varphi^2 + 4AB\psi^2) + g\varphi + h, \qquad (3)$$
$$\psi''_{tt} = [2bf\varphi + g + (-1)^n ab^n]\psi. \qquad (4)$$

On expressing φ from (4) in terms of ψ and substituting the result into (3), one arrives at a fourth-order nonlinear equation for ψ; if f, g, $h = \text{const}$, this equation is autonomous and hence admits reduction of order.

3°. Exact solution (c is an arbitrary constant):
$$w(x,t) = \varphi(t) + \psi(t)\cos(x\sqrt{b} + c), \qquad b > 0,$$

where the functions $\varphi(t)$ and $\psi(t)$ are determined by solving the following system of second-order ordinary differential equations with variable coefficients:

$$\varphi''_{tt} = bf(\varphi^2 + \psi^2) + g\varphi + h,$$
$$\psi''_{tt} = [2bf\varphi + g + (-1)^n ab^n]\psi.$$

7. $\dfrac{\partial^2 w}{\partial t^2} = a\dfrac{\partial^n w}{\partial x^n} + f\!\left(x, \dfrac{\partial w}{\partial x}\right) + g(t).$

Exact solution in additive form:
$$w(x,t) = \tfrac{1}{2}At^2 + Bt + C + \int_0^t (t-\tau)g(\tau)\,d\tau + \varphi(x).$$

Here, A, B, and C are arbitrary constants and the function $\varphi(x)$ is determined by the nonlinear ordinary differential equation
$$a\varphi_x^{(n)} + f(x, \varphi'_x) - A = 0.$$

8. $\dfrac{\partial^2 w}{\partial t^2} = a\dfrac{\partial^n w}{\partial x^n} + f\!\left(x, \dfrac{\partial w}{\partial x}\right) + bw + g(t).$

Exact solution in additive form:
$$w(x,t) = \varphi(t) + \psi(x).$$

The functions $\varphi(t)$ and $\psi(x)$ are determined by the ordinary differential equations
$$\varphi''_{tt} - b\varphi - g(t) = 0,$$
$$a\psi^{(n)}_x + f\!\left(x, \psi'_x\right) + b\psi = 0.$$

The general solution of the first equation is given by
$$\varphi(t) = C_1 \cosh(kx) + C_2 \sinh(kx) + \dfrac{1}{k}\int_0^t g(\tau)\sinh[k(t-\tau)]\,d\tau \quad \text{for} \quad b = k^2 > 0,$$
$$\varphi(t) = C_1 \cos(kx) + C_2 \sin(kx) + \dfrac{1}{k}\int_0^t g(\tau)\sin[k(t-\tau)]\,d\tau \quad \text{for} \quad b = -k^2 < 0,$$

where C_1 and C_2 are arbitrary constants.

9. $\dfrac{\partial^2 w}{\partial t^2} = a\dfrac{\partial^n w}{\partial x^n} + wf\!\left(t, \dfrac{1}{w}\dfrac{\partial w}{\partial x}\right).$

Exact solution in multiplicative form:
$$w(x,t) = e^{\lambda x}\varphi(t),$$

where λ is an arbitrary constant and the function $\varphi(t)$ is determined from the second-order linear ordinary differential equation
$$\varphi''_{tt} = \left[a\lambda^n + f(t,\lambda)\right]\varphi.$$

10. $\dfrac{\partial^2 w}{\partial t^2} = aw\dfrac{\partial^n w}{\partial x^n} + f(t)w + g(t).$

Exact solution:
$$w(x,t) = \varphi(t)\left(A_n x^n + \cdots + A_1 x\right) + \psi(t),$$

where $A_1, \ldots, A_n$ are arbitrary constants and the functions $\varphi(t)$ and $\psi(t)$ are determined from the second-order ordinary differential equations
$$\varphi''_{tt} = A_n a n!\,\varphi^2 + f(t)\varphi,$$
$$\psi''_{tt} = A_n a n!\,\varphi\psi + f(t)\psi + g(t).$$

11. $\dfrac{\partial^2 w}{\partial t^2} = aw\dfrac{\partial^n w}{\partial x^n} + bw^2 + f(t)w + g(t).$

Exact solution:
$$w(x,t) = \varphi(t)\Theta(x) + \psi(t),$$

where the functions $\varphi(t)$ and $\psi(t)$ are determined by the following system of second-order ordinary differential equations (C is an arbitrary constant):
$$\varphi''_{tt} = C\varphi^2 + b\varphi\psi + f(t)\varphi,$$
$$\psi''_{tt} = C\varphi\psi + b\psi^2 + f(t)\psi + g(t).$$

The function $\Theta(x)$ satisfies the nth-order linear ordinary differential equation
$$a\Theta^{(n)}_x + b\Theta = C.$$

12. $\dfrac{\partial^2 w}{\partial t^2} = aw\dfrac{\partial^n w}{\partial x^n} + f(x)w\dfrac{\partial w}{\partial x} + g(t)w + h(t).$

Exact solution:
$$w(x,t) = \varphi(t)\Theta(x) + \psi(t),$$

where the functions $\varphi(t)$, $\psi(t)$, and $\Theta(x)$ are determined by the ordinary differential equations

$$\varphi''_{tt} = C\varphi^2 + g(t)\varphi,$$
$$\psi''_{tt} = [C\varphi + g(t)]\psi + h(t),$$
$$a\Theta^{(n)}_x + f(x)\Theta'_x = C,$$

where C is an arbitrary constant.

B.8.3. Other Equations

1. $\dfrac{\partial w}{\partial y}\dfrac{\partial^2 w}{\partial x \partial y} - \dfrac{\partial w}{\partial x}\dfrac{\partial^2 w}{\partial y^2} = f(x)\dfrac{\partial^n w}{\partial y^n}.$

This is a special case of equation B.8.3.3.

Exact solution:
$$w(x,y) = \varphi(x)e^{\lambda y} - \lambda^{n-2}\int f(x)\,dx + C,$$

where $\varphi(x)$ is an arbitrary function; C and λ are arbitrary constants.

2. $\dfrac{\partial w}{\partial y}\dfrac{\partial^2 w}{\partial x \partial y} - \dfrac{\partial w}{\partial x}\dfrac{\partial^2 w}{\partial y^2} = f(x)\dfrac{\partial^{2n} w}{\partial y^{2n}} + g(x).$

This is a special case of equation B.8.3.3.

Exact solution:
$$w(x,y) = \varphi(x)e^{\lambda y} - \dfrac{1}{2\lambda^2 \varphi(x)}\left[\int g(x)\,dx + C_1\right]e^{-\lambda y} - \lambda^{2n-2}\int f(x)\,dx + C_2,$$

where $\varphi(x)$ is an arbitrary function and C_1, C_2, and λ are arbitrary parameters.

3. $\dfrac{\partial w}{\partial y}\dfrac{\partial^2 w}{\partial x \partial y} - \dfrac{\partial w}{\partial x}\dfrac{\partial^2 w}{\partial y^2} = F\left(x, w, \dfrac{\partial w}{\partial y}, \ldots, \dfrac{\partial^n w}{\partial y^n}\right).$

1°. If $w(x,y)$ is a solution of the equation in question, then the function

$$w_1(x,y) = w(x,\, y + \varphi(x)),$$

where $\varphi(x)$ is an arbitrary function, is also a solution of the equation.

2°. Let the right-hand side of the equation be independent of x explicitly. Then there are exact solutions of the form

$$w = w(z), \quad z = y + \varphi(x),$$

where $\varphi(x)$ is an arbitrary function and $w(z)$ is a solution of the ordinary differential equation $F\left(w, w'_z, \ldots, w^{(n)}_z\right) = 0$.

3°. Let the right-hand side of the equation be independent of x and w explicitly. Then there are exact solutions of the form

$$w = Cx + g(z), \quad z = y + \varphi(x),$$

where $\varphi(x)$ is an arbitrary function, C is an arbitrary constant, and $g(z)$ is a solution of the ordinary differential equation $F\left(g'_z, \ldots, g^{(n)}_z\right) + Cg''_{zz} = 0$.

4. $\dfrac{\partial w}{\partial t} = wF(t, \zeta_0, \zeta_1, \ldots, \zeta_n)$, $\zeta_k = \sum\limits_{i=k}^{n} \dfrac{(-1)^{i+k}}{k!\,(i-k)!} x^{i-k} \dfrac{\partial^i w}{\partial x^i}$, $k = 0, 1, \ldots, n$.

Exact solution in multiplicative form:
$$w(x, t) = (C_0 + C_1 x + \cdots + C_n x^n)\varphi(t),$$
where $C_0, C_1, \ldots, C_n$ are arbitrary constants and the function $\varphi = \varphi(t)$ satisfies the ordinary differential equation
$$\varphi'_t = \varphi F(t, C_0\varphi, C_1\varphi, \ldots, C_n\varphi).$$

⊙ *Reference*: Ph. W. Doyle (1996), the case $\partial_t F \equiv 0$ was considered.

5. $\dfrac{\partial^2 w}{\partial t^2} = wF(t, \zeta_0, \zeta_1, \ldots, \zeta_n)$, $\zeta_k = \sum\limits_{i=k}^{n} \dfrac{(-1)^{i+k}}{k!\,(i-k)!} x^{i-k} \dfrac{\partial^i w}{\partial x^i}$, $k = 0, 1, \ldots, n$.

Exact solution in multiplicative form:
$$w(x, t) = (C_0 + C_1 x + \cdots + C_n x^n)\varphi(t),$$
where $C_0, C_1, \ldots, C_n$ are arbitrary constants and the function $\varphi = \varphi(t)$ satisfies the ordinary differential equation
$$\varphi''_{tt} = \varphi F(t, C_0\varphi, C_1\varphi, \ldots, C_n\varphi).$$

6. $F\!\left(x, \dfrac{1}{w}\dfrac{\partial w}{\partial x}, \ldots, \dfrac{1}{w}\dfrac{\partial^n w}{\partial x^n};\, \dfrac{1}{w}\dfrac{\partial w}{\partial y}, \ldots, \dfrac{1}{w}\dfrac{\partial^m w}{\partial y^m}\right) = 0.$

Exact solution in multiplicative form:
$$w(x, y) = A e^{\lambda y} \varphi(x),$$
where A and λ are arbitrary constants and the function $\varphi(x)$ is determined from the nth-order ordinary differential equation
$$F\!\left(x, \varphi'_x/\varphi, \ldots, \varphi^{(n)}_x/\varphi;\, \lambda, \ldots, \lambda^m\right) = 0.$$

7. $F\!\left(x, \dfrac{1}{w}\dfrac{\partial w}{\partial x}, \ldots, \dfrac{1}{w}\dfrac{\partial^n w}{\partial x^n};\, \dfrac{1}{w}\dfrac{\partial^2 w}{\partial y^2}, \ldots, \dfrac{1}{w}\dfrac{\partial^{2m} w}{\partial y^{2m}}\right) = 0.$

1°. Exact solution:
$$w(x, y) = \big[A\cosh(\lambda y) + B\sinh(\lambda y)\big]\varphi(x),$$
where A, B, and λ are arbitrary constants and the function $\varphi(x)$ is determined by solving the nth-order ordinary differential equation
$$F\!\left(x, \varphi'_x/\varphi, \ldots, \varphi^{(n)}_x/\varphi;\, \lambda^2, \ldots, \lambda^{2m}\right) = 0.$$

2°. Exact solution:
$$w(x, y) = \big[A\cos(\lambda y) + B\sin(\lambda y)\big]\varphi(x),$$
where A, B, and λ are arbitrary constants and the function $\varphi(x)$ is determined by solving the nth-order ordinary differential equation
$$F\!\left(x, \varphi'_x/\varphi, \ldots, \varphi^{(n)}_x/\varphi;\, -\lambda^2, \ldots, (-1)^m\lambda^{2m}\right) = 0.$$

8. $f_1\!\left(x, \dfrac{\partial w}{\partial x}, \ldots, \dfrac{\partial^n w}{\partial x^n}\right) + f_2\!\left(y, \dfrac{\partial w}{\partial y}, \ldots, \dfrac{\partial^m w}{\partial y^m}\right) = kw.$

Exact solution in additive form:
$$w(x, y) = \varphi(x) + \psi(y).$$

The functions $\varphi(x)$ and $\psi(y)$ are determined by the ordinary differential equations
$$f_1\!\left(x, \varphi'_x, \ldots, \varphi^{(n)}_x\right) - k\varphi = C,$$
$$f_2\!\left(y, \psi'_y, \ldots, \psi^{(m)}_y\right) - k\psi = -C,$$
where C is an arbitrary constant.

9. $f_1\left(x, \dfrac{1}{w}\dfrac{\partial w}{\partial x}, \ldots, \dfrac{1}{w}\dfrac{\partial^n w}{\partial x^n}\right) + w^k f_2\left(y, \dfrac{1}{w}\dfrac{\partial w}{\partial y}, \ldots, \dfrac{1}{w}\dfrac{\partial^m w}{\partial y^m}\right) = 0.$

Exact solution in multiplicative form:

$$w(x,y) = \varphi(x)\psi(y).$$

The functions $\varphi(x)$ and $\psi(y)$ are determined by the ordinary differential equations

$$\varphi^{-k} f_1\left(x, \varphi'_x/\varphi, \ldots, \varphi^{(n)}_x/\varphi\right) = C,$$
$$\psi^k f_2\left(y, \psi'_y/\psi, \ldots, \psi^{(m)}_y/\psi\right) = -C,$$

where C is an arbitrary constant.

10. $f_1\left(x, \dfrac{\partial w}{\partial x}, \ldots, \dfrac{\partial^n w}{\partial x^n}\right) + e^{\lambda w} f_2\left(y, \dfrac{\partial w}{\partial y}, \ldots, \dfrac{\partial^m w}{\partial y^m}\right) = 0.$

Exact solution in additive form:

$$w(x,y) = \varphi(x) + \psi(y).$$

The functions $\varphi(x)$ and $\psi(y)$ are determined by the ordinary differential equations

$$e^{-\lambda\varphi} f_1\left(x, \varphi'_x, \ldots, \varphi^{(n)}_x\right) = C,$$
$$e^{\lambda\psi} f_2\left(y, \psi'_y, \ldots, \psi^{(m)}_y\right) = -C,$$

where C is an arbitrary constant.

11. $f_1\left(x, \dfrac{1}{w}\dfrac{\partial w}{\partial x}, \ldots, \dfrac{1}{w}\dfrac{\partial^n w}{\partial x^n}\right) + f_2\left(y, \dfrac{1}{w}\dfrac{\partial w}{\partial y}, \ldots, \dfrac{1}{w}\dfrac{\partial^m w}{\partial y^m}\right) = k\ln w.$

Exact solution in multiplicative form:

$$w(x,y) = \varphi(x)\psi(y).$$

The functions $\varphi(x)$ and $\psi(y)$ are determined by the ordinary differential equations

$$f_1\left(x, \varphi'_x/\varphi, \ldots, \varphi^{(n)}_x/\varphi\right) - k\ln\varphi = C,$$
$$f_2\left(y, \psi'_y/\psi, \ldots, \psi^{(m)}_y/\psi\right) - k\ln\psi = -C,$$

where C is an arbitrary constant.

REFERENCES

Abramowitz, M. and Stegun, I. A. (Editors), *Handbook of Mathematical Functions with Formulas, Graphs and Mathematical Tables*, National Bureau of Standards Applied Mathematics, Washington, 1964.

Acrivos, A., *A note of the rate of heat or mass transfer from a small sphere freely suspended in linear shear field*, J. Fluid Mech., Vol. 98, No. 2, pp. 299–304, 1980.

Aksenov, A. V., *Linear differential relations between solutions of the equations of Euler–Poisson–Darboux class*, Mechanics of Solids, Vol. 36, No. 1, pp. 11–15, 2001.

Akulenko, L. D. and Nesterov, S. V., *Determination of the frequencies and forms of oscillations of non-uniform distributed systems with boundary conditions of the third kind*, Appl. Math. Mech. (PMM), Vol. 61, No. 4, p. 531–538, 1997.

Akulenko, L. D. and Nesterov, S. V., *Vibration of an nonhomogeneous membrane*, Mechanics of Solids, Vol. 34, No. 6, pp. 112–121, 1999.

Akulenko, L. D. and Nesterov, S. V., *Free vibrations of a homogeneous elliptic membrane*, Mechanics of Solids, Vol. 35, No. 1, pp. 153–162, 2000.

Akulenko, L. D., Nesterov, S. V., and Popov, A. L., *Natural frequencies of an elliptic plate with clamped edge*, Mechanics of Solids, Vol. 36, No. 1, pp. 143–148, 2001.

Andreev, V. K., Kaptsov, O. V., Pukhnachov, V. V., and Rodionov, A. A., *Applications of Group-Theoretical Methods in Hydrodynamics*, Nauka, Moscow, 1994. (English translation: Kluwer, Dordrecht, 1999.)

Appell, P., *Traité de Mécanique Rationnelle, T. 1: Statique. Dinamyque du Point (Ed. 6)*, Gauthier-Villars, Paris, 1953.

Arscott, F., *Periodic Differential Equations*, Macmillan (Pergamon), New York, 1964.

Arscott, F., *The Whittaker–Hill equation and the wave equation in paraboloidal coordinates*, Proc. Roy. Soc. Edinburg, Vol. A67, pp. 265–276, 1967.

Babich, V. M., Kapilevich, M. B., Mikhlin, S. G., et al., *Linear Equations of Mathematical Physics* [in Russian], Nauka, Moscow, 1964.

Batchelor, G. K., *Mass transfer from a particle suspended in fluid with a steady linear ambient velocity distribution*, J. Fluid Mech., Vol. 95, No. 2, pp. 369–400, 1979.

Bateman, H. and Erdélyi, A., *Higher Transcendental Functions, Vol. 1 and Vol. 2*, McGraw-Hill, New York, 1953.

Bateman, H. and Erdélyi, A., *Higher Transcendental Functions, Vol. 3*, McGraw-Hill, New York, 1955.

Bateman, H. and Erdélyi, A., *Tables of Integral Transforms, Vol. 1 and Vol. 2*, McGraw-Hill, New York, 1954.

Belotserkovskii, O. M., and Oparin, A. A., *Numerical Experiment in Turbulence*, Nauka, Moscow, 2000.

Beyer, W. H., *CRC Standard Mathematical Tables and Formulae*, CRC Press, Boca Raton, 1991.

Bitsadze, A. V. and Kalinichenko, D. F., *Collection of Problems on Mathematical Physics Equations* [in Russian], Nauka, Moscow, 1985.

Bolotin V. V. (Editor), *Vibration in Engineering: a Handbook. Vol. 1. Vibration of Linear Systems* [in Russian], Mashinostroenie, Moscow, 1978.

Borzykh, A. A. and Cherepanov, G. P., *A plane problem of the theory of convective heat transfer and mass exchange,* PMM [Applied Mathematics and Mechanics], Vol. 42, No. 5, pp. 848–855, 1978.

Boyer, C., *The maximal kinematical invariance group for an arbitrary potential,* Helv. Phys. Acta, Vol. 47, pp. 589–605, 1974.

Boyer, C., *Lie theory and separation of variables for equation* $iU_t + \Delta_2 U - (\alpha/x_1^2 + \beta/x_2^2)U = 0$, SIAM J. Math. Anal., Vol. 7, pp. 230–263, 1976.

Bôcher, M., *Die Reihenentwickelungen der Potentialtheory,* Leipzig, 1894.

Brenner, H., *Forced convection-heat and mass transfer at small Peclet numbers from particle of arbitrary shape,* Chem. Eng. Sci., Vol. 18, No. 2, pp. 109–122, 1963.

Brychkov, Yu. A. and Prudnikov, A. P., *Integral Transforms of Generalized Functions,* Gordon & Breach Sci. Publ., New York, 1989.

Budak, B. M., Samarskii, A. A., and Tikhonov, A. N., *Collection of Problems on Mathematical Physics* [in Russian], Nauka, Moscow, 1980.

Burde, G. I., *The construction of special explicit solutions of the boundary-layer equations. Steady flows,* Q. J. Mech. Appl. Math., Vol. 47, No. 2, pp. 247–260, 1994.

Butkov, E., *Mathematical Physics,* Addison-Wesley, Reading, Mass., 1968.

Butkovskiy, A. G., *Characteristics of Systems with Distributed Parameters* [in Russian], Nauka, Moscow, 1979.

Butkovskiy, A. G., *Green's Functions and Transfer Functions Handbook,* Halstead Press–John Wiley & Sons, New York, 1982.

Carslaw, H. S. and Jaeger, J. C., *Conduction of Heat in Solids,* Clarendon Press, Oxford, 1984.

Clarkson, P. A and Kruskal, M. D., *New similarity reductions of the Boussinesq equation,* J. Math. Phys., Vol. 30, No. 10, pp. 2201–2213, 1989.

Colton, D., *Partial Differential Equations. An Introduction,* Random House, New York, 1988.

Courant, R. and Hilbert, D., *Methods of Mathematical Physics, Vol. 2,* Wiley–Interscience Publ., New York, 1989.

Crank, J., *The Mathematics of Diffusion,* Clarendon Press, Oxford, 1975.

Davis, B., *Integral Transforms and Their Applications,* Springer-Verlag, New York, 1978.

Davis, E. J., *Exact solutions for a class of heat and mass transfer problems,* Can. J. Chem. Eng., Vol. 51, No. 5, pp. 562–572, 1973.

Deavours, C. A., *An exact solution for the temperature distribution in parallel plate Poiseuille flow,* Trans. ASME, J. Heat Transfer, Vol. 96, No. 4, 1974.

Dezin, A. A., *Partial Differential Equations. An Introduction to a General Theory of Linear Boundary Value Problems,* Springer-Verlag, Berlin-New York, 1987.

Ditkin, V. A. and Prudnikov, A. P., *Integral Transforms and Operational Calculus,* Pergamon Press, New York, 1965.

Doyle, Ph. W., *Separation of variables for scalar evolution equations in one space dimension,* J. Phys. A: Math. Gen., Vol. 29, pp. 7581–7595, 1996.

Doyle, Ph. W. and Vassiliou, P. J., *Separation of variables for the 1-dimensional non-linear diffusion equation,* Int. J. Non-Linear Mech., Vol. 33, No. 2, pp. 315–326, 1998.

Elrick, D. E., *Source functions for diffusion in uniform shear flows,* Australian J. Phys., Vol. 15, No. 3, p. 283–288, 1962.

Faddeev, L. D. (Editor), *Mathematical Physics. Encyclopedia* [in Russian], Bol'shaya Rossiiskaya Entsyklopediya, Moscow, 1998.

Faminskii, A. V., *On mixed problems for the Corteveg–de Vries equation with irregular boundary data,* Doklady Mathematics, Vol. 59, No. 3, pp. 366–367, 1999.

Farlow, S. J., *Partial Differential Equations for Scientists and Engineers,* John Wiley & Sons, New York, 1982.

Galaktionov, V. A., *Invariant subspace and new explicit solutions to evolution equations with quadratic nonlinearities,* Proc. Roy. Soc. Edinburgh, Vol. 125A, No. 2, pp. 225–448, 1995.

Galaktionov, V. A. and Posashkov, S. A., *On new exact solutions of parabolic equations with quadratic nonlinearities,* Zh. Vych. Matem. i Mat. Fiziki, Vol. 29, No. 4, pp. 497–506, 1989.

Galaktionov, V. A. and Posashkov, S. A., *Exact solutions and invariant subspace for nonlinear gradient-diffusion equations,* Zh. Vych. Matem. i Mat. Fiziki, Vol. 34, No. 3, pp. 374–383, 1994.

Galaktionov, V. A., Posashkov, S. A., and Svirshchevskii, S. R., *Generalized separation of variables for differential equations with polynomial right-hand sides,* Dif. Uravneniya, Vol. 31, No. 2, pp. 253–261, 1995.

Gel'fand, I. M. and Shilov, G. E., *Distributions and Operations on Them* [in Russian], Fizmatlit, Moscow, 1959.

Gradshteyn, I. S. and Ryzhik, I. M., *Tables of Integrals, Series, and Products,* Academic Press, Orlando, 2000.

Graetz, L., *Über die Warmeleitungsfähigkeit von Flüssigkeiten,* Annln. Phys., Bd. 18, S. 79–84, 1883.

Grundland, A. M. and Infeld, E., *A family of nonlinear Klein–Gordon equations and their solutions,* J. Math. Phys., Vol. 33, No. 7, pp. 2498–2503, 1992.

Guenther, R. B. and Lee, J. W., *Partial Differential Equations of Mathematical Physics and Integral Equations,* Dover Publ., Mineola, 1996.

Gupalo, Yu. P., Polyanin, A. D., and Ryazantsev, Yu. S., *Mass Exchange of Reacting Particles with Flow* [in Russian], Nauka, Moscow, 1985.

Gupalo, Yu. P. and Ryazantsev, Yu. S., *Mass and heat transfer from a sphere in a laminar flow,* Chem. Eng. Sci., Vol. 27, pp. 61–68, 1972.

Haberman, R., *Elementary Applied Partial Differential Equations with Fourier Series and Boundary Value Problems,* Prentice-Hall, Englewood Cliffs, 1987.

Hanna, J. R. and Rowland, J. H. *Fourier Series, Transforms, and Boundary Value Problems,* Wiley-Interscience Publ., New York, 1990.

Happel, J. and Brenner, H., *Low Reynolds Number Hydrodynamics,* Prentice-Hall, Englewood Cliffs, 1965.

Hörmander, L., *The Analysis of Linear Partial Differential Operators. II. Differential Operators with Constant Coefficients,* Springer-Verlag, Berlin New York, 1983.

Hörmander, L., *The Analysis of Linear Partial Differential Operators. I. Distribution Theory and Fourier Analysis,* Springer-Verlag, Berlin, 1990.

Ibragimov N. H. (Editor), *CRC Handbook of Lie Group to Differential Equations, Vol. 1,* CRC Press, Boca Raton, 1994.

Ignatovich, N. V., *Invariant irreducible, partially invariant solutions of stationary boundary layer equations,* Mat. Zametki, Vol. 53, No. 1, pp. 140–143, 1993.

Ivanov, V. I., and Trubetskov, M. K., *Handbook of Conformal Mapping with Computer-Aided Visualization,* Boca Raton, CRC Press, 1994.

John, F., *Partial Differential Equations,* Springer-Verlag, New York, 1982.

Kalnins, E., *On the separation of variables for the Laplace equation in two- and three-dimensional Minkowski space,* SIAM J. Math. Anal., Hung., Vol. 6, pp. 340–373, 1975.

Kalnins, E. and Miller, W. (Jr.), *Lie theory and separation of variables, 5: The equations* $iU_t + U_{xx} = 0$ *and* $iU_t + U_{xx} - \frac{c}{x^2}U = 0$, J. Math. Phys., Vol. 15, pp. 1728–1737, 1974.

Kalnins, E. and Miller, W. (Jr.), *Lie theory and separation of variables, 8: Semisubgroup coordinates for* $\Psi_{tt} - \Delta_2 \Psi = 0$, J. Math. Phys., Vol. 16, pp. 2507–2516, 1975.

Kalnins, E. and Miller, W. (Jr.), *Lie theory and separation of variables, 9: Orthogonal R-separable coordinate systems for the wave equation* $\Psi_{tt} - \Delta_2 \Psi = 0$, J. Math. Phys., Vol. 17, pp. 331–335, 1976.

Kalnins, E. and Miller, W. (Jr.), *Lie theory and separation of variables, 10: Nonorthogonal R-separable solutions of the wave equation* $\Psi_{tt} - \Delta_2 \Psi = 0$, J. Math. Phys., Vol. 17, pp. 356–368, 1976.

Kamke, E., *Differentialgleichungen: Lösungsmethoden und Lösungen, I, Gewöhnliche Differentialgleichungen*, B. G. Teubner, Leipzig, 1977.

Kamke, E., *Differentialgleichungen: Lösungsmethoden und Lösungen, II, Partielle Differentialgleichungen Erster Ordnung für eine gesuchte Funktion*, Akad. Verlagsgesellschaft Geest & Portig, Leipzig, 1965.

Kanwal, R. P., *Generalized Functions. Theory and Technique*, Academic Press, Orlando, 1983.

Korn, G. A. and Korn, T. M., *Mathematical Handbook for Scientists and Engineers*, McGraw-Hill, New York, 1968.

Koshlyakov, N. S., Gliner, E. B., and Smirnov, M. M., *Partial Differential Equations of Mathematical Physics* [in Russian], Vysshaya Shkola, Moscow, 1970.

Krein, S. G. (Editor), *Functional Analysis* [in Russian], Nauka, Moscow, 1972.

Krylov, A. N., *Collected Works: III Mathematics, Pt. 2* [in Russian], Izd-vo AN SSSR, Moscow, 1949.

Lamb, H., *Hydrodynamics*, Dover Publ., New York, 1945.

Lavrent'ev, M. A. and Shabat B. V., *Methods of Complex Variable Theory* [in Russian], Nauka, Moscow, 1973.

Lavrik, V. I. and Savenkov, V. N., *Handbook of Conformal Mappings* [in Russian], Naukova Dumka, Kiev, 1970.

Landau, L. D. and Lifshits, E. M., *Quantum Mechanics. Nonrelativistic Theory* [in Russian], Nauka, Moscow, 1974.

Lebedev, N. N., Skal'skaya, I. P., and Uflyand, Ya. S., *Collection of Problems on Mathematical Physics* [in Russian], Gostekhizdat, Moscow, 1955.

Leis, R., *Initial-Boundary Value Problems in Mathematical Physics*, John Wiley & Sons, Chichester, 1986.

Levich, V. G., *Physicochemical Hydrodynamics*, Prentice-Hall, Englewood Cliffs, New Jersey, 1962.

Levitan, B. M. and Sargsyan, I. S., *Sturm–Liouville and Dirac Operators* [in Russian], Nauka, Moscow, 1988.

Loitsyanskiy, L. G., *Mechanics of Liquids and Gases*, Begell House, New York, 1996.

Lykov, A. V., *Theory of Heat Conduction* [in Russian], Vysshaya Shkola, Moscow, 1967.

Mackie, A. G., *Boundary Value Problems*, Scottish Academic Press, Edinburgh, 1989.

Makarov, A., Smorodinsky, J., Valiev, K., and Winternitz, P., *A systematic search for nonrelativistic systems with dynamical symmetries. Part I: The integrals of motion*, Nuovo Cimento, Vol. 52A, pp. 1061–1084, 1967.

Marchenko, V. A., *Sturm–Liouville Operators and Applications*, Birkhauser Verlag, Basel-Boston, 1986.

Markeev, A. P., *Theoretical Mechanics* [in Russian], Nauka, Moscow, 1990.

Mathematical Encyclopedia [in Russian], Sovetskaya Entsiklopediya, Moscow, 1977.

McLachlan, N. W., *Theory and Application of Mathieu Functions*, Clarendon Press, Oxford, 1947.

Meixner, J. and Schäfke, F., *Mathieusche Funktionen und Sphäroidfunktionnen*, Springer-Verlag, Berlin, 1965.

Mikhlin, S. G., *Variational Methods in Mathematical Physics* [in Russian], Nauka, Moscow, 1970.

Miles, J. W., *Integral Transforms in Applied Mathematics*, Cambridge Univ. Press, Cambridge, 1971.

Miller, W. (Jr.), *Symmetry and Separation of Variables*, Addison-Wesley, London, 1977.

Miller, J. (Jr.) and Rubel, L. A., *Functional separation of variables for Laplace equations in two dimensions*, J. Phys. A, Vol. 26, No. 8, pp. 1901–1913, 1993.

Moon, P. and Spencer, D., *Field Theory Handbook*, Springer-Verlag, Berlin, 1961.

Morse, P. M. and Feshbach, H., *Methods of Theoretical Physics, Vols. 1–2*, McGraw-Hill, New York, 1953.

Murphy, G.M., *Ordinary Differential Equations and Their Solutions*, D. Van Nostrand, New York, 1960.

Myint-U, T. and Debnath, L., *Partial differential equations for scientists and engineers*, North-Holland Publ., New York, 1987.

Naimark, M. A., *Linear Differential Operators* [in Russian], Nauka, Moscow, 1969.

Niederer, U., *The maximal kinematical invariance group of the harmonic oscillator*, Helv. Phys. Acta, Vol. 46, pp. 191–200, 1973.

Nikiforov, A. F. and Uvarov, V. B., *Special Functions of Mathematical Physics. A Unified Introduction with Applications*, Birkhauser Verlag, Basel-Boston, 1988.

Novikov, E. A., *Concerning turbulent diffusion in a stream with a transverse gradient of velosity*, Appl. Math. Mech. (PMM), Vol. 22, No. 3, p. 412–414, 1958.

Nusselt, W., *Abhängigkeit der Wärmeübergangzahl con der Rohränge*, VDI Zeitschrift, Bd. 54, No. 28, S. 1154–1158, 1910.

Olver, P. J., *Application of Lie Groups to Differential Equations*, Springer-Verlag, New York, 1986.

Ovsiannikov, L. V., *Group Analysis of Differential Equations*, Academic Press, New York, 1982.

Pavlovskii, Yu. N., *Analysis of some invariant solutions of boundary layer equations*, Zh. Bych. Matem. i Mat. Fiziki, Vol. 1, No. 2, pp. 280–294, 1961.

Petrovsky, I. G., *Lectures on Partial Differential Equations*, Dover Publ., New York, 1991.

Pinsky, M. A., *Introduction to Partial Differential Equations with Applications*, McGraw-Hill, New York, 1984.

Polozhii, G. N., *Mathematical Physics Equations* [in Russian], Vysshaya Shkola, Moscow, 1964.

Polyanin, A. D., *The structure of solutions of linear nonstationary boundary value problems of mechanics and mathematical physics*, Doklady Physics, Vol. 45, No. 8, pp. 415–418, 2000a.

Polyanin, A. D., *Partial separation of variables in unsteady problems of mechanics and mathematical physics*, Doklady Physics, Vol. 45, No. 12, pp. 680–684, 2000b.

Polyanin, A. D., *Linear problems of heat and mass transfer: general relations and results*, Theor. Found. Chem. Eng., Vol. 34, No. 6, pp. 509–520, 2000c.

Polyanin, A. D., *Handbook of Linear Mathematical Physics Equations* [in Russian], Fizmatlit, Moscow, 2001a.

Polyanin, A. D., *Transformations and exact solutions of boundary layer equations with arbitrary functions*, Doklady AN, Vol. 379, No. 3, 2001b.

Polyanin, A. D., *Exact solutions and transformations of the equations of a stationary laminar boundary layer*, Theor. Found. Chem. Eng., Vol. 35, No. 4, pp. 319–328, 2001c.

Polyanin, A. D., *Generalized separable solutions of Navier–Stokes equations*, Doklady AN, Vol. 380, No. 4, 2001d.

Polyanin, A. D. and Dilman, V. V., *Methods of Modeling Equations and Analogies in Chemical Engineering*, CRC Press, Boca Raton, 1994.

Polyanin, A. D., Kutepov, A. M., Vyazmin, A. V., and Kazenin, D. A., *Hydrodynamics, Mass and Heat Transfer in Chemical Engineering*, Gordon & Breach Sci. Publ., London, 2001.

Polyanin, A. D. and Manzhirov, A. V., *Handbook of Integral Equations*, CRC Press, Boca Raton, 1998.

Polyanin, A. D., Vyazmin, A. V., Zhurov, A. I., and Kazenin, D. A., *Handbook of Exact Solutions of Heat and Mass Transfer Equations* [in Russian], Faktorial, Moscow, 1998.

Polyanin, A. D. and Zaitsev, V. F., *Handbook of Exact Solutions for Ordinary Differential Equations*, CRC Press, Boca Raton, 1995.

Polyanin, A. D., Zaitsev, V. F., and Moussiaux, A., *Handbook of First Order Partial Differential Equations*, Gordon & Breach, London, 2001.

Polyanin, A. D. and Zhurov, A. I., *Exact solutions to nonlinear equations of mechanics and mathematical physics*, Doklady Physics, Vol. 43, No. 6, pp. 381–385, 1998.

Polyanin, A. D., Zhurov, A. I., and Vyazmin, A. V., *Generalized separation of variables in nonlinear heat and mass transfer equations*, J. Non-Equilibrium Thermodynamics, Vol. 25, No. 3/4, pp. 251–267, 2000.

Prudnikov, A. P., Brychkov, Yu. A., and Marichev, O. I., *Integrals and Series, Vol. 1, Elementary Functions*, Gordon & Breach Sci. Publ., New York, 1986.

Prudnikov, A. P., Brychkov, Yu. A., and Marichev, O. I., *Integrals and Series, Vol. 2, Special Functions*, Gordon & Breach Sci. Publ., New York, 1986.

Pukhnachev, V. V., *Group properties of the Navier–Stokes equations in the plane case*, Zh. Prikl. Mekh. i Tekhn. Fiziki, No. 1, pp. 83–90, 1960.

Rimmer, P. L., *Heat transfer from a sphere in a stream of small Reynolds number*, J. Fluid Mech., Vol. 32, No. 1, pp. 1–7, 1968.

Rotem, Z., and Neilson, J. E., *Exact solution for diffusion to flow down an incline*, Can. J. Chem. Engng., Vol. 47, pp. 341–346, 1966.

Schlichting, H., *Boundary Layer Theory*, McGraw-Hill, New York, 1981.

Sedov, L. I., *Plane Problems of Hydrodynamics and Airdynamics* [in Russian], Nauka, Moscow, 1980.

Shilov, G. E., *Mathematical Analysis: A Second Special Course* [in Russian], Nauka, Moscow, 1965.

Smirnov, V. I., *A Course of Higher Mathematics. Vols. 2–3* [in Russian], Nauka, Moscow, 1974.

Smirnov, M. M., *Second Order Partial Differential Equations* [in Russian], Nauka, Moscow, 1964.

Smirnov, M. M., *Problems on Mathematical Physics Equations* [in Russian], Nauka, Moscow, 1975.

Sneddon, I., *Fourier Transformations*, McGraw-Hill, New York, 1951.

Stakgold, I., *Boundary Value Problems of Mathematical Physics. Vols. I, II*, SIAM, Philadelphia, 2000.

Strauss, W. A., *Partial Differential Equations. An Introduction*, John Wiley & Sons, New York, 1992.

Sutton, W. G. L., *On the equation of diffusion in a turbulent medium*, Proc. Poy. Soc., Ser. A, Vol. 138, No. 988, pp. 48–75, 1943.

Svirshchevskii, S. R., *Lie–Bäcklund symmetries of linear ODEs and generalized separation of variables in nonlinear equations*, Phys. Letters A, Vol. 199, pp. 344–348, 1995.

Taylor, M., *Partial Differential Equations, Vol. 3*, Springer-Verlag, New York, 1996.

Temme, N. M., *Special Functions. An Introduction to the Classical Functions of Mathematical Physics,* Wiley-Interscience Publ., New York, 1996.

Tikhonov, A. N. and Samarskii, A. A., *Equations of Mathematical Physics*, Dover Publ., New York, 1990.

Thomas, H. C., *Heterogeneous ion exchange in a flowing system*, J. Amer. Chem. Soc., Vol. 66, pp. 1664–1666, 1944.

Tomotika, S. and Tamada, K., *Studies on two-dimensional transonic flows of compressible fluid, Part 1*, Quart. Appl. Math., Vol. 7, p. 381, 1950.

Urvin, K. and Arscott, F., *Theory of the Whittaker–Hill equation*, Proc. Roy. Soc., Vol. A69, pp. 28–44, 1970.

Vereshchagina, L. I., *Group fibering of the spatial nonstationary boundary layer equations*, Vestnik LGU, Vol. 13, No. 3, pp. 82–86, 1973.

Vladimirov, V. S., Mikhailov, V. P., Vasharin A. A., et al., *Collection of Problems on Mathematical Physics Equations* [in Russian], Nauka, Moscow, 1974.

Vladimirov, V. S., *Mathematical Physics Equations* [in Russian], Nauka, Moscow, 1988.

Vvedensky, D., *Partial Differential Equations*, Addison-Wesley, Wakingham, 1993.

Whittaker, E. T. and Watson, G. N., *A Course of Modern Analysis, Vols. 1–2*, Cambridge Univ. Press, Cambridge, 1952.

Zachmanoglou, E. C. and Thoe, D. W., *Introduction to Partial Differential Equations with Applications,* Dover Publ., New York, 1986.

Zaitsev, V. F. and Polyanin, A. D., *Handbook of Partial Differential Equations: Exact Solutions* [in Russian], MP Obrazovaniya, Moscow, 1996.

Zauderer, E., *Partial Differential Equations of Applied Mathematics*, Wiley–Interscience Publ., New York, 1989.

Zhdanov, R. Z., *Separation of variables in the non-linear wave equation*, J. Phys. A, Vol. 27, pp. L291–L297, 1994.

Zwillinger, D., *Handbook of Differential Equations*, Academic Press, San Diego, 1998.

INDEX

A

adjoint boundary conditions, 647
adjoint equation, 646
Airy equation, 668
Airy function, 163, 601
    first kind, 668
    second kind, 668
Airy stress function, 621
analytic function, 477
Appell transformation, 44
associated Legendre functions, 250–253, 408, 674

B

Bessel equation, 97, 311, 661
Bessel functions
    first kind, 661
    orthogonality properties, 664
    second kind, 661
    third kind, 665
    zeros, 664
Bessel's formula, 663
beta function, 660
    incomplete, 661
biharmonic equation, 621
binomial coefficients, 655
bipolar coordinates, 476
boundary conditions, 4
    adjoint, 647
    homogeneous, 4–7
boundary value problem
    first, 6, 26, 29
    homogeneous, 6, 490, 561
    mixed, 7, 26
    nonhomogeneous, 25, 28
    nonstationary, 15, 16, 25, 27, 28
    second, 6, 26, 29
    third, 7, 26
Boussinesq solution, 605

C

Cauchy problem, 5
    for equations of hyperbolic type, 24
    for equations of parabolic type, 23
    initial conditions, 5, 637
Cauchy–Riemann conditions, 468
Chebyshev functions, 679
Chebyshev polynomials, 679
classical solution, 7
classification of second-order PDE, 1
compatibility conditions, 16
complementary error function, 44, 106
complementary probability integral, 44
conditions
    boundary, 4–7, 12
    Cauchy–Riemann, 468
    initial, 5, 6, 637
    radiation, 493, 494, 564
    Sommerfeld, 493, 494, 564
conjugate heat and mass transfer problems, 50
coordinates
    bipolar, 476, 622, 632
    cylindrical, see Basic Notation
    polar, 622
    spherical, see Basic Notation
constant coefficient equations, 9–11, 43, 279
cosine integral, 658
cylinder function, 661

D

D'Alembert's formula, 280
Darboux equation, 331, 465
Dean's solution, 623
degenerate hypergeometric equation, 97, 669
degenerate hypergeometric functions, 86, 97, 99, 105, 669
    Wronskian, 670
descent method, 460
differential equation, see equation
diffusion equation, 161
Dini integral, 474
Dirac delta function, 11, 26, 203, 393, 633
Dirichlet problem, 467, 533, 477
Duhamel's principles, 38, 39

E

eigenfunctions, 14, 491, 561
eigenvalues, 14, 491, 561
ellipsoidal wave functions, 593
elliptic equation, 1, 4, 635
    problems, 30
    with three space variables, 533
    with two space variables, 467
elliptic operator, 635
equation
    adjoint, 646
    Bessel, 97, 311, 661
    biharmonic, 621
    Darboux, 331, 465
    degenerate hypergeometric, 97, 669
    diffusion, 161
    diffusion boundary layer, 155
    Euler–Poisson–Darboux, 314
    heat 1, 43, 65, 161, 188, 205, 254
    heat transfer in anisotropic media, 518
    homogeneous, 5, 6
    Klein–Gordon, 287, 302, 305, 376, 381, 414, 434
    Laplace, 1, 467, 533, 597

linear, 1, 5–7, 10
linearized Corteveg–de Vries, 601
mass transfer in anisotropic media, 518
mass transfer in liquid film, 105
Mathieu, 288, 675
modified Bessel, 666
modified Mathieu, 288, 677
nonhomogeneous biharmonic, 625
nonhomogeneous polyharmonic, 642
normal hyperbolic, 4
nonhomogeneous wave, 355, 412, 457
one-dimensional vibration with central symmetry, 298
Poisson, 478, 539, 598
polyharmonic, 638, 642, 643
Schrödinger, 85, 157, 200, 263, 510, 593
spheroidal wave, 591
Stokes, 630
telegraph, 321, 376, 386, 434, 438, 448
thermal boundary layer, 155
Tricomi, 526
turbulent diffusion, 265
ultrahyperbolic, 4
variance, 9
vibration of string, 279
wave, 1, 279, 341, 393, 456
wave, with axial symmetry, 295
Whittaker, 672
Whittaker–Hill, 162, 593
equations
   constant coefficient, 43, 279, 633
   elliptic, 1, 3, 4, 6, 635
   fourth-order, 602–633
   higher-order, 633–653
   hyperbolic, 1, 2, 4, 5, 637
   nonstationary, 613
   parabolic, 1, 2, 4, 5, 43, 161, 205
   regular, 637
   second-order, 43–600
   third-order, 601, 602
error function, 44, 656, 669
   complementary, 656
Euler constant, 660, 662, 670
Euler formula, 659
Euler–Poisson–Darboux equation, 314
   elliptic analogue, 528
exponential integral, 656
extremum principle, 467, 533

F

factorial, 655
first boundary value problem, 6, 29, 29
forced vibration of elastic rods, 606
formal solution, 7
formula
   Bessel, 663
   D'Alembert, 280
   Euler, 659
   Kirchhoff, 394
   Krylov, 606

Poisson, 343, 663
Stirling, 659
Fourier cosine transform, 18
Fourier sine transform, 18
Fourier transform, 18, 21
   inverse, 21
fourth-order equations
   nonstationary, 602
   stationary, 621
Fresnel integrals, 658
function
   Airy, 163, 601
   Airy, first kind, 668
   Airy, second kind, 668
   Airy stress, 621
   analytic, 477
   associated Legendre, 250, 252, 253, 674
   Bessel, first kind, 661
   Bessel, modified, 666
   Bessel, second kind, 661
   Bessel, third kind, 665
   beta, 660
   beta, incomplete, 661
   Chebyshev, 679
   cylinder, 661
   degenerate hypergeometric, 86, 97, 99, 105, 669
   Dirac delta, 11
   ellipsoidal wave, 593
   error, 44, 656, 669
   error, complementary, 44
   gamma, 83, 102, 659
   gamma, incomplete, 82, 83, 660
   Gegenbauer, 631, 632
   Green, 17, 25–37, 644, 646
   Hankel, 665
   harmonic, 467, 533, 597
   Hermite, 680
   hypergeometric, 672
   Legendre, 631, 674
   Macdonald, 666
   Mathieu, 675, 676
   Mathieu, modified, 677
   Neumann, 661
   parabolic cylinder, 45, 287, 675
   special, 655
   spheroidal wave, 591
   stream, 473, 477, 517, 621
   Whittaker, 672
fundamental solution, 11, 633, 636
   Cauchy problem, 23, 24, 634

G

gamma function, 83, 102, 659
   incomplete, 82, 660
   logarithmic derivative, 662, 666
Gaussian hypergeometric equation, 672
Gegenbauer functions, 631, 632
Goursat problem, 6, 284
Graetz–Nusselt equation, 514
Green's function, 17, 25–37, 644, 646

H

Hankel functions, 665
Hankel transform, 18
harmonic function, 467, 533, 597
heat equation, 1, 43
    n-dimensional, 268
    one-dimensional, 43
    three-dimensional, 205
    two-dimensional, 161
    with axial symmetry, 65, 70
    with central symmetry, 73, 79
    with source, 188, 254
Helmholtz equation, 490, 561, 599
Hermite functions, 680
Hermite polynomials, 669, 679
higher-order PDE, 633–653
homogeneous boundary condition, 4–7
homogeneous boundary value problem, 6, 490, 561
homogeneous equation, 5, 6
homogeneous initial condition, 5, 6
hyperbolic differential operator, 637
hyperbolic equation, 1, 2, 4, 5, 637
    first canonical form, 3
    problems, 27, 29
    second canonical form, 3
    with n variables, 456–466
    with one space variable, 279–340
    with three space variables, 393–455
    with two space variables, 341–391
hypergeometric equation, 672
hypergeometric functions, 672
hypergeometric series, 672

I

incomplete beta function, 661
incomplete gamma function, 82, 83, 660
incomplete separation of variables, 10, 24, 34, 203
initial conditions, 4
    for hyperbolic equations, 6
    for parabolic equations, 5
    homogeneous, 5, 6
integral
    complementary probability, 44
    Dini, 474
    Fresnel, 658
    Poisson, 473, 537, 598
    probability, 44
integral transforms, 17
inverse Fourier transform, 21
inverse transform, 17

J

Jacobi polynomials, 680

K

kernel of integral transform, 17
Kirchhoff's formula for wave equation, 394
Klein–Gordon equation
    one-dimensional, 287, 302, 305
    three-dimensional, 414
    two-dimensional, 362
    with axial symmetry, 302
    with central symmetry, 305
Krylov's formula, 606
Kummer series, 669
Kummer transformation, 670

L

Laguerre polynomials, 669, 678
    generalized, 678
Lamé equation, 540
Laplace equation, 1
    with n variables, 597
    with three variables, 533
    with two variables, 467
Laplace transform, 18
    inverse, 18
Legendre functions, 631, 674
    associated, 250–253, 408, 674
Legendre polynomials, 674
    zeros, 674
linear equation, 1, 5–7, 10
    homogeneous, 7
    nonhomogeneous, 10
linearized Corteveg–de Vries equation, 601
logarithmic derivative of gamma function, 662, 666
logarithmic integral, 657
Lorentz transformation, 280, 342, 394

M

Macdonald function, 666
mass transfer equation in liquid film, 105
Mathieu equation, 288, 675
Mathieu functions, 675, 676
Meijer transform, 18
method
    conformal mappings, 476, 489
    descent, 460
    integral transforms, 17
    separation of variables, 11
mixed boundary value problem, 7, 26
modified Bessel equation, 666
modified Bessel functions
    first kind, 666
    second kind, 666
modified Mathieu equation, 288, 677
modified Mathieu functions, 677

N

Neumann function, 661
Neumann problem, 467, 533
nonhomogeneous biharmonic equation, 625
nonhomogeneous polyharmonic equation, 642
nonhomogeneous wave equation, 355, 412, 457
nonstationary problems, 15, 16, 25, 27, 28, 38, 649
normal hyperbolic equation, 4

O

orthogonal polynomials, 677

P

parabolic cylinder functions, 675
parabolic equations, 1, 4, 16, 205
   in narrow sense, 4
   problems, 25, 28
   with n variables, 268–277
   with one space variable, 43–160
   with three space variables, 205–267
   with two space variables, 161–203
Pochhammer symbol, 656
Poisson equation, 478
   with n variables, 598
   with three variables, 539
   with two variables, 478
Poisson integral, 473, 537, 598
Poisson solution, 615
Poisson's formula, 663
   for wave equation, 343
polyharmonic equation, 638, 642, 643
polynomials
   Chebyshev, 679
   Hermite, 669, 679
   Jacobi, 680
   Laguerre, 669, 678
   Legendre, 674
   orthogonal, 677
principle
   Duhamel, 38, 39
   extremum, 467, 533
   linear superposition, 7, 15
   Rayleigh, 152
probability integral, 44
problem
   boundary value, 6, 7, 25, 28
   Cauchy, 5, 23
   conjugate, heat and mass transfer, 50
   diffusion boundary layer, 81, 84, 101, 155, 156
   Dirichlet, 467, 533
   first boundary value, 6
   for equations of elliptic type, 30
   for equations of hyperbolic type, 27, 29
   for equations of parabolic type, 25, 28
   Goursat, 6, 284
   mixed boundary value, 7, 26
   Neumann, 467, 533
   nonstationary, 15, 16, 25, 27, 28, 38, 649
   of axisymmetric straining flow about drop, 632
   of inflow of viscous fluid into half-plane through slit, 623
   of translational flow about drop/bubble, 632
   of transverse vibration of elastic rod in resisting medium, 612
   second boundary value, 6, 26, 29
   stationary, 31
   Sturm–Liouville, 151, 267, 276, 334, 454, 595
   third boundary value, 7, 26
   without initial conditions, 49

R

radiation conditions, 493, 494, 564
Rayleigh principle, 152
regular equations, 637
regular linear differential operator, 640
regularity index, 637

S

Schrödinger equation, 85, 593
   n-dimensional, 271
   one-dimensional, 157
   stationary, 510
   three-dimensional, 263
   two-dimensional, 200
second boundary value problem, 6, 26, 29
separable solutions, 9
sine integral, 657
solution
   Boussinesq, 605
   classical, 7
   Dean, 623
   formal, 7
   fundamental, 11, 633, 636
   fundamental for Cauchy problem, 23, 24, 634
   Poisson, 615
   separable, 9
   Stokes, 631
   Taylor, 632
Sommerfeld conditions, 493, 494, 564
special functions, 655
spectrum of boundary value problem, 491, 562
spheroidal wave functions, 591
stationary boundary value problems, 31
stationary Schrödinger equation, 510
Stirling formula, 659
Stokes equation, 630
Stokes solution, 631
stream function, 474, 477, 517, 621
   in spherical coordinates, 631
Sturm–Liouville problem, 151, 267, 276, 334, 454, 595

T

Taylor solution, 632
telegraph equation, 321
    in axisymmetric case, 386
    n-dimensional, 463
    three-dimensional, 434, 438, 448
    two-dimensional, 376, 381
third boundary value problem, 7, 26
third-order PDE, 601, 602
three-dimensional Laplace equation, 533
three-dimensional Poisson equation, 539
transform
    Fourier, 18, 21
    Fourier, inverse, 21
    Fourier cosine, 18
    Fourier sine, 18
    Hankel, 18
    integral, 17
    integral, inverse, 17
    Laplace, 18
    Laplace, inverse, 18
    Meijer, 18
    Mellin, 18
transformation
    Appell, 44
    Kummer, 670
    Lorentz, 280, 342, 394
Tricomi equation, 526

U

ultrahyperbolic equation, 4

V

variance equation, 9
vibration equation with axial symmetry, 295

W

wave equation, 1, 279
    n-dimensional, 456
    one-dimensional, 279
    three-dimensional, 393
    two-dimensional, 341
    with axial symmetry, 295
    with central symmetry, 298
Whittaker equation, 672
Whittaker functions, 672
Whittaker–Hill equation, 162, 593

Z

zeros of Bessel functions, 664
zeros of Legendre polynomials, 674